The Physics of Radiotherapy X-Rays and Electrons

The Physics of Radiotherapy X-Rays and Electrons

by Peter Metcalfe, Tomas Kron, and Peter Hoban

An updated successor to
The Physics of Radiotherapy X-Rays from Linear Accelerators

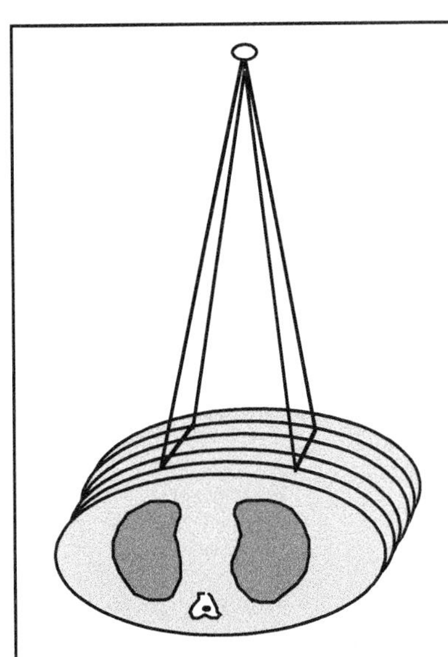

Medical Physics Publishing
Madison, Wisconsin

An updated successor to The Physics of Radiotherapy X-Rays from Linear Accelerators.

ISBN: 978-1-930524-36-1 (softcover)

Third printing softcover 2017

Library of Congress Cataloging-in-Publication Data:

Metcalfe, Peter.
The physics of radiotherapy x-rays and electrons / by Peter Metcalfe, Tomas Kron, and Peter Hoban.
p. cm.
Rev. ed. : The physics of radiotherapy X-rays from linear accelerators. 1997.
Includes bibliographical references and index.
ISBN 978-1-930524-35-4 (hc : alk. paper) -- ISBN 978-1-930524-36-1 (sc : alk. paper)
1. Medical physics. 2. Linear accelerators in medicine. 3. Cancer--Radiotherapy. 4. Radiation dosimetry. I. Kron, Tomas. II. Hoban, Peter. III. Metcalfe, Peter Physics of radiotherapy X-rays from linear accelerators. IV. Title.
R895.M48 2007
610.1'53—dc22

2007061228

Medical Physics Publishing
4555 Helgesen Drive
Madison, WI 53718
(608) 224-4508
E-mail: mpp@medicalphysics.org
Web: www.medicalphysics.org

Dedication

John Cameron (1923–2005), the founder of Medical Physics Publishing, has been an inspiration to many.

We wish to dedicate this book to John who gave us the opportunity to publish the first edition.

His contribution to the field of medical physics will always be appreciated.

Contents

Preface

"The increase in the average age of the population (i.e., 'age hardening'), improved cancer detection and better case reporting will produce a net increase in the incidence of cancer . . . towards a doubling of radiotherapy caseload by the year 2010."

(Jerry Battista 1996)

The long-term survival of cancer patients treated with radiotherapy alone or with combined-modality treatments such as surgery with chemotherapy is impressive for several common cancer types.

Meta-analysis of 78 randomized trials shows radiotherapy in combination with other modalities reduces the incidence of local breast cancer recurrence and has a potential impact on overall survival over 15 years (Abe et al. 2005). Another study of 36 trials shows adjuvant radiotherapy combined with mastectomy in advanced breast cancer leads to a small but significant increase in survival over 10 years (Gebski et al. 2006).

There have been several dose escalation studies showing the incidence in local recurrence in prostate cancer patients is reduced with dose escalation using conformal radiotherapy (e.g., Pollack et al. 2002) while rectal toxicity is maintained at acceptable levels. There has been one dose escalation cohort study showing significantly lower grade II rectal toxicity using intensity-modulated radiotherapy to 81 Gy versus the conformal radiotherapy (2% versus 14%) cohort arm to 81 Gy (Zelefsky et al. 2001).

There have also been gains in local control by combining adjuvant androgen deprivation therapy with radiotherapy, in particular with regard to advanced-stage prostate cancer patients; a 20% higher survival over 8 years has been shown in some subgroups when five Radiation Therapy Oncology Group (RTOG) prospective trials were analyzed (Roach et al. 2000).

Meta-analysis of randomized trials of head and neck cancers has shown cisplatin chemotherapy combined with radiotherapy decreases local recurrence by about 10% (Rosenthal and Ang 2004). More recently, a phase

III trial focusing on advanced stage head and neck cancer patients combining high-dose radiotherapy with cetuximab, a monoclonal antibody inhibiting the epidermal growth factor receptor. The trial showed improved loco-regional control without increase in severe mucositis associated with other chemotherapy agents, in essence allowing for better application of the radiotherapy component in the combined treatment (Bonner et al. 2006).

X-ray and electron beams produced by medical linear accelerators (linacs) are used to treat these cancer patients who have benefited from the advances in modern radiotherapy, a few of which have been mentioned above. The physical properties and uses of these beams are discussed in this book. The book is designed primarily as a useful reference for radiation oncology physicists, whether in training or established in their careers. The material is also intended to be accessible to radiation oncologists, dosimetrists, and radiation therapists who wish to gain a deeper understanding of the physical principles behind the technology they interact with daily.

Many simple empirical equations that the medical physicist can use to predict dose deposition in radiotherapy patients are introduced in this text. When the dose prediction process becomes too complex for manual calculations, computers are required. This text provides explanations of models used by radiotherapy treatment planning computers to calculate dose in patients from megavoltage x-ray and electron beams. Emphasis is placed on the methods for dose calculations in and around tissue inhomogeneities. Once x-rays or electrons penetrate the patient, there are a number of ways to predict dose modifications due to irregularities such as air cavities, lung, and bone. Photon beam inhomogeneity correction models currently used in monitor unit (MU) check programs are discussed. The methods studied include effective pathlength and more sophisticated models that include photon scatter and electron transport.

There are a number of areas in which standard x-ray beam correction methods fail to predict dose accurately. Therefore, two especially accurate beam modeling methods are discussed. These are the Monte Carlo and superposition/convolution methods. These methods provide us with new and exciting insights into x-ray beam interactions with the human body. With electron beam modeling there have also been advances from pencil-beam methods to Macro Monte Carlo methods. Both these approaches are discussed.

As the aim of radiotherapy is to kill all the tumor clonogens, the text also includes an explanation of radiobiological models that outline the principles behind tumor control. These models are also being used as planning objective tools and can help in comparing alternative treatment plans.

Since the first edition of this text there have been advances in shaping dose distributions around target volumes using intensity-modulated radiation therapy (IMRT) and in visualizing the target volumes with respect to the anatomy using image-guided radiation therapy (IGRT) techniques. Hence, both these areas are examined in the second edition. It was also decided that to make this a more complete text for teaching purposes, the physics of electron beams should be added and also that a series of questions and answers be added at the end of each chapter; some of these questions are from *The Q Book* and some new questions have been included.

Chapter one includes an outline of how electrons and x-rays are produced by linacs. The design principles behind microwave production devices, e.g., klystrons and magnetrons, are discussed. The principles of electron acceleration in standing and traveling waveguide accelerators are also outlined. Delivery of a clinically useful beam by means of a bending magnet, target or scattering foil, monitor ionization chambers, and collimators are also described in this chapter.

Interactions of radiotherapy x-ray and electron beams are discussed in chapter two. The processes involve the interactions of photons with electrons. These interaction processes include photoelectric effect, Compton scattering, and pair production. Such interactions cause electrons to be set in motion. Electrons then interact with other electrons, and this leads to energy absorption in the medium. The collisional and radiative energy loss mechanisms of electrons are also discussed. Included in this chapter is information about electron disequilibrium, as this is essential to the discussion of beam models that follows.

In chapter three ionization detectors and other modes of x-ray and electron detection such as film, semiconductor, and thermoluminescent dosimeters are discussed. Dosimetry phantoms such as water tanks and solid water/solid lung/solid bone slabs are discussed, as are anthropomorphic designs.

Chapter four includes discussion of buildup, depth dose, and dose profiles of x-ray beams ranging from kV to MV energies. Linac x-ray beam data are shown in a series of tables and graphs. This is typical of the data that are issued to the dosimetrist or physicist for treatment planning calculation checks. The data presented are also similar to the data that needs to be collected during the linac initial commissioning process and for radiotherapy treatment planning computer input.

Electron beam properties are dealt with in a new chapter five, including aspects of beam quality with emphasis on properties which impact on the use of electron beams in clinical practice.

The radiotherapy treatment planning process is discussed in chapter six. The use of patient x-ray simulators and CT scanners to provide planning information is included, as are other planning aids such as nuclear magnetic resonance (NMR) and positron emission tomography (PET) images. The hardware configuration of a computer planning system is also discussed. The planning system is equipped with a number of dose visualization tools that aid the planner in assessing the dose to the target volume. The volumes that require treatment are specified in this chapter, as are the dose visualization tools.

Chapter seven covers special procedures, including stereotactic radiosurgery, IMRT, IGRT, and tomotherapy. These new approaches and devices have revolutionized our ability to conform dose to the target volume and to ensure this dose is accurately placed relative to the anatomy on each treatment day. Inverse planning is an integral part of IMRT treatment planning; hence optimization of dose to the tumor using different inverse planning techniques is dealt with in this chapter.

Beam calibration protocols are discussed in chapter eight. While presenting the most current protocols, other historically important protocols are also discussed. This chapter is not to be used as a calibration protocol—its focus is on showing the basic concepts underlying absolute dosimetry with ionization chambers. The principles of operation of ionization chambers are outlined in chapter three (dosimetry) and it will help the reader to first be familiar with its contents.

In radiotherapy treatment planning it is necessary to calculate x-ray dose distributions with accuracy in order that the predicted dose is realized at treatment time. Incorrect predictions lead to suboptimum treatments, which can result in radiation-induced complications and loss of tumor control. In chapter nine, examples are shown of situations in which errors can occur, and rationale is given for these potential deviations from the plan. Chapter nine also includes a review of historically used photon beam models that are still currently used in MU checking software. The inhomogeneity correction methods studied in this chapter include effective depth and equivalent tissue air ratio corrections. This chapter also reviews electron pencil-beam models.

Photon beam methods based on convolution/superposition models are now used in most commercial radiotherapy treatment planning computers. These methods are described in chapter ten. Monte Carlo methods are also discussed in chapter ten, as these methods account for charged particle transport and are being increasingly used in the medical physics laboratory for dose accuracy benchmark tests. As predicted in the first edition improvements in hardware speed and optimization of code have led to hybrid Macro Monte Carlo codes being used for some clinical calculations for electron beams (Fippel et al. 1997) and x-ray beams (Laub et al. 2000). These methods are also discussed in chapter ten.

Once beam data have been collected from a linac during the commissioning process, care is taken to ensure that the beam properties stay the same. A series of quality assurance (QA) tests have been reported in many documents to ensure this occurs. A practical guide to QA tests for linacs and high-technology ancillary devices is given in chapter eleven. QA processes for multileaf collimators (MLCs), electronic portal imaging devices (EPIDs), linac-mounted imaging devices, CT simulators, and radiotherapy treatment planning computers are also discussed. The last section in chapter eleven includes discussion of in vivo dosimetry, potentially the ultimate QA tool for accurate dose delivery.

The tremendous advances in dose delivery afforded by techniques such as IMRT, stereotactic delivery, and tomotherapy also require similar efforts in ensuring best possible patient positioning during treatment planning and delivery. This includes immobilization and, increasingly, image guidance during treatment. These topics will be discussed in chapter twelve.

Environmental radiation protection principles that relate to linac room design and radiation safety tests are outlined in chapter thirteen. This chapter includes information about environmental regulations for occupational exposure and barrier calculation methods for linac bunkers.

Chapter fourteen summarizes some of the radiobiological models that have been designed to explain tumor control by radiotherapy. The linear quadratic model is used to describe the impact of tumor dose homogeneity and dose fractionation on tumor control and normal tissue damage. The principles of effective uniform dose are introduced. Evidence-based medicine is the cornerstone upon which new technology is being adopted; hence the basic framework upon which clinical trials are based is also briefly discussed in this chapter.

It is hoped that readers find the material useful. The authors are grateful for any comments, suggestions, and feedback.

References

Abe O., Abe R., Enomoto K., Koyama H. et al. (2005) Effects of radiotherapy and of differences in the extent of surgery for early breast cancer on local recurrance and 15-year survival: An overview of randomized trials. *Lancet* 366: 2087–2106.

Battista J.J. (1996) Radiotherapy in the next millennium. *Proceedings of AAPM Summer School—Teletherapy: Present and Future.* T.R. Mackie and J.R. Palta, Editors. Advanced Medical Publishing, Madison, Wisconsin, pp. 805–817.

Bonner J.A., Harari P.M., Giralt J., Azarnia N. (2006) Radiotherapy plus cetuximab for squamous-cell carcinoma of head and neck. *N. Eng. J. Med.* 354: 567–578.

Fippel M., Kawrakow I., Friedich K. (1997) Electron beam dose calculations with VMC algorithm and the verification of NCI working group. *Phys. Med. Biol.* 42: 501–520.

Gebski V., Lagleva M., Keech A., Simes J., Langlands A.O. (2006) Survival effects of postmastectomy adjuvant radiation therapy using biological equivalent doses: A clinical perspective. *J. Natl. Cancer Inst.* 98: 26–38.

Laub W., Alber M., Birkner M., Nusslin F. (2000) Monte Carlo dose computations for IMRT optimization. *Phys. Med. Biol.* 45: 1741–1754.

Pollack A., Zagars K., Starkshall G., Antolak J. et al. (2002) Prostate cancer radiation dose response: results of the M.D. Anderson phase III randomized trial. *Int. J. Radiat. Oncol. Biol. Phys.* 53: 1097–1105.

Roach M., Lu J., Pilepich M.V., Asbell S.O. (2000) Predicting long-term survival, and the need for hormonal therapy: A meta-analysis of RTOG prostate cancer trials. *Int. J. Radiat. Oncol. Biol. Phys.* 47: 617–627.

Rosenthal A., Ang K. (2004) Altered radiation therapy fractionation, chemoradiation, and patient selection for treatment of head and neck squamous cell carcinoma. *Semin. Oncol.* 14: 153–166.

Zelefsky M., Fuks Z., Hunt M., Lee H., et al. (2001) High dose radiation delivered by intensity modulated conformal radiotherapy improves the outcome of localized prostate cancer. *J. Urology* 166: 876–881.

Peter Metcalfe, PhD
Cancer Institute, NSW Chair,
Radation Oncology Physics
Centre for Medical Radiation Physics
University of Wollongong
Wollongong, New South Wales, Australia

Tomas Kron, PhD
Principal Research Physicist
Peter MacCallum Cancer Institute
Melbourne, Victoria, Australia

Dr. Peter Hoban
Senior Product Manager
Director, Stereotactic Radiosurgery
TomoTherapy Inc.
Madison, Wisconsin, USA

May 2007

Acknowledgments

Special Acknowledgment

> *The support of our families throughout the production of this manuscript has been incredible. Thanks to Tracy, Adam, and Jemma Metcalfe; Robyn-Woodward Kron, Alessa and Edmund Kron; Linda, Tilly, and Daisy Hoban.*

Many personnel contributed to the first and second editions:

Thanks to Betsey Phelps for editing and producing the final version of the second edition text. Thanks to Elizabeth Seaman for editing and producing the final version of the first edition text.

In particular, Patricia Ostwald whose study into electron beams with Tomas Kron helped greatly with the electron sections in the textbook. Matthew Williams whose advice on the manuscript was an inspirational help to Peter Metcalfe. Jake Van Dyk who gave advice and information on calibration and radiation protection to the authors.

Other personnel and friends who selflessly and generously gave information and data for the first and second editions include: Anatoly Rosenfeld (MOSFETs), Martin Carolan (dynamic wedges), Martin Butson (surface dose), Nick Hardcastle (Monte Carlo), Jo McNamara (gating), Anthony Arnold (immobilization), Iwan Cornelius (Monte Carlo), Dean Cutajar (brachytherapy), Lois Holloway (radiobiology), Bob Smee (radiosurgery), Clive Baldock (gel dosimetry), Melissa Siroky (kV x-rays), Michael Currie (beam models), Wayne Beckham (MRI), Paul Keall (convolution and gating), Martin Ebert (radiobiology), Timothy Zhu (beam models), Lisa Duggan (thermoluminescence dosimetry), C. Kleinschmidt (EGS-ray); Tony Wong, Lloyd Kelly, and John Farelly (linear accelerators).

Thanks to all the companies who supplied photographs for the text.

1 Medical Linear Accelerators

"A boy surfing on a water wave provides a useful traveling wave analogy."

(Karzmark and Morton 1981)

1.1 Introduction

For the treatment of deep-seated tumors, high-energy x-rays with penetrating characteristics are required. The medical *linear accelerator* (linac) is currently the most popular device for this application. To ensure a full understanding of linac beam properties it is important to review first the mechanism of x-ray beam production by these types of devices.

Early historically significant methods of producing x-ray beams for external beam radiotherapy (teletherapy) consisted of conventional x-ray tubes (with anode and cathode) producing x-rays of energy up to about 300 kV. These superficial (up to 150 kV) and orthovoltage (up to 300 kV) machines are still used very effectively for treatment of skin cancer and for palliation, respectively (Niroomand-Rad et al. 1987, Scrimger et al. 1986, Johns et al. 1952a, Oliver et al. 1949, Podgorsak et al. 1990, Butson et al. 1995, Biggs et al. 1999). They do not, however, provide skin sparing properties, and the beams are not penetrating enough to treat deep-seated tumors effectively.

The search for a more penetrating beam led to the development of cobalt-60 machines (Johns et al. 1952b, 1959). These became more popular than other teletherapy sources (e.g., cesium-137) because a large source activity of several thousand Curies of cobalt-60 could be packed into a very small cylinder of about 10 mm diameter (Johns et al. 1983). The cobalt-60 machine is still used in radiotherapy applications today. A cobalt-60 beam spectrum has two photo-peaks, at 1.17 and 1.33 MeV, giving a mean photon energy of about 1.25 MeV. Despite their rather low photon energy compared with that of a linac, they still provide reasonable skin sparing properties as maximum dose is achieved at 0.5 cm depth. Historically, most cobalt machines operate at 80 cm *source surface distance* (SSD). However, now some cobalt-60 designs do produce a 2 gray (Gy) per minute dose rate at 100 cm SSD. With this improvement, discussion as to the future role of cobalt units has arisen (Glasgow 1999, Glasgow et al. 1996, Van Dyk and Battista 1996). These units are also generally very reliable due to their relatively simple design. The units are also less susceptible to electric power fluctuations. They may continue to play a supporting role to linacs in some busy radiotherapy departments. The question of cobalt-60 radioactive source transportation and disposal are issues that need careful consideration with these devices.

The evolution of the linac was a direct result of radar development work that culminated in the production of microwave generators in the form of magnetrons and the klystrons. These devices are capable of establishing intense electromagnetic fields in microwave cavities. This enables the acceleration of electrons to relativistic velocities when incorporated with suitable waveguide structures. Linacs are currently the modality of choice for the production of high-energy x-rays in radiotherapy applications. These units are more popular than cobalt-60 teletherapy units because they have the following features:

i) Multiple electron and photon energies are available that allow the physician to tailor treatment to the required treatment depth. A modern linac is usually capable of producing at least two different photon and five different electron energies.
ii) The dose rates (1 to 10 Gy per minute) are higher than those of isotope machines. This enables shorter treatment times.
iii) Most linac x-ray beams, including 4 and 6 MV, have a sharper dose fall-off at the beam edge than cobalt beams. This is because linacs have a smaller virtual source size of about 2 mm in diameter (Munro et al. 1988). The exception to this rule is very high-energy linacs, which have significant lateral electron range that increases the width of their penumbra (see chapter two).

Other machines used for accelerating electrons for direct treatment application or to produce radiotherapy x-ray beams include betatrons, microtrons, and cyclotrons. A review of these machines can be found elsewhere (Khan 1984).

Alternate radiotherapy treatment beams include *fast* neutrons that are used to treat cancer patients in just a few centers. They have a very high linear energy transfer (LET) compared with photons. Hence most of the damage is direct. This leads to better destruction of hypoxic cells. Because the cell survival curves are not very curved (see chapter thirteen), the ability to spare late-responding tissues via fractionation is compromised compared with x-rays. Clinical trials show promise for these beams for treating slowly proliferative tumors (Maughan 1996). Epithermal boron neutron capture therapy (BNCT) is an interesting research area that uses the high-capture cross section of boron to thermal neutrons (Allen et al. 1992, Soloway et al. 1993). One of the tumor types this binary modality is being targeted toward is glioblastoma multiforme (GBM), a type of brain tumor that does not respond very well to conventional x-ray treatment. Some melanomas have also been treated using BNCT. Several treatment sites are at clinical trial stage and a few hundred patients have now been treated with BNCT (Coderre et al. 2003).

The most popular alternative to using linac x-ray beams seems to be *protons*. These are currently used to treat cancer patients at about 30 facilities worldwide (Moyers 1999) with several new centers currently under construction. Building a synchrotron or cyclotron to produce protons is expensive, even though the beam line may be shared by several (usually three) treatment delivery portals. The gantries may be of the order of 100 tons and complex magnetic beam steering is required. The cost of these facilities varies, but the capital expenditure for such a facility is typically between ten to twenty times more expensive than for setting up an x-ray linac facility.

Synchrotrons can also be used to deliver heavier particles (such as carbon-12. There are currently a handful of carbon-12 beam lines with associated gantries being used to treat cancer patients. Carbon-12 is particularly interesting, as it seems to have a direct ionizing damage effect (like neutrons) that makes it suitable for treatment of hypoxic tumors such as prostate (Ishikawa et al. 2006). Synchrotron designs exist which can combine beam lines to provide carbon-12 to one gantry and protons to another within the same facility.

One very favorable property of protons and carbon-12 is their well-defined Bragg peak. Due to this, the beam reaches a peak of dose deposition at significant depth and then a rapid fall-off beyond this depth. Proton beams of energy 250 MeV are used to treat deep-seated tumors. For further information about proton radiotherapy see Hall (1988), Miller (1996), Suit (2002), and Lomax (2003). While randomized trials comparing protons with x-ray are limited (Shipley et al. 1995), there are a number of planning studies which point to advantages of protons in sparing normal tissue by delivering much lower dose to these regions in particular (Lomax et al. 1999).

1.2 Principles of Operation

1.2.1 External and Geometric Features

Figures 1.1a to 1.1c show photographs of the external features of three commonly used linacs. Common features of all these linacs are: the patient is placed on the treatment couch, which has vertical, horizontal, and rotational movements in the horizontal plane for patient positioning; the linac beam production devices are mounted in the stand and gantry; the stand is fixed and holds the gantry in position; the gantry is able to rotate within a 360 degree arc around the patient to enable different beam angles to be aimed at the patient; the collimators are internally mounted at the end of the gantry; and the size and angle of the collimators are also adjustable.

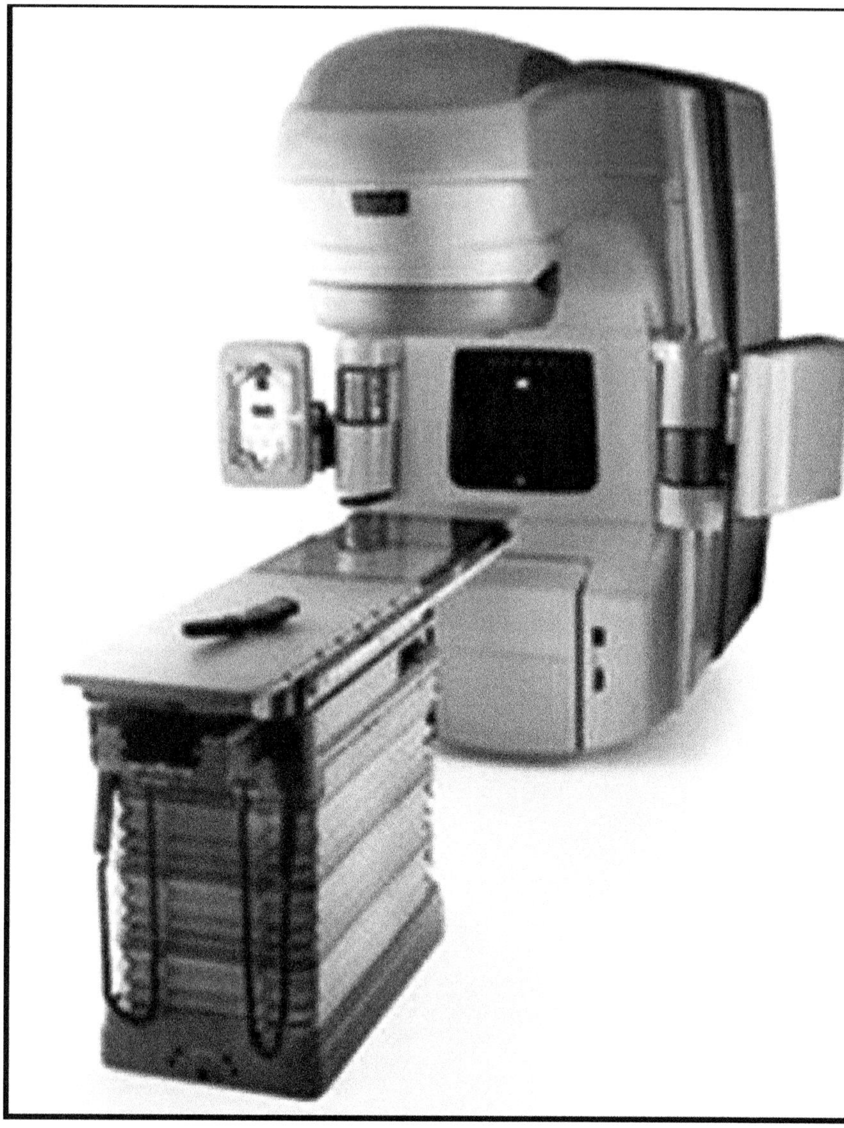

(a)

Figure 1.1: Medical linear accelerators (linacs): All these linacs are capable of producing two x-ray energies (popular combinations of 4/10 MV, 6/10 MV, 6/18 MV) and at least five electron energies ranging from 6 to 23 MeV. (a) Varian Clinac® accelerator with On-Board Imager® device for image-guided radiation therapy. (Reprinted with permission from: © Varian Medical Systems, Inc.) (b) Siemens Oncor linac (Reprinted with permission from: © Siemens Medical Systems, Inc.) (c) Elekta Synergy linac (Reprinted with permission from: © Elekta Inc.)

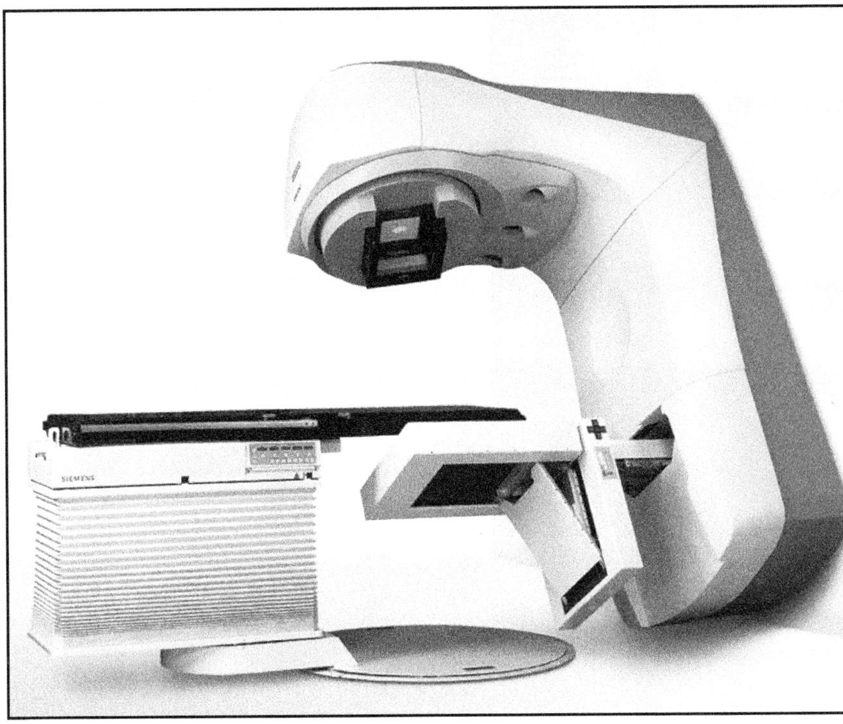

(b)

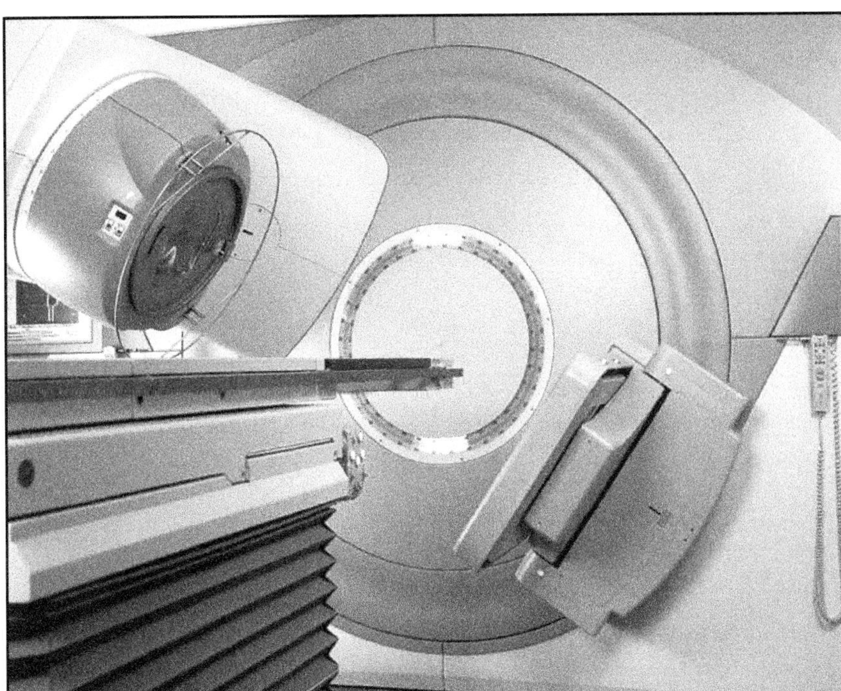

(c)

1.2.2 Operational Systems and Auxiliary Systems

A schematic diagram of the linac components that are housed in the gantry is given in figure 1.2a. This is the layout used by the Varian high-energy dual modality family of linacs, which are capable of producing x-ray and electron beams. All Varian dual modality linacs use klystrons, not shown in this figure, for microwave generation. The klystron is located in the stand, and microwaves are transmitted via a four port circulator in the stand through a rotary joint and up to the accelerating structure via a transmission waveguide.

Some low-energy linacs have no bending magnet and instead have a short guide mounted in the same axis as the beam, as shown in figure 1.2b. This is the method used by the single x-ray energy Varian linacs. Note that there is no bending magnet. The target is fixed to save space (i.e., no carousel). Despite this efficiency of space there is, however, a small recess required in the floor to allow 360 degree rotation of this design. The floor recess automatically opens and closes only when required. This ensures that the linac can be kept at a reasonably low height.

The Siemens family of linacs use a horizontal guide and bending magnet. They are all housed within a common gantry design. They have both magnetron and klystron driven machines, as shown in figure 1.2c.

As shown in figure 1.2d, the Elekta (previous vendor Philips) linacs use a slalom traveling waveguide. This consists of three separate magnets mounted at various locations along the guide. The final bending magnet bends electrons slightly more than 90 degrees toward the patient.

A detailed schematic showing a generic linac waveguide and beam path with beam monitoring and collimation systems is shown in figure 1.3. This figure shows the main components of the linac that are housed in the gantry represented in a two-dimensional schematic.

The linac accelerates electrons linearly to high velocity and energy using high-power microwaves. The heart of the linac is the accelerating waveguide, where acceleration of electrons occurs. The microwave energy required to accelerate the electrons is delivered to the accelerating structure in the form of short duration pulses (5 microseconds in duration) from the klystron or magnetron via a rectangular transmission microwave waveguide.

An electron gun injects low-energy electrons at one end of the accelerating waveguide. The timing of the electron injections into the guide is controlled by a gun driver system. The electrons are accelerated along the guide to speeds approaching the speed of light. The electrons then enter a bending magnet assembly whereby they are redirected toward the center of the beam's gantry axis of rotation, known as isocenter.

When a linac is operated in x-ray mode, the electron beam hits a target and x-rays are produced by the bremsstrahlung production process. When the linac is used in electron mode, the electrons generally strike a scattering foil prior to exiting the linac.

There are other important component systems and auxiliary systems that are critical to linac function, shown in schematic form in figure 1.4. The auxiliary support systems of the linac consist of (Greene 1986):

i) A water cooling system to regulate the temperature of linac components, e.g., the bending magnet.

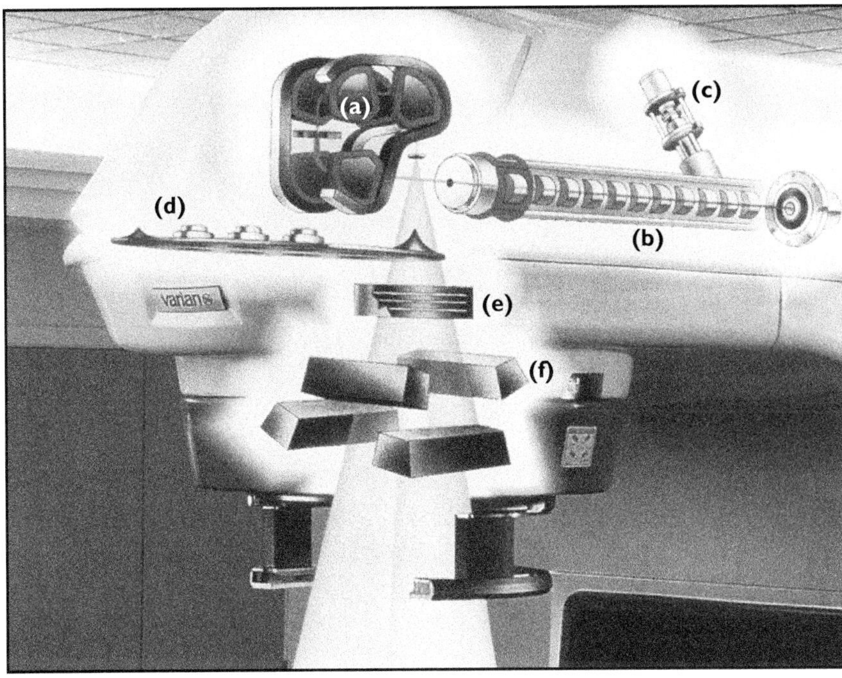

(a)

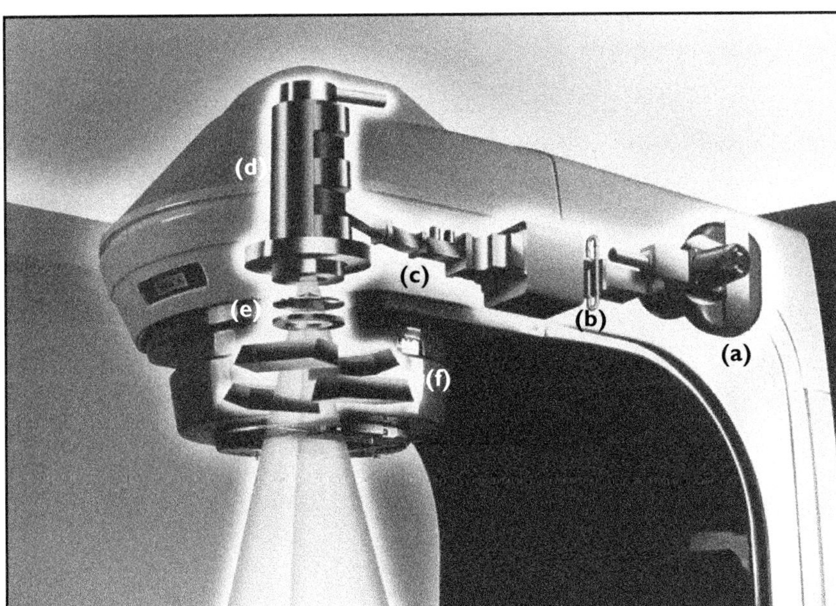

(b)

Figure 1.2: Different linac beam delivery systems. (a) Standing wave design for Varian 2100C, 2300C, and the newer 21eX and 21iX linacs. a = bending magnet, b = accelerating waveguide, c = energy switch, d = carousel, e = primary collimators, f = jaws. Note that this machine is klystron driven but the klystron is not shown in this schematic. (Reprinted with permission from: © Varian Medical Systems, Inc.) (b) Standing wave vertical guide design for a Varian 600C linac. a = magnetron, b = circulator, c = transmission waveguide, d = vertical accelerating waveguide, e = target, f = jaws. (Reprinted with permission from: © Varian Medical Systems, Inc.)

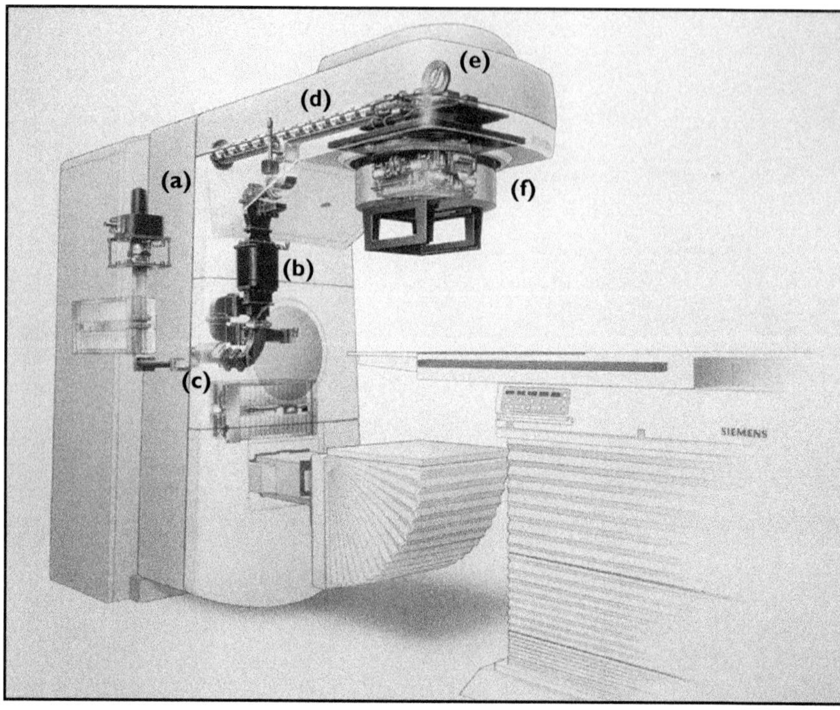

(c)

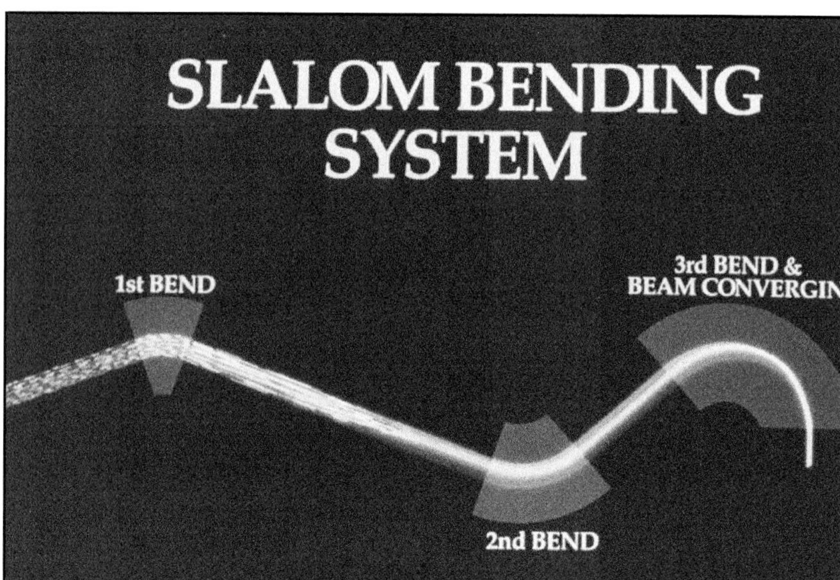

(d)

Figure 1.2: (Continued) Different linac beam delivery systems: (c) Standing wave design for the Siemens Mevatron family of linacs. This is the Siemens M type with the magnetron in the stand. The KD types use the same gantry, but the klystron is mounted in a room remote from the stand and is connected via a transmission waveguide. a = magnetron, b = circulator, c = transmission guide, d = accelerating waveguide, e = bending magnet, f = collimator assembly. (Reprinted with permission from: © Siemens Medical Systems, Oncology Care Systems Group.) (d) Traveling wave slalom guide design of Philips SL linacs. This design enables the height of the machine to be lower due to the final 90 degree bending magnet; the isocentric height is 125 cm from the floor. (Reprinted with permission from: © Philips Medical Systems (now Elekta.)

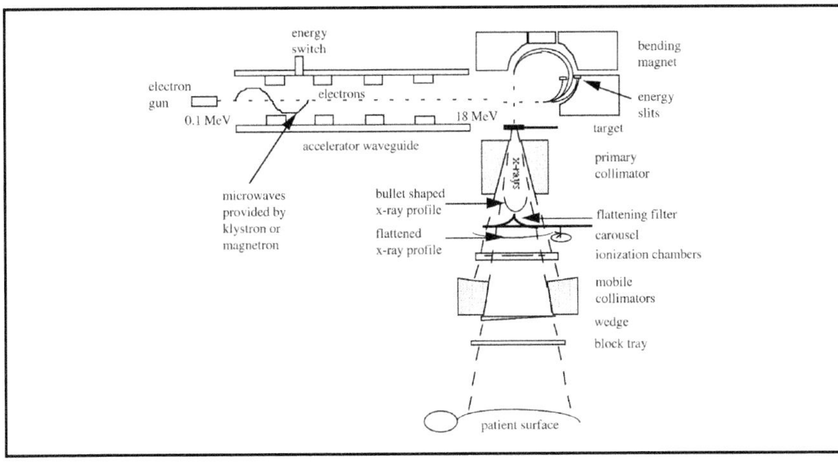

Figure 1.3: Linac beam delivery system in two dimensions including gun, guide, bending magnet, target, flattening filter, and collimation systems.

ii) A vacuum ion pump system to provide vacuum for the accelerating waveguide.
iii) An air pressure system for pneumatic drives, e.g., the target is pneumatically driven.
iv) A gas system to improve the dielectric strength of the transmission waveguide.

Description in this text is intended to give the reader an understanding of the principles of operation of the main components of the linac. Detailed descriptions of technical design variations of components such as waveguide structures and bending magnets can be obtained elsewhere (Greening 1981;

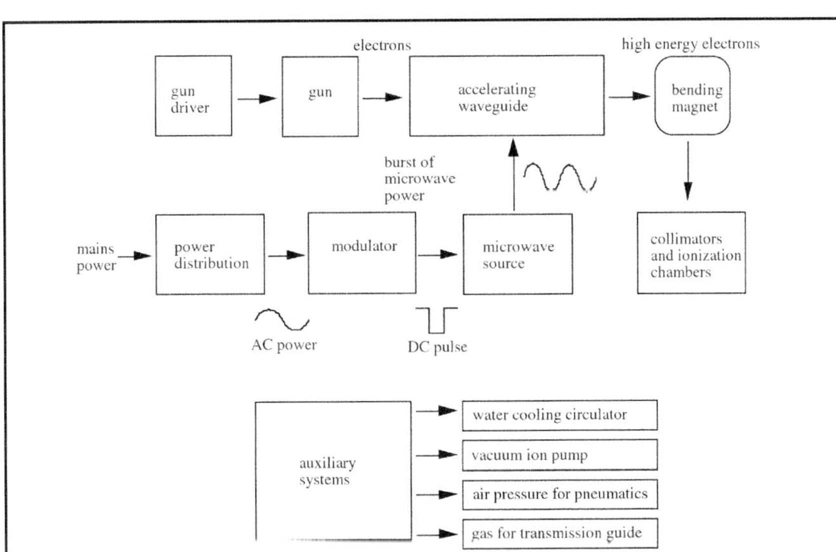

Figure 1.4: Block diagram showing linac support and auxiliary system.

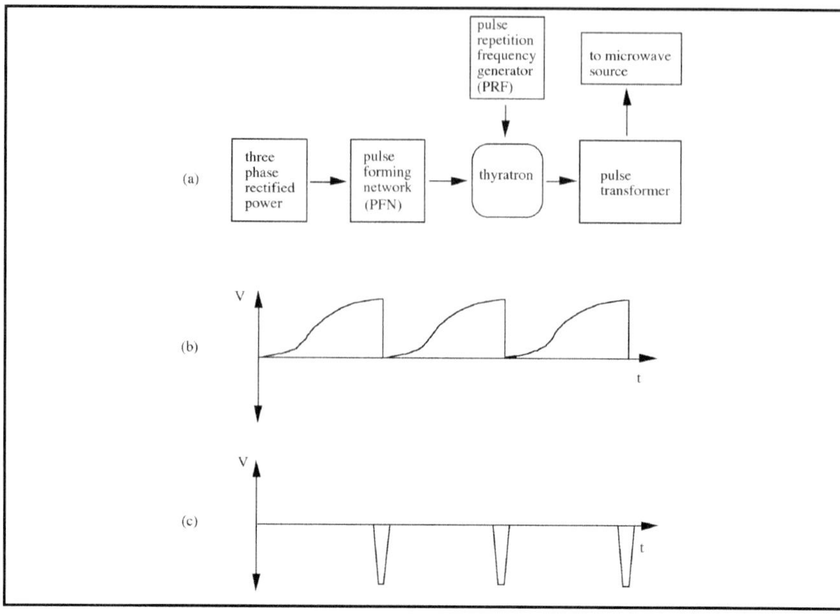

Figure 1.5: Modulator—block diagram of main components. (a) Pulse forming network (PFN) and thyratron. (b) PFN pulse voltage waveform as it is charged through diodes and discharged through the thyratron. (c) The negative voltage pulse at the high voltage end of the pulse transformer, this pulse is fed to the microwave generator/amplifier. (Adapted from: Greene 1986.)

Slater 1948; Karzmark 1984, Karzmark et al. 1993, 1989, 1981, 1973; Podgorsak et al. 1999).

1.2.3 The Modulator

The main components of the modulator are shown in figure 1.5a. The modulator circuit supplies high-voltage pulses to the cathode of the microwave generator (e.g., klystron). To do this a three-phase full wave rectifier and solid-state diodes are used to deliver about 10 kV to the *pulse forming network* (PFN). The PFN is an inductor/capacitor diode circuit that charges up and, when the hydrogen filled thyratron is fired, acts like a switch, discharging the voltage from the PFN. The peak current through the thyratron is about 500 amps, hence this is the only viable switching device. The resulting current passes through the primary windings of a pulse transformer. This is an autotransformer with multiple parallel windings; the low-voltage end is connected to earth, and the high-voltage end is connected to the cathode of the microwave generator. The voltage across the PFN follows a charge-hold-discharge cycle as shown in figure 1.5b. The discharge is determined by the PFN, and the frequency is determined by the pulses applied to the thyratron to make it conduct (Greene 1986). The frequency of the thyratron switching is controlled by a pulse repetition frequency (PRF) generator. The microwave source is then provided with short duration pulses from the pulse transformer (see figure 1.5c).

1.3 Klystrons and Magnetrons

Linear accelerators generally employ either a klystron amplifier or a magnetron generator. These are special kinds of electron tubes that supply amplified *microwave* power to the accelerating waveguide structure. Microwaves are similar to radio waves (up to 450 MHz) but have a much higher operating frequency (in the range of 3 GHz to 30 GHz). For linac operation the microwaves that are generally used have a frequency of about 3 billion cycles per second (3000 MHz = 3 GHz). As with all electromagnetic radiation the electric field vector, E, associated with microwaves changes sinusoidally in magnitude and direction with time. The wavelengths of microwaves are in the range of several centimeters. (3 GHz equates to a wavelength of about 10 cm.) Microwave devices use resonant microwave cavities that are generally cylindrical or rectangular in cross section and several cm in diameter and length.

Figure 1.6a is a schematic of a single microwave cavity with a circular hole cut in each end to allow electrons to pass through. This cavity is used in klystron and waveguide operation. The spatial relation between the E field and the magnetic field, H, are shown in the schematic. In this case, the orientation and magnitude of the magnetic field due to the Lorentz force does not significantly affect the motion of the electrons. It is the E field that accelerates the electrons. Electrons are attracted by the E field of opposite sign at each half cycle of the microwave pulse, hence energy is transferred from the E field to the electrons at this time. This occurs in an accelerating waveguide. If the electron already has significant speed and is slowed, some energy is transferred from the electrons to the E field. This occurs in the klystron catcher cavity.

An electron traveling at high velocity in the opposite direction to an electric field can transfer much of its kinetic energy to the E field, and the energy can be expended in the form of amplified microwaves. A two-cavity klystron is shown in figure 1.6b. The two-cavity resonators act as bunching and catching cavities connected by a drift tube. The electrons that are leading are traveling at a slower rate than those following; hence while traveling along the drift tube, the electrons are progressively bunched en route to the catcher cavity.

Electrons are accelerated from the cathode in the first buncher cavity. The buncher cavity is energized with low-power microwaves from a small microwave oscillator. This has the effect of setting up alternating E fields between the cavity walls. Note that the E field in the buncher cavity varies with time, as shown in figure 1.6c. Electrons in position between a and b are slowed down as they encounter a retarding E field, while those between b and c are sped up by an accelerating E field. The process is known as velocity modulation, and it causes bunches of electrons to form.

A catcher cavity is resonant at the arrival frequency of the bunches. As the electrons traverse the catcher cavity they encounter a retarding E field and are slowed. During this process much of the kinetic energy lost to the electron is gained by the E field and, therefore, the microwaves.

The low-power microwaves that are fed into the input of the klystron are usually provided by a solid state oscillator at about 50 watts. This is

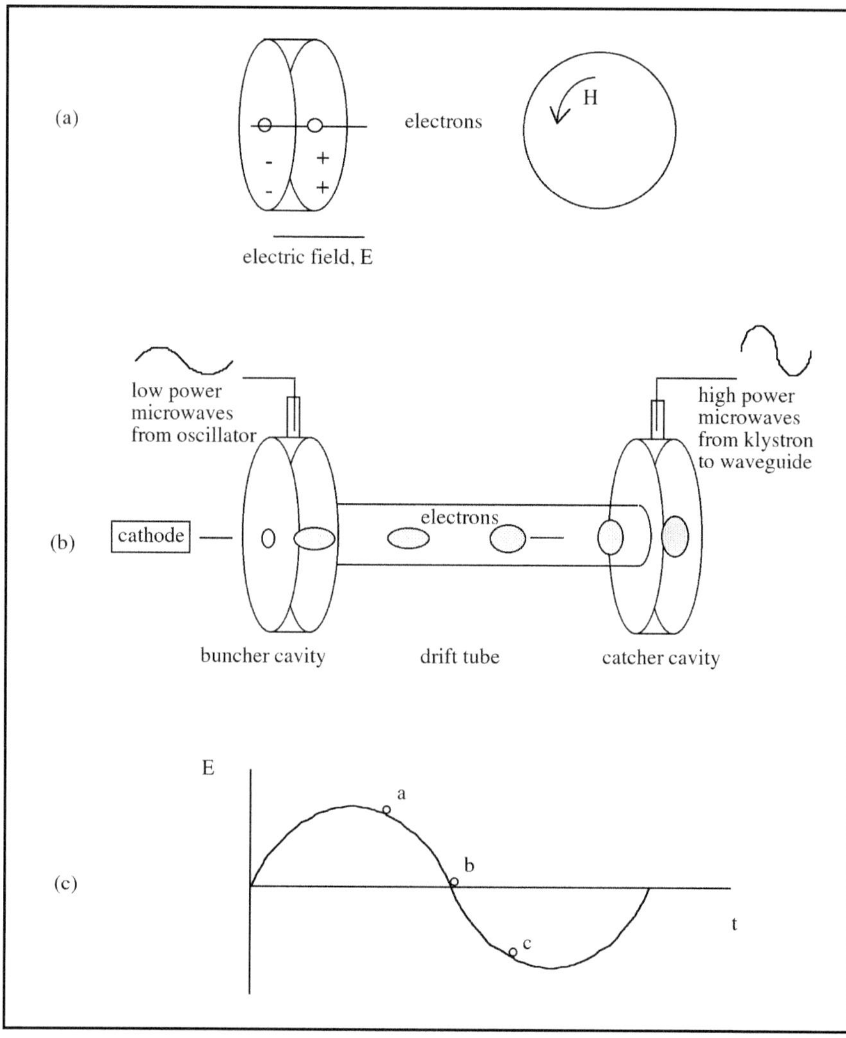

Figure 1.6: Microwave cavities and klystrons. (a) The electric and magnetic field orientations at an instant of time in a microwave cavity. (b) Schematic of a two-cavity klystron. (c) Timing diagram for the E field in the buncher cavity.

amplified by the klystron to an output in the order of megawatts, which is used to energize the accelerating waveguide. Klystrons used in linacs usually have five cavities to improve high-current bunching and increase amplification. An amplification of microwave power in the order of 10^{11} is typically achieved by this process. Typically 7 megawatts of peak power are produced by klystrons.

An alternative way to generate microwaves of sufficient energy is with a magnetron. These are generally used to power lower energy linacs, but this is not always the case. The magnetron is generally physically smaller and lighter than the klystron, hence klystrons are normally mounted in or behind the stand, whereas magnetrons can be mounted in the gantry. As

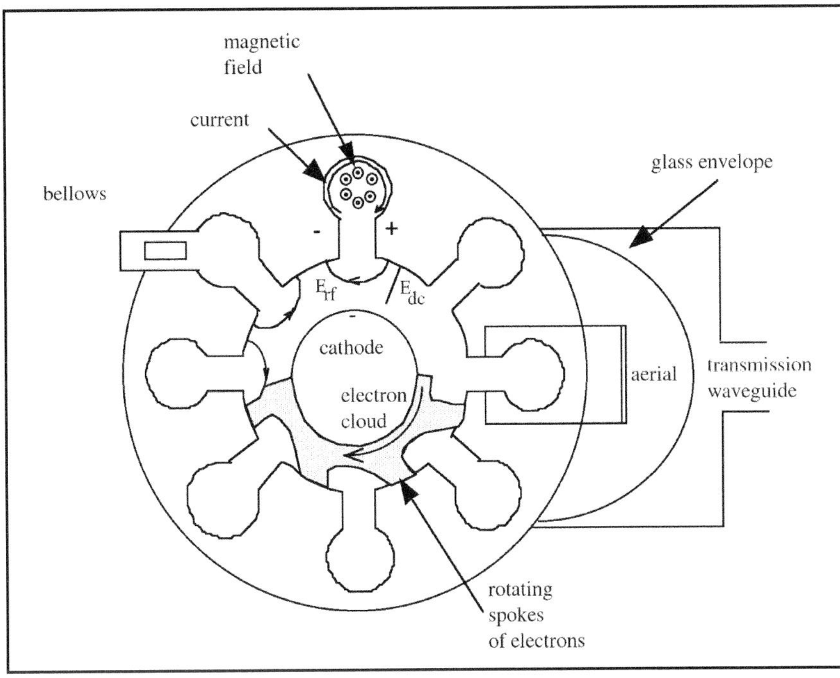

Figure 1.7: A magnetron used to generate microwaves. Electrons move in a coiled path under the influence of two electric fields: a static field E_{dc} and a pulsed electric field E_{rf} applied between anode and cathode. Because the electrons are moving in coiled paths at varying speeds, bunching occurs and microwave energy is given off. Electrons move along the negative direction of the field lines, some being accelerated and some retarded due to their phase relative to the rf field. The electrons cluster in a spoke formation that rotates around the cathode at the same velocity as the rf wave.

shown in figure 1.7, electrons emitted from the cathode are accelerated by a pulsed electric field, E_{rf}, toward the anode. The electrons produce an additional charge distribution that is shown on the anode poles and induces an electric field, E_{dc}, of microwave frequency between each segment of the anode. The electrons travel in spirals under the combined influence of the two E fields and the accompanying H field components. Most of the electron path deflection is due to the H fields. In the process the electrons are slowed, and most of the electron beam energy is converted to microwave power. An output aerial is inserted into one of the cavities to couple the microwave power from the magnetron to the waveguide.

Magnetrons generally supply about 2 megawatts of peak power. They are referred to as high-power oscillators and therefore originators of microwave power. Klystrons require a low-power oscillator to get them started, after which feedback may be employed. This is why klystrons are generally referred to as microwave amplifiers.

As the magnetron is a self oscillator, it is tuned to the frequency of the accelerator structure by a small rf feedback signal from the electromagnetic field in the waveguide, which activates a motorized plunger in the magnetron cavity array. The feedback mechanism is known as the automatic frequency control (AFC).

Microwave power is conveyed from the microwave source (klystron or magnetron) through a transmission waveguide (not an accelerating waveguide), which is filled with pressurized gas (sulfur hexafluoride, SF_6, to prevent arcing), to the accelerating waveguide structure. Two microwave windows (ceramic or beryllium) that are transparent to microwaves separate the pressurized gas from the vacuum maintained in the klystron/magnetron and the vacuum maintained in the accelerating waveguide.

1.4 Electron Gun

The electron gun is the source of electrons for the accelerating waveguide. It consists of an anode and cathode and may also contain a gun grid. In its simplest form the gun consists of a directly heated cathode in the form of a spiral tungsten filament that boils off electrons at a high temperature. The cathode is surrounded by a cone-shaped focusing electrode. Electrons are accelerated to the anode and escape through a port into the accelerator guide. The gun is usually pulsed by the same pulses that drive the microwave source.

In many modern linacs the reliability of the gun is enhanced by using an indirectly heated cathode known as a dispenser cathode. These are usually made from sintered tungsten impregnated with barium oxide to enhance electron emission (see figure 1.8). Because of the enhanced emission this type of cathode can be operated at lower voltage (nominal 10 kV, versus 45 kV). During gun operation the cathode is heated to temperatures of between 1100 and 1200 degrees Celsius. Gun lifetime is important because replacing the gun usually requires the linac to be out of operation for several hours while the vacuum system is reconditioned. Some designs even require the gun and guide to be replaced at the same time.

Voltages of between –150 and +180 are applied to a small grid in front of the cathode to control gun current; the more positive the voltage on the grid with respect to the cathode, the higher the gun current. The gun grid

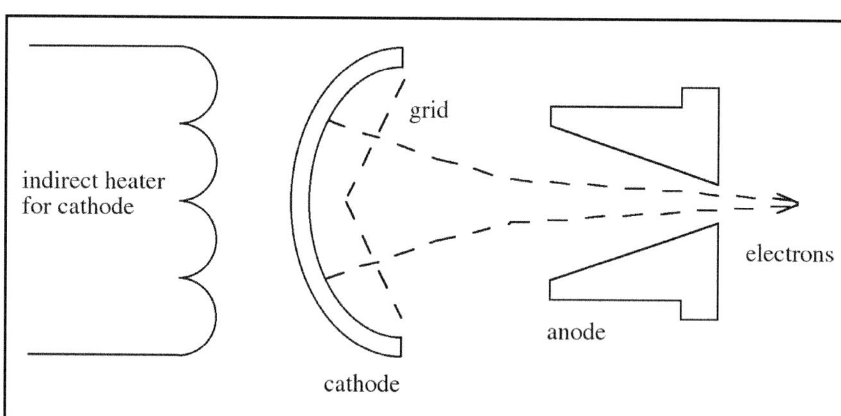

Figure 1.8: A grided dispenser type gun commonly used in modern linacs.

controls the flow of electrons from the cathode to the anode by altering the amplitude of the grid pulses. This is usually called gun current adjustment. If the grid is set at –150 V there is no electron flow and therefore no gun current. Electrons are injected simultaneously with the high-power microwaves into the accelerating guide.

The electron gun is controlled by a gun driver subsystem. Its functions are to:

i) Supply a source of power to heat the gun filament.
ii) Supply a source of high voltage for the gun cathode.
iii) Supply a program pulse of correct phase and wave shape for the gun grid.

1.5 Accelerator Waveguide

In the linac waveguide (which gave the linear accelerator its name), the cavity principle is applied toward an opposite goal than in the klystron, i.e., energy from the cavity's E field is used to accelerate electrons.

A hollow pipe such as the microwave transmission guide would allow microwaves to travel too fast for electrons to catch up. Hence the guide is loaded by introducing cavities. Electrons are captured and bunched on the moving E field by traveling in step with the advancing E field wave.

The first few cavities increase in length, or pitch, to give the microwaves a phase velocity that matches the initial low velocity of the injected electrons. This also has the effect of bunching the electrons and is called the bunching section of the guide. The later cavities are of uniform pitch. Electrons quickly gain energy and reach a velocity close to the speed of light in the first few cavities. From then on, the electrons continue to gain energy according to Einstein's special theory of relativity, but they only gain a little extra velocity because most of the energy gain goes into increasing their mass.

Accelerating guides are predominantly made from copper because this material has a high electrical conductivity at microwave frequencies. This results in low power loss when microwaves are reflected off the walls of the cavities. The accelerating guide is maintained at high vacuum, typically 10^{-6} torr (1 torr = 133.3 pascal), to prevent electron loss and arcing.

To ensure that electrons are retained in focus, direction, and position within the guide, a series of buncher steering coils and focusing solenoid coils are housed around the guides. For further details see Karzmark et al. (1993).

Accelerator waveguides are of two types: the traveling waveguide type and the standing waveguide type. The principles of operation of the traveling waveguides and standing waveguides are explained as follows.

1.5.1 Traveling Waveguide Accelerator

The microwave power for a traveling waveguide is fed in at the proximal (gun) end of the guide, and any remaining microwave power at the other end of the guide is absorbed by a microwave power load; none of the power is reflected.

Electrons injected by the electron gun are captured and bunched on the traveling E field, and they gain energy by traveling in phase with the advancing E field. Figure 1.9 shows a typical traveling wave E field moving to the right in the accelerating waveguide at three sequential instants of time t_1, t_2, and t_3. These time sequences are separated by one quarter cycle of the microwave pulse in time.

Figure 1.10 shows the spatial relation of the traveling wave E field at one instant in time. The electron bunches as shown are pushed by the negative charges and pulled by the positive charges along the waveguide.

There are four cavities per microwave wavelength, and each cavity is several centimeters long, which requires a rather long guide. In general,

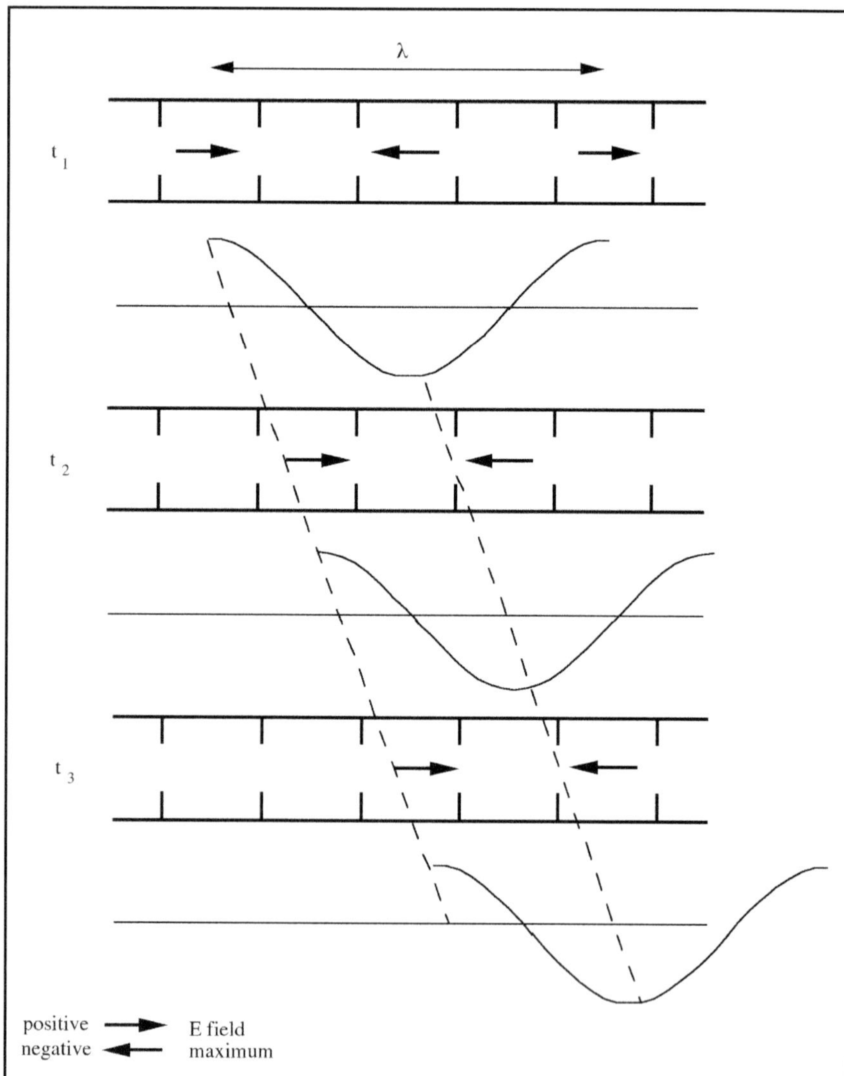

Figure 1.9: Traveling wave E field at times t_1, t_2, t_3. (Reprinted with permission from: Karzmark and Morton 1981.)

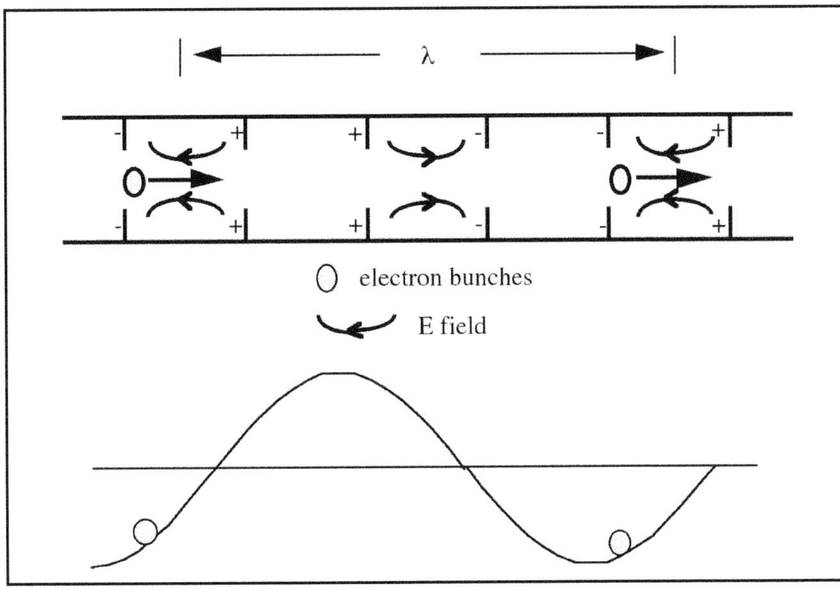

Figure 1.10: Charge distribution in a traveling wave accelerator in which electrons are accelerated. (Reprinted with permission from: Karzmark and Morton 1981.)

traveling waveguides are longer than standing waveguides that generate electrons of the same energy.

1.5.2 Standing Waveguide Accelerator

In contrast to traveling waveguide accelerators, the microwave power in the standing waveguide accelerator is not absorbed at the end of the waveguide but is instead reflected back down the guide. This gives rise to standing waves that do not travel down the guide but oscillate in magnitude with time. Standing waveguides require a device that deals with unwanted reflected power in the guide. Microwave power is fed into the guide (at any location along its length) via a device called a circulator (usually three or four ports). The circulator is a directional coupler designed to introduce the power at resonant frequency and also to act as a channel for the rejected energy via one of its ports to a water load.

Shown in figure 1.11 is the E field distribution for a standing wave accelerator. The upper arrow in each cavity represents the direction of the E field in the incident wave, and the lower arrow represents the direction of the E field from the reflected wave. The effective E field in each cavity is the sum of the incident and reflected E field. Note that every second cavity has a forward and backward E field component, which when summed gives a resultant zero E field in the intervening cavity. However, these cavities that contain no E field are essential to couple power between the accelerating cavities.

To reduce the length of the accelerating waveguide these cavities can be moved to the side of the accelerating structure away from the accelerating electron axis, as shown in figure 1.12. Most linac waveguides employ this

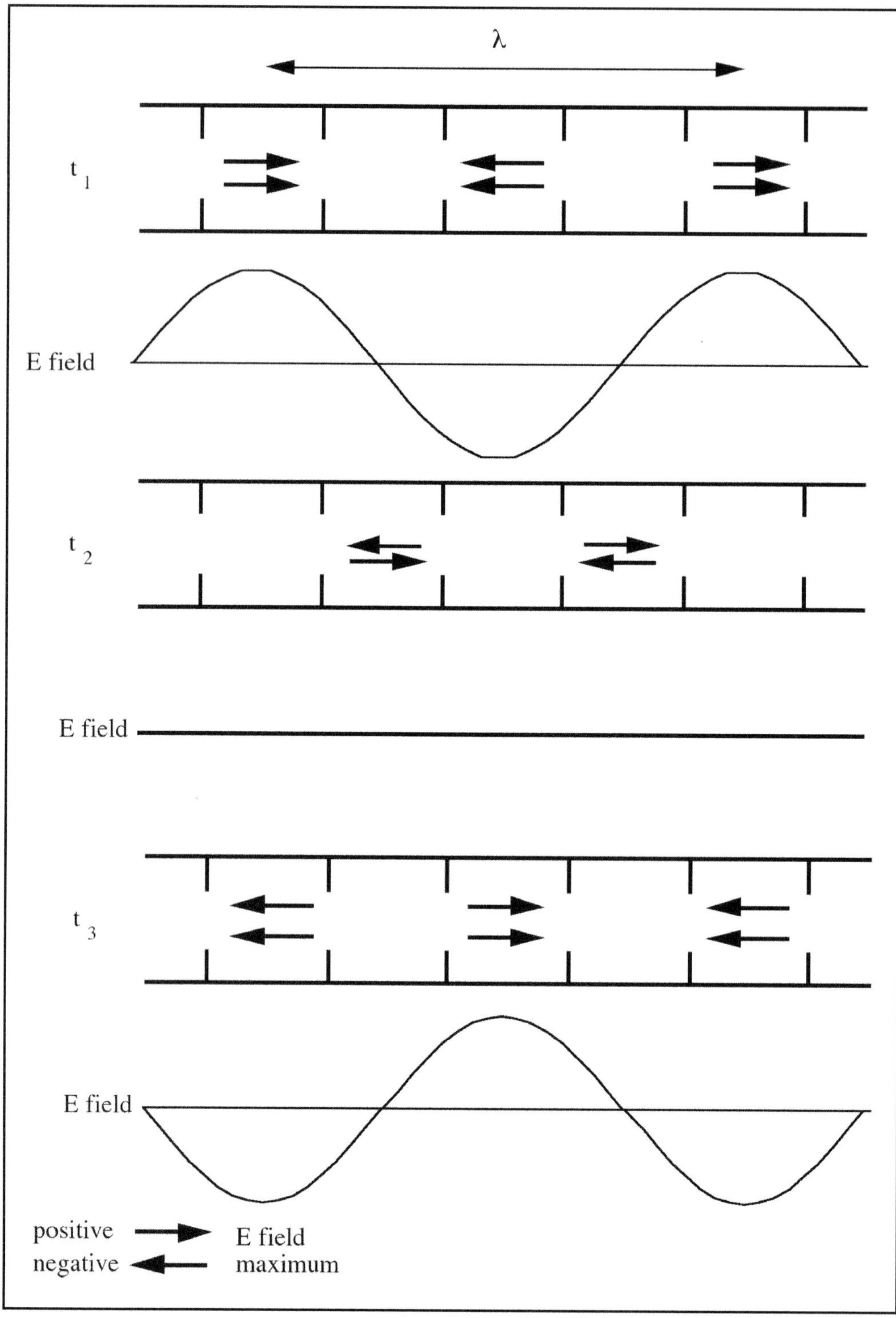

Figure 1.11: Standing waveguide E field distribution at times t_1, t_2, t_3. (Reprinted with permission from: Karzmark and Morton 1981.)

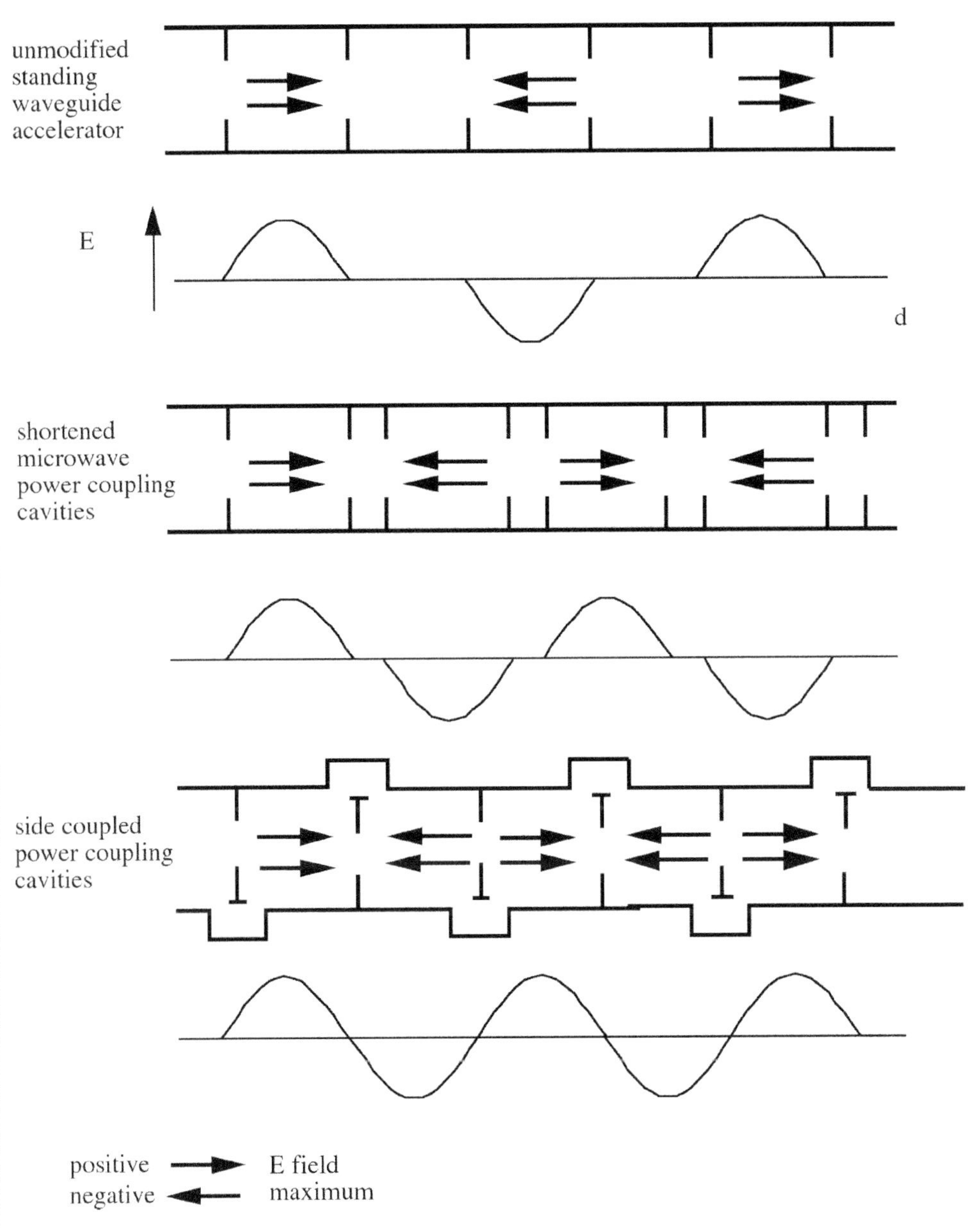

Figure 1.12: The evolution of the side-coupled standing waveguide design. (Reprinted with permission from: Karzmark and Morton 1981.)

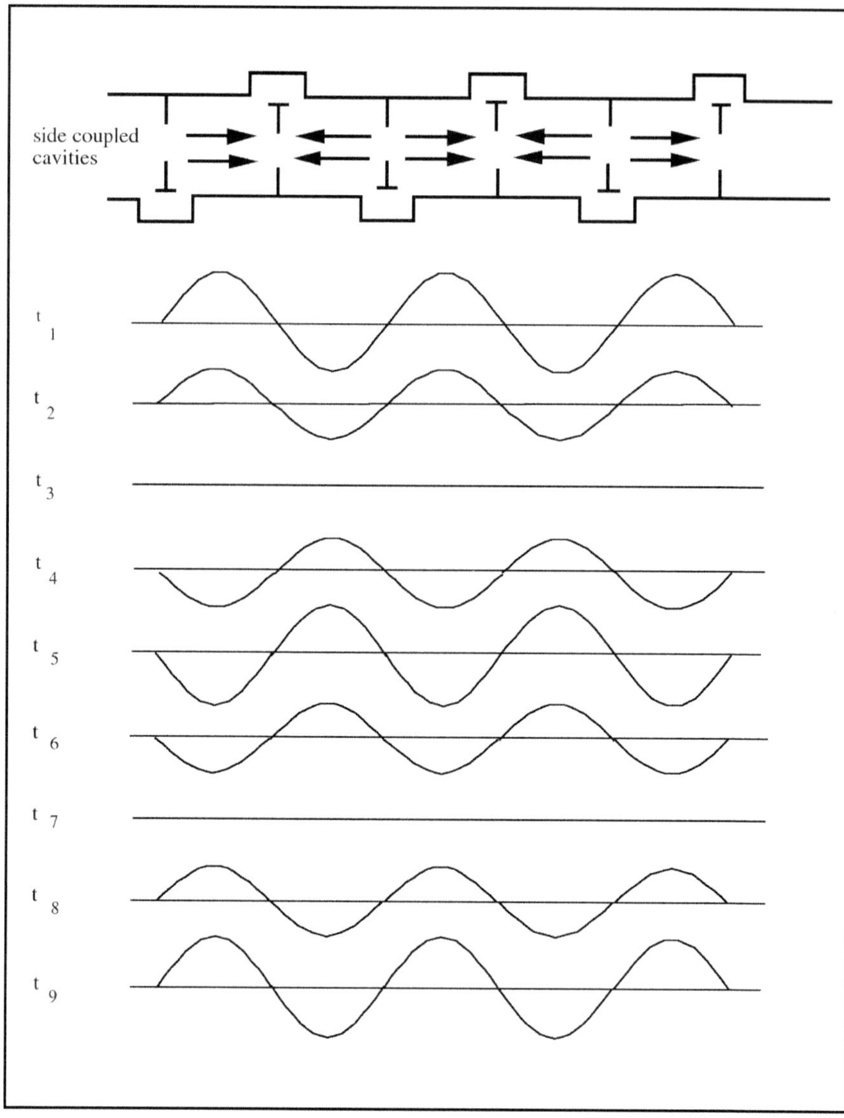

Figure 1.13: Standing wave E field pattern for one full microwave cycle shown at nine discrete time intervals. (Reprinted with permission from: Karzmark and Morton 1981.)

array of side-coupled cavities in their standing waveguide design. Using this design, a stronger electric field is obtained than with a traveling waveguide, and a shorter guide can be used to achieve the same electron energy.

Figure 1.13 shows a sequential look at the E field from times t_1 to t_9 over a complete microwave cycle for a standing wave accelerator. Injected electrons from the gun are captured, bunched, and accelerated in the first few cavities. They then pass through the next cavities in a negative E field and are accelerated. At this time the next cavity has a positive E field and no electrons can be accelerated in it, which doesn't matter because at this time

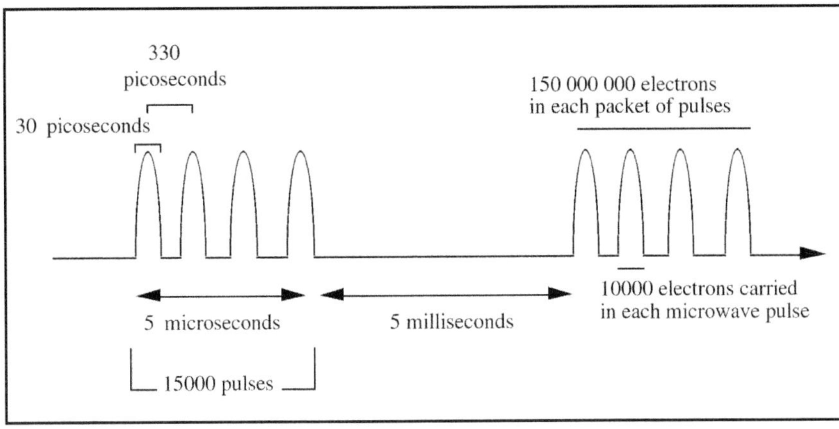

Figure 1.14: Typical microwave pulse sequences involved in linac operation. (Reprinted with permission from: Krieger and Petzold 1989.)

the electrons have not reached this cavity. As the electrons cross into the next cavity the E field in this cavity starts its negative excursion, and the electrons are further accelerated.

Figure 1.14 shows a typical pulse sequence and the number of electrons carried during each pulse sequence (Krieger et al. 1989). Note that there are actually thousands of pulses in each sequence. The microwave frequency is 3 GHz with a pulse micro spacing of 330 ps and a micro pulse length of 30 ps. The magnetron or klystron receives 5 ms pulses from the PFN. The repetition frequency is usually about 200 Hz for 18 MV x-ray production and 400 Hz for 6 MV x-ray production.

Example 1.1

Question: A linac is delivering a dose rate of 2 Gy/minute to a calibration point within a water medium. The radiation is in 5 microsecond pulses (assuming the pulses are continuous), and the pulse period is 5 milliseconds. Calculate the instantaneous dose rate during a pulse.

Answer: The dose rate is 2 Gy/minute, which is 2/60 Gy/second. The pulses have a 5 millisecond period, which means there are 200 pulses per second.

$$\text{Therefore the dose per pulse} = \frac{2}{60 \times 200} = 1.67 \times 10^{-4}\,\text{Gy}.$$

Therefore the dose rate in the pulse per minute equals the above value multiplied by 60 to get to dose per minute and divided by 5×10^{-6} as this is the pulse period. This gives an instantaneous dose rate within the pulse of 2000 Gy/minute.

Table 1.1: Features of traveling and standing wave accelerator structures used for linacs

Parameter	*Traveling Wave*	*Standing Wave*
electron acceptance	high	low
electric field strength	low	high
length of guide	longer	shorter
microwave power feed	at gun end	anywhere
acceleration	continuous	pulsed
frequency dependence	low	high
phase dependence	low	high
guide loading	low	high
vacuum requirements	lower	higher

1.5.3 Energy Selection

In electron mode the linac can be tuned for different electron energies by adjusting the microwave power in the waveguide and the electron current. Increasing the beam current progressively extracts an increasing fraction of the rf power from the guide; this is known as beam loading. When the gun current levels are low for electron beams, the amount of rf power being extracted is small, which means the load line is reasonably flat. Tuning for different electron energies using this method is effective.

However, for x-ray production to ensure sufficient bremsstrahlung production through the target, the gun current is much higher. This is typically 600 mA for 6 MV x-rays versus 150 mA for 18 MeV electrons. At these levels of gun current the load line is very steep, and retuning for different energies is more challenging for standing waveguide machines; a number of methods for creating different energies are discussed elsewhere (Karzmark et al. 1993). One practical method for producing low- and high-energy x-rays in a standing waveguide that does not compromise dose rate is the use of an energy switch. The switch drives into part of the accelerator structure, decoupling the second part of the waveguide so that only the first part is used. The electrons then reach the end of the guide without further acceleration. The whole length of the guide can be used to produce high-energy x-rays. In traveling waveguides the second portion of the rf field can be tapered down to produce low-energy x-rays (Karzmark et al. 1993). Table 1.1 is a summary of the different properties of traveling and standing waveguide designs.

1.5.4 Electron Energy

When electrons are accelerated to relativistic velocities, the mass, m, and energy, E, are related as postulated by Einstein's theory of relativity (1905), such that

$$E = mc^2 \tag{1.1}$$

where c is the speed of light.

Hence the kinetic energy, E_k, given to an electron accelerated in a linear accelerator waveguide is

$$E_k = mc^2 - m_0c^2 \tag{1.2}$$

where m_0 is the electron rest mass (9.11×10^{-31} kg). By rearranging equation 1.2 then

$$m = m_0 + \frac{E_k}{c^2}. \tag{1.3}$$

The particle's mass, m, is related to m_0 by

$$m = \frac{m_0}{\left(1 - \frac{v^2}{c^2}\right)^{\frac{1}{2}}}, \tag{1.4}$$

where v is the velocity of the particle. By rearranging equation 1.4 then

$$v = c\left(1 - \frac{m_0^2}{m^2}\right)^{\frac{1}{2}}. \tag{1.5}$$

Example 1.2

To acquire a feeling for the quantities involved when dealing with linear accelerators, let us apply the above equations to a typical electron energy. By applying equation 1.3 to a 6 MeV electron beam (1 MeV = 1.602×10^{-13} joules), the result is that the electron acquires a mass $m = 12.7m_0$. By applying equation 1.5, the velocity of the electron v = 0.994c, which is close to the velocity of light.

1.6 Beam Delivery

1.6.1 Bending Magnet Assembly

For low-energy standing wave linacs (generally up to about 6 MV) the waveguide is short enough (about 0.4 meters) to be mounted vertically in the gantry while still enabling full 360 degree gantry rotation with only a small floor inset to ensure a low couch position at isocenter. For these short waveguide linacs, electrons are accelerated toward the patient at isocenter with no further major beam steering required.

However, for virtually all traveling wave accelerators and for higher energy standing wave accelerators (those that produce x-ray energies greater than 6 MV), the guide is longer (e.g., an 18 MV standing waveguide is 1.5 meters long). This has to be mounted horizontally within the gantry. In this orientation, the accelerated electrons at the end of the guide have to be bent toward the patient.

In theory, only a 90 degree bend is required. Some linacs employ the 90 degree design as it produces a more compact bending magnet and generally the machine can be mounted at a lower height. This makes it more convenient for operators to place accessories on the machine.

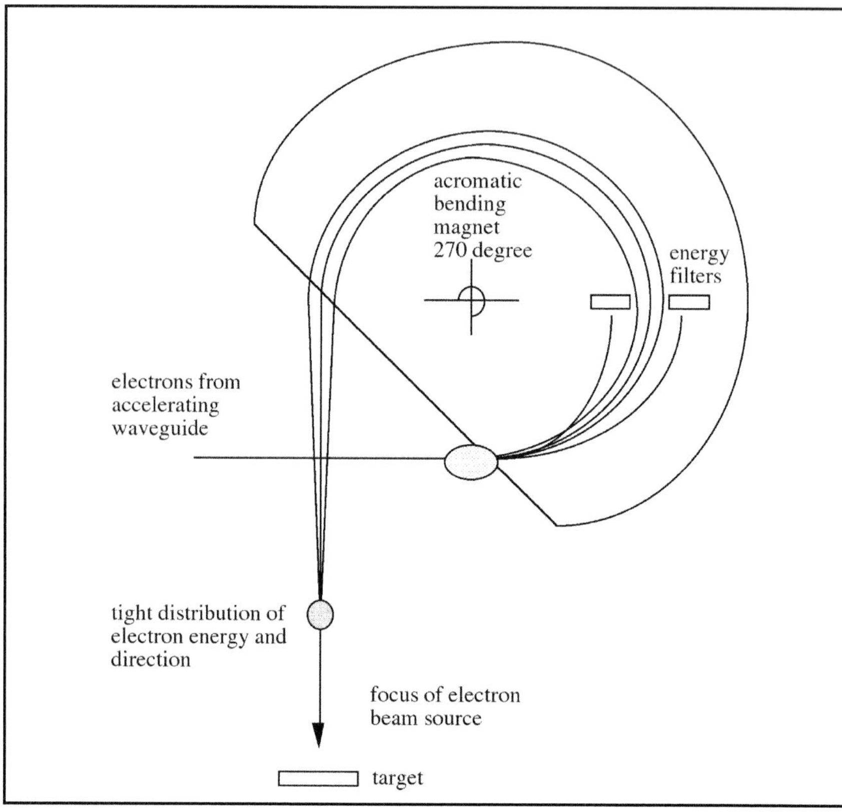

Figure 1.15: Linac bending magnet and energy slits showing a 270 degree bending magnet.

In practice, however, many linacs use a 270 degree bending magnet, as shown in figure 1.15. This is because the electrons get a long exposure to the magnetic field; at this time electron spectral spread or directional spread can be refocused by the bending magnet. The magnet is said to be achromatic because different energy electrons are drawn through a different path length and refocused at the desired position. This focus of electrons is needed to maintain a small virtual beam source size. Most of these magnet designs are fitted with an energy slit that removes electrons that are not within 5% of the nominal peak accelerated electron energy. The radius of curvature of less or more energetic electrons is different; therefore, the energy slit acts as a physical barrier and removes them from the beam path. In practice three 90 degree magnets are usually used, connected by short drift tubes. This in effect reduces the height of the machine isocenter because the target can be mounted at a location closer to where the electrons enter the bending magnet.

The Elekta (previous vendor Philips) traveling wave accelerators use magnets to deflect electrons by small angles within the guide with the final deflection being slightly more than 90 degrees (see figure 1.2d). This is known as a slalom waveguide and has three perceived advantages: it reduces the length of the guide, tightens the energy spectrum of the electrons, and lowers the machine height (Karzmark et al. 1993).

1.6.2 Target and Flattening Filter

The narrow electron beam is converted to a broad x-ray beam by bremsstrahlung production in a target (usually made of copper). The target may be a pneumatically driven device. Different thicknesses of copper are used. Lower energy photons require less thickness of copper, as shown in figure 1.16a where the deeper recess is for the lower energy x-ray target and the shallow recess is for the higher energy target. The high-energy x-rays emerging from the target are forward peaked in a bullet-shaped lobe that is further collimated by a fixed primary collimator.

To create a uniform beam, the x-ray dose profile is flattened by a flattening filter that is shaped like a cone with the point of the cone facing the target. This is made of tungsten or steel or a lead/steel combination. Two flattening filters are shown in figure 1.16b.

(a)

Figure 1.16: Photographs of components within a Varian linear accelerator treatment head. (a) Target from a linear accelerator with two different thicknesses of material for bremsstrahlung production 6 MV (left) and 10 MV (right). The target is water cooled by a thin pipe through the copper taking chilled water. This target resides inside the guide vacuum, hence the seals and bellows. It is pneumatically driven into one of three positions: out (electrons), 6 MV, and 10 MV.

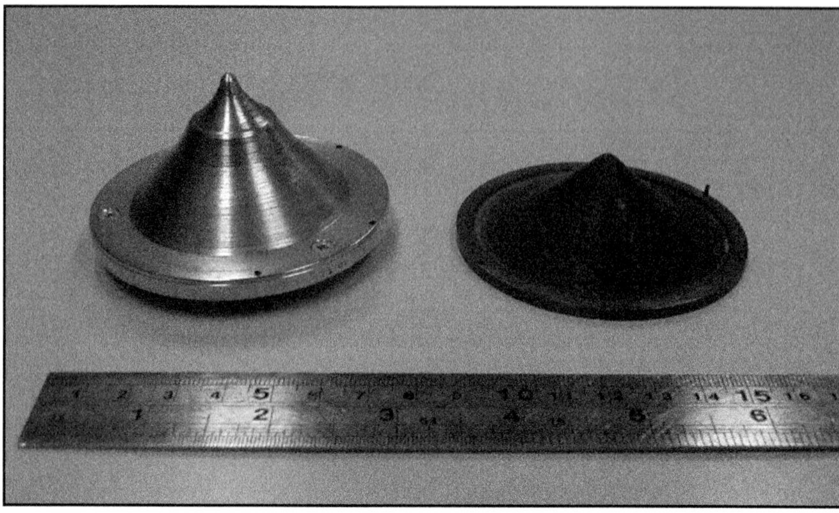

(b)

Figure 1.16: (Continued). Photographs of components within a Varian linear accelerator treatment head: (b) Two flattening filters from a linear accelerator: 18 MV (left) and 6 MV (right). (c) Monitor unit ionization chamber assembly from a linac. (d) Schematic of a monitor unit ionization chamber assemby from a linac. The orientation of the ionization chambers mounted in the treatment head used to monitor dose output, symmetry, and flatness is shown. AB and CD ionization chambers are used for integration of charge to give dose output in monitor units. The A to B and C to D current ratios are compared for feedback to angular beam steering in the radial and transverse planes. The E to F and G to H current ratios are compared for feedback to the position beam steering in the radial and transverse planes.

The cross beam profiles of large fields generally have a slightly higher dose near the edges than in the central axis; these high-dose regions are referred to as dose *horns*. Dose horns arise because the flattening filter is a compromise design to ensure a nearly flat beam profile at all field sizes. The horns contain low-energy components that may be attenuated more rapidly through the medium (Mohan et al. 1985). This is because the beam flattening filter is narrower at the edges than in the beam central axis. The horn effect tends to dissipate with depth, and by about 10 cm depth the horns are generally less than they are near the incident surface.

If electrons, instead of x-rays, are to be used to treat the patient, the target and flattening filters are generally replaced by an electron scattering foil. During a change of mode from x-rays to electrons, the target is retracted, the flattening filter is driven out of the beam and the foil is placed in the beam path, and the beam gun current is also significantly reduced. All these operations are monitored and interlocked to ensure that the correct programming is achieved. The x-ray flattening filter and electron foils usually are mounted below the target on a circular mechanism, which is referred to as the carousel.

The target is usually pneumatically driven into position, and there are several interlocks to ensure the machine can only operate in the selected mode. The target, foils, and flattening filter all have fault-detecting interlocks

(c)

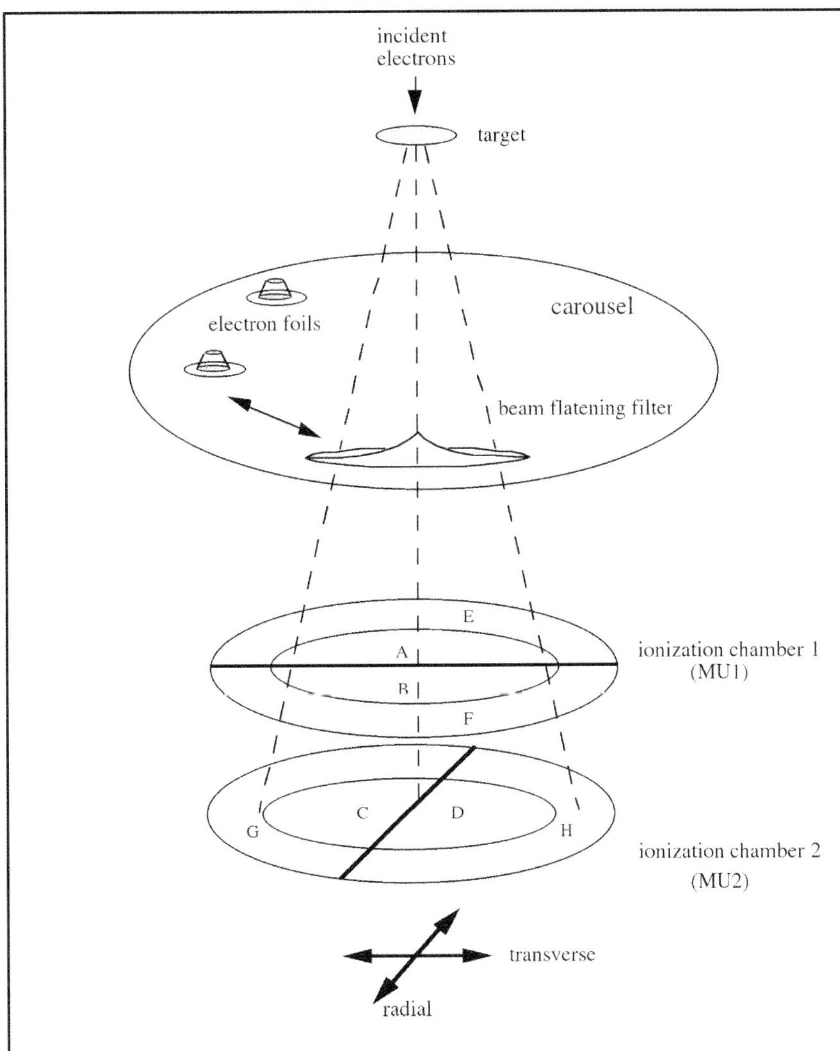
incident
electrons
target
carousel
electron foils
beam flatening filter
E
A
B
F
ionization chamber 1
(MU1)
C
D
G
H
ionization chamber 2
(MU2)
transverse
radial

(d)

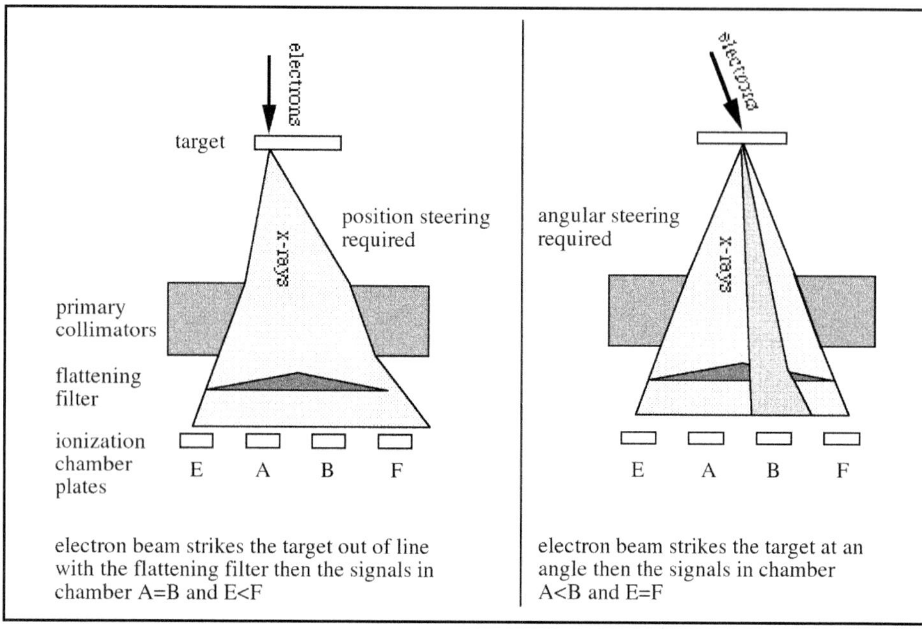

(e)

Figure 1.16: (Continued). (e) The x-rays' incident position and angle onto the target and flattening filter affect the beam symmetry as recorded by the charge collected in the ion chambers. Feedback of the current signal from the monitor ionization chambers to the electronic beam steering control circuit in the bending magnet is used to tune the beam for symmetry.

and back-ups to ensure correct operation. Linacs with vertical guides generally have a fixed target/flattening filter, which reduces the space and hence the height of the machine but precludes the use of electrons.

1.6.3 Monitor Unit Ionization Chambers

A photograph of a monitor unit ionization chamber used in a linac treatment head is shown in figure 1.16c. After the flattening filter, the beam traverses two parallel plate multichannel ionization chambers, as shown schematically in figure 1.16d. These are referred to as monitor unit ionization chambers because the monitor units are accrued using these chambers. The ionization chambers are usually made of Kapton®. Mica was previously used but because of a shortage of this material, Kapton is now more common. The bremsstrahlung contamination from Kapton chambers in electron mode is also less.

The mica chambers were usually filled with nitrogen, whereas the Kapton chambers are filled with oxygen-enriched air. Both these types are sealed to avoid the necessity of correcting for changes of gas density within the volume due to ambient pressure and temperature variation. Some linacs use chambers that are not sealed; these generally have an accurate electronic temperature pressure compensation device.

The monitor unit ionization chamber plates are mounted so that one is rotated 90 degrees on the beam axis from the other. This enables them to also monitor beam symmetry and flatness in the beam's radial and transverse planes, respectively. (Beam flatness and symmetry are defined in chapter four.) Beam flatness and symmetry are controlled by feedback circuits that run from the ion chambers to the bending magnet's beam steering coils. Figure 1.16e shows how the dose to the chamber plates will alter due to positional and angular deviations in beam steering.

The prescribed dose needs to be delivered reproducibly for each patient treatment. To achieve this routinely, one set of the monitor ionization chambers' plates (the inner plates) are used to monitor dose output. The units of dose recorded by these are referred to as monitor units (MU); one MU of dose has been delivered when the monitor chambers have detected a preset dose. Because these chambers are located above the final beam collimation system, an MU setting is calibrated to a standard dose for a standard field setting. For example, 1 MU may be calibrated to equal 1 cGy at 100 cm SSD for a 10×10 cm^2 field size at d_{max} in water. However, 1 MU will equate to a different dose at a different field size. This is why field size-dependent dose output tables are issued for dose calculations (see chapter four).

The monitor chambers are interlocked to stop beam production at an operator preset dose level. As mentioned for the symmetry monitoring, there are two sets of monitor ionization chambers. They are linked to two independent calibration circuits. One is a back-up to ensure that if the primary channel fails the other channel will terminate dose. The back-up is set to terminate dose at a small margin beyond the primary interlock.

Further dose regulation, apart from the monitor ionization chambers, includes a timer that terminates dose at a preset time. This means the treatment will terminate due to the time interlock if both ion chambers fail. Care is taken to ensure that the time is set slightly longer than the time needed to deliver the dose provided the preset dose rate is achieved. If a significantly lower than expected dose rate is being given, the treatment may also terminate on the time interlock first.

Of course the operator can terminate treatment at any time by using the beam-off button on the control console. This termination usually occurs if patient movement is detected within the room via the in-room video monitors. If this occurs the treatment is suspended, and the patient is realigned into the correct position prior to completing delivery of the prescribed dose. A two-way audio monitor is also provided for patient-operator communication.

1.6.4 Electron Beam Delivery

Modern linacs are also capable of producing clinically useful electron beams. In this case the electron beam is not directed onto a target but used directly. Therefore the production efficiency of electrons is much higher than for photons and, in principle, the dose rate for an electron beam can be much higher for electrons than for x-rays. However, in practice this is only used in special procedures such as total body electron skin irradiation (compare chapter 7).

For safety and dosimetric reasons typically the electron gun current is reduced and the power of the magnetron or klystron may also be reduced. This makes electron beams a good choice for linear accelerator warm up procedures such as morning start-ups.

Most linacs producing electron beams employ a bending magnet to ensure a relatively monoenergetic beam. The energy selection slit is typically set to allow electrons with a window of some 3% around the nominal energy through the bending magnet. Modern linacs typically have five or more electron energies available.

After the bending magnet the electron beam has a diameter of only a few millimeters. In order to create a useful clinical beam, the electron beam must be spread out. Two methods have been used in different linacs:

i) A magnetic field can be used to scan the electron beam across the area to be irradiated. This is similar to a television screen and has been used historically in some linac systems, and is currently used in the racetrack microtron. This technique produces a superior beam in terms of energy definition, i.e., the beam spectrum has a very small spread; photon contamination is also reduced as none is introduced by the absent scattering foil. However, the beam steering requires additional quality assurance and the high instantaneous dose rate in the scanned beam results in dosimetric difficulties. Therefore, most commercially available linacs employ a scattering foil.

ii) A scattering foil is employed in most current linac designs; most linacs use a dual scattering foil as shown in figure 1.17a. This foil is for a Varian 18 MeV electron beam. The foils are different for different energies and constitute a compromise between producing a large enough useful beam and limiting both energy degradation and bremsstrahlung production.

Because the electrons have not interacted with a target, the beam spectrum is closer to a monoenergetic beam of the peak energy as shown in figure 1.17b. Electrons as charged particles do interact more with air than photons and this leads to some energy degradation as the electrons traverse away from the scattering foil. There is also an increase in angular scattering. In order to achieve a relatively well-defined beam edge, it is therefore necessary to collimate the beam as closely as possible to the patient. This is achieved using an electron applicator as shown in figure 1.17c. The applicator has several planes in which the beam is collimated. The final collimation plane allows placement of an insert made from low melting alloy (LMA) in the applicator. This insert allows customization of the final collimation to individual patients. More details about electron beams are given in chapter five. Figure 1.17d illustrates schematically the role the applicator and other collimators play in the setup used on a linear accelerator to generate and transfer a useful clinical electron beam to a patient.

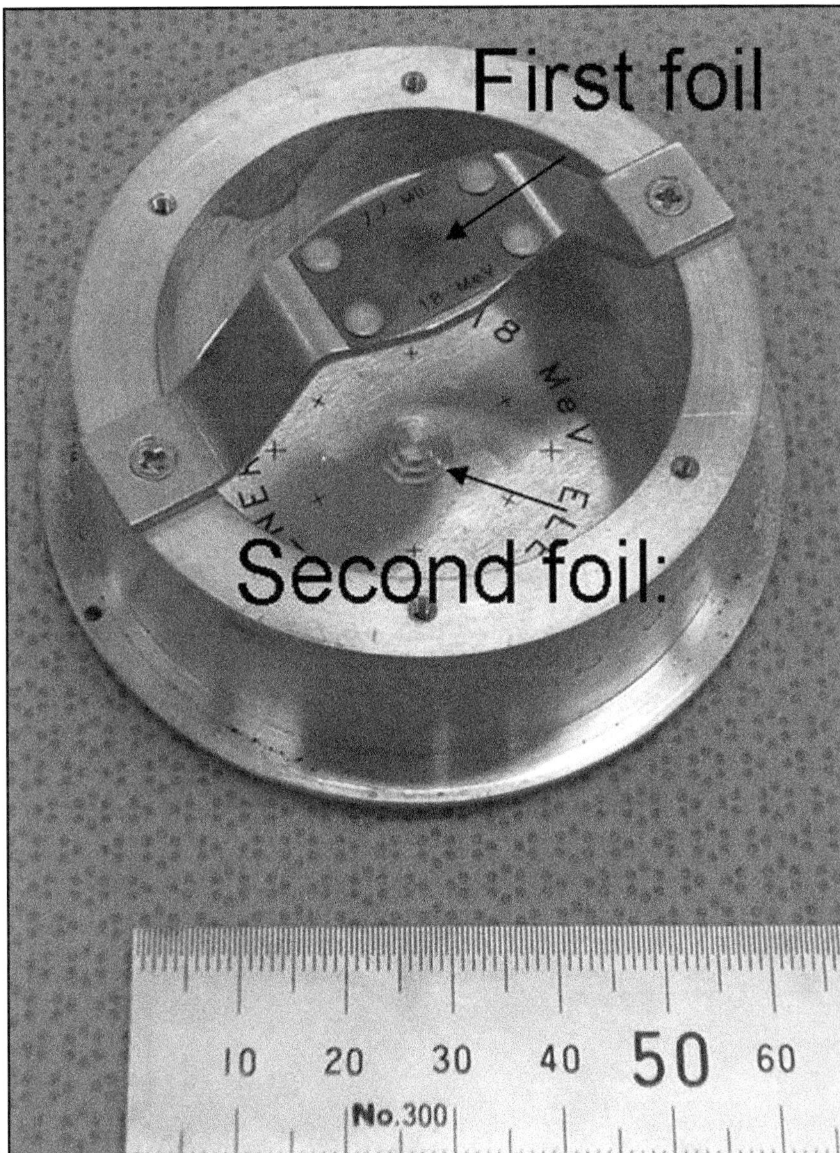

(a)

Figure 1.17: Electron scattering foils, spectral energy spread, and applicators. (a) Two-stage 18 MeV electron scattering foil from a Varian linac.

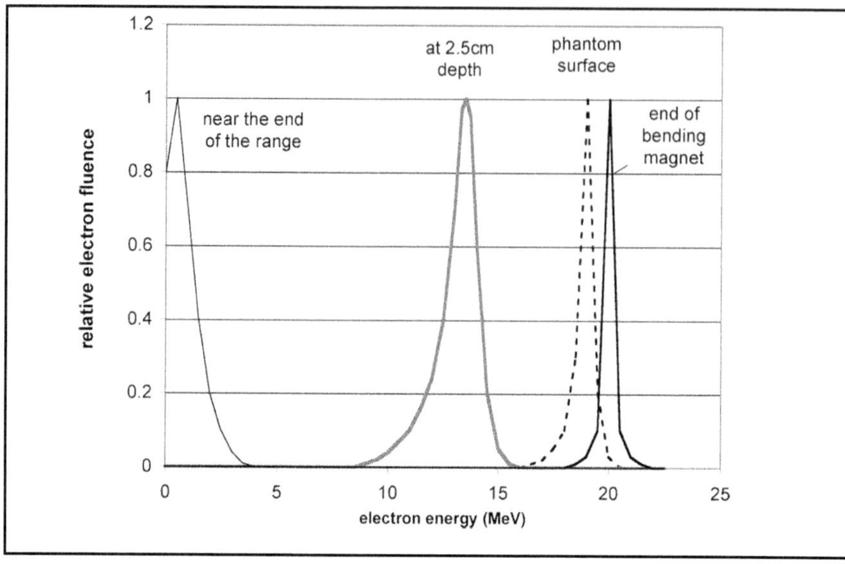

(b)

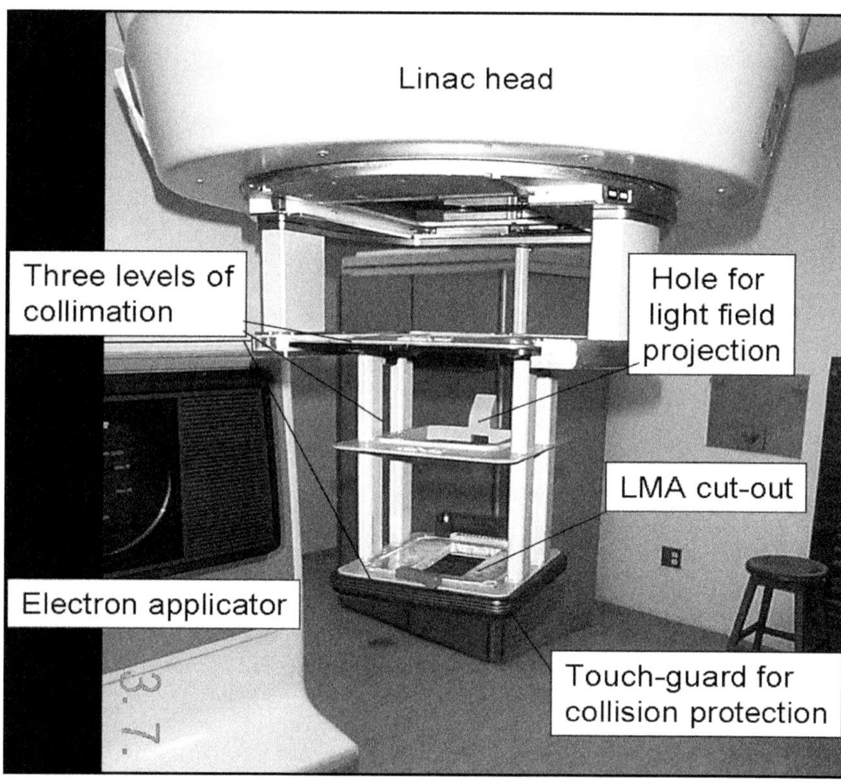

(c)

Figure 1.17: (Continued). Electron scattering foils, spectral energy spread, and applicators: (b) Typical spectral energy spread for an electron beam showing how the energy changes as the electrons move from the scattering foil to the patient surface. (c) Applicator used to improve scattering properties when electrons are employed. This applicator is placed on the block tray holder during electron mode and, when used at 100 cm SSD, the applicator is offset from patient skin surface by 5 cm. An LMA insert is placed at the bottom of the applicator to create irregular-shaped electron fields.

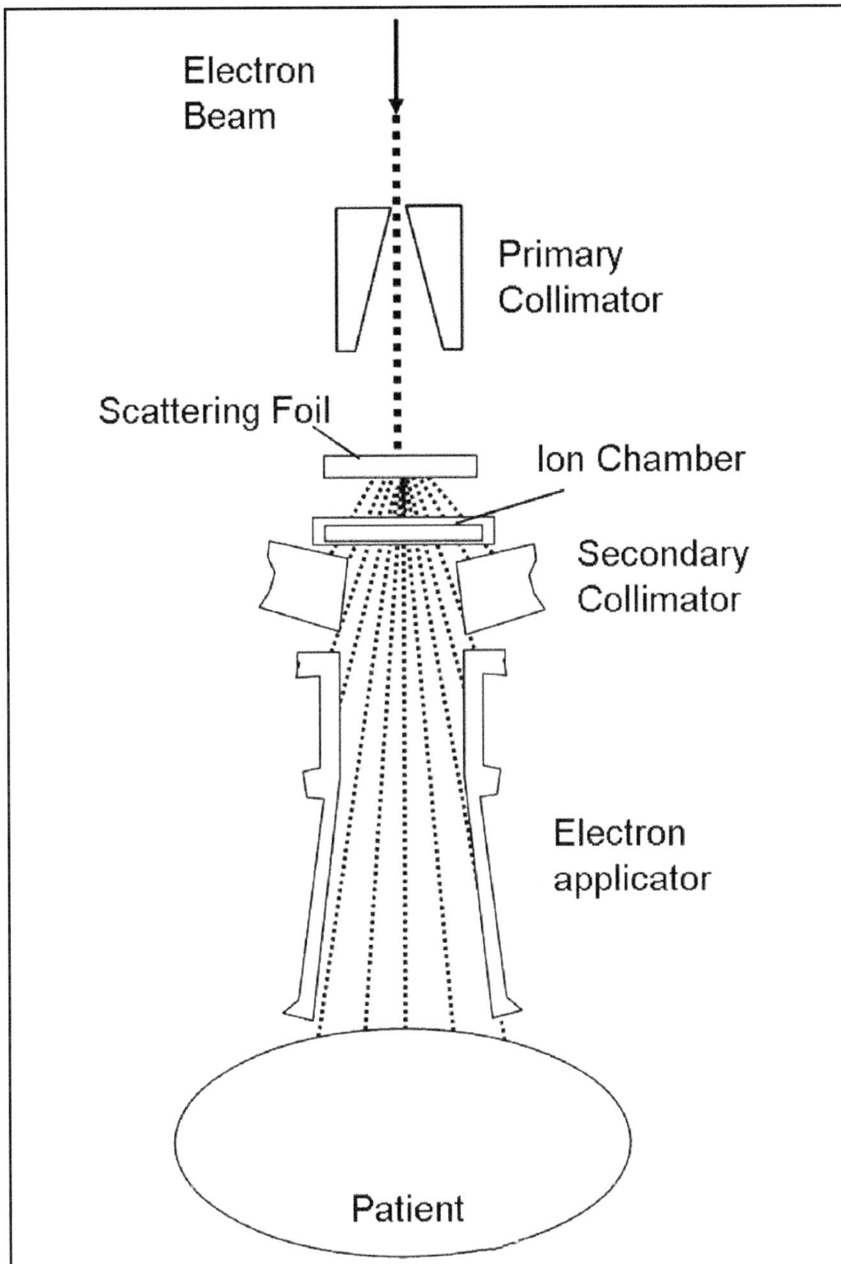

(d)

Figure 1.17: (Continued) Electron scattering foils, spectral energy spread, and applicators: (d) Schematic showing the electron applicator design in relation to the linac head components.

1.7 Collimation

Various devices are used to collimate and modify the intensity of the x-ray beam. These devices are reviewed in the following sections.

1.7.1 Primary Fixed Collimator

A primary fixed collimator (generally made from tungsten) is mounted below the target and above the flattening filter. Shaped like a slightly diverging cone open at both ends, this device allows only forward scattered x-rays to escape the linac (see figure 1.3). This collimator helps prevent head leakage, e.g., scattered photons escaping from the treatment head. The dimensions of this collimator are generally such that a circular beam of approximately 50 cm diameter would be incident at 100 cm SSD if the secondary collimators were not present.

1.7.2 Secondary Collimators (Jaws)

A secondary collimation system consists of two pairs of metal blocks usually made from tungsten or lead alloy of about 8 cm thickness. The common names given to these devices are collimators or jaws. The transmission through the jaws is about 0.4% of dose due to the unshielded incident beam. The collimators can be adjusted to produce different rectangular field definition from 0 cm to 40 cm for each jaw incident at the patient. For reasons of dose accuracy and adequate tumor coverage, fields smaller than 3×3 cm^2 are rarely used, except for some radiosurgical applications. The collimators are also able to rotate around the beam's central axis; this is referred to as collimator rotation.

The collimators are mounted at a source collimator distance (SCD) of about 40 cm. Because of the thickness of the collimators one set is mounted above the other set. This appears to have negligible effect on beam dose profiles. The beam profiles in the radial and transverse planes of a correctly operating linac are usually virtually identical. However, the jaw orientation does have a small effect on beam output (see chapter four for more detail).

The jaws are designed so that they drive at an angle to the beam that closely matches the angle of beam divergence at different field sizes. This ensures that the full jaw face is parallel to the edge of the x-ray beam. If the jaws were not so aligned this would lead to a broader beam penumbra due to partial jaw transmission. A combination of couch, collimator, and gantry angle can provide beam incidence at virtually any required treatment angle.

When used in electron mode, the secondary collimators are usually opened wider than the required field, and final collimation is provided by an electron applicator. This is a demountable device that attaches to an accessory mount and provides collimation at about 5 cm from the patient surface. This is required because electron scattering in air produces a wide beam penumbra if the collimators are further away, and a bremsstrahlung edge is created if placed directly on the skin. If the collimators are set too wide in relation to the electron cone size this can cause leakage through the sides of the cones (Perec et al. 1990). However, new foil and applicator designs on modern linacs have minimized this problem.

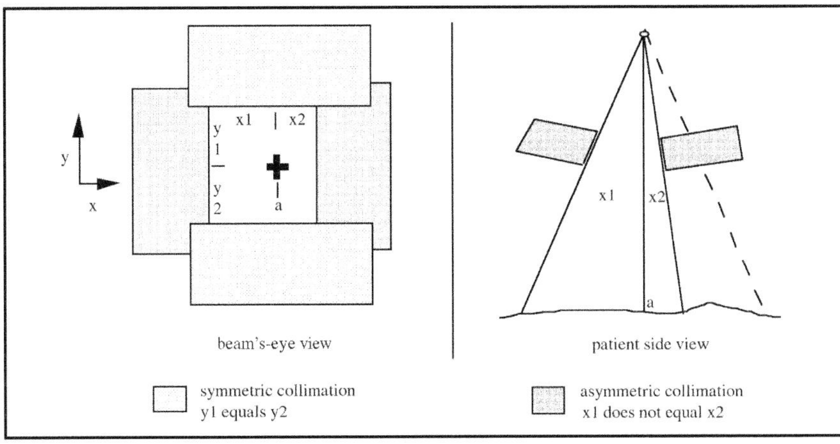

Figure 1.18: Collimators from a beam's-eye view and side view perspective. The jaws can be set asymmetrically or symmetrically about their axis.

A visible light field defines the geometric center and edges of the radiation field. On most modern linacs the light field is produced by a projector-type light source reflected from a thin nonretracting mylar mirror set in the head of the machine just above the collimators. Variations in design include retracting mirrors and lights placed on a slide in the target mount. The center of the field is defined by a crosshair etched onto a thin Mylar window mounted below the jaws, and the edge of the light field is defined by the jaws. The optics should be calibrated so that light field divergence and radiation field divergence agree over a range of clinically used SSDs.

1.7.3 Asymmetric Collimators

Modern linacs have independent drives for each set of jaws, which enables definition of asymmetric fields. For example, instead of having a 10×10 cm^2 field with symmetry on the central axis, one of the jaws can be driven to the center of the field and the other jaw can be driven to 10 cm from the central axis. This creates a 10×10 cm^2 asymmetric field with one edge of the field providing a nondiverging edge while the other has a more diverging edge (see figure 1.18). Detailed analysis of the x-ray fields produced by asymmetric collimators is given in chapter four. One pair of jaws can typically travel 2 cm beyond the central axis, while the other set can drive 10 cm beyond central axis. The extra overtravel of one set of jaws is to accommodate the use of this set of jaws in dynamic wedge mode (also see chapter four).

1.7.4 Blocks

The two sets of jaws can only provide rectangular field shapes on the patient surface. Therefore, other devices are used to create irregular field shapes so that delicate structures such as heart, lung, or spine may be shielded from the ray.

Lead blocks (see figure 1.19) are commonly used to modify the x ray treatment field dose distribution. These are mounted on trays known as

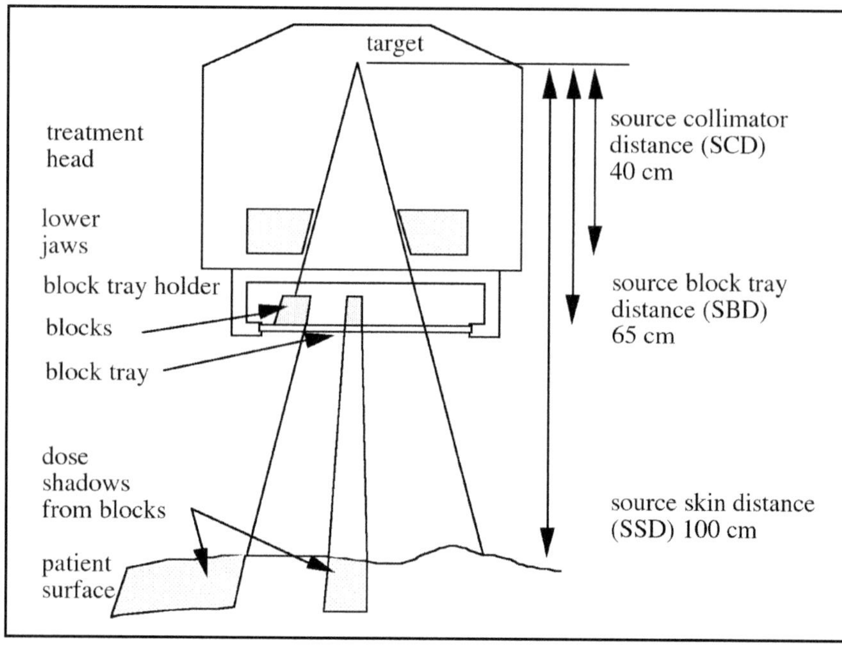

Figure 1.19: Blocks or beam attenuators that may be placed in the x-ray beam.

block trays that slide into a removable accessory mount just beyond the secondary collimators. The lead blocks are generally at a *source block distance* (SBD) of about 65 cm. Block thickness is generally sufficient to provide at least 5 *half-value layers* (HVLs) of shielding. This topic is discussed further in chapter two and in Khan (1984).

Each linac usually has a set of lead blocks in standard shapes, in particular a set of long narrow blocks for spinal cord shielding. Low melting point alloy (LMA) is also used in radiotherapy departments to shape blocks for irregular-field treatment. These are more convenient to mold than lead because of the low temperature involved. Lipowitz (Cerrobend) is one such alloy that has a low melting point of about 70 degrees Celsius. It consists of 50% bismuth, 26.7% lead, 13.3% tin, 10% cadmium (Khan 1984). Other LMA materials that do not contain cadmium, and are hence less hazardous for human long-term exposure, are also now available and are used in some clinics. Their melting point is generally slightly higher, about 120 degrees Celsius (McCullough et al. 1981, DeMeyer et al. 1986). The HVL of LMA is not quite as good as lead. The density is 8.3 g cm^{-3} versus 11.3 g cm^{-3}, and hence a slightly greater thickness of this material is required.

These blocks are usually produced by casting LMA in a polystyrene former. The polystyrene former is cut to shape using a hot wire cutter with a divergent edge that matches the angle of beam divergence. Fields defined by these blocks have a well-defined penumbral edge because there is no partial transmission through the edge of the block. For more reproducible positioning the blocks are often anchored to an individual block tray (with screws) for the duration of all patient dose fractions.

Because most radiotherapy department mold rooms have specialized equipment to produce these blocks, the compounds and processes are used in the creation of irregularly shaped electron inserts. Electron inserts are placed 5 cm from the patient surface in an electron applicator recess. The skin offset distance of 5 cm is to reduce bremsstrahlung contamination from the insert edge reaching the patient.

For x-ray treatments the blocks are suspended on block trays. Block trays may be acrylic or polycarbonate (typical dimensions 30 cm × 30 cm × 0.6 cm). As the light field shines through these materials, they are suitable for field alignment with patient skin marks. Blocks are often fixed to the tray by a thread and a wing nut to prevent block movement if gantry rotation is required.

Block trays create extra electron contamination in the beam, thus increasing patient surface dose. They also attenuate the beam, and a block tray correction factor for beam output is applied when they are used. Blocks are generally mounted on trays at least 15 cm from the patient to allow wide angle scattering of electrons to reduce the magnitude of patient skin dose due to electron contamination. Detailed discussion of block tray contribution to surface dose is given in chapter four.

1.7.5 Multileaf Collimators

Multileaf collimators (MLCs), as shown in figure 1.20a, have recently become so popular as block replacement devices that most new linacs are installed with an MLC attached. The leaves are now commonly used to replace lead blocks. These devices consist of a set of interleaved collimators.

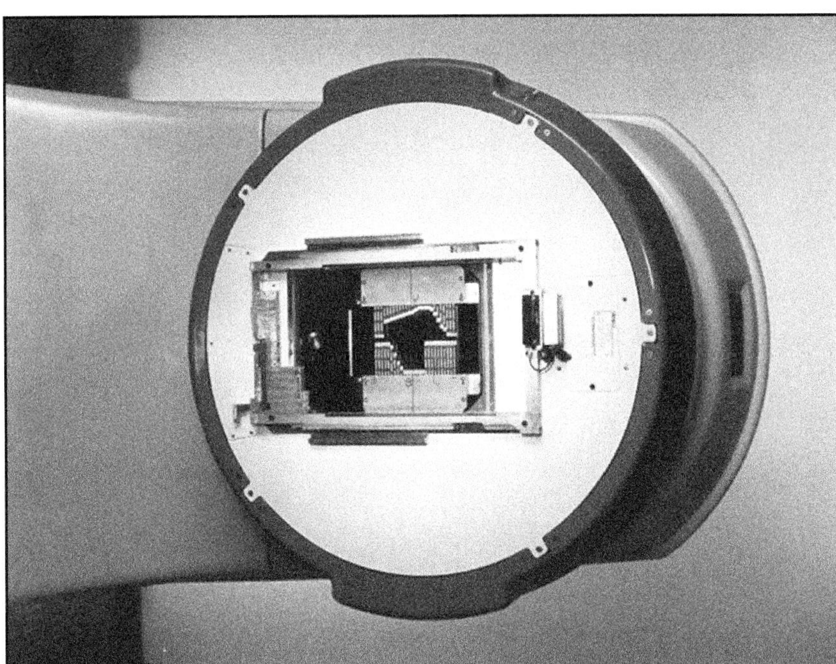

(a)

Figure 1.20: (a) A set of multileaf collimators. (Reprinted with permission from: © Varian Medical Systems, Inc.)

Each MLC leaf is capable of driving several centimeters beyond the geometric center of the field (leaf overtravel). The leaves can also be used in conjunction with a block tray on the rare occasion that blocking and MLCs need to be used in combination. Each leaf produces a jagged-shaped dose distribution at the field edge as each leaf produces a nominal 0.5 to 1 cm wide shadow at isocenter. This effect tends to smear out at depth (Galvin et al. 1993). Despite the discrete steps at the edge of the field formed by the leaves, the majority of patient field shapes can be attained satisfactorily with these devices (LoSasso et al. 1993, Powlis et al. 1993). Radiation therapy treatment planning software usually allows the flexibility to drive either the leaf's most protruding edge or each leaf center to the edge of the radiation oncologist's defined target volume.

Computer-controlled MLCs provide rapid field shape changes when multiple fields are employed; thus these devices are particularly useful for multiple field conformal radiotherapy applications (Leibel et al. 1991, Mohan et al. 1992). By using multiple fields (usually five to six) with MLCs the tumor dose can be increased while more normal tissue is spared by tightly conforming the treatment fields to the tumor volume (see chapter six for more detail). Cancer of the prostate is the site where this technique was first applied. This is because this site requires a high and focused dose to the tumor, and nearby delicate structures need to be given minimum dose. Note that similar conformal treatment can be achieved with blocks, but the treatments can be achieved much faster with a computerized MLC. The use of MLC to produce conformal fields of an irregular shape combined with three-dimensional (3-D) treatment planning is commonly known as 3-D conformal radiation therapy (3DCRT). This combined planning and treatment is now employed in virtually all cancer treatment sites (e.g., breast, head and neck, prostate). Due to the vastly reduced volume of normal tissue treated using 3DCRT, this is currently the accepted best practice technique with radiation therapy for most cancer treatment sites.

MLCs allow flexible field shaping without blocks by using leaves of tungsten that block the beam. There are typically 60 or more leaves to cover the whole photon radiation field. From a beam's eye view perspective they shape the beam into any irregular shape, hence blocking out normal tissue structures. The leaves, usually made from machined tungsten, travel in two banks as opposing pairs. Each leaf is driven by a small rotating electric motor connected to a thread or some other precision-drive device. Precise leaf position feedback is important when blocking the field edge and becomes even more important when in-field segmentation is employed (see more details in chapter seven). MLCs have typical dose leakage values of 1% with an extra 0.5% leakage between the leaves. This is better than a 5 HVL block transmission figure of 3.125%. Note the actual dose is still affected by in-patient scatter, and hence dose to critical structures, e.g., spinal cord, may still be several percent.

MLCs have virtually replaced lead and LMA blocks for 3DCRT. MLCs are more efficient to use in the clinic for the following reasons:

i) Automated positioning of MLCs to required shape reduces the time for setup and makes multiple shaped fields more viable.

ii) Using MLCs negates the time and hazard of lifting heavy blocks and mounting them above the patient.
iii) Significant block preparation time spent shaping molds and then pouring LMA and mounting the blocks on block trays is replaced with MLC.
iv) Time spent attaching blocks to block trays is avoided with MLCs.

There are still some situations where blocks are used instead of MLCs. These include:

i) Island blocking.
ii) Spinal cord shielding to avoid MLC end leaf leakage.
iii) Electron applicators still require LMA inserts.

Sets of collimators (jaws) are usually maintained on the machines so they can be driven to the field margins thereby minimizing leaf transmission, interleaf leakage due to the tongue and groove, and end leaf leakage (up to 30% for rounded leaf end designs).

Different manufacturers employ different MLC designs and positions within the linac treatment head.

Varian linacs retain two sets of jaws, hence the MLC is a tertiary collimator. The clearance from the bottom of the leaves to isocenter is 42 cm. A close-up of the Varian Millennium™ MLC is shown in figure 1.20b. This design has 60 pairs of leaves, i.e., 120 leaves in total, which are affixed to the machine just beyond the rectangular collimators and in front of the accessory holder. A single-focus leaf design is employed. This means the leaf sides diverge to align with beam divergence in one dimension. But because the leaves drive horizontally using a lead threaded screw design, a rounded leaf end is used to account for beam divergence in the other dimension. Leaf width at isocenter is 0.5 cm out to 20×20 cm^2 field size, then 1 cm leaves from 20 to 40 cm^2 field size.

The Siemens MLC retains a conventional upper jaw and the lower jaw is replaced by MLC leaves, shown in figure 1.20c. This provides good patient to treatment head clearance. The Siemens MLC employs a double-focused leaf design. This means the leaf end aligns with beam divergence, as does the side of the leaf. This is achieved by a different drive mechanism that employs a spur drive, enabling the leaf ends to tilt as they drive out. Leaf width at isocenter is 1 cm for the inner leaves.

The Elekta MLC, as shown in figure 1.20d, is mounted closer to the target than other vendor designs. The MLC is mounted above the two sets of jaws. One jaw is thinner and acts more as an extra MLC end leaf leakage attenuator. The MLC leaf pairs cannot drive closer than 1 cm. Leaf width at isocenter is 1 cm. The Elekta MLCs also employ a single-focused design with a rounded leaf end.

There is more clearance between the linac treatment head and the patient with the Siemens and the Elekta MLC designs, but they, in turn, need to have narrower physical leaf widths because they are mounted closer to the target. They both provide 1 cm leaf widths at isocenter. Therefore, the Varian 0.5 cm leaf width, which is obtainable because of its mount position further

(b)

(c)

Figure 1.20: (Continued) (b) Varian Millennium™ 120-leaf multileaf collimator. (Reprinted with permission from: © Varian Medical Systems, Inc.) (c) Siemens dual-focused MLC. (Reprinted with permission from: © Siemens Medical Systems, Inc.)

from the target, is currently the MLC treatment width benchmark (excluding niche mini multileaf designs).

Micro MLCs are niche products that are available from various companies (e.g., Radionics and BrainLAB). These typically have very small leaf widths of 3 to 4 mm at isocenter. They can generally only be used when small fields are employed (typically less than 12 cm) and are mainly used for stereotactic brain treatments. They are usually fixed to the end of the linac head beyond conventional collimators, and these MLC designs can usually be mounted and demounted as required. There is one specialized linac based on a Varian 6 MV design that is specifically designed for radiosurgery

(d)

Figure 1.20: (Continued). (d) Elekta MLC. (Reprinted with permission from: © Elekta Inc.)

applications with a fixed micro MLC; this device is known as a Novalis® linac (BrainLAB).

The dosimetric properties of MLCs when used as modulators in intensity-modulated radiation therapy beams are outlined further in chapter seven.

1.8 Wedges

Wedges are variable thickness absorbers that are placed in the beam and cause a progressive decrease in dose intensity across the beam, resulting in tilted dose profiles. They are designed to give an angled isodose curve at a predetermined isodose level. It may be the 50% isodose curve, or it may simply be determined at a reference depth such as 10 cm (Khan 1984).

1.8.1 Physical Wedges

Physical wedges are generally made from brass, lead, or steel and are used at a distance from the source of about 40 cm. The effect of a wedge on the dose profile is shown in figure 1.21a. A standard set of physical wedges generally provides profile angles of 15, 30, 45, and 60 degrees.

The introduction of a wedge produces a reduction in beam transmission (attenuation). This is accounted for by using wedge transmission factors to correct the dose delivered when a wedge is introduced into the x-ray beam. Because physical wedges attenuate low-energy photons, the beam is slightly harder than the open field. This results in depth dose curves for wedge fields being slightly shallower at depth. However, this effect tends to be very small, e.g., less than 2% difference for 6 MV at 30 cm depth. This effect does require separate central axis depth dose collection and input for modeling in radiation therapy treatment planning computers.

1.8.2 Flying Wedges

One manufacturer, Elekta, uses a steep physical wedge driven into the field for part of the treatment, sometimes called a flying wedge. These wedges are generally mounted internally in the gantry and are sometimes referred to as internal wedges. The dynamic movement of these devices produces resultant dose distributions that are the same as a physical fixed wedge. The final dose distribution is determined by how long the wedge, relative to the open beam, is retained in the beam. Therefore, a multiple number of beam wedge angles can be created by the use of this single wedge. For example, if the wedge is 60 degrees and maintained in the field for a short time with the rest of the dose delivered as an open field, the resultant wedged shape dose distribution will be some angle between 0 and 60 degrees (see figure 1.21b).

1.8.3 Dynamic Wedges

The other two linac vendors, Varian and Siemens, employ dynamic wedges (Kijewski et al. 1978, Levitt et al. 1990). This is actually a dynamic jaw that sweeps across the beam to produce a dose distribution that is similar to a

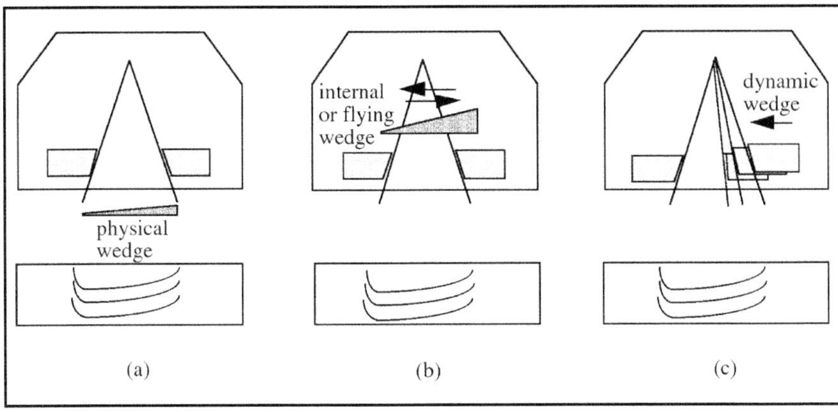

Figure 1.21: Wedges. (a) Physical wedge. (b) Flying wedge. (c) Dynamic wedge.

physical wedge's dose distribution (see figure 1.21c). Unlike physical and flying wedges there is no apparent beam hardening when dynamic wedges are employed. Hence the central axis depth dose for an open field is the same as for a dynamic wedge field of the same dimensions.

Any number of wedge angles are potentially available without physical wedge placement. The angles are not restricted, but in the interests of data collection efficiency the wedge angles commonly released for dynamic wedges are 10, 15, 20, 25, 30, 45, and 60 degrees (see chapter four).

Using a similar principle to the flying wedge, dynamic wedges have the jaws open and a large proportion of the dose is given as a flat dose profile. Then the jaws are closed or opened at progressive intervals during which a different preset dose is given. The preset dose per interval is obtained from a look-up table and is sufficient at each interval to give the steepest angled dose distribution available (i.e., 60 degrees). Then the other wedged angles are simply provided by altering the weight of the open field component of the beam relative to the weight of the steep wedged dose component of the beam.

As the resultant dose distribution is accrued over the time of the treatment, ratio ionization chamber measurements of field to reference chamber are of limited use (see chapter three). To get an accurate assessment of dose at a point using a conventional ionization chamber charge over time, measurements are required over the complete wedge treatment cycle. For collection of sufficient dose profile data, point-by-point acquisition is time consuming. Therefore, multiple linear detector arrays are used for collection and analysis of dynamic wedge dose profile data.

1.9 Compensators

Compensators are specialized devices designed to correct for missing tissue in two dimensions. As shown in figure 1.22, compensators are used where the patient outline is so irregular that normal wedges are insufficient. This is particularly relevant in the neck region, where rapid changes in patient contour

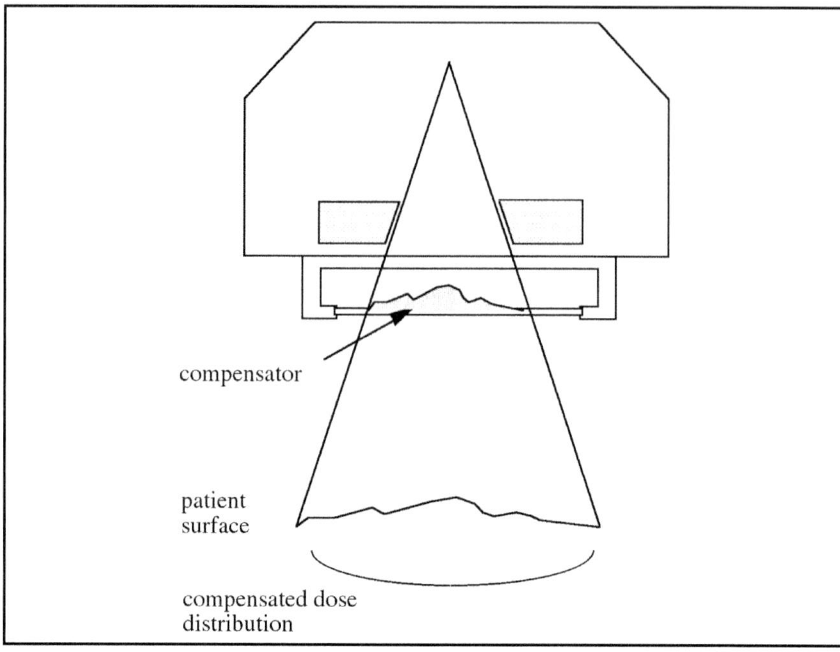

Figure 1.22: A retracted tissue compensator. This is a two-dimensional view—note that the compensator will account for contour changes in the third dimension. The compensator may be placed in the wedge slot or on the block tray. Some centers use compensators placed face up as shown in this schematic, and other centers mount them with the material face down. The most important thing to ensure is that the orientation is correct with respect to the patient.

are apparent. This leads, for example, to a high dose at the tip of the neck compared with the rest of the treatment region.

Compensators are generally mounted in the block tray or wedge slot. Compensators placed directly on the skin are known as bolus; they produce high skin dose. Hence, retracted compensators are more commonly used (Ellis et al. 1959).

These devices may be made from lead, wax, or some other material that provides an effective attenuation equivalent to compensate for missing tissue (Constantinou et al. 1989, Bagne et al. 1990). The thickness and shape of the compensator depends on the amount of missing tissue. Compensators can be designed to account for external contour changes (Khan et al. 1970, El-Khatib et al. 1986, Boyer 1982, Leung et al. 1974). Some designs also account for inhomogeneities such as lung (Ellis et al. 1973, Ulso et al. 1982). Most methods of compensation account for primary attenuation of the beam only. However, some work has been done on accounting for the scatter components (Robinson et al. 1989, 1990). Compensators have been successfully employed as solid intensity-modulated radiation therapy devices and some centers still use them for this purpose (Chang et al. 2004). It is more common however to use MLCs driven in a series of segment shapes to act as surrogate compensators. The advantage is that time for positioning

MLC segments is faster than placing several compensators onto the linac treatment head (see chapter seven).

1.10 Electronic Portal Imaging Devices

Most modern linacs now come with an electronic portal imaging device (EPID) as an option. EPIDs consist of a flat-plate detector array placed on the opposite side of the rotating gantry from the treatment beam portal. They are mounted on a retractable arm (see figure 1.23a) and designed to replace the historical method of treatment position validation, which is to place a film in this position (known as a port film). Because the port film or EPID image is produced using megavoltage (MV) beam energy, the image does not have as much contrast resolution as simulator images (diagnostic tube). Enough bony landmarks are often visible to enable an alignment comparison between EPID images/port films and simulator images or, indeed, digitally reconstructed radiographs (DRRs), which are images reconstructed from CT data. An EPID image is compared with a simulator image in figure 1.23b.

Most EPIDs consist of an amorphous silicon (a-Si) detector array overlaid with a fluorescent layer, usually gadolinium oxysulphide (GdOS) or cesium iodide (CsI). There are also fluorescent image-intensifier designs and liquid ion chamber designs. More detail about EPIDs is given in chapter twelve.

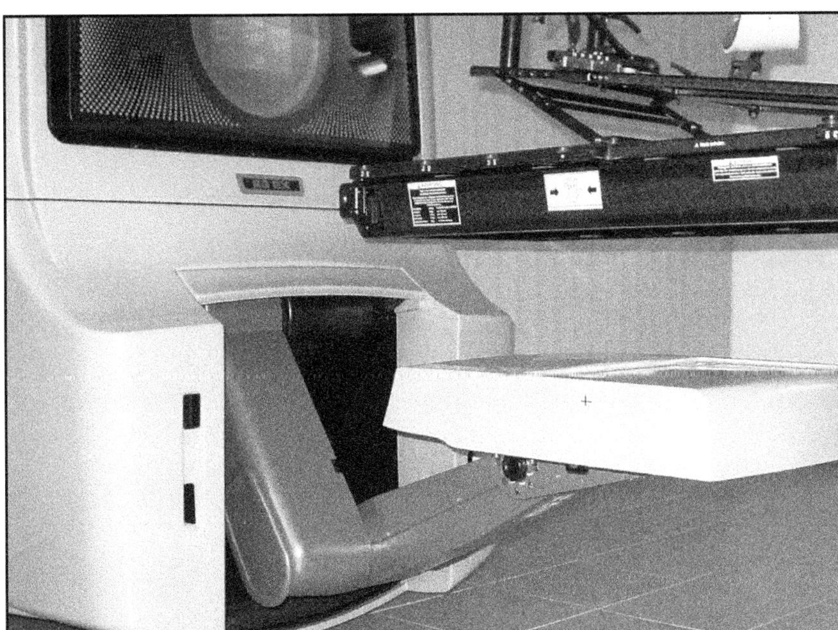

(a)

Figure 1.23: Electronic portal imaging device (EPID) and image. (a) EPID mounted on a mechanical arm so that the device can be retracted away during routine treatment. (Reprinted with permission from: © Varian Medical Systems, Inc.)

(b)
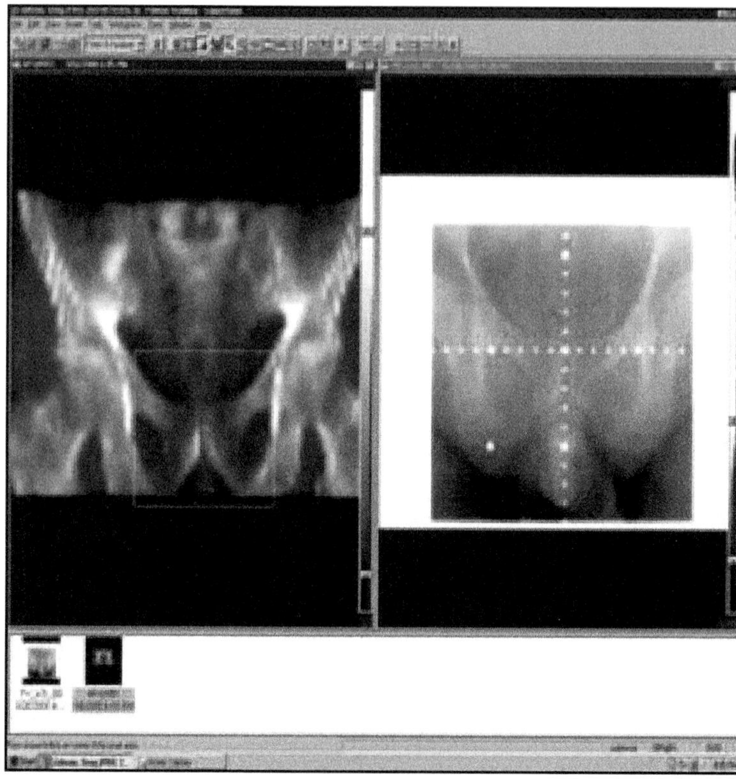

Figure 1.23: (Continued) Electronic portal imaging device (EPID) and image: (b) An EPID (MV) image (right) compared to a simulator image (kV) (left). (Reprinted with permission from: © Varian Medical Systems, Inc.)

1.11 On-Board Imaging Devices

An on-board imaging (OBI) device consists of a diagnostic kilovoltage (kV) x-ray tube and flat-panel detector device (FPD). As with an EPID, the FPD usually consists of an a-Si detector array and fluorescent layer. The OBI is placed on the linac gantry at 90 degrees to the linac treatment head. An OBI deployed on a Varian linac is shown in figure 1.24a and and OBI deployed on an Elekta linac is shown in figure 1.24b. Instead of providing a kV OBI, the Siemens linac uses the existing MV beam with an existing EPID to reconstruct MV cone-beam images. Both the x-ray tube and FPD are mounted on retractable arms and can be deployed when a high-quality image of patient position is required. Both can provide fixed beam's eye view type orthogonal x-ray images or can be rotated around the patient to provide cone-beam CT images of patient anatomy (see figure 1.24c).

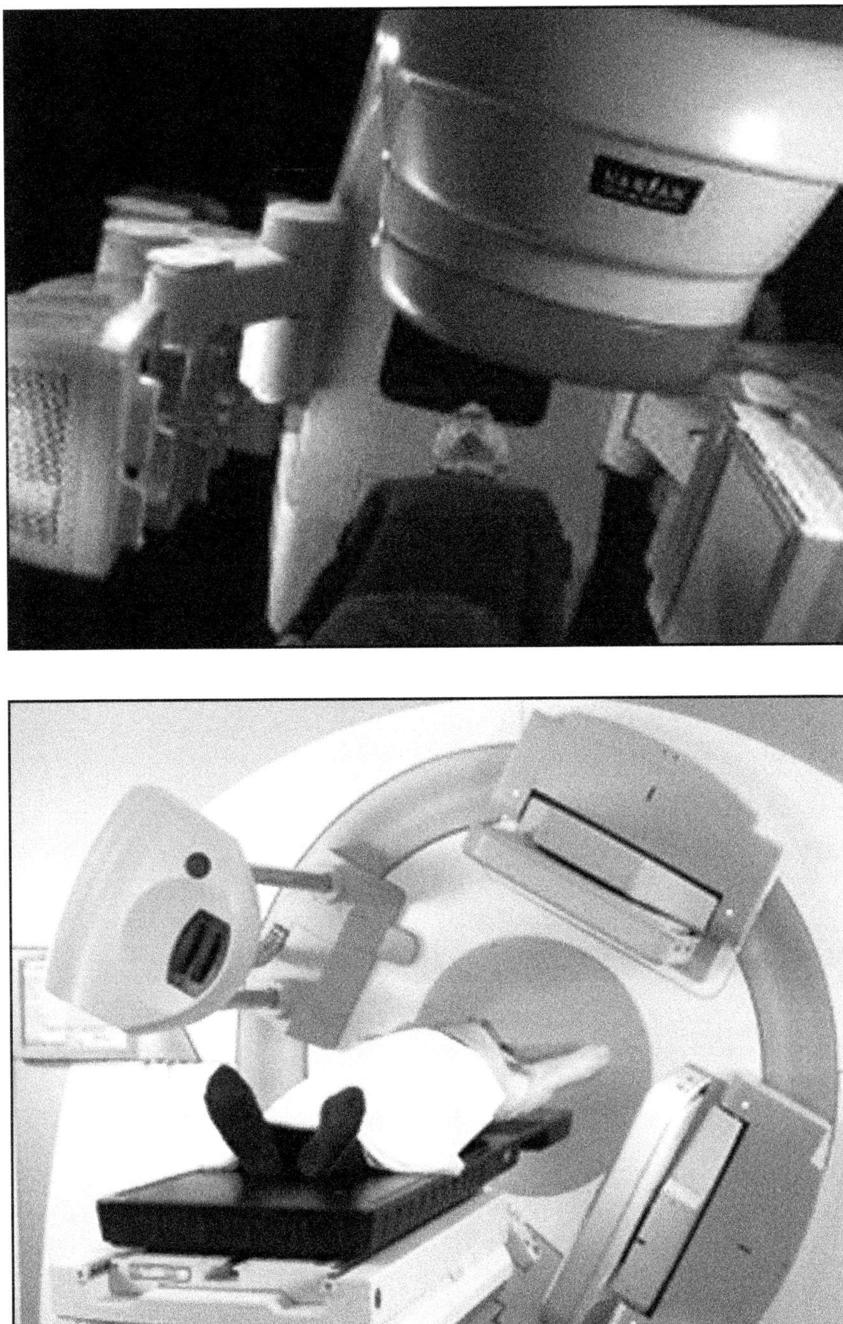

(a)

(b)

Figure 1.24: An On-Board Imager® (OBI) device mounted on the linac gantry. (a) Varian's OBI includes a diagnostic x-ray tube (left) and an a-Si flat-panel x-ray image detector (right). (Reprinted with permission from: © Varian Medical Systems, Inc.) (b) The Elekta OBI including a diagnostic x-ray tube (left) and an FPD (right) consisting of an a-Si detector array and a CsI fluorescent combination. (Reprinted with permission from: © Elekta Inc.)

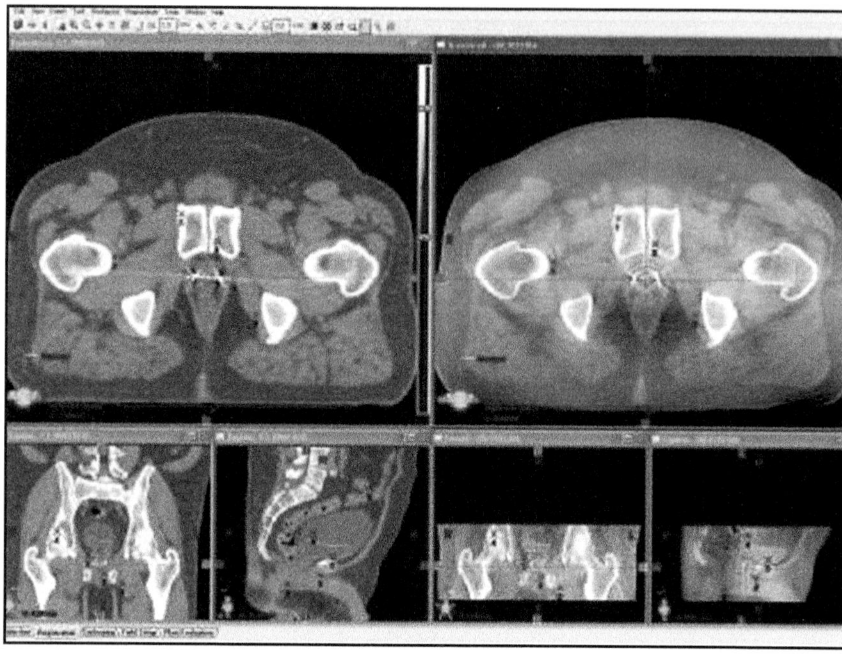

(c)

Figure 1.24: (Continued). (c) Cone-beam CT image from an OBI device (on right) as compared with the CT image used for radiotherapy treatment planning (on left); in this case for a prostrate treatment. (Reprinted with permission from: © Varian Medical Systems, Inc.)

1.12 Helical Tomotherapy

First proposed by Mackie (Mackie et al. 1993) helical tomotherapy linacs employ a linear accelerator with all ancillary equipment, including the modulator, mounted on a closed CT-type slip ring gantry as shown in figure 1.25. A fan beam treatment is delivered from multiple angles during the 360-degree gantry rotation. Beam modulation is achieved using a pneumatically driven binary MLC. Modulation in the superior-inferior dimension is also achieved by delivering the treatment using helical treatment delivery as the patient treatment couch moves through the gantry. Variable combinations of gantry rotation speed and table drive speed provide different pitches at which the treatment can be delivered. By reducing the pitch of beam delivery, beam modulation in the patient superior-inferior dimension is of extremely fine resolution (more details about tomotherapy are provided in chapter seven).

1.13 Gamma Knife® and CyberKnife®

Despite the requirement for multiple radioactive source changes and the relatively high cost, the Gamma Knife® does provide an excellent isocentric precision (less than 0.5 mm) and efficient treatment, particularly in the case

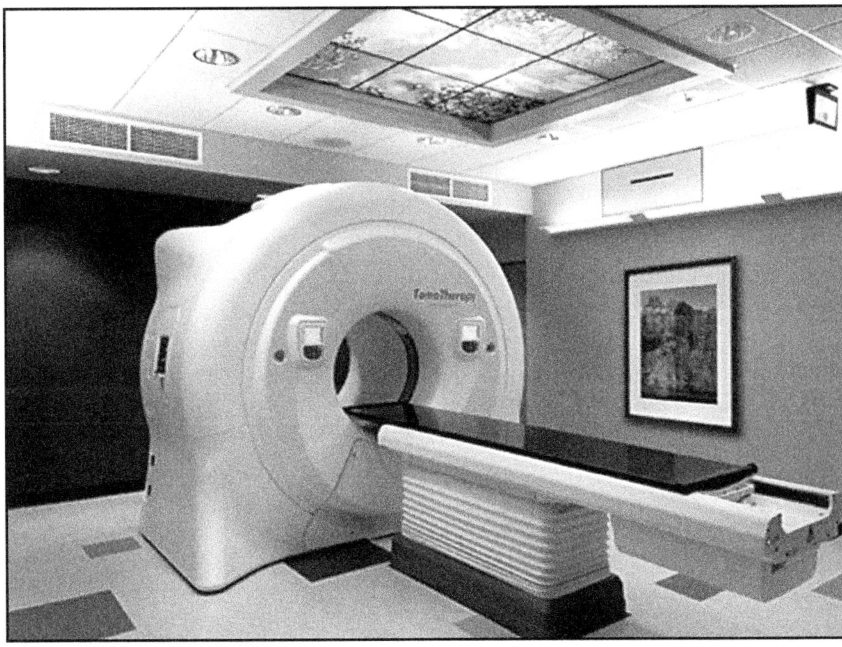

Figure 1.25: External view of a TomoTherapy Hi-Art System®. (Reprinted with permission from: © TomoTherapy Inc.)

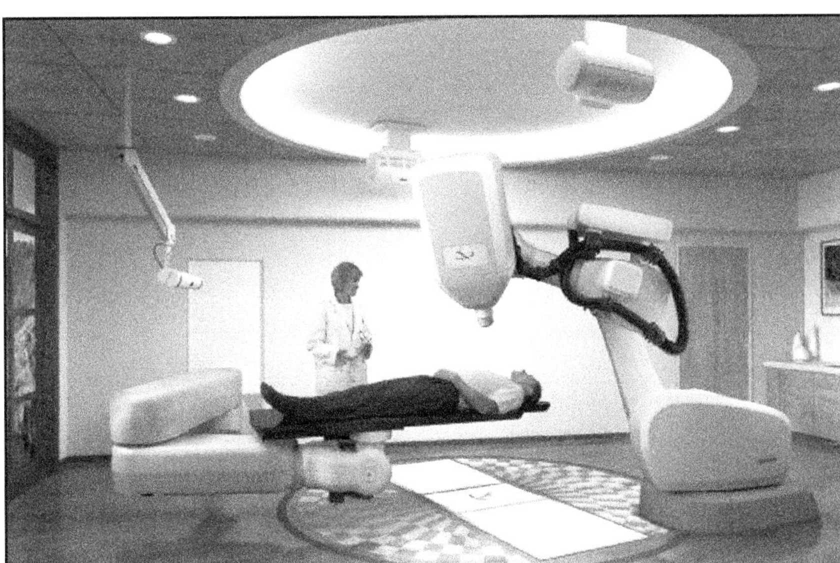

Figure 1.26: External view of a CyberKnife® linac mounted on a robotic arm (Image used with permission from Accuray Incorporated.)

of anterior venous malformations (AVMs), provided sufficient patients are available to be treated on this specialized device.

The CyberKnife® is a linac mounted on a robotic arm that can produce radiotherapy x-ray beams from multiple non-coplanar directions (see figure 1.26). The device can provide frameless stereotactic radiosurgery (SRS) using an image correlation algorithm for target localization (Adler et al. 1997). It is also used in combination with diagnostic x-ray tubes and flat-panel images in order to obtain frameless stereotactic images. This device was originally designed to give similar dose distributions to that provided by the Gamma Knife.

Despite the CyberKnife having a restricted maximum beam dimension of 5 cm diameter, it is also being used in some other parts of the body due to its image-guidance capability. It has been used for tracking radiotherapy lung treatment by a combination of lung position prediction and tracking of in-situ markers (Whyte et al. 2003). A review comparing three stereotactic radiosurgery systems including Gamma Knife, CyberKnife, and the Novalis radiosurgery linac systems has been reported by Andrews et al. (2006) and more details are provided in chapter seven.

References

Adler J.R. Jr., Chang S.D., Murphy M.J., Doty J., Geis P., Hancock S.L. (1997) The CyberKnife: a frameless robotic system for radiotherapy. *Stereotact. Funct. Neurosurg.* 69(1-4 Pt2): 124–128.

Allen B.J., Moore D.E., Harrington B.V., editors. (1992) *Progress in Neutron Capture Therapy for Cancer.* Plenum Publishing Corp., New York.

Andrews D.W., Bednarz G., Evans J.J., Downes B. (2006) A review of 3 current radiosurgery systems. *Surg. Neurol.* 66(6): 559–564.

Bagne F.R., Samsami N., Hoke S.W., Bron D.G. (1990) A study of effective attenuation coefficient for calculating tissue compensator thickness. *Med. Phys.* 17: 117–121.

Biggs P., Ma C.-M., Doppke K., Niroomand-Rad A., Beatty J. (1999). Kilovoltage x-rays, Chapter 9 in *The Modern Technology of Radiation Oncology*. Edited by J. Van Dyk. Medical Physics Publishing, Madison, Wisconsin, pp. 287–312.

Boyer A.L. (1982) Compensating filters for high energy x-rays. *Med. Phys.* 9: 429–433.

Butson M.J., Mathur J.M., Metcalfe P.E. (1995) Dose characteristics of a new 300 kVp orthovoltage machine. *Australas. Phys. Eng. Sci. Med.* 18: 133–138.

Chang S.X., Cullip T. J., Deschesne K.M., Miller E.P., Rosenman J.G. (2004) Compensators: an alternative IMRT delivery technique. *J Appl. Clin. Med. Phys.* 5(3): 15–36.

Coderre J.A., Turcotte J.C., Riley K.J., Binns P.J., Harling O.K., Kiger W.S. 3rd. (2003) Boron neutron capture therapy: cellular targeting of high linear energy transfer radiation. *Technol. Cancer Res. Treat.* 2(5): 355–375.

Constantinou C., Harrington J.C. (1989) Tissue compensators made of solid water or lead for megavoltage x-ray radiotherapy. *Med. Dosim.* 14: 41–47.

DeMeyer C.L., Whitehead L.W., Jacobson A.P., Brown D.G. (1986) Potential exposure to metal fumes, particulates, and organic vapors during radiotherapy shielding block fabrication. *Med. Phys.* 13: 748–750.

Einstein A. (1905) On the electrodynamics of moving bodies. *Ann. Physik.* 17: 891.

El-Khatib E.E., Podgorsak E.B., Pla C. (1986) The effect of lead attenuators on dose in homogeneous phantoms. *Med. Phys.* 13: 928–935.

Ellis F., Lescrenier C. (1973) Combined compensation for contours and heterogeneity. *Radiology* 106: 191–194.
Ellis F., Hall E.J., Oliver R. (1959) A compensator for variations in tissue thickness for high energy beams. *Br. J. Radiol.* 32: 421–422.
Galvin J.M., Smith A.R., Lally B. (1993) Characteristics of a multileaf collimator system. *Int. J. Radiat. Oncol. Biol. Phys.* 25: 181–192.
Glasgow P. (1999). Cobalt-60 Teletherapy, Chapter 10 in *The Modern Technology of Radiation Oncology*. Edited by J. Van Dyk. Medical Physics Publishing, Madison, Wisconsin, pp. 313–348.
Glasgow G.P., Kurup R.G., Leybovich L., Wang S., Fatyga M. (1996) Dosimetry and use of cobalt-60 100 cm source-axis-distance teletherapy units. *Curr. Oncol.* 3: 42–50.
Greene D. (1986) *Linear Accelerators for Radiation Therapy,* Medical Physics Handbook 17. Adam Hilger, Bristol, UK.
Greening J.R. (1981) *Fundamentals of Radiation Dosimetry,* first edition. Adam Hilger, Bristol, UK.
Hall E. (1988) *Radiobiology for the Radiologist*, third edition. J.B. Lippincott Co., Philadelphia, Pennsylvania. pp. 261–291.
Ishikawa H., Tsuji H., Kamada T., Hirasawa N., Yanagi T., Mizoe J.E., Akakura K., Suzuki H., Shimazaki J., Tsuijii H. (2006) Risk factors of late rectal bleeding after carbon ion therapy for prostate cancer. *Int. J. Radiat. Oncol. Biol. Phys.* 66(4): 1084–1091.
Johns H.E., Cunningham J.R. (1983) *The Physics of Radiology,* fourth edition. Charles C Thomas, Springfield, Illinois.
Johns H.E., Cunningham J.R. (1959) A precision cobalt-60 unit for fixed and rotational therapy. *Am. J. Roentgenol.* 81: 4–6.
Johns H.E., Fedoruk S.O., Kornelson R.O., Epp E.R., Darby E.K. (1952a) Depth dose data, 150 kVp to 400 kVp. *Br. J. Radiol.* 25: 542–549.
Johns H.E., Epp E.R., Cormack D.V., Fedoruk S.O. (1952b) 1000 Curie cobalt units for radiation therapy. *Br. J. Radiol.* 25: 296.
Karzmark C.J. (1984) Advances in linear accelerator design for radiotherapy. *Med. Phys.* 11: 105–128.
Karzmark C.J., Nunan C.S., Tanabe E. (1993) *Medical Electron Accelerators*. McGraw-Hill, New York.
Karzmark C.J., Morton R.J. (1989) *A Primer on the Theory and Operation of Linear Accelerators in Radiation Therapy.* Medical Physics Publishing, Madison, Wisconsin.
Karzmark C.J., Morton R.J. (1981) *A Primer on the Theory and Operation of Linear Accelerators in Radiation Therapy.* Bureau of Radiological Health, Rockville, Maryland.
Karzmark C.J., Pering N.C. (1973) Electron linear accelerators for radiation therapy: History, principles and developments. *Phys. Med. Biol.* 18: 321–354.
Khan F.M. (1984) *The Physics of Radiation Therapy,* first edition. Williams & Wilkins, Baltimore, Maryland.
Khan F.M., Moore V.C., Burns D.J. (1970) The construction of compensators for cobalt teletherapy. *Radiology* 96: 187–192.
Kijewski P.K., Chin L.M., Bjarngard B.E. (1978) Wedge-shaped dose distributions by computer-controlled collimator motion. *Med. Phys.* 5: 426–429.
Krieger H., Petzold W. (1989) *Strahlenphysik, Dosimetrie und Strahlenschutz,* B.G. Teubner, Stuttgart, Germany.
Leibel S.A., Kutcher G.J., Harrison L.B., Fass D.E., Burman C.M., Hunt M.A., Brewster M.S., Br. Ling C.C., Fuks Z. (1991) Improved dose distributions for 3-D conformal boost treatments in carcinoma of the nasopharynx. *Int. J. Radiat. Oncol. Biol. Phys.* 20: 823–833.
Leung P.M.K., Van Dyk J., Robins J. (1974) A method of large irregular field compensation. *Br. J. Radiol.* 47: 805–810.
Levitt D.D., Martin M., Moeller J.H., Lee W.L. (1990) Dynamic wedge field technique through computer controlled collimator motion and dose delivery. *Med. Phys.* 17: 87–91.

Lomax A.J. (2003) Compensated and intensity-modulated proton therapy. In: *Intensity-Modulated Radiation Therapy: The State of the Art*. Edited by J.R. Palta and T.R. Mackie. AAPM Medical Physics Monograph No. 29. Proceedings of the AAPM 2003 Summer School. Medical Physics Publishing, Madison, Wisconsin, pp. 787–827.

Lomax A.J., Bortfeld T., Goitein G., Debus J., Dykstra C., Tercier P.A., Coucke P.A., Mirimanoff R.O. (1999) A treatment planning inter-comparison of proton and intensity modulated photon radiotherapy. *Radiother. Oncol.* 51(3): 257–271.

LoSasso T., Chui C.S., Kutcher G.J., Leibel S.A., Fuks Z., Ling C.C. (1993) The use of multi-leaf collimators for radiotherapy of carcinomas of the prostate and nasopharynx. *Int. J. Radiat. Oncol. Biol. Phys.* 25: 161–171.

Mackie T.R., Holmes T., Swerdloff S., Reckwerdt P., Deasy J.O., Yang J., Paliwal B., Kinsella T. (1993). Tomotherapy: a new concept for the delivery of dynamic conformal radiotherapy. *Med. Phys.* 20(6): 1709–1719.

Maughan R.L. (1996) An update on neutron radiotherapy. *Proceedings of AAPM 1996 Summer School—Teletherapy: Present and Future.* Edited by T.R. Mackie and J.R. Palta, Advanced Medical Publishing, Madison, Wisconsin, pp. 723–759.

McCullough E.C., Senjem D.H. (1981) Airborne concentrations of toxic metals from the use of low melting point alloys to construct radiotherapy shielding. *Med. Phys.* 8: 111–114.

Miller D.W. (1996) Update on proton radiotherapy. *Proceedings of AAPM 1996 Summer School—Teletherapy: Present and Future.* Edited by T.R. Mackie and J.R. Palta, Advanced Medical Publishing, Madison, Wisconsin, pp. 591–610.

Mohan R., Mageras G.S., Baldwin B., Brewster L.J., Kutcher G.J., Leibel S., Burman C.M., Ling C.C., Fuks Z. (1992) Clinically relevant optimization of 3-D conformal treatments. *Med. Phys.* 19: 933–944.

Mohan R., Chui C. (1985) Energy and angular distributions of photons from medical linear accelerators. *Med. Phys.* 12(5): 592–597.

Moyers M.F. (1999) Proton therapy, Chapter 20 in *The Modern Technology of Radiation Oncology*. Edited by J. Van Dyk. Medical Physics Publishing, Madison, Wisconsin, pp. 823–869.

Munro P., Rawlinson J.A., Fenster A. (1988) Therapy imaging: Source size of radiotherapy beams. *Med. Phys.* 15: 517–524.

Niroomand-Rad A., Gillin M.T., Lopez F., Grimm G.F. (1987) Performance characteristics of an orthovoltage x-ray therapy machine. *Med. Phys.* 14: 874–878.

Oliver R., Kemp A.W. (1949) An investigation into factors affecting x-ray dose distribution and its measurement. *Br. J. Radiol.* 22: 33–43.

Perec A., Kubo H. (1990) Radiation leakage through electron applicators on Clinac–1800 accelerators. *Med. Phys.* 17: 715–719.

Podgorsak M.B., Schreiner L.J., Podgorsak E.B. (1990) Surface dose in intracavitary orthovoltage radiotherapy. *Med. Phys.* 17(4): 635–640.

Podgorsak E.B., Van Dyk J., Metcalfe P. (1999) Linear accelerators, Chapter 11 in *The Modern Technology of Radiation Oncology*. Edited by J. Van Dyk. Medical Physics Publishing, Madison, Wisconsin, pp. 349–435.

Powlis W.D., Smith A.R., Cheng E., Galvin J.M., Villari F., Bloch P., Kilgerman M.M. (1993) Initiation of multi-leaf collimator conformal radiation therapy. *Int. J. Radiat. Oncol. Biol. Phys.* 25: 171–181.

Robinson D.M., Scrimger J.W. (1990) Optimised tissue compensators. *Med. Phys.* 17: 391–396.

Robinson D.M., Scrimger J.W. (1989) Limitations of retracted missing tissue compensators: An experimental analysis. *Med. Dosim.* 14: 49–54.

Scrimger J.W., Connors S.G. (1986) Performance characteristics of a widely used orthovoltage x-ray machine. *Med. Phys.* 13: 267–269.

Shipley W.U., Verhey L.J., Munzenrider J.E., Suit H.D., Urie M.M., McManus P.L., Young R.H., Shipley J.W., Zietman A.L., Biggs P.J., et al. (1995) Advanced prostate cancer: the results of a randomized comparative trial of high dose irradiation boosting with conformal protons compared with conventional dose irradiation using photons alone. *Int. J. Radiat. Oncol. Biol. Phys.* 32(1): 3–12.

Slater J.C. (1948) The design of linear accelerators. *Rev. Mod. Phys.* 20: 473–518.

Soloway A.H., Barth R.F., Carpenter D.E., editors. (1993) *Advances in Neutron Capture Therapy for Cancer.* Plenum Publishing Corp., New York.

Suit H.D. (2002) The Gray Lecture; coming technical advances in radiation oncology. *Int. J. Radiat. Oncol. Biol. Phys.* 53(4): 798–809.

Ulso N., Christenson J. (1982) Optimization with compensating filters. *Proceedings of the World Congress of Medical Physics—Hamburg.*

Van Dyk J., Battista J.J. (1996) Cobalt-60: An old modality, a renewed challenge. *Curr. Oncol.* In print.

Whyte R.I., Crownover R., Murphy M.J., Martin D. P., Rice T. W., DeCamp M.M. Jr., Rodebaugh R., Weinhous M.S., Le Q.T. (2003) Stereotactic radiosurgery for lung tumors: preliminary report of a phase I trial. *Ann. Thorac. Surg.* 75(4): 1097–1101.

Questions

1.1 Explain the main difference between traveling and standing waveguide accelerators. (section 1.5)

1.2 With the aid of a sketch graph that shows how an increase in gun current can lead to a decrease in beam energy, explain the term "load line." (section 1.5)

1.3 If a 6 MeV electron beam needs to travel in a radius of curvature of 15 cm to traverse around the bend magnet, what magnitude of magnetic field is required if one assumes a constant magnetic field?
(section 1.6)

1.4 Name three advantages of multileaf collimators (MLCs) over shielding blocks.
section 1.7)

1.5 What are the practical advantages of using dynamic wedges instead of conventional wedges in combination with MLCs? Name one practical disadvantage of dynamic wedges.
(section 1.8)

2 Interaction Properties of X-Rays and Electrons

Wilhelm Conrad Röntgen discovered x-rays on the 8th of November 1895.

2.1 Introduction

The processes involved in the interaction of radiotherapy photon and electron beams with matter are discussed in this chapter, with an emphasis on the relevance of these processes to the commonly observed characteristics of photon and electron beam dose distributions. The most important phenomena are (i) attenuation of photons by scattering and absorption, (ii) energy transfer to charged particles, (iii) transport of charged particles, and (iv) energy deposition by charged particles.

The basic quantities governing the degree of attenuation are interaction cross sections, which are defined as the area that each interaction center presents to the incident photon. Cross sections and their relationship to attenuation coefficients are described in section 2.2. The different photon interaction processes that are dominant at radiotherapy x-ray beam energies are discussed in section 2.3. These interactions are photoelectric absorption, Compton scattering, and pair production. Attenuation is the removal of photons from a beam by these processes and is quantified by the *linear attenuation coefficient*, μ. This is defined as the probability per unit pathlength that a photon will interact or, equally, the fraction of photons in a beam that interact per unit pathlength. Note that here *removal* of photons means either that they have been absorbed (in photoelectric or pair production processes) or that they have been scattered (in Compton interactions).

In each interaction, some or all of the incident photon's energy is transferred to kinetic energy of charged particles. This means that there is a fraction of the *energy* in a beam lost per unit pathlength, and this is known as the *linear energy transfer coefficient*, μ_{tr}. Note that μ_{tr} must always be less than or equal to μ, since the proportion of energy transferred cannot be greater than the proportion of photons that interact! Of the energy transferred, some is absorbed in charged particle collisions and some is radiated away as *bremsstrahlung* photons. The fraction of energy lost per unit pathlength *that is actually deposited in the medium* is the *linear energy absorption coefficient*, μ_{ab}. Here also, μ_{ab} must be less than or equal to μ_{tr}, since there cannot be more energy absorbed than is transferred.

X-ray interactions cause charged particles to be set in motion. These then undergo collisions with orbital electrons (depositing energy), are decelerated by the electric fields of nuclei (giving rise to bremsstrahlung), and are elastically scattered by nuclei and (to a small degree) orbital electrons. The collisional and radiative energy loss mechanisms of electrons are described in section 2.5. It is important to note that charged particles may travel a considerable distance before running out of energy and coming to rest. This means that the distribution of energy *absorbed* in the medium is not the same as the distribution of energy *transferred* by photons (especially at high energies) and can lead to a phenomenon known as electronic disequilibrium (see sections 2.4 and 2.6).

2.2 Cross Sections and Attenuation Coefficients

2.2.1 Total and Differential Cross Section

The scattering cross section σ represents the area that an interaction center (an electron or nucleus) presents to the incident photon and is measured in units of cm^2 or barns (1 barn = 10^{-28} m^2). If the photon is incident on this region, an interaction will occur. Most importantly, the cross section of an interaction center is the probability that a photon incident somewhere within a unit area will interact with that interaction center. The total likelihood of interaction increases, of course, with the cross section and number of interaction centers. Also, the number of photons crossing a sphere of unit cross-sectional area is denoted by the fluence, Φ (units cm^{-2}). Now for a fluence Φ and a scattering cross section σ, the expected number of interactions as these photons cross a unit volume containing one scattering center is

$$\langle n \rangle = \sigma \Phi . \tag{2.1}$$

The situation of a photon fluence Φ incident on a unit volume containing an atom with cross section σ is depicted in figure 2.1.

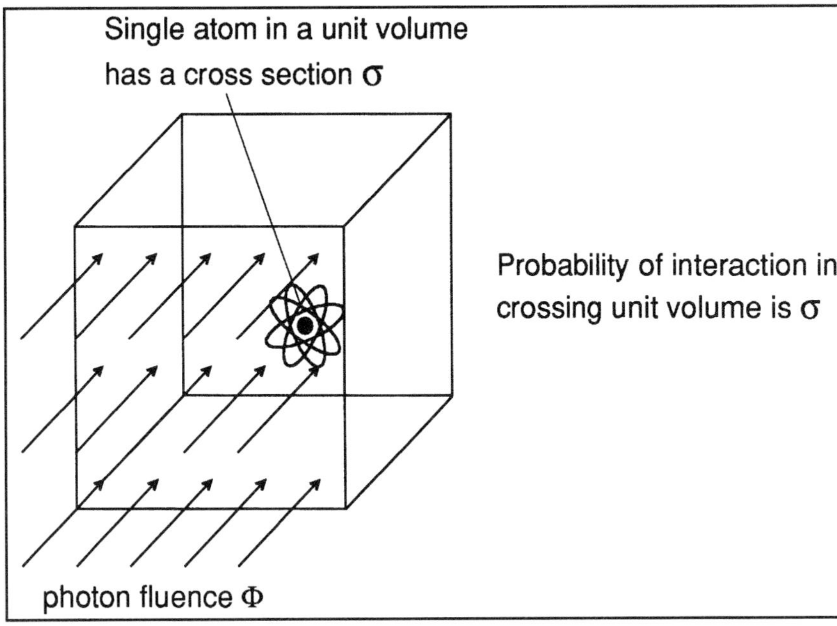

Figure 2.1: An atom with cross section σ is shown in a unit volume. There is a fluence Φ of photons incident on the unit area front face of the volume. The fraction of the area of the front face that the atom represents is $\sigma/1$, so the probability of interaction for any photon crossing the volume is σ.

The expected number of interactions in a unit volume, where there are ρ_{scat} scattering centers in the unit volume, is thus

$$\langle N \rangle = \rho_{scat}\,\sigma\,\Phi. \tag{2.2}$$

Note that this equation also gives the expected number of interactions when Φ photons incident on a unit area cross a unit pathlength. The scattering centers can be either electrons or nuclei. When electrons are responsible for scattering (as in Compton interactions), ρ_{scat} can refer to the density of either electrons or atoms, where the cross section will then correspond to the area presented per electron or per atom respectively—the product $\rho_{scat}\sigma$ is the same in each case. For interactions with nuclei, such as in photoelectric or pair production processes, the density and cross section refer to nuclei and thus to atoms. It is usual to express the cross section for Compton interactions on a per atom basis also, as the cross sections for each process can then be directly added. For the discussion following, the cross section refers to the total cross section per atom, and ρ_{scat} is the number of atoms per unit volume.

Differential Cross Section

The cross section σ is the total area presented by a scattering center to the incident photon, but the deflection of the photon (if it is not absorbed)

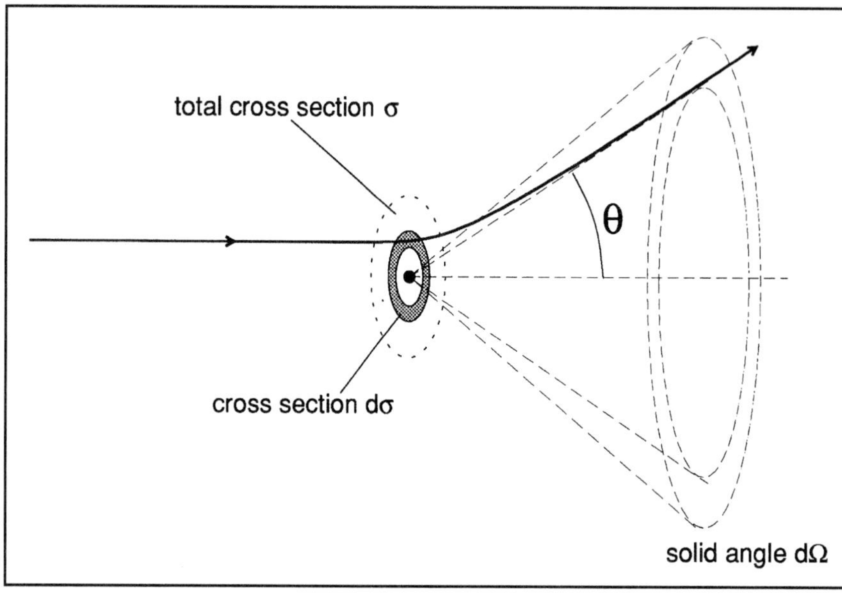

Figure 2.2: The area through which photons must pass in order to be scattered within a solid angle $d\Omega$ at deflection angle θ is $d\sigma$. The differential cross section per unit solid angle is thus $d\sigma/d\Omega$.

depends on the distance from the interaction center at which it is incident. The annular area $d\sigma$ within which the deflection will be within a solid angle $d\Omega$, centered on deflection angle θ, is the differential cross section $d\sigma/d\Omega$. The relationship between an increment in θ and an increment in Ω is, from simple geometry,

$$d\Omega = 2\pi \sin\theta \, d\theta, \tag{2.3}$$

so

$$\frac{d\sigma}{d\Omega} = \frac{1}{2\pi \sin\theta} \frac{d\sigma}{d\theta}. \tag{2.4}$$

The differential cross section concept is illustrated in figure 2.2.

2.2.2 Linear and Mass Attenuation Coefficient

Equation 2.2 is also the *loss* of fluence per unit pathlength; thus we can write the reduction per pathlength dz as

$$d\Phi = \rho_{scat} \sigma \Phi dz. \tag{2.5}$$

Now from equation 2.2, the fraction of photons lost from a beam in a unit pathlength is given by

$$\mu = \frac{\langle N \rangle}{\Phi} = \rho_{scat} \sigma, \tag{2.6}$$

which is the linear attenuation coefficient and has units of cm^{-1}. Inserting μ in equation 2.5, the fraction of photons lost in a pathlength dz is thus

$$\frac{d\Phi}{\Phi} = \rho_{scat}\sigma dz = \mu\, dz. \tag{2.7}$$

This differential equation can be solved to yield an equation for the reduction of photon fluence with depth z:

$$\Phi = \Phi_0 e^{-\mu z}. \tag{2.8}$$

This is the familiar equation for exponential attenuation of the primary fluence, where the fluence is Φ_0 at the medium surface. This exponential behavior results from the fact that a fixed proportion (rather than number) of photons remaining in the primary beam are removed in each depth increment. It is best, however, to think of the linear attenuation coefficient as a probability of photon interaction per unit pathlength rather than just as a component of an exponential equation, since it has other uses apart from describing exponential attenuation (as, for example, in calculating the *terma;* see section 2.4).

In equation 2.6, the attenuation coefficient is written as the number of atoms per unit volume times the cross section per atom. It is also useful to write the electron density in terms of mass density, ρ. The relationship between ρ_{scat} and ρ is easily deduced from the fact that the number of atoms per volume is the number of atoms per mass times the mass per volume. In turn, the number of atoms per mass is the atoms per mole times the number of moles per mass. We have then,

$$\frac{\text{atoms}}{\text{volume}} = \frac{\text{atoms}}{\text{mole}} \times \frac{\text{mole}}{\text{mass}} \times \frac{\text{mass}}{\text{volume}} \tag{2.9}$$

so

$$\rho_{scat} = N_A \times \frac{1}{A} \times \rho\,, \tag{2.10}$$

where N_A is Avogadro's number and A is the atomic mass number. This of course pertains in this simple form to elements, but a weighted sum over the elemental composition is possible for compounds (see below). Inserting equation 2.10 in equation 2.6, and dividing by the density then gives the mass attenuation coefficient:

$$\frac{\mu}{\rho} = \frac{N_A}{A}\sigma, \tag{2.11}$$

which is of course independent of density. The value for compounds is easily obtained by taking a weighted sum over each element:

$$\frac{\mu}{\rho} = \sum_i f_i \left(\frac{\mu}{\rho}\right)_i \tag{2.12}$$

where f_i is the fraction by weight of element i. This can be further expanded using the cross sections for the different interaction types, for each element:

$$\frac{\mu}{\rho} = \sum_i f_i \left[\left(\frac{\mu}{\rho}\right)_{i,pe} + \left(\frac{\mu}{\rho}\right)_{i,incoh} + \left(\frac{\mu}{\rho}\right)_{i,pair} \right]$$
$$= N_A \sum_i f_i \frac{1}{A_i} \left[\sigma_{i,pe} + \sigma_{i,incoh} + \sigma_{i,pair} \right] \tag{2.13}$$

where the subscripts *pe*, *incoh*, and *pair* refer to photoelectric, incoherent (Compton), and pair production processes, respectively. The relative importance of the cross sections above depends on the element and also on the energy, as will be discussed in the next section.

At this stage it is useful to write the contribution to the mass attenuation coefficient due to incoherent (Compton) scattering in terms of the cross section per electron, σ_e, and electron density, ρ_e. The atomic cross section for Compton scattering is equal to Z times the cross section per electron, as indicated in figure 2.3.

The electron density is of course a factor Z higher than the density of atoms, so we can write the coefficient for incoherent scattering as

$$\left(\frac{\mu_{incoh}}{\rho}\right) = N_A \left(\frac{Z}{A}\right) \sigma_e . \tag{2.14}$$

The reason why the mass attenuation coefficient for Compton interactions is best written like this is that the cross section per electron is independent of the element; therefore, (μ_{incoh}/ρ) can be directly related to the ratio of atomic to mass number. Furthermore, since Compton interactions overwhelmingly dominate the types of interactions undergone by megavoltage photons, equation 2.14 makes it simple to estimate the relative total mass attenuation coefficient in different materials. Note that because Z/A is approximately ½ for most elements in the body (except hydrogen), the mass attenuation coefficient does not vary greatly with material composition. It is interesting that the mass attenuation coefficient of water is higher than for most other materials because the ratio Z/A is 1 for hydrogen, which constitutes a significant fraction (1/9) of the mass of a water molecule. The mass attenuation coefficient of other materials is easily calculated (if the elemental composition is known) using a weighted sum of the ratio *Z*/*A* for each element:

$$\left(\frac{\mu_{incoh}}{\rho}\right) = N_A \left[\sum_i f_i (Z/A)_i \right] \sigma_e . \tag{2.15}$$

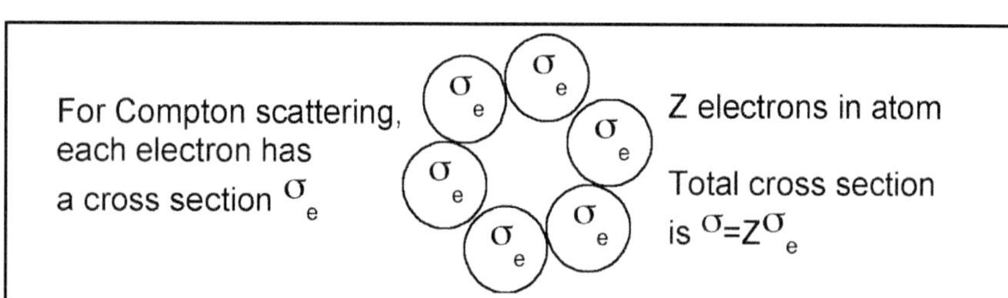

Figure 2.3: Total cross section per atom for Compton scattering, σ, is equal to Z times the cross section per electron σ_e.

Note that by including N_A and σ_e in the summation, a weighted sum over mass attenuation coefficients is obtained, as in the first line of equation 2.13. The summation in equation 2.15 is known as the *effective ratio of atomic to mass number*, $\langle Z/A \rangle$, which for water is easily shown to be equal to 10/18 = 0.555. For any substance *s*, the mass attenuation coefficient can be expressed in terms of the value for water by

$$\left(\frac{\mu_{incoh}}{\rho}\right)_s = \left(\frac{\mu_{incoh}}{\rho}\right)_w \frac{\langle Z/A \rangle_s}{\langle Z/A \rangle_w}, \tag{2.16}$$

and the mass attenuation coefficient relative to water is given then by

$$\left(\frac{\mu_{incoh}}{\rho}\right)_w^s = 1.8\langle Z/A \rangle_s = \frac{\langle Z/A \rangle_s}{0.555}, \tag{2.17}$$

which is also the number of electrons per *gram* relative to water. The relative number of electrons per *volume* is simply given by multiplying the number per mass by the mass per volume:

$$\left(\rho_e^w\right)_s = \frac{\langle Z/A \rangle_s}{0.555}\rho_s \tag{2.18}$$

where ρ_s is the density of substance *s*. It is obvious that for substances where there is no hydrogen, and thus Z/A for each element is at most ½, the relative electron density will not exceed 0.9. Also, the relative number of electrons per unit gram will not be above 0.9. There is some hydrogen in all body tissues, and *Z/A* = ½ for most other elements in these tissues, so the relative number of electrons per gram will be between 0.9 and 1.0. In muscle, the relative abundance of hydrogen is similar (though a little less) to that in water, so the relative number of electrons per gram approaches unity. Bone has a lower proportion of hydrogen, and also some heavier elements, so the figure is lower than for muscle. For heavier elements, *Z/A*<½, so the figure will be less than 0.9 for a high atomic number element. The electron density parameters (absolute and relative) for a few substances are shown in table 2.1. Note that the relative number of electrons per gram in carbon is exactly 0.9, since *Z/A* = ½.

Table 2.1: Electron densities of various substances, both per gram and per cm³. Substances containing higher atomic numbers have fewer electrons per gram because of their lower effective ratio of atomic to mass number <Z/A>.

	Density (g/cm)	*Elec./g (×10²¹)*	*Elec./cm³ (×10²¹)*	*⟨Z/A⟩*	*Rel. elec./g*	*Rel. elec./cm³*
Water	1.00	3.343	3.343	0.555	1.000	1.000
Muscle	1.04	3.312	3.444	0.550	0.991	1.030
Bone	1.65	3.192	5.267	0.530	0.955	1.575
Carbon	2.25	3.008	6.768	0.500	0.900	2.025
Aluminum	2.70	2.902	7.835	0.482	0.868	2.344
Lead	11.36	2.383	27.070	0.396	0.713	8.098

The relatively high number of electrons per gram in water and muscle is important as it means that the number of photon (and electron) interactions per unit mass is quite high. This corresponds to a high value of σ_{incoh}/A in equation 2.13. Consequently, the dose in water and soft tissue tends to be higher than in other materials such as bone, at least when Compton interactions dominate. This is not the case at low photon energies, however, where the nuclei of high atomic number elements give rise to a large cross section for photoelectric interactions.

As an example of the calculation of the mass attenuation coefficient due to Compton interactions for water, consider 3 MeV photons for which the cross section per electron is $\sigma_e = 0.1151$. Using equation 2.15, where the summation is replaced by $\langle Z/A \rangle = 0.555$,

$$\begin{aligned}\left(\frac{\mu_{incoh}}{\rho}\right) &= N_A \langle Z/A \rangle \sigma_e \\ &= 6.022\times10^{23}\,\text{mol}^{-1}\times0.555\,\text{mol g}^{-1}\times0.1151\times10^{-25}\,\text{cm}^2 \\ &= 0.0385\,\text{cm}^2\text{g}^{-1}. \end{aligned} \tag{2.19}$$

Compare this with the total mass attenuation coefficient for 3.0 MeV photons in water, which is 0.0397 cm^2g^{-1}. The 3% difference is due to the contribution from pair production interactions. At 1.0 MeV, there is virtually no difference in the total and incoherent mass attenuation coefficients (0.0707 cm^2g^{-1}) because the energy is low enough for pair production to be insignificant and high enough for photoelectric interactions to be virtually nonexistent.

2.2.3 Mass Energy Transfer and Absorption Coefficient

Just as there is a certain fraction of photons that interact in a unit pathlength, there is a certain fraction of the *energy* of these photons that will be lost in that distance. If the photons have an energy E and an average energy $\overline{E}_{tr}$ is transferred to the kinetic energy of charged particles in each interaction, the fraction of the total energy of the photons that is transferred in the unit pathlength will be

$$\mu_{tr} = \mu\left(\overline{E}_{tr} / E\right). \tag{2.20}$$

The fraction $\overline{E}_{tr}/E$ is very close to unity for photoelectric processes, since all the photon energy is given to the ejected electron, except that required to eject the electron from its shell. For Compton interactions, the fraction is always less than unity since some energy is always retained by the scattered photon. In pair production, the energy transferred is the incident photon energy less the rest mass energy of the created electron and positron. The energy transferred is thus an average over the probabilities of all interaction types, as well as over the different energy losses possible for each type. Also, similarly to the mass attenuation coefficient, the mass energy transfer coefficient is a weighted average of the elements in the irradiated material; so, in analogy to equation 2.12,

$$\left(\frac{\mu_{tr}}{\rho}\right)=\sum_i f_i\left(\frac{\mu_{tr}}{\rho}\right)_i. \tag{2.21}$$

Perhaps a more important quantity is the proportion of photon energy that is transferred to and subsequently *absorbed* by the medium. This quantity is represented by the mass energy absorption coefficient (μ_{en}/ρ) and is fraction g less than $(\mu_{\tau\rho}/\rho)$ where g is the mean proportion of energy transferred to charged particles that is *not* deposited in collisions with orbital electrons. This energy mainly includes that used in the production of bremsstrahlung photons (see section 2.5) and a small component from annihilation of positrons (produced in pair-production interactions) before they have lost all of their kinetic energy. We have then,

$$\left(\frac{\mu_{en}}{\rho}\right)=\left(\frac{\mu_{tr}}{\rho}\right)(1-g). \tag{2.22}$$

Note that sometimes (μ_{en}/ρ) is written as (μ_{ab}/ρ), but they are the same quantity. The mass energy transfer and absorption coefficients are used in the calculation of kerma and collision kerma, as will be discussed in section 2.4. It is the variation in (μ_{en}/ρ) with medium composition that is largely responsible for sharp changes in absorbed dose between two different adjacent materials. Furthermore, the relative difference between the two materials depends on the photon energy. For example, the (μ_{en}/ρ) values for soft tissue and bone differ more at low energies (where photoelectric interactions are important) than at energies where Compton scattering dominates.

As an example, again consider a 3.0 MeV photon in water, for which $\mu/\rho = 0.0397\ \text{cm}^2\text{g}^{-1}$. The mean energy transferred in each photon interaction is 1.74 MeV, so

$$\begin{aligned}\left(\frac{\mu_{tr}}{\rho}\right)&=\left(\frac{\mu}{\rho}\right)\frac{\overline{E}_{tr}}{E}\\&=0.0397\ \text{cm}^2 g^{-1}\times\frac{1.74}{3.00}\\&=0.0230\ \text{cm}^2 g^{-1}.\end{aligned} \tag{2.23}$$

Also, $(\mu_{en}/\rho) = 0.0228\ \text{cm}^2\text{g}^{-1}$ for this energy, so less than 1% of energy transferred is converted to bremsstrahlung. A much higher proportion of charged particle energy is converted to bremsstrahlung for higher atomic number materials.

In figure 2.4, the different coefficients for water are plotted as a function of energy. The sharp increase in all quantities can be seen at low energies where the photoelectric effect has a very large cross section. Furthermore, it can be seen that below approximately 0.01 MeV (10 keV), the energy coefficients approach the attenuation coefficient. This is because in photoelectric interactions almost all photon energy is given to kinetic energy of the ejected

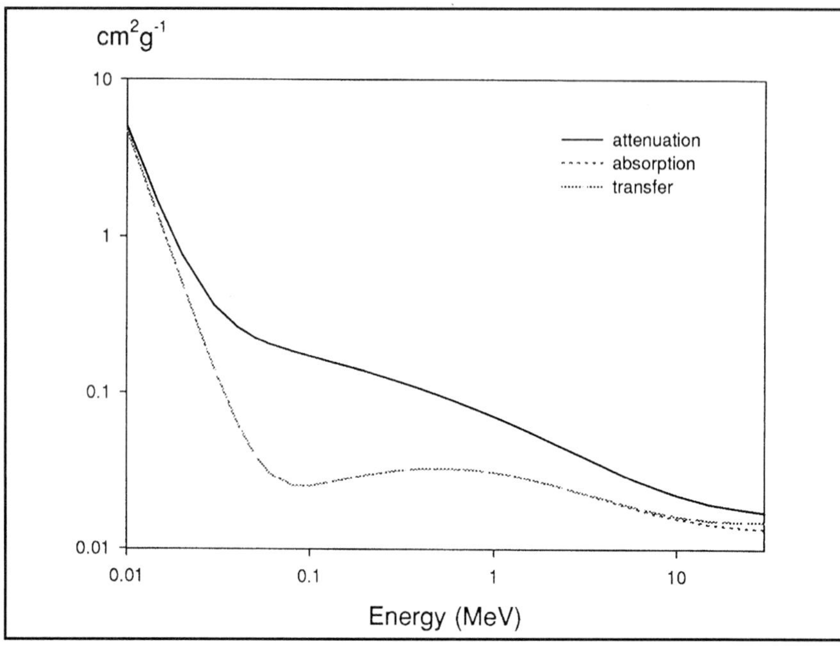

Figure 2.4: Mass attenuation, energy transfer, and absorption coefficients for water. Note the separation of transfer and absorption coefficients at high energies where bremsstrahlung becomes significant.

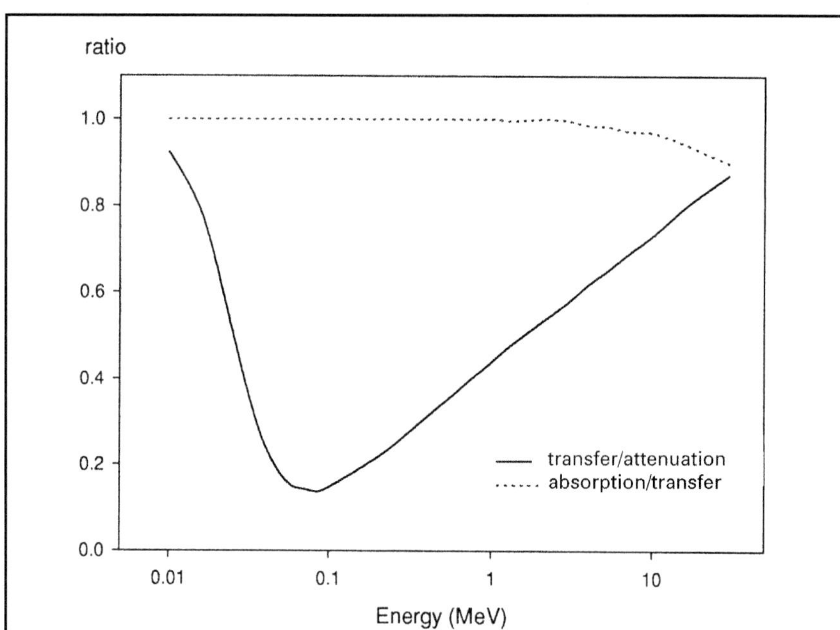

Figure 2.5: Ratio of mass energy transfer to mass attenuation coefficients as a function of photon energy. A low value indicates a small amount of energy transferred to secondary electrons per photon interaction. This in turn implies that a beam containing photon energies in this range will produce a high proportion of scatter dose.

electron. It is only at energies above approximately 10 MeV that a separation of (μ_{en}/ρ) and (μ_{tr}/ρ) due to bremsstrahlung production can be seen.

It is interesting to plot the ratio of transfer to attenuation coefficients, as in figure 2.5. This ratio is the mean fractional energy transfer per collision and is thus inversely proportional to the energy retained by the incident photon. This energy retained is zero for photoelectric and pair production processes, and the energy transferred in Compton collisions increases with energy. There is a minimum in the ratio, at which the energy retained by the scattered photon is a maximum. At this energy, which is at a little less than 100 keV from figure 2.5, we thus expect a maximum proportion of scatter to primary dose. Also plotted in figure 2.5 is the ratio of mass energy absorption to mass energy transfer coefficients, which is seen to be unity until near the high end of the energy range where bremsstrahlung becomes important.

2.3 Photon Interaction Processes

Now that the concepts of attenuation, energy transfer, and absorption coefficients have been introduced, it is time to examine the physics of each interaction process and their dependencies on energy and material composition.

Once electrons have been accelerated to relativistic velocities within the linear accelerator waveguide, they strike a target, and a broad energy spectrum of photons is emitted due to bremsstrahlung production. The resultant photon beam is a narrow bullet-shaped lobe of photons that is reshaped by a beam flattening filter to produce a uniform photon beam at the linear accelerator exit port. These photons then traverse air in a well-defined beam and strike the patient. On entering the patient, any one of the interactions discussed below may occur. The three photon interaction types of relevance at energies used in radiotherapy are photoelectric, Compton, and pair production processes. Coherent scattering is also discussed briefly here, although its cross section is small at radiotherapy energies and, furthermore, it does not contribute to energy deposition by ionization or excitation.

2.3.1 Photoelectric Absorption

Figure 2.6 is a description of the photoelectric process using the Böhr model of the atom. The photon strikes an outer or inner shell electron, and all of its energy is absorbed by the electron that escapes its shell. If an inner shell electron is involved, it is replaced by an outer shell electron. This electron in turn emits some energy in the form of a photon whose energy is characteristic of the energy difference between the two shells. This radiation is known as characteristic radiation. However, for the low atomic number elements in tissue, this energy difference is so small (for carbon it is 0.3 keV) that characteristic radiation is absorbed locally in closely neighboring atoms. In the photoelectric process, almost all of the photon's energy is transferred to the kinetic energy, E_k, of a bound electron such that (Meredith et al. 1977):

$$E_k = E_p - E_b, \tag{2.24}$$

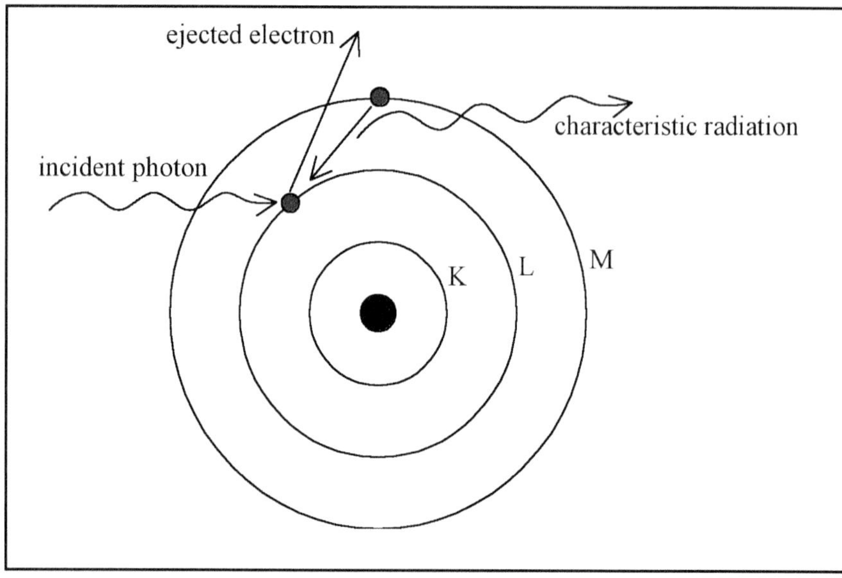

Figure 2.6: Illustration of photoelectric absorption, showing an electron being ejected from the L-shell. The photon is totally absorbed and gives up all its energy to the ejected electron (minus the binding energy). An electron from the M-shell replaces the ejected electron, giving rise to characteristic radiation.

where E_p is the initial photon energy and E_b is the electron binding energy of the ejected electron. The photoelectric component of the atomic cross section varies with atomic number as $Z^{3.8}$ for Z greater than or equal to 16 and varies as $Z^{3.0}$ for Z less than or equal to 15. It also varies as E_p^{-3}, thus the photoelectric component is most important in high Z materials and at low energies. Recalling the definition of the mass attenuation coefficient for elements in equation 2.11, we can write (for low atomic number materials)

$$\sigma_{pe} \propto Z^3 E_p^{-3}$$
$$\Rightarrow \left(\frac{\mu}{\rho}\right)_{pe} \propto \frac{Z^3}{A} E_p^{-3}. \qquad (2.25)$$

Note that this really implies a dependence on Z^2, since Z/A is approximately constant. The photoelectric effect is most likely to occur for photon energies just larger than the binding energy, hence the large decrease in cross section with energy. There are discontinuities in the cross section at energies corresponding to the binding energies of the shells from which electrons can be ejected, since an electron cannot be ejected if photon energy is less than the binding energy.

2.3.2 Compton (Incoherent) Scattering

The most important photon interaction process for the beam energies used in radiotherapy is Compton scattering. In fact from 200 keV to 2 MeV it is the only interaction of importance in soft tissues. As shown in figure 2.7, an

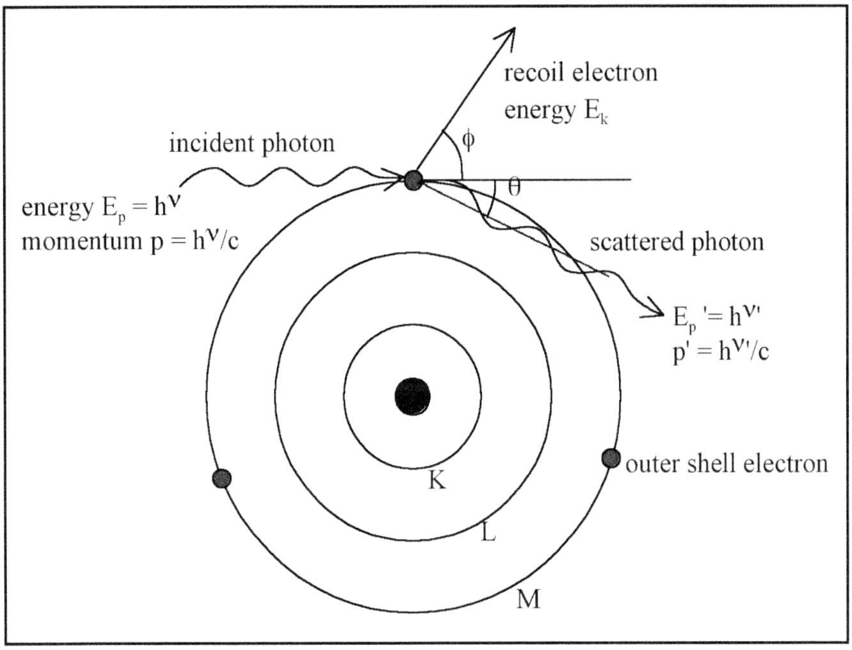

Figure 2.7: Compton scattering process. An incident photon with energy $E = h\nu$ and momentum of magnitude $p = h\nu/c$ interacts with a free electron, giving it an energy E_k and momentum q. The recoil electron is scattered at an angle ϕ and escapes its orbit. The incident photon gives up some energy to the recoil electron, and the rest of the energy goes to the scattered photon which has energy $E = h\nu'$ and momentum $p' = h\nu'/c$. The amount of energy given to the electron and recoil photon depends on the angle of photon scatter q. (Adapted from Fig. 2.14, p. 61, Johns and Cunningham, *The Physics of Radiology*, 4th ed., 1983. Courtesy of Charles C Thomas, Publisher, Ltd., Springfield., Illinois.)

incident photon with energy $E = E_p$ interacts with a free electron in an outer shell that is ejected with an energy E_k. Since the binding energy of the electron is negligible, by conservation of energy, the scattered photon retains an energy

$$E_p' = E_p - E_k. \tag{2.26}$$

Momentum is also conserved, since the collision is elastic. Hence if the momentum (a vector quantity) of the incident photon is p and that of the ejected electron is q, the momentum of the scattered photon is $p' = p - q$.

The ejected electron has a kinetic energy equal to its total energy, E, minus its rest mass energy, E_0 ($E_k = E - E_0$). Since the electron may have a velocity approaching the speed of light, c, then by using the relativistic expression for kinetic energy,

$$E_k = m_0 c^2 \left[\frac{1}{\left(1 - \frac{v^2}{c^2}\right)^{\frac{1}{2}}} - 1 \right]. \tag{2.27}$$

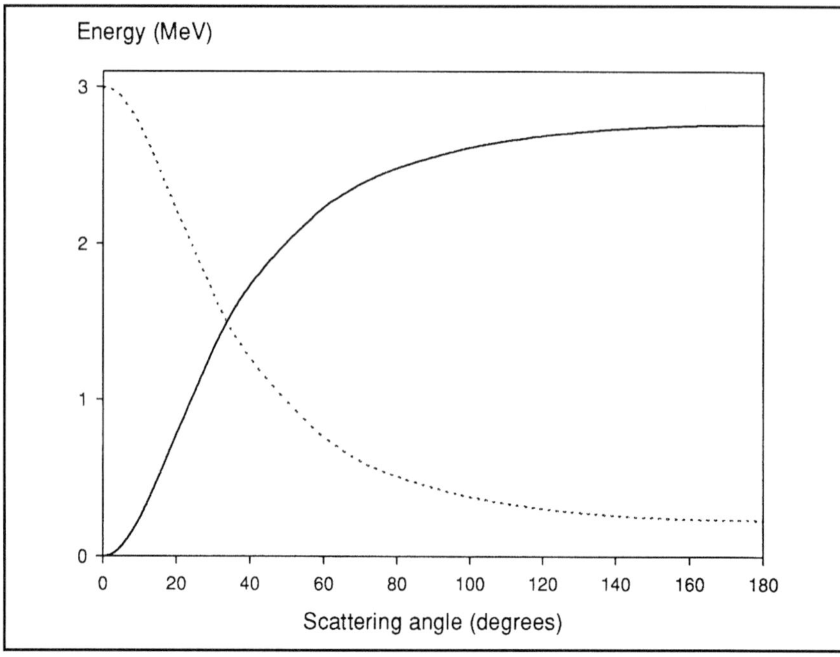

Figure 2.8: The amount of energy given to an electron and photon from an incident 3 MeV photon. The energy depends on the angle of photon scatter θ. For example, from data shown in the graph, the interaction of an incident photon with an electron that causes the photon to scatter at $\theta = 45°$ will produce a scattered photon of energy $h\nu' = 1.104$ MeV and a recoil electron of energy $E_k = 1.896$ MeV. The horizontal axis units are degrees and the vertical axis units are MeV.

Note that for small values of v, equation 2.27 simplifies to $E_k = -m_0v^2$.

By vector analysis, Johns and Cunningham (1983) give the following equations, which show that the kinetic energy given to the recoil electron and the energy retained by the scattered photon are dependent on θ, the angle of recoil photon scatter, as follows:

$$E_k = E_p\left[\frac{\alpha(1-\cos\theta)}{1+\alpha(1-\cos\theta)}\right], \tag{2.28}$$

where $\alpha = E_p/m_0c^2$, the ratio of incident photon energy to electron rest mass energy (0.511 MeV). Now using equation 2.26,

$$E_p' = E_p\left[\frac{1}{1+\alpha(1-\cos\theta)}\right]. \tag{2.29}$$

Figure 2.8 shows the energy given to the recoil electron and scattered photon resulting from a 3 MeV incident photon as a function of scattering angle θ using equations 2.28 and 2.29.

Compton Differential Cross Section

Klein and Nishina (1929) determined by quantum mechanical analysis that the differential cross section for Compton interactions is given by (Heitler 1960):

$$\frac{d\sigma_e}{d\Omega} = \frac{d\sigma_0}{d\Omega} F_{KN}, \tag{2.30}$$

where the subscript *e* is used to indicate that this is a cross section per *electron*. The term $d\sigma_0/d\Omega$ is the classical scattering expression (Johns et al. 1983):

$$\frac{d\sigma_0}{d\Omega} = \frac{r_0^2}{2}\left(1+\cos^2\theta\right), \tag{2.31}$$

where r_0 is the classical electron radius. The term F_{KN} is the Klein-Nishina factor and is given by (Klein et al. 1929):

$$F_{KN} = \left(\frac{1}{1+\alpha(1-\cos\theta)}\right)^2 \left(1+\frac{\alpha^2(1-\cos\theta)^2}{(1+\alpha(1-\cos\theta))\left(1+\cos^2\theta\right)}\right). \tag{2.32}$$

The differential cross section for energy transfer to electrons represents the fraction of incident energy transferred by photons that are scattered into a solid angle dΩ. This is of course given by

$$\frac{d\sigma_{tr}}{d\Omega} = \frac{d\sigma_e}{d\Omega}\frac{E_k}{E_p}, \tag{2.33}$$

where E_k is obtained from equation 2.28.

In figure 2.9, for the case of an incident 3 MeV photon beam, the classical and Klein-Nishina differential cross sections are shown along with the energy transfer differential cross section. The probability of scattering through an angle θ can be seen to decrease as θ increases. Furthermore, the preferential scattering angle becomes smaller as incident photon energy increases—high-energy photons are deflected less.

Note that all the previous equations assume that electrons are free, whereas they are in fact bound. An approximate solution to this problem can be obtained by including an incoherent scattering function that accounts for the electrons that do not escape due to their binding energy. Tables of this information have been produced by Hubbell et al. (1975). However, for energies above 500 keV where Compton interactions dominate, this is a negligible effect (Johns et al. 1983).

The total cross section per electron is obtained by integrating the differential cross section over all Ω:

$$\sigma_e = \int_\Omega \frac{d\sigma_e}{d\Omega}\, d\Omega, \tag{2.34}$$

where the cross section has an approximate energy dependence of $E^{-1/2}$. Similarly, the total energy transfer cross section is

$$\sigma_{tr} = \int_\Omega \frac{d\sigma_{tr}}{d\Omega}\, d\Omega = \int_\Omega \frac{d\sigma_e}{d\Omega}\frac{E_k}{E_p}\, d\Omega, \tag{2.35}$$

which is the fraction of incident energy transferred per electron as photons in a beam of unit cross-sectional area traverse a unit pathlength.

Most importantly with Compton interactions, there is no dependence on the charge of the nucleus (apart from the fact that the number of electrons per

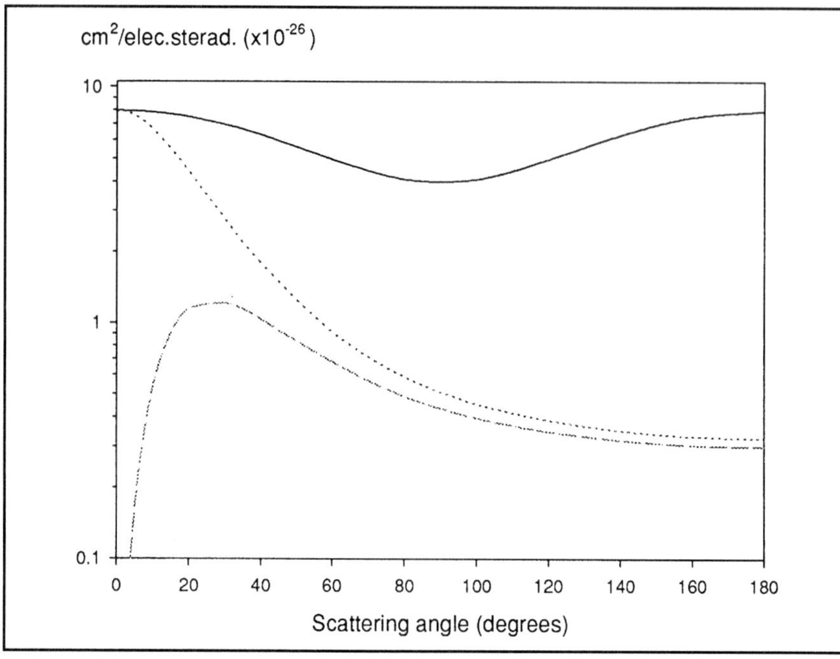

Figure 2.9: The differential cross section $d\sigma$ per unit solid angle $d\Omega$ plotted against angle of photon scatter θ. Data shown in the graph include the classical differential cross section per unit solid angle $d\sigma_0/d\Omega$, the Compton scattered photon differential cross section per unit solid angle $d\sigma/d\Omega$, and the Compton recoil electrons differential cross section per unit solid angle $d\sigma_{tr}/d\Omega$. The horizontal axis units are degrees, and the vertical axis units are (cm²/electron steradian)×10⁻²⁶.

atom is equal to the atomic number) because the interactions are with free electrons. Now recalling equation (2.14), and using the $E^{-\frac{1}{2}}$ dependence of σ_e,

$$\left(\frac{\mu}{\rho}\right)_{incoh} \propto \frac{Z}{A} E_p^{-\frac{1}{2}}. \tag{2.36}$$

So, the Compton mass attenuation coefficient is proportional to the number of electrons per unit mass ($\propto Z/A$) and inversely proportional to the square root of the energy.

2.3.3 Pair Production

When the energy of an incident photon is greater than the rest mass energy of an electron and a positron ($2m_0c^2 = 1.022$ MeV), the photon may be absorbed through the mechanism of pair production. Here, the photon passes very close to the nucleus, and its energy is used to create an electron-positron pair. The excess energy (E_p–1.022 MeV) is shared between the electron and positron as kinetic energy.

It is easily shown that the rest mass energy of an electron or positron is 0.511 MeV using Einstein's equation, $E = mc^2$. The mass m_0 of the electron is 9.11×10^{-31} kg and the speed of light is approximately 3.00×10^8 ms^{-1}, thus

$$\begin{aligned} E &= m_0 c^2 \\ &= 9.109\times10^{-31}\,kg\times\left(2.998\times10^{8}\,ms^{-1}\right) \\ &= 8.187\times10^{-14}\,joules \end{aligned} \tag{2.37}$$

and by converting from joules to MeV,

$$\begin{aligned} E &= \frac{8.187\times10^{-14}\,joules}{1.602\times10^{-13}\,joules\;MeV^{-1}} \\ &= 0.511\,MeV. \end{aligned} \tag{2.38}$$

Thus the energy equivalent of one electronic mass unit is 0.511 MeV. As two charged particles are formed, the threshold energy required for pair production by the photon is 1.022 MeV. The mechanism is described in figure 2.10, which shows the incident photon colliding with the nucleus and a electron-positron pair being formed. The positron interacts with an electron after losing its kinetic energy, in which process both particles are annihilated. In this process two photons each of 0.511 MeV are emitted in the conversion of mass to energy. This radiation is called annihilation radiation. Also shown in figure 2.10 is triplet production. This is a similar interaction to pair production except that the interaction occurs in the field of an electron rather than in the field of the nucleus. Three particles are involved—the original electron plus the created electron and positron. The threshold for this interaction is 2.044 MeV. The cross section for triplet production is usually small compared with pair production and is usually included in the pair production cross sections.

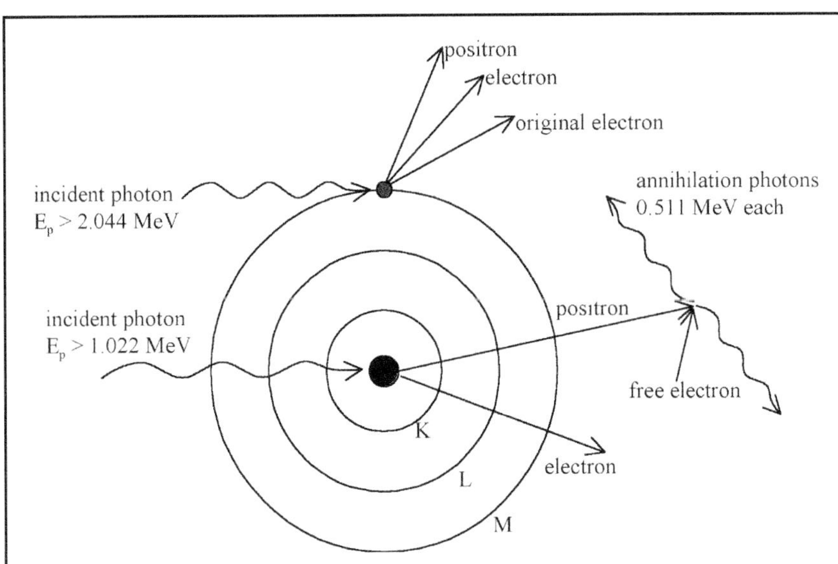

Figure 2.10: The absorption of photons by the pair and triplet production processes. The incident photons require an energy of 1.022 and 2.044 MeV, respectively. In each case a positron-electron pair is emitted. The positron then interacts with a free electron to produce two 0.511 MeV photons.

Note that energy and charge are conserved between the photon, electron, and positron, but momentum is not conserved between the three particles because some momentum is given to the nucleus. As the nucleus attains some of the momentum, it is difficult to analyze the collision in detail because the exact angular deviation of the positron and electron are unknown. The differential cross section for the production of an electron-positron pair has been shown by Heitler (1960) to be approximately equal to

$$\frac{d\sigma_{pair}}{d\Omega} = \frac{Z^2}{137}\frac{r_0^2}{2\pi} m_0^2 c^4 F_{pair}. \tag{2.39}$$

The term F_{pair} is a complicated function of momentum, energy, and angle of projection of the positron and the electron, and $d\Omega$ is the solid angle into which the photon is projected.

The pair production cross section also increases with energy, which means a high-energy beam is less penetrating than a low-energy beam when the main attenuation process is pair production. This is in contrast to the Compton and photoelectric processes, whose probability of occurrence decreases with increasing energy.

Now using equation 2.11, and the fact that the pair production cross section is proportional to Z^2 and to the log of the energy, we can write

$$\begin{aligned} &\sigma_{pair} \propto Z^2 \log E_p \\ &\Rightarrow \left(\frac{\mu}{\rho}\right)_{pair} \propto \frac{Z^2}{A} \log E_p, \end{aligned} \tag{2.40}$$

which gives a dependence on Z, since Z/A is approximately constant.

2.3.4 Coherent Scattering

This process is not of much interest in radiotherapy since it is insignificant at energies above approximately 100 keV in soft tissues, and, furthermore, no energy is transferred to charged particles. This means that coherent scattering is only important in the deflection of low-energy photons and is not directly responsible for energy deposition. Coherent scattering occurs when the electromagnetic wave associated with a photon causes the atomic electrons to momentarily oscillate (at the same frequency, ν, as the photon, where the energy of the photon is $h\nu$). This vibration causes electromagnetic radiation to be emitted from *all* electrons in the atom, hence the radiation is termed *coherent.*

The cross section for coherent scattering is approximately proportional to $Z^{2.5}$ and to E_p^{-2}, so we have

$$\begin{aligned} &\sigma_{coh} \propto Z^{2.5} \log E_p^{-2} \\ &\Rightarrow \left(\frac{\mu}{\rho}\right)_{coh} \propto \frac{Z^{2.5}}{A} \log E_p^{-2}, \end{aligned} \tag{2.41}$$

which gives an approximate dependence on $Z^{1.5}$ and an energy dependence which is similar to that for the photoelectric effect.

2.3.5 The Relative Importance of Each Process

The total mass attenuation coefficient is the sum of the attenuation coefficients due to photoelectric, incoherent (Compton) scattering, pair production, and coherent scattering processes:

$$\frac{\mu}{\rho}=\left(\frac{\mu}{\rho}\right)_{pe}+\left(\frac{\mu}{\rho}\right)_{incoh}+\left(\frac{\mu}{\rho}\right)_{pair}+\left(\frac{\mu}{\rho}\right)_{coh}. \tag{2.42}$$

The components of the mass attenuation coefficient (along with the total coefficient) for the element oxygen are plotted against energy in figure 2.11(a). It can be seen that Compton interactions are dominant from about

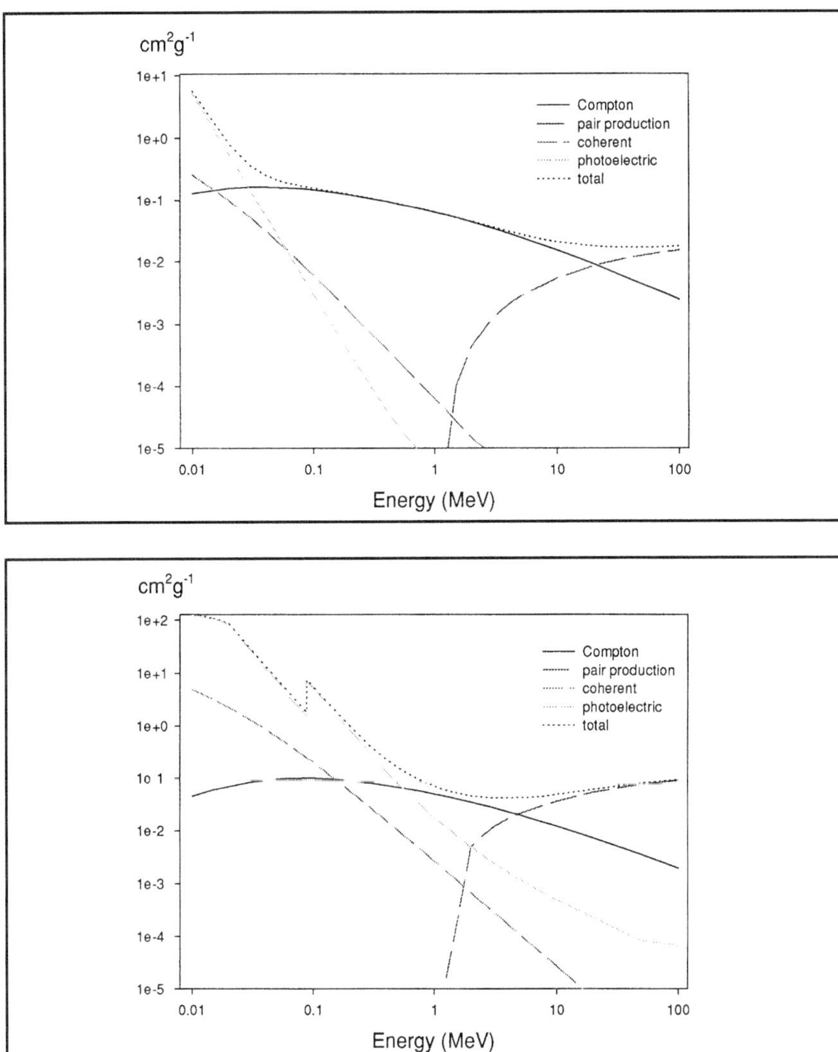

Figure 2.11: (a) Mass attenuation coefficients for oxygen as a function of energy, showing the components due to Compton, pair production, coherent, and photoelectric processes. The values are calculated using equation 2.11, with atomic cross section values from Johns and Cunningham (1983). (b) The comparison repeated for lead.

30 keV to 20 MeV. In fact, for much of the energy range of interest in radiotherapy, the Compton cross section is orders of magnitude greater than the other cross sections—the Compton and total curves are seen to coincide from approximately 0.15 to 2 MeV. This comparison is repeated for lead in figure 2.11b, where it can be seen that both the photoelectric and pair production processes are much more significant than in the case of oxygen and that the energy range dominated by Compton interactions is much narrower. Moreover, the total and Compton curves do not coincide. In the curve for the photoelectric component in lead, the K-shell edge can be seen, where the cross section increases as the energy becomes high enough to eject an electron from the K-shell.

The consequence of high photoelectric and pair production cross sections for high atomic number materials is that at energies below and above the range where Compton interactions dominate (and this range decreases as atomic number increases), the total mass attenuation coefficient is higher for high atomic number materials. The mass attenuation and energy absorption coefficients are shown for water and lead in figure 2.12, where the sharp increase (due to photoelectric absorption) in both coefficients for lead as energy is decreased can be seen. There is also an increase for high energies (due to pair production) that is not seen for water. Notice that between approximately 1–3 MeV, the coefficients are larger for water than for lead.

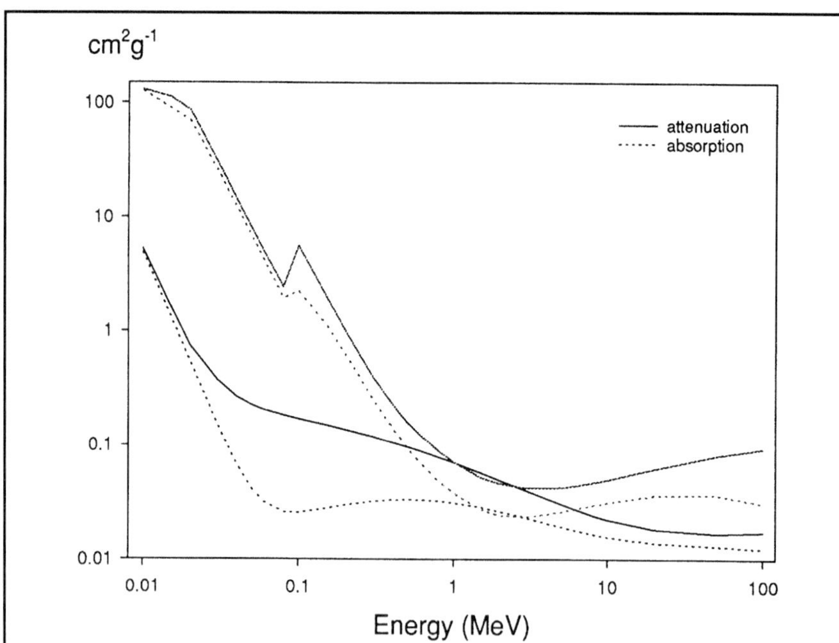

Figure 2.12: Mass attenuation and mass energy absorption coefficients for water and lead. The large photoelectric cross section for high atomic numbers causes the high values for lead compared with water at low energies. In the middle of the energy range, where Compton scattering dominates, the coefficients are larger for water than for lead due to a higher number of electrons per gram. At higher energies, the high pair production cross section of lead causes the coefficients to be higher than for water.

This is because Compton interactions dominate in this region, and the number of electrons per gram is larger for water than for lead. It is also interesting to note that for lead, the mass energy absorption coefficient at lower energies is closer to the mass attenuation coefficient than is the case for water. This is because of the dominance of photoelectric interactions in the case of lead, where all photon energy is transferred to the electron.

2.4 Photon Beam Transport

In this section, some of the interesting properties of photon beam transport and dose distribution will be examined in light of the characteristics of the interaction and absorption coefficients discussed above. Most beam characteristics, such as the distributions of fluence, energy fluence, kerma, and dose, are dependent on the energy and material dependence of these coefficients. These quantities will be discussed and calculated for a range of *polyenergetic* beams, each of which contains a spectrum of energies. Beams used as examples include 250 kVp x-rays (used in orthovoltage therapy) as well as 6 MV and 24 MV linear accelerator beams. The broad energy spectrum contained within each beam makes the calculation of parameters more difficult but also more interesting, since each energy component is attenuated and absorbed differently. Furthermore, the different energy ranges of each beam mean that (i) the relative variations and magnitudes of parameters characterizing the beams are different and (ii) different considerations are important when the beams are used for treatment.

Very little measured information about high-energy spectra is available, mainly due to difficulties with the high photon fluence saturating detecting systems. Some measurements have been attempted using spectral analysis (Levy et al. 1976, Nath et al. 1976) and attenuation analysis (Huang et al. 1982a, 1982b). The devices and computations employed are complicated and computationally intensive, respectively. The spectra used here are from data distributed with the EGS4 Monte Carlo code (Nelson et al. 1985). These data have been produced using Monte Carlo simulations combined with geometric information about the treatment heads of particular accelerators (Mohan et al. 1985).

2.4.1 Fluence and Attenuation of Polyenergetic Beams

Beam Spectrum

Photons in a beam containing a spectrum of energies are attenuated at different rates that depend on their energy as well as the density and composition of the absorbing material. The total fluence of a polyenergetic beam at the surface ($z = 0$) of the medium is given by the sum of the fluence of each component:

$$\Phi(0) = \int_0^{E_{max}} \frac{\partial\Phi(E,0)}{\partial E} dE, \tag{2.43}$$

where $\partial\Phi(E,0)/\partial E$ is the fluence per unit energy interval (per MeV) at energy E and depth 0. E_{max} is the maximum energy in the beam. The function describing the fluence per energy interval can be normalized such that the value of the integral is unity. In N discrete energy intervals of (possibly varying) width ΔE_n, the fluence at the surface can be written

$$\Phi(0) = \sum_{n=1}^{N} \Phi_n(0)\Delta E_n, \tag{2.44}$$

where $\Phi_n(0)$ is the fluence per MeV at the surface for energy interval n, which is centered on energy E_n. The spectrum of a megavoltage photon beam is characteristic of bremsstrahlung emission from a thick target, where the maximum photon energy is equal to the maximum incident electron energy. The photon fluence increases with decreasing energy, typically reaching a peak at an energy of about one-third of the maximum. Photon fluence drops off below the energy of the peak because of preferential attenuation of low-energy photons by the target material. (In bremsstrahlung from a thin target, such self-absorption is less and the reduction in fluence at low energies is correspondingly reduced.)

Beam Hardening

The attenuation of a polyenergetic beam can be calculated on a component-by-component basis, using the energy-dependent mass attenuation coefficient $(\mu/\rho)_n$. Using equation 2.8, the fluence of component n at depth z is

$$\Phi_n(z) = \Phi_n(0)\exp\left(-\left(\frac{\mu}{\rho}\right)_n z\right) \tag{2.45}$$

and the total fluence is

$$\Phi(z) = \sum_{n=1}^{N} \Phi_n(z), \tag{2.46}$$

where for our purposes the reduction in fluence due to beam divergence from a point source is ignored. Now, since the degree of attenuation is energy dependent, the relative heights of each component in the spectrum will vary with depth. At radiotherapy energies in soft tissue, the mass attenuation coefficient decreases with energy (see figures 2.11a and 2.12), so the attenuation of low-energy components is greatest, giving an increase in mean energy with depth and a shift of the spectrum plot to the right. This phenomenon is known as *beam hardening* and is illustrated for the 250 kVp, 6 MV, and 24 MV beams in figure 2.13, where the spectra at the surface and at 20 cm deep are shown (both normalized such that the total fluence is unity).

Note that the discussion above refers to the primary beam only. The actual energy distribution at any depth will include scattered photons, which have a lower average energy than the primary photons. In figure 2.13b, for the 6 MV beam, the spectrum of *all* photons at 20 cm deep is shown along with the primary spectrum. This total spectrum is seen to include a considerable low-energy component, below about 0.5 MeV. The total spectrum

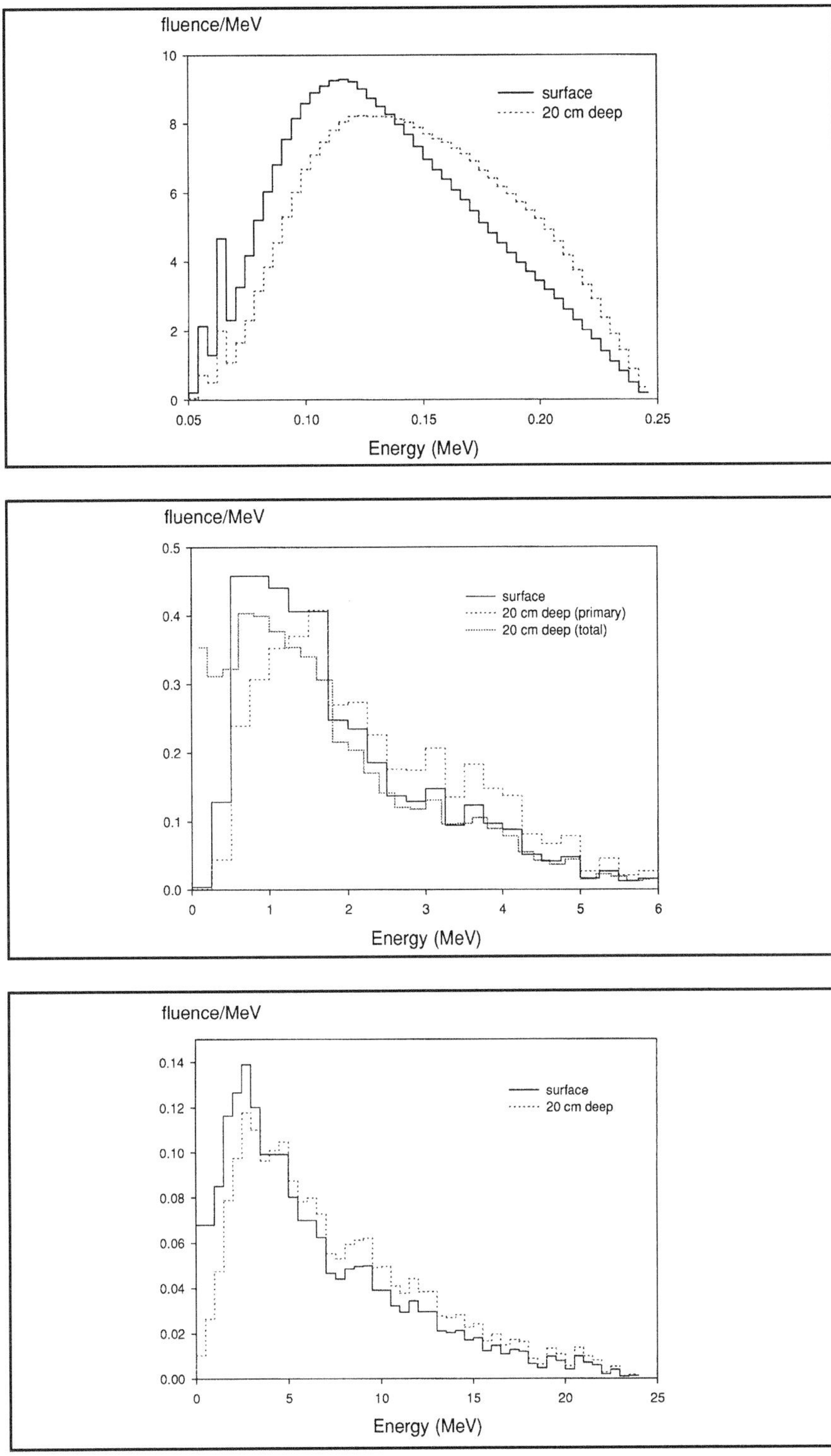

Figure 2.13: Primary spectra at the surface and at 20 cm deep for (a) 250 kVp, (b) 6 MV, and (c) 24 MV photon beams. For 6 MV, the spectrum of all photons including scatter is also shown (obtained from a Monte Carlo simulation). Incident spectra are taken from EGS4 Monte Carlo data files.

shown here was generated with an EGS4 Monte Carlo simulation, where the energies of all photons at a range of depths were recorded and assigned to the appropriate energy interval. Interestingly, the mean energy of the incident spectrum and the total spectrum at 20 cm deep is the same (1.80 MeV), while the mean of the primary photons at 20 cm deep is 1.91 MeV.

Energy Fluence

Often the spectrum of a polyenergetic beam is defined in terms of the *energy fluence* of each component. If the fluence of photons of energy E at some point is Φ, then the energy fluence is simply $\Psi = \Phi E$. Thus, the energy fluence of component n at depth z is

$$\begin{aligned}\Psi_n(z) &= \Phi_n(z)E_n \\ &= \Phi_n(0)E_n \exp\left(-\left(\frac{\mu}{\rho}\right)_n z\right) \\ &= \Psi_n(0)\exp\left(-\left(\frac{\mu}{\rho}\right)_n z\right).\end{aligned} \tag{2.47}$$

and the total energy fluence is, of course,

$$\Psi(z) = \sum_{n=1}^{N} \Psi_n(z). \tag{2.48}$$

Now by definition, the ratio $\Psi(z)/\Phi(z)$ is the mean energy at depth z, since this is a fluence-weighted sum of the energy components. It is interesting, therefore, to calculate both fluence and energy fluence so that we may obtain the mean energy versus depth. Note that the fluence and energy fluence must not be renormalized for this calculation. Plots of the total fluence and energy fluence, both normalized to 1.0 at the surface, are shown for the three beams in figure 2.14. The shallower slope of the energy fluence curve compared with the fluence curve indicates an increase in mean energy with depth, while the ratio of the two normalized curves gives the relative increase in mean energy. Also shown in figure 2.14 is the mean energy versus depth, which is seen to be linearly increasing. This linear increase is a consequence of the relatively small change in mean energy and the near-linear variation in the mass attenuation coefficient over that range.

It is interesting that the increase in mean energy seen in figure 2.14 is larger for the 6 MV and 24 MV beams than for the lower energy. This can be explained by the fact that the variation in mass attenuation coefficient throughout the energy range of each beam is larger and, hence, there is more differential attenuation of the components. Mass attenuation coefficient is shown plotted against energy (which is normalized to the maximum in each spectrum) in figure 2.15.

Again, it should be emphasized that the hardening shown here is for the *primary* component of the beam only. When scattered photons are included, the mean energy varies in a more complicated way that depends on the field

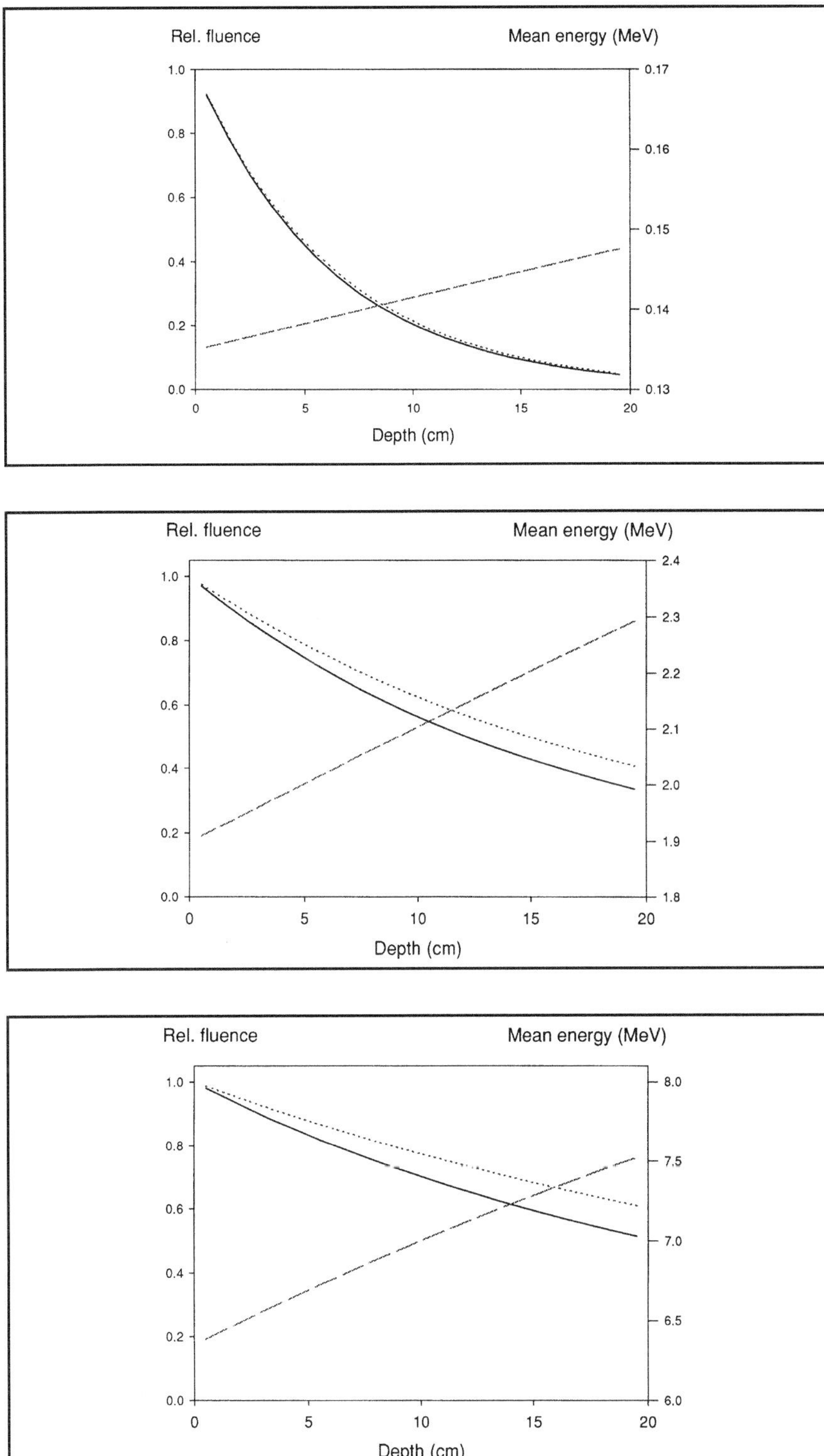

Figure 2.14: Total fluence and energy fluence for (a) 250 kVp, (b) 6 MV, and (c) 24 MV beams normalized to unity at the surface. Also shown is the mean energy variation with depth.

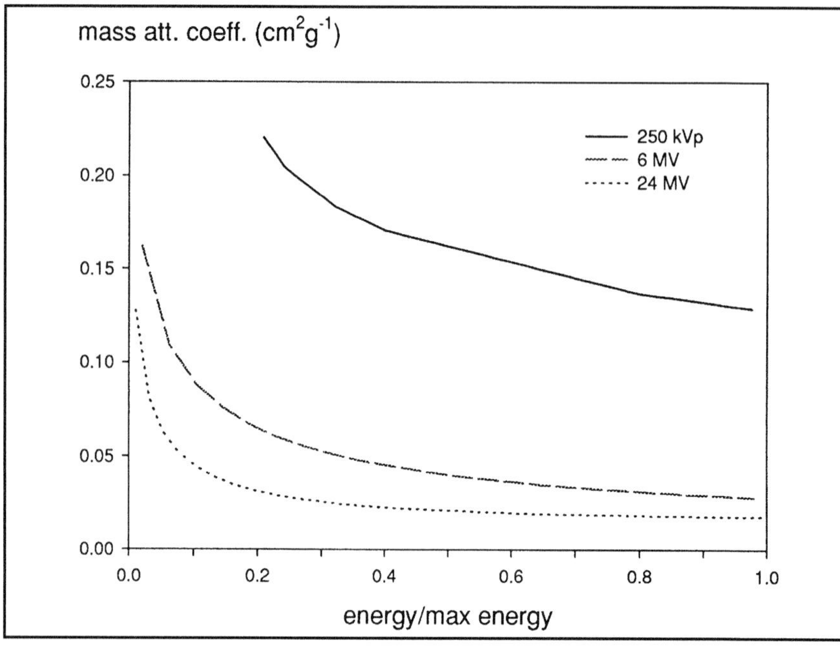

Figure 2.15: Variation in mass attenuation coefficient for 250 kVp, 6 MV, and 24 MV spectra versus energy, where energy is normalized to the maximum in each spectrum.

size. A larger field gives more scatter and hence a larger contribution to the mean energy from low-energy photons. The mean energy of all photons actually decreases initially with depth because there is a rapid increase in scatter, then begins to increase as hardening of the primary becomes significant. The rate of the initial reduction, depth at which mean energy begins to rise, and the rate of the rise depend on field size.

2.4.2 Kerma, Collision Kerma, and Terma

These are the *middlemen* between fluence or energy fluence and the absorbed dose, since their distributions are characterized by the distribution of photon fluence, but they represent an energy imparted to the medium and have the same units as dose.

Kerma is the energy imparted by photons to the medium, which may subsequently be deposited and give rise to dose. Kerma stands for kinetic energy released per unit mass and is expressed in Gy. If there is an energy $d\overline{E}_{tr}$ transferred to a mass dm, the kerma at that point is

$$K = \frac{d\overline{E}_{tr}}{dm}. \tag{2.49}$$

At a point where there is a fluence Φ of photons with energy E, the kerma is

$$K = \Phi E\left(\frac{\mu_{tr}}{\rho}\right) = \Psi\left(\frac{\mu_{tr}}{\rho}\right). \tag{2.50}$$

This equation obviously gives energy imparted per unit mass, since there is an energy Ψ incident on the unit mass (a region of unit cross-sectional area and length $1/\rho$ cm) and a fraction μ_{tr}/ρ of this energy is imparted in that volume. It is simple to calculate the kerma for each energy component in the spectrum at depth z from the energy fluence of the component

$$K_n(z) = \Psi_n(z)\left(\frac{\mu_{tr}}{\rho}\right)_n \tag{2.51}$$

and the total kerma is the sum of each component.

The kerma includes all energy transferred to charged particles but, as discussed in section 2.2.3, some energy transferred is subsequently radiated away as bremsstrahlung and thus does not contribute to the locally absorbed dose. The kerma is thus partitioned into collisional and radiative parts, where the collision kerma is equal to the absorbed dose except for the slight difference caused by the transport of charged particles away from the site of energy transfer. Similarly to the kerma, the collision kerma is given by

$$K_c = \frac{d\bar{E}_{en}}{dm} = \frac{d\bar{E}_{tr}}{dm}(1-g). \tag{2.52}$$

The *collision kerma* for component n of an energy spectrum is then equal to

$$K_{c,n}(z) = \Psi_n(z)\left(\frac{\mu_{en}}{\rho}\right)_n = K_n(z)\frac{\left(\frac{\mu_{en}}{\rho}\right)_n}{\left(\frac{\mu_{tr}}{\rho}\right)_n} = K_n(z)(1-g_n). \tag{2.53}$$

Since the fraction, g, of energy transferred that is converted to bremsstrahlung is low at low energies, the kerma and collision kerma are only different at higher energies (just as is the case for mass energy transfer and absorption coefficients). Note that if equation 2.48 is used to calculate the energy fluence at depth z, the above expressions give the *primary* components of the kerma and collision kerma.

Collision kerma, and thus the dose, obviously increases with the mass energy absorption coefficient. We therefore expect that in two adjacent regions (of different composition) where the energy fluence can be considered equal, the dose will be higher in the region where the mass energy absorption coefficient is higher (on average over the spectrum). A good example of this is illustrated in figure 2.16a, where an EGS4 Monte Carlo simulation has been performed for the three beams considered here, all incident on a water medium containing a 1 cm thick slab of cortical bone. In the case of the 6 and 24 MV beams, the dose dips in the bone, whereas the dose rises sharply for the 250 kVp beam. This behavior is explained by the fact that Compton interactions dominate in water and bone for the higher energies and therefore the effective bone/water ratio of mass energy absorption

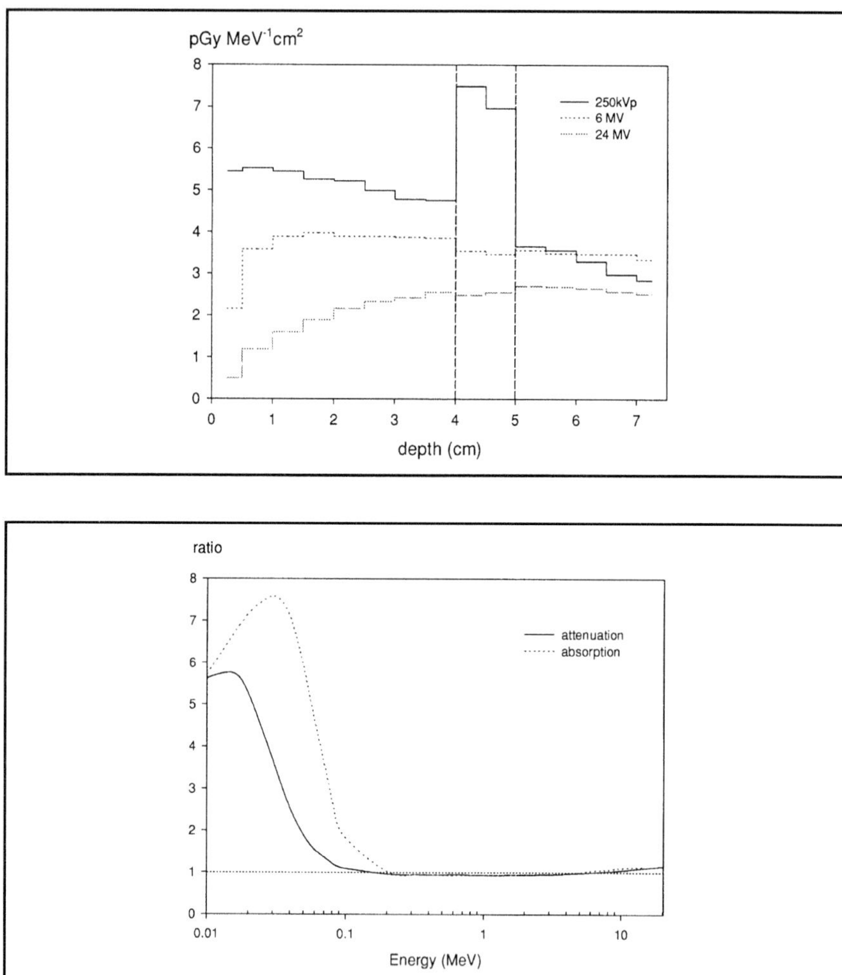

Figure 2.16: (a) Dose calculated using EGS4 Monte Carlo for a water medium containing a 1 cm bone slab for the 250 kVp, 6 MV, and 24 MV beams. Dose in bone drops for the higher energies because of the lower number of electrons per gram and increases for the low energy because of the increased photoelectric cross section. (b) Bone/water ratio of mass attenuation and mass energy absorption coefficients as a function of energy. The high ratio at low energies (due to photoelectric interactions) gives rise to the increase in dose in bone for 250 kVp in (a). The ratio of less than unity for energies between 0.2 and 5 MeV gives the reduction in dose in bone for the higher energies.

coefficients is less than unity (since there are fewer electrons per gram in bone than in water). On the other hand, there is a large enough proportion of photoelectric interactions occurring in the lower energy beam (for which the cross section is significantly higher for bone) that the ratio is considerably greater than unity. The bone/water ratio of mass energy absorption and mass attenuation coefficient is shown for a large energy range in figure 2.16b, where values of less than 1.0 for both ratios are seen between approximately 0.2 and 5 MeV. It is interesting that it is the absorption ratio that rises most rapidly with decreasing energy. This is because not only are photoelectric interactions becoming more prevalent but also because in each interaction *all* photon energy is then transferred.

The *terma* is the *total* energy released per unit mass and includes the energy retained by the scattered photon. It is thus the energy *removed* from the primary beam and is given for component n by

$$T_n(z) = \Psi_n(z)\left(\frac{\mu}{\rho}\right)_n . \tag{2.54}$$

Terma is of course always larger than the kerma by a factor μ/μ_{tr}. It is in general not as useful as the kerma or collision kerma because it does not approximate the primary or total dose. (This is useful since it is always larger than the primary, and energy carried away from a point by scattered photons is deposited a long distance away, if at all, in the irradiated medium.) It is, however, normally used as the factor modulating the *energy deposition kernel* in the *convolution* method of photon beam dose calculation (see chapter seven) and so is an important quantity to understand. Another way of expressing the terma is in terms of the *divergence* of the energy fluence (the energy carried out of a small region minus the energy carried in). The terma is actually the negative of this quantity divided by the density (Ahnesjö 1989):

$$T(z) = -\frac{1}{\rho(z)}\nabla\cdot\Psi(z). \tag{2.55}$$

2.4.3 Kerma and Dose for Different Energies

If it could be assumed that charged particles ejected in photon interactions traveled an insignificant distance from the interaction site, the collision kerma would exactly give the dose. In fact, the kerma would then also give the dose, since the energy of the ejected electrons would be so low that bremsstrahlung would be insignificant or nonexistent. For megavoltage beams, however, the electron range can be several centimeters, especially when a low-density material such as lung is irradiated. The dose distribution therefore resembles a *smeared out* collision kerma distribution, where the amount of smearing depends on the electron range. A good example of this smearing is the convolution of a kerma (or collision kerma) distribution with an energy deposition kernel in the convolution method of photon beam dose computation. Although the terma is normally used in convolution calculations to calculate primary and scatter dose, primary kerma or collision kerma can also be used along with kernels representing the energy deposition due to ejected electrons (at least for calculation of primary dose). Since electrons are predominantly ejected in a forward direction, it is expected that, on average, the energy imparted to electrons in primary photon collisions will be deposited slightly downstream of the primary interaction site. Thus, the primary dose curve is expected to resemble the primary collision kerma curve, displaced to a slightly greater depth. The scatter dose can also be calculated accurately using a convolution calculation, where the terma is convolved with a kernel representing the spreading out of scattered photons from the primary interaction site.

We can write an approximate expression for the relationship of primary dose to primary collision kerma that includes a displacement term. If dose resembles the collision kerma displaced a distance $\overline{z}$ downstream, then

$$D_p(z) = K_c(z - \overline{z}), \tag{2.56}$$

where $\overline{z}$ can be considered the effective longitudinal range of ejected electrons. Now, using the approximate exponential attenuation of the primary collision kerma (with an effective attenuation coefficient μ').

$$K_c(z) = K_c(0)\exp(-\mu' z) \tag{2.57}$$

and making the approximation that $\exp(-\mu' z) \approx 1 + \mu' z$, we have

$$D_p(z) \approx K_c(z)(1 + \mu'\overline{z}), \tag{2.58}$$

Also, an approximate expression using the primary kerma instead of the primary collision kerma is

$$D_p(z) \approx K(z)(1 + \mu'\overline{z})\frac{\mu'_{en}}{\mu'_{tr}}, \tag{2.59}$$

where μ_{en}' and μ_{tr}' are effective energy absorption and energy transfer coefficients for the spectrum of photon energies at depth z.

From equations 2.58 and 2.59 it is expected that primary dose will be slightly larger than primary *collision* kerma, but it may be smaller than the primary kerma depending on the relative magnitudes of $1 + \mu' z$ and μ_{en}'/μ_{tr}.

It is important to note that the above approximations are only valid under conditions *of lateral electronic equilibrium*, which is discussed in section 2.6. If this condition does not exist, the assumption that primary dose is equal to collision kerma slightly upstream is not valid and primary dose can then be much lower than the collision kerma. This is due to a disequilibrium situation in which electrons are displaced laterally to regions outside the beam boundaries, from which no ejected electrons are originating.

In this section, calculations of primary kerma and collision kerma (using the methods in the last section) are compared with Monte Carlo calculations of primary scatter and total dose, for the 250 kVp, 6 MV, and 24 MV beams. Particular note should be taken of the relative magnitudes of primary dose, kerma, and collision kerma. Also interesting is the variation in the relative abundance of scatter dose as energy and field size change.

6 MV Beam

This beam is examined first because it is perhaps the least interesting. The scatter dose contribution is relatively small, bremsstrahlung production by ejected electrons is negligible, and electron range is also small. Figure 2.17a shows the Monte Carlo-calculated primary, scatter, and total dose along the beam axis compared with the primary kerma and collision kerma for a 5 cm diameter beam. Also included is a Monte Carlo calculation of primary kerma, as a check that the analytical model and the Monte Carlo simulations are consistent. It can be seen that calculated kerma and collision kerma are almost identical (since the ratio μ_{en}'/μ_{tr}' is approximately unity) and agree very well with the Monte Carlo-generated kerma. Primary dose coincides with the kerma data, indicating that the product $\mu'\bar{z}$ is negligible and the effective longitudinal range of ejected electrons is small (at least for unit density material). Scatter dose is only 10–20% of the total dose in this case. In the case of the 20 cm beam in figure 2.17b, however, scatter dose is considerably increased due to the larger volume in which primary photon interactions occur. Primary dose and kerma, of course, remain the same as for the smaller beam.

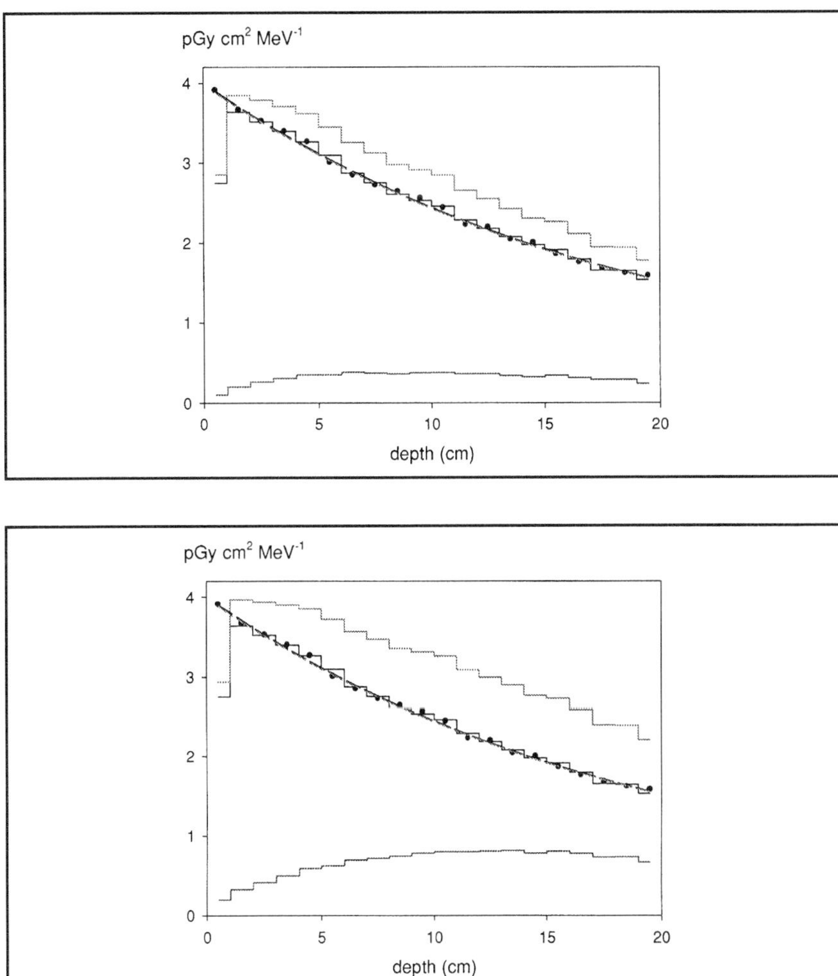

Figure 2.17: Monte Carlo-generated primary scatter and total dose (histograms) compared with calculated kerma and collision kerma (broken lines) for a 6 MV beam. Also shown for comparison is the Monte Carlo-generated kerma (dots). (a) shows the results for a 5 cm diameter beam and (b) for a 20 cm diameter beam.

250 kVp Beam

Again, Monte Carlo simulations were performed for 5 and 20 cm diameter beams, and the primary scatter and total doses versus depth were recorded along with the kerma. Results are plotted for the 5 cm beam in figure 2.18a, in which the calculated kerma and collision kerma are overlaid. Bremsstrahlung emission by the ejected electrons is nonexistent at such low energies, and thus kerma and collision kerma are equivalent. Electron range is of course also negligible and primary dose thus coincides with kerma. The interesting aspect of this graph is the large proportion of scatter dose compared with the 6 MV beam. This is a consequence of the small proportion of energy transferred to electrons in each primary photon interaction (see figure 2.5), where the relatively low values of μ_{tr}/μ for the energy range in the 250 kVp beam are obvious. The data for the 20 cm beam in figure 2.18b show a very large scatter dose component, now at least twice the primary component for most of the depth

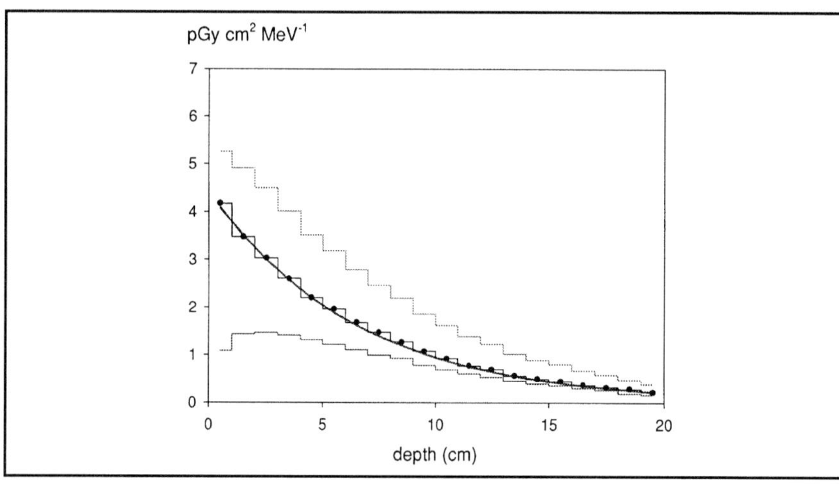

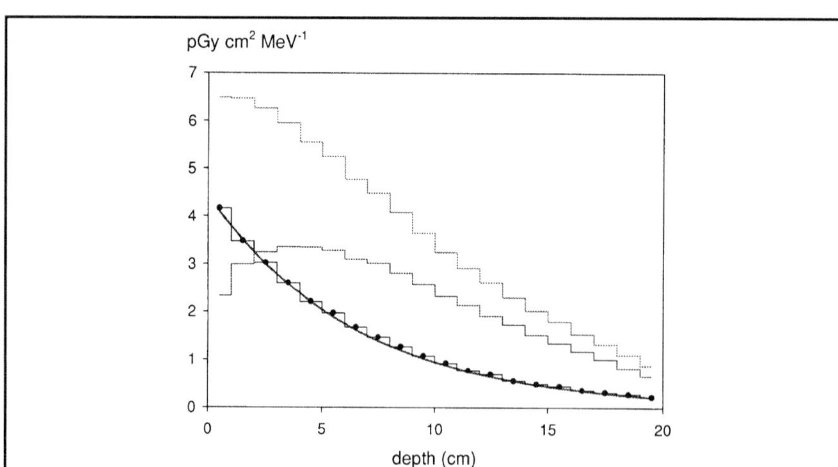

Figure 2.18: Monte Carlo-generated primary scatter and total dose (histograms) as well as kerma (dots) compared with calculated kerma and collision kerma (broken lines) for a 250 kVp beam. (a) shows the results for a 5 cm diameter beam and (b) is for a 20 cm diameter beam.

range. Note, however, that this trend of more scatter at lower energies does not continue—as in photoelectric interactions all energy is transferred to the electron, and no scatter dose results. (Again refer to figure 2.5 where the increase in μ_{tr}/μ at low energies can be seen.)

24 MV Beam

This beam gives the most interesting results, since bremsstrahlung is now appreciable, as is the effective longitudinal electron range. The field sizes chosen for the 24 MV beam were 2 and 20 cm diameters; the small diameter serves to illustrate the effect of lateral electronic disequilibrium. Looking first at the 20 cm beam (figure 2.19a), the collision kerma is now considerably less than the kerma because the ratio μ_{en}'/μ_{tr} is significantly less than unity. Also, primary dose is larger than collision kerma, due to the forward

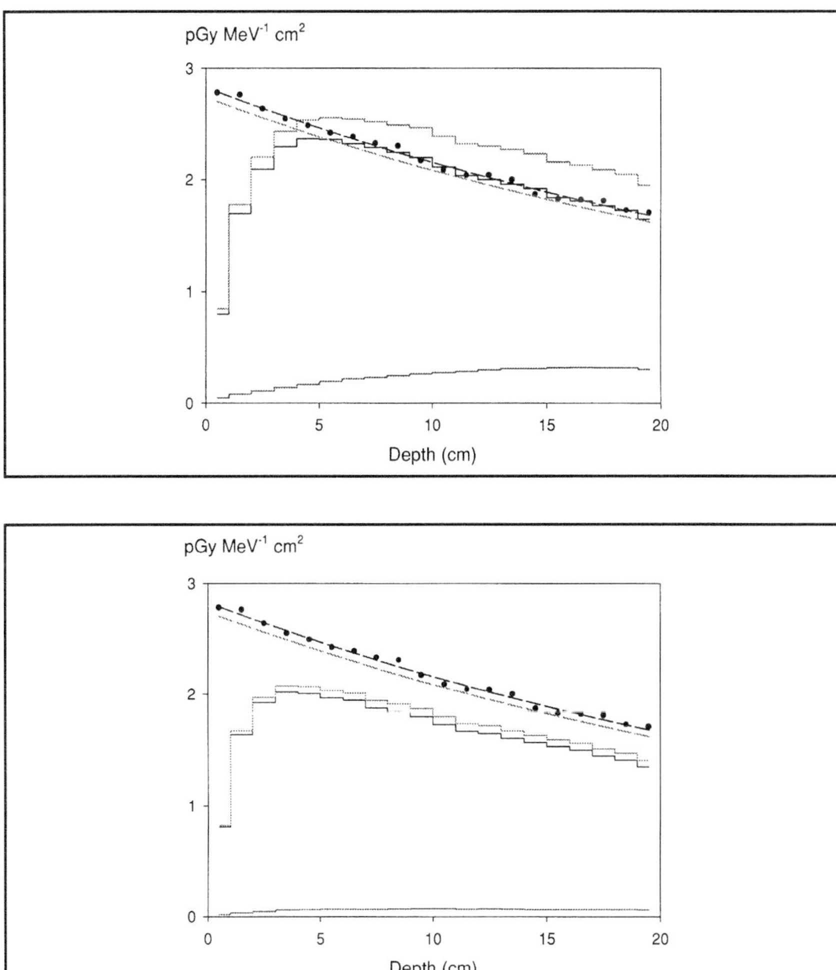

Figure 2.19: Monte Carlo-generated primary scatter and total dose (histograms) as well as kerma (dots) compared with calculated kerma and collision kerma (broken lines) for a 24 MV beam. (a) Shows the results for a 20 cm diameter beam and (b) for a 2 cm diameter beam (where lateral electronic disequilibrium exists).

transport of ejected electrons. By observation, the displacement of the primary dose curve from the collision kerma curve (and hence the effective longitudinal electron range) is about 1 cm. By coincidence, the primary dose is approximately equal to the kerma, indicating that the values of $1 + \mu' z$ and μ_{en}'/μ_{tr} are similar. Comparing the scatter dose in this graph to that for the 20 cm diameter 6 MV beam in figure 2.17a, the reduction in scatter dose for higher energies is apparent (corresponding to the increase in μ_{tr}/μ in figure 2.5). Results for the 2 cm beam in figure 2.19b show the primary and total dose to be lower than the kerma curves. This is because of lateral electronic disequilibrium, which occurs at the central axis of the beam when the lateral electron range exceeds the beam radius. This phenomenon is described below and is again discussed in relation to other forms of charged particle disequilibrium in section 2.6.

The reason for lateral electronic disequilibrium is illustrated in figure 2.20, where ejected electrons and their ranges are indicated. For the larger diameter beam in (a), electrons ejected at the edge of the beam are seen not to reach the central axis: the beam is thus wide enough that any further increase in diameter will not result in an increase in primary dose at the beam axis. In (b), however, electrons originating at the beam edge have a larger lateral range than half the beam diameter and an increase in diameter *would* cause an increase in primary dose on the axis. Lateral electronic disequilibrium is said to exist on the central axis for the smaller beam because electrons ejected and displaced laterally from the beam axis are not replaced

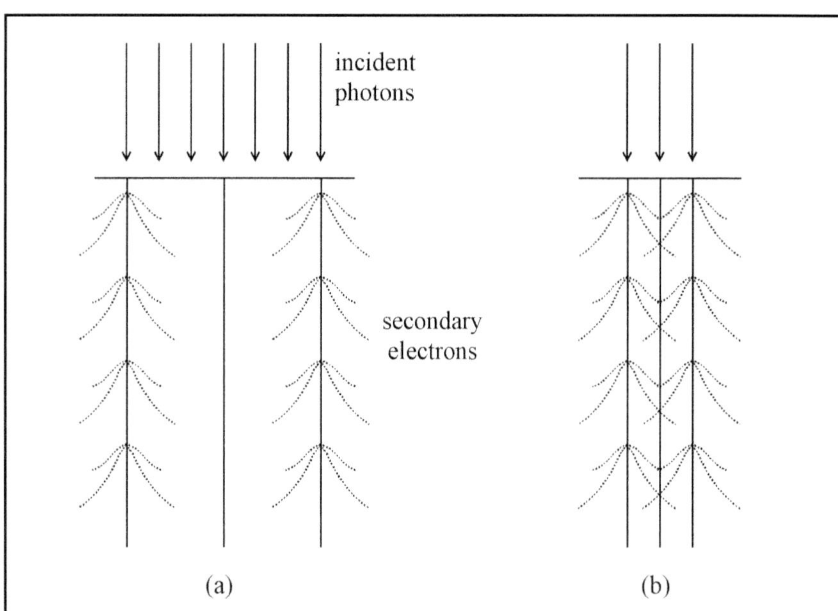

Figure 2.20: Schematic representation of lateral electronic disequilibrium. For the beam on the left (a), electrons displaced laterally from the central axis are replaced by equal numbers from elsewhere, since the lateral electron range does not exceed half the field width. Also, a further increase in field size would not cause an increase in central axis primary dose. For the smaller diameter beam on the right (b), lateral electronic disequilibrium exists because the lateral electron range exceeds half the field width.

by equal numbers displaced laterally from elsewhere. This can be seen to be the case in (b) because electrons ejected from the axis are able to reach further than the beam edge, outside of which there are no electrons ejected.

Further insight into the appearance of the primary and scatter dose curves for the three energies above is obtained by observing the secondary electron tracks produced by a narrow photon *pencil beam* as in figure 2.21. Each of the

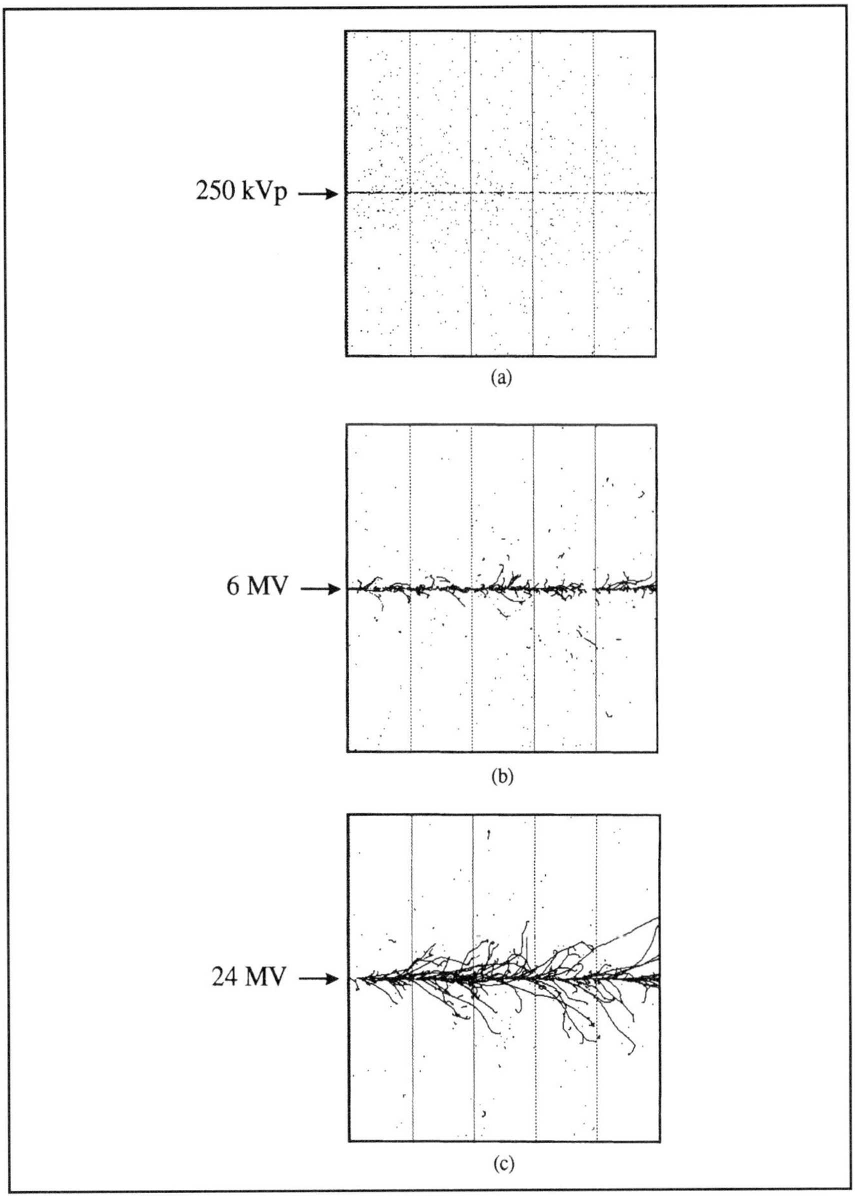

Figure 2.21: EGS4-generated charged particle tracks due to pencil beams of (a) 250 kVp, (b) 6 MV, and (c) 24 MV photons. Tracks originating at the beam axis are due to electrons ejected by primary photons, whereas isolated tracks and points indicate electrons set in motion by scattered photons. Electron range increases with energy and the number of electrons set in motion by scattered photons decreases.

EGS4 Monte Carlo-generated sets of tracks represents a projection of points at which energy is deposited by charged particles ejected by both primary and scattered photons. Tracks originating at the pencil beam axis thus represent primary dose, and isolated short tracks and points represent scatter dose. The vertical lines in the plots are separated by 2 cm. First, for the 250 kVp beam it can be seen that the secondary electron range is insignificant, and because of the large number of points away from the axis, scatter dose will be large. This is expected from the small ratio of energy transfer to attenuation coefficient for this energy range (see figure 2.5) and is confirmed by the large scatter dose in figure 2.18. For the 6 MV beam, the proportion of dose deposition points away from the axis is decreased, indicating less scatter dose, and the range of secondary electrons is, of course, increased. The trend continues for the 24 MV beam where dose deposition points due to scattered photons is reduced further and the maximum electron range is now several centimeters. (The lateral electron range is considerably larger than 1 cm, hence the large degree of lateral electronic disequilibrium for the 2 cm beam in figure 2.19b.)

2.5 Electron Interactions

When an x-ray photon interacts with a medium by either a photoelectric, Compton, or pair production process, charged particles are set in motion and deposit energy until they run out of energy and stop. In the case of pair production, an electron and a positron are first created and then transported; in other cases, a single electron is ejected. Since the interaction properties of electrons and positrons are virtually identical, the term *electron* will be used to refer to both unless otherwise stated. The ejected electron from a photon interaction is sometimes referred to as a *secondary electron*, since it is not incident in the primary beam. The process leading to secondary electron production in photon beams is shown in figure 2.22a.

For clinical use of electron beams the electrons accelerated in the linac are spread out using a scattering foil. It is these primary electrons which enter the patient. The primary electrons can produce other secondary electrons by interactions which are described in the next section. The interaction processes as the electrons enter the body and interact with other electrons are shown in figure 2.22b.

The important mechanisms of electron transport are energy loss and scattering; the rates at which both occur depend on the electron energy as well as the composition of the absorbing medium. A higher initial electron energy and/or a slower rate of energy loss gives rise to a larger electron range. The processes whereby electrons lose energy include collisions with atomic electrons (termed *collision energy loss*) and interactions with the electric fields of nuclei (*radiative energy loss*) in which a bremsstrahlung photon is emitted. Both of these interaction types are inelastic; hence, energy is lost by the incident electron. When the electron passes by a nucleus at a larger distance, the interaction with the electric field is elastic and the electron is deflected with no energy loss. This process is known as *multiple coulomb scattering*, since the electron attains a net angular deflection after a very large number of coulomb interactions. It is collision energy loss that leads to the ionization and excitation processes responsible for biological damage to the cell.

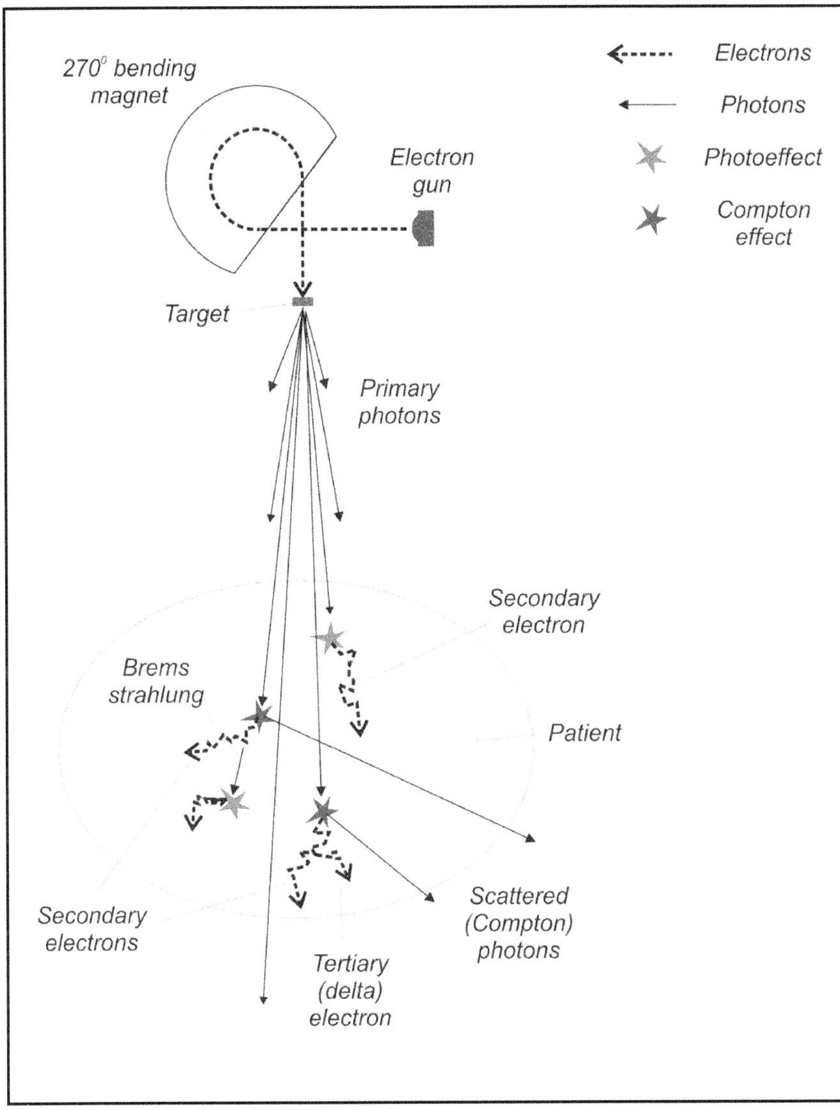

(a)

Figure 2.22 Electron and photon interactions. (a) when linac in x-ray beam mode; (b) when linac in electron beam mode.

2.5.1 Collisional Energy Loss

Collisions that cause electron energy loss are of major interest, as these events give rise to deposition of energy. Collisional energy loss occurs when an electron passes close enough to an atomic electron to eject it from its shell either permanently or temporarily. If the electron is permanently removed, the atom is ionized; if it is temporarily elevated to a higher energy level, excitation of the atom occurs. Collisional energy losses, therefore, include ionization or excitation of the electron within the atom. In general, this process is closely modeled by a free electron collision because the outer

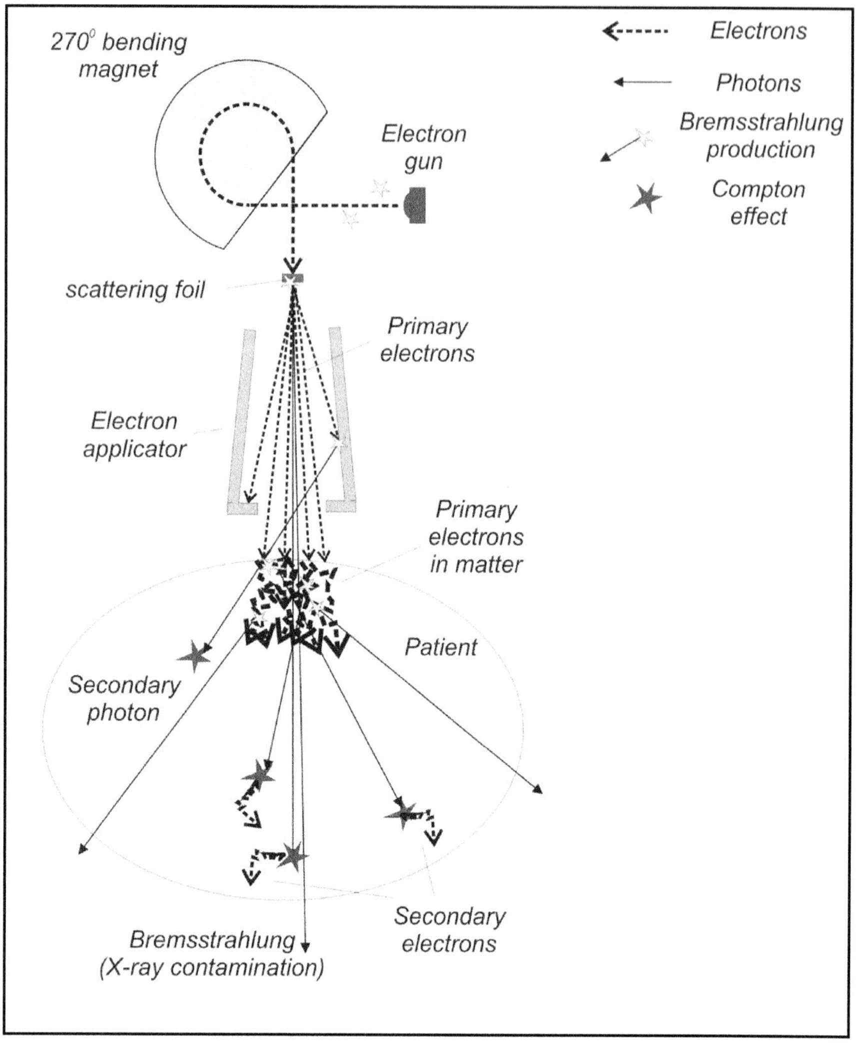

(b)

Figure 2.22 Electron and photon interactions (continued). (b) when linac in electron beam mode.

shell electrons are loosely bound. The excitation interaction is shown in figure 2.23a, whereby temporary excitation of an electron into another shell can cause the emission of a visible photon, and the ionization interaction is shown in figure 2.23b. It is also possible for a collisional event to lead to characteristic radiation to be emitted during the event.

Collision energy losses in which the electron loses a small amount of energy are very frequent. The rate of energy loss by this mechanism depends on electron energy, number of atomic electrons per unit volume, and slightly on the ionization energy of the atoms in the medium. Less frequently, large energy losses occur when a much higher proportion of the secondary electron's energy is transferred to an orbital electron. This is called a knock-on collision, and the ejected electron is referred to as a δ-ray, which itself causes

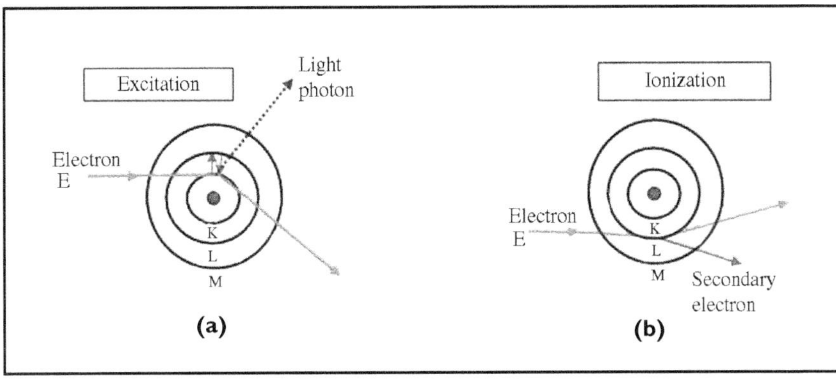

Fig 2.23: Electron collisional energy loss events. (a) Excitation; (b) Ionization.

ionization and excitation. Such collisions are called *Møller* scattering events. When the δ-ray is produced by a positron, the process is *Bhabha* scattering.

2.5.2 Radiative Energy Loss (Bremsstrahlung Production)

When an electron passes close to the nucleus of an atom, it experiences an electromagnetic force that results in an energy loss as shown in figure 2.24. Bremsstrahlung (from the German words *bremsen,* which means decelerate, and *strahlung,* which means radiation) radiation is emitted by conservation of energy. The probability of such an interaction increases as the distance of the electron from the nucleus decreases. The maximum energy of the

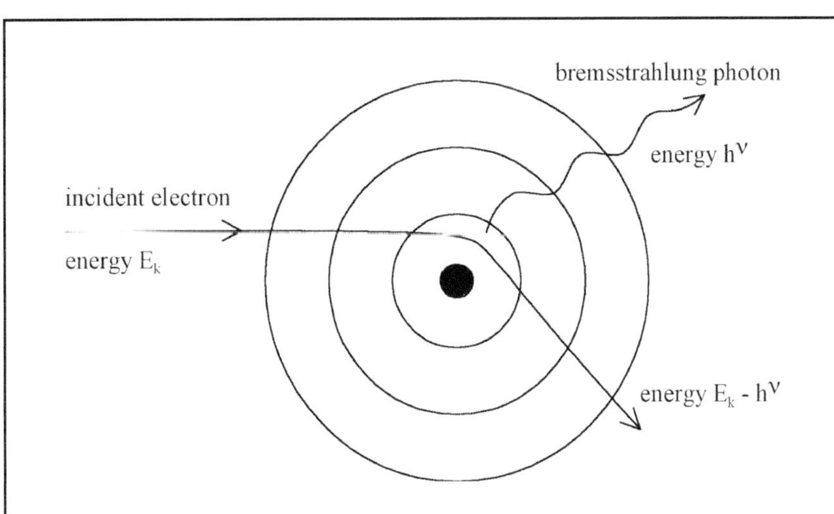

Fig 2.24: Bremsstrahlung photons are emitted when an electron interacts with the coulomb field of the nucleus. The electron continues with its energy reduced to E_K–*hv* where *hv* is the energy lost to the photon.

bremsstrahlung photon can be no more than the incident electron energy, and a spectrum of photon energies below this value is produced. Because the energy lost is converted to a bremsstrahlung photon, this is referred to as a radiative energy loss. There is also a small radiative energy loss component due to electron-electron bremsstrahlung, in which the coulomb field of an orbital electron decelerates the secondary electron.

2.5.3 Electron Stopping Powers

Energy transferred from the electron to the medium by collisional or radiative processes is quantified by the use of stopping powers. The rate of energy loss per unit path length, dE/dz, is termed the *stopping power* and has components from collision and radiative losses. As with the attenuation and absorption coefficients for photons, the stopping powers are usually tabulated as density-independent quantities called the *mass collision stopping power* and *mass radiative stopping power*. The total mass stopping power has units of MeV cm^2g^{-1} and is denoted by

$$\left(\frac{S}{\rho}\right)_{tot} = \left(\frac{S}{\rho}\right)_{col} + \left(\frac{S}{\rho}\right)_{rad}. \tag{2.60}$$

Tables of stopping powers are available (Berger et al. 1966).

Collision Stopping Power

Berger and Seltzer's expression for the mass collision stopping power of an element with atomic number Z and mass number A is

$$\left(\frac{S}{\rho}\right)_{col} = \frac{2\pi r_0^2 m_0 c^2 N_A Z}{\beta^2 A}\left[\ln\left(\frac{\tau^2(\tau+2)}{2(I/m_0c^2)^2}\right) + F^-(\tau) - \delta\right], \tag{2.61}$$

where β is the velocity relative to the speed of light c, τ is the ratio of electron kinetic energy to rest mass energy, I is the mean excitation energy, δ is the density effect correction, and $F(\tau) = (1 - \beta^2)(1 + \tau^2/8 - (2\tau+1)\ln 2)$ accounts for Møller scattering. For positrons, $F(\tau)$ is replaced by $F^+(\tau)$ to account for Bhabha scattering. The mass collision stopping power decreases sharply with increasing energy at first because the electron spends less time in the electric field of the orbital electron and is less likely to cause ionization or excitation (there is an initial increase in β^2 in the denominator of equation 2.61). There is a minimum for body tissues at about 1.5 MeV, whereafter there is a rise due to relativistic effects (see the explanation of the *density effect* below). Since the ratio Z/A (which is proportional to the number of electrons per gram) is relatively constant, the mass collision stopping power is not very material dependent (similarly to the Compton component of the mass attenuation coefficient).

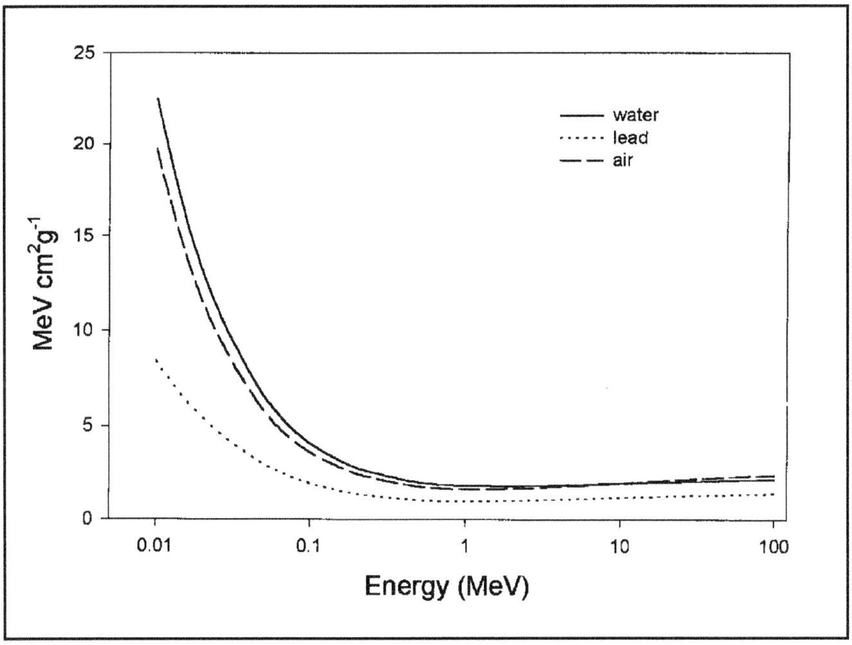

Figure 2.25: Mass collision stopping powers for water, air, and lead. Note the relatively low values for lead because of its *Z/A*. The slope of the relativistic rise for water can be seen to be less than that for air. This is due to the density effect.

Density Effect

As the electron energy increases, the electric fields of orbital electrons at an increasing distance are felt because of relativistic effects (Lorentz contraction of the coulomb field). This causes an increase in the mass collision stopping power known as the *relativistic rise*. However, the effect of distant electrons is screened by intervening atoms, which reduces the slope of the relativistic rise from what it otherwise would be. For higher density materials, the number of intervening atoms is greater and thus the reduction in slope is larger. Therefore, as energy increases, the increase in collision stopping power with energy is smaller for higher density materials; hence the name *density effect*. This phenomenon is most important in dosimetry in the ratio of mass collision stopping powers of water to air: The ratio decreases as energy increases because of much greater screening in water (the value of δ in equation 2.61 is larger for water).

The mass collision stopping powers for water, air, and lead are plotted in figure 2.25. Note that the values for lead are lower than for the other materials because of the lower value of *Z/A*. The relativistic rise can be seen for all materials, but note the reduced gradient for water compared with air due to the density effect.

Restricted Stopping Power

For dosimetry, the mass collision stopping power is often represented by a special case called the *restricted mass stopping power*, L/ρ, which excludes energy losses above a certain value (say 100 keV). This is because in calculating the energy deposited at a point, the energy given to δ-rays that have a significant range should not be counted, as it will be deposited remotely. The actual energy deposited at a point is, therefore, defined as the energy transferred by secondary electrons in collisional energy losses less than a specific value Δ. The linear restricted stopping power, L_Δ, also known as the *linear energy transfer* (LET), is therefore the energy loss per unit pathlength in which the energy loss per collision is less than Δ. Note that if there is no restriction placed on the energy loss, ($L_\Delta = L_\infty = S_{coll}$). The restricted mass stopping power is therefore defined as

$$\left(\frac{L}{\rho}\right)_\Delta = \left(\frac{S}{\rho}\right)_{col} \quad [\text{energy losses less than } \Delta], \tag{2.62}$$

where the parameter Δ is typically set at 100 keV because the range of electrons in water at such a low energy is less than 1 mm.

Radiative Stopping Power

Radiative energy losses are less frequent than ionizations and excitations, but each event can include a large energy loss. In soft tissues, for the energy range of secondary electrons in radiotherapy, the radiative stopping power is to the order of 1% of the collision stopping power. The relative proportion however increases for higher atomic numbers and higher energies. An approximate expression for the mass radiative stopping power is

$$\left(\frac{S}{\rho}\right)_{rad} = \frac{4N_A Z(Z+1)}{A} r_0^2 E \frac{183}{Z^{1/3}}. \tag{2.63}$$

Note that the numerator of this equation includes a factor $Z^2 + Z$. The two terms represent electron-nucleus and electron-electron bremsstrahlung, respectively, so it is obvious that the former is more significant by a factor of Z. It is important to note that the mass radiative stopping power is approximately proportional to Z (since Z/A is approximately constant) and to energy E, whereas the mass collision stopping power is almost independent of both for energies larger than that where the minimum occurs. A very approximate expression for the ratio of radiative to collision stopping power is

$$\frac{S_{rad}}{S_{col}} \approx \frac{E(Z+1.2)}{800}, \tag{2.64}$$

which predicts that the two stopping powers will be equal when the energy is approximately equal to $800/Z$ MeV.

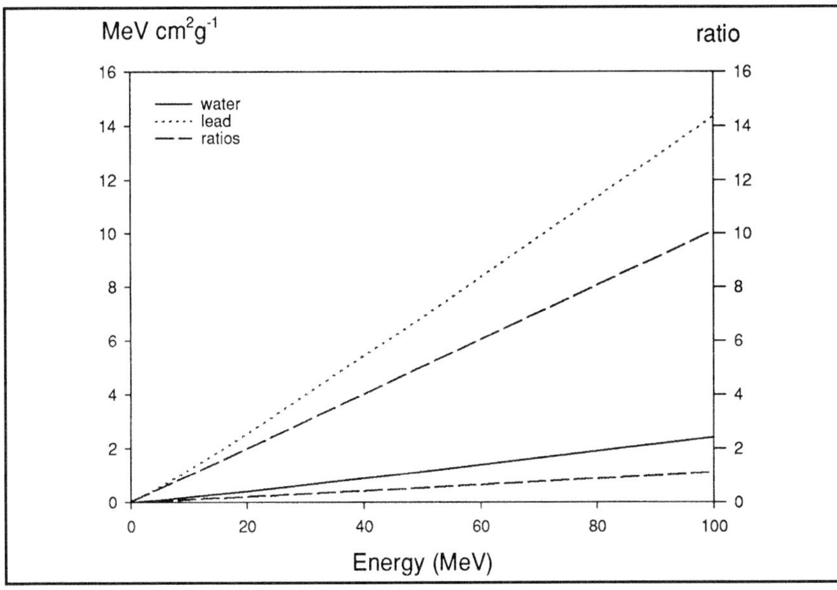

Figure 2.26: Mass radiative stopping powers for water (solid line) and lead (dotted line), along with the ratios of radiative to collision stopping powers for both (dashed lines—upper is lead, lower is water). The mass radiative stopping power, as well as the ratio, are much higher for lead because of the approximate Z dependence.

In figure 2.26, the mass radiative stopping powers for water and lead are plotted along with the ratios of radiative to collision stopping powers for both materials. In contrast to the mass collision stopping power, the values for lead are much larger than for water in this case due to the approximate Z dependence of S_{rad}/ρ. The gradient of the radiative/collision stopping power ratio is also larger for lead and agrees quite well with that predicted by equation 2.64 (approximately $Z/800$).

2.5.4 Electron Scattering Powers

Electrons are continuously deflected by electromagnetic interactions with nuclei. These are predominantly frequent, small-angle deflections, but less frequently large-angle scattering also occurs. The net angular deflection that an electron undergoes in traversing a certain thickness of material, due to small-angle scattering, is a statistical quantity that can be calculated using multiple coulomb scattering theory.

In this theory the mean square angular deflection per unit pathlength is called the scattering power $d\bar{\theta}/dz$, which defines a Gaussian distribution of likely scattering angles after an electron has traversed a unit pathlength. Scattering power decreases with electron energy and increases with the atomic number of the medium. Multiple coulomb scattering theory has been used successfully to model the scattering of electrons in electron beam treatment-planning algorithms and is also important in modeling the transport of high-energy secondary electrons. The scattering power can be used to calculate the approximate angular and lateral distributions of an electron beam after penetrating a medium. In its most basic application, it can be used to determine the standard deviation of the angular deflection for an electron penetrating a short distance Δz. According to the Fermi-Eyges theory (Klevenhagen 1985),

$$\sigma_\theta = \sqrt{\frac{d\overline{\theta}}{dz}\Delta z} \qquad (2.65)$$

and the x-component of lateral standard deviation after penetrating a distance z is

$$\sigma_\theta = \sqrt{\frac{1}{2}\int_0^z \frac{d\overline{\theta}}{dz}\left(z - z'\right)^2 dz}, \qquad (2.66)$$

where the scattering power can be a function of position z'. This is the sort of calculation used in obtaining the lateral standard deviation due to multiple coulomb scattering in the widely used Hogstrom pencil beam method for electron beam planning (Hogstrom et al. 1981). As with the stopping power, the scattering power is usually tabulated as a *mass scattering power* $d\overline{\theta}/\rho dz$, which is approximately proportional to atomic number Z and inversely proportional to the square of the energy:

$$\frac{d\overline{\theta^2}}{\rho dz} \propto \frac{Z(Z+1)}{A}\frac{1}{E^2}. \qquad (2.67)$$

The $Z^2 + Z$ term takes account of electron-nucleus and electron-electron scattering in a similar manner to the situation with radiative energy loss. Note that the approximate proportionality of Z to A reduces the dependence on atomic number to the first power. The $1/E^2$ dependence means that scattering increases rapidly near the end of the electron range, causing the trajectory to be very tortuous just before the electron stops.

The Gaussian distributions predicted by the Fermi-Eyges theory are not rigorous because of the neglect of large-angle scattering. The true angular distribution is in fact more peaked in the forward direction with longer tails at large angles. More rigorous is the Molière multiple scattering theory (Molière 1948) where the entire angular distribution is represented by an infinite series. This theory is used by the EGS4 Monte Carlo system to obtain angular distributions from which to sample the change in direction of an electron after a transport step. Even more accurate scattering distributions, produced by the Goudsmit-Saunderson theory (Goudsmit et al. 1940), are used by the ETRAN Monte Carlo code.

2.6 Absorbed Dose

As discussed in section 2.4.3, absorbed dose is very close to the collision kerma. This is true for both the primary and scatter components of these quantities. The difference in collision kerma and dose is that energy is carried away from the site of photon-electron energy transfer by the ejected electron. As seen in section 2.4.3, this transport results in a dose curve resembling the collision kerma curve shifted slightly downstream. The shift is very small, however, except for high photon energies where the secondary electron range is significant. Absorbed dose is the quantity of most interest in radiotherapy, and its measurement via detection of charge liberated in ionization is covered in detail in chapter three. The relationship of absorbed dose to kerma and collision kerma will be briefly discussed here along with the concept of charged particle equilibrium (also known as electronic equilibrium).

2.6.1 Dose, Kerma, and Charged Particle Equilibrium

In this section we will consider four examples of energy transfer/deposition in a small volume and the relationships of dose to kerma, with and without the constraint of charged particle equilibrium. For an excellent and detailed discussion of the concepts introduced in this section, see Attix (1979).

First, the energy deposited in a volume is equal to the difference in the energy entering the volume and the energy exiting; this energy can be in the form of photons or charged particles. This net energy is thus given by

$$\varepsilon = \sum \varepsilon_{in} - \sum \varepsilon_{out}, \tag{2.68}$$

which can be expanded to contain separate terms for photons and charged particles:

$$\varepsilon = \sum (\varepsilon_{in})_p - \sum (\varepsilon_{out})_p + \sum (\varepsilon_{in})_c - \sum (\varepsilon_{out})_c . \tag{2.69}$$

The condition of charged particle equilibrium (CPE) means that the energy carried in and out of the volume by charged particles is equal, so

$$\sum (\varepsilon_i)_c = \sum (\varepsilon_{out})_c \qquad \text{for CPE.} \tag{2.70}$$

For small numbers of particles, ε is a *stochastic* quantity—its value varies discontinuously with space and time. On averaging successive values of ε for the same volume, a nonstochastic quantity is obtained from which the dose can be defined. If the mean value of ε is $\bar{\varepsilon}$, the dose at a point is defined as

$$D = \frac{d\bar{\varepsilon}}{dm}. \tag{2.71}$$

Now referring to figure 2.27, consider a photon of energy $h\nu$ entering a volume, transferring an energy T_0 to an electron, and exiting the volume with an energy $h\nu' = h\nu - T_0$. The electron stops within the volume in this case.

Equation 2.69 for this situation reduces to

$$\begin{aligned} \varepsilon &= h\nu - h\nu' \\ &- T_0 \end{aligned} \tag{2.72}$$

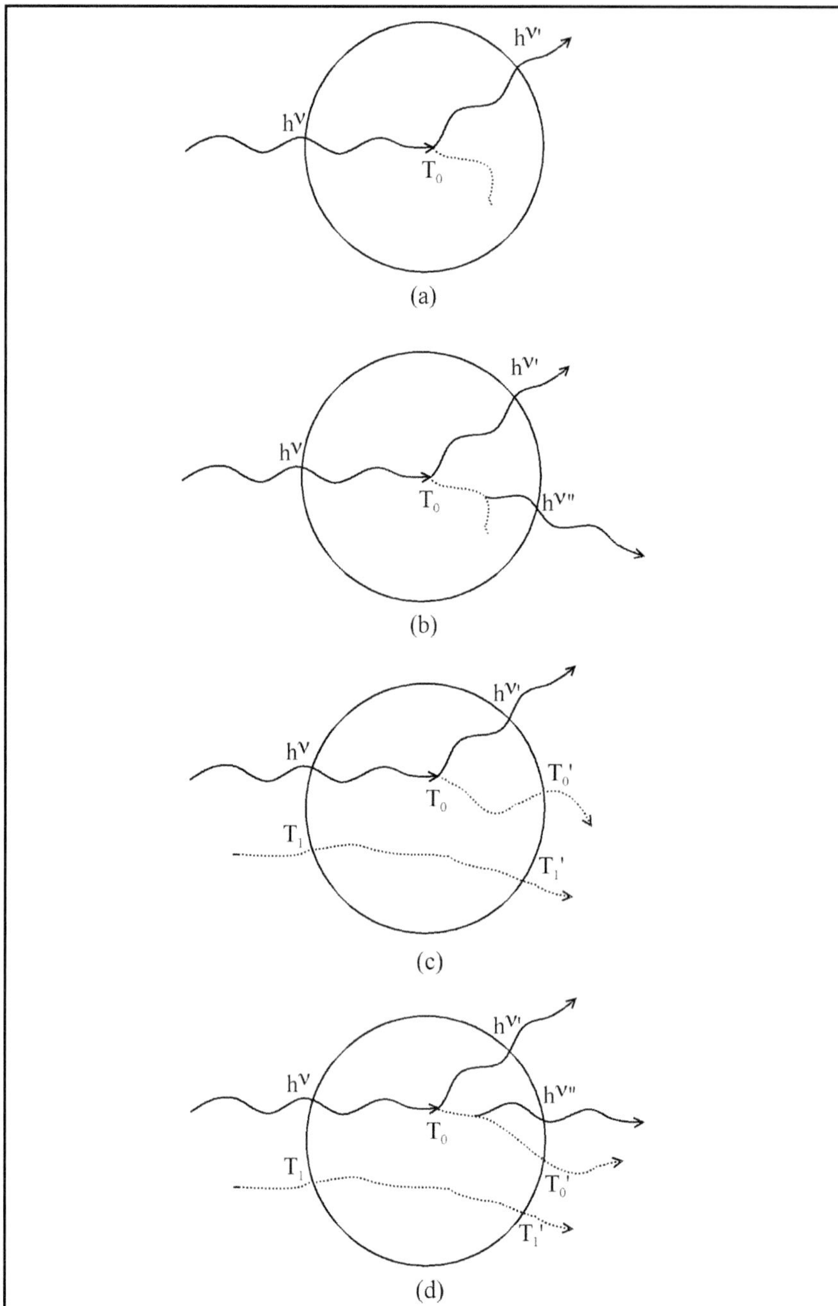

Figure 2.27: Energy imparted to a small volume due to charged and uncharged particles entering and exiting the region. (a) A photon enters and transfers an energy T_0 to a charged particle; no bremsstrahlung photons are produced. (b) As in (a) but a bremsstrahlung photon is produced by the ejected electron. (c) The ejected electron exits the volume and another electron enters and exits. (d) As in (c) but a bremsstrahlung photon is produced by the ejected electron. In (a) and (c), dose is equal to kerma and collision kerma. In (b) and (d), dose is equal to collision kerma but less than kerma.

and, of course, CPE exists as there are no charged particles entering or exiting the volume. This energy deposited is also equal to the energy transferred, ε_{tr}, since there is no bremsstrahlung produced. The dose is thus equal to the kerma and collision kerma in this case ($D = K = K_c$).

Now consider the situation in which the charged particle gives rise to a bremsstrahlung photon of energy $h\nu''$ as in figure 2.27b. The energy deposited is now given by

$$\begin{aligned}\varepsilon &= h\nu - h\nu' - h\nu'' \\ &= T_0 - h\nu''\end{aligned} \tag{2.73}$$

and the energy transferred is equal to T_0 as before. CPE still exists, since no charged particles enter or exit the volume; dose is equal to collision kerma but is now smaller than the kerma ($D = K_c$ and $D < K$).

With a charged particle entering the volume and a secondary charged particle exiting the volume, dose is not equal to collision kerma unless CPE exists. In figure 2.27c, a photon of energy $h\nu$ enters along with a charged particle of kinetic energy T_1. The photon transfers an energy T_0 to a charged particle that exits the volume with an energy T_0', and the photon exits with energy $h\nu' = h\nu - T_0$ as before. Now equation 2.69 becomes

$$\begin{aligned}\varepsilon &= h\nu - h\nu' + T_1 - T_1' - T_0' \\ &= T_0 - T_0' + T_1 - T_1'.\end{aligned} \tag{2.74}$$

With CPE ($T_1 = T_1' + T_0'$) it reduces to simply

$$\varepsilon = T_0, \tag{2.75}$$

as in the simple case of figure 2.27b. Thus, as long as we have CPE, dose is equal to kerma and collision kerma in the case of no bremsstrahlung ($D = K = K_c$).

Finally, consider that the ejected charged particle produces a bremsstrahlung photon as in figure 2.27d. The energy deposited in the volume is now

$$\begin{aligned}\varepsilon &= h\nu - h\nu' - h\nu'' + T_1 - T_1' - T_0' \\ &= T_0 - T_0' + T_1 - T_1' - h\nu'',\end{aligned} \tag{2.76}$$

which with CPE reduces to

$$\varepsilon = T_0 - h\nu'', \tag{2.77}$$

as in the case of figure 2.27b with no incident or exiting charged particles. As in that case, dose is equal to collision kerma and less than the kerma ($D = K_c$ and $D < K$).

These examples show that under conditions of CPE, dose is equal to collision kerma, regardless of whether or not any bremsstrahlung photons are produced.

Transient Charged Particle Equilibrium

In reality, CPE does not exist at any point, since the varying photon fluence causes an imbalance in total charged particle energy entering and exiting a region. In particular, primary photon fluence decreases with depth, thus reducing the number of charged particles entering successive regions as depth increases. This concept is illustrated in figure 2.28, which displays the number of electrons crossing into each region along with the equal number of photon interactions in the preceding region. This represents the equivalence of dose to the collision kerma slightly upstream. In this situation, a state of *transient* CPE is said to exist, a state in which dose is a constant proportion of the collision kerma, as in the examples of section 2.4.3. Note that if photon fluence were constant, true equilibrium would exist at all points past the depth of maximum dose, and dose would exactly equal collision kerma. The phenomenon of transient longitudinal electron equilibrium has been discussed in detail by Dutriex et al. (1965, 1966).

CP Disequilibrium in the Build-Up Region

An obvious case of charged particle disequilibrium occurs in the build-up region, where dose increases before decreasing exponentially. Here, the charged particle energy entering successive regions increases with depth until approximately the maximum longitudinal range of the secondary electrons. The number of electrons crossing into each region accumulates with depth. (See figure 2.29.) This accumulation is eventually balanced by secondary electrons running out of energy and by attenuation of photon fluence.

Lateral CP Disequilibrium

Lateral CP or electronic disequilibrium was discussed and illustrated in section 2.4.3 in relation to the 24 MV beam. It was seen that dose at central axis

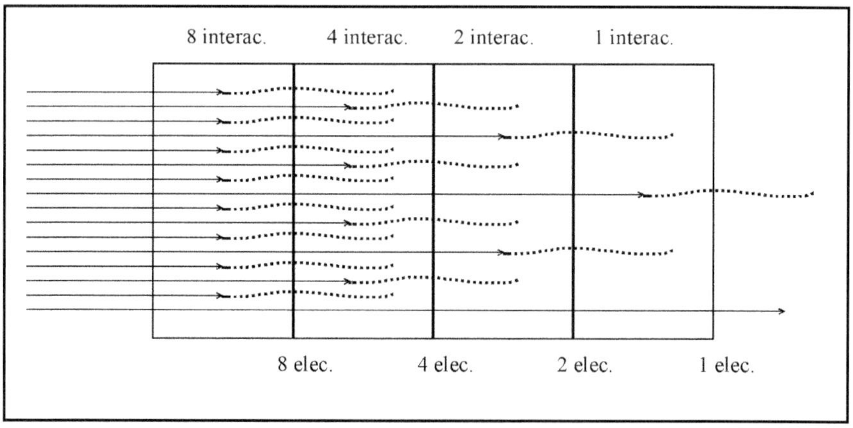

Figure 2.28: Transient CPE. Photons enter successive regions, ejecting electrons in the forward direction (for simplicity). The number of photons entering each region decreases with depth due to attenuation, so the total charged particle energy entering each region also decreases. This gives rise to disequilibrium, but since dose is proportional to kerma, a state of transient charged particle equilibrium is said to exist.

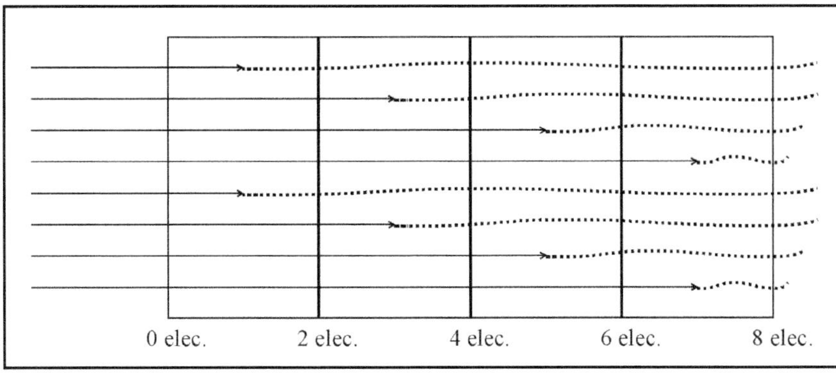

Figure 2.29: CP disequilibrium in the build-up region. The number of charged particles crossing into successive regions accumulates with depth until a balance is reached due to (i) secondary electrons running out of energy and (ii) photon attenuation.

is reduced for small field sizes due to the larger lateral range of secondary electrons compared with the beam radius. To unify the concept of lateral CP disequilibrium with the examples of longitudinal CP disequilibrium above, a situation in which photons are entering adjacent regions and ejecting charged particles is illustrated in figure 2.30. There it can be seen that secondary electrons ejected laterally from the innermost volumes are replaced by electrons from adjacent regions. The outer volumes however suffer from disequilibrium, since this replacement does not occur. Lateral CP disequilibrium thus always occurs at the beam edges and exists at distances from a beam edge up to the maximum lateral secondary electron range. The phenomenon of lateral electron disequilibrium has been experimentally investigated by other researchers (Kornelson et al. 1982, Young et al. 1983, Mackie et al. 1985).

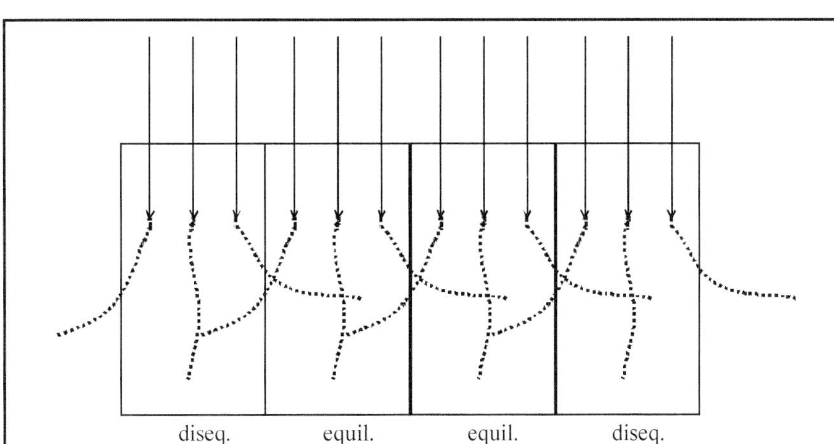

Figure 2.30: Lateral CP disequilibrium exists in the outer regions shown because charged particles ejected laterally out of the volumes are not replaced by equal numbers from adjacent regions (as is the case with the inner regions).

Significant lateral electron disequilibrium regions, which can affect radiotherapy dose distributions, may occur in the following situations:

i) High photon beam energies, where electrons ejected from throughout the irradiated medium may range outside the beam boundaries.
ii) Small field sizes, such as in stereotactic radiosurgery, in which a large proportion of the irradiated region is in a state of CP disequilibrium even though the beam energy is typically only 6 MV.
iii) Low density media, since electron range varies inversely with density. For example, in lung of density 0.25 gcm^{-3}, disequilibrium will occur at field sizes four times larger than in water (Mackie et al. 1985, Metcalfe et al. 1988).
iv) Air/tissue interfaces where disequilibrium exists in the air and also in the tissue until it is re-established after a few millimeters (Kornelson et al. 1982, Wong et al. 1992).

Because of lateral electron disequilibrium, the dose in these regions can be lower than expected (a dose void occurs). This means that empirical dose calculation models (which assume CPE) often overestimate the dose (see chapter nine).

Situations of CP disequilibrium are nicely shown by the EGS4 Monte Carlo-generated electron tracks in figure 2.31. Here, a small number of photons from a 15 MV energy spectrum were incident in beams of dimensions 0.5×0.2 cm, 1.0×0.2 cm, and 2.0×0.2 cm, and the incident fluence was the same in all cases. Only the charged particle tracks are shown for clarity. For all beams, a build-up region is obvious near the surface where the density of tracks is less than at greater depths. The most interesting observation, however, is the increase in track density (after build-up) as field size increases. This is due to increases in lateral CP equilibrium. Of course, this condition does not truly exist for any of these cases, since the beam dimension perpendicular to the page is only 0.2 cm.

2.6.2 Absorbed Dose from Exposure

So far dose has been related to the energy transferred to secondary electrons via the kerma and collision kerma. However, if it is to be *measured*, we must use a relationship to some quantity that instruments can detect. This quantity is the *exposure*, X, which is the charge released in ionizations per unit mass of the detector material. The detector material is typically the air in an ionization chamber.

The quantity W (*ev* per ion pair) is the mean energy expended in the gas per ion pair produced. This quantity is constant in dry air for electron energies above a few keV. By dividing W by e (the charge of the electron), the energy deposited in air per coulomb of charge released is obtained:

$$\left(\frac{W}{e}\right)_{air} = 33.97 \text{ J/C.} \tag{2.78}$$

Note also (e/W) gives the coulombs of charge released per joule of energy.

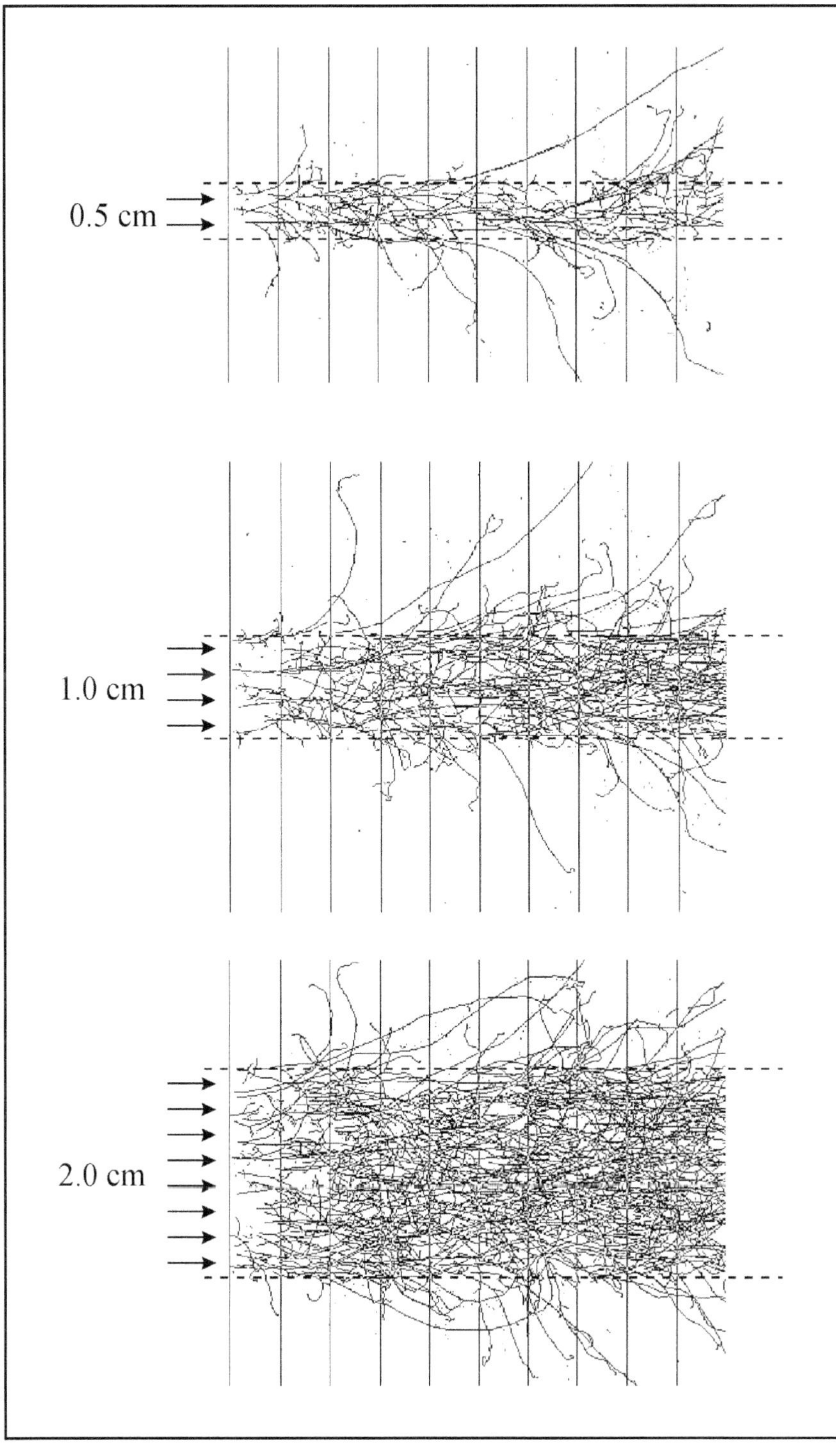

Figure 2.31: EGS4 Monte Carlo-generated charged particle tracks for 15 MV photon beams of dimensions 0.5×0.2 cm, 1.0×0.2 cm, and 2.0×0.2 cm. The reduced charged particle track density in the build-up region is seen, along with the increase in track density as field size increases due to increases in lateral CP equilibrium.

Exposure, X, is the absolute value of charge produced by ions (of one sign) in dry air when all the electrons liberated in air of mass dm are completely stopped:

$$X = \frac{dQ}{dm}. \tag{2.79}$$

The absorbed dose in air is then given by the charge produced per unit mass multiplied by the energy required to produce a unit charge:

$$D_{air} = X\left(\frac{W}{e}\right)_{air}. \tag{2.80}$$

Note that this is also the collision kerma in the air under conditions of CPE.

Now to convert the dose in air to the dose in the absorbing medium, it must be multiplied by the ratio of mass collision stopping powers. In fact, the ratio of restricted mass collision stopping powers is used (see section 2.5.3) because we do not want to include the energy lost to δ-rays that exit the air volume. The dose to the medium is thus given by

$$D_{med} = D_{air}\left(\frac{L}{\rho}\right)_{air}^{med}. \tag{2.81}$$

Details of the protocols used in the measurement of dose by ionization chamber dosimetry are given in chapter three.

References

Ahnesjö A. (1989) Collapsed cone convolution of radiant energy for photon dose calculation in heterogeneous media. *Med. Phys.* 16(4): 554–577.

Attix F.H. (1979) The partition of KERMA to account for bremsstrahlung. *Health Phys.* 36: 347–354.

Berger M.J., Seltzer S.M. (1966) *Stopping Power Tables.* NASA SP3036. Washington D.C.

Dutriex J., Bernard M. (1966) Dosimetry at interfaces for high energy x and gamma rays. *Br. J. Radiol.* 39: 205–210.

Dutriex J., Dutriex A., Tubana M. (1965) Electronic equilibrium and transition stages. *Phys. Med. Biol.* 10(2): 177–190.

Goudsmit S., Saunderson J.L. (1940) Multiple scattering of electrons. *Phys. Rev.* 57: 24–29.

Heitler W. (1960) *The Quantum Theory of Radiation,* third edition. Clarendon Press, Oxford, 215.

Hogstrom K.R., Mills M.D., Almond P.R. (1981) Electron beam dose calculations. *Phys. Med. Biol.* 26: 445–59.

Hubbell J.H., Veigele W.J., Briggs E.A., Brown R.T., Cromer D.T., Howerton R.J. (1975) Atomic form factors, incoherent scattering functions and photon scattering cross sections. *J. Phys. Chem. Ref. Data.* 4: 471.

Huang P.H., Kase K.R., Bjarngard B.E. (1982a) Simulation studies of 4 MV x-ray spectra by numerical analysis of transmission data. *Med. Phys.* 9(5): 695–703.

Huang P.H., Kase K.R., Bjarngard B.E. (1982b) Reconstruction of 4 MV bremsstrahlung spectra from measured transmission data. *Med. Phys.* 10(6): 778–785.

Johns H.E., Cunningham J.R. (1983) *The Physics of Radiology,* fourth edition. Charles C Thomas, Springfield, Illinois.

Klein O., Nishina Y. (1929) Uber die streuung von Strahlung durch freie Electronen nach der neuen relativistischen Quanten dynamic von Dirac: Translation—On the scattering of radiation by free electrons based on the relativistic quantum dynamics of Dirac. *Zs. f. Phys.* 52: 853–869.

Klevenhagen S.C. (1985) *Physics of Electron Beam Therapy,* first edition. Adam Hilger Ltd., Bristol, UK.

Kornelson R.O., Young M.E.J. (1982) Changes in the dose-profile of a 10 MV x-ray beam within and beyond low density material. *Med. Phys.* 9: 114–116.

Levy L.B., Waggener R.G., Wright A.E. (1976) Measurement of primary bremsstrahlung spectrum from an 8 MeV linear accelerator. *Med. Phys.* 3 (5): 173–175.

Mackie T.R., Scrimger J.W., Battista J.J. (1985) A convolution method of calculating dose for 15 MV x-rays. *Med. Phys.* 12: 188–196.

Meredith W.J., Massey J.B. (1977) *Fundamental Physics of Radiology,* third edition. Wright and Sons Publishers, Bristol, UK, 69.

Metcalfe P.E., Battista J.J. (1988) Accuracy of inhomogeneity corrections in lung irradiated with high energy x-rays. *Australas. Phys. Eng. Sci. Med.* 11(2): 67–75.

Mohan R., Chui C. (1985) Energy and angular distributions of photons from medical linear accelerators. *Med. Phys.* 12 (5): 592–597.

Molière G. (1948) Theorie die Strueuung schneller geladener Teilchen II: Hehrfach and Vielfachstreuung. *Z. Naturforsch.* 3a: 78–79.

Nath R., Schultz R.J. (1976) Determination of high energy x-ray spectra by photoactivation. *Med. Phys.* 3(3): 133–141.

Nelson W.R., Hirayama H., Rogers D.W.O. (1985) *The EGS4 Code System.* Stanford Linear Accelerator Center Publication, 265.

Wong T.P.Y., Metcalfe P.E., Kron T., Emeleus T.G. (1992) Radiotherapy x-ray dose distribution beyond air cavities. *Australas. Phys. Eng. Sci. Med.* 15 (3): 138–146.

Young M.E.J., Kornelson R.O. (1983) Dose corrections for low-density tissue inhomogeneities and air channels for 10 MV x-rays. *Med. Phys.* 10: 450–455.

Questions

2.1 Explain why the interaction cross section is both an area and a probability. (section 2.2)

2.2 Calculate the conversion factor from cross section σ to linear μ and mass attenuation coefficient μ/ρ for oxygen given that Avogadro's number (6.011×10^{-23} mol^{-1}), ρ is the mass density (for oxygen $\rho = 1.332\times10^{-3}$ gcm^{-3}), and A is the mass number (for oxygen A = 16.)
(section 2.2)

2.3 Define the mathematical relationship which describes the differential cross section per unit solid angle for Compton interactions. Use this to calculate $d\sigma/d\Omega$ for a 3 MeV photon scattered by 10 degrees.
(sections 2.2 and 2.3)

2.4 (i) Given that the Compton (incoherent) scattering component of the mass attenuation coefficient in water for 3 MeV photons is 0.0385 cm^{2}g^{-1}, what is the corresponding value for lead?
(ii) Why does water have a larger mass attenuation coefficient than most other substances?
(section 2.3)

2.5 At radiotherapy energies, the mean energy of primary photons in a beam incident on soft tissue increases with depth. Why does this phenomenon occur and what is it called?
(section 2.4)

3 Dosimetry of Megavoltage X-Rays and Electrons

"Mysterious in the open day, Nature lets no man of veil bereave her. What to thy mind herself will not betray, Thou canst not from her wrest with screw and lever."

(J.W. Goethe, Faust)

3.1 Introduction: Why Measure Dose?

Ionizing radiation and its effects can be described by a number of different quantities (see e.g., ICRU 1976, ICRP 1975) such as flux, fluence, exposure, kerma, and absorbed dose. In general, one can distinguish three different sets of quantities:

i) Quantities that describe the source of radiation, such as activity or source strength. These quantities are most commonly used in connection with radioactive isotopes.
ii) Quantities that relate to the radiation beam and its particles, such as fluence and energy fluence. These are useful to describe the primary photon beam, and they are often used in diagnostic radiology (a field in which photons that do not interact carry the most important information).
iii) Quantities that measure the effect of ionizing radiation in matter such as exposure and dose. These quantities are of most importance in radiotherapy (a field in which the effect of ionizing radiation is of greatest concern).

In any case, it is important to note that there is not necessarily a straightforward correlation between these quantities.

The important feature of ionizing radiation used for radiotherapy is the destruction of tumor cells. For photons, this biological effect seems to be related not to the number of photons or particles passing through tissue, the fluence, but to the amount of energy deposited in a certain mass of tissue, the absorbed dose. More details of biological effects of ionizing radiation can be found in chapter eight. The absorbed dose is measured in grays (Louis H. Gray, British radiation physicist, 1905–1960; 1 Gy = 1 J/kg). The aim of dosimetry in radiotherapy must therefore be to measure or predict the absorbed dose in various tissues, neoplastic or normal, of a patient undergoing radiotherapy. This typically happens in two steps. First one assesses the deposited dose by means of a radiation detector in a suitable and well-defined environment, the *phantom*. Then one uses this *beam data* and patient data, such as tumor location and patient contours, to predict the

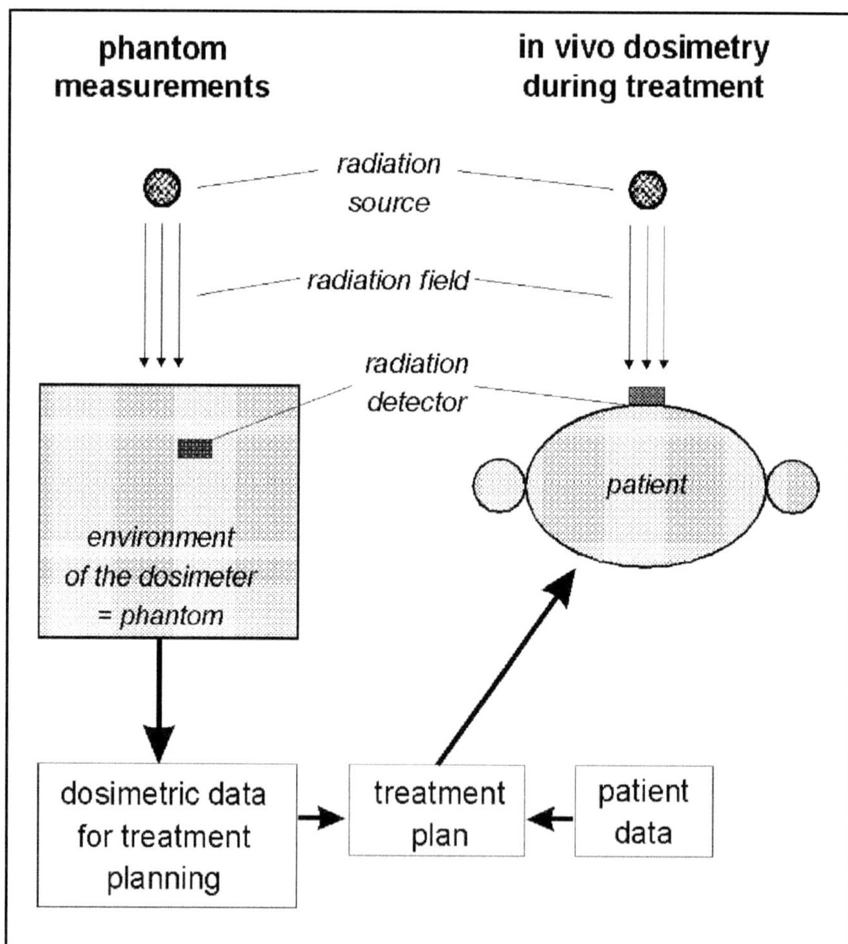

Figure 3.1: The roles of dosimetry in external beam radiotherapy.

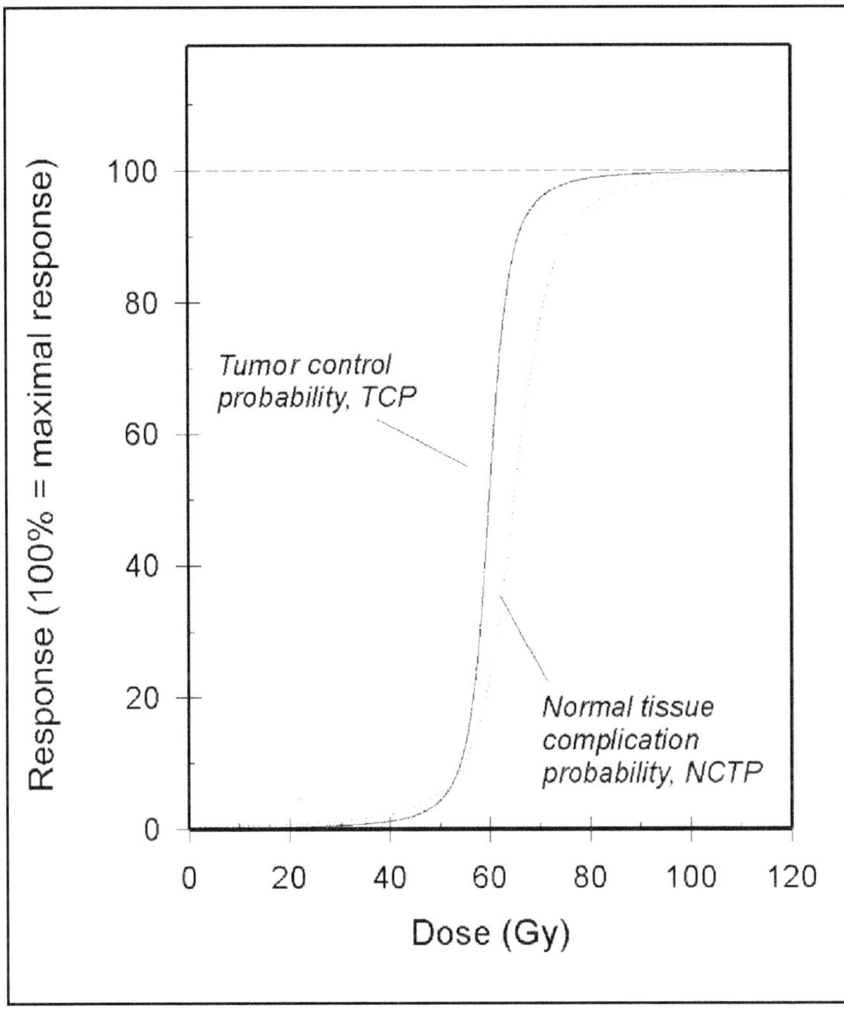

Figure 3.2: Dose-response of tumor control probability (TCP) and normal tissue complication probability (NTCP). The dose on the x-rays refers to the tumor dose in the given treatment.

dose at any point in the patient by means of appropriate mathematical algorithms. The latter is described in more detail in chapters nine and ten.

Figure 3.1 illustrates this task of dosimetry. Dose is measured in a phantom (water, slab, or anthropomorphic [see also section 3.3]), and the data are used to help create a treatment plan. The figure also shows a second important role of dosimetry in radiotherapy. A radiation detector can be used on the patient during the treatment to verify the dose given to the patient in vivo. This is discussed in more detail in chapter eleven on quality assurance (QA).

A typical relation between total dose and effect (i.e., effect in this context would be tumor control probability, or TCP) is shown in figure 3.2. This figure also includes a dose-response curve plotted for normal tissue; full effect would represent intolerable side effects for the patient. The aim of radiotherapy is to maximize the effect of radiation on the tumor while minimizing normal tissue reactions. This point can be determined by finding the maximum of the difference between the two dose-effect curves as shown in figure 3.2. The difference between the tumor control curve and the normal tissue complication probability is called the therapeutic ratio. Dose-response curves depend on a number of parameters, such as tumor or tissue cell type and radiosensitivity, which cannot be determined for each individual patient appropriately. It is also difficult to define what constitutes an intolerable side effect. For some patients the development of a cataract is a small price to pay for a higher cure rate, but others could consider this as an intolerable side effect. Therefore, published dose-response curves usually reflect the tumor response of a whole patient population; this averaging tends to make the curves shallower.

The shape of dose-response curves can give an estimate of the accuracy of dose delivery required for radiotherapy and therefore for dosimetry. It is generally assumed that a difference of 5% in the delivered dose to a radiotherapy patient does make a clinically detectable difference in treatment outcome (e.g., ICRU 1976, Brahme 1984). An underdosage of the tumor by 5% may reduce the tumor control probability by more than 15% for a typical tumor control probability (Brahme 1984). Similarly, an increase of 5% in the dose delivered to the normal surrounding tissues of particular organs at risk can cause intolerable side effects. This has also been confirmed in clinical observations of head and neck disease radiotherapy (Skladowski et al. 1994, Chappell and Fowler 1994). It is noteworthy that for an individual patient even smaller differences in dose can make a difference in the treatment outcome. However, due to patient variability this is very difficult to demonstrate clinically.

The demand of only 5% uncertainty in the absorbed dose to tissues is not a minor task. E. Hassenstein estimates the overall uncertainty as 7.8% in clinical routine (Hassenstein and Nuesslin 1985), and Karzmark (1993) confirmed these estimates (1993). Most of these error margins are already taken up by uncertainties in planning calculation procedures and in the definition of the target volume (i.e., tumor outline, patient geometry, tissue inhomogeneities, and patient movement). Therefore, the dosimetry should be as accurate as possible to make sure that the same dose is given every day to patients at any one radiotherapy center and that the dose delivered at one center is the same as the dose given at any other center. A wide variety of different detector types have been developed to quantify the dose for different applications and requirements. Before these are discussed in more detail in the following chapters, the ideal dosimeter will be described in section 3.2. In this section general characteristics of dosimeters will be discussed, including a number of definitions that will be useful for the discussion of dosimeter types in the following sections.

Whatever effect or signal is used to quantify radiation dose, it will depend on the materials employed in the measurements. Therefore, section 3.3 is devoted to the discussion of radiotherapy phantoms, including geometrical and material properties of a number of typically used models. As ionization chambers are the most important tool for day-to-day dosimetry, they are the recommended standard dosimeter for calibration purposes. In sections 3.4 and 3.5 the theory of ionization chambers is introduced and several practical examples are discussed. More information on ionization chambers is discussed in the context of calibration in chapter eight.

A number of other techniques used for x-ray dosimetry are described in sections 3.6–3.8. The most important techniques, semiconductors, thermoluminescence dosimetry (TLD), and various two-dimensional (2-D) dosimeters, such as film, are dealt with in separate sections.

Other dosimetric techniques are discussed in section 3.9. Some have quite a long historical background, such as calorimetry and biological dosimetry. Others are relatively new developments like diamond detectors and gel dosimetry, which have just started to establish their place in routine radiotherapy dosimetry.

The final section in chapter three presents an overview of different dosimetric methods. The aim of this section is to help the reader choose a suitable dosimeter for a particular application.

3.2 The Ideal Dosimeter: Some Definitions

From a physics point of view a radiation detector for radiotherapy should be designed to measure absorbed dose—energy absorbed per unit mass. However, except for calorimetry (see section 3.11) no other dosimeter detects absorbed energy directly, and other effects of radiation such as ionization and chemical changes are used to assess the impact of radiation on tissues. The quantity of the measured effect has to be converted into absorbed dose. In practice, the latter can be a very difficult task involving a number of factors and assumptions. Chapter eight will provide more information on this topic. However, there are also a nmber of other characteristics that need to be considered before using a certain dosimeter.

The ideal dosimeter has a number of features, defined below. The real dosimeters discussed in the following sections have to be compared with these characteristics.

Accuracy

The most important feature of any dosimeter is the ability to indicate a physical dose correctly, i.e., accuracy. The accuracy may be limited by systematic and stochastic errors. Typical systematic errors are electrometer leakage (see section 3.5) or fogged film (see section 3.8), both of which shift measurement results systematically in one direction, and repeated measurements will not improve the accuracy of the result. Stochastic errors, on the other hand, can be reduced by multiple measurements as they result from random variations which may go in all directions.

Precision

The precision of a measurement technique specifies the reproducibility of its results under similar conditions. This definition excludes systematic errors and therefore does not allow any conclusions about the correctness of the result. However, precision is an important measure, for example, in consistency checks. The precision is usually expressed as 1 or 2 standard deviations of the fluctuations of the measurement scores around the mean. In the context of this book 2 standard deviations will be used (if not stated otherwise), which means that more than 95% of all measurement scores will fall within the mean value ± precision. Figure 3.3 illustrates the concepts of accuracy and precision.

Detection Limit

The lowest dose detectable with a certain dosimeter type is called the detection limit. This dose is usually determined by fluctuations of the background (natural radiation and instrument noise only) readings. It can be distinguished between a qualitative ("there is some dose, but it is not possible to quantify how much") and a quantitative detection limit. Typically, the latter is quoted as three times the standard deviation of the background fluctuations. For radiotherapy, which uses relatively high doses, the detection limit is usually not a major concern (with the exception of radiation safety surveys).

Measurement Range

The dosimeter should be able to detect radiation from very low doses to very high doses. The range between the lowest and highest detectable dose is called the *usable range*. Dosimeters are often limied at high doses by saturation.

Dose Response

The reading of the dosimeter should be linearly proportional to the given dose, i.e., the reading for a certain additional dose is independent of the dose already registered. Strictly speaking, this is not required as long as dose response is reproducible and predictable. However, a linear dose response reduces the likelihood of errors and enables faster processing.

Dose Rate Dependence

The ideal dosimeter should be independent of the dose rate. This is particularly relevant to the radiation produced by linear accelerators that deliver very high doses in very short pulses (see chapter one).

Variation of Response with Radiation Quality (Tissue Equivalence)

The ultimate aim of a dose measurement in radiotherapy is to determine the dose in tissue of a patient. In this context, tissue could be all tissues, such as lung, muscle, or bone. In practice, however, this usually refers to water as a reasonable substitute for a variety of human tissues. This would be achieved if the detector were interactive with the radiation, like tissue, i.e., be tissue equivalent. Unfortunately, most detectors are complex in their construction and not made from water and a correction must be applied. To complicate matters, the magnitude of the correction does depend on the radiation quality.

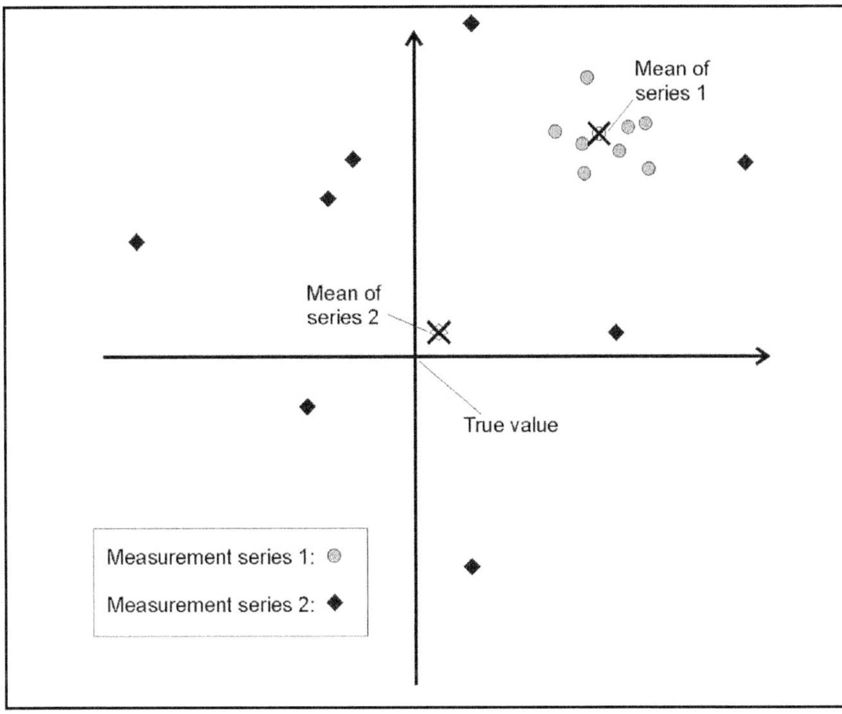

Figure 3.3: Illustration of the concepts of accuracy and precision: The measurements of series 1 have high precision (reproducibility) while the measurements of series 2 are more accurate. (Adapted from Van Dyk 2005.)

Therefore, it is important to test how the radiation response of the dosimeter changes with radiation quality. The requirement of minimal change in dose response with radiation energy usually implies an effective atomic number of the dosimeter that is very close to that of the tissue under investigation. In practice, tissue equivalence is a very important issue when choosing an appropriate dosimeter because it is difficult to make predictions for a complex detector system.

Spatial Resolution

The dosimeter should allow the determination of the dose in a very small volume (point dose). The location and size of this "point" should be well defined. However, in practice the size of the volume is limited by the number of events required to create a signal and the stochastic mechanism of dose deposition in microscopic dimensions. The energy, dE, deposited in a mass element, dm, is shown schematically in figure 3.4. It shows z = dE/dm as a function of dm. The energy deposition is a random event at the microscopic level. The dose D can be calculated from the mean value of this random unit <dE/dm> according to:

$$D = \lim < \frac{dE}{dm} > \text{ for } dm \rightarrow 0, \tag{3.1}$$

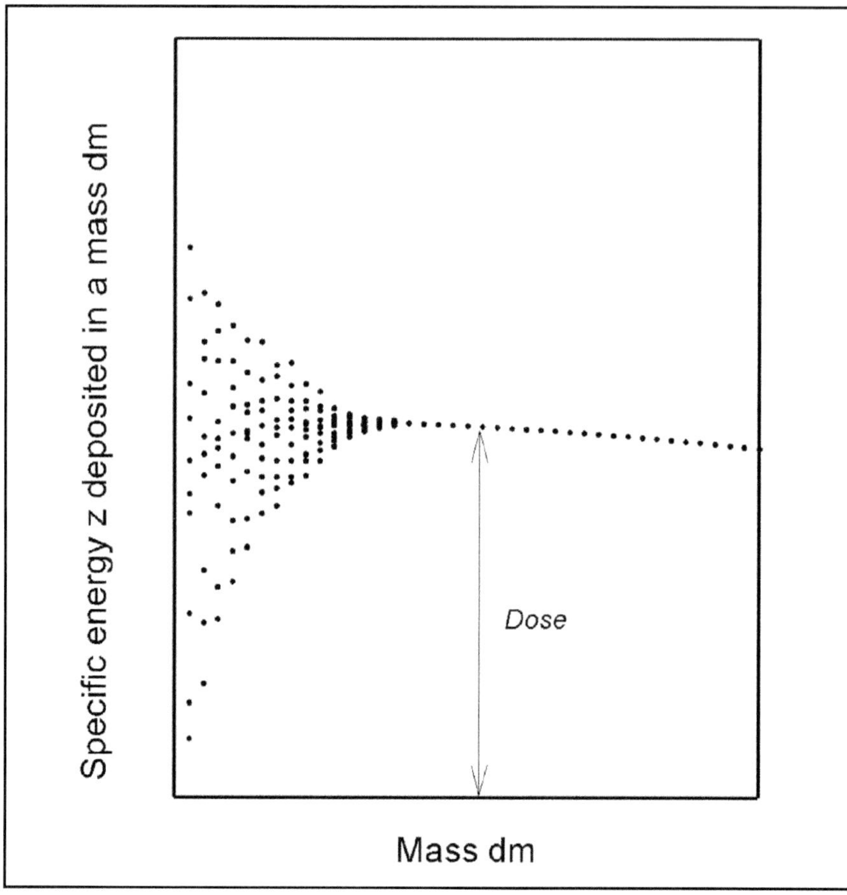

Figure 3.4: Illustration of the concept of microdosimetry. Each point represents the result of a single measurement. (Adapted from Rossi 1968.)

For practical considerations, when measurement time is restricted, the size of the volume for dose measurements is limited. A reasonable limit for the dimensions would be 10 μm, the typical diameter of a cell. However, in practice this is difficult to achieve with detectors of dimensions that have a sufficient signal-to-noise ratio. Interestingly, for low LET (linear energy transfer) radiation such as photons and electrons, a dose of less than 0.2 mGy will result on average in one radiation track per cell nucleus of 8 μm diameter (Fry 1996). On average, a 1 MeV electron will lose some 200 eV per μm in 6 events along this track (Tubiana et al. 1990). As such, a typical radiotherapy dose of 2 Gy will deposit energy in thousands of events in a volume of 10 μm diameter. This ensures that therapeutic doses are still well defined on this microscopic scale and not affected significantly by the random nature of the energy deposition events. A similar discussion can be made for an air-filled ionization chamber (see sections 3.4 and 3.5) where the decreased medium density is well compensated for by larger size. More information on microdosimetry can be obtained from the International Commission on Radiation Units and Measurements (1983) and Rossi and Zaider (1991).

Since the ideal dosimeter does not exist, compromises must be made in dosimetric practice. To minimize the risk of serious errors due to these compromises, it is good practice to repeat instrument calibration at regular intervals. Also, comparisons with dosimeters from different institutions can help avoid mistakes. Alternatively, dosimeters can be checked internally by comparison with a reference source of radiation, typically a source of radioactive ^{90}Sr. These checks cannot replace regular calibrations, but they do monitor the performance of dosimeters and help to detect possible defects.

For many situations it can also be useful to compare the dosimetric results of different techniques and dosimeter types. All this effort can make one feel more confident that the "correct" dose is administered to patients undergoing radiotherapy.

3.3 Radiotherapy Dosimetry Phantoms

3.3.1 General Considerations

In radiotherapy the term *phantom* is used to describe a material and structure that models the radiation absorption and scattering properties of the human tissue of interest. *Phantom* also refers to the environment of a radiation detector as shown in figure 3.1. To appreciate the quality of phantoms and their materials, it is necessary to review briefly the properties of human tissues themselves.

In general, two different phantom types are available: *geometrical phantoms,* which mimic the dosimetric properties of human tissues but reduce the complexity of the human anatomy to simple square or circular geometries (see sections 3.3.3 and 3.3.4), and *anthropomorphic phantoms,* which will be discussed in section 3.3.5.

The use of radiotherapy phantoms requires a number of considerations to be kept in mind:

i) Phantoms are only designed to mimic the properties of human tissue for specific radiation qualities; a phantom designed for diagnostic x-rays (e.g., a CT-phantom) is not necessarily suitable for megavoltage dosimetry.
ii) Phantoms may be designed for use with a particular detector type only; or, in reverse, some detectors such as ionization chambers employ a certain wall material that requires a particular phantom material.
iii) Phantoms are an essential part of dosimetric procedures; they should be the object of an appropriate quality assurance program. Often neglected are features such as:
    - Does a material dry out and change physical density or does it shrink? This is common for gels and may occur in human skeleton incorporated in some anthropomorphic phantoms.

- Does a phantom material accumulate charge, thereby altering the dose distribution from charged primary or secondary radiation? This has been reported for PMMA in electron beams (Galbraith et al. 1984).
- Does the surface of a slab phantom stay smooth and flat? Inappropriate storage may lead to deformations of some materials, which can result in air gaps between slabs of phantom material. Similarly, lung equivalent phantom material may be lung equivalent inside, but the surface could be sealed, making it unsuitable for interface dosimetry.

iv) For practical reasons, all phantoms should be allowed to equilibrate with the ambient air temperature for sufficient time. In the case of a large water phantom this can take several hours.

3.3.2 Human Tissues and Phantom Materials

Phantom material is designed to mimic certain properties of human tissues. A number of different reference texts are available listing the body composition of human tissues (ICRP 1975, Iyengar et al. 1978). The elemental composition that determines radiation absorption and scattering properties can vary considerably between different organs. However, for most purposes in radiotherapy it is sufficient to distinguish between muscle tissue, lung, and bone only. Most other organs of interest, such as kidney, liver, brain, and intestines are considered to be equivalent to muscle. This is not always correct, and for low-energy x-rays the difference between dose given to muscle and fat can be substantial. However, as discussed in chapter two, Compton effect is predominant for the megavoltage x-rays, and the difference between tissues reduces to the difference between their electron densities. Fat and adipose tissue have a relative electron density of close to 0.9. Therefore, lung and bone constitute the extremes of tissue density, and their difference cannot be neglected in dose calculations.

In certain cases it can be necessary to include air as a constituent of cavities and air passages. Air is part of the human anatomy in cavities, and it is commonly used as the surrounding medium for air kerma measurements, which are recommended by many protocols as calibration procedures for x-rays with energies below 3 MV. Even though all linear accelerators produce x-rays of higher peak energies, air as a phantom material will be included in the discussion because many national standards laboratories provide an in-air exposure measurement for ^{60}Co as reference for the calibration of higher energy x-rays.

Table 3.1 lists the elemental composition of muscle, lung, and bone. Similar data can also be found as part of many Monte Carlo calculation programs as they require this information for dose calculations. Also included are air and water as well as some commercially available tissue substitutes.

Table 3.1: Elemental composition of human tissues, water, air, and some tissue substitutes. The tissue composition is given in percentage of mass composition. (Data from White 1997, White et al. 1986, Constantinou et al. 1982, Woodard and White 1986)

Material	*Elemental Composition %fi (%)*
Water	H (12) O (88)
Solid water	H (8.09) C (67.22) N (2.4) O (19.84) Ca (0.13)
Air	N (78.09) O (20.97) Ar (0.93) C (0.01) Ne, He, Kr (traces)
Muscle	H (10.2) C (14.3) N (3.4) O (71) K (0.4) Na (0.1) P, S, Cl (traces)
Muscle analog	H (9.15) C (70.66) N (4.28) O (13.34) Cl (0.11) Ca (2.46)
Fat analog (RMI 453)	H (8.5) C (72.88) N (2.24) O (16.25) Cl (0.13)
Bone	H (8.67) C (13) N (3.6) O (66.4) S (0.46) P (2.43) Ca (4.96) Na, Mg, K (trace)
Bone analog (RMI 456: "Inner bone")	H (7.9) C (63.79) N (4.23) O (9.88) Cl (14.2)
Lung	H (10.3) C (10.1) N (2.9) O (75.5) Na (0.2) P (0.2) S (0.3) Cl (0.2)
Lung analog (RMI 455)	H (8.36) C (60.41) N (1.67) O (17.33) Cl (0.15) Si (0.72) Mg (11.36)
Perspex™ (PMMA Lucite™, Plexiglas™)	H (2.13) C (63.83) O (34.04)
Polystyrene	H (7.69) C (92.31)

3.3.3 Water Phantoms

Water is one of the most convenient phantom materials for physicists. It has a simple chemical formula, it is transparent, liquid, cheap, and readily available. Water is one of the main constituents of many human tissues, and its effective atomic number ($Z_{eff} = 7.51$) (Johns et al. 1983) and its physical density ($\rho = 1$ g/cm^3) are close to muscle as a tissue equivalent. Therefore, many dosimetric procedures are performed in water phantoms.

A very versatile phantom is the so-called water tank. It is the workhorse in many radiotherapy departments. It consists of a PMMA cuboid basin with typical dimensions of at least $50 \times 50 \times 50$ cm^3. It is open at the top and can be filled with (distilled) water. A precision carriage mounted within the tank can be moved in three dimensions; this allows for the placement of detectors such as ionization chambers or semiconductors at any position in the tank. However, it is advisable to leave at least 5 cm of water between the detector and any of the walls to provide enough scatter material for dose measurements. Depth dose, profile, and isodose track scans can be produced using this carriage mechanism within the water tank. Figure 3.5 shows a photograph of a commercially available water phantom.

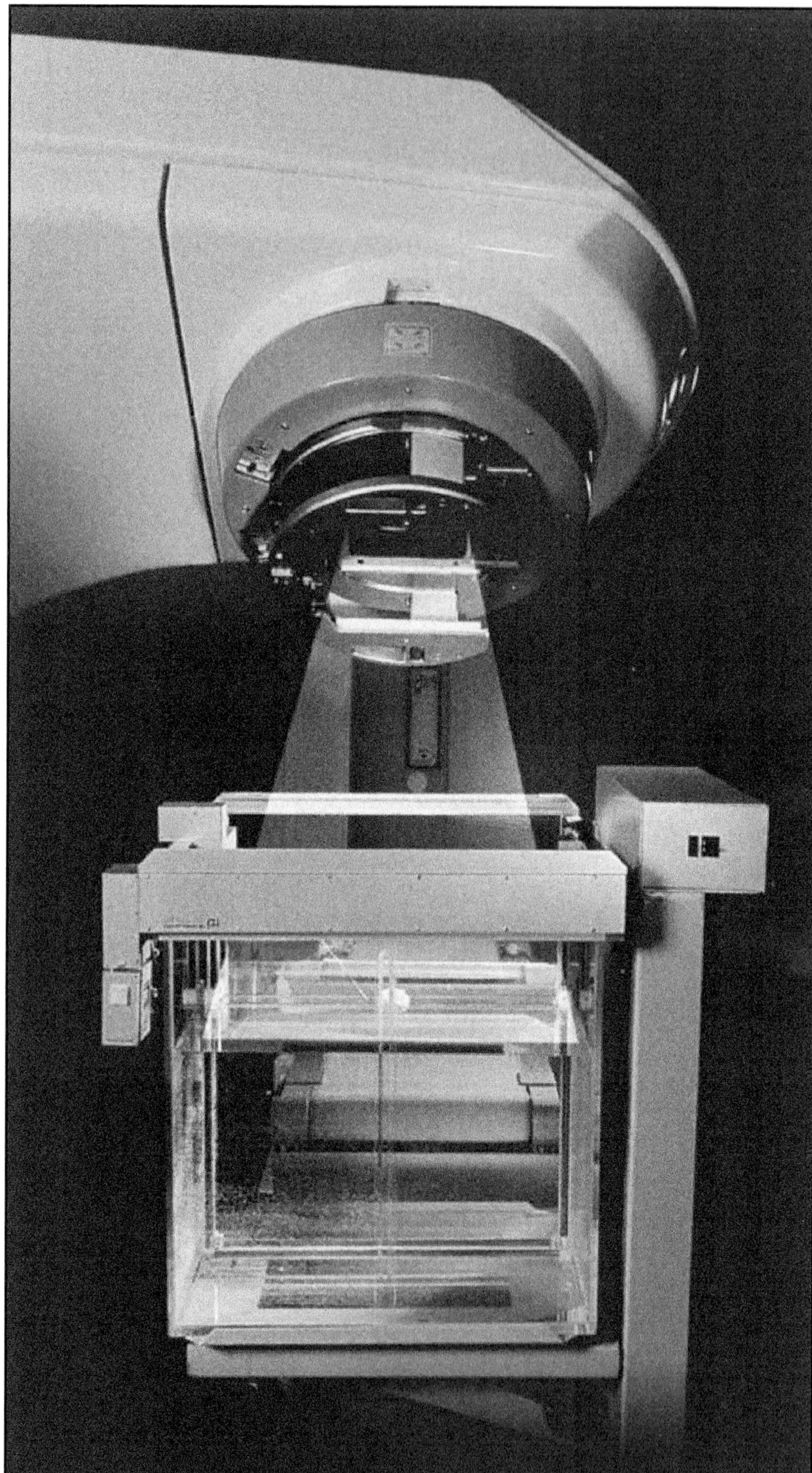

Figure 3.5: A commercial water tank phantom, (Courtesy of Scanditronix).

If scans are to be measured for radiation with varying output such as pulsed radiation produced by a linear accelerator, a second detector is used as reference dosimeter in a fixed position in the radiation beam. This can be mounted anywhere (within the radiation beam) but is usually placed above the water surface in air. The signal from the dosimeter on the movable carriage in water is then compared with the reference signal from the stationary reference detector, thus avoiding fluctuations in the scans due to the pulsed nature or instability of the radiation. If the second detector is mounted to the water tank itself, one needs to take care that the reference detector does not move if the filling of the tank is altered such as for TMR measurements (the weight of the water may deform the tank and turn the mounting). Alternatively, the reference chamber can be mounted to the head of the linear accelerator.

Figure 3.6 shows a water tank system schematically. Most commonly, ionization chambers are used as detectors in the tank, but semiconductors and diamond detectors are also available. More recently, linear arrays of semiconductors and ionization chambers have become available for rapid beam profile data acquisition. The acquisition of data using such an array of detectors can be very useful in assessing dose distributions during dynamic beam delivery such as dynamic wedging or arc therapy. A waterproof ionization chamber array is shown in figure 3.7.

The setup of water tank phantoms such as shown in figure 3.4 takes some time. About 20 minutes or more are needed to align and level the tank, fill the tank, set the right water height, and align the detector. However, water tank measurements are very important. They are required for commissioning of new accelerators (IPSM 1988, Constantinou 1994, Kutcher et al. 1994) and are used for the data acquisition of most computerized planning systems. For quality assurance, water tanks can be used periodically to check beam profiles and depth dose curves of the radiation.

Water tanks can also be useful for clinical measurements such as output factor determination for complicated blocked fields. The dose can be measured not only in the central axis of the beam but at any position (e.g., behind the thin blocks used for larynx shielding) and at any depth. This is important, for example, in mantle treatments.

A limitation of all open water tank systems is that measurements can only be performed with the linac in vertical gantry position. This restricts the checking of radiation properties in other gantry positions and does not allow the use of the tank at very extended focus-to-surface distances (FSDs). To overcome this problem, some manufacturers have put a narrow, thin window in the side of the tank to allow measurements to be taken in gantry positions 90° and 270° at least.

An interesting variation of a water phantom is shown in figure 3.8. This 'buddelship' ("boat in a bottle") consists of a fully closed water tank that allows 2-D scanning of a radiation beam. One scanning axis is aligned with the beam direction to assess depth dose data while the other can acquire a profile. Other profiles can be assessed by rotating the collimator, and deeper depths can be reached by adding PMMA blocks. The buddelship can be attached to the head of the linac. As such, it allows acquisition of data at all gantry angles, a process which is particularly useful during beam commissioning.

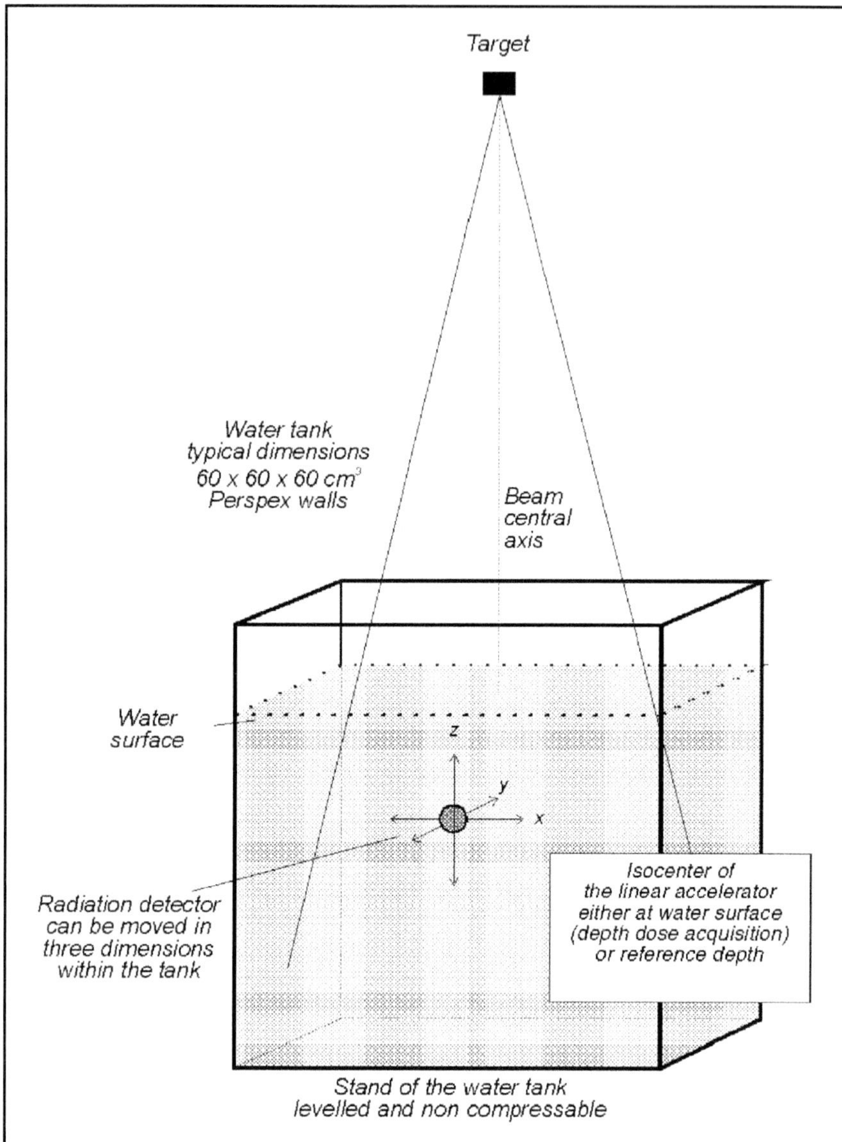

Figure 3.6: Schematic drawing of a scanning water phantom. Many modern systems also allow an automatic adjustment of the water surface for tisue-maximum ratio (TMR) or tissue-phantom ratio (TPR) measurements.

Many water tank systems also offer the purchase of other beam scanning devices that use similar hardware and software. Examples are film densitometers (see section 3.8) and in-air scanners. The latter devices attach to the treatment head and rotate with the collimator and gantry. They allow the 2-D scanning of the radiation beam at a certain specified distance, which is usually smaller than the standard FSD. An in-air scanner typically employs a reference and a field radiation detector and allows a PMMA sheet of variable thickness to be used as build-up. In-air scanners are very well suited for profile measurements as a quality assurance measure or for beam steering,

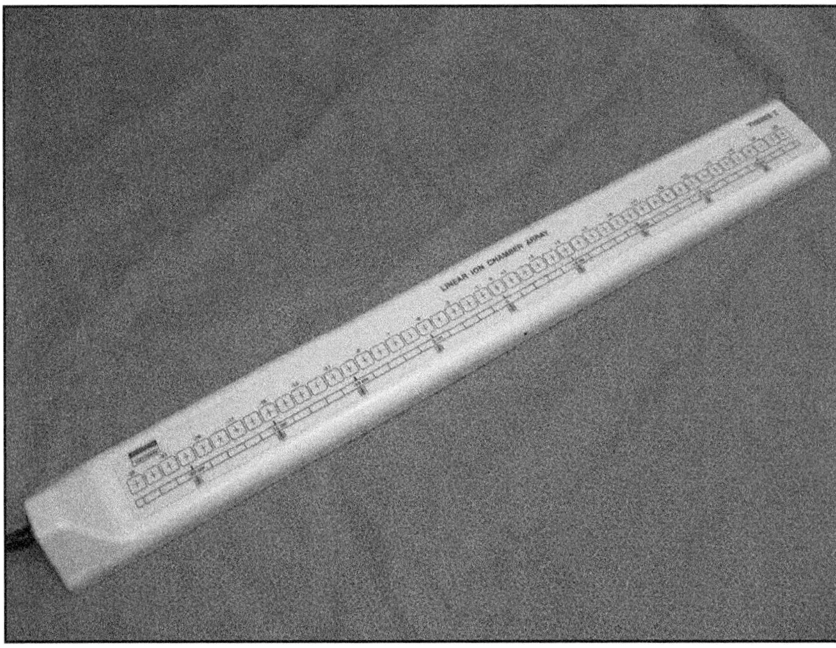

Figure 3.7: Photograph of a waterproof linear array of ionization chambers.

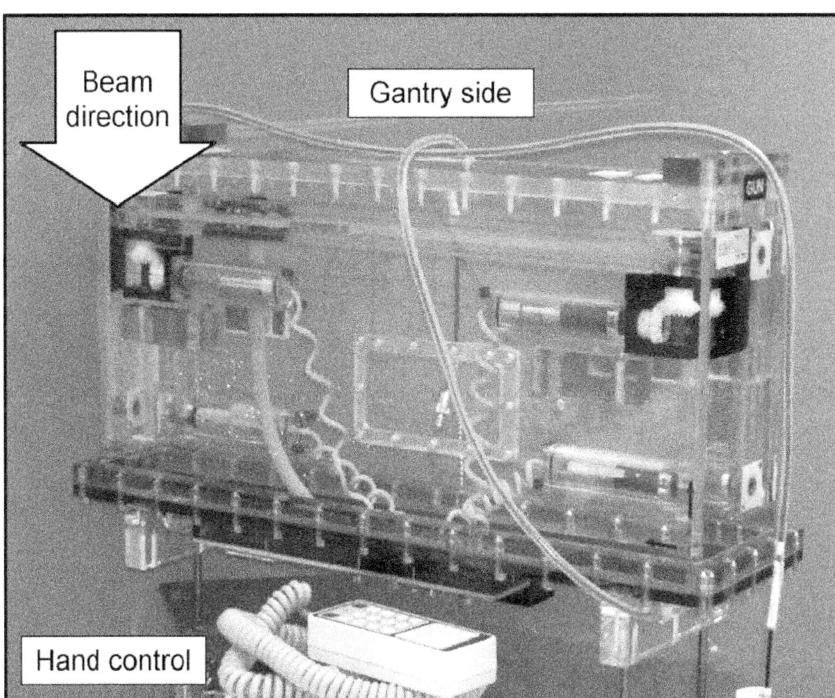

Figure 3.8: A water phantom with two-dimensional scanning mechanism. The phantom is designed to attach to the gantry head of a linac.

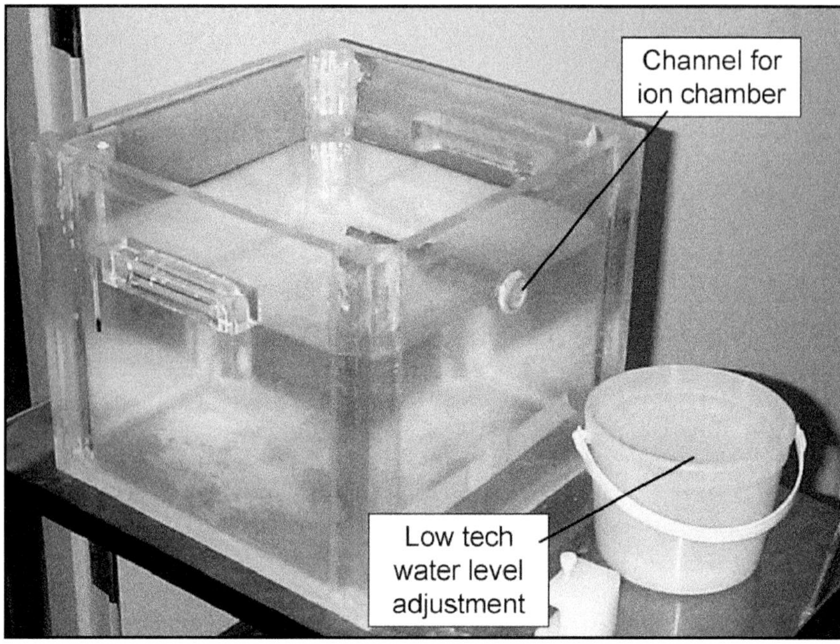

Figure 3.9: A simple water phantom with a single water-protected holder for a Farmer-type chamber..

since they set up faster than a water tank phantom. As the flatness of an x-ray beam also depends strongly on the energy (due to the flatness filter employed), a profile scan can indicate problems with the effective energy of an x-ray beam.

Water tank phantoms play an important role in the calibration of the radiation beam. The use of calibrated ionization chambers in the water tank is recommended by most protocols for the primary calibration of the linac (compare chapter eight). In these cases great care has to be taken to waterproof the chamber without restricting the free ventilation of vented chambers. A smaller type of water phantom is commercially available (or can be easily constructed) for calibration purposes only. These phantoms have a fixed space for an ionization chamber (typically a 0.6 cc Farmer-type chamber). The hole for the chamber is surrounded by a thin layer of PMMA or polystyrene and the rest of the phantom is filled with water. These calibration phantoms are available in a closed (sealed with a thin PMMA cover) and open version(shown in figure 3.9). Both are useful for rapid calibration checks and offer typically at least one calibration depth of 5 cm in water.

3.3.4 Solid Phantom Materials

For routine checks of the output consistency of the linac it is too time consuming to set up the large water tank. In these situations solid phantoms are a useful alternative. Solid stack phantoms are usually cross-calibrated with liquid water phantoms to enable rapid routine tests for beam output consistency or beam energy (by comparing the reading at two different depths). These phantoms are often more practical for some other specialized tests, such as inhomogeneity stack phantoms and phantoms for surface dose measurements (no meniscus effect with a parallel plate ionization chamber). They also allow the use of radiation detector types such as film or TLDs without the need for waterproofing.

A variety of solid materials are employed as radiotherapy dosimetry phantoms. The increasing quality of these water or tissue substitutes makes them more and more applicable for a variety of tasks in a radiotherapy department, replacing the time-consuming setup of a water tank. The importance of the materials can also be seen from the number of papers, conferences, and reviews dedicated to that subject in the 1980s and 1990s (e.g., HPA 1977, White 1977, 1978, Baird and White 1982, ICRU 1989, Kleck et al. 1990). Phantom materials include "off the shelf" materials such as polystyrene and PMMA (also known as acrylic or Perspex). More recently epoxy resin–based analog materials such as solid water (Constantinou et al. 1982) have become commercially available. The elemental composition of solid water, PMMA, and polystyrene is given in table 3.1.

Some materials such as PMMA and polystyrene are very good insulators. In electron beams this can cause the material to accumulate charge, which will alter the radiation dose. Charge accumulation in megavoltage x-ray beams is of no concern, since the net amount of charges transported into and out of a certain volume is approximately equal (except for the build-up region).

For many compounds of different atomic number (polystyrene, acrylic, and Delrin®) the electron densities are calculable since the chemical formula and mass density are available. Electron densities and effective atomic numbers were calculated for such compounds, using a formula of McCullough and Holmes (1985). The results of calculations for some phantom materials are provided in columns three and four of table 3.2.

Table 3.2: Mass density, ρ, electron density relative to water, ρ_e^w, effective atomic number, Z_{eff}, and R_σ of some materials and tissue substitutes. Also listed are the effective atomic number relative to water, Z_{eff}^w, and the Hounsfield CT number. The calculation of ρ_e^w, Z_{eff}, and R_σ is shown in chapter six.

Material	ρ	ρ_e^w	Z_{eff}	Z_{eff}^w	R_σ	*CT Number*
Water	1	1	7.4	1		0
Air	0.0012	0.0012	7.55	1.02		−1000
Solid water (WT/SR6)	1.04	1.02	7.54	1.02		17
PMMA ($C_5H_4O_2$)	1.19	1.16	7.55	1.02		133
Polystyrene (C_8H_8)	1.04	1.01	5.7	0.77		8
Delrin (CH_2O)	1.41	1.36	6.96	0.94		348
Muscle analog (MS10)	1.03	1.02	7.39	1.0	0.9860	14
Lung analog (LN 10/75)	0.3	0.28	7.51	1.01	0.9879	-677
Fat analog (RMI 453)	0.92	0.895				
Breast tissue substitute (RMI 454)	0.99	0.98				
Soft bone analog (IB1)	1.16	1.14	9.26	1.25	0.9130	228
Hard bone analog (SB3)	1.84	1.71	13.27	1.79		1283
BC-400 plastic scintillator	1.032					

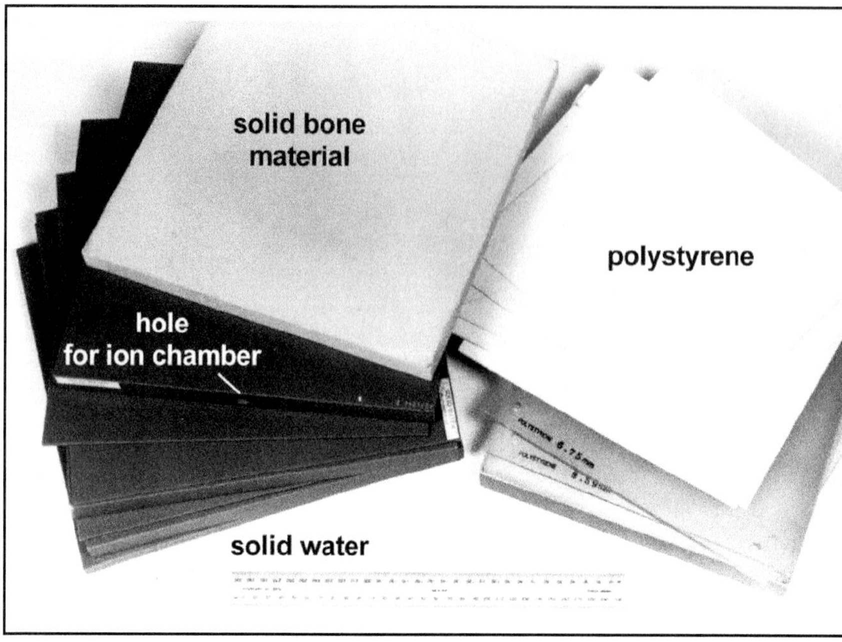

Figure 3.10a: Phantom from slabs of material: left, solid water and bone; right, white and clear polystyrene.

Figure 3.10b: Solid water phantom consisting of vertical slabs of solid water which allow sandwiching of prepacked film for QA of IMRT dose distributions. (Courtesy of E. Wong.)

Solid water, bone, muscle, and lung epoxy resin–based analogs are commercially available as dosimetry phantom material. Figure 3.10 shows a solid water phantom consisting of many slabs of solid water with various thicknesses. The typical size of these phantoms is 30×30 cm^2. This provides enough scattering material for most field sizes used in radiotherapy, as one typically would require 5 cm material out of the field under investigation. The size is a compromise between allowing for enough side scatter material and not being too heavy (a $30 \times 30 \times 4$ cm^3 slab of solid water weighs about 3.6 kg).

Solid phantom materials are available as substitutes for a variety of tissues such as muscle, fat, lung, and bone. The recipes and methods of D. White et al. (1977, 1986) can be followed in the manufacture of solid phantom materials which mimic a variety of substances. A method has been developed by White (1977, 1978) to show that the radiation properties of these analogs match closely those of their respective equivalent human tissues. The mass stopping powers, S/ρ, and mass attenuation coefficients, μ/ρ, of these substances match those of their human tissue equivalents to within 3% over the x-ray energy range of 0.1 to 10 MeV (White 1977, White et al. 1986). The solid water recipe of Constantinou et al. (1982) manufactured by a company (RMI) is now commonly used as a standard dosimetry phantom material for absolute calibration and other beam tests. Some information about commonly used phantom materials is listed in table 3.2.

Slab phantoms are available in many forms for many purposes. Figure 3.10b shows a slab phantom that is used for verification of dose distributions in complex radiation deliveries such as intensity-modulated radiation therapy (IMRT). It can accommodate film (as shown) as well as ionization chambers, metal oxide semiconductors-field effect transistors (MOSFETs), and thermoluminescent dosimeters. Also, cylindrical slab phantoms are available, which match the external contours of the human body better than a square slab phantom. As such, they are often used for the dosimetry of dynamic procedures such as rotation or arc treatments.

Only in a few dosimetry techniques are the phantom material and the detector identical. An example is gel dosimetry, in which a chemical dosimetric solution is embedded in agarose or gelatin gels. Here the advantages of tissue equivalence are combined with three-dimensional (3-D) assessment of dose distributions (compare section 3.9.5).

3.3.5 Anthropomorphic Radiotherapy Phantoms

The ideal radiotherapy phantom not only matches the attenuation and scattering properties of human tissues but also mimics the external and internal contours of the patient. Anthropomorphic phantoms are available for these purposes (Kleck et al. 1990, ICRU 1992). However, these phantoms are also only a compromise, since all patients differ in stature, internal structures, and outline.

A simple approach to an anthropomorphic phantom is wax shaped according to the outlines of a patient. This can be sufficient to investigate areas such as the skin dose in complicated geometries and has the advantage of being individualized for a particular patient. A typical example is shown in figure 3.11. The cost of this type of anthropomorphic phantom is comparatively low. Wax can easily be cut or molded as desired to accommodate radiation detectors. Similarly, it is possible to incorporate other structures such as metallic implants, bone equivalent, or cartilage (Ostwald et al. 1996).

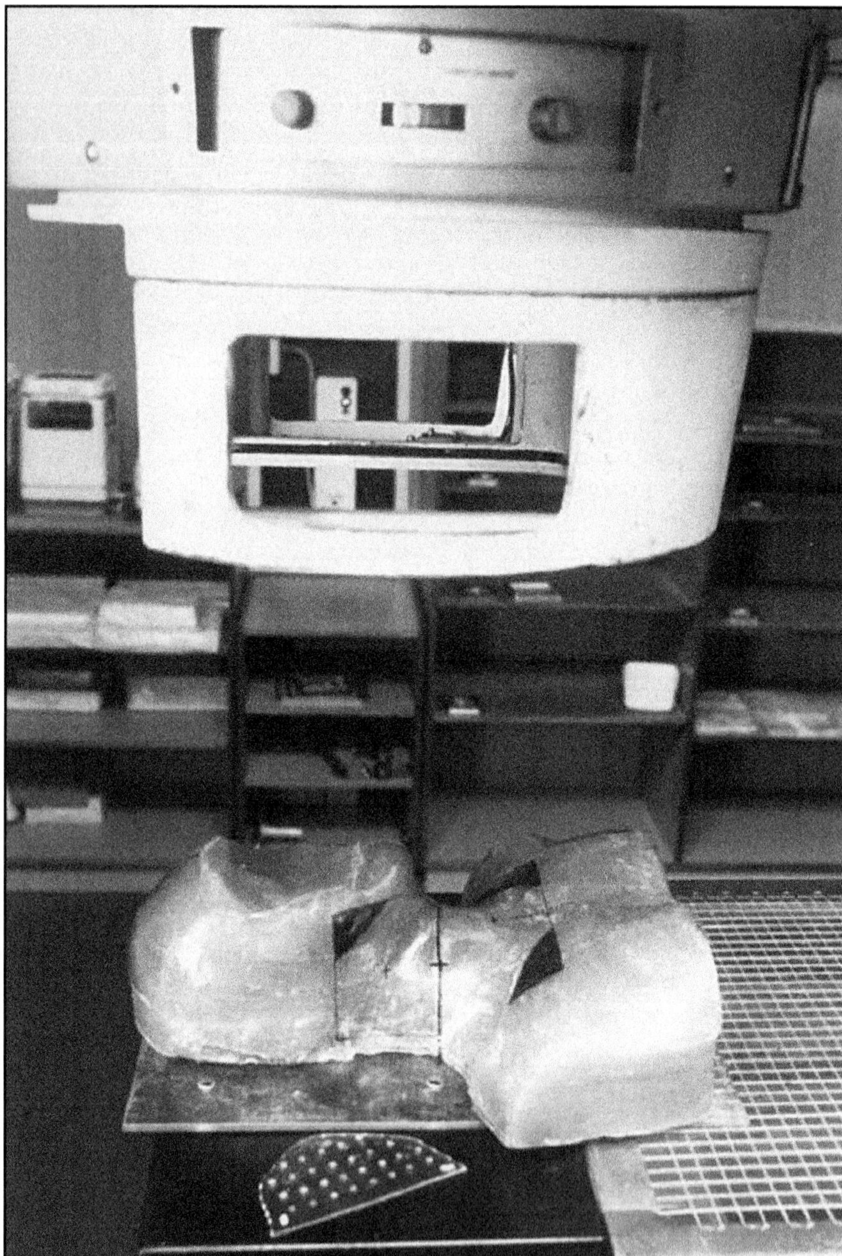

Figure 3.11: Anthropomorphic wax phantom in treatment position. The present phantom is modeled from a volunteer and was used in a study on dosimetry at the junction of photon fields in head and neck treatment. (Reprinted from Zhu et al. ©1997 with permission from Elsevier.)

However, wax is not a very good tissue substitute at lower x-ray energies, as its atomic number is closer to fat than to tissue. It is also impractical to include extensive bone structures or lung. Therefore, more sophisticated phantoms are commercially available. Most of these phantoms are based on the ARL (Alderson Research Laboratories) phantom, which was introduced in 1960 (Kleck et al. 1990). Figure 3.12 shows a typical example.

Older types of these phantoms were built around a real human skeleton or a bone analog skeleton. Real skeletons tend to dry out over a long period of time and give unrealistic bone densities. Analog skeletons on the other hand often do not model the difference between hard and soft bone adequately and, therefore, only give a compromise bone density. Tissue and lung equivalent materials are used to complete the whole body, which is available in a male and/or female version. Major air cavities, such as the trachea, are drilled out to improve the match between the phantom and a human patient.

Most anthropomorphic phantoms are cut in the axial plane in slabs of typically 2 to 3 cm thickness, as can be seen in figures 3.12 and 3.13. Figure 3.12 shows two completely assembled phantoms that are available in male or female versions, while figure 3.13 shows the cross section of two different

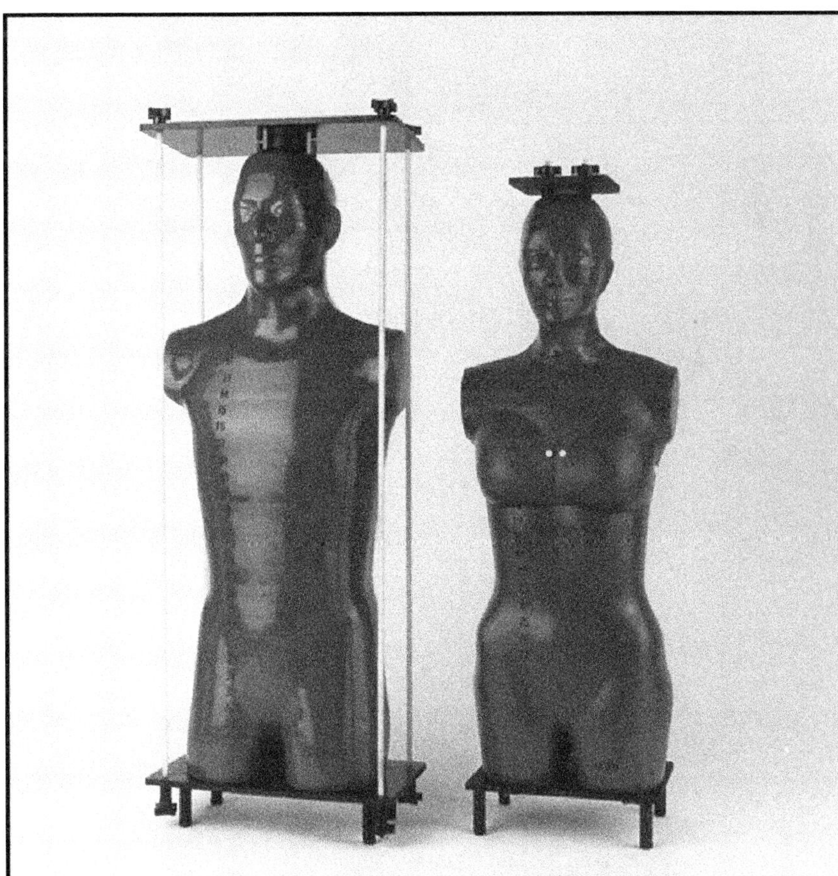

Figure 3.12: Male and female ART phantoms. (Courtesy of Radiation Services Australia.)

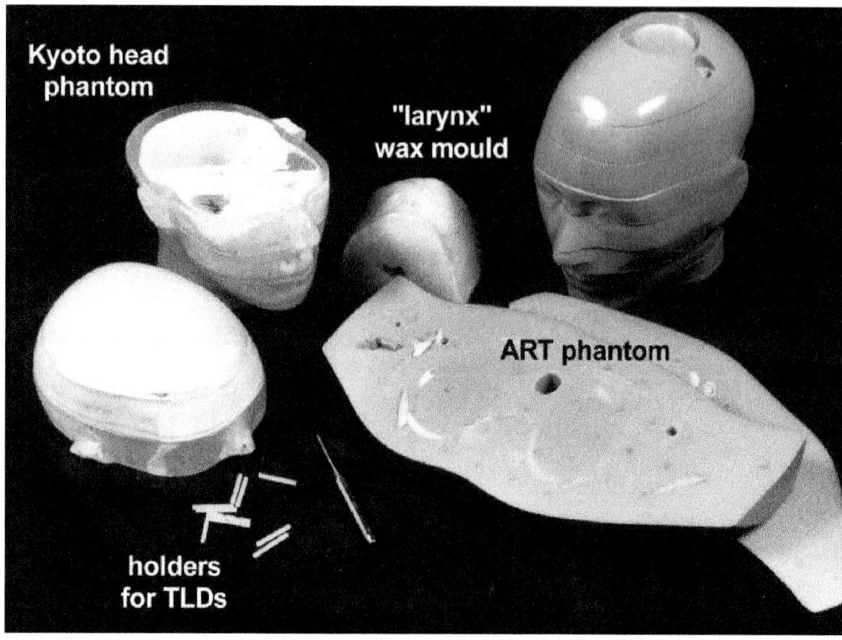

Figure 3.13: Parts of several anthropomorphic phantoms: left, the "Kyoto head phantom" with cavities drilled out, respectively; middle, top of a wax phantom of a larynx cavity; and right, the ART (adaptive radiotherapy) phantom.

phantoms. Holes are drilled in the slabs in a regular pattern to accommodate TL dosimeters. If the holes are filled with phantom material, film can be sandwiched between slabs. If the holes are located in bone or lung, care should be taken to fill them with the appropriate phantom material. A bracket is used to keep the phantom together and put equal pressure on the film to avoid air cavities around it. If an anthropomorphic phantom is to be used for film dosimetry, it should be opaque to avoid exposing the film to light. Film can then be cut to the appropriate shape, sandwiched within the phantom and covered on the edges with black tape.

A more recent development are phantoms that can mimic human motion. These can consist of an anthropomorphic phantom, which is just mounted on a movable platform (Kanagaki 2006, Liu 2007) or involve a phantom designed with moving (Dietrich 2005) or deformable parts (Nioutsikou 2006). Moving anthropomorphic phantoms are now also commercially available from a few manufacturers. Most of the motion phantoms are designed to mimic breathing motion, as this is the largest source of regular intrafraction patient motion. There are also now many radiotherapy centers in the world commencing treatment protocols which include either breath-holding or gating for compensation of breathing motion in lung, liver or breast cancer patients (Keall 2006a). Moving phantoms are essential for the study of motion and its effect on a particular treatment approach—this is particularly important for dynamic deliveries such as arc therapy and IMRT (Keall 2006b). In addition to this, moving phantoms can verify that the motion compensation is performing as expected and allows for appropriate QA of all devices involved in the delivery (compare also chapter twelve).

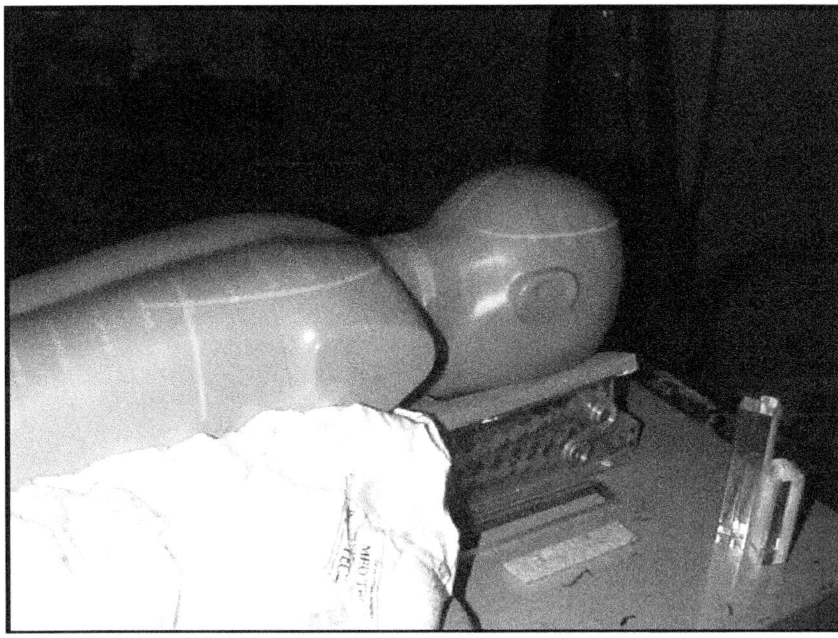

Figure 3.14: Anthropomorphic phantom of a 5-year-old child in prone position.

Even though many anthropomorphic phantoms are primarily designed for radiology (which is the larger market), they are a very useful tool in any radiotherapy department for teaching, testing new treatment techniques, verifying planning algorithms, evaluating dose distributions within a realistic patient (Dunscombe et al. 1996), and allowing dosimetric intercomparisons in realistic scenarios (Kron et al. 2002a). It is important to keep in mind that most anthropomorphic phantoms represent a healthy person. In practice, difficulties often arise for the nonstandard patient, such as an obese person. Children are another group of patient who requires different phantoms, and recently anthropomorphic phantoms of children have become available; an example for a 5-year-old child is shown in figure 3.14 (Hood et al. 2005).

3.4 Theory of Ionization Chamber Dosimetry

Dosimetry of ionizing radiation with ionization chambers is the most common type of dosimetry in radiotherapy departments. The general principle is well understood and measurements can be easily and reliably performed in most clinically relevant situations. Dosimetry of radiation with ionization chambers has a long history, and one of the most successful concepts in radiation dosimetry, exposure and the unit "roentgen," was devised to suit ionization chamber measurements. Even if the roentgen itself is no longer to be used as a unit and the absorbed dose has replaced exposure as the most fundamental quantity in dosimetry (IAEA 1987), the determination of ionization in air (or any other medium) due to ionizing radiation remains the standard of all dosimetric techniques.

The following four sections are dedicated to the theory and practice of ionization chamber measurements. After a general introduction to the concept of ionometric measurements (section 3.4), a number of ionization chamber types relevant to clinical radiotherapy dosimetry are discussed in section 3.5. This section also includes a brief discussion of commercially available electrometers. One of the most important tasks in radiotherapy dosimetry is the calibration of the output of a radiation source—virtually exclusively done using ionization chambers. The protocols available for this are discussed and compared in chapter eight during discussion of complementary material.

There is a wealth of literature available on ionization chambers used in radiotherapy dosimetry. Some useful standard textbooks and review articles are Attix et al. (1966, 1968, 1969), Greening (1981), Hendee et al. (2004), Johns and Cunningham (1983), Khan (2003), Krieger and Petzold (1989), Rajan (1992), Reich (1990), Van Dyk (1999), and Wiliams and Thwaites (2000).

3.4.1 General Considerations

The ionometric determination of x-rays is a three-step process, as illustrated in figure 3.15.

The first step can be linked to the quantity kerma (kinetic energy released per unit mass), K. In homogenous materials, the kerma is directly proportional to the photon energy fluence (see chapter two). If a reference material with known radiation properties, such as air, is used, this is a good measure of the strength of a radiation source. It should be noted that the secondary electrons generated in this step have a spectrum of energies determined by the original photon spectrum and the probability of a certain energy transfer during Compton interactions (compare chapter two). For dosimetry of primary electron radiation the following steps are also relevant —the difference is that one starts the process with a monoenergetic electron distribution.

The second step in figure 3.15 refers to the quantity collision kerma, which is only the kinetic energy released and actually deposited in the material. Some electrons set in motion in the first step lose energy due to radiative losses such as bremsstrahlung. As outlined in chapter two this fraction of energy is called g, and only the rest of the energy that is deposited in the material via electron collisions $(1 - g)$ contributes to the collision kerma.

Most ionization chambers use air as the active medium in which ions are to be generated. In this case the collision kerma is related to the exposure, X. The exposure is measured in C/kg air (the old unit was the roentgen: $1R = 2.58\times10^{-4}$ C/kg) and is a measure of the ability of photons to generate charge (i.e., ions) in air. It is interesting to note that the concept underlying the definition of exposure has been different in different countries (Reich 1990). In the United States and most English-speaking countries, exposure was related to the number of electrons generated in a mass element, dm, while in German-speaking countries exposure was related to the energy transferred from secondary electrons to a mass element, dm. The location of these mass elements is different in regard to the first interaction of megavoltage photons.

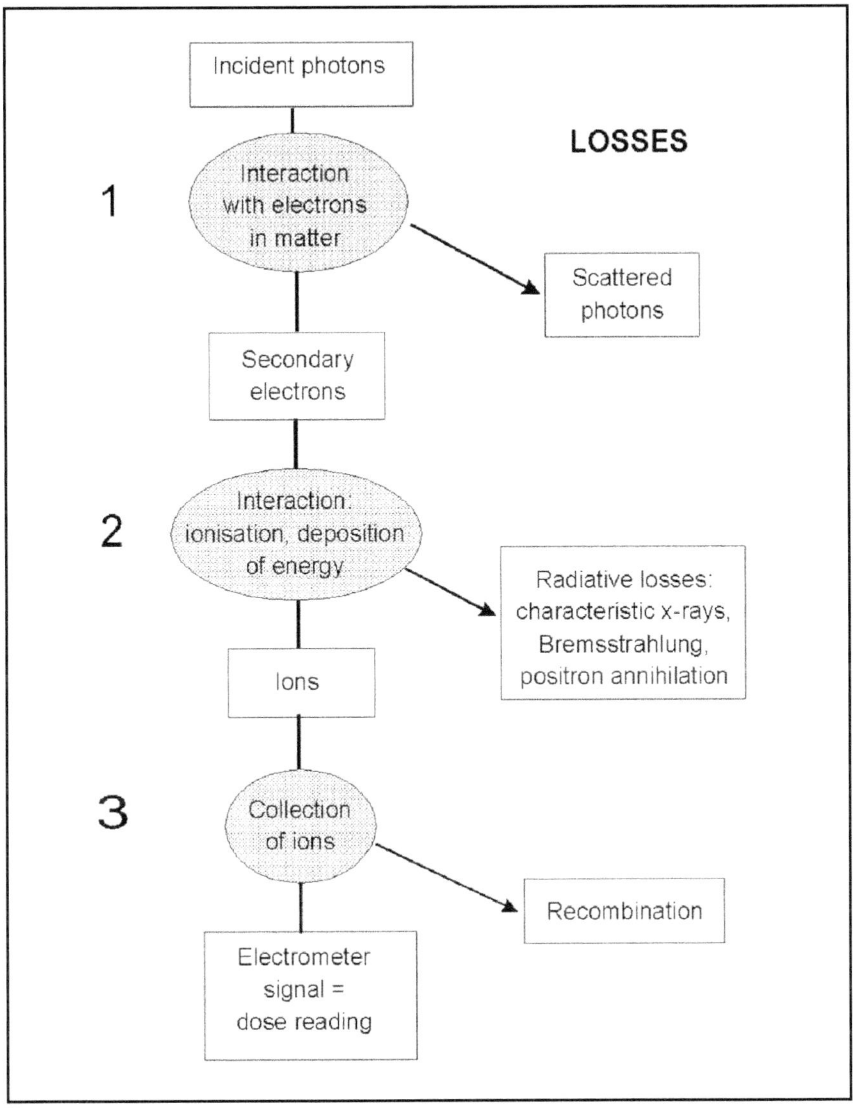

Figure 3.15: Schematic diagram of the steps involved in determination of photons using ionization chambers.

Exposure (defined in the first, Anglo-Saxon way) is linked to the energy-related quantity collision kerma via the mean energy required to produce one ion pair, W/e. It is noteworthy that the exposure refers to the point where the electrons are set in motion, not where they deposit their energy. Thus, the definition of exposure requires air twice: As medium in which the electrons are produced and as medium in which they cause ionization. This is not necessarily the same location in space (ICRU 1962). The quantity that refers to the location where the electrons produce ions is called the mass *ionization*. Exposure and mass ionization share the same unit; however, they refer to different concepts of measurement.

The generated ions are collected using an electric field and the charge is measured in the third step shown in the figure, typically by using an electrometer (see section 3.5). The number of ions produced in the chamber (and therefore the recorded charge) depends on the electric field and, therefore, on the applied high voltage. Figure 3.16 shows the number of ions produced as a function of the applied voltage U—the characteristic curve—of a typical ionization chamber. The figure assumes a cylindrical ionization chamber of about 1 cm^3 volume. As the electric field strength ($E = U/d$, d = distance) is proportional to the acceleration of the ions, the absolute value of the applied voltage can change depending on chamber design and volume. After an initial range ($U < 100V$) in which most ions recombine before they reach the electrodes, a plateau is reached when just all ions produced reach the electrodes. This range ($100V < U < 400V$) is used for

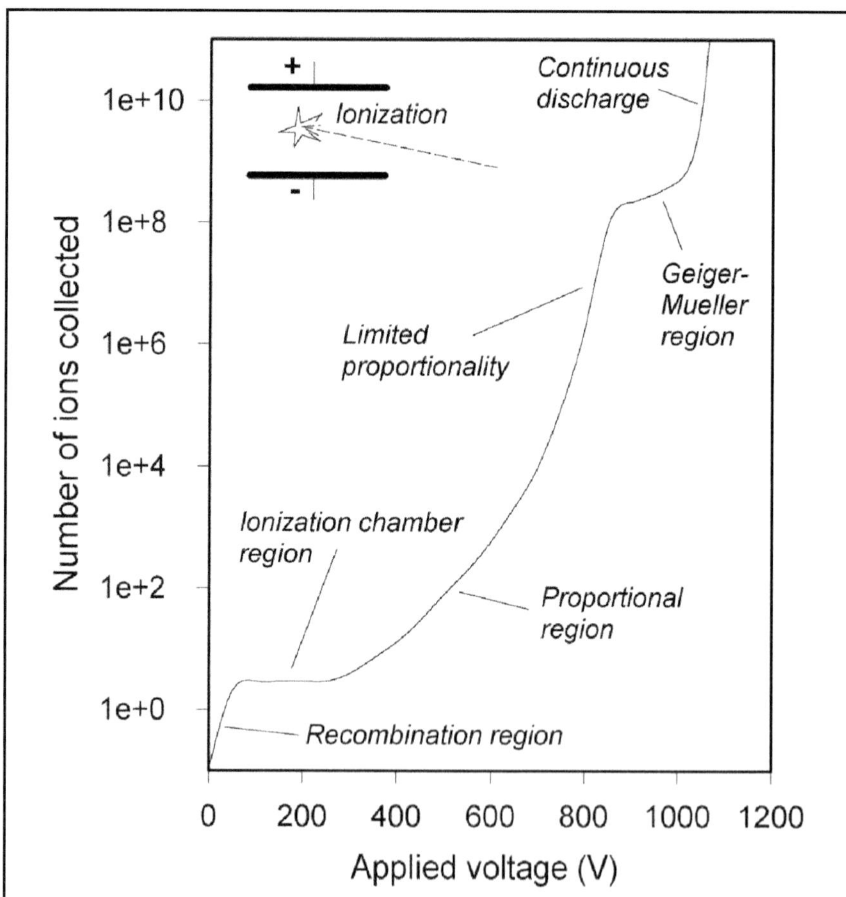

Figure 3.16: Characteristic curve of an ionization chamber. The actual voltage required for a particular effect depends on the chamber design.

dosimetry with ionization chambers. If the voltage is increased further, the ions gain enough energy in the electric field to cause secondary ionization. This increases the recorded number of ions in the so-called "proportionality range." With very high voltages (typically U > 800V), the number of secondary ions does not increase, and one records just one avalanche of ions per primary photon. This is called the Geiger-Müller range, and many survey meters operate under these conditions, as many more ions are created per radiation event. In the low radiation exposures typically encountered in radiation safety surveys, this results in a much larger signal than the one obtained from an ionization chamber.

It is intuitively evident that the charge collected in an ionization chamber should also be a measure of the energy that is absorbed in the air or in any medium replaced by the chamber. To convert from measured charge (in C) to energy (in J) per unit mass, one has to know the mean energy required to produce 1 ion pair, W/e. For dry air this is 33.97 ± 0.15 J/C (CCEMRI 1985). This factor converts the charge-related quantities into energy-related quantities, as indicated in figure 3.17.

Under certain conditions, the collected charge is proportional to the dose that would be delivered at the same location in space. Then the readout of an ionization chamber in coulombs can be converted to J/kg and used for the determination of absorbed dose. There are two theoretical situations where this is possible. These situations will be described further following a discussion of energy balance given by Harder (1966) and described by Krieger et al. (1989).

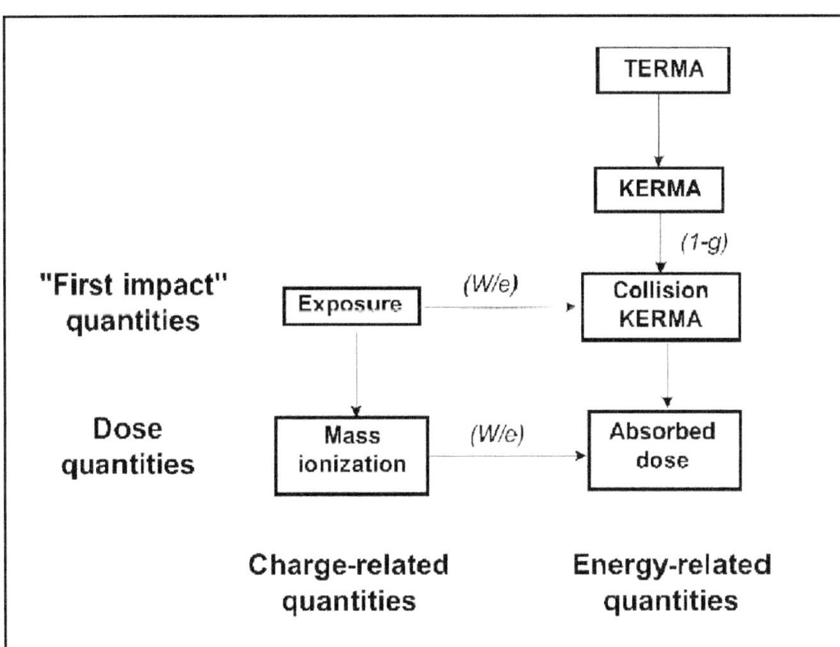

Figure 3.17: Quantities used for determination of x-rays using ionization chambers.

Any radiation detector can be regarded as shown in figure 3.18. The energy deposited in the detector, and therefore the dose, D, can be calculated from a simple energy balance using the following equation:

$$D = \frac{1}{m}\left(E^{\gamma}_{in} - E^{\gamma}_{ex} - E^{\gamma,e}_{ex} + E^{e}_{in} - E^{e}_{ex} - E^{e,\delta}_{ex} + E^{\delta}_{in} - E^{\delta}_{ex}\right), \tag{3.2}$$

with m as the mass, which is penetrated by each energy, E, with the index "in" and from which all energy contributions indexed with "ex" are escaping. The γ indicates photons, e refers to secondary electrons, and δ to delta electrons (see chapter two). The mixed high indices refer to interactions within the mass element, m. Therefore, γ,e, refers to electrons generated by a photon in m, and e,δ shows delta electrons generated by electrons in m.

This energy balance can be used to characterize two important conditions in ionometric dosimetry: secondary electron equilibrium and Bragg-Gray conditions (Krieger et al. 1989).

3.4.2 Secondary Electron Equilibrium

If there is a balance between the energy transported into and out of the detector by electrons (including delta particles), the last six terms in equation 3.2 cancel each other. This is called electron equilibrium, and the dose in the detector is only dependent on the energy transported into and out of the detector by photons. A less restrictive case would be to allow delta particle disequilibrium. In this case:

$$-E^{\gamma,e}_{ex} + E^{e}_{in} - E^{e}_{ex} = 0\,, \tag{3.3}$$

with no restrictions toward delta electrons. The condition described in equation 3.3 is called secondary electron equilibrium. It ensures that the energy transported into and out of the detector by secondary electrons is equal. Therefore, secondary electrons do not deposit energy in the detector, and the recorded signal is only due to photons (and delta particles).

In practice, electron equilibrium cannot be completely achieved in an ionization chamber with finite dimensions. This is due to the difference in photon flux at the entrance and exit side of the chamber, which leads to the production of different amounts of secondary electrons in different positions in the chamber. The correct term for "real" measurement conditions is therefore *transient electron equilibrium*. However, in megavoltage x-ray beams from linear accelerators with chamber dimensions small compared with the target to the chamber distance, this effect is negligible for dosimetry.

The detector used for measurements under electron equilibrium conditions has to fulfill certain requirements:

i) The wall of the detector must be made from the same material as the detector itself to ensure that spectrum and angular distribution of electrons generated in the wall equals the spectrum and the distribution of electrons generated in the detector. There are no specific requirements for the density of the material, which can differ between wall and detector. This will be the case particularly for ionization chambers, where the air equivalent wall will have about 1000

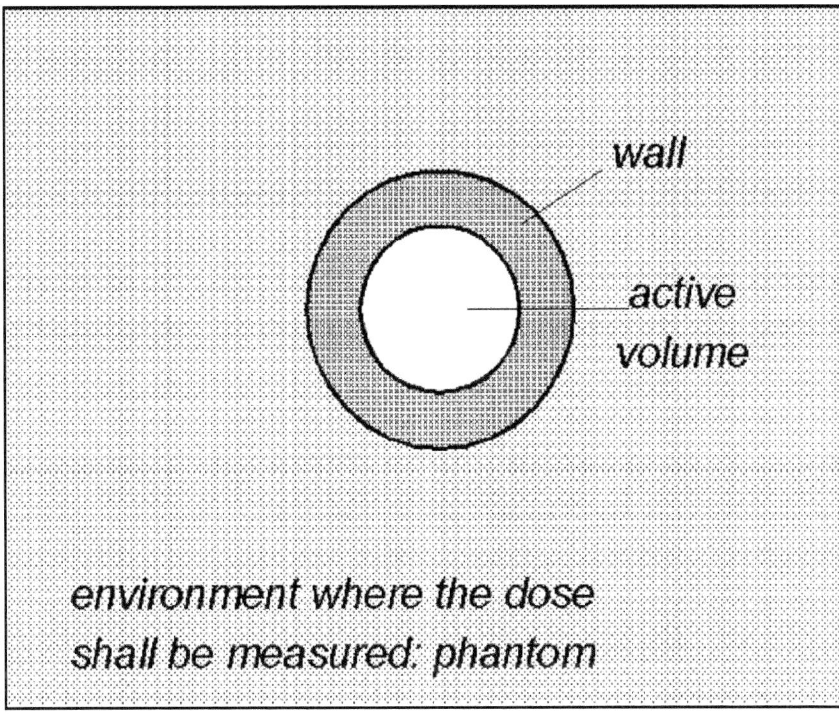

Figure 3.18: Schematic drawing of an ionization chamber.

times the density of the air in the chamber. The situation is covered by Fano's theorem (Fano 1954, Reich 1990), which states that: "*If a given material is irradiated with radiation of constant flux then the secondary particle flux will also be constant independent of density and density fluctuations.*"

This theorem is only valid for regions of secondary electron equilibrium. In photon radiation it also breaks down above energies of 1 MeV due to the so-called density effect. This is a result of polarization of matter due to the electrons, which is denser in matter of higher density. Therefore, the gaseous state will exhibit a (slightly) higher mass stopping power.

ii) The detector wall must be at least as thick as the range of the secondary electrons. This ensures that no secondary electrons from the outside medium reach the detector. In practice, this sets a limit for the usefulness of electron equilibrium conditions in high-energy photon beams because the build-up cap used to ensure the electron equilibrium must be very thick (e.g., for 18 MV x-rays it would need to be about 7 cm in diameter). In this case the radiation flux will vary throughout the detector (and wall) volume.

As a rule of thumb, dosimetric measurements using electron equilibrium conditions should not be made in photon beams of effective energies above 3 MeV. However, it has recently been shown that reasonably accurate *in-air* measurements can be performed even in photon beams of 25 MV if the correct equipment (mini-phantoms) and some empirical corrections are used (van Gasteren et al. 1991). This is important for the measurement of collimator factors, for example.

The *free-air* chamber shown in figure 3.19 is an ideal example of an ionization chamber using electron equilibrium conditions. A small collimated x-ray beam traverses a large air volume. The active volume shaded in figure 3.19 is surrounded from all sides by air in dimensions larger than the range of the secondary electrons. This fulfills the second requirement listed above, and since the "wall" material itself is air also, the first condition is satisfied.

In practice, free-air chambers are limited to x-rays with peak energies below about 400 kV. The limiting factors above this energy are the required chamber dimensions (60 cm for 400 kV) and the high voltage needed for chamber saturation (10 kV for 400 kV) (Reich 1990). Experiments have been performed with high-pressure chambers to extend the useful range of free-in-air chambers (Khan 2003). However, these devices can only extend the useful range to about 2 MV x-ray energy. Due to their large plate separation, free-air ionization chambers may have problems with electric field inhomogeneities and increased recombination of ions. In general, they are very difficult to set up and to use, and most of them are located in large institutions such as National Standards Laboratories. There, they are very useful primary standards for exposure measurements.

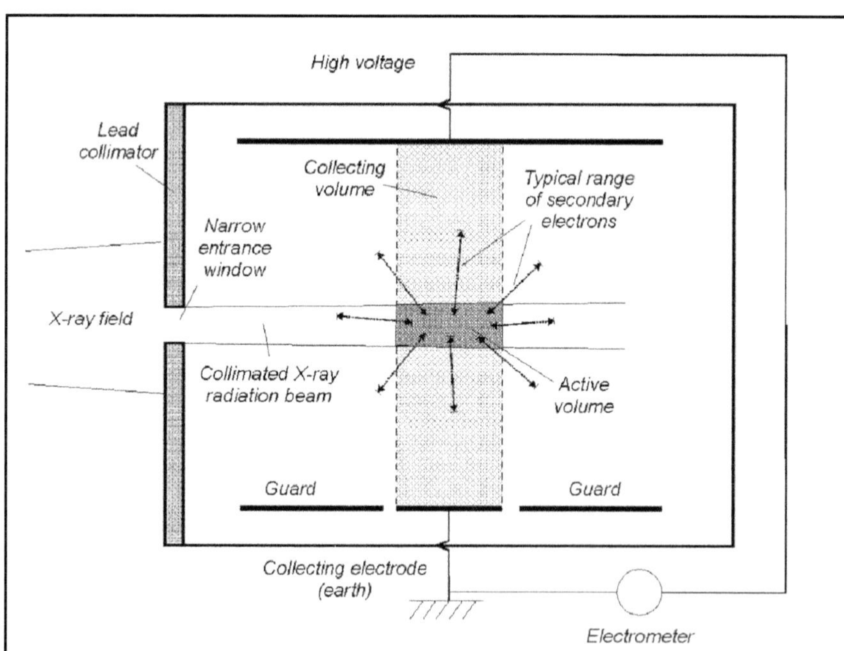

Figure 3.19: Schematic drawing of a free-air ionization chamber.

3.4.3 Bragg-Gray Cavity Theory

Under Bragg-Gray conditions only the secondary electrons are measured with the detector. This is the other extreme compared with electron equilibrium measurements, which determine only energy transferred from photons in the detector.

If the detector is used in a medium and its dimensions are small, the contributions of photons to the energy deposited in the detector are very small compared with the energy deposited by electrons. If the chamber can be constructed in a way to allow also delta electron equilibrium, then:

$$E^{\gamma}_{in} - E^{\gamma}_{ex} - E^{\gamma,e}_{ex} - E^{e,\delta}_{ex} + E^{\delta}_{in} - E^{\delta}_{ex} = 0. \tag{3.4}$$

Equation 3.4 characterizes Bragg-Gray conditions where neither photons nor delta electrons contribute to the energy deposited in the detector. Clearly then, all the signal recorded in the chamber is due to secondary electrons.

The ideal chamber to be used under Bragg-Gray conditions is very small to avoid disturbance of the secondary electron flux. It also has a very thin wall to make sure that (nearly) all electrons recorded in the detector have been generated in the phantom medium. For this reason, Bragg-Gray cavity theory is only applicable to photon energies above about 300 keV where the range of the secondary electrons is large enough to pass through a thin chamber wall. Figure 3.20 shows the range of the secondary electrons as a function of the energy of the incident photon energy. The figure also shows the range of dimensions and x-ray energies in which measurements under Bragg-Gray conditions may be performed.

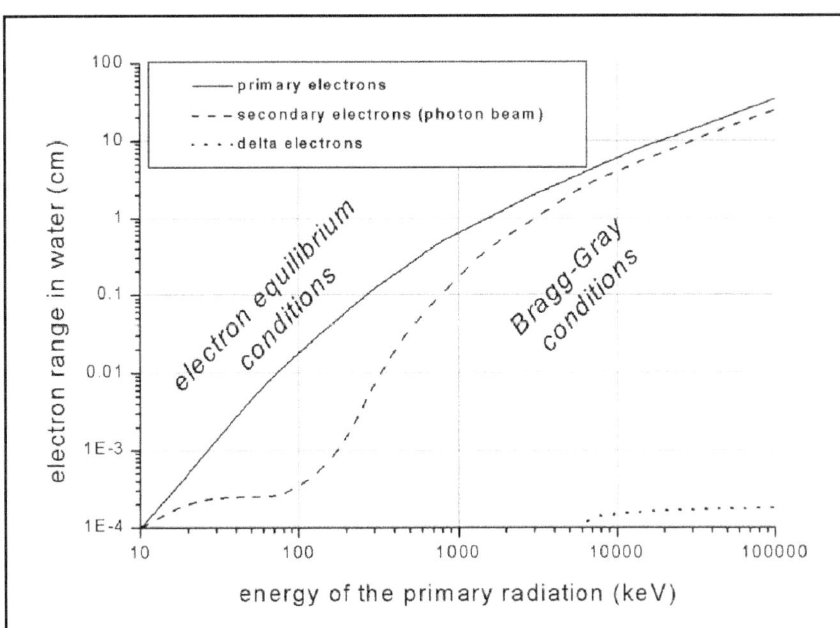

Figure 3.20: Range of primary and secondary electrons in water. Also shown are the typical track lengths for delta electrons. (Adapted from Van Dyk 2005.)

The condition that no delta electrons from the medium can reach the chamber (and the other way around) requires the wall of the chamber to be of finite thickness to stop all delta particles within the medium before they can reach the chamber. This is usually fulfilled, since the range of delta particles for photon energies below 10 MeV is less than 1 mm, as can be seen in figure 3.14.

If these conditions are fulfilled, the (mass) ionization determined in the chamber can be related directly to the dose in the chamber using W/e, as shown in figure 3.17. Under Bragg-Gray conditions it is possible to convert this "dose (to water as all secondary electrons have been generated in the medium) measured in air," D_{air}, to "dose to water measured in water," D_w, using the stopping power ratio between water and air, saw as shown in figure 3.21:

$$D_w = D_{air}\, s_{aw} = Q/m\ W/e\ s_{aw}\ , \tag{3.5}$$

with the charge, Q, collected in the chamber with a mass of air, m. This is also discussed in chapter eight in the context of the AAPM TG-21 protocol (AAPM 1983).

It is important to note that the stopping power ratio depends on the radiation quality. In photon beams it is generally assumed that s_{aw} is independent of the point of measurement in a phantom because the energy spectrum of the secondary electrons is in first approximation independent of the location. However, in electron beams the depth of measurement will determine the effective electron energy (compare chapter five) and s_{aw} will depend on the location of the measurement in a phantom.

Spencer-Attix Extension of Bragg-Gray Cavity Theory

The discussion so far has neglected the fact that electrons not only cause local ionization and bremsstrahlung along their track through matter, but they can also generate higher-energy electrons that may transport energy away from the original particle track. These electrons are called *delta electrons* and they are second-generation electrons in directly ionizing radiation (megavoltage electrons) and third-generation particles in indirectly ionizing radiation (megavoltage photons). Per definition, the maximum energy of a delta electron is half the energy of the original electron as the electron with

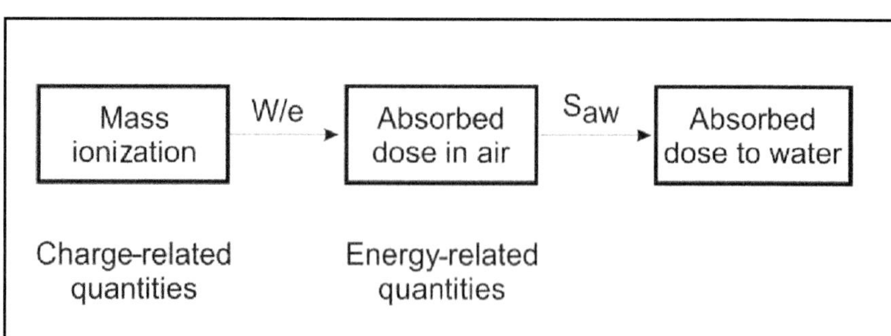

Figure 3.21: Quantities used for determination of dose under Bragg-Gray conditions using ionization chambers.

the lower energy after the interaction is automatically referred to as the delta electron. In megavoltage radiation the range of delta electrons is too large to ensure that equation (3.4) is fulfilled, and the effect of delta electrons can be considered following an extension to the Bragg-Gray cavity theory developed by Spencer and Attix (Spencer 1955).

Delta electrons are physically identical to other electrons and cannot be distinguished by measurement. Therefore, it is convenient for dosimetric purposes to consider the delta electron as part of the original electron beam. This could be "primary" electrons in the case of megavoltage electrons or "secondary" electrons generated by primary photons. Including delta electrons as part of the original electron beam will change the slowing down spectrum of the electrons and therefore a modification to the stopping power is required. The restricted collision stopping power, L_Δ, is introduced, which only takes into account locally relevant energy losses in the stopping power. Energy losses larger than a cut-off energy Δ are ignored as they generate delta electrons with an energy large enough to exit the volume element of interest. These delta electrons are then accounted for as part of the original electron spectrum. The magnitude of the cut-off energy depends on the volume of interest and is chosen typically such that an electron of the energy Δ has an expected range approximately equal to the size of the volume of interest. In case of a thimble ionization chamber of the Farmer type the cut-off energy is usually chosen as 10 keV (AAPM 1983, IAEA 1987. Klevenhagen 1993). Fortunately, the restricted stopping power ratios that are of importance for calibration do not vary dramatically with the choice of the cut-off energy in practice.

Two comments made by Klevenhagen (1993) summarize the use of restricted stopping powers well:

i) "...the Bragg-Gray cavity theory uses unrestricted stopping powers averaged over the slowing-down spectrum of only the primary (*or secondary in the case of photons*) electrons, the Spencer-Attix formulation uses restricted stopping powers averaged over the slowing-down spectrum of all generations of electrons;
ii) the Bragg-Gray theory is independent of cavity size, the Spencer Attix formulation takes account of the finite size of practical ionization chambers by choosing a suitable value of the cut-off energy Δ...."

In this context, it is also interesting to note that the linear energy transfer (LET) used in radiobiology to determine radiobiological effectiveness equals the restricted stopping power as the quantity of concern is the locally deposited energy.

Thus, the ratio of stopping powers seen in equation 3.5 is typically the ratio of restricted stopping powers. One could calculate the stopping power ratio according to the original Bragg-Gray theory using the unrestricted stopping power if "delta electron equilibrium" exists. In this case, the signal generated in the chamber would be solely due to the primary electrons in electron beams and secondary electrons in the case of photon irradiation. Delta electron equilibrium can, in principle, be achieved by using an air-equivalent wall, similarly to the discussion for electron equilibrium conditions above. In this case, the air-equivalent wall must be thick enough to ensure delta electron equilibrium and thin enough to make the chamber small enough to avoid perturbation of the radiation field. In practice, the use of restricted stopping powers using the Spencer-Attix formulation of the Bragg-Gray cavity theory is more commonly used. s_{aw} can be calculated as:

$$S_{aw} = \frac{\left\{\left[\int_{\Delta}^{E_{max}} \Phi_{E,w} (L/\rho)_{\Delta,w}\, dE\right] + \left[(\Phi_E(\Delta))_w (S(\Delta)/\rho)_w \Delta\right]\right\}}{\left\{\left[\int_{\Delta}^{E_{max}} \Phi_{E,w} (L/\rho)_{\Delta,air}\, dE\right] + \left[(\Phi_E(\Delta))_{air} (S(\Delta)/\rho)_{air} \Delta\right]\right\}}, \quad (3.6)$$

with

E_{max}	the maximum electron energy
$\Phi_{E,w}$	the spectral electron fluence for water (the presence of the cavity is ignored)
L	the restricted stopping power
S	the unrestricted stopping power
Δ	the cut-off energy (usually set to 10 keV to match the dimensions of a Farmer-type chamber.

The integral starts at the cut-off energy to exclude electrons with an energy below Δ. The second term in numerator and denominator is the product of the cut-off energy Δ, the spectral electron fluence $\Phi E(\Delta)$ and the unrestricted stopping power $S(\Delta)/\rho$, the latter two only at the energy Δ.The stopping power ratios for water and air calculated using the Spencer-Attix formulation are typically less than 0.5% higher than the ones calculated using the original Bragg-Gray formulation (Reich 1990).

In photon beams it is assumed that the stopping power ratio is constant at all depths in the phantom, provided the measurement location is deeper than the depth of maximum dose. This is a reasonable assumption, as the spectrum of secondary electrons in photon beams does not alter considerably with depth in most circumstances. In extreme cases such as very small fields where beam hardening is substantial or in very large fields where build-up of scattered photon radiation can be significant, this assumption may not be correct. Calibration procedures stay away from these "extreme" cases on purpose. In electron beams it is necessary to correct for changes in stopping power ratio with depth in the phantom, as the electrons lose energy along their path.

3.4.4 Chamber Correction Factors

Absolute dosimetry, in particular the calibration of linear accelerator output, requires a number of correction factors. Some of them can be taken from tables in the literature; others have to be determined for a particular chamber-electrometer combination and a particular radiation quality. Following different protocols a different set a correction factors is used, as discussed in more detail in chapter eight. However, all these sets correct for the same physical phenomena in ionization chamber dosimetry. The corrections that need to be made to allow for the chamber response itself are dealt with here. Other corrections, depending on the choice of phantom and the interaction between the chamber and its environment, are discussed in chapter eight.

Issues that require correction before an ionization chamber reading can be used for absolute dosimetry are:

i) Temperature and pressure: In ionization chambers that are open to the atmosphere, the number of air molecules varies with temperature, T, and pressure, p. This leads to a different number of ions for a certain dose in the active volume, V. Since typical measurement conditions do not vary considerably from the reference pressure, p_r, and temperature, T_r, a first-order correction can be made:

$$p_{p,T} = \left(\frac{p_r}{p}\right) \times \left(\frac{T}{T_r}\right). \tag{3.7}$$

Typical values for T_r and p_r are 293.2 K (20 °C) and 101.3 kPa. However, the reference conditions can differ from this (e.g., AAPM TG-21 protocol: T_{ref} = 295.2 K = 22 °C [1983]), and it is important to apply the same reference values as those used by the respective National Standards Laboratory.

While one would probably sense a temperature difference that results in a measurement problem of 1% (about ±3 °C), this is not as easy for air pressure determinations. It is therefore advisable to check the local device for the air pressure measurement in regular intervals by comparing it to the reading of the nearest meteorological service (commonly available at airports—caution is to be taken to ensure that the value of the meteorological station is not corrected for sea level).

The application of equation (3.7) assumes the chamber is well vented. Therefore, it is important that exchange of air from inside the chamber to the outside environment must be easily possible. If the ionization chamber is to be submerged in a water phantom, it is essential to ensure ventilation through the sheath used to waterproof the cable.

It should also be noted that temperature may cause component expansion; hence the physical dimensions of the chamber may change slightly, altering the active air volume. Temperature may also have some influence on the electronic components of the electrometer used. However, these effects are generally small and difficult to account for (Reich 1990). Therefore, they are usually not corrected for in practice. Finally, one needs to be aware that room temperature may affect the measurement of pressure in some devices (such as a mercury barometer where the mercury expands with temperature).

ii) Humidity: Ideally all measurements should be done at very low humidity; the W/e factor used in dosimetry actually refers to dry air only. Typical measurement conditions are under humidity of 20% to 70%. At room temperature the influence of this on the W/e is still negligible (< ±0.1%) (ICRU 1979). However, at higher humidity there could be additional chamber and electrometer leakage due to moisture on the surfaces. Therefore, leakage should be carefully watched at these conditions, in particular if the chamber is constructed from hygroscopic material, such as A-150 plastic or nylon (IAEA 1987). In general, all therapeutic dosimetry equipment should be constructed to keep humidity as a source of error well below 1% (Reich 1990). A correction for humidity can therefore usually be omitted, in particular if the calibration at the national standards facility uses 50% relative humidity. However, if the calibration reference is to dry air, the IAEA (1987) recommends the use of a correction factor $p_h = 0.997$ for ^{60}Co radiation.

iii) Polarity: The number of primary and secondary electrons entering and leaving the measuring electrode is not equal. Also, if the electrical field is parallel to the direction of the radiation, the secondary charged particles of different polarities will gain different energies and speeds on their way to the collecting electrodes. This results in the so-called polarity effect: The direction of the polarity of an ionization chamber can affect its reading. The polarity effect can be corrected for by measuring with both polarities (positive and negative high-voltage applied) and taking the average as the corrected reading. Polarity effects—mainly a problem using plane parallel chambers in electron beams (Gerbi and Khan 1987, Havercroft and Klevenhagen 1994, Kron et al. 2006b)—can affect the dosimetry of high-energy photons in particular in the build-up region or the penumbra, where electron disequilibrium prevails (Butson et al. 2003b). The effect depends on chamber construction and radiation conditions. Therefore, it has to be investigated for each particular measurement situation.

iv) Recombination: Even with the recommended high voltage used for any particular ionization chamber there will be a certain number of ions that will recombine before reaching the electrodes. This loss of signal is typically very small for continuous radiation (such as ^{60}Co) but can amount to more than 1% in pulsed radiation from linear accelerators. Recombination can be accounted for by the *two-voltage method* (Boag and Currant 1980, Boag 1982, Weinhous and Meli 1984). This technique uses two voltages, U_1 and U_2, with U_2, as the normal operation voltage and U_1 a smaller voltage. U_1 should ideally be at least three times smaller than U_2 (Reich 1990). However, in practice many electrometers offer the option of *half voltage* only. From the ratio of the two respective readings one can obtain the value for the recombination correction, p_r, from published graphs (Boag 1982, Reich 1990) or published formulas and tables (Weinhous et al. 1984).

In liquid-filled ionization chambers it is impossible to achieve saturation as outlined in articles by Wickman (1974), Wickman and Holmstroem (1992) and Bydder et al. (1988). However, for air-filled ionization chambers, ion saturation is nearly achieved provided a bias voltage of above about 200 volts is applied. As a first approximation the correction factor for pulsed radiation from linear accelerators can be calculated according to Reich (1990) as:

$$p_r = \frac{\left(\frac{U_2}{U_1} - 1\right)}{\left(\frac{U_2}{U_1} - \frac{M_2}{M_1}\right)}. \qquad (3.8a)$$

This approximation is based on conduction of positive and negative ions only and does not account for electrons that are liberated from oxygen and reach the electrodes. This requires a correction that is particularly noticeable for large doses per pulses and decreases the effect of recombination (Boag et al. 1996). However, equation 3.6 is commonly applied and valid in the case of a larger interval time between pulses than the collection time for the ions in the electrometer (i.e., relatively small ionization chambers) and for corrections up to about $p_r = 1.05$ (Reich 1990). It is also the method recommended in the calibration protocol of Task Group 51 of the AAPM (1999) discussed in more depth in chapter eight.

A slightly different approach is taken in the protocols of the IAEA (1987, 2004) where the polarity correction is determined using:

$$P_{ion} = a_0 + a_1\,(M_1 / M_2) + a_2\,(M_1 / M_2)_2 \qquad (3.8b)$$

with the constants a_1 given by the protocol as a function of the ratio of the two voltages V_1 and V_2 used. The constants are different for continuous, pulsed, and pulsed-scanned radiation beams because the instantaneous dose rate is considerably different in these cases. Consequently, the number of charges created in a very short time in the cavity varies and the chance of recombination increases with dose rate.

While p_T, p_p, and p_h are dependent only on environmental conditions, p_{pol} and p_r are dependent on the radiation quality. They have to be determined, therefore, for each radiation beam of interest individually.

With all ionization chambers it is advisable to check the reading from time to time with a reference radioactive source. These checks typically employ a beta-emitting $^{90}S/^{90}Y$ source of several hundred MBq housed in appropriate shielding. A "strontium" source for consistency checks of cylindrical chambers is depicted in figure 3.15. The ionization chamber can be exposed in close (and reproducible) vicinity to the source, and after correction for radioactive decay, the reading can be compared with previous readings. If temperature and pressure corrections are applied, the reading for a good ionization chamber and electrometer combination can be expected to be well within 1% of the baseline reading.

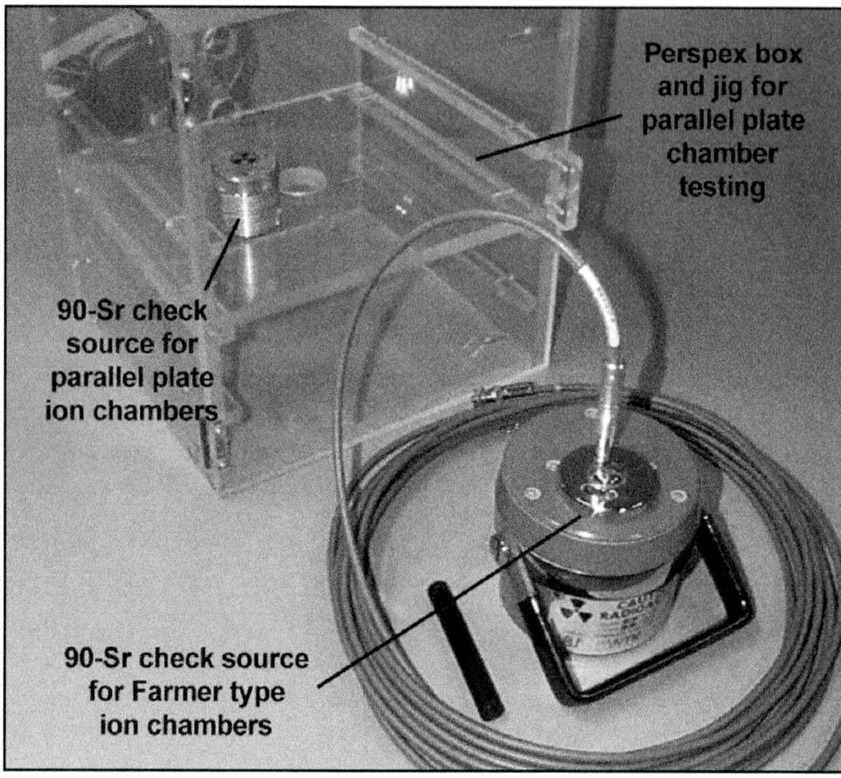

Figure 3.22: Photograph of two check sources for consistency tests of ionization chambers. (Adapted from Van Dyk 2005.)

3.5 Ionization Chambers: Commercial Chambers and Electrometers

3.5.1 General Considerations

In addition to the free-in-air chamber, which has been discussed in the previous section, a multitude of different ionization chambers have been developed for different applications. They all employ one medium for radiation detection (typically air) and a second one as the wall. The main ways in which the chambers can differ are:

1. Geometry.
2. Wall (and electrode) material and thickness.
3. (Air) volume.

The following discussion is divided into the two main types of chamber geometries in use in radiotherapy: cylindrical (thimble type) and plane parallel

Table 3.3: Ionization chamber volume and typical applications (Adapted from Petzold et al. 1988)

Chamber Volume (cm^3)	*Application*	*Typical Dose Rate*	*Typical Ionization Current (pA)*
> 10000	environmental monitoring	0.1 μSv/hr	0.001 – 0.1
100 – 1000	health physics/radiation protection	0.01 – 1 mSv/hr	0.1 – 10
1 – 5	diagnostic radiology	0.1 Gy/min	10
0.1 – 1	megavoltage radiotherapy: range recommended for MV beam calibration	1 Gy/min	100
0.03 – 0.3	superficial and deep x-ray radiotherapy	10 Gy/min	100

(pancake type) chambers. Thimble chambers typically have a cylinder height larger than or equal to the diameter and are irradiated through the cylinder mantle. Pancake chambers have a smaller cylinder height than diameter, and they are irradiated with the primary beam through one of the end plates. The extrapolation chamber dealt with in section 3.5.4 is a special type of plane parallel chamber.

Other types of ionization chamber geometries, such as a spherical (compact) chamber, are available (see e.g., Petzold and Krieger 1988, Reich 1990) but not commonly used for the dosimetry of x-rays produced by linacs. However, the transmission chamber is worth a mention because of its relevance to modern linear accelerators. It was discussed in chapter one and for some trouble shooting with these monitor chambers it is useful to keep the physics behind the correction factors discussed in section 3.4.5 in mind.

The various materials used for chamber construction reflect the need to have air equivalent walls for measurements in electron equilibrium and water/tissue equivalent walls for measurements under Bragg-Gray conditions. In addition to this, parts of the chamber (e.g., electrodes, guard rings) need to be conductive, and others have to be very effective insulators to prevent leakage. Some of the materials used for ionization chamber construction are also discussed in section 3.3.

The volume of an air-filled ionization chamber depends on the application sought. It is always a compromise between minimizing the volume to achieve the best spatial resolution and maximizing the volume to improve the signal-to-noise ratio. When increasing the size of the ionization chamber, one also needs to consider that the homogeneity of the electric field gets worse and higher voltages may be required. Table 3.3 gives a brief summary of chamber volume and typical applications.

3.5.2 Thimble Ionization Chambers

Thimble type ionization chambers are the most frequently used dosimeters in medical radiation physics at present. This is because of their ease of use combined with a wealth of information about their properties. They were originally constructed as a modification of the free-air chamber discussed in the previous section. Therefore, the wall of the chamber is typically air equivalent and practically condenses the required volume of air into a solid material. Thimble type chambers are of cylindrical geometry, as can be seen in figure 3.23. The outer shell of the cylinder is normally operated at ground potential, and a central conducting rod conducts the applied voltage. In this case a field that varies inversely with radius is created in the air space between the electrodes.

The wall of thimble chambers is relatively thin to make the chamber suitable for measurements under Bragg-Gray conditions. As it is sometimes made from graphite, it can be fragile and care has to be taken in handling it. If the chamber is to be used for exposure measurements under electron equilibrium conditions in x-ray beams with an energy > 0.5 MV, an appropriate build-up cap has to be used. A Farmer type chamber and a range of build-up caps for different photon energies are depicted in figure 3.24. The thickness required for a build-up cap increases with the photon energy. For ^{137}Cs (E_K = 661 keV) a radius of 0.2 cm is required; for ^{60}Co (E_K = 1.25 MeV) 0.5 cm and for 6 MV x-rays, it would be 1.5 cm. Clearly, the concept of localized exposure measurements with build-up caps breaks down for higher

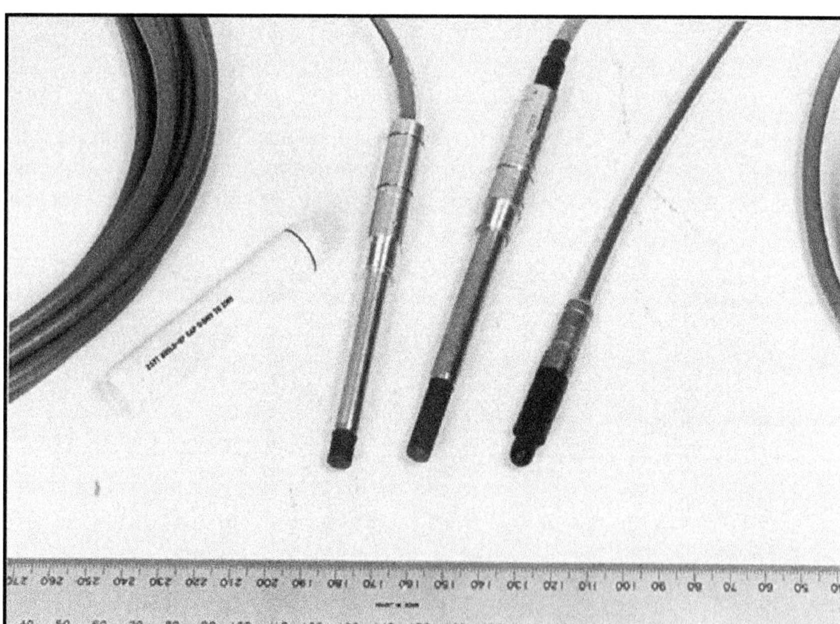

Figure 3.23: Cylindrical ionization chambers. Depicted are from left to right: NE 2577 volume 0.2 cc with delrin build-up cap, Farmer type NE 2505/3 volume 0.6 cc, and Wellhöffer IC 10 chamber for water tank measurements.

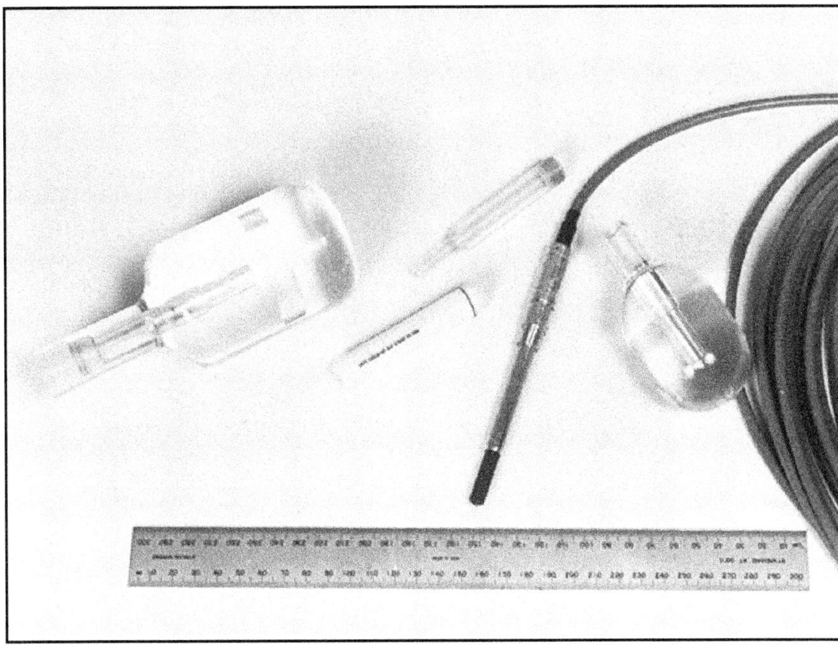

Figure 3.24: Farmer type ionization chamber with four different build-up/protective caps. From left clockwise they are: Perspex build-up cap for 6 MV photons, delrin build-up cap for 0.3 to 2 MV photons, Perspex build-up cap for 10 MV photons, and thin (0.5 mm) protective Perspex cap for waterproofing of a Farmer type chamber for beam calibration.

energy x-rays, such as for an 18 MV x-ray beam where the diameter of the build-up cap exceeds 6 cm. For the spectral beam of linear accelerators one should, strictly speaking, use a customized build-up cap for every beam. They can be designed by using a large PMMA cylinder around the chamber in air and shaving the mantle of the cylinder back until the dosimeter reading decreases. This indicates that the thickness of the build-up cap is just a bit too small to allow for full secondary electron build-up conditions. A suitable cap can then be made to give just the maximum dosimeter reading. In practice, the use of miniphantoms seems to be more appropriate for these measurements. Their design is discussed by J. van Gasteren et al. (1991).

Figure 3.25a shows the cross section of a Farmer-type chamber (Aird and Farmer 1972), which is one of the most commonly used thimble chambers and is particularly recommended for absolute dose measurements. In the Farmer chamber the inner electrode was originally made from aluminum. This material with a relatively high atomic number of 13 was used purposely to compensate for the carbon wall with a low atomic number of 6 as compared to tissue (effective atomic number 7.5). As such, this design flattens the energy response of the chamber for photon energies below 100 keV. For absolute dosimetry purposes, corrections have to be applied for this as discussed in chapter eight.

The original Farmer chamber has a nominal volume of 0.6 cm^3 with a length of 24 mm, which makes it a rather large chamber. As can be expected from its asymmetrical design and stem, the dose response will also vary with

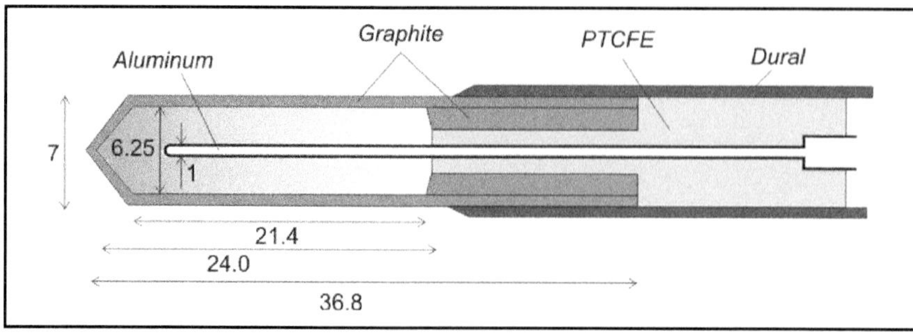

Figure 3.25a: Cross section through a Farmer-type ionization chamber; all dimensions are in millimeters. (Adapted from Aird and Farmer 1972).

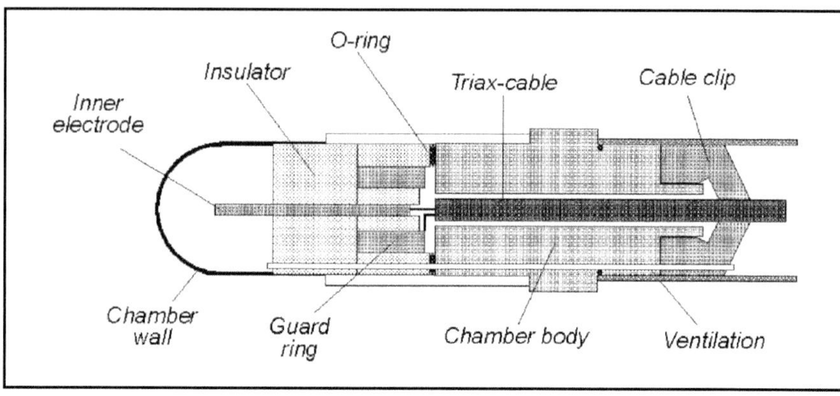

Figure 3.25b: Cross section through a cylindrical ionization chamber designed for scanning in a water phantom. Please note the dimensions in axial and cross-sectional directions are identical.

the incident angle of radiation to its cylinder axis. This is particularly noticeable at superficial and orthovoltage radiation energies.

A modification of the original Farmer chamber with an active volume of 0.2 cm^3 and a shorter length is available and better suited for the measurement of field size factors for small fields. However, this chamber is also asymmetric, being longer than its diameter. For measurements in a water phantom where dose profiles will be scanned in two dimensions, it is advantageous and convenient to employ a symmetrical chamber. The broadening of steep dose gradients due to the detector response function (Dawson et al. 1984, 1986) is then at least equal in both dimensions. Most of these thimble chambers designed for beam scanning are also waterproof to be easily adapted for water tank measurements. However, it is advisable not to leave chambers submersed in water if not required. Figure 3.25b shows an example of this chamber type.

With improvements in electrometers and chamber design it is now possible to produce smaller ionization chambers with volumes less than 0.1 cc. While these are recommended to be used for calibration (compare chapter eight), there is considerable scope for their application in small fields such as used for stereotactic radiotherapy (Vahc et al. 2001, Mack et al. 2002) or IMRT (Martens et al. 2000, Laub and Wong 2003, McNiven and Kron 2004).

Table 3.4 lists a number of commercially available thimble type ionization chambers. It includes a so-called "pinpoint" chamber specifically designed for determination of dose in small fields. Data on other types of ionization chambers can be obtained from AAPM Task Groups 21 and 51 (AAPM 1983, 1999), Gastorf et al. (1986), and Rogers (1992), and the IAEA protocol TRS 398 (2004).

3.5.3 Parallel Plate Ionization Chambers

As can be seen in table 3.4, the typical diameter of a thimble chamber is of the order of 6 mm. In steep dose gradients this is too large to allow meaningful dose measurements. Therefore, there is a need for a better localized (at least in one dimension) volume that still contains enough air to collect enough charge at clinically used dose rates. The plane parallel ionization chamber was developed for that purpose. Figure 3.26 and figure 3.27 show typical plane parallel ionization chambers.

Plane parallel ionization chambers were primarily designed for low-energy x-rays and electron beams that feature steep depth dose gradients along the central axis of the beam. However, they are also very useful for measurements in the dose build-up region of megavoltage x-ray beams produced by linear accelerators (Gerbi and Khan 1990, Rawlinson et al. 1992). Even exit doses have been measured in phantoms using plane parallel ionization chambers (Purdy 1986, Klein and Purdy 1993, Kron and Ostwald 1995). For both entry and exit dose measurements, chambers with a very thin entrance window are desirable. This thickness and other properties of some commercially available plane parallel ionization chambers can be seen in table 3.5. If the plate separation is not too large, one generally assumes that the effective point of measurement is just behind the front plate of the chamber. This point refers to the geometrical location in a phantom at which the dose equals the one measured with the chamber. For chambers with relatively large dimensions, such as the ones used for very accurate exposure measurements, this assumption breaks down and the effective point of measurement depends on plate separation and thickness of the front window. A record of this for a Kemp-Barber ionization chamber was recently given by Kotler and Webb (1996).

In contrast to thimble type chambers, plane parallel chambers are generally not well suited for measurements in a water phantom because of the following reasons:

i) They create disturbances in the water during movement due to their shape.
ii) It is difficult to set the water surface level accurately due to adhesion to the water meniscus.
iii) Most of them employ a considerable mass of (nonwater) material in the walls and the exit side of the chamber.

A wide variety of materials has been used for the construction of plane parallel ionization chambers as can be seen in table 3.5. Some chambers have the ability to be waterproofed for water tank measurements (e.g.,

Table 3.4: Comparison of some commercially available thimble type ionization chambers. The selection does not imply any recommendations. It is advisable to check the figures with the supplier before ordering.

	Farmer Type Chambers			*Other Thimble Chambers*					
	NE 2505/3A	*NE 2561*	*NE 2571*	*IC 10*	*RK*	*PTW "Normal"*	*Exradin Spokas*	*Pnpoint*	*Exradin A 1 SL*
Manufacturer	NE	NE	NE	Wellhöffer	Scanditronix	PTW	Exradin	PTW	Standard Imaging
Volume (cm^3)	0.6	0.325	0.6	0.147	0.12	0.3	0.5	0.015	0.056
Length (mm)	24	9.2	24	6.3	10	1.8	0.97	5	4.1
Internal radius (mm)	3.15	3.7	3.15	3	2	2.5	4.7	1	2.1
Wall material	graphite	graphite	graphite	Shonka C552	graphite/epoxy, polystyrene	Perspex	AE-plastic	Perspex and graphite	Shonka plastic
Wall thickness (mm; g/cm^2)	0.36; 0.065	0.5; 0.09	0.36; 0.065	0.4	1.5	0.5; 0.059	1.02; 0.182	0.66; 0.085	1.1
Remarks	reference chamber	reference chamber	reference chamber	waterproof	waterproof			Very small for stereo-tactic beams	Used for tomotherapy (Balog et al. (2006)

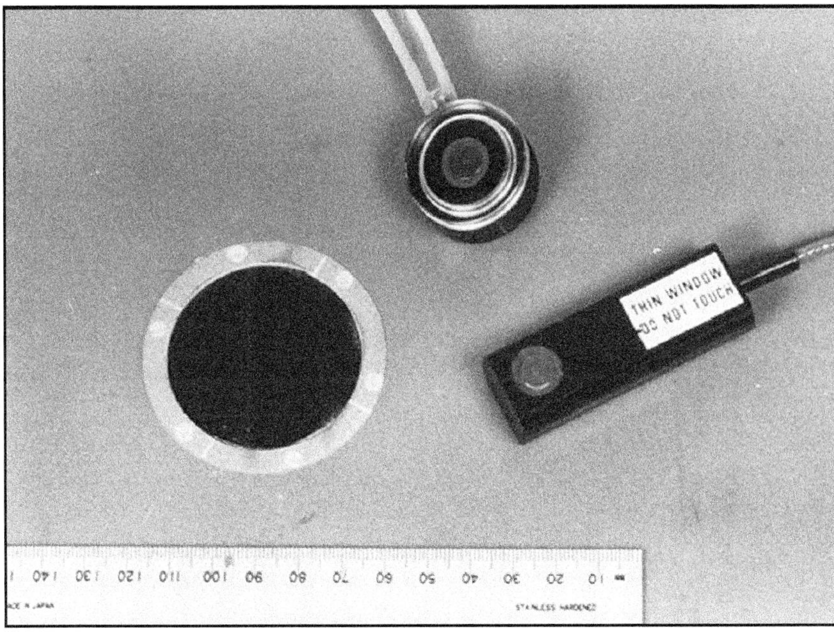

Figure 3.26: Plane parallel chambers. From left to right are shown: RMI Attix chamber embedded in a solid water slab, PTW Markus chamber, and NE thin window chamber, volume 0.02 cc, particularly designed for measurements in superficial x-ray beams.

Markus chamber, NACP chamber). This can be of importance during, for example, the calibration of an electron beam.

Plane parallel chambers are more commonly used in combination with a solid slab phantom. It is advisable to use a chamber and a phantom constructed of the same material in order to avoid interface problems. Measurements in inhomogeneous slabs are difficult, as plane parallel chambers are typically mounted on substantial slabs of material (such as acrylic for the Markus chamber and solid water for the Attix chamber); whereas ideally the chamber should be housed in a slab of similar material to that of the rest of

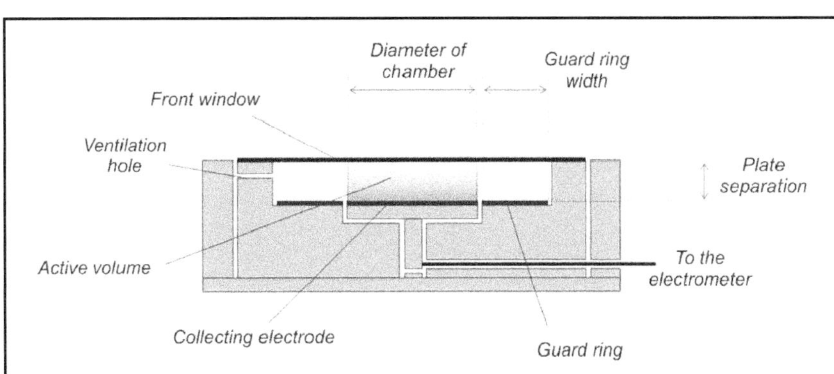

Figure 3.27: Schematic cross section through a plane parallel ionization chamber.

Table 3.5: Comparison of some commercially available plane parallel ionization chambers. The selection does not imply any recommendations.

	Holt Chamber	*Markus Chamber*	*NACP Chamber*	*Attix Chamber*	*Capintec PS-033*	*Exradin A, P, or T-11*
Manufacturer		PTW or NA	Scanditronix	RMI	Capintec	Standard Imaging
Phantom material,	polystyrene chamber body	Perspex	graphite	solid water	C552	A air; P polystyrene; or T tissue equivalent plastic
Wall material	graphited polystyrene	Perspex	graphite	solid water		as above
Window material		graphited polyethylene	graphite	conductive Kapton	aluminized Mylar	as above
Electrode material	graphited polystyrene	graphited polystyrene	graphite	colloidal (dag) on polyethylene insulator	C552	as above
Electrode diameter (mm)	25	5.4	10	12.7	14.9	20
Cavity depth (mm)	2	2	2	1	2.4	2
Guard ring width (mm)	5	0.2 advanced Markus 2	3.2	13.5	2.5	2
Window thickness (mm; mg/cm^2)	4; 420	0.022, 2.3	0.5	0.025, 4.8	0.025, 0.5	1, 118
Active volume (cm^3)	1	0.04	0.15	0.125	0.5	
Cavity diameter (mm)	35	6	16.4	40	21	45
Maximum/recommended polarizing voltage (V)		300		300		1000
Remarks	robust	thin window, waterproof	waterproof	wide guard ring, thin window	angled wall	waterproof different materials

the phantom. Maucari and Kase (1987) have addressed this problem by mounting chambers in suitable lung analog slabs.

As shown in figure 3.27, a guard ring is an essential ingredient for the construction of a plane parallel chamber. It has the same potential as the collecting electrode and creates therefore a homogenous electric field between the two electrodes. A wide guard ring is desirable not only for field homogeneity but also to avoid secondary electrons scattered from the wall being counted in the chamber. It should be at least several millimeters wide (Reich 1990). This has created some problems and discussion with the Markus type chambers (Gerbi and Khan 1990, Mellenberg 1990) and an improved version of the Markus chamberis mow commercially available (Pearce et al. 2006). In addition to this it can be advantageous to have the side walls of the chamber angled to avoid contributions to the reading from secondary electrons from the wall (Nilsson and Montelius 1986). It should be noted that the inclusion of a guard can be of value also for cylindrical chambers providing additional shielding between the two electrodes and therefore reducing leakage. The guard wire leading to the chamber provides shielding of the cable, which can significantly reduce noise, pick-up of unwanted signals, and leakage.

One of the most important applications for plane parallel ionization chambers in which the above-mentioned issues have to be considered is the determination of dose in the build-up region of megavoltage photon beams. Several empirical methods have been developed to correct for imperfections of plane parallel ionization chambers in these measurements (Velkley et al. 1975, Gerbi and Khan 1990, Rawlinson et al. 1992). As the "Rawlinson correction" seems to be the most rigorous one, it shall be briefly outlined here. The true percentage depth dose at depth d in a photon beam of energy E, D(d,E) can be derived from the measured dose D´ as follows (Rawlinson et al. 1992):

$$D(d,E) = D'(d,E,G) - \zeta\,(d,E,G), \tag{3.9}$$

with G identifying the geometry of the chamber and ζ being an empirical correction. At the surface of the phantom (d = 0) ζ can be calculated as

$$\zeta(0,E,G) = c(E)\left(\frac{s_{el}}{d_w}\right)\rho^{0.8}, \tag{3.10a}$$

with the correction c = 27% for 6 MV x-rays and 15% for 18 MV x-rays, s_{el} the separation of the electrodes, d_w the wall diameter, and ρ the density of the chamber walls. At other depths in the build-up region ζ can be found from

$$\zeta(d) = \zeta(0)\exp\left(\frac{-\alpha d}{d_{max}}\right), \tag{3.10b}$$

with α an empirical factor and d_{max} the depth of maximum dose for the respective photon energy. A comparison of corrected and uncorrected measurement results can be seen in figure 4.2 in the next chapter.

The electrical field lines of plane parallel ionization chambers are parallel to the direction of the radiation. As such, secondary electrons that are predominantly forward directed gain energy if the entrance window is negatively

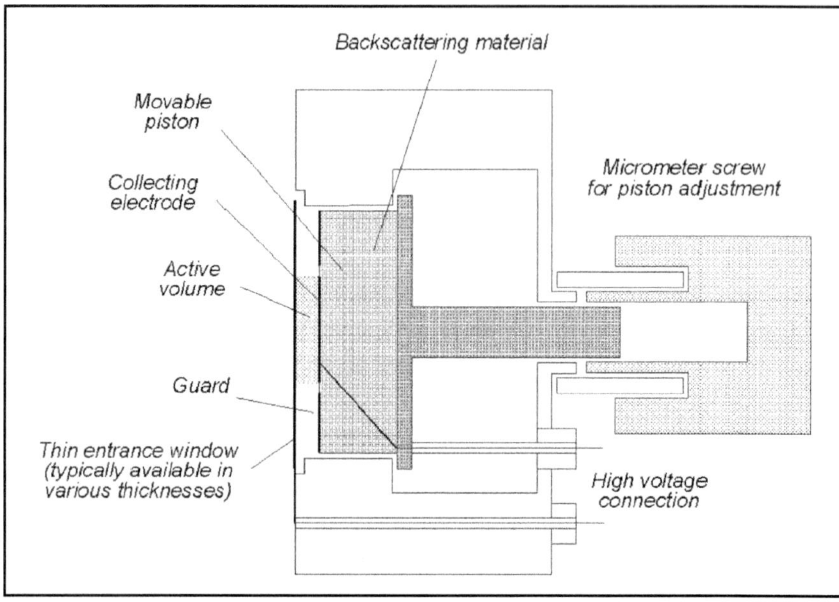

Figure 3.28: Schematic drawing of an extrapolation chamber.

and the collecting electrode positively charged. This may lead to more ionization events than in the opposite polarity. Therefore, plane parallel ionization chambers are more prone to polarity effects than cylindrical chambers. This should be evaluated before a chamber is used for a particular application.

3.5.4 Extrapolation Chambers

A special type of plane parallel ionization chamber is the extrapolation chamber (Failla 1937, Nilsson and Montelius 1986). These chambers were originally designed for beta dosimetry (Boehm 1986), but they are also very useful for megavoltage x-ray dosimetry in the build-up region (Gagnon and Grant 1975, Velkley et al. 1975, Mellenberg 1990). Shown in figure 3.28, the extrapolation chamber employs a piston to vary the distance between the two electrodes, which changes the active air volume. By decreasing the distance between the two electrodes, one can extrapolate the ionization reading to an infinitesimal thin chamber. If different entrance windows can also be employed with the chamber (e.g., PTW 2535A and B chamber) the actual dose at the very surface can be determined through extrapolation.

3.5.5 Liquid-Filled Ionization Chambers

The volume of the ionization chamber is a compromise between minimizing the volume to obtain well-localized measurements and maximizing it to get maximum signal and the best signal-to-noise ratio. As such, liquid-filled ionization chambers have the advantage over gas-filled chambers of an active material of higher density. However, the design and use of liquid-filled ioniza-

Figure 3.29: Photograph of two transmission chambers used in linacs. (Courtesy of G. Sorell.) The coin is 10c Australian.

tion chambers is difficult because they require a relatively high voltage (approximately 1000 V), are never completely free of recombination, and their dose response may not be completely linear. Nevertheless, the work of Wickman since the 1970s in Sweden has also demonstrated some significant advantages of liquid-filled chambers for clinical use (Wickman 1974, Wickman and Holmstroem 1992, Wickman and Nystroem 1992, Johansson et al. 1995). In addition to their high sensitivity and small physical size, they offer negligible perturbation effects and a relatively constant stopping power ratio to water over a wide range of photon energies. A typical liquid ionization chamber employs graphite electrodes in a plane parallel plate chamber geometry with a plate separation of 0.3 mm (Wickman and Nystroem 1992). This is filled with a dielectric liquid, such as isooctane or tetramethylsiline.

In addition to some research applications, liquid-filled ionization chambers are employed in portal imaging devices (Meertens et al. 1985, van Herk and Meertens 1988, van Herk 1991). These devices are now commercially available and clinically used for the verification of treatment field positioning (Meertens et al. 1990, Michalski et al. 1993). More recently they also have been used for in vivo dose verification on patients (Ying et al. 1990, Wong et al. 1990). This is discussed in more detail in chapter eight.

3.5.6 Transmission Chambers for Dose Monitoring in Linear Accelerators

A very important part in the dose delivery of linear accelerators is the monitor chamber assembly. The design is discussed in more detail in chapter one. These ionization chambers can be used as closed or open systems. Figure 3.29 shows two chambers: the one on the left is sealed (Varian) while the one on the right is an open transmission chamber (Philips). Since the accuracy and reliability of these chambers has to be very high, even small temperature and pressure changes in the environment have to be corrected for in open systems. Some manufacturers of linear accelerators employ an automatic temperature and pressure correction with the appropriate sensors.

A different approach is to use the monitor chambers in a sealed version. The chambers were generally nitrogen- or oxygenated air–filled with Mylar or Kapton foils on both sides. The use of Kapton is advantageous particularly for electron beams where it reduces the bremsstrahlung production. In closed transmission chambers the active gas volume does not change with ambient temperature and pressure changes, which eliminates the need to correct for temperature and pressure. The disadvantage of closed chambers is that they require thicker walls, which introduce additional beam scattering and degradation. In megavoltage photon beams these problems are negligible; however, in electron mode the construction of the chamber influences the spectrum considerably.

3.5.7 Electrometers

To collect the charge generated in any ionization chamber, a charge measuring device called an electrometer is required. To get a feeling for the detection limits and accuracy required in an electrometer used for dosimetry of therapeutic x-rays, the following considerations are useful (Petzold and Krieger 1988). The typical dose one would like to detect in radiotherapy is of the order of 1 Gy with an accuracy of at least 1 mGy. If one assumes the volume of an ionization chamber to be 0.1 cm^3 at standard conditions, this results in a sensitive mass of 1.293×10^{-7} kg. The energy deposited in this chamber by 1 Gy is therefore:

$$E = D\,m = 1\ \text{Gy} \times 1.293 \times 10^{-7}\ \text{kg} = 1.293 \times 10^{-7}\ \text{J}. \tag{3.11}$$

This energy is used to create charged particles with W/e = 33.97 J/C as the mean energy required to produce 1 C of charge of one polarity. The charge created by 1 Gy in 0.1 cm^3 of air is therefore:

$$Q = \frac{E}{\left(\frac{W}{e}\right)} = \frac{1.293 \times 10^{-7}\ \text{J}}{33.97\ \frac{\text{J}}{\text{C}}} = 3.8 \times 10^{-9}\ \text{C} = 3.8\ \text{nC}. \tag{3.12}$$

This charge consists of 2.3×10^{10} elemental charges of 1.6×10^{-19}C or 2.3×10^{10} ions. If a resolution of 1 mGy is required, this leads to an electrometer that can measure charge in the pC range accurately. If a dose of 1 Gy is delivered in 15 s (typical linear accelerator dose rate at isocenter 4 Gy/min), it is necessary that the electrometer can detect a current of:

$$I = \left(\frac{Q}{t}\right) = \frac{3.8\ \text{nC}}{15\ \text{s}} = 250\ \text{pA} \tag{3.13}$$

or less at depth in a phantom or at extended source to surface distance (SSD).

In a typical clinical measurement situation it follows that the leakage current in the instrument has to be well below 1 pA to avoid false dose readings. Therefore, in practice a galvanometer that has a typical detection limit of 10^{-9} A does not suffice.

The schematics of a modern electrometer is shown in figure 3.30. The voltage measured across the capacitor, C, is proportional to the collected charge in the ionization chamber. To avoid false readings, the leakage resistance, R_{lk}, has to be very high, i.e., of the order of 10^{15} A. The measurement setup illustrated in figure 3.30 is called *charge measurement mode.* This is typically done in feedback mode using an operational amplifier to amplify the small charges involved (Rajan 1992).

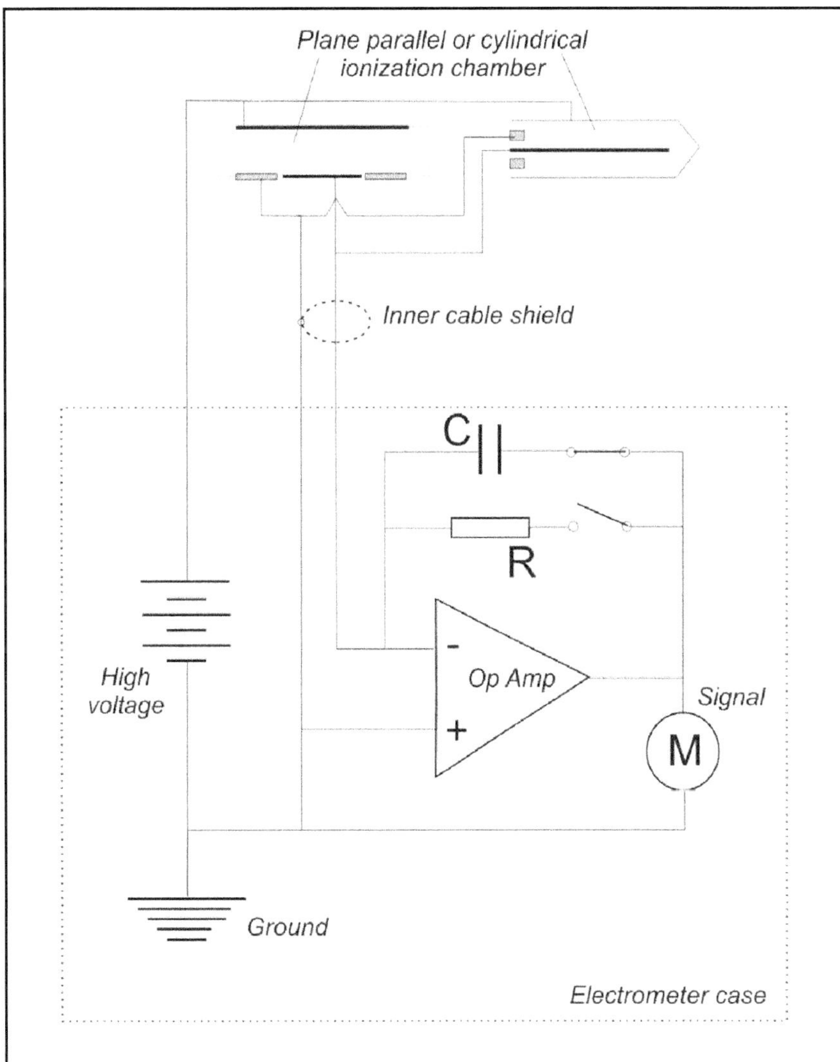

Figure 3.30: Schematic drawing of an electrometer. (Adapted from Van Dyk 2005.)

There are quite stringent criteria for the selection of an appropriate electrometer for clinical measurements. In addition to the requirements of accuracy and high sensitivity, the electrometer readings need to be stable over time. Typically, calibrations are performed on an ionization chamber/electrometer combination once every 2 years at a national standards laboratory. The electrometer used as a local standard should not change in this time period by more than 0.25%. A consistency check of the electrometer using a radioactive source and several different ionization chambers may be useful in detecting variations in electrometer response.

A wide range of electrometers is commercially available. Figure 3.31a shows a typical electrometer (Keithley model 614). Some electrometers were specifically developed for ionometric dosimetry; others are a modification of routinely used multimeters with current (dose rate) and charge (dose) measurement facilities. Additional features of electrometers used for clinical dosimetry may include the following:

i) Even though the ionization chamber is used in saturation, a stable voltage supply is required. The voltage should be of the order of 200–400 V depending on the ionization chamber used. If measurements with an extrapolation chamber are to be attempted, a variable voltage source will allow one to maintain the same electrical field strength throughout the chamber at different plate separations.
ii) The electrometer should offer polarity change. The polarity of the high voltage applied to the chamber has to be altered to measure polarity effects. A no polarity (zero voltage) option should also be available as no voltage should be applied when chambers are exchanged or connectors checked. This can be seen in figure 3.31b, which shows the back of a commercially available electrometer.
iii) The electrometer should have the facility to apply half or third values of full voltages for both polarities. If the high voltage is 300 V, the electrometer should also be able to supply 150 V or 100 V for the evaluation of recombination effects in the ionization chamber. This feature is also available in the electrometer depicted in figure 3.31b.
iv) Some consideration should be given to the connector type used. The present standard is a screw on or bayonet type triax connector; however, some older chambers have different connectors. Since adapters are not only expensive but also increase the chance of leakage, the number of connections between the electrometer and the chamber should be kept to a minimum.
v) For measurements in superficial, orthovoltage, and isotope beams that are programmed for a set time of radiation, it is advantageous to have an internal timer in the electrometer. However, since the dose rate of linear accelerators is not quite as stable and their dose delivery is controlled by monitor chambers, this is not an essential requirement for electrometers used for linac dosimetry.
vi) The instrument should read in units of charge and current (typically nC and pA) to avoid confusion if ionization chambers with different volumes are used (NACP 1980).

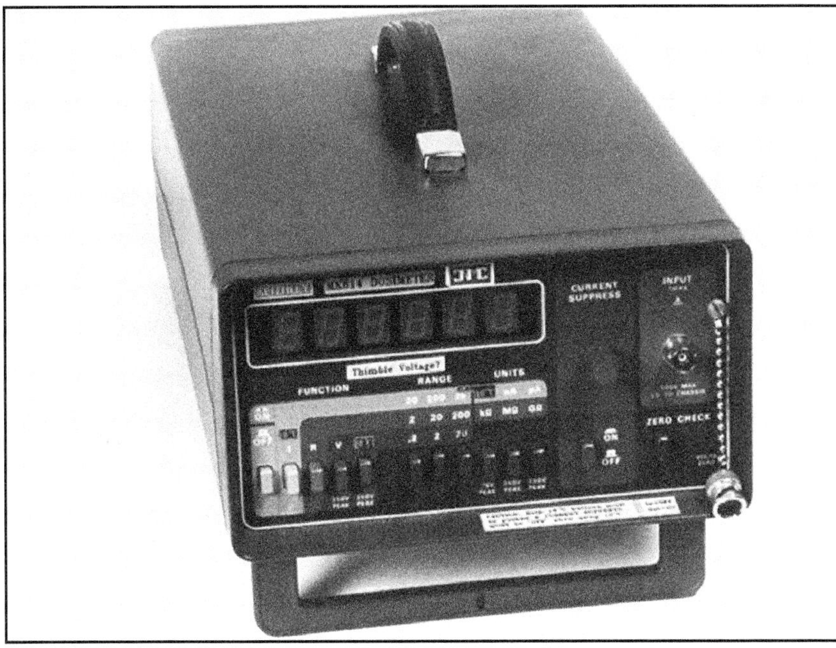

Figure 3.31a: Front cover of a commercially available electrometer Keithley 614; note the triax bayonet connector at the right side.

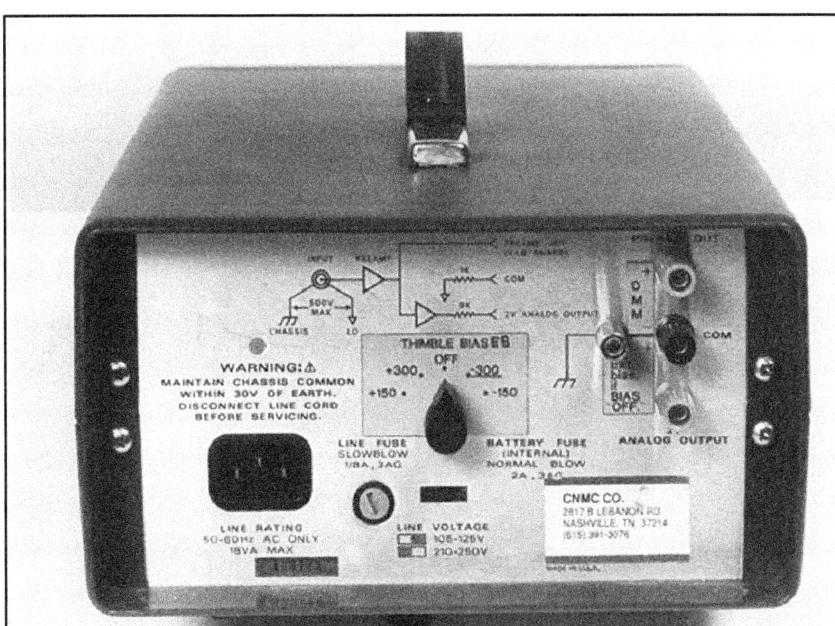

Figure 3.31b: Back cover of a commercially available electrometer Keithley 614; note the switch to vary the high-voltage bias for the determination of recombination and polarity correction.

Table 3.6: Comparison of some commercially available electrometers. The selection does not imply any recommendation. Some data, such as the stability, are taken from the manufacturers' specifications—it is advisable to check the figures with the supplier before ordering.

Electrometer	*Manufacturer*	*Resolution*	*Linearity*	*Channels*	*Bias*	*Stability*
CMNC 11	CMNC			1	±150V, ±300V	± 0.5% 2 yr
Keithley 35040	Keithley	10fC	±0.1%	1	adjustable 0–500V	± 0.1% 5 yr
Keithley 617	Keithley	100fC		1	0, ±150V, ±300V	±0.4% 1yr
Model 530	Victoreen	100fC	±0.3%	1	internal, adjustable 0–500V	±0.5% 1yr
Farmer 2570	NE			1	±250V	
Unidos	PTW	10fC	±0.5%	1	adjustable 0 – ±400V	±0.5% 1yr
Ionex Dosemaster	NE	30fC	±0.01%	10	±20V – ±1000V	±0.2% 1yr
Janus	Precitron			2	±100V – ±400V	
Max 4000	Standard Imaging	10fC	±0.6%	1	0, ±150V, ±300V	±0.2%

Table 3.6 shows some commercially available electrometers and gives some of their technical details. In 2000, the Institute of Physics and Engineering in Medicine (IPEM) in UK published a recommendation on specification for electrometers to be used for calibration (Morgan et al. 2000). Some of these specifications are reproduced in chapter eight in table 8.4.

For absolute dosimetry it is important to note that most electrometers do not have a charge collection efficiency of 100%. The efficiency can be tested with reference charge sources and may be part of the calibration report sent out by a national standards laboratory. The error is typically in the order of less than 0.2%.

For day-to-day use a number of other requirements should be followed. Most electrometers need some warming up before they can be used. This time is of the order of 30 minutes but can vary from instrument to instrument. Therefore, it is advisable to place and leave the electrometer where it will be used (pressure, temperature, and humidity equilibration) and turn it on some time before its intended use. In addition to this, several Gy should be given to the ionization chamber connected to the electrometer prior to recording the first readings. It is necessary to repeat this procedure and to wait for a couple of minutes after each change of high voltage (value or polarity).

All sophisticated electronic equipment should be protected from excess humidity. If the environmental conditions are very humid, it is advisable to place the electrometer in an airtight box with desiccator to avoid moisture build-up in electronic parts and connectors. This box can at the same time be used to store ionization chambers and cables to protect them from humidity.

3.6 Semiconductor Detectors

For dosimetry of ionizing radiation several types of semiconductor devices can be used (Rosenfeld 2006). Compared to ionization chambers they typically have a small active volume, as the energy required to create an ion pair is some 10 times smaller and a solid rather than a gaseous detector material is used. Most of them are classified as active radiation detectors, as they allow immediate dose readout. This includes also larger detectors with volumes of 100 cm^3 or more used for spectroscopic measurements (e.g., lithium-drifted silicon crystals or pure germanium crystals). These detectors are rather expensive and their applications in radiotherapy are limited. Semiconductor detectors may also be complex memory chips where the amount of information lost is proportional (within limits) to the dose of radiation. Also, these detectors are not commonly used in radiotherapy. Therefore the discussion in the following will be restricted to:

i) silicon diodes.
ii) Metal oxide semiconductors–field effect transistors (MOSFETs).

3.6.1 Silicon Diodes

Most semiconductor radiation detectors are silicon based. Silicon itself is a very poor conductor. However, the silicon matrix can be doped with impurities

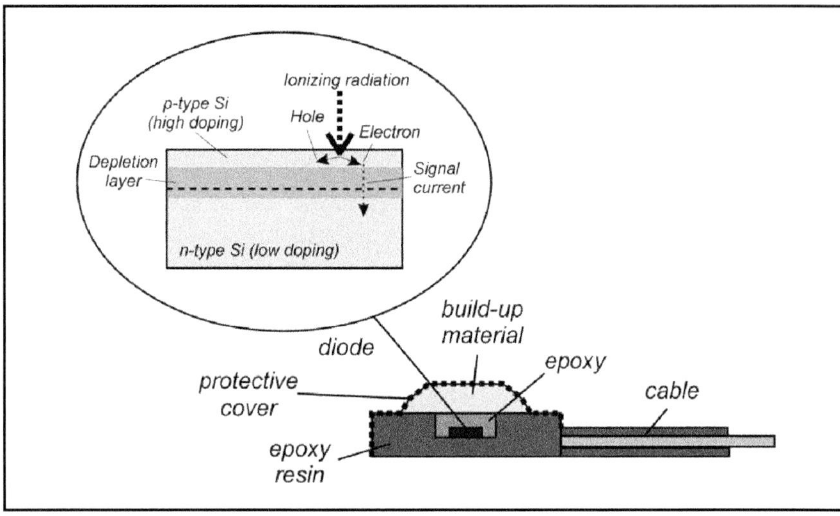

Figure 3.32: Schematic illustration of a diode detector for in vivo dosimetry.

of electron donors (e.g., P, As, Se) to produce n-type semiconductors or with electron acceptor atoms (e.g., Ga, In) to produce a p-type semiconductor. The radiation sensitive diode is formed by a thin layer of one type of semiconductor on top of a substrate from the other type. The insert in figure 3.30 shows the setup schematically as part of the cross section through a diode detector designed for in vivo dosimetry.

The zone between the two materials, called a 'p-n junction,' is depleted of charged particles and poses a barrier for current. When the semiconductor is irradiated, charged particles such as electrons and electron holes are set free, which allows a signal current to flow.

In principle, the ionizing radiation is not only recorded by the diode but also damages the semiconductor. This leads to a change of radiation sensitivity with radiation history in diode detectors. It has been shown by Rikner and Grusell (Rikner 1986, Rikner and Grussell 1983, 1987) that p-type semiconductors are more suitable for radiation detection. After pre-irradiation with several kGy of ionizing radiation, they demonstrate less-pronounced radiation damage leading to a slower change in sensitivity with dose.

Diode semiconductors can be operated in two different modes: (a) with bias and (b) without bias. In the first case, the resistance of the diode is taken as a measure of the intensity of the radiation. The current is measured as a function of the bias using an ampere meter. In the second mode no bias is required, and the diode works similarly to a solar cell. In this case the voltage generated by the diode is proportional to the radiation intensity (or dose rate). This voltage leads to a current, and the charge collected from the diode can be taken as a measure of dose. For radiation applications the diode is usually used in this photovoltaic mode. In this mode the junction electrodes are left open and no external bias is applied. The diode biases itself in the forward direction, so that if no radiation is applied the net current is zero and an open circuit voltage is developed.

The electric field that exists in a semiconductor junction causes any electrons that are created in or near the junction to be swept toward the n-type material and any holes are swept toward the p-type material. The region is then called a depletion region because the number of electrons and holes is decreased. Hence any electron-hole pairs that are created in this region by ionizing radiation are swept from the region by an electric field; their motion causes a current to flow. The resulting charge is monitored by an electrometer.

Typical Applications

In radiation detector applications requiring extremely good spatial resolution, the semiconductor detector has advantages because the solid medium of the junction is approximately 1000 times more dense than the gaseous medium found in ionization chambers. This and the small mean energy required to produce a charge carrier (approximately 3 eV) means that the detector volume can be extremely small and still produce sufficient ion pairs for an adequate electric current to flow. This makes diodes a method of choice when performing measurements in steep dose gradients (e.g., beam penumbra) and small fields (e.g., stereotactic radiotherapy) (Griessbach et al. 2005, McNiven and Kron 2006).

In addition to this, the stopping power ratio of silicon and water is more or less constant over the range of electron energies of interest in radiotherapy. This is in contrast to the air-to-water stopping power ratio, which is required to correct ionization chamber response in electron beams (compare chapters five and eight). Therefore, diodes may be used for depth dose measurements in electron beams without the need for an energy-dependent "ionization to dose" correction (compare chapter eight). However, other influences on the dose response such as dose rate dependence and directional response still need to be considered (see below) as they could lead to considerable differences between diode and ionization chamber measurements in electron beams (Song et al. 2006).

Due to these features and the fact that diodes are relatively cheap, diodes are used in a variety of commercially available dose measurement devices as illustrated in figure 3.33. It shows diodes used for measurements in a water phantom (figure 3.33a), diodes used for in vivo dosimetry (Auckett 1991, Heukelom et al. 1991b, Howlett et al. 1999) (figure 3.33b), a diode device for daily output consistency checks (figure 3.33c), and a dual array for profile, symmetry, and flatness checks in megavoltage radiotherapy beams (3.33d). The small size of diodes allows incorporating many of them into an array such as the one shown in figure 3.33d. The detector spacing of 0.4 cm allows the determination of flatness, symmetry, and beam profiles. In particular for commissioning and quality assurance of dynamic wedges, these diode arrays are of value.

Limitations

When using semiconductor detectors one has to be aware of a number of limitations concerning their use.

i) Generally, semiconductors show a temperature dependence. This of course also applies to radiation detection. In particular for in vivo

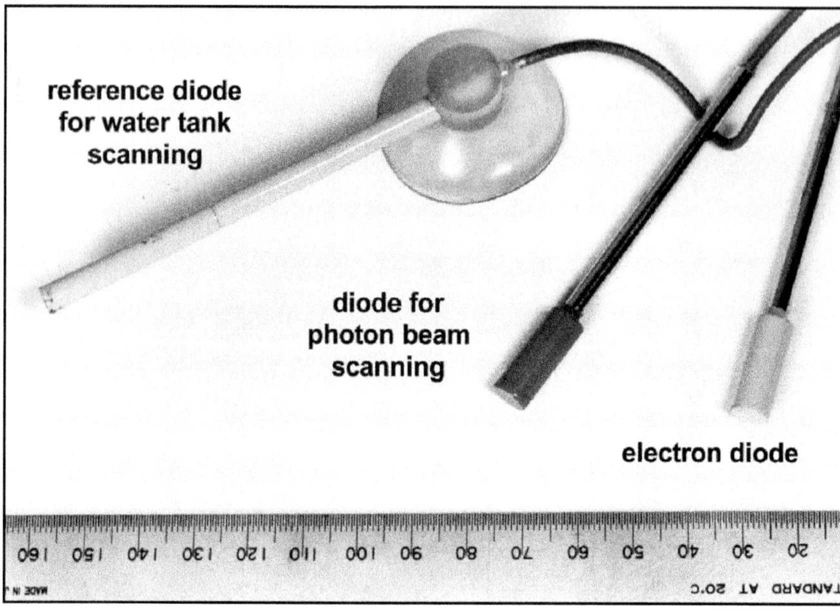

Figure 3.33a: Waterproof diodes used for beam scanning in a water phantom (Scanditronix).

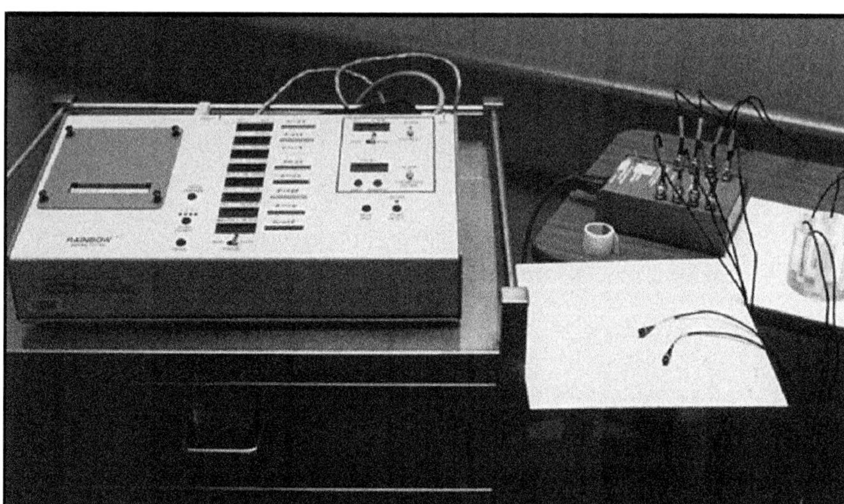

Figure 3.33b: Commercial diode system for in vivo dosimetry (Nuclear Associates Rainbow model 37-701). Two of the eight diodes available are shown on a sheet of white polystyrene. Their active volume is cylindrical to minimize directional variations of the dose response. A metal build-up that is approximately equivalent to 1 cm of water is incorporated in the waterproof encapsulation. The electrometer shown on the left side of the photo has eight separate read-outs and a built-in printer to produce a permanent record of the dose given to a patient.

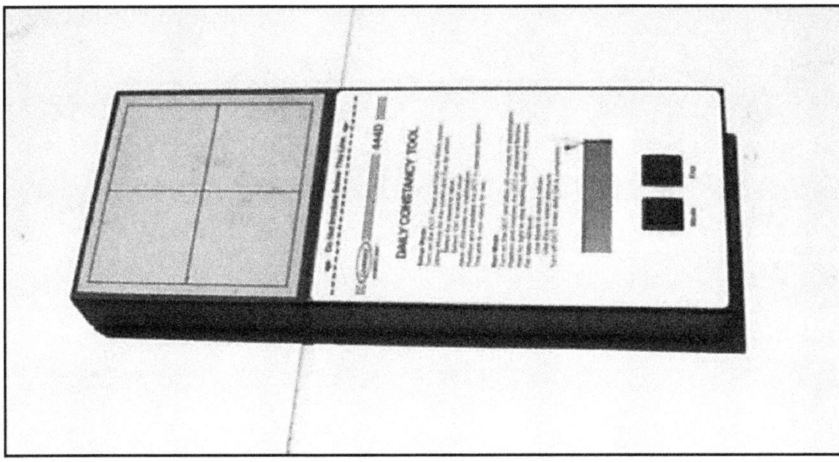

Figure 3.33c: A simple battery-operated device for daily consistency checks (RMI 444). The memory in the device maintains a record of up to 99 exposures, which allows convenient acquisition of data for several radiation qualities and field arrangements without entering the room.

dosimetry where the diodes are in contact with the patient and not in temperature equilibrium, this effect has to be thoroughly checked (Grusell and Rikner 1986).

ii) Diodes display a dose-rate dependence of their reading (VanDam et al. 1990). This can be particularly of concern in pulsed radiation

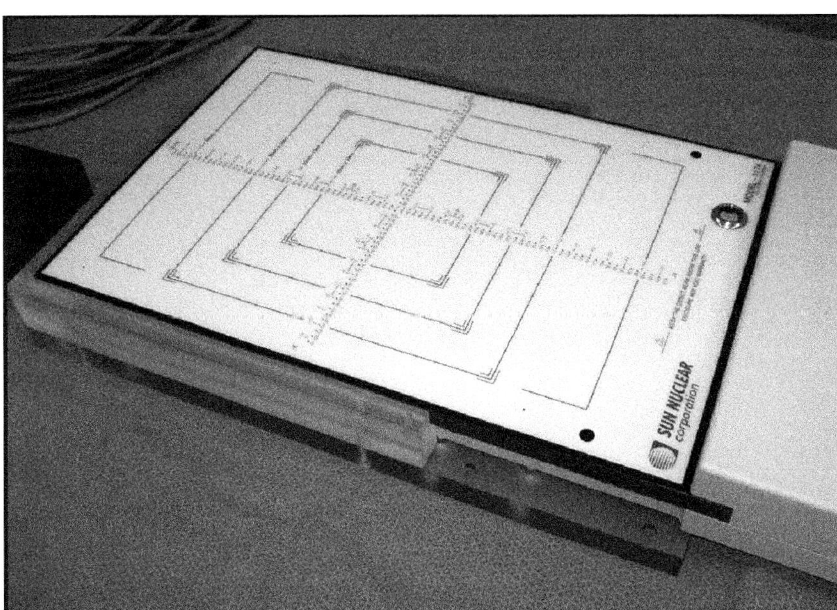

Figure 3.33d: A diode array for beam profile measurements (Sun Nuclear Profiler II). The device contains two orthogonal rows of diodes with 4 mm spacing (81 and 57 diodes, respectively).

beams produced by medical linear accelerators and may distort dose measurements.

iii) The shape of the p-n junction and its geometry determines the directional response of semiconductors. In any case there will be a directional dependence of the dose reading that is enhanced by the presence of the cable connection. Figure 3.34a shows how the dose response varies with different irradiation directions in a 6 MV photon beam. The measurements shown were taken using a commercially available diode system for in vivo patient dosimetry (Nuclear Associates, Rainbow diode system).

Rainbow diodes have a cylindrical shape. As a result of this there is no directional dependence of the dose reading if the diodes are irradiated from different angles through the mantle. However, there is a clear dependence of the reading when the diode is irradi-

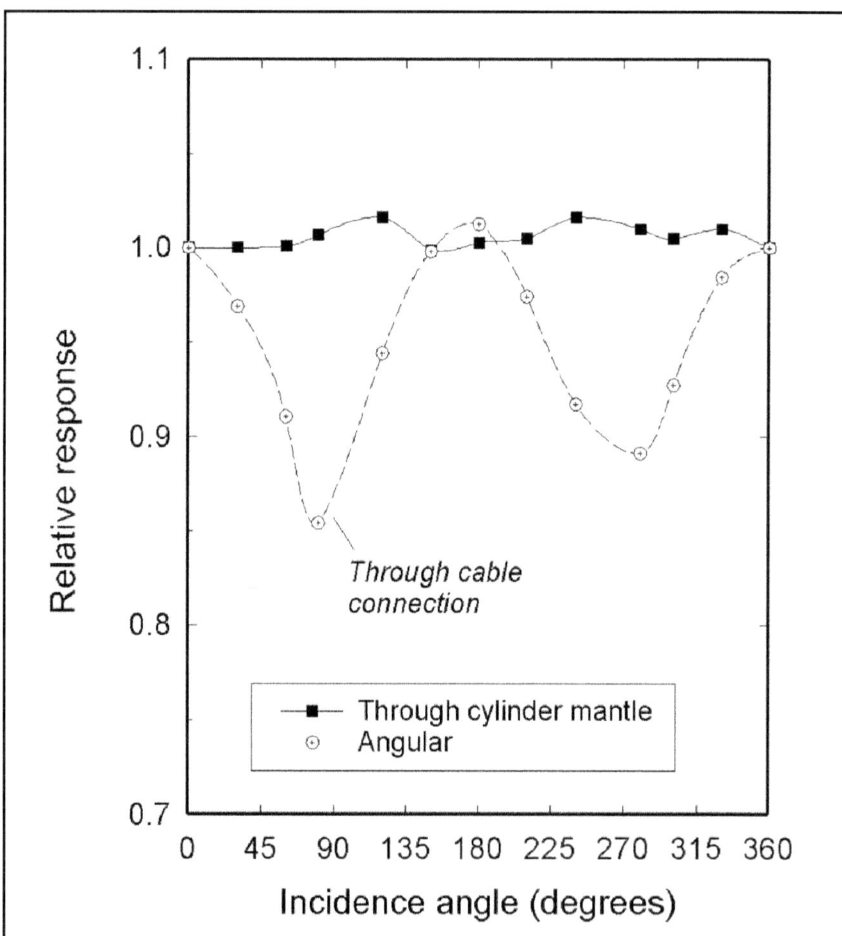

Figure 3.34a: Variation of dose response with incidence angle onto diodes used for in vivo dosimetry (NA Rainbow system).

ated through the top and bottom of the cylinder as demonstrated in figure 3.34a ("angular").

iv) Many commercially available diodes are encapsulated. McKerracher and Thwaites (2006) recently demonstrated this very nicely by taking x-rays of a variety of solid-state detectors. In particular, diodes designed for in vivo dosimetry often also include build-up. This build-up, which is schematically illustrated in figure 3.32, is typically made

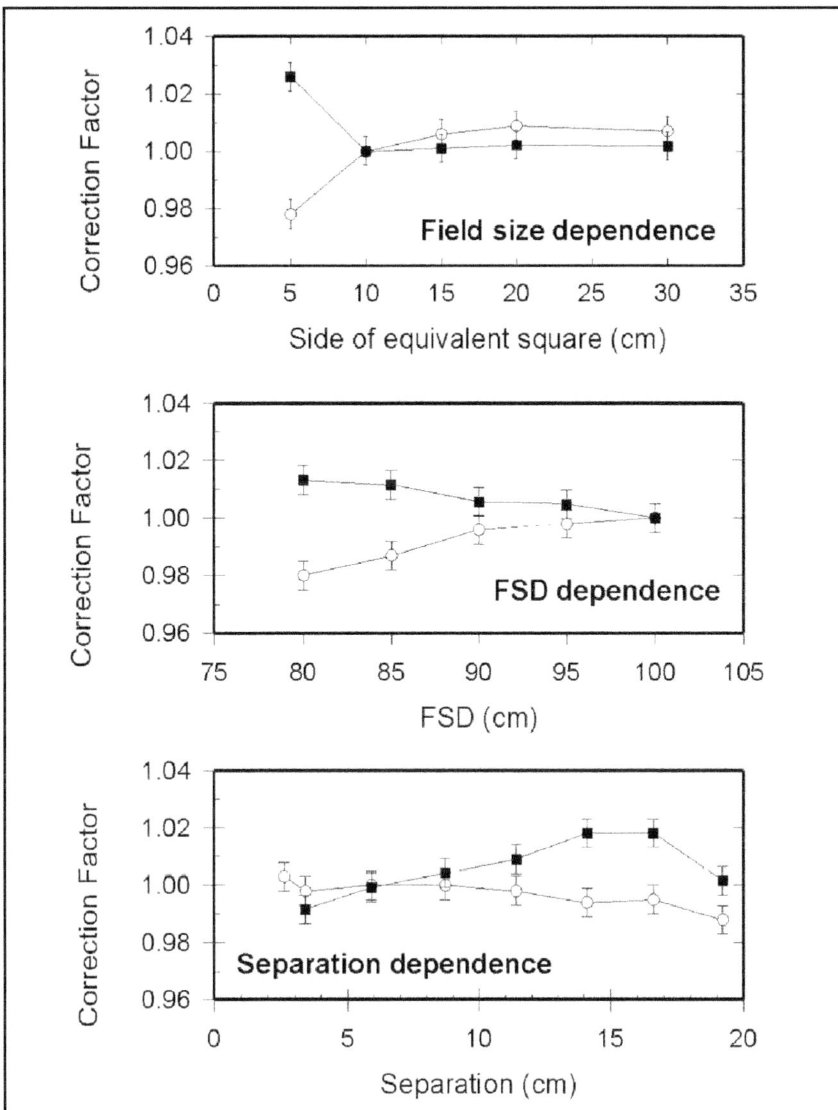

Figure 3.34b: Example for entrance and exit dose correction factors for diodes used for in vivo dosimetry: (top) for variations in field size; (middle) for variations in focus-to-detector distance, and (bottom) for variations in patient/phantom thickness. Open symbols: entrance dose measurement. Closed symbols: exit dose measurement.

from steel or other metals to keep the favorable small dimensions of the detector. The build-up is designed for a particular type of radiation, and it must be checked to ensure that it is still suitable under altered radiation conditions. This has to be taken into account by using an appropriate correction factor (which may be field size dependent) or by adding additional build up with thin wax sheets if the diode is designed for a photon energy lower than the one under investigation.

v) The most subtle problem is the relatively high atomic number in the active zone. This makes the detector susceptible to slight variations in the spectral composition of radiation beams (Yin et al. 2004). The diode depletion region contains silicon, which has a high atomic number ($Z = 14$) compared to air ($Z = 7.78$). This causes an enhanced response to photoelectric beam components, but fortunately the effect is only significant at low energies where the diode overresponds to the photoelectric beam component. Most low-energy photons in megavoltage beams are from scattered radiation, which generally increases with depth of measurement. Therefore, special backscatter filters have been designed for commercial semiconductors for photon dosimetry to help reduce the effect (Rikner and Grusell 1983, 1987, Grusell and Rikner 1986). Although higher energy beams do have low-energy components, these are not significant enough to affect the diodes' depth dose response by more than 2% in the range of energies typically produced by linear accelerators (1.5 to 15 MeV effective energy).

As the effective photon spectrum is affected by a variety of parameters, a number of correction factors may be required if diodes are used for in vivo dosimetry. These include corrections for field size, focus-to-detector distance (FDD), and patient/phantom thickness (Heukelom et al. 1991a). Particular problems may also be encountered at interfaces. A set of correction factors measured for a commercially available diode system (Nuclear Associates, Rainbow diode system) is illustrated in figure 3.34b.

Due to the nature of the photon spectrum at entrance and exit position, the factors required are quite different for exit and entrance dose measurements, as can be seen in figure 3.34.

3.6.2 MOSFET Dosimeters

A relatively new development for dosimetry of ionizing radiation are metal oxide semiconductors–field effect transistors (MOSFETs). They became available for dosimetry in the early 1980s and have the unique feature of integrating dose information as well as allowing an immediate dose readout (Thomson et al. 1984). Figure 3.35 shows a schematic diagram of a MOSFET detector.

The measurement is based on the fact that ionizing radiation alters the threshold voltage, V_T, at the gate. This voltage variation is most commonly evaluated by maintaining a constant current (e.g., 50 μA) between source and drain and measuring the required gate voltage. Change of the threshold voltage is proportional to the absorbed radiation. Changes in the voltage,

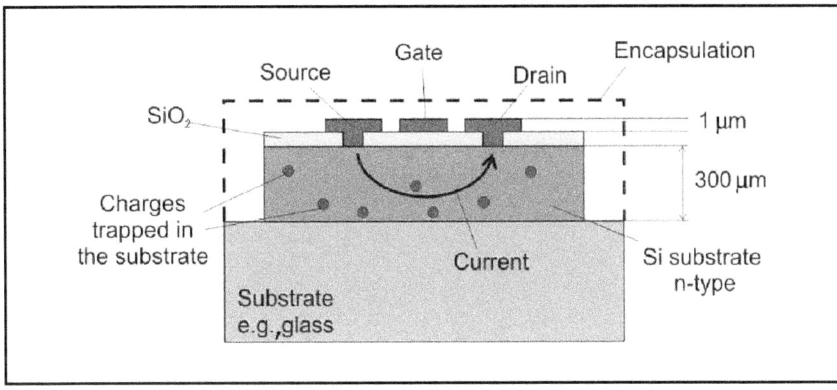

Figure 3.35: Schematic diagram of a MOSFET detector.

required to maintain a constant current, are due to charge carriers (electrons or holes) that are generated in the insulating silicon oxide layer. Positive charges (holes) are moving toward the Si-SiO_2 interface and are trapped leading to a charge build-up in the gate oxide SiO_2 of the MOSFET. In many applications (Butson et al. 1996a, Rosenfeld et al. 1995), for dosimetry p-MOSFET is used, i.e., silicon substrate is n-type Si. MOSFET detectors can be used in a passive or an active mode. In an active mode the gate is biased positively relative to source during irradiation. The electrical field in the gate oxide reduces the recombination of electron-hole pair induced by ionization leading to more holes being trapped on the Si-SiO_2 interface.

The threshold voltage change depends on gate bias during irradiation and the thickness of the insulating SiO_2 layer. To achieve acceptable sensitivity, the layer needs to be relatively "thick" (0.5 to 1 µm is a typical thickness). Also, a high gate bias during irradiation increases the sensitivity, and with a bias of 100 V, a sensitivity of 5 V per Gy can be achieved (Litovchenko et al. 1990). However, as a high sensitivity also leads to a fast saturation of the detector, a bias voltage of 3 V to 6 V is typically applied (Butson et al. 1996a). With 6 V bias, the detectors are linear with dose within 5% up to approximately 50 Gy.

The main advantage of MOSFET detectors is their extremely small physical size—the gate can be as small as 10 µm wide with a sensitive layer of 1 µm thickness or less. An experimental MOSFET system (University of Wollongong Centre for Medical Radiation) is shown in figure 3.36.

There is an increasing number of reports on clinical applications of MOSFETs. These include surface dose measurements in photon beams (Butson et al. 1996a), multileaf collimator (MLC) leakage assessment (Kron et al. 2002c), and in vivo dosimetry (Ramani et al. 1997). Due to their small physical size they may particularly be useful for measurements in brachytherapy and microbeam radiation therapy (Kaplan et al. 2000, Rosenfeld et al. 2001, Brauer-Krisch et al. 2003). Of some interest are also measurements in mixed radiation fields (Rosenfeld et al. 1995). A summary of applications of MOSFET detectors in radiation therapy has been published by Rosenfeld (2002). The manufacturer Thomson and Nielsen provides a large number of relevant references on their webpage (http://www.thomson-elec.com/; accessed 4/2007).

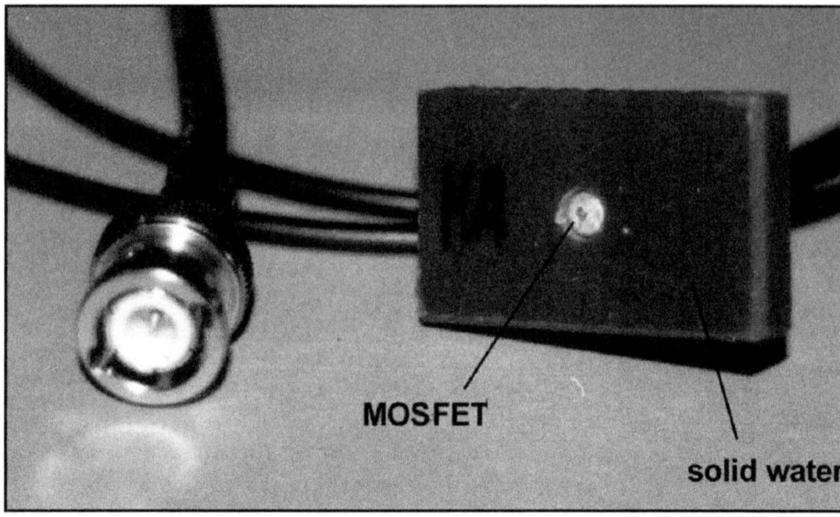

Figure 3.36: Photograph of a MOSFET radiation detector used for experimental measurements.

When using MOSFETs the following points should be considered (Soubra et al. 1994):

i) As all semiconductors, MOSFETs exhibit a temperature dependence.

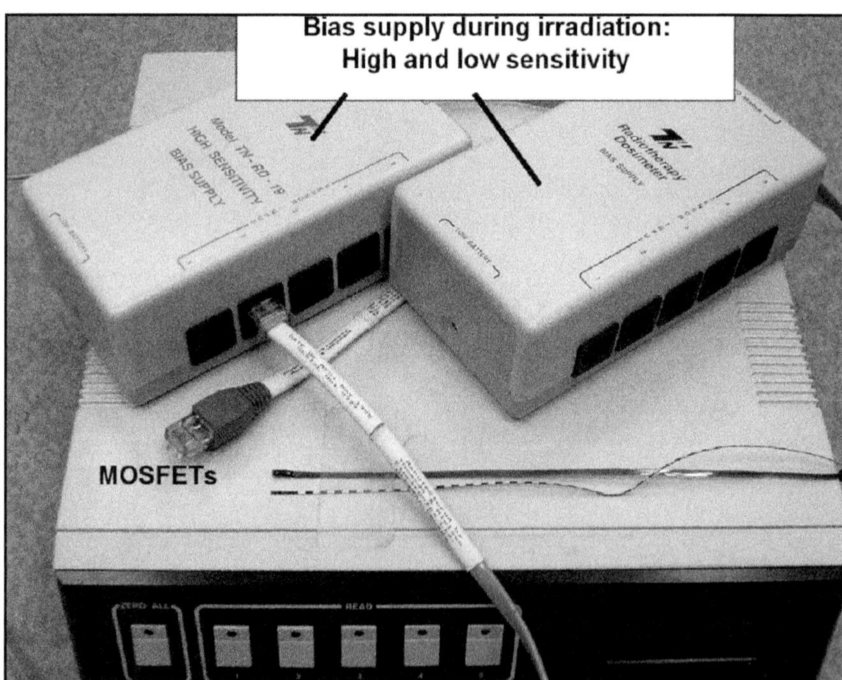

Figure 3.37a: Photograph of a commercial MOSFET system used for in vivo dosimetry (Thomson and Nielsen).

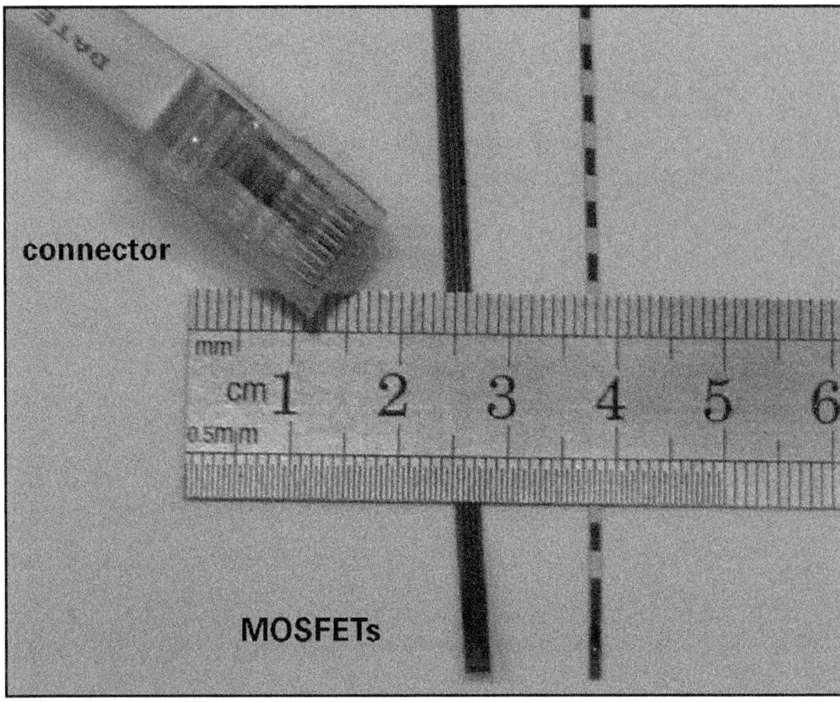

Figure 3.37b: Photograph of two commercial MOSFETs for in vivo dosimetry (Thomson and Nielsen). The left detector is standard size, while the right one is a miniaturized MOSFET suitable for insertion into brachytherapy catheters.

ii) The dose response of individual MOSFETs is not linear (Kron et al. 1998). It decreases with total absorbed dose and this needs to be corrected for when using single detectors.

iii) The reading of MOSFETs drifts slightly after irradiation as more charges are trapped and some released. Therefore, the reading should be taken at a well-specified time after irradiation.

iv) The detectors have a limited life.

v) The sensitivity of the detector can change with bias voltage during irradiation; this voltage is typically supplied using batteries and must be stable.

vi) MOSFETs can be used as integrative dosimeters, and it has been suggested to use them to monitor the whole fractionated treatment of a patient. In this case, fading must be checked for.

vii) MOSFETs are not tissue equivalent and their dose response varies with radiation quality (Edwards et al. 1997, Kron et al. 1998). As the active detector volume is quite small, the packaging of the detector will affect its response characteristics substantially (Rosenfeld et al. 1995).

MOSFET detectors are commercially available for radiotherapy dosimetry (Thomson and Nielson, Ottawa, Canada). These detectors are of dual bias, dual MOSFET design, which employs two identical MOSFETs with different gate bias (Soubra et al 1994). Figure 3.37 shows such a MOSFET system designed for in vivo dosimetry. By taking a differential reading of the two detectors, one can overcome the first two problems mentioned above. Both temperature dependence and dose linearity are substantially improved in these dosimeters (Soubra et al. 1994).

Very interesting recent developments are implantable MOSFET detectors for in vivo dosimetry (Black et al. 2005). These devices make use of the small physical size, which allows the production of a miniaturized detector system with radio transmitter. This can be implanted into the target tissue or other structures of interest, providing the ultimate in vivo dosimeter (compare also chapter eleven).

3.7 Thermoluminescence Dosimeters

Thermoluminescence dosimetry (TLD) is a relatively old dosimetric technique that was first applied some 100 years ago (Kron 1995). A whole variety of TL materials and forms are available, some of which are shown in figure 3.38. They all share the same advantages of small size and no attached cables. However, TLD has often been regarded as a "black art" that only the "initiated" can use to obtain reliable results. The following section addresses some of the problems with TLD by first introducing a simplified theory of TL and then illustrating the characteristics of a TLD measuring system using the example of LiF:Mg,Ti—the most commonly used TL material in radiotherapy dosimetry.

3.7.1 Basic Theory of TLD

The theory of thermoluminescence is complex. It is beyond the scope of the present section to cover it in detail, and the theory will be only discussed as far as relevant for an understanding of TLD measurements in radiotherapy. For more details one should refer to the articles of Horowitz (1981) and Kron (1994), the monographs of Cameron (Cameron et al. 1968) and McKeever (McKeever et al. 1995), the three-volume handbook by Horowitz (1984c), or the recet summary of microdosimetry responses by the same author (Horowitz 2006a,b).

TL dosimeters consist of crystals that are in general nonconducting—i.e., the conduction band in the band model of crystals is empty. At room temperature all electrons are confined to the valence band. If the crystal is irradiated with ionizing radiation, some of the electrons gain enough energy to get into the conductivity band (see figure 3.39). In TL materials there are a number of imperfections in the crystal that can trap electrons from the conduction band at an energy state that is between the conduction and the valence band. The energy gap between the conduction band and the trap is of the order of a few eV, and the number of electrons trapped is a function of the incident radiation intensity. Typically less than 1% of the electrons in the conduction band are trapped (Horowitz 1984c). The probability that the electrons can gain enough energy to escape back to the conduction band from which they can fall back to the lowest energy state in the valence band depends on the depth of these traps and the temperature. Figure 3.39 simpli-

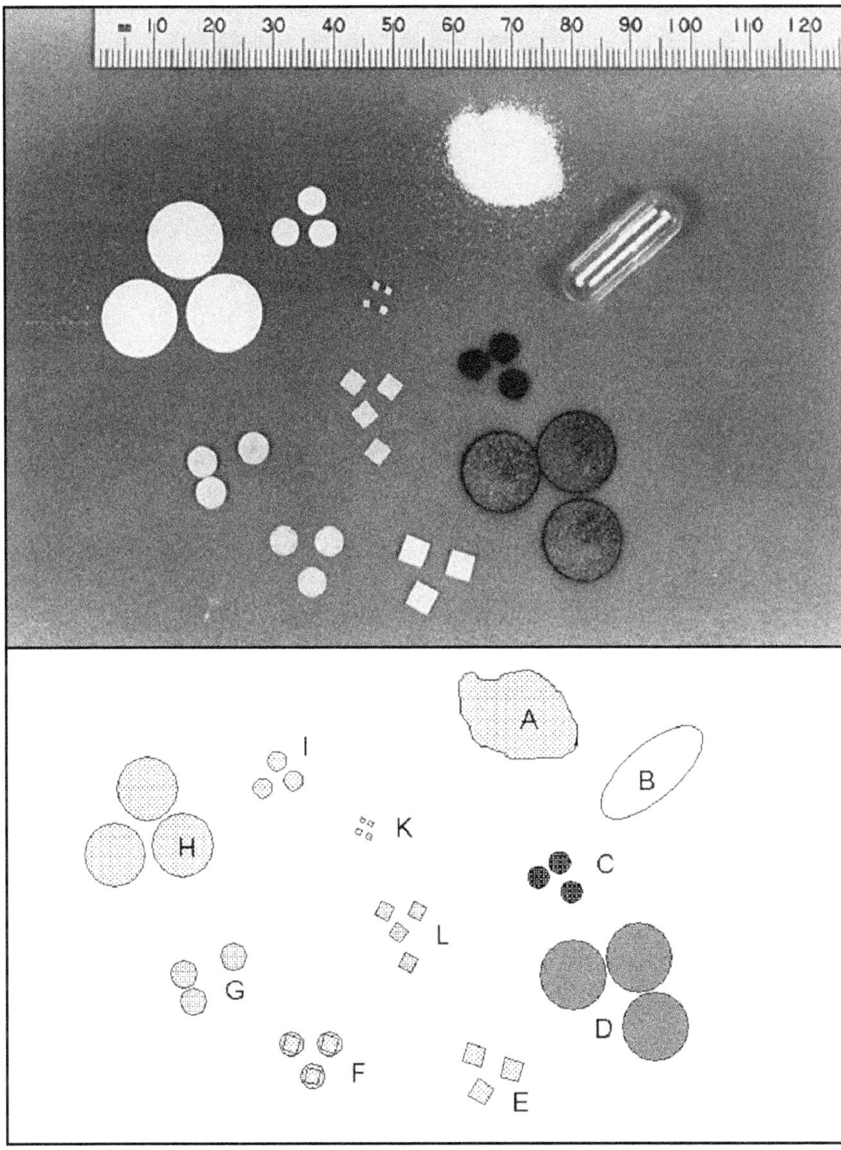

Figure 3.38: Variety of thermoluminescence (TLD) detectors depicted are: A: powder; B: capsule used to hold TLD powder; C: carbon loaded LiF disks; D: as C but in Teflon matrx; E: LiF:Mg,Cu,P chips; F: EXTRAD chips (thin active layer); G: LiF:Mg,Ti in Teflon matrix; H: LiF:Mg,Ti in a Teflon matrix; I: LiF:Mg,Ti disks; K: small TLD cubes; L: standard TLD-100 ribbons.

fies the whole process by showing only the electrons. Positively charged holes created by the ionizing radiation in the crystal structure behave in a similar way. With rising temperature, the probability increases that electrons have enough energy to be raised to the conduction band and release energy in the form of light when they recombine with a positive hole in the valence

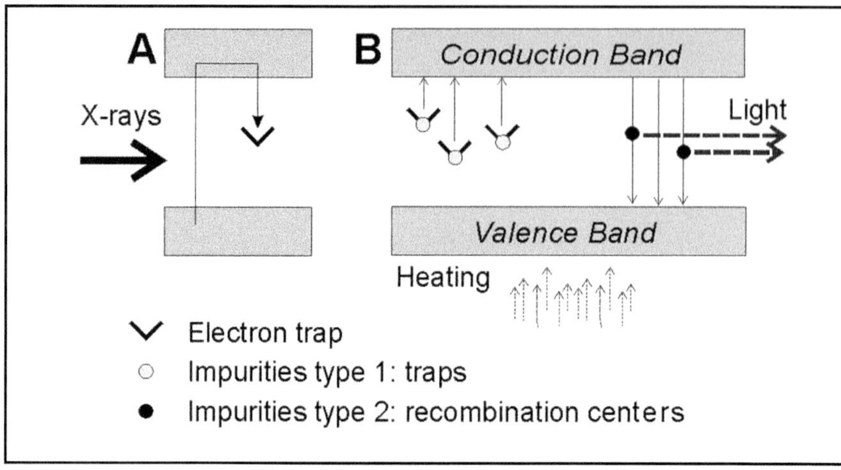

Figure 3.39: Schematic drawing of the TLD process.

band. Impurities again play an important role. They create the centers where the transition of the electron from conduction to valence band results in the emission of light (the so-called "color centers"). The amount of light emitted is dependent on the temperature and the number of electrons trapped and therefore on the absorbed dose in the crystal.

Figure 3.39 illustrates the principle of thermo (*"one applies heat"*) luminescence (*"and the crystal emits light"*) dosimetry *("of which the intensity is related to the dose of ionizing radiation absorbed by the crystal prior to heating"*). In most TL materials there is more than one trap type. These traps have different energy gaps to the conduction band and will therefore empty at different temperatures. In principle, all traps shown in figure 3.39 are filled with different numbers of electrons (or holes). The multitude of traps (e.g., eight or more in LiF:Mg,Ti) leads to a complicated function of light intensity versus temperature as displayed in figure 3.40. This is called *glow curves*.

In order to describe the glow curves mathematically, one has to consider that all the traps shown in figure 3.39 will be filled with different numbers of electrons (or holes). There will also be a dependence on the so-called frequency function, s, which is a trap-specific constant related to crystal lattice vibrations (Burch 1968, Horowitz 1984a). For a constant temperature, T, and a trap with an energy gap of E_g, the number of electrons, n, released after a time, t, can be calculated as:

$$n = n_0 \exp\left(-s\,t\,\exp\left(\frac{-E_g}{kT}\right)\right), \tag{3.14}$$

with n_0 as the number of electrons (and/or positrons) trapped, s the frequency factor or "attempt to escape" frequency associated with a particular defect, and k the Boltzmann constant.

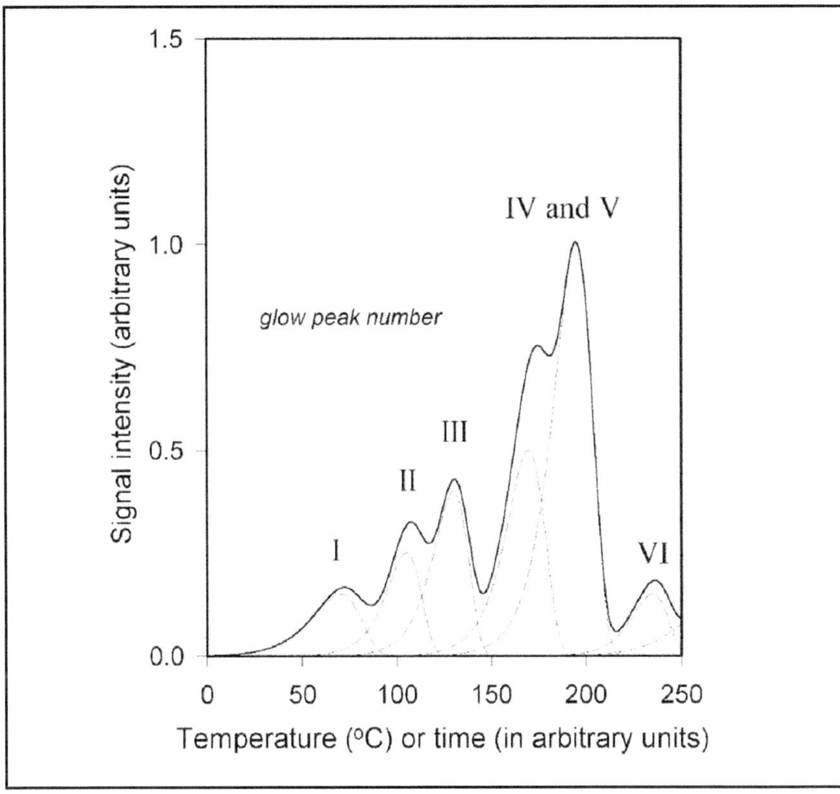

Figure 3.40: Schematic drawing of the glow curve of Lif:Mg,Ti. the shape of the individual peaks was calculated using equaton 3.16.

However, in practice the desired temperature, T, is not reached immediately. This leads to the observation of glow curves similar to the one depicted in figure 3.40. The figure shows the light emitted from a LiF:Mg,Ti crystal as a function of its temperature. The shapes of individual peaks in the glow curve that are associated to a particular trap are shown as a dotted line. They can be calculated using the Randall-Wilkins theory of thermoluminescence (Randall and Wilkins 1945). As a first approximation, the crystal in most TLD readers is exposed to a constant heating rate R = dT/dt = constant. Assuming first order kinetics (i.e., once released into the conduction band the retrapping of electrons is neglected) any peak in the glow curve can be described as follows (McKinlay 1981b, Chen 1984):

$$I(T) = n_0\, s \exp\left(\frac{-E_g}{kT}\right) \exp\left(\int_0^T \left(\frac{-s}{R} \exp\left(\frac{-E_g}{kT'}\right)\right) dT'\right) \tag{3.15}$$

At low temperatures the first term in equation 3.14 dominates, describing a rapid initial rise of the TL signal I. The second term becomes more important at higher temperatures, where it leads to a fall in signal. The function given in equation 3.15 is too complicated to be used directly for glow

Table 3.7: Characteristics of the glow peaks of LiF:Mg,Ti. The table was compiled using references (Horowitz 1984c and 1990, McKinlay 1981, Khan 2003, Goldfinch et al. 1993, McKeever et al. 1995). Shown are only the peaks observed in typical radiotherapy applications.

Peak Number[1]	*Approximate Energy Gap (eV)*	*Readout Temperature (°C)*	*Approximate Contribution to Readout*[1]	*Half-Life at Room Temperature*
I	1.04	70	0	10 min
II	1.07	105	0.2	20 hr
III	1.05	130	0.3	6 months
IV	1.53	170	0.5	10 yr
V	2.21 (1.61 for peak Va)[2]	195	1	80 yr
VI	1.70	235	0.1	>100 yr
VII	1.80	260		
VIII	1.96	285		

[1]Readout after irradiation with 6 MV photons from a linear accelerator and a 10 minute pre-read anneal on 100 °C, normalized to the contribution of peak V. As pre-read anneal, waiting times, and variation in doping alter the contribution of the peaks, this can only be seen as a rough guideline.

[2]The frequency factor for peak V is assumed to be approximately 10^7 times larger than for other peaks (Horowitz 1984a, Fairchild et al. 1978). Peak Va has been reported only for some batches of LiF:Mg,Ti (Bos et al. 1993a).

curve fitting and analysis. However, a simplified expression can be used to describe the light, I, emitted from the TL detector as a function of the temperature (Podgorsak et al. 1972, Delgado et al. 1992).

$$I(T) = I_{max} \exp\left(1 + W (T - T_{max}) - \exp\left(W (T - T_{max})\right)\right), \quad (3.16)$$

with I_{max} the light intensity at the temperature of the glow curve maximum T_{max}, and $W = E_g/k\,T_{max}^2$. In this case the information on the trapping parameters, s and n_0, is lost, which is of no importance to most dosimetric applications. Equation 3.16 or similar approaches (Bos et al. 1993b, Delgado et al. 1995) can be used for mathematical glow curve analysis (Delgado et al. 1992, Horowitz and Yossian 1995). The glow curve in figure 3.40 was created using equation 3.16 for the seven first peaks of LiF:Mg,Ti and the parameters given in table 3.7. Provided the heating rate is constant, temperature, T, as the label of the ordinate in figure 3.40 is interchangeable with time, t. It should be noted that figure 3.40 illustrates a schematic curve. Most clinical readers employ a relatively large heating rate to keep the overall readout time short, which leads to a collapse of two and more peaks into one.

3.7.2 Characteristics of LiF:Mg,Ti

The characteristics of TL dosimeters of importance for dosimetry in radiotherapy will be discussed using the most commonly used material, lithium fluoride, doped with magnesium and titanium (LiF:Mg,Ti) as an example. A good, recent review of the underlying physics was given by Horowitz (2006b).

In practice, LiF:Mg,Ti comprises a whole subgroup of materials with different doping and different isotopic compositions such as of ^{6}Li and ^{7}Li. The following discussion will apply in general to all these materials. However, some differences may occur even between batches of nominally identical material, and all parameters should be verified for a particular material in use.

Fading

Even with relatively large energy gaps, Eg, between the conduction band and the traps, there is a finite probability that an electron gains enough energy at room temperature to escape from the trap. The smaller the energy difference, the higher the probability of spontaneous emission of light, which diminishes the population of the traps and therefore reduces the TL signal. This process is called fading. The half-lives for the different traps in LiF:Mg,Ti at room temperature range from a few minutes to several years. They are listed in table 3.7. As can be expected, they increase with increasing gap width (i.e., increasing readout temperature).

The integral under peaks IV and V, the most commonly used peaks for dosimetry in LiF:Mg,Ti, do not exhibit pronounced fading, as can be seen in table 3.7. It is of the order of 5% over 12 weeks. The extent of fading for other materials is listed in table 3.8.

Variation of the TL Signal with Dose

Figure 3.41 shows the dose response of a typical TL dosimeter to low LET radiation such as photons and electrons. Four ranges can be distinguished. Between points A and B the dose reading levels out for very small doses. The detection limit, which is of the order of 0.1 mGy for a LiF:Mg,Ti detector of 100 mg (Hirning 1992) is close to point B. Between points B and C the dose response is linear to the absorbed dose. This is the most useful range for measurements. At dose levels above point C, the TL readings are actually overestimating the given dose. The sensitivity of the dosimeters is increased. This phenomenon is called supralinearity, and it has to be considered for many applications in radiotherapy. At even higher radiation doses of about 300 Gy, the relative sensitivity decreases again due to radiation damage to the detector.

Supralinearity

In radiotherapy, where typically doses of more than 1 Gy have to be assessed, supralinearity of TL dosimeters has to be accounted for. Many theories have been developed to explain the behavior and to correct for it (Delgado et al. 1992, Attix 1975, Lakshmanan and Bhatt 1981, Feist 1988). A relatively simple model is given by Claffy (Attix 1975)—the track interaction model. It assumes that the filled traps and activated recombination centers, both created by the

Table 3.8: TLD materials commonly used in medical applications and their characteristics

TLD material	*LiF: Mg,Ti (TLD 100)*	*LiF: Mg,Cu,P*	*$Li_2B_4O_7$:Mn (TLD 800)*	*$CaSO_4$:Mn*	*$CaSO_4$:Dy (TLD 900)*
Physical density (g/cm^3)	2.64	2.64	2.3	2.61	2.61
Effective atomic number	8.2	8.2	7.4	15.3	15.3
Sensitivity to ^{60}Co radiation (relative to LiF:Mg,Ti)	1	about 30	0.3	70	about 15
Energy response 30 keV/1.25 MeV	1.7	1.25	0.9	12	12
Temperature of main glow peak (°C)	195	210	200	110	220, 250
Maximum wavelength of emitted light (nm)	400	380	600	500	480, 570
Fading of main glow peak at 20 °C	<10% per year		10% per month	50% per day	6% in 6 months
Typical annealing procedure	1 hr: 400 °C 20 hr: 80 °C	¼ hr: 250 °C 2 hr: 100 °C	½ hr: 300 °C	½ hr: 400 °C	½ hr: 400 °C
Useful dose range (Gy)	5×10^{-5} to 10^3	$<10^{-6}$ to 100	10^{-4} to 10^4	10^{-7} to 100	10^{-6} to 10^3
Available physical forms	powder, rods, crystal ribbons, bulbs, cards, Teflon-based chips	ribbons, powder	powder, Teflon-based chips, cards, ribbons	powder	ribbons, powder, Teflon-based chips
Toxicity	high if ingested	high if ingested	high if ingested	low	low
Principle dosimetric applications	personal, radiotherapy TLD 600: neutrons		diagnostic, radiotherapy	environmental, low dose	environmental, personal
Remarks	complex glow curve, most common TLD, available as ^{6}Li and ^{7}Li	good signal to noise, no supralinearity, low-temperature annealing	tissue equivalent	low dose TLD, high sensitivity	complex glow curve, high sensitivity

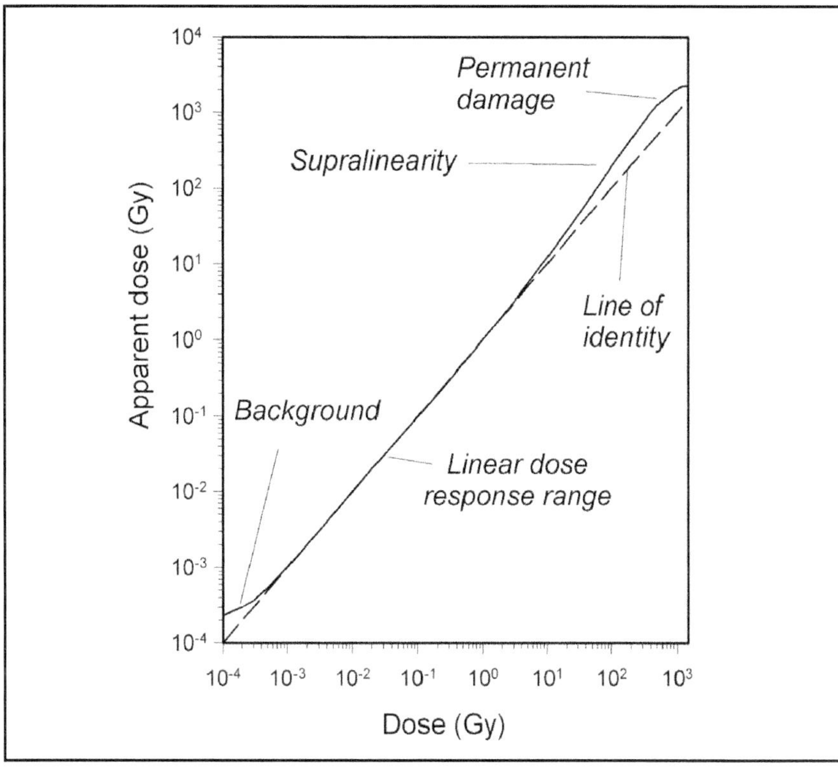

Figure 3.41: Dose response of Lif:Mg,Ti over 7 orders of magnitude in dose.

radiation, are closer together for high absorbed doses. This would lead to a higher probability of recombination under light emission and therefore a larger TL signal. Accordingly, LET (linear energy transfer) will also influence supralinear behavior (Horowitz 1984b).

The supralinear behavior of LiF:Mg,Ti depends on the chemical composition of the material. In particular the concentration of titanium determines the onset of supralinear behavior. Higher titanium concentrations shift the supralinearity threshold to higher doses. For the practice of TL dosimetry it is important to note that supralinearity can be influenced by the glow peak under investigation. Different peaks in the glow curve can exhibit different supralinear behavior (Busoli 1981). In LiF, glow peaks of higher number exhibit a stronger supralinear behavior. If higher peaks such as peak VII are included in the evaluation, this may lead to the onset of supralinearity at doses well below 1 Gy (Horowitz 1993). As thermal history influences the weighting of the different glow peaks in the curve, it has an influence on supralinearity. Annealing temperatures and the cool down rate have also been shown to alter the supralinearity in LiF (Horowitz 1984b).

In clinical practice several formulas have been proposed to correct for the effect of supralinearity on the reading (Feist 1988, Mayles et al. 1993). They typically contain a correction of the reading with a higher order of the dose received by the TL material, D. For example, Feist proposes a correction as follows (1988):

$$\alpha D^2 + (1 - \alpha) D = K M, \tag{3.17}$$

with M the TL reading, K the calibration factor obtained at a dose of 1 Gy, and α the percentage supralinearity per Gy. In Feist's measurements, α varies between 0.002 and 0.04 for different TL dosimeters and irradiation conditions. Bearing in mind the complexity of the supralinear behavior of TLD, it is important to verify the supralinearity correction used for all particular irradiation and readout conditions. A way to minimize errors due to supralinearity is to expose some TL detectors parallel to the measurement to a known dose of the radiation in question. If this dose is as close as can be predicted to the dose to be measured, one can correct for supralinearity.

For radiotherapy measurements a complete annealing cycle is required after a high dose exposure to reset the supralinearity (Busoli 1981, Krieger and Petzold 1989). The simple and short anneal in the TLD reader that is sometimes performed may not be sufficient to restore the original TLD sensitivity.

Sensitivity

The sensitivity of LiF:Mg,Ti depends on the material composition and the thermal history. It can vary between different batches of nominally similar material and even between different dosimeters from one batch without correlation to variations in weight. This may be due to different surface conditions that influence how much light can escape the TL detector. A very important (and often neglected) influence on sensitivity is the heating rate during readout and the cooling rate after the high-temperature annealing step (Bos et al. 1992). Regulla showed that the TL output of LiF:Mg,Ti can be increased by at least a factor of three if the cooling down rate is increased from 2 °C to 2000 °C per minute (Regulla 1981). This process, called *thermal quenching*, may lead to a larger contribution of peaks II and III to the TL signal, and the sensitivity of the dosimeters may vary with time (Driscoll et al. 1986). A second low-temperature annealing (LiF:Mg,Ti either 80 °C for 24 hours or 100 °C for 2 hours) is recommended after the use of thermal quenching. The sensitivity of chips decreases with time and usage. Horowitz states that there is a loss of about 1.5% in sensitivity per 10 Gy absorbed dose (Horowitz 1993), which is in general agreement with other reports in the literature (Kron et al. 1993d).

Dose Rate Dependence

No dose rate dependence has been observed for LiF:Mg,Ti up to dose rates of 10^8 Gy/s (Karzmark et al. 1964, Tochilin et al. 1966), which seems to be equally valid for other types of TL dosimeters.

Variation of the TL Signal with Radiation Quality

Generally, the variation of dose response with radiation quality depends on the glow peak under consideration (which is related to the type of trap) and the LET of the radiation (which determines the spatial distribution of activated traps and recombination centers). Both will determine the response to a particular radiation quality and also influence the degree of supralinearity observed.

The stopping power ratio for electrons of LiF to water is approximately 0.8 and constant over a wide range of electron energies (10 keV to 20 MeV, Mayles et al. 1993). In spite of this the response characteristics of LiF:Mg,Ti to electrons of low energies is complex and cavity theory is not easily applicable as TLDs typically cannot be regarded as small compared to the electron tracks (Mobit and Sandison 2006). Depending on electron energy and detector size the response was found to be either stable or to decrease by up to 15% compared to Co-60 radiation (Horowitz 1984). A good summary of response of 1 mm thick LiF:Mg,Ti detectors with radiation energy is given by Mobit et al. (1996), who compared experimental results with Monte Carlo calculations. Duggan et al. (2004) discussed the impact of variations in energy response of LiF detectors for clinical dosimetry. Considering the complexities of TLD response with radiation quality it is generally advisable to calibrate TL dosimeters in a radiation beam with similar radiation quality as the radiation that will be investigated.

When measuring low- to medium-energy x-rays, the following additional influences have to be considered:

i) The effective atomic number of LiF (8.14 compared to 7.51 for water.
ii) Supralinearity can be different for different radiation qualities. It is advisable to restrict high-precision measurements therefore to doses below about 1 Gy.
iii) Attenuation of radiation in the TL material itself. This effect is most pronounced for larger TL detectors and low radiation energies. In the commonly used LiF ribbons with a thickness of 0.9 mm, about 70% of the intensity of a 10 keV x-ray beam is absorbed in the crystal.
iv) The readout process. The over-response to low-energy x-rays in LiF:Mg,Ti varies for different glow peaks. In general, the over-response increases with the temperature of the glow peak. For example, peak VII has been reported to over-respond by as much as a factor of 2 (Horowitz 1993).
v) Thermal history. Since the thermal history influences the contribution of different glow peaks, it will also influence the energy response of LiF:Mg,Ti.
vi) If the TL material is embedded in a different matrix such as Teflon, the physical properties of the matrix and the grain size of the TL material can also contribute to variations of the dose response with energy. This leads to a complex function of dose response with effective x-ray energy, which should be tested for each detector type (or even for each batch of similar detectors) if measurements in low to medium x-ray energies are to be performed.

Figure 3.42 illustrates the variation of dose response with photon energy for three comonly used TLD materials.

Accuracy/Precision

TLD is a relative dosimetric technique because there are too many factors that relate the dose to the emitted light, and they are difficult to control. The accuracy of the measurement depends therefore only on the reproducibility (i.e., the precision) of the TL reading of the respective dosimeter and the standard TL dosimeter used to define its sensitivity. Depending on the effort (and funds) put into the setup of a TLD system, very different levels of precision can be achieved. For LiF:Mg,Ti this ranges from ±10% (Julius 1981) down to 0.3% (Feist 1988) or even 0.2% (1 SD) (Martensson 1969). The latter has been achieved with a pre-annealing temperature below 0 °C and a dedicated irradiation facility for uniformity checks and calibration. With reasonable effort, reproducibilities of ±2% or better can be achieved for a single ribbon or rod in the therapeutic dose range. If no correction is made for individual detectors, the reproducibility is typically worse (e.g., ±4%) (Mansfield and Suntharalingam 1976).

The best accuracy can be achieved using *calibration detector groups* (Julius 1981, Kron et al. 1993a). If all detectors in a set of, e.g., 50 are individually calibrated and their sensitivity is known, several dosimeters of a given set can be exposed to a known dose at approximately the same time as the actual measurement with other detectors from the same set. If the reference detectors are exposed to a dose of the same order as expected for the measurement using the same radiation quality at the same time, this procedure overcomes problems with supralinearity, energy response, and fading. Many other influences on the reproducibility are also minimized, since measurement and reference detectors share the same thermal history. In clinical measurements typically more than one dosimeter is used for any one measurement point (Mansfield and Suntharalingam 1976, Wolf et al. 1989, Kron et al. 1990, Kron et al. 1993e). Assuming two dosimeters per measurement point and four used as standard, an accuracy of better than 5% on a 95% confidence level can be achieved. This is generally regarded as adequate for in vivo dosimetry. In order to improve a TLD system beyond that level to an accuracy of better than 3% (2 SD), the reproducibility of each single chip readout must be controlled to be within 1% (Toivonen 1993).

Other Influences on the TLD Reading

A number of other parameters influence the TL reading of LiF:Mg,Ti and other materials. The best way to avoid their impact on the validity of measurements is to use identical dosimeters for measurement and calibration and use the same irradiation conditions for both. Among these problems are:

1. Size: The size of the TL detector will influence its specific dose response (response per unit mass of detector) (Kron et al. 1993d). This is a result of the high physical density of LiF of 2.7 g/cm^3, which leads to an increased number of secondary electrons released in LiF

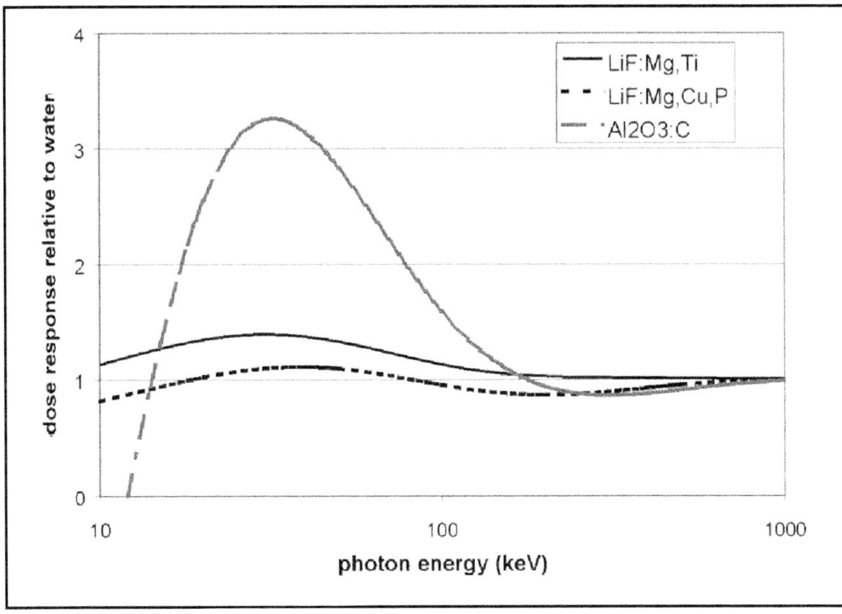

Figure 3.42: Variation of dose response for three commonly used TL materials with photon energy in the low- to medium-energy x-ray range normalized to the response to Co-60. Above approximately 1 MeV, Compton effect is predominant and all materials have a stable dose response.

as compared with tissue. This effect can also affect the reading of TL dosimeters in close proximity to each other where an increase of the reading by more than 1% has been observed due to neighboring detectors (Kron et al. 1993d).

An opposite effect of size that decreases the effective reading is absorption in LiF. This can be attenuation of low-energy radiation leading to inhomogeneous dose distribution throughout the detector. Alternatively, it can be the absorption of light in the TL material itself before it can be registered by the photomultiplier "self-absorption" (Nariyama et al. 1993, Kron et al. 1993a, Mukherjee and Vana 1984).

2. Directional response: While this may affect the TLD reading in low-energy x-rays, it is not a problem in megavoltage photon beams (Bagne 1977).
3. Light sensitivity: Many TL materials are sensitive to visible and UV light. In addition to optical fading, light can cause a false signal (Julius 1981). Even though this effect is typically small and can be neglected in therapeutic radiation doses, it is advisable to keep all TL detectors in a lightproof container when not in use.
4. Heating rate during readout (Bos et al. 1992): If the area under the curve of the glow curve is evaluated, the heating rate will influence the integral reading. The optimum heating rate is different for different TL materials and is of the order of 100 °C per minute for LiF:Mg,Ti (Julius 1981).

The Importance of the Thermal History

Following the discussion of influences on the TLD reading it is evident that the thermal history is of outstanding importance for good performance of a TLD system. It is important to note that thermal history does not just start after the first irradiation. It is recommended to initialize TL detectors prior to their first use with a heating program that can be quite extensive. Driscoll lists recommendations for annealing procedures and initialization cycles for different TL materials (Driscoll et al. 1986). For LiF:Mg,Ti it is recommended that a sequence of four annealing procedures be applied prior to the first measurement. A typical usage cycle of a LiF:Mg,Ti detector set is shown in figure 3.43. The steps that involve temperature are shaded. Similarly, thermal history does not end after a complete annealing procedure. If TL detectors are stored after annealing for more than a week, their sensitivity may already be reduced (Marshall 1984). The time between complete anneal and reuse should therefore be short. In other words, a busy TLD department will have more reproducible results. Thermal history has to include the heating and cooling rates (Bos et al. 1992, Julius 1981). Together this will determine background, detection limit, sensitivity, supralinearity, and energy response of the detectors. The typical requirement for the absolute temperature of the high-temperature anneal of LiF:Mg,Ti is 400 ±5 °C. However, it seems that a considerably broader range of temperatures between 360 °C and 440 °C may be used without major impact on the sensitivity, but the accuracy of the low temperature in the second phase of the annealing procedure seems to be more important (Julius 1981, Horowitz 1990). In general, as long as one temperature cycle is exactly equal to the next, all parameters of a particular detector batch have to be evaluated only once for each radiation quality.

3.7.3 TLD Materials

Nature provides us with more than 1000 different TL materials. However, only a few of them are suitable for dosimetry of ionizing radiation. These include lithium fluoride (LiF), lithium borate (Li_2B_4O), and beryllium oxide (BeO), as dosimeters with approximately tissue equivalent atomic number, and calcium sulfate ($CaSO_4$), calcium fluoride (CaF_2), and aluminum oxide (Al_2O_3), as dosimeters with high atomic number. Since most materials require doping for the creation of traps and color centers, the number of TL materials increases because different dopants can be used. Two examples of this (LiF, and $CaSO_4$) are given in table 3.8, which lists the properties of some commonly used TL materials. Also listed are other parameters of interest for clinical photon dosimetry, such as density ρ, effective atomic number Z_{eff}, and the usable range.

In clinical dosimetry lithium fluoride is the most commonly used TL material. It was first produced by Harshaw Chemicals doped with impurities of magnesium (Mg 300 ppm) and titanium (Ti 15 ppm). This type of material is commonly referred to as LiF:Mg, Ti or TLD 100. Therefore, the sensitivity values of different materials are given relative to LiF:Mg,Ti in table 3.8. It should be noted, however, that the given responses can be partially attributed

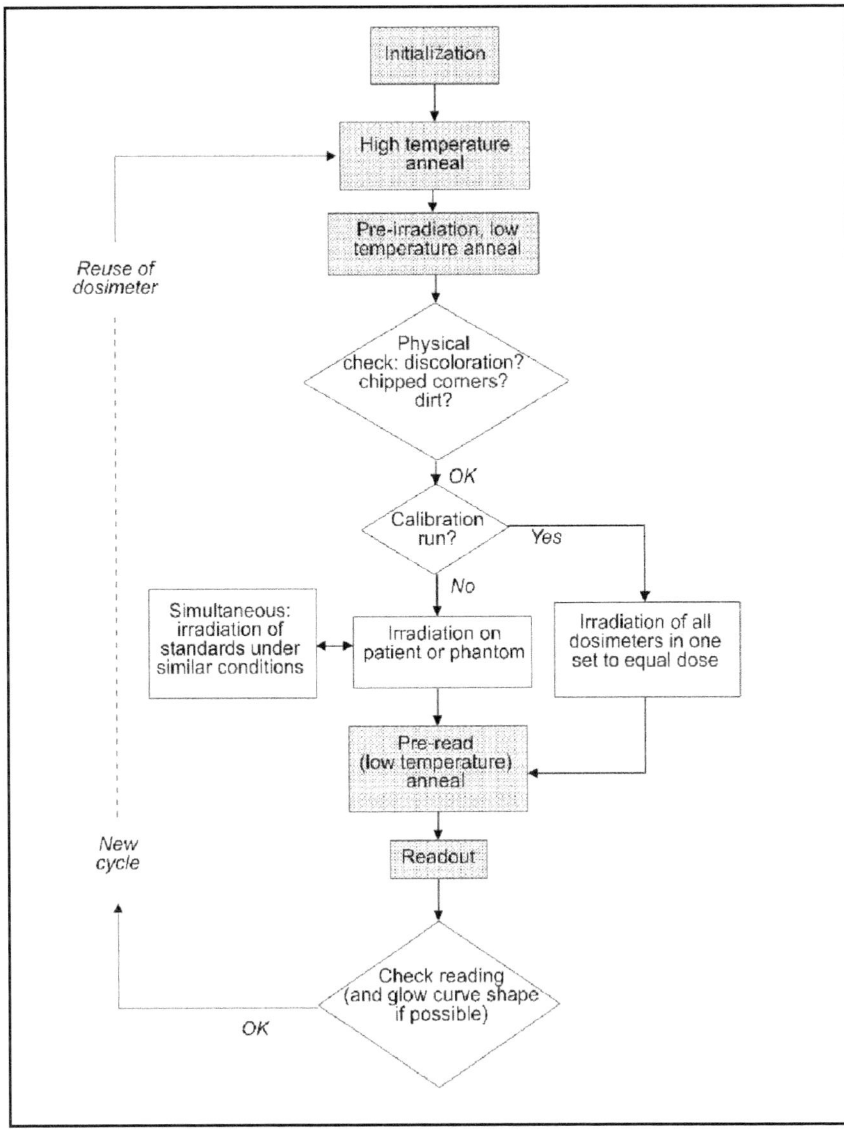

Figure 3.43: Flow chart illustrating the use of a TL dosimeter such as LiF:Mg,Ti. All steps that involve temperature are shaded.

to the reduced sensitivity at wavelengths other than 400 nm with readers optimized for LiF: Mg,Ti.

For absorbed dose experiments a variety of TLD material configurations are available. Table 3.9 lists some configurations and geometries in which LiF:Mg,Ti and other materials are commercially available. Powder is the most convenient form provided a constant sample volume can be dispensed because only one calibration is required for the whole batch of powder. However, the use of powder requires a dispenser and a microbalance to determine accurately the amount of LiF used for a particular dosimetric

Table 3.9: Commercially available forms of LiF:Mg,Ti. The materials marked with a hatch (#) are also available with enriched ^{6}Li and/or ^{7}Li for neutron dosimetry. Please note that this list is not complete and should be used as a guideline only.

Detector Form	*Dimensions*	*Remarks*
(#) Powder	grain size 3–200 μm	flexible use, requires other handling equipment
(#) Extruded ribbons	$3.1 \times 3.1 \times 0.9$ mm^3	the "standard" TL detector form
Rods	length 6 mm, diameter 1 mm	available with round or square cross section
(#) Thin ribbons	$3.1 \times 3.1 \times 0.39$ mm^3	
Extra thin ribbons	$3.1 \times 3.1 \times 0.14$ mm^3	fragile and difficult to handle
Small cubes	$1 \times 1 \times 1$ mm^3	close to a point detector
(#) Hot sintered disks	diameter 4.5 mm, thickness 0.9 mm	
Teflon-based disks	diameter 9.5 mm, thickness 0.4 or 0.13 mm	Teflon-based material cannot be annealed at 400 °C
Ultra-thin Teflon-based disks	diameter 6 mm, thickness 0.02 mm	for interface measurements, cut from a rod with a microtome
Carbon loaded material ("black TLDs")	several disk types are available	useful for surface and interface dosimetry
Cards with multiple elements	different sizes and materials for filtration	for personnel dosimetry
TLD arrays	1.5 mm diameter LiF material in 3 mm spacing	available for laser readout for dose mapping
Doped active LiF material on undoped substrate	like extruded ribbons but with only the surface doped	useful for surface dose measurements

application. Dosimetry with powder is affected by the grain size of the powder, which should be as uniform as possible. Since typically only milligram amounts are used, the results also depend on the accuracy of the balance involved. Due to these disadvantages an increasing number of centers use LiF in solid form as ribbons, rods, or disks. These can be easily packed in small packs encapsulated by thin foil or can be placed inside phantoms in appropriate recesses.

The most commonly used form of TLDs is the square ribbon (3.15×3.15 mm^2 area), which is available in various thicknesses (Kron et al. 1993b) ranging from 0.14 mm to 0.9 mm. Rods are also popular; they typically have a diameter of 1 mm and a length of 6 mm. These are particularly well suited for in vivo dosimetry in cavities because they easily fit into catheters.

LiF:Mg,Ti is also available in various forms in a Teflon matrix (Nash and Attix 1971, Webb 1967). This has the advantage of easy handling, and the possibility of obtaining dosimeters in a wide variety of shapes including ultra-thin disks, as can be seen in table 3.9. However, Teflon-based material cannot be annealed at the recommended temperature of 400 °C. Typically this high-temperature anneal is substituted by thermal treatment at 300 °C for one hour. Teflon-based chips also have the disadvantage of poor light transmission (Gooden and Brickner 1972). This reduces the signal, particularly if the material thickness is 1 mm or more.

For surface dose measurements LiF:Mg,Ti is available as carbon-loaded TL dosimeters (Thomas and Palmer 1989, Ostwald et al. 1995). In these dosimeters carbon is mixed with the thermoluminescent LiF. While the carbon does not affect the initial dose distribution significantly, it affects the amount of light that can leave the TL detector and can therefore be detected by the photomultiplier. The recorded light is emitted only from a shallow surface layer of the chip and, therefore, will reflect only the dose absorbed in this layer. The carbon loading is typically adjusted to achieve an effective depth of measurement of 0.07 mm, which is the recommended depth for skin dose measurements by the ICRU (1985). It reflects the average depth of the basal cell layer, the primary target for skin reactions and the development of skin cancer. Other means of surface dose measurements with TLD include TLD extrapolation (Kron et al. 1993b), the placement of a thin layer of TLD powder on Kapton foil (Harvey and Felstead 1979), or the use of thin active layers of TLD material on an inactive (not doped and potentially carbon loaded) LiF substrate (Horowitz 1993).

Most TL materials have some disadvantages, and the search for the ideal material is ongoing. A most promising material is LiF: Mg,Cu,P (DeWerd and Cameron 1984, Horowitz 1993, Olko et al. 1994, Pradhan 1995, Duggan and Kron 1999, Duggan et al. 2004). LiF:Mg,Cu,P is approximately 30 to 50 times more sensitive than LiF:Mg,Ti without losing its close tissue equivalence. The variation of dose response with photon energy is different from the one of LiF:Mg,Ti. LiF:Mg,Cu,P has an anomalous photon energy response with a minimum relative sensitivity at approximately 80 keV. This does not reflect the fact that the atomic number of LiF (8.3) is higher than that of water, which should result in an over-response at lower energies.

LiF:Mg,Cu,P has as complex a glow curve structure as LiF:Mg,Ti does. Readout is at approximately 240 °C, and a short anneal is recommended at temperatures not exceeding 270 °C. The latter is of paramount importance as LiF:Mg,Cu,P loses its dosimetric properties entirely when heated above 270 °C. The annealing cycle is shorter than in LiF:Mg,Ti, but the reproducibility of all thermal treatments is just as important for good results. An additional advantage of LiF:Mg,Cu,P is the late onset of supralinearity, which has been reported to be in excess of 10 Gy (Horowitz 1993).

3.7.4 TLD Handling and Equipment

Handling of TL Dosimeters

Whatever material is used in a TLD system, great care should be taken when handling the TL detectors. Scratches or contamination of the chip (or rod) surface should be carefully avoided because they reduce the light emitted from the chip. Therefore, all TL material should be handled with vacuum tweezers. These do not need to be complicated and expensive: figure 3.44 shows a simple set of vacuum tweezers with manual suction.

In many TLD applications a large number of TL detectors are grouped together as one set (Gooden and Brickner 1972, Mansfield and Suntharalingam 1976, Mayles et al. 1993, Kron et al. 1993e). Figure 3.45 shows a tray for a set of 50 TL dosimeters made from aluminum. A set should be kept together in a dedicated holder. These holders should have at least 50 recesses to allow the setup of large enough sets of TL dosimeters. Pure aluminum is a good material for annealing trays because its excellent thermal conductivity allows equal temperature distribution to all chips. A lid to the tray is useful to protect the chips from UV radiation, dust, and other environmental influences (e.g., spillage).

If used for in vivo dosimetry, TLD crystals must be packed to avoid contamination and scratching of their surface and to allow labeling. TL dosimeter holders for rods and ribbons are commercially available for most anthropomorphic phantoms. Other phantom measurements do not pose a particular problem for the packing of the dosimeters, since the appropriate recesses can be milled in phantom materials. For use on patients the packaging is more critical. Several methods have been proposed ranging from commercially available capsules to shrink-wrap packaging of single chips (Spanne 1984, Kron et al. 1993a, Mayles et al. 1993). A useful method is the use of small paper snips on sticky tape. They should have about the size of the detector in use, as the detectors are placed on them. The whole arrangement can then be covered by a thin plastic foil that is attached at all sides around the detector to the sticky tape. (See figure 3.46.) The advantages are that individual detectors can be labeled on the paper snips while a thin window facing the radiation in question is maintained.

The TLD Reader

The typical TLD reader consists of two separate components: the heating element and the light detection system. A commercially available reader is depicted in figure 3.47. The carousel can accommodate 50 detectors at any one time. In a busy radiotherapy department, the additional expense for an automated reader is easily made up for by the increasing throughput. A sample changer also encourages more consistency in measurements and quality assurance activities.

Different methods of heating have been employed (Julius 1981, Spanne 1984). The most common technique is the use of a heating planchet employing ohmic heating. The advantage of contact heating is its simplicity and the ease with which temperatures can be varied. However, it has the disadvantage

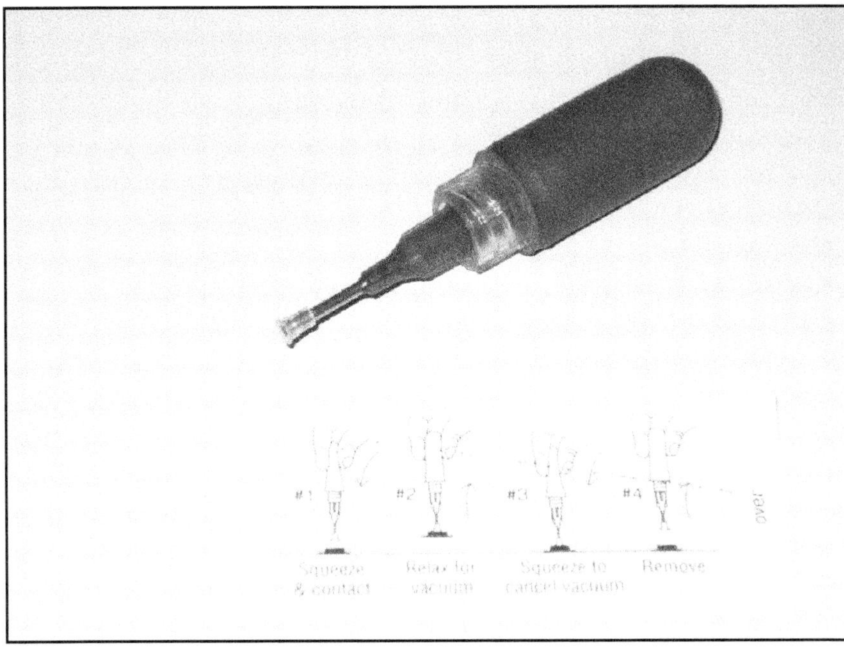

Figure 3.44: A "manual" set of vacuum tweezers. (Courtesy of K. Harrison.)

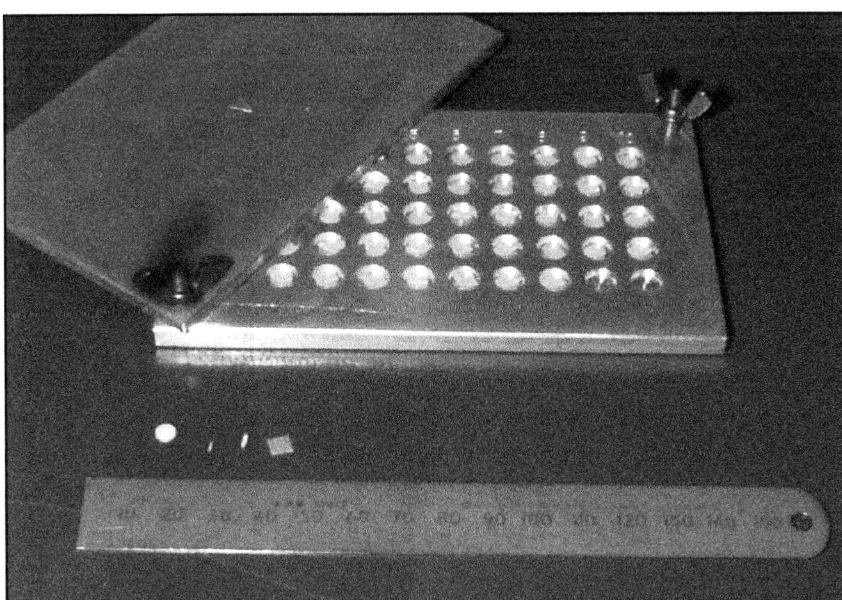

Figure 3.45: Aluminum tray accommodating 50 TL detectors. The tray has a lid and can be used for storage as well as for annealing.

that the thermal contact between planchet and TL material is crucial (and difficult to control) and that there may be a problem of surface contamination of the dosimeter. Most modern TLD readers work with the TL material in an inert gas atmosphere (typically nitrogen) during the readout process. This reduces spurious light emission and therefore the background reading.

The heating cycle for TL detectors can also be provided by a hot gas stream (Botter-Jensen 1978). The reader shown in figure 3.47 employs hot gas read-out. Although this technique is not suitable for TL powder, it has a number of advantages over contact heating, e.g., less danger of contamination and a more uniform heating of crystals.

Most modern TLD readers employ photomultipliers to detect the small quantities of light emitted during TLD evaluation (Spanne 1984). These are usually optimized for a light emission of about 400 nm, which equals the predominant spectral emission of LiF:Mg,Ti. Other TL materials may show a reduced dose response in the same reader, which could primarily be due to a reduced efficiency of the photomultiplier. The PMT tube should have a low infrared sensitivity to avoid an increased background signal due to higher readout temperatures. TLD readers often also employ an infrared filter to minimize the influence of thermal radiation on the photomultiplier reading.

Most modern TLD readers have a built-in reference light source. This reference light can be either placed manually in the position of the TL dosimeter to check the photomultiplier when required or the photomultiplier may be exposed to the light source each time a new TL dosimeter is loaded.

Annealing

In principle, all thermal treatments for TL dosimeters can be done in the TLD reader. However, this practice has a number of important disadvantages:

i) Only one detector can be treated at any one time. In addition to being time consuming, this practice leads to (minor) differences between the thermal history of TL dosimeters in one set.
ii) No long thermal treatments can be performed (e.g., for LiF:Mg,Ti the time recommended for a pre-irradiation anneal at 80 °C is 20 hours (Driscoll et al. 1986)).
iii) Typically temperatures above 400 °C cannot be achieved because of the risk of damage to the photomultiplier.
iv) The reproducibility of the temperature is worse than in a dedicated annealing oven because the thermal capacity of the planchet is far smaller than that of a typical oven.

As the thermal history is crucial for the performance of any TLD system, a dedicated annealing oven is an important part of a TLD setup (McKinlay 1981, Driscoll et al. 1986, Horowitz 1990, Kron 1994). The TLD oven should be able to keep a prescribed temperature within well-specified margins. However, the reproducibility of a temperature cycle is usually of greater importance than the absolute accuracy of the temperatures reached. This includes the heating and cooling rates. In particular, the reproducibility

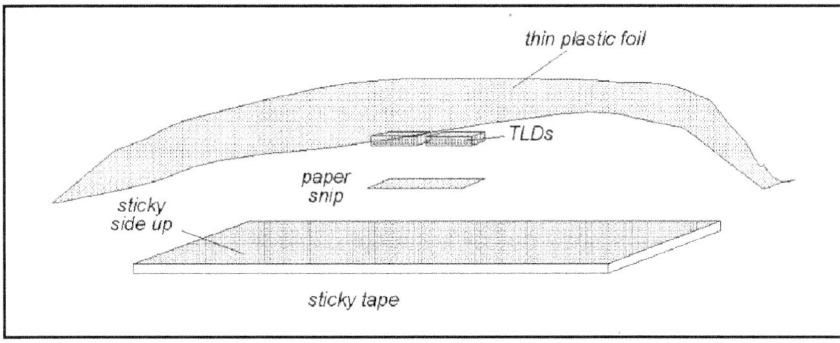

Figure 3.46: Schematic diagram of a method to assemble two detector ribbons for in vivo dosimetry.

of the latter is of great importance for the performance of a TLD system based on LiF:Mg,Ti.

Evaluation/Computing

A real-time display of the TL glow curve is a very useful quality assurance tool. It may be facilitated in the TLD reader itself or in a computer connected to it. The complete glow curve has to be stored to allow glow curve analysis at a later stage. It has been shown that the integral under a glow peak is a more reproducible means of evaluating TL readings than measuring the peak height. For a first evaluation of the TL readings it is also useful

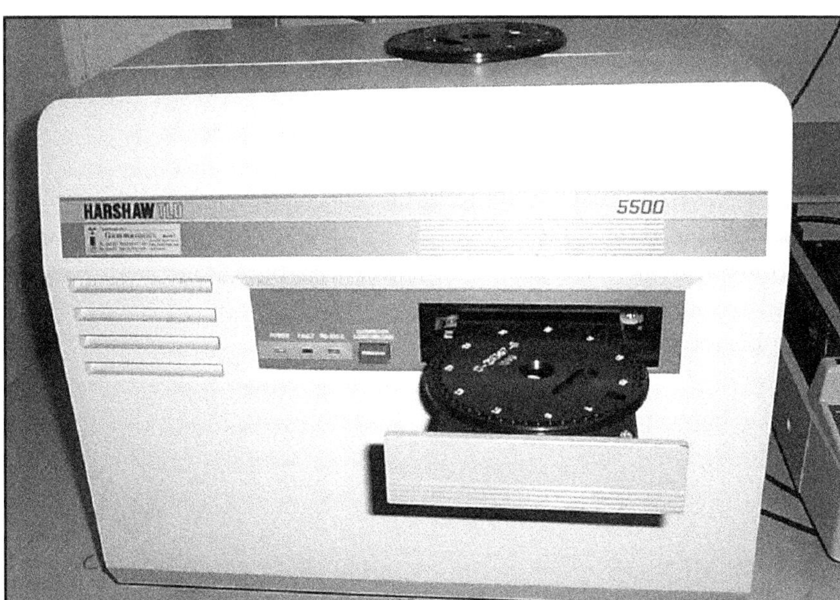

Figure 3.47: An automatic TLD reader (Harshaw 5500).

to be able to define regions of interest around the glow peaks of interest. The total signal in these regions can then be used for the TL analysis.

The great number of TL dosimeters typically used in clinical applications makes a computerized evaluation desirable. For this purpose it is useful to have a direct data transfer from the TLD reader to a (personal) computer. Ideally the whole glow curve should be transferred. However, it may be sufficient to transfer only the integrated light signal under the glow peaks of interest. Even the latter allows the direct evaluation of dosimeters using the reading of detectors exposed to a known dose (standards). This could be done with software provided by the manufacturer of the reader or with a spreadsheet (Wolf et al. 1989). In any case, automatic background subtraction should be made. Since many TLD measurements are repeated to improve the precision of the measurements, a spreadsheet can also be used for statistical evaluation. In any case, this is of importance to quantify the performance of TL dosimeters in regular background and standard exposure measurements.

If the whole glow curve can be transferred to a computer, glow curve analysis can be performed. Different levels of sophistication are possible, ranging from a curve fit according to equation 3.15 for all peaks to an iterative "removal" of peaks reconstructed from their first signal rise. This improves the evaluation of most TL dosimeters by increasing the precision of a single reading. Glow curve analysis is of particular interest in diagnostic radiology or radiation safety where small doses have to be assessed (Horowitz 1995).

Table 3.10 summarizes the considerations regarding the setup of a clinical TLD system based on solid forms of TLD material (crystals, ribbons, rods, or Teflon-based material). In practice, the main advantages of TLD are the small size of the detectors and the fact that no cables are attached to the detector during irradiation. As such, the main applications of TL dosimeters s in radiotherapy are for in vivo dosimetry (Kron et al. 1993) and measurements in anthropomorphic phantoms. Not surprisingly, most anthropomorphic phantoms such as the one shown in figure 3.13 have recesses specifically designed for TLD. TLD is also widely used for dosimetric intercomparisons (Kron et al. 2006) such as verification of calibration performed by the International Atomic Energy Agency (IAEA) (Izewska and Andreo 2000, Izewska et al. 2002) and the Radiological Physics Center in Houston, Texas (Aguirre et al. 2002; compare also http://rpc.mdanderson.org/rpc/; accessed 4/2007). An intercomparison can also be combined with the use of an anthropomorphic phantom to perform a level III dosimetric intercomparison that verifies the whole treatment chain from imaging for treatment planning to the delivery of radiation (Kron et al. 2002a). Due to the fact that TL detectors can integrate low doses over a long period of time, they are also commonly used in dose monitoring for radiation protection. This will be discussed in chapter thirteen.

Table 3.10: Considerations for the setup of a clinical TLD system using solid TLD detectors

	Equipment	*Comments*
TL dosimeters	LiF chips, rods	at least 50 dosimeters of any one type should be used together; more should be purchased if possible to obtain dosimeters from one production batch; while individual dosimeters are evaluated they should not be marked—they can be identified by their position in the annealing tray
Handling	trays	should have about 50 recesses to accommodate dosimeters; preferable material: ceramics or Pyrex, but also pure aluminum may be used; for storage the trays need marks for chip positions and a lid
	vacuum tweezers	essential to avoid scratching of the dosimeter surface; a plastic or Teflon tip should be used
	other equipment	"tissue equivalent" holder with 50 or more recesses for uniformity checks; scissors, scalpel, shrink-wrap, and double-sided tape are useful for preparation of TLD packs for use on patients
	cleaning agents	recommendations vary—best if not required; alcohol may be used for LiF
Reader	heater	contactless heating in a hot nitrogen stream produces better results for chips and rods than contact planchet heating—careful adjustment of the dosimeter position in the stream is necessary; in general, contactless heating is not as versatile as planchet heaters in regard to altering temperature profiles and accommodating different TLD materials
	photomultiplier	voltage as low as compatible with application; should be left on all the time; a reference light source for photomultiplier check is recommended
	electronics	real-time visualization of the glow curve is very useful for QA (as is simultaneous display of the temperature profile); a parallel port should be provided to transfer data to and from a personal computer
Evaluation	computer	compatible with other computers to allow data transfer in other programs—e.g., graphics or spread sheet; hard copy of results (glow curves and integrated readings) should be possible for records
Annealing	dedicated oven	pre-irradiation and preread cycles have to be available; oven should accommodate several annealing trays; should preferably allow free temperature programming without compromising reproducibility; reproducibility more important than accuracy; for certain materials an inert gas atmosphere may be required

3.7.5 Optically Stimulated Luminescence Dosimetry

Optically stimulated luminescence (OSL) is based on similar physical phenomena as TLD. Instead of heat, light can be used to provide the energy required to release trapped electrons back into the conduction band as seen in figure 3.39 (Rao 1993). Potential advantages of OSL over TL are the speed of the read-out process and the fact that light can be applied localized in a crystal, thereby stimulating only parts of it and providing potentially for high spatial resolution. It has also been shown that OSL can be used for real-time dosimetry (Aznar et al. 2004).

The traditional setup of OSL includes the excitation of the crystal with a light source and the recording of the luminescence light under 90 degrees using heavy filtration to avoid any influence of the stimulating light on the recorded light. This technique has primarily been employed for dating and has not yet found a lot of applications in radiotherapy. In a newer development, pulsed OSL (or POSL) (McKeever et al. 1996), the sample is excited using pulses from an Ar-ion laser. After an intense light pulse of some 30 ms, a shutter isolates the laser and another one exposes the photomultiplier to the luminescent light from the crystal. The most promising material for POSL is Al_2O_3:C, which is also as a highly sensitive TLD material. McKeever et al. (1996) found that the light emitted by the same crystal using POSL was more than a factor of 3 times more intense than after thermal heating. Therefore, POSL with aluminum oxide is currently one of the most sensitive solid-state dosimetry techniques available. At present, the most important commercial application is in personnel dosimetry.

3.8 Film Dosimetry

The use of radiation-sensitive film is one of the earliest detection methods of x-rays (see e.g., Trevert 1896). Probably the first documented effect of x-rays was the "strange" picture of two coins on a radiographic plate observed by Professor Goodspeed in Pennsylvania in 1890 (Webster 1995). However, Prof. Goodspeed did not realize what had exposed his radiographic plate, and it was left to Wilhelm Röntgen to discover x-rays in November 1895 (Röntgen 1895). Incidentally, Röntgen also used film extensively to document his discovery and included a copy of an x-ray film taken of his wife's hand with the first reprints of his famous paper on the discovery of x-rays sent to colleagues and friends on January 1, 1896 (Foelsing 1995).

Film dosimetry is currently undergoing a dramatic change. This is prompted largely by diagnostic radiology phasing out film for imaging in favor of digital radiography. Therefore, manufacturers reduce their film production and film processing becomes less available. This and the availability of new 2-D dosimetric media such as radiochromic film and computed radiography cassettes make it likely that conventional film will not be available in the future. Nevertheless, radiographic film will be discussed first in this section because of its historic significance and the fact that it has been widely used as a detector for commissioning of dynamic wedges (Klein 1996, Beavis et al. 1997), dynamic treatments (Kubo and Wang 2000) stereotactic radiotherapy (Paschalis et al. 2006) and, more recently, IMRT (Bayouth et al. 2003, Yan et al. 2005).

However, many radiotherapy departments are at present in the process of replacing radiochromic film. In the context of portal imaging this has already widely occurred with the introduction of electronic portal imaging (EPI) (Langmack 2001, Antonuk 2002, Herman 2005). These devices have also found limited applications in dosimetry, mainly for relative dosimetry in IMRT field verification (e.g., Renner et al. 2003, Warkentin et al. 2003, Vieira et al. 2006). However, also for other QA applications, EPI may be useful. Examples are MLC QA (Baker et al. 2005, Parent et al. 2006, Mohammadi and Bezak 2007), dynamic wedge (Greer and Barnes 2007), and isocentricity (Liu et al. 2004). The main replacement for dosimetry with radiographic film though has been radiochromic film, which will be discussed in section 3.8.2. Other options may arise from developments in diagnostic radiology such as computed radiography (CR) screens. This will be briefly presented at the end of this section.

3.8.1 Radiographic Film

Radiographic film typically consists of a clear polyester base that is coated on one or both sides with a radiosensitive emulsion. Figure 3.48 shows a cross section of a double emulsion x-ray film. The sensitive layer normally consists of silver halide (most commonly bromide) crystals (diameter 0.2–2 μm) embedded in gelatin. The absorption of ionizing radiation causes the following reaction (simplified):

$$\text{photon} + Br^- \rightarrow Br + e^-$$
$$e^- + Ag^+ \rightarrow Ag.$$

The elemental silver is black and produces the latent image. During the development of the film, other silver ions are reduced in the presence of silver atoms. Therefore, if one silver ion is reduced in a silver bromide crystal, all silver in this crystal (or grain) will be reduced during development. The rest of the silver halide (in undeveloped grains) is then washed away from

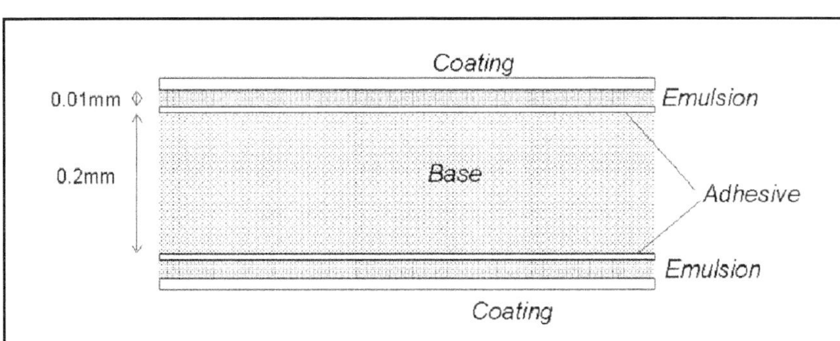

Figure 3.48: Cross section of radiographic film.

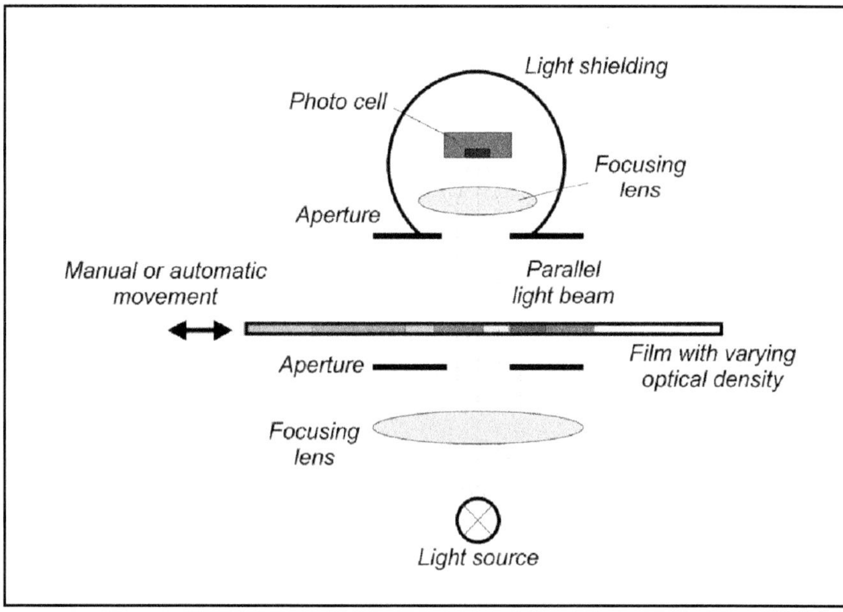

Figure 3.43: Schematic diagram of a film densitometer. (Adapted from Van Dyk 1999.)

the film during fixation, and only the areas of the film which were hit by radiation appear black.

The blackening of the film can be quantitatively evaluated using optical densitometry. The measurement setup of a densitometer is shown in figure 3.49. The light used in densitometers is commonly in the infrared range to avoid interference with the room lighting. While film has theoretically a very good spatial resolution, in practice this is limited by the aperture of the light-collecting device in the densitometer. This is typically of the order of 2 mm diameter. As a result, film is now mostly evaluated using film scanners such as the one shown in figure 3.50. These devices digitize the whole film (or specified section thereof). While earlier scanners had some limitations for dosimetry due to the limited grey scale resolution of 8 bits ($2^8 = 256$ density levels), there are now scanners available which offer 16-bit contrast resolution. This is more than adequate for film dosimetry.

The optical density, OD, can be calculated from the ratio between the intensity of the light emitted from the photodiode Φ_0 and the one detected by the photodetector on the other side of the film Φ according to:

$$\mathrm{OD} = \log_{10}\left(\frac{\Phi_0}{\Phi}\right). \tag{3.18}$$

In clinical practice, the optical densities are of the order one to four. The advantage of this definition is that optical densities are additive. Two layers of film (or two exposures of the same film) produce a sum of optical densities:

$$\mathrm{OD} = \mathrm{OD}_1 + \mathrm{OD}_2 \tag{3.19}$$

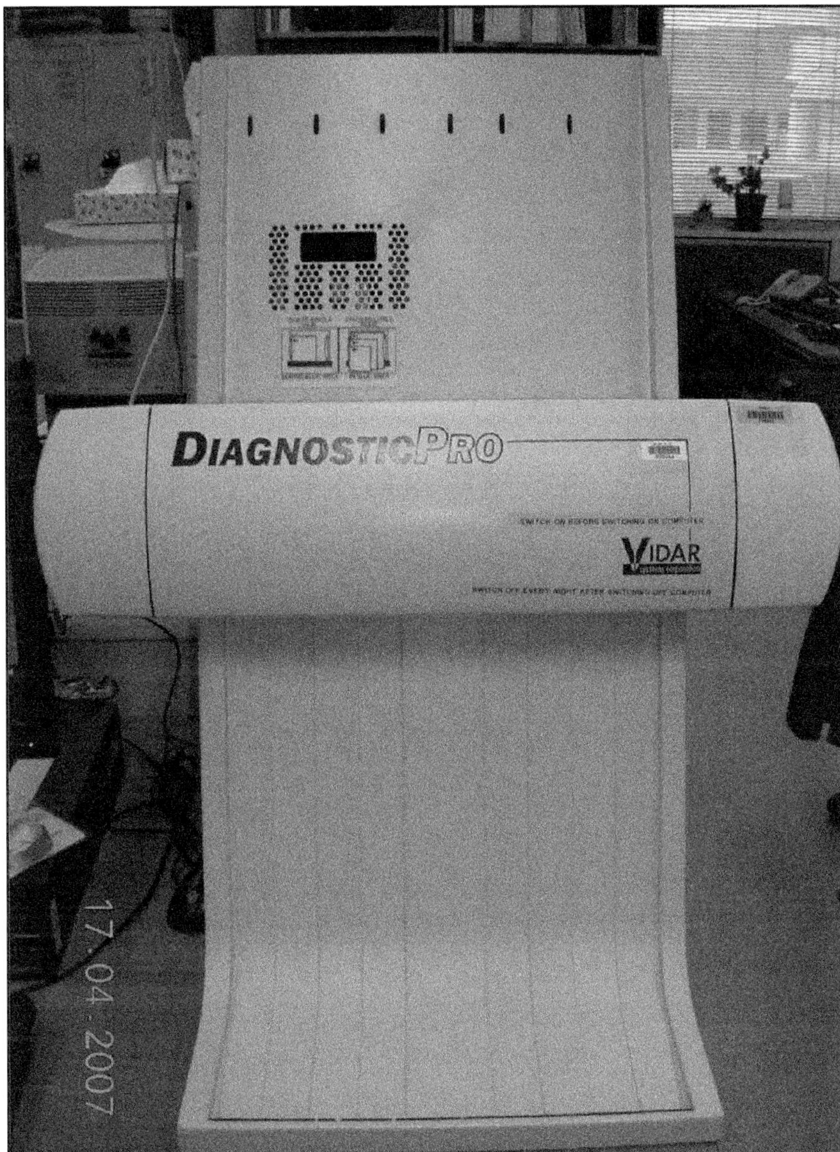

Figure 3.50: Photograph of a commercial film scanner for dosimetry (VIDAR).

One has to take into account that the nonexposed film will already have a finite optical density. This is due to some spontaneous reduction of silver and background radiation (a contribution called *fog*) and light absorption in the film base (called *base*). Using optimal processing conditions can reduce the fog, but some remaining background density (base + fog) has always to be subtracted from the optical density readings after film exposure. The optical density of any given film depends on the processing conditions. Therefore, it is important to control the processing conditions, in particular the temperature of the developer and the time, very accurately. With a typical automated processor available in a radiotherapy department, one can expect to achieve a reproducibility of about 3% over a period of one day. An interesting complicating factor may be the Schwartzschild effect (Djouguela et al. 2006) where the overall optical density is reduced when the dose is delivered in protracted fractions. In clinical practice these effects are typically small enough to be ignored.

For accurate quantitative measurements it is advisable to process a whole batch of films at one time. This could be done with a fog and a film exposed to a standard dose developed every three films to control processor variations. It may also be beneficial to develop several old and unusable films prior to the series of films of interest to ensure that the chemicals are replenished and well mixed.

Figure 3.51 shows the optical density of radiographic film as a function of the radiation dose. This is a double logarithmical plot taking into account the definition of optical density. The optimum measurement conditions are in the "linear part" of the curve. The extent of the linear part of the curve depends on the film type, with more uniform silver bromide grain sizes resulting in better dose response. For a limited time a type of radiographic film, CEA film, had been available (Cheng and Das 1996), which featured uniform silver halide grain size. More expensive than traditional radiographic film, CEA film was packaged in a shiny, vacuum-sealed polyester envelope that could be submersed in water. All this made the film highly suitable for radiotherapy dosimetry (Cadman 1998).

However, the film type that had the most success in radiotherapy has been Kodak EDR2 film, which combines low sensitivity and a relatively linear dose response (Olch 2002, Ahmad 2006). The low sensitivity allows the use of conventional radiotherapy dose fraction sizes (2 Gy and more) on phantoms to verify IMRT dose distributions (Georg et al. 2003, Tangboonduangjit et al. 2004, Yan et al. 2005).

However, despite the more linear dose response relationship, dose linearization programs are commonly employed to extend the linear part of the curve (Mota et al. 1990, Ebert and Kron 1994). This technique makes use of the observation that dose-response curves for a film type are similar in shape and referred to as density-to-dose curves (DDC). This is also approximately valid for different irradiation and processing conditions. One can then fit these curves using a polynomial. Alternatively, a look-up table, and density-to-dose conversion can be implemented over the dose range of interest.

Film is available in different sensitivities to radiation. This is called the speed of film, with a fast film requiring a lot less dose than a slow one to develop to the same optical density. While in diagnostic radiology a fast film is preferable to avoid unnecessary exposure to the patient, film used for the dosimetry of therapeutic x-rays is usually very slow. In a radiation field where the smallest unit equals about 1 cGy, one clearly requires film that develops into the optimum optical density region after exposure to some tens of cGy. Films are commercially available for radiotherapy dosimetry from most manufacturers such as Kodak, Dupont, and Agfa. These films are individually prepackaged in lightproof paper for convenient use without the need for cassette loading. A prepacked film is shown in figure 3.52. It can be

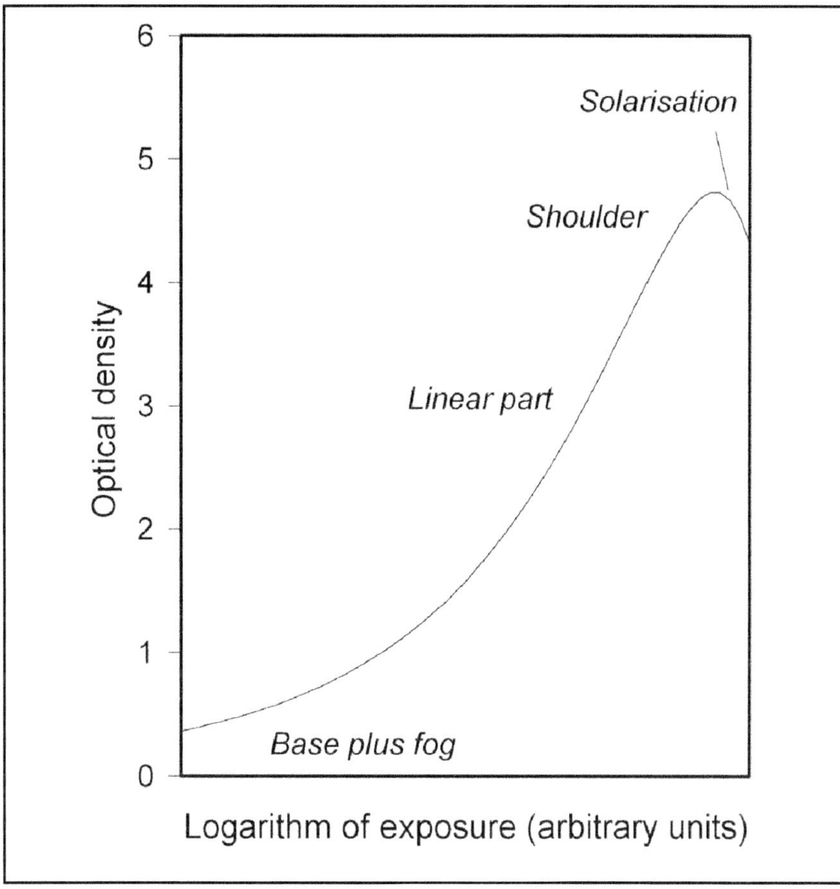

Figure 3.51: Schematic plot of optical density versus dose in radiographic film.

used directly in slab phantoms without darkening the room or using red light. As prepackaged film cannot conform to irregularly shaped surfaces such as encountered in anthropomorphic phantoms, it may sometimes be necessary to cut the film in the darkroom to the required shape. The edges can then be sealed using thick black tape. The edges can then be sealed using thick black tape.

When prepackaged film is used, care should be taken to avoid air bubbles within the packet. It is good practice to pierce the film with a pin (a personal dosimeter or an identity card with a pin may be handy) at one side and slide the phantom material over it from the other side pressing out the air. The prick holes can also later be used for identification of the film orientation. (compare figure 3.54).

A problem of film for photon dosimetry is the high atomic number of the active material. This leads to a distinct over-response to low-energy x-rays. Film is approximately 20 times more sensitive to 40 keV x-rays than to 1 MeV x-rays, and the ratio of mass energy absorption coefficients for silver bromide to water is 140:1. Because of this and the many parameters that

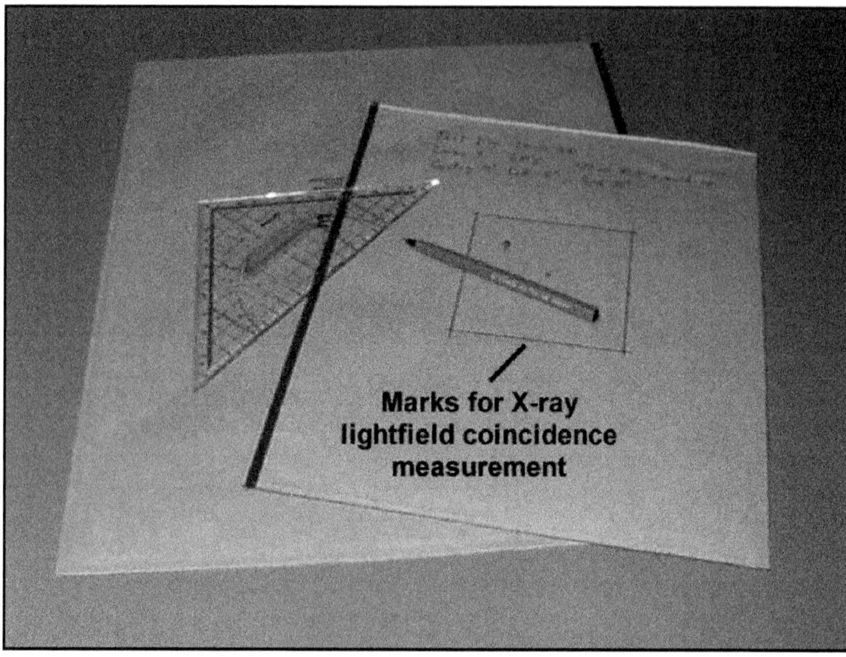

Figure 3.52: Prepacked radiographic film in two different sizes. Shown are marks for an x-ray light field coincidence measurement. (Adapted from Van Dyk 1999.)

influence film blackening during development, film is not well-suited for absolute dosimetry. On the other hand, film is well suited for dosimetry of megavoltage electrons since the stopping power ratio between film and water is relatively constant. Therefore, film is particularly of interest for electron dosimetry (Shui et al. 1989, van Battum and Huizenga 1990), where it has also been used in inhomogeneous physical phantoms (el-Khatib et al. 1992).

The main advantages of film over other types of dosimeters are:

i) The spatial resolution, which can be significantly better than 1 mm if an appropriate film scanner is used. Theoretically, only the grain size (about 1 μm) sets a lower limit for the resolution.
ii) The availability of a 2-D dose distribution map. An example can be seen in figure 3.53 that shows the depth dose distribution in a 6 MV x-ray beam. From one exposure it is possible to obtain depth dose curves as well as profiles at any desired depth.
iii) The integrative nature of the dose acquisition. This is an essential feature for the dosimetry of dynamic treatment modalities such as dynamic wedges, arc, and rotation therapy. The strong energy dependence of film, however, means that care must to be taken when interpreting the optical density-to-dose conversion. Any change in the beam spectrum with depth or outside the edges of the beam (where scattered radiation of low energy may be present) can lead to distortion of results in this region.

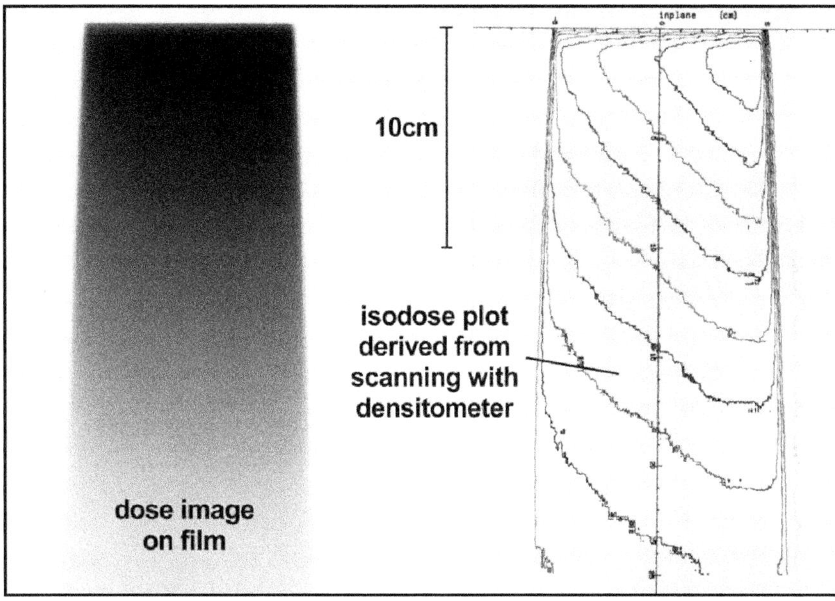

Figure 3.53: Dose distribution in a wedged x-ray field (6 MV, 45-degree wedge, 10 × 10 cm^2 field size). The film on the left was exposed "edge on" with the field axis parallel to the film plane. The dose distribution on the right was derived from a densitometer scan of the film after correction for dose-to-density variations with dose.

Collection of central axis depth dose data and off-axis profiles using ionization chambers is quite laborious, as after each discrete reading is taken the probe is repositioned within the phantom and another reading is recorded. This becomes particularly laborious (or even impossible) when the beam is changing dynamically over the course of delivery, such as with treatment arcs or dynamic wedges. An alternative technique is to sandwich a radiation-sensitive film between phantom material and expose this parallel or perpendicular to the beam central axis. In this way a 2-D profile of the x-ray dose distribution is recorded on the film. This has become a very important method to verify dose distribution in IMRT fields (Budgell et al. 2005). An example for this is shown in figure 3.54. The dose distribution is from an intensity modulated arc therapy treatment of a prostate cancer case (Wong et al. 2002, 2005).

The perturbation of the radiation flux by the high atomic number silver film emulsion is small due to the very thin layer of this material. However, to avoid any problems with fluence perturbation when exposing film parallel to the beam direction (Suchowerska et al. 2001), one can angle the film by about 2° against the plain of the depth dose. This is more a problem in electron dosimetry and will affect the depth dose distribution only slightly (and can be corrected for using trigonometry). It avoids any significant perturbation of the fluence.

In the context of radiographic film it may also be useful to discuss the difference between "quantitative" and "qualitative" dosimetry. This is not to be confused with absolute and relative dosimetry; these both are quantitative

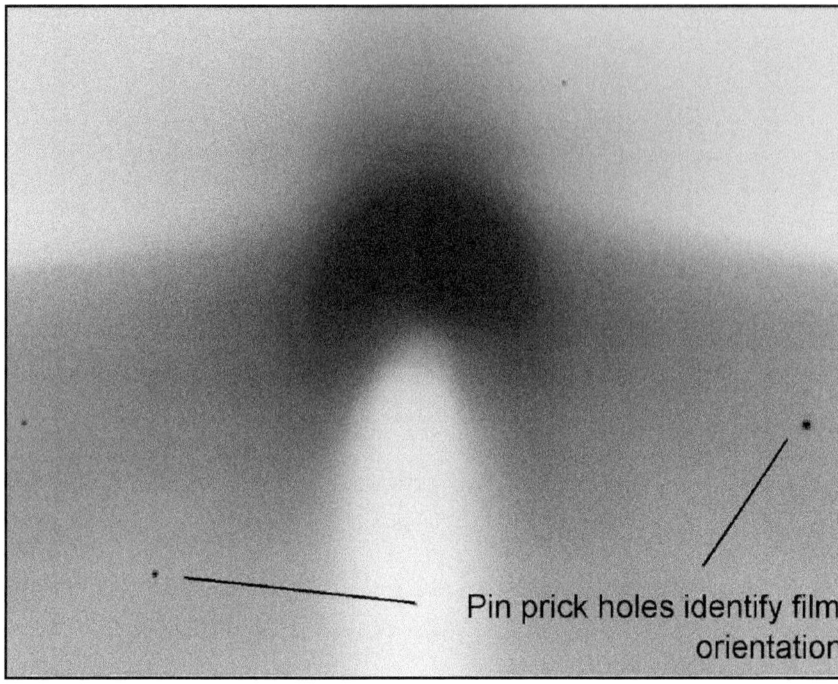

Figure 3.54: Dose distribution from an IMRT prostate treatment recorded on film in a phantom such as one shown in figure 3.10b. (Courtesy of E. Wong.)

applications of dosimetry. Strictly speaking, qualitative dosimetry is a contradiction in terms—dosimetry always intends to quantify the dose. However, the term is sometimes used to describe the quantification of spatial relationship between high-dose and low-dose regions. In the context of film this is using the spatial resolution of film without actually evaluating the dose information.

Many applications of film in radiotherapy are of this nature. Examples are:

i) Port films that determine the position of an x-ray film in relation to the patient's anatomy during treatment.
ii) The check of x-ray/light field coincidence is a very important part of quality assurance on a linear accelerator. This can be performed by marking the light field edges (and the center of the collimator rotation) on a prepackaged film by impressing the edges of the light field with a pen or ballpoint pen on the emulsion (compare figure 3.52). If the film is then exposed to x-rays, one can see x-ray (dark areas) and light field (white lines where the emulsion has been compressed) in one film. Alternatively, a template with radiopaque markers that are aligned with the light field can be used.
iii) Film is ideally suited to determine the center of the x-ray field by exposing spoke patterns. For this technique one of the collimator

jaws is completely opened, and the other one is nearly completely closed. With collimator or gantry rotation the film can be exposed to this narrow x-ray field under many different angles, revealing the actual center of the x-ray field (Khan 2003), as also discussed in chapter eleven.

iv) Film can easily be cut to any desired shape that can then be exposed (with appropriate light exposure prevention) in an anthropomorphic phantom. This gives a useful assessment of the dose distribution achieved in a particular treatment technique. The films can be evaluated either quantitatively or qualitatively.

3.8.2 Radiochromic Film

A more recent development for dosimetry of high radiation doses is radiochromic film. The importance which radiochromic film has achieved over the last years is demonstrated by the fact that the American Association of Physicists in Medicine (AAPM) has convened a task group to report on the issue. The findings of AAPM Task Group 55 were recently published and provide a good starting point for dosimetry using radiochromic film (Niroomand-Rad et al. 1998). Another excellent reference is a review article of Butson et al. (2003).

The most commonly used type of radiochromic film, Gafchromic film, is the trademark of GAF Chemicals Corporation, Wayne, New Jersey, now part of International Specialty Products (ISP) Technologies. However, many more different types of film are available and several new developments feature film with higher sensitivity (Butson et al. 2002, Dini et al. 2005, Soares 2006), some of them also suitable for dosimetry in diagnostic radiology (Butson et al. 2006, Watanabe et al. 2006). Over the years a variety of different films have been marketed and described. The web page of ISP provides details on many of the films and gives links to references (http://www.ispcorp.com/products/dosimetry/content/gafchromic/index.html; accessed April 2007). While sheets of radiochromic film have traditionally been relatively small, as a "film" replacement, radiochromic film has now also become available in larger sheets (Gafchromic EBT, up to dimensions of 36 × 42 cm). This makes it an interesting option for film replacement despite its higher costs.

Gafchromic MD-55 exists in two versions, as illustrated in figure 3.55. MD-55-1, the original film consists of a 15 μm thick radiochromic layer on a 67 μm thick polyester base. As the emulsion was directly exposed to environmental influences, it was prone to variations in thickness and scratches. Therefore, the newer MD-55-2 film employs a dual-emulsion structure that is entirely sealed from the environment, as can be seen in the cross section of figure 3.55. Without irradiation radiochromic film appears clear, transparent, and colorless. On irradiation it produces a very fine-grained blue color due to a solid-state polymerization process. The color change has two maxima at wavelengths of some 610 and 670 nm, which can be evaluated using a densitometer. However, as the absorption peak may shift slightly with radiation dose, the wavelength has to be carefully monitored to obtain a reproducible dose response curve (Klassen et al. 1997).

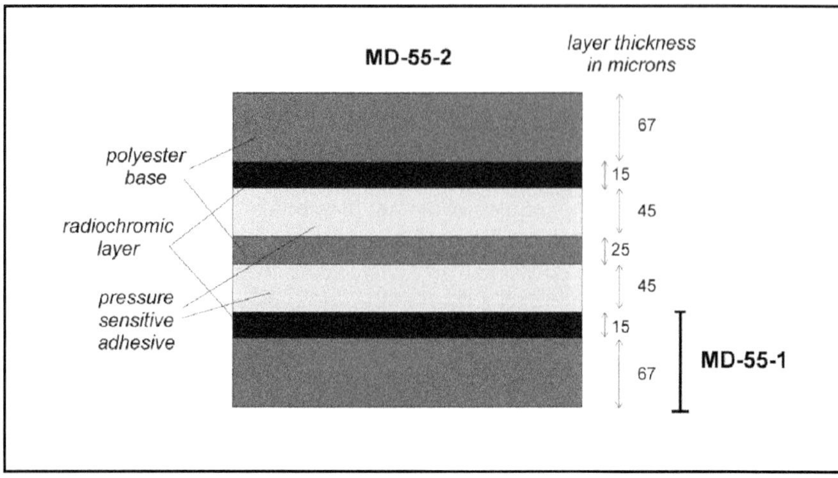

Figure 3.55: Cross section through radiochromic film (ISP: Gafchromic Md-55). The figure indicates dimensions and composition of two variants of the film: a single- and a dual-emulsion type. (Adapted from Van Dyk 1999.)

When using radiochromic film a number of points should be considered:

i) Radiochromic film is self-developing: no developer or fixer is required. However, it takes several hours for the color change to stabilize sufficiently for evaluation (Chu et al. 1990, Butson et al. 1996b). For MD-55-2 film the waiting period must be at least 1 hour to ensure reasonable stability of the optical density (Butson et al. 2003a). The variation of optical density with time is also temperature dependent and typically shows an increase in optical density that can be described by a logarithmic function (Klassen et al. 1997).
ii) Excess humidity at temperatures above 25 °C should be avoided.
iii) Ultraviolet light may cause a color change without exposure to ionizing radiation; this has to be taken into account when light is used for evaluation (Butson et al. 2003a).
iv) The measured optical density in all radiochromic films depends on the wavelength of the light used for the evaluation (Butson et al. 2003a).
v) The measured optical density in MD-55-2 film depends on the polarization of the light used for evaluation (Klassen et al. 1997). Therefore, orientation and alignment of the film should be identical for calibration and measurement. In addition to this, interference patterns due to multiple reflections as a result of changes in refractive index along the light's path have been observed (Newton's rings, Butson et al. 2003a).
vi) The thickness of the radiosensitive emulsion may vary within a sheet of film. Due to the manufacturing process this variation is particularly noticeable in one direction only (Butson et al. 2003a).

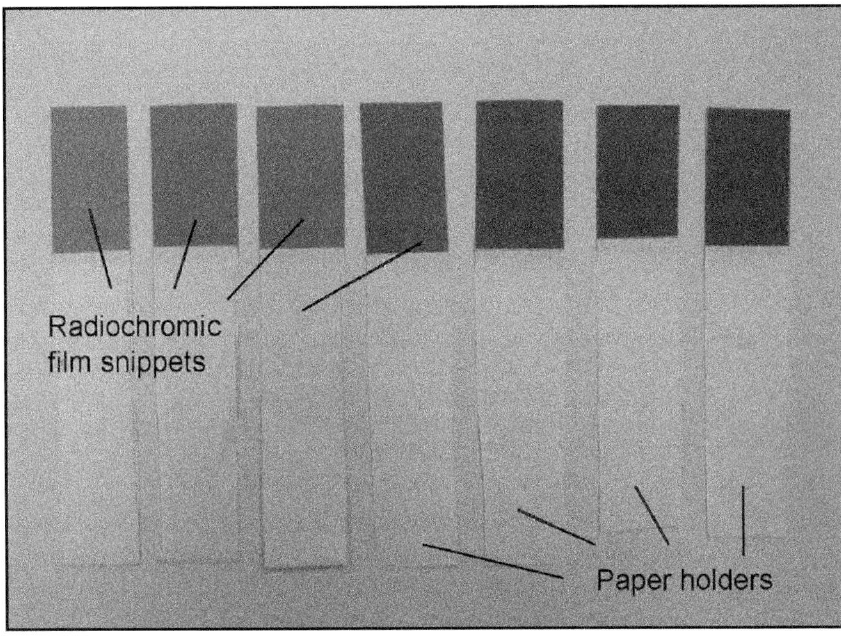

Figure 3.56a: International Specialty Products (ISP) Technologies Inc., Gafchromic EBT QD+ film exposed left to right to 10, 30, 50, 70, 90, 110, and 130 cGy.

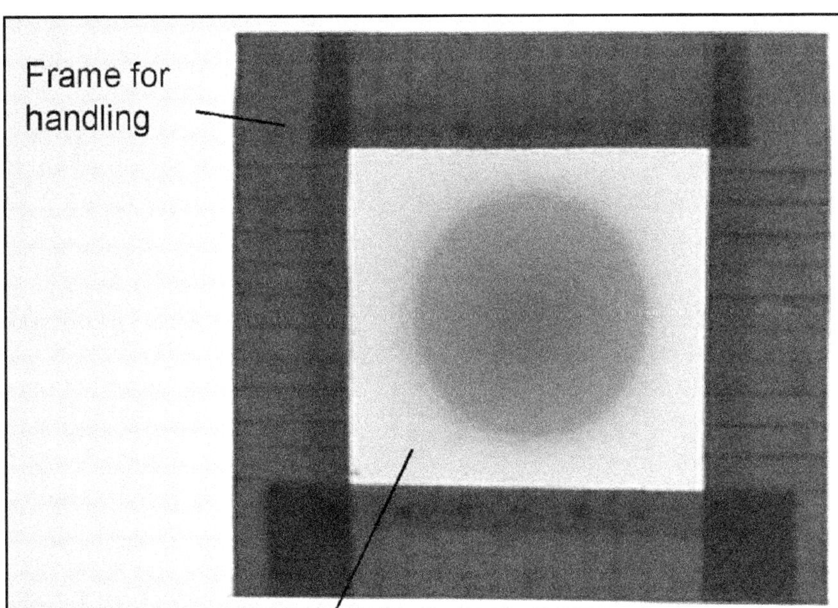

Figure 3.56b: Circular dose distribution from a brachytherapy source acquired with radiochromic film. (Reprinted from Kron 2002b with permission of IOP Publishing.)

vii) Radiochromic film is a relative dosimeter and dose standards must be included in the evaluation. A double irradiation method (Klassen et al. 1997, Niroomand-Rad et al. 1998) seems to be well suited for this. In this type of exposure the film is scanned prior and after exposure to a uniform dose. The difference between the two films provides a measure of the dose response in each pixel of the film. After this, the film is exposed to the radiation field to be measured. The resultant optical density map is evaluated using the dose response determined from the uniform exposure.

viii) The dose response is slightly curved with reduced sensitivity at higher doses. A single exponential fit may be employed to correct for this.

ix) No dose rate or fractionation dependence has been reported (Butson et al. 2003a).

x) Gafchromic film is not entirely tissue equivalent. At low photon energies around 25 keV its dose response is only about half the one of water (Muench et al. 1991, Kron et al. 1998).

xi) Sensitivity can be increased by using multiple layers of film (Butson et al. 2003a).

The first application identified for radiochromic film was the monitoring of dose in food irradiation where doses of up to 30,000 Gy are delivered (Saylor et al. 1988). Consistently with this, the first type of Gafchromic film produced had a low radiation sensitivity and a dose of some 1000 Gy was required to produce an optical density of 1 (Muench et al. 1991). The dose response of Gafchromic film has improved considerably since then and an optical density of 1 in MD-55-2 film can be achieved using a dose of 10 Gy (Klassen et al. 1997). Provided considerable care is taken with environmental conditions, it is possible to achieve a reproducibility of 1% using a double irradiation method (Klassen et al. 1997). This makes the film well suited for applications where high spatial resolution and tissue equivalence are required. Examples for this are brachytherapy dose measurements (Pai et al. 1998), surface/skin dose measurements in megavoltage photon beams (Butson et al. 1996b, 1998), stereotactic radiotherapy fields (Lee et al. 2006) and of course IMRT dose verification (Tangboonduangjit et al. 2004, Watanabe 2006, Zeidan et al. 2006).

3.8.3 Computed Radiography

An interesting recent development is computed radiography (CR). CR should not be confused with digital (or direct readout) radiography where the image is directly acquired using a silicon-based device, similar to modern electronic portal imaging devices (EPIDs) (Neitzel 2005). CR technology utilizes photostimulating storage phosphor in an imaging plate as film replacement (Rowlands 2002). The image plate consists of a barium-based phosphor that stores the energy deposited by radiation and can be read out using optical stimulation by a focused laser (Olch 2005). It is as light sensitive as radiographic film and is usually handled in a cassette similar to film. Figure

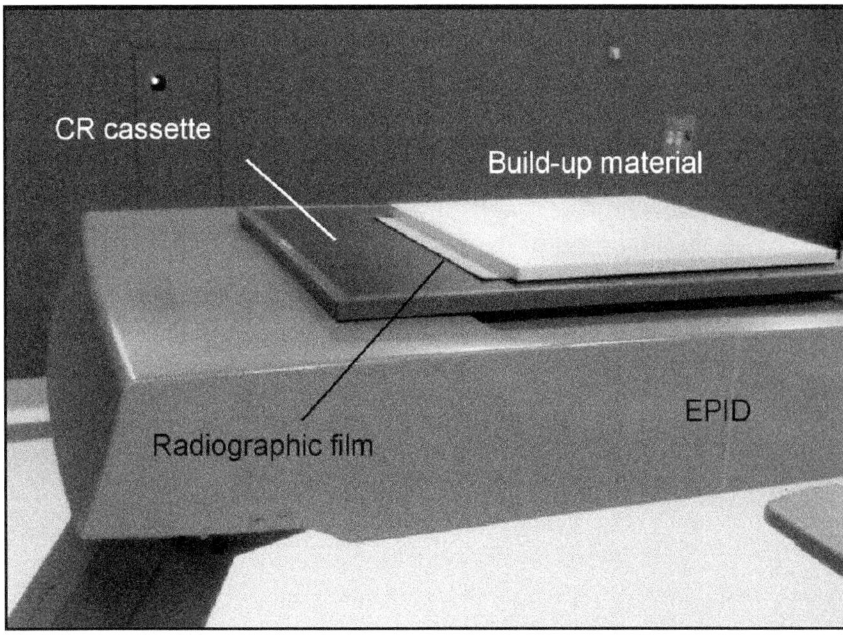

Figure 3.57: Setup for an intercomparison of EPI, CR, and radiochromic film.

3.57 shows such a cassette during testing. While cassettes were originally developed for diagnostic radiology, several manufacturers now offer cassettes for radiotherapy applications. These are similar to port-film cassettes lined with a high-density material such as lead or tungsten to improve the detection of megavoltage photon quanta.

The image plate is read out using a laser scanner in some respects similar to the OSL detectors discussed in section 3.7.5. The readout produces a digital image with similar features as radiographic film. The advantages are typically faster readout (1 minute or less for a 14 × 17 in. plate) and adjustable contrast and sensitivity. The spatial resolution is still less than radiographic film; however this is of no concern in megavoltage dosimetry where the secondary or primary electrons have considerable range anyway. Most readers, such as the one shown in figure 3.58. are designed to provide images that resemble the diagnostic film used by the radiologists. As such, the raw image information (point-by-point light output from the image plate as the laser scans across) is commonly already processed before being displayed. This must be taken into account when attempting to use the information even for relative dosimetry.

There are now a few groups that study the usefulness of CR for a variety of applications in radiotherapy including portal imaging (Whittington et al. 2002, Yamada and Murase 2005, Wu et al. 2005) to quality assurance and dosimetry (Olch 2005, Li 2007).

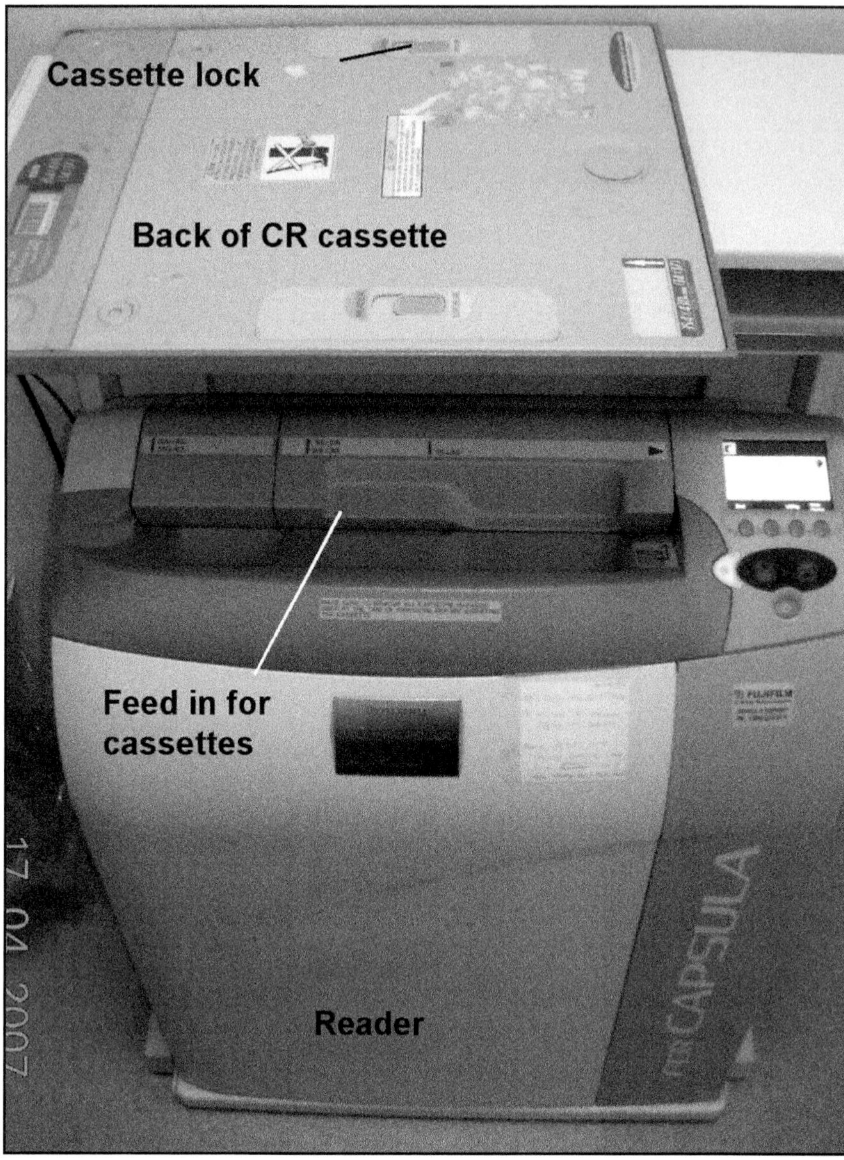

Figure 3.58: A CR reader (Fuji). Also shown is a CR cassette on top of the reader. Visible is the lock, which ensures good contact between the screen and the CR sheet during exposure.

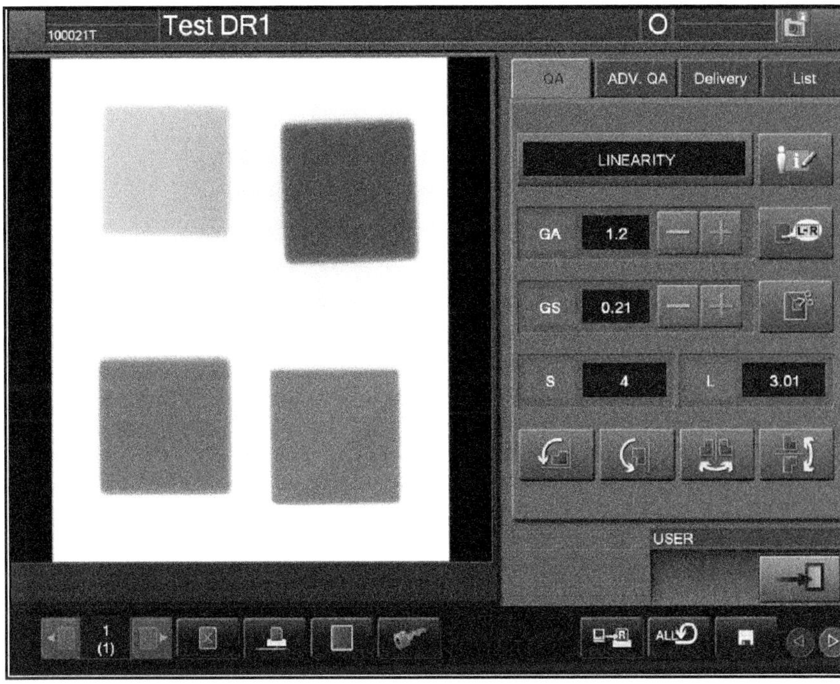

Figure 3.59: Screen shot of a CR reader. Shown is the image of four exposures of a CR cassette with different dose levels. The right side of the panel shows tools for digital image enhancement.

3.9 Other Means of Dosimetry

3.9.1 Calorimetry

If the measurement of absorbed dose is required, calorimetry is the most fundamental of all dosimetry techniques. The cross section of a calorimeter is shown in figure 3.60. It measures the change in temperature of a test medium (usually water or graphite) due to ionizing radiation. This is a direct measure of the absorbed energy in the medium if the heat defect can be accounted for:

$$D = \frac{dH}{dm} + \frac{dE_C}{dm}. \tag{3.20}$$

H is the heat and E_C the part of the deposited energy used for chemical changes, the heat defect. The heat defect can be positive or negative (Khan 2003). It is of the order of –2% for water (Ross et al. 1989) and can be assumed to be 0 for graphite (Reich 1990). Unfortunately, the heat defect is difficult to ascertain and depends on minor impurities in the medium. A good review of water calorimetry and the influence of the heat defect is given by Ross and Klassen (1996).

However, if the heat defect for a certain medium is known, the heat, H, can be directly assessed via temperature changes in the medium:

$$-T = \frac{H}{c_{p,m}}, \qquad (3.21)$$

with $c_{p,m}$ as the specific heat of the substance. The rise of temperature is extremely small (about 0.00024 K for 1 Gy absorbed dose in water) and therefore difficult to measure.

Calorimeters can be operated in three different modes (Rajan 1992): adiabatic (–T is measured directly), isothermal (an external heater compensates for H), and quasi-adiabatic. In the latter mode one accepts a difference of temperature between core and ambient to start with and corrects –T for this constant shift. Most present calorimeters operate in this mode with several jackets (see figure 3.60) insulating the core from the temperature fluctuations of the environment.

The temperature and temperature changes can be detected by means of thermistors evaluated via electrical bridges. The ideal material in a calorimeter should be tissue equivalent in terms of the radiation and should be a good thermal conductor to avoid temperature disequilibrium throughout the sample. Therefore, water and graphite are the most commonly used materials.

Figure 3.60 shows a graphite calorimeter. The experimental difficulties of providing appropriate thermal insulation and ensuring an accurate temperature assessment in these instruments commonly limit the use of calorimetry to NSLs where the technique is used for absolute dosimetry. However, a different type of water calorimeter that does not require thermal isolation (Ross and Klassen 1996) is an interesting development. If the water is perfectly still and no convection takes place, its thermal conductivity is rather small. As such, temperature changes can be assessed in a small

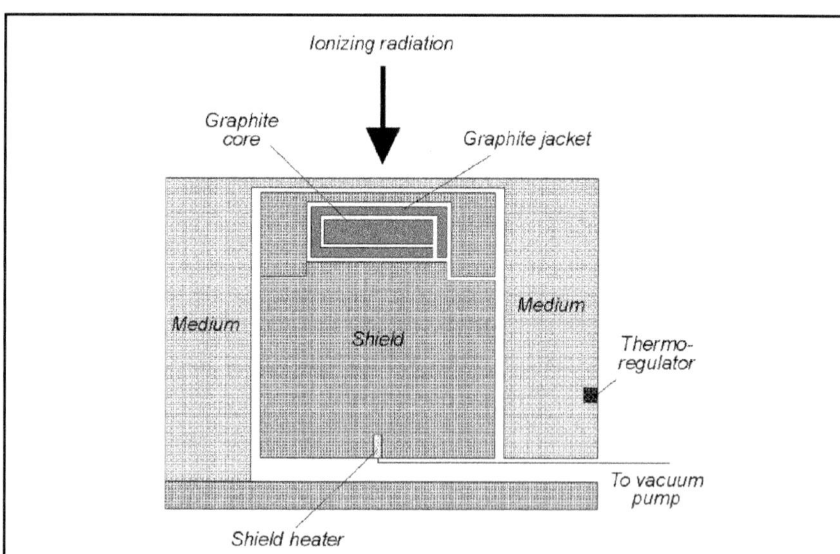

Figure 3.60: Schematic drawing of a calorimeter. (Adapted from Domen et al. 1974.)

subvolume of a large water reservoir. Convection can be further reduced by keeping the water at 4 °C (Ross and Klassen 1996). Calorimeters of this type were first proposed and realized by Domen in 1982.

Of interest for clinical physicists may be the development of a relatively small portable calorimeter by McEwen (McEwen and Duane 2000). This is a design for absolute dosimetry in therapy beams, with the promise to achieve an accuracy of better than 1% "in the field."

3.9.2 Diamond Detectors

Carbon has a long tradition as graphite in calorimetry and as wall material in ionization chambers. The other crystal modification of carbon, the diamond, has only recently become available for dosimetry of megavoltage radiation (Planskoy 1980, Heydarian et al. 1993, Hoban et al. 1994, Rustgi 1995).

A schematic diagram of a diamond detector is shown in figure 3.61. The dimensions in the figure refer to a commercially available detector (PTW, Freiburg, FRG). A natural diamond of 1.4 mm^3 volume is sealed in a cylindrical polystyrene housing. Bias is applied through two thin gold contacts. Simplified, a diamond detector acts as a resistive element with the resistance inversely proportional to the dose rate of the radiation administered. At a given dose rate the current is proportional to the applied voltage, which is typically of the order of 100 V.

In a more rigorous treatment, one needs to consider dose rate effects and impurities in the crystal (Hoban et al. 1994). Recombination of electrons and holes created by the radiation will reduce the measurement signal. However, as the recombination probability increases with the number of charge carriers present, one can expect a pronounced dose rate dependence of the detector. This can be counteracted by the presence of impurities, which can trap electrons. If the number of trapped electrons is large, the number of holes will always be large compared with the number of free

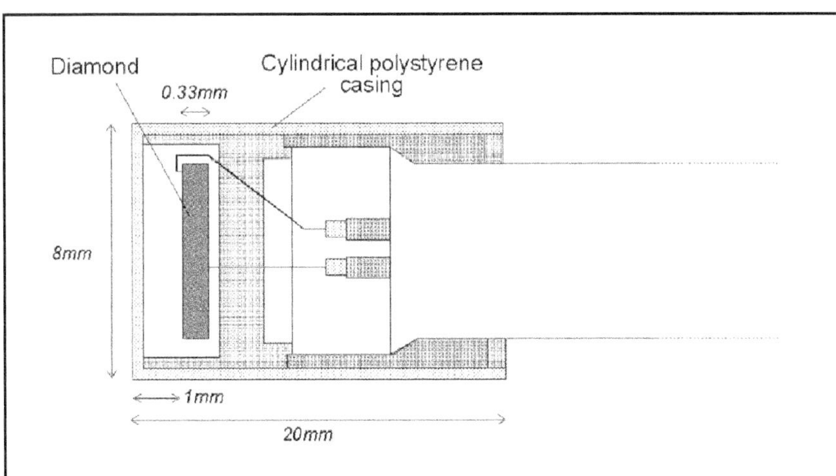

Figure 3.61: Schematic drawing of a diamond detector. (Adapted from Heydarian et al. 1993.)

electrons. As such, a variation of dose rate and therefore number of electrons will not result in a significant change of recombination rates. While this renders the diamond detector independent of dose rate, it also reduces the overall sensitivity. Even more problematic is the charging up ("polarization") of the detector due to the trapped electrons. Polarization will reduce the sensitivity with increased total absorbed dose, and it was recommended to pre-irradiate the diamond detectors with a dose of approximately 5 Gy to stabilize its dose response.

However, with a good compromise in impurity concentration, diamond detectors can be useful tools for the dosimetry of megavoltage photons from linear accelerators. The stopping power ratio of carbon to water is virtually constant for all energies of interest. Diamond detectors have negligible directional variation in dose response (Rustgi 1995), and their small physical size and tissue equivalence makes them well-suited for measurements in the beam penumbra. Similarly, they have been successfully used in small photon fields (Westermark et al. 2000) employed for stereotactic radiosurgery (Rustgi and Frye 1995, Mack et al. 2002) and IMRT (Laub and Wong 2003, Bucciolini et al. 2003). In depth dose measurements one has to take the dose rate dependence into account. At typical dose rates delivered by linear accelerators the dose response reduces by about 3% from 0.8 Gy/min to 4.8 Gy/min. Therefore, the dose at the tail of the depth dose curve may be overestimated by some 1% (Heydarian et al. 1996). A correction for these effects has been suggested by Fidanzio (Fidanzio et al. 2000).

A recent development are diamonds made by chemical vapor deposition (Marczewska et al. 2002). These CVD diamonds are used as TL dosimeters and feature excellent tissue equivalence and high spatial resolution. Several laboratories produce these detectors now (Marczewska et al. 2006, De Angelis et al. 2006, Benabdesselam et al. 2006) and it will be interesting to see if they will find a place among other detectors in radiotherapy dosimetry.

3.9.3 Scintillation Detectors

Small plastic scintillation detectors are a relatively new development for radiotherapy dosimetry that occurred in two laboratories in Madison, Wisconsin (Beddar et al. 1992a,b) and Dortmund, Germany (Fluhs et al. 1996). While scintillators are common for radiation dosimetry in health physics, they have not yet established a role for themselves in routine clinical applications. However, plastic scintillators have some important advantages such as tissue equivalence and small physical size, which make it a particularly promising detector for brachytherapy dosimetry (Fluhs et al. 1996, Quast et al. 1998). A typical detector based on the design of Beddar (1992) is shown in figure 3.62.

A problem is that light emitted by the scintillator has to be transmitted via optical fiber to a light detector, which should be removed from the primary radiation beam. However, any material with a refraction index larger than one will be prone to the production of Cerenkov radiation in megavoltage beams, which would affect the signal. Several methods to correct for this have been suggested (Clift et al. 2000), and the design of the detector in figure 3.62 includes two identical light fibers, one of them connected to the scintillator and the other used as a background Cerenkov radiation detector.

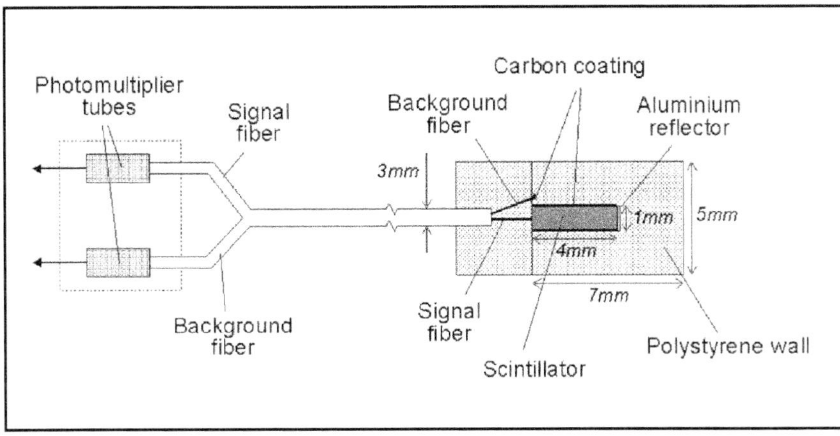

Figure 3.62: Schematic drawing of a scintillation counter. (Adapted from Beddar et al. 1992a.)

The difference of the two signals is proportional to the light emitted by the scintillator.

The plastic chosen by Beddar and associates was BC-400 with a physical density of 1.032, a relative (to water) electron density of 0.979, and an effective atomic number close to 6 (Beddar et al. 1992a). This detector proved to be:

i) Independent of dose rate.
ii) Prone to a very small temperature dependence of less than 1% between 20 °C and 50 °C.
iii) Not significantly affected by radiation damage up to doses of 10 kGy. However, radiation damage that will lower the light yield should be monitored (Aoyama et al. 1995).
iv) Reproducible in their dose reading by better than 0.2%.

The dimensions of the detector indicated in figure 3.62 are small compared to ionization chambers, and detectors of a volume of 0.2 mm^3 have been reported (Quast et al. 1998). Attempts have been made to reduce the size of the scintillator even further and theoretically it is possible to achieve detectors of a diameter of a few microns only. Combined with their tissue equivalence this makes scintillators particularly useful for applications such as brachytherapy dosimetry (Lambert et al. 2006, Sliski et al. 2006) but the technology may also find applications in other areas such as IMRT (Petric et al. 2006), in particular if multiple detectors are combined.

3.9.4 Chemical Dosimetry

A wide range of different chemical compounds undergoes changes after irradiation that can be used to quantify radiation effects. Radiographic film, as discussed in section 3.10.1, is one example. Since chemical dosimetry directly assesses radiation effects based on ionization, it can be taken as a direct measure of dose and may be considered for absolute dosimetry (Angerstein et al. 1987). Reviews of chemical dosimetry with solutions can be found in Fricke and Hart (1966) and Spinks and Woods (1964).

The principle of most chemical dosimeters is that metal ions in aqueous solution are transferred into a different oxidization stage by ionizing radiation. The yield of this reaction is quantified by the so-called G-factor:

$$G = \frac{n}{E}, \tag{3.22}$$

with n the relative number of molecules showing the desired chemical reaction, and E the energy absorbed in the medium due to radiation. Typical G values are of the order of several micromoles per Joule. For the measurement of 1 Gy of absorbed dose, this requires the detection of micromoles of a substance in 1 kg of medium, which may contain some 10^{10} µmol of substance. In general, chemical dosimetry is therefore not a very sensitive radiation detection method, and its use is limited to high-dose applications.

The chemical changes in dosimetric solutions are normally evaluated spectrophotometrically. This is based on the fact that many chemical dosimeters display a color change associated with the transition of a metal ion in a different oxidization state.

The most commonly used chemical dosimeter is based on the work of Fricke and Morse (Fricke and Morse 1927, Fricke and Hart 1966). Radiation oxidizes Fe^{2+} in a ferrous sulfate solution to Fe^{3+}, ferric iron. The iron concentration used is approximately 1 mM. Since the medium for the solution is water with about 0.1 M sulfuric acid, it is close to water/tissue equivalent. The basic reaction can be written as (simplified):

$$Fe^{2+} + OH + \text{radiation} \rightarrow Fe^{3+} + OH^{-}.$$

This is followed and accompanied by a series of other reactions converting Fe^{2+} into Fe^{3+}. The overall reaction is well-studied and quantified. The G value for Fricke solution is of the order of 1.6 µmol/J. It varies slightly with radiation energy. However, the G values are well-documented in various textbooks (Spinks and Woods 1964). In Fricke dosimetry the iron transition can be detected using a spectrophotometer in the red range of the visible light spectrum. As most chemical dosimetry, Fricke solution is not very radiation sensitive and attempts have been made to increase its yield. A doubling of sensitivity is possible by adding benzoic acid; however, other organic agents have also been used. For example, agarose, used also for gel dosimetry as discussed below, increases the G value of the Fricke solution.

The main application of Fricke dosimetery in radiation oncology is as a standard dosimeter. Fricke solution is water based and therefore virtually tissue equivalent. It can easily be placed in sealed containers or cuvettes for dosimetry. The containers can be shaped in any way required for dosimetry and—most importantly—they can be easily transported and mailed. Therefore, some national standards laboratories utilize Fricke solution as basis of a dosimetric intercomparison service (e,g., PTB, Germany). Similarly, Fricke solution can be handled like a solution containing cells (e.g., placed in the same containers or Petrie dishes). Therefore, it also plays an important role in radiobiological cell experiments. Reviews of chemical dosimetry with solutions can be found in Fricke and Hart (1966) and Spinks and Woods (1964).

3.9.5 Gel Dosimetry

One of the most important boosts for interest in Fricke dosimetry is its inclusion into a gel matrix. Presently, gel dosimetry can be divided into two branches: Fricke-based gels (Chu 2001, Schreiner 2004) and polymer gels (McJury et al. 2000, McAuley 2006). Many reviews of the topic are available (Day 1990, De Deene 2002, MacDougal et al. 2002, Oldham 2004) and an AAPM task group report is awaited. Also, the proceedings of the regular DOSGEL meetings provide an excellent update on this fast developing field —the last two are available online as Vol 3 and Vol 56 of the Institute of Physics (UK) Journal of Physics Conference Series (www.iop.org). While magnetic resonance imaging (MRI) was the first technique used for the evaluation of the gel dosimeters (Olsson et al. 1990), other methods such as ultrasound (Mather et al. 2002), x-ray CT (Hilts et al. 2000, Hill et al. 2005, Hilts 2004) and Raman spectroscopy (Baldock et al. 1998a,b, Jirasek et al. 2001, Rintoul et al. 2003) have been employed. Optical methods have particularly been successful and will be discussed below.

Fricke-based Gels

In 1984, Gore et al. (1984) pointed out that the chemical transition from ferrous to ferric iron can also be detected by nuclear magnetic resonance (NMR). It is beyond the scope of the present section to discuss the physics of NMR and MRI. For a good introduction into the principles of MRI the reader may refer to an article by Leach (1992). Reviews of MRI as a tool for gel dosimeter evaluation have been provided by De Deene (2004) and Lepage (2006) as part of the international conferences on radiotherapy gel dosimetry (DOSGEL). The two different ionic states of iron have a different magnetic moment, different ionic radii, and correlation times which influence the proton relaxation times T1 and T2 in an NMR experiment. If the spatial position of the iron in a Fricke dosimeter is fixed by embedding it in a gel matrix, the effect can be used to obtain a 3-D image of the dose distribution in the gel using spatially resolved MR imaging (Day 1990, Olsson et al. 1990). This type of dosimetry has three major advantages:

i) The three-dimensional nature of the phantom and the MR imaging technology.
ii) Detector and phantom are identical and virtually tissue equivalent for ionizing radiation (Kron et al. 1993c).
iii) Phantoms can be shaped in any desirable form, and other different materials (such as bone or lung substitutes) can be incorporated in them.

Figure 3.63 shows the image of a 2 cm thick slab of iron-doped agarose gel that was irradiated with five superficial radiation beams to approximately 20 Gy each. The dose was delivered from the top of the sample and the dose fall-off and the divergence of the radiation beam with depth can be

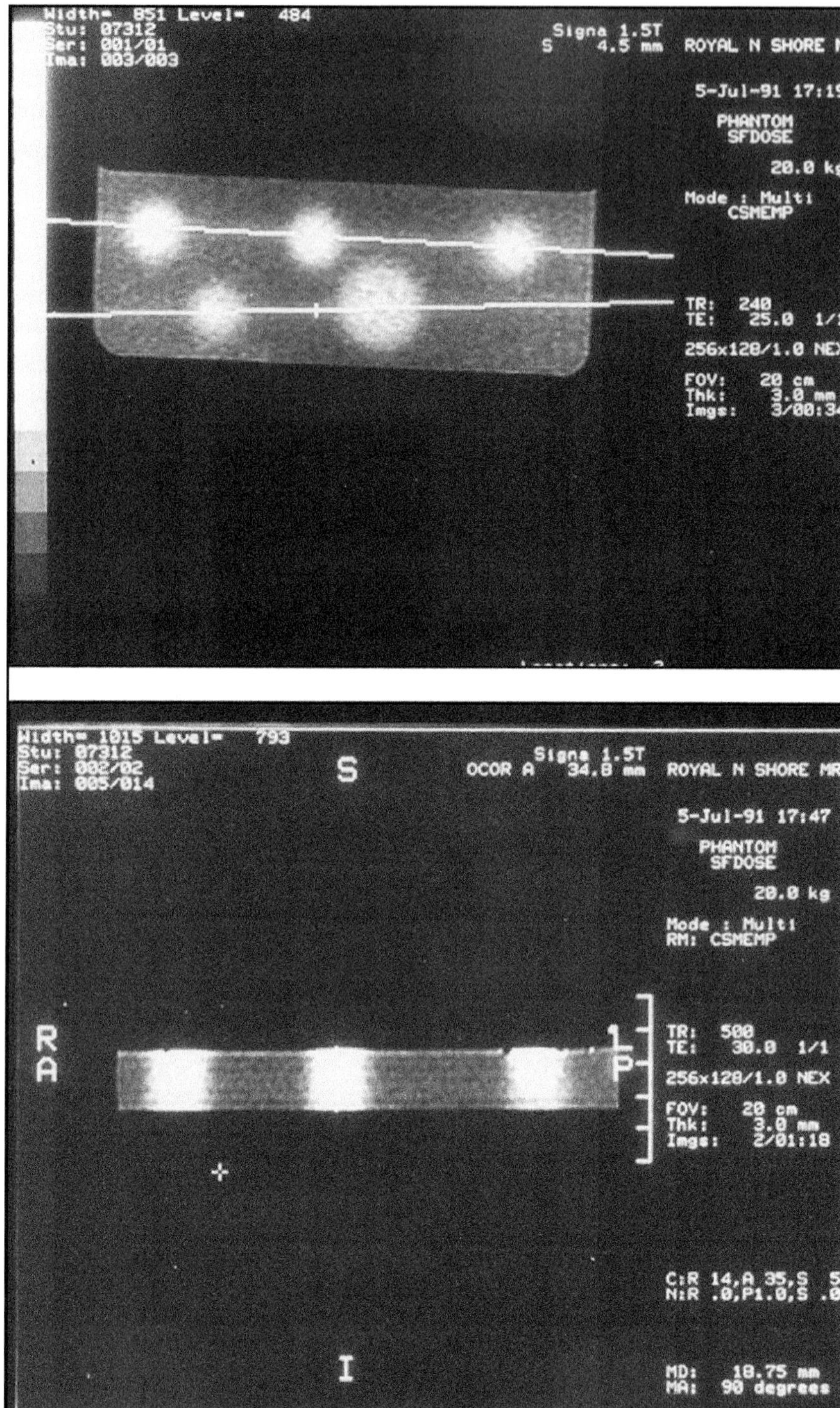

Figure 3.63: T_1 weighted image of an agarose gel irradiated with five different radiation fields: (a) top image: transverse slice; (b) bottom image: "coronal" slice.

seen. Using a relatively short repetition time (TR 500 ms) for acquisition of the images, the areas where the radiation dose was absorbed are clearly visible as bright spots because of their shorter T_1. In this regard Fe^{3+} has the same effect as gadolinium-based contrast agent, commonly used in diagnostic MRI. Regions of interest can be used to evaluate the relaxation times in particular parts of the image.

The basic principle of NMR dosimetry is illustrated in figure 3.64. It depicts the change of T_1 with the concentration of ammonium ferrous (Fe^{2+}) and ferric (Fe^{3+}) sulfate. The linear relationship between $1/T_1$ and the concentration of iron c is described by Podgorsak and Schreiner (1992):

$$1/T_1 = 1/T_{1,0} + R_1 \, c_{ion} \tag{3.23}$$

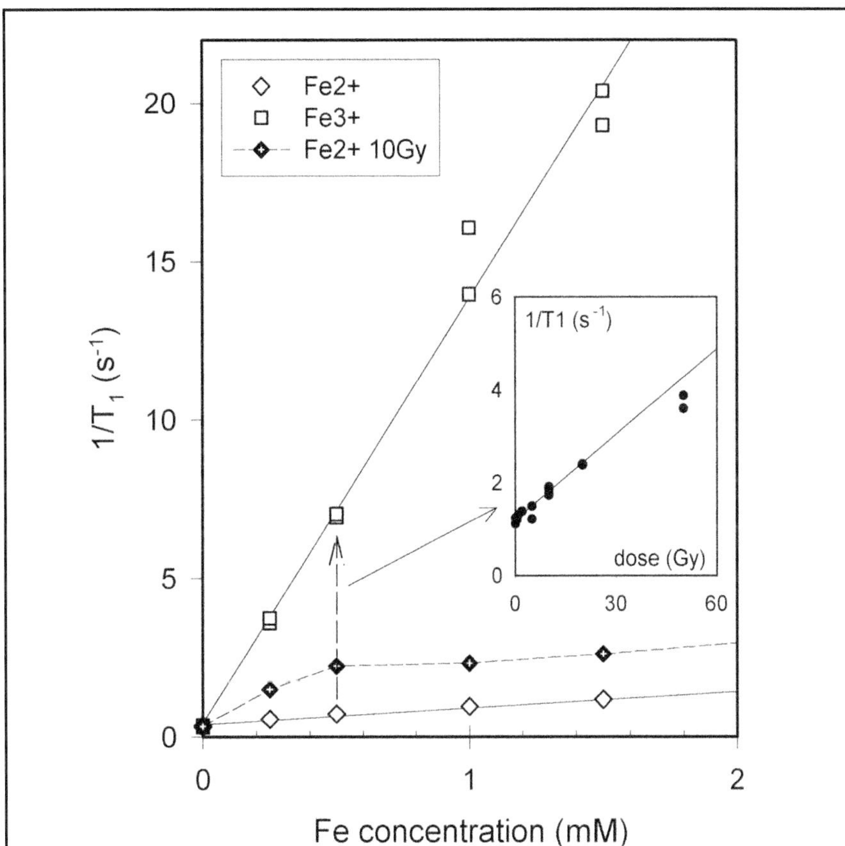

Figure 3.64: Variation of $1/T_1$ as a function of the iron concentration in three gels Fe^{3+} (ferric), Fe^{2+} (ferrous), and a gel doped with Fe^{2+} and irradiated with 10 Gy causing partial conversion of ferrous to ferric iron. The insert depicts a dose response curve for a gel doped with 0.5 mmol Fe^{2+}.

with $T_{1,0}$ as the spin lattice relaxation time of the sample without any ion present, and R_1 the T_1 relaxivity of the ion. A similar equation is valid for the spin-spin relaxation time T_2, and both T_1 and T_2 changes have been employed for NMR dosimetry. While T_2 changes seem to be more sensitive to radiation (Prasad et al. 1991, Gambarini et al. 1994, Duzenli et al. 1994), T_1 changes can be assessed more accurately and they are therefore also often used in Fricke gel dosimetry (Hazle et al. 1991, Olsson et al. 1992, Kron and Pope 1994). The relaxivity (the slope of the curves for Fe^{2+} and Fe^{3+} in figure 3.64) for Fe^{2+} is about 20 times smaller than the one of Fe^{3+},a difference which can be used for dosimetry. This is illustrated in the figure where $1/T_1$ of the Fe^{2+} samples are shown after irradiation with 10 Gy (symbols with a cross). It can be seen that T_1 is shortened after irradiation due to oxidization of Fe^{2+} to Fe^{3+}. If the iron concentration is sufficient, the variation of $1/T_1$ is independent of the iron concentration, as the total amount of iron converted is proportional to the dose.

The change of the relaxation time with dose can also be displayed as a dose response curve. This is shown in the insert of figure 3.64, which depicts the change of $1/T_1$ as a function of dose in 1.5% agarose gels doped with 0.5 mM ammonium ferrous sulfate and 0.125 M sulfuric acid using a 4.7T superconducting magnet (Kron and Pope 1994).

The dose response in Fricke gel systems varies considerably (Kron et al. 1997). Due to different experimental conditions, such as oxygenation (Schulz et al. 1990) and various additives, dose responses from 0.01 to 0.12 s^{-1} Gy^{-1} have been reported in the literature (Schulz et al. 1990, Olsson et al. 1990, Rae et al. 1996, Kron et al. 1997). These can be compared to a dose response of 0.2 s^{-1} Gy^{-1} achieved by T2 measurements recently (Gambarini et al. 1994) in the same Fricke based gel systems and 0.25 s^{-1} Gy^{-1} observed in the polymer gels discussed below (Maryanski et al. 1994a,b). The dose response is also related to the detection limit, which is presently of the order of 0.5 Gy.

Due to its great potential, many applications have been identified for gel dosimetry. They range from electron beams (Appleby 1987) to superficial x-rays (Kron and Pope 1994), the verification of stereotactic radiosurgery (Olsson et al. 1992, Schulz et al. 1993, Rousseau et al. 1994), and of course IMRT (De Deene 2002, Saur et al. 2005). Figure 3.65 shows the dose distribution in a Fricke gel doped with xylenol orange, a method to make chemical change visible (see below). Considerable interest has also been in the area of brachytherapy dosimetry, where tissue equivalence and 3-D approaches are particularly valuable (Schreiner et al. 1994, Olsen et al. 1994). However, while showing great promise, MRI-based Fricke gel dosimetry is complex and its wide spread implementation is challenged by:

i) High costs and problems of obtaining access to an MRI scanner.
ii) Reproducibility of the gel manufacturing process: even under well-controlled conditions it is difficult to produce gels with equal radiation sensitivity in different batches. As the reproducibility is typically poorer than 10%, it is necessary to run standards of each gel batch with the measurement. This is illustrated for a polymer gel in figure 3.66.

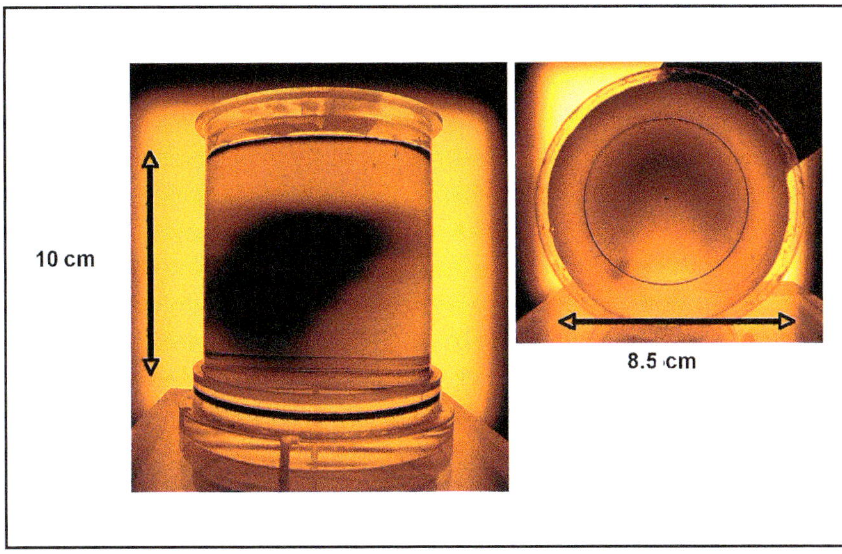

Figure 3.65: Dose distribution from an IMRT prostate treatment as determined using a radiochromic Fricke gel. (Courtesy of E. Wong, K. Jordan, and J. Battista.)

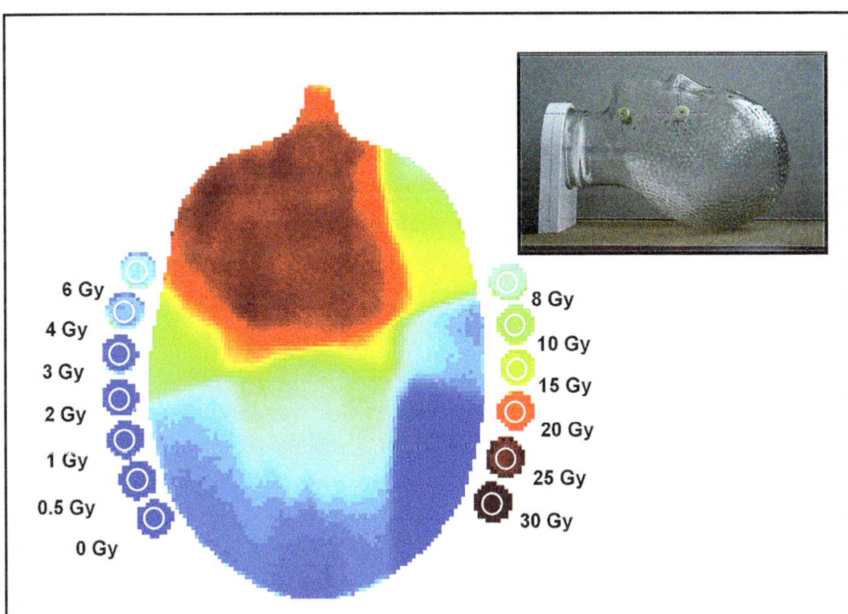

Figure 3.66: Dose distribution in a head phantom treated for a paranasal sinus cancer using IMRT. The head phantom shown in the insert was filled with polymer gel. The dose distribution was determined using MRI localized T_2 measurements. A set of gels irradiated to known doses was positioned around the phantom during the imaging process. A set of these reference gels is shown in figure 3.67. (Courtesy of A. Venning and K. Vergote.)

iii) Fricke gels exhibit spontaneous oxidization; this compromises the long-term stability of the gels and it is advisable to prepare a fresh batch of gels prior to each measurement series.

iv) Due to field inhomogeneities and variations in tip angle throughout the field of view in a typical MR scanner there may be differences in dose response in different parts of a gel phantom. It may be necessary to map these prior to quantitative analysis.

v) MR-based gel dosimetry is time consuming. Typically a couple of measurements are required to obtain a map of relaxation times. Fast imaging (Kron et al. 1997) or magnetization to dose (M&D) calibration (Schreiner et al. 1995) have been suggested to overcome this.

vi) The diffusion of the radiation product, Fe^{3+}, requires the gels to be evaluated soon after irradiation to avoid a blurring of the dose image. This appears to be the most important problem in Fricke gels and a variety of approaches have been suggested to overcome it. They include fast imaging (Kron et al. 1997), mathematical methods (Harris et al. 1996), and the inclusion of chelating agents such as xylenol orange in the gels (Rae et al. 1996). The latter has the additional advantage, that a visible color change can be observed as a function of the absorbed radiation dose (Appleby and Leghrouz 1991).

Polymer Gels

Many of the problems listed above have been addressed by the use of polymer gels (Maryanski et al. 1994a,b, 1996, Audet 1996, Audet et al. 1996, Baldock et al. 1996, 1998, McAuley 2006). A typical composition of these gels is 3% acrylamide, 3% N,N' methylene-bisacrylamide, and 5% gelatine. They are variously referred to as PoGels, PAGs (PolyAcrylamide Gels), or BANG® (Bis Acrylamide Nitrogen Gelatine - BANG is a trademark of MGS Research, CT) gels and exist in different slightly different chemical compositions (Maryanski et al. 1996, Baldock et al. 1996). Polymer gels rely on radiation-induced polymerization and cross-linking of acrylic monomers. These radiation "products" are part of the gel matrix and cannot diffuse within the gel.

The cross-linking of chains also alters the optical properties of the gels. This is shown in figure 3.67 by a set of test vessels irradiated with different radiation doses. With increasing dose the gels get more opaque because of a significant increase in scattering of light within the gel. Nowadays there are many groups all over the world who have successfully manufactured polymer gels and a few groups have reported successful clinical implementation (Ibbott et al. 1997, De Deene 2002). The long-term stability of PAG gels is of particular interest for dosimetric intercomparison. MGS Research offers a "QA by mail" service (http://www.mgsresearch.com/custom.asp?id=63396&page=9&showsite=true) and the Radiological Physics Center in Houston, Texas, has used gels for dose verifications in stereotactic radiotherapy and IMRT (Ibbott 2006).

When evaluated with MRI, one typically measures T_2 and a detailed model for the changes in T_2 with dose is provided by Audet (1996). This is also the case for the phantom measurement shown in figure 3.66.

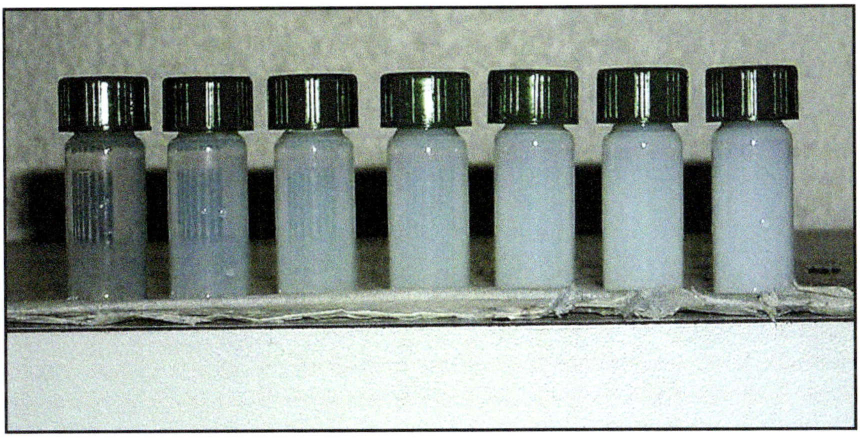

Figure 3.67: Polymer gels irradiated to doses of 1, 2, 3, 4, 5, 10, and 15 Gy, respectively.

Compared to Fricke gels the polymer gels have also a number of disadvantages: they are more expensive and difficult to produce, they require an airtight environment to protect them from oxygen, and some of the chemicals employed are toxic. Several attempts have been made to improve the ease of polymer gel preparation by using less toxic ingredients and adding oxygen scavengers to lower the susceptibility of the gels to oxygen damage. Of note are the MAGIC gels (methacrylic and ascorbic acid in gelatin initiated by copper) (Fong 2001, Mather 2002, Gustavsson 2003) and MAGAT gels (methacrylic acid, gelatin and tetrakis [hydroxymethyl] phosphonium chloride) (Hurley et al. 2005, Bayreder et al. 2006). Both are "normoxic" gel formulation, i.e., they can be handled in normal atmosphere. Normoxic gels can be evaluated using both MRI or optical methods (Fong 2001) and have been applied for a variety of applications in dosimetry ranging from CT scanners (Hill et al. 2005) to IMRT (Gustavsson et al. 2003) and stereotactic radiosurgery (Scheib et al. 2006).

Radiation dose distributions in polymer gels can also be evaluated using x-ray CT (Hilts et al. 2000, Trapp et al. 2001, Hilts 2004). The relatively small radiation dose delivered during the imaging process adds negligible additional dose in radiotherapy applications. CT evaluation of gels is not very sensitive and limited in its application as it is based on a change in physical density of the gel (Trapp et al. 2002). However, it is very convenient (every radiotherapy center has access to a CT scanner), fast, and the spatial accuracy is excellent.

Optical Evaluation of Gels

Most gel types can also be evaluated using optical tomography (Maryanski et al. 1996, Tarte et al. 1996, Kelly et al. 1998). Optical methods were originally based on the use of a dye, xylenol orange, which forms a colored complex with Fe^{3+} in Fricke-based solution (Gupta et al. 1978, Appleby and Leghrouz 1991); however, the change in optical absorption due to cross-linking in the polymer gels can also be used. While 2-D images of thin slabs of gel were originally acquired, the technique can be extended to 3-D objects using a technique similar to CT scanning (Kelly et al. 1998). Optical CT as a means for gel dosimetry evaluation has been reviewed recently by Jordan (2004) and Oldham (Oldham 2004, 2006, Oldham et al. 2005).

The aim is to acquire 3-D dose distributions. As such, it must be possible to acquire several slices or a full volume. A variety of different approaches have been taken, all trying to strike a compromise between rapid acquisition of data and the need to reduce scatter from the sample. The designs range from a moving laser (Oldham et al. 2001, Maryanski et al. 1996, Islam et al. 2003) to a cone beam setup with an array of light-emitting diodes (LEDs) (Doran et al. 2001, Jordan 2004). Figure 3.68 shows a schematic drawing of an optical CT scanner of the latter design.

Optical CT is limited to samples of cylindrical shape like the one shown in figure 3.65; an anthropomorphic shape as shown in figure 3.66 would need to be embedded into a suitable cylindrical container. It seems easier to use MRI in this case. For optical CT the cylindrical sample must have a thin wall to avoid refraction of the light. The container is typically immersed in a rectangular tank filled with fluid optically matched to the gel. Another limitation is a restriction of sample sizes (typically not larger than 10 cm diameter) because of the maximum light intensity of the source. However, optical evaluation of gels is characterized by excellent spatial resolution (Oldham et al. 2001) and despite the limitations discussed above, optical CT has made gel dosimetry more readily available for many users.

A commercial optical CT scanner is shown in figure 3.69 (Modus Medical Devices, London, Ontario). Compared to MRI, optical evaluation of gels reduces costs, thereby making 3-D gel dosimetry accessible for more radiotherapy centers. Figure 3.70 shows a gel that was irradiated using two fan beams from a helical tomotherapy unit (Hi·Art®, TomoTherapy Inc.,

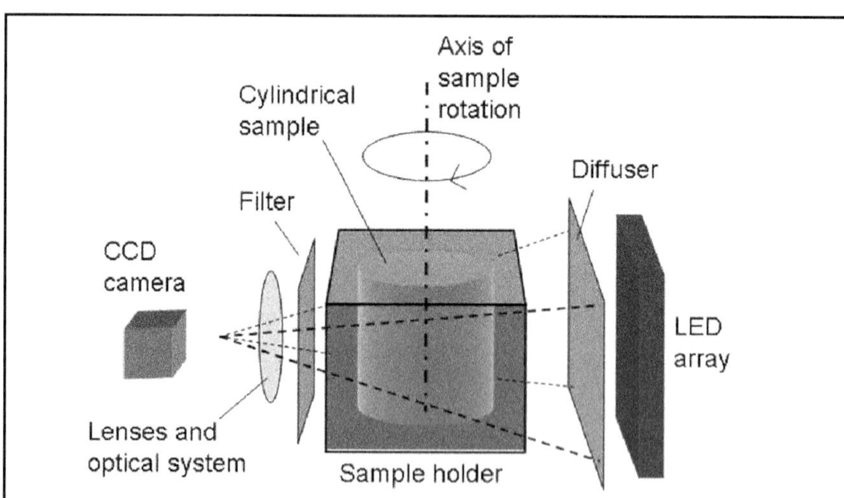

Figure 3.68: Schematic drawing of an optical cone beam CT scanner for 3-D gel dosimetry.

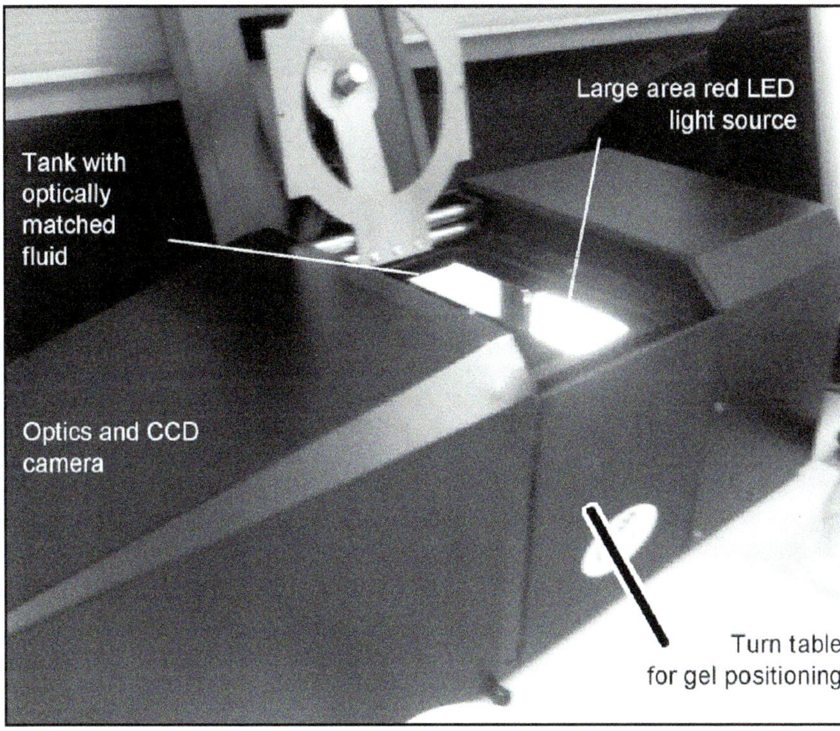

Figure 3.69: Optical CT scanner for 3-D evaluation of gels in dosimetry (Modus Medical).

Madison, Wisconsin). In each beam every second leaf was shut, resulting in a comblike dose distribution. Putting two of these "combs" together results in the dose pattern seen in the cylindrical gel sample. Both attenuation and

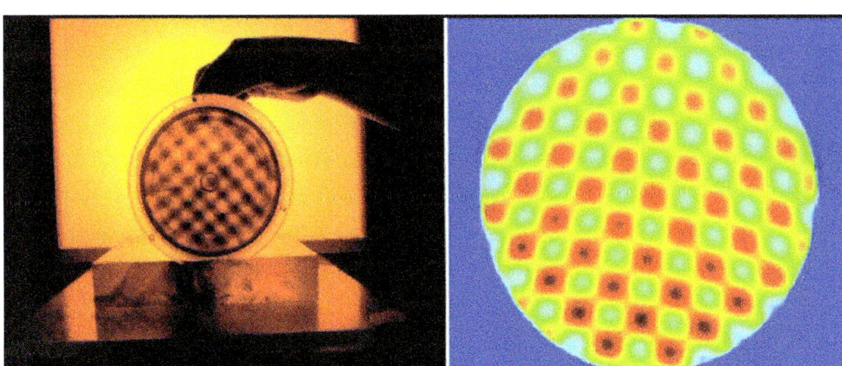

Figure 3.70: Xylenol orange doped Fricke gel irradiated with a cross fire of 6 MV x-rays from a helical tomotherapy unit (compare chapter seven) with every second leaf in the MLC closed. The picture on the right shows a color plot of the dose distribution as acquired using an optical CT scanner (shown on the cover of the *International Journal of Radiation Oncology, Biology and Physics*, Volume 57, Issue 4, © 2003.) (Reprinted with permission of Elsevier.)

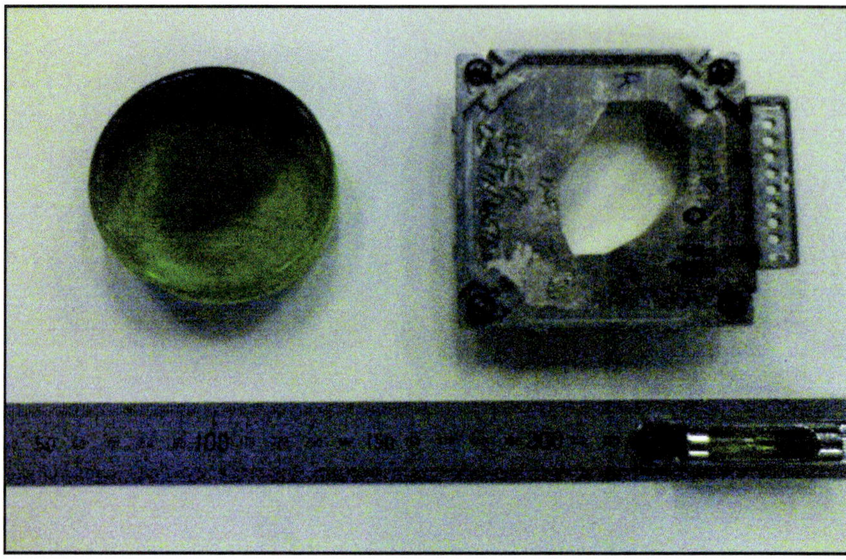

Figure 3.71: Electron dose distribution in a "pressage" dosimeter (Modus Medical).

divergence of the beams can be seen. The gel was evaluated in an optical CT scanner and the resulting dose distribution is shown in the figure. It can be seen that the optical evaluation results in an excellent representation of the qualitative dose distribution.

Another interesting recent development for 3-D dosimetry using optical evaluation are "pressage" dosimeters (Adamovics et al. 2006a,b). Pressage dosimeters are solid, transparent plastics that are radiation sensitive. They contain leuco dyes that change color with radiation. Pressage dosimeters have a cylindrical cross section suitable for optical scanning. Figure 3.71 shows such a sample used for verification of the dose distribution in an irregular electron field. The cutout (compare chapter five) is shown adjacent to the dosimeter.

3.9.6 Biological Dosimetry

The endpoint in radiotherapy is a biological one—the killing of all clonogenic tumor cells. Similarly, the unwanted side effects are biological reactions to radiation. Therefore, biological dosimetry is assessing and quantifying the very effect on which radiotherapy is based. However, biological effects are difficult to quantify, and considerable care has to be taken to interpret results in such a complex "dosimeter" as a biological organism.

Two types of biological dosimetry do play a role in radiotherapy and will be briefly discussed here: erythema and chromosome aberration assays. Other indicators have been suggested and are discussed in more detail in the extensive literature on health physics and nuclear accidents. Some attention should also be drawn to the fact that some predictive assays that try to predict the likelihood of uncomplicated tumor control for an individual patient may employ the assessment of biological endpoints as a function of radiation dose.

Erythema

The first biological effect of ionizing radiation observed was a skin reddening similar to a sunburn. This effect could then be related to the probability of tumor cure in patients undergoing radiotherapy. Therefore, lacking a better quantification of dose, erythema became the first dose measure in radiation oncology. The erythema dose, ED, was defined as the dose required to achieve a clearly detectable reddening of the skin. Skin reactions are still used for clinical judgment of an individual's radiation sensitivity.

There have been several attempts to quantify the degree of redness by using color tables. A more recent development is the reflectance spectrophotometer (Feather et al. 1988, 1989, Denham et al. 1995). This device analyses the light reflected from the skin. Since no human observer is involved, this reading can be used as the basis of a quantitative redness evaluation, provided individual baseline differences (skin type, part of the body) and environmental factors (patient stress, temperature) are accounted for.

Chromosome Aberrations

In this technique a blood sample is taken from a person, and the number of chromosome aberrations studied using a microscope. This can be used to quantify the radiation given to a person starting from a total body dose of approximately 0.1 Sv (Reich 1990). For partial body irradiations the threshold can be a lot higher. The accuracy of the technique is limited, and false positive results for relatively low exposures are possible due to spontaneous chromosome aberrations. To establish a baseline for a particular individual, a second blood sample should be investigated about four weeks after the radiation exposure.

In general, chromosome aberration analysis is a useful technique for major radiation accidents. It is typically only available in bigger centers like national laboratories.

3.10 Dosimeter Overview

In the history of the medical application of x-rays virtually all effects due to ionizing radiation have been used for dosimetric purposes. Table 3.11 lists some of the effects and the dosimetry techniques which make use of a particular effect.

According to the scope of this chapter only effects were included that can be employed for the dosimetry of therapeutic x-rays. For detectors used for other types of ionizing radiation, such as electrons, neutrons, and heavy ions, one can refer to many other sources (Attix et al. 1966, 1968, 1969, 1972, Reich 1990, Khan 2003).

All dosimetric techniques have their respective advantages and disadvantages, as has been discussed in detail in this chapter. Table 3.12 lists in the first four rows dosimetric techniques that can be expected to be available in every radiotherapy department: ionization chambers, semiconductors, thermoluminescence dosimetry, and radiographic film. The latter is an example of the fast changes experienced in dosimetry. Due to the replacement of film in

Table 3.17: Effects of ionizing radiation and their use in dosimetry

Radiation Effect	*Dosimetric Method*
ionization in gases	ionization chamber
ionization in liquids	liquid-filled ionization chamber
ionization in solids	semiconductors diamond detectors
luminescence	thermoluminescence dosimetry
fluorescence/phosphorescence	scintillation detectors screens for film or CR
chemical transitions	radiographic film chemical dosimetry gel dosimetry
heat	calorimetry
biological effects	erythema chromosome damage

diagnostic radiology, it is likely that radiotherapy departments will not have access to film dosimetry in the future. Section 8 of this chapter discusses this issue and suggests some alternative dosimeters. The most important 2-D detector systems in linac-based radiotherapy are electronic portal imaging devices (EPIDs) (Boyer et al. 1992, Shalev 1995). As they are a fixture on most linacs, they are readily available and their geometry is *per se* well defined. In addition to treatment verification, EPIDs can be used as a QA tool for dynamic treatments and for in vivo exit dosimetry (Wong et al. 1990, Essers et al. 1996, Boellaard et al. 1998, Warkentin et al. 2003, Baker et al. 2005, Greer and Barnes 2007, McDermott et al. 2007).

Table 3.12 also states typical areas of usage. This is only a selection of applications and should not imply that a particular technique can only be used for a certain type of measurement. The second column lists the number of the section in this chapter where more details of the methods can be found.

A number of other dosimetric techniques are employed to study megavoltage x-rays and electrons and their dose distribution in matter. Some of them are also contained in table 3.12. These techniques are not used in many radiotherapy departments on a routine basis. However, they have specific advantages and demonstrate certain features of dosimetry in general, which makes it worthwhile to discuss them in the context of this chapter. Some of them, like MOSFETs and gel dosimetry, are also recently available in a commercial form, and it can be expected that their use will increase. A wide range of other techniques for x-ray dosimetry is available. However, the discussion of all of them would be beyond the scope of the present book. This leads to a somewhat arbitrary selection of "other" x-ray dosimetry techniques.

Table 3.12: X-ray dosimeters—advantages, disadvantages and applications

Dosimetric Technique	*Section*	*Advantages*	*Possible Disadvantages*	*Important Applications*
Air-filled ionization chamber	3.4, 3.5	high precision, wide variety of chambers available (cylindrical vs. parallel plate). real time measurement, standard, well-documented and investigated, no dose rate dependence	high voltage required, relatively large volume, variation of stopping power ratio of water/air with electron energy, often fragile	calibration, output factor measurements, water tank measurements, acquisition of data for planning, QA, surface dose with thin window pp chamber, absolute dose point in IMRT verification.
Semiconductors	3.6	small size, ease of use, real time measurement, arrays are available, robust, relatively cheap	energy, temperature, dose rate, directional dependence, recalibration required	water tank measurements (penumbra, electrons) in vivo dosimetry, small field dosimetry, detector arrays
Thermoluminescence detectors	3.7	small size, standalone measurement, various materials and forms available, mailable, large number can be used	delayed readout, usually not as precise as ionization chambers, "black art"	in vivo dosimetry, measurements in anthropomorphic and slab phantoms, intercomparisons between centers, personal dosimetry
Radiographic film	3.3.1	measurement of dose distributions, integrative, very good spatial resolution	requires development, variability, energy and dose dependence, not reusable, typically darkroom required, film processors are phased out	measurements in solid and anthropomorphic phantoms, qualitative dosimetry—QA, portal imaging, IMRT, QA

Table 3.12: Continued

Dosimetric Technique	*Section*	*Advantages*	*Possible Disadvantages*	*Important Applications*
Radiochromic film	3.8.2	self-developing, high spatial resolution, easy to use	low sensitivity, not stable after irradiation, light sensitive, expensive	starts to replace film in a number of dosimetric appplications
Calorimetry	3.9.1	absolute determination of energy deposited in matter	insensitive, bulky, difficult to set up	reference dosimeter at standards laboratory
Diamond detectors	3.9.2	small physical size, tissue equivalence, real time measurement	cost, may require pre-irradiation	water tank measurements, small field dosimetry
Chemical dosimetry	3.9.4	different compounds available, versatile, tissue equivalent	insensitive, requires chemical laboratory	reference dosimeter, cell culture measurements, inter-comparisons between centers
Gel dosimetry	3.9.5	tissue equivalence, 3-D, integrative	complex evaluation. limited accuracy and reproducibility	IMRT QA, intercomparisons
Biological dosimetry	3.9.6	ultimate in vivo dosimetry, refers to individual patient	low accuracy and precision	accident dosimetry, individual evaluation of treatment response

References

AAPM (American Association of Physicists in Medicine) (1999) AAPM's TG-51 protocol for clinical reference dosimetry of high-energy photon and electron beams. *Med. Phys.* 26: 1847–1870.

AAPM (American Association of Physicists in Medicine) (1983) A protocol for the determination of absorbed dose from high energy photon and electron beams. Report of Task Group 21. *Med. Phys.* 10: 741–771.

Adamovics J., Jordan K., Dietrich J. (2006a) Pressage TM: Development and optimization studies of a 3D radiochromic plastic dosimeter: Part I. *J. Phys. Conf. Ser.* 56: 172–175.

Adamovics J., Guo P., Burgess D., Manzoor A., Oldham M. (2006b) Pressage TM: Development and optimization studies of a 3D radiochromic plastic dosimeter: Part II. *J. Phys. Conf. Ser.* 56: 176–178.

Aguirre J.F., Tailor R., Ibbott G., Stovall M., Hanson W. (2002) Thermoluminescence dosimetry as a tool for remote verification of output for radiotherapy beams: 25 years experience. *Standards and Codes of Practice in Medical Radiation Dosimetry*, Proceedings of an International Symposium, November 25–28, 2002. IAEA, Vienna, pp. 191–199.

Ahmad M., Chen Z., Song H., Deng J., Nath R. (2006) Evaluation of the EDR-2 film for relative dosimetry of high-energy photon and electron beams. *Radiat. Prot. Dosimetry* 120: 159–162.

Aird E., Farmer E. (1972) The design of a thimble chamber for the Farmer dosemeter. *Phys. Med. Biol.* 17: 169–174.

Andreo P., Nahum A., Hohlfeld K., et al. (1992) Status report on technical report series No. 277 of the IAEA.

Andreo P., Brahme A. (1986) Stopping power data for high energy photon beams. *Phys. Med. Biol.* 31: 839–858.

Angerstein W., Gursky S., Hegewald H. (1987) *Strahlenphysik und radiologische Technik in der Medizin*. Leipzig: VEB Thieme.

Antonuk L.E. (2002) Electronic portal imaging devices: A review and historical perspective of contemporary technologies and research. *Phys. Med. Biol.* 47: R31–R65.

Aoyama T., Maekoshi H., Tsuzaka M., Koyama S. (1995) A scintillating fiber beam-energy monitor for electron beam therapy. *Med. Phys.* 22: 2101–2102.

Appleby A., Leghrouz A. (1991) Imaging of radiation dose by visible color development in ferrous-agarose-xylenol orange gels. *Med. Phys.* 18: 309–312.

Appleby A., Christman E.A., Leghrouz A. (1987) Imaging of spatial radiation dose distribution in agarose gels using magnetic resonance. *Med. Phys.* 14: 382–384.

Attix H. (1975) Further considerations on the track interaction model for thermoluminescence in LIF (TLD100). *J. Appl. Phys.* 47: 81–88.

Attix H., Roesch W., Tochilin E. (1972) *Radiation Dosimetry.* Supplement 1: Topics in radiation dosimetry. Academic Press, London.

Attix H., Roesch W., Tochilin E. (1969) *Radiation Dosimetry.* Vol. III: Sources, fields, measurements and applications. Academic Press, London.

Attix H., Roesch W., Tochilin E. (1968) *Radiation Dosimetry.* Vol. I: Fundamentals. Academic Press, London.

Attix H., Roesch W., Tochilin E. (1966) *Radiation Dosimetry.* Vol. II: Instrumentation. Academic Press, London.

Audet C., Duzenli C., Kwa W., Tsang V., Mackay A. (1996) An example of MRI polymer gel dosimetry applied to 3D conformal radiotherapy. *Med. Phys.* 23: 803.

Aukett R.J. (1991) A comparison of semiconductor and thermoluminescent dosimeters for in vivo dosimetry. *Br. J. Radiol.* 64: 947–952.

Aukett R. (1991) A comparison of semiconductor and thermoluminescent dosimeters for in vivo dosimetry. *Br. J. Radiol.* 64: 947–952.

Aznar M.C., Andersen C.E., Botter-Jensen L., Back S.A., Mattsson S., Kjaer-Kristoffersen F., Medin J. (2004) Real-time optical-fibre luminescence dosimetry for radiotherapy: Physical characteristics and applications in photon beams. *Phys. Med. Biol.* 49: 1655–1669.

Bagne F. (1977) A comprehensive study of LiF TL response to high energy photons and electrons. *Radiology* 123: 753–760.
Baird L., White D.R. (1982) A survey of tissue substitutes and phantoms in 98 US medical centres. *Med. Phys.* 9: 434–435.
Baker S.J., Budgell G.J., MacKay R. I. (2005) Use of an amorphous silicon electronic portal imaging device for multileaf collimator quality control and calibration. *Phys. Med. Biol.* 50: 1377–1392.
Baldock C., Murry P., Kron T. (1999) Uncertainty analysis in polymer gel dosimetry. *Phys. Med. Biol.* 44: N243–N246.
Baldock C., Burford R.P., Billingham N., Wagner G. S., Patval S., Badawi R.D., Keevil S.F. (1998a) Experimental procedure for the manufacture and calibration of polyacrylamide gel (PAG) for magnetic resonance imaging (MRI) radiation dosimetry. *Phys. Med. Biol.* 43: 695–702.
Baldock C., Rintoul L., Keevil S.F., Pope J.M., George G.A. (1998b) Fourier transform Raman spectroscopy of polyacrylamide gels (PAGs) for radiation dosimetry. *Phys. Med. Biol.* 43: 3617–3627.
Baldock C., Burford R., Billingham N., Cohen D., Keevil S. (1996) Polymer gel composition in magnetic resonance imaging dosimetry. *Med. Phys.* 23: 1070.
Balog J., Holmes T., Vaden R. (2006) A helical tomotherapy dynamic quality assurance. *Med. Phys.* 33: 3939–3950.
Bayouth J.E., Wendt D., Morrill S.M. (2003) MLC quality assurance techniques for IMRT applications. *Med. Phys.* 30: 743–750.
Bayreder C., Georg D., Moser E., Berg A. (2006) Basic investigations on the performance of a normoxic polymer gel with tetrakis-hydroxy-methyl-phosphonium chloride as an oxygen scavenger: Reproducibility, accuracy, stability, and dose rate dependence. *Med. Phys.* 33: 2506–2518.
Beavis A.W., Weston S.J., Whitton V.J. (1996) Implementation of the Varian EDW into a commercial RTP system. *Phys. Med. Biol.* 41: 1691–1704.
Beddar A., Mackie T.R., Attix F.H. (1992a) Water-equivalent plastic scintillation detectors for high energy beam dosimetry: I Physical characteristics and theoretical considerations. *Phys. Med. Biol.* 37: 1883–1900.
Beddar A., Mackie T.R., Attix F.H. (1992b) Water-equivalent plastic scintillation detectors for high energy beam dosimetry: II. Properties and measurements. *Phys. Med. Biol.* 37: 1901–1913.
Benabdesselam M., Serrano B., Iacconi P., Wrobel F., Lapraz D., Herault J., Butler J.E. (2006) Thermoluminescence properties of CVD diamond for clinical dosimetry use. *Radiat. Prot. Dosimetry* 120: 87–90.
Black R.D., Scarantino C.W., Mann G.G., Anscher M.S., Ornitz R.D., Nelms B.E. (2005) An analysis of an implantable dosimeter system for external beam therapy. *Int. J. Radiat. Oncol. Biol. Phys.* 63: 290–300.
Boag J.W. (1982) The recombination correction for an ionisation chamber exposed to pulsed radiation in a 'swept beam' technique. *Phys. Med. Biol.* 27: 201–211.
Boag J., Currant J. (1980) Current collection and ionic recombination in small cylindrical ionisation chambers exposed to pulsed radiation. *Br. J. Radiol.* 53: 471–478.
Boag J., Hochhaeuser E., Balk O. (1996) The effect of free-electron collection on the recombination correction to ionization measurements of pulsed radiation. *Phys. Med. Biol.* 41: 885–897.
Boehm J. (1986) The national primary standard of the PTB for realising the unit of absorbed dose rate to tissue for beta dosimetry. *PTB Berichte* DOS 13.
Boellaard R., Essers M., van Herk M., Mijnheer B.J. (1998) New method to obtain the midplane dose using portal in vivo dosimetry. *Int. J. Radiat. Oncol. Biol. Phys.* 41: 465–474.
Bos A., Pithers T. (1993a) Success and failure of the Randall-Wilkins model for thermoluminescence in LiF(TLD). *Radiat. Protect. Dosim.* 47: 41–47.
Bos A., Pithers T., Gomez Roz J., Delgado A. (1993b) An intercomparison of glow curve analysis computer programs: I. synthetic glow curves. *Radiat. Protect. Dosim.* 47: 473–477.

Bos A., Vijverberg R., Pithers T., McKeever S. (1992) Effects of cooling and heating rate on trapping parameters in LiF: Mg,Ti crystals. *J. Phys. D.* D25: 1249–1257.

Botter-Jensen L. (1978) A simple hot N2 gas TL reader incorporating a post-irradiation annealing facility. *Nucl. Instr. Meth.* 153: 413–418.

Boyer A., Antonuk L., Fenster A., van Herk M., Meertens H., Munro P., Reinstein L., Wong J. (1992) A review of electronic portal imaging devices (EPIDs). *Med. Phys.* 19: 1–16.

Brahme A. (1984) Dosimetric precision requirements in radiation therapy. *Acta Radiol. Oncol.* 23: 379–391.

Brauer-Krisch E., Bravina A., Lerch M., Rosenfeld A., Stepanek J., Di Michiel M., Laissue J.A. (2003) MOSFET dosimetry for microbeam radiation therapy at the European Synchrotron Radiation Facility. *Med. Phys.* 30: 583–589.

Bucciolini M., Buonamici F. B., Mazzocchi S., De Angelis C., Onori S., Cirrone G.A. (2003) Diamond detector versus silicon diode and ion chamber in photon beams of different energy and field size. *Med. Phys.* 30: 2149–2154.

Budgell G.J., Perrin B.A., Mott J.H., Fairfoul J., Mackay R.I. (2005) Quantitative analysis of patient-specific dosimetric IMRT verification. *Phys. Med. Biol.* 50: 103–119.

Burch W.M. (1968) Two level isothermal readout for high precision thermoluminescence dosimetry with LiF. *Phys. Med. Biol.* 13: 627–634.

Busoli G. (1981) General characteristics of TLD materials. *Applied Thermoluminescence Dosimetry,* Edited by M. Oberhofer and A. Scharmann. Adam Hilger, Bristol, UK, 83–96.

Butson M.J., Cheung T., Yu P.K. (2006) Measurement of energy dependence for XRCT radiochromic film. *Med. Phys.* 33: 2923–2925.

Butson M., Yu P., Cheung T., Metcalfe P. (2003a) Radiochromic film for medical radiation dosimetry. *Material Sci. Eng.* R41: 61–120.

Butson M.J., Yu P.K., Cheung T. (2003b) Polarity effect on surface dose measurement for an attix parallel plate ionisation chamber. *Australas. Phys. Eng. Sci. Med.* 26: 85–87.

Butson M.J., Yu P.K., Cheung T., Metcalfe P. (2002) High sensitivity radiochromic film dose comparisons. *Phys. Med. Biol.* 47: N291–N295.

Butson M.J., Yu P.K., Metcalfe P.E. (1998) Measurement of off-axis and peripheral skin dose using radiochromic film. *Phys. Med. Biol.* 43: 2647–2650.

Butson M.J., Rozenfeld A., Mathur J.N., Carolan M., Wong T.P.Y., Metcalfe P.E. (1996a) New radiotherapy surface dose detector: The MOSFET. *Med. Phys.* 23(5): 655–658.

Butson M., Mathur J., Metcalfe P. (1996b) Radiochromic film as a radiotherapy surface-dose detector. *Phys. Med. Biol.* 41: 1073–1078.

Bydder E., Crouch A., Metcalfe P. (1988) In house detector systems for a Clinac 18 linear accelerator. *Australas. Phys. Eng. Sci. Med.* 11: 27–35.

Cadman P. (1998) Use of CEA TVS film for measuring high energy photon beam dose distributions. *Med. Phys.* 25: 1435–1437.

Cameron J., Suntharalingam N., Kenny G. (1968) *Thermoluminescence Dosimetry*, University of Wisconsin Press, Madison.

CCEMRI (1985) Effect of a change of stopping power values on the W value recommended by the ICRU for electrons in dry air. Bureau International de Poids et Measures Report 85–8. CCEMRI, Sevres.

Chappell R., Fowler F. (1994) Steepness of dose-response curve for larynx cancer. *Radiother. Oncol.* 30: 90.

Chen R. (1984) Kinetics of thermoluminescence glow peaks. *Thermoluminescence and Thermoluminescent Dosimetry.* By Y. Horowitz. CRC Press, Boca Raton, Florida, 50–88.

Cheng C.-W., Das I.J. (1996) Dosimetry of high energy photon and electron beams with CEA films. *Med. Phys.* 23: 1225–1232.

Chu W.C. (2001) Radiation dosimetry using Fricke-infused gels and magnetic resonance imaging. *Proc. Natl. Sci. Counc. Repub. China* B 25: 1–11.

Chu R., Van Dyk G., Lewis D., O'Hara K., Buckland B., Dinelle F. (1990) Gafchromic dosimetry media: a high dose, thin film routine dosimeter and dose mapping tool. *Radiat. Phys. Chem.* 35: 767–773.

Clift M.A., Sutton R.A., Webb D.V. (2000) Dealing with Cerenkov radiation generated in organic scintillator dosimeters by bremsstrahlung beams. *Phys. Med. Biol.* 45: 1165–1182.

Constantinou C. (1994) *Protocol and Procedures for Quality Assurance of Linear Accelerators.* Constantinou, Brockton, Massachusetts.

Constantinou C., Attix F.H., Paliwal B.R. (1982) A solid water phantom material for radiotherapy x-ray and gamma ray beam calibrations. *Med. Phys.* 9: 436–451.

Dawson D., Schroeder N., Hoya J. (1985) Penumbral measurements in water for high energy x-rays. *Med. Phys.* 13: 101–104.

Dawson D., Harper J., Akinradewo A. (1984) Analysis of physical parameters associated with the measurement of high energy x-ray penumbra. *Med. Phys.* 11: 491–497.

Day M.J. (1990) Radiation dosimetry using nuclear magnetic resonance: An introduction review. *Phys. Med. Biol.* 35: 1605–1609.

De Angelis C., Bucciolini M., Casati M., Lovik I., Bruzzi M., Lagomarsino S., Sciortino S., Onori S. (2006) Improvements in CVD diamond properties for radiotherapy dosimetry. *Radiat. Prot. Dosimetry* 120: 38–42.

De Deene Y. (2004) Fundamentals of MRI measurements for gel dosimetry. *J. Phys. Conf. Ser.* 3: 87–114.

De Deene Y. (2002) Gel dosimetry for the dose verification of intensity modulated radiotherapy treatments. *Z. Med. Phys.* 12: 77–88.

Delgado A., Gomez Ros J.M., Muniz J.L. (1995) Computerised analysis of LiF GR-200 TL signals: Application to dose measurements in the uGy range. *Radiat. Protect. Dosim.* 60(2): 147–53.

Delgado A., Gomez Roz J., Muniz J., Portillo J. (1992) Application of glow curve analysis methods to improve TLD-100 dose reassessment performance. *Health Phys.* 62: 228–234.

Denham J.W., Hamilton C.S., Simpson S.A., Ostwald P.M., O'Brien M., Kron T., Joseph D.J., Dear K.B. (1995) Factors influencing the degree of erythematous skin reactions in humans. *Radiother. Oncol.* 36: 107–120.

DeWerd L.A., Cameron J.R. (1984) Characteristics of a new dosimeter material: LiF(Mg,Cu,P). *Radiat. Protect. Dosim.* 6(1–4): 350–352.

Dietrich L., Tucking T., Nill S., Oelfke U. (2005) Compensation for respiratory motion by gated radiotherapy: An experimental study. *Phys. Med. Biol.* 50: 2405–2414.

Dini S.A., Koona R.A., Ashburn J.R., Meigoonia A.S. (2005) Dosimetric evaluation of GAFCHROMIC XR type T and XR type R films. *J. Appl. Clin. Med. Phys.* 6: 114–134.

Djouguela A., Kollhoff R., Ruhmann A., Willborn K.C., Harder D., Poppe B. (2006) Physical mechanism of the Schwarzschild effect in film dosimetry--theoretical model and comparison with experiments. *Phys. Med. Biol.* 51: 4345–4356.

Domen S. (1982) An absorbed water calorimeter: Theory design and performance. *J. Res. NBS* 87: 211–235. (Quoted from Ross and Klassen 1996.)

Doran S.J., Koerkamp K.K., Bero M.A., Jenneson P., Morton E.J., Gilboy W B. (2001) A CCD-based optical CT scanner for high-resolution 3D imaging of radiation dose distributions: Equipment specifications, optical simulations and preliminary results. *Phys. Med. Biol.* 46: 3191–3213.

Driscoll C., Barthe J., Oberhofer M., Busoli G., Hickman C. (1986) Annealing procedures for commonly used radiothermoluminescent materials. *Radiat. Protect. Dosim.* 14: 17–32.

Duggan L. Kron, T. (1999) Glow curve analysis of long-term stability of LiF:Mg,Cu,P as compared to LiF:Mg,Ti. *Radiat. Prot. Dosimetry* 85: 213–216.

Duggan L., Hood C., Warren-Forward H., Haque M., Kron T. (2004) Variations in dose response with x-ray energy of LiF:Mg,Cu,P thermoluminescence dosimeters: Implications for clinical dosimetry. *Phys. Med. Biol.* 49: 3831–3845.

Dunscombe P., McGhee P., Lederer E. (1996) Anthropomorphic phantom measurements for the validation of a treatment planning system. *Phys. Med. Biol.* 41: 399–411.

Duzenli C., Sloboda R., Robinson D. (1994) A spin-spin relaxation rate investigation of the gelatin ferrous sulphate NMR dosimeter. *Phys. Med. Biol.* 39: 1577–1592.

Ebert M., Kron T. (1994) Portal films for relative exit dose determination. *Australas. Phys. Eng. Sci. Med.* 17: 71–78.

Edwards C., Green S., Palethorpe J., Mountford P. (1997) The response of a MOSFET, ptype semiconductor and LiF TLD to quasi-monoenergetic x-rays. *Phys. Med. Biol.* 42:2383–2392.

el-Khatib E., Antolak J., Scrimger J. (1992) Evaluation of film and thermoluminescent dosimetry of high-energy electron beams in heterogeneous phantoms. *Med. Phys.* 19: 317–323.

Essers M., Boellaard R., van Herk M., Lanson J., Mijnheer B. (1996) Transmission dosimetry with a liquid filled electronic portal imaging device. *Int. J. Radiat. Oncol. Biol. Phys.* 34: 931–941.

Failla G. (1937) The measurement of tissue dose in terms of the same unit for all ionizing radiations. *Radiology* 29: 202.

Fairchild R., Mattern P., Lengweiler K., Levy P. (1978) Thermoluminescence of LiF TLD-100: Glow curve kinetics. *J. Appl. Phys*. 49: 4523–4533.

Fano U. (1954) Note on the Bragg-Gray cavity principle for measuring energy dissipation. *Radiat. Res.* 1: 237–240.

Feather J., Hajizadeh-Saffar M., Leslie G, Dawson J. (1989) A portable scanning reflectance spectrophotometer using visible wavelengths for the rapid measurement of skin pigments. *Phys. Med. Biol.* 34: 807–820.

Feather J., Ellis D., Leslie G. (1988) A portable reflectometer for the rapid quantification of cutaneous haemoglobin and melanin. *Phys. Med. Biol.* 33: 711–722.

Feist H. (1988) Einfluss des Regenerier- und Auswerteverfahrens auf das supralineare Verhalten von LiF-Thermolumineszenzdosimetern. *Strahlenther. Onkol.* 164: 223–227.

Fidanzio A., Azario L., Miceli R., Russo A., Piermattei A. (2000) PTW-diamond detector: dose rate and particle type dependence. *Med. Phys.* 27: 2589–2593.

Fluhs D., Heintz M., Indenkampen F., Wieczorek C. (1996) Direct reading measurement of absorbed dose with plastic scintillators—the general concept and applications to ophthalmic plaque dosimetry. *Med. Phys.* 23: 427–304.

Foelsing A. (1995) *Wilhelm Conrad Roentgen*. Hanser, Muenchen.

Fong P.M., Keil D.C., Does M.D., Gore J.C. (2001) Polymer gels for magnetic resonance imaging of radiation dose distributions at normal room atmosphere. *Phys. Med. Biol.* 46: 3105–3113.

Fricke H., Hart J. (1966) Chemical dosimetry. *Radiation Dosimetry.* Vol. II: Instrumentation. Edited by H. Attix, W. Roesch, and E. Tochilin, Academic Press, London, 167–239.

Fricke H., Morse S. (1927) The chemical action of roentgen rays on dilute ferrous sulfate solutions as a measure of dose. *Am. J. Roentgerol. Radiat. Ther.* 18: 430–432.

Fry, R.J.M. (1996) Effects of low doses of radiation. *Health Phys.* 70: 823–827.

Gagnon W. (1979) Measurement of surface dose. *Int. J. Radiat. Oncol. Biol. Phys.* 5: 449–450.

Gagnon W., Grant III W. (1975) Surface dose from megavoltage therapy machines. *Radiology* 117: 705–708.

Galbraith D., Rawlinson J., Munro P. (1984) Dose errors due to charge storage in electron irradiated plastic phantoms. *Med. Phys.* 11: 197–204.

Gambarini G., Arrigoni S., Cantone M., Mohlo N., Facchielli L., Sichirollo A. (1994) Dose-response curve slope improvement and result reproducibility of ferrous-sulphate-doped gels analysed by NMR imaging. *Phys. Med. Biol.* 39: 703–717.

Gastorf R., Humphries L., Rozenfeld M. (1986) Cylindrical chamber dimensions and the corresponding values of A_{wall} and $N_{gas}/(N_xA_{ion})$. *Med. Phys.* 13: 751–754.

Georg D., Kroupa B., Winkler P., Potter R. (2003) Normalized sensitometric curves for the verification of hybrid IMRT treatment plans with multiple energies. *Med. Phys.* 30: 1142–1150.

Gerbi B., Khan F. (1990) Measurement of dose in the build-up region using fixed separation plane parallel ionisation chambers. *Med. Phys.* 17: 17–26.

Gerbi B.J., Khan F.M. (1987) The polarity effect for commercially available plane-parallel ionization chambers. *Med. Phys.* 14: 210–215.

Goldfinch E., McKeever S., Scharman A. (1993) *Solid State Dosimetry*—proceedings of the 10th international conference on solid state dosimetry, Washington D.C. July 13–17, 1992. Nuclear Technology Publishing, Ashford, U.K.

Gooden D., Brickner T. (1972) The routine use of thermoluminescence dosimetry for radiation therapy. *Radiology* 102: 685–689.

Gore J.C., Kang Y., Schulz R.J. (1984) Measurement of radiation dose distribution by nuclear magnetic resonance (NMR) imaging. *Phys. Med. Biol.* 29: 1189–1197.

Greening J. (1981) *Fundamentals of Radiation Dosimetry,* HPA Medical Physics Handbook 6, Adam Hilger, Bristol, UK.

Greer P.B., Barnes M.P. (2007) Investigation of an amorphous silicon EPID for measurement and quality assurance of enhanced dynamic wedge. *Phys. Med. Biol.* 52: 1075–1087.

Griessbach I., Lapp M., Bohsung J., Gademann G., Harder D. (2005) Dosimetric characteristics of a new unshielded silicon diode and its application in clinical photon and electron beams. *Med. Phys.* 32: 3750–3754.

Grusell E., Rikner G. (1986) Evaluation of temperature effects in p-type silicon detectors. *Phys. Med. Biol.* 31: 527.

Gustavsson H., Karlsson A., Back S.A., Olsson L.E., Haraldsson P., Engstrom P., Nystrom H. (2003) MAGIC-type polymer gel for three-dimensional dosimetry: Intensity-modulated radiation therapy verification. *Med. Phys.* 30: 1264–1271.

Harder D. (1966) Physikalische Grundlagen der Dosimetrie. *Strahlentherapie,* Sonderband 62: 254–274 (in German).

Harris P., Piercy A., Baldock C. (1996) A method for determining the diffusion coefficient in Fe(II/III) radiation dosimetry gels using finite elements. *Phys. Med. Biol.* 41: 1745–1753.

Harvey J., Felstead S. (1979) Thin layer thermoluminescent dosemeters based on high temperature self-adhesive tape. *Phys. Med. Biol.* 24: 1250–1257.

Hassenstein E., Nuesslin F. (1985) Medizinische und physikalische Aspekte der Qualitaetssicherung in der Radioonkologie. *Strahlentherapie* 161: 685–693.

Havercroft J.M., Klevenhagen S.C. (1994) Polarity effect of plane-parallel ionization chambers in electron radiation. *Phys. Med. Biol.* 39: 299–304.

Hazle J., Hefner L., Nyerick C., Wilson L., Boyer A. (1991) Dose response characteristics of a ferrous-sulphate-doped gelatin system for determining radiation absorbed dose distributions by magnetic resonance imaging (Fe MRI). *Phys. Med. Biol.* 36: 1117–1125.

Hendee W., Ibbott G., Hendee E. (2004) *Radiation Therapy Physics*. 3rd edition. Wiley-Liss, New York.

Herman M.G. (2005) Clinical use of electronic portal imaging. *Semin. Radiat. Oncol.* 15: 157–167.

Heukelom S., Lanson J., Mijnheer B. (1991a) Comparison of entrance and exit dose measurements using ionisation chambers and silicon diodes. *Phys. Med. Biol.* 36: 47–59.

Heukelom S., Lanson J., van Tienhoven G., Mijnheer B. (1991b) In vivo dosimetry during tangential breast treatment. *Radiother. Oncol.* 22: 269–279.

Heydarian M., Hoban P., Beddoe A. (1996) A comparison of dosimetric techniques in stereotactic radiosurgery. *Phys. Med. Biol.* 41: 93–110.

Heydarian M., Hoban P., Beckham W., Borchardt I., Beddoe A. (1993) Evaluation of a PTW diamond detector for electron beam measurements. *Phys. Med. Biol.* 38: 1035–1042.

Higgins P.D., Sibata C.H., Siskind L., Sohn J.W. (1995) Deconvolution of detector size effect for small field measurement. *Med. Phys.* 22: 1663–1666.

Hill B., Venning A.J., Baldock C. (2005) A preliminary study of the novel application of normoxic polymer gel dosimeters for the measurement of CTDI on diagnostic x-ray CT scanners. *Med. Phys.* 32: 1589–1597.

Hilts M., Jirasek A., Duzenli, C. (2004) Effects of gel composition on the radiation induced density change in PAG polymer gel dosimeters: A model and experimental investigations. *Phys. Med. Biol.* 49: 2477–2490.

Hilts M., Audet C., Duzenli C., Jirasek A. (2000) Polymer gel dosimetry using x-ray computed tomography: a feasibility study. *Phys. Med. Biol.* 45: 2559–2571.

Hirning R. (1992) Detection and determination limits for thermoluminescence dosimetry. *Health Phys.* 62: 223–227.

Hoban P., Heydarian M., Beckham W., Beddoe A. (1994) Dose rate dependence of a PTW diamond detector in the dosimetry of a 6MV photon beam. *Phys. Med. Biol.* 39: 1219–1229.

Hood C., Kron T., Hamilton C., Callan S., Howlett S., Alvaro F., Back M. (2005) Correlation of 3D-planned and measured dosimetry of photon and electron craniospinal radiation in a pediatric anthropomorphic phantom. *Radiother. Oncol.* 77: 111–116.

Horowitz Y. (2006a) *Microdosimetric Response of Physical and Biological Systems to Low and High LET Radiations.* Elsevier, Amsterdam.

Horowitz Y. (2006b) A unified and comprehensive theory of the dose response of thermoluminescent systems applied to LiF:Mg,Ti. Chapter 2 in *Microdosimetric Response of Physical and Biological Systems to Low and High LET Radiations.* Y. Horowitz (ed.). Elsevier, Amsterdam.

Horowitz Y. (1993) LiF: Mg,Ti versus LiF: Mg,Cu,P: The competition heats up. *Radiat. Protect. Dosim.* 47(1): 135–141.

Horowitz Y. (1990) The annealing characteristics of LiF: Mg,Ti. *Radiat. Protect. Dosim.* 20: 219–230.

Horowitz Y. (1984a) General characteristics of TLD materials. *Thermoluminescence and Thermoluminescent Dosimetry.* CRC Press, Boca Raton, Florida, Vol. I: 89–172.

Horowitz Y. (1984b) TL dose response. *Thermoluminescence and Thermoluminescent Dosimetry*. CRC Press Boca Raton, Florida, Vol. II: 1–41.

Horowitz, Y. (1984c) *Thermoluminescence and Thermoluminescent Dosimetry,* Volumes I, II, and III. CRC Press, Boca Raton, Florida.

Horowitz Y. (1981) The theoretical and microdosimetric basis of thermoluminescence and applications to dosimetry. *Phys. Med. Biol.* 26: 765.

Horowitz Y., Yossian D. (1995) *Computerised Glow Curve Deconvolution: Application to Thermoluminescence Dosimetry.* Nuclear Technology Publishing, Ashford, UK.

Howlett S., Duggan L., Bazley S., Kron T. (1999) Selective in vivo dosimetry in radiotherapy using P-type semiconductor diodes: A reliable quality assurance procedure. *Med. Dosim.* 24: 53–56.

HPA (Hospital Physicists Association) (1977) *Phantom Materials for Photons and Electrons*. London.

Hurkmans C.W., Scheepers E., Springorum B.G., Uiterwaal H. (2005) Influence of radiotherapy on the latest generation of pacemakers. *Radiother. Oncol.* 76: 93–98.

Hurley C., Venning A., Baldock C. (2005) A study of a normoxic polymer gel dosimeter comprising methacrylic acid, gelatin and tetrakis (hydroxymethyl) phosphonium chloride (MAGAT). *Appl. Radiat. Isot.* 63: 443–456.

IAEA (International Atomic Energy Agency) (2004) An international protocol for absorbed dose determination. IAEA technical report series TRS-398 (V.11b). Vienna, Austria.

IAEA (International Atomic Energy Agency) (1987) *Absorbed Dose Determination in Photon and Electron Beams,* Technical Report Series N277. Vienna.

Ibbott G. (2006) Clinical applications of gel dosimeters. *J. Phys. Conf. Ser.* 56: 108–131.

ICRP (International Commission on Radiological Protection) (1975) *Report of the Task Group on Reference Man,* report 23. Pergamon Press, Oxford.

ICRU (International Commission on Radiation Units and Measurements) (1992) *Phantoms and Computational Models in Therapy, Diagnosis and Protection,* ICRU report 48. Bethesda, Maryland.

ICRU (International Commission on Radiation Units and Measurements) (1989) *Tissue Substitutes in Radiation Dosimetry and Measurements,* ICRU report 44. Bethesda, Maryland.

ICRU (International Commission on Radiation Units and Measurements) (1985) *Determination of Dose Equivalent Resulting from External Sources,* ICRU report 39. Bethesda, Maryland.

ICRU (International Commission on Radiation Units and Measurements) (1983) *Microdosimetry*, ICRU report 36. Bethesda, Maryland.

ICRU (International Commission on Radiation Units and Measurements) (1979) *Average Energy Required to Produce an Ion Pair,* ICRU report 31. Bethesda, Maryland.

ICRU (International Commission on Radiation Units and Measurements) (1976) *Determination of Absorbed Dose in a Patient Irradiated by Beams of X or Gamma Rays in Radiotherapy Procedures,* ICRU report 24. Bethesda, Maryland.

ICRU (International Commission on Radiation Units and Measurements) (1962) *Radiation Quantities and Units,* ICRU report 10A. Bethesda, Maryland.

IPSM (Institute of Physical Sciences in Medicine) (1988) *Commissioning and Quality Assurance of Linear Accelerators,* IPSM report 54, York.

Islam K.T., Dempsey J.F., Ranade M.K., Maryanski M.J., Low D.A. (2003) Initial evaluation of commercial optical CT-based 3D gel dosimeter. *Med. Phys.* 30: 2159–2168.

Iyengar G., Kollmer W., Bowen H. (1978) *The Elemental Composition of Human Tissues and Body Fluids.* Verlag Chemie, Weinheim.

Izewska J., Andreo P. (2000) The IAEA/WHO TLD postal programme for radiotherapy hospitals. *Radiother. Oncol.* 54: 65–72.

Izewska J., Bera P., Vatnitsky S. (2002) IAEA/WHO TLD postal dose audit service and high precision measurements for radiotherapy level dosimetry. International Atomic Energy Agency/World Health Organization. *Radiat. Prot. Dosimetry* 101: 387–392.

Jirasek A.I., Duzenli C., Audet C., Eldridge J. (2001) Characterization of monomer/crosslinker consumption and polymer formation observed in FT-Raman spectra of irradiated polyacrylamide gels. *Phys. Med. Biol.* 46: 151–165.

Johansson B., Wickman G., Holmstroem T. (1995) Properties of liquid ionization chambers at LDR brachytherapy dose rates. *Phys. Med. Biol.* 40: 575–587.

Johns H.E., Cunningham J.R. (1983) *The Physics of Radiology.* Charles C Thomas, Springfield, Illinois.

Jordan K. (2004) Advances in optical CT scanning for gel dosimetry. *J. Phys. Conf. Ser.* 3: 115–121.

Julius H. (1981) Instrumentation. *Applied Thermoluminescence Dosimetry*. By M. Oberhofer and A. Scharmann, Adam Hilger, Bristol, UK, 39–66.

Kanagaki B., Read P.W., Molloy J.A., Larner J.M., Sheng K. (2007) A motion phantom study on helical tomotherapy: The dosimetric impacts of delivery technique and motion. *Phys. Med. Biol.* 52: 243-255.

Kaplan G.I., Rosenfeld A.B., Allen B.J., Booth J.T., Carolan M.G., Holmes-Siedle A. (2000) Improved spatial resolution by MOSFET dosimetry of an x-ray microbeam. *Med. Phys.* 27: 239–244.

Karzmark C. (1993) *Medical Linear Accelerators.* McGraw-Hill, Inc., New York.

Karzmark C., White J., Fowler J. (1964) Lithium fluoride thermoluminescence dosimetry. *Phys. Med. Biol.* 9: 273–286.

Keall P.J., Mageras G.S., Balter J.M., Emery R.S., Forster K.M., Jiang S.B., Kapatoes J.M., Low D.A., Murphy M.J., Murray B.R., Ramsey C.R., van Herk M.B., Vedam S S., Wong J.W., Yorke E. (2006a) The management of respiratory motion in radiation oncology. Report of AAPM Task Group 76. *Med. Phys*. 33: 3874–3900.

Keall P., Vedam S.S., George R., Bartee C., Siebers J., Lerma F., Weiss E., Chung T. (2006) The clinical implementation of respiratory-gated intensity-modulated radiotherapy. *Med. Dosim.* 31: 152–162.

Kelly R.G., Jordan K.J., Battista J.J. (1998) Optical CT reconstruction of 3D dose distributions using the ferrous-benzoic-xylenol (FBX) gel dosimeter. *Med. Phys.* 25: 1741–1750.

Khan F. (2003) *The Physics of Radiation Therapy.* 3rd edition. Williams & Wilkins, Baltimore, Maryland.

Klassen N., van der Zwan L., Cygler J. (1997) GafChromic MD-55: Investigation as a precision dosimeter. *Med. Phys.* 24: 1924–1934.

Kleck J., Smathers J., Holly F., Myers L. (1990) Anthropomorphic radiation therapy phantoms: A quantitative assessment of tissue substitutes. *Med. Phys.* 17: 800–806.

Klein E., Purdy J. (1993) Entrance and exit dose regions for a Clinac 2100C. *Int. J. Radiat. Oncol. Biol. Phys.* 27: 429–435.

Klein E.E., Low D.A., Maag D., Purdy J.A. (1996) A quality assurance program for ancillary high technology devices on a dual-energy accelerator. *Radiother. Oncol.* 38: 51–60.

Klevenhagen S. (1993) *Physics and Dosimetry of Therapy Electron Beams*. Medical Physics Publishing, Madison, Wisconsin.

Kotler L., Webb D. (1996) Experimental determination of the effective point of measurement of a parallel plate ionisation chamber. *Australas. Phys. Eng. Sci. Med.* 19: 1–10.

Krieger H., Petzold W. (1989) *Strahlenphysik, Dosimetrie und Strahlenschutz,* Band 2. Teubner, Stuttgart (in German).

Kron T. (2005) Calibration of megavoltage radiation beams. Chapter 9 in *The Modern Technology of Radiation Oncology Volume 2*. J. Van Dyk (ed.). Medical Physics Publishing, Madison, Wisconsin, pp. 301–385.

Kron T. (1999) Dosimetric tools. Chapter 19 in *The Modern Technology of Radiation Oncology*. J. Van Dyk (ed.). Medical Physics Publishing, Madison, Wisconsin, pp. 753–821.

Kron T. (1995) Thermoluminescence dosimetry and its applications in medicine—part 2: history and applications. *Australas. Phys. Eng. Sci. Med.* 18(1): 1–25.

Kron T. (1994) Thermoluminescence dosimetry and its applications in medicine—part 1: physics, materials and equipment. *Australas. Phys. Eng. Sci. Med.* 17: 175–199.

Kron T., Ostwald P. (1995) Skin exit dose in megavoltage x-ray beams determined by means of a plane parallel ionisation chamber (Attix chamber). *Med. Phys.* 22: 129–137.

Kron T., Pope J. (1994) Dose distribution measurements in superficial X-ray beams using NMR dosimetry. *Phys. Med. Biol.* 39: 1337–1349.

Kron T., McNiven A., Witruk B., Kenny M., Battista J. (2006) An experimental study of recombination and polarity effect in a set of customized plane parallel ionization chambers. *Australas. Phys. Eng. Sci. Med.* 29: 291–299.

Kron T., Jordan K., Battista J. (2003) Tomotherapy Images from London. Canadian Organization of Medical Physics Newsletter. *Interactions* 49: 97.

Kron T., Hamilton C., Roff M., Denham J. (2002a) Dosimetric intercomparison for two Australasian clinical trials using an anthropomorphic phantom. *Int. J. Radiat. Oncol. Biol. Phys.* 52: 566–579.

Kron T., Haque M., Foulke K., Jeraj R. (2002b) A flattening filter for brachytherapy skin irradiation. *Phys. Med. Biol.* 47: 713–722.

Kron T., Rosenfeld A., Lerch M., Bazley S. (2002c) Measurements in radiotherapy beams using on-line MOSFET detectors. *Radiat. Prot. Dosimetry* 101: 445–448.

Kron T., Duggan L., Smith T., Rosenfeld A., Butson M., Kaplan G., Howlett S., Hyodo K. (1998) Dose response of various radiation detectors to synchrotron radiation. *Phys. Med. Biol.* 43: 3235–3259.

Kron T., Jonas D., Pope J.M. (1997) Fast T_1 imaging of dual gel samples for diffusion measurements in NMR dosimetry gels. *Magn. Reson.Imaging* 15: 211–221.

Kron T., Butson M., Hunt F., Denham J. (1996) Skin dose measurements in breast cancer patients using TLD extrapolation. *Radiother. Oncol.* 41: 119–123.

Kron T., Murray A., Metcalfe P. (1994) Addendum to the penumbra of a 6-MV x-ray beam as measured by thermoluminescent dosimetry and evaluated using an inverse square root function [*Med. Phys.* 20: 1429–1438 (1993)]. *Med. Phys.* 21: 1261.

Kron T., Butson M., Wong T., Metcalfe P. (1993a) Readout of thermoluminescence dosimetry chips using a contact planchet heater. *Australas. Phys. Eng. Sci. Med.* 16: 137–142.

Kron T., Elliott A., Wong T., Showell G., Clubb B., Metcalfe P. (1993b) X-ray surface dose measurements using TLD-extrapolation. *Med. Phys.* 20: 703–711.

Kron T., Metcalfe P., Pope J.M. (1993d) Investigation of the tissue equivalence of gels used for NMR dosimetry. *Phys. Med. Biol.* 38: 139–150.

Kron T., Metcalfe P., Wong T. (1993e) Thermoluminescence dosimetry of therapeutic x-rays with LiF ribbons and rods. *Phys. Med. Biol.* 38: 833–845.

Kron T., Schneider M., Murray A., Mameghan H. (1993f) Clinical thermoluminescence dosimetry: How do expectations and results compare? *Radiother. Oncol.* 26: 151–161.

Kron T., Schmiedeberg D., Schneider M., Oliver L. (1990) Thermoluminescence dosimetry in the Prince of Wales Hospital, Sydney. *Australas. Phys. Eng. Sci. Med.* 13: 192–196.

Kubo H.D., Wang L. (2000) Compatibility of Varian 2100C gated operations with enhanced dynamic wedge and IMRT dose delivery. *Med. Phys.* 27: 1732–1738.

Kutcher G.J., Coia L., Gillin M., Hanson W., Leibel S., Morton R.J., Palta J., Purdy J., Reinstein L., Svensson G., et al. (1994) Comprehensive QA for radiation oncology: Report of AAPM therapy committee task group 40. *Med. Phys.* 21: 581–618. AAPM Report No. 46.

Lakshmanan A., Bhatt R. (1981) Supralinearity and LET dependence of thermoluminescence dosimeters. *Phys. Med. Biol.* 26: 1157–1163.

Lambert J., McKenzie D.R., Law S., Elsey J., Suchowerska N. (2006) A plastic scintillation dosimeter for high dose rate brachytherapy. *Phys. Med. Biol.* 51: 5505–5516.

Langmack K.A. (2001) Portal imaging. *Br. J. Radiol.* 74: 789–804.

Laub W.U., Wong T. (2003) The volume effect of detectors in the dosimetry of small fields used in IMRT. *Med. Phys.* 30: 341–347.

Leach M. (1992) Spatially localised nuclear magnetic resonance. *The Physics of Medical Imaging.* Edited by S. Webb. IOP Publishing, Bristol, UK, pp. 389–397.

Lee K.Y., Fung K.K., Kwok C.S. (2006) Application of high-resolution radiochromic film dosimetry in verifying a small-field stereotactic radiosurgery plan. *Appl. Radiat. Isot.* 64: 934–939.

Lepage M. (2006) Magnetic resonance in polymer gel dosimetry: Techniques and optimisation. J. *Phys. Conf. Ser.* 56: 86–94.

Litovchenko P., Barabash L., Rosenfeld A., Khivrich V., Zinets O., Kuts V., Marusan I., Petrov V., Sluchenkov G., Koval G., et al. (1990) MOS structure for emergency gamma and proton dosimetry. *Radiat. Prot. Dosimetry* 33: 179–182.

Li H.H., Gonzalez A.L., Ji H., Duggan D.M. (2007) Dose response of BaFBrl: Eu2+ storage phosphor plates exposed to megavoltage photon beams. *Med. Phys.* 34: 103–111.

Liu Q., McDermott P., Burmeister J. (2007) Effect of respiratory motion on the delivery of breast radiotherapy using SMLC intensity modulation. *Med. Phys.* 34: 347–351.

Liu G., van Doorn T., Bezak E. (2004) The linear accelerator mechanical and radiation isocentre assessment with an electronic portal imaging device (EPID). *Australas. Phys. Eng. Sci. Med.* 27: 111–117.

MacDougall N.D., Pitchford W.G., Smith M.A. (2002) A systematic review of the precision and accuracy of dose measurements in photon radiotherapy using polymer and Fricke MRI gel dosimetry. *Phys. Med. Biol.* 47: R107–R121.

Mack A., Scheib S.G., Major J., Gianolini S., Pazmandi G., Feist H., Czempiel H., Kreiner H.J. (2002) Precision dosimetry for narrow photon beams used in radiosurgery-determination of Gamma Knife output factors. *Med. Phys.* 29: 2080–2089.

Mansfield C., Suntharalingam N. (1976) Thermoluminescence dosimetry in radiation oncology. *Applied Radiology* 2: 43–48.

Marczewska B., Bilski P., Olko P., Nesladek M., Rebisz M., Guerrero M.J. (2006) On the dose response of some CVD diamond thermoluminescent detectors. *Radiat. Prot. Dosimetry* 119: 319–322.

Marczewska B., Olko P., Nesladek M., Waligorski M.P., Kerremans Y. (2002) CVD diamonds as thermoluminescent detectors for medical applications. *Radiat. Prot. Dosimetry* 101: 485–488.

Marshall T. (1984) Accuracy and precision in thermoluminescence dosimetry. *Practical Aspects of Thermoluminescence Dosimetry.* By A. Hufton. Hospital Physicists Association, London, 12–22.

Martens C., De Wagter C., De Neve W. (2000) The value of the PinPoint ion chamber for characterization of small field segments used in intensity-modulated radiotherapy. *Phys. Med. Biol.* 45: 2519–2530.

Martensson B. (1969) Thermoluminescence of LiF: a statistical analysis of the influence of pre-annealing on the precision of measurements. *Phys. Med. Biol.* 14: 119–130.

Maryanski M., Ibbot G., Eastman P., Schulz R.J., Gore J.C. (1996a) Radiation therapy dosimetry using magnetic resonance imaging and polymer gels. *Med. Phys.* 23: 699–705.

Maryanski M., Zastavker Y., Gore J. (1996b) Radiation dose distributions in three-dimensions from tomographic optical density scanning of polymer gels: II. Optical properties of the BANG polymer gel. *Phys. Med. Biol.* 41: 2705–2717.

Maryanski M., Schulz R., Ibbott G., Gatenby J., Xie J., Horton D., Gore J.C. (1994a) Magnetic resonance imaging of radiation dose distributions using a polymer-gel dosimeter. *Phys. Med. Biol.* 39: 1437–1455.

Maryanski M., Gore J., Kennan B., Schulz R. (1994b) NMR relaxation enhancement in gels polymerized and cross-linked by ionizing radiations: A new approach to 3-D dosimetry by MRI. *Magn. Reson. Imaging* 11: 253–258.

Mather M.L., De Deene Y., Whittaker A.K., Simon G.P., Rutgers R., Baldock C. (2002) Investigation of ultrasonic properties of PAG and MAGIC polymer gel dosimeters. *Phys. Med. Biol.* 47: 4397–4409.

Maucari T., Kase K. (1987) Effects of ionization chamber construction on dose measurements in a heterogeneity. *Med. Phys.* 14: 653–656.

Mayles W., Heisig S., Mayles H. (1993) Treatment verification and in vivo dosimetry. *Radiotherapy Physics.* By J. Williams and D. Thwaites. Oxford University Press, Oxford, pp. 227–251.

McAuley K. (2006) Fundamentals of polymer gel dosimeters. *J. Phys. Conf. Ser.* 56: 35–44.

McCullough E., Holmes T. (1985)Acceptance testing computerised radiation therapy treatment planning systems: Direct utilization of CT scan data. *Med. Phys.* 12: 237–242.

McDermott L.N., Wendling M., Sonke J.J., van Herk M., Mijnheer B.J. (2007) Replacing pretreatment verification with in vivo EPID dosimetry for prostate IMRT. *Int. J. Radiat. Oncol. Biol. Phys.* 67: 1568–1577.

McEwen M.R., Duane S. (2000) A portable calorimeter for measuring absorbed dose in the radiotherapy clinic. *Phys. Med. Biol.* 45: 3675–3691.

McJury M., Oldham M., Cosgrove V.P., Murphy P.S., Doran S., Leach M.O., Webb S. (2000) Radiation dosimetry using polymer gels: methods and applications. *Br. J. Radiol.* 73: 919–929.

McKeever S., Moscovitch M., Townsend P. (1995) *Thermoluminescence Dosimetry Materials: Properties and Uses.* Nuclear Technology Publishing, Ashford, UK.

McKeever S., Akselrod M., Markey B. (1996) Pulsed optically stimulated luminescnece dosimetry using -Al_2O_3:C. *Radiat. Prot. Dosimetry* 65: 267–272.

McKerracher C., Thwaites D.I. (2006) Notes on the construction of solid-state detectors. *Radiother. Oncol.* 79: 348–351.

McKinlay A. F. (1981) *Thermoluminescence Dosimetry*, Medical Physics Handbook 5. Adam Hilger, Bristol, UK.

McNiven A.L., Mulligan M., Kron T., Battista J.J. (2006) The response of prototype plane-parallel ionization chambers in small megavoltage x-ray fields. *Med. Phys.* 33: 3997–4004.

McNiven A., Kron T. (2004) Interpolation and extrapolation of dose measurements with different detector sizes to improve the spatial resolution of radiotherapy dosimetry as demonstrated for helical tomotherapy. *Phys. Med. Biol.* 49: 3665–3674.

Meertens H., van Herk M., Bijhold J., Bartelink H. (1990) First clinical experience with a newly developed electronic portal imaging device. *Int. J. Radiat. Oncol. Biol. Phys.* 18: 1173–1181.

Meertens H., van Herk M., Weeda J. (1985) A liquid ionisation detector for digital radiography of therapeutic megavoltage photon beams. *Phys. Med. Biol.* 30: 313–321.

Mellenberg D. (1990) Determination of build-up overresponse corrections for a Markus type chamber. *Med. Phys.* 17: 1041–1044.

Michalski J., Wong J., Gerber R., Yan D., Cheng A., Graham M., Renna M., Sawyer P., Perez C. (1993) The use of on-line image verification to estimate the variation in radiation therapy dose delivery. *Int. J. Radiat. Oncol. Biol. Phys.* 27: 707–716.

Mobit P., Sandison G. (2006) Cavity theory. Chapter 5 in *Microdosimetric Response of Physical and Biological Systems to Low and High LET Radiations*. Y. Horowitz (ed.). Elsevier, Amsterdam, pp. 331–366.

Mobit P., Mayles P., Nahum A. (1996) The quality dependence of LiF TLD in megavoltage photon beams: Monte Carlo simulation and experiments. *Phys. Med. Biol.* 41: 387–398.

Mohammadi M., Bezak E. (2007) Evaluation of MLC leaf positioning using a scanning liquid ionization chamber EPID. *Phys. Med. Biol.* 52: N21–N33.

Morgan A.M., Aird E.G., Aukett R.J., Duane S., Jenkins N.H., Mayles W.P., Moretti C., Thwaites D.I. (2000) IPEM guidelines on dosimeter systems for use as transfer instruments between the UK primary dosimetry standards laboratory (NPL) and radiotherapy centres. *Phys. Med. Biol.* 45: 2445–2457.

Mota H., Sibata C., Roberts W., Higgins P. (1990) Film dosimetry: Linearisation of dose response for relative measurements of dose distribution. *Phys. Med. Biol.* 35: 565–569.

Mouton J., Haug R., Bridier A., Dodinot B., Eschwege F. (2002) Influence of high-energy photon beam irradiation on pacemaker operation. *Phys. Med. Biol.* 47: 2879–2893.

Muench P., Meigooni A., Nath R., McLaughlin W. (1991) Photon energy dependence of the sensitivity of radiochromic film and comparison with silver halide and LiF TLDs used for brachytherapy dosimetry. *Med. Phys.* 18: 769–775.

Mukherjee B., Vana N. (1984) Self-absorption effect in LiF thermoluminescent dosimeters. *Nucl. Instr. Meth.* 226: 572–573.

NACP (Nordic Association of Clinical Physicists) (1980) Procedures in external radiation therapy dosimetry with electron and photon beams with maximum energies between 1 and 50 MeV. *Acta Radiol. Oncol.* 19: 55–79.

Nariyama N., Tanaka S., Nakane Y., Asano,Y., Hirayama H., Ban S., Nakaskima H., Namito Y. (1997) Responses and glow curves of $Li_2B_4O_7$:Cu, BeO and $CaSO_4$:Tm TLDs to 10–40 keV monoenergetic photons from synchrotron radiation. *Radiat. Prot. Dosimetry* 74: 155–161.

Nariyama N., Tanaka S., Yoshizawa M., Hirayama H., Ban S., Nakaskima H., Namito Y., Nakane N. (1993) Responses of LiF TLDs to 10-40 keV monoenergetic photons from synchrotron radiation. *Radiat. Prot. Dosimetry* 49: 451–457.

Nash A., Attix F. (1971) Use of LiF teflon discs in laminated identification cards for personnel accident dosimetry. *Health Phys.* 21: 435–439.

Neitzel U. (2005) Status and prospects of digital detector technology for CR and DR. *Radiat. Prot. Dosimetry* 114: 32–38.

Nilsson B., Montelius A. (1986) Fluence perturbation in photon beams under non-equilibrium conditions. *Med. Phys.* 13: 191–195.

Nioutsikou E., Richard N. Symonds-Tayler J., Bedford, J.L., Webb S. (2006) Quantifying the effect of respiratory motion on lung tumour dosimetry with the aid of a breathing phantom with deforming lungs. *Phys. Med. Biol.* 51: 3359–3374.

Niroomand-Rad A., Blackwell C.R., Coursey B.M., Gall K.P., Galvin J.M., McLaughlin W.L., Meigooni A.S., Nath R., Rodgers J.E., Soares C.G. (1998) Radiochromic film dosimetry: Recommendations of AAPM Radiation Therapy Committee Task Group 55. *Med. Phys.* 25:2093–2115. AAPM Report No. 63.

Olch A.J. (2005) Evaluation of a computed radiography system for megavoltage photon beam dosimetry. *Med. Phys.* 32: 2987–2999.
Olch A.J. (2002) Dosimetric performance of an enhanced dose range radiographic film for intensity-modulated radiation therapy quality assurance. *Med. Phys.* 29: 2159–2168.
Oldham M. (2006) 3D dosimetry by optical CT scanning. *J. Phys. Conf. Ser.* 56: 58–71.
Oldham M. (2004) Optical CT scanning for of polymer gels. J. Phys. Conf. Ser. 3: 122–135.
Oldham M., Kim L., Hugo G. (2005) Optical-CT imaging of complex 3D dose distributions. *J. Phys.* 5745: 138–146.
Oldham M., Siewerdsen J.H., Shetty A., Jaffray D.A. (2001) High resolution gel-dosimetry by optical-CT and MR scanning. *Med. Phys.* 28: 1436–1445.
Olko P., Bilski P., Michalik V.M. (1994) Microdosimetric analysis of the response of LiF thermoluminescent detectors for radiations of different qualities. *Radiat. Protect. Dosim.* 52(1–4): 405–408.
Olsen D.R., Hellesnes J. (1994) Absorbed dose distribution measurements in brachytherapy using ferrous sulphate gel and magnetic resonance imaging. *Br. J. Radiol.* 67: 1121–1126.
Olsson L., Arndt J., Fransson A., Nordell B. (1992) Three dimensional dose mapping from gamma-knife treatment using a dosimeter gel and MR-imaging. *Radiother. Oncol.* 24: 82–86.
Olsson L., Fransson A., Ericsson A., Matsson S. (1990) MR imaging of absorbed dose distributions for radiotherapy using ferrous sulphate gels. *Phys. Med. Biol.* 35: 1623–1632.
Ostwald P., Kron T., Hamilton C. (1996) Assessment of mucosal underdosing in larynx irradiation. *Int. J. Radiat. Oncol. Biol. Phys.* 36: 181–187.
Pai S., Reinstein L.E., Gluckman G., Xu Z., Weiss T. (1998) The use of improved radiochromic film for in vivo quality assurance of high dose rate brachytherapy. *Med. Phys.* 25: 1217–1221.
Parent L., Seco J., Evans P.M., Dance D.R., Fielding A. (2006) Evaluation of two methods of predicting MLC leaf positions using EPID measurements. *Med. Phys.* 33: 3174–3182.
Paschalis T., Sandilos P., Tatsis E., Karaiskos P., Antypa, C., Chatzigiannis C., Dardoufas K., Georgiou E., Vlachos L. (2006) Dosimetric evaluation of a new collimator insert system for stereotactic radiotherapy. *Br. J. Radiol.* 2006 Dec 6 [Epub ahead of print].
Pearce J., Thomas R., Dusautoy A. (2006) The characterization of the Advanced Markus ionization chamber for use in reference electron dosimetry in the UK. *Phys. Med. Biol.* 51: 473–483.
Petric M.P., Robar J.L., Clark B.G. (2006) Development and characterization of a tissue equivalent plastic scintillator based dosimetry system. *Med. Phys.* 33: 96–105.
Petzold W., Krieger H. (1988) *Strahlenphysik, Dosimetrie und Strahlenschutz,* Band 1. Teubner, Stuttgart (in German).
Planskoy B. (1980) Evaluation of diamond radiation dosimeters. *Phys. Med. Biol.* 25: 519 532.
Podgorsak M., Schreiner L.J. (1992) Nuclear magnetic relaxation characterisation of irradiated Fricke solution. *Med. Phys.* 19: 87–95.
Podgorsak E., Morgan P., Cameron J. (1972) Interpretation of resolved glow curve shapes in LiF(TLD100) from 100K to 500K. *Proceedings of the 3rd International Conference on Luminescence Dosimetry 1971* Oct. 11–14, 1971. Danish Atomic Energy Commission, Risoe, Denmark, 1–8.
Pradhan A.S. (1995) Influence of heating rate on the TL response of LiF TLD-700, LiF: Mg,Cu,P and Al2O3: C. *Radiat. Protect. Dosim.* 58(3): 205–209.
Prasad P., Nalcioglu O., Rabbabi B. (1991) Measurement of three dimensional radiation dose distributions using MRI. *Radiat. Res.* 128: 1–13.
Purdy J. (1986) Buildup/surface dose and exit dose measurements for a 6MV linear accelerator. *Med. Phys.* 13: 259–262.
Quast U., Fluhs D., Bambynek M. (1998) Endovascular brachytherapy—treatment planning and radiation protection. *Herz* 23: 337–346.

Rae W., Willemse C., Loetter M., Engelbrecht J., Swarts J. (1996) Chelator effect on ion diffusion in ferrous-sulfate-doped gelatin gel dosimeters as analyzed by MRI. *Med. Phys.* 23(1): 15–23.

Rajan G. (1992) *Advanced Medical Radiation Dosimetry.* Prentice Hall of India, New Delhi.

Ramani R., Russell S., O'Brien P. (1997) Clinical dosimetry using MOS-FETs. *Int. J. Radiat. Oncol. Biol. Phys.* 37: 959–964.

Randall J., Wilkins M. (1945) Phosphorescence and electron traps II. The interpretation of long period phosphorescence. *Proceedings of the Royal Society of London, Series A: Mathematical and Physical Sciences,* A184: 390–407.

Rao R.P. (1993) Luminescence in medical sciences: An introductory review. *Physica Medica* 9: 1–12.

Rawlinson J., Arlen D., Newcombe D. (1992) Design of parallel plate ionisation chambers for build-up measurements in megavoltage photon beams. *Med. Phys.* 19: 641–648.

Regulla D. (1981) Operational aspects. Applied Thermoluminescence Dosimetry. By M. Oberhofer and A. Scharmann: Adam Hilger, Bristol, UK, 123–141.

Reich H. (1990) *Dosimetrie ionisierender Strahlung.* Teubner, Stuttgart (in German).

Renner W.D., Sarfaraz M., Earl M.A., Yu C.X. (2003) A dose delivery verification method for conventional and intensity-modulated radiation therapy using measured field fluence distributions. *Med. Phys.* 30: 2996–3005.

Rikner G. (1986) Silicon diodes as detectors in relative dosimetry of photon, electron and proton radiation fields PhD thesis, University of Uppsala, Uppsala.

Rikner G., Grusell E. (1987) General specifications for silicon diodes for use in radiation dosimetry. *Phys. Med. Biol.* 32: 1109–1117.

Rikner G., Grusell E. (1983) Effects of radiation damage on p-type silicon detectors. *Phys. Med. Biol.* 28: 1261–1267.

Rintoul L., Lepage M., Baldock C. (2003) Radiation dose distribution in polymer gels by Raman spectroscopy. Appl. Spectrosc. 57: 51–57.

Rogers D. (1992) The advantages of absorbed dose calibration factors. *Med. Phys.* 19: 1227–1239.

Röntgen W. (1895) Ueber eine neue Art von Strahlen. *Sitzungsberichte Der Wuerzburger Medizinisch Physikalischen Gesellschaft:* 137–147.

Rosenfeld A.B. (2002) MOSFET dosimetry in modern radiation oncology modalities. *Radiat. Prot. Dosimetry* 101: 393–398.

Rosenfeld A.B. (2006) Semiconductor radiation detectors in modern radiation therapy. Chapter 6 in *Microdosimetric Response of Physical and Biological Systems to Low and High LET Radiations.* Y. Horowitz (ed.). Elsevier, Amsterdam, pp. 367–410.

Rosenfeld A.B., Lerch M.L.F, Kron T., Brauer-Krisch E., Bravin A., Holmes-Siedle A. (2001) Feasibility study of on-line, high spatial resolution MOSFET dosimetry in static and pulsed X-ray radiation fields. *IEEE Trans. on Nucl. Sci.* 48: 2061–2068.

Rosenfeld A.B., Carolan M.G., Kaplan G.I., Allen B.J., Khivrich V.I. (1995) MOSFET dosimeters: The role of encapsulation on dosimetric characteristics in mixed gamma-neutron and megavoltage x-ray fields. *IEEE Trans. Nucl. Sci.* 42(6): 1870–1876.

Ross C., Klassen N. (1996) Water calorimetry for radiation dosimetry. *Phys. Med. Biol.* 41: 1–29.

Ross C., Klassen N., Short K., Smith G. (1989) A direct intercomparison of water calorimetry and Fricke dosimetry. *Phys. Med. Biol.* 34: 23–42.

Rossi H. (1968) Microscopic energy distribution in matter. *Radiation dosimetry.* Vol. I: Fundamentals. Edited by H. Attix, W. Roesch, and E. Tochilin. Academic Press, London, 43–92.

Rossi H., Zaider M. (1991) Elements of microdosimetry. *Med. Phys.* 18: 1085–1092.

Rousseau J., Gibon D., Sarrazin T., Doukhan N., Marchandise X. (1994) Technical Note: Magnetic resonance imaging of an agarose gel phantom for assessment of three-dimensional dose distribution in linac radiosurgery. *Br. J. Radiol.* 67: 646–648.

Rowlands J.A. (2002) The physics of computed radiography. *Phys. Med. Biol.* 47:R123–R166.

Rustgi S. (1995) Evaluation of the dosimetric characteristics of a diamond detector for photon beam measurements. *Med. Phys.* 22: 567–570.

Rustgi S., Frye D. (1995) Dosimetric characterization of radiosurgical beams with a diamond detector. *Med. Phys.* 22: 2117–2121.

Saur S., Strickert T., Wasboe E., Frengen J. (2005) Fricke gel as a tool for dose distribution verification: Optimization and characterization. *Phys. Med. Biol.* 50: 5251–5261.

Saylor M., Tamargo T., McLaughlin W.L., Khan H.M., Lewis D.F., Schenfele R.D. (1988) A thin film recording medium for use in food irradiation. *Radiat. Phys. Chem.* 31: 529–536.

Scheib S., Crescenti R., Vogelsanger W., Schenkel Y., Gianolini S., Lazaridis E., Mack, A. (2006) [Application of normoxic polymer gels in 3D-dosimetry for radiosurgery]. *Z. Med. Phys.* 16: 180–187.

Schreiner L.J. (2004) Review of Fricke gel dosimeters. *J. Phys. Conf. Ser.* 3: 9–21.

Schreiner L.J., Parker W., Henri C., Evans M.D.C., Podgorsak E.B. (1995) 3-D MRI radiation dosimetry using the magnetization to dose (M&D) calibration technique. *Med. Phys.* 22: 951–952. (abstract).

Schreiner L.J., Crooks I., Evans M.D.C., Keller B.M., Parker W.A. (1994) Imaging of HDR brachytherapy dose distributions using NMR Fricke-gelatin dosimetry. *Magn. Reson. Imaging* 12: 901–907.

Schulz R., Maryanski M., Ibbot G., Bond J. (1993) Assessment of the accuracy of stereotactic radiosurgery using Fricke-infused gels and MRI. *Med. Phys.* 20: 1731–1735.

Schulz R., deGuzman A., Nguyen D., Gore J. (1990) Dose response curves for Fricke infused gels as obtained by nuclear magnetic resonance. *Phys. Med. Biol.* 35: 1611–1620.

Shalev S. (1995) Treatment verification using digital imaging. *Radiation Therapy Physics*. A. Smith (ed.). Springer, Berlin, pp. 155–173.

Shui A., Otte V., Hogstrom K. (1989) Measurement of dose distributions using film in therapeutic electron beams. *Med. Phys.* 16: 911–915.

Skladowski K., Law M., Maciejewski B., Steel G. (1994) Planned and unplanned gaps in radiotherapy: The importance of gap position and gap duration. *Radiother. Oncol.* 30: 109–120.

Sliski A., Soares C., Mitch M.G. (2006) A fibre optic scintillator dosemeter for absorbed dose measurements of low-energy x-ray-emitting brachytherapy sources. *Radiat. Prot. Dosimetry* 120: 24–27.

Smith A., Sha G.A., Kron T. (1998) Variation of patient dose in head CT. *Br. J. Radiol.* 71: 1296–1301.

Soares C.G. (2006) New developments in radiochromic film dosimetry. *Radiat. Prot. Dosimetry* 120: 100–106.

Solan A.N., Solan M.J., Bednarz G., Goodkin M.B. (2004) Treatment of patients with cardiac pacemakers and implantable cardioverter-defibrillators during radiotherapy. *Int. J. Radiat. Oncol. Biol. Phys.* 59: 897–904.

Song H., Ahmad M., Deng J., Chen Z., Yue N.J., Nath R. (2006) Limitations of silicon diodes for clinical electron dosimetry. *Radiat. Prot. Dosimetry* 120: 56–59.

Soubra M., Cygler J., Mackay G. (1994) Evaluation of a dual bias dual metal oxide-silicon semiconductor field effect transistor detector as radiation dosimeter. *Med. Phys.* 21: 567–572.

Spanne P. (1984) TL readout instrumentation. *Thermoluminescence and Thermoluminescent Dosimetry.* By Y. Horowitz. CRC Press, Boca Raton, Florida, pp. 1–47.

Spinks J., Woods R. (1964) *An Introduction to Radiation Chemistry.* John Wiley & Sons, Inc., New York.

Suchowerska N., Hoban P., Butson M., Davison A., Metcalfe P. (2001) Directional dependence in film dosimetry: Radiographic and radiochromic film. *Phys. Med. Biol.* 46: 1391–1397.

Tangboonduangjit P., Metcalfe P., Butson M., Quach K.Y., Rosenfeld A. (2004) Matchline dosimetry in step and shoot IMRT fields: A film study. *Phys. Med. Biol.* 49: N287–N292.

Tangboonduangjit P., Wu I., Butson M., Rosenfeld A., Metcalfe P. (2003) Intensity modulated radiation therapy: flm verifcation of planar dose maps. *Australas. Phys. Eng. Sci. Med.* 26: 194–199.

Tarte B.J., Jardine P.A., van Doorn T. (1996) Laser-scanned agarose gel sections for radiation field mapping. *Int. J. Radiat. Oncol. Biol. Phys.* 36: 175–179.

Thomas S., Palmer N. (1989) The use of carbon-loaded thermoluminescent dosimeters for the measurement of surface dose in megavoltage x-ray beams. *Med. Phys.* 16: 902–904.

Thomson I., Thomas R., Berndt L. (1984) Radiation dosimetry with MOS sensors. *Radiat. Protect. Dosim.* 6: 121–124.

Tochilin E., Goldstein N. (1966) Dose rate and spectral measurements from pulsed x-ray generators. *Health Phys*. 12: 1705–1713.

Toivonen M. (1993) Improving the accuracy of TLD systems in clinical applications. *Radiat. Protect. Dosim.* 47: 497–503.

Trapp J.V., Michael G., de Deene Y., Baldock C. (2002) Attenuation of diagnostic energy photons by polymer gel dosimeters. *Phys. Med. Biol.* 47: 4247–4258.

Trapp J.V., Back S.A., Lepage M., Michael G., Baldock C. (2001) An experimental study of the dose response of polymer gel dosimeters imaged with x-ray computed tomography. *Phys. Med. Biol.* 46: 2939–2951.

Trevert E. (1896) *Something about X Rays for Everybody.* Bubier, Lynn, Massachusetts. Reprinted by Medical Physics Publishing, Madison, Wisconsin, 1988.

Tubiana M., Dutreix J., Wambersie A. (1990) *Introduction to Radiobiology.* Taylor & Francis, London.

Vahc Y.W., Chung W.K., Park K.R., Lee J.Y., Lee Y.H., Kwon O., Kim S. (2001) The properties of the ultramicrocylindrical ionization chamber for small field used in stereotactic radiosurgery. *Med. Phys.* 28: 303–309.

van Battum L., Huizenga H. (1990) Film dosimetry of clinical electron beams. *Int. J. Radiat. Oncol. Biol. Phys.* 18: 69–76.

van Dam J., Leunens G., Dutreix A. (1990) Correlation between temperature and dose-rate dependence of semiconductor response—influence of accumulated dose. *Radiother. Oncol.* 19: 345–351.

Van Dyk J. (ed.) (2005) *The Modern Technology of Radiation Oncology, Volume 2.* Medical Physics Publishing, Madison, Wisconsin.

Van Dyk J. (ed.) (1999) *The Modern Technology of Radiation Oncology.* Medical Physics Publishing, Madison, Wisconsin.

van Gasteren J., Heukelom S., van Kleffens H., van der Laarse R., Venselaar J., Westerman C. (1991) The determination of phantom and collimator scatter components of the output of megavoltage photon beams. *Radiother. Oncol.* 20: 250–257.

van Herk M. (1991) Physical aspects of a liquid-filled ionization chamber with pulse polarizing voltage. *Med. Phys.* 18: 692–702.

van Herk M., Meertens H. (1988) A matrix ionization chamber imaging device for on-line patient setup verification during radiotherapy. *Radiother. Oncol.* 11: 369–378.

Velkley D., Manson D., Purdy J., Oliver G. (1975) Build-up region of megavoltage photon radiation sources. *Med. Phys.* 2: 14–19.

Vieira S.C., Bolt R.A., Dirkx M.L., Visser A.G., Heijmen B.J. (2006) Fast, daily linac verification for segmented IMRT using electronic portal imaging. *Radiother. Oncol.* 80: 86–92.

Warkentin B., Steciw S., Rathee S., Fallone B.G. (2003) Dosimetric IMRT verification with a flat-panel EPID. *Med. Phys.* 30: 3143–3155.

Watanabe Y., Patel G.N., Patel P. (2006) Evaluation of a new self-developing instant film for imaging and dosimetry. *Radiat. Prot. Dosimetry* 120: 121–124.

Webb G. (1967) Thermal annealing effects in the thermoluminescence of lithium fluoride Teflon dosimeters. *Health Phys.* 13: 814–816.

Webster E.W. (1995) X-rays in diagnostic radiology. *Health Phys.* 69: 610–635.

Weinhous M., Meli J. (1984) Determination of Pion, the correction factor for recombination losses in an ionisation chamber. *Med. Phys.* 11: 846–849.

Westermark M., Arndt J., Nilsson B., Brahme A. (2000) Comparative dosimetry in narrow high-energy photon beams. *Phys. Med. Biol.* 45: 685–702.

White D.R. (1978) Tissue substitutes in experimental radiation physics. *Med. Phys.* 5: 467–479.

White D.R. (1977) The formulation of tissue substitute materials using basic interaction data. *Phys. Med. Biol.* 22: 889–899.

White D.R., Constantinou C, Martin R. (1986) Foamed epoxy resin-based lung substitutes. *Br. J. Radiol.* 59: 787.

Whittington R., Bloch P., Hutchinson D., Bjärngard B.E. (2002) Verification of prostate treatment setup using computed radiography for portal imaging. *J. Appl. Clin. Med. Phys.* 3: 88–96.

Wickman G. (1974) Radiation quality independent liquid ionisation chamber for dosimetry of electron radiation from medical accelerators. *Acta Radiol.* 13: 37–47.

Wickman G., Holmstroem T. (1992) Polarity effect in plane parallel ionisation chambers using air or dielectric liquid as ionization medium. *Med. Phys.* 19: 637–640.

Wickman G., Nystroem H. (1992) The use of liquids in ionization chambers for high precision radiotherapy dosimetry. *Phys. Med. Biol.* 37: 1789–1812.

Williams J., Thwaites D. (1993) *Radiotherapy Physics.* Oxford University Press, Oxford.

Wolf D., Dean D., Cytacki E., Evans R. (1989) The use of a spreadsheet computer program to expedite TLD reporting. *Med. Dosim.* 14: 125–127.

Wong E., D'Souza D.P., Chen J.Z., Lock M., Rodrigues G., Coad T., Trenka K., Mulligan M., Bauman G S. (2005) Intensity-modulated arc therapy for treatment of high-risk endometrial malignancies. *Int. J. Radiat. Oncol. Biol. Phys.* 61: 830–841.

Wong E., Chen J.Z., Greenland J. (2002) Intensity-modulated arc therapy simplified. *Int. J. Radiat. Oncol. Biol. Phys.* 53: 222–235.

Wong J., Slessinger E., Hermes R., Offutt C., Roy T., Vannier M. (1990) Portal dose images I: Quantitative treatment plan verification. *Int. J. Radiat. Oncol. Biol. Phys.* 18: 1455–1463.

Woodard H., White D.R. (1986) The composition of body tissue. *Br. J. Radiol.* 59: 1209.

Wu S., Jin X., Xie C., Cao G. (2005) Optimal steel thickness combined with computed radiography for portal imaging of nasopharyngeal cancer patients. *Med. Phys.* 32: 3112–3116.

Yamada S., Murase K. (2005) Effectiveness of flexible noise control image processing for digital portal images using computed radiography. *Br. J. Radiol.* 78: 519–527.

Yan Y., Papanikolaou N., Weng X., Penagaricano J., Ratanatharathorn V. (2005) Fast radiographic film calibration procedure for helical tomotherapy intensity modulated radiation therapy dose verification. *Med. Phys.* 32: 1566–1570.

Yin Z., Hugtenburg R.P., Beddoe A.H. (2004) Response corrections for solid-state detectors in megavoltage photon dosimetry. *Phys. Med. Biol.* 49: 3691–3702.

Ying X., Geer L., Wong J. (1990) Portal dose images II: patient dose estimation. *Int. J. Radiat. Oncol. Biol. Phys.* 18: 1465–1475.

Zeidan O.A., Stephenson S.A., Meeks S.L., Wagner T.H., Willoughby T.R., Kupelian P.A., Langen K.M. (2006) Characterization and use of EBT radiochromic film for IMRT dose verification. *Med. Phys.* 33: 4064–4072.

Zhu L., Kron T., Barnes K., Johannson S., O'Brien P. (1997) Junctioning of lateral and antrior fields in head and neck cancer: A dosimetric assessment of the monoisocentric technique (including reproducibility). *Int. J. Radiot. Oncol. Biol. Phys.* 41: 227–232.

Questions

3.1 Most dosimeters consist of many different materials (wall, structural support, electrodes, and connections). Please list dosimeters which consist (or could consist) of only one material. Please discuss briefly their characteristics (Hint: You should consider at least two dosimeter types). (sections 3.7 and 3.9)

3.2 Please complete the following table:

	Typical Diameter (mm)	*Typical Height (mm)*	*Advantages and Disadvantages for Measurements in a Water Phantom*	*Typical Applications (Please list two per chamber type)*
Parallel plate ionization chamber				
Cylindrical/ thimble ionization chamber				

3.3 J.A. Rawlinson et al. suggested in 1992 a design for a parallel plate ionization chamber with a fixed separation for surface dose measurements in megavoltage beams. The design of this chamber is redrawn in the figure below. Please discuss the features of this chamber design.

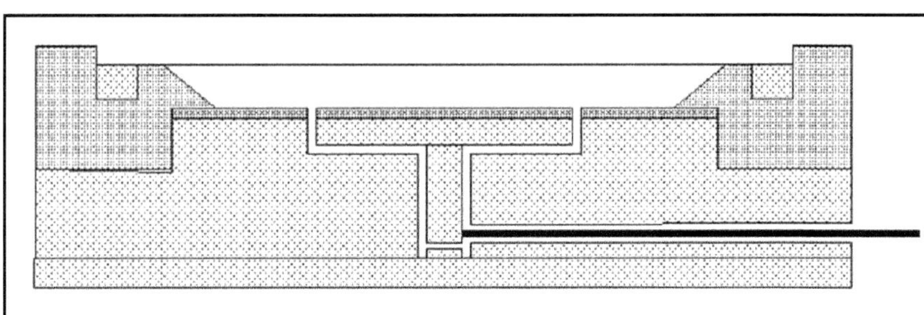

(section 3.5.3)

3.4 Please list some advantages of semiconductor diode detectors for radiotherapy dosimetry.
(section 3.6.1)

3.5 What does the term "glow curve" mean?
(section 3.7)

3.6 Please compare the dosimetric features of radiographic and radiochromic film.
(section 3.8)

3.7 Please state the typical spatial resolution for the following detectors:
- NE 2571 Farmer type ionization chamber
- TLD100 ribbon
- Radiographic film evaluated with a typical densitometer
- A silicon diode detector for water tank measurements (e.g., Scanditronix)

(section 3.4)

3.8 Using a water tank computerized beam-scanning system, a beam profile in a 6 MV x-ray beam is measured using a cylindrical ionization chamber of 0.2 cc volume. The scanning system is moving stepwise and it is adjusted to measure for 0.2 seconds in each location before moving to the next measurement position. What charge is approximately collected for each measurement point when acquiring a profile at 10 cm depths?

If one were to determine the collimator jaw transmission (which is of the order of 0.5%) in this way, what charge needs to be detectable by the electrometer?
(section 3.3)

3.9 A radiographic film has been exposed to a beam of 6 MV photons with the developed film shown schematically in the figure below. A simple object has been placed in the center of the film during exposure. Can you give a plausible explanation for the resultant film pattern and explain what the object is?
(section 3.4)

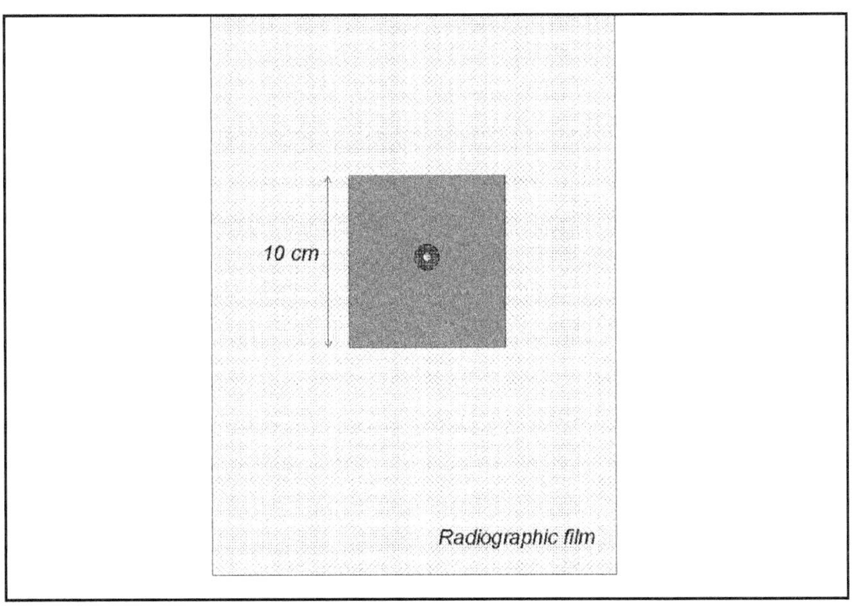

3.10 You are called to a give advice on the treatment of a patient who has a pacemaker. The patient is treated for lung cancer and the margins of the proposed treatment field are just 1 cm away from the pacemaker. The policy in the radiotherapy center is to:

- take no action if the total dose is below 2 Gy.
- calculate and verify the dose if the dose is expected to be between 2 Gy and 10 Gy.
- explant the pacemaker if the total dose is expected to exceed 10 Gy.

The treating clinician expects the dose to the pacemaker to be between 2 Gy and 10 Gy and asks for verification. What methods would you consider to assess this problem? (Hint: Both in phantom and in vivo measurements should be considered.)
(section 3.4)

4 X-Ray Beam Properties

> *"Scatter-air ratios are empirical quantities derived from tissue-air ratios for use in calculating dose from scattered radiation. . . ."*
>
> *(Jack Cunningham 1972)*

4.1 Introduction

To ascertain the dose that patients receive during radiotherapy treatment a series of manual or computerized dose assessment methods are used. These methods employ basic beam data that has been collected in various phantoms using the detection methods outlined in chapter three. The data collected can be presented in various forms such as depth dose curves and dose profiles, which are presented and discussed in this chapter. Because each linac has its own unique depth dose and beam profile properties, the data presented here are not identical to other like machines. However, the data provide a valuable guide for comparison with other machine data (BJR 1996).

All linacs are calibrated according to one of the dose calibration protocols outlined in chapter three. As the dose monitor chambers of the linac are placed above the collimators, the calibration dose is only valid for one field size. Hence the effect of field size changes on dose output is accounted for by the use of a field factor as outlined in section 4.2.

In megavoltage x-ray beams, some skin sparing is achieved because of the range of electrons set in motion prior to dose deposition. This leads to a dose build-up region where the beam reaches maximum dose (D_{max}) at a depth (d_{max}), e.g., for a 6 MV x-ray beam d_{max} is 1.5 cm. Detailed central axis dose build-up data is presented in section 4.3. The impact of electron contamination on surface dose is discussed, as are the effects of field size and block tray on the magnitude of surface dose.

After the x-ray beam has penetrated to the d_{max} position there is a gradual fall-off in dose deposited. The magnitude of this fall-off in dose is very important when considering how much dose will be delivered at depth. In section 4.4 the depth dose fall-off characteristics of x-ray beams are outlined. Examples of depth dose curves and how they are affected by energy, source-surface distance (SSD), and field size are summarized. Depth dose curves vary with SSD. Two related quantities that are SSD independent are introduced in section 4.5; these are known as the tissue-phantom ratio (TPR) and tissue-air ratio (TAR). These quantities are particularly useful for fixed source-axis distance (SAD) treatments that include stationary and arc isocentric treatment techniques.

The term dose profile is used to describe dose data collected by scanning a dosimeter perpendicular to the beam axis. The data are graphically presented with dose (vertical axis) versus off-axis distance (horizontal axis). Dose profile data are usually collected across the beam in line with the crosshairs or across the beam diagonals. Dose profiles can be collected at any specified depth and are unique to the depth of collection.

The dose profile is relatively flat to within about 1 cm of the field edge; this region is known as the inter umbral region. Near the edges of the field there is a rapid fall-off in the dose profile, which is known as the penumbral region. In section 4.6 dose profiles in the inter umbral and penumbral regions are discussed. Dose profiles vary depending on whether symmetric or asymmetric beams are used. There is a rapid dose gradient in the penumbra. Dose in this region may be modeled by penumbral forming functions, several of which are discussed in section 4.6.4.

Beam attenuation and modulation using blocks, MLCs, wedges, and dynamic wedges is discussed in section 4.7. MLCs and dynamic wedges are having an increasing impact on efficient patient treatments as these devices can be activated without need to enter the room between treatment fields. Because these devices are being used more frequently, the understanding of how physics properties of the x-ray beam are affected by MLC and dynamic wedges is discussed in detail. Different manufacturers have produced different MLC designs and the impact of this on beam properties is also discussed in section 4.7.

Kilovoltage and orthovoltage x-rays are commonly used in most radiotherapy clinics for treatment of skin cancers and for palliation. These beams show interesting and contrasting properties to megavoltage beams in the build-up region, fall-off region, and in their interactions with high Z materials; these aspects are discussed in section 4.8.

4.2 Calibration Field Factors

Ionization chambers in the head of the linac monitor the dose output. Feedback circuits from these chambers have a calibration adjustment at the console control that can be adjusted to match a calibrated dose rate. The units recorded at the control console from the linac ionization chambers are known as monitor units (MUs). For example, 1 MU may be calibrated to equal 1 cGy at 100 cm SSD for a 10×10 cm^2 field size at d_{max} in water. This is achieved following a calibration protocol (e.g., AAPM 1999, IAEA 2004) as described in chapter three. However 1 MU equals a different dose at different field sizes. The monitor ionization chambers are located above the final beam collimation system; the wider the collimators are set, the more dose will be delivered per MU, due to an increase in scatter. To compensate for this effect, field factors are applied in the planning dose calculation (see table 4.1a). When these factors are applied to the 1 cGy per MU for a 10×10 cm^2 field, this gives the dose applied at d_{max} for the new field size.

Note that the field factor is collimator dependent. The historical explanation is that extra scatter was collected by the monitor chambers because the elongated upper collimators are closer. This caused faster monitor chamber dose accumulation and hence dose termination at a lower dose level recorded at the patient/phantom position. Therefore, the field factor was less in comparison to elongated lower collimators. A more recent explanation is that scatter from the flattening filter is the component that affects the jaw-dependent output. As the jaws are at different distances from the flattening filter, the area of flattening filter exposed to the measuring volume is jaw dependent due to beam divergence. If the equivalent square of the jaws exposed to the flattening filter is plotted against head scatter factor, the jaw dependence is resolved (Lam et al. 1996). If the magnitude of the effect is significant, collimator dependent field factors may be issued for clinical radiotherapy plan calculations.

To measure *field factors* an ionization chamber may be placed at d_{max} in water with 100 cm SSD to the top of the water. The field size is altered, and the difference in charge output is given as a ratio to the charge for a standard field size. This is defined in this text as the field factor and includes dose increase due to scatter from within the phantom and outside (the collimators). An alternative approach to measuring this factor is to place the detector at d_5, measure the charge, and convert this reading back to dose at d_{max} by including the percentage depth dose reading for the respective field size. There are two arguments for using the second approach:

i) The position of d_{max} changes slightly with field size.
ii) The calibration protocols call for calibration at a reasonable depth in the dose fall-off region (e.g., d_5). This leads to field factors being derived at the same depth. Note that field factors refer in this text to dose at a reference point with full scatter. Hence they could equally be measured at 100 cm SAD, provided this is specified in the field factor table.

Output factor information may be measured in a different way. For those centers that choose to separate collimator dose information from in-phantom scatter information, an ionization chamber is usually placed *in air* at isocenter with a build-up cap. The collimator component of the dose output taken as a ratio to the dose for a standard field size is defined here as the collimator scatter factor (S_c). This is what Khan defines as the *output factor* (Khan 1984). (See table 4.1b.) The in-phantom scatter factor (S_p) can then be ascertained from the difference between in-air and in-phantom data. The combination of these two quantities as a product is sometimes called the *area factor*. For further description of how to achieve this, the reader is referred to the following articles: Ten Haken et al. (1996), Bjärngard (1996), and van Gasteren (1991). If the area factor is measured at 100 cm SAD and the treatment is to the patient at 100 cm SSD, an *inverse square* conversion (INVSQ) is required where

$$INVSQ = \left(\frac{SAD}{SSD + d_{max}} \right)^2 . \tag{4.1}$$

The output factor calculation can also be affected by the introduction of beam modifiers, such as wedge factors and block tray factors, which tend to attenuate the beam. Examples of factors introduced to correct for these devices are given in table 4.1c and 4.1d, respectively.

4.3 Build-Up Region

4.3.1 Surface Dose

Electrons that are set in motion by high-energy x-rays interacting near the patient surface have a range of several millimeters (mm). Therefore maximum dose is achieved at depth beneath the patient skin surface, and skin sparing is achieved when high-energy x-rays are used for radiotherapy treatment. This means that doses that exceed skin tolerance dose at d_{max} can be administered because the skin dose is less. However, the surface dose and near surface dose are not negligible, as electrons are produced from x-rays striking beam modifying devices such as collimators, blocks, and block trays. An illustration of this phenomenon is shown in figure 4.1. These electrons have a long range in air, and the lowest energy contamination electrons produce dose deposition at the surface. The remainder of the incident dose at the surface is due to electrons from backscattered photon interactions in the patient and from electrons produced in the air volume between the linac head and the patient. The angle of incidence of the beam to the skin and the exit dose also contribute significantly to patient skin dose.

The skin is prone to early damage in the basal cell layer, which lies at about 0.07 mm depth (Orton and Seibert 1972). The dose gradient in this region is steep, hence surface dose does not give an exact estimate of skin dose. Late radiation effects originate within the dermis at 0.05 mm to 3 mm where blood vessel damage may occur (ICRP 1991). Therefore dose at the surface and down to about 3 mm depth should preferably be assessed for clinical situations.

The use of head immobilization devices (head masks) can also add significantly to skin dose. The standard cellulose acetate heat mold material varies in thickness, but for example assume it is 3 mm thick (density 1.2). This in effect puts the skin surface at about 3.6 mm depth on the build-up curve. The sensitive skin layer is then between 3.6 and 6.6 mm on the build-up curve with an associated increase in dose to the skin.

Table 4.1: Data tables showing factors that account for linac dose output changes. Data measured for a 6 MV x-ray beam from a Varian 2100C linac

(a) Field factors measured against field size. These need to be applied as a ratio to alter the monitor units set in order to deliver 1 cGy/MU at 100 cm SSD in a phantom.

Field Size (cm^2)	*Output (MU/100 cGy)*	*Field factor (rel. 10×10 cm^2)*
5×5	106	0.940
10×10	100	1.000
15×15	97	1.031
20 × 20	94	1.056

(b) Field factors measured against field size and showing the influence of field size.

Field Size (cm^2)	*Field Factor++*	*Collimator Scatter Factor (S_c) or Output Factor+++*
5×5	0.943	0.960
10×10	1.000	1.000
20×20	1.056	1.033
30×30	1.086	1.050
40×40	1.104	1.063
5L×40U+	1.000	0.976
40L×5U+	1.027	1.011

+ Note: The use of the equivalent square (ESQ) method for calculating field factors can be inaccurate for elongated fields. Field factors also vary with collimator orientation. Hence, data issued for clinical use generally include collimator orientation.

++ With ionization chamber in water at 100 cm SSD.

+++ With ionization chamber and build-up cap in air at 100 cm SAD.

(c) Wedge factors for a 10×10 cm^2 field.

Wedge Angle (degrees)	*Output (MU/100 cGy)*	*Wedge Factor*
15	128	0.78
30	159	0.63
45	204	0.49
60	244	0.41

(d) Block tray factors and block transmission.

Block Tray	*Factor*
metal (honeycomb)	0.995
PMMA (6 mm)	0.970
PMMA (10 mm)	0.951

Block	*Transmission*
10 cm lead	0.7%
10 cm LMA+	3.8%

+Low melting point alloy (LMA).

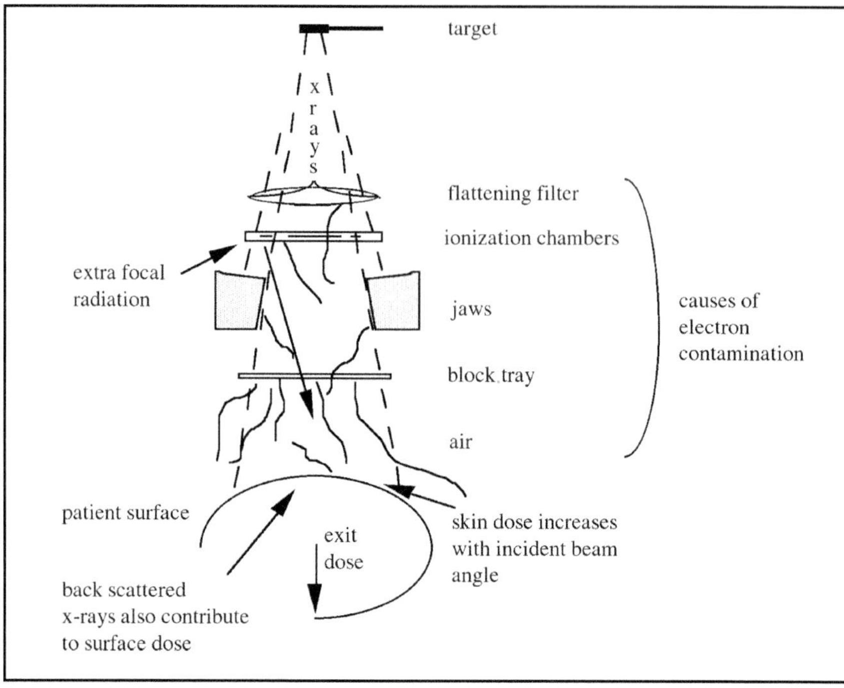

Figure 4.1: Contributions to patient skin dose from x-ray beams.

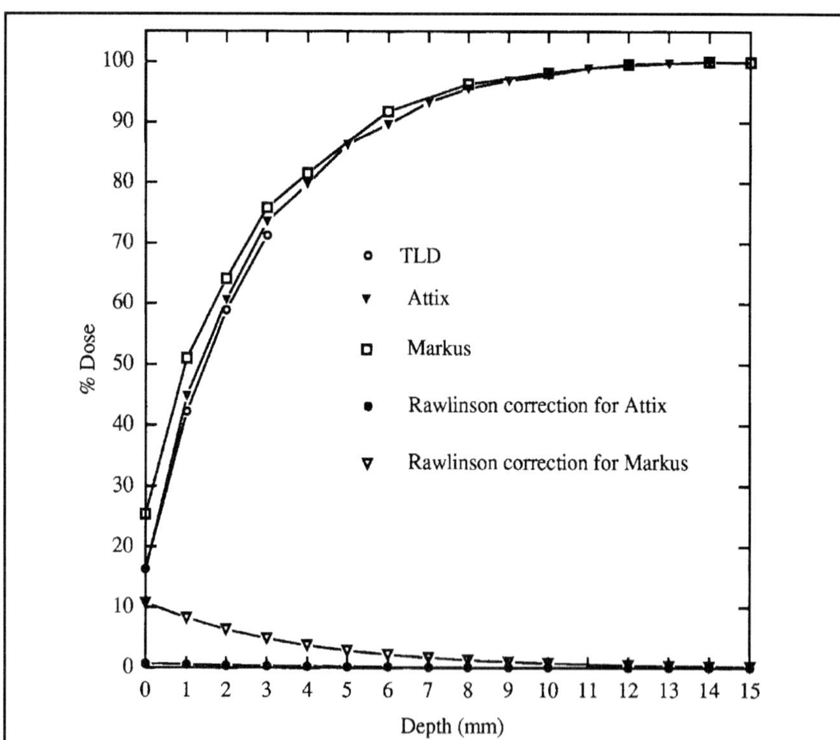

Figure 4.2: The build-up curve in the central axis for 6 MV x-rays as measured using Attix and Markus chambers and thin TLDs. Curves showing the magnitude of the Rawlinson correction for electron in-scattering for both the Attix and Markus chambers are also shown.

Total removal of head cast material in the field is usually not practical as marks for field crosshair alignment need to be maintained. The physical integrity of the mask also needs to be maintained in order to avoid excessive patient movement. Many centers randomly mill large gaps in about half of the shell area of the mask that falls within the field. This has the effect of sparing at least some of the skin in the field and helps basal cell repopulation from nearby batches. Note that some new mask materials have a mesh of heat sensitive plastic (for molding to shape) with large air gaps between the mesh. This adds to the skin sparing in these open areas for the patient (Fiorino et al. 1992).

4.3.2 Dose Build-Up

The dose deposition sharply increases in the build-up region over the first few millimeters. This is predominantly due to dose deposition by electrons generated in the medium. Electrons (not x-rays) deposit dose, and electrons have a finite longitudinal and lateral range (Metcalfe et al. 1990).

Accurate measurement of dose in the build-up region requires the use of different detector systems, usually parallel plate chambers and TLD techniques. The overresponse of some parallel plate chambers has been discussed in detail elsewhere in chapter three.

A comparison of Markus chamber results (corrected and uncorrected) with Attix and thin TLD extrapolation measurements in the dose build-up region are shown in figure 4.2. Note that in such a steep dose gradient region large differences in dose relate to small spatial differences in depth. Note also that the Attix chamber, which has a wide guard ring, is the more ideal parallel plate chamber for surface dose measurements because the in-scattering correction is minimal. The chamber has a wide diameter, however, and small field sizes still need to be monitored with other chambers (e.g., Markus).

Table 4.2 is a review of x-ray beam surface dose measurements obtained by various authors for 6 MV x-rays using various detection methods on different linacs. The results agree well and appear to be relatively independent of accelerator type.

A comparison of dose build-up at three different energies is shown in figure 4.3. The energies compared are 1.25 MeV (cobalt 60), 4 MV, and 6 MV. It is interesting that the surface doses are similar, but the doses between 2 mm and 3 mm are very different. This data is relevant to breast and head/neck treatment as there is a requirement to keep near surface dose reasonably high to reduce the risk of superficial recurrences.

The surface dose also increases with field size; this is probably due to the larger surface area of the jaws and the larger air volume contributing extra contamination components (along with some extra backscatter). Shown in figure 4.4 are the dose build-up curves for 6 MV x-rays versus field size; the greater the field size, the higher the x-ray surface dose. As beam energy increases, the depth of d_{max} increases. For example, the build-up depth is 15 mm for 6 MV, compared with 33 mm for 18 MV. This is because electrons produced in the medium are also of higher energy, and their range is longer then d_{max} is at greater depth. The position of d_{max} is described as the position of transient longitudinal equilibrium. For a detailed discussion see Attix

Table 4.2: Summary of surface dose measurements for 6 MV x-rays including various surface dose measurements. Block tray measurements are for a 6 mm PMMA block tray.

Detector (Reference)	*Accelerator Type*	*10×10 cm² Open (Tray)*	*20×20 cm² Open (Tray)*
TLD extrapolation (Kron et al. 1993)	Clinac 2100C	16.3 (17.7)	26.9 (34.2)
Markus—uncorrected (Kron et al. 1993)	Clinac 2100C	25.4 (27.8)	36.6 (43.4)
Markus—corrected (Kron et al. 1993)	Clinac 2100C	17.1 (25.9)	25.9 (32.7)
extrapolation chamber (Mallenberg et al. 1990)	Clinac 6/100	12.5	23.6
extrapolation chamber (Gagnon et al. 1975)	Clinac 6/100	11.2 (13.0)	23.6 (32.0)
parallel plate (other) (Purdy 1986)	Clinac 6/100	12.5	23.6
parallel plate (Attix) (Cross 1993)	Siemens Mevatron	13.5	23.3
parallel plate (Attix) (Butson et al. 1994)	Clinac 2100C	15.2 (17.1)	26.4 (31.8)

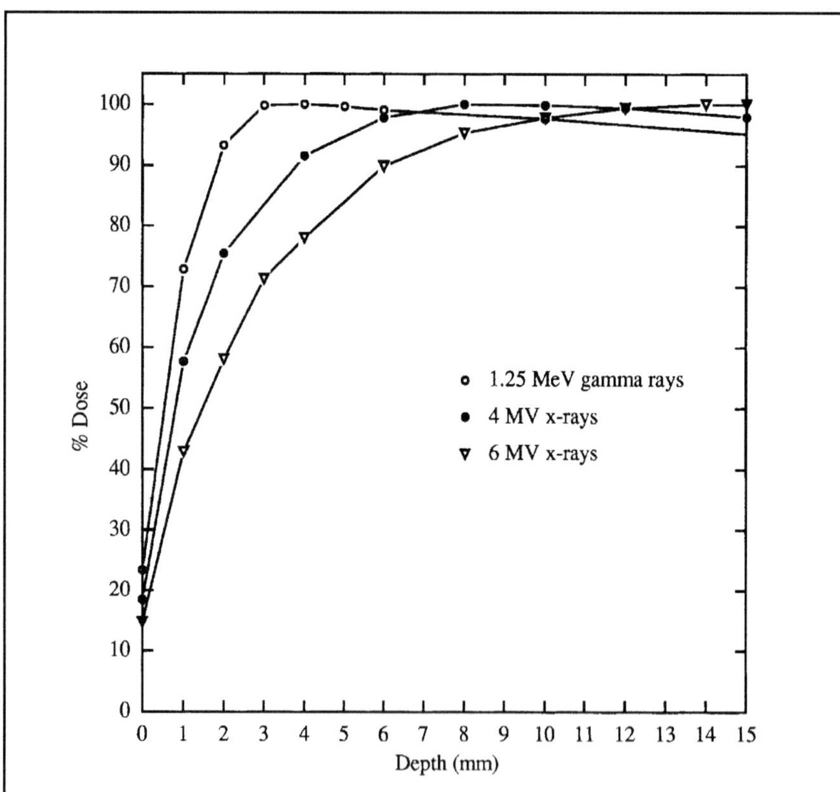

Figure 4.3: Build-up curves for Cobalt 60 (Theratronics 780), 4 MV (Varian 4/80), and 6 MV (Varian Clinac 2100C) x-rays of 10×10 cm² open field size. Showing the change in surface dose and build-up curves when beam energy is different.

(1979) and Dutreix et al. (1965, 1966). The position of d_{max} is shallower at small field sizes, then reaches a maximum depth at about 5×5 cm^2 field size, then gets progressively shallower again. This has been shown to occur for energies ranging from 6 to 18 MV (Sixel and Podgorsak 1994).

As shown in figure 4.5, the surface dose is increased when electron contamination is increased by electrons striking the collimators, block tray holder, or the air column. Block trays, usually constructed from 6 mm polymethylmethacrylate (PMMA), are commonly used for practical reasons (i.e., they are transparent). Polycarbonate is also sometimes used instead of PMMA; these trays are very strong but tend to flex a little more. The contamination and x-ray transmission properties of polycarbonate block tray holders are similar to PMMA.

Steel grid block tray holders are also available, made from a honeycomb-shaped steel mesh. Their extensive use in the clinic has some practical constraints. It is more difficult to align blocks, as the light field produces a grid-shaped shadow on the patient. The material also tends to bend and flex. However, electron contamination levels are lower using this block tray than those from block trays of solid construction. Electron contamination is similar to open field levels. (See figure 4.5.) This is because the device consists of large air gaps due to its honeycomb construction and steel, which self-absorbs most of the contamination it produces.

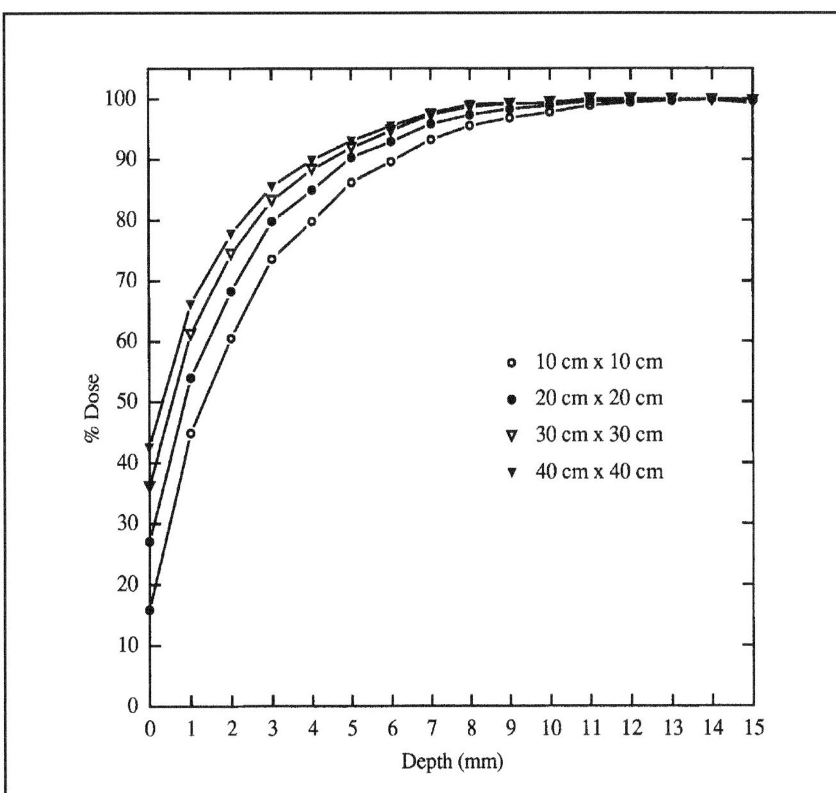

Figure 4.4: Build-up curves for 6 MV x-ray open fields of size 10×10 cm^2, 20×20 cm^2, 30×30 cm^2, and 40×40 cm^2. Note the increase in surface dose when field size is increased. Data collected using Attix chamber and corrected using Rawlinson correction.

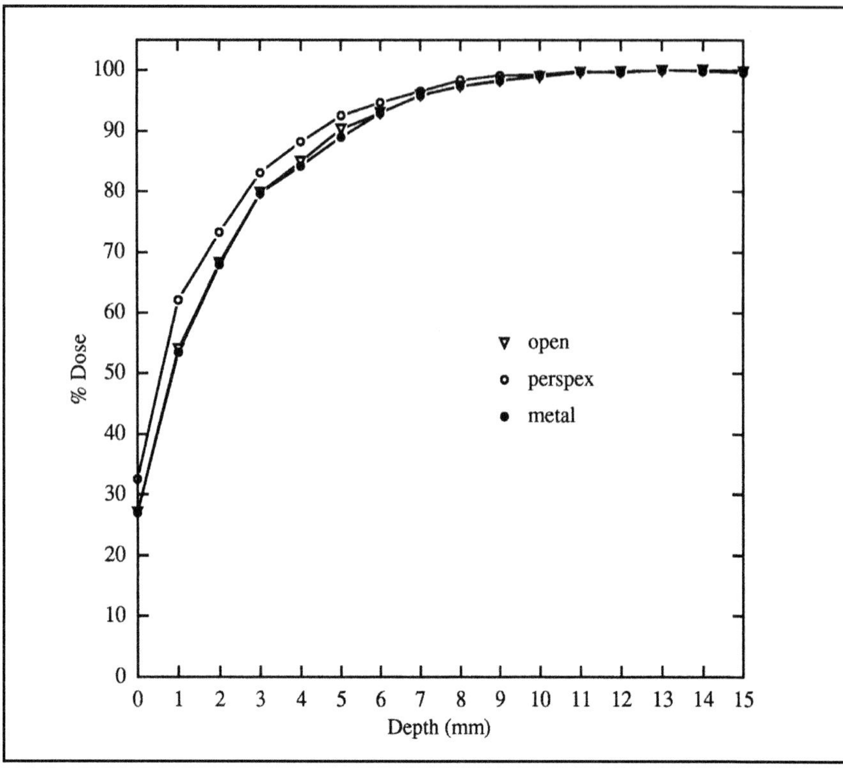

Figure 4.5: Build-up curves for a 6 MV x-ray beam measured with Attix chamber for 20×20 cm^2 field size. The steel honeycomb gives virtually the same results as for an open field whereas the PMMA tray significantly increases surface dose.

The contamination component can be further reduced by the use of filters (such as thin layers of lead) placed beneath the block tray to absorb the contamination (Biggs and Ling 1979, Parthasaradhi et al. 1989, Rao et al. 1988). It is not common to use lead filters in the clinic as the lead attenuates the beam and another tray factor has to be added to the calculation along with an interlock to ensure that the device is in place. Because physical wedges are made from high-density, high atomic number media they also self-absorb electron contamination. Therefore, surface dose with wedges relative to D_{max} is similar to that of open field levels.

The contamination component can also be reduced by the use of electromagnets to sweep the contamination away. These are not incorporated on commercial linacs due to the weight and extra complexity of such devices (Ling and Biggs 1979). Lightweight fixed ceramic magnets are at an early stage of development for this purpose (Butson et al. 1996, Jursinic and Mackie 1996, Damrongkijudom et al. 2006). Note that neither the magnet nor filters reduces the reintroduction of electron contamination due to the air space between decontaminator and patient; the extent of this component may be significant (McParland 1991).

The surface dose also increases as the field size increases, as shown in figure 4.4. This is due to extra in-phantom scatter with increasing field size and also due to an increase in jaw surface area and air volume adding to the electron contamination.

Shown in figure 4.6 is the build-up curve for a 10×10 cm^2 6 MV x-ray beam. Also shown is the result of a contamination-free Monte Carlo computer beam simulation. Contamination was not included in the Monte Carlo calculation as the simulation was set up with x-rays set incident at the patient surface. Hence the difference in results between the Monte Carlo electron contamination-free data and the experimental electron contaminated data gives an indication of the total contamination component. Note that a small component due to extra focal x-rays will also be included in these data. The process has been repeated for different field sizes, and the results are shown plotted in figure 4.7.

As there is a steep dose gradient in the beam build-up region, radiotherapy treatment planning (RTP) computer systems approximate the dose distributions in these regions by empirical models. To avoid storing data in this region, the GE Target system, for example, uses the following approximate method. For the dose points between surface and d_{max}, linear interpolation is

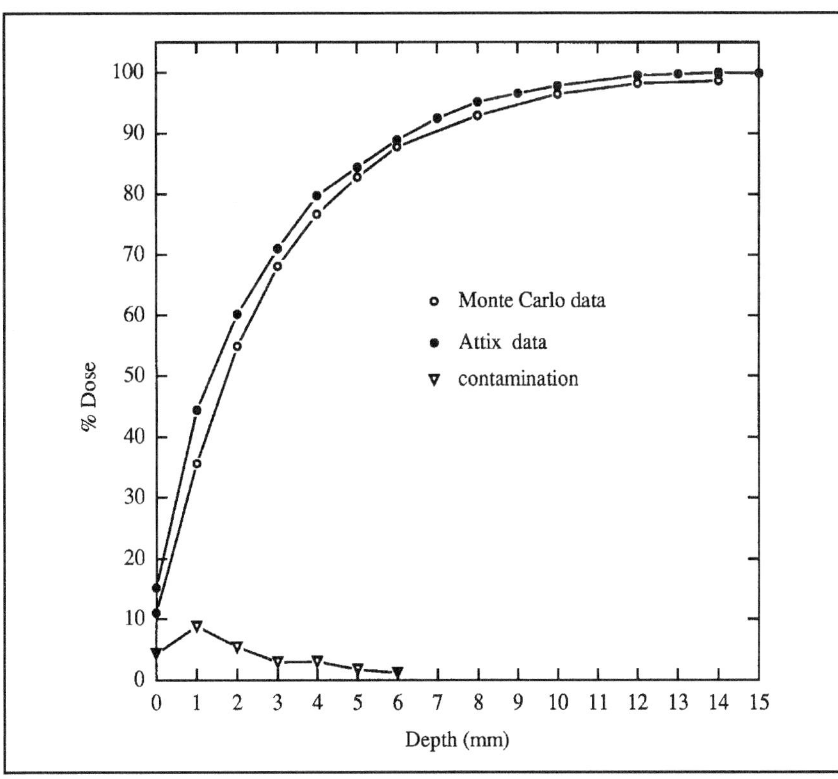

Figure 4.6: Experimental and contamination-free Monte Carlo-computed dose build-up curves, 6 MV x-rays, 10×10 cm^2 field size.

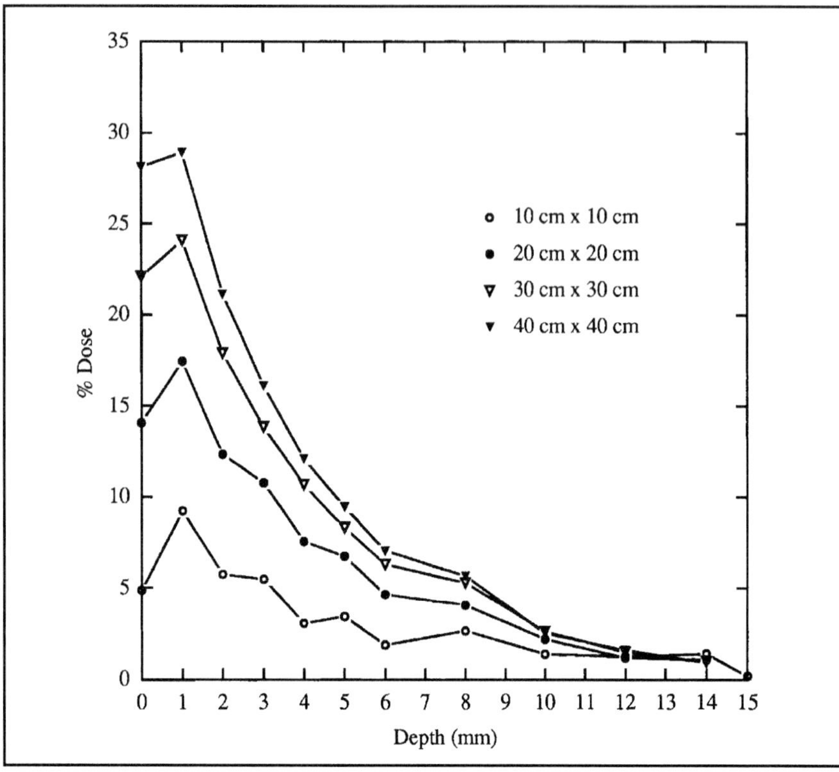

Figure 4.7: Electron contamination components for different field sizes 10×10 cm^2, 20×20 cm^2, 30×30 cm^2, and 40×40 cm^2. Determined by subtraction of Monte Carlo data from experimental data in the build-up region of a 6 MV x-ray beam.

used. Then the curve is given its shape by adjusting d_{ref} to a position d_{ref}' and then looking up the dose on the linear part of the line joining surface dose with dose at d_{max} (Target 1990).

The distance d_{ref}' is given by

$$d_{ref}' = d_{max} - \frac{\left(d_{max} - d_{ref}\right)^3}{d_{max}^2}. \tag{4.2}$$

Figure 4.8 shows the result of applying this approximation for the 6 MV beam; we call this the cubic fit. Note that the large differences in dose between the fit and experiment relate to small differences in spatial terms of about 1 mm. This is due to the steep dose gradient.

An empirical fit that separates the two components of build-up dose, phantom build-up and electron contamination, has been derived with field size dependence. A comprehensive empirical method of predicting central axis dose that accounts for field size, block trays, and beam angle has been presented by Butson et al. (1996). In the interest of brevity a simplified version is presented here that excludes the block trays and beam angle (note that both effects are significant). Phantom build-up dose is modeled using an exponential fit, such that

$$\%D = 100 - (100 - C(s))e^{-\alpha z}, \tag{4.3}$$

where C(s) is a field size dependent surface dose value, z is the depth in water (mm), and α is the empirical value that defines the curvature. For a 6 MV x-ray beam, $\alpha = 0.34$ and the surface dose is modeled by

$$C(s) = \beta - \beta e^{-\gamma s'}, \tag{4.3a}$$

where $\beta = 14.3$, $\gamma = 0.148$ and s' is the square field size between 0×0 cm^2 and 40×40 cm^2.

The contamination part of the x-ray build-up dose also follows an exponential shape. Contamination is modeled by the empirical formula,

$$\%D = D(s)e^{-\delta z}, \tag{4.3b}$$

where $\delta = 0.235$, z is the depth in water (mm), $D(s)$ is the field size dependent electron contamination surface dose, and δ is an empirical value defin-

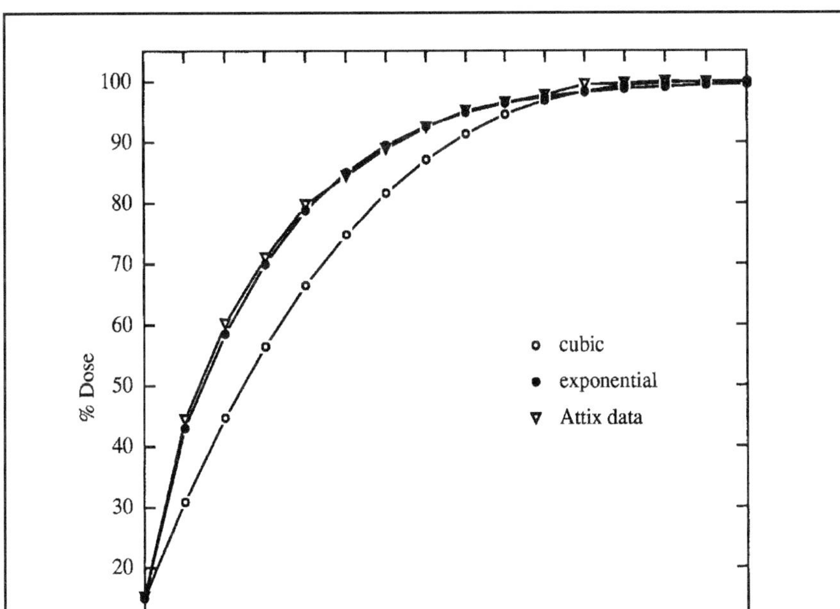

Figure 4.8: Match between experimental data and analytical fits to the data in the build-up region for 6 MV x-rays, 10×10 cm^2 field size, open field, 100 cm SSD, cubic and exponential fits to the data are shown.

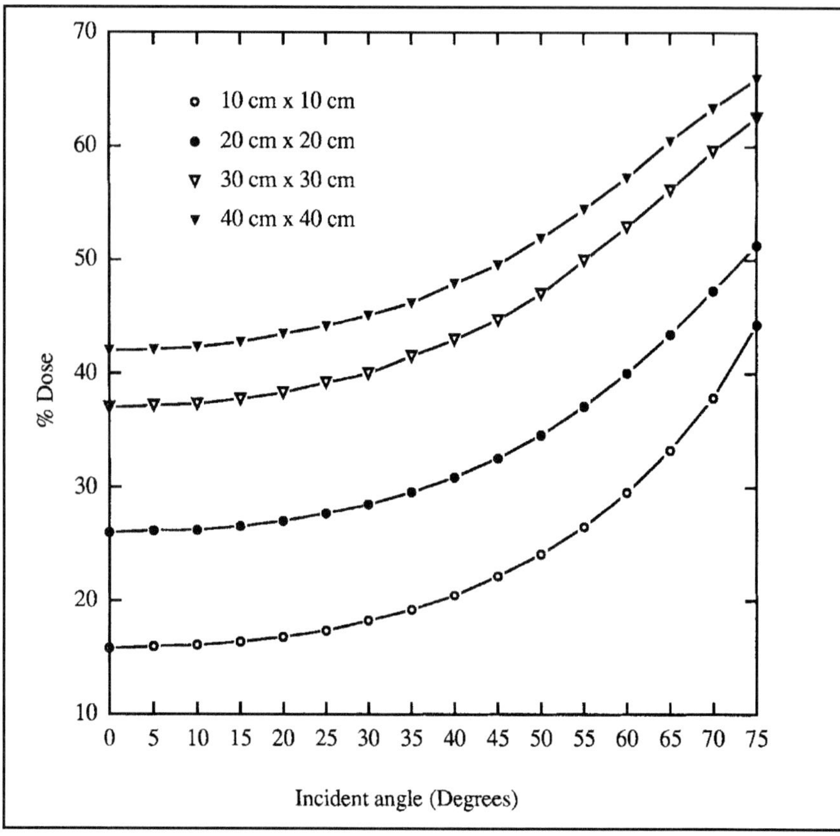

Figure 4.9: Change in surface dose with incident beam angle, 6 MV measured with an Attix chamber.

ing the curvature (for 6 MV $\delta = 0.235$). Surface dose due to electron contamination is best modeled by a third order polynomial such that

$$D(s) = \varepsilon s' + \zeta (s')^2 + \eta (s')^3, \qquad (4.3c)$$

where s' is the field size and the empirical values, ε, ζ, and η (0.057, 0.04, and -6.14×10^{-4}, respectively) relate to the fitting parameters for a 6 MV x-ray beam.

The results of using the above exponential and the previously mentioned cubic approximation are shown in figure 4.8. Note that the results show a better fit to the data using the exponential method.

The Pinnacle[3] treatment planning system uses an exponential fall-off parameter modified near the surface with a linear fall-off parameter to directly model the electron contamination dose component D_{ec}, such that (Pinnacle 2007)

$$D_{ec}(F,r,d) = D_{ec}(F,d) \times OA_{effect}, \qquad (4.4)$$

where

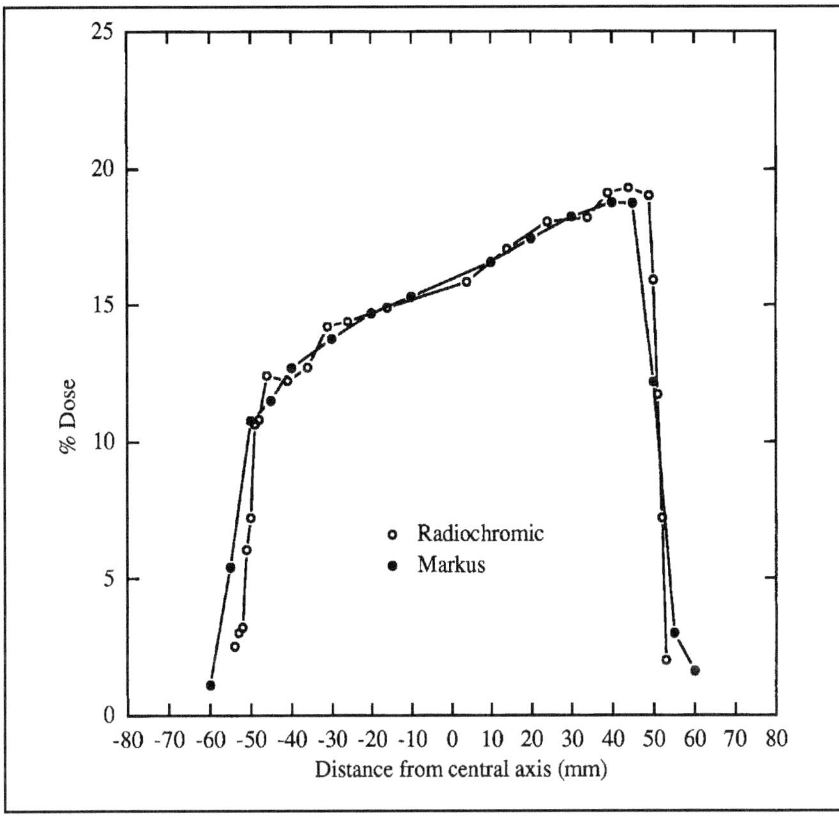

Figure 4.10: Profile of surface dose off-axis collected at 6 MV using radiochromic film and a Markus chamber. (Reprinted with permission from: Butson et al. 1996 and IOP Publishing.)

$$D_{ec}(F,d) = P(F) \times \frac{e^{-Kd} - e^{-kD_{\max}}}{1 - e^{-kD_{\max}}}, \tag{4.5}$$

where $P(F)$ is a polynomial function accounting for field size changes in contamination and the off-axis reductionin contamination is modeled by a gaussian parameter such that

$$OA_{effect} = e^{Br^2}. \tag{4.6}$$

The incidence of beam entry angle on the patient surface has a significant effect on surface dose; the more oblique the beam angle the higher the surface dose. Once again this has relevance to the surface dose for any oblique beam entry point such as head/neck and breast sites (Kron et al. 1993b, McParland 1991, Orton et al. 1972, Jackson 1971, Gerbi et al. 1987). In figure 4.9 we show the change in surface dose with incident beam angle, as measured with an Attix chamber.

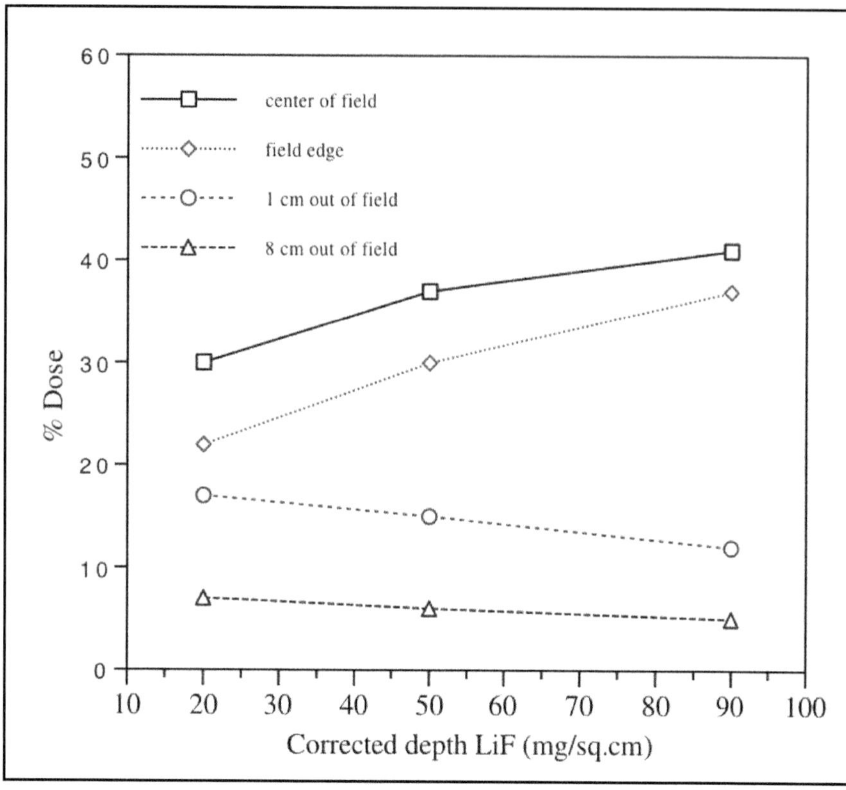

Figure 4.11: Surface and build-up dose characteristics, central axis versus off-axis, 6 MV x-rays. (Reprinted with permission from: Kron et al. 1993b and AAPM .)

It has been demonstrated elsewhere that the electron contamination component reduces outside the field boundary. A broad profile is produced which falls off more slowly than the collimated penumbral profile at depth (Harper et al. 1991). The change in surface dose off-axis, for a wedged field, is shown in figure 4.10. Note also that the penetration characteristics of electron contamination off-axis also change (Kron et al. 1993b), as shown in figure 4.11. This implies a change in the energy and/or angular distribution of the electron contamination component off-axis.

If the fields are opposed, the beam exit dose becomes the most significant contribution to surface dose. As a rule of thumb, about 85% of the full-scatter dose just prior to exit applies to the skin. This means that if the depth dose in full scatter at a depth is x%, then the skin dose at exit due to lack of scatter is about x multiplied by 0.85. Kron et al. (1996) have shown that for breast fields with 2 Gy target dose per fraction, the total surface dose is about 1 Gy with two-thirds of this contribution coming from the exit beam dose and only one-third from the entry beam.

4.4 Percentage Depth Dose

4.4.1 Definitions

Absorbed dose, D, is the mean energy absorbed, E_{ab}, in a medium per unit mass, dm, due to electron interactions. The dose can be calculated at any point using the dosimetry protocols outlined in chapter three (e.g., AAPM 1999, IAEA 2004).

A reference dose is usually calculated at the depth of maximum dose $d_{\max}$, where the percentage relative dose is normalized to $D(d_{\max}, f, S, E)$ or $D_{\max}$. The dose within the body is then considered for computer treatment planning calculations in terms of percentage depth dose, $\%D$, which is the ratio, expressed as a percentage, of the absorbed dose at any point to the dose at $D_{\max}$ by

$$\%D(d_{ref}, f, S, E) = \frac{D(d_{ref}, f, S, E)}{D(d_{\max}, f, S, E)} \times 100. \tag{4.7}$$

The relationship is functionally related to any reference depth, d_{ref}, in the patient or phantom, field size at the surface, S, SSD f, and the beam energy, E, (see also figure 4.12 for further explanation of percentage depth dose).

If the value of absorbed dose at a point, for example $D_{\max}$, can be calculated and the percentage depth dose is known, then the absorbed dose to any other central axis point can be calculated as

$$D(d_{ref}, f, S, E) = \frac{\%D(d_{ref}, f, S, E)}{100} \times D(d_{\max}, f, S, E). \tag{4.8}$$

The central axis percentage dose is often plotted against depth as this indicates the penetration properties of the beam as shown in figure 4.12b. This type of curve is commonly referred to as a percentage depth dose curve.

4.4.2 Beam Energy

The depth dose curves for different nominal beam energies are shown in figure 4.13. Note that as the energy increases, the depth of penetration increases from 250 kV (orthovoltage), 662 keV (cesium-137), 1.25 MeV (cobalt-60) to 6 MV and 18 MV x-rays produced by linacs.

Linacs with a higher nominal peak energy have depth dose curves that maintain a higher dose at the same depth in the depth dose fall-off region. Depth dose data is machine specific and depends on the target and wave-guide specification of each machine, which in turn affects the beam spectral distribution of energy, E. It is important to realize that the specification of beam energy by its nominal peak gives no indication of the shape of the beam spectrum produced by the linear accelerator.

As discussed in chapter three a convention that enables a beam penetration comparison is to divide dose at 20 cm depth by the dose at 10 cm depth at 100 cm SSD ($D_{20/10}$). This ratio is used for looking up the nominal beam energy parameters, which are in turn used to select chamber calibration factor data from tables in dosimetry protocols such as IAEA (1987). Table 4.3 includes typical values of interest at these depths for some nominal linac beam energies.

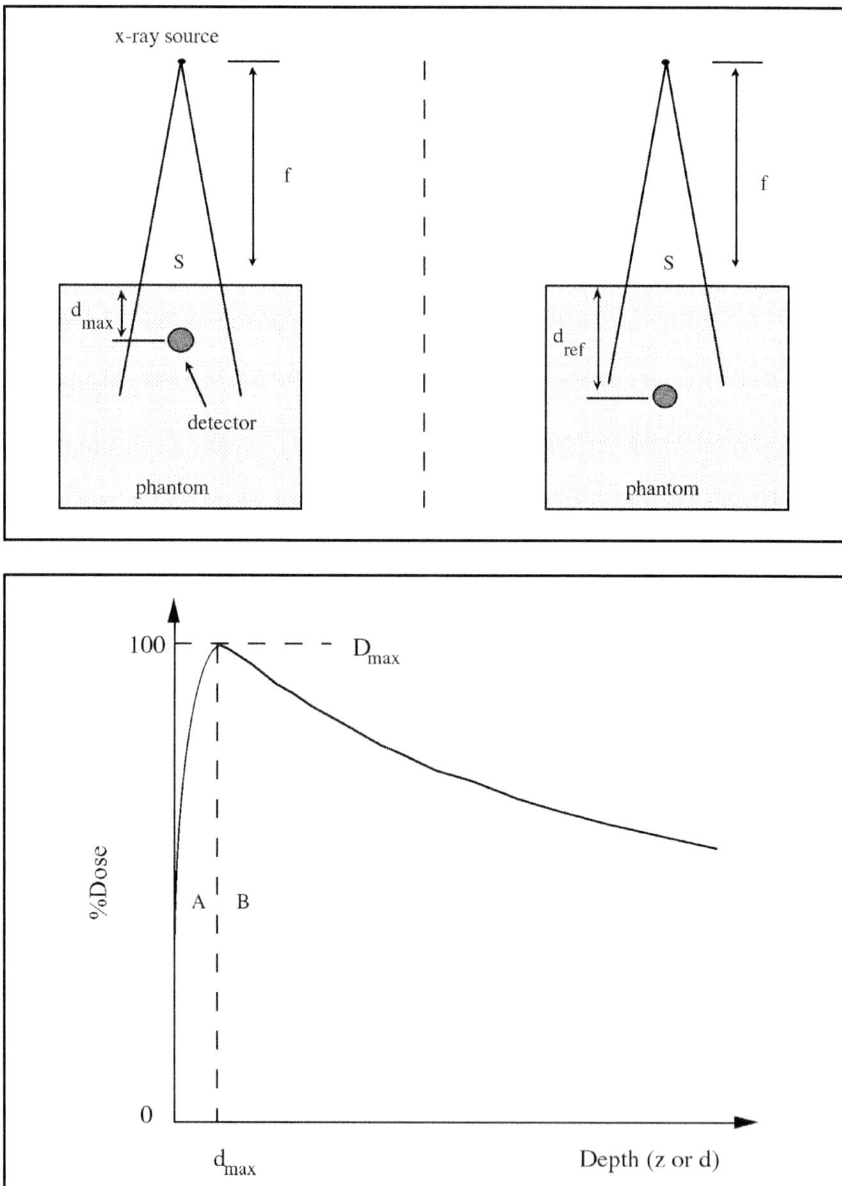

Figure 4.12: Percentage depth dose. (a) The SSD is constant, but the source detector distance changes. Percentage depth dose varies depending on beam energy, SSD, and phantom material. The phantom material is usually water, but it may also be a solid slab phantom. (b) A is the dose build-up region, and B is the dose fall-off region.

Table 4.3: Position of d_{max} and the percentage depth dose values at d_{10} and d_{20} for different linear accelerator energies. The values labeled + are derived from BJR (1983) and the other values are from measurements on a Varian 2100C linear accelerator.

Energy (MV)	*4*	*6*	*10*	*18*
d_{max} (cm)	1.3+	1.5	2.5+	3.3
D_{10} (%)	62.7+	66.9	75.3+	78.7
D_{20} (%)	34.0+	38.6	48.2+	52.0

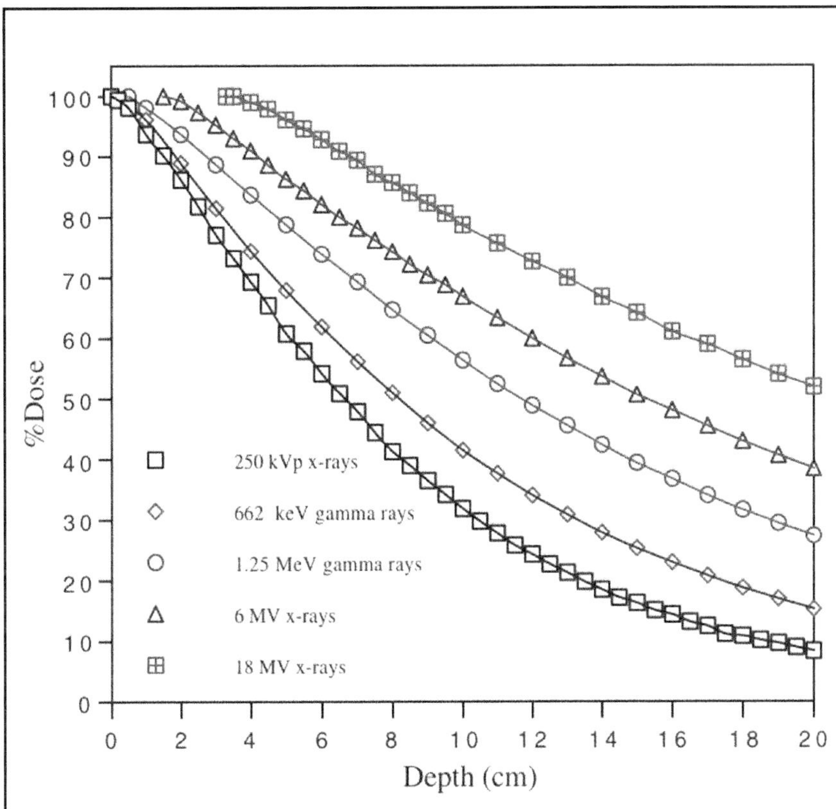

Figure 4.13: Comparison of depth dose curves for different radiotherapy beams. Orthovoltage 250 kVp x-rays 50 cm SSD, cesium-137 40 cm SSD, cobalt-60 80 cm SSD, 6 MV x-rays 100 cm SSD, 18 MV x-rays 100 cm SSD. Note—all fields 10×10 cm^2.

4.4.3 Field Size

Shown in figure 4.14 and table 4.4 are the percentage depth dose curves and data for different field sizes. Notice how the depth dose increases with field size, S, due to the increase in scatter at larger field sizes. Note that the depth dose for a 40×40 cm² field is similar to a 30×30 cm² field because the mean free path of photon scatter is of this order of magnitude and lateral photon scatter equilibrium is being approached at both these large field sizes.

It is not obvious from this data that the position of d_{max} may sometimes be observed to change slightly when field size is changed. It usually becomes shallower by a very small amount (about 1 mm) when field size is increased.

It is useful to refer to typical data for commonly used linacs for comparison with data collected for a specific machine. This acts as a check on the data collected. Because depth dose and profile data vary slightly from machine to machine, it is important to collect machine-specific data for patient planning calculations. Linac vendors now routinely supply beam matching. This means different linacs are tuned to have very similar depth dose and dose profile properties within a department. While the physical design of targets and flattening filters cannot be altered, the beam properties can be tuned by small adjustments to gun, beam current, and bend magnet settings. This is very useful if one machine has a lengthy malfunction because patients can be transferred and treated with the other machine without the need to generate a new computer treatment plan (Marshall 1993).

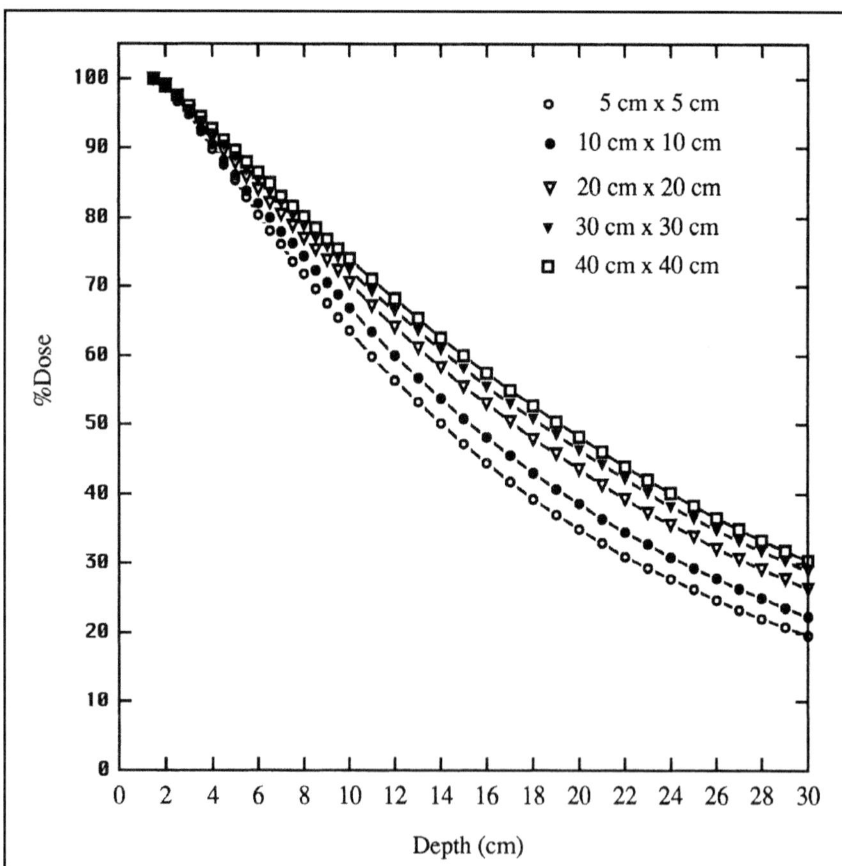

Figure 4.14: Depth dose curves for a 6 MV x-ray beam from a Varian 2100C linac at different field sizes 5×5 cm², 10×10 cm², 20×20 cm², 30×30 cm², 40×40 cm².

Table 4.4: Clinac 2100C 6 MV x-ray beam percentage depth dose (%D) data at 100 cm SSD

	Field size (cm^2)				
z (cm)	*5×5*	*10×10*	*20×20*	*30×30*	*40×40*
1.5	100.0	100.0	100.0	100.0	100.0
2.0	99.2	99.2	99.2	99.2	99.2
3.0	94.8	95.2	95.2	95.5	96.1
4.0	89.8	91.0	91.4	92.1	92.8
5.0	85.4	86.3	87.8	88.7	89.6
6.0	80.4	82.1	84.1	85.3	86.5
7.0	76.1	78.2	80.5	82.1	83.1
8.0	71.7	74.3	77.0	78.8	80.2
9.0	67.5	70.4	73.7	75.7	76.9
10.0	63.6	66.9	70.5	72.3	73.9
11.0	59.9	63.4	67.1	69.2	70.9
12.0	56.3	60.0	64.1	66.4	68.1
13.0	53.2	56.7	61.1	63.6	65.3
14.0	50.1	53.7	58.3	60.8	62.5
15.0	47.1	50.7	55.4	58.1	60.0
16.0	44.4	48.1	53.0	55.4	57.4
17.0	41.7	45.5	50.4	52.9	54.9
18.0	39.3	43.0	47.8	50.7	52.7
19.0	37.0	40.7	45.7	48.5	50.3
20.0	34.9	38.6	43.5	46.3	48.2
21.0	32.9	36.4	41.3	44.2	46.0
22.0	30.9	34.4	39.3	42.2	43.9
23.0	29.3	32.7	37.3	40.2	42.0
24.0	27.7	30.8	35.6	38.2	40.2
25.0	26.2	29.3	33.9	36.5	38.3
26.0	24.7	27.8	32.1	34.8	36.6
27.0	23.3	26.3	30.6	33.2	34.9
28.0	22.0	25.0	29.1	31.7	33.3
29.0	20.8	23.6	27.7	30.3	31.8
30.0	19.6	22.3	26.3	28.8	30.4

4.4.4 Source Surface Distance

Another factor that affects percentage depth dose is source surface distance. The depth dose actually falls off faster at a shorter SSD. This can be explained by the inverse square law (ISL) principle, which states that radiation intensity (fluence) is reduced with the square of the distance from the radiation source.

Figure 4.15 shows the geometric changes that occur when converting from one SSD to another. When converting from one SSD at f to another at f', there is an associated change in field size at the patient surface from S to a field size S/F where F is given by

$$F = \left(\frac{f + d_{ref}}{f}\right)\left(\frac{f'}{f' + d_{ref}}\right). \tag{4.9}$$

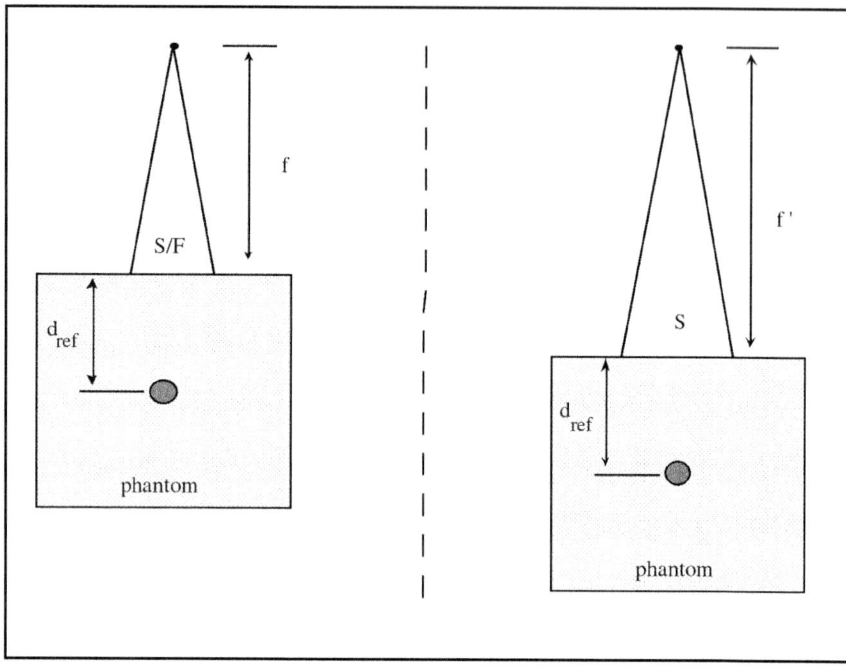

Figure 4.15: The effect of change in SSD on depth dose measurement parameters.

The change in absolute dose at d_{max} can be calculated using the inverse square fall-off principle as

$$D(d_{max}, f', S, E) = D(d_{max}, f, \frac{S}{F}, E) \times \left(\frac{f + d_{max}}{f' + d_{max}} \right)^2, \qquad (4.10)$$

where $D(d_{max}, f, S/F, E)$ is the dose at one SSD and $D(d_{max}, f', S, E)$ is the dose at the other SSD.

The percentage depth dose, $\%D(d_{ref}, f, S/F, E)$, can be converted to the dose $\%D$ at d_{ref} at a different SSD f' as:

$$\%D(d_{ref}, f', S, E) = \%D(d_{ref}, f, \frac{S}{F}, E) \times \frac{PSF\left(\frac{S}{F}\right)}{PSF(S)} \times M, \qquad (4.11)$$

where M is the Mayneord factor (Khan 1984), such that

$$M = \left(\frac{f + d_{ref}}{f + d_{max}} \right)^2 \times \left(\frac{f' + d_{max}}{f' + d_{ref}} \right)^2. \qquad (4.12)$$

The peak scatter factor (PSF) is the backscatter component at d_{max}, and the ratio of PSFs is a small correction that is introduced to account for the scatter increase due to field size increase by beam divergence from one SSD to another (BJR 1996). M gives the inverse square factor conversion for the

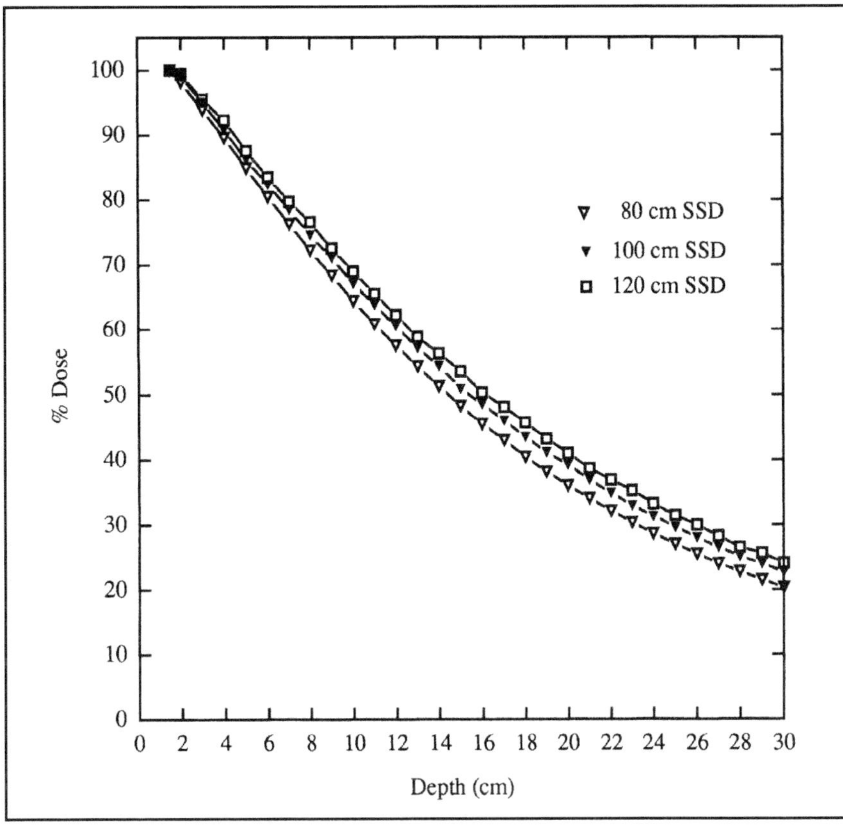

Figure 4.16: Depth dose curves for a 6 MV x-ray beam from a Varian 2100C linac at 80, 100, and 120 cm SSD.

change of SSD. Figure 4.16 shows the percentage depth dose curves for a 6 MV x-ray beam at three different SSDs. Note that the curves are shallower with increasing SSD as predicted by equation 4.11.

4.5 Isocentric Dose Ratios

4.5.1 Tissue-Phantom Ratio

The tissue-phantom ratio (TPR) is defined as the ratio of absorbed dose at any given point to the absorbed dose at the same distance from the source but with the surface of the phantom moved so that this point is at a specified reference depth d_{ref} (see figure 4.17):

$$TPR(d_{ref2}, S, E) = \frac{D_{tissue}(d_{ref2}, S, E)}{D_{tissue}(d_{ref1}, S, E)}. \tag{4.13}$$

This is a useful quantity particularly because isocentric data can be referenced to the depth of calibration (e.g., $d_{\text{ref1}} = 5$ cm for 6 MV calibration).

The tissue-maximum ratio (TMR) is a special case of the TPR where the specified reference depth is at d_{max} (BJR 1983):

$$TMR(d_{ref}, S, E) = \frac{D_{tissue}(d_{ref2}, S, E)}{D_{tissue}(d_{max}, S, E)}. \tag{4.14}$$

The majority of radiotherapy x-ray treatment plans use a common point of origin within the patient. This point of origin is known as the *isocenter* and is defined in more detail in section 4.9.1. This form of planning is known as isocentric planning, and the beams are referred to as isocentric beams, which have a common point of rotation. The source axis distance (SAD) is usually 100 cm to isocenter. The main advantage of isocentric treatment is the practicality of setup; the patient can be set up and treated with one field, and the machine can be rotated to the next angle for the next field of treatment without any need to reposition the patient. The TPR and TMR concepts are useful when planning with isocentric beams as dose ratios at different depths are provided directly, and these quantities have been shown to be relatively independent of SSD (Johns et al. 1958, Gupta and Cunningham 1966). Table 4.5 shows a series of TMR data at different depths for a selection of field sizes for a 6 MV x-ray beam.

4.5.2 Tissue-Air Ratio and Scatter-Air Ratio

The ratio of the dose at a point in tissue to the dose at the same point in air is known as the tissue-air ratio. The measurement setup is shown in figure 4.18. A TAR versus depth curve is produced by measuring the dose in air with an ionization chamber and then measuring the dose at that same point with varying thickness of overlying tissue. The ratios of the dose for each

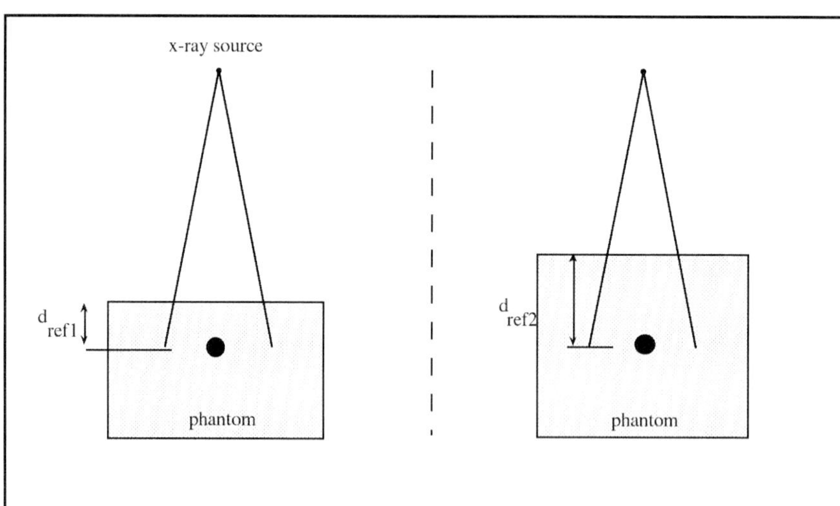

Figure 4.17: Tissue-phantom ratio (TPR) and tissue-maximum ratio (TMR). The dose in a phantom at depth d is taken as a ratio to the dose in a phantom at a reference depth d_{ref} at a fixed SAD. For TMR the reference depth d_{ref} is d_{max}. The phantom generally consists of water or solid water.

Table 4.5: Clinac 2100C 6 MV x-ray tissue-maximum ratio (TMR) data

	Field Size (cm²)				
z (cm)	*5×5*	*10×10*	*20×20*	*30×30*	*40×40*
1.5	1.000	1.000	1.000	1.000	1.000
2	1.002	1.002	1.002	1.002	1.002
3	0.976	0.980	0.980	0.982	0.990
4	0.942	0.955	0.959	0.966	0.974
5	0.911	0.923	0.938	0.947	0.957
6	0.875	0.894	0.916	0.928	0.941
7	0.841	0.866	0.891	0.908	0.922
8	0.807	0.838	0.870	0.885	0.902
9	0.773	0.807	0.846	0.864	0.882
10	0.740	0.780	0.819	0.843	0.863
11	0.709	0.750	0.795	0.821	0.842
12	0.678	0.722	0.775	0.800	0.822
13	0.649	0.693	0.749	0.778	0.800
14	0.621	0.668	0.722	0.755	0.778
15	0.594	0.642	0.698	0.734	0.758
16	0.569	0.613	0.675	0.711	0.738
17	0.544	0.589	0.652	0.692	0.715
18	0.520	0.568	0.631	0.670	0.696
19	0.498	0.545	0.609	0.647	0.674
20	0.476	0.524	0.586	0.626	0.655
21	0.456	0.502	0.565	0.607	0.636
22	0.436	0.483	0.545	0.587	0.617
23	0.416	0.462	0.525	0.568	0.597
24	0.400	0.443	0.504	0.550	0.579
25	0.383	0.429	0.487	0.529	0.561
26	0.368	0.409	0.470	0.511	0.541
27	0.350	0.391	0.451	0.494	0.524
28	0.337	0.376	0.435	0.478	0.508
29	0.323	0.361	0.419	0.461	0.491
30	0.309	0.346	0.402	0.444	0.474

depth to the dose in air then gives a TAR. A TAR versus depth curve is similar to a depth dose curve, except that since there is no distance related (ISL) fall-off in dose with depth, the fall-off in dose is less. As with TPRs, TARs are independent of distance from the source for distances greater than approximately 50 cm (Johns and Cunningham 1983). At shorter SSDs than this the approximation breaks down, as the changes in scatter become significant due to beam divergence.

TARs are useful for determining the variation in dose with field size as well as with depth—there is an increase in the TAR with field size. This increase occurs because more photons are scattered by the medium to the point of dose measurement in tissue as field size increases, and there is no such increase at the same point in air. For a particular depth in tissue, a TAR versus field size curve can be plotted. An illustration of such a curve is given in figure 4.19. Measured TARs are indicated by the points on the curve. If

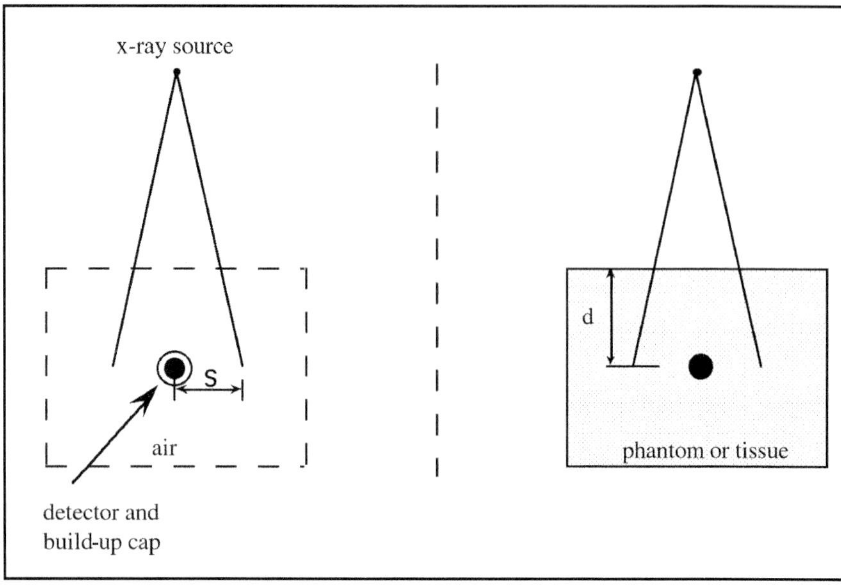

Figure 4.18: Tissue-air ratio (TAR) and scatter-air ratio (SAR). The dose reading taken in air is divided by the dose taken in full scatter conditions at depth in water.

the TAR values are taken at d_{max} these have the special term, peak scatter factors (PSF), such that

$$PSF(S) \approx TAR(d_{max}, S), \tag{4.15}$$

with the exception that PSFs are traditionally measured at 100 cm SSD and a TAR(d_{max}, S), while SSD independent would traditionally be measured at 100 cm SAD.

If all the $PSF(S)$ values are renormalized so that $PSF(10\times10\ \text{cm}^2) = 1$, these values are called normalized peak scatter factors (NPSF). This is nearly equal to the in-phantom scatter factor, S_p, which as previously stated can be measured directly with a build-up cap and does not require the extrapolation step. The only subtle difference is that NPSF is usually measured at 100 cm SSD whereas S_p is usually measured at 100 cm SAD.

For high-energy photon beams, there is a lower limit on the field sizes for which dose can be measured accurately by ion chamber dosimetry. This limit is based on the secondary electron range. For the 10 MV beam the smallest field size for which dose is normally measured is approximately 5×5 cm^2. For field sizes smaller than this limit, TAR values are found by extrapolation of the measured points. If the curve is extrapolated back to a field size of zero, a zero-area TAR is obtained. This quantity represents the TAR when no scattered photons contribute to the dose in tissue and is indicated as TAR(d, 0).The *extra* dose contributed by scattered photons is represented by the difference in the TAR and TAR(d, 0) and is called the scatter-air ratio (SAR). Thus, for a beam of field size S, the TAR at a depth d can be expressed as the sum of the zero area TAR and the SAR:

$$TAR(d,S) = TAR(d,0) + SAR(d,S). \tag{4.16}$$

Note that S can represent a field of an equivalent square or an equivalent radius. The increase of scatter-air ratio (SAR) for a unit increase in beam diameter is referred to as the differential scatter-air ratio (dSAR). (See chapter six for more detail.)

The points marked on the curve in figure 4.19 indicate measured TARs. The zero-area TAR, TAR(d, 0), is given by the intersection of the curve with the TAR axis. The difference in the TAR for a particular field size and TAR(d, 0) is the SAR(d,S) for that field size. The TAR and SAR for a field size S are indicated. Because this TAR data was measured at d_{max} this is equivalent to PSF data. If the PSF data is normalized to a 10×10 cm^2 field then this is NPSF data as shown on the graph.

TARs can be measured or can be derived from depth dose data (BJR 1996) such that:

$$TAR(d,S) = \frac{\%D\left(d, f, \dfrac{Sf}{(f+d)}\right)}{100} \times PSF\left(\frac{Sf}{f+d}\right) \times \left(\frac{f+d}{f+d_{max}}\right)^2 . \quad (4.17a)$$

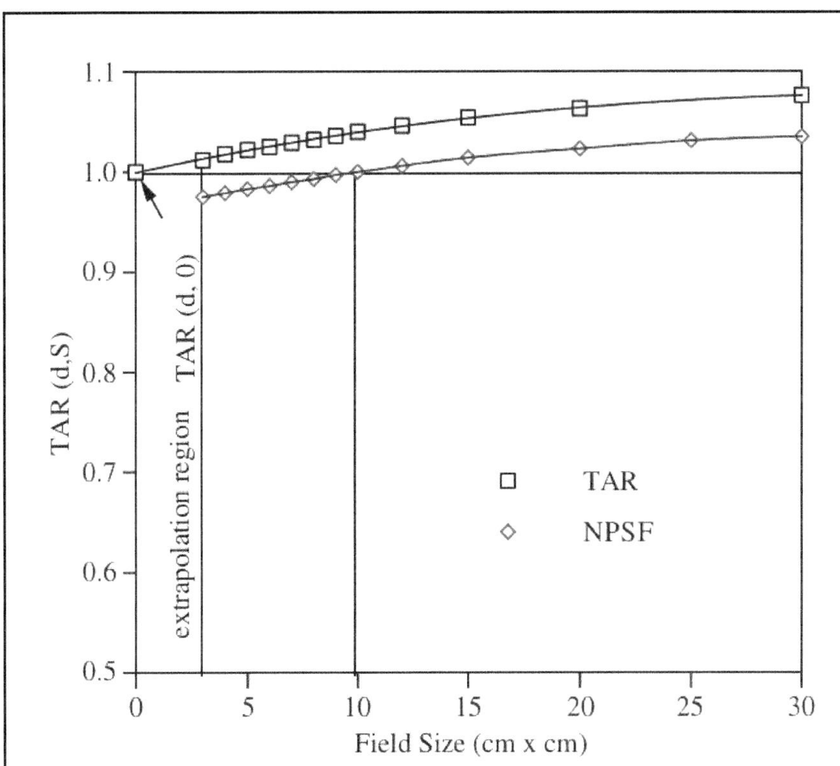

Figure 4.19: Tissue-air ratio (TAR) and normalized peak scatter factor (NPSF) versus field size. Measured TARs are indicated by the points on the curve. The zero area TAR, TAR(0), is given by the intersection of the curve with the TAR axis.

A series of TAR data at different depth for a selection of field sizes is shown for a 6 MV x-ray beam in table 4.6.

When TPR data or TMR is recalculated from depth dose data, the equation is given by:

$$TPR(d,S)=\frac{\%D\left(d,f,\frac{Sf}{(f+d)}\right)}{\%D\left(d,f,\frac{Sf}{(f+d_{ref})}\right)}\times\frac{PSF\left(\frac{Sf}{(f+d)}\right)}{PSF\left(\frac{Sf}{(f+d_{ref})}\right)}\times\left(\frac{f+d}{f+d_{ref}}\right)^2, \tag{4.17b}$$

where d is the depth in the phantom and d_{ref} is the reference depth. Note that because the PSFs are expressed as a ratio, the ratio of NPSFs can also be used. Because TMR is a special case of TPR, when $d_{ref} = d_{max}$ the following equation applies:

$$TMR(d,S)=\frac{\%D\left(d,f,\frac{Sf}{(f+d)}\right)}{100}\times\frac{PSF\left(\frac{Sf}{(f+d)}\right)}{PSF\left(\frac{Sf}{(f+d_{ref})}\right)}\times\left(\frac{f+d}{f+d_{max}}\right)^2. \tag{4.17c}$$

4.6 Beam Dose Profiles

4.6.1 Profile Shape

A megavoltage x-ray beam dose profile consists of two distinct regions, as indicated in figure 4.20. The inter umbral region, A, is where the beam is unaffected by the collimators, and the penumbral region, B, is where the beam is affected by the field defining collimators. There is no exact definition for the geometric position where transition between inter umbra and penumbra occurs on the beam profile. A nominal position 1 to 1.5 cm inside the geometric field edge (50% dose level) is generally accepted as the approximate location where this occurs.

The penumbral region is characterized by a sharp dose gradient fall-off where the primary photons are shielded by the collimators. However, there is a finite width of a few millimeters due to the finite size of the virtual x-ray source and lateral electron disequilibrium. The fall-off is sigmoid in shape. (Extra focal irradiation may also have a small impact.) Then there is a long shallow fall off-in dose. This is known as the penumbral tail region where there is a small component of dose due to transmission through the jaws and a significant component due to in patient x-ray scatter. This component, and hence the penumbral dose in the tail, increases in magnitude with depth and field size.

Table 4.6: Clinac 2100C 6 MV x-ray tissue-air ratio (TAR) data

	Field Size (cm^2)				
z (cm)	*5*	*10*	*20*	*30*	*40*
1.5	1.022	1.039	1.063	1.076	1.079
2	1.023	1.041	1.064	1.078	1.081
3	0.997	1.018	1.041	1.056	1.068
4	0.962	0.992	1.019	1.039	1.051
5	0.931	0.959	0.997	1.018	1.032
6	0.894	0.929	0.973	0.999	1.015
7	0.860	0.900	0.947	0.976	0.995
8	0.824	0.871	0.925	0.952	0.974
9	0.789	0.839	0.899	0.930	0.952
10	0.756	0.811	0.871	0.907	0.931
11	0.725	0.779	0.845	0.884	0.909
12	0.692	0.751	0.824	0.860	0.887
13	0.663	0.721	0.795	0.837	0.864
14	0.634	0.695	0.767	0.812	0.840
15	0.606	0.667	0.742	0.790	0.818
16	0.581	0.637	0.717	0.765	0.796
17	0.556	0.612	0.692	0.744	0.772
18	0.531	0.590	0.670	0.721	0.751
19	0.509	0.567	0.647	0.696	0.728
20	0.486	0.545	0.623	0.673	0.707
21	0.466	0.522	0.601	0.653	0.686
22	0.445	0.502	0.579	0.631	0.665
23	0.425	0.480	0.558	0.611	0.644
24	0.409	0.461	0.536	0.591	0.625
25	0.392	0.446	0.518	0.569	0.605
26	0.376	0.425	0.499	0.550	0.584
27	0.357	0.407	0.480	0.531	0.566
28	0.345	0.391	0.462	0.514	0.548
29	0.330	0.375	0.445	0.496	0.529
30	0.316	0.360	0.427	0.478	0.511

X-ray dose profile uniformity is generally measured by a scan along the center of both the major beam axes. As the primary collimator and flattening filter are circular, diagonal scans further test the radial symmetry of construction and placement of these devices. Two parameters that quantify field uniformity are calculated. These are field flatness and symmetry.

To introduce the concept, a simple descriptive method (IPSM 1988) is first outlined here. Symmetry is a measure of whether the beam's dose is consistent on each side of the beam profile. Beam symmetry may be established by measuring the dose levels equidistant from the central axis of the beam. For example, the symmetry ratio in figure 4.20 is defined as C1/C2. The flatness is established by comparing the maximum dose within the profile with the minimum dose within the profile. The flatness ratio in figure 4.20 is defined as D1/D2.

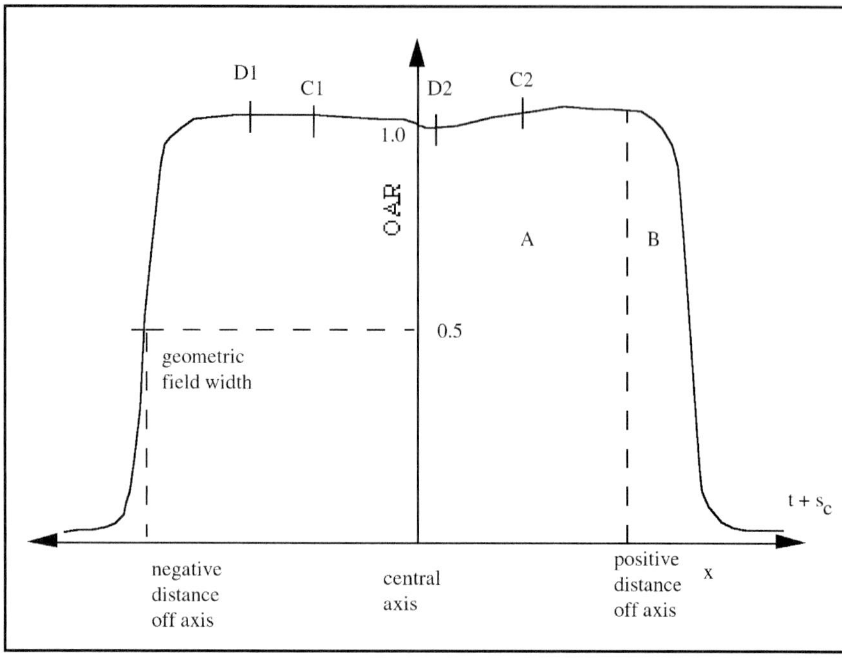

Figure 4.20: Profile of an x-ray beam taken by scanning across the x or y axis of the field, region A defines the inter umbral region, and B is the penumbral region. The geometric field width usually coincides with the 50% dose level of the profile.

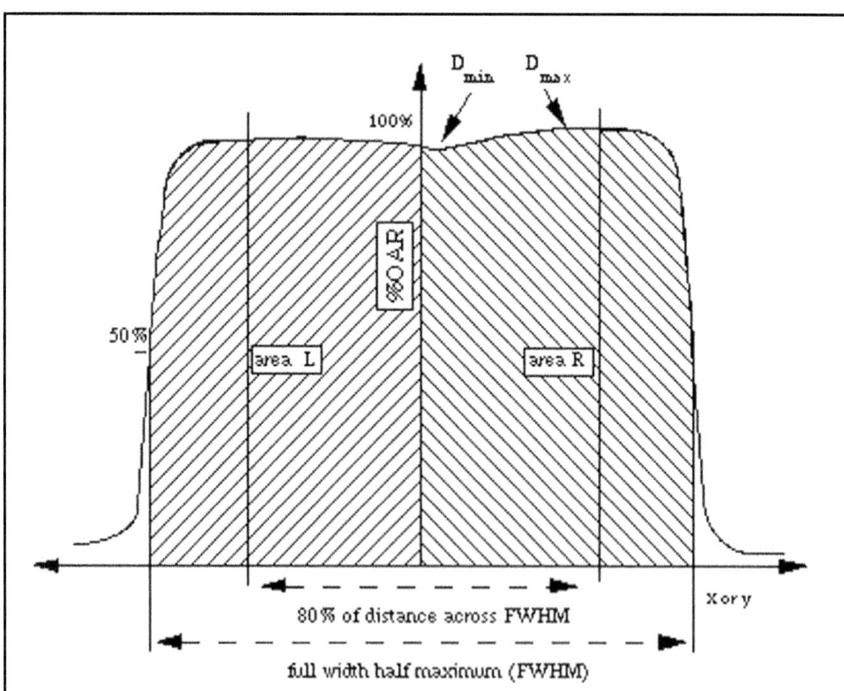

Figure 4.21: Profile scan definition of beam flatness and beam symmetry following the IEC 1987 guideline (IEC 1989b).

The most common quantitative measures of symmetry and flatness follow the IEC (1989) guideline. The percentage flatness, $\%F$, is assessed by finding the maximum, D_{max}, and minimum, D_{min}, dose point values on the profile within 80% of the beam width (see figure 4.21),

$$\%F = \left(\frac{D_{max} - D_{min}}{D_{max} + D_{min}} \right) \times 100, \tag{4.18}$$

Dose profiles are taken to test flatness at a typical treatment depth, generally at a depth of 10 cm where the horns are reduced and the flatness is generally better than 5%. The scan at depth is used to avoid the horn effect, which is greatest at d_{max}. Dose *horn* is the term given to describe the profile shape if the dose at the beam off axis position exceeds the dose in central axis (see figure 4.20).

Symmetry may be assessed at d_{max}, which is the most sensitive position to test symmetry. The beam symmetry, %S, is assessed by comparing the area under each side of the profile, as shown in figure 4.21, such that

$$\%S = \left(\frac{Area_L - Area_R}{Area_L + Area_R} \right) \times 100. \tag{4.19}$$

The area under the profiles can often be obtained using an automatic option on the water tank scanning device. If this is not available, counting squares on graph paper with a hard copy of the profile is an alternative. With current linac beam steering, the beam profiles should be able to be tuned close to symmetry in both the transverse and radial axis. The symmetry should be within ±2%.

4.6.2 The Inter Umbra

When a detector is scanned at depth d_{ref} along either the x or y axis, a beam profile is formed. These cross plot profiles are generally normalized to unity at the center of the field, thus obtaining an off-axis ratio (OAR) for each off-axis distance x. Dose at any point on the profile can be calculated using the depth dose at central axis and the OAR. The dose at a point $D(d_{ref}, x, y)$ is given by

$$D(d_{ref}, f, S, x, y) = OAR(d_{ref}, f, S, x, y) \times \frac{\%D(d_{ref}, f, S, 0, 0)}{100} \times D(d_{max}, f, S, 0, 0). \tag{4.20}$$

Figure 4.22 shows the dose profile from 6 MV x-rays for a 20×20 cm^2 field at d_{max}. The horn effect is typical for large fields and is due to a necessary compromise in the flattening filter design. Note that the 20×20 cm^2 field profile at 10 cm depth is flatter; this implies that the outer distribution of softer x-rays has been absorbed by 10 cm of water. Remember that these

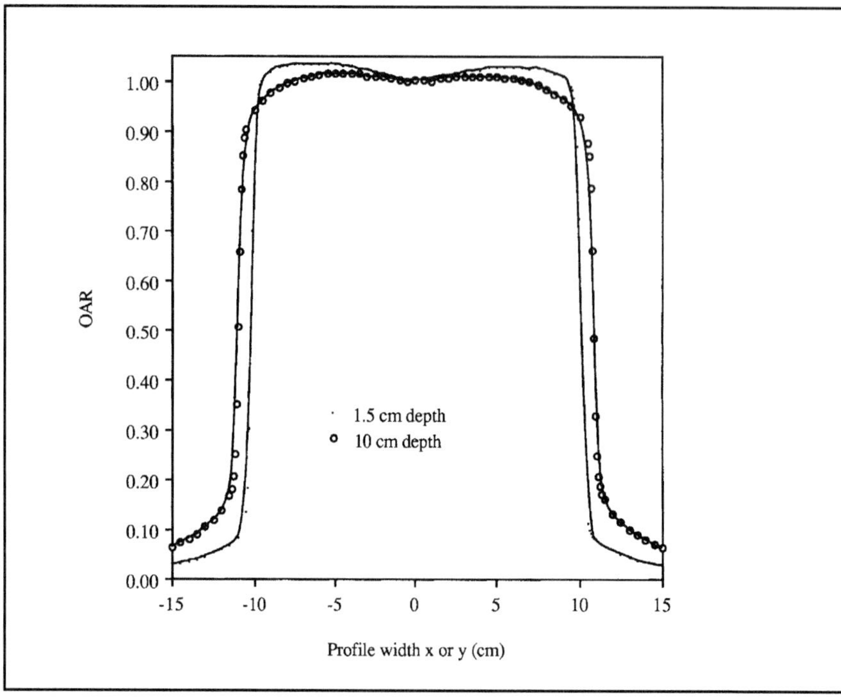

Figure 4.22: Dose profile for a 20×20 cm^2 6 MV x-ray beam at d_{max} and at 10 cm depth d_{10}. The profiles were scanned in the transverse axis, but a scan in the radial axis of the linac should produce virtually an identical result, provided the linac beam steering is tuned correctly.

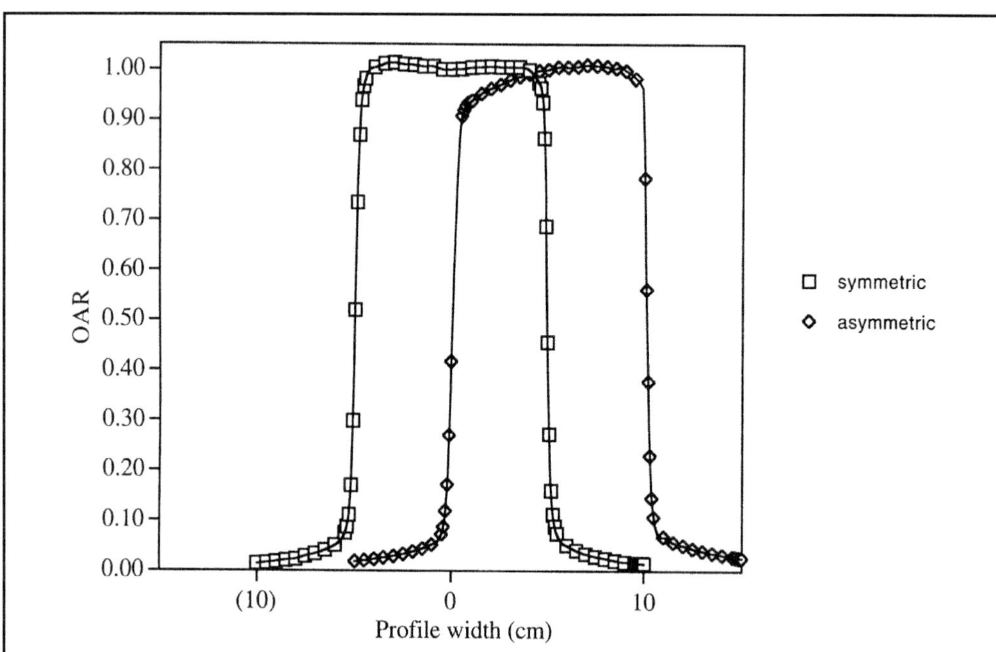

Figure 4.23: Dose profile for a 10×10 cm^2 symmetric field 6 MV x-ray beam at d_{max} and a 10×10 cm^2 asymmetric field at d_{max}.

x-rays have traversed a thinner part of the flattening filter, and therefore more low-energy components are likely to be included (Mohan et al. 1985). The increased x-ray scatter component at depth may also act to smear out the horns.

As detailed in chapter one an asymmetric x-ray collimator system has a set of collimators that can move independently of one another (commonly referred to as independent jaws). This means that a portion of the field from one side can be blocked without affecting the opposite jaw setting.

When one set of asymmetric collimators is driven to the center of a symmetric field, as previously explained in section 1.7.3, divergence is removed from one beam edge. The clinical advantages of using the nondivergent edge near delicate body structures and for field matching has been shown elsewhere (Woo et al. 1989, Conte et al. 1988, Rosenow et al. 1990).

Figure 4.23 shows a 10×10 cm^2 symmetric field compared with a 10×10 cm^2 asymmetric field. There is a nonsymmetric dose distribution across the inter umbral region of the asymmetric field. In contrast, the 10×10 cm^2 field shows a relatively flat and symmetric dose distribution. The calculation of dose distributions in the inter umbral region of asymmetric fields has been examined in detail elsewhere (Khan et al. 1986, Klemp et al. 1988, Dean and Davis 1992, Chui and Mohan 1988, Chui et al. 1988, Loshek and Keller 1988, Slessinger et al. 1993).

For an asymmetric isocentric treatment field the dose can be expressed as (Sohn et al. 1995)

$$D(d_{ref}, y, s) = \left(\frac{D}{MU}\right)_{std,SAD,Air} \times S_c(s) \times S_p(s) \times MU \times TMR(d_{ref}, s) \times OAR_{dref}(y), \tag{4.21}$$

where $(D/MU)_{\text{std,SAD,Air}}$ includes the dose (e.g., cGy/MU) for a symmetric field of standard dimension 10×10 cm^2 in air, S_c and S_p are the collimator and in-phantom scatter factors for a field size (s×s cm^2), MU is the monitor units, TMR is for the symmetric field, and OAR is the off-axis ratio at depth. Note that y is the distance off-axis to the pseudocenter of the asymmetric field. If the linac is calibrated at 100 cm SSD and the treatment is at fixed SSD,

$$D\left(d_{ref}, y, S\right) = \left(\frac{D}{MU}\right)_{sym,SSD,Water} \times F \times MU \times \%D\left(d, S\right) \times OAR_{dref}\left(y\right) \times M, \tag{4.22}$$

where $(D/MU)_{\text{sym,SSD,Water}}$ is the dose per monitor unit (e.g., cGy/MU) measured at 100 cm SSD for a 10×10 cm^2 field in water. F is the field factor and M is the Mayneord factor defined in equation 4.12.

Head scatter factors can be affected by the position of the collimators. For example in the Varian Clinac 21 family of accelerators the distance from the bottom of the upper set of jaws to the bottom of the lower set of jaws is about 8 cm. This is simply to accommodate the movement of the two sets of collimators due to their thickness. This means that the amount of flattening filter seen from a detector's eye view along the beam axis toward the target is jaw dependent. The collimator field factor S_c can be measured for a series of jaw-dependent field openings directly, or an empirical correction to equivalent-square field sizes can be applied based on the geometry of the jaw offsets. Kim et al. (1997) have introduced a simple approach to this and the method has been applied to asymmetric fields by Shih et al. (1999) whereby the head scatter factor at any off-axis point can be calculated for any asymmetric jaw offset configuration.

4.6.3 The Penumbra

Measured penumbral dose data are used in radiotherapy computer planning systems that calculate radiotherapy patient dose distributions. Some systems use the data directly to formulate penumbral shape (Milan and Bentley 1974). Other systems use the data as the basis for penumbral forming functions to calculate penumbral shape (Johns and Cunningham 1983).

Because the dose in the penumbra changes rapidly, a detector with good spatial resolution is required. Experimental comparisons are shown in figure 4.23 for different detector types. They show that the p-type diode produces the sharpest profile; this is because it has the best spatial resolution of these detectors. The measurement in the penumbra with ionization chambers of finite diameter produces a pronounced rounding of the profile shape. It is possible to correct for this using various deconvolution calculation methods (Sibata et al. 1991, Dawson et al. 1984 and 1986, Metcalfe et al. 1993, Higgins et al. 1995, Laub and Wong 2003).

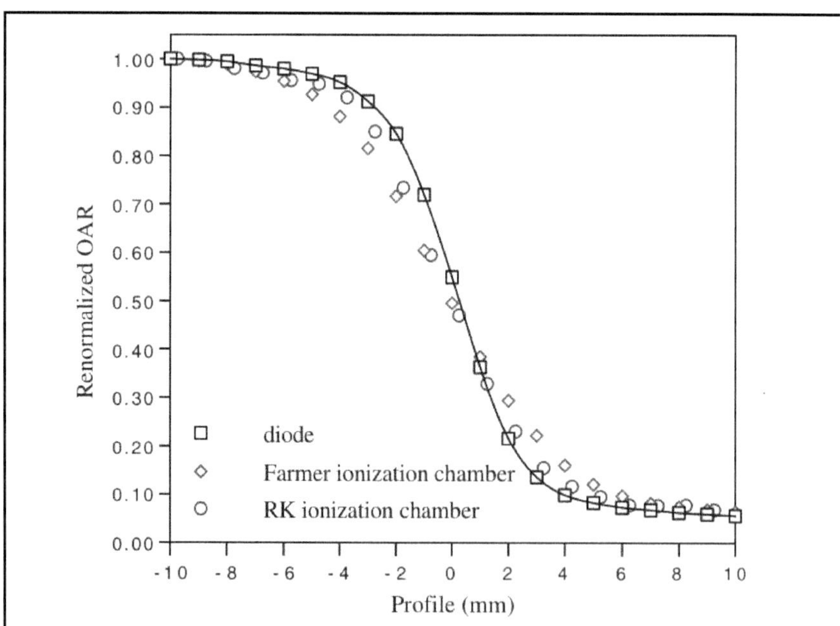

Figure 4.24: Comparison of penumbral dose profile shape as measured by different detectors at d_{max} in a 6 MV x-ray field. Dose for this graph has been normalized to 1 at a position 1 cm inside the 50% geometric penumbral edge.

The penumbral width can be quantified by defining the distance measured horizontally across the profile width between the 80%–20% dose level as p′ (with reference to p′ in figure 4.24). The dose fall-off is relatively linear in this region; hence if the tangent line joining the 80%–20% is extended, $p = p' \times 10/6$. Other regions such as the 90%–10% may be useful in some applications (Biggs et al. 1991). This region may be much wider than the 80%–20% at depth because the scatter component in the shallow fall-off part of the penumbra may exceed 10% but rarely exceeds 20%.

Table 4.7 shows a summary of the p′ values obtained using a combination of detector systems and detector correction methods. The 80%–20% results obtained with the diode show close agreement with TLD and film results, and these detectors show good spatial resolution.

Comparison of symmetric and asymmetric profiles in the penumbral region are shown in figure 4.25. All penumbral profiles were normalized to be equal to one at a position 10 mm inside the geometric edge of the field. Also shown on this graph are the standard ways of measuring p′ and p for the forming function calculation outlined in the next section.

Data for table 4.8 have been drawn from graphs similar to figure 4.24, taken at different depths with the diode detector. This table shows a summary of the 80%–20% penumbral widths, p′, obtained for symmetric and asymmetric fields at different depths. The differences in p′ for symmetric and asymmetric fields is less than 0.2 mm at all depths tested. The slight increase in traversal across the beam edge by the detector due to beam divergence of symmetric beams is a negligible effect. The slope from the 80%–20% in the beam profile is often misinterpreted as being due to beam divergence.

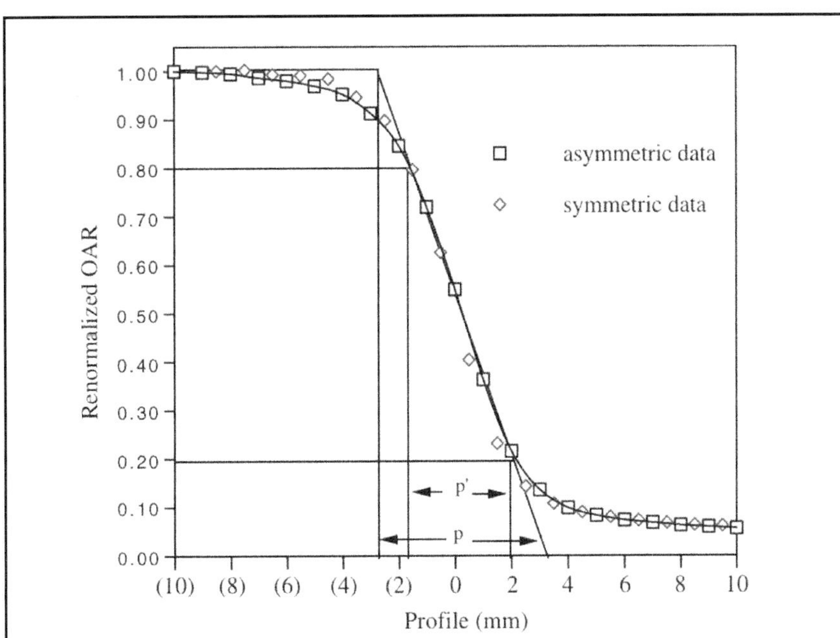

Figure 4.25: Comparison of 6 MV x-ray asymmetric and symmetric beam penumbra at 1.5 cm depth; also included is the way p is measured. For this graph dose has been normalized to 1 at a position 1 cm inside the 50% geometric penumbral edge.

Table 4.7: A comparison of 80%–20% penumbral widths for an asymmetric field using the diode and other detectors at 100 cm SSD. (Data adapted from Metcalfe et al. 1993)

Measurement	*Asymmetric 10×10 cm² 80%–20% width p′ (mm)*
Shielded diode (Scanditronics)	3.6
Point diode detector (calculated)	3.4
TLDs (Harshaw TLD 100 rods)	3.4
Film (Kodak XTL)	3.6
Ionization chamber (NE2571)	6.0
Ionization chamber (RK)	5.0
Ionization chamber (extrapolated)	3.1

4.6.4 Penumbral Forming Functions

Some commercial radiotherapy computer planning systems use collected beam data in a fan beam geometry. Unless the data points are spaced closer together in the penumbra, there are only a limited number of data points to describe a high dose gradient region. For this reason several systems use empirical penumbral forming functions to calculate the dose in the penumbral region. Some systems use both; for example, one system uses fan beam data for rectangular fields but uses a forming function only when beam blocks are employed (Target manual 1990). Another system uses a penumbral forming function for both rectangular field and blocked beam (irregular field) computations (Theraplan manual 1989). Three forming functions are outlined as follows.

Double Exponential Forming Function

This function has curvature tuning constants that can be altered to adjust the upper profile curve shape (α_1) and the lower profile curve shape (α_2). With reference to parameters defined in figure 4.26, the dose relative to the central axis dose at a particular depth central axis (i.e., the OAR) defined here as D_{pen} can be calculated from a fit produced by the following empirical equations (Johns and Cunningham 1983),

$$D_{pen}(d_{ref}, x) = 1 - \frac{1}{2}\exp\left(-\frac{\alpha_1}{p}\right)\left(\left(\frac{W_d}{2}\right) - |x|\right) \quad for \quad |x| \le \frac{W_d}{2} \tag{4.23}$$

and

$$D_{pen}(d_{ref}, x) = (t + s_c)\left(\frac{1}{2} - (t + s_c)\right)\exp\left(-\frac{\alpha_2}{p}\right)\left(|x| - \left(\frac{W_d}{2}\right)\right), \quad \text{for} \quad |x| \ge \frac{W_d}{2} \tag{4.24}$$

Table 4.8: A comparison of 80%–20% penumbral widths for symmetric and asymmetric fields using the diode at different depths for 6 MV x-rays on a Varian 2100C linac, 100 cm SSD

Depth of Profile Scan (cm)	*Symmetric 10×10 cm² 80%–20 % Width p′ (mm)*	*Asymmetric 10×10 cm² 80%–20% Width p′ (mm)*
1.5	3.4	3.6
5.0	4.1	4.2
10.0	4.8	4.9
20.0	6.8	6.7
30.0	11.2	10.7

The profile is assumed to have symmetry about the central axis, so the absolute value of the off-axis distance x is used. $W_d/2$ is the half width of the beam at depth d_{ref}, and t is the collimator transmission factor.

The model of Johns and Cunningham (1983) was originally designed with cobalt-60 beams in mind. It has since been adapted for linear accelerators as follows. The parameter, p, can be measured directly, which essentially

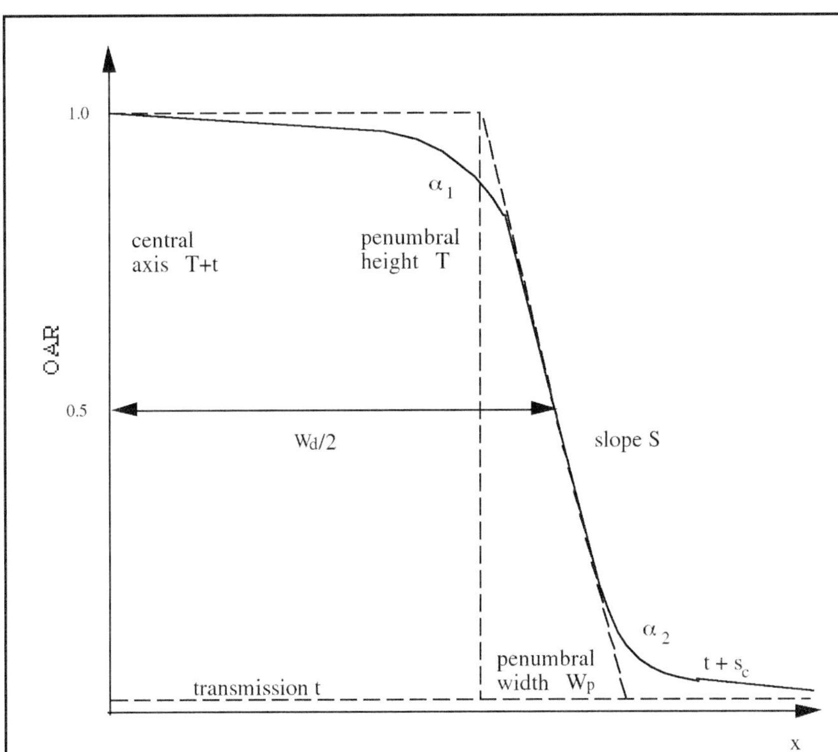

Figure 4.26: Empirical methods used to model the shape of the profiles. The lower the values of α in equations 4.23 and 4.24, the more rounded the curves become. Also includes definition of terms used in square root forming function described in equation 4.27.

Table 4.9: Suitable exponential forming function parameters that fit asymmetric field profile data collected with a shielded diode. Beam = 6 MV x-rays, machine = Varian 2100C linac, field size = 10×10 cm², SSD = 100 cm

Depth of Profile (cm)	α_1	α_2	*p (mm)*	$t + s_c$ *(%)*
1.5	3.1	3.3	6.0	5.4
5.0	3.5	3.5	7.0	7.5
10.0	3.8	3.9	8.2	11.0
10.0+	4.3	4.0	8.3	11.0
10.0++	3.6	6.8	8.2	15.5
20.0	4.9	4.9	11.1	15.6
30.0	7.2	7.7	17.8	19.1
Uncertainty	±0.2	±0.2	±0.2	±0.5

+ Data for a 10×10 cm² asymmetric field at 10 cm depth. One jaw has been driven 5 cm beyond the central axis of the field crosshairs. The data show very small differences from the other asymmetric field data where the jaw has been driven to the center of the crosshairs.

++ Data for a 20×20 cm² asymmetric field at 10 cm depth. This shows an increase in scatter causing $t + s_c$ to go up from 11.0% to 15.5%. This also has an impact on the shape of the lower profile. Note the value of α_2 for the 20×20 cm² field compared with α_2 for the 10×10 cm² field.

removes the reliance of the model on the source size feature where p is calculated from source size s' (Johns and Cunningham 1983).

The original authors used the model for *dose due to primary* and subtracted the scatter component from the penumbra tail. As we are dealing with profiles that include scatter, we have included the measured transmission, t, and scatter, s_c, outside the field. Note that in the Theraplan implementation, s_c is subtracted at all depths and it is assumed that the α values can be set at one depth and will not alter as this scatter component is removed. Shown in table 4.9 is a list of parameters at different depths that best fit the analytical equations to beam data for a 6 MV asymmetric x-ray field.

The main advantage of this method is that the small asymmetries between the upper and lower profile curvature can be accounted for because α_1 and α_2 are separate. One disadvantage of the method is that two equations are employed and discontinuities can occur where the functions match. For example, this has bearing on applications where Fourier deconvolution in the beam penumbra is attempted (Chui et al. 1988, Murray 1994). No attempt is made here to adjust the penumbral shape due to penumbral flaring in low-density media; however some progress has been made in this regard elsewhere (Hoban et al. 1992).

Cosine Forming Function

Another empirical penumbral forming function uses a sinusoidal intensity function such that (Target manual 1991):

$$D_{pen}(d_{ref},x)=\frac{\cos\left(\frac{d_{pen}}{s'}\times 180\right)+1}{2}\times\left(1-(t+s_c)\right)+(t+s_c), \quad (4.25)$$

where d_{pen} is the distance across the penumbra from the unblocked field edge, and t is the block transmission factor.

Note that the virtual source size for linear accelerators is about 2 mm (Munro et al. 1988), but the use of this value produces a profile that is much too narrow in the penumbra. Because the Varian lower asymmetric collimators are at a source collimator distance of f_c = 36 cm, the quasipenumbral source size that was found by using (Target manual 1991):

$$s'=\left(\frac{f_c}{f+d_{ref}}\right)\times p'', \quad (4.26)$$

where $p''=2.4\times p'$, p' is the penumbral width measured from the 80% to the 20% level at d_{ref}, and 2.4 is an empirical constant that spreads the width to account for electron ranging and perhaps some extra focal radiation effects. For a more correct empirical model the constant 2.4 would vary with energy. In this case a d_{ref} of 1.5 cm was chosen, and p' was measured to be 4 mm for the nondivergent edge of a 10×10 cm^2 asymmetric field. Applying equation 4.25, this relates to an s' of 3.4 mm.

Square Root Forming Function

A single continuous forming function that shows a good fit to experimental data is based on an inverse square root function (Kron et al. 1993a). The function is described as follows:

$$D_{pen}(s)=\frac{1-(t+s_c)}{2}\times\left(1-\frac{s}{\sqrt{s^2+n}}\right)+(t+s_c), \quad (4.27)$$

where s is the distance from the field edge that is located at $s = 0$. Therefore, $s = x - W_d/2$. Note that $1 - (t + s_c)$ is the maximum height of the penumbra at the reference position, and n is an empirical parameter (typical value for 6 MV is at d_{max} $n = 4.5$) that gives the curvature of the function—the greater the magnitude of n the steeper the penumbral curvature.

It can be shown also that

$$p'=1.5\sqrt{n}, \quad (4.28)$$

which is a useful way to estimate penumbral width (Kron et al. 1993a). The disadvantage of the square root and cosine functions is that symmetry is assumed in the upper and lower curved parts of the penumbral profile. The results of applying the three different penumbral forming functions to a 6 MV penumbral profile at d_{max} are shown in figure 4.27. The inverse square root and exponential methods both show good fits to the data. The cosine method does not; the slope of the curve is correct, but the curved edge of the penumbra at the bottom and top of the penumbra is not a good match to the data.

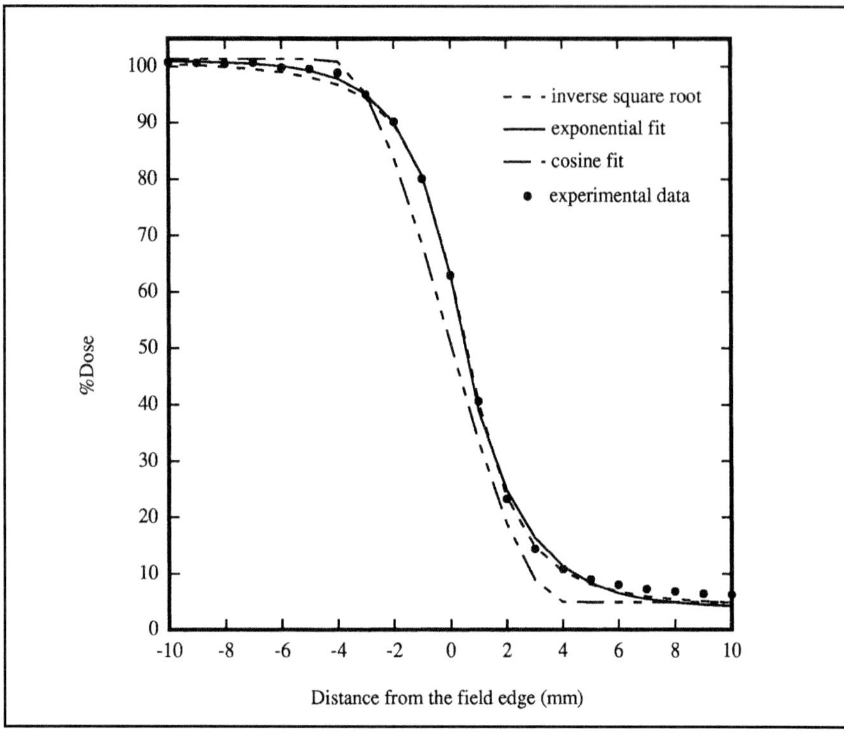

Figure 4.27: Exponential, cosine, and square root forming function fits to 6 MV profile data at 1.5 cm depth. Exponential forming function details: The alphas are set to 3.1 and 3.3 respectively, p is 0.6 cm, and t + s is 5.4%. Dose is normalized to 1 at a position 1 cm inside the 50% geometric penumbral edge. Square root forming function details: $T = (1 - (t + S_c)) = 48.8$, t is 3.6, s is −0.528, and n is 4.34.

4.7 Attenuation and Shielding

4.7.1 Compensators

Compensators are used to account for the effect of missing tissue on the dose distribution. Frank Ellis realized in 1951 that by retracting compensating material away from the patient surface, skin sparing could be maintained. Retracted compensators are generally mounted in the block tray or wedge slot. The thickness and shape of the compensator depends on the amount of missing tissue. Lead or lead alloy provides an efficient compensation material (i.e., significant attenuation in a compact thickness). All results are shown here for layers of lead, although various other materials have been used (Boge et al. 1974, Jackson 1970, Constantinou and Harrington 1989, Robinson and Scrimger 1990, Chang et al. 2004).

In order to estimate the equivalent amount of retracted compensator material needed to replace that of tissue we define a transmission curve as the graph representing percentage dose transmission versus thickness of attenuating material. These data were originally used to compensate manually for missing tissue with an equivalent amount of retracted material (Ellis 1960, Hall and Oliver 1961).

Transmission curve data can also be used to calculate effective *attenuation coefficients* as used in computer plan calculations. This data is obtained by stacking layers of attenuating material (e.g., lead) in the retracted position (e.g., wedge or block tray). An ionization chamber is then placed in air or in phantom to measure charge readings with different layers of attenuator. As the transmission curve approximates an exponential, the assumption is usually made that an effective attenuation coefficient, μ, can be calculated for each layer using the exponential equation.

$$\Phi = \Phi_0 \exp(-\mu x), \tag{4.29}$$

where Φ is the intensity with the attenuator present in the beam, Φ_o is the intensity without the attenuator, μ is the measured linear attenuation coefficient, and x is the thickness of the attenuator. Note that the assumptions implied by this approximation are:

i) The beam is infinitely narrow. (It actually has finite width).
ii) The beam is monoenergetic; in reality it is polyenergetic.
iii) There is no self-absorption by the attenuator.
iv) μ is depth independent. (This is an approximation.)
v) μ is not field size dependent. (It is due to increasing scatter.)
vi) The attenuator is placed at the same retracted source to compensator distance each time (stacking material causes a small offset in position).

Figure 4.28 shows a percentage transmission curve for a 10 × 10 cm² field at a depth of 10 cm in water. Somewhat surprisingly, comparisons of

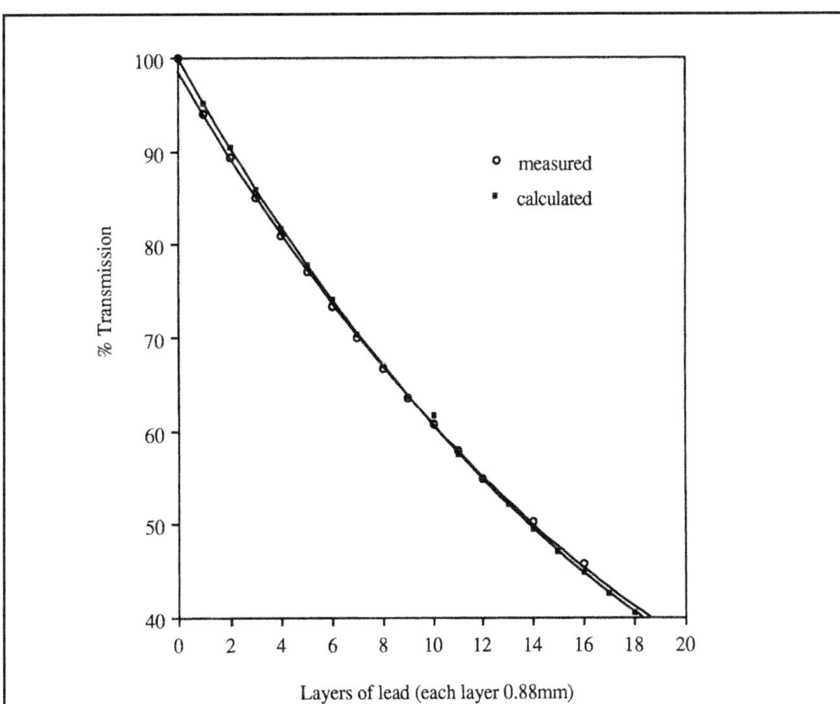

Figure 4.28: Percentage transmission of dose versus thickness of retracted lead compensator. Also included is the transmission curve generated by using μ = 0.57 cm⁻¹.

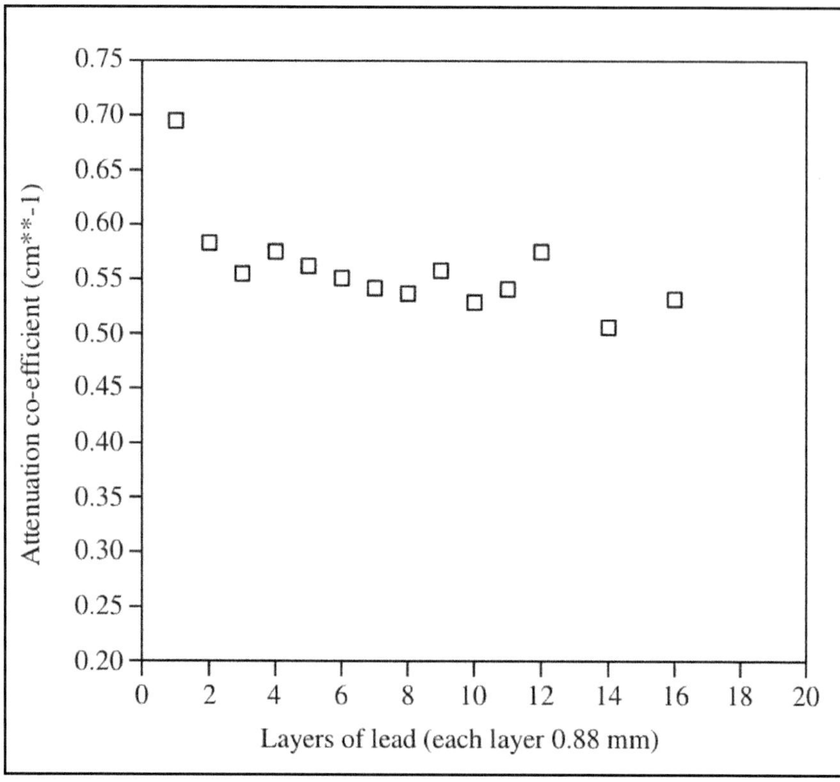

Figure 4.29: Different values of μ derived from the transmission data shown in figure 4.28.

transmission curves at d_{max} in water and in air are very similar. As μ is depth dependent, d=10 cm is presented here, which is a typical clinical depth of interest.

The results of calculating different μ are shown in figure 4.29 for different layers of lead. The large variation in these values immediately shows the need for an averaging method. Calculating a single μ value is therefore difficult for such a broad-spectrum photon beam. However, a specific value can be chosen so that the reconstructed transmission curve closely matches the measured one in most regions.

It is relatively easy to match the middle region of both curves once the photon beam has transmitted through several layers. This is because the lower energy components of the beam have been removed by the first few layers of lead; the rest of the transmitted curve almost follows the exponential shape.

One way to produce a reasonable match to the transmission data is to calculate μ by obtaining the half-value thickness (HVT) from the graph and substituting it in the following equation.

$$\mu = \frac{\ln 2}{HVT}, \tag{4.30}$$

With this method an average value of μ for all the lead layers up to the thickness of HVT is given. It deweights the high proportional absorption of the first few layers of lead.

Another successful approach is to start with μ selected for five layers, layers 2 to 6, 3 to 7, 4 to 8, 5 to 9, and so on, and obtain an average. Then iteratively correct μ until a best match is found for transmission curves. The first layer of lead is usually ignored due to extra attenuation of low-energy components by this layer.

The position of the attenuator does have some effect on the attenuation properties; in general, less material is required as the source attenuator distance is reduced. The ratio of the thickness required at shorter SSD to reduce the intensity to a known amount to the thickness required at longer SSD to reduce the intensity to the same amount is commonly called the *retraction ratio*. For MV linacs this value can be as low as one and as high as two.

4.7.2 Shielding Blocks

When blocks (lead or LMA) are placed in the beam, they are used to shield critical organs. In this position they only partially cover the field. Hence though the block transmission is usually less than five half-value layers (HVLs), there is more dose under the block than this, due mainly to in-phantom scatter. Figure 4.30 shows the results of depth dose scans taken

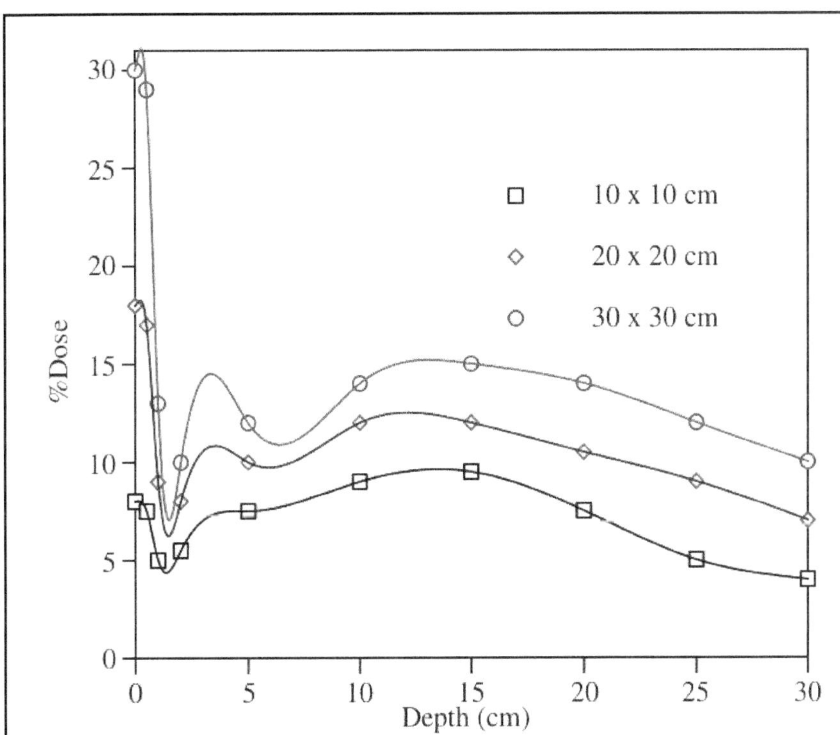

Figure 4.30: Depth dose curves under a 2 cm wide block with field sizes 10 × 10 cm^2, 20 × 20 cm^2, and 30 × 30 cm^2, respectively. The larger fields produce more dose under the block due to increased scatter. Note that the initial dose is high due to electron contamination, followed by a rapid reduction in dose followed by a slow climb in dose to a plateau at about 15 cm.

under a block that was placed over the central axis of a 6 MV x-ray beam. The lead block was 8 cm thick by 2 cm wide, extending beyond maximum field length. To demonstrate the scatter effect, three different field sizes were chosen. The maximum dose under the block occurs at the maximum field size. The curves show a maximum dose at the surface due to electron contamination and then a rapid decrease in dose that gradually increases and reaches a plateau at about 15 cm depth. It is the in-phantom scatter that causes this characteristic shape beyond d_{max}. Some standard lead blocks are manufactured with perpendicular sides; as they may be placed anywhere in the field, these blocks produce a wider penumbra than the jaw when they are placed off-axis because of partial transmission effect. Custom LMA blocks whose molds are manufactured using a divergent cutting device that mimics the beam source focus do have a tighter penumbra edge similar to jaws.

4.7.3 Multileaf Collimators (MLCs)

MLC Properties and Types

A multileaf collimator (MLC) consists of multiple adjacent vanes or leaves of high-density attenuating material, with each leaf capable of individual motion in and out of the radiation field. The primary purpose of an MLC is to shield regions of the patient from radiation in a similar fashion to shielding blocks. For an MLC to have broad application in radiotherapy it must be capable of achieving a variety of aperture shapes across a large range of radiation field sizes up to 40×40 cm^2. The majority of MLCs can have their leaves incremented across a range of positions; the single largest range is that of the Elekta MLC (Elekta, Stockholm, Sweden; formerly Philips), which can span 32.5 cm.

The properties of several commercial MLCs are listed in table 4.10. This table includes current commercial MLC used in general radiotherapy applications and micro-MLC designs used in stereotactic radiosurgery applications. The table does not include binary MLC designs. The binary MLC designs, commercially available versions of which include the MIMiC MLC (NOMOS Corporation, Sewickley, PA) and tomotherapy (TomoTherapy Inc., Madison, WI), are discussed in chapter seven.

Definitions

The AAPM Task Group 50 Report 72 (AAPM 2001a) definitions have been adopted to describe the MLC leaf properties and orientations. With reference to figure 4.31, the *width* of a leaf is the dimension of the leaf perpendicular to the direction of propagation of the x-ray beam and perpendicular to the direction of leaf motion. The *length* of each leaf refers to the leaf dimension parallel to the direction of leaf motion. The leading surface of each leaf that inserts into the field is the *leaf end*. The leaf surfaces in contact with adjacent leaves are the *leaf sides*. The *leaf height* refers to the dimension of the leaf along the direction of the primary x-ray beam. The leaf height extends from the surface near the x-ray source to the bottom of the leaf nearest the isocenter. The height of the leaf combined with its leaf density determines its attenuation properties.

Table 4.10: Summary of MLC parameters for various linear accelerators (all units cm). (Some parameters adapted from AAPM 2001a)

	Manufacturer						
	Siemens		*Varian*		*Elekta*	*BrainLAB*	*Radionics*
MLC name	*3-D MLC*	*OptiFocus*	*Standard MLC-80*	*Millennium MLC-120*	*MLCi*	*BrainLAB m3*	MMLC
Number of leaves	2 x 27	2 x 41	2 x 40	2 x 60	2 x 40	2 x 26	2 x 31
Leaf width at Isocenter	10 mm with outermost 6.5 cm	10 mm with outermost 5 cm	10 mm	Central 20 cm: 5 mm, Outer 20 cm: 10 mm	10 mm	3.0 mm, 4.5 mm, and 6.0 mm	4.0 mm
Leaf thickness material	76 mm WHA (tungsten heavy alloy metal)	76 mm WHA	55.3 mm WHA	55.3 mm WHA	75 mm WHA	60 mm tungsten with coating	73 mm WHA
Focusing	Double	Double	One round and one focused	One round and one focused	One round and one focused	One round and one focused	Single
Tongue and groove	Yes	Yes	Yes	Yes	Yes	Yes	Yes
Replacement of jaw	Yes	Yes	No	No	Yes (backup diaphragm)	No	No
Maximum field size	27x40 cm^2	40x40 cm^2	40x40 cm^2	40x40 cm^2	40x40 cm^2	10.2x10 cm^2	12x10cm^2
Overtravel	10 cm	10 cm	16.0 cm	19.7 cm	12.5 cm	4.7 cm	~5 cm
Maximum single leaf extension	30 cm	30 cm	15 cm	15 cm	32.5 cm	9.7 cm	~10 cm
Interdigit	No	No	Yes	Yes	No	Yes	Yes

Configurations

Nearly all MLCs are constructed from tungsten heavy alloys (WHA); these are high-density metal alloys. Pure tungsten is brittle and difficult to machine. By adding varying amounts of nickel, iron, and copper the machinability can be improved, which is fortunate as physical leaf side dimensions as thin as 1 mm are being manufactured. Pure tungsten has a density of 19.3 g/cm^3, and the alloys have densities that range from 17.0 to 18.5 g/cm^3.

The amount of radiation that passes through an MLC leaf is defined as the *intraleaf transmission*, and the radiation passing between the leaves is the *interleaf transmission*. These transmissions are sometimes referred to as *leaf*

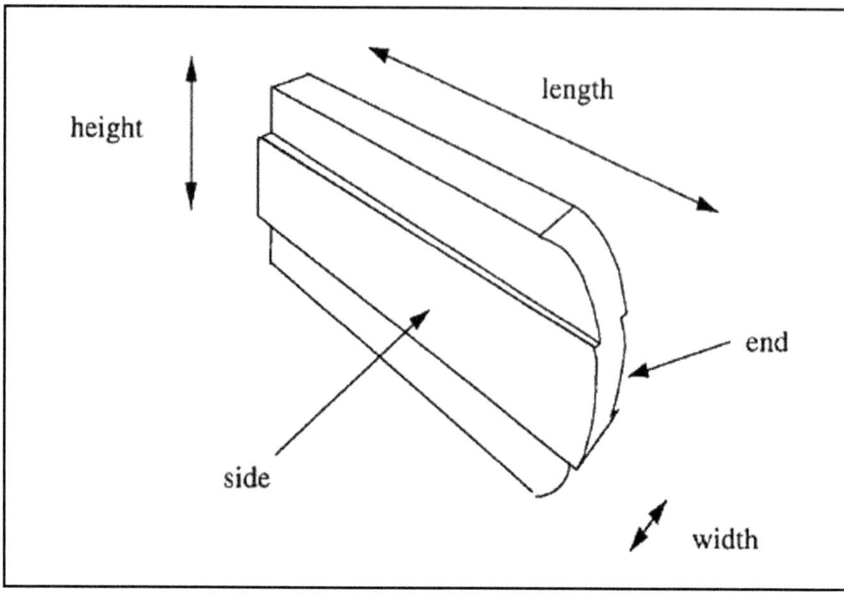

Figure 4.31: Schematics of a generic MLC showing the terminology used to describe various parts of each leaf. (Reprinted from AAPM (2001a) with permission from American Association of Physicists in Medicine (AAPM).)

leakage. When an MLC is positioned as a secondary collimation device, it replaces either the upper or lower secondary jaws (e.g., Siemens and Elekta). Tertiary collimators sit below both of the secondary collimators (e.g., Varian). When used as a secondary collimator the MLC is required to attenuate the beam as per the jaws they have replaced to within 2% of the original intensity, and when acting as a tertiary collimator the transmission should not exceed that produced by shielding blocks (generally between 4 and 5 HVL); hence 5% transmission is accepted (AAPM 2001a). To comply with the transmission requirements, MLCs that act as secondary collimators are generally constructed thicker in the leaf height dimension than MLCs that are designed as tertiary collimators. The overriding rationale behind these somewhat arbitrary leakage limits is to limit dose to normal tissues outside the target volume.

Focus Designs

The ability of the MLC to conform to a target shape in sections of curvature is dependent on the leaf width. A narrower leaf width provides better conformity to steeply curved aperture shapes. The space required to house the various components associated with the positioning (e.g., motors and lead screws) and verification (e.g., potentiometers) of each individual MLC leaf restricts the number of MLC leaves that can fit inside the treatment head. A reduced leaf width means more leaves are required to cover the same area, but the number of leaves is limited also by the amount of space available.

Penumbra width is minimized by presenting the beam with a focused diverging side leaf construction. This is achieved by designing the MLC leaves with a divergent or focused cross section. The two leaf sides are constructed with a divergence focused on the x-ray source. The majority of MLCs are focused in this manner, as shown in figure 4.32. To aid in efficiency of manufacture and replacement, small sets of neighboring leaves may have the same divergence, then the next set of leaves diverge a little more as the sets progress away from the central axis.

With the divergent design of the MLC leaf sides corresponding to beam divergence, the slightest separation between adjacent leaves could result in an unhindered path of radiation from the source to patient. To prevent this occurrence, the sides of MLC leaves are designed with an interlocking tongue and groove (see also figure 4.32). This design limits the amount of radiation that leaks between adjacent leaves by providing a physical barrier, the tongue, to attenuate the beam. Because the leaf sides still diverge, this is known as a single-focus MLC.

The leaf can also be focused in the direction of leaf motion. This is achieved by maintaining the leaf end parallel to the radiation beam, independent of the leaf position across the field. The Siemens MLC achieves double (or dual) focusing by combining a diverging MLC cross section with an arced path of motion for each leaf in direction of leaf travel; see figure 4.33.

A simpler method to approximate focusing in the direction of leaf motion is to use a rounded leaf end (see figure 4.33b). In this method the leaf moves in a plane perpendicular to the central axis of the beam. The angle of curvature of the leaf end is designed so that the leaf end is always tangential to the divergent beam. The position of the leaf tip and the light field projected is also different, depending on off-axis position. This is accounted for by use of an MLC look-up position calibration table that corrects for this effect (Vial et al. 2006). As a consequence of the curved leaf design the transmission through the curved surface varies, depending on the off-axis position of the leaf. This results in an offset between the projected

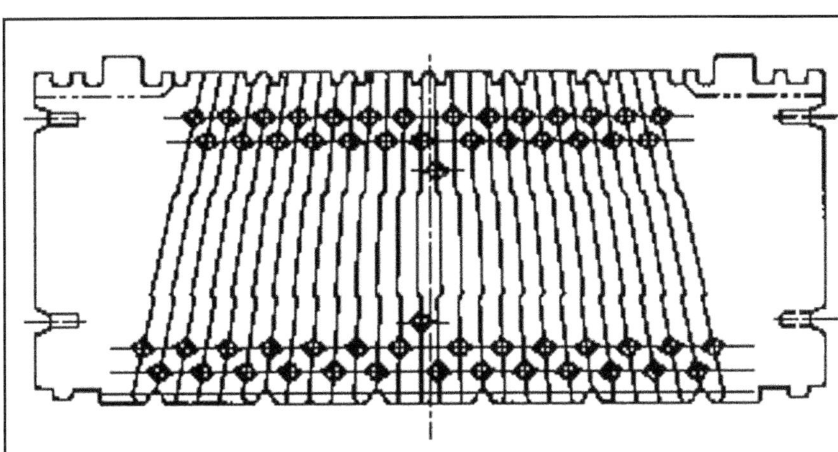

Figure 4.32: Schematic showing MLC divergence and tongue and groove effect for a Siemens MLC with rounded leaf end. (Reprinted from AAPM (2001a) with permission from American Association of Physicists in Medicine (AAPM).)

light field and the radiation field as defined by the 50% penumbra. The radiation field penumbra is also slightly broader than for jaw transmission. In figure 4.34 the difference in field dose profile under a rounded leaf end 10 × 10 cm^2 MLC versus a 10 × 10 cm^2 jaw at d_{max} in a 6 MV photon field is shown.

The Varian MLC has a physical fuse as its weakest link so that in the event of a leaf pair collision, major damage is avoided. The calibration ensures that at nominal 0 leaf gap, the gap is actually 0.05 mm. An extra safety margin can be set in the planning computer also (typically 1 mm); however this does lead to increased end leaf leakage in the event that the jaws are not shielding the end leaf opening. A dosimetry film showing end leaf leakage is shown in figure 4.35. The curved leaf compromise with diverging MLC cross-section focus maintained is known as a single-focus design. These designs do enable the use of threaded screws, which provide very accurate placement of leaves to their designated placement positions. The Varian MLC has positional placement to within ±0.1 mm.

Leaf Extension

The distance of the single leaf extension is the maximum difference in position by which two adjacent MLC leaves can be separated. Overtravel is the maximum distance past the central axis that a leaf can travel. The field-shaping limits of the Siemens and Elekta MLCs are shown in figure 4.36a. The Siemens MLC can extend 20 cm to the center of the field and then overtravel a further 10 cm across the centerline. This gives a maximum leaf extension for this collimator system of 30 cm. The Elekta MLC has 12.5 cm overtravel, with a maximum single leaf extension of 32.5 cm.

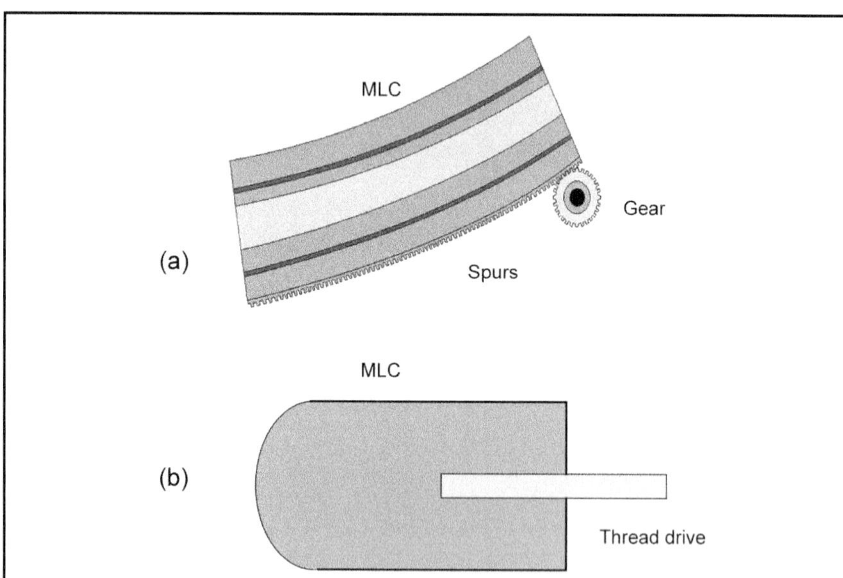

Figure 4.33: Schematic showing different MLC designs: (a) dual-focus MLC with spur gear; (b) single-focus MLC with rounded leaf end.

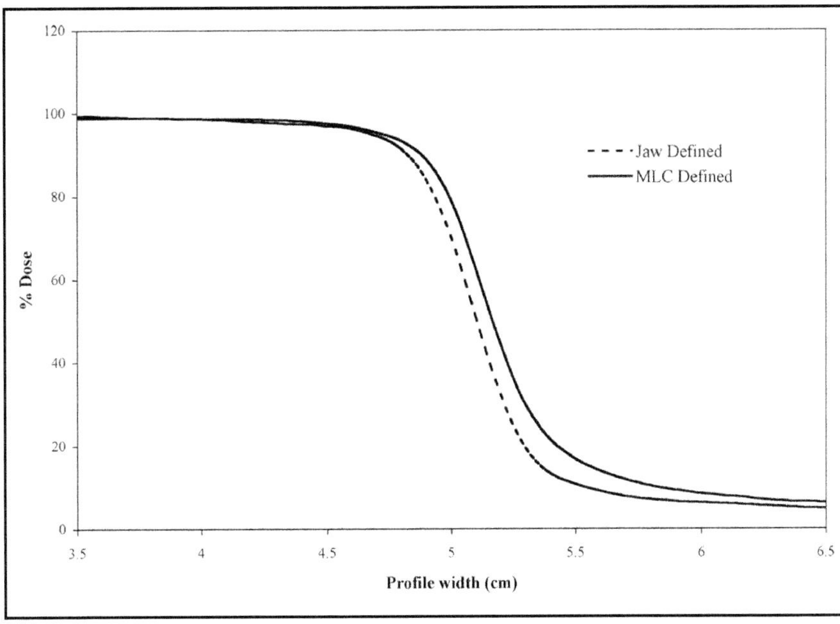

Figure 4.34: Dose profile for 10 × 10 cm^2 MLC compared with a collimated 10 × 10 cm^2 field across a Varian rounded leaf end MLC at d_{max} in a 6 MV photon field. Note how the MLC projects a wider profile and a broader penumbra due to rounded end leaf transmission. The leakage in the tail of the MLC penumbra is also higher than that under the jaw. Note only half of the symmetric profile is shown; also, the jaws were set at 30 × 30 cm^2 for data collection when the MLC was set at 10 × 10 cm^2.

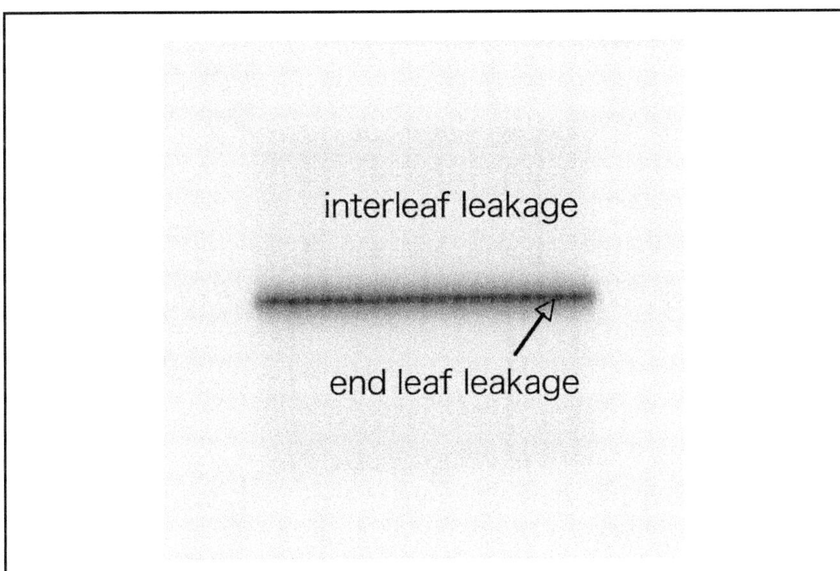

Figure 4.35: Dosimetry film showing end leaf leakage under a Varian MLC; note this leakage is about 30%, but in clinical treatments this is usually avoided by jaw shielding.

An MLC that has a relatively small single-leaf extension distance can achieve a useful overtravel distance by mounting the MLC leaves on a movable carriage (e.g., Varian). By moving the carriage the maximum leaf extension distance is maintained, but the overtravel distance that can be achieved is increased. The Varian MLC is the only design that has a movable MLC carriage. This extends the range of MLC positioning across the field, as shown in figure 4.36b. The maximum leaf extension is only 14.5 cm for the Varian MLC; however the carriage can be moved to positions anywhere from outside the maximum field boundary to 2 cm beyond the field center. Note the carriage is interlocked to remain stationary during beam on. This means the Varian MLC can move further across central axis than any of the other systems. For single-field block replacement, the Varian tertiary configuration is more limited than the other systems for fields that are large in the leaf length dimension. On the other hand, this configuration lends itself to large-field applications for intensity-modulated radiation therapy (IMRT) if overlapping fields are used (Metcalfe et al. 2004). In these practical applications the jaw shielding and carriage are usually driven 2 cm short of central axis, and by combining two overlapping fields a 25 cm combined field width is achievable.

The Elekta and Siemens leaf designs do not allow leaves on one side of a field to interdigitate with neighboring leaves on the opposing leaf bank; the Varian MLC can interdigitate. The ends of odd-numbered leaves from the right-hand bank can be driven past the ends of even-numbered leaves from the left-hand bank. While this may not be important for most block replacement applications, interdigitation can be useful when segments are required for IMRT applications. For example, if superimposed beam segments at a fixed gantry angle are used for intensity modulation, unusual beam shapes that employ leaf interdigitation can improve the efficiency of dose delivery (Galvin et al. 1993a,b).

4.7.4 Wedges and Dynamic Wedges

Physical Wedges

Wedges differentially attenuate the beam across the dose profile; they are generally used to compensate dose changes due to axial contour changes. They are also useful in reducing hot dose regions within a target when orthogonal beams are employed. The wedge angles quoted relate to the isodose angle produced by the wedge specified at a nominated depth, e.g., 10 cm (Khan 1984). This is preferred to selecting the 50% isodose level to assess the angle, as this is generally deeper than the typical patient half separation depth. Another reasonable level to choose for wedge angle assessment at energies such as 6 MV is the 80% isodose level. The wedge angle definition of IEC (1989a) also addresses the curvature of the wedge profile as shown in figure 4.37. Note that physical wedges produce some beam hardening off-axis; hence beam horn effects tend to appear less in dose profiles collected under wedged beams in the non-wedge dimension than profiles collected at equivalent depths in open fields.

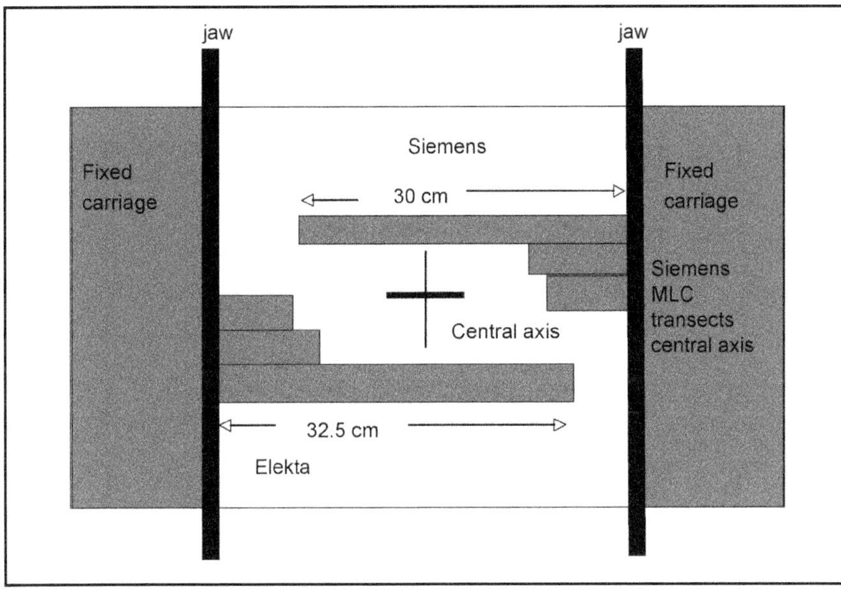

Figure 4.36(a): Leaf span and over-center travel of Siemens and Elekta MLC with fixed carriage.

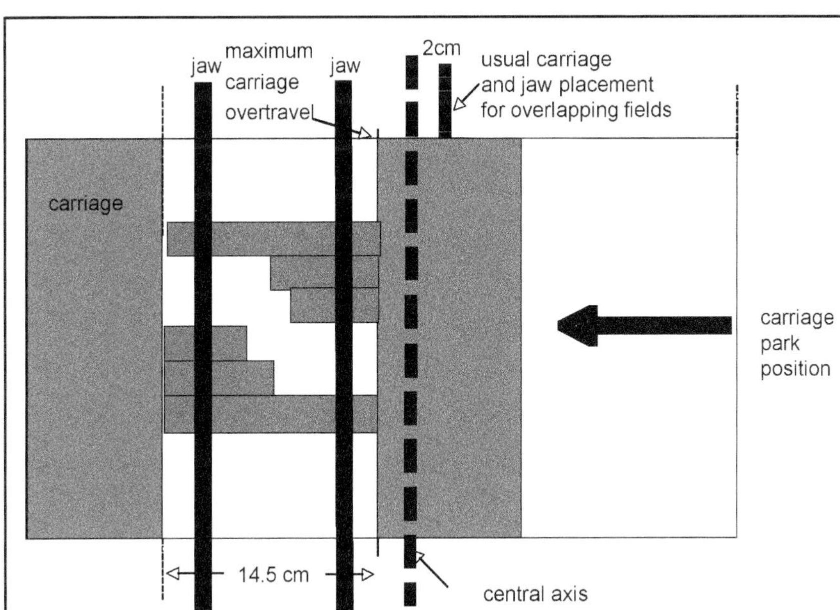

Figure 4.36(b): Leaf span and over-center travel of Varian MLC with movable carriage.

Physical wedges attenuate the beam. To account for this, wedge factors are determined as the ratio of dose at d_{max} with the wedge to the dose at d_{max} for an open field. To account for shift in d_{max}, due to changes in beam hardening and electron contamination, the measurements may however be done at a greater depth with the inclusion of an appropriate depth dose conversion back to d_{max}. Block tray holders attenuate the beam also, and therefore block tray factors are measured in the same manner as wedge factors. Typical wedge factors and block tray factors are listed in table 4.1. Wedge factors should be tested for different field sizes and can be issued as field size dependent if the dependence of wedge factor on field size is considered significant.

Wedge profiles and wedge factors should be collected and assessed for consistency with the wedge placed in different orientations (i.e., with respect to the collimators and with respect to collimator rotation) to ensure the wedge is aligned correctly. Note that the volume of wedge material within elongated rectangular fields depends on its orientation to the collimators within the field. The effect of this on the wedge factor should also be tested.

Depth dose beneath the wedge should be tested. There is generally a small amount of beam hardening under physical wedges that leads to a slightly more penetrating depth dose than the open field equivalent. Dosimetry of solid wedged fields with asymmetric collimation is dealt with by Khan (1993). Elekta linacs use a 60-degree physical wedge mounted internally (a flying wedge) in the treatment head and this is driven in the field for a varying proportion of the beam weight to obtain different beam angles. These devices display similar characteristics to physical wedges.

Dynamic Wedges

Physical wedges require operators to enter the room and mount these devices when required for each treatment field. Dynamic wedges are essentially a series of preset jaw moves that can be programmed outside the room at the linac control console for each field. The time efficiency gains of dynamic wedges compared with physical wedges are significant. The improvement in efficiency using MLC and dynamic wedge has been quantified by Delaney et al. (2005).

Dynamic wedges usually only operate using the jaw set which has the most overtravel. The Varian linac dynamic wedges are called *enhanced dynamic wedge* (*EDW*). The term evolved due to some improvements in wedge characteristics since the original release of dynamic wedges (Leavitt et al. 1990). The motion of the Varian overtravel jaw set coincides with MLC leaf travel (i.e., along the MLC length direction). Because of this, collimators and the MLC rotate together. This occasionally precludes the use of dynamic wedge with MLC shaping of some concave target shapes. It is also not possible to use EDW with arc fields.

Dynamic wedge angles are usually assessed using isodose data reconstructed from multi-diode or multi-ionization chamber array data. For dynamic wedges the difference in depth dose between open field and wedged appears to be imperceptible (Leavitt et al. 1990). Beam hardening is also not evident in dynamic wedge field profiles. EDW dose profiles kick up

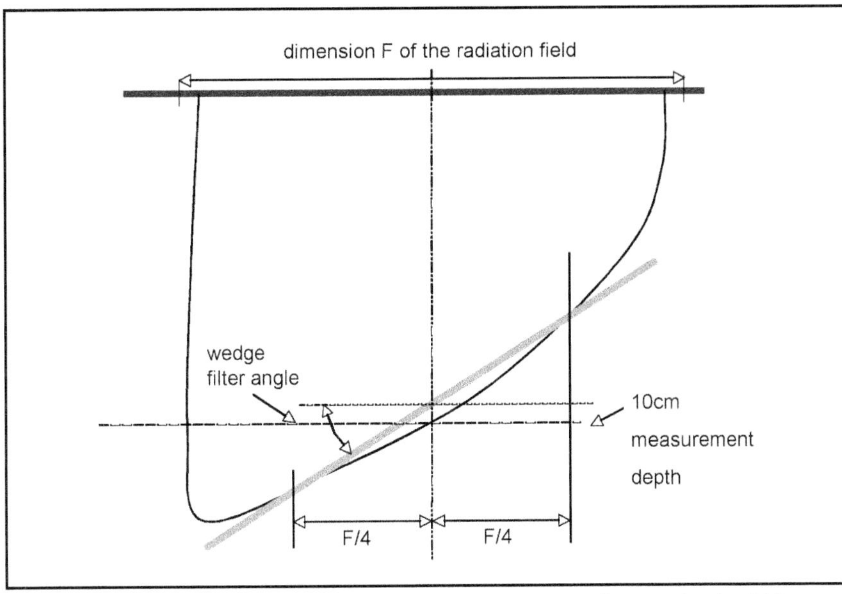

Figure 4.37: Wedge angle convention which allows for curved isodoses at reference depth of 10 cm.

more than physical wedge profiles in the wedge direction and beam horns are maintained in the cross EDW direction whereas they seem to be flatter for physical wedges.

For dynamic wedges, the wedge factor is field size dependent, and hence wedge factor tables versus field size have to be issued. *Enhanced dynamic wedge factors* (*EDWFs*) are essentially a composite of open field and sweeping collimator effects. The Varian EDW operates based on scanning one of the Y jaws with the beam on and modifying the rate of dose delivery in order to deliver a wedged dose distribution. The tabulation of how many monitor units (MUs) have been delivered at each point during the travel of the jaw is called the *golden segmented treatment table* (*GSTT*). On current model Varian linacs the valid range of travel for a EDW jaw is from 20 cm off-axis to a point 10 cm beyond central axis. The GSTT is defined such that a 60-degree wedged dose distribution results from the linac jaws moving with dose delivery conforming to this table. In order to obtain smaller wedge angles the GSTT is combined with an open field. The proportion of MUs delivered to the open field prior to the initiation of jaw motion determines the effective wedge angle of the field. Each energy has a unique GSTT.

Gibbons (1998) has presented a way of predicting the EDWFs using an analytic expression based on an exponential fit to the GSTT data. Note that the GSTT is the output from the maximum wedge angle and field size applied. It provides the relationship between MUs and moving Y jaw position. Gibbons' method represents an extension of the MU fraction model presented by Waldron (Waldron et al. 1994). The methods are based on a wedge angle being determined as a combination of weighted beams from open field and the maximum wedge angle field. The weights $W_{0°}$ and $W_{60°}$ are computed for the selected wedge angle θ where (Petti and Siddon 1985):

$$W_{0^\circ} = \frac{(\tan 60^\circ - \tan\theta^\circ)}{\tan 60^\circ}, \tag{4.31}$$

and

$$W_{0^\circ} = \frac{\tan\theta}{\tan 60^\circ}. \tag{4.32}$$

Because there are a number of effects such as S_c and dose due to jaw transmission, a number of publications present various ways to cater to these effects. We have given but a few of the many references available (Beavis et al. 1996, Yu 2002, Papatheodorou et al. 1999, Sethi et al. 2000, Kuperman 2004, 2005).

One simple empirical approach to calculating EDWFs by Yu (2002) is based on the observation that dose is accumulated by a detector at a point Y_0 in the middle of the field until the moving jaw passes that point in the field. The number of MUs delivered up to this position $GS(Y_0)$ divided by the MUs for the whole field sweep $GS(Y_F)$ gives the EDWF. A small empirical correction α is included to account for the fact that the moving jaw stops 0.5 cm short of the opposing jaw at the end of its travel. This correction factor could also account for scatter or non-ideal behavior of the jaw sweep and, as suggested by Ansbacher and Neath (2003), may yield more accurate results if made a function of field size. Hence the EDWF is given by

$$EDWF = \frac{GS(Y_0 - \alpha)}{GS(Y_F)}. \tag{4.33}$$

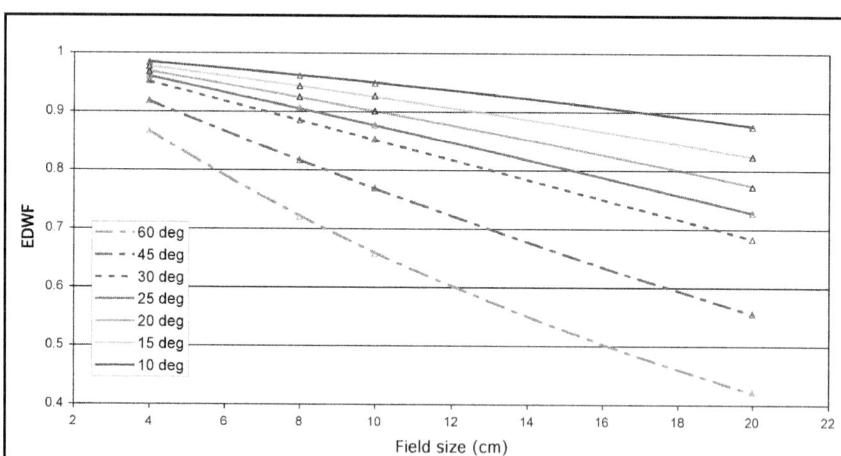

Figure 4.38: Enhanced dynamic wedge factors (EDWFs) for a 6 MV x-ray beam (linac type: Varian 21eX).

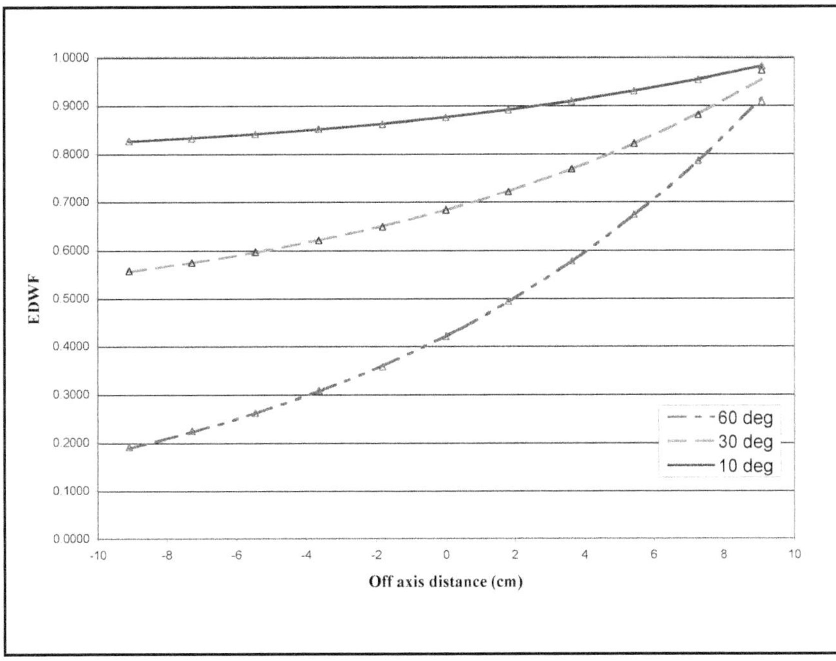

Figure 4.39: Enhanced dynamic wedge (EDW) off-axis factors for a 6 MV x-ray beam (linac type: Varian 21eX).

Enhanced dynamic wedge output factors for Varian linacs are field width dependent but relatively independent of field length. The output factors versus field size for three wedge angles are shown in figure 4.38. The change of output factor as the mid-field point moves away from the linac central axis to mid-field axis for asymmetric fields is shown in figure 4.39.

Since the photon energy spectrum in EDW fields corresponds closely to the energy in open fields, the percentage depth dose curves are very similar. Therefore, as Kuperman (2005) has observed, there is no significant dependence of central-axis EDWF on depth from 5 cm to 20 cm or for SSDs from 80 cm to 100 cm. Expressions for EDWF calculation at points off the central axis for symmetric and asymmetric fields have been presented by Prado et al. (2003) and Kuperman (2005).

The Siemens linac version of dynamic wedge is called a virtual wedge (VW) (McGhee et al. 1997). The implementation includes an internal MU compensation method to set the wedge factor to one for all field sizes. The method seems to deviate by no more than 2% for most field sizes (Rathee et al. 1999) and with a maximum deviation of 3.5% for large field sizes and wedge angles (van Santvoort 1999).

4.8 A Few Comments on kV X-Rays

Orthovoltage and superficial x-ray units pre-dated linacs for cancer treatment. Because these units are still used in radiotherapy clinics, a small section is provided here outlining beam properties. Figure 4.40 shows a combined

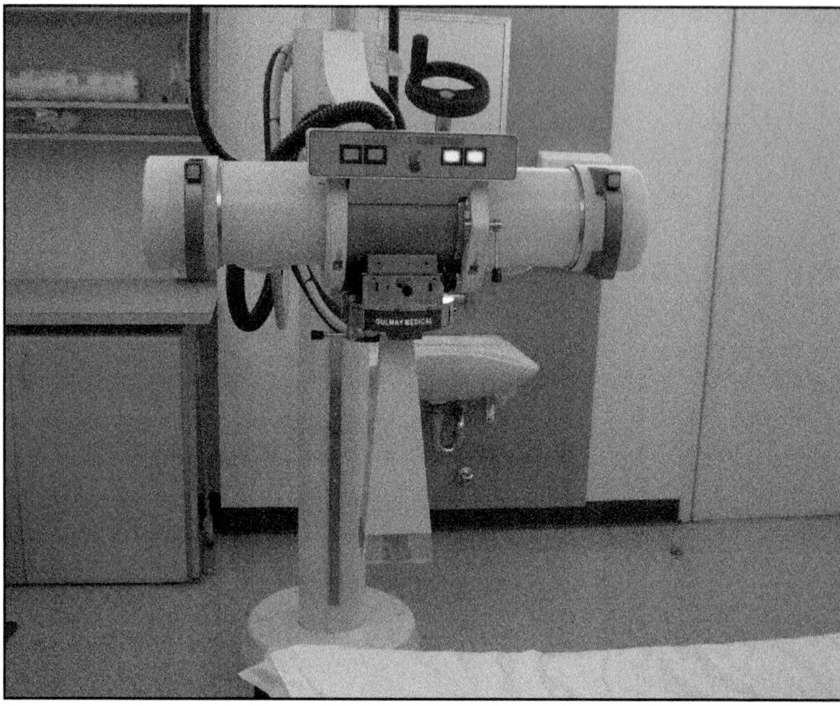

Figure 4.40: Photograph of a combined orthovoltage/superficial x-ray machine (Gulmay D3300) with an orthovoltage applicator cone attached. The machine would operate at energies 150 to 300 kVp using this type of applicator.

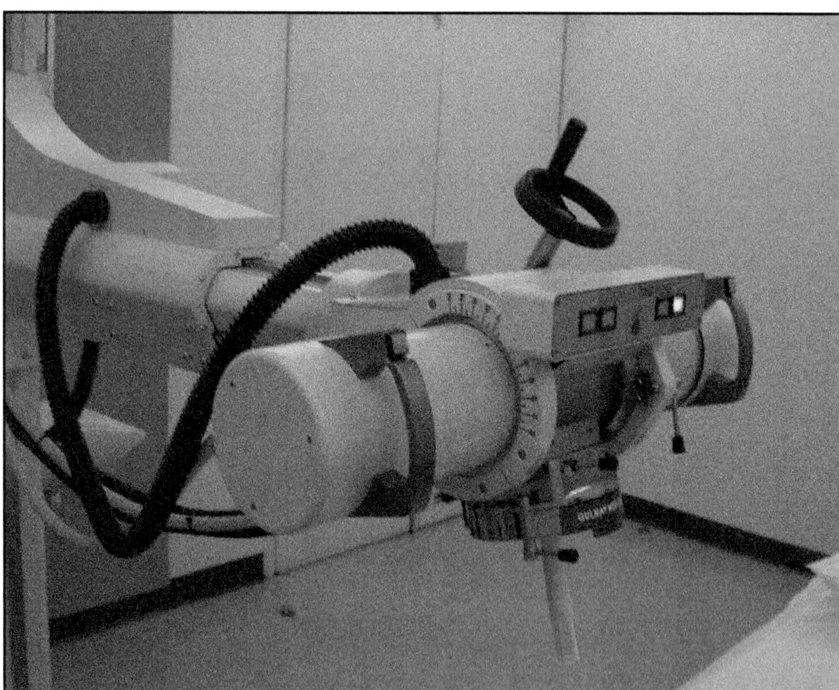

Figure 4.41: Photograph of an orthovoltage/superficial x-ray machine (Gulmay D3300) with a superficial applicator cone attached. The machine would operate at energies 50 to 120 kVp using this type of applicator.

orthovoltage/superficial x-ray unit with a large 10×10 cm^2 cone mounted for orthovoltage treatment. Figure 4.41 shows the same unit with a small, 2 cm diameter cone used for superficial treatment. The cones are abutted to the patient skin surface; sometimes an irregular lead shield is also mounted between the cone and the patient skin surface.

Machines can also be purchased which do superficial treatments only. These typically have a smaller, more maneuverable tube and can generate beams up to 150 kVp. Historically, there was a machine called a Philips RT 100 which had a very small tube and a very high dose output (it is a pity this machine was discontinued). The orthovoltage cone is at 50 cm SSD; this helps with depth dose and beam divergence. The superficial cone is at 30 cm SSD, which gives a higher dose rate.

Superficial therapy applies to treatment with x-rays ranging from 50 to 150 kV while orthovoltage therapy describes a range from 150 to 500 kV. Shown in figure 4.42 are the depth dose curves for the kV x-ray range of 50 to 250 kV. The penetration properties are not as good as megavoltage beams and there is no skin sparing.

Clinically, superficial therapy is used for very superficial, small tumors such as skin cancers. These machines are excellent for this as they give high skin dose where skin tumors often reside and very small inserts can be easily maneuvered into difficult positions. Shielding of sensitive structures such as eyes is often more easily achieved than with electron beams.

Orthovoltage energies are sometimes deployed for palliative treatment. As mentioned, the machine is very maneuverable, so palliative treatment with these machines may be warranted. An overview of the uses of superficial and orthovoltage therapy machines in the clinical situation and various measurements can be found in Johns and Cunningham (1983), Bomford and Kunkler

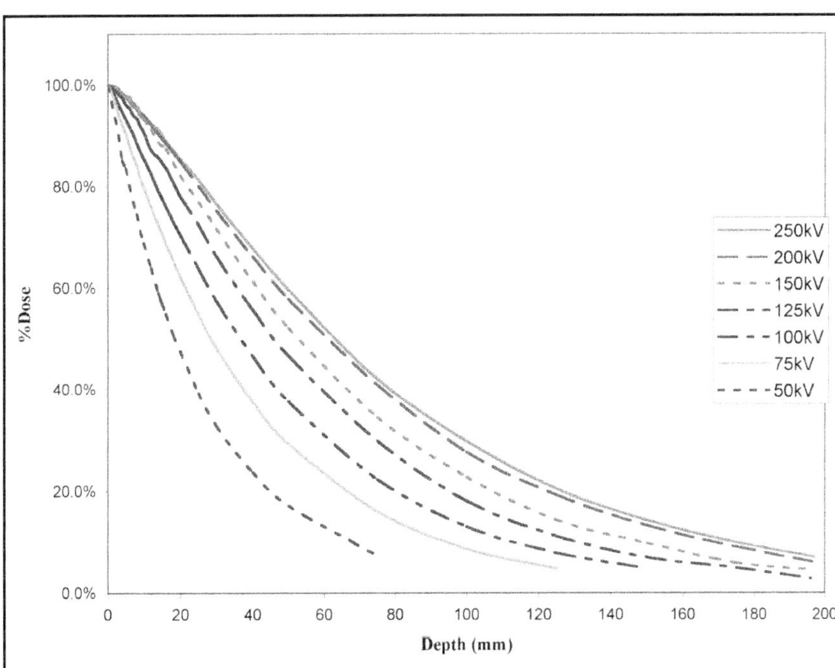

Figure 4.42: Depth dose curves for a kilovoltage x-ray machine from 50 kVp to 250 kVp.

(2003), and Biggs et al. (1999). For an in-depth description on measurements and calibration see IPEMB (1996). The calibration of these devices is dealt with in various protocols (e.g., AAPM 2001b). One of the historical changes has been a move from calibration in air to calibration in water at some kV energies. The use of solid water for some output factor applications has been characterized (Hill et al. 2005). The x-ray backscatter factor (BSF) at kV energies represents a significant component of dose and is field size dependent. Tables of BSFs are available in various reference sources (e.g., BJR 1996, Klevenhagen 1989, AAPM 2001a).

When dealing with superficial and orthovoltage therapy machines, the nominal energy in kilovolts does not exactly describe the penetration characteristics because varying thickness of filtration are added to harden the beam to a desired amount. The filtration affects the spectrum. It is traditional for the energy surrogate for kV beams to be expressed as the half-value layer (HVL). As defined earlier in this chapter, this is the thickness of a specified material that when introduced into the path of the beam below the filtration reduces the exposure rate by one-half.

The HVL is determined by measuring the intensity of the beam transmitted through various thicknesses of the chosen material and interpolating to find the thickness that would transmit 50% of the intensity measured with zero thickness of material added. This is done using the experimental arrangement shown in figure 4.43.

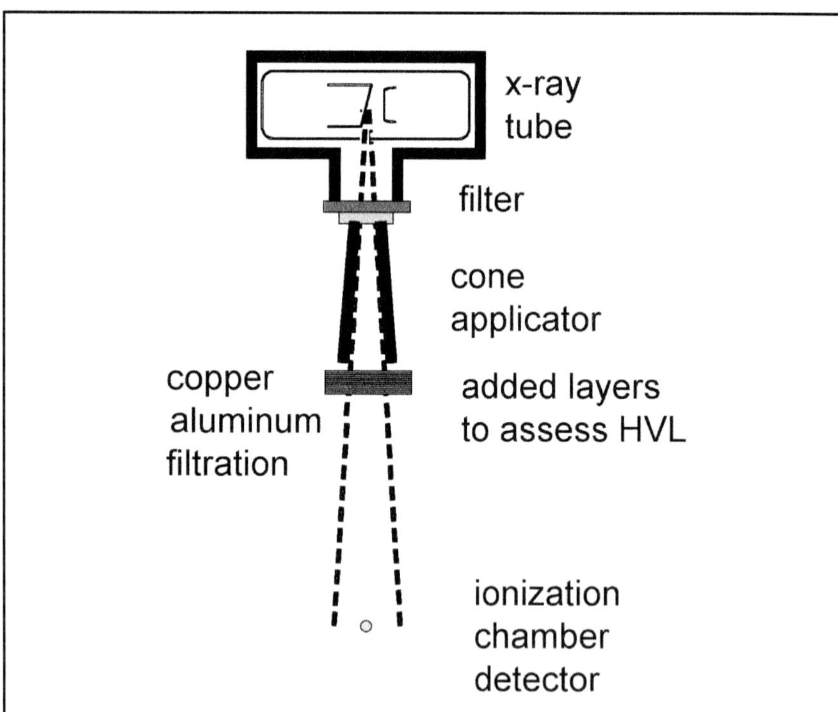

Figure 4.43: The layout for the measurement of HVL under an orthovoltage or a superficial x-ray machine.

Table 4.11: Half-value layers (HVLs) for different kV x-ray beams

Energy (kVp)	*Material*	*HVL (mm)*
50	Aluminum	1.40
75	Aluminum	2.48
100	Aluminum	3.72
125	Aluminum	6.40
125	Copper	0.29
150	Copper	0.63
200	Copper	1.35
250	Copper	2.24

To measure the HVL layer for kV therapy machines, it is important that the detector used, normally an ionization chamber, have an air equivalent wall over the range of photon energies being measured. The detector should be at least 50 cm from the end of the treatment cone or beam-defining system of the unit so that radiation scatter from the added material is minimized. The minimum field size that just covers the detector should also be employed to avoid scatter. Additionally, the detector should be positioned at least 50 cm from the floor, wall, and any other scattering objects. The machines have a slot into which the tube filter is positioned. A clamp should be arranged to hold additional attenuators at the end of the applicator for HVL measurements (layers of aluminum or copper are used). A graph of dose versus attenuator thickness should then be plotted and the HVL determined. Table 4.11 shows some HVL values for kV beams ranging from 50 to 250 kV.

Superficial and orthovoltage machine dose profiles exhibit a *heel effect.* The heel effect describes the marked variation in intensity (I) along the anode-cathode axis of the useful x-ray beam. This variation is the result of the x-rays produced attenuating in the target as they pass out of the anode material. The attenuation is greater in the anode direction than in the cathode direction because of the difference in the path length within the target (figure 4.44). The variation in x-ray intensity perpendicular to the anode-cathode direction can be seen in the dose profile shown in figure 4.45. The heel effect shown in this figure becomes more pronounced with the steeper angles encountered with large applicators. The heel effect also diminishes at depth as the beam traversing more distance through the target is harder.

Bone

As discussed in section 2.3, in bone at orthovoltage energies there is a significant increase in dose deposited due to increased electron fluence from the increased photoelectric absorption in this higher than tissue Z material (Spiers 1951, Johns and Cunningham 1983, Khan 2003). This occurs because the photoelectric effect is dominant for low kV beam energies and as the photoelectric component of the mass energy absorption coefficient is approximately proportional to the atomic number (Z) of a material to the power of 3, this causes the high absorption of x-ray energy from photoelectric electrons in bone.

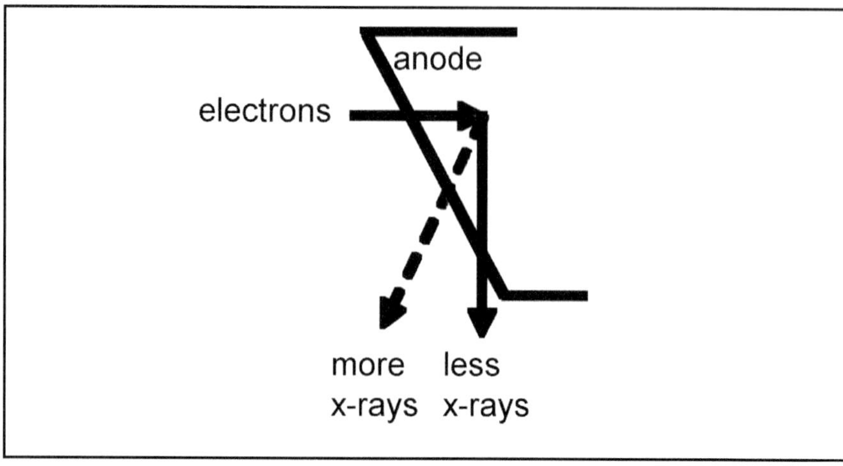

Figure 4.44: Interactions of electrons at the anode demonstrating the heel effect.

The ratio of extra dose absorbed can be calculated using the mass energy absorption coefficients of bone as a ratio to that of tissue, such that

$$\frac{D_B}{D_T} = \frac{\left(\frac{\mu_{en}}{\rho}\right)_{air}^{bone}}{\left(\frac{\mu_{en}}{\rho}\right)_{air}^{tissue}}. \tag{4.34}$$

Example 4.1: Dose Enhancement in Bone

Shown in table 4.12 are the values for 100 keV x-rays. Assume this is equivalent to the dominant energy within the spectral distribution of a 300 kVp beam, then calculate the dose enhancement in bone using equation (4.34).

Table 4.12: Mass energy absorption coefficients (μ_{en}/ρ) at different photon energies for air, water, muscle, and bone. Units are in g/cm². (Adapted from Khan 2003)

Photon energy	*Air* $\rho = 1.205 \times 10^{-2}$ g/cm³ Z=7.78	*Water* $\rho = 1.000$ g/cm³ Z=7.51	*Muscle* $\rho = 1.040$ g/cm³ Z=7.64	*Bone* $\rho = 1.850$ g/cm³ Z=12.31
100 keV	2.318×10^{-2}	2.539×10^{-2}	2.538×10^{-2}	3.838×10^{-2}
200 keV	2.672×10^{-2}	2.966×10^{-2}	2.942×10^{-2}	2.994×10^{-2}
1 MeV	2.787×10^{-2}	3.106×10^{-2}	3.072×10^{-2}	2.959×10^{-2}
2 MeV	2.342×10^{-2}	2.604×10^{-2}	2.580×10^{-2}	2.487×10^{-2}
6 MeV	1.644×10^{-2}	$1,804 \times 10^{-2}$	1.785×10^{-2}	1.784×10^{-2}
20 Mev	1.308×10^{-2}	1.386×10^{-2}	1.367×10^{-2}	1.474×10^{-2}

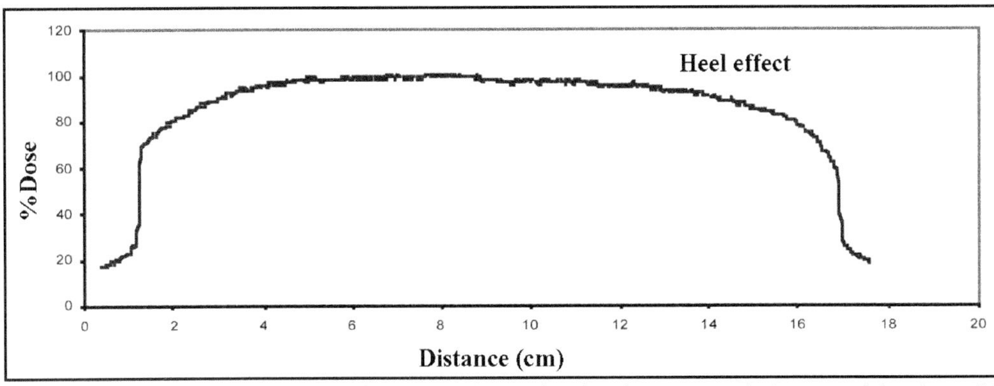

Figure 4.45: Dose profiles for a 250 kVp x-ray beam showing the asymmetrical nature of the dose with off-axis distance known as the heel effect.

$$\frac{D_B}{D_T} = \frac{\left(\frac{\mu_{en}}{\rho}\right)_{air}^{bone}}{\left(\frac{\mu_{en}}{\rho}\right)_{air}^{tissue}} = \left(\frac{3.838\times10^{-2} / 2.318\times10^{-2}}{2.538\times10^{-2} / 2.318\times10^{-2}}\right) = 1.5 \qquad (4.35)$$

By applying the same calculation to the 2 MeV data in table 4.12, the relation is slightly reversed. This means that for a 6 MV photon beam the dose deposited in tissue (assumed to be muscle for these calculations) is just slightly greater than that deposited in bone. Then the bone-to-tissue ratio increases slightly again as pair production occurs at higher energies. The dose to soft tissue imbedded in bone (bone marrow) is perhaps even more important. This has been estimated be greater than the dose to bone because the electrons per gram are greater in soft tissue than in bone. The ratio enhancement (γ) defined by Khan (2003) has been estimated as:

$$\gamma = \frac{D_{bone}}{D_{tissue}} = \left(\frac{\mu_{en}}{\rho}\right)_{tissue}^{bone} \times \left(\frac{S}{\rho}\right)_{bone}^{tissue}. \qquad (4.36)$$

Example 4.2: Dose Enhancement in Tissue beyond Bone

Assume the dominant energy for a 6 MV x-ray beam is 2 MeV; then calculate the dose enhancement in tissue close to bone using equation (4.36) and the mass energy absorption coefficient data from table 4.12 and the stopping power ratio data from table 4.13, then

$$\gamma = \frac{D_{bone}}{D_{tissue}} = \left(\frac{\mu_{en}}{\rho}\right)_{tissue}^{bone} \times \left(\frac{S}{\rho}\right)_{bone}^{tissue} = \left(\frac{2.487\times10^{-2}}{2.580\times10^{-2}}\right) \times \left(\frac{1.804}{1.684}\right) = 1.03. \qquad (4.37)$$

Table 4.13: Collision mass stopping powers (S/ρ) at different electron energies. Units are in MeV g/cm^2. (Adapted from Khan 2003)

Electron energy	*Air* $\rho = 1.205 \times 10^{-2}$ g/cm^3 Z=7.78	*Water* $\rho = 1.000$ g/cm^3 Z=7.51	*Muscle* $\rho = 1.040$ g/cm^3 Z=7.64	*Bone* $\rho = 1.850$ g/cm^3 Z=12.31
100 keV	3.633	4.115	4.008	3.820
200 keV	2.470	2.793	2.769	2.599
1 MeV	1.661	1.844	1.830	1.705
2 MeV	1.684	1.825	1.804	1.684
6 MeV	1.870	1.919	1.892	1.778
20 MeV	2.134	2.051	2.026	1.909

There is a nominal 8% increase in dose occurring just in front of bone at MV beam energies due to backscattered electrons at just in front of bone (Khan 2003). The effect is short range, typically less than 2 mm and is usually not shown by beam displays on radiotherapy treatment planning systems, as dose voxels are usually 2 mm or larger. The same effect has also been quantified from high-density hip prostheses in MV x-ray beams (Carolan et al. 2000). Electron backscatter from bone at orthovoltage energies also occurs but over a very short range.

High Z Materials

Because lead is often used for internal shielding for kV beams (e.g., mouth cavity), any dose enhancement in these situations is important. There is a significant dose enhancement, and backscattered electron dose enhancement for kV beams at short range has been reported by Huq et al. (1992), Lanzon and Sorell (1993), Mitchell et al. (1998), and Verhaegen (2002). However the dose upstream from backscattered x-rays is actually reduced. Das and Chopra (1995) quantify the short-range dose enhancement as about a factor of 10 for superficial energies; however the range of this dose enhancement is only 0.1 mm. They also point out that there is a very small reduction in backscatter component of dose between 0.1 and 100 mm, the reduction being 0.95 for a 300 kV beam. They explain this reduction as the coherent and fluorescent photons scattering from the high Z material contributing less to dose than Compton-scattered photons from the low Z material. To avoid the high-dose enhancement close to lead shields, a layer of wax on both sides of the lead is recommended. Das and Chopra (1995) suggest 1 mm thickness is sufficient.

We have shown that kV beams show higher dose in the build-up region and a faster depth dose fall-off than MV beams. These properties can sometimes be useful for treating superficial tumors such as skin cancers. The maneuverability of these devices combined with a sharp penumbra (between 30% and 90%) also lends them to some usefulness near critical structures such as eyes. Orthovoltage beams are occasionally also useful for palliation.

References

AAPM (American Association of Physicists in Medicine) (2001a) Basic applications of multileaf collimators: Report of the AAPM Radiation Therapy Committee Task Group 50. Report 72. Medical Physics Publishing, Madison, Wisconsin.

AAPM (American Association of Physicists in Medicine) (2001b) AAPM protocol for 40-300 kV x-ray beam dosimetry in radiation therapy. *Med. Phys.* 28: 868–893. Also printed as Report 76. Medical Physics Publishing, Madison, Wisconsin.

AAPM (American Association of Physicists in Medicine) (1999) AAPM's TG-51 protocol for clinical reference dosimetry of high-energy photon and electron beams. *Med. Phys.* 26: 1847–1870.

Ansbacher W., Neath C. (2003) Comment on "Analytical representation of enhanced dynamic wedge factors for symmetric and asymmetric fields." Letter to the editor. *Med. Phys.* 30: 722–723.

Attix F.H. (1979) The partition of KERMA to account for bremsstrahlung. *Health Phys.* 36: 347–354.

Beavis A.W., Weston S. J., Whitton V. J. (1996) Implementation of the Varian EDW into a commercial RTP system. *Phys. Med Biol.* 41: 1691–1704.

Biggs P., Ma C.-M., Doppke K., Niroomand-Rad A., Beatty J. (1999) Kilovoltage x-rays. Chapter 9 in *The Modern Technology of Radiation Oncology*. J. Van Dyk (Editor). Medical Physics Publishing, Madison, Wisconsin, pp. 287–312.

Biggs P., Capalucci J., Russell M. (1991) Comparison of the penumbra between focused and nondivergent blocks—implications for multileaf collimators. *Med. Phys.* 18: 753–758.

Biggs P.J., Ling C.C. (1979) Electrons as the cause of the observed d_{max} shift with field size in high energy photon beams. *Med. Phys.* 6: 291–295.

Bjärngard B. (1996) Meter-set calculations. Proceedings of 1996 AAPM Summer School—*Teletherapy: Present and Future.* T.R. Mackie and J.R. Palta (Editors). Advanced Medical Publishing, Madison, Wisconsin, pp. 221–233.

BJR (British Journal of Radiology) Supplement 25. (1996) *Central axis depth dose data for use in radiotherapy departments.* British Institute of Radiology, London, U.K.

Boge J.R., Edland R.W., Matthes D.C. (1974) Tissue compensators for megavoltage radiotherapy fabricated from hollow styrofoam filled with wax. *Radiology* 111:193–198.

Bomford C.K., Kunkler I.H. (2003) *Walter and Miller's Textbook of Radiotherapy.* Sixth Edition, Churchill Livingstone, London, U.K.

Butson M.J., Perez M.D., Mathur J.N., Metcalfe P.E. (1996) 6 MV x-ray dose in the build up region: Empirical model and the incident angle effect. *Australas. Phys. Eng. Sci. Med.* 19: 74–82.

Butson M.J., Wong T.P., Kron T., Mathur J., Clubb B., Metcalfe P.E. (1994) Surface doses from combined electron/photon fields in a radiotherapy. *Australas. Phys. Eng. Sci. Med.* 17: 14–22.

Carolan M., Dao P., Fox C., Metcalfe P. (2000) Effect of hip prostheses on radiotherapy dose. *Australas. Radiol.* 44: 290–295.

Chang S.X., Cullip T.J., Deschesne K.M., Miller E.P., Rosenman J.G. (2004) Compensators: an alternative IMRT delivery technique. *J. Appl. Clin. Med. Phys.* 5: 15–36.

Chui C.S., Mohan R. (1988) Extraction of pencil beam kernels by the deconvolution method. *Med. Phys.* 15: 138–144.

Chui C.S., Mohan R., Fontenla D. (1988) Dose computations for asymmetric fields defined by independent jaws. *Med. Phys.* 15: 92–95.

Chui C.S., Mohan R. (1986) Off-center ratios for three-dimensional dose calculations. *Med. Phys.* 13: 409–412.

Constantinou C., Harrington J.C. (1989) Tissue compensators made of solid water or lead for megavoltage x-ray radiotherapy. *Med. Dosim.* 14: 41–47.

Conte G., Nascimben O., Turcato G., Polico R., Idi M.B., Belleri L.M., Bergoglio F., Simonato F., Stea L., Bugin F., Bartot N. (1988) Three-field isocentric technique for breast irradiation using individualized shielding blocks. *Int. J. Radiat. Oncol. Biol. Phys.* 14: 1299–1305.

Cross P. (1992) The use of parallel plate ion chambers to determine surface dose of a 6 MV photon beam. *Aust. Phys. Eng. Sci. Med.* 15: 208–213.

Cunningham J.R. (1972) Scatter-air ratios. *Phys. Med. Biol.* 17: 42–51.

Damrongkijudom N., Oborn B., Butson M., Rosenfeld A. (2006) Measurement of magnetic fields produced by a "magnetic deflector" for the removal of electron contamination in radiotherapy. *Australas. Phys. Eng. Sci. Med.* 29: 321–327.

Das I.J., Chopra K.L. (1995) Backscatter dose perturbation in kilovoltage photon beams at high atomic number interfaces. *Med. Phys.* 22:767–773.

Dawson D.J., Schroeder N.J., Hoya J.D. (1986) Penumbral measurements in water for high-energy x rays. *Med. Phys.* 13: 101–104.

Dawson D.J., Harper J.M., Akinradewo A.C. (1984) Analysis of physical parameters associated with the measurement of high-energy x-ray penumbra. *Med. Phys.* 11: 491–496.

Dean E.M., Davis J.B. (1992) Output factors for blocked fields. *Br. J. Radiol.* 65: 517–522.

Delaney G.P., Shafiq R.J., Jalaludin B.B., Barton M.B. (2005) Technology enhancements and changes in radiotherapy throughput in New South Wales. *Clin. Oncol.* 17: 305–310.

Dutreix J., Bernard M. (1966) Dosimetry at interfaces for high energy X and gamma rays. *Br. J. Radiol.* 39: 205–210.

Dutreix J., Dutriex A., Tubana M. (1965) Electronic equilibrium and transition stages. *Phys. Med. Biol.* 10: 177–190.

Ellis F. (1960) Accuracy in compensation for tissue heterogeneity in treatment by x-rays, supervoltage x-rays and electron beams. *Br. J. Radiol.* 33: 404–405.

Fiorino C., Cattaneo G.M., del Vecchio A., Longobardi B., Signorotto P., Calandrino R., Fossati V., Volterrani F. (1992) Skin dose measurements for head and neck radiotherapy. *Med. Phys.* 19: 1263–1266.

Gagnon W.F., Grant W. 3rd (1975) Surface dose from megavoltage therapy machines. *Radiology* 117: 705–708.

Galvin J.M., Smith A.R., Lally B. (1993a) Characterization of a multi-leaf collimator system. *Int. J. Radiat. Oncol. Biol. Phys.* 25: 181–192.

Galvin J., Chen X., Smith R. (1993b) Combining multileaf fields to modulate fluence distributions. *Int. J. Radiat. Oncol. Biol. Phys.* 27: 697–705.

Gerbi B.J., Meigooni A.S., Khan F.M. (1987) Dose buildup for obliquely incident photon beams. *Med. Phys.* 14: 393–399.

Gibbons J.P. (1998) Calculation of enhanced dynamic wedge factors for symmetric and asymmetric photon fields. *Med. Phys.* 25: 1411–1418.

Greene D. (1986) *Linear Accelerators for Radiation Therapy,* Medical Physics Handbook 17, Adam Hilger, Bristol, U.K.

Gupta S.K., Cunningham J.R. (1966) Measurement of tissue-air ratios and scatter functions for large field sizes, for cobalt 60 gamma radiation. *Br. J. Radiol.* 39: 7–11.

Hall E.J., Oliver R. (1961) The use of standard isodose distributions with high energy radiation beams—the accuracy of a compensator technique in correcting for body contours. *Br. J. Radiol.* 34: 43–52.

Harper N.R., Metcalfe P.E., Hoban P.W., Round W.H. (1991) Electron contamination from 4 MV and 10 MV radiotherapy x-ray beams. *Australas. Eng. Phys. Sci. Med.* 14: 141–145.

Higgins P., Sibata C., Siskind L., Sohn J. (1995) Convolution of detector size effect for small field measurement. *Med. Phys.* 22: 1663–1666.

Hill R., Holloway L., Baldock C. (2005) A dosimetric evaluation of water equivalent phantoms for kilovoltage x-ray beams. *Phys. Med. Biol.* 50: N331–344.

Hoban P.W., Keall P.J., Round W.H. (1992) The effect of density on the 10MV photon beam. *Australas. Eng. Phys. Sci. Med.* 15: 113–123.

Huq, M., Venkataramanan N., Meli J.A. (1992) The effect on dose of kilovoltage x-rays backscattered from lead. *Int. J. Radiat. Oncol. Biol. Phys.* 24: 171–175.

IAEA (International Atomic Energy Agency) (2004) An international protocol for absorbed dose determination. IAEA technical report series TRS-398 (V.11b). Vienna, Austria.

ICRP (International Commission on Radiological Protection) (1991) Publication 60. Recommendations of the International Commission on Radiological Protection. *Annals of the ICRP* 21: 149–163.

IEC (International Electrotechnical Commission) (1989a) Document 976. Medical electron accelerators—functional performance characteristics.

IEC (International Electrotechnical Commission) (1989b) Document 977. Medical electron accelerators in the range 1 MeV to 50 MeV—guidelines for functional performance characteristics.

IPEMB (Institution of Physics and Engineering in Medicine and Biology) (1996) The IPEMB code of practice for the determination of absorbed dose for x-rays below 300 kV generating potential (0.035 mm Al – 4 mm Cu HVL; 10–300 kV generating potential). *Phys. Med. Biol.* 41: 2605–2625.

IPSM (The Institute of Physical Sciences in Medicine) (1988) Report 54. Commissioning and quality assurance of linear accelerators. IPSM, York, England.

Jackson W. (1971) Surface effects of high-energy x rays at oblique incidence. *Br. J. Radiol.* 44: 109–115.

Jackson W. (1970) Wax retraction as a technique for compensating the effect of surface irregularities in high-energy radiotherapy. *Br. J. Radiol.* 43: 859–867.

Johns H.E., Cunningham J.R. (1983) *The Physics of Radiology,* fourth edition. Charles C Thomas, Springfield, Illinois.

Johns H.E., Bruce W.R., Reid W.B. (1958) The dependence of depth dose on focal skin distance. *Br. J. Radiol.* 31: 254–260.

Jursinic P.A., Mackie T.R. (1996) Characteristics of secondary electrons produced by 6, 10 and 24 MV x-ray beams. *Phys. Med. Biol.* 41: 1499–1509.

Khan F.M. (2003) *The Physics of Radiation Therapy,* Third edition. Lippincott Williams and Wilkins, Baltimore, Maryland.

Khan F.M. (1993) Dosimetry of wedged fields with asymmetric collimation. *Med. Phys.* 20: 1447–1451.

Khan F.M. (1984) *The Physics of Radiation Therapy,* First edition. Williams and Wilkins, Baltimore, Maryland.

Khan F.M., Gerbi B.J., Deibel F.C. (1986) Dosimetry of asymmetric x-ray collimators. *Med. Phys.* 13: 936–941.

Kim S., Zhu T.C., Palta J.R. (1997) An equivalent square field formula for determining head scatter factors of rectangular fields. *Med. Phys.* 24: 1770–1774.

Klemp P.F., Perry A.M., Hedland-Thomas B., Stevenson S., Zucaro D. (1988) Commissioning of a linear accelerator with independent jaws: computerised data collection and transfer to a planning computer. *Phys. Med. Biol.* 33: 865–871.

Klevenhagen S.C. (1989) Experimentally determined backscatter factors for x-rays generated at voltages between 16 and 140 kV. *Phys. Med. Biol.* 34:1871–1882.

Kron T., Butson M., Hunt F., Denham J. (1996) TLD extrapolation for skin dose determination in vivo. *Radiother. Oncol.* 41: 119–123.

Kron T., Elliott A., Metcalfe P. (1993a) The penumbra of a 6-MV x-ray beam measured by means of thermoluminescent dosimetry and evaluated using an inverse square root function. *Med. Phys.* 20: 1429–1438.

Kron T., Elliott A., Wong T., Showell G., Clubb B., Metcalfe P.E. (1993b) X-ray surface dose measurements using TLD extrapolation. *Med. Phys.* 20: 703–711.

Kuperman V.Y. (2005) Analytical representation for Varian EDW factors at off-center points. *Med. Phys.* 32(5): 1256–1261.

Kuperman V.Y. (2004) A new analytical model for Varian enhanced dynamic wedge factors. *Phys. Med. Biol.* 49: 2841–2851.

Lam K.L., Muthuswamy M.S., Ten Haken R.K. (1996) Flattening-filter-based empirical methods to parameterize the head scatter factor. *Med. Phys.* 23: 343–352.

Lanzon P.J., Sorell G.C. (1993) The effect of lead underlying water on the backscatter of x-rays of beam qualities 0.5 mm to 8 mm Al. *Phys. Med. Biol.* 38: 1137–1144.

Laub W., Wong T. (2003) The volume effect of detectors in the dosimetry of small fields used in IMRT. *Med. Phys.* 30: 341–347.

Leavitt D.D., Martin M., Moeller J.H., Lee W.L. (1990) Dynamic wedge field techniques through computer-controlled collimator motion and dose delivery. *Med. Phys.* 17: 87–91.

Ling C.C., Biggs P.J. (1979) Improving the buildup and depth-dose characteristics of high energy photon beams by using electron filters. *Med. Phys.* 6: 296–301.

Loshek D.D., Keller K.A. (1988) Beam profile generator for asymmetric fields. *Med. Phys.* 15: 604–610.

Marshall M.G. (1993) Matching the 6-MV photon beam characteristics of two dissimilar linear accelerators. *Med. Phys.* 20: 1743–1746.

McGhee P., Chu T., Leszczynski K., Dunscombe P. (1997) The Siemens virtual wedge. *Med. Dosim.* 22: 39–41.

McParland B.J. (1991) The effects of a universal wedge and beam obliquity upon the central axis buildup for 6-MV x rays. *Med. Phys.* 18: 740–743.

McShan D.L., Fraass B.A. (1996) Computer controlled linacs. *Proceedings of AAPM Summer School—Teletherapy: Present and Future.* T.R. Mackie and J.R. Palta (Editors). Advanced Medical Publishing, Madison, Wisconsin, pp. 379–402.

Mellenberg D.E. Jr. (1990) Determination of build-up region over-response corrections for a Markus-type chamber. *Med. Phys.* 17: 1041–1044.

Metcalfe P., Tangboonduangjit P., White P. (2004) Intensity-modulated radiation therapy: overlapping co-axial modulated fields. *Phys. Med. Biol.* 49: 3629–3637.

Metcalfe P.E., Kron T., Elliott A., Wong T., Hoban P. (1993) Dosimetry of 6-MV x-ray beam penumbra. *Med. Phys*. 20: 1439–1445.

Metcalfe P.E., Hoban P.W., Murray D.C., Round W.H. (1990) Beam hardening of 10 MV radiotherapy x-rays: analysis using a convolution/superposition method. *Phys. Med. Biol.* 35: 1533–1549.

Milan J., Bentley R.E. (1974) The storage and manipulation of radiation dose data in a small digital computer. *Br. J. Radiol.* 47: 115–121.

Mitchell G., Kron T., Back M. (1998) High dose behind inhomogeneities during medium-energy x-ray irradiation. *Phys. Med. Biol.* 43:1343–1350.

Mohan R., Chui C., Lidofsky L. (1986) Differential pencil beam dose computation model for photons. *Med. Phys.* 13: 64–73.

Mohan R., Chui C., Lidofsky L. (1985) Energy and angular distributions of photons from medical linear accelerators. *Med. Phys.* 12: 592–597.

Munro P., Rawlinson J.A., Fenster A. (1988) Therapy imaging: Source sizes of radiotherapy beams. *Med. Phys.* 15: 517–524.

Murray A. (1994) Fourier analysis of asymmetric beam penumbra, M.Sc. Thesis, University of Adelaide, South Australia.

Orton C.G., Seibert J.B. (1972) Depth dose in skin for obliquely incident ^{60}Co radiation. *Br. J. Radiol.* 45: 271–275.

Papatheodorou S., Zefkili S., Rosenwald J.C. (1999) The "equivalent wedge" implementation of the Varian enhanced dynamic wedge into a treatment planning system. *Phys. Med. Biol.* 44: 509–524.

Parthasaradhi K., Prasad S.G., Rao B.M., Lee Y., Ruparel R., Garces R. (1989) Investigation on the reduction of electron contamination with a 6-MV x-ray beam. *Med. Phys.* 16:123–125.

Petti P.L., Siddon R.L. (1985) Effective wedge angles with a universal wedge. *Phys. Med. Biol.* 30: 985–991.

Pinnacle® Radiotherapy Treatment Planning Physics Manual (2007) software version 8.0.

Prado K.L., Kirsner S.M., Kudchadker R.J., Steadham R.E., Lane R.G. (2003) Enhanced dynamic wedge factors at off-axis points in asymmetric fields. *J. Appl. Clin. Med. Phys.* 4: 75–84.

Purdy J.A. (1986) Buildup/surface dose and exit dose measurements for a 6-MV linear accelerator *Med. Phys.* 13: 259–262.

Rao B.M., Prasad G.S., Parthasaradhi K., Lee Y., Ruparel R., Garces R. (1988) Investigations on the near surface dose for three 10-MV x-ray beam accelerators with emphasis on the reduction of electron contamination. *Med. Phys.* 15: 246–249.

Rathee S., Kwok C.B., MacGillivray C., Mirzaei M. (1999) Commissioning, clinical implementation and quality assurance of a Seimen's Virtual Wedge. *Med. Dosim.* 24: 145–153.

Robinson D.M., Scrimger J.W. (1990) Optimized tissue compensators. *Med. Phys.* 17: 391–396.

Rosenow U.F., Valentine E.S., Davis L.W. (1990) A technique for treating local breast cancer using a single set-up point and asymmetric collimation. *Int. J. Radiat. Oncol. Biol. Phys.* 19: 183–188.

Sethi A., Leybovich L.B., Dogan N., Glasgow G.P. (2000) Elimination of field size dependence of enhanced dynamic wedge factors. *Phys. Med. Biol.* 45: 3359–3365.

Shih R., Li X.A., Chu J.C., Hsu W.L. (1999) Calculation of head scatter factors at isocenter or at center of field for any arbitrary jaw setting. *Med. Phys.* 26: 506–511.

Sibata C.H., Mota H.C., Beddar A.S., Higgins P.D., Shin K.H. (1991) Influence of detector size in photon beam profile measurements. *Phys. Med. Biol.* 36: 621–629.

Sixel K.E., Podgorsak E.B. (1994) Buildup region and depth of dose maximum of megavoltage x-ray beams. *Med. Phys.* 21: 411–416.

Slessinger E.D., Gerber R.L., Harms W.B., Klein E.E., Purdy J.A. (1993) Independent collimator dosimetry for a dual photon energy linear accelerator. *Int. J. Radiat. Oncol. Biol. Phys.* 27: 681–687.

Sohn J.W., Suh J.H., Pohar S. (1995) A method for delivering accurate and uniform dosages to the head and neck with asymmetric collimators and a single isocentre. *Int. J. Radiat. Oncol. Biol. Phys.* 32: 809–813.

Spiers (1951) Radiotherapeutic physics II: Dosage in irradiated soft tissue and bone. *Br. J. Radiol.* 24: 365–369.

Target Series 11 (1990) Physics reference manual first edition, GE Medical Systems, B1-27.

Theraplan (1989) Photon beam calculations, Theraplan users manual appendix 1.

Ten Haken R.K., Fraass B.A., Lam K.L. (1996) Dosimetry and data acquisition. Proceedings of 1996 AAPM Summer School—Teletherapy: Present and Future. T.R. Mackie and J.R. Palta (Editors). Advanced Medical Publishing, Madison, Wisconsin, pp. 191–219.

van Gasteren J.J., Heukelom S., van Kleffens H.J., van der Laarse R., Venselaar J.L., Westermann C.F. (1991) The determination of phantom and collimator scatter components of the output of megavoltage photon beams: Measurement of the collimator scatter part with a beam-coaxial narrow cylindrical phantom. *Radiother. Oncol.* 20: 250–257.

van Santvoort J. (1998) Dosimetric evaluation of the Siemens virtual wedge. *Phys. Med. Biol.* 43: 2651–2663.

Verhaegen F. (2002) Evaluation of the EGSnrc Monte Carlo code for interface dosimetry near high-Z media exposed to kilovoltage and ^{60}Co photons. *Phys. Med. Biol.* 47: 1691–1705.

Vial P., Oliver L., Greer P.B., Baldock C. (2006) An experimental investigation into the radiation field offset of a dynamic multileaf collimator. *Phys. Med. Biol.* 51: 5517–5538.

Waldron T.J., Boyer A.L., Wells N.H., Otte V.E. (1994) Calculation of dynamically-wedged isodose distributions from segmented treatment tables and open-field measurements. *Med. Phys.* 21: 873.

Woo S.Y., Donaldson S.S., Heck R.J., Nielson K.L., Shostak C. (1989) Minimizing and measuring lens dose when giving cranial irradiation. *Radiother. Oncol.* 16: 183–188.

Yu M.K. (2002) Analytical representation of enhanced dynamic wedge factors for symmetric and asymmetric photon fields. *Med. Phys.* 29: 2606–2610.

Questions

4.1 Identify the different sources of electron contamination in x-ray beams. (section 4.3)

4.2 Define the following terms:
(i) tissue-maximum ratio (TMR)
(ii) tissue-air ratio (TAR)
(iii) peak scatter factor (PSF)
(section 4.5)

4.3 Convert the %D value for a 22×22 cm^2 field size at 10 cm depth from table 4.6 to a TAR for a 20×20 cm^2 field size at the same depth. (section 4.5)

4.4 Convert the %D value for a 22×22 cm^2 field size at 10 cm depth from table 4.6 to a TMR for a 20×20 cm^2 field size at the same depth. Given that 200 cGy is to be delivered at this depth and the linac is calibrated $(D/MU)_{std,SSD,water} = 1$ cGy/MU, calculate the MU to be set given $S_c \times S_p$ for this field is 1.033. (section 4.5)

4.5 Define the terms "beam symmetry" and "beam flatness". Describe how these two parameters are calculated from linac beam profiles. (section 4.6)

5 Linear Accelerator Electron Beam Properties

5.1 Introduction

As outlined in chapter two, electrons are directly ionizing radiation. Therefore, the average energy of any electron beam reduces with depth in tissue as the electrons continuously lose small amounts of energy in many interactions (mainly ionization and excitation). The typical energy deposited in each interaction is small (several electron volts [eV]) and the direction of the electron's motion is only altered slightly. It takes an electron of 10 MeV more than a million interactions before all its energy is deposited in matter. This is illustrated in figure 5.1a, which shows the tracks of ten 10 MeV electrons calculated using EGS4 Monte Carlo code implemented in the software package EGS-ray(http://www.uni-koeln.de/~airm7/EGSRay/index.html; last accessed May 2007) developed by Kleinschmidt of the University of Cologne (Kleinschmidt 2001).

The continuous loss of energy of an electron beam in matter results in two significant differences to photon beams:

i) The difference in energy in different depths results in slightly different interaction probabilities at different depths and a change in stopping power. If the variation in stopping power ratio with energy is not identical between a detector and the tissue of interest (often water), a correction must be made which depends on the electron energy.

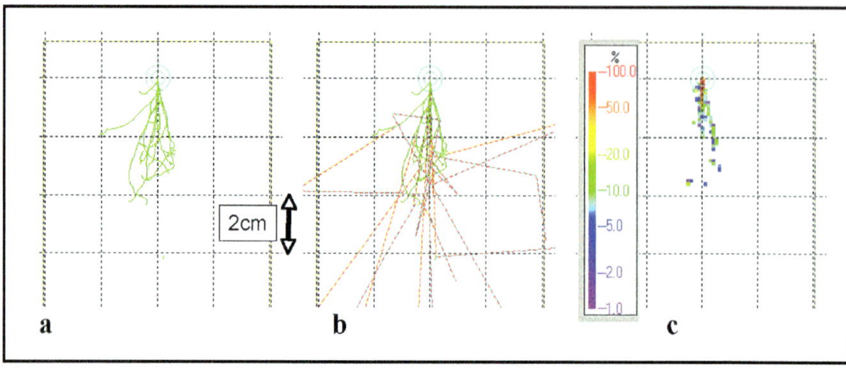

Figure 5.1: Monte Carlo calculated electron tracks for electrons of 10 MeV energy in water: (a) 10 electron tracks from a point source; (b) as in (a), but shown with secondary photon tracks; (c) dose deposition from the electron tracks; the scale is logarithmic and normalized to 100% for maximum dose. Dose levels below 1% are not shown. Scoring voxel size 1 × 1 × 1 mm^3. The figure was generated using EGS-ray.

ii) When the incident energy of the electron is used up, the electron stops and no dose is delivered beyond the end of the electron track. This allows sparing of organs behind a target structure and generally results in a smaller integral dose for the patient.

The latter is even more pronounced in proton beams, as will be discussed at the end of the chapter. For electrons the two features above mean that in spite of generally more complicated dosimetry, electrons will always be of interest if the integral dose shall be kept low. We will discuss this briefly in section 5.9.

The chapter starts with a description of typical electron beam properties (energy, depth dose, and profiles) and then discusses the features of typical clinical beams. Following this, we will discuss some of the multiple aids and gadgets used to deliver electron beams in practice. In order to use electron beams clinically, one must understand their dose distribution in the presence of contour changes (section 5.6). Particular emphasis will also be on tissue inhomogeneities as they will affect electron beam much more than photons. The chapter builds on chapter two (Interactions) and will be complemented by chapters nine and ten where beam models are presented.

Recently, there has been an increasing interest in proton radiotherapy. This is to a large extent due to the fact that the integral dose to the patient's body is lower in charged particle irradiation such as electrons and protons. While proton radiotherapy is not directly related to the emphasis of the present book (linac-based radiotherapy), some features of charged particle radiotherapy become more obvious when using protons rather than electrons. As such, the chapter will conclude with some remarks on proton radiotherapy as another example for charged particle irradiation.

The reader who would like to obtain more information is directed to two publications by the late Stan Klevenhagen (Klevenhagen 1993, Klevenhagen 1988) and excellent review by two of the pioneers of electron radiation physics, Kenneth Hogstrom and Peter Almond (Hogstrom and Almond 2006). Two publications on dosimetry, which will be discussed in more detail in chapter eight, are worth reading when considering electrons: The report of the AAPM Task Group 25 (TG 25) (AAPM 1991) and the technical report of the IAEA 381 which is dedicated to the use of plane parallel ionization chambers (IAEA 1997). Good book chapters on electrons can also be found in the textbooks of Khan (1994) and Williams and Thwaites (1993).

5.2 Characterization of a Clinical Electron Beam

Electron beams are available from linear accelerators. As discussed in chapter one, section 1.6 electron beams are generated most commonly using scattering foils and transported to the patient via an applicator that ensures collimation of the beam as closely as possible to the patient. This is illustrated in figure 5.2. Only few older linacs and the race track microtron unit by Scanditronix

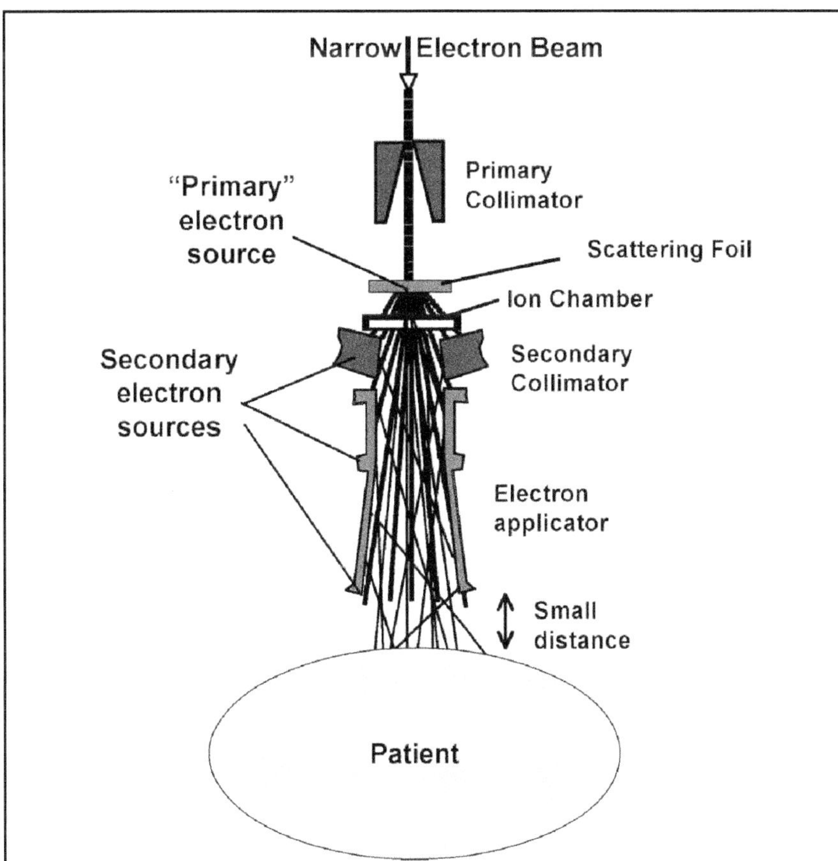

Figure 5.2: Illustration of the delivery system of electron beams generated by a linear accelerator.

employ scanning electron beams in order to create large clinically useful field sizes (Ertan et al. 1984, Karlsson et al. 1992).

As discussed above, electrons are charged particles, which lose their energy in matter in many small steps. This is a random process, which is accompanied also by small changes of direction. Therefore the average energy, the energy distribution, and the average direction of the electron beam vary with depth in a phantom. This is shown in figure 5.3. Two parameters that are useful for the characterization of electron beams are also shown: the mean energy at a certain depth and the most probable one. This is schematically shown for the surface and a depth of 2.5 cm. The mean energy is reduced faster with depth.

In order to quantify the effect of this energy and angular spread on characteristics of a clinical electron beam, it is helpful to extend the Monte Carlo simulations of figure 5.1. Figure 5.4a shows the dose distribution for a pencil beam of 10 MeV incident energy in water. As one can see, a significant amount of dose is deposited laterally of the beam. Therefore, adjacent pencil beams will increase the dose in the direct path of their neighbors. This has two important consequences:

i) The variation of dose with depth is different in an extended beam as shown in figure 5.4b. The figure shows a Monte Carlo calculation for the same 10 MeV electrons as in figure 5.4a but spread out over a field size of $4 \times 4\ cm^2$.
ii) For small radiation fields the lateral range of the electrons may exceed the field size. In this case, the dose at central axis will be affected at depth and the beam will be less penetrating than expected from measurements in larger fields.

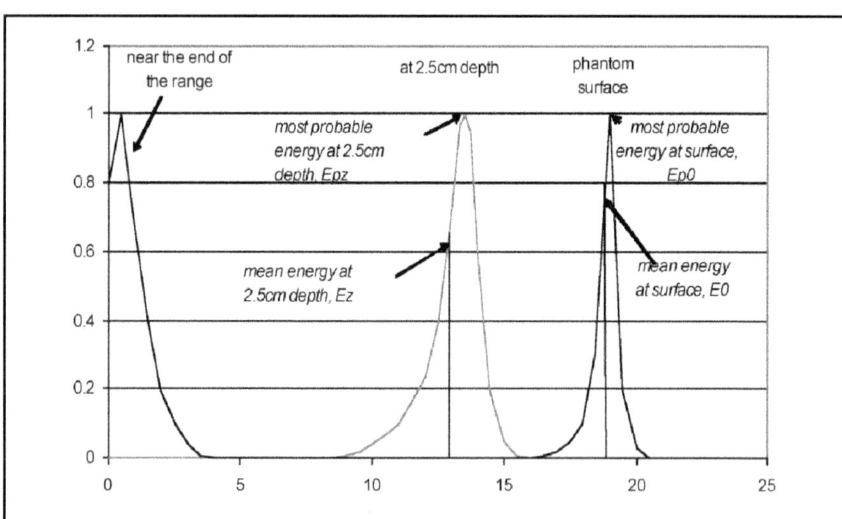

Figure 5.3: Electron spectrum at different depths in a water phantom. Each of the three spectra shown is normalized to 100%.

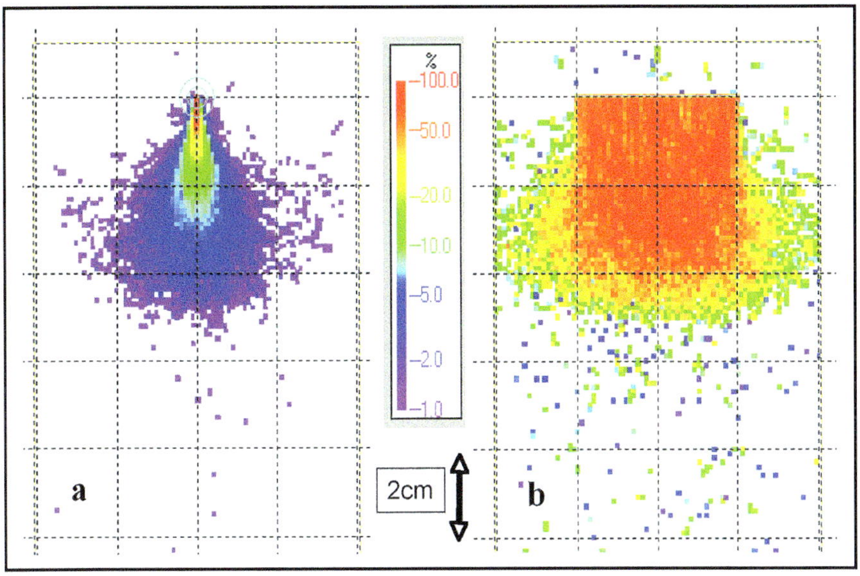

Figure 5.4: Monte Carlo calculated dose distribution in water for 10,000 histories of 10 MeV electrons: (a) point source; (b) a square-shaped source of 4 x 4 cm^2. Both plots are normalized to 100% at the maximum dose level in each figure, respectively. The figure was generated using EGS-ray.

There are several ways to approximate the variation of energy with depth in a clinical electron beam. They are illustrated in figure 5.5, adapted from the AAPM TG 21 protocol (AAPM 1983). The most simple approximation is a linear reduction of energy with depth as proposed by Harder (1966) and adopted by the AAPM TG 21 for their calibration protocol published in

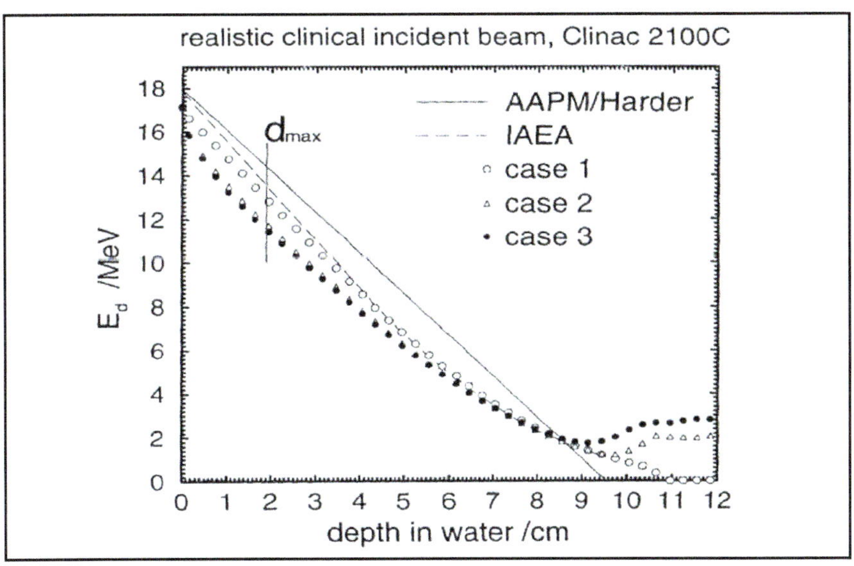

Figure 5.5: Variation of electron energy with depth in water. The figure is adapted from Ding et al. (1996) and illustrates a 20 MeV clinical electron beam.

1983. The IAEA provides a more realistic approximation, which is given in its protocol TRS 277 in tabular format (IAEA 1987). An even better prediction can be achieved by Monte Carlo calculations. The rise of average energy beyond the range of the electron beam shown in figure 5.5 is a result of secondary electrons to photons generated by the electron beam. While the average energy becomes higher, the fluence of electrons is significantly reduced. It is noteworthy that the mean electron energy may also change across the field at a given depth (Andreo and Brahme 1981). However, these variations in effective electron energy are only large enough to cause significant problems with dosimetry close to the field edge.

Typically, the electron beam is characterized by a nominal energy. Unfortunately, the energy specified by the manufacturer is not necessarily a good indication of penetration (Followill et al. 2004). This group from the Radiological Physics Center in Houston, Texas, studied more than 2000 electron beams from linacs of all manufacturers represented in the United States as part of their provision of dosimetric services. They found that in particular for high-energy electron beams the values for depth of maximum dose, d_{max}, and half maximum dose, d_{50}, varied by as much as 20 mm for nominally identical electron energies.

From a physics perspective the electron energy could be assessed, for example, by deflection of the electrons in a known magnetic field. However, the most common way to determine the average energy at the surface is to acquire a depth dose in a field sufficiently large to ensure lateral electron equilibrium (Faddegon and Blevis 2000). If reduced to a few measurement points, this is the same approach as in other radiation types such as megavoltage photons where the percentage depth dose at 10 cm depth, D_{10X},or the ratio of tissue-maximum ratios (TMRs) between depths of 20 cm and 10 cm, TMR_{10}^{20} , is used for specification of radiation quality (compare chapter eight). Depth dose curves for five clinical electron beams are shown in figure 5.6. The curves are typical for a clinical electron beam that has been spread out using a scattering foil. A scanning beam would typically have a lower surface dose and sharper dose fall-off at the end of the range (Wittkamper and Mijnheer 1993).

Figure 5.6 shows the parameters necessary to characterize the electron beam spectrum (adapted from Kron 2005).

The mean energy at the surface of the phantom can be calculated from the depth of half maximum dose, d_{50}, as (IAEA 1987):

$$E_0 = 2.33\, d_{50}\ \text{MeV}. \tag{5.1}$$

In this equation, the half-value depth of the beam in water, d_{50}, must be specified in centimeters. Similarly, the most probable energy at the surface is calculated from the practical range of the electron beam, R_p, as

$$E_{p,0} = C_1 + C_2 R_p + C_3 R_p^2\ \text{MeV}, \tag{5.2}$$

with $C_{1,2,3}$ as constants given by the protocol (IAEA 1987). It is important to note that these equations are only applicable for field sizes that are large enough to ensure lateral electron equilibrium. In practice, this is 10×10 cm^2 for energies below 20 MeV (16 MeV in IAEA TRS 398 [2000]) and larger

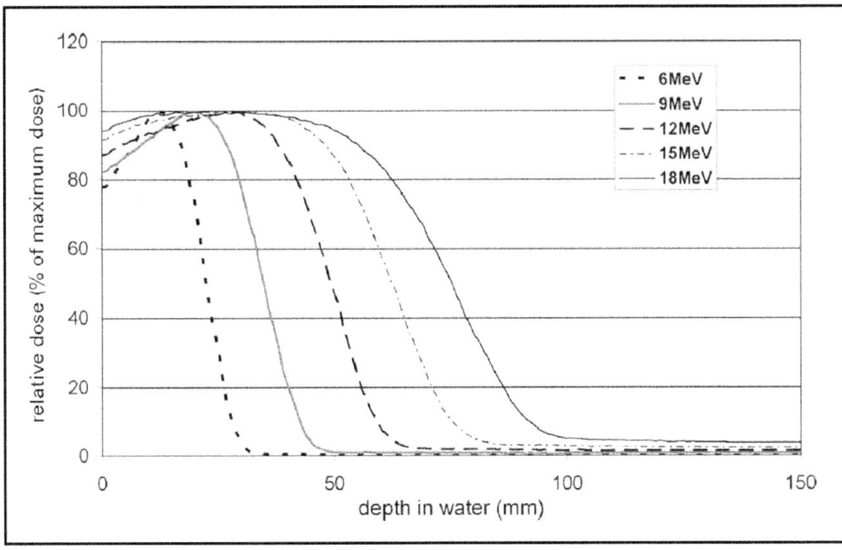

Figure 5.6: Percentage depth dose curve for five different electron energies.

above this energy. It is also important to note that other formulas have been proposed to convert the information in a depth dose curve into a nominal energy (compare e.g., Klevenhagen 1993, p. 134). The variations obtained in this fashion from the same data sets can be substantial and may exceed 1 MeV. As such, it is important to specify the method used to calculate the electron energy and/or obtain the original depth dose data. In practice, this also applies to the specifications provided by manufacturers.

5.3 Depth Doses and Profiles

In order to provide clinicians (and planning systems) with appropriate information on when and how to use a particular electron beam, 3-D dose distributions must be available. These are typically acquired and saved as percentage depth dose curves and profiles.

Figures 5.6 and 5.7 show percentage depth dose curves for a variety of electron beams. The important parameters to note are illustrated in figure 5.7:

i) The maximum dose, D_{max}, normalized to 100% in a percentage depth dose curve.
ii) The depth of maximum dose, d_{max}. It is important to note that this is not necessarily identical for maximum ionization, which is the quantity often measured. The difference and conversion methods are discussed in more detail in chapter eight.
iii) The surface dose: this typically increases with beam energy, which is opposite to photon beam surface dose.
iv) The depth of half maximum dose, d_{50}, which is used as per equation 5.1 to calculate the incident electron beam energy.

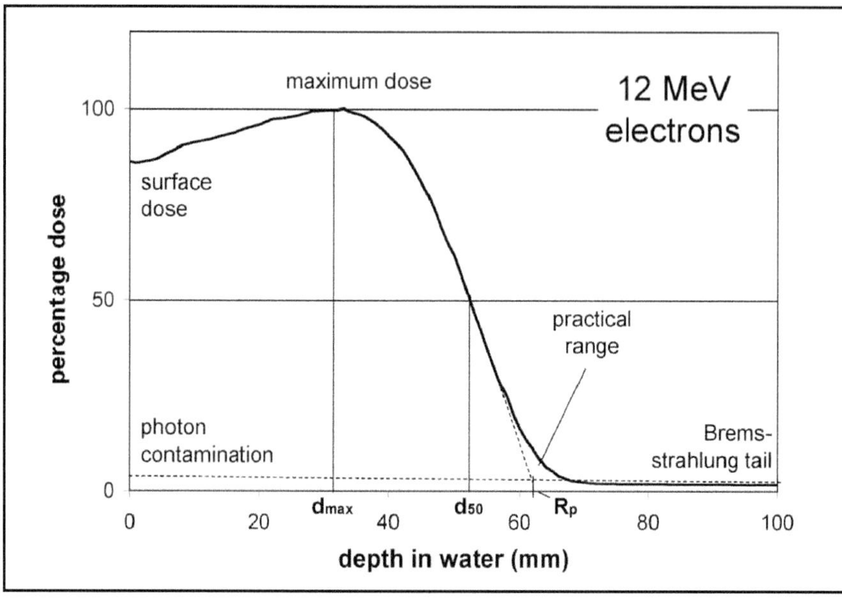

Figure 5.7: Important parameters used to characterize a clinical electron beam from the depth dose distribution in a radiation field of sufficient size for lateral electron equilibrium.

v) The photon contamination: every electron beam has a small component of photons in it. They are produced mainly in the treatment head through bremsstrahlung production (Zhu et al. 2001). Careful selection of materials for scatter foils and transmission chambers can reduce this contamination (el-Khatib et al. 1991). However, high-energy electrons hitting other media will also generate x-rays. For example, it was shown that in small fields up to 30% of the bremsstrahlung may come from the cut-out (Zhu et al. 2001). The bremsstrahlung "tail" extends beyond the range of the electron beam and the "photon contamination" of the electron beam can be determined by extrapolating the bremsstrahlung tail back to the phantom surface. The higher the electron energy and the higher the atomic number of the material, the higher the yield for photon production. This process, which is identical to x-ray production in the target, was discussed in more detail in chapter two. As can be seen in figure 5.6 and table 5.1, the photon contamination increases with radiation energy.

vi) The practical range, R_p, is defined by the slope of the dose fall-off. The intersection between the tangential to the steepest part in the fall-off of the depth dose curve and the photon contamination part is defined as practical range (Klevenhagen 1993).

vii) For clinical purposes also, other depths related to certain dose values may be of interest. As electron prescriptions were often made to the 90% or even 80% isodose line of the maximum dose; the corresponding depths d_{80} and d_{90} are also often quoted.

A useful relation between the practical range and the half-value depth of the electron beam has recently been given by Rahim (2003):

$$R_p = 1.193\, d_{50} + 0.154. \quad (5.3)$$

This formula described the depth dose characteristics in 19 accelerators from different manufacturers with high accuracy.

Table 5.1 lists these parameters for a number of photon beams from a typical Varian linac of the 21 series. The data is representative of a dual scattering foil.

Of particular interest is the "build-up" region. The surface dose is much higher than in photon beams. Unlike in photon beams it also increases with electron energy. The build-up in electron beams is more difficult to understand than the one in photon beams. In addition to a greater interaction probability with decreasing energy at depth, the angular distribution of the electrons plays a role. This is illustrated in figure 5.8. With increasing depth the mean angle of an electron track with the incident direction increases. This means that the actual length of the electron track and therefore the number of interactions in a given depth is larger.

Figure 5.9 shows a color plot of the dose distribution in a plane parallel to the incidence direction of a 4×4 cm^2 electron beam. This is shown for three different electron energies based on Monte Carlo calculations. One can easily see the difference in penetration but also in lateral electron spread at depth. At larger depths the 20 MeV beam is affected even at central axis by lateral electron disequilibrium. This effect can also be seen in table 5.1 where depth dose

Table 5.1: Example data for electron radiation beams; data given for a field size of 10×10 cm^2. The data are typical for a linac with dual scattering foil and the percentages are of the maximum dose in the respective field. Please note the data are purposely not taken from the same linac as figure 5.6 (Adapted from Ostwald and Kron 1996b.) These data are NOT intended for clinical use.

Nominal energy (MeV)	*4*	*6*	*9*	*12*	*12*	*16*	*20*	*20*
Applicator size (side of square field: cm)	15	15	15	6	15	15	6	15
Surface dose (%)	75	75	80	83	85	90	92	92
Depth of 90% in the build-up region	6	8	10	9	6	na	na	na
Depth of maximum dose (mm)	8	14	17	26	25	26	15	23
Depth of 90% dose (mm)	12	19	25	38	38	50	49	58
Depth of 80% dose (mm)	13	21	31	44	43	58	70	71
Depth of 50% dose d_{50} (mm)	16	25	34	50	50	67	81	83
Practical range (mm)	19	31	43	60	61	83	99	99
Photon contamination (%)	<1	0.6	1.6	3.3	3.3	4.9	5.9	6.5

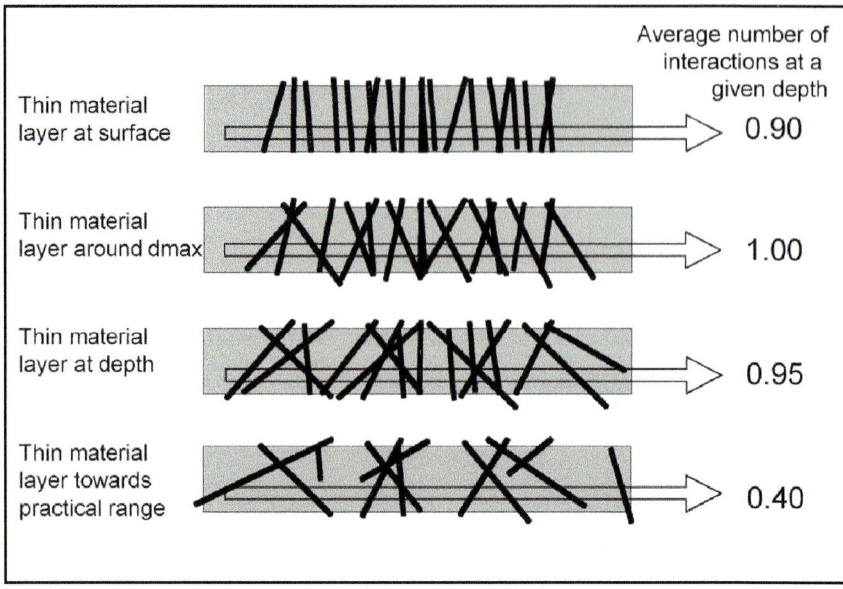

Figure 5.8: Illustration of one of the main contributors to the build-up effect in electron beams. Each black line represents a section of the electron track. In first approximation, the number of interactions per millimeter of track, and therefore the dose deposited, can be assumed to be identical at all depths. In practice, the interaction probability increases with decreasing energy at depth. This also contributes to a build-up of dose.

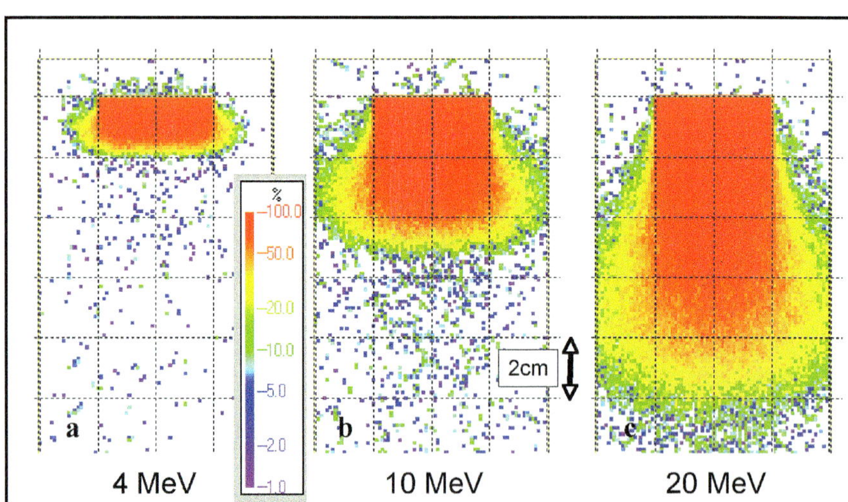

Figure 5.9: Monte Carlo calculated dose distribution in water for 106 histories of electrons with different nominal energies: (a) 4 MeV; (b) 10 MeV; and (c) 20 MeV. The scale is logarithmic, which makes it difficult to appreciate the build-up effect. The figure was generated using EGS-ray software.

parameters are shown for different electron energies. Beam data are given for the 12 and 20 MeV for a 6 × 6 and 15 × 15 cm^2 applicator. The 15 × 15 cm^2 field can be assumed to achieve lateral electron equilibrium along the central axis for all energies as it is unlikely that electrons even with an incident energy of 20 MeV travel 7.5 cm orthogonally to their incidence direction. While the parameters for different applicator sizes are not too different for the 12 MeV beam, the depth of 90% dose is clearly affected at central axis for the 20 MeV beam.

Figure 5.10 shows a schematic representation of the isodose lines in an electron beam. The figure is adapted from a publication of the International Electrotechnical Commission (IEC standard 976 1989). The IEC defines the standard measurement depth as the plane orthogonal to the incidence direction containing the depth of maximum dose while the base depth is parallel to this plane containing the 90% point on central axis. A typical relation between the isodoses and the nominal field size (which will be projected by the light field) is indicated. The figure also shows the cross section through the beam with isodose lines at different depths.

The measured penumbra at different depths in a 20 MeV electron beam produced by a 15 × 15 cm^2 electron applicator is shown in figure 5.11. It can

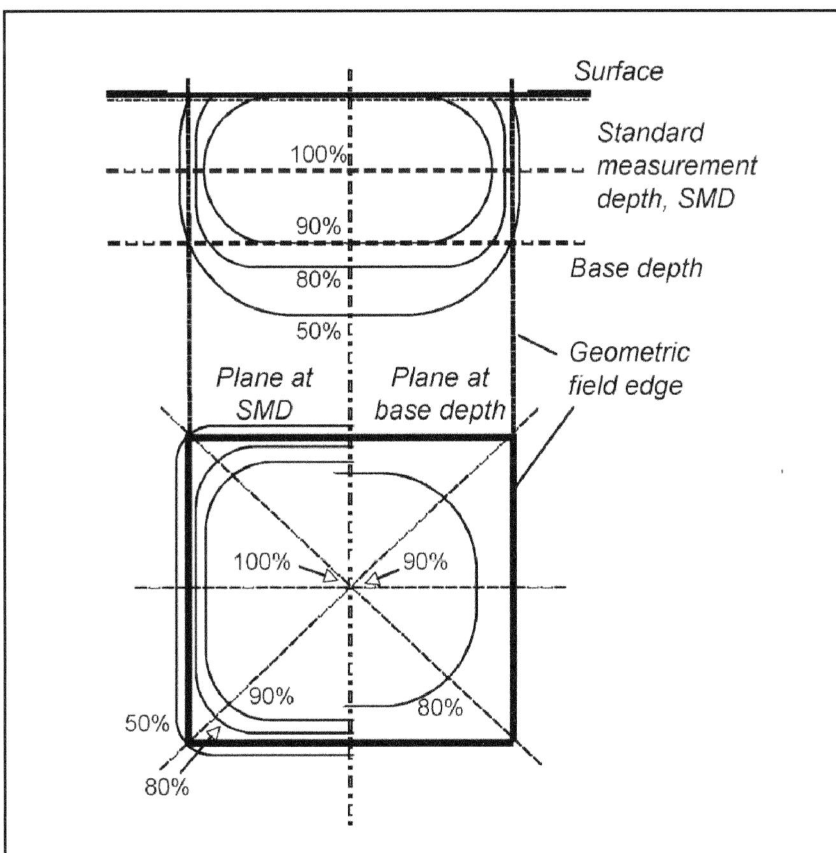

Figure 5.10: Isodose profile of an electron beam (adopted from IEC standard 976, 1989). (a) Profiles parallel to incident beam direction in water; (b) cross section at two different depths: left half at standard measurement depth, right half at base depth.

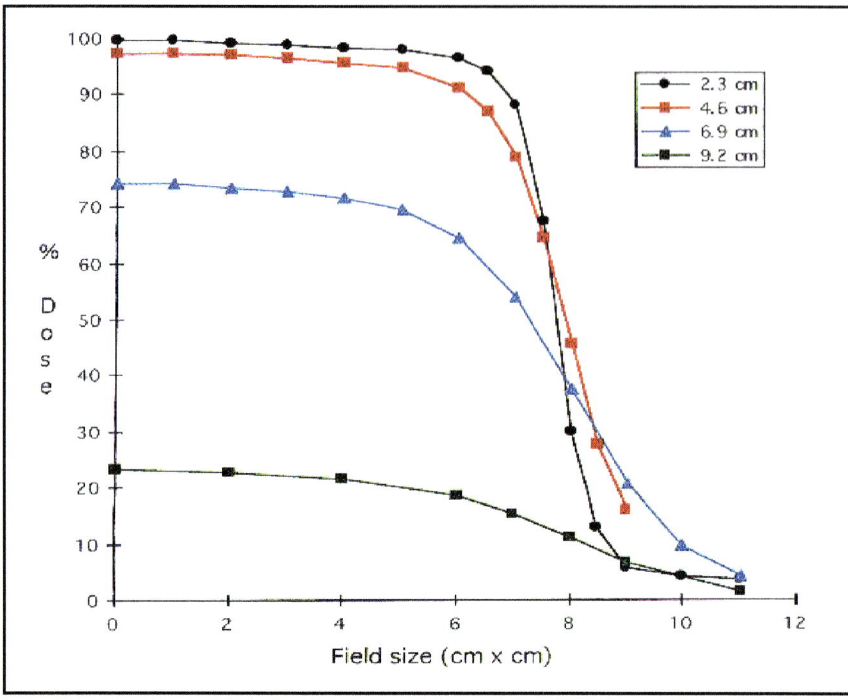

Figure 5.11: Typical profiles at various depths in water for a 20 MeV electron beam, including depth of maximum dose (2.3 cm).

be seen how the penumbra widens with depth due to lateral electron transport. The distance from the central axis to the position of half the central axis dose follows more or less the divergence of the radiation beam. From a clinical perspective the distance between the 90% and 50% dose locations is of importance: when prescribing the dose to an electron beam to 90% of the maximum dose, this distance will determine how far from the field edge an adequate dose will be delivered.

Table 5.2 shows this penumbra width for a variety of nominal electron energies and different applicator sizes. For all but the highest electron energies,

Table 5.2: Distance in millimeters between 90% and 50% isodose line for different nominal incident electron energies measured at depth of maximum dose. The 90% isodose line is often the prescription dose, while the 50% line is identical with the field edge. These data are typical for a linac with dual scattering foil and NOT intended for clinical use

Nominal energy (MeV	*6*	*9*	*12*	*16*	*20*
6 x 6 cm^2 applicator	8	8	8	7	4
10 x 10 cm^2 applicator	8	8	8	7	4
15 x 15 cm^2 applicator	8	7	7	7	3
20 x 20 cm^2 applicator	7	6	6	5	3
25 x 25 cm^2 applicator	7	6	5	4	2

the distance from the field edge to the 90% isodose line is more than 5 mm. This is larger than the equivalent distance in x-ray beams and adds to the field size required for treatment.

5.4 Electron Applicators and Cut-outs

Electron radiation is typically delivered to patients in a fixed SSD (source to skin (or surface) distance) arrangement at 100 cm SSD. As pointed out above, it is important to provide collimation for an electron beam as close as possible to the patient's skin. This will ensure relatively sharp field edges and a flat radiation beam. Electron applicators such as the one shown in figure 5.12 provide this collimation. They are available in many sizes as shown in figure 5.13. The multiple planes of collimation help to produce the desired narrow beam penumbra and a flat radiation beam. The jaws of the linac used to define photon beams are set to a field size slightly larger than the nominal

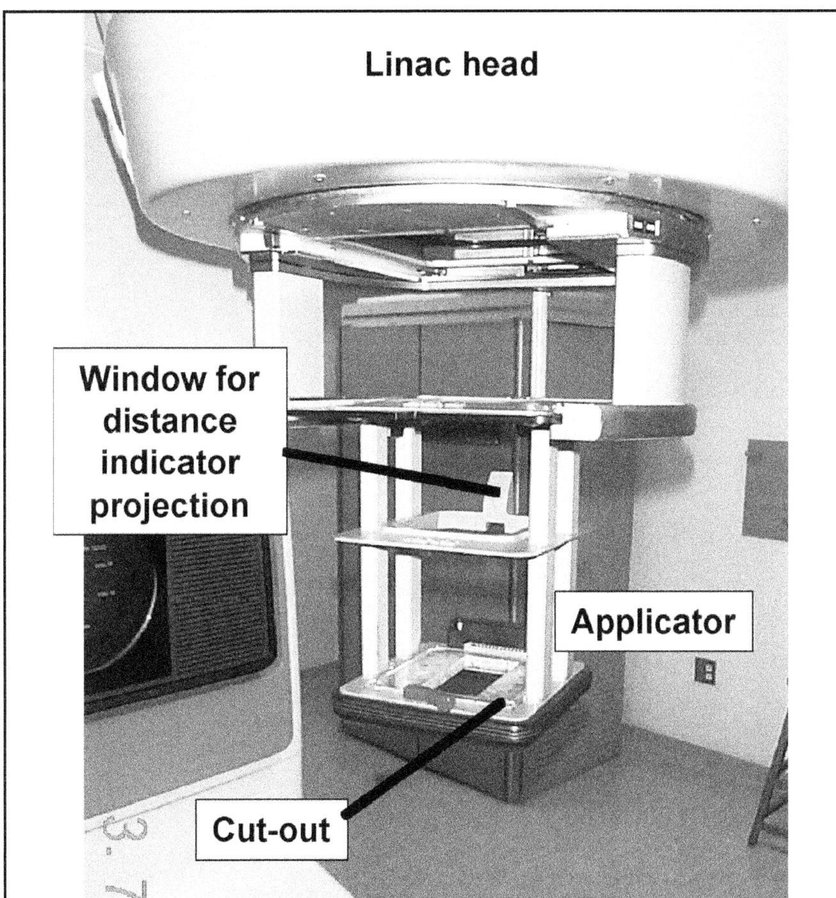

Figure 5.12: Modern linac with electron applicator in place.

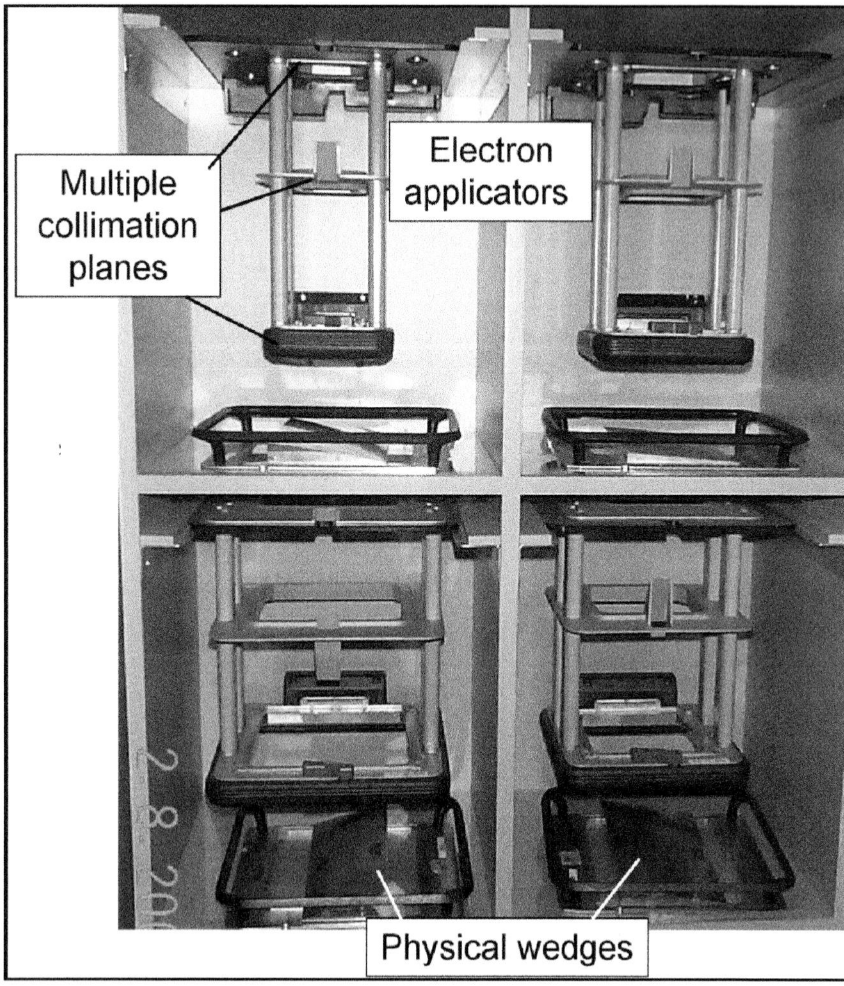

Figure 5.13: Shelf for storage of electron applicators in the treatment room. The shelf does also store the physical wedges.

size of the electron applicator. These collimator settings are chosen individually for each applicator size and electron energy to optimize output and electron beam flatness. They vary from applicator to applicator and can differ for different electron energies. While the required collimator positions are typically coded into the machine settings and automatically adjusted once a particular applicator and electron energy is chosen, it is important to verify the settings regularly.

Most electron applicators provide the final collimation 5 cm from the patient's skin. The choice of applicator will also affect the output of the machine. Unlike in photon beams there is usually no uniform correlation between the output at central axis and applicator size as the dose in the field depends not only on the primary electrons but also on the photon jaw settings and the scatter from the applicator. Therefore, output factors must be established for each radiation energy during commissioning.

Figure 5.14 shows an electron applicator with a final collimator ("cut-out") for physics measurements in place. This cut-out is made from high-density material and provides collimation to standard sizes such as the 10×10 cm^2 field size shown in the figure. A microswitch is activated when a cut-out is in place and ensures that no treatment goes ahead without an appropriate cut-out. The applicator also features a little slide (or window) which allows projection of the optical distance indicator (ODI) into the light-field within the applicator. The slide or pusher is open when setting up the patient and closed during treatment to ensure symmetric collimation. As the applicator is also relatively close to the patient (typically 5 cm), it has a touch guard, which disables all linac motions when activated.

The physics cut-out can be replaced by cut-outs which are customized for individual patients. This is shown in figure 5.15. The rectangular cut-out on the right side of figure 5.15a was made from low melting point alloy (LMA). The thickness of material required to ensure adequate attenuation of all electrons up to a maximum electron energy of 20 MeV is of the order of 10 g/cm^2. Using LMA with a density of 8 g/cm^3 or higher, the typical thickness of a cut-out is 15 mm. Many modern linacs have a coding for cut-outs, which can be verified in the record and verify (R&V) system. This is seen in both figures 5.15 and 5.16. In this way, only the cut-out designed for a specific field for a particular patient can be used for this treatment.

During commissioning of the electron radiation beam a variety of cut-outs in geometric shapes can be used to determine field-size factors and off-axis factors. Examples for these cut-outs are given in figure 5.15b. While it is possible to determine the output for each cut-out by measurement, it is

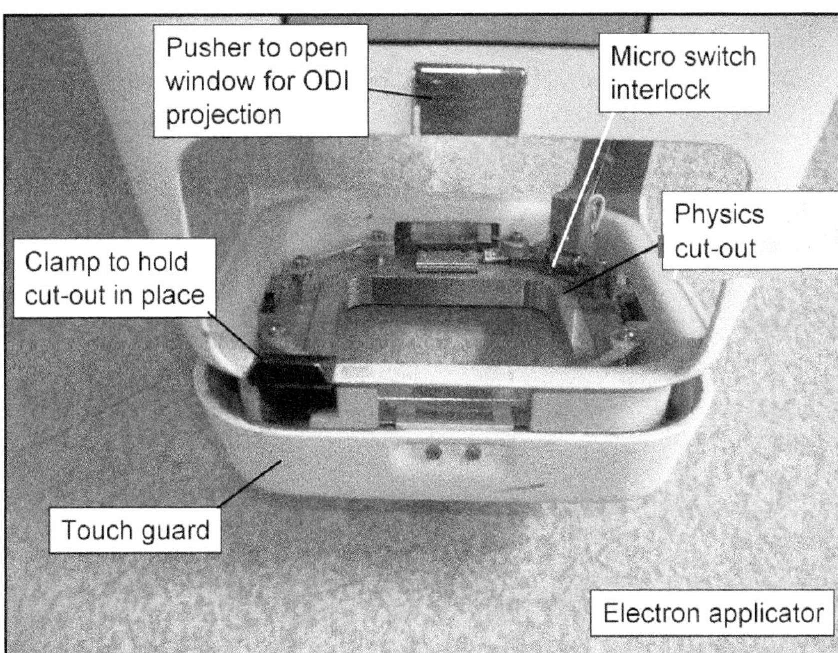

Figure 5.14: View into the final collimator ("cut-out") of a 10×10 cm^2 electron applicator.

(a)

(b)

Figure 5.15: (a) Frames to hold customized electron cut-outs. On the left is an empty frame, while on the right a rectangular cut-out is mounted. (b) LMA cut-outs for physics measurements: on the left side are shown six cut-outs with different field shapes, sizes, and locations for determination of out-put factors. On the right side is a cut-out without any opening for transmission and leakage requirements.

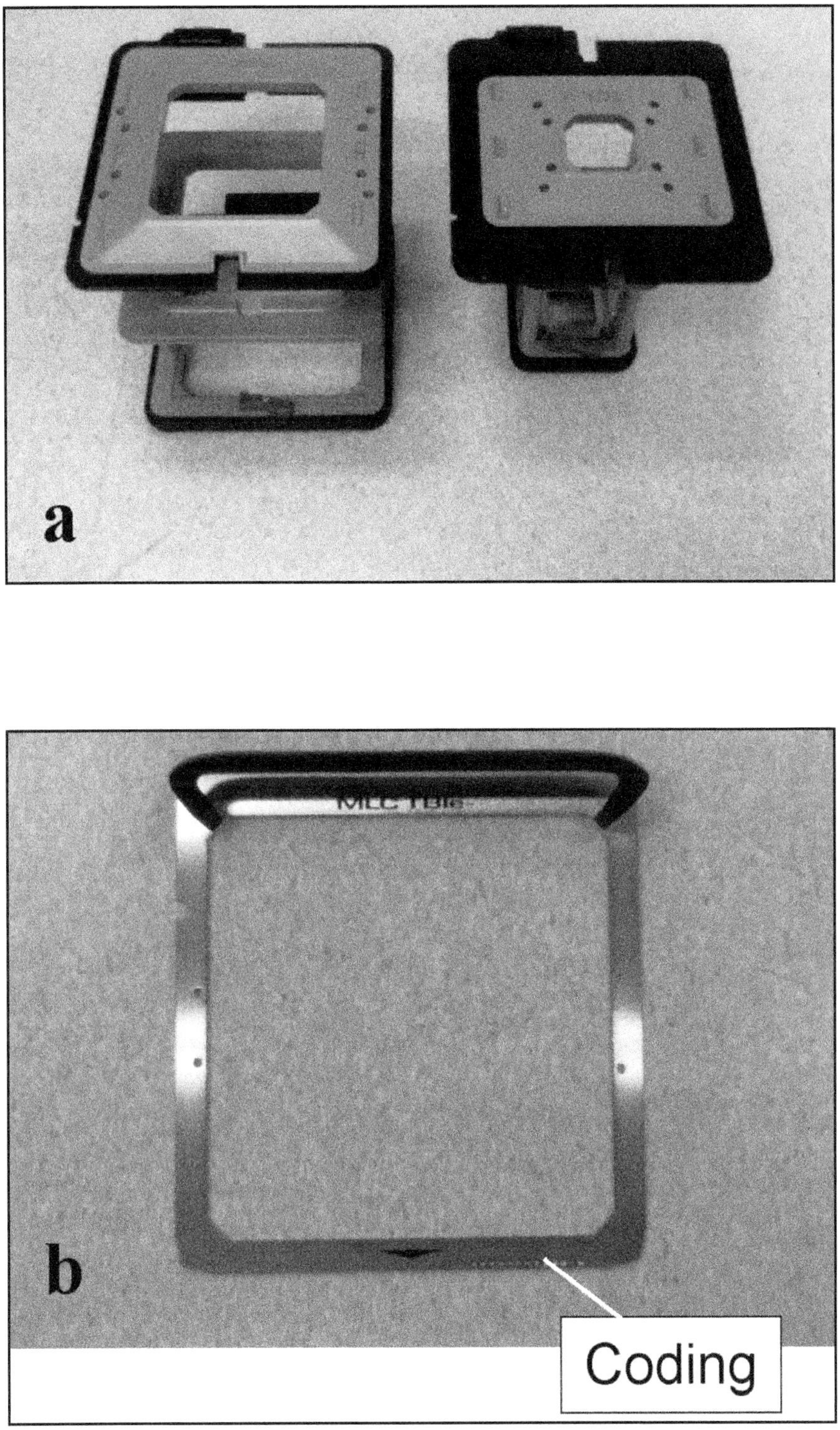

Figure 5.16: Photographs of three different electron applicators: (a) 6 × 6 and 20 × 20 cm^2 applicators. Please note that the collimation in all planes is used to optimize the beam characteristics. (b) "Applicator to produce a large open field such as needed for total body irradiation.

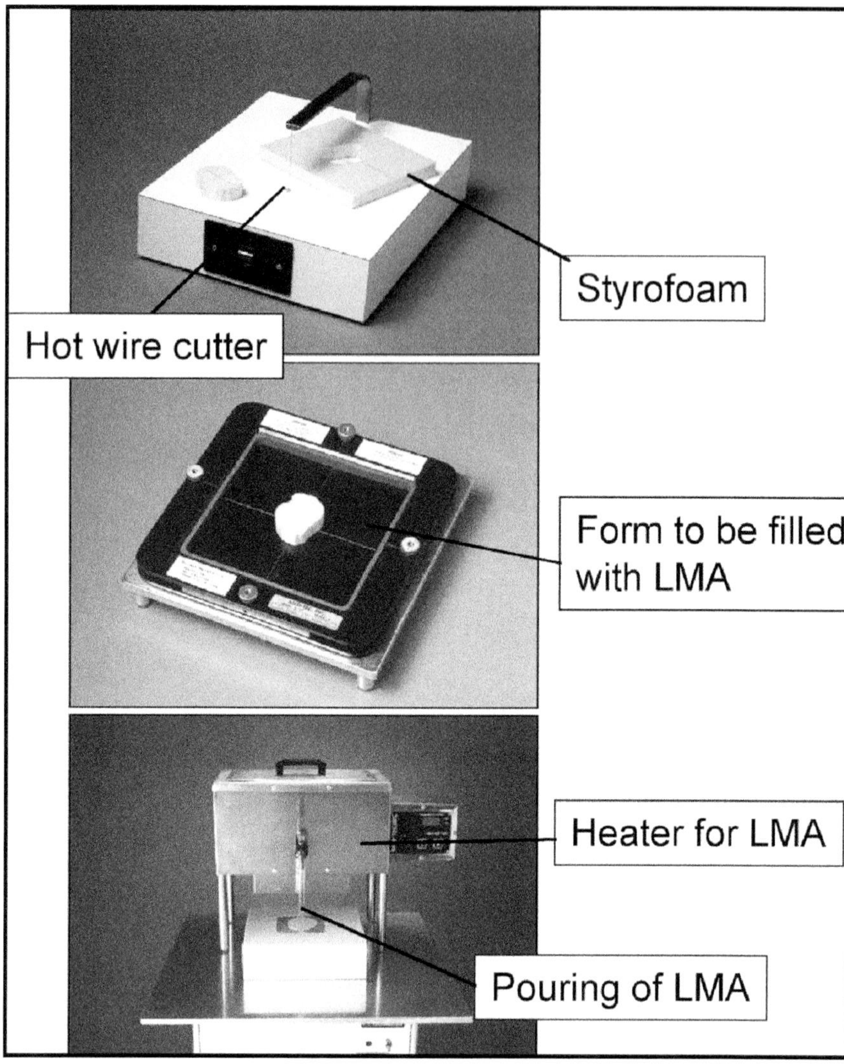

Figure 5.17: Process of manufacturing a customized electron cut-out: (a) Photo of a hot wire cutter. The desired cut-out shape is cut into a piece of Styrofoam™. (b) The foam with the desired outline of the opening is placed in a tray. The tray has the outer shape of a cut-out and the inner shape (opening) is defined by the foam block. (c) Low melting point alloy (LMA) is poured in the tray in a similar fashion to the photon block being poured in this picture.

typically possible to generate a "library" of cut-out factors which can be used to estimate the output in most fields. A set of regularly shaped cut-outs such as the ones shown in figure 5.15b can be used to generate this library during commissioning of the linac. Also, software has been developed (often in house) to calculate output factors and/or percentage depth dose in electron fields (Choi et al. 2003).

A customized cut-out is most commonly made using a hot wire cutter and polystyrene foam, as illustrated in figure 5.17. The outline of the desired field is marked by a clinician. The size is then uniformly increased by a few millimeters to account for the difference in location between prescription isodose line and the 50% isodose line that indicates the field edge. Table 5.2 lists typical distances for this. The new outline is then scaled to the appropriate distance of the collimator from the skin (typically 95 cm/100 cm). The outline is then cut out of a sheet of foam. This is placed in a tray with the dimensions of the applicator end and the area around the foam filled with hot LMA. After cooling, the cut-out can be used in the applicator.

Electron applicators are designed to be used at the nominal SSD. This is possible when the patient's contours are relatively flat or convex. Examples for this are boosts for breast radiotherapy and many skin treatments with low-energy electrons. However, it gets more complicated for treatments such as a lateral neck field where the applicator does not always fit above the shoulder of the patient. In this case, the treatment will not proceed at standard SSD, but a correction must be made for the distance between the nominal and actual SSD as discussed in section 5.6.

5.5 Small and Irregular Fields

Cut-outs allow the delivery of arbitrarily shaped fields after taking the distance from the geometric field edge to a high isodose line into account (compare table 5.2). This however, may lead to complex dosimetric problems as lateral electron disequilibrium may affect dose distribution and output. Lateral electron disequilibrium is defined as an imbalance between electrons entering and exiting a volume element laterally. This is illustrated in figure 5.18. The figure shows the lack of lateral electron equilibrium in

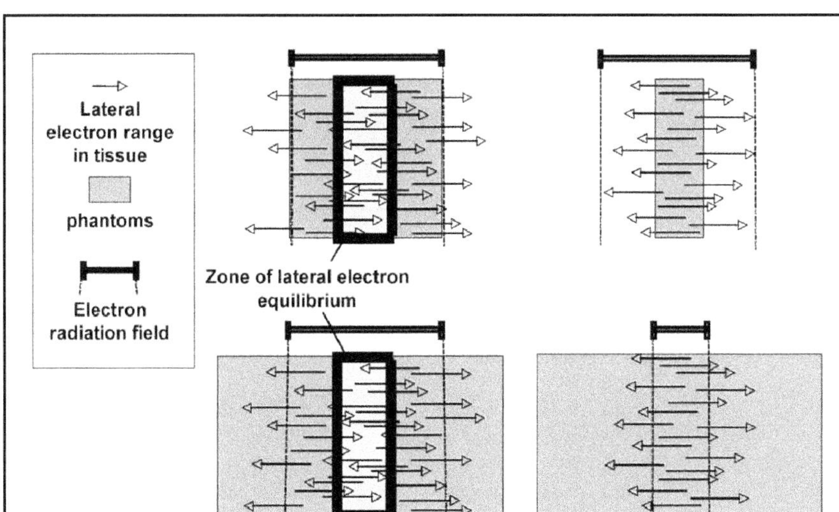

Figure 5.18: Illustration of the concept of lateral electron equilibrium. The figure only shows the lateral component of the electron transport. Electrons only can be set in motion where a radiation field interacts with the phantom. Only where the number of electrons scattered in and out of the volume is the same is there lateral electron equilibrium..

two distinct scenarios: lack of scattering material in the beam (top illustrations) and small electron beams (bottom row). In the illustrations on the right side of the figure no lateral electron equilibrium exists because of lack of electrons scattered back into the field towards central axis. This is either due to lack of scattering material or lack of primary radiation.

As discussed above and shown in figure 5.4, electrons will travel laterally. The actual distance traveled laterally by electrons is a random distribution that depends on:

i) electron energy (lower energy electrons scatter more),
ii) scattering power of the material, and,
iii) penetration depth in the material (the larger the depth in matter, the more scattering events have occurred already and the lower the energy of the electrons).

In addition to the many small angular deflections due to excitations and ionization, there are also some large angle scattering events as discussed in chapter two. They play an important role in determining accuracy of Monte Carlo codes for dose deposition as discussed in chapters nine and ten.

In general, lateral electron disequilibrium is more likely to extend across the whole field in small fields and high-energy electron beams at depth. The lateral range of the electrons and therefore the dimensions of disequilibrium will also increase in low-density material such as lung. This will be discussed in more detail in section 5.7. In very high-energy electron beams (such as 20 MeV), the field size may exceed 10×10 cm^2, which led the IAEA to recommend calibration of high-energy electron beams with a half-value depth in water of 7 cm or more to be performed using at least a field size of 20×20 cm^2 (IAEA TRS 398 2000). (Compare also chapter eight).

The discussion above applies obviously not only to small electron fields but also to all narrow sections in larger fields. As such, it is typically recommended to avoid using narrow sections in electron radiotherapy. It will affect not only the maximum dose in the middle of the section but also the penetration of the beam. This is illustrated in figure 5.19, which shows an electron cut-out shape. While it may be clinically important to extend a radiation field, e.g., along a scar, the dose distribution may be adversely affected in a narrow extension of the field in exactly the same way as a small radiation field in the first place.

The effect of field size of penetration can be seen in figure 5.20, which shows the depth dose distribution for an 18 MeV electron beam for several field sizes. Already at a field size of 8×8 cm^2 one can see a very small reduction of the penetration of the 18 MeV beam compared to a 20×20 cm^2 beam that ensures full lateral electron equilibrium at central axis. For a very small field of 2×2 cm^2 the penetration is nearly halved.

It is important to note that not only field size and beam energy will affect the dose distribution but also the position of the field opening within a cut-out. This has recently been demonstrated for small electron fields (Chow and Grigorov 2007).

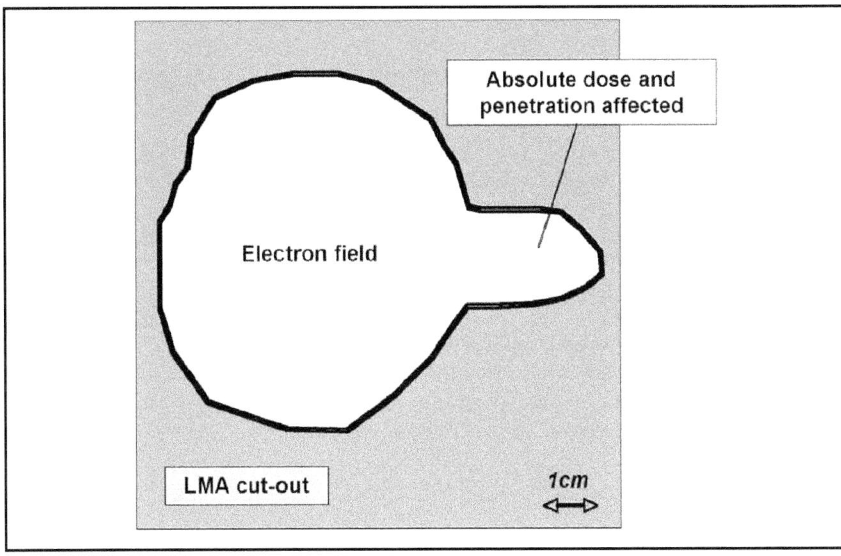

Figure 5.19: Illustration of an electron field in which the right part would be affected by lateral electron equilibrium.

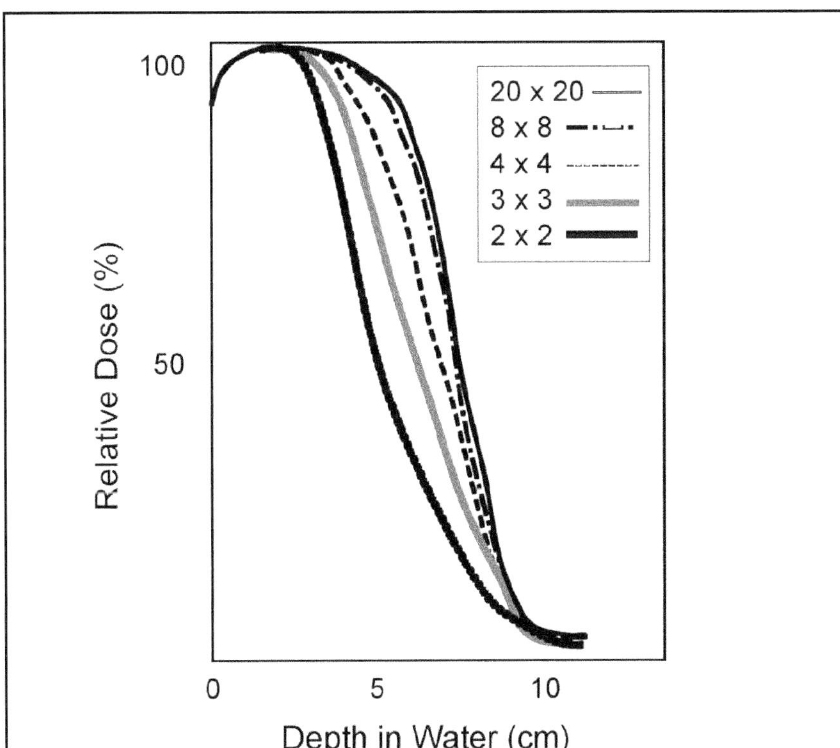

Figure 5.20: Measured field-size dependence of depth doses in water for an 18 MeV beam. (Redrawn from A. Nahum The computation of dose distributions in electron beam radiotherapy. Thesis. Umea: Umea University, 1985. Courtesy of P.M. Ostwald.).

5.6 Variations in Geometry

The properties of electron beams discussed above will also affect their behavior in nonstandard geometries. Important examples for this are extended source to SSD and oblique incidence. Both are encountered frequently in clinical patient treatment scenarios.

5.6.1 Extended SSD

Extending the SSD of electron beams can increase their effective field size. However, the most common reason for using a larger SSD than normal is that the applicator physically cannot be placed close enough to the patient to allow treatment at 100 cm SSD. An example for this is lateral radiation fields to the neck to treat lymph nodes in head and neck cancer. The use of electrons will ensure avoidance of the spinal cord; however, often the applicator does not fit past the shoulder of the patient. In this case, a larger SSD must be accepted.

Extension of the SSD will result in two effects:

i) A reduction in dose.
ii) The increase in field size.

Both are obviously related and both can be used to find the effective source location using the inverse square law or simple geometry according to the following simple equations:

i) The effective source position can be determined from decrease of dose D with distance from the nominal source-to-surface distance (SSD) of 100 cm:

$$D(100) = D(\text{SSD})\ (100 / \text{ESSD} + d)^2. \tag{5.4}$$

with ESSD as the effective SSD of the source and d the distance from nominal SSD location.

ii) The virtual source position is derived from the change in full-width at half maximum (FWHM = distance between the points of 50% dose on both sides of the field) with distance as:

$$\text{FWHM}(100) = \text{FWHM}(\text{SSD})\ (100 / \text{VSSD} + d)^2. \tag{5.5}$$

with VSSD as the virtual SSD of the source.

Both parameters are specific to a particular electron beam energy and collimator size. They are both useful and clinically relevant but usually not identical. The effective SSD is the more commonly used, as it directly relates to the number of monitor units required to deliver a prescribed dose. Table 5.3 shows some sample data for effective SSDs for different nominal beam energies and field sizes. At large field sizes and high energies the ESSD is

Table 5.3: Example data for effective SSD for various electron radiation beams. The data are typical for a linac with dual scattering foil and electron applicators. The typical uncertainty is ±3 cm. (Data reproduced from Ostwald and Kron 1996). These data are NOT intended for clinical use

Nominal energy (MeV) *Field size (cm²)*	*4*	*6*	*9*	*12*	*16*
4 x 4	27	33	43	60	60
6 x 6	43	50	55	67	65
10 x 10	62	65	66	74	74
15 x 15	68	71	76	80	82
20 x 20	71	75	82	85	85
25 x 25	77	81	83	86	87

nearly as large as the photon SSD. However with shrinking field size and lower electron energy the effective field size is reduced: the output of a small field of low-energy electrons drops dramatically with distance from the applicator.

Due to the use of an applicator and the relatively large number of scattering events for electrons prior to reaching the patient, the flatness of the beam is also affected as the air gap between the patient and the final collimating aperture increases. This is illustrated in figure 5.21. The figure also illustrates why the variation of output and field size with distance is not a simple relation for electron beams. The electrons reaching the patient originate from different sources: "primary electrons" originate from the scattering foil and only undergoing scattering in air, other electrons have scattered from the collimator or applicator closer to the patient. These electrons would therefore exhibit a different dose fall-off with distance as the "reference" distance for the inverse square law fall-off is much smaller. The behavior of electron beams with distance from the source depends on the mix of the electrons originating effectively at different distances from the patient. This is difficult to predict; hence ESSD should be measured during commissioning of any linac with electron capability to yield data sets similar to the one shown in table 5.3.

While the electron profiles and output vary with distance from the applicator surface, the change in effective energy at central axis is relatively small because there are only few scattering events in the air above the patient. Therefore, the percentage depth dose curve is only affected slightly as the distance between the applicator and the patient changes. This can be seen in figure 5.22. The data in the figure are normalized to 100% at depth of maximum dose. The maximum dose rate changes with distance from the applicator. Therefore the figure has to be seen in connection with the absolute dose fall-off described by the effective SSD shown in the next figure. The most important feature observable in figure 5.22 is a reduction of surface dose as less low energy electrons scattered from the applicator reach the surface.

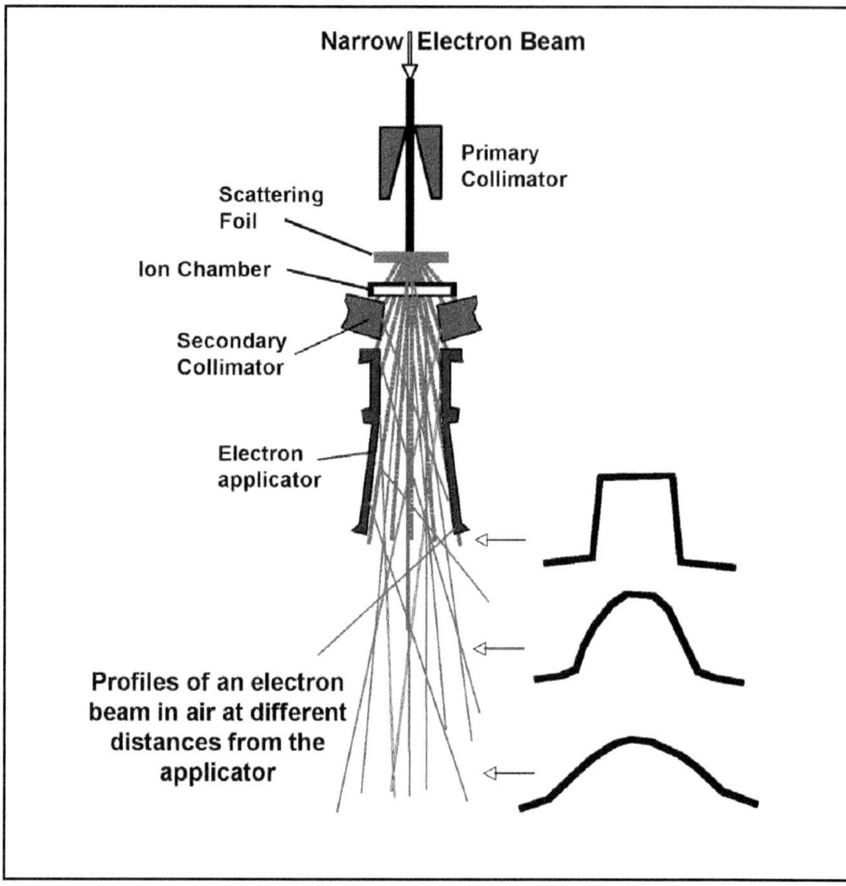

Figure 5.21: Illustration of the change in profile of electron beams in air at different distances from the electron applicator.

The additional shift of the depth of maximum dose also reflects an overall electron energy spectrum, which features more high-energy electrons because the number of lower-energy electrons scattered from the applicator are reduced. This is also illustrated in figure 5.23, which shows the variation of effective SSD with nominal electron energy and applicator size (Ostwald and Kron 1996a,b). The data were acquired using an Attix-type plane parallel ionization chamber with thin front window (compare chapter three). As can be expected, the effective SSD decreases with decreasing electron energy and field size. Figure 5.23a shows measurements performed at the surface of a slab phantom. As can be expected, the effective SSD is generally smaller than when measured at depth of maximum dose as shown in figure 5.23b. This is a result of a fast reduction of lower energy electrons scattered from the electron applicator with distance from the applicator as discussed in the previous paragraph.

This can be appreciated also in figure 5.24, which shows the variation of effective SSD with depth of measurement in a 12 MeV electron beam with a nominal field size of 15×15 cm^2 at 100 cm SSD (Ostwald and Kron 1996a,b). The data were acquired with an applicator reproducing the results of figure

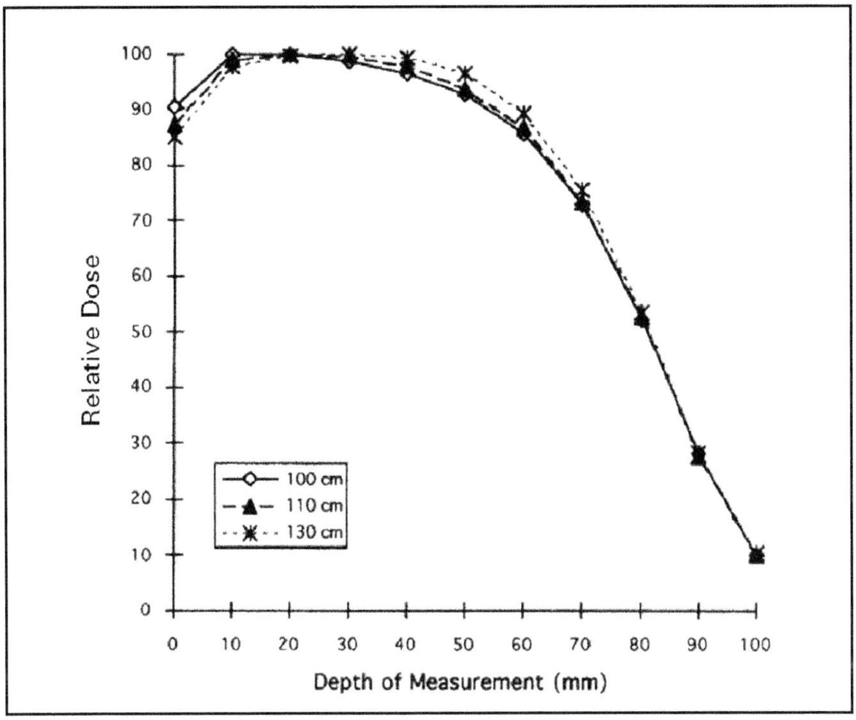

Figure 5.22: Depth dose curves plotted for a 20 MeV electron beam at three different nominal SSDs (100, 110, and 130 cm). The data were measured for an applicator size of 15 x 15 cm^2 using an Attix chamber and connected using the method provided by Khan et al. (1991). (Courtesy of P.M. Ostwald.)

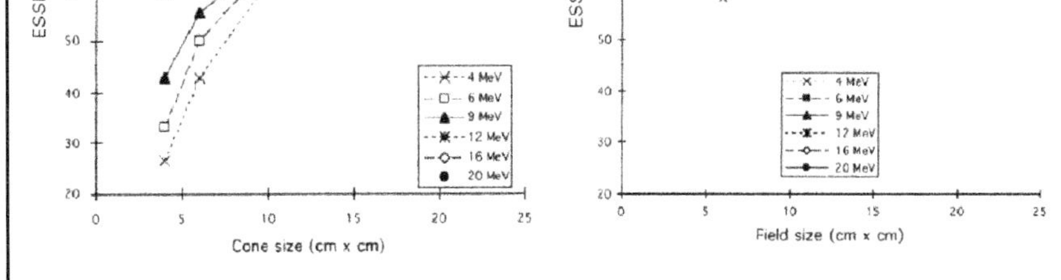

Figure 5.23: Variation of effective SSD for different electron energies with applicator cone size. The data are similar to the data in table 5.3. Figure 5.23a shows the effective SSD measured at the surface of a phantom while 5.23b shows the same measurements performed at depth of maximum dose. (Adapted from Ostwald and Kron 1996b).

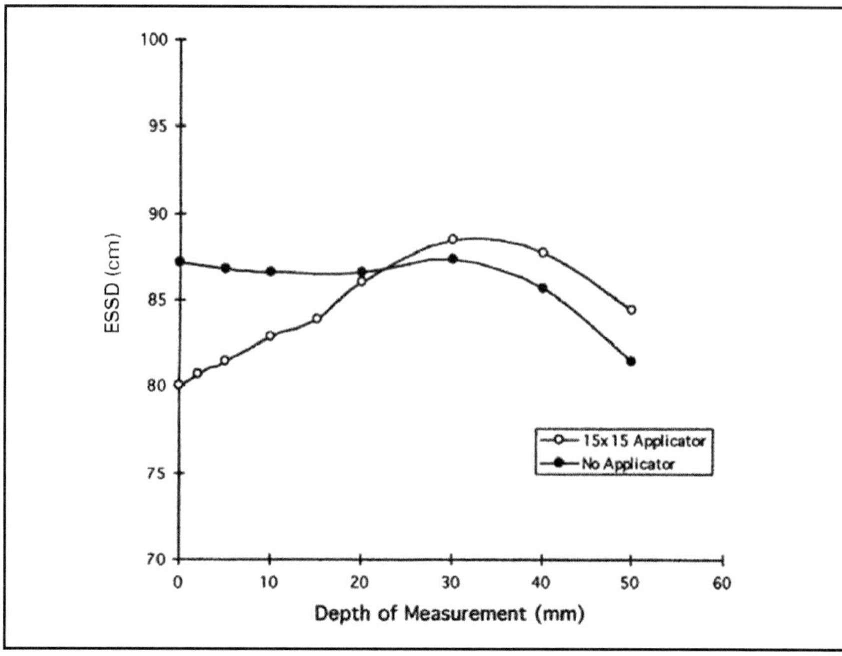

Figure 5.24: Variation of effective SSD as a function of the depth of measurement in a 12 MeV electron beam. (Adapted from Ostwald and Kron 1996b).

5.23 and table 5.3. The measurements were repeated without an applicator by collimating the beam using the adjustable jaws that are used for x-ray delivery. In this case, there is an increase of effective SSD specifically at the surface because of the lack of electrons scattered from the applicator.

It is important to note that the discussion above pertains to square regular fields. In irregularly shaped fields the fall-off in dose with distance may be more complicated and may vary in different parts of the field. A single value cannot adequately describe the complex dose fall-off pattern in an irregularly shaped elongated field and this often can only be determined by measurements.

5.6.2 Surface Contour Changes and Bolus

As can be seen in the depth dose curves shown in figures 5.6, 5.20, and 5.24, the dose at surface in electron beams is lower than the dose at depth of maximum dose. In particular for low-energy electron beams, which are used to treat lesions at the skin, this is undesirable as the dose to some of the target structures in the skin can be substantially lower than the prescription dose. It is therefore common practice to use a bolus such as the one displayed in figure 5.25 to increase the skin dose where needed. It is useful to have different thicknesses of bolus available for different energies.

It is obvious that the application of bolus also alters the penetration of the electron beam; this can be used to adjust the depth in circumstances where the distal part of the target volume is at varying depth from the surface. Examples

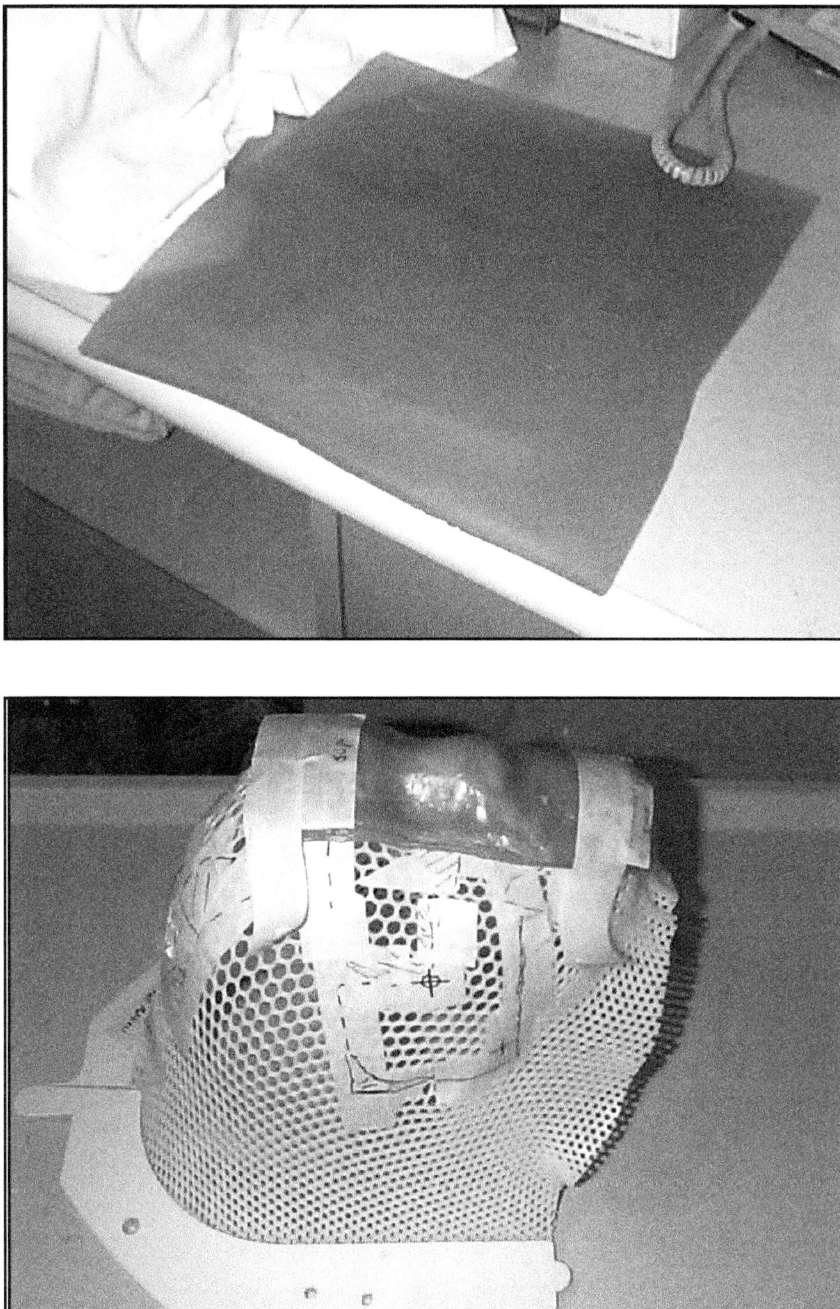

(a)

(b)

Figure 5.25: (a) Picture of a flexible sheet of bolus used for electron beam radiotherapy on the patient's skin. (b) Picture of a sheet of wax bolus mounted over the nose part of the immobilization device of a patient..

would be parotid treatments and, potentially, spinal irradiation in pediatric patients (Hood et al. 2005). In these cases a customized bolus can be manufactured, which features different thicknesses in different parts of the beam. This is in some respect similar to compensators in photon treatment (compare chapter four). However, there are two important differences:

i) Attenuating material in photon beams changes the number of photons (fluence) and therefore the dose. The penetration pattern is only marginally affected by small changes in the spectrum due to preferential attenuation of low-energy photons. In electron beams the number of particles behind the bolus is not significantly affected, however, the energy (and angular distribution) and therefore the penetration depth is reduced.
ii) A photon compensator is employed at a distance from the patient to avoid scatter from the compensator reaching the patient, while an electron bolus is typically placed directly on the patient's skin.

Bolus material may also be used to reduce the impact of sharp contour changes on the patient. This can be seen in figure 5.26. On the left side in figure 5.26a an electron beam with a nominal energy of 20 MeV is incident on a surface that includes a step of 2 cm height. The significantly lower level of interaction in air, compared with the medium, means that scattered electrons from the higher surface edge are not deposited in the air, but into the medium further below (creating a hot spot). The same lack of interaction in air means a lack of electrons scattered into the cold spot adjacent to the lower edge. In this case, the dose variation in the "treatment" field exceeds 20%, ranging from a cold spot of less than 90% to a hot spot in excess of 110%. Both would be unacceptable in clinical practice.

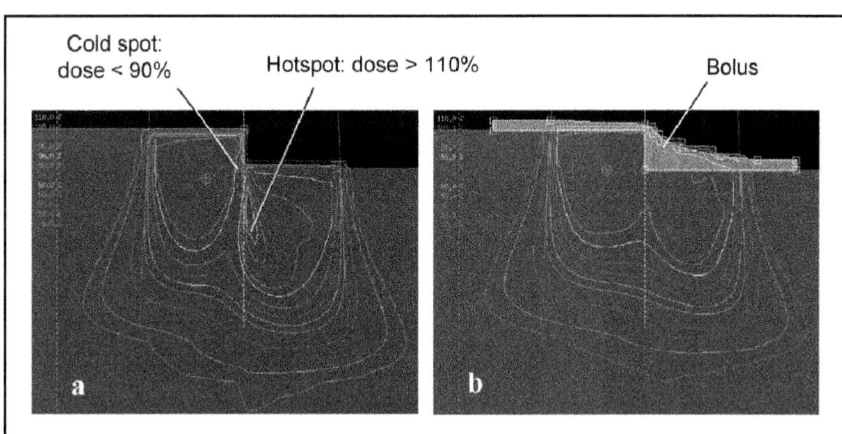

Figure 5.26: Isodose lines of an electron beam of 20 MeV incident on a surface with a sharp edge of 2 cm height. (a) Hot and cold areas due to the edge. (b) Improvement of the dose distribution by smoothing the surface using bolus. Both pictures were created using ADAC Pinnacle planning system with 3-D pencil beam algorithm (compare chapters nine and ten).

The problem can be addressed by smoothing the surface contour, e.g., by using bolus as shown in figure 5.26b. As can be seen from the dose distribution, the dose inhomogeneity has been reduced to less than 10%. However, care must be taken to avoid any edges of the bolus itself being within the field. If the edge is sharp and the bolus exceeds some 5 mm in thickness, the hot spots shown in figure 5.26 will also occur in the patient (Nygaard et al. 2005).

5.6.3 Oblique Incidence

An important issue with electron beams is oblique radiation incidence. As there are only a few large, flat surfaces on the human shape, some degree of obliquity is always likely in a large electron field. Figure 5.27 shows a Monte Carlo calculation of the dose distribution in an electron beam with an area of 4×4 cm^2 on the phantom (patient) surface. The figure shows the data normalized to the respective maximum dose in a linear dose scale. The most important feature of the dose distribution is a shift of the distal part of the dose cloud to shallower depths.

This needs to be seen in conjunction with an effective reduction of field size, as a field of 2.6 cm width casts a 4 cm "foot print" on a surface under 45° to the incidence beam. The typical convention is to specify the incident angle as the angle from normal incidence. The integral dose in both scenarios is therefore not identical as the fields in figure 5.27 employ a different field size. Similarly, the two figures do not display the difference in absolute dose as both are normalized to the dose maximum in the respective dose distribution. Figure 5.28 addresses both issues by showing the isodose lines for 20 MeV electron beam incident onto the surface under 60° from normal incidence (left side) and normally (right side).

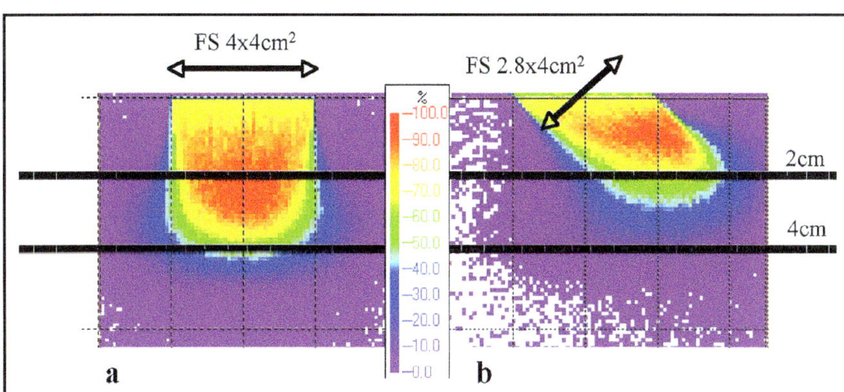

Figure 5.27: Monte Carlo calculated dose distribution in water for 10 MeV electrons: (a) a square shaped source of 4×4 cm^2 incident orthogonally to the phantom surface. (b) The same irradiated area on the phantom surface but with electrons incident under 45° to the normal incidence. This is equivalent to a field size of 2.8×4 cm^2. Both plots are normalized to 100% at the maximum dose level in each figure, respectively. The figure was generated using EGS-ray.

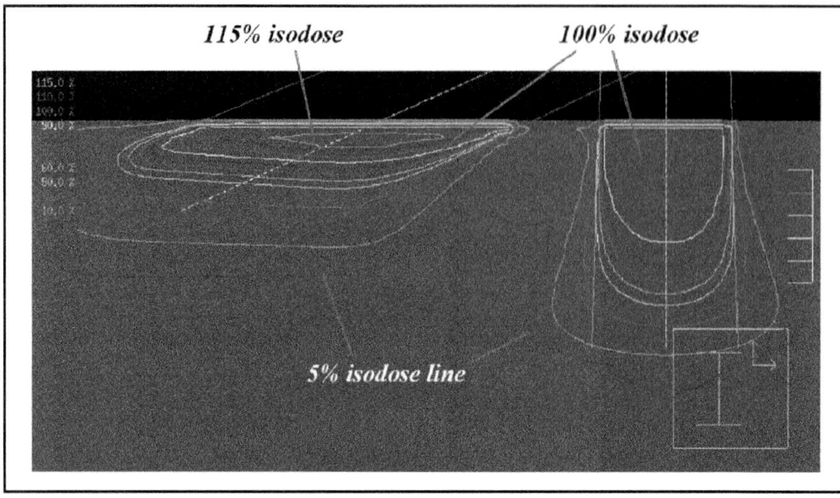

Figure 5.28: Isodose lines of an electron beam of 20 MeV incident on a surface. On the right is a beam with normal incidence to the surface while the isodose contours on the left are from the same field incident under 60°. Both isodose contours were created using ADAC Pinnacle planning system with 3-D pencil beam algorithm (compare chapter nine).

Two important features can immediately be identified:

i) The geometric field size at the skin is increased while the penetration is reduced. The overall dose deposited (J per kg) is identical in both scenarios; however their geometric distribution is different.
ii) The maximum dose in the oblique incidence case exceeds the maximum dose in the normal incidence scenario by more than 15%. This can seriously affect the prescription dose if not taken into consideration.

Figure 5.29 shows the depth dose curves for a 9 MeV electron beam incident onto a surface under different angles. This is a commonly used electron beam energy in breast or head and neck cancer treatments where these oblique incidences may occur. The maximum dose in a beam incident under 60° is nearly 20% higher than the maximum dose in the beam with normal incidence (incident angle 0).

Figure 5.30 provides a simplified geometric explanation for the increase in dose at oblique incidence angles. One can consider the electron radiation field to be a sum of many "pencil beams" as shown in the top of the figure and previously illustrated in figure 5.3. Each of these pencils has a pear-shaped dose distribution, which broadens with depth due to the increased angular struggling with depth. There is also an effective increase in dose at depth due to the relatively longer path of electron tracks at a particular depth and an increased stopping power with reducing electron energy at depth. This was discussed in section 5.3. Under oblique incidence the broad parts of the dose distribution overlap resulting in an increased dose at depth.

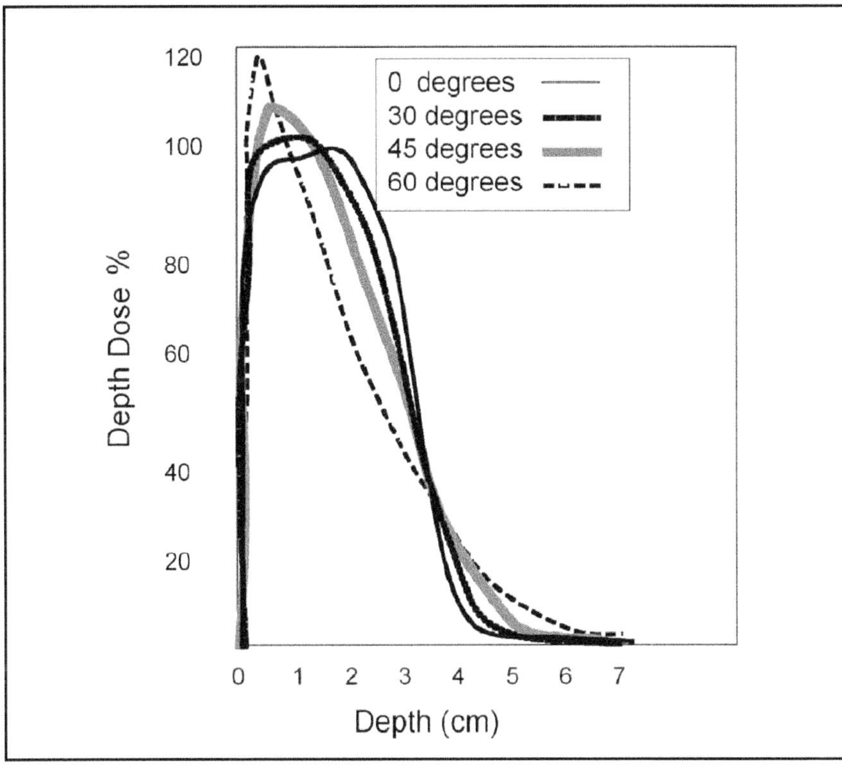

Figure 5.29: Depth dose distribution for a 9 MeV electron beam incident onto a surface under an oblique angle.

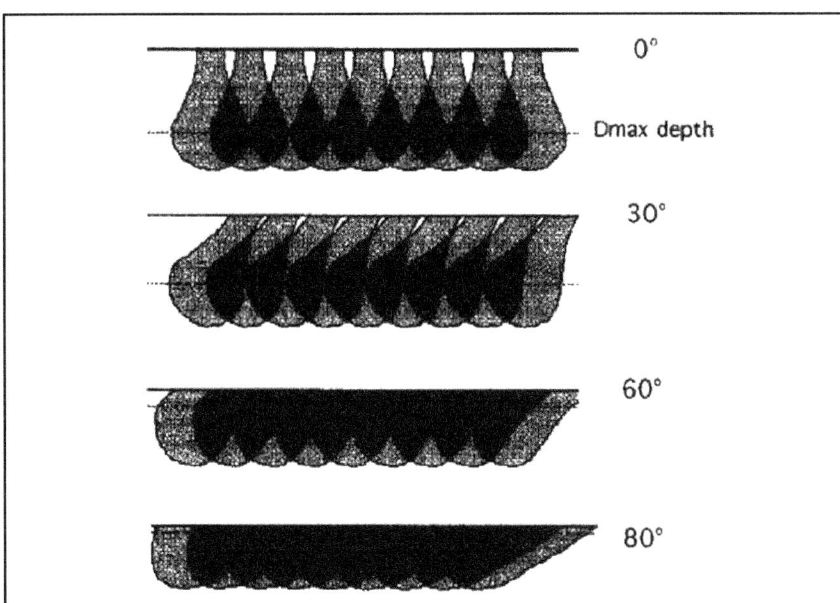

Figure 5.30: A simple geometric interpretation of the dose distribution in electron beams with incidence oblique to a surface using "pencil beams." (Adapted from Ostwald and Kron 1996a.)

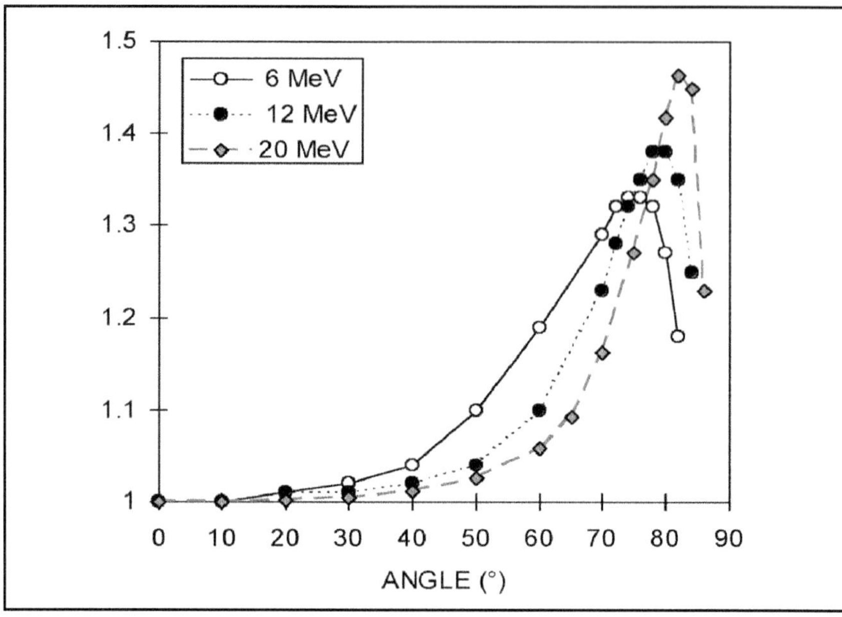

Figure 5.31: Measured variation of maximum dose with nominal energy and obliquity of the incident electron beam. (Adapted from Ostwald and Kron 1996a).

Depending on beam energy and incidence the increase of dose in an oblique beam can be dramatic. In general, the dose maximum increases with electron energy and incidence angle as shown in figure 5.31. Under glancing incidence of a high-energy electron beam a dose enhancement of more than 40% can be observed. The prediction of this dose increase is complicated and not all computerized treatment planning systems may be able to predict it accurately (Lax 1987, Ostwald et al. 1994a,b).

5.7 Inhomogeneities

An *inhomogeneity* is any variation in material properties (in practice, density, stopping power, and scattering power) from tissue equivalence. In the case of charged particle beams such as electrons and protons the density of electrons in the material is the most important component of the inhomogeneity as most interactions between the incident radiation and matter occur via charged particle interaction (compare also chapter two). In the case of electrons the inhomogeneity will affect dose deposition, energy and angular distribution of the electron beam. As such, its impact on the resulting dose distribution is typically more significant than in photon beams.

Figure 5.32 shows Monte Carlo generated electron tracks in media. The tracks are for monoenergetic electrons of 10 MeV and each picture shows the tracks of 100 histories. The left side of the figure shows the tracks in a homogenous medium (water). This is compared with the same tracks when a 1 cm thick lead (middle) or a 3 cm thick air (right side) inhomogeneity is introduced. As can be expected, 1 cm lead attenuates all electrons. The range of the electrons that reach the lead sheet with about 6 MeV rest energy is only of the order of 2 mm in lead. The same electrons traverse the air cavity and produce a dose distribution at the other side of the cavity which is similar to the dose distribution of the electron beam in homogenous medium at a depth greater than 2 cm. The main difference is a less well-defined field edge and a slightly larger effective field size due to the path of the electrons through the air cavity.

The lower part of figure 5.32 shows the tracks of secondary photons in addition to the electron tracks. The most important feature is the increased photon production due to bremsstrahlung with the lead present (compare chapter two). While most photons are forward directed and absorbed in the lead sheet, there is a considerable number of backscattered photons. If the lead sheet had been thinner, it would have acted similarly to a target in a linac as discussed in chapters one and two.

The dose distributions around a lead and air inhomogeneity are shown in figure 5.33. The profound effect of the inhomogeneity on the dose distribution can be easily seen. In the case of the lead sheet the maximum dose is directly before the lead. This is due to backscattered electrons from the lead, which add to the dose from the incident electron beam. An illustration of the dose enhancement in front of the inhomogeneity decreases with distance from the high-density material as can be seen in figure 5.34.

In the case of the air cavity, there is still dose deposition beyond the cavity. By comparing figure 5.33b with 5.27a one can see that the beam now reaches a larger depth. Also, the effect of lateral electron disequilibrium within the cavity can be clearly seen. Due to the large range of electrons in

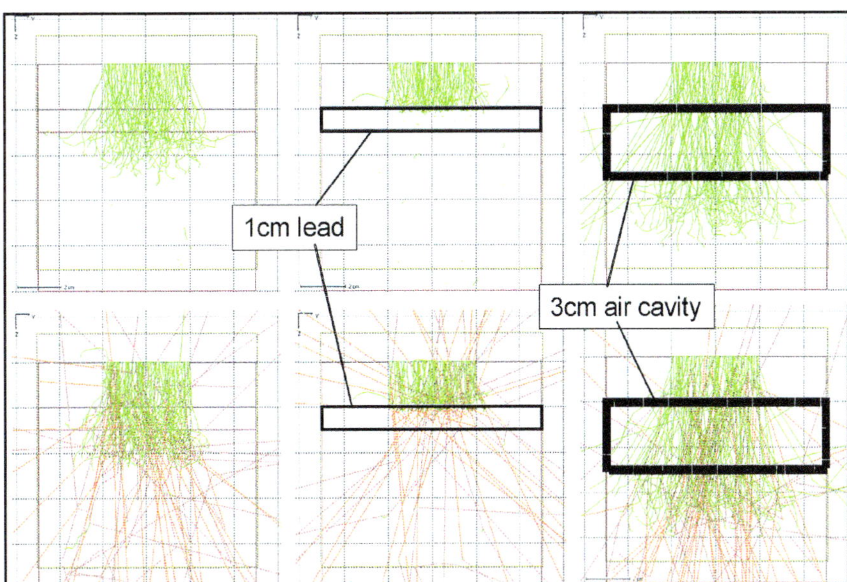

Figure 5.32: Monte Carlo calculated tracks of 100 10-MeV electrons incident over an area of 4×4 cm^2. On the left are shown tracks in homogeneous medium (water) while in the middle and on the right side a lead and an air inhomogeneity are introduced, respectively. Top row: electrons (green) tracks only. Bottom row: electron and photon (red) tracks. The figure was generated using EGS-ray.

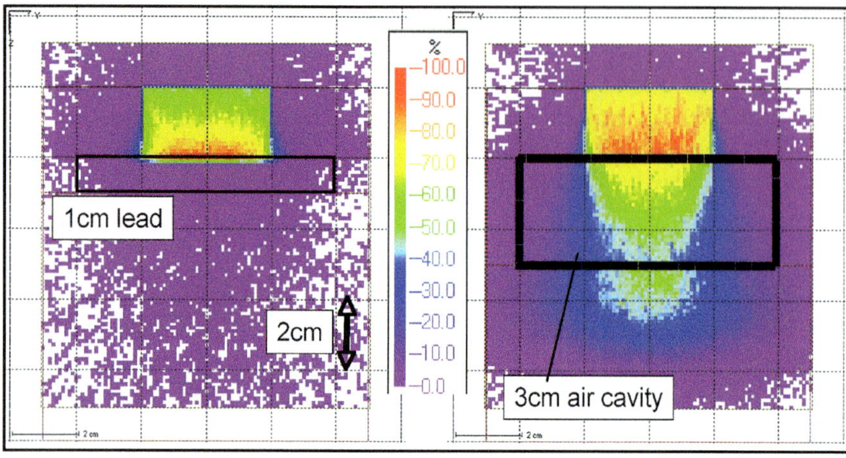

Figure 5.33: Monte Carlo calculated dose distribution for a 4 x 4 cm^2 beam of 10 MeV electrons: (a) a 1 cm thick lead sheet at 2 cm depth in water; (b) a 3 cm air cavity at 2 cm depth in water. Both figures are normalized to 100% at the maximum dose level in each figure, respectively. The figure was generated using EGS-ray.

air, lateral electron equilibrium cannot exist and the dose at central axis will fall off more rapidly with depth.

The lack of lateral electron equilibrium has consequences in clinically relevant scenarios. This is illustrated in figure 5.35, which shows the dose

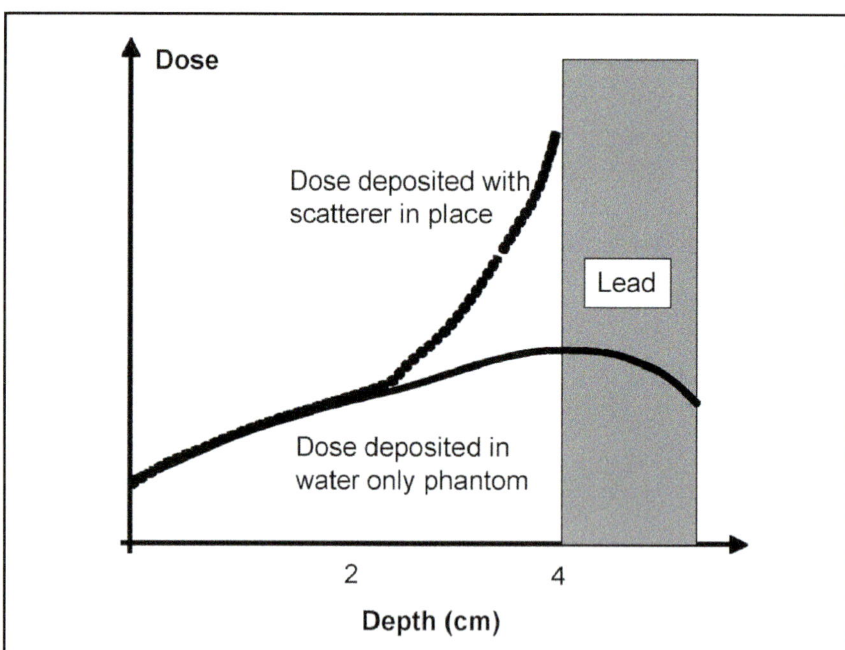

Figure 5.34: Schematic illustration of the dose enhancement in electron beams in front of a high-density material due to backscatter. (Courtesy of P. M. Ostwald.)

distribution of a 10 MeV electron beam of 4 × 4 cm² field size in water and lung. In the case of lung there is 1 cm of water overlaying the lung material. Both pictures are normalized to 100% for their respective dose maximum. The larger volume of high dose for the lung scenario should not be surprising as the dose (energy per unit mass) is spread wider in low-density material. The lateral electron disequilibrium results in two clinically important consequences:

i) The dose distribution spreads out wider at depth. This would be significant in the case of electrons reaching the lung (e.g., due to chest wall irradiation) (Prasad et al. 1983). If accurate treatment planning is not available, a set of measured inhomogeneity correction factors may be useful (Molden and el-Khatib 1989).
ii) The depth dose curve at central axis falls off faster than one would expect from a direct comparison of radiological depth.

As can be seen in figure 5.35, there is no significant variation in dose across the interface between water and lung 1 cm below the surface. This is reflected also in the data shown in figure 5.36. The figure depicts the backscatter from an inhomogeneity in an electron beam as a function of the inhomogeneity material and the effective electron energy at the interface. Lower-energy electrons and high-density materials result in larger back scatter factors (BSFs). The BSF is the ratio of dose at the interface with the inhomogeneity in place to the one in the same location without the inhomo-

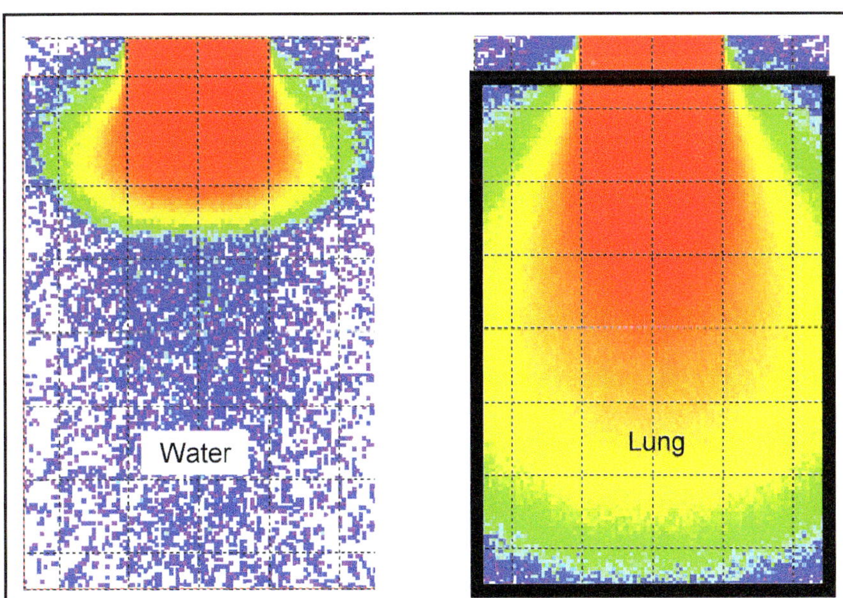

Figure 5.35: Comparison of dose distribution from a 10 MeV electron beam with a field size of 4 × 4 cm² in water and lung. The lung density was chosen to be 1/3 of water. Both figures are normalized to 100% at the maxmum dose level in each figure, respectively. The figures were generated using EGS-ray.

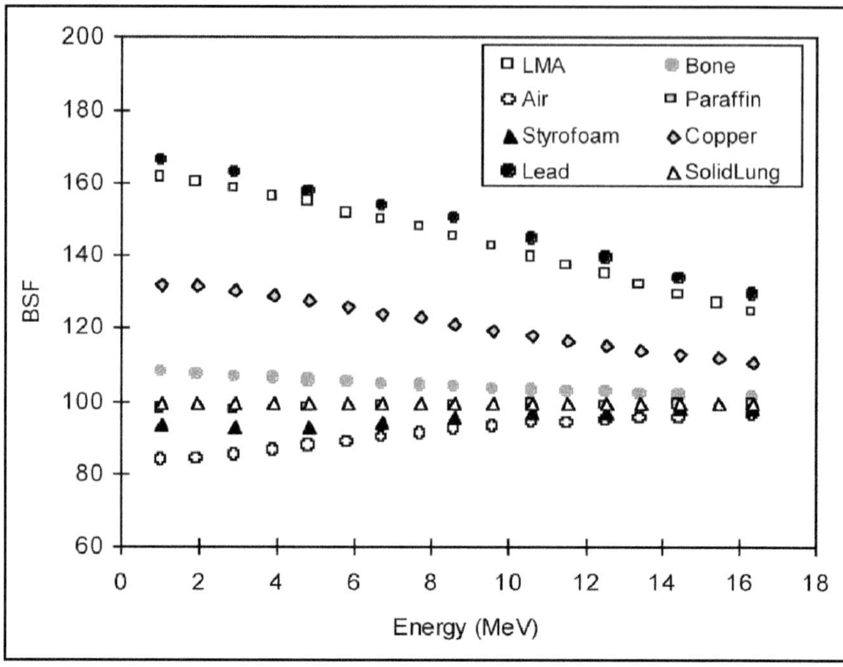

Figure 5.36: Variation of backscatter factor (BSF) as a function of material and electron energy at the interface. (Courtesy of P. M. Ostwald.)

geneity. Note that the fall-off of dose with distance from the inhomogeneity depends also on the electron energy. At low energies there are more electrons backscattered, but their remaining range is less than that of higher-energy electrons.

5.8 Typical Electron Beam Applications

Electron radiation beams have always been considered more complicated to use for radiotherapy than photons. This is in part due to the fact that there is less clinical experience with electrons because cobalt-60 units and early linacs did not produce clinically useful electron beams. Early experience with electrons in radiotherapy was gained mostly using betatrons (Turano et al. 1959). However, the complexities of the dose distribution in small fields or presence of inhomogeneities and oblique incidence also discussed above resulted in more careful application of electrons. The present section aims to illustrate some of the complexities of the use of electrons for radiotherapy in the context of a few clinical scenarios.

5.8.1 Contour Variations

The potential of unexpected hot and cold spots below sharp contour changes has already been discussed in the context of figure 5.26. Figure 5.37 shows another example of these problems. It shows an electron beam of 10 MeV energy incident into a corner. The left picture shows the dose distribution in a logarithmic scale illustrating the low-dose levels while the right picture displays it in a linear scale emphasizing variation in the high-dose region.

The dose distribution at depth is not just a reflection of the external contour. It features a significant hot spot under the corner due to the lack of attenuation in air and a contribution of electrons forward scattered from the walls. This is sometime referred to as a "funneling effect." Clinically, it may be significant when treating the perineum or some areas in the head and neck region. It may particularly be a concern when treating the ear (Morrison et al. 1995) and also may cause problems when irradiating breast and including skin folds. The use of bolus as shown in figure 5.26 can avoid or reduce the effect (Dubois et al. 1996).

The actual extent and location of the hot spot depends on electron energy and geometry. Pencil beam dose calculation algorithms (Hogstrom et al. 1981, Ostwald et al. 1994a,b) as discussed in chapters nine and ten may provide an indication of the problem but may fail to predict the magnitude of the hot spot. Prior to using a treatment technique clinically that may result in "electron funneling" it may be advisable to study the dose distribution in a phantom.

As discussed above, electron beams are particularly useful in circumstances where structures at the distal side of the target have to be spared. This may be the case in skin irradiation (Griep et al. 1995), the delivery of a boost to the primary site of breast cancer (Verney and Morgan 2001, Benda et al. 2003, Vass and Bairati 2005), or even if the spinal canal is to be treated in children (Hood et al. 2005). Unfortunately, the distal contour of the target

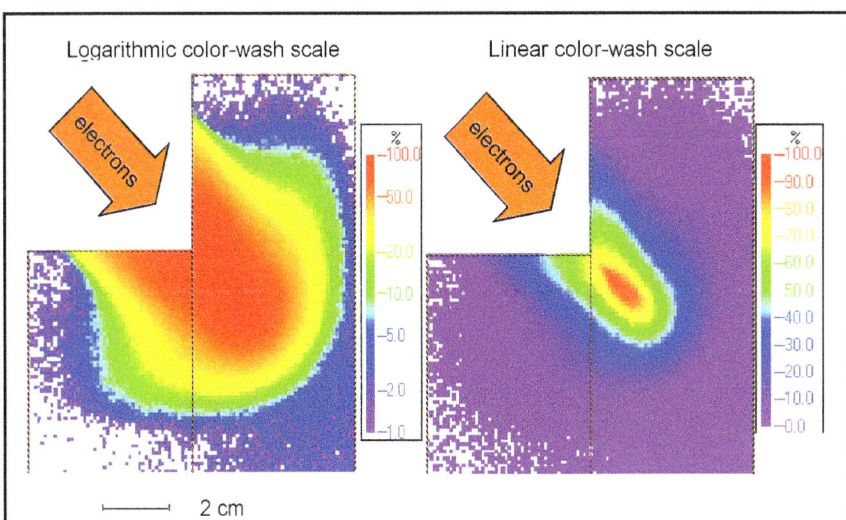

Figure 5.37: Monte Carlo calculated dose distributions for a 4 x 4 cm^2 beam of 10 MeV electrons incident on a corner of material. The pictures are normalized to 100% at the maximum dose level. The left picture shows the dose distribution in a logarithmic scale, illustrating the low-dose levels, while the right picture displays it in a linear scale, emphasizing variation in the high-dose region. The figures were generated using EGS-ray.

does not always conform to the skin. As such, different electron energies may be used and fields junctioned (see next section) or customized bolus employed. The group at M.D. Anderson Cancer Center has designed an elegant method to do this for a variety of applications ranging from breast (Perkins et al. 2001) to head and neck treatments (Kudchadker et al. 2002, 2003). The flexibility that can be achieved using customized bolus and judicious choice of electron beam energy is so great that the technique can benefit from inverse treatment planning and has been referred to as electron IMRT (e-IMRT) (Kudchadker et al. 2002, Gentry et al. 2006).

The importance of accounting for obliquity of the beam has been discussed in section 5.6. Clinically, this may be important in head and neck and breast cancer.

5.8.2 Junctioning of Electron Beams

When treating large areas it may be necessary to abut several electron beams (Zavgorodni et al. 1996, McKenzie 1998). Due to the variations of penumbra with depth this is often a complicated process and a match of the 50% dose lines of both fields at the surface may not be ideal. As can be seen in figure 5.38 there is a hot spot at depth between two 20 MeV electron beams matched at the 50% dose line at the surface. If the beams are matched at depth to avoid the hot spot, there is a potential for underdosage at the surface.

The situation gets even more complicated on curved surfaces. This is a more likely scenario for large field electron beam matching and the hot spot can be of the order of 50% as shown in figure 5.39a. A way to improve the situation with electron beam matching would be to employ a moving junction similar to photon beams. A moving junction refers to having the junction at

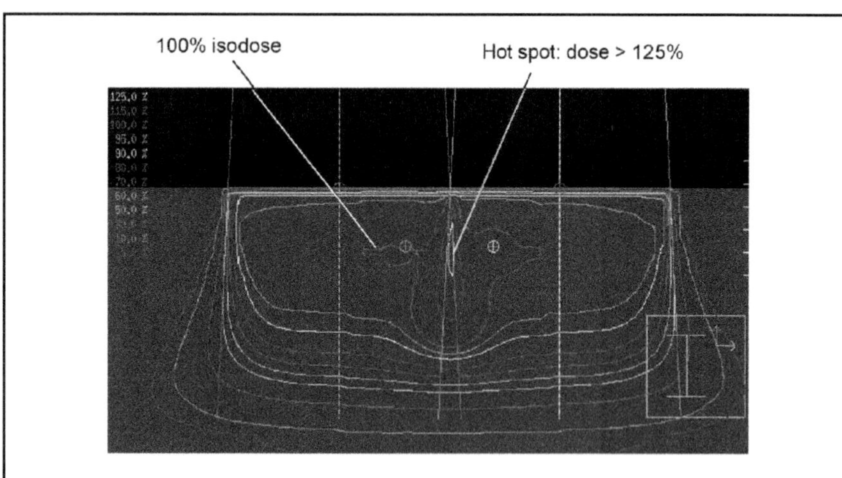

Figure 5.38: Isodose lines of two electron beams of 20 MeV incident on a flat water surface. The two beams were matched at the 50% dose position in the penumbra at the surface. The picture was created using ADAC Pinnacle planning system with 3-D pencil beam algorithm (compare chapter nine.)

two or more different locations on different treatment days. The distance between these positions is typically of the order of 1 cm, and a moving junction ensures that any dosimetric problems that may affect the junction are not occurring at the same location within the patient on all treatment days. In the case of electrons this would require the fabrication of two (or more) electron cut-outs per field. A less labor-intensive possibility is to employ a penumbra spoiler. This is a thin sheet of tissue-equivalent material placed on the electron cut-out. It will decrease the effective electron energy, increase the skin dose, and most importantly widen the penumbra of the electron beam. This allows easier matching and reduces hot (or cold) spots. The dose

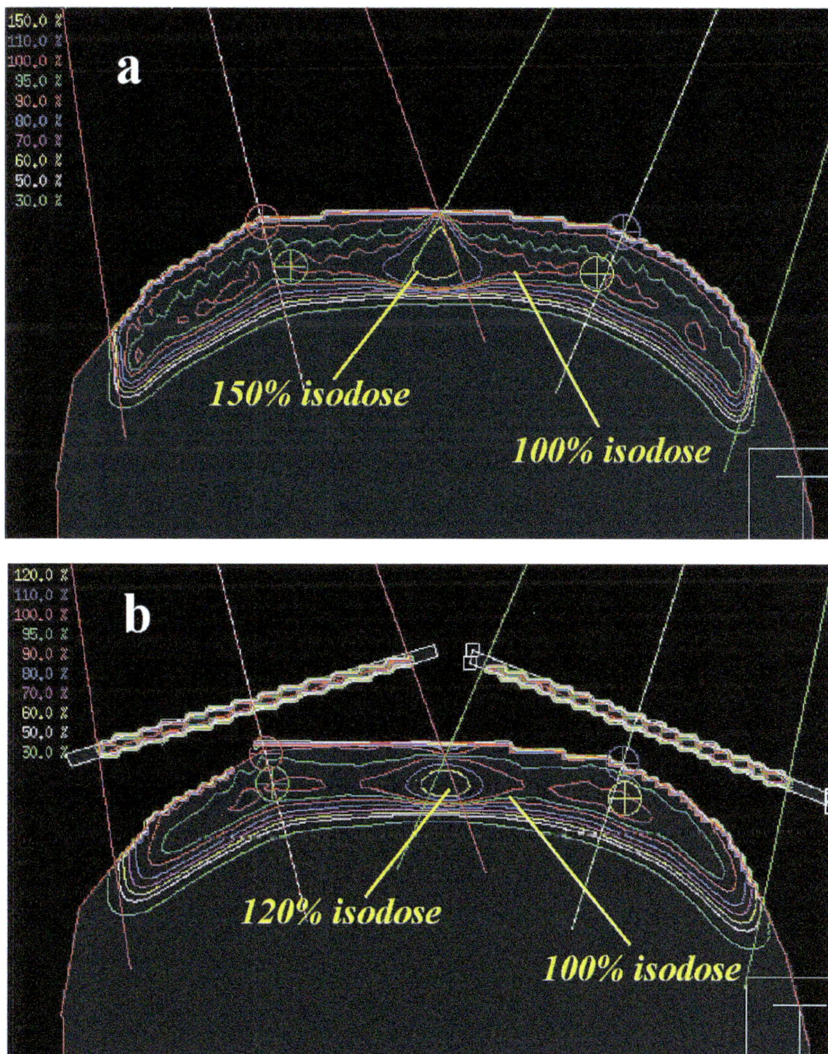

Figure 5.39: Isodose lines of two electron beams of 20 MeV incident on a curved water-equivalent surface. (a) The two beams were matched at the 50% dose position in the penumbra at the surface. (b) As in (a) but with two 5 mm thick water-equivalent "spoilers" added. The pictures were created using an ADAC Pinnacle planning system. (Courtesy of P. M. Ostwald.).

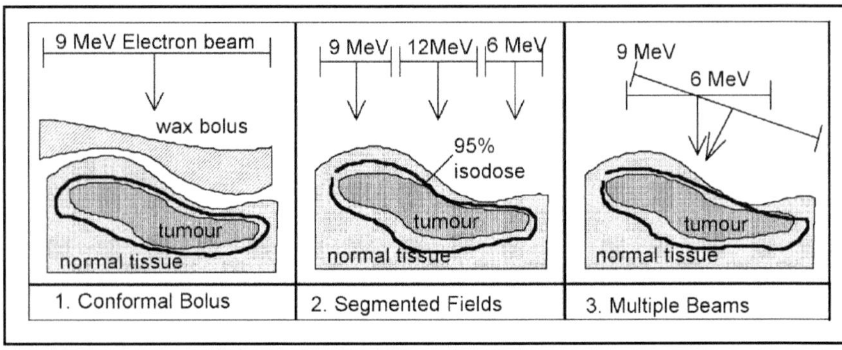

Figure 5.40: Three different methods to employ electrons from conformal radiotherapy. (Courtesy of P. M. Ostwald.)

distribution for the situation shown in figure 5.39a is shown for the use of penumbra spoilers in 5.39b. For similar dose coverage the hot spot between the two beams is reduced to 120%. Penumbra spoilers have also been used for electron photon matching (Papiez et al. 1992).

Provided bolus and junctioning of electron beams are done with care (and possibly experimental verification), very conformal dose distributions can be achieved. This is illustrated for a superficially located tumor in figure 5.40, which shows three different methods to provide good tumor coverage while maintaining the excellent sparing of deeper-lying normal tissues.

5.8.3 Electron Photon Combination

The problems encountered when matching electron beams also affect the combination of adjacent electron and photon beams (Sun et al. 1998). The difference in penumbra between the beams and the variation of penumbra of the electron beam with depth make a perfect abutment impossible. This may even be more complex in the case of photon fields delivered using IMRT (Dogan et al. 2002). However, in some clinical scenarios such as breast and head and neck, deeper-lying targets are adjacent to more superficial ones. In this case, it may be advantageous to combine electron and photon beams in a single treatment. An advantage is often that in clinical scenarios where photon beams are used, the skin dose usually does not need to be high. As such, matching at depth is a more successful strategy.

In any case where an electron field will be matched either with other electron beams or photon field, it is worthwhile to investigate if better matching and simpler patient setup can be achieved using the same central axis for the abutting fields (Hawkins et al. 1997, Yorke et al. 1998).

Electron and photon beams can also be used together in the same field (Butson et al. 1994, Vass and Bairati 2005). This electron/photon mix maintains some of the favorable characteristics of both radiation modalities: the skin sparing of photon beams and the finite penetration of electrons. Figure 5.41 illustrates the concept for a 4 MV photon beam and a 12 MeV electron beam, both of a field size of 10×10 cm^2. Different weightings for the mix are possible and the choice of different photon and electron energies allows tailoring the dose distribution to the clinical problem.

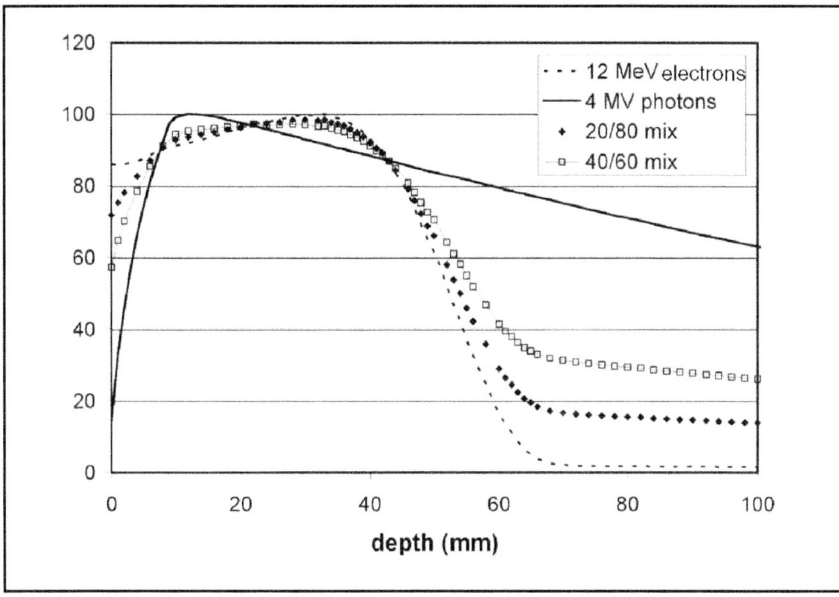

Figure 5.41: Percentage depth dose curves for different mixes of 12 MeV electrons and 4 MV photons.

5.8.4 Electron Arcs

When treating large, curved surfaces such as limbs, chest wall, or ribs, electron arcs may be an option to overcome the problems faced when junctioning several electron fields (Khan et al. 1977, Lam et al. 1987, et al. Tobler 1995). Most clinical experience has been gained for radiotherapy of breast cancer patients after mastectomy (Gaffney et al. 2001). For example, for chest wall volumes that extend posteriorly beyond the mid-axillary line, the conventional tangential field techniques irradiate too much lung, and electron arc therapy may be well suited. The electron arc treatment is typically performed without a conventional electron applicator and the isocenter is placed several centimeters within the patient. It should be well beyond the practical range of the electron beam used. Collimation is achieved by the photon jaws and a secondary collimator at the height of the block tray. However, MLCs have also been found useful for this (Klein et al. 1996). The typical (geometric) field width at isocenter is 4 to 8 cm and the length is variable. The actual field width is larger due to the distance between the applicator and the skin (compare figure 5.21) and the penumbra is relatively wide. Therefore, it is common to delineate the desired field borders by lead placed on the skin. This is shown in figure 5.42, which shows the dose distribution in an electron arc for the same surface shown in figure 5.39. The dose distribution is clearly superior in the arc when compared with field junctions.

Electron arcs need to be specially commissioned and care must be taken to determine the photon contamination (Kase and Bjärngard 1979, Pla et al. 1989). While no electrons reach the isocenter, the photon contamination is focused on this point. Therefore, it is also important during planning to use an electron algorithm, which estimates well the photon

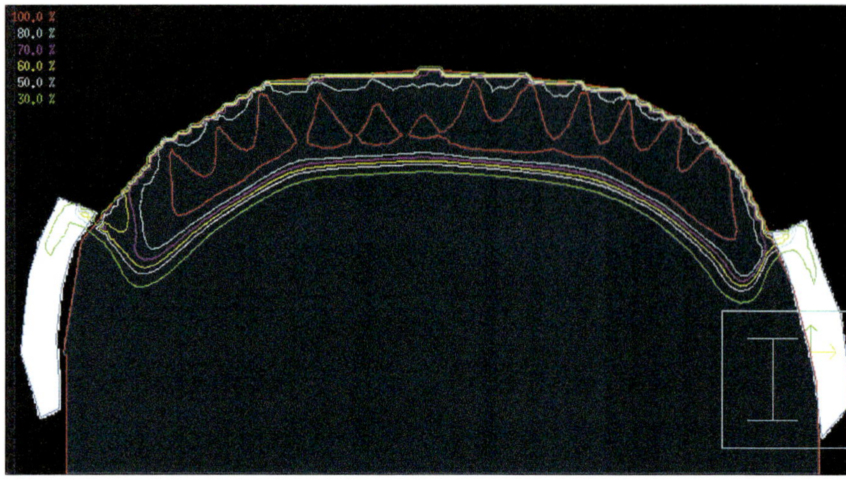

Figure 5.42: Isodose lines of a 20 MeV electron arc to treat the same curved surface as shown in figure 5.39. The dose distribution is clearly superior. Sharp field edges are produced by lead strips on the patient's skin (shown in white). The picture was generated using ADAC Pinnacle planning system with 3-D pencil beam algorithm.

contamination, as this will be more significant in arc therapy than for a single electron field.

5.8.5 Total Body Skin Irradiation (TBSI)

A special case of rotational electron therapy is total body skin irradiation (TBSI). In this treatment technique, which is used to treat a skin condition called mycosis fungoides (Maingon et al. 2000), the gantry does not rotate around the patient but the patient rotates in the electron beam. Detailed description of TBSI is beyond the scope of the present text. The reader is refereed to AAPM Report 23 (AAPM 1987) for more details. TBSI typically employs a low-energy electron beam at a large distance. Figure 5.43 shows the setup schematically. The patient stands on a small pedestal, which also could be a rotating platform. He/she is typically standing with one leg in front of the other, alternating one arm up and one arm down. The patient is supported by a frame (not shown) as the treatment can take more than half an hour at an extended distance of 3 or more meters. However, if the patient cannot stand for the required period of time a treatment technique with a reclined patient may be considered (Gerbi et al. 1989).

In order to reduce the treatment time, several manufacturers now offer high-dose rate electron beam options (typically only for the lowest energy beams). In this mode the linac delivers 1000 or more monitor units per minute. Great care must be taken to commission such a beam as the high-dose rate can lead to recombination losses in the monitor chambers (compare chapter three).

The treatment is performed without an applicator and the photon jaws set to maximum size. A steel frame applicator with large opening, such as the one shown in figure 5.16b, may be used to provide the coding for the

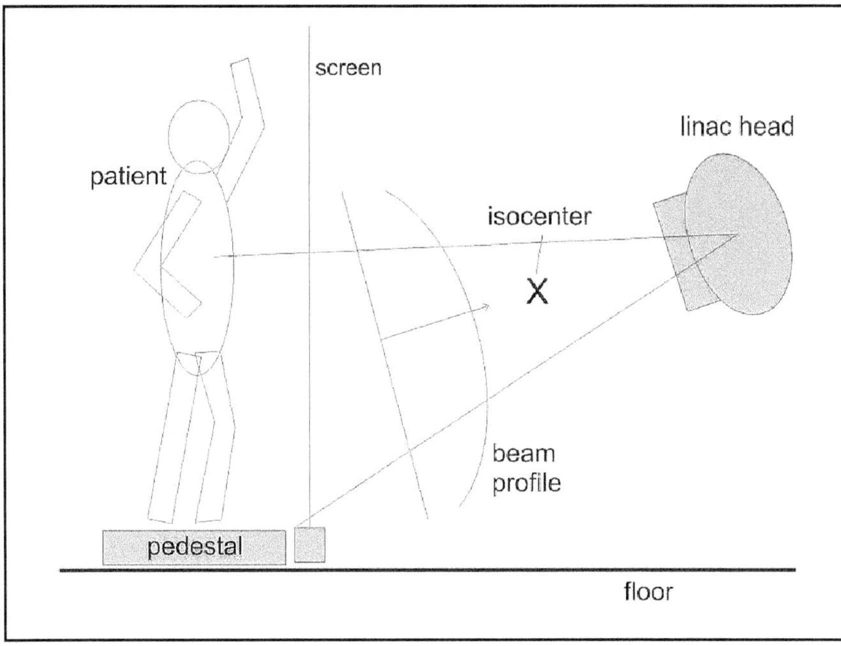

Figure 5.43: Schematic setup of a total body skin irradiation (TBS). Shown is only one gantry angle of two.

computerized linac operation. However, even at the largest distance from the isocenter possible in typical treatment rooms, the field size is not large enough to treat a standing patient in a single field. Therefore usually two gantry angles are used and the fields matched at the 50% dose line at the treatment distance (Chen et al. 2004). Figure 5.43 shows the gantry pointing down. A similar field is treated with the gantry pointing up. This setup is maintained for a rotational treatment or the use of several static fields. In both cases, usually a thin Perspex™ screen is placed some 20 cm from the patient. This screen, sometimes referred to as a decelerant, allows energy moderation of the electron beam and will provide for better electron coverage at all surfaces.

If static field directions are chosen, the patient must be rotated between fields. Figure 5.44 shows a six-direction arrangement. It is common that in this case not all fields are treated every day; for example, fields 1, 3, and 5 can be treated on one day and the remaining fields on the next. From the figure it can be appreciated that oblique beam incidence must be considered when determining the dosimetry. Other areas of consideration are the dose to skin, which is "shielded" by the patient him/herself. Even if the leg and arm positions are varied between fields (e.g., left/right foot forward, arm up/down), there will be skin between the legs and under the arms that may be underdosed. In vivo dosimetry can identify these areas and an electron boost may be given. Similarly, the skin on top of the head is typically not covered by the irradiation technique described above and requires boosting. On the other hand, it is appropriate to shield eyes and finger and toe nails to reduce likelihood of cataract formation and nail loss.

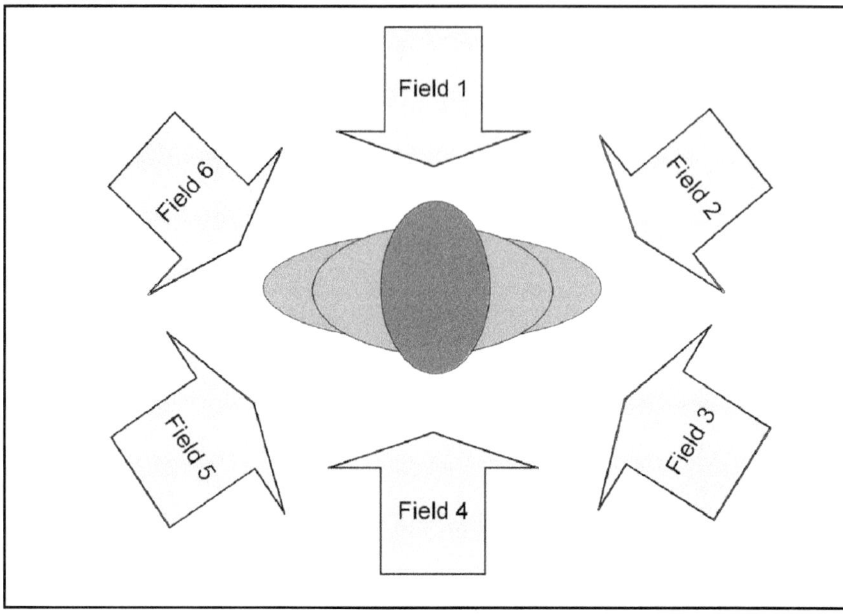

Figure 5.44: Six-field "Stanford" technique for TBSI. The six fields are given once with the gantry angled down as shown in figure 5.43 and once with the gantry up..

Given all the complexities of TBSI it is essential that the setup be carefully commissioned. It is also advisable to perform in vivo dosimetry at least for some of the treatment fractions. TL dosimeters are well suited for this (Weaver et al. 1995).

5.8.6 Inhomogeneities

Many clinical issues arising from inhomogeneities in electron radiotherapy have been discussed in section 5.7. The scenario depicted in figure 5.34 is not uncommon in clinical practice. When treating superficial lesions in the head and neck region it is common to employ internal shielding in the oral cavity, the nostrils, or the ear canal. In the latter this also avoids the problems of "electron funneling" discussed above under contour variations. However, care must be taken to reduce the impact of backscatter from the shield on the overlaying skin or mucosa (Klevenhagen et al. 1982, Perez-Caladayut et al. 2000). Therefore, it is common to use a coating of wax or aluminum (Weidlich et al. 1996) over the shield to absorb the backscattered low-energy electrons.

Figure 5.45 shows the dose distribution for a small air cavity in a 20 MeV electron beam. The considerable perturbation caused by the cavity can be clearly seen. This may be of concern when using an *en face* medium-energy electron beam for treatment of a laryngeal cancer (Yorke et al. 2001).

The dose to lung may be an issue in breast cancer radiotherapy with electrons. In principle, the use of electrons avoids or reduces dose to structures below the target. This has been demonstrated to work well in patients

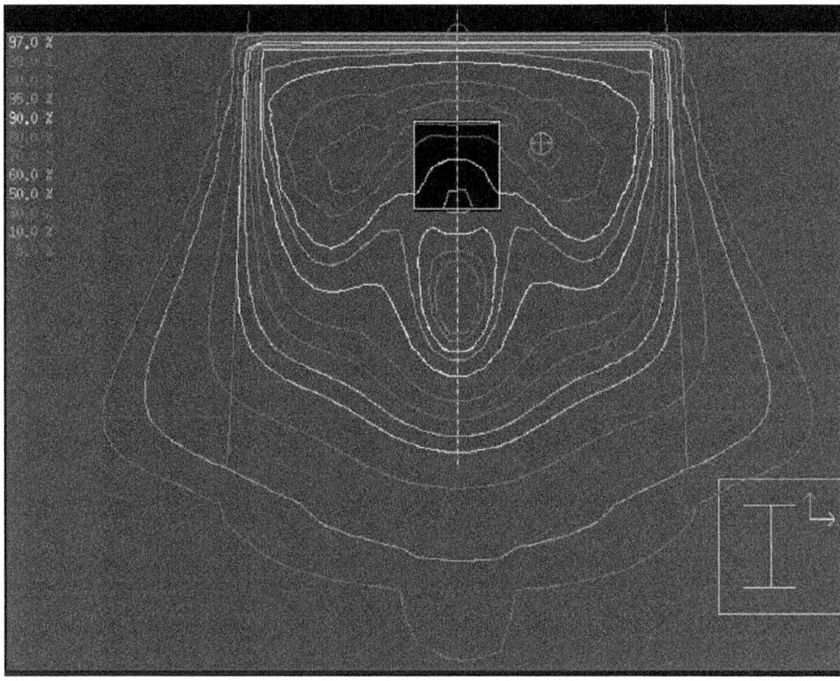

Figure 5.45: Isodose lines of an electron beam of 20 MeV incident on flat phantom including a 2 × 2 cm^2 air cavity (indefinite length in third dimension). The pictures were created using ADAC Pinnacle planning system with 3-D pencil beam algorithm. (Couresy of P. M. Ostwald.)

after mastectomy (min-Zimmerman et al. 2005) and has recently been included in planning of partial breast irradiation (Taghian et al. 2006). However, the dose to lung may not be accurately predicted using pencil beam algorithms (Seuntjens et al. 1994) and careful selection of beam energy and bolus is essential.

5.9 A Few Comments on Protons

Similar to electrons protons are charged particles. As such, their interaction with matter is similar: energy is deposited in many small interactions along the track of the particle. Towards the end of the track the proton has lost most of its energy and has slowed down. Therefore its interaction probability increases and more dose is deposited. This increase of dose towards the end of the track is called the *Bragg peak*, named after Sir William Henry Bragg who discovered it in 1903. The difference between protons and electrons (in addition to their opposite charge) is that a proton is approximately 2000 times heavier than an electron. Therefore, the angular straggling is smaller and the Bragg peak appears at the end of a relatively straight track. As such, the overall depth dose curve displays a distinct dose increase at depth as shown in figure 5.46. The figure shows the percentage depth dose curve of a 6 MeV electron beam compared to the one of a 70 MeV proton

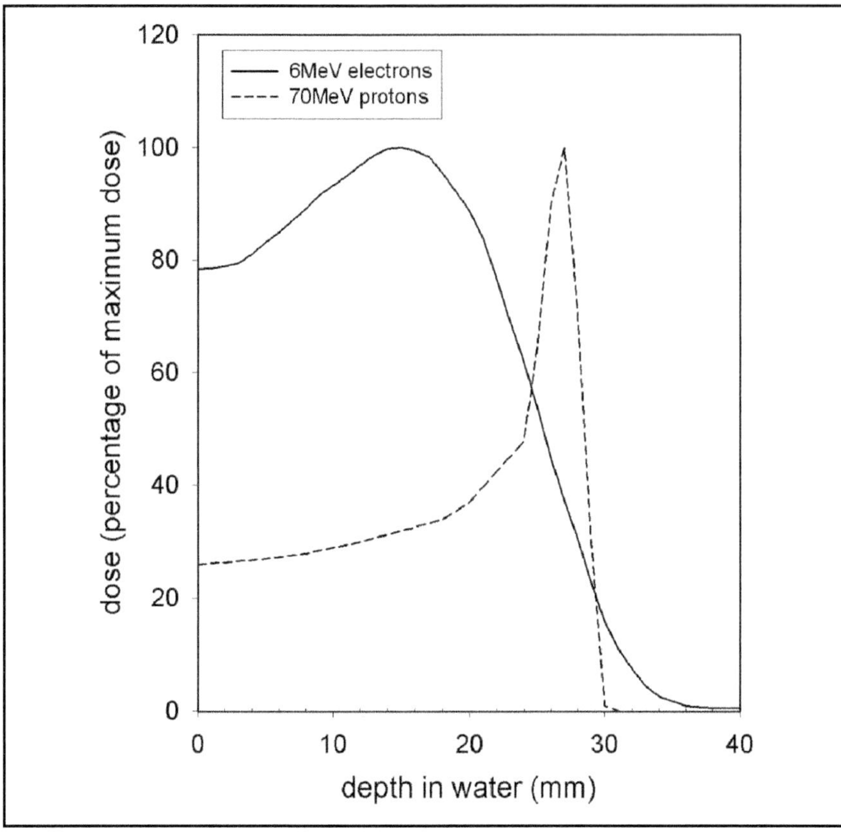

Figure 5.46: Comparison of the percentage depth dose curve for a 6 MeV electron beam and a 70 MeV proton beam.

beam. It becomes immediately obvious that a dose distribution such as the one afforded by the proton beam will be useful for radiotherapy of deeper-lying structures. Good reviews of the physics of proton radiotherapy are given by Miller (1995), Moyers (1999), and Smith (2006). The recent interest in the topic of protons for radiotherapy can be seen from a number of recent point/counterpoint discussions on proton issues published in *Medical Physics*, the journal of the American Association of Physicists in Medicine (AAPM) (Moyers et al. 2007, Schulz et al. 2007, Ma and Maughan 2006).

The penetration of the beam will depend on its energy. Figure 5.47a shows the Bragg peak of an ideal 250 MeV proton beam. The beam that will reach the patient has undergone some scattering and attenuation in the beam delivery pathway due to vacuum windows, scattering foils, and transmission ionization chambers. Therefore, the realistic proton beam features a slightly smaller penetration and an increased energy spread, resulting in a broader peak. As can be seen from the figure, a proton beam of 250 MeV has a penetration of around 30 cm in water and, as such, would be suitable for most radiotherapy treatment scenarios involving deep-seated tumors.

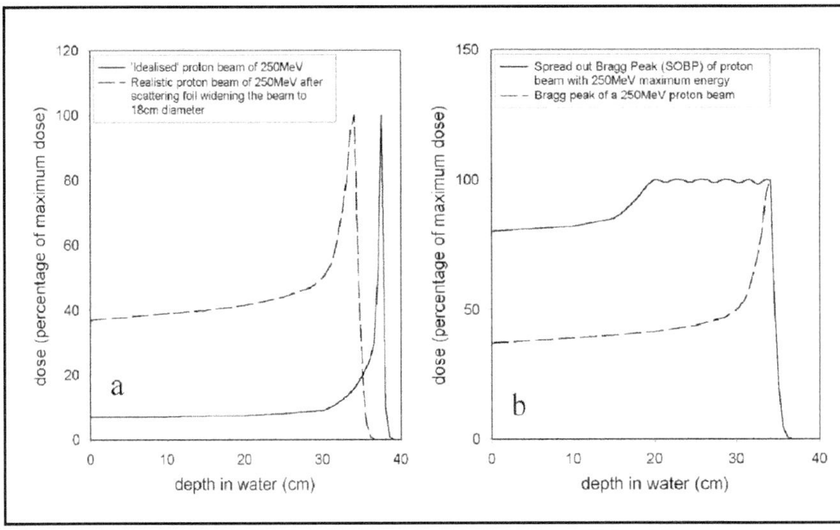

Figure 5.47: A "realistic" proton beam: (a) Beam degradation due to scattering foil, dose monitoring system, and collimation; (b) Spread out Bragg peak.

Electron tracks also feature a Bragg peak. However, as electrons straggle more, the Bragg peak of different electrons occurs at different depths in tissue; therefore the depth dose does not feature a dose enhancement at depth.

Beyond the Bragg peak there is no primary dose deposition in both electron and proton beams. This makes both of them well suited for the treatment of a target while sparing a critical structure directly behind the target. However, the low-dose region and large Bragg peak make protons more attractive. Protons also have a larger linear energy transfer (LET) than electrons. Therefore, a larger initial energy is required per particle to reach the same depth in tissue. However, the LET is still not large enough to make a significant difference in radiobiological effectiveness (RBE) in most of the proton track. Only towards the end of the range at the distal end of the Bragg peak, the protons have slowed so much that their LET increases and the RBE goes up. This must be taken into account when planning dose delivery with protons to a target close to a critical structure (Wilkens and Oelfke 2005).

5.9.1 Proton Beam Production

Protons for radiotherapy are generated either by synchrotrons or cyclotrons, both being ring accelerators with a diameter of several meters. Cyclotrons accelerate the particles on a spiral path in a constant magnetic field to a fixed energy, while synchrotrons accelerate a pulsed beam on a constant orbit of a pulsed magnetic ring structure and deliver an ion beam, the energy of which can be varied in less than a second without additional losses. Most proton treatment facilities at present utilize cyclotrons, which are somewhat easier to control and produce higher beam currents. Synchrotrons require more sophisticated beam steering but may be more flexible, e.g., for fast changes of energy.

A proton beam from an accelerator has a diameter of only several millimeters and the Bragg peak has a FWHM of around 1 cm. Therefore, in most clinical scenarios it will be necessary to spread out the proton beam in three dimensions (including the spreading out of the Bragg peak) to treat a typical target. This can be done in active and passive modes.

Lateral Beam Spreading

The proton beam can be spread out using a scattering foil similar to an electron beam. This passive scattering is typically achieved using one or more high atomic number foils that widen the beam. In order to achieve beam uniformity across the useful field of better than 5%, only the central part of the beam can be used and dual scattering foils of complex design are needed (Grusell et al. 1994).

A different method to spread the dose distribution laterally is to scan the original pencil beam in two dimensions across the target (dynamic/active scanning) (Pedroni et al. 1999, Lomax et al. 2004). This method makes more effective use of the accelerator beam current but is more difficult to control. There are also safety and dosimetric considerations for a scanning beam, as the instantaneous dose rate of the pencil beam is very high.

Longitudinal Beam Spreading

The spreading of the Bragg peak in beam direction can be achieved passively by using a rotating filter. The filter consists of material of different thicknesses. Each filter part will reduce the proton energy relative to its

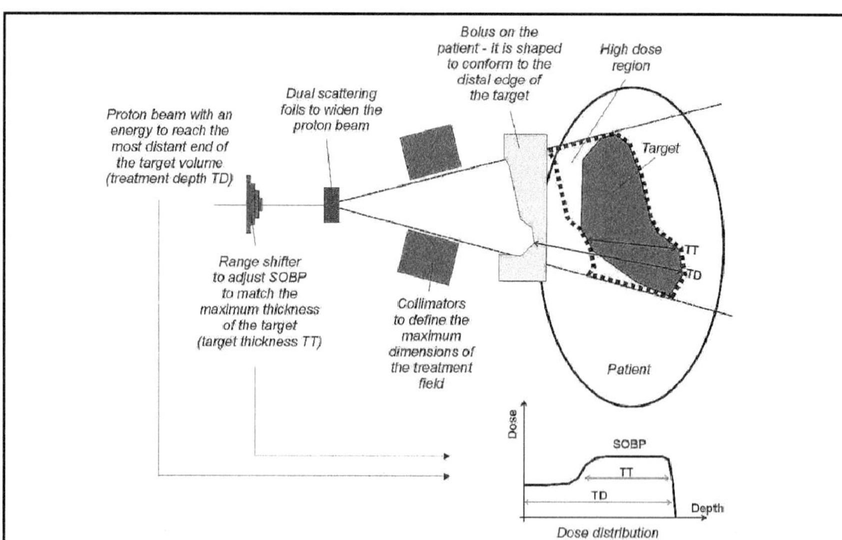

Figure 5.48: Schematic illustration of passive proton beam generation.

thickness, thereby altering the depth of the Bragg peak in the patient. If several different material thicknesses are brought into the beam by rotating the filter, the peak is spread out as shown in figure 5.47b. This is called the *Spread Out Bragg Peak* (*SOBP*).

The depth of the Bragg peak can also be changed "actively" by altering the energy of the proton beam in the accelerator. In a modern synchrotron this can be achieved in less than a second. Again, the active method preserves the advantageous characteristics of the beam better than the passive method, as no additional attenuation and scattering is required before the beam enters the patient. However, the active method requires a more sophisticated accelerator and is more difficult to control.

Figure 5.48 illustrates a proton beam delivery using a passive system. This mode of beam delivery is typically combined with a compensator. The compensator allows for differences in tissue overlaying the distal edge of the tumor and modifies the depth of proton penetration at different locations in the patient to conform to the distal end of the target. As can be seen in the figure, this process leads to conformity of the high-dose volume to the distal edge of the tumor. However, it also overdoses normal tissues in front of thin parts of the target volume. The passive beam delivery using a spread out Bragg peak can be improved by delivering the radiation dose to the target via multiple beams.

Figure 5.49 shows the active beam delivery using energy modulation and dynamic beam scanning. This technique requires advanced accelerator technique and excellent beam delivery verification. Therefore, the passive beam delivery is typically cheaper and at present more widely used. There are also a variety of "hybrid" systems available where the beam is only scanned magnetically in one dimension and the other dimension is covered by moving the couch with the patient through the beam.

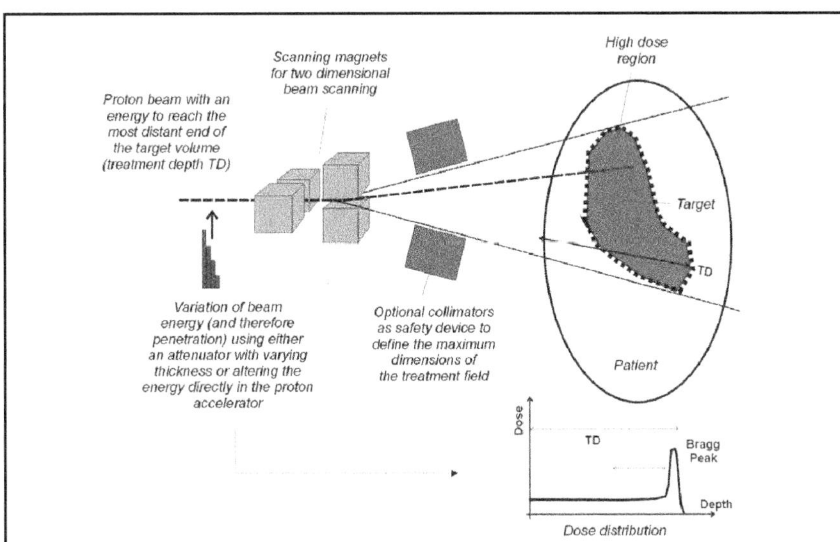

Figure 5.49: Schematic illustration of active proton beam generation.

In summary, a variety of beam delivery modes are possible. However, it is fair to say that active spot scanning techniques make better use of the advantageous physical characteristics of protons for radiotherapy. There has also been some recent discussion about the radiobiological consequences of neutron production in proton treatment facilities (Hall 2006). While the actual neutron levels in proton radiotherapy facilities are still debated, the active delivery method reduces neutron production compared to passive systems and as such optimizes the dose to normal tissues more effectively.

5.9.2 Proton Beam Delivery

In order to make best use of an expensive accelerator, most clinical proton treatment facilities direct the proton beam into multiple treatment rooms. This is illustrated for a fictitious proton treatment facility in figure 5.50. After acceleration (and energy definition) the beam is transported in an evacuated beam line to one or more treatment rooms with gantries or fixed beam lines. Magnetic dipoles and quadrupoles are used to deflect and to focus the proton beam. The beam-line is designed to deliver the same proton beam parameters at the entrance of each treatment room, with the ability to switch the beam between treatment rooms in less than 1 min. A fast switching magnet allows, in case of a safety problem, to direct the proton beam into a beam dump in less than 100 μs.

There are several options for proton beam delivery to the patient. The cheapest is a stationary port that would, for example, be perfectly suitable for eye treatment using a relatively low-energy proton beam. Also, for many other targets in the human body one lateral port possibly combined with a vertical beam would be adequate to deliver three radiation fields. Due to the favorable internal dose distribution of protons this may be sufficient for some treatments. However in other cases, horizontal and vertical beams may not be sufficient to optimize the delivery of radiation. In this case, a rotating

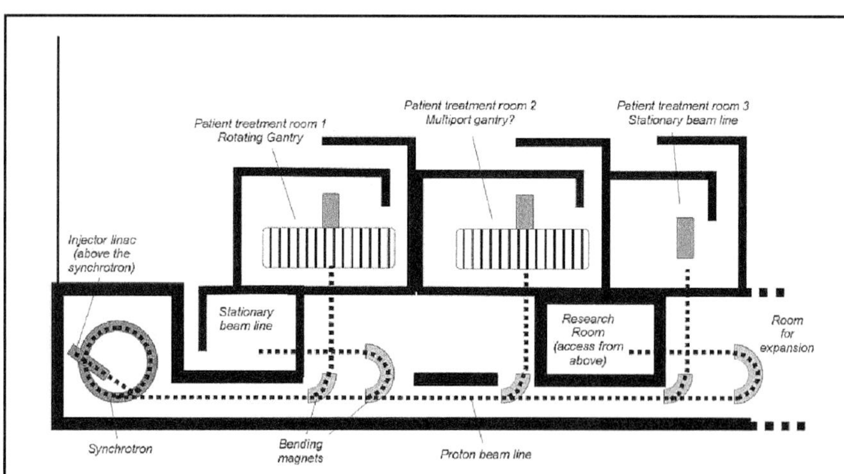

Figure 5.50: Schematic floorplan of a proton treatment facility.

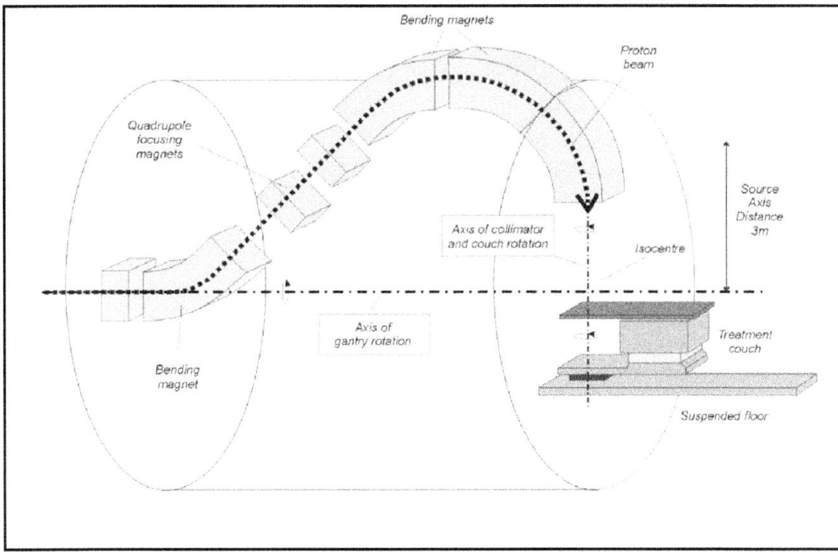

Figure 5.51: Schematic drawing of a rotating gantry for proton radiotherapy.

proton gantry would be ideal as it matches the clinical experience from isocentric treatment with x-rays from linacs.

However, proton gantries are large, drum-shaped structures. Due to the higher energy and particle weight, it requires significantly more force to bend a proton beam than the electron beam used in a linac. Therefore, the rotating gantry employs typically only a 90° bending magnet, and even then the bending radius is of the order of 3 m. This is illustrated in figure 5.51. However, if constructed well, the isocentricity of proton gantries can match the one from conventional linacs despite their different size (Moyers and Lesyna 2004).

5.9.3 Clinical Experience

Protons have been used for cancer treatment since 1954. Much of the early work was done in the United States, Russia, and Japan. However, more recently other countries have also commenced proton treatment. Most facilities operating to date are not dedicated to radiotherapy. They often are an add-on to existing nuclear physics research installations and/or share facilities with other users. This results in a large variety of different beam qualities and physical characteristics of the facilities, which makes it difficult to compare treatment outcomes. It also means that in most institutions only a small number of patients were treated in often-suboptimal environments (and with suboptimal beam configurations). As such, it is not surprising that there are only a few well-conducted clinical trials that demonstrate indications for proton radiotherapy.

Over the last 50 years most patients who received proton radiotherapy were treated for ocular cancers where only a stationary beam line of relatively low-energy (70 MeV) is required (Bonnett et al. 1993, Weber et al. 2005a, Rethfeldt et al. 2006). Of significance is the Loma Linda University

Medical Center (California), where a hospital-based proton treatment facility opened in the early 1990s (Nelson et al. 2001, Slater 2006). Most of the patients treated there over the last 15 years were treated for prostate cancer. However, more hospital-based treatment facilities are being rapidly developed. Nearly 10 of the 30 proton treatment centers are in a clinical environment at present. The most important treatment indications are brain tumors particularly of the skull base (Huq and Slater 1999, 2000, Weber 2005b) and pediatric cancers (Archimbeau et al. 1992, McAllister et al. 1997, Wilson et al. 2005, Timmermann et al. 2007) where the significantly reduced integral dose in proton beams is most relevant (Mu et al. 2005).

It is interesting to note that the challenges for proton radiotherapy are very similar to other radiotherapy approaches. Intensity-modulated proton radiotherapy (IMPT) is introduced (Lomax et al. 2003, Oelfke and Bortfeld 2004) and most centers are capable of delivering image-guided radiation therapy (IGRT) (Schulte et al. 2005, Schulte and Li 2006).

Useful information and updates on proton treatment facilities can be found on the page of the Proton Therapy Co-Operative Group PTCOG (http://ptcog.web.psi.ch/).

References

AAPM (American Association of Physicists in Medicine) (1991) Clinical electron-beam dosimetry. Report of Task Group 25. *Med. Phys.* 18: 73–109. Also available as AAPM Report No. 32.

AAPM (American Association of Physicists in Medicine) (1987) Report No. 23. Total skin electron therapy: technique and dosimetry. Report of Task Group 30. American Institute of Physics, New York.

AAPM (American Association of Physicists in Medicine) (1983) A protocol for the determination of absorbed dose from high energy photon and electron beams. Report of Task Group 21. *Med. Phys*. 10: 741–771.

Andreo P., Brahme, A. (1981) Mean energy in electron beams. Med. Phys. 8: 682-687.

Archambeau J.O., Slater J.D., Slater J.M., Tangeman R. (1992) Role for proton beam irradiation in treatment of pediatric CNS malignancies. *Int. J. Radiat. Oncol. Biol. Phys.* 22: 287–294.

Benda R.K., Yasuda G., Sethi A., Gabram S.G., Hinerman R.W., Mendenhall N.P. (2003) Breast boost: Are we missing the target? *Cancer* 97: 905–909.

Bonnett D.E., Kacperek A., Sheen M.A., Goodall R., Saxton T.E. (1993) The 62 MeV proton beam for the treatment of ocular melanoma at Clatterbridge. *Br. J. Radiol*. 66: 907–914.

Butson M.J., Wong T.P., Kron T., Mathur J., Clubb B., Metcalfe P.E. (1994) Surface doses from combined electron/photon fields in a radiotherapy. *Australas. Phys. Eng. Sci. Med.* 17: 14–22.

Chaney E., Cullip T., Gabriel T. (1994) A Monte Carlo study of accelerator head scatter. *Med. Phys*. 21: 1383–1390.

Chen Z., Agostinelli A.G., Wilson L.D., Nath R. (2004) Matching the dosimetry characteristics of a dual-field Stanford technique to a customized single-field Stanford technique for total skin electron therapy. *Int. J. Radiat. Oncol. Biol. Phys*. 59: 872–885.

Choi D.R., Mobit P.N., Breitman K.E. (2003) The clinical implementation of a method for calculating the output factor and per cent depth dose for an electron beam. *Phys. Med. Biol.* 48: 899–908.

Chow J.C., Grigorov G.N. (2007) Electron radiotherapy: A study on dosimetric uncertainty using small cutouts. *Phys. Med. Biol*. 52: N1–11.

Ding G., Rogers D.W.O., Mackie T., (1996) Mean energy, energy-range relationships and depth scaling factors for clinical electron beams. *Med. Phys*. 23: 361–376.

Dogan N., Leybovich L.B., Sethi A., Emami B. (2002) Improvement of dose distributions in abutment regions of intensity modulated radiation therapy and electron fields. *Med. Phys.* 29: 38–44.

Dubois D., Bice W., Bradford B., Schneid T., Engelmeier R. (1996) Moldable tissue equivalent bolus for high-energy photon and electron therapy. *Med. Phys.* 23: 1547–1549.

el-Khatib E.E., Scrimger J., Murray B. (1991) Reduction of the bremsstrahlung component of clinical electron beams: Implications for electron arc therapy and total skin electron irradiation. *Phys. Med. Biol.* 36: 111–118.

Ertan E., Muller-Sievers K., Riehl G. (1984) A new approach to overcome the inconveniences in electron dosimetry associated with the beam scanning technique in linacs. *Phys. Med. Biol.*29: 789–796.

Faddegon B.A., Blevis I. (2000) Electron spectra derived from depth dose distributions. *Med. Phys.* 27: 514–526.

Followill D., Davis D., Ibbott G. (2004) Comparison of electron beam characteristics from multiple accelerators. *Int. J. Radiat. Oncol. Biol. Phys*. 59: 905–910.

Gaffney D.K., Leavitt D.D., Tsodikov A., Smith L., Watson G., Patton G., Gibbs F.A., Stewart J.R. (2001) Electron arc irradiation of the postmastectomy chest wall with CT treatment planning: 20-year experience. *Int. J. Radiat. Oncol. Biol. Phys.* 51: 994–1001.

Gentry J.R., Steeves R., Paliwal B.A. (2006) Inverse planning of energy-modulated electron beams in radiotherapy. *Med. Dosim.* 31: 259–268.

Gerbi B.J., Khan F.M., Deibel F.C., Kim T.H. (1989) Total skin electron arc irradiation using a reclined patient position. *Int. J. Radiat. Oncol. Biol. Phys.* 17: 397–404.

Griep C., Davelaar J., Scholten A.N., Chin A., Leer J.W. (1995) Electron beam therapy is not inferior to superficial x-ray therapy in the treatment of skin carcinoma. *Int. J. Radiat. Oncol. Biol. Phys.* 32: 1347–1350.

Grusell E., Montelius A., Brahme A., Rikner G., Russell K. (1994) A general solu tion to charged particle beam flattening using an optimized dual-scattering-foil technique, with application to proton therapy beams. *Phys. Med. Biol.* 39: 2201–2216.

Hall E.J. (2006) Intensity-modulated radiation therapy, protons, and the risk of second cancers. *Int. J. Radiat. Oncol. Biol. Phys.* 65: 1–7.

Harder D. (1966) Physikalische Grundlagen der Dosimetrie. *Strahlentherapie Sonderband* 62: 254–274.

Hawkins R.B., Pullen B.P., Wilenzick R.M. (1997) Treatment of adjoining field areas with electron beams of different energy using a single central axis. *Med. Dosim*. 22: 87–90.

Hogstrom K.R., Almond P.R. (2006) Review of electron beam therapy physics. *Phys. Med. Biol.* 51: R455–R489.

Hogstrom K.R., Mills M.D., Almond P.R. (1981) Electron beam dose calculations. *Phys. Med. Biol.* 26: 445–459.

Hood C., Kron T., Hamilton C., Callan S., Howlett S., Alvaro F., Back M. (2005) Correlation of 3D-planned and measured dosimetry of photon and electron craniospinal radiation in a pediatric anthropomorphic phantom. *Radiother. Oncol.* 77: 111–116.

Hounsell A. (1998) Monitor chamber backscatter for intensity modulated radiation therapy using multileaf collimators. *Phys. Med. Biol.* 43: 445–454.

Huq, E.B., Slater J.D. (1999) Proton radiation therapy for pediatric malignancies: Status report. *Strahlenther. Onkol.* 175 Suppl 2: 89–91.

Huq E.B., Slater J.D. (2000) Proton radiation therapy for chordomas and chondrosarcomas of the skull base. *Neurosurg. Clin. N. Am*. 11: 627–638.

IAEA (International Atomic Energy Agency) (2000) Absorbed dose determination in external beam radiotherapy. IAEA Technical Report Series N398. IAEA, Vienna, Austria.

IAEA (International Atomic Energy Agency) (1997) The use of plane parallel ionization chambers in high energy electron and photon beams. IAEA Technical Report Series N381. IAEA, Vienna, Austria.

IAEA (International Atomic Energy Agency) (1987) Absorbed dose determination in photon and electron beams. IAEA Technical Report Series N277. IAEA, Vienna, Austria.

IEC (International Electrotechnical Commission) (1989) Medical electron accelerators - functional performance characteristics. International Standard 976. Bureau Central de la Commission Electrotechnique Internationale, Geneva, Switzerland.

Karlsson M., Nystrom H., Svensson H. (1992) Electron beam characteristics of the 50-MeV racetrack microtron. *Med. Phys.* 19: 307–315.

Kase K.R., Bjärngard B.E. (1979) Bremsstrahlung dose to patients in rotational electron therapy. *Radiology* 133: 531–532.

Khan F. (1994) *The Physics of Radiation Therapy*. Williams and Wilkins, Baltimore, Maryland.

Khan F.M., Fullerton G.D., Lee J.M., Moore V.C., Levitt S.H. (1977) Physical aspects of electron-beam arc therapy. *Radiology* 124: 497–500.

Klein E.E., Li Z., Low D.A. (1996) Feasibility study of multileaf collimated electrons with a scattering foil based accelerator. *Radiother. Oncol.* 41: 189–196.

Kleinschmidt C. (2001) [EGs-Ray, a program for visualization of Monte Carlo calculations in radiation physics]. *Z. Med. Phys.* 11: 119–123.

Klevenhagen S. (1988) Current status of electron beam therapy. *Br. J. Radiol.* Supplement 22.

Klevenhagen S. (1993) *Physics and Dosimetry of Therapy Electron Beams*. Medical Physics Publishing, Madison, Wisconsin.

Klevenhagen S.C., Lambert G.D., Arbabi A. (1982) Backscattering in electron beam therapy for energies between 3 and 35 MeV. *Phys. Med. Biol.* 27: 363–373.

Kron T. (2005) Calibration of megavoltage radiation beams. Chapter 9 in *The Modern Technology of Radiation Oncology*, Volume 2. J. Van Dyk (ed.). Medical Physics Publishing, Madison, Wisconsin, pp. 301–385.

Kudchadker R.J., Antolak J.A., Morrison W.H., Wong P.F., Hogstrom K.R. (2003) Utilization of custom electron bolus in head and neck radiotherapy. *J. Appl. Clin. Med. Phys.* 4: 321–333.

Kudchadker R.J., Hogstrom K.R., Garden A.S., McNeese M.D., Boyd R.A., Antolak J.A. (2002) Electron conformal radiotherapy using bolus and intensity modulation. *Int. J. Radiat. Oncol. Biol. Phys.* 53: 1023–1037.

Lam K.S., Lam W.C., O'Neill M.J., Lee D.J., Zinreich E. (1987) Electron arc therapy: Beam data requirements and treatment planning. *Clin. Radiol.* 38: 379–383.

Lax I. (1987) Accuracy in clinical electron beam dose planning using pencil beam algorithms. *Radiother. Oncol.* 10: 307–319.

Lomax A.J., Bohringer T., Bolsi A., Coray D., Emert F., Goitein G., Jermann M., Lin S., Pedroni E., Rutz H., Stadelmann O., Timmermann B., Verwey J., Weber D.C. (2004) Treatment planning and verification of proton therapy using spot scanning: Initial experiences. *Med. Phys.* 31: 3150–3157.

Lomax A.J., Goitein M., Adams J. (2003) Intensity modulation in radiotherapy: Photons versus protons in the paranasal sinus. *Radiother. Oncol.* 66: 11–18.

Ma C.-M., Maughan R.L. (2006) Within the next decade conventional cyclotrons for proton radiotherapy will become obsolete and replaced by far less expensive machines using compact laser systems for the acceleration of the protons. *Med. Phys.* 33: 571–573.

Maingon P., Truc G., Dalac S., Barillot I., Lambert D., Petrella T., Naudy S., Horiot J.C. (2000) Radiotherapy of advanced mycosis fungoides: indications and results of total skin electron beam and photon beam irradiation. *Radiother. Oncol.* 54: 73–78.

McAllister B., Archambeau J.O., Nguyen M.C., Slater J.D., Loredo L., Schulte R., Alvarez O., Bedros A A., Kaleita T., Moyers M., Miller D., Slater J.M. (1997) Proton therapy for pediatric cranial tumors: Preliminary report on treatment and disease-related morbidities. *Int. J. Radiat. Oncol. Biol. Phys.* 39: 455–460.

McKenzie A.L. (1998) A simple method for matching electron beams in radiotherapy. *Phys. Med. Biol.* 43: 3465–3478.

Miller D.W. (1995) A review of proton beam radiation therapy. *Med. Phys.* 22: 1943–1954.

min-Zimmerman F., Paris K., Minor G.I., Spanos W. (2005) Postmastectomy chest wall radiation with electron-beam therapy: Outcomes and complications at the University of Louisville. *Cancer J.* 11: 204–208.

Moldon C., el-Khatib E. (1989) Measured inhomogeneity correction factors for lung in electron beam treatments of the chest wall. *Med. Dosim.* 14: 89–98.

Morrison W.H., Wong P.F., Starkschall G., Garden A.S., Childress C., Hogstrom K.R., Peters L.J. (1995) Water bolus for electron irradiation of the ear canal. *Int. J. Radiat. Oncol. Biol. Phys.* 33: 479–483.

Moyers M. (1999) Proton radiotherapy. Chapter 20 in *The Modern Technology of Radiation Oncology.* J. Van Dyk (ed.). Medical Physics Publishing, Madison, Wisconsin, pp. 823–869.

Moyers M.F. Lesyna W. (2004) Isocenter characteristics of an external ring proton gantry. *Int. J. Radiat. Oncol. Biol. Phys.* 60: 1622–1630.

Moyers M.F., Pouliot J., Orton C.G. (2007) Point/Counterpoint. Proton therapy is the best radiation treatment modality for prostate cancer. *Med. Phys.* 34: 375–378.

Mu X., Bjork-Eriksson T., Nill S., Oelfke U., Johansson K.A., Gagliardi G., Johansson L., Karlsson M., Zackrisson D.B. (2005) Does electron and proton therapy reduce the risk of radiation induced cancer after spinal irradiation for childhood medulloblastoma? A comparative treatment planning study. *Acta Oncol.* 44: 554–562.

Nafstadius P., Brahme A., Nordell B. (1984) Computer assisted dosimetry of scanned electron and photon beams for radiation therapy. *Radiother. Oncol.* 2: 261–269.

Nahum A. The computation of dose distributions in electron beam radiotherapy. Thesis. Umea: Umea University, 1985.

Nelson G.A., Green L.M., Gridley D.S., Archambeau J.O., Slater J.M. (2001) Research activities at the Loma Linda University and Proton Treatment Facility--An overview. *Phys. Med.* 17 Suppl 1: 30–32.

Nygaard K., Odland O.H., Kvinnsland Y., Nygaard B., Heggdal J., Muren L.P. (2005) Measurements and treatment planning calculations of electron dose distributions below bolus edges. *Radiother. Oncol.* 74: 217–220.

Oelfke U., Bortfeld T. (2003) Optimization of physical dose distributions with hadron beams: Comparing photon IMRT with IMPT. *Technol. Cancer Res. Treat.* 2: 401–412.

Ostwald P.M., Kron T. (1996a) Surface dose measurements for highly oblique electron beams. *Med. Phys.* 23: 1413–1420.

Ostwald P.M. Kron, T. (1996b) Variation in calculated effective source-surface distances with depth. *Phys. Med. Biol.* 41: 2067–2078.

Ostwald P.M., Cooper S.G., Denham J.W., Hamilton C.S. (1994a) Dosimetry of high energy electron therapy to the parotid region. *Radiother. Oncol.* 33: 148–156.

Ostwald P.M., Metcalfe P.E., Denham J.W., Hamilton C.S. (1994b) A comparison of three electron planning algorithms for a 16 MeV electron beam. *Int. J. Radiat. Oncol. Biol. Phys.* 28: 731–740.

Papiez E., Dunscombe P.B., Malaker K. (1992) Matching photon and electron fields in the treatment of head and neck tumors. *Med. Phys.* 19: 335–341.

Pedroni E., Bohringer T., Coray A., Egger E., Grossmann M., Lin S., Lomax A., Goitein G., Roser W., Schaffner B. (1999) Initial experience of using an active beam delivery technique at PSI. *Strahlenther. Onkol.* 175 Suppl 2: 18–20.

Perez-Calatayudt J., Ballester F., Serrano M.A., Lluch J.L., Casal E., Carmona V. (2000) Dosimetric characteristics of backscattered electrons in lead. *Phys. Med. Biol.* 45: 1841–1849.

Perkins G.H., McNeese M.D., Antolak J.A., Buchholz T.A., Strom E.A., Hogstrom K.R. (2001) A custom three-dimensional electron bolus technique for optimization of postmastectomy irradiation. *Int. J. Radiat. Oncol. Biol. Phys.* 51: 1142–1151.

Pla M., Podgorsak E.B., Pla C. (1989) Electron dose rate and photon contamination in electron arc therapy. *Med. Phys.* 16: 692–697.
Prasad S.C., Bedwinek J.M., Gerber R.L. (1983) Lung dose in electron beam therapy of chest wall. *Acta Radiol. Oncol.* 22: 91–95.
Rahim H. (2003) R_{50} und R_p in der Elektronendosimetrie. *Z. Med. Phys.* 13: 49–53.
Rethfeldt C., Fuchs H., Gardey K.U. (2006) Dose distributions of a proton beam for eye tumor therapy: Hybrid pencil-beam ray-tracing calculations. *Med. Phys.* 33: 782–791.
Schulte R.W., Li T. (2006) Innovative strategies for image-guided proton treatment of prostate cancer. *Technol. Cancer Res. Treat.* 5: 91–100.
Schulte R.W., Bashkirov V., Klock M.C., Li T., Wroe A.J., Evseev I., Williams D.C., Satogata T. (2005) Density resolution of proton computed tomography. *Med. Phys.* 32: 1035–1046.
Schulz R., Smith A.R., Orton C.G. (2007) Proton therapy is too expensive for the minimal potential improvements in outcome claimed. *Med. Phys.* 34: 1135–1137.
Seuntjens J., Van der Plaetsen A., Thierens H., Piessens M. (1994) Comparison of measured and calculated dose distributions in lung after electron beam treatment of the chest wall. *Med. Phys.* 21: 1959–1968.
Slater J.D. (2006) Clinical applications of proton radiation treatment at Loma Linda University: review of a fifteen-year experience. *Technol. Cancer Res. Treat.* 5: 81–89.
Smith A.R. (2006) Proton therapy. *Phys. Med. Biol.* 51: R491–R504.
Sun C., Cheng C.W., Shimm D.S., Cassady J.R. (1998) Dose profiles in the region of abutting photon and electron fields in the irradiation of head and neck tumors. *Med. Dosim.* 23: 5–10.
Taghian A.G., Kozak K R., Doppke K.P., Katz A., Smith B.L., Gadd M., Specht M., Hughes K., Braaten K., Kachnic L.A., Recht A., Powell S.N. (2006) Initial dosimetric experience using simple three-dimensional conformal external-beam accelerated partial-breast irradiation. *Int. J. Radiat. Oncol. Biol. Phys.* 64: 1092–1099.
Timmermann B., Schuck A., Niggli F., Weiss M., Lomax A.J., Pedroni E., Coray A., Jermann M., Rutz H.P., Goitein G. (2007) Spot-scanning proton therapy for malignant soft tissue tumors in childhood: First experiences at the Paul Scherrer Institute. *Int. J. Radiat. Oncol. Biol. Phys.* 67: 497–504.
Tobler M., Leavitt D.D., Hayes J.K. (1995) Dosimetry and treatment planning of complex electron arc therapy. *Med. Dosim.* 20: 229–235.
Turano L., Biagini C., Bompiani C., Paleani-Vettori P.G. (1959) [Radiobiological, dosimetric and clinical principles of therapy with fast electrons of a 15-MeV-betatron.]. *Strahlentherapie* 109: 489–504.
Van Dyk J. (ed.) (1999) *The Modern Technology of Radiation Oncology*. Medical Physics Publishing, Madison, Wisconsin.
Vass S., Bairati I. (2005) A cosmetic evaluation of breast cancer treatment: A randomized study of radiotherapy boost technique. *Int. J. Radiat. Oncol. Biol. Phys.* 62: 1274–1282.
Verney J.N., Morgan A.M. (2001) Evaluation of in vivo dose measurements for patients undergoing electron boost treatments. *Radiother. Oncol.* 59: 293–296.
Weaver R.D., Gerbi B.J., Dusenbery K.E. (1995) Evaluation of dose variation during total skin electron irradiation using thermoluminescent dosimeters. *Int. J. Radiat. Oncol. Biol. Phys.* 33: 475–478.
Weber D.C., Bogner J., Verwey J., Georg D., Dieckmann K., Escude L., Caro M., Potter R., Goitein G., Lomax A J., Miralbell R. (2005) Proton beam radiotherapy versus fractionated stereotactic radiotherapy for uveal melanomas: A comparative study. *Int. J. Radiat. Oncol. Biol. Phys.* 63: 373–384.
Weber D.C., Rutz H.P., Pedroni E.S., Bolsi A., Timmermann B., Verwey J., Lomax A.J., Goitein G. (2005) Results of spot-scanning proton radiation therapy for chordoma and chondrosarcoma of the skull base: the Paul Scherrer Institute experience. *Int. J. Radiat. Oncol. Biol. Phys.* 63: 401–409.
Weidlich G.A., Nuesch C.E., Fuery J. J. (1996) Added aluminum shielding to attenuate back scatter electrons from intra-oral lead shields. *Med. Dosim.* 21: 165-167.

Wilkens J.J., Oelfke U. (2005) Optimization of radiobiological effects in intensity modulated proton therapy. *Med. Phys.* 32: 455–465.
Williams J., Thwaites D. (1993) *Radiotherapy Physics*. Oxford University Press, Oxford, UK.
Wilson V.C., McDonough J., Tochner Z. (2005) Proton beam irradiation in pediatric oncology: An overview. *J. Pediatr. Hematol. Oncol.* 27: 444–448.
Wittkamper F.W., Mijnheer B.J. (1993) Dose intercomparison at the radiotherapy centers in The Netherlands. 3. Characteristics of electron beams. *Radiother. Oncol.* 27: 156–163.
Yorke E.D., Kassaee A., Doyle T., Loevner L.A., Rosenthal D.I. (2001) The dose distribution of medium energy electron boosts to the laryngectomy stoma. *J. Appl. Clin. Med. Phys.* 2: 9–20.
Yorke E.D., Kassaee A., Lin L.C., Rosenthal D.I. (1998) Dosimetric comparison of centered and off-centered posterior neck electron fields. *Med. Dosim.* 23: 284–287.
Zavgorodni S.F., Beckham W.A., Roos D.E. (1996) A method for producing uniform dose distributions in the junction regions of large hinge angle electron fields. *Int. J. Radiat. Oncol. Biol. Phys.* 35: 765–770.
Zhu T.C., Das I.J., Bjärngard B.E. (2001) Characteristics of bremsstrahlung in electron beams. *Med. Phys.* 28: 1352–1358.

Questions

5.1 Please compare the skin sparing in an electron radiation beam with the one in a photon beam.

5.2 The International Atomic Energy Agency does recommend performing calibration of electron beams with a half-value depth of $R_{50} > 7$ g/cm^2 to be done in a radiation beam with a field size of at least 20×20 cm^2 at the surface of the phantom.
i) Why?
ii) What nominal electron energy would $R_{50} = 7$ g/cm^2 refer to?

5.3 An electron boost to the neck using 9 MeV electrons and a field size of 7×5 cm^2, must be treated at 115 cm SSD since the applicator cannot fit close to the skin near the patient's shoulder. The standard treatment SSD is 100 cm. The equation to obtain an Inverse Square Law correction factor is:

$$ISL = ((oldSSD + d_{max})/(newSSD + d_{max}))^2$$

where newSSD = oldSSD + 15 cm.

Please give an estimate for "oldSSD".

5.4 Please consider figure 5.18. It is assumed that all electrons in the radiation beam travel the same distance laterally. Why is this not correct?

5.5 An electron beam has half-value dose (d_{50}) of 5 cm; please calculate the nominal energy of the beam at the patient surface. Where would you expect the practical range to be? Can you estimate its nominal energy?

6 Radiotherapy Treatment Planning: X-Rays

"There is no region of regret when the dose you see is the dose you get."
(Peter Metcalfe 1994)

6.1 Introduction

Radiotherapy treatment planning (RTP) is the process whereby the therapeutic strategy of the radiation oncologist is realized as a set of treatment instructions together with a physical description of the distribution of dose in the patient. The aims of treatment planning can be summarized as follows (Williams et al. 1993):

i) Localize the tumor volume in the patient and define the target volume for treatment.
ii) Measure the outline of the patient and mark the target volume and other anatomical structures that may affect the dose distribution, or for which dose constraints may be necessary.
iii) Determine the optimum treatment configuration required to irradiate the target volume to the specified dose within particular clinical constraints.
iv) Calculate the resultant dose distribution in the patient.
v) Prepare an unambiguous set of treatment instructions for other radiation therapists.

The treatment planning process consists of a number of steps which can be summarized as follows (AAPM TG 53 1998):

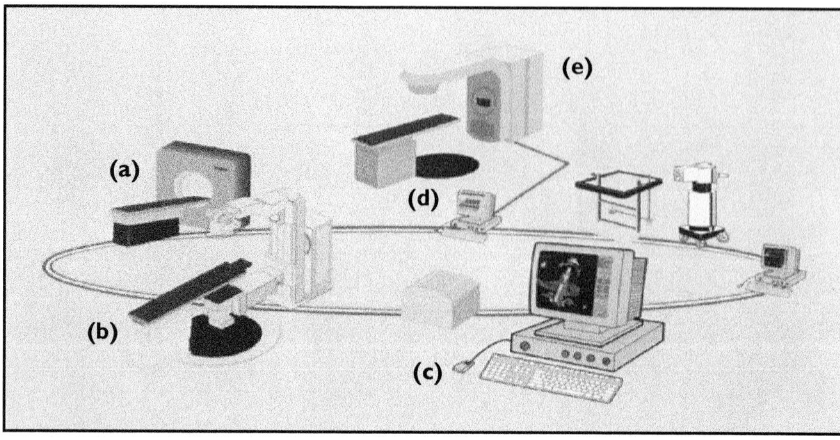

Figure 6.1: The radiotherapy treatment planning process. (a) CT scanner, (b) simulator, (c) planning computer, (d) linac control, (e) linac. (Reprinted with permission from: © Nucletron-Odelft.)

i) Patient positioning and immobilization.
ii) Image acquisition and input (e.g., computed tomography [CT] data input).
iii) Anatomy defnition (defning target and normal tissue contours).
iv) Beam technique (e.g., beam energy, beam angles, and feld shapes).
v) Dose calculations (select and implement prescription and calculation).
vi) Plan evaluation (analyze dose displays and dose-volume histograms [DVHs]; compare plan trials).
vii) Plan implementation (calculate monitor units [MUs]; transfer plan data to R&V [record and verify]).
viii) Plan review (review all aspects of plan including independent MU check).
ix) If all checks are within specifcation, then treat patient.

Dose distributions are calculated to provide the clinician with a physical description of the treatment prescribed and to provide a record of that treatment. Formerly, all treatment plans were produced laboriously by hand. Hand planning limited the opportunity for optimization of the dose distribution and limits the degree of sophistication to which patients could be treated. RTP computer systems are now almost universally used for radiotherapy planning and this has been made possible by the decreasing cost of hardware and the expanding market in software, much of it developed by physicists working in the clinical environment. The most common two-dimensional (2-D) display tool used by RTP computers is the isodose distribution, as outlined in section 6.2.

The final product of treatment planning is the treatment sheet. This includes the prescription and provides a set of instructions for the radiation therapist, enabling the treatment to be delivered as described by the dose distribution. The instructions must be both unambiguous and accurate. Accuracy has to be ensured by a series of checks through the whole planning process, and these should form a part of a fully documented quality assurance system.

The main features of different radiotherapy processes and machines used to plan patients prior to radiotherapy treatment are reviewed in this chapter. Figure 6.1 shows a summary of the treatment planning process. The patient is likely to go through a CT and/or simulation procedure where patient-specific data are collected. Then calculations are usually performed on the RTP computer. Once a treatment sheet is produced and all appropriate checks have been undertaken, then treatment of the patient on the linac. can proceed.

Calculation of linac machine MUs is required to deliver a specified dose to the patient. MU calculations are undertaken by the RTP computer. Independent validation of these MU calculations is essential to the checking process. The calculation methods used to undertake these checks are dealt with in section 6.3. Because calibration methods and normalization procedures vary, it is advisable that a physicist who has detailed knowledge of these calculations be involved in the plan checking process.

Prior to planning dose distributions on a RTP computer, various patient-specific data related to linac orientations are simu;ated. The process has evolved from the use of classical simulators to the more frequent use of CT scanners with virtual simulation tools. Hence both simulation options are described in section 6.4.

CT scans provide the main source of patient contour data for the RTP process. This is currently the modality of choice because geometric distortions are negligible and the physics is well understood. In particular, the relationship between CT number and relative electron density remains consistently reproducible. This is important because relative electron density is the parameter used in inhomogeneity correction algorithms to calculate effective pathlength through a medium using ray tracing methods (see chapter nine). Hence the relationship between CT number and density is outlined in detail in section 6.5.

In section 6.6 other modalities that can provide useful data for the RTP process are discussed. These include nuclear magnetic resonance imaging (MRI). In some cases, MRI provides better tumor volume definition than CT. However, an accurate electron density map is not obtained using this method; electron density may be approximated from proton density scans by assumptions about water content. The accuracy of this method is currently being investigated. To achieve an optimum image containing tumor and electron density information, currently an overlay between CT and MRI is ideal in these situations. Positron emission tomography (PET) is another diagnostic imaging modality that provides valuable tumor staging information. The potential for this modality to aid in outlining tumor margins is also discussed. Image data from modalities such as MRI and PET can be fused with CT data sets to provide images that show enhanced tumor definition and provide valuable tumor stage information. The process of image registration and fusion is also disccssed in section 6.6.

In section 6.7, the requirements of RTP computer hardware and peripherals are outlined. These are the calculation computers that are used to calculate and display dose distributions prior to patient treatment. In section 6.8 display tools that are available on various RTP systems are discussed. These range from 2-D isodose displays to various exotic three-dimensional (3-D) views. DVHs are discussed in section 6.9. These are currently the mainstay tool for analyzing 3-D dose distributions because of the quantitative dose information they convey.

It is critical that any dose-volume tool is applied to target volumes that are consistently defined, as this ensures consistent doses are delivered to patient cohorts. These definitions are important in providing consistent treatment between patients within a clinic. Consistent volume definition is also important for multicenter clinical trial dose reporting. The definitions used to describe volumes and margins by the recognized protocols are dealt with in section 6.10.

6.2 Isodose Curves

The most common way to display the dose distribution is with isodose curves. When points of equal dose are joined together, an isodose curve is produced. These are the curves used for visualizing dose distributions on cross-sectional planes of the patient. This is necessary because the dose to many points in a patient needs to be assessed and compared with the dose at a reference point (usually d_{max}). A series of isodose curves for a range of different doses, such as %D of 90, 70, 50, 30, and 10%, can be combined to form an isodose chart or isodose distribution. This form of display is convenient, since the distribution gives a map of the dose contours in one plane.

Figure 6.2 shows a series of isodose distributions. Figures 6.2a and b represent isodose distributions from a 6 MV and an 18 MV x-ray beam produced by a linac at 100 cm source-surface distance (SSD). Note that the light field and an extension of the 50% isodose curve to the surface should be coincident; this line represents the beam's divergent field edge. Comparing these figures shows that the penetration of the 18 MV beam is greater, with d_{max} being at 3.3 cm compared with 1.5 cm and all other doses being greater at the same depth. The 18 MV beam also has a slightly wider penumbra. This is due to the greater lateral range of electrons from the 18 MV beam causing a marginally wider zone of lateral electron disequilibrium.

Figure 6.2c illustrates the isodose distribution for a 12 MeV electron beam produced by a linac. The penumbra width is also significant due to lateral electron scattering. The penetration properties of this beam are much less than for the x-ray beams. For this reason (and also because electrons give a high skin dose, approximately 90% of D_{max}) electron beams are predominantly used for boost fields. Electron fields usually either add dose to superficial regions, e.g., breast radiotherapy boost, or add dose to regions where the spine may have reached tolerance, e.g., head and neck off-cord boost. As an aside, electrons are less forgiving of density changes. As such, treatment near air cavities and bone should be undertaken with due regard of this (Cygler et al. 1987, Jette 1984). Because electron boost fields generally contribute less than 20% to the total dose, the accuracy of electron dose calculations is less critical than x-ray dose calculations in terms of total dose delivered. As this chapter concentrates on x-ray beams, the reader is referred to the following texts and papers which outline the fascinating physics models behind the dose distributions created by radiotherapy electron beams (Klevenhagen 1985 and 1993, Hogstrom et al. 1981).

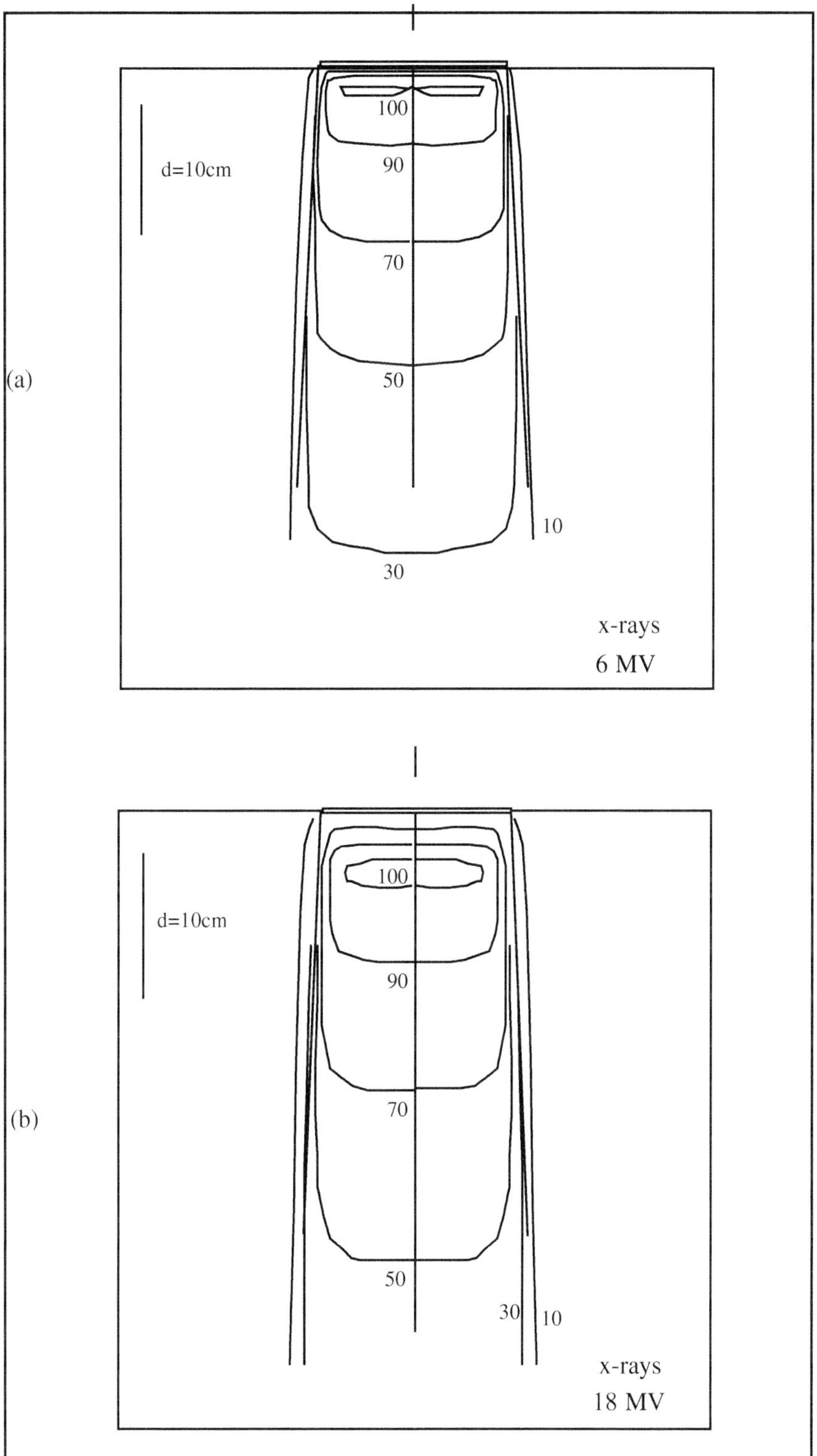

Figure 6.2: Isodose curves for different energy beams. (a) 6 MV x-ray beam from a Varian 2100C linac, 10×10 cm^2 field at 100 cm SSD. (b) 18 MV x-ray beam from a Varian 2100C linac, 10×10 cm^2 field at 100 cm SSD.

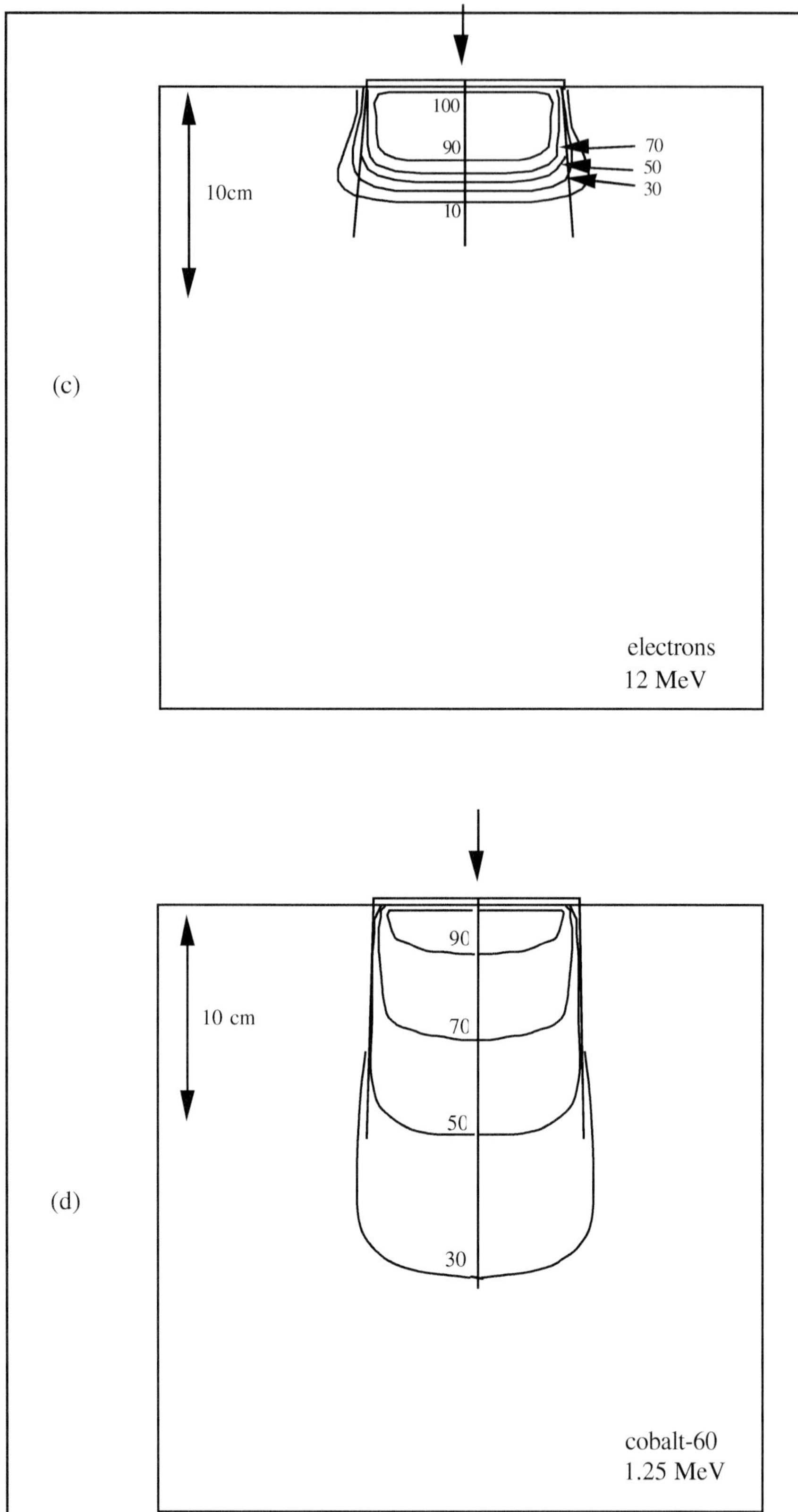

Figure 6.2: (Continued) (c) 12 MV electron beam from a Varian 2100C linac, 10×10 cm^2 field at 100 cm SSD. (d) Cobalt-60 (1250 keV) gamma ray beam from a Theratron 80, 10×10 cm^2 field at 80 cm SSD.

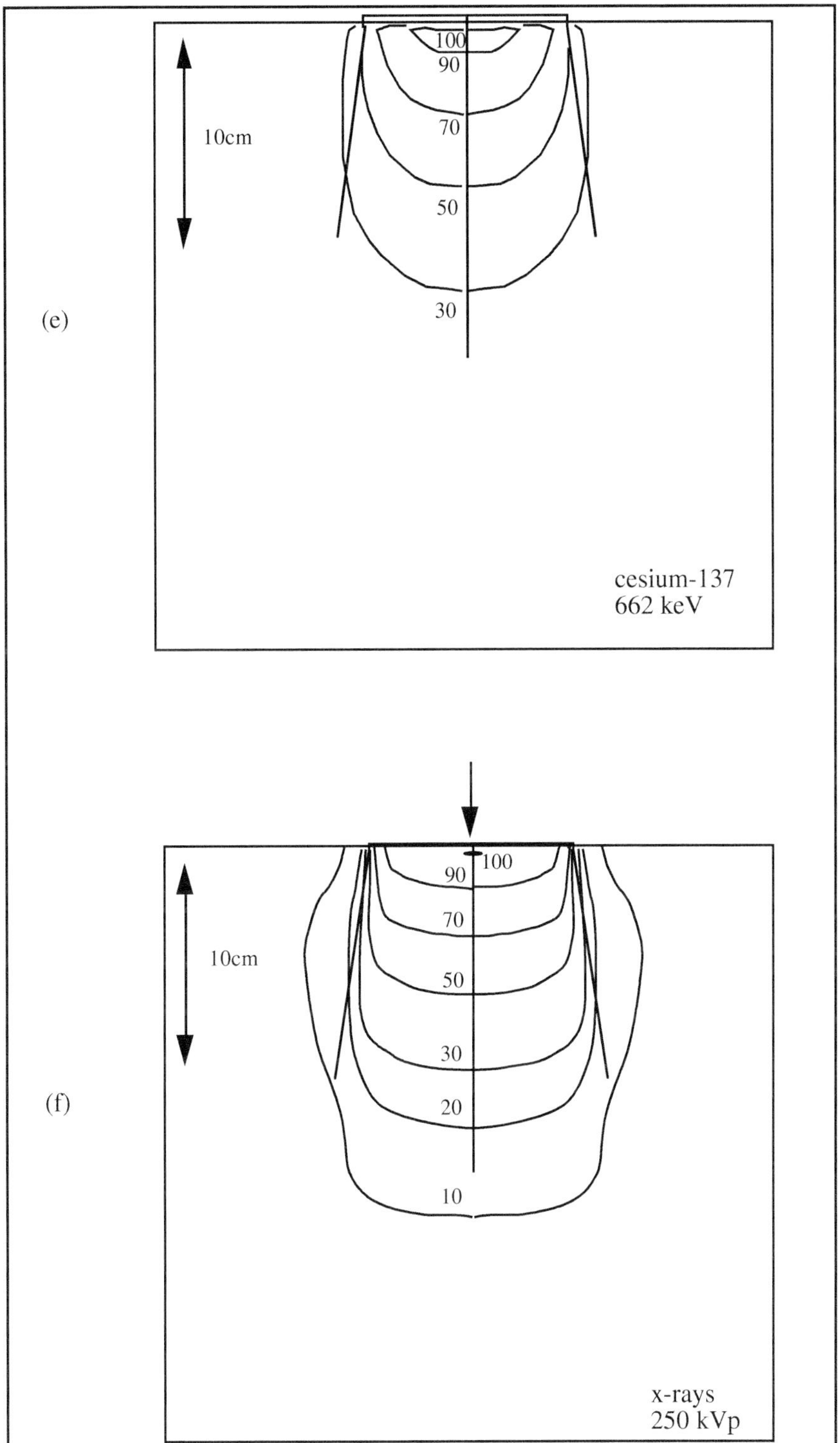

Figure 6.2: (Continued) (e) Cesium-137 (662 keV) gamma ray beam from a caesa Gammatron 80, 10×10 cm^2 field at 40 cm SSD. (f) Orthovoltage (250 kVp) x-ray beam from a Pantac Therapax DXT 300, 10×10 cm^2 field at 50 cm SSD. (e) 250 kVp 50 cm SSD (Pantac Orthovoltage); note the extra beam divergence due to short SSD.

Figures 6.1d, e, and f show the isodose curves produced by cobalt-60 (nominal 1250 keV), cesium-137 (662 keV), and orthovoltage (250 kV) beams, respectively. The cobalt beam shows less penetration than the linac beams and a slightly wider penumbra (cobalt has a 5 mm source diameter compared to a linas's 1 to 2 mm). The cesium beam has less depth penetration and a very wide penumbra due to its very large source size. The 250 kV beam has even less penetration and virtually no skin sparing properties; the penumbra is moderately wide due to the source size and photon scatter. The isodose levels from about 30% to 70% are very close together for 250 kV beams; this is because the electron range is short. The penumbra spreads out in the low-dose region due to photon scatter dominating in the penumbral tail.

6.3 Monitor Unit Calculations

6.3.1 Calibration Position

The monitor unit settings on the linac can be calibrated in different ways to achieve an output of 1 cGy per MU for a standard field size (usually 10×10 cm^2). The three common scenarios that are shown in figure 6.3 are:

i) $(D/MU)_{std,SAD,Air}$ = 1 cGy/MU at 100 cm SAD in air.
ii) $(D/MU)_{std,SAD,Water}$ = 1 cGy/MU at 100 cm SAD in a water phantom at d_{max}.
iii) $(D/MU)_{std,SSD,Water}$ = 1 cGy/MU at 100 cm SSD in a water phantom at d_{max}.

The calibration method has an impact on the MU calculation. The first two methods have been designed for easy use with isocentric calculations using isocentric weighting. The third method can be used for SSD and isocentric calculations directly using applied weighting. An inverse square correction is required if isocentric weighting is used in combination with method (iii) because the outputs are not measured at isocenter (see equation 4.1). The output, area, or field factors are then all derived as ratios of charge measurements for different field sizes to the charge measurement for the standard field size in the matching geometric configuration.

The *beam weight* of a treatment beam is its relative contribution to the treatment dose. Most current computer planning systems allow the choice of two weighting options. For treatments at fixed SSDs the weight represents a multiplying factor for the dose at the depth of dose maximum, usually expressed as a percentage. In example 6.2, which follows, two opposed fields each have a weight of 100%. However, other beam weights could have been selected, which in this simple case would represent the amount of dose one beam would deliver relative to the other beam's dose.

For isocentric treatments the field weight is defined as either the relative contribution of the beam to dose at d_{max}, applied weight = applied weight, or the relative contribution of the beam to the dose at isocenter. For the second

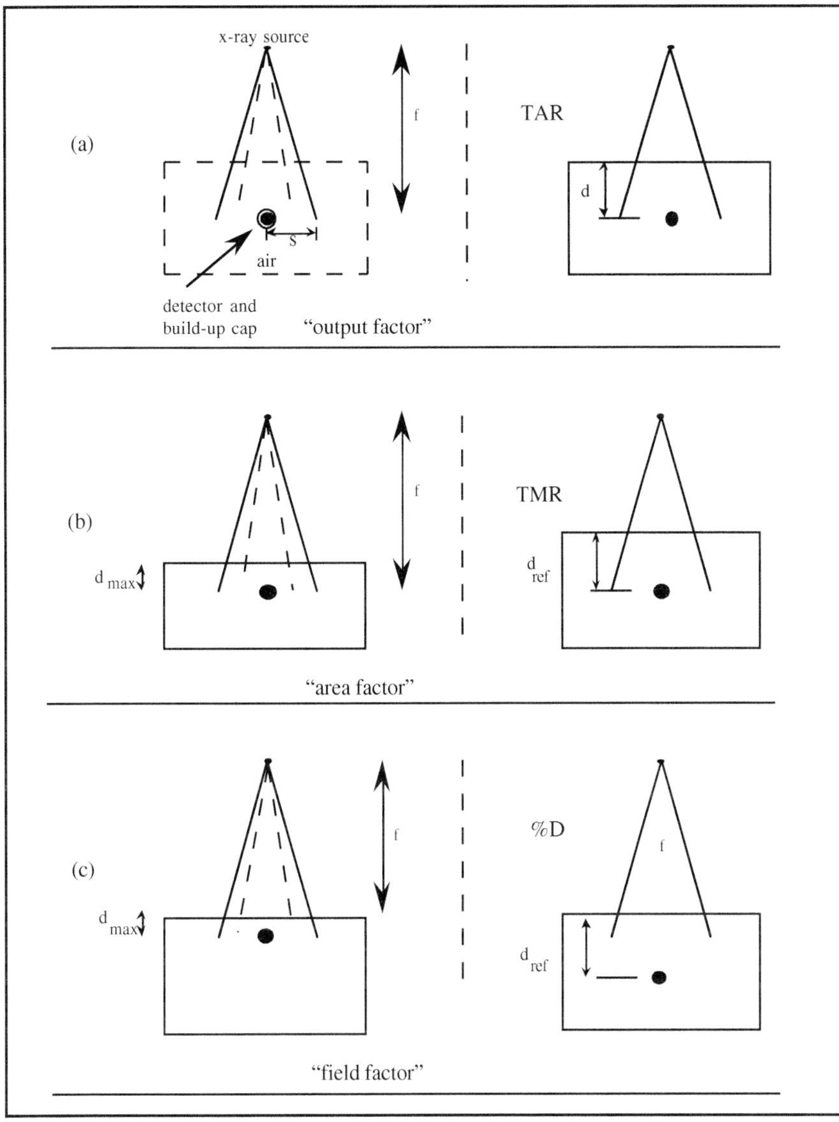

Figure 6.3: Calibration for 1 cGy/MU. The *factor* is the ratio of charge readings of a field size to the charge readings for a 10×10 cm^2 field. (a) Output factor and TAR. (b) Area factor and TMR. (c) Field factor and depth dose.

scenario if calibrated at isocenter, applied weight = isocenter weight×TMR. If calibrated at SSD, applied weight = isocenter weight×TMR×(INVSQ).

There are also three common ways of normalizing planned dose distributions. Combined beams can be displayed if applied weighting is used:

i) Don't renormalize.
ii) Renormalized to 100%.
iii) Renormalized to represent total dose at isocenter.

With isocentric weighted beams the weights can be altered to achieve 100% at isocenter. This is done by applying the weights as a reciprocal of the number of beams. Weights can also be altered for applied weighted beams to achieve the required normalization at isocenter; however, the procedure is not so straightforward. In the following section we apply some calculations for applied weighted and isocentric weighted beams; these examples serve to illustrate the different types of calculations involved. It is becoming increasingly common for plans to be normalized with an actual dose prescription and fractionation to isocenter. These plan prescriptions can either show Gy per fraction or total dose in Gy isodose lines. The MU calculated per fraction for each beam to deliver this dose to isocenter is recorded in the beam data sheet.

6.3.2 Fixed SSD Beam Calculations

Single-field treatment is a term applied to those situations in which the irradiated volume does not overlap any other volume treated concurrently. In photon therapy, treatments with a single field are generally used only for palliative purposes, a typical example being irradiation of the spine to relieve pain in patients who have spinal metastases.

For these palliative treatments a computer plan is not always generated and instead depth dose tables may be acceptable (see table 4.4). Alternately if a plan is generated, a suitable isodose level may be selected.

Example 6.1

The following illustrates how to calculate the MUs that need to be set for a single fraction of 6 MV x-rays at 100 cm SSD. The prescription is 30 Gy in 15 fractions at a depth of 12 cm.

$$MU = \frac{D \times 100}{\left(\frac{D}{MU}\right)_{std,SSD,Water} \times F \times \%D} = \frac{200 \times 100}{1 \times 1.031 \times 62.2} = 312 \quad MU. \tag{6.1a}$$

Table 6.1: Parameters used in example 6.1

Parameter	*Symbol*	*Magnitude*	*Units*	*Table or (Figure)*
Dose per fraction	D	200	cGy	
Target depth	d	12	cm	
Field size	S	15 ×15	cm^2	
SSD	f	100	cm	
Percent depth dose	%D	62.2	%	4.4 (5.4a)
Field factor	F	1.031		4.1
Calibration	$(D/MU)_{std,SSD,Water}$	1	cGy/MU	

Example 6.2

Two opposed coaxial beams can be used to irradiate volumes of tissue in which the tumor cannot be accurately defined or in which the intention is to give a relatively low dose for palliation. Many tumor sites can be irradiated in this way, and the combination of two beams in this configuration is commonly referred to as a parallel pair or parallel-opposed fields.

The following illustrates how to calculate the machine MUs to be set for each field for two opposed fields of 6 MV x-rays at 100 cm SSD. All other parameters are the same as example 6.1 except %D equals twice the above, i.e., 2×62.2. The percent depth dose could also be derived from figure 6.4b.

$$
\begin{aligned}
MU &= \left(\frac{D \times 100}{\left(\frac{D}{MU}\right)_{std,SSD,Water} \times F \times \%D} \right) \\
&= \left(\frac{200 \times 100}{1 \times 1.031 \times 124.4} \right) \qquad (6.1b) \\
&= 156 \quad MU\ per\ field \quad .
\end{aligned}
$$

6.3.3 Isocentric Beam Planning

In examples 6.1 and 6.2 the calculations are for treatment at a fixed SSD. In most centers there is a preference for performing these treatments isocentrically. Isocentric irradiation involves positioning the patient so that the center of the target volume is coincident with the center of rotation of the treatment unit (isocenter). When the first beam has been set up and treated, it is then only necessary to rotate the gantry to the next angle before treating the second and subsequent fields. This has the advantage of minimizing patient setup time for the second and subsequent fields, if we assume the machine was still calibrated at $(D/MU)_{std,SSD,Water}$ = 1 cGy/MU. Since treatment is carried out at a nonstandard SSD, the output and depth dose charts cannot be used directly and corrections can be made as follows:

Example 6.3

Assume that instead of treating the single field at 100 cm SSD, the target point is placed at isocenter, which means that the SSD is now 100 – 12 cm = 88 cm. Calculate the MUs required for the same dose for a single field.

(a) The percentage depth dose is adjusted from 62.2% to 60.7% using equation 4.11 with Peak Scatter Factor (PSF) values derived from figure 4.19.

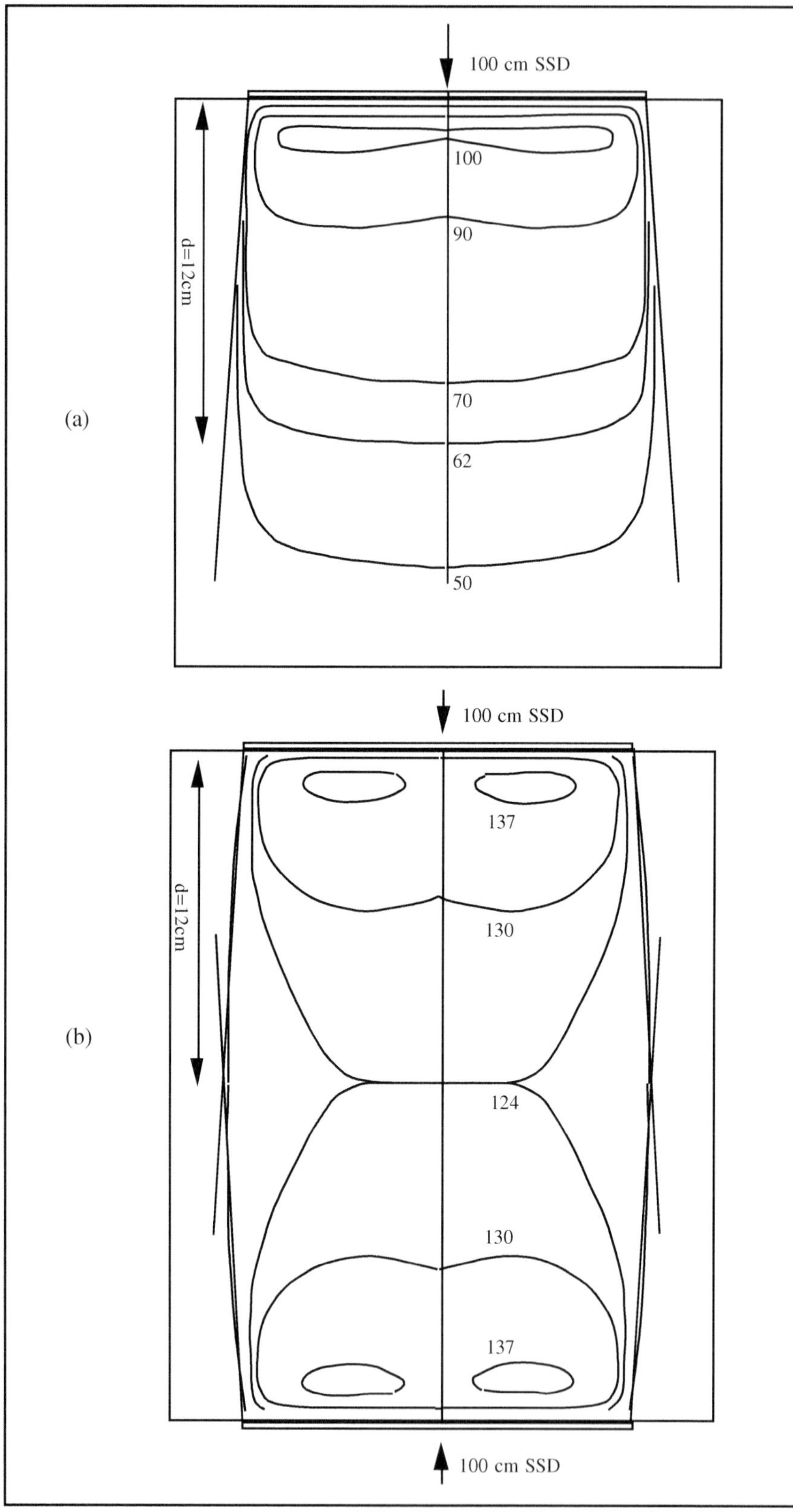

Figure 6.4: Isodose curves using monitor unit calculations for 6 MV x-rays, field size 15 × 15 cm². (a) Single field 100 cm SSD. (b) Two opposed fields 100 cm SSD.

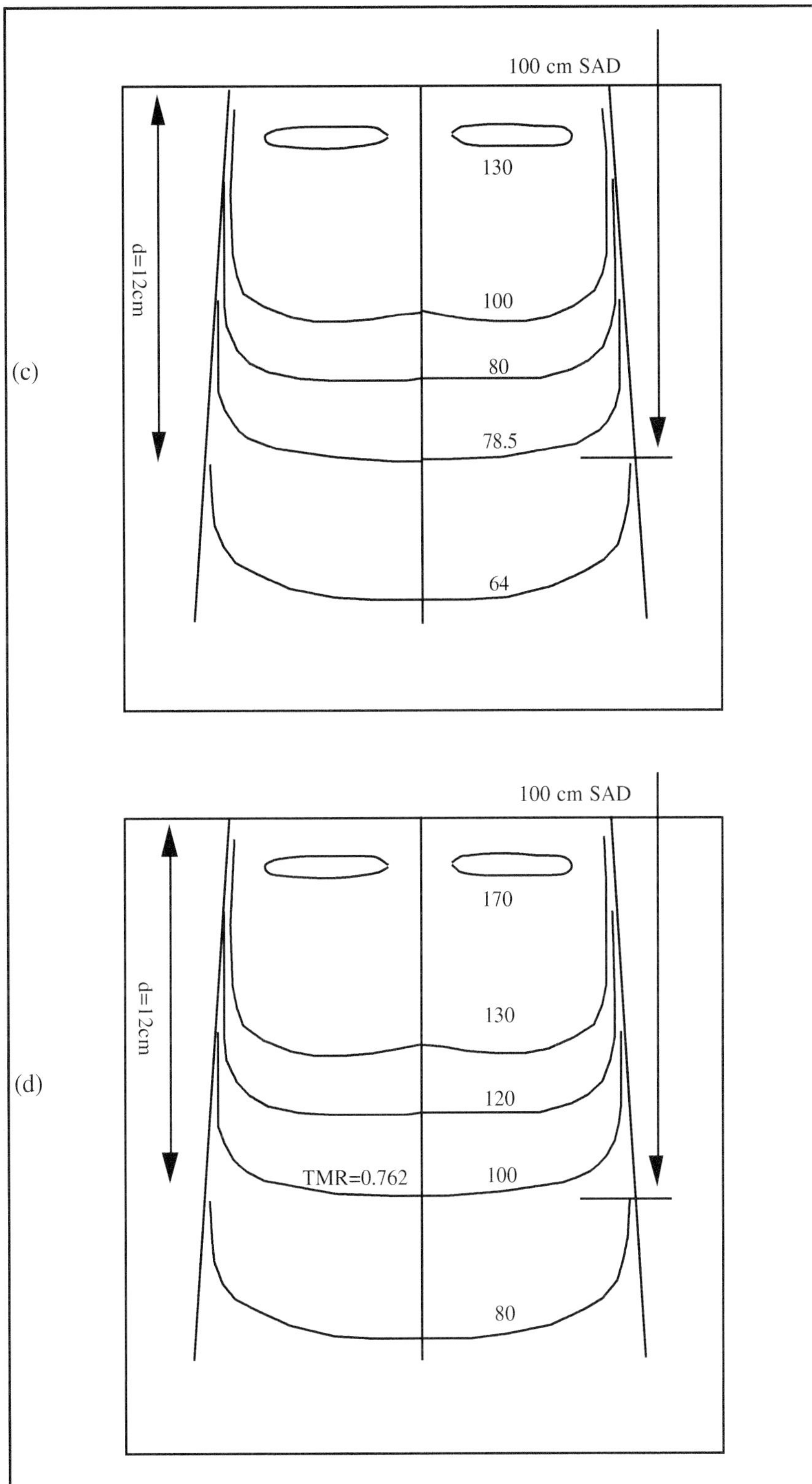

Figure 6.4: (Continued) (c) Single field applied weight 100 cm SAD. (d) Single field isocentric weight 100 cm SAD.

(b) The field factor is adjusted by:

$$
\begin{aligned}
F(f') &= F(f)\times\left(\frac{f+d_{max}}{f'+d_{max}}\right)^2 \times\left(\frac{PSF(S)}{PSF\left(\frac{f'}{f}\right)\times(S)}\right)\\
&=(1.031)\times\left(\frac{100+1.5}{80+1.5}\right)^2\times\left(\frac{1.040}{1.036}\right)\\
&=1.333
\end{aligned}
\tag{6.2}
$$

then

$$
\begin{aligned}
MU &= \left(\frac{D\times 100}{\left(\frac{D}{MU}\right)_{std,SSD,Water}\times F(f')\times\%D(f')}\right)\\
&=\left(\frac{200\times 100}{1\times 1.333\times 60.7}\right)\\
&=247 \quad MU.
\end{aligned}
\tag{6.3}
$$

Note that because ratios are involved NPSF values could have been used instead of PSFs to achieve the same results.

Example 6.4

Revisiting example 6.3 but this time using the applied weight with the isodose plan normalized to d_{max} at 100 cm SSD as shown in figure 6.4c:

$$
\begin{aligned}
MU &= \left(\frac{D\times 100}{\left(\frac{D}{MU}\right)_{std,SSD,Water}\times F\times\%D}\right)\\
&=\left(\frac{200\times 100}{1\times 1.031\times 78.5}\right)\\
&=247 \quad MU.
\end{aligned}
\tag{6.4}
$$

An alternative method for the calculation of the dose to the isocenter is to use isocentric normalized beam weighting combined with a tissue-maximum ratio (TMR). The advantage of this parameter is that it is independent of SSD so that a single table of TMRs as a function of field size at the isocenter and depth, d, is required.

Example 6.5

Using the isodose plan normalized to 100% at 100 cm SAD as shown in figure 6.4d:

$$
\begin{aligned}
MU &= \left(\frac{D \times INVSQ}{\left(\frac{D}{MU}\right)_{std.SSD.Water} \times F \times TMR} \right) \\
&= \left(\frac{200 \times 0.97}{1 \times 1.031 \times 0.763} \right) \qquad (6.5) \\
&= 247 \quad MU.
\end{aligned}
$$

The inverse square term corrects for reposition of field factor from 100 cm SSD to 100 cm SAD.

If the beam includes a wedge or a block tray, an output adjustment is required that accounts for the reduction in beam dose output at d_{max} central axis. Table 4.11 lists wedge factors for different wedge angles. Note that the beam attenuation is significant!

The data in tables 4.1 and 4.4 are normally tabulated for square fields. If the field is rectangular, an equivalent square field size that would have the same output and depth dose characteristics is used. That is, we are trying to emulate the same scattering properties as an equivalent square field. It has been shown that these data can sometimes be adequately represented by the following relationship:

$$
S = \left(\frac{2 \times x \times y}{x + y} \right), \qquad (6.6)
$$

where S is the side of the equivalent square, and x and y are the sides of the rectangular field.

If the field has blocks included and then becomes an irregular shaped field, the following may be used:

$$
S = \left(\frac{4 \times Area\ of\ field}{Perimeter\ of\ field} \right). \qquad (6.7)
$$

Note that these are empirical relationships that should be used with caution. Several tests have shown that these methods are reasonable for calculating the change in %D. However, the method is less accurate for calculating most irregular field output factors. The method generally gives a good estimate of in-phantom scatter contributions. Hence, when backscatter factors (BSFs), derived from ESQs (equivalent squares) for two different field dimensions, are ratioed, this gives an estimate of how much the in-phantom scatter dose component changes. Taking the same ratio for extra focal scatter contributions does not give a good estimate for this component.

Equivalent squares were historically used for direct output calculation adjustment of cobalt beams that used timers, not ion chamber monitors. This tends to imply that backscatter into the ion chambers, which does not occur with blocked fields because blocks are further away, may be the cause of the difference. This point has been argued in the literature with conflicting views. Another contributing factor may be extra focal radiation; this is the radiation scattered from the flattening filter and collimators (Tatcher and Bjärngard 1994, Chaney et al. 1994).

Different centers have different protocols (developed with experimental validation) to cope with blocked fields. A typical planning protocol for a department might follow these rules of thumb:

i) If shielding less than 30% of field area, use unblocked field for calculation because the ratio of PSFs is negligible when field size differences are small.
ii) If shielding greater than 30%, use blocked equivalent square for %D tables and use ratio of PSF for the blocked field size to the PSF for the unblocked field size to adjust output factor (note that a ratio of NPSFs is just as valid).

Correction method (ii) simply corrects for in-phantom scatter effects, but because the MU is retained for the unblocked field size, collimator effects are removed. An example of this calculation is given as follows.

Example 6.6

One way of adjusting MUs for a blocked field using a block correction factor (BCF) for in-phantom scatter is as follows. An open 20×20 cm^2 field is blocked and the ESQ for the blocked field using equation 6.8 is calculated as 12×12 cm^2. The block tray is 6 mm PMMA, hence the block tray factor (BF) = 0.96. The PSFs were derived from figure 4.19 by assuming that they were approximately equal to the TAR(d_{max},S). Assume the dose required is 200 cGy at 12 cm depth and find the %D for a 12×12 cm^2 field size by interpolation from table 4.4.

$F(20 \times 20\ cm^2) = 1.064$
$F(12 \times 12\ cm^2) = 1.020$

$$BCF = \left(\frac{PSF(12cm)}{PSF(20cm)}\right) = \left(\frac{1.046}{1.063}\right) = 0.984 \tag{6.8}$$

$$MU = \frac{D \times 100}{\left(\frac{D}{MU}\right)_{std,SSD,Water} \times F\left(20\,cm^2\right) \times BCF \times BF \times \%D\left(12\ cm^2\right)} = \frac{200 \times 100}{1 \times 1.064 \times 0.984 \times 0.96 \times 60.8} = 327\ MU. \tag{6.9}$$

Note that applying the ESQ for the small field to directly alter the MU gives a higher value of 335 MU. Experiments show that this gives a higher dose than that prescribed.

Output factors have been shown to be slightly dependent on collimator orientation (i.e., upper, U, and lower, L, jaw). Therefore, dose output factors may be issued for jaw-dependent rectangular fields. This increases the amount of output factor data to be collected significantly, but it also improves the output calculation accuracy by up to about 2% for very elongated fields (e.g., 5U cm × 40L cm field versus 40U cm × 5L cm field). The BCF correction can still be applied to the collimator dependent output factors directly.

If the collimator and phantom scatter factors are separated, then the BCF used in example 6.6 is very similar to applying S_c for the open field size and S_p for the blocked field ESQ. For isocentric beams calibrated at isocenter, the following applies (Khan 1984, Purdy et al. 1992):

$$MU = \frac{D}{\left(\frac{D}{MU}\right)_{std,SAD,Water} \times TMR\left(d, s_{phant}\right) \times S_c\left(s_{col}\right) \times S_p\left(s_{phant}\right) \times WF \times BF} \tag{6.10a}$$

where WF is the wedge factor, and BF the block tray factor, $S_c(s_{col})$ is the collimator scatter factor applied to the open collimated field size, and $S_p(s_{phant})$ is the phantom scatter factor applied to the blocked field size. Equation 6.10a is a generic representation of the check calculation most commonly used in most clinics. Most MU calculations checks within most clinics use this or a slight variation of this equation.

Example 6.7

A linac is calibrated in air at 100 cm SAD. There is a 15 degree wedge added to the beam and the field is 20 × 20 cm² blocked down to 10 × 10 cm². The blocks are placed on a 6 mm PMMA block tray. Calculate the MU setting required to give 200 cGy at 12 cm depth.

$$\begin{aligned} MU &= \frac{D}{\left(\frac{D}{MU}\right)_{std,SAD,Air} \times TMR(d, s_{phant}) \times S_c\left(s_{col}\right) \times S_p\left(s_{phant}\right) \times WF \times BF} \\ &= \frac{200}{1 \times 0.722 \times 1.033 \times 1 \times 0.78 \times 0.97} \\ &= 354 \quad MU. \end{aligned}$$

(6.10b)

Table 6.2: Parameters used in example 6.7

Parameter	*Symbol*	*Magnitude*	*Units*	*Table or (Figure)*
Dose per fraction	D	200	cGy	
Target depth	d	12	cm	
Open field	S_c	1.033		4.1
Blocked field	S_p	1.000		(4.19)
Tissue-max. ratio	TMR	0.722		4.5
Calibration	$(D/MU)_{std,SAD,Air}$	1	cGy/MU	
Block tray factor	BF	0.96		
Wedge factor	WF	0.78		

6.4 Classical Simulation

A *classical* simulator (shown in figure 6.5a) consists of a diagnostic x-ray tube (generally of peak energy 120 kVp and nominal mean energy of 40 keV). It is mounted on a gantry arrangement that mimics the movements of the linear accelerator (hence the term "simulator"). As shown in figure 6.5b, mobile wires in the head of the machine are projected onto the patient (by light projection and x-ray projection). These show a beam's eye view (BEV) of the proposed treatment fields in relation to the anatomy. An outer set of collimators beyond the wires are provided to restrict the x-ray field.

Simulators historically used an image intensifier with an additional film cassette holder. In the most recent designs the *image intensifier* has been replaced by a *flat-panel detector* (*FPD*) (e.g., amorphous silicon) plate as shown in figure 6.5a. This reduces the bulk of the detector system. The new FPDs provide much more clearance from patient to floor than the image intensifiers. As with other open gantries in radiotherapy, rotation is regulated to achieve one rotation in one minute.

The x-ray simulator is used by planning radiation therapists and radiation oncologists prior to patient treatment. It provides the clinician with an x-ray BEV of the treatment volume and surrounding structures. This means that the clinician can view the coverage of the target from the beam's perspective so that accurate treatment of the target can be achieved, and surrounding normal tissues and radiation-sensitive structures are spared any unnecessary dose.

There are three main advantages of the classical simulation technique:

i) A target area is clearly defined in BEV in relation to sensitive structures within and beyond the treatment field margins.
ii) Treatment machine parameters such as gantry, collimator, and couch angle, as well as jaw sizes can be preset for use in calculations by the radiotherapy planning computer and ultimately for use on the linac.
iii) Treatment strategy can be devised prior to placement of the patient on the linac, thus minimizing valuable setup time on the linac.

There were deficiencies in the historical simulation method, such as measurements of patient external and internal contour were not provided and no density information was available, e.g., densities in different regions in lung vary greatly (Van Dyk et al. 1982).

However these systems are now also capable of cone beam CT, which can provide contour and some density information. The consistency of the density information for RTP calculations is not yet as well characterized as for a conventional CT. One such example of a modern classical simulator is the Acquity simulator (see figure 6.5a), which can acquire data with one gantry rotation and cone beam CT scans can be reconstructed (scan field of view along the patient is 20 cm). Similar technology is also used on linac-mounted imaging devices, as discussed in chapter one and chapter eleven of this text and in detail elsewhere (Siewerdsen and Jaffray 2000, Jaffray et al. 2002, Groh et al. 2002).

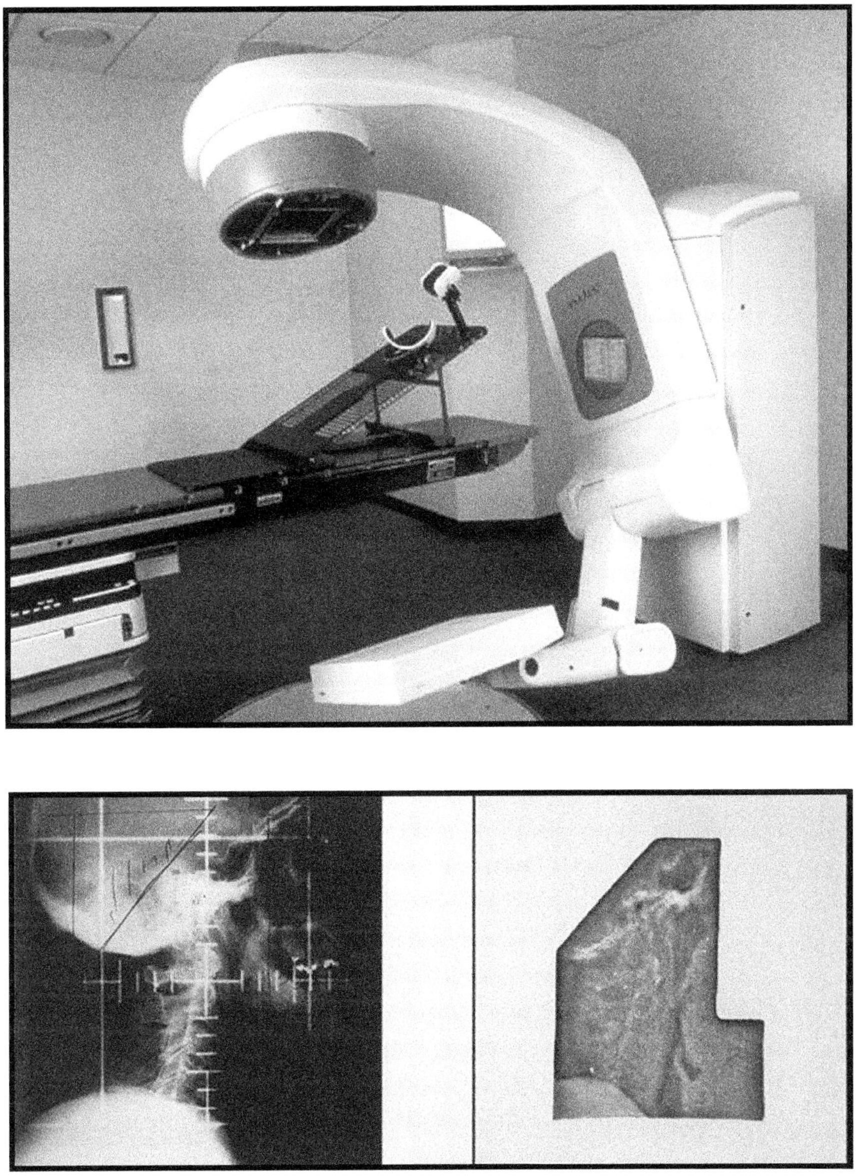

(a)

(b)

Figure 6.5: (a) A modern simulator (Varian Acquity), including a kV x-ray tube and flat-panel detector mounted on an open C-arm gantry. This simulator is capable of planar x-ray views and cone beam CT reconstruction. (b) X-ray beam's eye view from simulator and linac portals: (left) simulator film, (right) electronic portal image. (Reprinted with permission from: © Varian Associates, Inc.)

6.5 Computed Tomography Simulation

6.5.1 CT Scanners and Virtual Simulation

X-ray *computed tomography* (CT) scanners provide high-contrast x-ray images in the transverse axial patient plane (Hounsfield 1973, Brooks and Di Chiro 1975). This information is used to create a density map from which an accurate calculation of x-ray beam penetration is possible. CT is the modality of choice for radiotherapy planning in terms of providing accurate internal/external contour information, combined with an accurate map of electron density.

An excellent review of the use of CT in radiotherapy is given by Ling et al. (1983). This points to a number of studies that show how treatment plans are altered in a large number of cases when CT scan data is available (Van Dyk et al. 1980). It has even been shown that disease-free survival may be improved by the use of plans that involve CT data (Goitein 1979, Pilepich et al. 1980, Emami et al. 1978). These studies show the number of geographic misses is significantly reduced when CT information is used (Smith and Purdy 1991).

Most radiotherapy departments have their own in-house CT. One advantage of diagnostic CT devices is the speed of each scan, which takes only a few seconds. This means that multiple slices can be produced, which aids in 3-D plan visualization and 3-D dose calculation. The disadvantage of these devices is that the scout view modality does not account for beam divergence. A beam divergent view such as that provided by a simulator BEV is required. CT scanners that include software manipulation to emulate a BEV similar to that of a simulator are known as CT-simulators. The process used to carry out this image manipulation is also known as *virtual simulation* (Sherouse and Chaney 1991). The views are obtained by overlaying software-reconstructed beam portals on transverse patient views that are reconstructed from axial CT scan data. These reconstructions are known as *digitally reconstructed radiographs* (DRRs) (Sherouse et al. 1990).

A 2-D CT scan reconstruction through the axial plane is referred to as a CT cut or CT slice. The optimum number and spacing of CT slices for radiotherapy planning has not yet been the subject of detailed analysis. Historically, it was suggested that slice spacing between 0.3 to 0.5 cm in the head region and 0.5 to 1.0 cm in the body region might be optimum (Smith and Purdy 1991). With modern helical scanners full volumetric data sets are routinely collected. Scan reconstructions of 2 mm are routine in the head and neck region and out to 5 mm may be used in the pelvic region.

To allow rapid, safe scans, CTs have an enclosed doughnut gantry that restricts the table movement to less than that which can be obtained on treatment simulators and treatment machines. The enclosed gantry of a diagnostic CT is usually 70 cm in diameter and this also restricts patient arm position. This is a particular problem for breast cancer radiotherapy patients because an arm up position during treatment is favored, but scanning with arms down distorts the anatomy.

CT scanner designs are now available with radiotherapy treatment virtual simulation in mind. An example of such a CT is shown in figure 6.6. This scanner has an 82 cm physical gantry doughnut with a field of view out to a similar range, which is about 12 cm wider than a conventional CT. They also are installed with a laser alignment system and a flat couch top that mimics patient setup on the linac. Most radiotherapy patients can now be accommodated in these gantries and less classical simulation is being used.

The process whereby isocenter is established with reference to patient markers using a virtual simulator is outlined in a series of steps, shown in figure 6.7. In brief, the patient is placed on the CT couch and a reference position is marked using CT fiducials. These are aligned with a special set of lateral and overhead lasers that are placed at an exact known couch offset, usually 0.5 m from the CT scan start position. The patient is then CT scanned with reference to this exact offset.

Once the scan is completed and CT data are reconstructed, a target volume is then outlined, as are the reference fiducial markers. The center of this target volume usually becomes isocenter. The CT couch is then driven back out of the CT gantry to the original laser setup position. The virtual simulation software computes the movable laser offsets required to move the lasers by the same amount as the x,y,z shifts from the reference position fiducials to the isocenter position. The patient fiducial positions are then marked at these new positions so that the treatment beams will intersect at isocenter.

Another option is to have just one movable laser mounted on the ceiling for lateral shifts as the CT couch does not move laterally. Then the CT couch can be moved anterior/posterior and inferior/superior in relation to fixed wall lasers as instructed by the virtual simulator couch offsets. This one

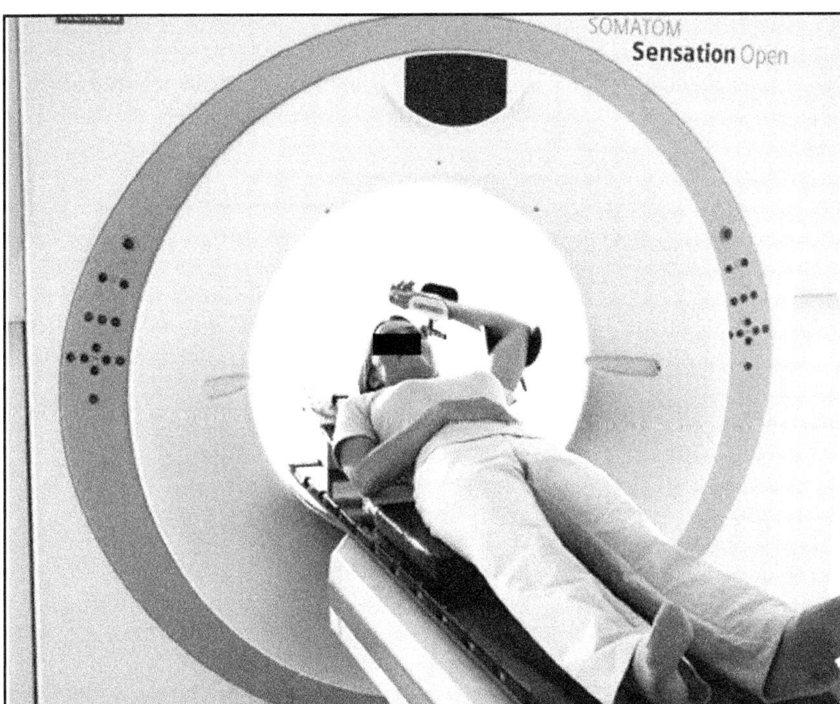

Figure 6.6: CT simulator used to generate CT data and DRRs for virtual simulation purposes. The CT simulator shown (Siemens Sensation Open CT) has a large doughnut opening of 82 cm diameter. This is particularly useful for breast simulation whereby the arm positoning breast boards require a large gantry opening. (Reprinted with permission from: © Siemens Medical Systems.)

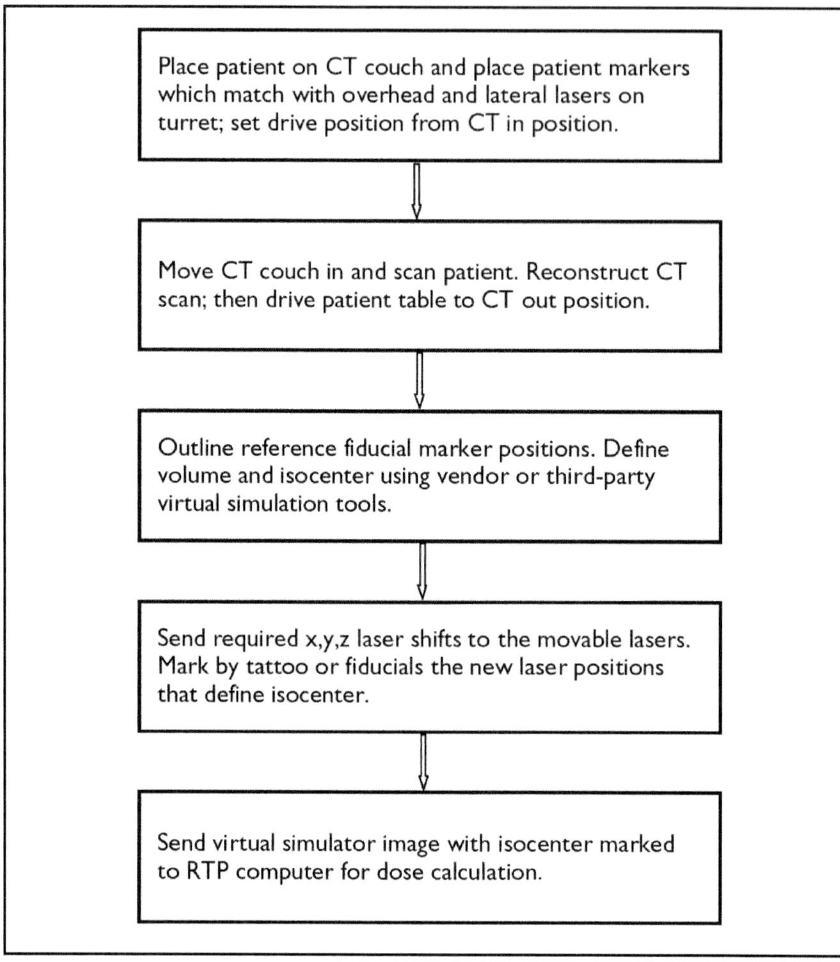

Figure 6.7: Schematic showing the process used to relate isocenter to patient skin marks using virtual simulation.

movable ceiling laser system with fixed wall lasers then achieves the same result as with the ceiling and wall movable laser system.

The CT data and beam isocenter positions may then be sent to the RTP workstation for dose calculation. Any subsequent change which might be required to get a dose coverage of the target volume may be accommodated by small changes in beam geometric field edges using the RTP computer.

With some sites (e.g., prostate) it may be possible to set isocenter before the scan using a standard protocol and not adjust the lasers. Instead, asymmetric jaw shifts are used to accommodate the asymmetry of the target from the isocenter marked.

If isocenter is moved after the scan and cannot be accommodated by jaw shifts, then a daily offset from the patient marks (e.g., tattoo) on the treatment machine may be required. This offset process on the linac treatment couch may be more prone to error on a day-to-day basis.

6.5.2 Obtaining CT Calibration Lines

RTP computers use a data table to compute electron density from CT number (N_{CT}). Some RTP computers require data for mass density versus CT number, and conversion to electron density is computed within the system. Care should be taken to ensure the correct data set is inserted as required by the specific RTP computer, i.e., mass density or relative electron density. In order to collect these data, the physicist should ascertain samples of analog material for bone, soft tissue, lung, and solid water (White et al. 1977 and 1986, Constantinou et al. 1982, Ho and Paliwal 1986). The analogs are generally shaped into cylinders of approximately 2 cm diameter and placed in a cylindrical phantom of approximately 30 cm diameter. This phantom with analogs inserted is then CT scanned.

The phantom should be cylindrical or anthropomorphic (of patient anatomy) in shape to avoid CT image distortions due to reconstruction artifacts (Brooks and di Chiro 1975). A suitable phantom made from solid water is shown in figure 6.8. It is important that the CT phantom have sufficient solid water or tissue equivalent plugs so that samples can be scanned individually, thus avoiding cross talk from one sample affecting the CT number recorded in the other sample position. It is also important to have sample positions across the scan plane to quantify the small deviation in CT number with position (Henson and Fox 1984). A N_{CT} versus density line is shown in figure 6.9a. It varies only slightly for different scanners from N_{CT} = –1000 to 0. However, because of the change in nominal tube energy between scanner types, the representative bone line depends on energy and is therefore scanner

Figure 6.8: CT calibration phantom (Gammex RMI 467) with inserted samples of known mass and electron density.

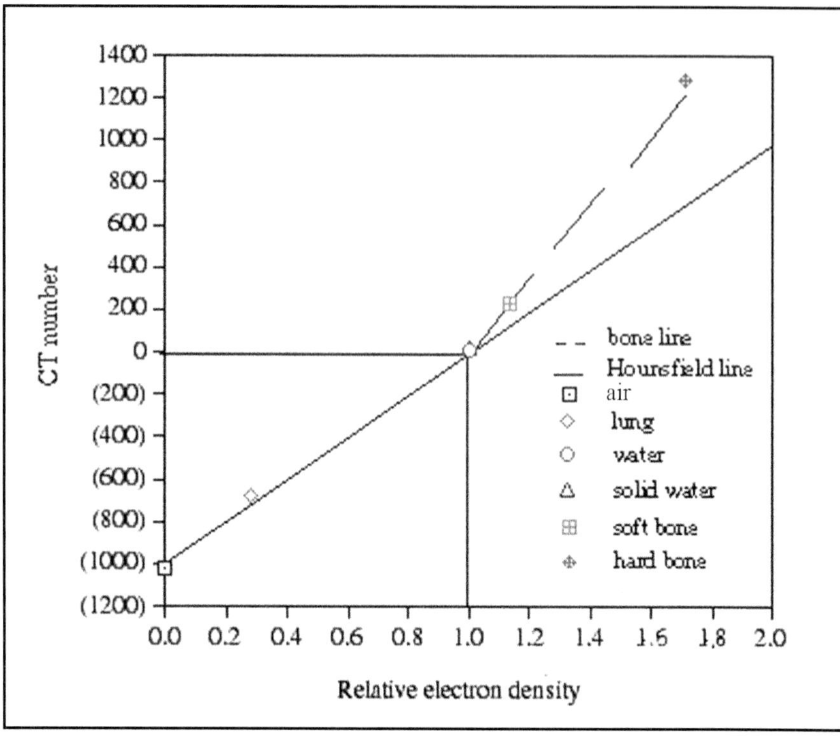

Figure 6.9a: CT number versus electron density. Measured on a Philips CT at 120 kVp, 71 keV.

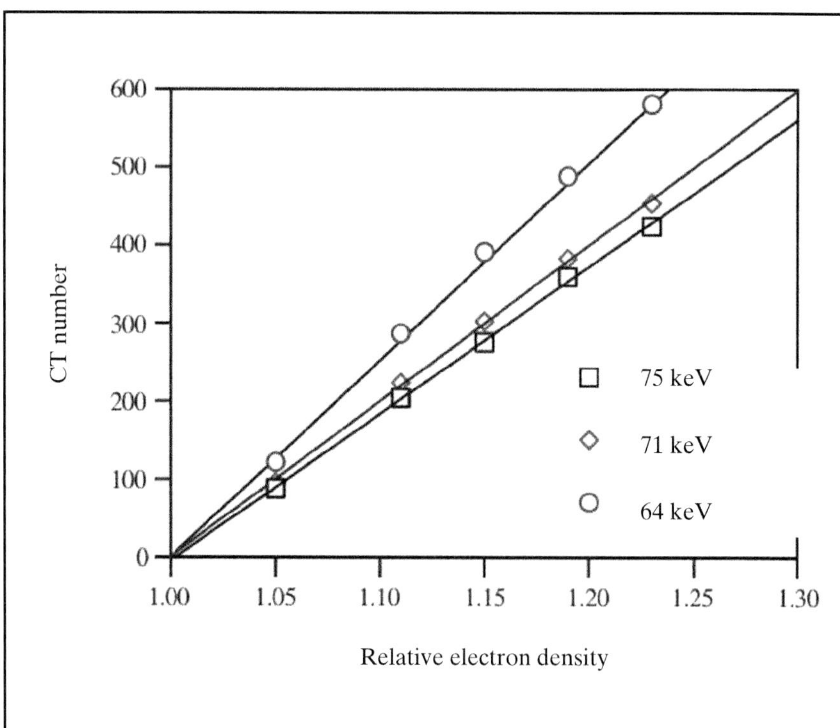

Figure 6.9b: CT number versus electron density. for liquid bone analog (K_2HPO_4) at three different keV. Measured on a Philips CT.

energy dependent as shown in figure 6.9b. To ensure all axis data input are positive numbers, for convenience some RTP computers require data input in CT number plus 1000 as an offset. Shown in figure 6.10 is a screen shot showing a CT calibration line used in an RTP computer for calculations. This shows the CT number plus 1000 on the x axis and mass density on the y axis. Note there is also a large range CT mass density data point put in to account for hip prostheses.

A detailed soft bone calibration line can be obtained using liquid bone analogs. Bone analogs have been investigated elsewhere using liquid solutions with various concentrations of solutions such as monohydrogen dipotassium phosphate (K_2HPO_4) (Huizenga and Storchi 1985). Data for the liquid analogs are limited to a maximum electron density level of $\rho_e^w = 1.23$. This is because the solutions become saturated above this level. Hence, it is impossible to use this solution to model hard bone, which is of higher density than this. A number of solid hard bone analog materials are available (White 1977, 1978, White et al. 1977, 1986) that can be used to extend the line into the hard bone region. Due to convenience and reproducibility solid analogs are now more commonly used than liquid mixtures.

The effective energy of the CT x-ray beam spectrum can be determined by the use of effective keV liquids (White and Speller 1980). This method involves the mixing of carbon tetrachloride with water in various concentrations. The particular concentration that matches that of a water CT number is energy dependent, and the test is reasonably easy to perform. Tests have revealed that CT tube energies of 120 kVp relate to a nominal average energy of about 70 keV (Metcalfe et al. 1988). Note that this is a "harder" beam for the same nominal peak energy than that produced by an x-ray simulator.

The change in x-ray CT spectra has little effect on the soft tissue analog calibration lines, but it has a more significant effect on the bone analog line gradients. This is attributed to the enhanced photoelectric absorption in

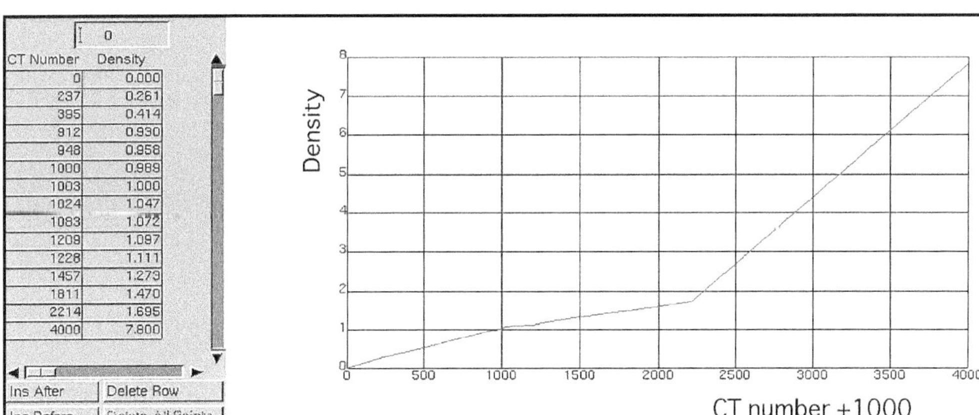

Figure 6.10: CT calibration data used in an RTP computer. This shows the CT number +1000 on the x axis and mass density on the y axis. The data points mass density versus CT number +1000 are also visible in a table (left). The data point (4000, 7.8) is included in an attempt to account for hip prosthesis. These data were collected by scanning the RMI CT phantom in the Siemens Sensation Open CT scanner. (Screen shot from a Philips Pinnacle RTP computer).

samples of greater atomic number when beam energy is reduced (Henson and Fox 1984).

6.5.3 Experimental Relation between CT Number and Electron Density

By fitting regression equations to the data shown in figure 6.9 this can be represented by

$$\rho_e^w = R_\gamma \left(\frac{1}{1000} N_{CT} \right) + 1, \tag{6.11}$$

where R_γ denotes the experimental regression line gradient for substances of different atomic number, Z. The regression equations are useful for radiotherapy planning purposes as these equations are used to convert from N_{CT} to relative electron density, ρ_e^w, for bone-type substances (Metcalfe abd Beckham 1988, McCullough and Holmes 1985). Typical values of R_γ are 1.0 for tissue type substances and 0.5 for bone type substances. An exact calibration tailored to each CT scanner is recommended because the CT number is dependent on the x-ray tube energy.

Before using CT data on RTP computers the scan data should be put into a calibration look-up table of relative electron density versus CT number. This is held in an RTP computer file to ensure that the correct electron density conversion from CT number occurs. Some equations are presented here that will help the reader understand the theoretical relation between these quantities; a protocol is presented that shows how correct calibration data can be obtained.

6.5.4 Theoretical Relation between CT Number and Electron Density

A theoretical relationship between N_{CT} and attenuation coefficients, cross sections lead to a way of predicting electron density from N_{CT}. This gives insight into the fundamental electron density data that can be obtained from CT scans. The theoretical relationship defined by Hounsfield (1973) between N_{CT} and linear attenuation coefficient μ is

$$N_{CT} = \frac{1000\,(\mu_s - \mu_w)}{\mu_w}, \tag{6.12}$$

where μ_s is the linear attenuation coefficient for a substance at a specific energy, and μ_w is the linear attenuation coefficient of water at this energy. The unit of CT number is called the *Hounsfield Unit* (*HU*).

The product of electron density, ρ_e (electrons per cm^3), and total electron cross section, σ_e (cm^2 per electron), is equal to μ (McCullough 1975, Geise and McCullough 1977):

$$\mu = \rho_e \sigma_e \tag{6.13}$$

where σ_e is the sum of component electron cross sections due to Rayleigh scattering, σ_R, photoelectric absorption, σ_τ, Compton scattering, σ_c, and pair production, σ_κ, such that

$$\sigma_e = \sigma_R + \sigma_\tau + \sigma_C + \sigma_\kappa . \tag{6.14}$$

The pair production term can be neglected as at diagnostic CT energies of less than approximately 70 keV there is no such contribution (since 1.022 MeV is required), but it is included here for completeness.

Now let ρ_{es} and ρ_{ew} be the electron densities of a substance and of water, respectively. Also defining σ_{es} and σ_{ew} as the total electron cross sections of a substance and of water, respectively, then substituting equation 6.13 into 6.12 gives

$$N_{CT} = \frac{1000(\rho_{es}\sigma_{es} - \rho_{ew}\sigma_{ew})}{\rho_{ew}\sigma_{ew}} , \tag{6.15}$$

and this is equivalent to

$$\frac{\rho_{es}}{\rho_{ew}} = \frac{\sigma_{ew}}{\sigma_{es}}\left(\frac{N_{CT}}{1000}+1\right) . \tag{6.16}$$

The term ρ_{es}/ρ_{ew} is the electron density of a substance relative to that of water, commonly called the relative electron density, ρ_e^w. The ratio σ_{ew}/σ_{es} has been defined by Battista and Bronskill (1981) as the ratio of electron cross sections, R_σ:

$$R_\sigma = \frac{\sigma_{ew}}{\sigma_{es}} = \frac{(\sigma_R + \sigma_\tau + \sigma_C)_w}{(\sigma_R + \sigma_\tau + \sigma_C)_S} . \tag{6.17}$$

Therefore equation 6.16 can be written as

$$\rho_e^w = R_\sigma\left(\frac{N_{CT}}{1000}+1\right) \tag{6.18}$$

or

$$N_{CT} = \frac{1000}{R_\sigma}(\rho_e^w - R_\sigma), \tag{6.19}$$

Using known elemental composition data, R_σ can be calculated as follows (Metcalfe et al. 1989).

$$R_\sigma = \frac{\left[\sum_{i=1}^{N} f_i \left(\frac{\mu}{\rho}\right)_i \left(\frac{A}{Z}\right)_i\right]_w}{\left[\sum_{j=1}^{N} f_j \left(\frac{\mu}{\rho}\right)_j \left(\frac{A}{Z}\right)_j\right]_s} . \tag{6.20}$$

Data for the fractional elemental compositions f_i for bone analog is given in table 3.1. Included in this table for bone analog are values of $(\mu/\rho)_i$, which were estimated at 71 keV using data from Jackson and Hawkes (1981), and values of $(A/Z)_i$, which were obtained from the periodic table for chemical elements. Values of R_σ calculated for bone analog are calculated using equation 6.20 as shown in example 6.8.

By referring to equation 2.18 it can be seen that from ρ and the atomic composition, ρ_ε^w can be accurately calculated using the relation (Henson 1989):

$$\rho_e^w = \frac{N_A \rho}{\rho_{ew}} \left[\sum_{j=1}^{N} f_j \left(\frac{Z}{A} \right)_j \right]_S , \tag{6.21}$$

where N_A is Avogadro's number (6.023×10^{23} mol^{-1}) and ρ_{ew} is 3.33 $\times 10^{23}$ cm^{-3}. Substituting values for N_A and ρ_{ew} into 6.21 gives

$$\rho_e^w = 1.81 \rho \left[\sum_{j=1}^{N} f_j \left(\frac{Z}{A} \right)_j \right]_S . \tag{6.22}$$

Values of ρ_e^w for the body analogs for their equivalent human tissues are presented in table 3.2. Another quantity of interest in assessing the radiation properties of analogs is the effective atomic number. This is assessed by using an adaptation of the methods developed by McCullough and others (McCullough 1975, McCullough and Holmes 1985, Geise and McCullough 1977) as follows.

A very approximate relation is

$$\rho_e^w \approx \rho . \tag{6.23}$$

This is heavily dependent on the amount of hydrogen in the compound. This can be seen by studying equation 6.22; if (Z/A) equals approximately 0.5 as it does for most compounds, the above equation holds. Note however for hydrogen (Z/A) equals 1 and this tends to cause changes in the above approximation. For most substances that have less hydrogen than water the term in brackets in equation 6.22 is less than (1/1.81), and therefore ρ_e^w is less than ρ for most substances.

The effective atomic number Z_{eff} is another important parameter as this affects the cross section; it can be derived from McCullough and Holmes (1985):

$$Z_{eff} = \left[\sum_{i=1}^{q} \alpha_i (Z_i)^a \right]^{\frac{1}{a}} , \tag{6.24}$$

where the exponent "a" accounts for photoelectric effects at CT energies. The value of $a = 2.94$ is used in the calculations to conform with other publications (McCullough and Holmes 1985). The coefficient α_i in equation 6.24 can be calculated by (McCullough and Holmes 1985):

$$\alpha_i = \frac{n_i Z_i}{\sum_{j=1}^{q} (n_j Z_j)} . \qquad (6.25)$$

The quantity n_i is the relative number of the ith element in the substance (i.e., from the empirical formula). But to use equation 6.25 directly from values of constituent fraction by mass f_i, α_i needs to be determined in terms of f_i. It has been shown that (Metcalfe et al. 1990)

$$\alpha_i = \frac{f_i \left(\frac{Z}{A} \right)_i}{\sum_{j=1}^{q} \left(\frac{Z}{A} \right)_j} . \qquad (6.26)$$

Using equation 6.24 combined with equation 6.26, values of Z_{eff} have also been calculated for human tissue and analogs as shown in table 3.2.

Also shown in figure 5.9 is the *Hounsfield* regression line. Note all the bone analog data are above the Hounsfield line. The gradient of the bone line becomes steeper if the energy at which the CT scan is taken is reduced (Metcalfe and Beckham 1988). A number of liquid bone analogs have been scanned at different energies to show this effect. (See figure 6.10.) This is because the bone analogs have a higher Z_{eff} than water, and hence their cross section is greater. This leads to a higher CT number per electron density than with water-like substances. Because the cross section affects the linear attenuation coefficient and μ increases as energy is reduced, for a given electron density the linear attenuation coefficient of bone is higher than that of water-type substances. Because CT number varies linearly with μ (see equation 6.12), the CT number is greater also.

An example of how to calculate the values shown in table 3.2 is provided in the following example. This is also included for the convenience of readers who may wish to repeat the procedure with other analogs or in vivo human data.

Example 6.8

The following is a calculation of R_σ, ρ_e^w, and Z_{eff} for bone analog material. To calculate R_σ at 71 keV the mass attenuation coefficients at this energy for the composite elements are found in tables (Jackson and Hawkes 1981). The A/Z ratios are found in the periodic table. Percentage by mass of elements are from White et al. (1977).

To calculate R_σ then for water

$$\left[\sum_{i=1}^{N} f_i \left(\frac{\mu}{\rho} \right)_i \left(\frac{A}{Z} \right)_i \right]_w = 0.112 \times 1.008 \times 0.3165 + 0.888 \times 2.0 \times 0.1736$$

$$= 0.3440.$$

Table 6.3: CT numbers for analogs and human tissue. The calculations used data from table 3.2. These are compared with CT numbers measured on a Siemens DRH CT scanner at a nominal energy of 71 keV (column 3).

Sample	N_{CT} *(Calculated)*	N_{CT} *(Measured)*	ΔN_{CT}
Bone analog (RMI 456)	222	228	–6
Human bone	224	236[+]	–12
Muscle analog (MS 10)	34	14	20
Human muscle	44	30	1
Lung analog (RMI 455)	–716	–677	39
Human lung	–699	–711[+]	12

[+] The human bone data and human lung data were obtained by sampling five patients whose bone and lung regions were CT scanned. It is recognized that bone and lung density and composition vary greatly depending on site and patient.

Also for the bone analog substance

$$\left[\sum_{i=1}^{N} f_i \left(\frac{\mu}{\rho}\right)_i \left(\frac{A}{Z}\right)_i\right]_s = 0.079\times1.008\times0.3165+0.6379\times2.0\times0.1666 +0.0423\times2.0\times0.1736+0.0988\times2.0\times0.1736 = 0.3768.$$

Then from equation 6.20 the ratio of these two gives $R_\sigma = 0.9330$.

The mass density was measured as $\rho = 1.16$ then using equation 6.22 $\rho_e^w = 1.14$. Applying equation 6.19 gives $N_{CT} = 222$ HU.

The N_{CT} value calculated using the method outlined above is shown in table 6.3. In order to calculate Z_{eff} it is necessary to use equation 6.26. The denominator is equal to 0.5357. Therefore repeating the summation for the constituent elements as outlined in equation 6.23 and including Z_i with the exponent ($a = 2.94$), the numerical value of $Z_{eff} = 9.26$ is calculated.

It is important to note that the composition of real inner bone is used for the calculation of N_{CT} in table 6.3. These data are based on the estimate of White et al. (1977) that inner bone has an average composition of 22.4% hard bone and 77.6% red marrow.

As shown in table 6.3 samples of the analog as scanned by a Siemens DRH CT scanner give experimental N_{CT} to compare with those theoretically predicted. It is important that CT calibration data be collected for the RTP computer and that this calibration be checked at routine intervals to ensure that tube voltage and CT calibration has remained consistent. Note that inconsistencies have been shown to arise (Levi et al. 1982).

6.6 Other Treatment Planning Aids

6.6.1 Nuclear Magnetic Resonance Imaging

In contrast to CT scans, which use x-rays to reconstruct an image of the electron density distribution, nuclear magnetic resonance imaging (MRI) is sensitive to the magnetic moments of protons. In addition to the proton density, other properties of protons can be used to obtain contrast in MRI. In particular, the relaxation times T1 and T2 play an important role in the design of an appropriate MRI experiment. These relaxation times describe how energy and entropy are redistributed after an excitation pulse of radio frequency (RF) to the protons. In general the relaxation times of different tissue types are different; in particular there is a difference between neoplastic and normal tissue. This allows the design of a customized MRI pulse sequence to image a particular tumor. In general, MRI has a significant advantage over CT with regard to visualizing many tumor types and outlining the tumor margins (Boechat et al. 1988, Krestin et al. 1988, Jelinek et al. 1990, Brant-Zawadzki et al. 1984, Shuman et al. 1985, Henkelman 1992).

Acquisition of an MRI image usually takes a rather long time (several minutes). Therefore, the traditional strong points of MRI are the head/brain, the spine, and joints, which are not prone to a lot of internal movement which would blur the images. In these areas MRI allows the distinction between edema, cysts, and tumors, playing an important role not only in the diagnosis but also treatment planning (tumor outline) of tumors.

However, more recently fast imaging sequences have become available on clinical MRI scanners that allow the acquisition of an image in less than a minute or even shorter times (echo-planar imaging). The relatively poor signal-to-noise ratio in the latter can be improved with repeated images. Fast imaging allows the reduction of motion artifacts and opens other areas of the human body for MRI investigation. In particular, in the pelvic region it is expected that the superb soft tissue contrast of MRIs may contribute to improved tumor outlining and treatment planning. A number of studies have shown that this typically leads to a reduction of the field size used, which reduces the number of side effects in radiotherapy treatment.

Other recent developments, like MR angiography, allow the display of blood vessels without injecting contrast agents and can be helpful in diagnosis of arteriovenous malformations (AVMs) and obstruction of vessels by the tumor. In addition to that it may be possible in the future to assess the tumor blood flow and hence oxygenation in vivo thus allowing a possible superior rationale for RTP on a more radiobiological basis.

In general, MRI does not require the use of contrast agents. However, Gadolinium DTPA is widely used to enhance contrast in MR images. This substance is also able to cross the blood brain barrier at locations where it has been damaged by a growing tumor. Therefore, Gd DTPA is sometimes even referred to (falsely) as a tumor marker. It certainly plays an important role in the diagnosis of brain tumors and their distinction from other structures like edema.

MRI is a *3-D* technique allowing the acquisition of volume data as well as the information in any oblique slice that may be required for treatment planning. This can be achieved without repositioning the patient. The typical voxel size of MR images is smaller than one millimeter. Exact size depends on field of view, the number of pixels, and the imaging time. In addition, the resolution of images can be improved with customized surface coils.

Contrast between voxels in MRI is due to different physical processes than CT, which only images electron density variations. Note however that MRI scans do not provide electron density information. This means that accurate dose calculations that rely on electron density information cannot be obtained from MRI scans. This gives the dilemma of superior tumor definition to CT but inferior inhomogeneity correction than from CT. Current methods therefore involve overlaying the MRI tumor information onto the CT electron density image (Fraass et al. 1987, Kessler et al. 1987, Ten Haken et al. 1992).

The spatial accuracy of MRI scans can vary (Fraass et al. 1987, Gauvin 1992, Covell et al. 1986). This is because the spatial localization gradient magnetic fields may not always be perfectly linear. This appears to lead to small distortions of less than a millimeter for small field of view applications such as head scans for stereotactic radiosurgery planning (Henri et al. 1991). The geometric accuracy of any system should be checked on a routine basis to ensure the geometric accuracy of these devices.

6.6.2 Magnetic Resonance Spectroscopy

Magnetic resonance spectroscopy imaging (MRSI) using Proton (1H) mode can be used to image water, lipids, choline, citrate, lactate, and creatine. It has been hypothesized that an elevated choline level relative to that of citrate (choline/citrate ratio of 2.1 ± 1.3) is an indicator of active tumor (Kurhanewicz et al. 1995). For this technique an RF probe is inserted into the rectum, which distorts the prostate anatomy. This is an example where deformable object information is used to fuse the MRS image shape onto the CT data set used for radiotherapy planning and targeting (Ling et al. 2003). Based on this information feasibility studies into boosting intraprostatic lesions, while maintaining standard dose to the rest of the prostate have been reported (Xia et al. 2001).

Using MRSI information as possible target zones to *dose paint* has been suggested by Ling et al. (2000) by giving extra boost dose using IMRT to intratarget volumes. Being able to do this to structures such as the prostate relies on being able to precisely target the position of these intratarget structures on a fraction-by-fraction basis. Since MRI is completely noninvasive, these scans can be repeated as many times as desired (if the costs can be met). This can be useful for the early assessment of tumor response to radiotherapy by means of MRI and/or MRSI. One challenge for dose painting is that hypoxic regions may not stay in the same location over time.

6.6.3 Nuclear Medicine Bone and Lymph Node Scans

Nuclear medicine scans are often used to establish a diagnosis prior to radiotherapy. Most nuclear medicine techniques yield primarily physiological information rather than anatomical data. Radioactive markers label compounds that follow certain metabolic pathways or target a certain cell type. Targeting tumor cells can be useful for the detection of metastases. A special type of scan commonly used in oncology is a *bone scan*, which targets areas of rapid bone turnover. These scans are a good method of establishing that a patient does not have any bone metastases prior to the decision to offer radiotherapy.

Another example of the usefulness of nuclear medicine scans is the imaging of lymph nodes that are not visible with CT or simulation. Breast cancer in certain patients is now treated by breast conservation. Lumpectomy plus radiotherapy to the breast yields similar results in terms of survival and disease-free survival as mastectomy (Recht et al. 1988 and 1989).

One way of localizing the lymph nodes in the breast is to use a gamma camera anterior and lateral scan. The nodes are spatially positioned using a method devised by Yip and Ege (1985) and referred to elsewhere (Ege and Clark 1985, Collier et al. 1983, Siddon et al. 1982). The method is known as lymphoscintigraphy. The lymph node positions can then be placed onto CT scan data as regions of interest. (Note that lymph nodes do not show up well on CT scans.) The region of interest transfer from gamma camera to CT is required because the CT scan data are ideal for RTP and the gamma camera data are not.

There is a useful hand-held device called the *intraoperative gamma* probe (Zanzonico and Heller 2000) that detects radioactive tracer and helps surgeons locate the first (sentinel) node in a chain. This is used for breast cancer patients during surgery. Negative pathology on biopsy of the located sentinel node can sometimes allow the other nodes to stay in place, hence leading to less-invasive surgery.

Note also that treatment of thyroid cancer with sodium iodide (I-131) and bone palliation with strontium (Sr-89) may be undertaken in nuclear medicine or radiotherapy departments. Coordination between the radiation oncologist and the nuclear medicine physician is essential for the treatment of these patients. It is also proposed to extend the application of tumor (primary or metastases) radiotherapy using radioactive isotopes attached to monoclonal antibodies. This could be an important cancer treatment option in the future.

6.6.4 Positron Emission Tomography

The advent of better localization techniques such as single photon emission computed tomography (SPECT) and in particular positron emission tomography (PET) provides further scope for the improvement of radiotherapy planning. For a brief review of these applications refer to Smith and Purdy (1991).

Recently the use of fluorodeoxyglucose (FDG) PET images has become invaluable as a tool for cancer staging. These images show regions of increased glucose metabolism that point to enhanced uptake in tumor and malignant cells. Clinical studies of disease sites including brain, breast, head and neck, lung, and colorectal and ovarian cancer have shown PET imaging with FDG is useful in improving detection and in staging of tumors (Mac Manus et al. 2001). This, in turn, has led to better design of a suitable treatment course based on a more accurate assessment of tumor stage.

These FDG images may potentially also be useful in defining tumor volumes. Sites of particular interests include lung and head and neck tumors. These PET images may be registered and fused with CT anatomical images to help define the tumor location. With the combination of PET CT on a common gantry (Townsend and Cherry 2001), the fusion process becomes even more straightforward as the CT and PET are done in the one planning session. Currently, PET CT fused data are used as a beacon in terms of nodal recurrence and can then guide the clinician as to how much dose to give to various target regions based on node positive or node negative.

Due to varying intensity levels from PET scans the use of the data to set margins is still not fully resolved. The scan is often used to show up areas of irregular metabolic activity and margins may still be defined using CT anatomy.

The ideal concept is to have the tumor outlined using the PET information overlaid on the CT image; then a target volume is grown. There are several methods being developed to try to standardize the margins using PET scan data. One suggested method to standardize the intensity of PET scans uses a *standardized uptake volume* (SUV) defined as the decay corrected activity per tissue volume divided by the injected activity per body weight (Biehl et al. 2006). Attempts to match a percentage of the SUV to a CT intensity level have also been reported (Hong et al. 2007). The complexity of FDG uptake in tumors precludes both these methods working successfully for all patient cohorts and more research in this area is required. However, early dose escalation planning studies using PET CT to define target volumes in lung look promising (e.g., De Ruysscher et al. 2005).

There are some hypoxic imaging agents for PET, e.g., ^{18}F-fluoromisoidazole (known as F-miso) that may be used to image regions of high tumor hypoxia (Ling et al. 2003). A kinetic model using this tracer to analyze tumor hypoxia overcomes the problem that hyopoxic cells are spatially separated from well-perfused vessels and secondly the tracer uptake occurs only in viable hypoxic cells (Thorwarth et al. 2005). As with MRSI and dose painting mentioned earlier, there are issues with chronic versus acute hypoxic zones and their possible mobility within the tumor volume over time; this mobility aspect still needs further study.

6.6.5 Ultrasound

Ultrasound was historically used to augment conventional simulation information for some plans where it is not possible to undertake a CT scan. For example, the chest wall may be ultrasound scanned to ascertain chest wall separation (i.e., the distance between lungs and external body contour). This method does not provide patient-specific information about electron densities of these structures, and hence mean densities (derived from patient surveys) were assigned. As CT data have become more available for RTP, ultrasound scans to obtain chest wall thickness estimates have become obsolete.

Ultrasound has been successfully used to ascertain lens depth for treatment of retinoblastomas (Glatstein et al. 1985). It has been hypothesized that this may also be a useful modality for providing rectal wall position information for high-dose conformal prostate treatments (Ten Haken et al. 1991).

Since then, ultrasonic imaging systems have found more of a role in daily image-guided radiotherapy procedures. Devices such as the B-mode Acquisition and Targeting (BAT®) system (NOMOS Corporation) or Exac-Trac® (BrainLAB Inc.) are used in some centers to localize the prostate prior to irradiation.

The position of the prostate, as determined by the transabdominal ultrasound imaging, is compared to the previously CT-defined prostate location relative to isocenter. The position of the ultrasound transducer is referenced to the coordinate system of the linac's isocenter, either through physical connection to the gantry, as is the case for BAT, or via an infrared localizing system such as that used by ExacTrac. Typically, it is the interface regions between the bladder and prostate and the rectum and prostate, which show on ultrasound, that are used as reference landmarks. The locations of these interfaces relative to isocenter are compared to the planned location, and the necessary calculated couch shifts relocate the prostate to the correct position.

There have been several studies into the use of transabdominal imaging for tracking prostate motion (Jani et al. 2005, Trichter and Ennis 2003). A study by Lattanzi et al. (2000) of 54 prostate patients found that the variation in prostate position relative to the baseline (planning) CT was largest in the anterior/posterior direction, followed by the superior/inferior direction with standard deviations of 8.3 mm and 6.5 mm, respectively. At a minimum, transabdominal ultrasound imaging of the prostate can provide a better indication of appropriate margin sizes for target volume delineation. At best, it allows for the daily relocation of the prostate in an attempt to better cover the mobile prostate organ, leading to potential reduction in the target margin.

6.6.6 Image Registration and Fusion

Image registration refers to the coordinate transformation required to geometrically match one data set with another. Fusion is the overlaying of some data from one data set onto another. This can create an enhanced view of some tumor volumes with respect to a reference anatomy data set such as that provided by CT. In radiation oncology MRI fused with CT and PET fused with CT image data provide extra information about tumor stage, and potentially tumor location for some tumor sites. Image fusion of these images allows for a combination of functional (MRI or PET) and anatomical (CT) images for better diagnosis, staging, and therapy planning (Delbeke et al 2001). Subsequent MRI or PET patient images may be compared over time in order to monitor progress.

Image registration can be rigid or non-rigid, and can be performed automatically or manually. Rigid registration refers to fusing images using only translations, rotations, reflection, and linear scaling. Non-rigid registrations include affine transformations, which consist of uniform and non-uniform scaling and shearing. Another non-rigid transformation involves polynomial or curved transformations, which use warping to remove spatial distortions between images.

Manual registration refers to a software program that allows the user to physically align the images based on sighting by eye; this is, in effect, a manual rigid transformation. Automatic image registration is provided by a software program that performs the rigid or non-rigid fusion based solely on the equations in the program. Most image registration software used today involves an initial approximate manual alignment followed by an automatic fusion by the software (Weber and Ivanovic 1994).

Fusion applications started with registration to external fiducial markers, mounted on stereotactic head frames used in radiosurgery applications. The head frames are fixed to the skull for single fraction applications, or bite block frames can be used for multiple fraction applications. However, automatic registration to bony or other anatomical landmarks without external fiducial markers is now quite common in radiotherapy applications involving MRI or PET fusion onto CT images. Registration to internally inserted prostate fiducials is another option (Chung et al. 2004).

6.7 Radiotherapy Treatment Planning Hardware

A photograph of an RTP computer planning system is show in figure 6.11. This RTP computer employs features common to many other systems. Shown from left to right are a hard copy printer, control video display unit (VDU), display VDU, text printer, and a high-resolution hard copy printer. More modern workstations combine commands and display on one VDU. The printers are also often networked between several workstations.

Figure 6.11: Photograph of a radiotherapy treatment planning system. (a) = hard copy display, (b) = control VDU, (c) = display VDU, (d) = text printer, (e) = high resolution printer. (Reprinted with permission from: © Siemens Medical Systems, Oncology Care Systems Group.)

There is a continual trend toward faster processors, enabling rapid and accurate dose calculations and isodose display. RTP started on computers like the Rad 8, where 8 stands for 8 kilobytes of CPU random access memory (RAM), in 1973 (Milan et al. 1974). Now there are workstations of ever-increasing RAM, typically 8 gigabytes, with massive disk storage and ever-increasing computational speed. There was a major increase in the software capabilities of RTP computers in the 1990s as several clinics pursued RTP calculation and display solutions for 3DCRT (Smith and Purdy 1991).

Convolution calculations are now routinely used by several commercial vendors of radiotherapy planning computers as the basic tool for photon beam dose calculations. These calculations can typically be undertaken in less than half a minute on modern computer processors. Times of less than one minute have been achievable for some time (Murray et al. 1989).

Macro Monte Carlo methods are achieving acceptable speeds for routine electron calculations and routine macro Monte Carlo calculation for photon beams with times approaching 1 hour for complex IMRT plan calculations are now available on some commercial vendor planning systems (e.g., CMS VMXC Macro Monte Carlo calculation).

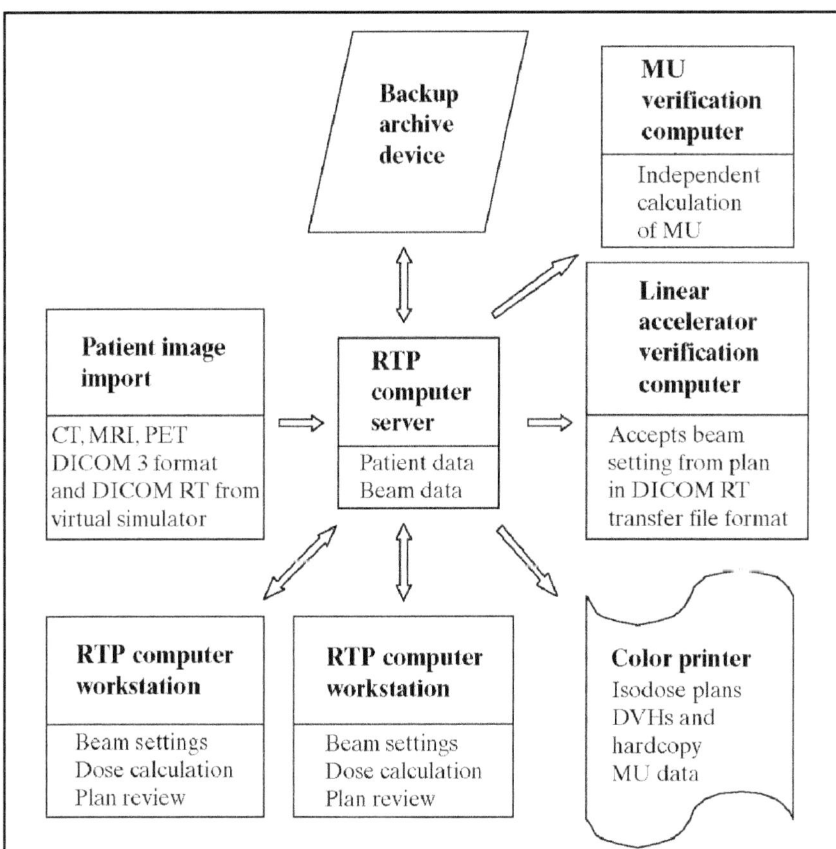

Figure 6.12: Schematic diagram showing the interactions and data flows into and out from a computerized radiotherapy treatment planning (RTP) computer.

Suffice it to say that hardware speeds are improving at about 30% per year and that the faster the processor the better, particularly for complex 3-D calculations and 3-D display.

Shown in figure 6.12 are common data flows to and from the RTP computer to other peripherals. The input of patient data is usually by network Ethernet line and the file format followed is currently the Digital Imaging and Communications in Medicine 3 (DICOM 3) standard. This avoids each vendor having to supply a software handler to convert different CT data formats. An extended DICOM RT file format for transfer is also available for transfer of CT image data information from a virtual simulator. These files include target volume information, beam placement, and beam geometry parameters.

The modern RTP computer system then generally has several workstations capable of beam manipulation and dose calculation. These workstations often have most of the tools available on the virtual simulation stations so that further volume outlining of critical structures can proceed if necessary. The line between virtual simulator workstations and RTP workstation is becoming somewhat blurred. Virtual simulator software is usually used for laser moves and for DRR production. Virtual simulator stations sometimes have more imaging tools and may have image fusion tools, while the only tool not usually resident on a virtual simulator workstation (that is always on a RTP computer) is the dose calculation tool.

A number of RTP computer workstations are usually networked so that plans for different patients can be worked on in parallel. Each RTP workstation is capable of treatment planning and beam calculations. Some workstations may have additional software licenses to enable inverse planning for IMRT and/or image fusion of MRI and PET images with CT images. Output of beam planning data is sometimes directly provided to an independent MU check computer running MU check code (e.g., RadCalc, LifeLine Software Inc., Bullard, TX, or MU Check (Oncology Data Systems, Oklahoma City, OK, or IMSure, Standard Imaging Inc., Middleton, WI).

The RTP computer workstations are usually linked to a hard-copy print-out device such as a color laser printer. Printouts of isodose distributions over-laid on CT scans, MU per fraction, DVHs, and machine/beam parameters can be printed out if required. Direct transfer of machine beam parameters such as field size, field angle, and MLC shapes can be provided via Ethernet using DICOMRT file format transfer to the linac verification computer. These parameter settings can then be acquired by the linac verification computer. Some of these parameter settings can in turn be acquired by the linac control computer if the linac is being used in auto field sequence (AFS) mode.

Backup of patient treatment plan data is required in case of computer failure and this may be provided by a backup server. RAID (redundant array of independent disks) arrays are another redundancy disk option. Archive of plans to be kept for posterity may be achieved using tape, DVD, optical disk, or a myriad of other storage media devices that are now available.

6.8 Dose Display

6.8.1 2-D Plans

A 2-D plan consists of an isodose distribution resulting from the chosen beams overlaid on the patient structures and clearly shows the dose to the target volume (see figure 6.13a). The radiation oncologist then reviews the plan and decides whether the chosen beams are providing the required dose distribution. This is all done prior to patient treatment to ensure that the dose distribution delivered to the patient is adequate. Superimposing isodose curves on a series of 2-D transverse image sections is the most common method of displaying dose distributions. 2-D isodose curves can also be viewed overlaid on sagittal or coronal views.

Dose distributions may also be displayed as color-wash overlays on image planes whereby each dose range is assigned a specific color. The CT image is displayed with the various dose range regions flooded with transparent color. Images of regret show regions of overdose to normal critical structures and underdose to the target volume. These may also be displayed as color-wash overlays on CT images, with color indicating over- or underdose. These displays highlight potential trouble spots without the need to search for them (Shalev et al. 1988).

6.8.2 3-D Plan Displays

As the contours in many regions vary from slice to slice, so too does the dose distribution. Displaying 2-D image sections side-by-side can be very difficult (see figure 6.13b). A plan may appear to be superior in one plane but may be totally unacceptable in another. Therefore, a number of different modes of displaying 3-D anatomy and dose distributions have been developed. These assessment tools have become essential with the advent of 3DCRT because now that irregular shapes can be defined with MLCs, it is important to be able to visualize the blocking of structures and tumor coverage in 3-D.

Historically, 3-D display of treatment plans has been developed using wire-frame, surface rendering, and BEV techniques (Mohan 1989, Goitein et al. 1983). This section defines dose visualization methods and outlines dose optimization tools. Several of the definitions are adapted from Mohan (1989).

Figure 6.13c is an example of a surface-rendered 3-D plan. The target volume is visible on the left side of the patient's neck surrounded by an isodose plane shown as a series of ribbons. The intensity is governed by the distance of the point from the viewer (depth queuing). The utility of information conveyed by surface dose displays depends on the assumption that the dose at the surface is an adequate indicator of dose distribution within the structure and that there are no sudden gradients in the volume. This should be the case in the vast majority of situations.

Isodose surfaces may be shown along with the surface displays of the tumor volume and anatomical structures. To distinguish the anatomy from

(a)

(b)

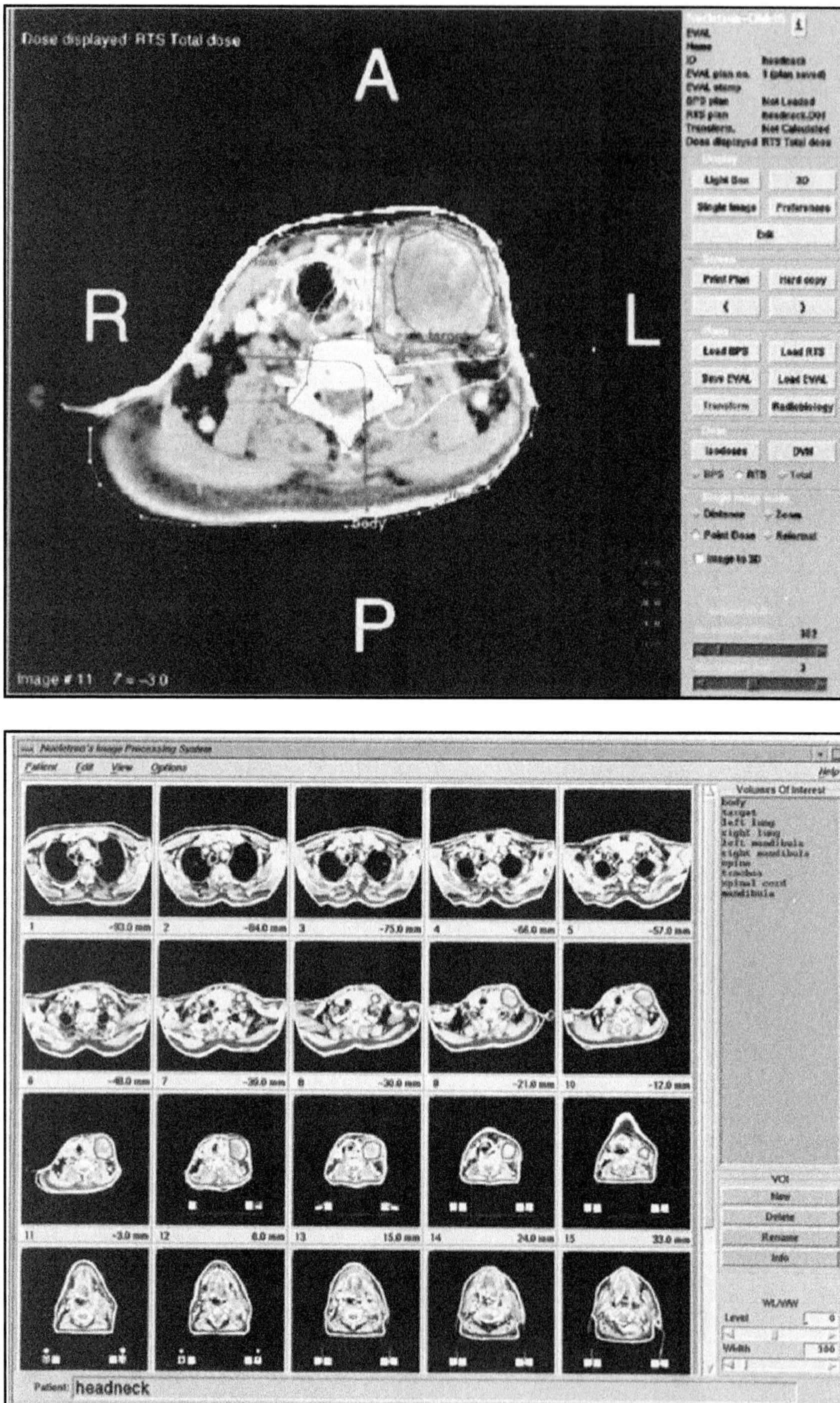

Figure 6.13: Dose display tools. (a) A 2-D head/neck plan showing isodoses overlayed on a CT image. (b) A series of 2-D CT slices with target drawn.

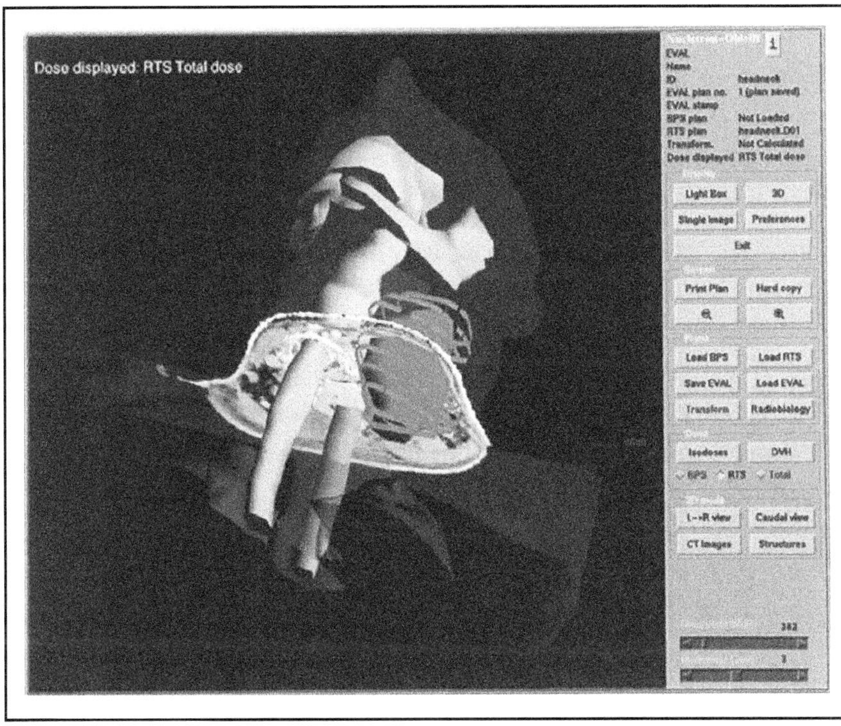

(c)

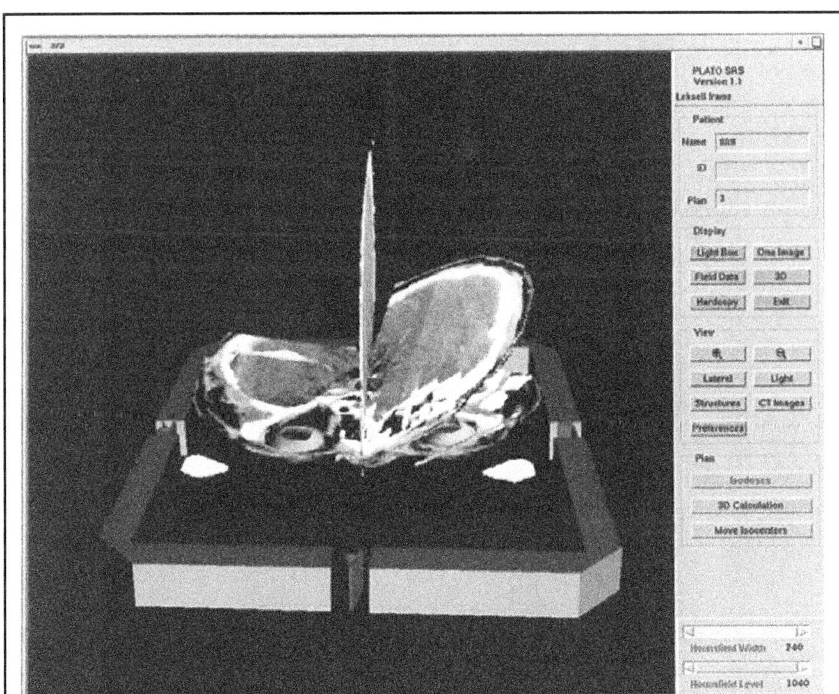

(d)

Figure 6.13: (Continued) (c) 3-D surface-rendered image showing target and isodose ribbons with respect to anatomical structures. (d) Clipping planes. (Images (a)-(d) reprinted with permission from: © Nucletron-Odelft.)

the dose distribution, different surface rendering techniques need to be employed. For example, the anatomy may be shown as a stack of contours, and the dose may be displayed as a mesh of points or as a solid transparent surface. Normally, it is not useful to display more than one isodose surface at a time.

A great deal of versatility has been achieved by displaying various combinations of gray-scale image planes, wire-frame surfaces, and solid shaded surfaces. *Axonometric* displays show several planes of gray-scale images registered together as they would be in 3-D. For example, a sagittal, coronal, and several transverse cuts may be shown simultaneously on different parts of the RTP computer display screen.

A variation on the combination of gray-scale images and solid surfaces is the use of cut-out or "watermelon" displays. An anatomical object (for example, the skin surface of the head) is shown as a solid, shaded object. A box-shaped section can then be scooped out to reveal the gray-scale image data within the head in the resulting cut planes. Isodose levels can be shown as contours on the gray-scale surfaces.

A useful feature is a clipping plane that can be freely moved in space to cut open various structure and dose surfaces. As shown in figure 6.13d, where three clipping planes have been selected, the plane at the level of the eyes also shows a stereotactic frame. While all these tools look impressive, due to the complexity of information presented and some subjectivity in their interpretation, they have not found a great deal of use in the clinic.

6.9 Dose Volume Histograms

6.9.1 Definitions

Visual examination of dose distribution data with respect to the anatomy allows the physician to evaluate a plan subjectively. Additional tools are required for further quantitative assessment of plans. *Dose volume histograms* (DVHs), for example, can graphically summarize large amounts of 3-D dose distribution information throughout the volume of the tumor or a normal anatomic structure. They present the volume (or a percentage of the total organ volume) that receives dose in a given range of dose levels.

Because of the complexity of designing plans that cover the treatment volume, new techniques have been established to quantify target coverage, such as DVHs (Chen et al. 1987, Niemierko and Goitein 1991). A DVH graphically demonstrates the dose that various organs receive. The dose to an entire volumetric structure is reduced to a single graph. The histogram shown is an integral DVH, which displays how much volume receives at least a certain dose threshold. The DVH tool combined with stepping through the 2-D isodose curves are the most common tools used for dose assessment in the clinic.

Differential DVHs are constructed by grouping the frequency of voxels within a dose distribution into discrete ranges, for example, as shown in figure 6.14, the number of voxels with a dose between 60 and 70 Gy is 9.

Integral or cumulative DVHs, which are the most commonly used DVHs, then take these frequency distributions and group each distribution as a

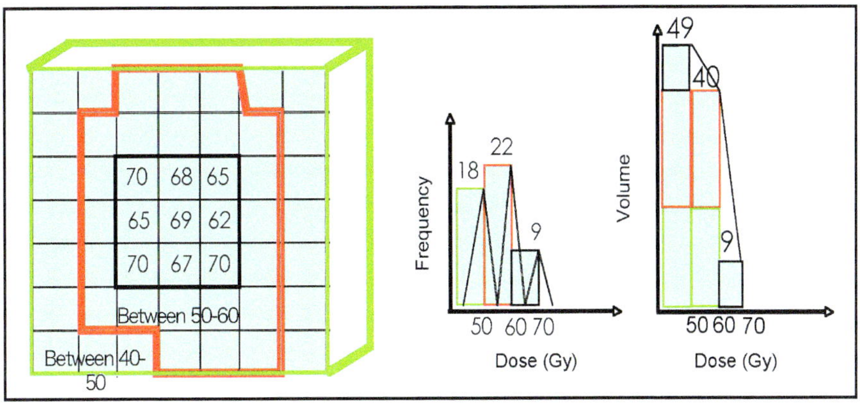

Figure 6.14: Description of how differential and integral dose-volume histograms (DVHs) are generated from dose voxel data.

cumulative dose more than a particular dose range. For example, in figure 6.14 there are 22 dose voxels between 50 and 60 Gy and 18 dose voxels with dose between 40 and 50 Gy, so there are 31 voxels with dose of 60 Gy or less, as shown on the cumulative DVH. Because the voxels are generally small, the lines joining the data points are generally smooth and the "histogram" graphical presentation is removed, but the name is retained, hence dose-volume *histogram*. The y axis volume in cm^3 can be user selected; however this is often also renormalized to 1.0 representing the maximum total target or normal tissue volume. The dose in the x axis may also be user selected as dose per fraction or, more commonly, total dose in Gy.

Note that the spatial relation of dose within a defined volume is lost when looking at DVH data, but a simple summary of dose levels within the volume is obtained. The position of dose cold spots or hot spots within the volume is best viewed using isodose displays.

Shown in figure 6.15 is the integral DVH within the target and to radiosensitive normal tissue such as spinal cord, respectively. The aim is to include the shape of the DVH within the target as squarely as possible so that most of the tumor volume receives a high dose. To normal tissue structures the aim is to have the DVH fall off as quickly as possible so that the minimum volume receives a high dose in this region. It is easier to achieve this if the normal tissue structures do not overlap the target volume.

DVHs can indicate potential trouble at a glance by highlighting hot and cold spots within a structure. However, they obscure information about the precise geometric position of these regions. Therefore, final judgment about the value of a plan must still be based on detailed dose distribution analysis. Dose-area histograms are similar to DVHs; however, they only display the area in a particular slice that is given a particular dose (i.e., only 2-D information).

DVHs are often used to compare two or more plans for the same patient, which are assigned as different plan *trials*. For example, a conformal plan using MLC blocking may be compared with an IMRT plan to see if dose to rectum and/or head of femur or bladder are reduced. Figure 6.16 shows a typical presentation of two plan trials whereby the target volume DVH curve is similar but because the volume of normal tissue receiving a

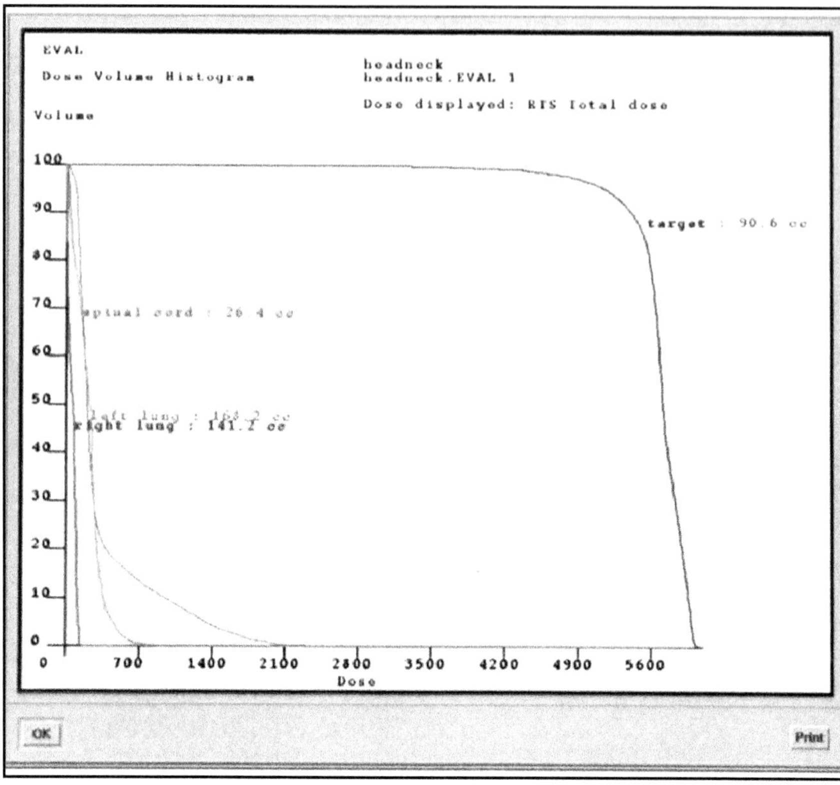

Figure 6.15: Dose-volume histogram. (Reprinted with permission from: © Nucletron-Odelft.)

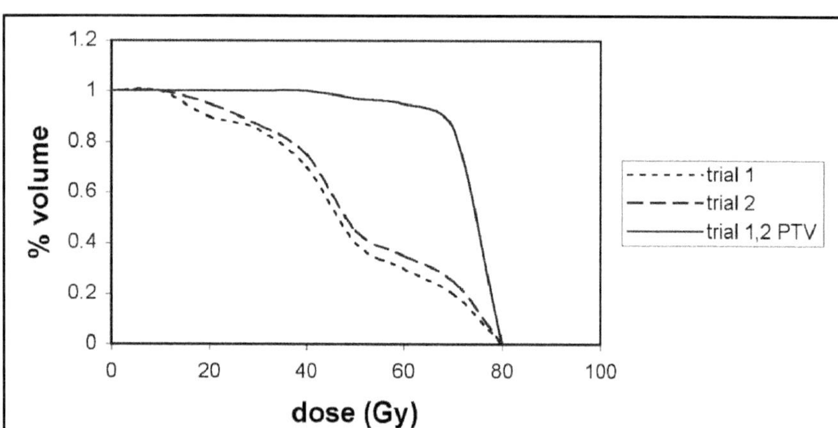

Figure 6.16: A typical presentation of two plan trials whereby the target volume DVH curve for plan trials 1 and 2 are similar but plan 2 presents a better DVH than plan 1 for the patient because the volume of normal tissue (e.g., rectal wall) receiving a particular dose is less.

particular dose at all dose points is less, trial B presents a better DVH than trial A for the patient.

6.9.2 DVH for Prostate and Rectum

The challenge with assessing DVHs for prostate tumors is the target volume and rectal tissue overlap. It is important to reduce the rectal DVH as much as possible while maintaining a realistic internal margin around the planning target volume (PTV). This is the reason why the rectal DVH does not fall off rapidly, unlike structures that have a clear margin between the PTV and normal tissue structure.

There are several ways to define the rectal volume of rectal tissue. This is important, in particular when prostate tumors are treated with radiation therapy. The rectal volume for the DVH can be defined by outlining the outer rectal surface and including the enclosed volume. Alternately, the rectal wall can be outlined and dose-wall histogram (DWH) data described. It is difficult to visualize and outline the rectal inner and outer surface; so as a surrogate a ring of 2 mm nominal thickness from the outer wall may be used (Garcia-Vincent et al. 2005). Alternately, a surface can be defined and dose-surface histogram (DSH) data presented (Lu et al. 1995). Another method is to account for the change in rectal thickness with bowel filling by defining the same number of surface dose elements at all contour levels to provide a normalized DSH (MacKay et al. 1997). While the actual dose-wall volume is typically less than half that of the solid rectal volume, there is a good correlation when relative DVH curves are compared with DWHs (Garcia-Vincent et al. 2005).

Different conventions are used for defining the length of the rectal tissue. One convention is to outline from 2 cm below the inferior-most aspect of the ischial tuberosities to a set length (e.g., 11 cm) or to the rectal arch. The volume of rectum defined significantly affects the normalized DVH. This can be demonstrated by contouring more bowel, as shown in figure 6.17. This affects the absolute amount of rectal volume in the absolute DVH shown in figure 6.18 and drives down the normalized DVH shown in figure 6.19. This demonstrates the importance of defining consistent rectal volumes when maximum dose thresholds are being determined to minimize toxicity and also for consistency within clinical trials.

Recommended dose thresholds exist above which it is not recommended to exceed for a certain volume of tissue. For rectal DVHs dose thresholds can usually be related back to a certain level of grade II rectal toxicity as found in randomized trials. For example, the cut points recommended from one dose escalation level III randomized trial comparing 70 Gy with 78 Gy of Pollack et al. (2002) as described by Huang et al. (2002) were:

i) Less than 60 Gy to 40.6% rectal volume.
ii) Less than 70 Gy to 26.2% rectal volume.
iii) Less than 76 Gy to 15.8% rectal volume.
iv) Less than 78 Gy to 5.1% rectal volume.

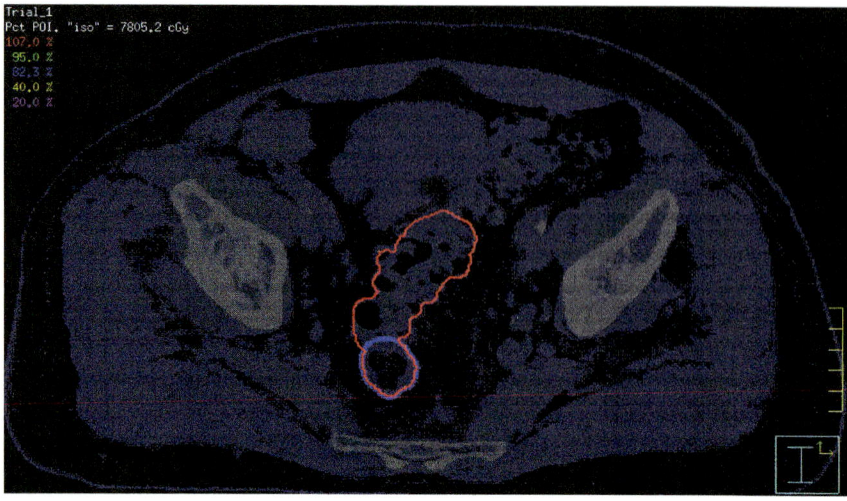

Figure 6.17: CT contour showing two contours of bowel (blue small and red large), hence giving different volume of bowel defined. (Screen shot from a Philips Pinnacle RTP).

The additional thresholds were:

i) Less than 75.6 Gy to 3.8 cm^3 absolute volume.
ii) Less than 78 Gy to 1.4 cm^3 absolute volume.

The less than 70 Gy to 26.2% rectal volume relates back to a correlated freedom from grade II late rectal toxicity rate of 54% for those with more than 26.2% of volume treated to 70 Gy, while the rate was only 13% for

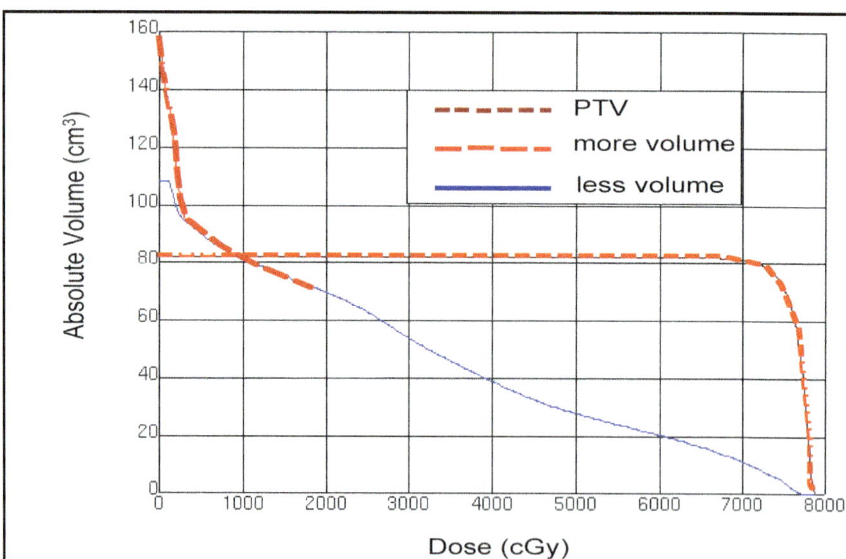

Figure 6.18: Absolute volume DVH (y axis cm^3) with different volume of bowel defined.

those with less than 26.2% of volume treated to 70 Gy. It is beneficial to stay as far below the dose threshold as possible in an attempt to further reduce the rate of complications, and this may sometimes be achievable if clinic gross target volume (GTV) to PTV margins are smaller, as validated by careful assessment, than the margins used in the original trial.

Figure 6.20 is an example of two plans under trial comparison, showing 2-D isodose distributions for conformal prostate (left) and IMRT prostate (right). The clinician assesses the isodose coverage for each trial at each CT slice (only one slice shown here) and also assesses the DVH data for the trial comparison (see figure 6.21). The inverse planning methods used to generate these IMRT plans are discussed further in chapter seven.

The use of a small rectal balloon placed in the rectum during each fraction of treatment has been reported to lower DWHs when both conformal and IMRT plans are employed. The effect may be even more significant on the rectal DWH if the seminal vesicles are included in the target volume (Patel et al. 2003). Reports of decreased rectal toxicity with use of rectal balloons has been reported (e.g., van Lin et al. 2007).

The rectal balloon probably helps in three ways:

i) Reduces the rectal volume treated to a high dose (Wachter et al. 2002).
ii) Reproduces the position of the rectum from day to day.
iii) May help immobilize the prostate (Patel et al. 2003).

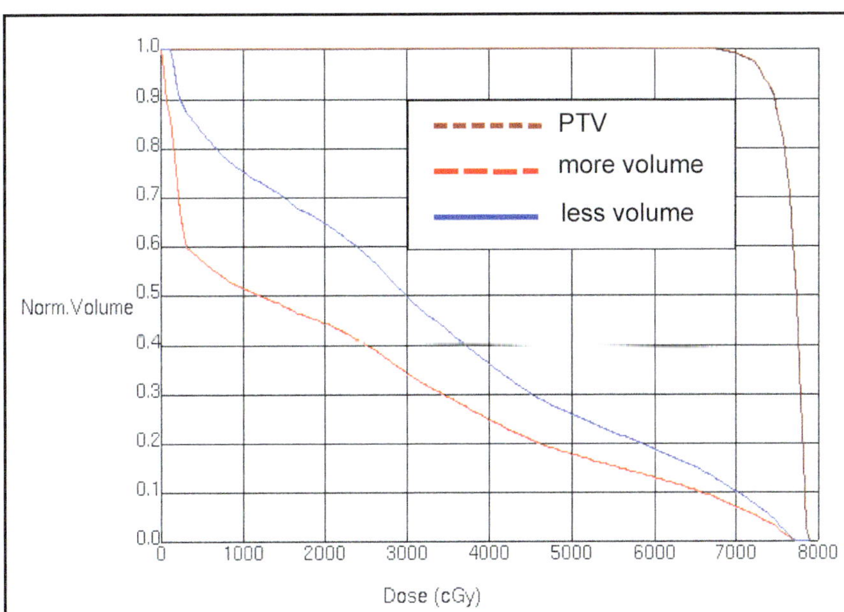

Figure 6.19: Normalized DVH with different bowel volume defined. This demonstrates the importance of consistent rectal volume when cut points are being determined to minimize toxicity and also for consistency within clinical trials.

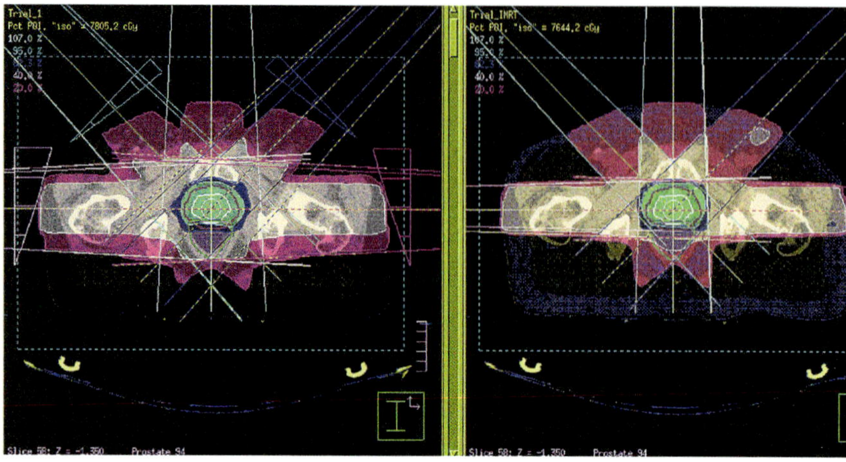

Figure 6.20: Isodose display for two plan trials comparisons: (i) conformal prostate (left) and (ii) IMRT prostate plan (right). Note: For these plans seminal vesicles were not included in volume (Screen shot from a Philips Pinnacle RTP).

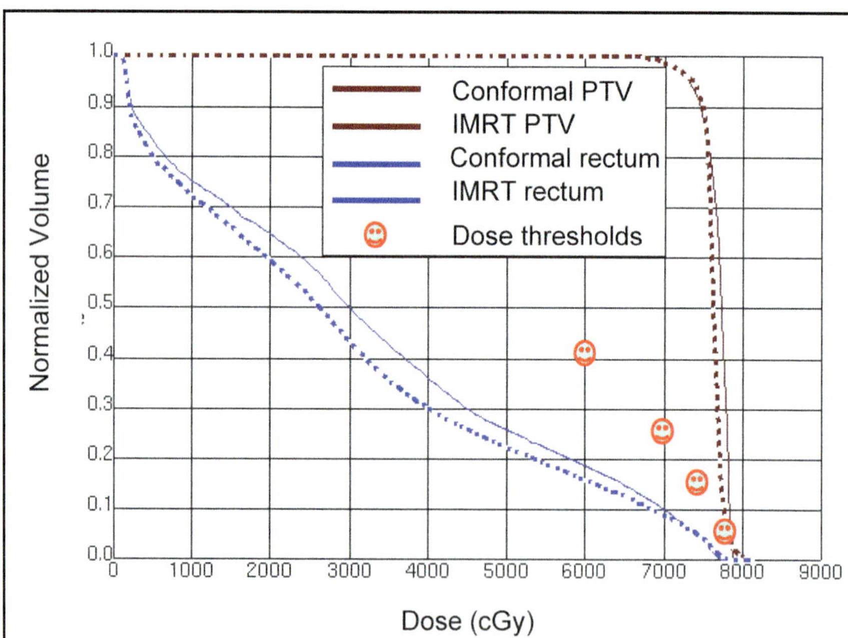

Figure 6.21: DVH curves for target and rectal volume for two plan trials comparisons: (i) conformal prostate target and rectal DVH and (ii) IMRT prostate target and rectal DVH. The dose threshold points from a phase III clinical trial are also shown. (Screen shot from a Philips Pinnacle RTP).

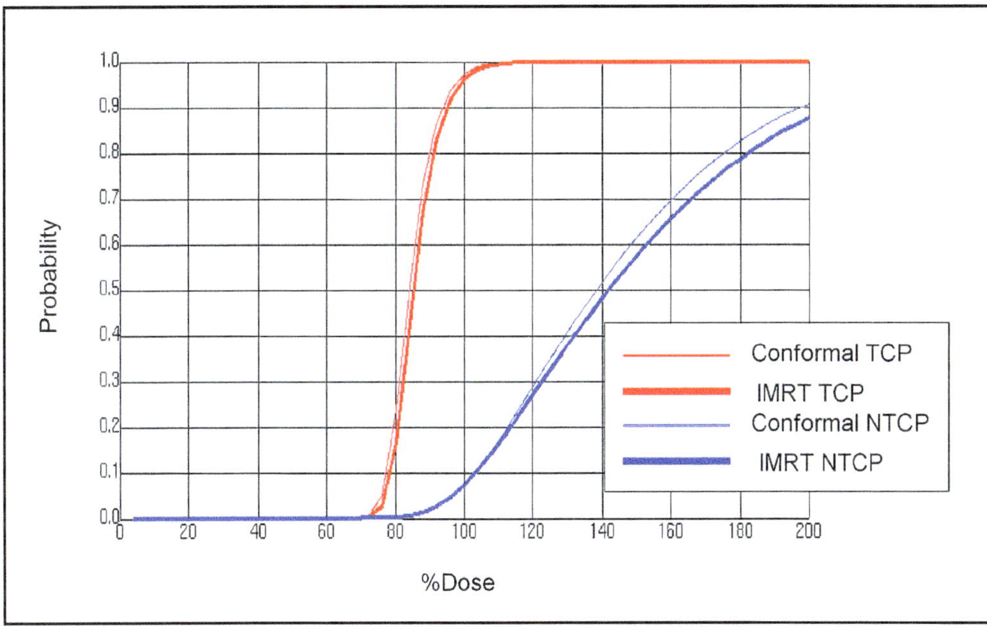

Figure 6.22: Tumor control probability (TCP) and normal tissue complication probability (NTCP) curves for (i) conformal prostate target TCP and rectal TCP and (ii) IMRT prostate target TCP and rectal NTCP. (Screen shot from a Philips Pinnacle RTP.)

6.9.3 Combining DVH Data with Radiological Models

Dose distribution 3-D data sets may be used in conjunction with observed biological responses of normal tissue complication probabilities (NTCPs) and tumor control probabilities (TCPs). Niemierko (1996) has combined fractionation data with DVH data to examine the impact of dose homogeneity on TCP. TCP and NTCP curves calculated for the prostate DVH data using the previous example for conformal versus IMRT are shown in figure 6.22. While these models are rarely used in the clinic to guide treatment, they give the clinician another piece of information when assessing what gains have been made by using one technique versus another. The radiobiological calculation models used to generate these curves are discussed in more detail in chapter fourteen.

6.9.4 Setting Dose Objectives on a DVH

Improvements in optimization techniques are also aiding in the choice of beam configurations that give the best planned dose distribution (Mohan et al. 1992). With these techniques, the operator sets a few parameters, such as number of fields and beam angles, and the computer determines the optimum beam parameters.

Inverse planning is the term used to describe computer-aided optimization of treatment plans and generally refers to setting dose objectives to various volumes and having the computer calculate the most suitable beam parameters.

The dose objective tool can often be used in combination with DVH graphs; the required dose objectives can usually be interactively positioned by the operator on a DVH graph prior to the dose objective process being attempted. Inverse planning has become critical to the success of IMRT because multiple complex segments could not be designed using forward planning methods. In fact, currently it is often only the segment shapes and segment weights that are inverse planned by RTP computer systems. The other parameters such as beam energy, number, and angles are forward planned. Inverse planning techniques are not restricted to calculating IMRT segments and weights. Inverse plans can be generated using conventional fields. In this case, the dose objective is best met by calculating the most suitable beam weights for each field. For more detail on inverse planning and setting dose objectives, refer to the IMRT section of chapter seven.

6.10 Defining Treatment Planning Volumes

The aim of *radical* treatment of malignant disease or curative radiotherapy is to decrease the number of tumor cells to a level that achieves permanent local tumor control. To achieve this, the volume that encompasses demonstrated tumor and subclinical spread must be included. If the tumor has been removed (e.g., mastectomy, hysterectomy), the remaining tissue may contain subclinical disease.

In planning radiation treatments the precise definition of the volume(s) to be treated is of utmost importance. Treatment volumes bear a complex relationship to the primary lesion, involved lymph nodes, and regions of suspected microscopic extensions. For the purposes of discussion, it is useful to adopt the following definitions of biological tumor volumes and PTVs following the convention established elsewhere (ICRU 1993).

Specification of volumes and doses is required for consistent prescription, recording, and reporting of dose. Complete, accurate, and explicit information on patient disease, volume/dose irradiated, radiation techniques, and treatment time/fractions needs to be reported to ensure patient results can be sensibly compared.

With reference to figure 6.23 the volumes are defined as follows:

i) The *gross tumor volume* (GTV) is the gross palpable or visible/demonstrable extent and location of malignant growth.

Whether a physician chooses to include certain sites in the tumor volume may depend upon the patient's current physical status, history prior to radiation treatment, prior and concurrent treatments with other modalities, etc. He or she will often weigh the probability of involvement of a particular site against the consequences of treating the site. The wide variety of factors to be considered and the associated uncertainties make the process of decision making difficult and subjective, leading to inconsistencies. This is an area for further development in which expert systems can be valuable to the physician for recommending the regions of suspected disease to be included in the target volume.

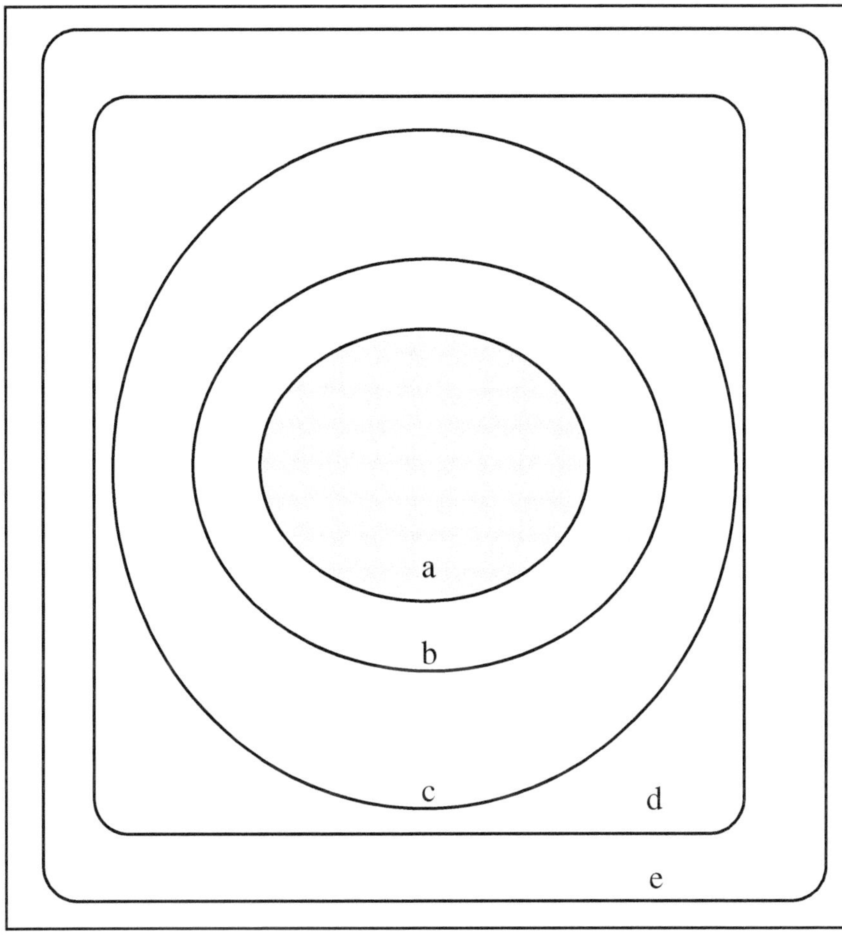

Figure 6.23: Definition of volumes: (a) gross tumor volume (GTV), (b) clinical target volume (CTV), (c) planning target volume (PTV), (d) treated volume (TV), (e) irradiated volume (IV).

While regions of gross disease may, sometimes, be distinguishable on CT scans, the extent of microscopic disease, nodal involvement, and distant spread of the disease are usually difficult to visualize and extract. However, for more accurate localization of the tumor volume, it may be necessary to incorporate information from other imaging modalities, such as MRI, PET, SPECT, lymphoscintigraphy, lymphangiography, radio-labeled monoclonal antibodies, or even simulation radiographs.

ii) The *clinical target volume* (CTV) is the tissue that contains the GTV and/or subclinical microscopic malignant disease that has to be eliminated. This volume must be treated adequately in order to achieve the aim of cure or palliation. If the tumor has been removed by surgery, the volume of potential microscopic malignancy is the CTV. Uncertainty regarding the extent of the disease currently limits the accuracy of this assessment.

iii) The *planning target volume* (PTV) was previously referred to in the literature as the *target volume.* It takes into account the net effect of all possible geometric variations to ensure that the prescribed dose is delivered to the CTV. This is the margin that usually lies beyond the CTV and accounts for movement of tissues (e.g., patient respiration), variations in tissues that contain CTV (e.g., full bladder), and variations in beam geometry.

In a subsequent report 62 (ICRU 1999) the margin between the CTV and the PTV has been further subdivided into two more clearly described volume subsets. These are the *internal margin* (IM), which is added to account for variations in position and/or shape and size of the CTV, and the *setup margin* (SM), which is added to account for all the variations in patient-beam positioning. Hence CTV+IM+SM defines the PTV. To be exact, because the IM and SM account for random errors the magnitude of the combined margin is actually the square root of the sum of squares of the individual margins. ICRU report 62 (1999) also included the concept of *region of interest* (ROI) margins for sensitive structures or *organs at risk* (OAR), such as spinal cord, when this may be the limiting dose organ. In these scenarios it makes sense to include a margin around these critical structures to account for setup.

The PTV is a static, geometric concept used for treatment planning. The PTV can be considered as a 3-D envelope in which the tumor and any macroscopic extensions reside. The GTV and CTV can move within this envelope but not through it (Purdy 1996). As shown in figure 6.24, these regions are shown on a lung plan in two planes.

Patient and organ motion are difficult uncertainties to quantify. On-line portal imaging is helping provide information to improve the quantification of margins to allow for this. For many sites patient motion and internal organ motion may be major sources of uncertainty, requiring us to leave large margins to ensure proper tumor coverage. For many situations, external custom-designed casts and other restraining devices have been found to be useful in keeping the patient still during treatments as well as in reproducing his or her position for the entire course of treatments. Development of systems for on-line, real-time imaging during treatment are expected to be of significant value in ensuring proper positioning of target volume and normal structures with respect to the radiation field.

Patient setup and alignment variations occur. Regardless of how careful one is, at least a small degree of uncertainty is introduced by variations in setup and alignment of a patient from one day to the next. These may be attributable to such factors as the finite size of marks and markers, resolution of imaging device, and basic human limitations.

Note that the PTV and the CTV may be very similar if there is little scope for patient movement during treatment (e.g., pituitary tumor including restriction of movement with a head mask). Or alternatively the PTV may be much larger than CTV (e.g., lung including patient respiration movement).

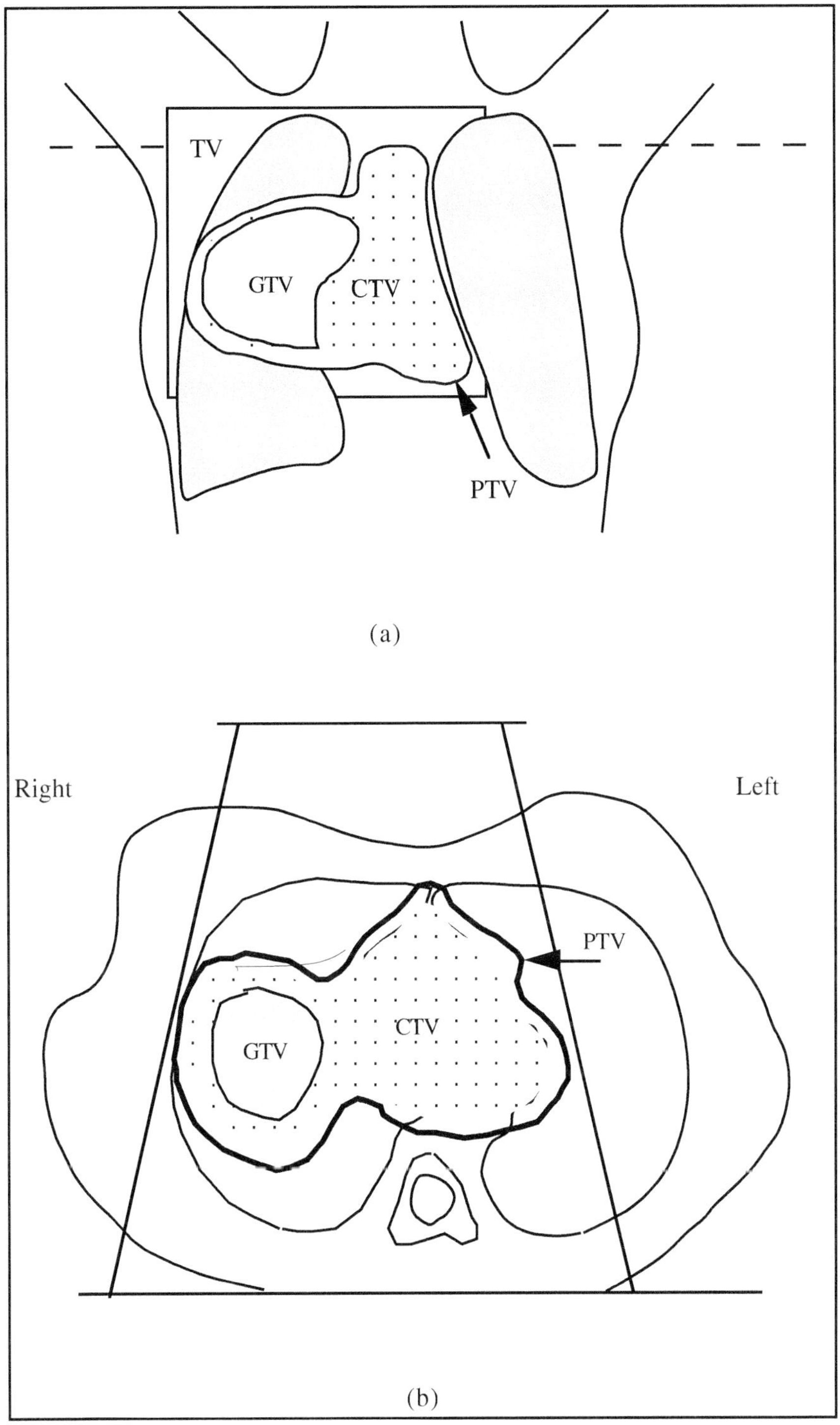

Figure 6.24: Lung plan showing typical margins for clinical and planning target volumes.

iv) The *treated volume* (TV) is the volume enclosed by an isodose surface, selected and specified by the radiation oncologist as being appropriate to achieve the purpose of treatment (e.g., tumor irradiation or palliation). The isodose level used to represent the TV is usually 95%. Ideally dose is delivered to the PTV. Due to radiation treatment technique variations, the isodose curve usually approximates this.

The reasons for identifying the treated volume are:

(i) The shape and size of TV relative to PTV is an important optimization tool. Dosimetrists use many different beam configurations to achieve the best target coverage.
(ii) Recurrence within TV but outside PTV is an in-field recurrence due to inadequate dose, but a marginal recurrence due to inadequate volume is not an in-field recurrence.

v) The *irradiated volume* (IV) is the tissue volume that receives a dose that is considered significant in relation to normal tissue tolerance. It is usually represented by the 20% isodose line. Defining this gives an indication of the volume of normal tissue treated due to the beam technique employed. Shown in figure 6.25 is an example of the different ratios of areas of tissues treated to the same dose. Note that this approach ignores the organs at risk whose radiation sensitivity may significantly influence treatment planning and/or prescribed dose. Hence plans with a good TV to IV ratio with relation to the PTV and low dose to sensitive organs are ultimately selected.

According to the ICRU (1993) a –5% to +7% variation in dose within the PTV is within acceptable limits. A higher dose heterogeneity may be acceptable if the hot dose area is in a high malignant cell population in the GTV.

The ICRU (1993) recommends dose be reported from three locations:

i) Minimum dose such as described by the isodose line enclosing the PTV.
ii) Maximum dose within the PTV. (It is suggested that diameter of maximum dose region be at least 15 mm.)
iii) The ICRU reference point, which is selected as a high tumor density region close to the center of the PTV (usually isocenter).

If the reference point cannot be located at the center or near the central axis of the beam in the middle of the PTV (e.g., asymmetric breast treatment), it should be specified in a potentially high tumor cell population region.

The application of these conventions for outlining tumors has been embraced by radiation oncologists as they see the benefit when comparing patient treatments in particular in multicenter randomized trials. The challenge remains for the radiation oncologists in defining the GTV and CTV using the current diagnostic imaging data available. This is a challenging process.

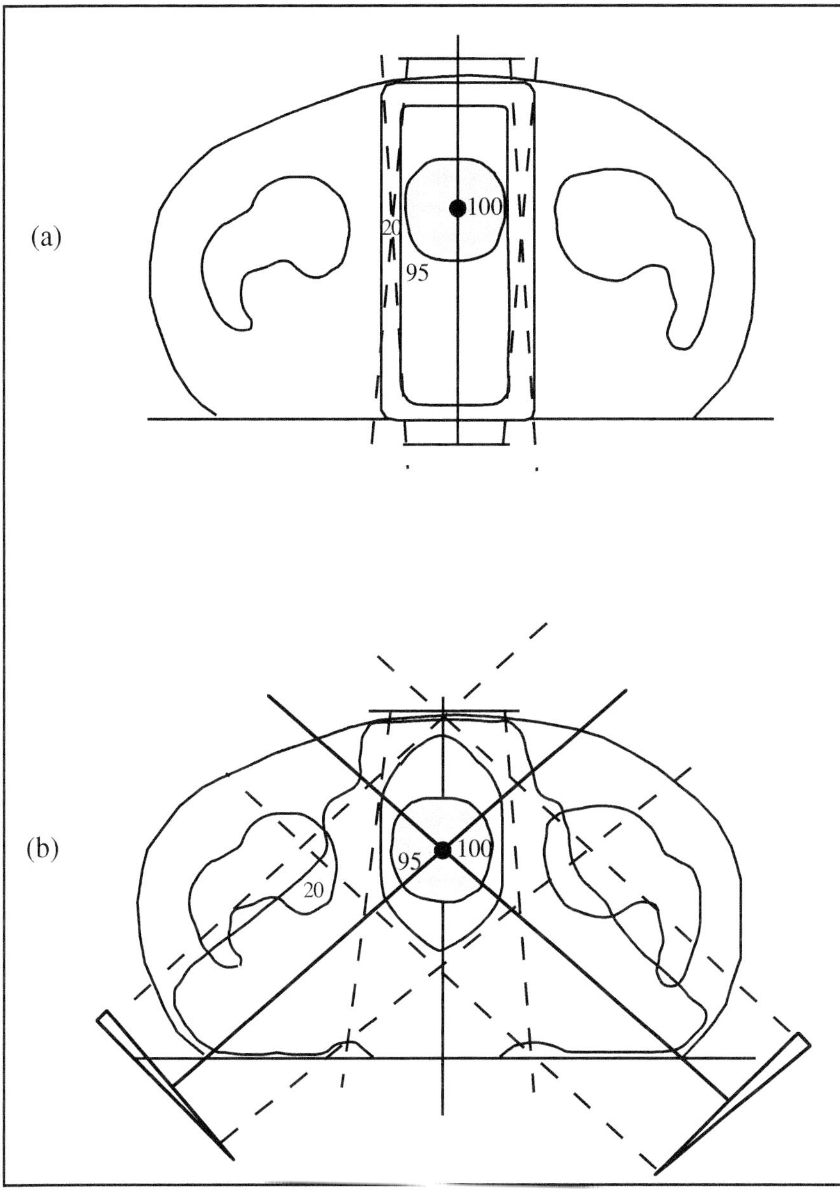

Figure 6.25: Ratio of treatment and irradiated volumes to the planning target volume for different plan techniques. Respective ratios of volume are PTV:TV:IV (a) 1.0:2.4:4.6 (b) 1.0:1.9:9.0 (c) 1.0:1.9:9.0 (d) 1.0:1.1:11.0 (two 130° arc fields).

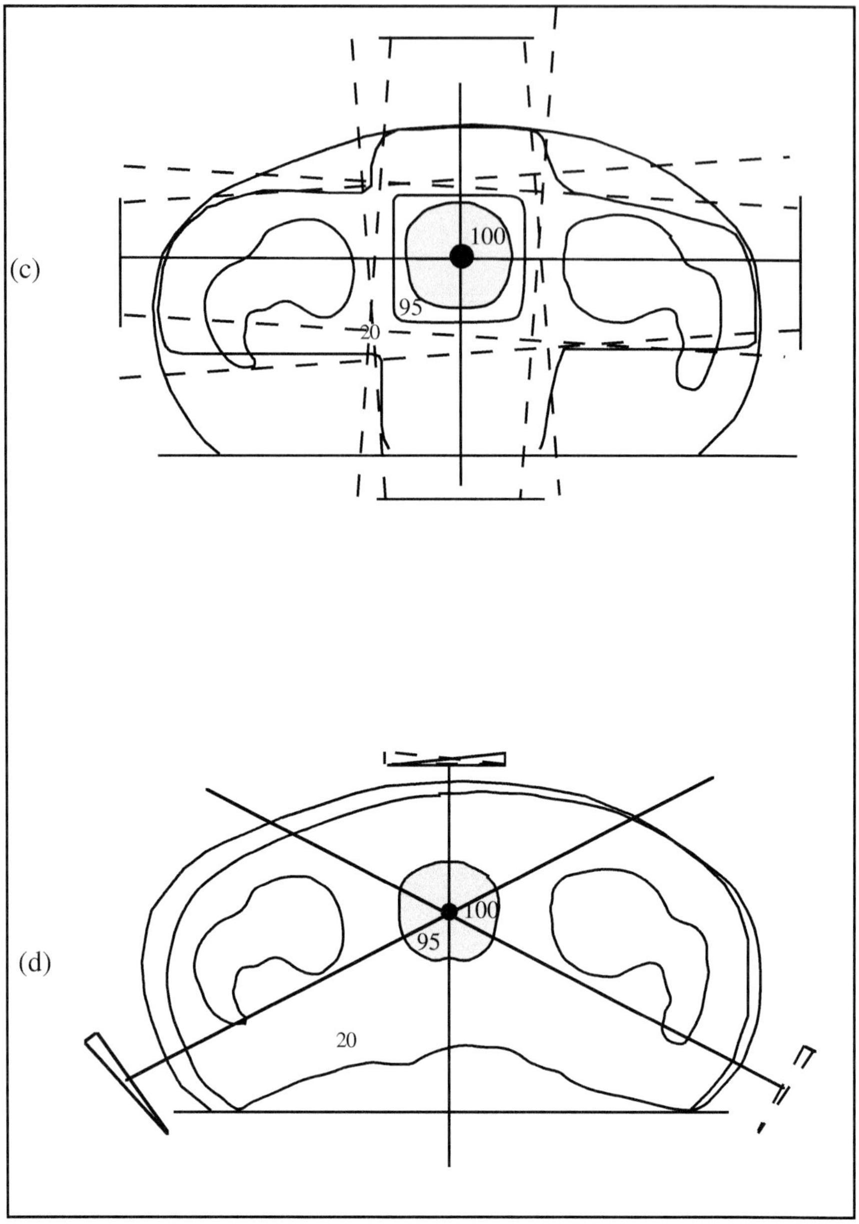

Figure 6.25: (Continued)

References

AAPM (American Association of Physicists in Medicine) (1998) Quality assurance for clinical radiotherapy treatment planning. Report of Task Group 53. *Med. Phys.* 25: 1173–1829. Also available as AAPM Report No. 62.

Battista J.J., Bronskill M.J. (1981) Compton scatter imaging of transverse sections: An overall appraisal and evaluation for radiotherapy planning. *Med. Phys. Biol.* 26(1): 89–99.

Biehl K.J., Kong F.-M., Dehdashti F., Jin J.Y., Mutic S., El Naqa I., Siegel B.A., Bradley J.D. (2006) 18F-FDG PET definition of gross tumor volume for radiotherapy of non-small cell lung cancer: Is a single standardized uptake value threshold approach appropriate? *J. Nucl. Med.* 47: 1808–1810.

Boechat M.I., Kangarloo H., Ortega J., Hall T., Feig S., Stanley P, Gilson V. (1988) Primary liver tumor in children: A comparison of CT and MR. *Imaging Radiol.* 169: 727–732.

Brant-Zawadski M., Norman D., Newton D., Newton T.H., Kelly W.M., Kjos B., Mills C.M., Dillon W., Sobel D., Crooks L.E. (1984) Magnetic resonance of the brain: The optimal screening technique. *Radiology* 152: 71–77.

Brooks R.A., Di Chiro G. (1975) Theory of image reconstruction in computed tomography. *Radiology* 117: 561–572.

Chaney E.L., Cullip T.J., Gabriel T.A. (1994) A Monte Carlo study of accelerator head scatter. *Med. Phys.* 21: 1383–1390.

Chen G.T.Y., Pelizzari C.A., Spelbring D.R., Awan A. (1987) Evaluation of treatment plans using dose volume histograms. *Radiat. Ther. Oncol.* 10: 44–55.

Chung P., Haycocks T., Brown T., Cambridge Z., Kelly V., Alasti H., Jaffray D.A., Catton C.N. (2004) On-line aSi portal imaging of implanted fiducial markers for the reduction of interfraction error during conformal radiotherapy of prostate carcinoma. *Int. J. Radiat. Oncol. Biol. Phys.* 60: 329–334.

Collier B.D., Palmer D.W., Wilson J.F., Greenberg M., Komaki R., Cox J.D., Lawson T.L., Lawlor P.M. (1983) Internal mammary lymphoscintigraphy in patients with breast cancer. *Radiology* 147: 845–848.

Constantinou C., Attix F.H., Paliwal B. (1982) A solid water phantom material for radiotherapy x-ray and gamma ray beam calibrations. *Med. Phys.* 9: 436–441

Covell M.M., Hearshen D.O., Carson P.L., Chenevert T.P., Shreve P., Aisen A.M., Bookstein F.L., Murphy B.W., Martel W. (1986) Automated analysis of multiple performance characteristics in magnetc resonance imaging systems. *Med. Phys.* 13: 815–823.

Cygler J., Battista J.J., Scrimger J.W., Mah E., Antolok J. (1987) Electron dose distributions in experimental phantoms: A comparison of 2-D pencil beam calculations. *Phys. Med. Biol.* 32: 1073–1078.

Delbeke D., Martin W.H., Patton J.A., Sandler M.P. (2001) Value of iterative reconstruction, attenuation correction, and image fusion in the interpretation of FDG PET images with an integrated dual-head coincidence camera and x-ray-based attenuation maps. *Radiology* 218: 163–171.

De Ruysscher D., Wanders S., Minken A., Lumens A., Schiffelers J. et al. (2005) Effects of radiotherapy planning with a dedicated combined PET-CT-simulator of patients with non-small cell lung cancer on dose limiting normal tissues and radiation dose-escalation: A planning study. *Radiother. Oncol.* 77: 5–10.

Ege G.N., Clark R.M. (1985) Internal mammary lymphoscintigraphy in the conservative management of breast carcinoma. An update and recommendations for a new TNM staging. *Clin. Radiol.* 36: 469–472.

Emami B., Melo A., Carter B.L., Munzenrider J.E., Piro A.J. (1978) Value of computed tomography in radiotherapy of lung cancer. *AJR. Am. J. Roentgenol.* 131: 63–67.

Fraass B.A., McShan D.L., Diaz R.F., Ten Haken R.R., Aisen A., Gerbaski S. (1987) Integration of magnetic resonance imaging into radiation therapy treatment planning. *Int. J. Radiat. Oncol. Biol. Phys.* 13: 1897–1908.

García-Vicente F., Zapatero A., Floriano A., Cruz-Conde A., Pérez L., Marín A., Minguez C., Torres J. (2005) Statistical analysis of dose-volume and dose-wall histograms for rectal toxicity following 3D-CRT in prostate cancer. *Med. Phys.* 32: 2503–2509

Gauvin A. (1992) Geometrical distortion of magnetic resonance images. M.Sc. Thesis. Department of Physics, McGill University, Montreal, Canada.

Geise R.A., McCullough E.C. (1977) The use of CT scanners in megavoltage photon beam therapy planning. *Radiology* 124: 133–141.

Glatstein E., Lichter A.S., Fraass B.A., Kelly B.A., Van De Geijn J. (1985) The imaging revolution and radiation oncology: Use of CT, ultrasound, and NMR for localization, treatment planning and treatment delivery. *Int. J. Radiat. Oncol. Biol. Phys.* 11: 299–314.

Goitein M. (1979) The utility of computed tomography in radiation therapy: An estimate of outcome. *Int. J. Radiat. Oncol. Biol. Phys.* 5: 1799–1807.

Goitein M., Abrams M., Rowell D., Pollari H., Wiles J. (1983). Multi-dimensional treatment planning: II. Beam's eye-view, back projection, and projection through CT sections. *Int. J. Radiat. Oncol. Biol. Phys.* 9: 789–797.

Groh B.A., Siewerdsen J.H., Drake D.G., Wong J.W., Jaffray D.A. (2002) A performance comparison of flat-panel imager-based MV and kV cone-beam CT. *Med. Phys.* 29: 967–975.

Henkelman R.M. (1992) New imaging technologies: Prospects for target definition. *Int. J. Radiat. Oncol. Biol. Phys.* 22: 251–257.

Henri C.J., Collins D.L., Peters T.M. (1991) Multimodality image integration for stereotactic surgical planning. *Med. Phys.* 18: 167–177.

Henson P.W. (1989) Determination of electron density, mass density and calcium fraction by mass of soft and osseous tissue by dual energy CT. *Australas. Eng. Phys. Sci. Med.* 12: 3–10.

Henson P.W., Fox R.A. (1984) The electron density of bone for inhomogeneity correction in radiotherapy planning using CT numbers. *Phys. Med. Biol.* 29: 351–359.

Ho A.K., Paliwal B.R. (1986) Stopping-power and mass energy-absorption coefficient ratios for solid water. *Med. Phys.* 13: 403–404.

Hogstrom K.R., Mills M.D., Almond P.R. (1981) Electron beam dose calculations. *Phys. Med. Biol.* 26: 445–459.

Hong R., Halama J., Bova D., Sethi A., Emami B. (2007) Correlation of PET standard uptake value and CT window-level thresholds for target delineation in CT-based radiation treatment planning. *Int. J. Radiat. Oncol. Biol. Phys.* 67: 720–726.

Hounsfield G.N. (1973) Computerised transverse axial scanning (tomography): Part 1 description of the system. *Br. J. Radiol.* 46: 1016–1022.

Huang E.H., Pollack A., Levy L., Starkschall G., Dong L., Rosen I., Kuban D.A. (2002) Late rectal toxicity: Dose-volume effects of conformal radiotherapy for prostate cancer. *Int J. Radiat. Oncol. Biol. Phys.* 54: 1314–1321.

Huizenga H., Storchi P.R.M. (1985) The use of computed tomography numbers in dose calculations for radiation therapy. *Acta Radiat. Oncol.* 24: 509–519.

ICRU (International Commission on Radiation Units and Measurements) Report 62 (1999) Prescribing, recording, and reporting photon beam therapy. (Supplement to ICRU report 50.) ICRU, Bethesda, Maryland.

ICRU (International Commission on Radiation Units and Measurements) Report 50 (1993) Prescribing, recording, and reporting photon beam therapy, ICRU, Bethesda, Maryland.

Jackson D.F., Hawkes D.T. (1981) Attenuation coefficient data tables. *Physics Reports* 70: 169–233.

Jaffray D., Siewerdsen J., Wong J., Martinez A. (2002) Flat-panel cone-beam computed tomography for image-guided radiation therapy. I*nt. J. Radiat. Oncol. Biol. Phys.* 53: 1337–1349.

Jani A.B., Gratzle J., Muresan E., Martel M.K. (2005) Impact on late toxicity of using transabdominal ultrasound for prostate cancer patients treated with intensity modulated radiotherapy. *Technol. Cancer Res. Treat.* 4: 115–120.

Jelinek J.S., Redmond J., Perry J.J., Burrell L.M., Benedikt R.A., Geyer C.A., Peller P.J., Wacks L.L., Wise B.J., Ghaed V.N. (1990) Small cell lung cancer: Staging with MR imaging. *Radiology* 177: 837–842.

Jette D. (1984) The problem of electron dose calculation: Inhomogeneities and beam shaping. *Am. J. Med. Dosim.* 9: 12–17.

Kessler M.L., Pitluck S., Chen G.T.Y. (1987) Techniques and applications of image correlation in radiotherapy treatment planning. *Front. Radiat. Ther. Oncol.* 21: 25–32.

Khan F. (1984) *The Physics of Radiation Therapy.* Williams and Wilkins, Baltimore, Maryland.

Klevenhagen S.C. (1993) *Physics and Dosimetry of Therapy Electron Beams.* Medical Physics Publishing, Madison, Wisconsin.

Klevenhagen S.C. (1985) *Physics of Electron Beam Therapy.* Adam Hilger, Bristol, U.K.

Krestin G.P., Steinbrich W., Friedman G. (1988) Recurrent rectal cancer: Diagnosis with MR imaging versus CT. *Radiology* 168: 307–311.

Kurhanewicz J., Vigneron D., Nelson S., Hricak H., MacDonald J.M., Konety B., Narayan P. (1995) Citrate as an in vivo marker to descriminate prostate cancer from benign prostatic hyperplasia and normal prostate peripheral zone: detection via localized proton spectroscopy. *Urology* 45: 459–466.

Lattanzi J., McNeeley S., Donnelly S., Palacio E., Hanlon A., Schultheiss T.E., Hanks G.E. (2000) Ultrasound-based stereotactic guidance in prostate cancer—quantification of organ motion and set-up errors in external beam radiation therapy. *Comput. Aided Surg.* 5: 289–295.

Levi C., McCullough E.C., Hattery R.R. (1982) The unreliability of CT numbers as absolute values. *Am. J. Roentgenol.* 139: 443–447.

Ling C.C., Leibel S., Fuks Z., Humm J., Koutcher J., Hricak H. (2003) Imaging for IMRT. Chapter 1 in *A Practical Guide to Intensity-Modulated Radiation Therapy*. Medical Physics Publishing, Madison, Wisconsin, pp. 1–13.

Ling C.C., Humm J., Larson S., Amols H., Fuks Z., Leibel S., Koutcher J.A. (2000) Towards multidimensional radiotherapy (MD-CRT): Biological imaging and biological conformality. *Int. J. Radiat. Oncol. Biol. Phys.* 47: 551–560.

Ling C.L., Rogers C.C., Morton R.J. (1983) *Computed Tomography in Radiation Therapy.* Raven Press, New York.

Lu Y., Li S., Spelbring D., Song P., Vijayakumar S., Pelizzari C., Chen G.T. (1995) Dose-surface histograms as treatment planning tool for prostate conformal therapy. *Med. Phys.* 22: 279–284.

MacKay R.I., Hendry J.H., Moore C.J., Williams P.C., Read G. (1997) Predicting late rectal complications following prostate conformal radiotherapy using biologically effective doses and normalized dose-surface histograms. *Br. J. Radiol.* 70: 517–526.

Mac Manus M.P., Hicks R.J., Ball D.L., Kalff V., Matthews J.P., Salminen E., Khaw P. et al. (2001) F-18 fluorodeoxyglucose positron emission tomography staging in radical radiotherapy candidates with nonsmall cell lung carcinoma: Powerful correlation with survival and high impact on treatment. *Cancer* 92: 886–895.

McCullough E.C. (1975) Photon attenuation in computed tomography. *Med. Phys.* 2: 307–320.

McCullough E.C., Holmes T.W. (1985) Acceptance testing computerized radiation therapy treatment planning systems: Direct utilization of CT scan data. *Med. Phys.* 12: 237–242.

Metcalfe P.E., Beckham W.A. (1988) Radiotherapy planning accuracy in terms of CT numbers and inhomogeneity correction techniques. *Australas. Radiol.* 23: 371–379.

Metcalfe P.E., Hoban P.W., Harper N.R., Murray D.C., Round W.H. (1990) The production of body analog materials for use in medical radiation physics. *Australas. Phys. Eng. Sci. Med.* 13: 117–128.

Metcalfe P.E., Beckham W.A., Long B.H., Battista J.J. (1988b) The effect of patient density variation on radiotherapy dose calculations. *Australas. Phys. Eng. Sci. Med.* 11: 107–115.

Milan J., Bentley R.E. (1974) The storage and manipulation of radiation dose data in a small digital computer. *Br. J. Radiol.* 45: 115–121.

Mohan R. (1989) Three-dimensional radiation treatment planning. *Australas. Phys. Eng. Sci. Med.* 12: 73–91.

Mohan R., Mageras G.S., Baldwin B., Brewster L.J., Kutcher G.J., Leibel S., Burman C.M., Ling C.C., Fuks Z. (1992) Clinically relevant optimization of 3-D conformal treatments. *Med. Phys.* 19: 933–944.

Murray D. C., Hoban P.W., Metcalfe P.E., Round W.H. (1989) Modelling polychromatic high energy photon beams by superposition. *Australas. Phys. Eng. Sci. Med.* 12: 128–138.

Niemierko A. (1996) Treatment evaluation. *Proceedings of the AAPM Summer School—Teletherapy: Present and Future.* Edited by T.R. Mackie and J.R. Palta. Advanced Medical Publishing, Madison, Wisconsin, pp. 175–191.

Niemierko A., Goitein M. (1991) Calculation of normal tissue complication probability and dose-volume histogram reduction schemes for tissue with a critical element architecture. *Radiother. Oncol.* 20: 166–176.

Patel R., Orton N., Tome W., Chappell R., Ritter M. (2003) Rectal dose sparing with a balloon catheter and ultrasound localization in conformal radiation therapy for prostate cancer. *Radiother. Oncol.* 67: 285–294.

Pilepich M.V., Perez C.A., Prasad S. (1980) Computed tomography in definitive radiotherapy of prostate carcinoma. *Int. J. Radiat. Oncol. Biol. Phys.* 6: 923–926.

Pollack A., Zagars G.K., Starkschall G., Antolak J.A., Lee J.J., Huang E., von Eschenbach A.C., Kuban D.A., Rosen I. (2002) Prostate cancer radiation dose response: Results of the M.D. Anderson phase III randomized trial. *Int. J. Radiat. Oncol. Biol. Phys.* 53: 1097–1105.

Purdy J.A. (1996) Volume and dose specification, treatment evaluation, and reporting for 3D conformal radiation therapy. *Proceedings of the AAPM Summer School—Teletherapy: Present and Future.* Edited by T.R. Mackie and J.R. Palta. Advanced Medical Publishing, Madison, Wisconsin, 235–251.

Purdy J.A., Williamson J.F., Wong J.W., Klein E.E. (1992) Clinical photon beam dosimetry. *Advances in Radiation Oncology Physics*, AAPM monograph 19, J.A. Purdy (Editor). American Institute of Physics, New York, pp. 258–288.

Recht A., Schnitt S.J., Connolly J. L., Rose M.A., Silber B., Come S., Henderson I.C., Slavin S., Harris J.R. (1989) Prognosis following local or regional recurrence after conservative surgery and radiotherapy for early stage breast carcinoma. *Int. J. Radiat. Oncol. Biol. Phys.* 16: 3–9.

Recht A., Silen W., Schnitt S.J., Connolly, L.J., Gelman R.S., Rose M.A., Silver B., Harris J.R. (1988a) Time course of local recurrence following conservative surgery and radiotherapy for early stage breast cancer. *Int. J. Radiat. Oncol. Biol. Phys.* 15: 255–261.

Shalev S., Bartel L., Therrien P., Hahn P., Carey M. (1988) The objective evaluation of alternative treatment plans I. Images of regret. *Int. J. Radiat. Oncol. Biol. Phys.* 15: 763–767.

Sherouse G.W., Chaney E.L. (1991) The portable virtual simulator. *Int. J. Radiat. Oncol. Biol. Phys.* 21: 475–482.

Sherouse G.W., Novins K.L., Chaney E.L. (1990) Computation of digitally reconstructed radiographs for use in radiotherapy treatment design. *Int. J. Radiat. Oncol. Biol. Phys.* 18: 651–658.

Shuman W.P., Griffin B.R., Haynor D.R., Johnson J.S., Jones D.C., Cromwell L.D., Moss A.A. (1985) MR imaging in radiotherapy planning. Work in progress. *Radiology* 156: 143–147.

Siddon R.L., Chin L.M., Zimmerman R.E., Mendel J.B., Kaplan W.D. (1982) Utilization of parasternal lymphoscintigraphy in radiation therapy of breast carcinoma. *Int. J. Radiat. Oncol. Biol. Phys.* 8: 1059–1063.

Siewerdsen J., Jaffray D. (2000) Optimization of x-ray imaging geometry with specific application to flat-panel cone-beam computed tomography. *Med Phys.* 27:1903–1914.

Smith A.R., Purdy J.A., eds. (1991) State-of-the-art of external photon beam radiation treatment planning: Photon treatment planning collaborative group. *Int. J. Radiat. Oncol. Biol. Phys.* 21: 9–23.

Tatcher M., Bjarngard B.E. (1994) Head-scatter factors in blocked photon fields. *Radiat. Ther. Oncol.* 33: 64–67.

Ten Haken R.K., Thornton A.F., Sandler H.M., La Vigne M.L., Quint D.J., Fraass B.A., Kessler M.L., McShan D.L. (1992) A quantitative assessment of the addition of MRI to CT based, 3-D treatment planning of brain. *Radiat. Ther. Oncol.* 25: 121–133.

Ten Haken R.K., Forman J.D., Heimberger D.K., Gerhardsson A., McShan D.L., Perez-Tamayo C., Schoeppel S.L., Lichter A.S. (1991) Treatment planning issues related to prostate movement in response to differential filling of the rectum and bladder. *Int. J. Radiat. Oncol. Biol. Phys.* 20: 1317–1325.

Thorwarth D., Eschmann S., Paulsen F., Alber M. (2005) A kinetic model for dynamic [18F]-Fmiso PET data to analyze tumor hypoxia. *Phys. Med. Biol.* 50: 2209–2224.

Townsend D., Cherry S. (2001) Combining anatomy and function: the path to true image fusion. *Eur. Radiol.* 11: 1968–1974.

Trichter F., Ennis R. (2003) Prostate localization using transabdominal ultrasound imaging. *Int. J. Radiat. Oncol. Biol. Phys.* 56: 1225–1233.

Van Dyk J., Keane T.J., Rider W.D. (1982) Lung density as measured by computer tomography: Implications for radiotherapy. *Int. J. Radiat. Oncol. Biol. Phys.* 8: 1363–1372.

Van Dyk J., Battista J.J., Cunningham J.R., Rider W.D., Sontag M.R. (1980b) On the impact of CT planning on radiation planning. *Comput. Tomogr.* 4: 55–65.

van Lin E.N., Kristinsson J., Philippens M.E., de Jong D., van der Vight L.P., Kaanders J.H., Leer J.W., Visser A.G. (2007) Reduced late rectal mucosal changes after prostate three-dimensional conformal radiotherapy with endorectal balloon as observed in repeated endoscopy. *Int. J. Radiat. Oncol. Biol. Phys.* 67: 799–811.

Wachter S., Gerstner N., Dorner D., Goldner G., Colotto A., Wambersie A., Potter R. (2002) The influence of a rectal balloon tube as internal immobilization device on variations of volumes and dose-volume histograms during treatment course of conformal radiotherapy for prostate cancer. *Int. J. Radiat. Oncol. Biol. Phys.* 52: 91–100.

Weber D., Ivanovic M. (1994) Correlative image registration. *Semin. Nucl. Med.* 24: 311-323.

White D.R. (1978) Tissue substitutes in experimental radiation physics. *Med. Phys.* 5: 467–479.

White D.R. (1977) The formulation of tissue substitute materials using basic interaction data. *Phys. Med. Biol.* 22: 889–899.

White D.R., Speller R.D. (1980) The measurement of effective photon energy and linearity in computerised tomography. *Br. J. Radiol.* 53: 5–11.

White D.R., Constantinou C., Martin R.J. (1986) Foamed epoxy resin-based lung substitutes. *Br. J. Radiol.* 59: 787–790.

White D.R., Martin R.J., Darlison R. (1977) Epoxy resin based tissue substitutes. *Br. J. Radiol.* 50: 814–821.

Williams J.R., Thwaites D.I. (1993) *Radiotherapy Physics in Practice.* Oxford University Press, Oxford, U.K.

Xia P., Pickett B., Vigneault E., Verhey L.J., Roach M 3rd. (2001) Forward or inversely planned segmental multileaf collimator IMRT and sequential tomotherapy to treat multiple dominant intraprostatic lesions of prostate cancer to 90 Gy. *Int. J. Radiat. Oncol. Biol. Phys.* 51: 244–254.

Yip T.C.K., Ege G.N. (1985) Determination of depth distribution of internal mammary lymph nodes on lateral lymphoscintigraphy. *Clin. Radiol.* 36: 149–152.

Zanzonico P., Heller S. (2000) The intraoperative gamma probe: Basic principles and choices available. *Semin. Nucl. Med.* 30: 33–48.

Questions

6.1 Calculate the monitor units per fraction that need to be set to deliver a dose of 50 Gy in 25 fractions using a single treatment field to a position at 10 cm depth. A single 6 MV beam of field size 20 x 20 cm^2 is employed.
i) Assume the patient is at 100 cm SSD.
ii) Assume the point is at isocenter and hence this is an SAD treatment.
(section 6.3 and data from sections 4.1 and 4.4)

6.2 A single field isocentric treatment with collimator settings of 30 x 30 cm^2 is blocked to 20 x 20 cm^2. Calculate the monitor units required if the linac is calibrated in water at 100 cm SAD and there is a 30 degree wedge and a 6 mm PMMA block tray in the beam.
(section 6.3 and data from tables 4.1 and 4.5)

6.3 Explain why the in-phantom component of scatter is adjusted with the blocked field size in the output factor calculation but the collimator scatter contribution is relatively unaffected by the block.
(section 6.3)

6.4 Draw a sketch graph of relative electron density (x axis) versus CT number (y axis) showing the lines that fit Hounsfield number data for tissue and bone. Place points in approximate positions on these lines to label lung, water, and soft bone (label these approximate positions on the graph). Derive the equation that experimentally relates CT number to relative electron density.
(section 6.5)

6.5 Given that the mass density of soft bone is 1.12 g/cm^3 and elemental compositions of body tissues are given in table 3.1, calculate:
i) The electron density.
ii) The expected CT number if the mean tube energy is 71 keV.
Assume the following approximation that (Z/A) = 1.0 for hydrogen and (Z/A) = 0.5 for all other elements.

Table Q6.1: The mass attenuation coefficients at 71 keV of elements

Element	H	C	N	O	S	P	C	Na	Mg	K
(μ/ρ)	0.3165	0.1666	0.1706	0.1732	0.3063	0,2706	0.4597	0.1956	0.2157	0.4034

(section 6.5)

7 Special Procedures: Stereotactic Radiosurgery and IMRT

7.1 The Rationale for Stereotactic Radiosurgery

Radiosurgery, as the term implies, is the use of radiation as a substitute for surgical excision. An anatomical target volume is defined during treatment planning within which all tissue is intended to be ablated, or rendered incapable of further growth or function. This volume should contain only tumor tissue and, as such, consideration is generally not given to sparing healthy tissue within the volume. Indeed, dose is traditionally delivered in a large single fraction and thus without the tissue-sparing fractionation effects used in conventional radiotherapy. It is obviously a central goal of radiosurgery that the high-dose region can be confined to the intended anatomy at *treatment time* as well as in the treatment plan. This requires both a planned dose distribution with confinement of the high dose to a region including little more than the planned target volume and highly accurate targeting at treatment time. The targeting aspect is where stereotactic techniques come into play and give rise to the term *stereotactic radiosurgery* (SRS).

Stereotactic targeting involves referencing an internal or external feature or object at treatment time that is also visible or at a known location within the treatment plan. For example, if a bony feature such as the skull can be seen on the treatment plan and is also at a known location at treatment time, the skull can be positioned during treatment such that the target volume is also at the correct location. Indeed, the skull is the most commonly used reference object in stereotactic targeting, whether used directly (as in x-ray or CT image-guided techniques) or indirectly (as with traditional stereotactic targeting frames). Both of these techniques will be discussed here.

Note that the radiosurgery paradigm is quite different to the radiotherapy paradigm. Treatment volumes defined for radiotherapy include a margin for positioning error, on top of a margin for sub-clinical disease (ICRU 1999). The effectiveness of radiotherapy is thus limited mostly by dose that can be safely delivered to a target volume that inevitably includes a significant amount of healthy tissue. Given that with radiosurgery the intent is to ablate all tissue within the target volume, under the assumption that this volume includes minimal normal tissue whose ablation will not cause complications, normal tissue tolerance is typically not the limiting factor in the effectiveness of radiosurgery. Instead, treatments are more likely to fail due to the prescribed dose geographically missing part of the diseased area, and this is one reason why accurate targeting is so important. Another reason is that the ablative dose must be restricted to the intended target volume to minimize potentially very harmful effects on normal tissue.

The earliest stereotactically directed radiation delivery was in the treatment of functional disorders, where a small region of the brain is irradiated to suppress neurological activity (for example in the treatment of Parkinson's disease). This was carried out by Swedish neurosurgeon Lars Leksell using orthovoltage x-rays (Leksell 1951). Probably the most significant technological advance in stereotactic radiosurgery was the development of the Leksell Gamma Knife, which in its original form used 179 focused cobalt-60 sources (Larsson et al. 1974). It was soon realized that the techniques used in irradiation of dysfunctional regions of the brain were equally applicable to the obliteration of intracranial tumors. The modern Gamma Knife has 201 sources which can be selectively blocked in order to spare critical structures. The patient's head is precisely located within a "helmet," such that the sources produce a near-spherical high-dose region centered on the chosen target. In order to treat nonspherical lesions, multiple targets are defined and a cluster of spherical high-dose regions are used to shape the dose distribution (this is known as a multiple isocenter treatment). The Gamma Knife is currently in wide use for the treatment of benign and malignant tumors, as well as in the correction of functional disorders.

While SRS has been practiced since the late 1960s using Gamma Knife units, linac-based SRS was first used in the 1980s (Winston and Lutz 1988). The technical basis for traditional SRS using a linac is to intersect multiple arced beams at one or more anatomical focal points, which at treatment time will coincide with the isocenter of the linac. The beams are collimated using small circular "cones," which are most suitable in creating a spheroid-shaped dose distribution at the intersection of the arcs.

7.2 Clinical Indications for SRS

Traditionally, SRS has been confined to the treatment of intracranial (brain) lesions, of both malignant and benign histology. The skull provides an excellent *frame of reference* for stereotactic localization and targeting. Specifically, a stereotactic *head ring* can be affixed to the skull via screw-in pins and used as the basis for a stereotactic coordinate system. Furthermore, brain tumors are often small, quite regular in shape and easily delineated using CT or MRI, making them suitable for an ablative treatment technique. Lesions treated with SRS are typically less than about 3 cm in diameter. Larger target volumes mean a larger volume of normal tissue irradiated to high-dose levels, mostly due to greater overlap between incident beams (or elements of a beam arc). Thus the opportunity for SRS instead of conventional radiotherapy rapidly diminishes as tumor size increases.

The most common application of SRS is in the treatment of brain metastases arising from primary tumors elsewhere in the body such as in the lung, breast, or prostate. Metastases are well suited to SRS because they are generally contained within a defined and regularly shaped volume. Often there are multiple lesions arising and at least a limited number can be treated in a single session. These comprise about 40% of cases in a typical clinical SRS program. SRS is often used in conjunction with whole brain radiotherapy (WBR) in the treatment of metastases. The rationale for WBR plus SRS is to kill the large number of malignant cells within radiologically visible lesions with SRS and the smaller number of invisible cells likely to exist elsewhere in the brain with WBR. Clinical studies have shown a benefit of this approach over WBR alone, at least for a small number of lesions (Andrews et al. 2004).

Another indication comprising a relatively large proportion of SRS treatments is benign brain tumors such as meningiomas, acoustic schwannomas (also known as acoustic neuromas), and pituitary tumors. These lesions are often very close to, or even abutting, critical structures such as the brainstem and optic chiasm. In these situations the precise targeting of stereotactic techniques must be accompanied by especially fast dose fall-off between the tumor and adjacent structures. Meningiomas, in particular, can be highly irregular in shape due to the fact that they arise in the lining of the brain (the *meninges*) and often follow bony surfaces. These lesions thus present an additional challenge in terms of requiring conformity of the prescription dose to a complex 3-D shape.

Arteriovenous malformations (AVMs) are another type of benign condition quite commonly treated using SRS. An AVM is a malformed knot of blood vessels that would otherwise have been a normal array of fine capillaries. There is a significant chance that at AVM will rupture, hence the need to remove it or stop the flow of blood through it. As with other lesions, SRS is typically used in the treatment of AVMs only when immediate removal of the lesion via surgery would present an unacceptable risk to the patient. Stereotactic radiosurgery for small (say <2 cm) AVMs achieves a high obliteration rate 2 years after treatment (Alexander et al. 1993). Treatment of larger lesions is not as successful because of a lower probability of obliteration of a larger *nidus* and the necessary reduction in dose to avoid irradiating a large volume of normal brain to a high dose.

SRS is sometimes used in the treatment of primary malignant brain tumors, such as gliomas and astrocytomas, in conjunction with other forms of treatment such as surgery, chemotherapy, and conventional radiotherapy. The role of SRS in conjunction with radiotherapy is to increase overall cell kill via an ablative dose to a limited region. In some cases, fractionated SRS will be used with the intent of reducing toxicity to healthy brain tissue (Cho et al. 2004).

7.3 Traditional Stereotactic Targeting

As with most radiation treatments, a central challenge of SRS is how to ensure correct aiming of the beams at a target that is usually invisible to the human eye. What is needed is a way of relating the location of the anatomical target, as defined in the coordinate system of a *planning image*, to a reference coordinate system that can in turn be related to the coordinate system of the treatment machine.

In modern radiotherapy and radiosurgery, the primary planning image is typically a CT scan, on which a target volume and other anatomical volumes of interest are drawn. If physical features defined in the reference coordinate system are also visible on the planning image, then a transformation from one coordinate system to the other can be deduced. Once this match has been established, the task of positioning the target defined in the planning image becomes one of positioning the apparatus that forms the basis of the reference coordinate system. This rather abstract explanation is in fact an equally valid high-level description of patient setup in conventional radiotherapy, SRS, and image-guided radiotherapy. The distinction is in how the reference coordinate systems are defined and used. In this section the use of a traditional stereotactic head ring as the basis of the reference coordinate system (*stereotactic coordinates*) for SRS targeting will be described.

SRS traditionally involves a head ring, rigidly affixed to the skull, as the platform for various pieces of apparatus used for *localization* and *targeting*. Localization is the process of establishing a transformation between planning image coordinates and stereotactic coordinates, or more specifically, finding the stereotactic coordinates of the anatomical point designated during planning as the treatment isocenter. Targeting is the process of positioning the patient, according to the stereotactic coordinate determined during localization, so that the treatment isocenter coincides with the machine isocenter.

Figure 7.1 illustrates the path via which anatomy defined on a planning image (in planning coordinates) is positioned correctly relative to the treatment machine (in machine, or beam, coordinates). The key to this process is the intermediate coordinate system established by the combination of a stereotactic head ring and stereotactic localizer.

7.3.1 The Head Ring

The position and orientation of a stereotactic head ring relative to the patient's anatomy must remain constant throughout the planning and treatment process. This is so that the transformation between coordinate systems

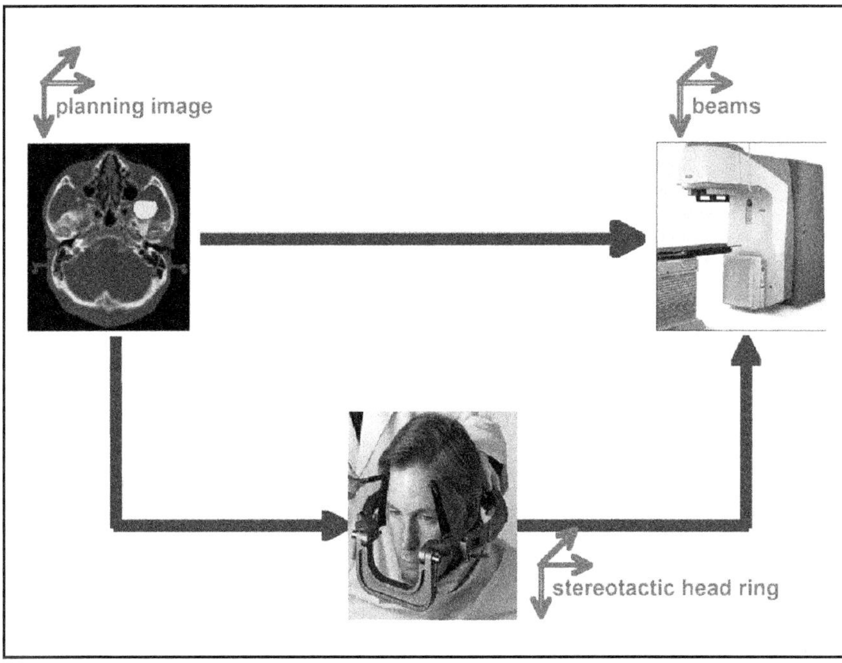

Figure 7.1: Coordinate systems corresponding to (i) patient anatomy as seen in the planning image; (ii) reference coordinate system (in this case defined by a stereotactic head ring); and (iii) the treatment beams. The reference coordinate system links the anatomical and beam coordinate systems in the patient positioning process.

established after the planning scan remains valid for targeting. The correspondence between a given anatomical point and a point in stereotactic coordinates should be within 1 mm for single fraction radiosurgery treatments. To keep the head ring fixed relative to the skull, sharp pins threaded through posts attached to the head ring itself are used. Figure 7.1 shows a head ring being fitted and forming the basis for the stereotactic coordinate system. Such an "invasive" fixation of a head ring is reserved for single fraction treatments as (i) the head ring cannot be left screwed to the skull for an extended period, and (ii) it is not clinically feasible to repeatedly attach the head ring in this way.

Head rings can be "noninvasively" affixed in various ways that allow repeated use for targeting after an initial localization session. A common method of noninvasive fixation is a bite block containing a substance that is molded to the upper dentition and palate. Between localization and the first treatment procedure, and between treatments, a noninvasive head ring is removed and stored. Noninvasive fixation is generally accepted to provide less stability in head ring location relative to the skull than invasive fixation, but targeting accuracy is still very high (typically 1 to 2 mm). Figure 7.2 is a picture of a patient set up for treatment using a bite block–based head ring. Note also that the patient's head is being supported by an occipital pad and straps over the top of the head are also used to assist in immobilization.

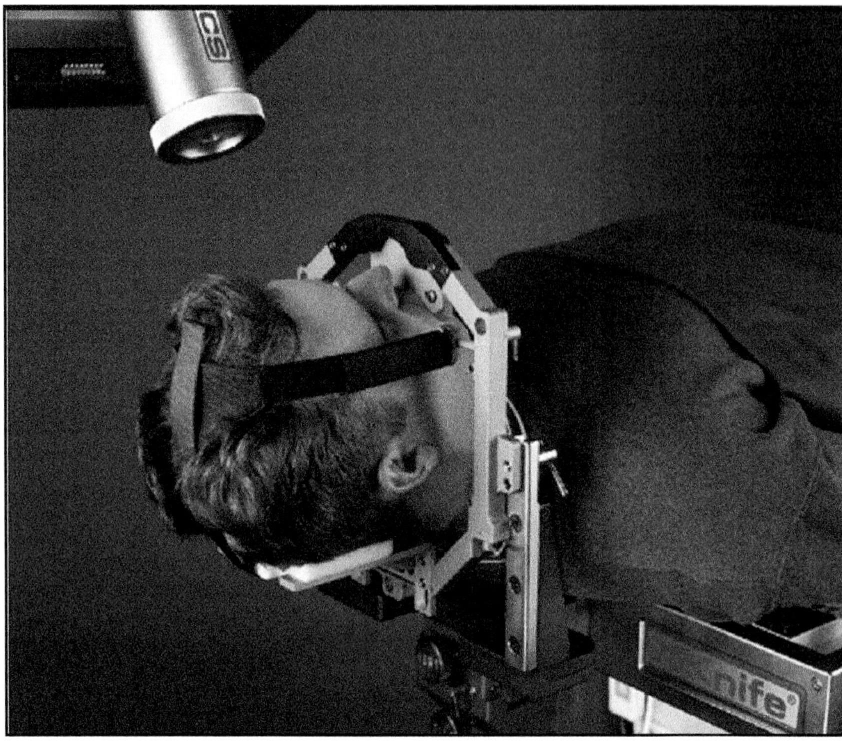

Figure 7.2: A patient set up for treatment using a bite block–based head ring. The patient's head is being supported by an occipital pad and straps over the top of the head are also used to assist in immobilization. (Courtesy of Radionics.)

7.3.2 The Localizer

The *localizer* is used to transform each pixel in the planning CT image to a stereotactic coordinate. It is a frame that connects securely to the head ring during imaging, in an orientation that is fixed and known to the stereotactic treatment planning system. Features of the localizer, typically a set of carbon fiber rods, are visible in the image slices. Since the geometry of the rods is known relative to the head ring, a transformation between CT and stereotactic coordinates can be deduced.

Figure 7.3a is a picture of a BRW stereotactic localizer. When the planning CT image is acquired, its superior-inferior limits will be set to be within the bounds of the localizer so that all image pixels can be assigned a stereotactic coordinate via the location of the rods as they appear on each image slice. With this localizer, there are nine rods, three of them diagonal and the rest perpendicular to the plane of the head ring. The diagonal rods appear progressively shifted as the axial plane becomes more superior or inferior. It is the locations of these diagonal rods on each axial CT slice that is used to derive a coordinate transformation between each CT pixel and the stereotactic coordinate system. Figure 7.3b is an axial CT image including the nine localizer rods. In this image, the rods have been detected in a particular order

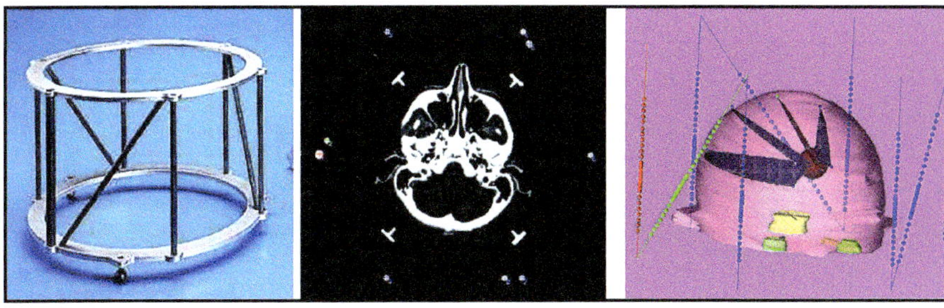

a b c

Figure 7.3: (a) BRW stereotactic localizer. This attaches to the head ring to establish a stereotactic coordinate system within the planning image. (b) An axial CT image including the nine localizer rods. (c) The localizer rods, after detection by the planning software (Radionics, Burlington, MA) superimposed on 3-D surfaces representing the patient anatomy. (Courtesy of Radionics.)

by the planning software. Rod 1 is the left-most in the image (on the patient's right) and the other rods are numbered in a clockwise fashion. Rods 2, 5, and 8 are the diagonal rods. As an example, as rod 2 appears progressively closer to rod 3 and further from rod 1, this indicates the CT slice is more superior. Mathematically, the imaged locations of the diagonal rods form three points at known stereotactic coordinates (calculated from their distances from the rods at either side) and also at known locations within the image. From these two sets of three points, a transformation from CT to stereotactic coordinates can be derived for that slice. Figure 7.3c shows the localizer rods, after detection by the planning software, superimposed on 3-D surfaces representing the patient anatomy.

The 3-D stereotactic coordinate axes are commonly denoted as AP (anterior-posterior, positive toward the patient's anterior), LAT (lateral, positive toward the patient's right), and VERT (vertical as the patient sits down, positive toward the patient's superior). After this transformation has been performed, the planner can draw contours on the CT image and these, too, are defined in stereotactic coordinates. 3-D volumes formed from the contours thus occupy a 3-D space within this coordinate system.

7.3.3 The Targeting Frame

During treatment, the patient must be set up so that the linac isocenter is at the planned stereotactic coordinate. This is normally achieved via a *targeting frame* that attaches to the head ring in the same manner as does the localizer. The targeting frame often has a scale on each of the three coordinate axes, whereby crosshairs are slid into the position as derived using the localizer and planning software. Figure 7.4a shows a patient affixed to the treatment couch via a head ring with a targeting frame attached. Lasers indicating the linac isocenter can be seen coinciding with the frame crosshairs. Note that

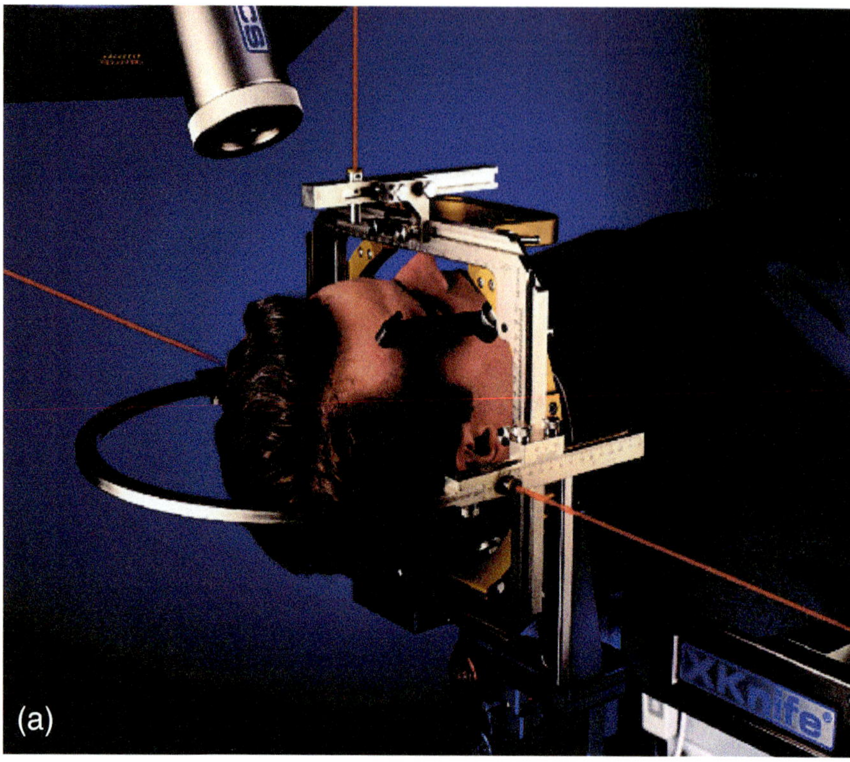

Figure 7.4: (a) Patient affixed to the treatment couch via a head ring with a targeting frame attached. Lasers indicating the linac isocenter can be seen coinciding with the frame crosshairs. (Courtesy of Radionics.)

this particular frame includes a bow-shaped component that can be used to ensure there is no head tilt; a coronal laser creating a horizontal sheet of light should align with the center of the bow throughout its circumference.

An alternative targeting frame is shown in figure 7.4b. Here instead of a mechanical scale there is a "targeting box" with overlay sheets on the lateral, anterior, and superior faces. Each sheet is a printout containing the planned isocenter coordinate corresponding to that face. This type of targeting device removes the chance of human error in setting coordinates. Another advantage is that the printout can include an outline of the beam projections as they should impinge on the face. Comparing these outlines with the shapes created by the field light can be used as an extra quality assurance step.

7.4 Frameless Stereotactic Targeting

7.4.1 Camera Based

With the targeting systems discussed so far, whether invasive or noninvasive, the processes of immobilization and localization/targeting are linked via the head ring. That is, the head ring provides immobilization by either pins or

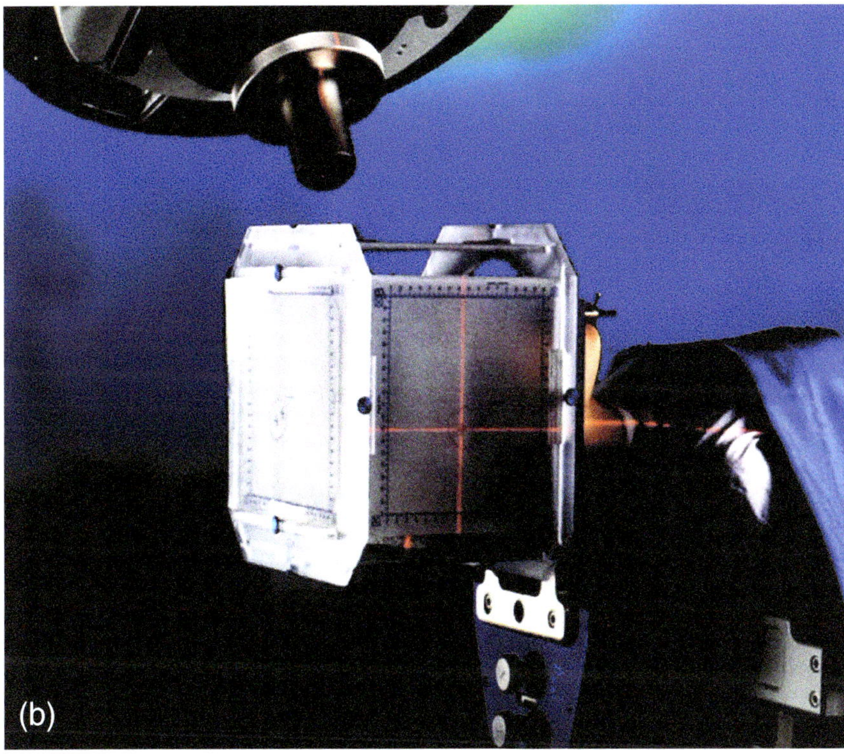

Figure 7.4: (Continued) (b) A "targeting box" with overlay sheets on the lateral, anterior, and superior faces to indicate the planned isocenter coordinate corresponding to that face. (Courtesy of Radionics.)

bite block and also localization and targeting via the localizer and targeting frame attachments. It is possible to decouple these two functions so that the role of the immobilization device is keeping the patient still during treatment, but does not *per se* need to retain precisely the same spatial relationship to the patient between planning and treatment (or from treatment to treatment). The immobilization device should however maintain the patient at a consistent orientation, even if actual position needs to be adjusted via the localization and targeting system.

Such systems are known as *frameless*, because a traditional head ring, localizer, and targeting frame are not used. One example of the frameless SRS system uses a mask to immobilize the patient's head and an array of reflectors attached to a bite block for localization and targeting. The mask is relatively comfortable and provides good immobilization during treatment but would not necessarily be able to be reproducibly fitted to the patient with enough accuracy and precision to act as the basis for a stereotactic coordinate system (as does a head ring). The reflector array is not attached to the mask but directly to the patient, again noninvasively via the bite block. The accuracy of this paradigm is dependent upon the array maintaining a reproducible and static relationship to the patient's anatomy between planning and each treatment session.

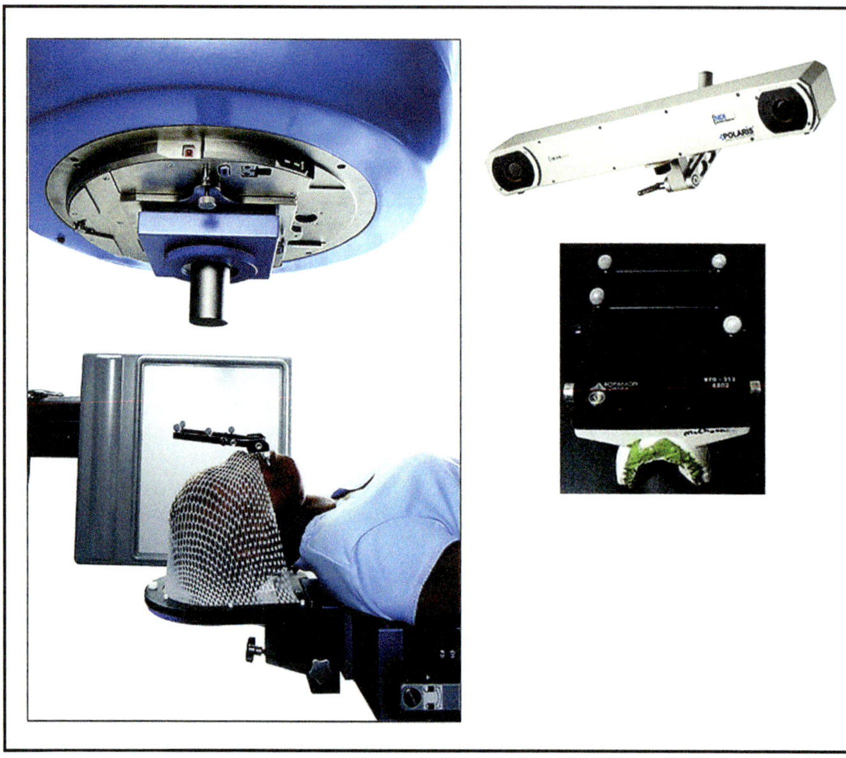

Figure 7.5: Components of a commercially available frameless setup. The reflectors mounted on the array are illuminated by infrared light from a camera system that also detects the 3-D position of each reflector. The array of reflectors is at a required position and orientation relative to the linac isocenter (according to the plan). (Courtesy of Varian Medical Systems.)

Figure 7.5 shows a picture of some of the components a of commercially available frameless setup. The reflectors mounted on the array are illuminated by infrared light from a camera system that also detects the 3-D position of each reflector. The camera's frame of reference is calibrated such that the location of the reflectors is known relative to the isocenter of the treatment machine. During treatment planning, the location of each reflector relative to the planned isocenter is determined, so that during treatment there is a required position of each reflector relative to the machine isocenter. With the four reflectors used here, the frameless targeting system can calculate the translations and rotations of the array, and thus the patient's head, needed to bring the array to its planned location and orientation in the room.

This type of frameless SRS system is used for both single-fraction and multiple-fraction treatments, thus providing a general procedure. A big operational advantage of such a technique over the use of an invasive head ring for single-fraction treatments is that all planning scans and treatment planning can be done prior to the day or treatment, as for multi-fraction treatments. The development of the original University of Florida frameless radiosurgery system is reported by Bova et al. (1997).

7.4.2 X-Ray Based

Just as a head ring provides the basis of a coordinate system within which a target can be defined, the skull itself can provide such a basis as long as we know where it is during treatment. That is, we define a target somewhere within an image of the anatomy that contains the skull, then at treatment time we determine the location of the skull relative to the linac. The means by which we can directly locate the skull at treatment time is 2-D or 3-D imaging. When we align the patient, the objective is to position the skull at the planned location relative to the isocenter and we can do this as long as we know the imaging system's geometry relative to the linac. This is taking the frameless approach a step further than with the reflective array discussed above—not only is there no frame but there is no surrogate localization/targeting device, as this role is fulfilled by the patient himself.

Following the same framework as figure 7.1, figure 7.6 gives an example of 2-D image-based stereotactic targeting. Such a strategy can be realized using two x-ray cameras to acquire stereoscopic images of the bony anatomy in the region of the target. These two images are compared with the images that *should* be seen considering the location of the x-ray sources and detector panels relative to the treatment machine's isocenter, as well as the planned anatomical location of the isocenter. Just as a head ring is something whose location we know relative to the treatment machine, so too are the images. Any difference between the actual acquired images and the

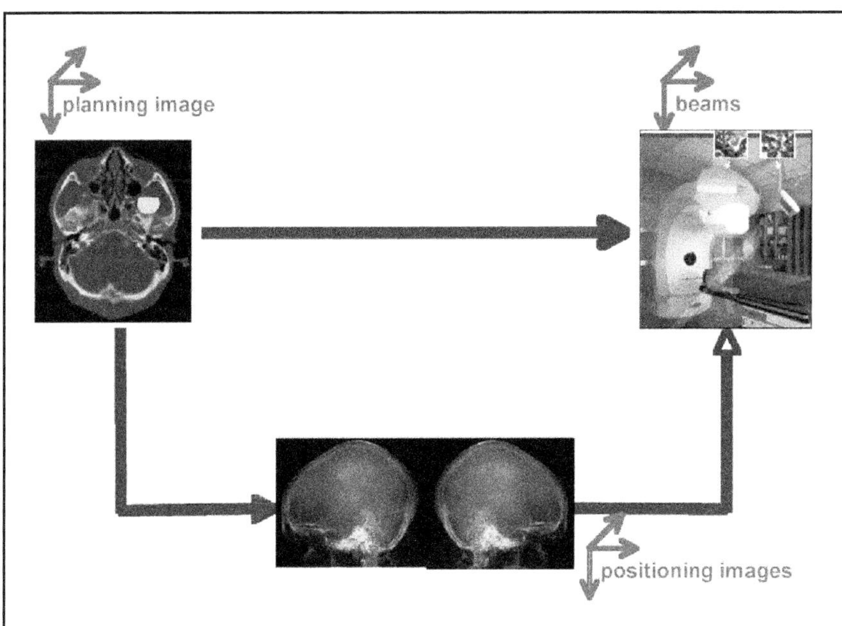

Figure 7.6: As with figure 7.1, but with the reference coordinate system provided by a pair of x-ray images taken at treatment time. The imaging geometry is known relative to the treatment beams.

images that are expected to be seen is interpreted via image-matching software as a required shift of the patient.

Similarly, 3-D images acquired at treatment time can be compared with 3-D planning images and a required shift calculated. Again, image-matching software is used to detect any misalignment of the anatomy relative to the treatment machine (in a very similar procedure as used for registration of CT and MR images during planning). An advantage of using 3-D images is that there is the potential to perform the alignment using not just images of the bony anatomy but of soft tissue and even of the tumor itself. Thus the actual tumor can become the basis for the stereotactic coordinate system!

7.5 SRS Treatment Planning

The treatment planning stage normally begins by transferring CT and MR image data to the planning computer. Fiducial markers on the CT and MR images are detected, where those on the MR images will be the same as on the CT images if image fusion has been used (the MR data set is thus treated as being equivalent to CT). When the fiducials have been detected, the image sets are thus stereotactically defined and contouring of the external volume, tumor, and critical structures can proceed. Critical structures typically outlined include the brainstem, eyes, optic nerves, and optic chiasm.

After contouring has been done and the structures are displayed as a 3-D image, the target coordinates can be set so that the intersection of all beams is at the center of the tumor. Usually the size of the currently chosen collimator (as projected to isocenter) will be represented as a sphere or circle and displayed at the isocenter. The collimator size is thus chosen so that the sphere covers the lesion. If the lesion is nonspherical in shape, more than one isocenter may be used, where the beam diameter can be different for each. The collimator size should be chosen to be the minimum that covers the lesion. Note however that the final choice of collimator should be made by assessing how well the high-dose region covers the tumor.

For treatment using multiple non-coplanar arcs, several couch angles are chosen and gantry start and stop angles are set for each. The choice of couch and gantry angles depends on the spatial relationship of the tumor to critical structures. A beam's-eye view of the lesion is useful for selecting these angles, where the gantry can be rotated, and start and stop angles chosen such that no critical structures are "seen" along the beam path, especially between the entry point and the tumor. Usually approximately five couch angles are used and gantry arcs typically range from 30 degrees up to 140 degrees. A total arc angle (over all couch angles) of at least 400 degrees is desirable to spread out the dose and to achieve good focusing of the dose distribution at the lesion. Another technique is to use a continuously rotating couch and gantry (Podgorsak et al. 1988), where the couch rotates from +75 to −75 degrees and the gantry from 30 to 330 degrees (with the convention that the couch angle is 0 degrees when along the direction of the gantry axis and the gantry angle is 180 degrees when the beam is directed at the floor).

Assessment of the dose distribution is normally via isodose lines, isodose surfaces, and dose volume histograms. Dose is usually normalized to 100% at the isocenter. Isodose contours should ideally wrap tightly around the target volume, such that a particular isodose level is at the edge of the tumor. If the high-dose region is too large, a smaller collimator can of course be chosen. If the lesion is irregular, two or more isocenters may need to be chosen so that the isodose contours are nonspherical. This may incur a "hot spot" where high-dose regions due to two adjacent isocenters overlap—this will cause a less uniform dose within the tumor but hopefully there will be less dose to surrounding normal tissue. An isodose level that coincides with the edge of the tumor should be used as the *prescription isodose*, to which the desired dose to the tumor should be equated. For example, it may be decided to deliver 2000 cGy to the 90% isodose, which will correspond to a dose at isocenter of 2000 cGy × 100%/90% = 2222 cGy.

Figure 7.7 shows two stereotactic "cone" collimators, one of which projects a beam of diameters 17.5 mm at the isocenter and the other a beam of diameters 20 mm. These cones typically range in diameter such that the circular field they produce projects a beam of diameter 10 to 40 mm at isocenter.

Figure 7.8 is a schematic of an arc configuration for a four-arc treatment. There is a 260-degree arc with the couch at 0 degrees and 100-degree arcs with the couch at 90, 45, and –45 degrees. This represents a total of 560 degrees of arc, which will result in a highly focused, near spherical, dose distribution at the isocenter. If one were to omit the arc with the couch at 90 degrees, for example, the dose distribution would be somewhat flattened. Similarly, a series of short arcs from similar couch angles will produce an elongated dose distribution. Also indicated in figure 7.8 are the stereotactic coordinate axes AP, LAT, and VERT.

Figure 7.7: Stereotactic collimators ("cones"). These particular collimators create circular beams with diameter 17.5 mm and 20.0 mm at isocenter.

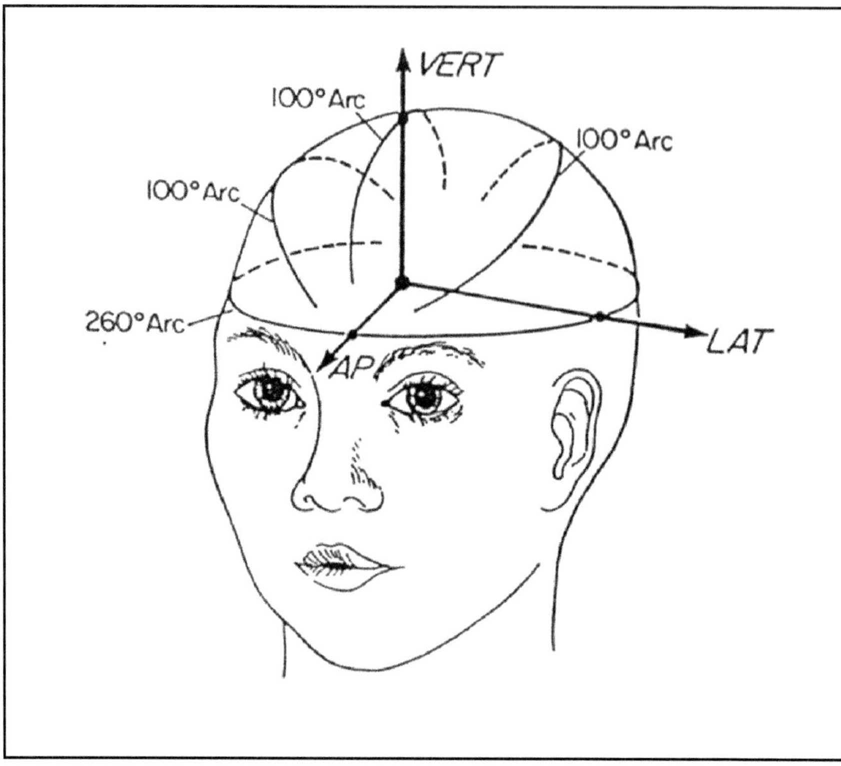

Figure 3.8: Schematic of an arc configuration for a four-arc treatment. There is a 260° arc with the couch at 0° and 100° arcs with the couch at 90°, 45°, and −45°. Also indicated are the stereotactic coordinate axes AP, LAT, and VERT.

In figure 7.9a, a single isocenter plan is shown with six arcs. The circular collimator chosen is represented by the red sphere, where the diameter of the sphere is the projected diameter of the field at isocenter. In figure 7.9b, a double isocenter plan is shown. Note that only the arcs for the currently selected (red) isocenter are shown. In a multiple isocenter treatment, the patient is repositioned each time for a new set of arcs to be delivered. Multiple isocenters are used when the dose distribution needs to be of a more complex shape than a sphere or near to a sphere (a single isocenter can produce an ellipsoidal dose distribution via strategic arc directions). Notice the nine rows of beads outside the head in figure 7.9b. These are the localizer rods used to create the stereotactic coordinate system, as discussed above.

An axial CT image, with target volume drawn in white and isodose lines for a simple, single isocenter, plan is shown in figure 7.10. The cone diameter is indicated in red. Note that the 80% isodose line (dose is normalized to 100% at isocenter) closely matches the cone diameter. At top left are the stereotactic coordinates (AP, LAT, and VERT) of the point indicated by the yellow cross and the dose at this point (86.3% of the dose at isocenter). The maximum dose in this slice is 101% as seen at bottom right.

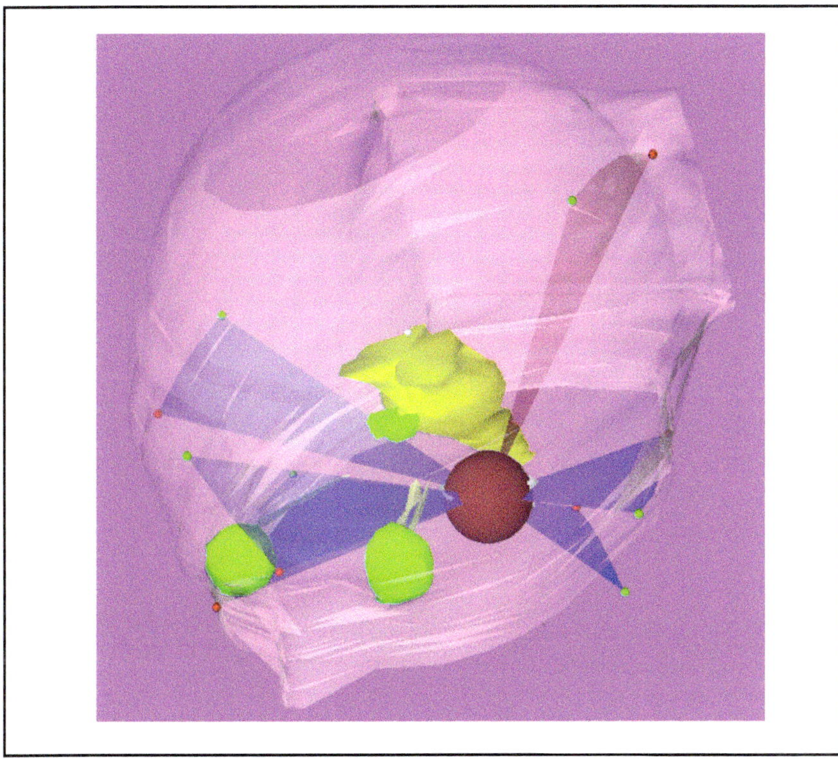

(a)

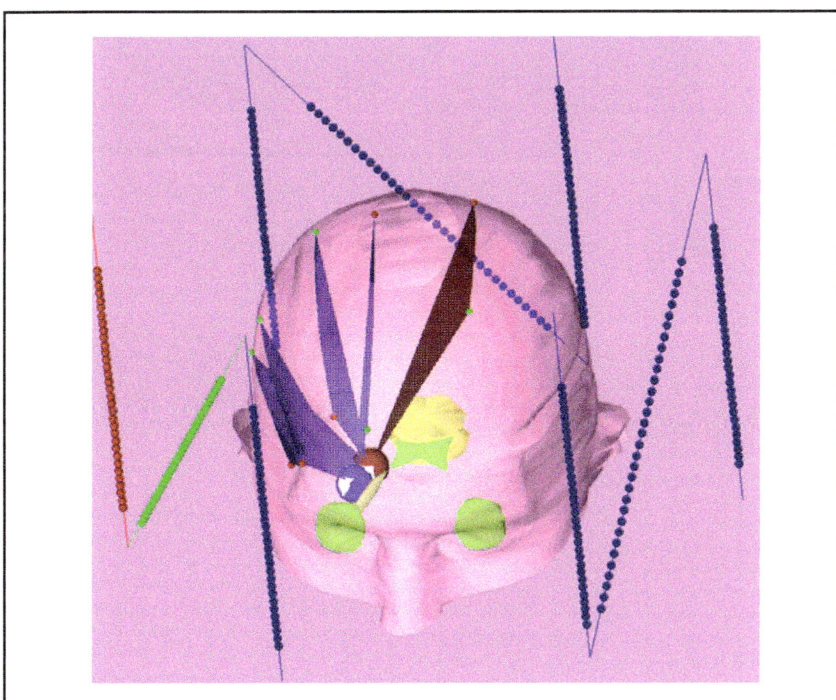

(b)

Figure 7.9: (a) A single isocenter plan with six arcs. The diameter of the circular collimator chosen is represented by the red sphere. (b) A double isocenter plan. Only arcs for the currently selected (red) isocenter are shown.

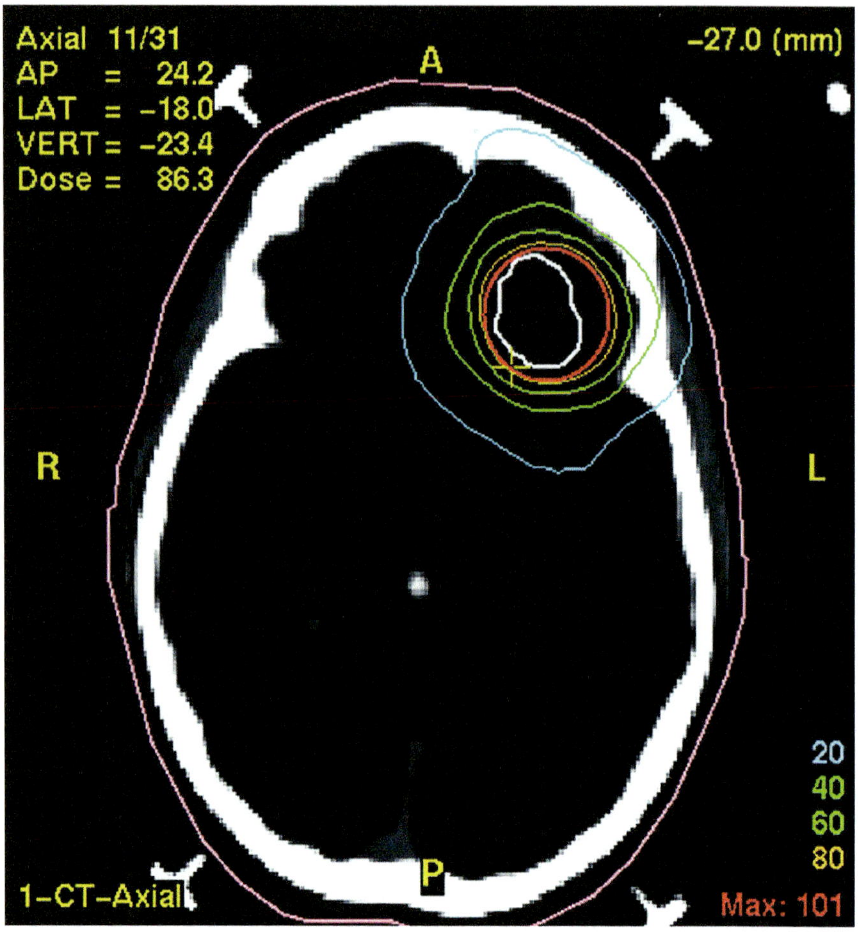

Figure 7.10: An axial CT image with target volume drawn in white and isodose lines for a simple, single isocenter. The cone diameter is indicated in red. Note that the 80% isodose line closely matches the cone diameter. (Courtesy of Radionics.)

7.5.1 Use of MRI via Image Fusion

Because MR scans generally show soft-tissue boundaries much more distinctly than do CT scans, it is highly desirable to be able to draw tumor and organ contours on the MR images. Of course for this to fit into the stereotactic process, the MR pixels also have to be able to be transformed into stereotactic coordinates. While it used to be quite common to attach a special MR-compatible localizer to a special MR-compatible head ring and thus transform the MR scans independently of CT, this is inconvenient and has been almost totally superseded by CT-MR image fusion. This process simply matches the 3-D MR image to the 3-D CT image and relies on the CT transformation to obtain the stereotactic coordinates of any pixel or contours on either set of scans. Specifically, any pixel in the MR image has a corresponding location in the CT image and this location has a corresponding stereotactic coordinate. Figure 7.11a is a diagram of the image fusion process: the MR scans are put into stereotactic coordinates via a two-step process of image fusion and then transformation of the CT scans. Figure 7.11b shows a common way of checking the match between CT and MR images, whereby a split-screen display (with scrollable boundary position) is used for assessment of whether the anatomy on either side of the CT-MR boundary is correctly aligned.

Image fusion is commonly performed using either bony anatomy as the common region to be matched (Kooy et al. 1993) or via a more general algorithm that takes into account image pixels throughout each image volume (Maes et al. 1997).

7.5.2 Dose Calculation for Arc Treatment

Here we will briefly look at a basic dose calculation scheme for cone/arc treatments. This is by no means the only scheme, but serves as an example and highlights some important factors.

For circular beams there are three main properties to consider in calculating the dose at some point under the surface:

i) *Output factor* (OF). This is the dose for a given cone at the depth of dose maximum, relative to the dose at the same depth, for a calibration field size (usually 10×10 cm^2). The output factor increases as the cone diameter increases because of a larger volume of tissue irradiated and thus more scattered x-rays.
ii) *Tissue maximum ratio* (TMR). This is the dose fall-off with depth that occurs because of x-ray attenuation. TMR is distinct from the similar, and perhaps more familiar, percentage depth dose (PDD or %dd) curve in that for a TMR the dose measurement point remains at the same distance from the beam source. The depth of measurement is increased by increasing the depth of water or tissue above the detector, rather than moving the detector to greater depth as with a PDD. The detector is placed at the isocenter, at a distance from the beam source known as the "source-axis distance" (SAD). The SAD is 100 cm for modern conventional linacs.
iii) *Off-axis ratio* (OAR). This is the lateral "profile" of the beam, giving the dose at an off-axis position relative to that on the central axis. It is generally assumed that these profiles are radially symmetric (dependent only on distance from the axis).

Figure 7.12 shows real examples of curves for each of these quantities. In figure 7.12a, output factors for 0.5, 1.0, 2.0, and 3.0 cm diameter cones are plotted. Note the sharp reduction in output for the 0.5 cm cone. This is difficult to measure accurately due to detector technology; hence dosimetry for such small fields is often somewhat uncertain.

Figure 7.12b shows a family of TMR curves for the same cone sizes. A larger cone diameter gives a slower fall-off of the curve. This is because the

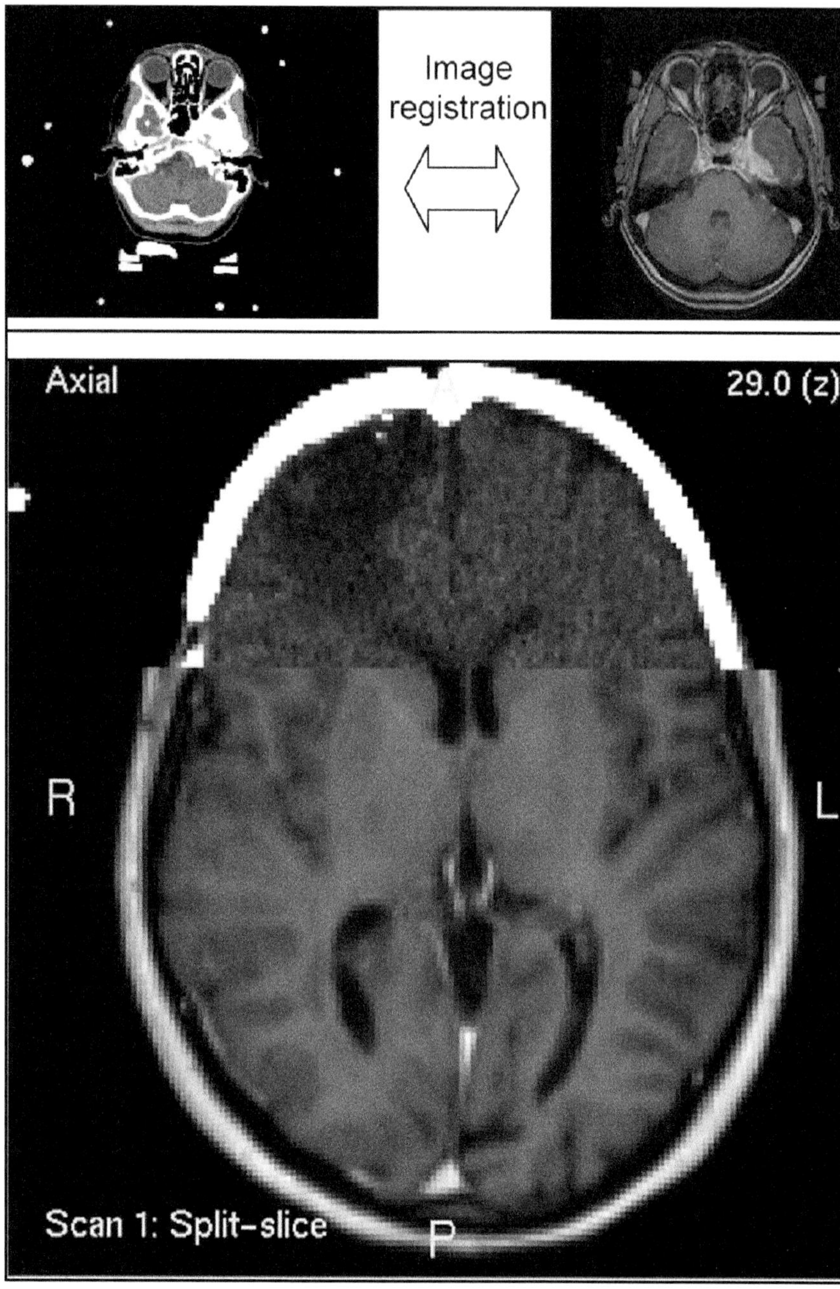

(a)

(b)

Figure 7.11: (a) Diagram of the CT-MR image fusion process. The MR scans are put into stereotactic coordinates via a two-step process of image fusion and then transformation of the CT scans via the localizer rods. (b) Checking the match between CT and MR images via a split-screen display with scrollable boundary position.

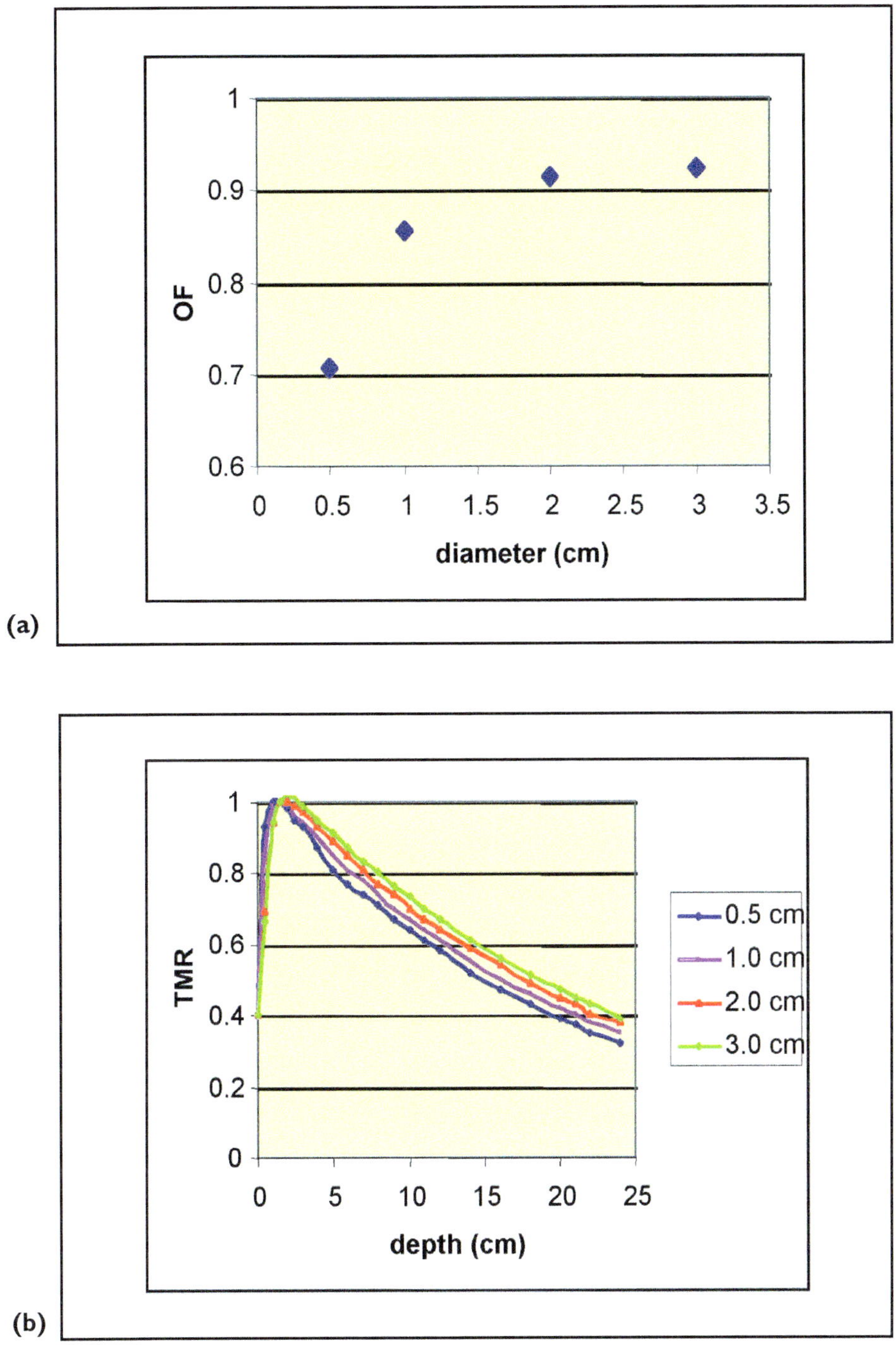

Figure 7.12: (a) Output factors for 0.5, 1.0, 2.0, and 3.0 cm diameter cones. Note the sharp reduction in output for the 0.5 cm cone. (b) Tissue maximum ratio (TMR) curves for the same cone sizes.

(c)

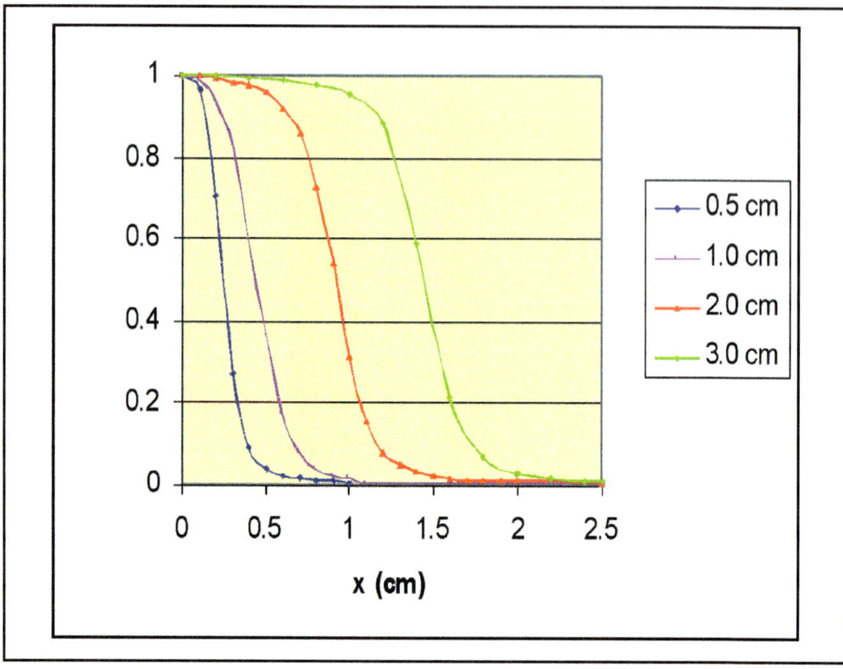

Figure 7.12: (Continued) (c) Off-axis ratio (OAR) curves for these collimator diameters.

increase in scatter with larger cones is more pronounced at greater depths, where the dose is therefore more bolstered than at the surface.

In figure 7.12c, the corresponding family of OAR curves is shown. There is typically a reasonably flat inner portion and a steep fall-off region known as the *penumbra*. Note that for the smaller cones, there is almost no flat portion since the entire profile is in the beam penumbra. One important point that should be made here is that when a complete arc, or a set of arcs, is delivered, the "3-D" penumbra resulting from the overlap of arc segments is much broader than the individual penumbral shapes as seen here. In fact, the 3-D penumbra is almost entirely characterized by the degree of overlap rather than by the shape of the static beam penumbra. A larger cone diameter means more overlap due to simple geometry and thus the 3-D penumbra becomes broader (larger cones give a less focused beam).

Figure 7.13 is a simple diagram of a divergent beam of diameter, s, indicating the isocenter point at depth, d. In this framework, the dose per monitor unit (MU) at the isocenter is simply:

$$D(d,s) / MU = OF(s)TMR(d,s). \tag{7.1}$$

For a beam arc using a given cone, the average dose per MU at the isocenter is calculated using the average TMR for the arc, where the average TMR is a function of the range of depths to the isocenter throughout the arc. Average dose per MU for a given arc (index n, in a set of N arcs, where a given arc has I sectors) is thus:

$$D_n^{arc}/MU = OF(s)\,AVE_{i=1,I}\left[TMR(d_i,s)\right]. \tag{7.2}$$

Importantly for programming the linac to perform the arc treatment, the number of MUs is then obtained from the required dose for the arc and the average dose per MU:

$$MU = \frac{D_n^{arc}}{OF(s)\,AVE_{i=1,I}\left[TMR(d_i,s)\right]}. \tag{7.3}$$

It is common for the contribution from each arc to the dose at isocenter to be such that the dose per degree of arc is equal for all beams. Thus if the total dose at isocenter is to be D_P and there are N arcs, where arc n covers an angle θ_n, then the dose due to arc n should be

$$D_n^{arc} = D_P \frac{\theta_n}{\sum_{i=1}^{N} \theta_i}. \tag{7.4}$$

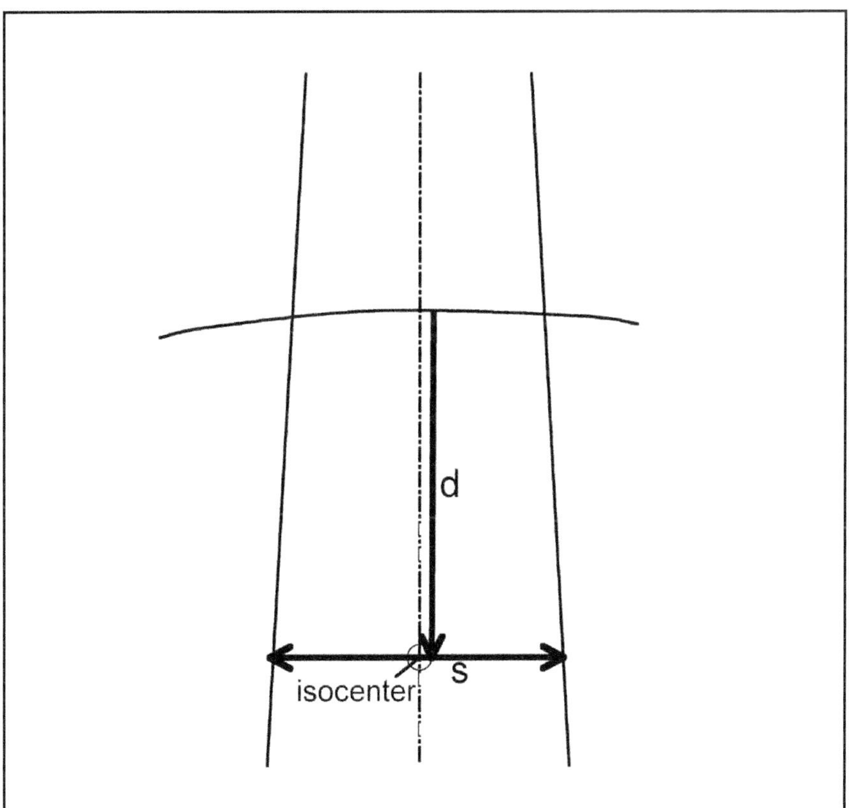

Figure 7.13: Simple diagram of a divergent beam of diameter s, indicating the isocenter point at depth *d*.

7.6 MMLC-based Radiosurgery Techniques

7.6.1 Overview

There are limitations, both fundamental and practical, to the shape of the dose distribution that can be achieved using a cone/arc technique. The most obvious limitation is that circular fields tend to form spherical or near-spherical dose distributions, which do not conform to irregular targets. So while dose will fall away rapidly outside of the high-dose volume formed by the intersecting arcs, the high-dose volume itself may include more normal tissue than is clinically desirable. While multiple isocenters expand the flexibility of cone/arc techniques, the more isocenters that are used, the more time-consuming is the treatment and also the planning phase. Also, overlap between the dose distributions formed around each isocenter means that the aggregate dose distribution can contain regions of unacceptably high dose, both inside and outside of the planned target volume. Regions of low dose adjacent to the overlap cause a further clinical problem and can further exacerbate the issue of a high maximum dose when a given minimum target dose must be maintained.

An alternative collimation technique is the mini-multileaf collimator (MMLC). These are a type of MLC with narrow leaves and a relatively small maximum field size, designed especially for targets that are larger and more irregular than are well suited to cone/arc treatments, but still smaller than those usually treated using a conventional MLC. The leaf width of MMLCs varies between manufacturers but is typically in the 3 to 4 mm range, with some having multiple leaf widths. This is in comparison to large field MLCs, the currently available commercial versions of which have a leaf width of either 5 or 10 mm. Narrower leaves means that the edges of a beam aperture formed can more resemble a continuous curve, rather than a "stair-step" shape and can be more irregular.

It should be noted that many, if not most, stereotactic procedures delivered using an MMLC are multiple-fraction treatments. This is because the choice of an MMLC over a convergent arc/cone treatment is usually (though not always) because of relatively large lesion size. Such treatments are commonly termed *stereotactic radiotherapy* (SRT) or *stereotactic body radiotherapy* (SBRT), as is discussed later in this chapter.

Figure 7.14a shows an MMLC mounted on the treatment head of a linac and figure 7.14b is a close-up of the leaves. This MMLC has 31 pairs of leaves, each projecting a width of 4 mm at the isocenter. Maximum field size is approximately 10 cm × 12 cm. In this case the MMLC is temporarily attached for a session of stereotactic treatments. In contrast, "dedicated" stereotactic linacs have a MMLC permanently attached. The advantages of a dedicated stereotactic linac are that (i) the MMLC does not need to be attached and removed, and (ii) the leaf control system is integrated with the linac control system. The disadvantage is that the linac cannot be used to treat targets larger than the maximum field size of the MMLC. Such a dedicated unit is that in the image guidance example in figure 7.6.

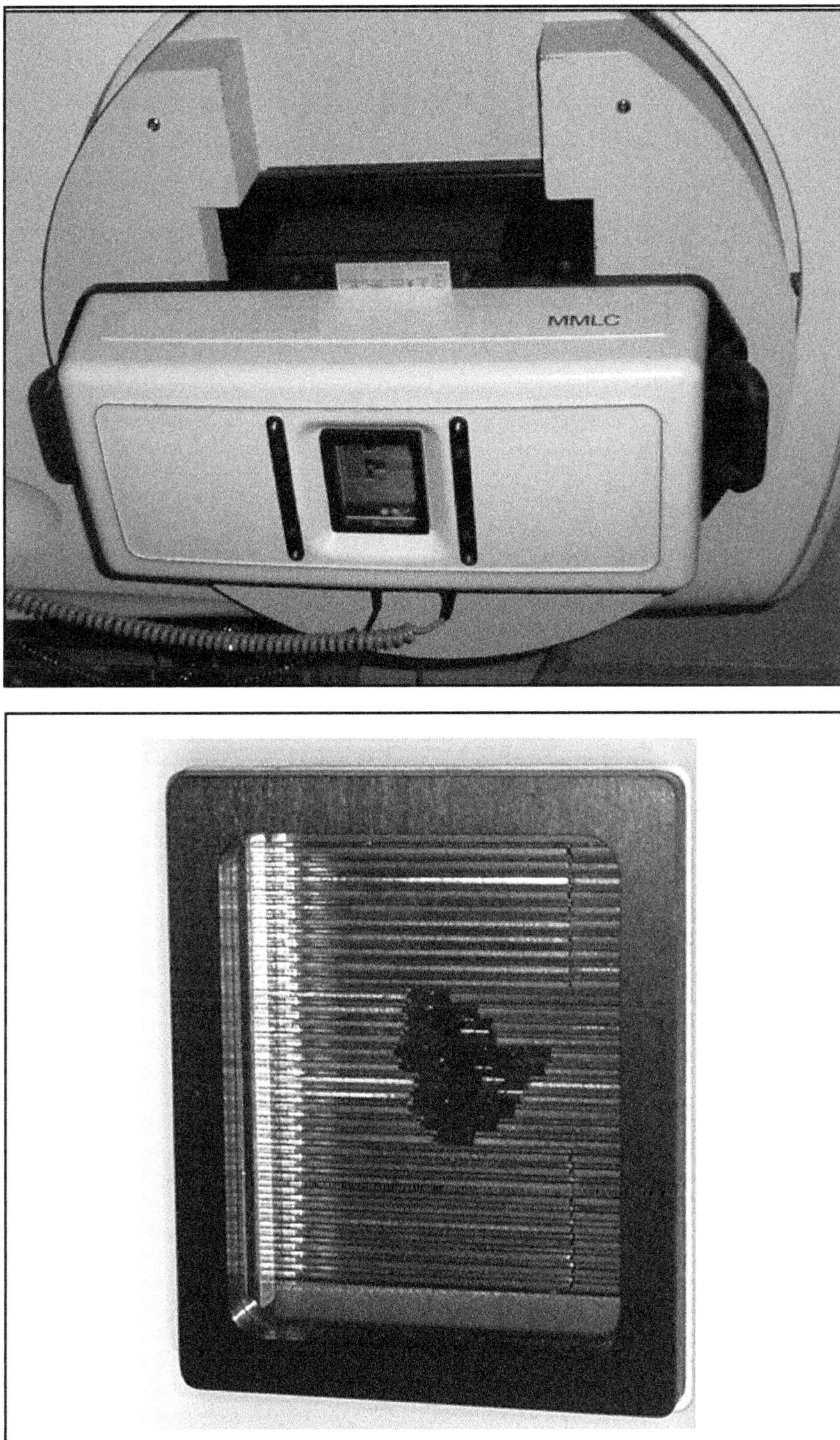

(a)

(b)

Figure 7.14: (a) A mini-multileaf collimator (MMLC) mounted on the treatment head of a linac. This MMLC has 31 pairs of leaves, each projecting a width of 4 mm at the isocenter. Maximum field size is approximately 10 cm by 12 cm. (b) A close-up picture of the levels.

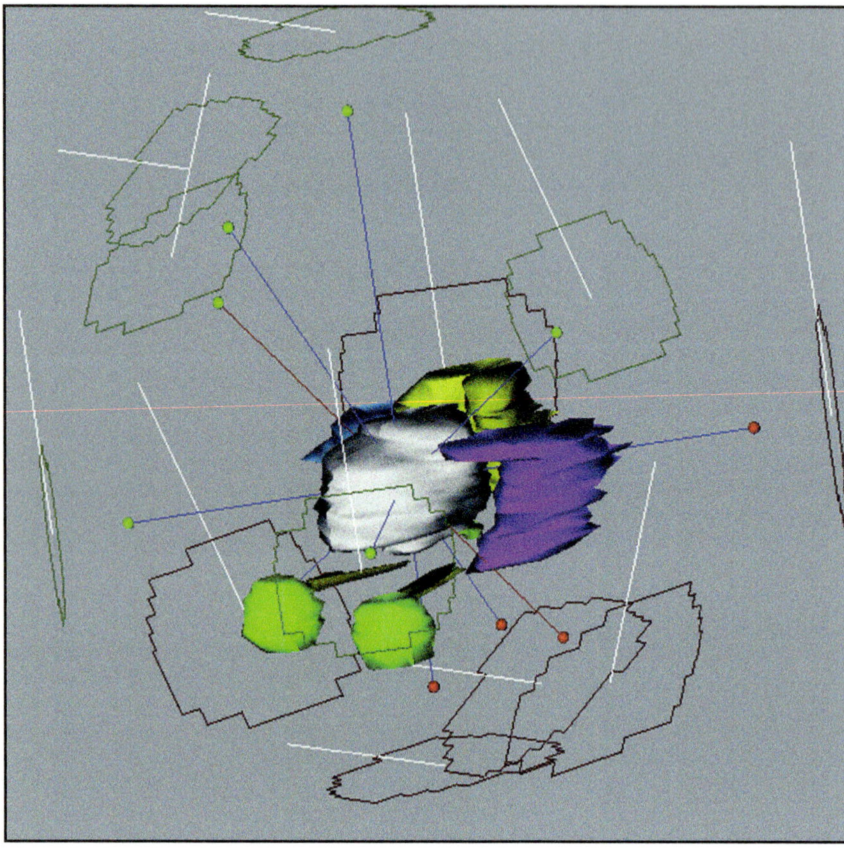

Figure 7.15: An example of a case where the use of an MMLC is advantageous over a cone/arc delivery. This is a six-beam configuration for an irregular target and will produce a more conformal and homogeneous dose distribution than a multiple-isocenter arc plan.

7.6.2 Fixed Conformal Fields

A good example of a case where the use of a MMLC is advantageous over a cone/arc delivery is shown in figure 7.15. This is a large (4 to 5 cm across) tumor close to the brain stem, where fractionated stereotactic radiotherapy is indicated. The target is nonspherical so an arc treatment requires two isocenters, for which the price to pay is a "hot spot" where the spheres representing the hot region from each isocenter overlap. Figure 7.15 shows a six-beam MMLC configuration. This is a good example of a situation where the high leaf resolution of a MMLC is beneficial and the lesion size is well within the allowed range. Note that most of the MMLC beams enter from the superior aspect and do not significantly intersect the outlined critical structures (brain stem and temporal lobes).

7.6.3 Dynamic Conformal Arc Fields

A new and very effective use of a MMLC is the so-called "dynamic conformal arc" technique. Here, arced beams are used but instead of using a simple circular cone for beam collimation, a MMLC with dynamically changing leaf patterns is used. During planning, the beam aperture is made to follow the "beam's-eye view" of the target, or some other user-defined pattern sequence, as the gantry rotates. This requires sophisticated, synchronized control of the MMLC, such that the aperture continually corresponds to the intended shape for the current gantry angle. Note that this technique retains the fundamental property of arc-based delivery—that of very rapid dose fall-off outside the prescription dose volume—making it more appropriate than fixed MMLC fields for single-fraction radiosurgery. This delivery process is depicted in figure 7.16. Although three distinct MMLC shapes are shown here, in fact the shape continually deforms as the gantry rotates. A portion of one arc is indicated here but several couch angles may be used to yield the additional benefits of a non-coplanar treatment. Note that this technique retains the

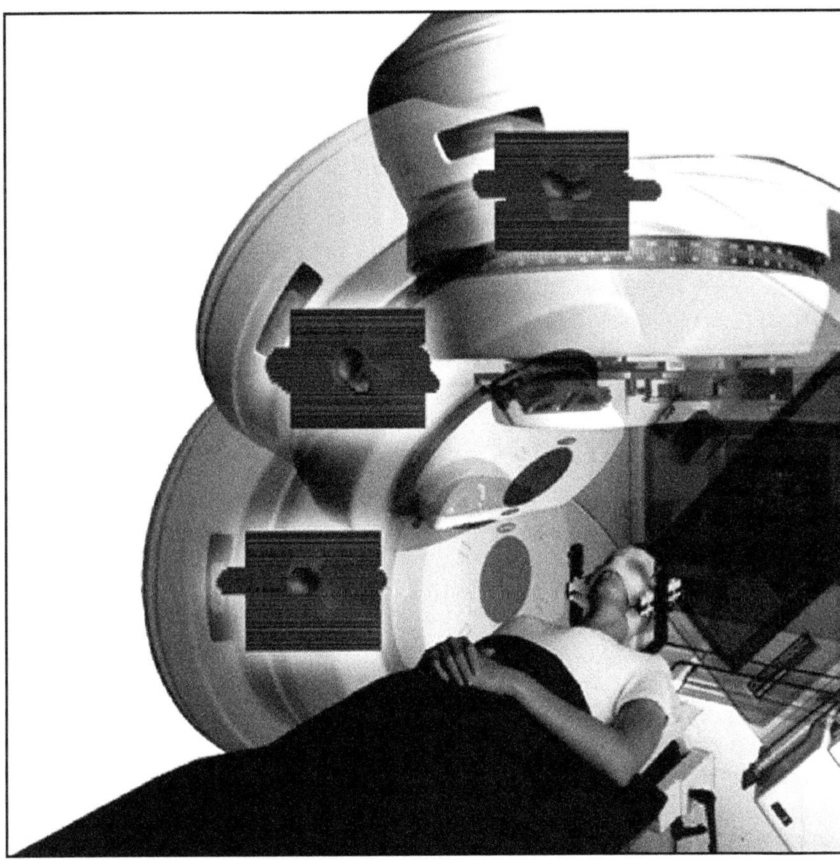

Figure 7.16: Depiction of the "dynamic conformal arc" technique. An MMLC with dynamically changing leaf patterns is used to create a "beam's-eye view" of the target throughout multiple gantry rotations for different couch angles. (Courtesy of BrainLAB.)

fundamental property of arc-based delivery—that of very rapid dose fall-off outside the prescription dose volume—making it more appropriate than fixed MMLC fields for single-fraction radiosurgery.

7.6.4 Intensity-Modulated Radiosurgery

The emergence of intensity-modulated radiation therapy (IMRT) is also having an impact on radiosurgery, especially as a form of SRT or SBRT. With modulated beams, we have the ability to sculpt the dose distribution around complex target shapes to a degree that is not possible using beams with uniform intensity. This is true even if the shapes of the uniform beams are customized to the target shape for a large number of beam directions. In particular, IMRT can create dose distributions with a concave shape, such as that required to treat, for example, a tumor that partially or completely encircles the brain stem or spinal cord. The degree of modulation, and hence dose distribution sculpting, that can be achieved depends somewhat on the spatial resolution of the intensity-modulated beams. Since a MMLC has narrower leaves than a MLC, beam apertures with more complex shapes can be created and thus higher spatial resolution intensity patterns can be created. MMLCs are thus very useful devices for IMRT treatments involving small targets with complex shapes, as often occur in the brain or spine. When an IMRT treatment is carried out in a single fraction, or very few fractions, it is normally referred to as *intensity-modulated radiosurgery* or *IMRS*.

A comparative study of conformal (fixed beam), dynamic conformal arc and intensity-modulated treatment delivery has been performed by Ding (Ding et al. 2006). Dynamic arc–based delivery was found to be a very good all-round modality, with dosimetric advantages over fixed conformal beam plans for all lesion types. IMRT was only found to be advantageous over dynamic conformal delivery for complex geometries where a large target overlapped with critical structures.

7.7 Robotic Radiosurgery: The CyberKnife®

The key elements of an SRS procedure—accurate targeting and focusing of beams within the target volume—are well suited to being carried out with a robot-based system. As mentioned earlier in this chapter, stereotactic treatments are not defined by, or limited to, a frame-based method of localization and targeting. Patient images acquired immediately prior to or during treatment provide a valid stereotactic coordinate system for aiming the beams. Automated image acquisition and comparison to reference images is an efficient method of determining required patient offsets and/or alterations in beam trajectory. Likewise, a fully automated sequence of treatment beams, each with a direction and intensity as determined by treatment plan optimization software, is an effective method of applying a large number of convergent beams to a tumor volume.

The CyberKnife® (Accuray Inc., Sunnyvale, CA) is such a system (Adler et al. 1999), developed specifically as a radiosurgery device that could both complement and compete with the Gamma Knife and linac-based systems. An image of the latest-model CyberKnife is shown in chapter one (figure 1.26). It is based on an industrial robot, normally used for such applications as car assembly, with a compact *X-band* (10 GHz) linac waveguide, target, and collimation system as the "tool" positioned and directed by the manipulator arm. This assembly is well within the 210 kg maximum payload for the robot and is thus able to be manipulated quickly and accurately. The CyberKnife's robotic platform has six axes of rotation, giving it six degrees of freedom (all translational and all rotational axes). A depiction of these "joints" is shown in figure 7.17.

CyberKnife has the key advantage of regular 2-D x-ray imaging *during* treatment to keep track of patient position. This is distinct from both pretreatment 2-D/3-D imaging and motion tracking using external surrogates such as infrared markers during treatment. Image guidance via matching of x-ray images with digitally reconstructed radiographs (DRRs) is known to be highly accurate, especially for lesions in the skull. Thus, an accurate anatomy-based stereotactic reference coordinate system is maintained throughout treatment and the need for rigid fixation is lessened. In particular, CyberKnife

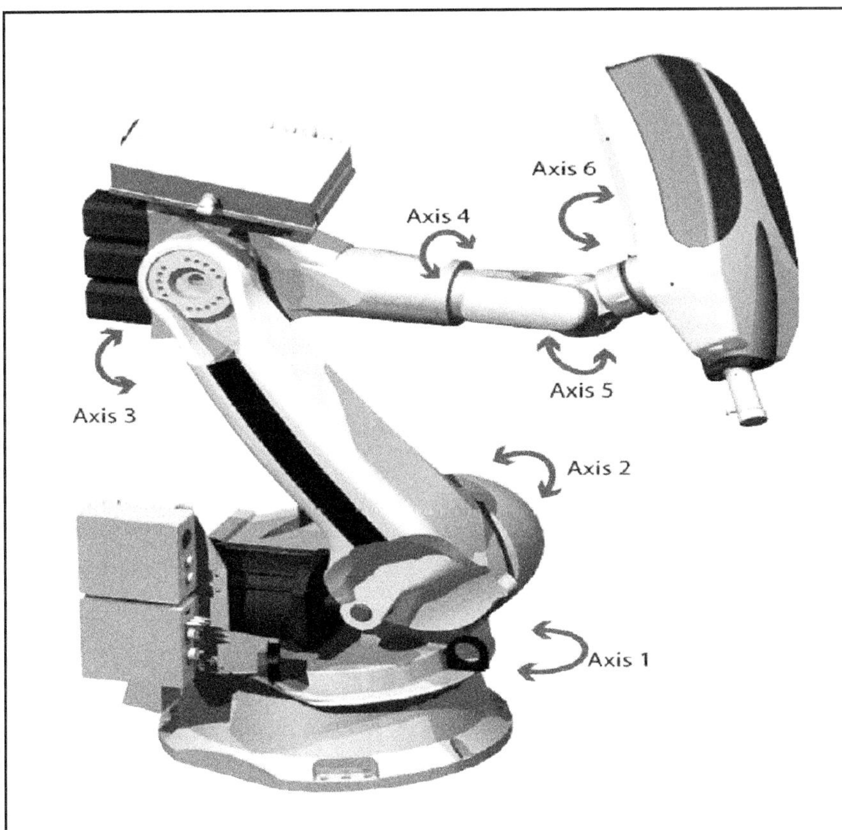

Figure 7.17: Six "joints" used by the CyberKnife's robotic platform to give the beam manipulating head three axes of rotation and three translational directions. (Courtesy of Accuray Incorporated.)

procedures do not include an invasive stereotactic head ring even for single fraction treatments. Comfortable, noninvasive (usually mask-based) immobilization is used to minimize but not necessarily eliminate motion of the target volume relative to the treatment unit.

7.7.1 CyberKnife Beam Delivery

Instead of multiple arcs as a means of beam focusing, the CyberKnife uses a large number of discrete, non-coplanar beam angles. This is achieved by precisely aiming the beams at the target volume from a series of up to approximately 100 "node" positions. Beams are of circular cross section, with a choice of diameters between 0.5 cm and 60 cm at 80 cm from the source. During treatment, the robot can "walk" the beam along one or more paths on the surface of a sphere or ellipse. Such a set of nodes is indicated by the points on the surface of a sphere in figure 7.18. At each node there can be a range of up to 12 angular offsets to direct the beam at points throughout the

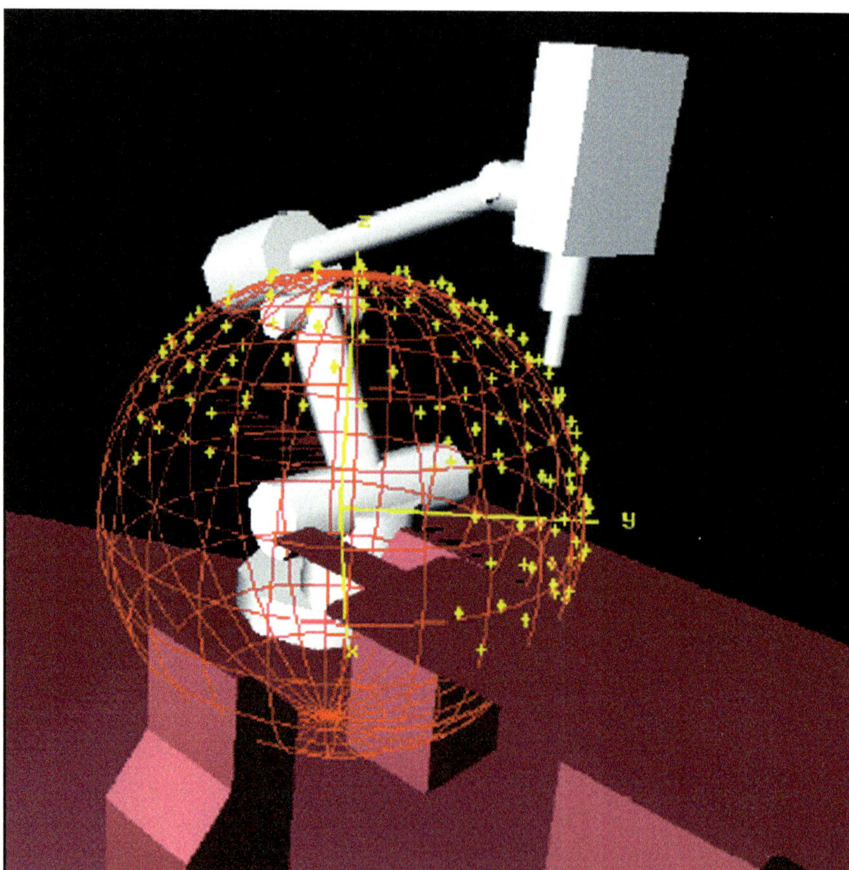

Figure 7.18: A set of nodes (yellow) on the surface of a sphere, indicating the paths along which the robotic manipulator can "walk" the beam nozzle. There are typically 50 to 300 of these nodes, depending on the case complexity. (Courtesy of Accuray Incorporated.)

target volume. Anywhere between about 50 and 300 total beam directions are used, with a larger number of beams increasing treatment time but also giving more flexibility in terms of creating a suitable dose distribution in terms of conformality, target coverage, and rate of fall-off outside the target.

For nonspherical target volumes, a smaller collimator diameter improves dose conformality but again at the expense of treatment time. Multiple beam diameters can be used in order to better achieve high dosimetric quality while minimizing treatment time. Because the dose distribution is built up of many overlapping circular beams, dose is typically less homogeneous within the target than for conformal beam delivery systems. However, as is typical of highly focused beam delivery techniques (including arc-based and Gamma Knife treatments), dose fall-off outside the target is very rapid and this is of primary importance for the ablative treatment paradigm.

The user can select one of three modes of delivery:

i) A single isocenter plan, where all beams pass through the same anatomical point. This mode is suitable when the target volume is close to spherical in shape and a near-spherical prescription dose volume gives good target coverage without including excessive normal tissue. Figure 7.19 indicates a large number of beam angles converging on a single isocenter.

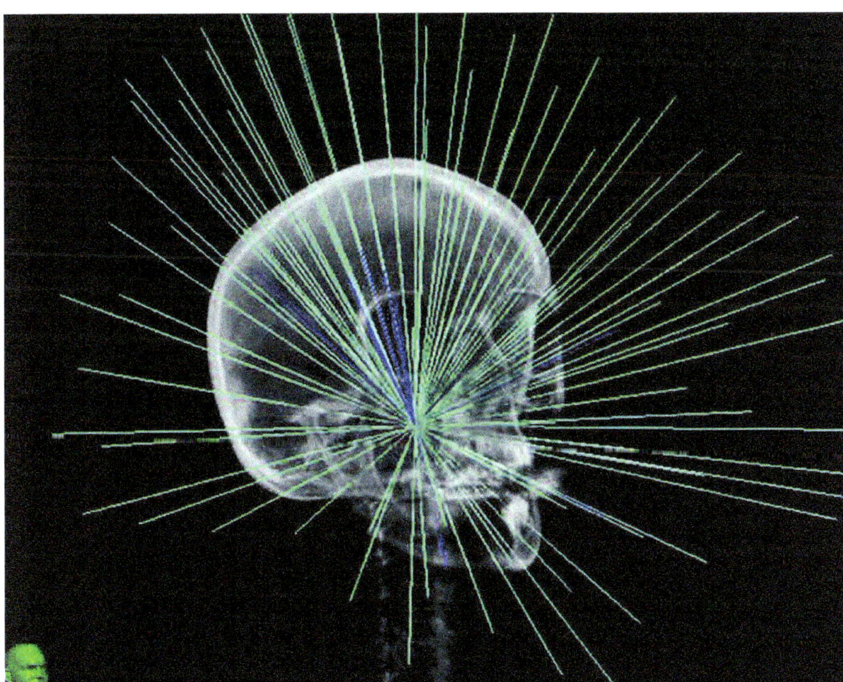

Figure 7.19: A large number of beam angles converging on a single isocenter. Note that multiple isocenter treatments can also be performed, as well as treatments where beams intersect the target as required to achieve a conformal dose. (Courtesy of Accuray Incorporated.)

ii) A multiple isocenter plan, where several target points, each at the focus of many beams are used. This process is used for the treatment of multiple near-spherical lesions or to improve conformality for an irregular lesion.
iii) A conformal plan, where multiple independent target points are distributed according to the geometry of the lesion. Each beam is targeted at a different point within the target volume rather than at a single or few focal points. This mode results in more conformal dose distributions.

Although a large number of beams are used to focus dose in the target volume and disperse dose outside the target volume, one issue with the CyberKnife system is its inability to use beams that enter from the posterior aspect of the patient (manipulator arm below the couch). For posterior targets such as spine lesions, angles that would correspond to the least depth to the target and thus the least required beam intensity are not available.

7.7.2 CyberKnife Image Guidance

The geometry of the robotic beam arm that directs the beam as well as the configuration of the pair of x-ray cameras used for image guidance is known to the CyberKnife's treatment planning system. Thus for a given placement and orientation of the patient's anatomy within the planning software's coordinate system, there is a uniquely corresponding *required* position and orientation of the patient relative to the system hardware. In particular, there is a known view of the anatomy that should be "seen" by each x-ray camera. These views are stored as DRRs, created from the planning CT image set, that are compared with the actual x-ray images acquired both before treatment begins and before each beam is delivered during treatment. In fact, DRRs for a range of small translational and rotational offsets from the required setup position and orientation are stored. These are searched to find the best match with the x-ray images and thus the geometry of the patient is determined relative to the required geometry. The robotic beam manipulator is able to adjust its aim (within limits) so that each beam will intersect the target at the planned location and orientation. This process corresponds to the image-based stereotactic targeting paradigm depicted by figure 7.6, with the process being performed repeatedly.

As well as compensation for patient motion via the robotic arm, the system's treatment table is able to adjust patient position and orientation. The latest robotic table can make adjustments with six degrees of freedom (three translational and three rotational axes). Table adjustments are used for initial setup and to readjust the patient (after a pause in the treatment sequence) if translation or rotation exceeds the limits able to be compensated for by the robotic arm.

The x-ray sources are mounted on the ceiling of the treatment room, directing beams orthogonal to each other and intersecting in the region where the anatomy will be placed for treatment. Flat-panel image detectors are embedded in the floor of the treatment room so as to be unobtrusive (earlier models used detectors mounted above the floor).

There are in fact four modes of tracking used by the CyberKnife:

i) *Skull tracking*, where x-ray and DRR images are matched via rigid-body image registration to obtain targeting information. The skull provides feature-rich DRRs and x-ray images, as well as being a rigid structure, so this relatively simple mode provides very accurate targeting.
ii) *Skeletal tracking*, where non-rigid registration of an optimal region of interest in the x-ray images is used. The displacement of a point of interest within the region of interest can thus be determined. This is mostly for spinal treatments, where non-rigid movements of the bony anatomy may occur.
iii) *Fiducial tracking*, where several implanted fiducial markers are used as high-contrast features. An algorithm is used to first identify the markers in each x-ray image and to compare their 3-D locations to their locations in the planning CT image set. This is used for low-contrast soft-tissue targets and for spine treatments if the skeletal tracking system is not available or effective.
iv) *Motion tracking*, where the imaged trajectories of implanted fiducials are first correlated with the trajectories of external reflective markers prior to treatment. Subsequent to this, the external markers can be continuously tracked throughout treatment and the motion of the internal markers inferred via the correlation. This mode is used for targets such as lung, liver, and pancreas tumors.

7.7.3 Clinical Applications

The CyberKnife system's key advantage of near real-time target tracking and the consequent lack of need for invasive patient fixation mean that it is suitable for treating tumors both inside and outside the head. Intracranial treatments of small lesions with a simple shape are generally the fastest and simplest, since often a single isocenter and single collimator size can be used, with skull tracking as the targeting mode. More complex lesions require either multiple isocenters or conformal treatment mode in order to maintain a high level of dose conformity. Many extracranial tumors are large and complex in shape, incurring an extended beam delivery time compared to other forms of treatment. Nevertheless, the range of sophisticated tracking modes available means that the CyberKnife system is a valuable tool for the difficult task of accurately targeting lesions throughout the body for treatment with a high dose of radiation per fraction.

Clinical indications for which the CyberKnife system is commonly used include:

- Brain metastases
- Benign brain lesions such as meningiomas and acoustic neuromas
- Spine metastases
- Cranial and spinal arteriovenous malformations (AVMs).

Emerging indications, consistent with the burgeoning field of SBRT and the frameless targeting abilities of the CyberKnife system are:

- Primary and metastatic lung lesions
- Primary and recurrent head and neck tumors
- Pancreatic lesions
- Liver lesions
- Prostate lesions.

The CyberKnife system is evolving into a useful tool for SBRT as well as SRS. In as much as SBRT is becoming an alternative to conventional radiotherapy, the use of the CyberKnife system as an alternative to traditional radiotherapy devices is likely to continue to grow.

Technical aspects of the current CyberKnife system's design, and its mode of beam delivery, mean that conventionally fractionated treatments are relatively less efficient than the 1 to 5 fraction treatments for which it is typically used. In particular, the use of circular beams to conformally treat an irregular target volume necessitates a large number of beams, each contributing a small proportion of the total dose. With a smaller dose per fraction, a larger proportion of treatment time is spent transitioning between beam angles, thus not reducing total time per fraction to a level that would make 30 to 40 fractions practical.

It is worth remembering that the CyberKnife system, as is the case with all SRS devices, is primarily an alternative to surgery. It has contributed to a shift from surgical excision to ablative radiation treatment for many indications. Conventional radiotherapy is often used as a "safe" alternative to surgery, but without the intended equivalence to surgical removal provided by an ablative treatment. With the increased assurance of accurate targeting provided by devices such as the CyberKnife system, it is likely that there will be a shift from conventional radiotherapy to more definitive radiosurgical treatments. Thus while the CyberKnife has traditionally not "cannibalized" conventional radiotherapy treatments, this is a likely scenario as the practice of SBRT grows.

7.8 Stereotactic Body Radiotherapy

Depending on the size and shape of the lesion, and proximity to critical structures, treatments with ablative intent are often delivered over several fractions. This allows a higher total dose for a given toxicity than a single fraction treatment—a fact that can be taken advantage of to increase probability of tumor control or to reduce probability of complications. Such precise treatments, to a small target anywhere in the body, where a small number of fractions are used to deliver a high dose, are most often referred to as stereotactic body radiotherapy (SBRT). A course of 1 to 5 fractions is typically used in SBRT (in fact, a treatment with more than five fractions does not qualify as SBRT according to some reimbursement coding schemes).

Note that in modern terminology the term *SBRT* is intended to include single- and multiple-fraction treatments outside the head and multiple -fraction treatments inside the head. SRS refers only to single-fraction intracranial treatments. The latter being the odd-man-out is consistent with the fact that it is typically only single-fraction intracranial treatments that use traditional invasive fixation devices.

Extracranial tumors are normally larger and more mobile, making ablative treatments a technical challenge. Tumor mobility makes accurate targeting more difficult due to the uncertainty in tumor location relative to a frame of reference such as the skeleton. Furthermore, the sorts of invasive stereotactic localization and targeting systems traditionally used for the head cannot be readily adapted to the rest of the body, or to the head for multi-fraction treatments. New and more versatile targeting techniques are required to translate ablative treatment strategies to areas such as the lung, liver, and prostate.

SBRT draws on techniques developed to increase the accuracy of radiotherapy, while retaining the ablative intent of radiosurgery. In particular, the targeting principles of image-guided radiotherapy (IGRT) have been very effectively applied to SBRT. The targeting techniques used by the CyberKnife system, discussed above, are a form of IGRT; in that case via 2-D x-ray images. IGRT is also practiced using 3-D CT images, as in the case of the TomoTherapy Hi-Art System™ discussed later in this chapter. Here, just the fundamental principle will be reiterated: the patient's anatomy (in particular the tumor) *should* be at a location relative to the treatment machine geometry as defined by the treatment plan; an image acquired during the treatment session can reveal the *actual* tumor location; and a shift can be applied. Just as a traditional stereotactic head ring establishes a coordinate system within which the tumor has a well-defined (but invisible) location, the patient's anatomy itself forms the reference coordinate system in IGRT. That is, using stereotactic targeting apparatus to position a head ring at a certain location relative to the treatment machine is the equivalent of a treatment-time image of the patient being used to correctly position the anatomy.

In fact, SBRT has been in clinical use for many years, using special stereotactic body frames. Blomgren and Lax at the Karolinska Institute (Stockholm, Sweden) pioneered "extracranial stereotactic radiosurgery," with treatments beginning in 1991 (Blomgren et al. 1995). The targeting technique employed by this group uses a stereotactic coordinate system within a "body frame" (a three-sided box) that the patient lies within. The body frame contains a vacuum pillow molded to the patient's contours to aid repositioning accuracy. Fiducial rods visible on CT scans are embedded within the walls of the frame. These are used to establish a stereotactic coordinate system within which to define a treatment isocenter. A targeting bar on which the isocenter coordinates are set is attached to the frame for patient setup. This setup is akin to a head ring for intracranial treatments, whereby the immobilization device acts as the basis for a stereotactic coordinate system that can in turn be easily related to the coordinate system of the treatment machine. In both cases, maintaining the spatial relationship between target anatomy and stereotactic coordinate system is crucial to targeting accuracy.

An excellent collection of articles on SBRT, from both a clinical and physical perspective, is provided in the book *Stereotactic Body Radiation Therapy* by Kavanagh and Timmerman (2005).

7.9 The Rationale for IMRT

.Conformality of the prescription dose level to the defined target volume is often an important clinical requirement, especially when sensitive structures are adjacent or nearby. However until the recent availability of intensity-modulated radiotherapy (IMRT), dose conformality to all but very simple target volumes was not possible, or impractical. This is especially true near concave surfaces of the target. If there is a sensitive structure within that concavity (i.e., the tumor "wraps around" the structure), not being able to keep high dose off this region may incur unacceptable complications and certainly limit the dose that can be delivered. Such situations include, for example, a tumor that is partially surrounding the spinal cord or brainstem. The situation is further exacerbated if previous irradiation has been given, in which case the availability of IMRT can mean the difference between radiotherapy being an option or not.

3-D conformal radiotherapy (3DCRT) is "conformal" mostly in terms of the way beam apertures are shaped via a *beam's-eye view* (BEV) of the target. Unfortunately, even the use of a very large number of such conformal beams does not imply that dose will conformally surround the target volume. The problem is that a BEV of the anatomy does not give any indication of the 3-D shape of a structure since all that is seen is a projection of the shape from each angle. For example, a series of such views of an egg cup, even from non-coplanar directions, will give the impression of a solid hemispherical object—no knowledge of the hollowed out nature of the cup will be gained.

Creating a true conformal dose distribution for complex objects firstly requires a 3-D view of the object (and surrounding objects) instead of simply a series of 2-D projections. Such a volumetric description of target and sensitive structure geometry is part of 3DCRT planning, but only with IMRT is there a means of making use of this geometry to design not just beam apertures but the non-uniform 2-D beam intensity profiles required to create a conformal dose. There are typically two steps in the IMRT treatment planning process:

i) Determining suitable beam intensity profiles via an *inverse planning* technique. This chapter will outline a typical process for such beam intensity optimization.
ii) Realizing these intensity profiles via a practical beam delivery mechanism. MLC-based delivery will be covered here, but it should be noted that other means are possible.

The development of IMRT and its adoption into widespread clinical practice represents one of the most significant advances in radiotherapy treatment planning and delivery technology. An intense research effort, particularly throughout the 1990s, and the commitment of equipment vendors (in conjunction with clinical research partners) to design reliable technology to implement the procedures developed has been required. As with many technological advances, the rapid increase in computer processing power over the past decade or so has been a critical enabler for IMRT, particularly for the beam optimization stage. Furthermore, treatment delivery mechanisms have had to be made much more sophisticated than those required to simply shape a beam, specifically in correlating the mechanical motion of MLC leaves with the time or monitor units for which radiation has been emitted.

An example is given here to reinforce the concept of the need for a computational method to produce a dose distribution which conforms to a concave shape. The situation of the cup-shaped tumor is shown schematically in 2-D in figure 7.20. Achieving conformality of a dose distribution to such a shape with unmodulated coplanar beams (left) is not possible (just as conforming to a 3-D cup shape with unmodulated coplanar or non-coplanar beams is not possible). Intensity-modulated beams are shown at right, conforming the dose distribution to the target. In this case, such a result is made possible by an optimization algorithm that "sees" the 2-D shape of the target volume from each direction instead of just seeing the projection of the shape as a line. In reality, the algorithm must see a 3-D shape.

The IMRT optimization process is often referred to as an *inverse planning* because of the fact that the beam shapes and profiles are derived via computer from the desired dose distribution. *Forward planning*, on the other hand, is the conventional process where beam apertures and weights are determined by a

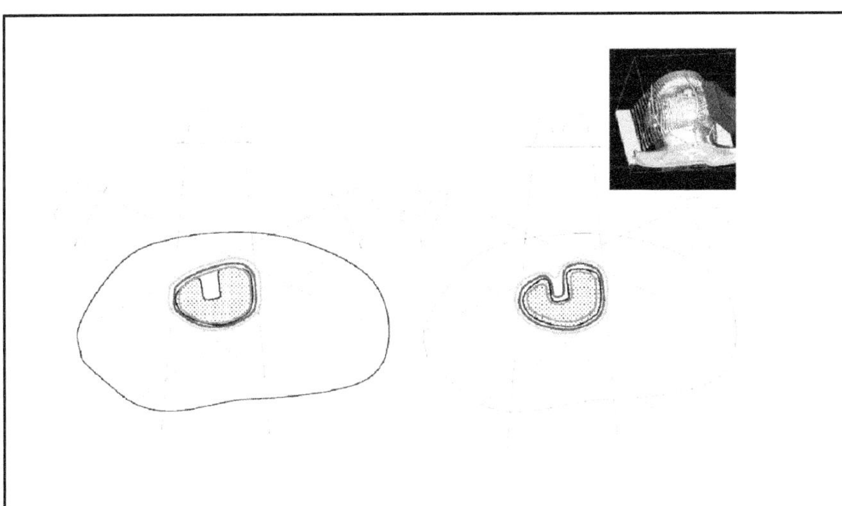

Figure 7.20: Conformality of dose to a target volume with a concave surface cannot be achieved via simple "beam's-eye view" planning as shown on the left. Achieving conformality of a dose distribution to such a shape with unmodulated coplanar beams (left) is not possible (just as conforming to a 3-D cup shape with unmodulated coplanar or non-coplanar beams is not possible). At right, intensity-modulated beams are shown, conforming the dose distribution to the target.

trial-and-error process of adjusting beam properties and examining the resulting dose distribution. This distinction is arguably an unfortunate one, since in practice even the computer performs many trials, or iterations, to come up with what it deems the best set of beam intensity profiles. The real difference lies in the fact that optimization is performed by computer rather than via the planner's judgment.

7.10 Computation of Optimal Intensity Patterns for IMRT

7.10.1 Overview

Early reports of *inverse planning* involved mathematical inversion techniques where the principle was to take the matrix representing the desired dose distribution and transform this via other matrices representing the deposition of dose via individual elements of each beam (often called beamlets) into a matrix representing the required beam intensity profiles (arrays of beamlet *weights*). This can also be thought of as a *deconvolution* procedure, where the beamlet matrices are being deconvolved from the desired dose matrix to obtain the intensity profiles (Brahme et al. 1982; Brahme 1988; Holmes et al. 1991). In the corresponding convolution operation, as performed for dose calculation, the beamlet matrices are convolved with the intensity profiles.

A major problem with deconvolution approaches is that we do not know beforehand whether the desired dose distribution is actually deliverable via the beamlets for any *achievable* set of beam intensity profiles. If we simply ask for a prescription dose within a target volume and no dose elsewhere, then this is most certainly not deliverable! Inevitably, deconvolution techniques lead to *negative* beamlet weights—that is, dose should be taken away by some elements of a beam intensity matrix to make up for excess dose deposited by another element. The need for positivity of beamlet weights means that any negative weights have to be eliminated (for example, assigned a weight of zero) and this, of course, leads to a dose distribution that differs from the one we wanted to achieve. The best approach is then to examine what dose distribution is achievable given the constraint of positive beamlet weights and then try again to correct the dose via another iteration of the matrix algebra procedure.

It has been found that a much more practical approach than deconvolution is an iterative weight adjustment based on simple differences between the prescribed dose and the dose resulting from a given set of beam weights. In particular, for each beamlet, the dose at all (or a subset of) voxels influenced by that beamlet's dose matrix can be assessed for their deviation from an applicable dosimetric goal. Typical dosimetric goals include:

- Minimum target dose (applies to target voxels whose dose is below the minimum)
- Maximum target dose (applies to target voxels whose dose is above the maximum)

- Maximum dose to a sensitive structure (applies to sensitive structure voxels whose dose is above the maximum)
- Maximum sensitive structure volume that should receive a given dose or more.

The last one of these goals is called a *DVH goal.* The way this is applied is typically the penalized voxels include those with a dose above the specified goal and below the dose that actually corresponds to the volume specified. This concept was reported by Bortfeld et al. (1997).

Figure 7.21a illustrates the maximum and minimum target dose goal in terms of a dose-volume histogram (DVH) display. Any voxel with dose above the maximum and below the minimum will be penalized. The hatched areas are truncated for clarity but should be interpreted as extending beyond the area shown.

Figure 7.21b illustrates the maximum sensitive structure dose goal and a sensitive structure DVH goal. For the DVH goal there is a coordinate (D_{max},V_{max}) at the origin of the hatched area indicating a region of the DVH that we ideally want to eliminate. That is, we want to reduce the dose in all voxels that currently have a dose between D_{max} and D′ to D_{max}.

The larger the deviation of a voxel's dose from the applicable dose goal, the more that voxel influences the change in beamlet weight. Some voxels may act to try and raise the beamlet weight (if their dose is too low) and others may have the opposite influence—it is a weighted average of the influence of all voxels sampled that determines the final alteration in beamlet weight. Note that many beamlets can influence the dose in a given voxel, as well as many voxels being influenced by a given beamlet, so the dose deviation in a particular voxel has a small effect on the weights of many

(a)

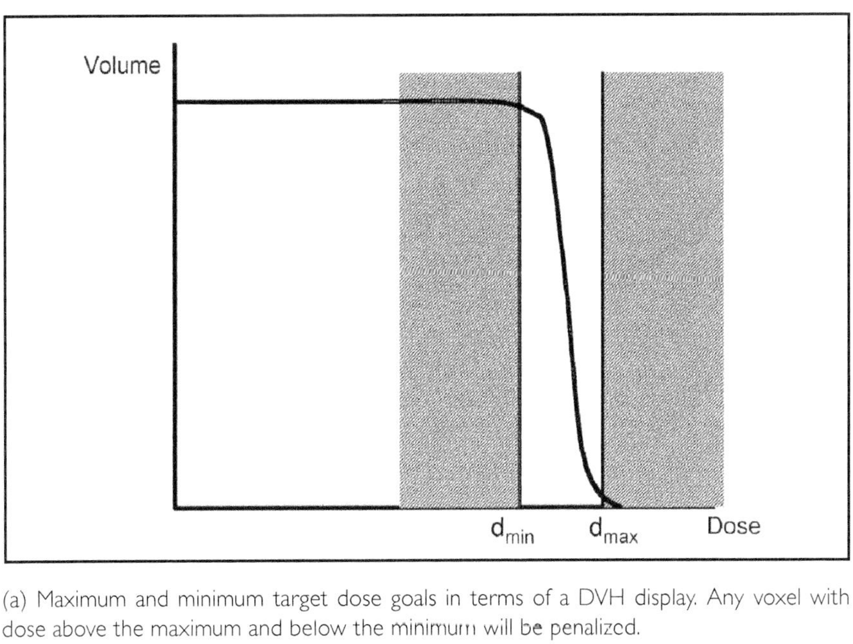

Figure 7.21: (a) Maximum and minimum target dose goals in terms of a DVH display. Any voxel with dose above the maximum and below the minimum will be penalized.

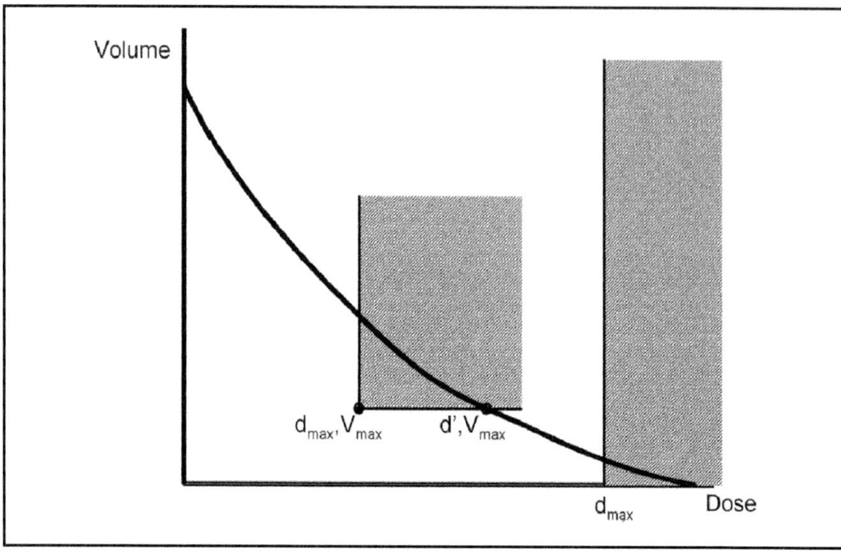

(b)

Figure 7.21: (Continued) (b) Maximum sensitive structure dose goal and a sensitive structure DVH goal. For the DVH goal there is a coordinate (D_{max}, V_{max}) that we want the DVH curve to pass through (see text for further explanation).

beamlets. As such, any update of beamlet weights during the optimization process is intended to steer the dose distribution towards a state that is closer to the overall goal. Each goal has an importance relative to other goals, so that the user has some control over which goals are more closely achieved in the event that a compromise must be made. A typical compromise is one between minimum target dose and maximum dose to an adjacent sensitive structure—it may be considered clinically prudent to allow minimum target dose to drop below the goal if protection of the sensitive structure is the highest priority (for example, in treatment of a spinal tumor where the spinal cord has already been irradiated to near its tolerance in a previous course of treatment).

7.10.2 Simple Optimization Mathematics

The Dose Calculation

If the matrix containing all of beamlet weights is w and the matrix of calculated doses in the patient is d, the dose distribution is obtained from the matrix multiplication $d = Dw$ where D gives the change in d for a unit increment in each beam weight. Note that d and w are single-column matrices and D is a rectangular matrix with the same number of rows as d and the same number of columns as w has rows. In fact, d has as many rows as there are dose calculation points and w has as many rows as there are beamlets in the problem.

Figure 7.22 shows a schematic of an intensity profile (the weights included in the w vector) and a dose distribution described by the d vector. An arrow is drawn (representing an element of the D matrix) indicating the influence of a beam element on a dose element.

Ideally, we would simply invert this matrix algebra expression and obtain w from $w = D^{-1}d$, but this is not a straightforward procedure for reasons such as those discussed above.

It is more instructive to write the dose calculation equation as a summation, where the dose in the ith voxel can be expressed as

$$d_i = \sum_{j=1}^{M} D_{ij} w_j. \tag{7.5}$$

Here j is the beamlet index and there are M beamlets over all beams used. Note that D_{ij} is the dose delivered to the ith voxel by the jth beamlet for a unit beam weight. A unit of beam weight may be a unit of beam fluence, energy fluence, opening density, exposure time, or any quantity that the D_{ij} matrix is normalized with respect to.

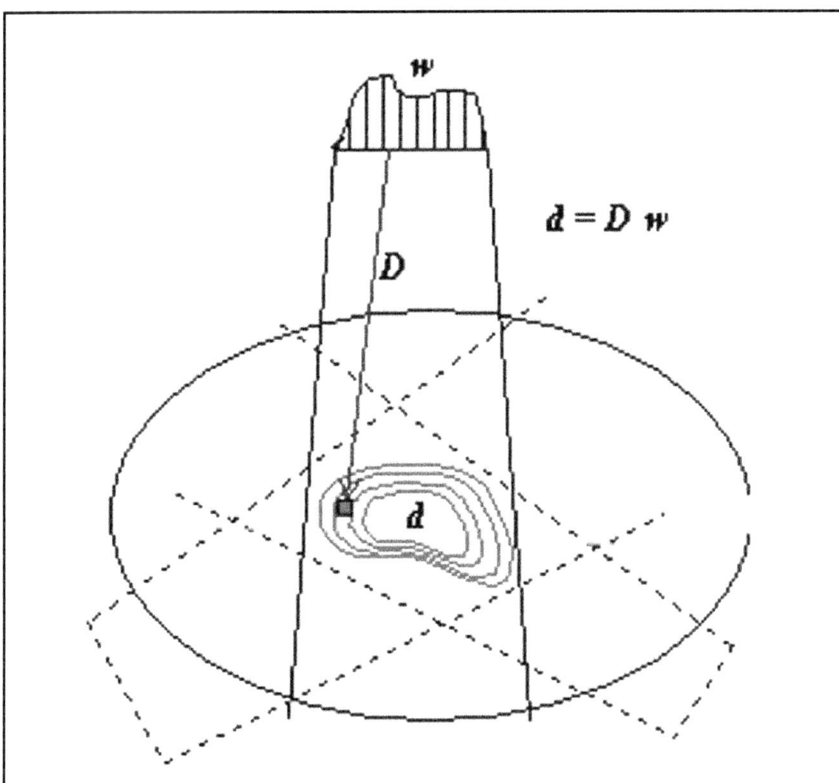

Figure 7.22: Schematic of an intensity profile (the weights included in the w vector) and a dose distribution described by the d vector. An arrow is drawn (representing an element of the D matrix) indicating the influence of a beam element on a dose element

The Objective Function

An optimization algorithm needs an *objective function* to measure how far the result currently is from the desired solution. For a dose optimization problem, the objective function needs to express an overall voxel-by-voxel deviation from the desired dose, including the importance, or penalty, for each of those deviations. Deviations need not include a subtraction of desired dose from actual dose, but this is a useful and obvious measure. In fact, a quadratic objective function including a weighted sum of the squared difference between desired and actual dose is often used. As mentioned above there may be several goal doses that apply to different voxels, and these goal doses act as the desired dose. If we denote the goal dose for the ith voxel as a prescription dose d_i^{pres} and the associated penalty for deviating from that goal as s_i, then the objective function can be defined as

$$F = \sum_{i=1}^{N} s_i \left(d_i - d_i^{pres}\right)^2 . \tag{7.6}$$

Here there are N voxels in which dose is to be considered, throughout the dose matrix and corresponding to both targets and sensitive structures. The penalty values are typically user-defined by type and target/structure. If a voxel i is in a given target and its dose value exceeds the maximum dose goal for that target, the prescription dose used for that voxel in evaluating the objective function will be the maximum dose goal for that target, along with its associated penalty value; similarly for target voxels with dose beneath a minimum dose goal and sensitive structure voxels exceeding the corresponding maximum dose goal.

Figure 7.23 shows the optimization panel from the KonRad inverse planning system (Preiser et al. 1997). Here maximum and minimum dose goals are set for the target structures clinical target volume (CTV) and planning

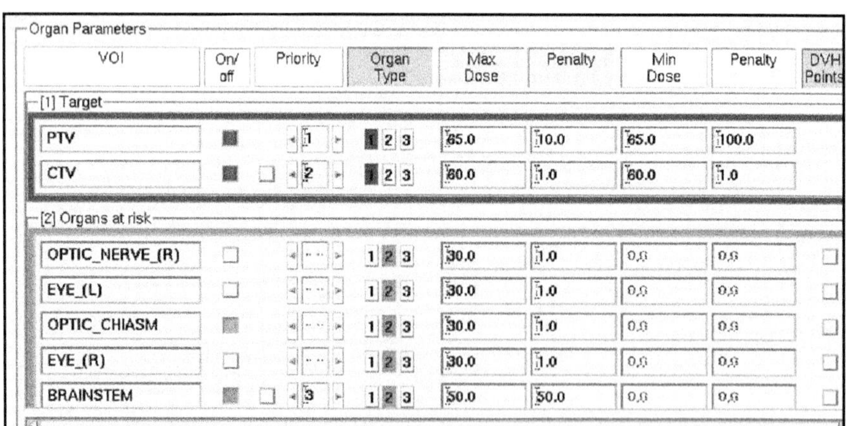

Figure 7.23: Optimization panel from the KonRad inverse planning system (Preiser et al. 1997). Maximum and minimum dose goals are set for the target structures CTV and PTV and maximum dose goals are set for several critical structures. Penalty values are set for each dose goal.

target volume (PTV) and maximum dose goals are set for several critical structures. Penalty values are set for each dose goal. Note that "Brainstem" is the sensitive structure with the highest penalty value in this case, indicating that keeping dose away from this organ is of paramount importance. Practically, penalty values are used as a way of manipulating the optimizer into giving a desired result as much as being based on any prioritization of how important a structure is. One feature of this interface that is common to most systems is the setting of "priority" values. Here, PTV has a priority of 1 and CTV has a priority of 2, meaning that all voxels common to both structures are considered part of PTV. In fact, since PTV encloses all of CTV, there are no voxels belonging to CTV and it is effectively ignored in the optimization.

Iterative Optimization of Beamlet Weights

In order to drive the dose to a solution that is "better" according to the objective function, the most efficient tactic is to alter beamlet weights in some *deterministic* way. One such method is a so-called *gradient technique*, where the idea is to reduce the objective function value via calculated changes in the beamlet weights until its rate of change is zero. That is, the objective function is minimized. This concept is illustrated in figure 7.24a, where as iterations continue, the gradient of the objective function change decreases until it reaches a plateau and the user (or system) would cease the optimization. Figure 7.24b is a plot from the KonRad system of the reduction in relative difference in the objective function from iteration to iteration. Here, the optimization is set to stop when the relative difference reaches 10%.

According to the commonly used quasi-Newton gradient technique, the calculated changes in beamlet weight can be arrived at via the derivative and second derivative of the objective function with respect to that particular weight. The change in weight of beamlet j for a given iteration will be:

$$\Delta w_j = -\alpha\left(\frac{dF}{dw_j}\Big/\frac{d^2F}{dw_j^2}\right). \tag{7.7}$$

The factor α is a damping term to ensure convergence to a solution. The proviso must be imposed that the adjusted weight cannot be less than zero (if it would become so, it is set to zero). Now using equation 7.7, the change in weight becomes

$$\Delta w_j = -\alpha\frac{\sum_{i=1}^{N} s_i\left(d_i - d_i^{pres}\right)D_{ij}}{\sum_{i=1}^{N} s_i D_{ij}^2}. \tag{7.8}$$

Note that the weight change for beamlet j is essentially a weighted average of the dose-differences for all voxels influenced by that beamlet, where the weights in the average are dependent on the penalty value appropriate to that voxel and on the dose per unit beamlet weight in that voxel. Thus, as should intuitively be the case, the influence of a particular voxel's dose-deviation on a given beamlet weight is larger when the dose per unit beamlet weight is higher for that voxel.

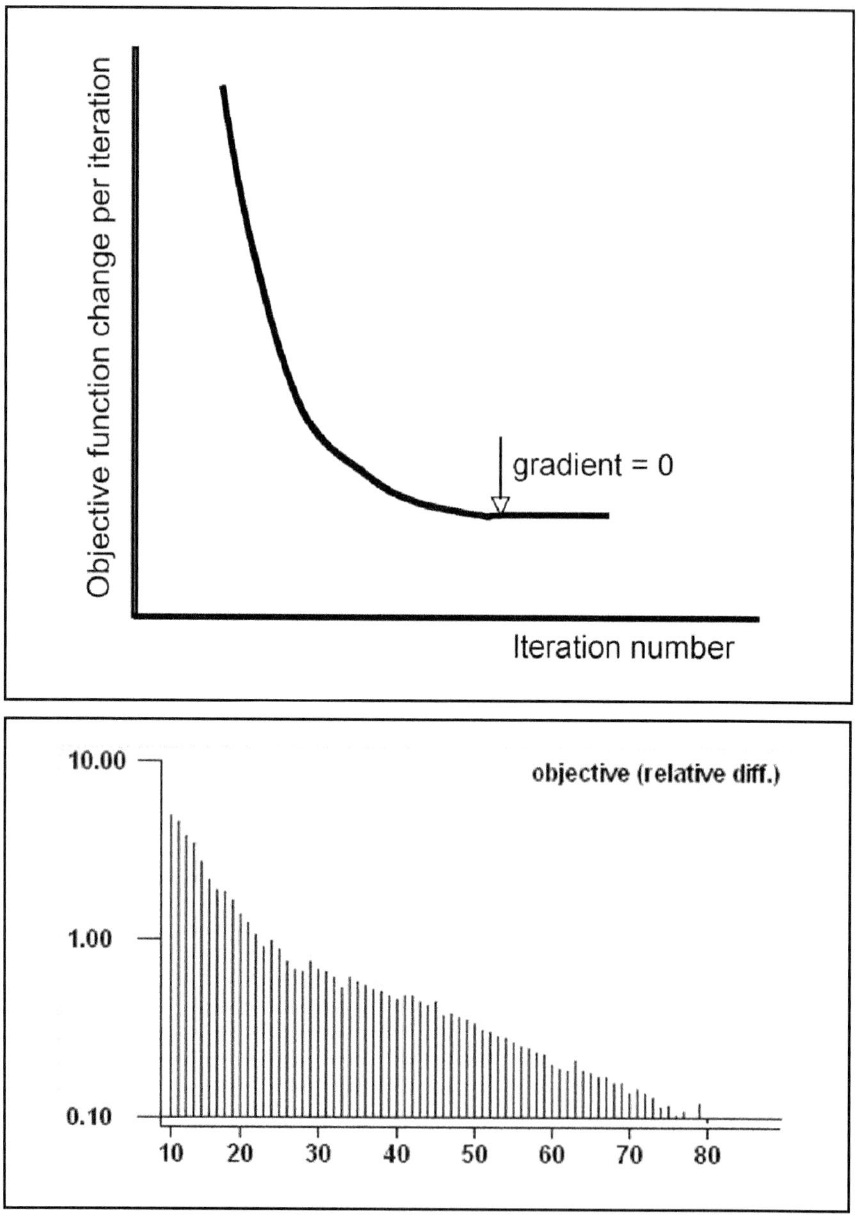

Figure 7.24: (a) Plot representing the change in objective function value as iteration progress. The aim is for the gradient to reach a plateau indicating the best achievable solution has been reached. (b) A plot from the KonRad sysem showing reduction in relative difference in the objective function value from iteration to iteration. Here, the optimization is set to stop when the relative difference reaches 10%.

Figure 7.25 illustrates a single beamlet weight w_j within a single intensity map, the corresponding beamlet doses D_{ij} and a few dose voxels d_i. In iteration, k, in dose voxels coinciding with each point in the D_{ij} matrix the difference in dose from the desired dose can be used in a weighted average to update beamlet weights as in equation 7.8.

The result of an optimization procedure using the KonRad system is shown in figure 7.26. The high-dose region (orange and yellow) can be seen to conform well to the PTV contours and the brain stem is largely spared from dose at this level. In the right-hand panel of this display is an optimized intensity map for one of the beams (anterior direction). In this case the intensity maps have been *discretized* to six levels in order to reduce the number of beam segments required to create the intensity map. Elements of an intensity map are sometimes referred to as *bixels*, as in a beam-pixel. Darker bixels indicate a higher intensity in this case.

It is useful to express beamlet intensity as the total number of monitor units (MU) for which radiation should be incident on the corresponding area, or sometimes on a single point. This is often referred to as opening density (OD), since it is a measure of total "openness" of the beam, as controlled by the motion of the MLC leaves, during the beam-on period (Mohan et al. 2000). Considering a point in the calculated intensity map, located in a plane passing through the isocenter, OD can ideally be considered the number of MU for which the point can "see" the radiation source in a view up through the MLC leaf openings during a beam delivery. In a real situation, transmission through the MLC leaves contributes to OD and must be taken into account during conversion of OD to a leaf sequence.

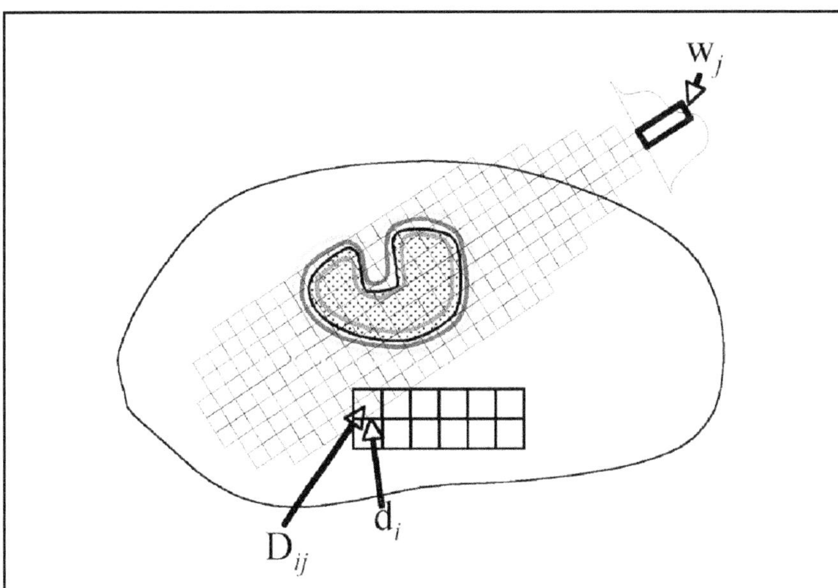

Figure 7.25: A single beamlet weight w_j within a single intensity map, the corresponding beamlet doses D_{ij} and a few dose voxels d_i. In iteration k, in dose voxels coinciding with each point in the D_{ij} matrix, the difference in dose from the desired dose can be used in a weighted average to update beamlet weights as in equation 7.8.

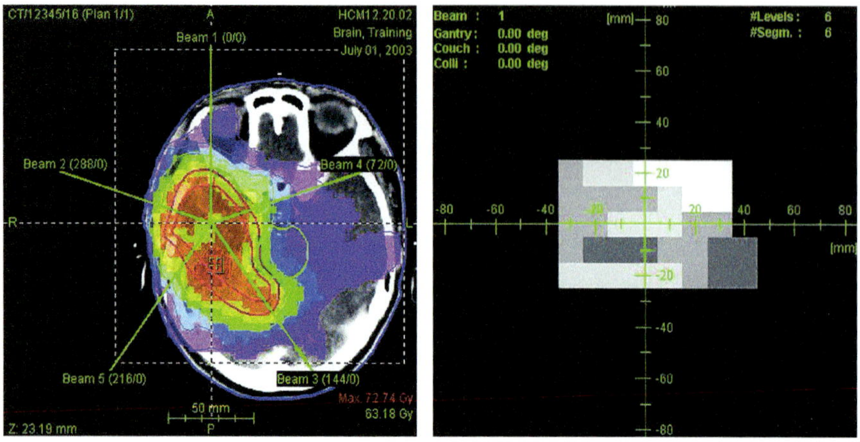

Figure 7.26: Result of an optimization procedure using the KonRad system. The high-dose region (orange and yellow) conforms well to the PTV and the brain stem is largely spared from high dose. Also shown is an optimized intensity map for one of the beams (anterior direction). Intensity maps have been discretized to six levels in order to reduce the number of beam segments required to create the intensity map.

7.11 MLC Leaf Sequences for IMRT

Once the beamlet intensity distributions have been determined for each beam, typically in the form of opening density maps, these must be converted to MLC leaf sequences that can as closely as possible reproduce the optimized maps.

In the *step-and-shoot* method of dose delivery the dose is only delivered when the leaves are stationary and there is no dose delivered as the leaves move to the next segment position. In figure 7.27, an example of a step-and-shoot sequence, overlaid on the corresponding OD map, is shown. A step-and-shoot sequence may be represented simply as a series of beams, each with a (usually) small MU setting. It may also be represented as a *control point sequence*, containing a series of machine states that include progression of MUs and of individual leaf positions. In the control point sequence for a step-and-shoot delivery there is an alternating sequence of "step" (leaves move) and "shoot" (MUs increment) stages where successive control points either express an increment in leaf positions or in MUs delivered.

A *dynamic* delivery is an alternative to step-and-shoot delivery. Here the leaves move while the beam remains on and the control point sequence contains successive states with both leaf position and MU value increments. Dynamic delivery sequences can be affected by mechanical limitations of the MLC—in particular whether the leaves can move to the next point listed in the control point sequence within the time that the corresponding MU increment will take. If necessary, the beam may need to be temporarily halted while the leaves "catch up." In fact, a step-and-shoot sequence is really a special case of a dynamic sequence where the increment in MUs for each increment in leaf positions is zero, thus presenting a situation where

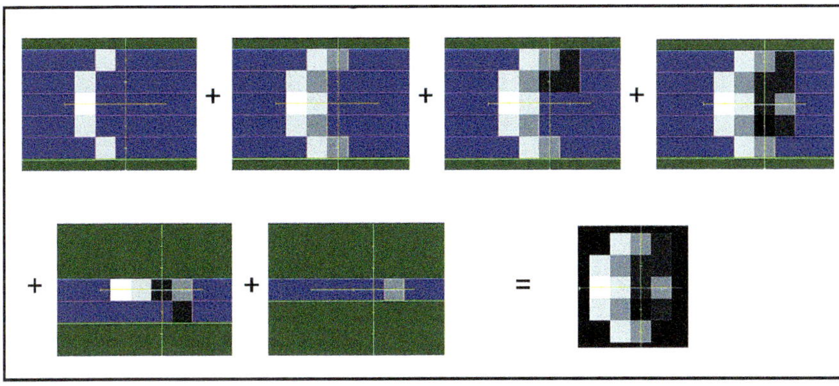

Figure 7.27: Example of a step-and-shoot sequence overlaid on the corresponding opening density map. In this case, seven MLC field segments have been superimposed.

the leaf speed would need to be infinite if the beam were to remain on! In this case, the beam is simply "held off" while the leaves move to their next control point, creating a step-and-shoot sequence.

7.11.1 Sliding Window Sequence for Step-and-Shoot Delivery

There is a very simple method for determining the required sequence of leaf positions for a single pair of opposing leaves to create a given row of an intensity map. This is illustrated in figure 7.28, where first the row of intensities (OD) is converted to a histogram of MU values. This OD map has in fact been discretized into 10 equally spaced levels of 5 MU—the MU values can only be 5, 10, 15,..., 50. Also notice in this histogram that there are 15 "up-steps" and 15 "down-steps." As was described by Bortfeld et al. (1994), each corresponding up and down step can be paired to obtain the sequence of paired leaf positions for a step-and-shoot delivery. That is, the left leaf in the pair for "shoot" 1 should be at the position of the first up-step and the right leaf should be at the position of the first down-step.

Figure 7.29 shows these leaf positions graphically, where the rods in the histogram have been "pushed" such that, if a rod is higher than the rod to its left, the bases are lined up and if the rod is lower than the rod to its left the tops are lined up (Siochi 1999). Once this "rod pushing" has been performed, pairs of leaves can be positioned starting at the bottom of the stack and proceeding to the top. The 15 up-steps in the OD histogram translate to 15 levels in this stack. Note that there are in fact only 5 *different* left-right leaf position pairs in this sequence. That is, there are 5 shoots. This is a unidirectional, left-to-right "sweep" of both leaves in each pair. It is, in fact, the method that yields the minimum number of MU in realizing a given OD map, though not necessarily with the least steps.

It is important to realize that fewer discrete levels in the OD histogram mean fewer discrete levels in the stack and thus the fewer distinctly different leaf pair positions. This is very important for some implementations,

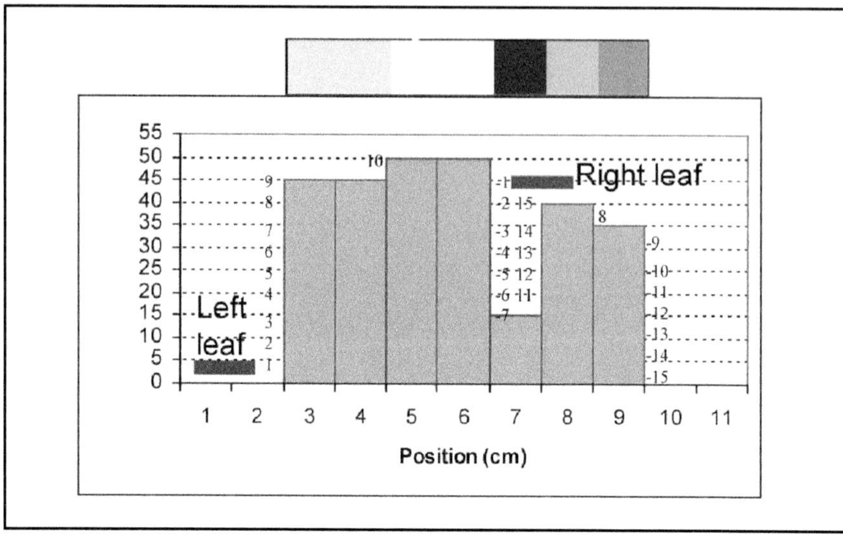

Figure 7.28: In a simple method for determining the required sequence of leaf positions for a single pair of opposing leaves the row of intensities (OD) is converted to a histogram of MU values. This map has first been discretized into 10 equally spaced levels of 5 MU. In this histogram there are 15 "up-steps" and 15 "down-steps," pairs of which can be used to obtain the sequence of paired leaf positions for step-and-shoot delivery.

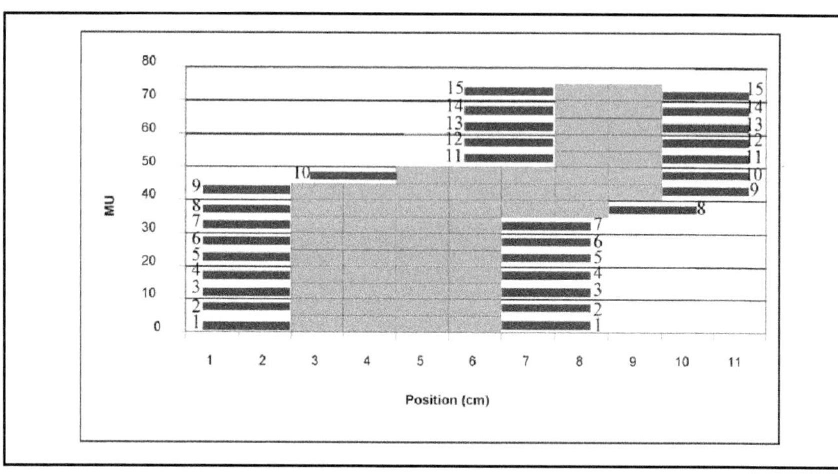

Figure 7.29: Rods in the histogram of figure 7.28 have been "pushed" such that if a rod is higher than the rod to its left, the bases are lined up and if the rod is lower than the rod to its left, the tops are lined up (Siochi 1999). Pairs of leaves can then be positioned starting at the bottom of the stack and proceeding to the top. The 15 up-steps in the OD histogram translate to 15 levels in this stack and 15 leaf-pair positions.

where there the time spent between shoots (the "steps") can represent the majority of the delivery time. If step time is very small, then number of levels is not such a significant factor in overall time.

OD map discretization can be simply via, for example, taking the maximum value in the map and dividing by the desired number of levels to obtain the difference between levels. An example of a more sophisticated algorithm is K-means clustering (Hartigan 1975, Wu et al. 2000a,b). Another series of algorithms designed for efficient step-and-shoot delivery, particularly in minimizing number of steps, are within the IMFAST system (Siochi 1999).

7.11.2 Sliding Window Sequence for Dynamic Delivery

The sliding window strategy described above for step-and-shoot delivery can be easily translated to a dynamic delivery, again with a left-right sweep of each leaf pair. There are typically a much larger number of steps within a dynamic leaf sequence than that within a step-and-shoot sequence. This is because there is essentially no penalty in terms of time since the beam is always on. In principle, therefore, more complex OD maps can be delivered in a dynamic sequence than in a step-and-shoot sequence.

The OD values derived from the optimization can either be treated as rods as above, or the values can be treated as points whose position is the center of the rod. The techniques for realizing each are slightly different; here we will examine the latter for brevity. Essentially, what is required is to create a leaf sequence such that the difference in MU value between when the right leaf "uncovers" a point (radiation begins to be delivered to that point) and the left leaf "closes off" radiation to that point is equal to the MU value required to create the optimized OD value for that point (Convery and Rosenbloom 1992).

Mathematically, if the required OD at a point x is $OD(x)$, $M_A(x)$ is the MU value at which the A (left) leaf reaches x and $M_B(x)$ is the MU value at which the B (right) leaf reaches x, then (for now ignoring leaf transmission)

$$OD(x) = M_A(x) - M_B(x). \tag{7.9}$$

To create an OD map in this way with a dynamic leaf sequence, the algorithm must be aware of the maximum leaf speed as well as the MU-rate. The control point sequence that drives the leaf positions as a function of MU will include these constraints in its trajectory for each leaf. Note that the MU rate typically stays constant while leaf speed varies up to its maximum in order to create the modulated intensity profile.

Figure 7.30 illustrates leaf trajectories for a dynamic sequence, where the goal is to create an OD map with values that follow the line segments drawn between midpoints of the OD histogram in figure 7.28. Ideally, to create the first (left-most) OD point in figure 7.30, the right leaf would move in zero time from its initial position by a distance equal to one bar width and the left leaf would follow at a slower controlled rate to allow the OD values to ramp up at the required rate (Stein et al. 1994). The right leaf has a finite maximum speed so there will be a small number of MUs elapsed when it

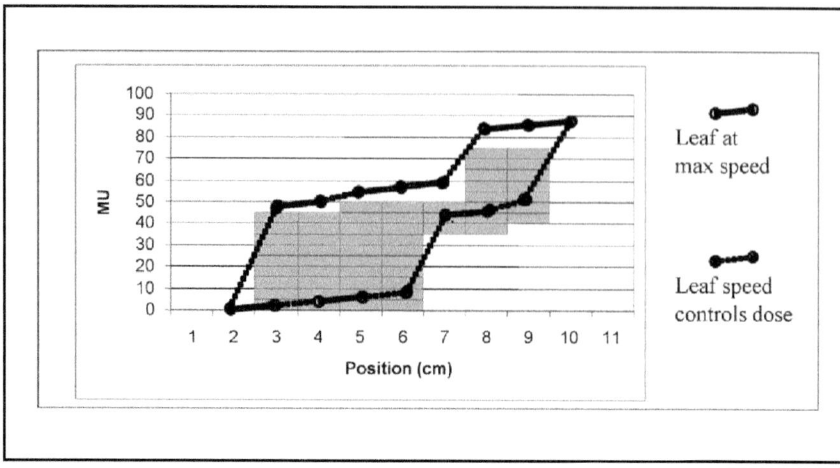

Figure 7.30: Leaf trajectories for a dynamic sequence, where the goal is to create an OD map with values that follow line segments drawn between midpoints of the OD histogram in figure 7.28. When the OD values are ramping up, the right leaf moves at maximum speed (with left leaf following at a slower rate as above) and when the OD values are ramping down, the right leaf moves at a slower controlled rate with the left leaf following at maximum speed.

reaches this point. This means that this MU value has to be added to the value for when the left leaf reaches the same point to retain the required MU difference. When the OD values are ramping up, the right leaf moves at maximum speed (with left leaf following at a slower rate as above) and when the OD values are ramping down, the right leaf moves at a slower controlled rate with the left leaf following at maximum speed. Figure 7.30 illustrates this point, with the dashed lines representing the leaf moving at the slower rate (or equal in the case of no intensity change between points).

With a lower MU-rate, this extra MU value will be less, and the total number of extra MU for the whole sequence will be less. This is an advantage in terms of having to account for radiation transmission through the leaves, but a disadvantage in terms of overall delivery time. Leaf transmission can be accounted for in designing a dynamic leaf sequence. This requires knowledge of the total MU delivered, which itself can be approximated (Mohan et al. 2000). Transmission adds to total OD, thus the MU difference between left leaf and right leaf reaching a point needs to be decreased.

There are a number of analyses that can be performed using leaf tracking software to ensure the speed and dose delivery of the MLC leaves are within tolerance. These tests are outlined in detail in the acceptance test documents and are the subject of current review by an AAPM task group (Ezzell et al. 2003).

While the leaf sequence may be designed to deliver the correct intensity profile, the actual dose delivered may be different due to a number of dosimetric complications due to nonidealities in the MLC design. In particular, the rounded leaf ends and leaf transmission lead to a dose increase above what is predicted. (LoSasso et al. 1998).

7.12 Dosimetric Properties of MLCs

IMRT delivery is a challenge for MLCs, which were originally designed for defining the larger fields used in conformal therapy. In particular, with a small gap between opposing leaves, any error in leaf position can have a big impact on field width and thus on the integrated dose deposited by that field. When a narrow field is sweeping across a point, the width of the field is directly proportional to the dose. Furthermore, details of leaf tip design are important to model correctly as the pattern of beam transmission though these tips, and thus beam penumbra, impacts the effective position of the end of the leaf. LoSasso et al. have analyzed these effects for a Varian MLC and determined potential errors (LoSasso et al. 1998).

Some other commonly known effects are outlined below.

Matchline Effects

When an MLC with rounded leaf tips is used for beam shielding in simple single segment beams, the rounded leaf ends may cause only a small difference in the penumbra at the beam edge; however if multiple segments are involved in step-and-shoot delivery, this small penumbral asymmetry may lead to high-dose lines between neighboring segments: these are known as *matchlines.*

While matchlines are not observed with sliding window delivery, the extra rounded leaf end transmission does lead to a higher dose in between the leaf gap than expected (Mohan et al. 2000).

The MLC design that leads to matchlines is shown by the schematic in figure 7.31. A planar film view of a series of matchlines produced by driving leaf banks in 2 cm steps with 2 cm gaps is shown on dosimetry film in figure 7.31 (bottom), and dosimetry film showing two matchlines produced from a random clinical leaf sequence is shown in figure 7.32. One approach to correct for this effect is to apply a small leaf gap offset of about 0.7 mm for each leaf (Cadman et al. 2002, Tangboonduangjit et al. 2004).

More elegant rounded leaf end models have recently been provided by the planning computer models, which provide a better solution as target dose uniformity is not adversely affected by correctly calculated matchlines at the planning stage (Cadman et al. 2005, Williams and Metcalfe 2006). Note there is a very small opening (usually about 0.05 to 0.25 mm) between the leaves also to avoid collision in their closed positions. It is assumed this also contributes somewhat to the matchlines observed in a clinical IMRT field.

Tongue and Groove Effect

Each MLC has a tongue and groove design so the leaves lock together while minimizing leakage between the leaves. For single segment deliveries this leads to a small extra leakage dose of about 0.5% in the tongue and groove region. In IMRT delivery, however, the extra small protrusion of each tongue can overlap with protrusions from adjacent leaves and cause an underdose along the "track" at the boundary between leaves (Siochi 1999). Leaf synchronization, keeping adjacent leaves at the same setting when

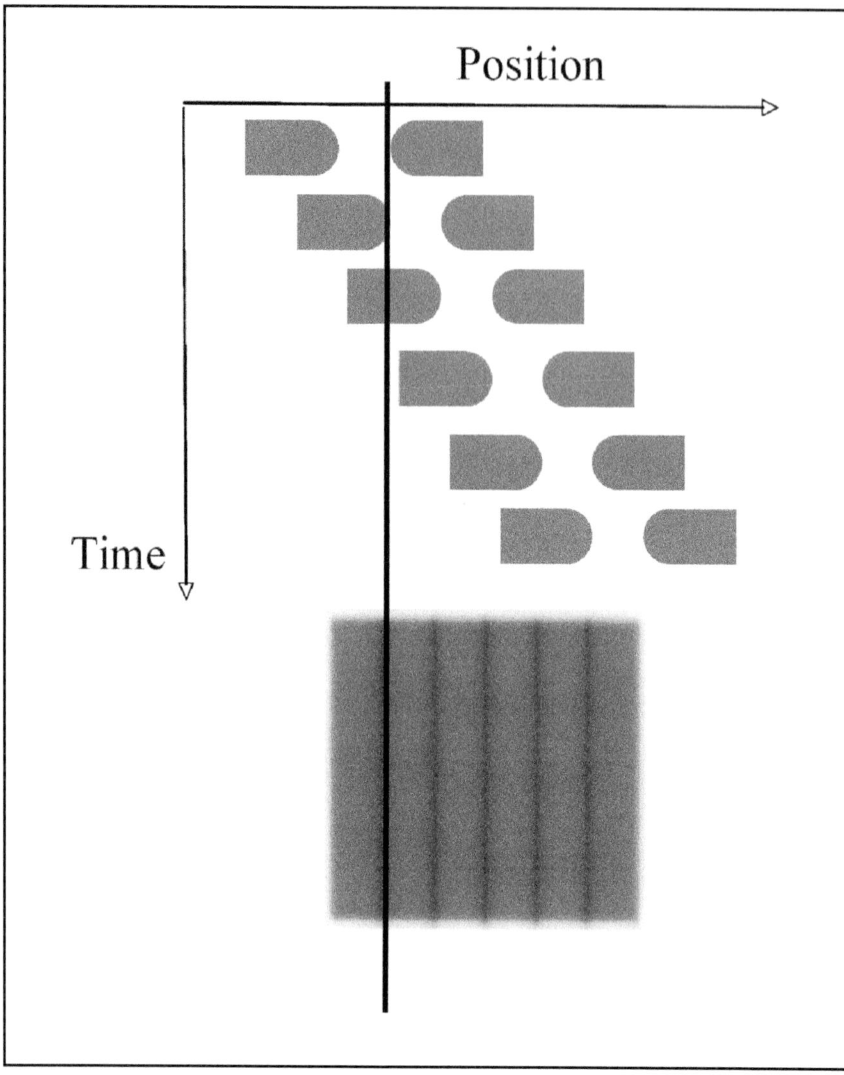

Figure 7.31: Schematic of rounded MLC leaf ends causing *matchlines* when dwell positions for left and right leaves are nominally the same in successive segments. The matchlines are indicated by dark lines on the dosimetry film at bottom.

necessary, can be performed to reduce this effect. The physical leaf design that leads to this effect has previously been described in chapter four, section 4.7, figure 4.32.

Planning software usually offers tongue and groove minimization at the leaf sequencing stage, so the effect should be rarely seen in clinical leaf sequences. It is worth noting that dosimetry problems due to these leaf nonidealities are rare in clinical fields, but vigilant quality assurance is still wise until one gains confidence that these are not affecting the delivered dose distribution.

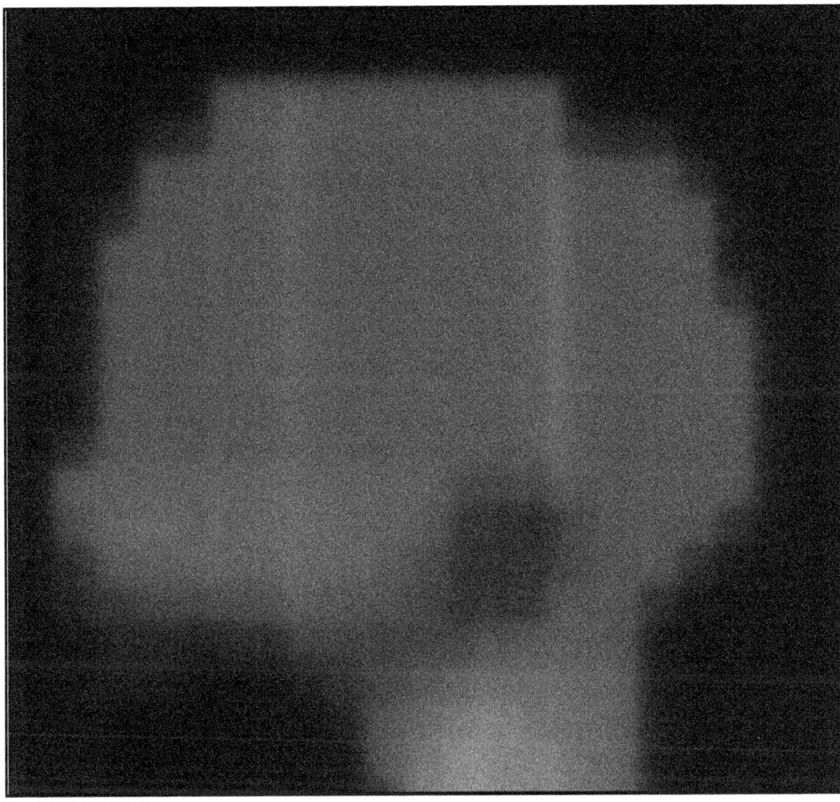

Figure 7.32: Planar dosimetry film of matchlines observed in a clinical IMRT treatment beam represented by the two light grey vertical lines.

Leaf Leakage

Leaf transmission (leakage) is typically 1.5%, and interleaf transmission (leakage) is typically only 0.5% of dose delivered in static mode. However the actual dose due to leakage depends on the MUs delivered to give the dose. In combined conformal field treatments it usually takes slightly more than 200 MU to deliver 200 cGy at isocenter; hence the MU to cGy dose ratio is close to unity. For multisegment IMRT treatments the MU to cGy dose ratio is typically between 3 and 5. This is because small leaf openings are often employed. Small leaf openings lead to the MU to dose ratio increasing and hence the leakage component of dose in the region between the jaws and irregularly MLC shaped field increases. Therefore the 2% leaf leakage is about 2 cGy for conformal and 10 cGy if the MU to dose ratio was 5.

End Leaf Leakage

End leaf leakage has been measured by film as 30% of the open field dose while interleaf leakage (also known as interleaf transmission) is typically 1.5% to 2%. End leaf leakage does not present a problem provided leaf pairs which have finished segment delivery are driven under a jaw while other leaf pairs are delivering dose. Computed leaf sequences should ensure unused leaf pairs are placed under a jaw for extra shielding when not in use. This process should also be pre-checked with leaf verification software. A planar dosimetry film showing interleaf and end leaf leakage is shown in chapter four, section 4.7.3, figure 4.35.

Wide Field IMRT

Wide IMRT fields are sometimes required, especially for head and neck cases. Establishing field widths wider than 14.5 cm with the Varian MLC presents a challenge because this is the distance the leaf can travel from the carriage. However the Varian leaf carriages can move so a pair of split overlapping fields can be used to extend the field size. Figure 4.36b in chapter four shows how the MLC carriages can be moved to 2 cm beyond central axis; however to protect the carriage electronics, a jaw needs to be moved. The dual overlapping fields (typically 2 to 4 cm overlap) can then be "feathered" so that contributions ramp down and up for the left and right components, respectively. Leaf modulation is used for this feathering (Metcalfe et al. 2004a, Wu et al. 2000a,b). Figure 7.33 shows an IMRT plan with overlapping fields that have been used to extend the field size in the leaf direction.

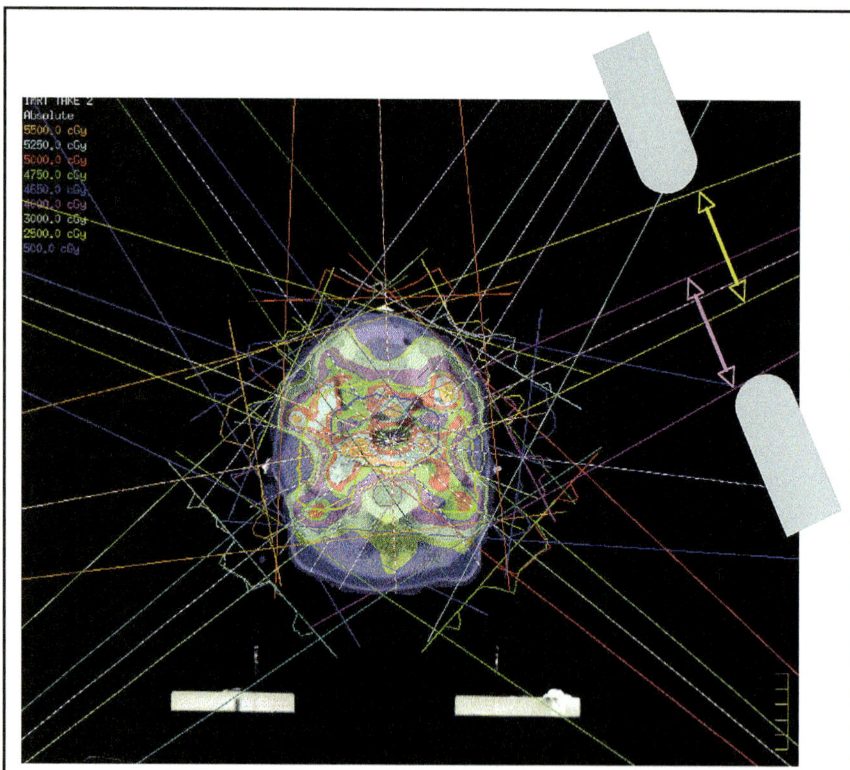

Figure 7.33: A 2-D axial view of a treatment plan using six "split beams" to ensure wide enough fields in the MLC leaf travel direction. (Plan generated on a Philips Pinnacle RTP computer.)

Entry Surface Dose

Skin dose from IMRT fields has been measured to give a surface dose between 15% and 25% by Price et al. (2006) with only a small increase in surface dose when a high MU to dose ratio is involved. This nonsignificant increase in surface dose is similar to that found by Dogan and Glasgow 2003.

There have however been incidences of excess skin dose to IMRT patients. The problem appears to arise when dose objectives are applied to the PTV and this volume is close to the skin surface. Then, optimizer solutions that contain high-intensity tangential beam segments close to the surface arise (Thomas and Hoole 2004). This can be avoided by setting the dose objective to the CTV to a higher importance or by setting a skin volume as a sensitive structure and giving it a low dose objective (Lee et al. 2003).

Bolus may also be applied to the area and although the skin dose will be high in this case, the prediction of skin dose will be more accurate. Something to consider is that for 3DCRT head and neck treatments there are often two opposing highly weighted beams, so entrance and exit dose add to sometimes cause severe skin reactions. With IMRT up to nine beam angles are generally employed, so the beam exit dose is nonopposing and each beam has a low weight. This means less incident surface dose per beam, but the volume of skin irradiated is greater. Another significant feature may be that the extra beams may traverse the head cast, and even the neck support, hence increasing the skin dose beyond that incurred by beams that just traverse the head cast material.

7.13 Independent Plan Dosimetry Verification

Because it is unlikely that the radiotherapy computer treatment planning calculation accounts for all the above effects, dosimetry comparisons between planned IMRT fields and delivered IMRT fields are carried out. Because there are a large number of beam segments delivered, a manual check of dose per field is very difficult. Hence these dosimetry tests become a suitable substitute for the traditional independent dose check calculation. Usually validation of planar and axial dose distributions is undertaken using comparison of film and ionization chamber readings. Sometimes thermoluminescent dosimeters or diode arrays may be employed, and recently progress in using electronic portal imaging devices (EPIDs) as dosimetry validation devices has been made (Greer and Popescu 2003). The IMRT QA dosimetry process undertaken for each patient is discussed in detail in the QA chapter of this book (chapter eleven, section 11.4). We briefly discuss the principle here.

Ionization Chamber Checks

Each individual planned IMRT beam is transferred to a phantom and the dose delivered to a planar check film is compared against the dose predicted using the radiotherapy planning computer. Then all beam parameters are transferred to a phantom, which may be a cylinder or a simple block of solid water; then the dose is delivered and dose checks are made using film placed parallel to the beam axis to get an axial dose comparison. Point dose checks may also be undertaken with an ionization chamber placed at isocenter and integrating dose over the entire treatment of combined multiple IMRT fields (Boyer et al. 1999). As an alternative an ionization chamber check can be done with all beams delivering dose with gantry at the same angle and dosimeter at a fixed depth, then a tissue-phantom ratio correction may be applied to each ionization chamber measurement, allowing for the different depths of each beam to isocenter (LoSasso et al. 2001).

Film Dosimetry Checks

Shown in figure 7.34 is a schematic of the plan validation dosimetry which is undertaken prior to treatment. Also previously shown in figure 7.32 was an IMRT planar film density map. This map can be converted to dose using a film calibration curve and, in turn, compared with planar dose maps produced using RTP computer planning tools.

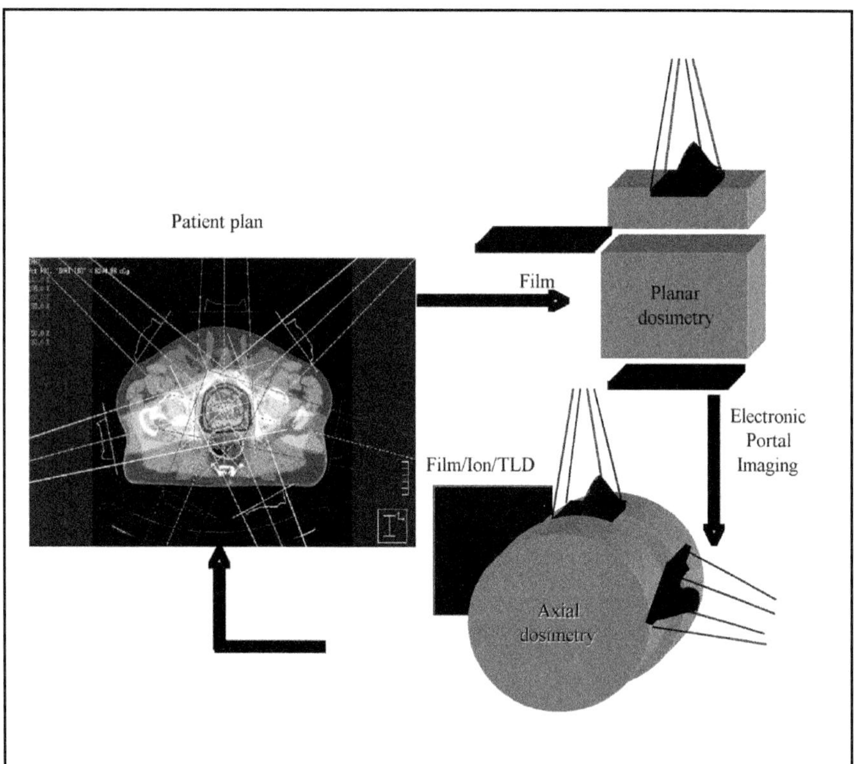

Figure 7.34: Schematic of the dosimetry verification loop required prior to IMRT treatment of a patient and the dosimetric options available including: ionization chambers, film, EPIDs, and thermo-luminiscent dosimeters.

EPID Dosimetry Checks

Rapid progress has been made in using the portal dosimeter to collect IMRT distributions and correct these to compare with planned dose distributions (Metcalfe et al. 2004a,b). Detector response kernels and off- axis profile correction can be applied, and currently one planning system (Eclipse) offers both dose and Gamma difference map (Low et al. 1998) comparison of EPID dose predicted by the RTP computer versus dose accrued by the EPID (Van Esch et al. 2004). The use of EPIDs for IMRT dose verification is discussed further in chapter eleven, section 11.4.

7.14 Clinical Treatment with IMRT

Head and Neck IMRT

The potential for parotid sparing and the removal of the requirement of electron boost fields for head and neck treatment make this site ideal for IMRT, particularly because immobilization with various head casts or stereotactic frames provides excellent fixation. The dose to the parotid glands (mean dose of less than 25 Gy) can reduce xerostomia, dry mouth due to lack of salivary production, and this can be achieved with IMRT as shown in figure 7.35 (Wu et al. 2000a, Dawson et al. 2000, Boyer et al. 1997).

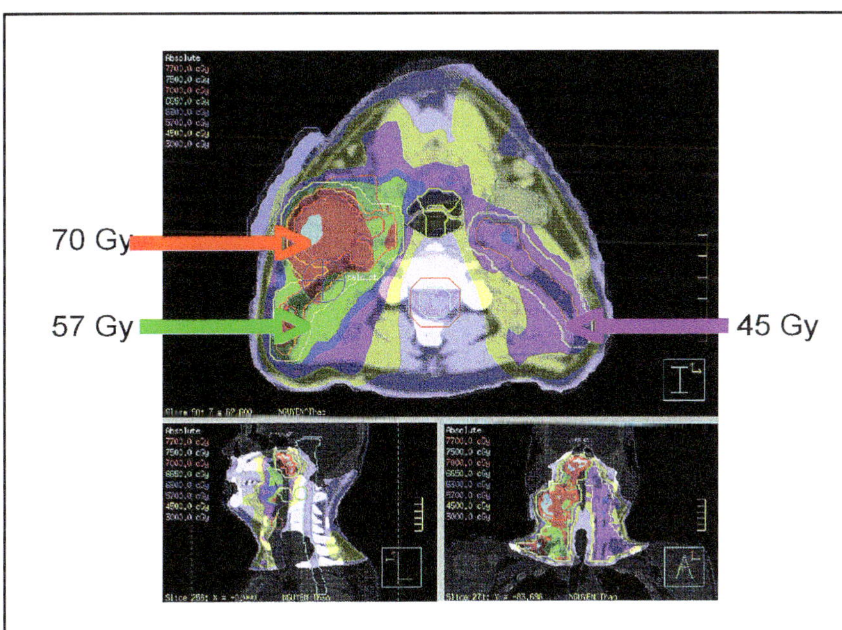

Figure 7.35: Concurrent boost using IMRT to three treatment volumes in a head and neck tumor. The prescription was 70 Gy in 2 Gy fractions to the GTV, 65 Gy in 1.85 fractions to CTV 1, and 57 Gy in 1.63 fractions to CTV 2 (node-negative on PET scan). (Plan generated on a Philips Pinnacle RTP computer.)

Because dose objectives in inverse planning can be set to defined volumes, it is relatively simple to provide concurrent (also known as concomitant) boost fields using IMRT as shown in figure 7.35. Because of mucositis from combined treatment with cisplatin, the concomitant boost schedules are typically not accelerated very much above 2 Gy per fraction at this point in time. With other agents such as Cetuximab, accelerated regimens may become more prevalent (see chapter fourteen). Concomitant boosts with IMRT are available, but not the only approach—boosting sequentially is still an option. The three-zone schedule shown in figure 7.35 was set at 2 Gy per fraction for the gross target volume (GTV) and other volumes received less than 2 Gy per fraction.

The potential for sparing other structures in the head and neck region is significant with subsequent improvements in patient quality of life. Another clinical example is sparing of the lacrimal gland to ensure tear production is maintained and a gritty dry eye is not encountered (Metcalfe et al. 2004b).

Breast

Forward planned IMRT or in-field segmentation is used to treat breast tangent fields (Kestin et al. 2000). Using a combination of forward planning, the planner segments off hot dose areas to produce three or four in-field beam segments. The planner then alters the beam weights by trial and error to come up with a plan with fewer hot-spots. This has the effect of acting as a coarse compensation technique. This has found favor in breast tangents as the target often reaches near the surface where a rapid dose gradient makes it difficult to set inverse plan dose objectives (Chui et al. 2003). Another advantage is that a few segments can be manually checked and the rigorous and time-consuming dosimetry checks required for inverse planned IMRT are not essential. As shown in figure 7.36 this has the effect of acting as a coarse compensation technique.

Prostate

IMRT has originally been used to treat concave tumor volumes that have either overlapping or very close normal tissue structures which require some sparing. Dose escalation studies have mainly focused on prostate treatment where escalation to over 80 Gy has been noted (Burman et al. 1997, Perez 1997, Ling et al. 2003). The long-term improvement in local tumor control may take a long time to evaluate because prostate is a slow-growing tumor. Nevertheless, normal tissue complication rates can be maintained at a low level while target dose is escalated with minimal increase in dose to critical structures, such as the rectum. An IMRT prostate treatment plan is shown in figure 7.37. This example has five fields and 50 step-and-shoot segments for all fields. Seven-field plans are also common with fewer segments required.

Prostate position verification and immobilization nevertheless present a major challenge for these techniques. Improvements in targeting the prostate with improved images from electronic portal imaging technology and on-board cone beam CT, as well as other methods (e.g., ultrasound positioning) are welcome additions to position verification technology.

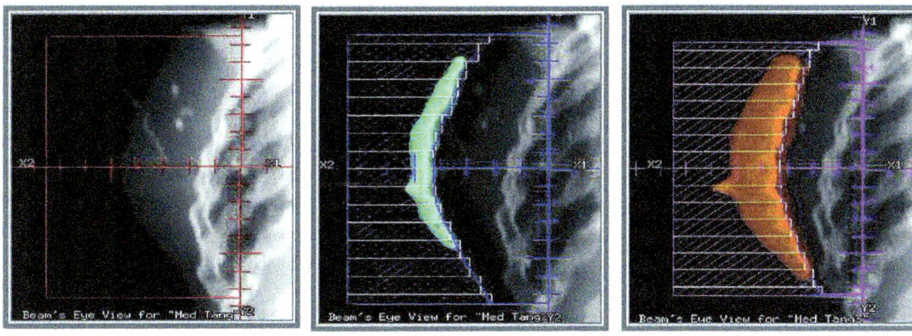

Figure 7.36: Radiotherapy treatment plan beam's-eye view of a three-segment breast tangent field. The first segment is an open field; then the other two segments block dose to the region, which would otherwise have a hot dose region. (Plan generated on a Philips Pinnacle RTP computer.)

IMRT treatment plans are usually characterized by a large number (7 to 9) of nonopposing beams. While this is a similar number of beams to that used for conformal prostate treatments, it is many more than used for conformal parallel-opposed head and neck treatments. This recipe is said to help give the optimization engine a dose gradient (no opposing beams) and

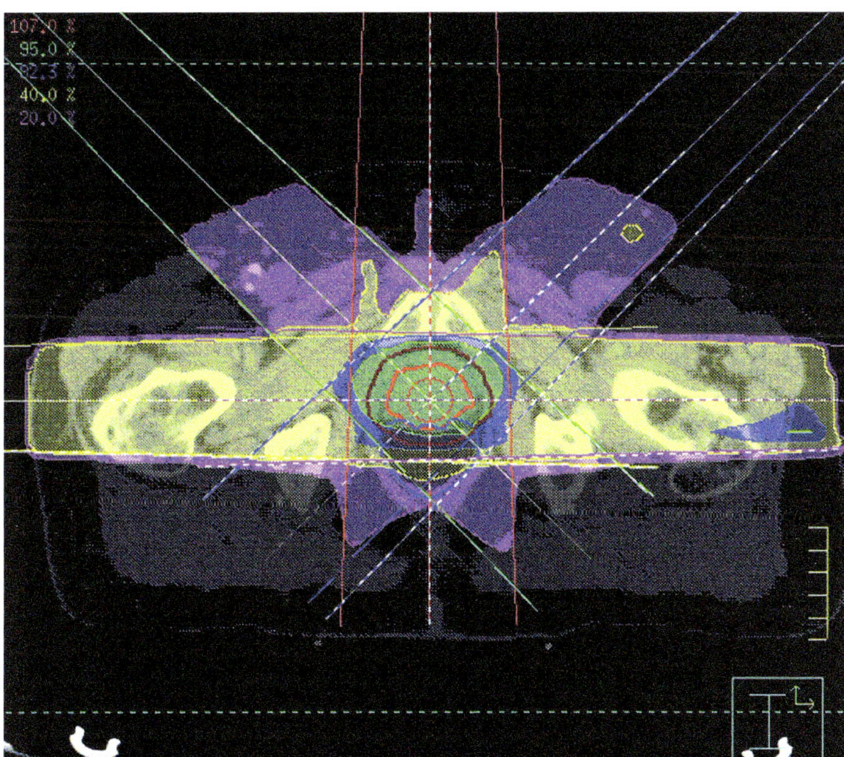

Figure 7.37: An IMRT prostate treatment plan showing a 2-D isodose display for one of the CT slices. This example has five fields and 50 step-and-shoot segments for all fields. Seven-field plans are also common. (Plan generated on a Philips Pinnacle RTP computer.)

more degrees of freedom due to several beam angles (Stein et al. 1997). This probably also helps provide a tight fall-off in dose around the target margin. Note that an inverse plan can be designed using open field beams; however a more optimum dose distribution is nearly always obtained when modulated beams are used.

As a deleterious side effect, IMRT can lead to a low-dose background irradiation. This is not dissimilar to dose encountered in arc therapy plans. The relatively large low-dose irradiated volume (IV) has contributed to discussions about advantages of proton therapy, which incurs a characteristically smaller low-dose volume than photon treatment (Lomax et al. 1999, Cozzi et al. 2001). In general though it is accepted that the improved conformality provided by IMRT (e.g., sparing of parotid glands) far outweighs the small extra low-dose burden.

7.15 Tomotherapy as an IMRT Modality

The generic process known as tomotherapy is a form of IMRT that combines rotational beam delivery with intensity modulation. In terms of beam delivery, tomotherapy is essentially equivalent to a scenario where a conventional linac gantry would rotate through a single 360-degree arc, while continually changing the modulated 2-D intensity profile of the beam. In practice, tomotherapy uses multiple rotations of a narrow "fan beam" with a continually changing one-dimensional profile. Tomotherapy literally means '"slice therapy," just as tomography refers to slice-based imaging. The mode of image acquisition used in CT scanning is in fact a close analogy to the way in which beams are optimized and delivered in the tomotherapy process. It has been shown that a larger number of modulated beams improves the conformality and homogeneity of the delivered dose that can be achieved (Shepard et al. 1999).

Tomotherapy has been implemented both as a sequence of rotational modulated beams (known as *serial tomotherapy*), with each rotation at a fixed couch position (Carol 1995; Grant and Woo 1999), or as a beam following a helical trajectory relative to the patient (Mackie et al. 1993). The latter is the paradigm used by the commercially available TomoTherapy Hi-Art System® (TomoTherapy Inc., Madison, WI), which will be used here as an example of a system employing the tomotherapy form of IMRT.

Although TomoTherapy Hi-Art systems have been in clinical use since only 2002 (Mackie 2006), several years after the first IMRT treatments with conventional MLCs, serial tomotherapy was the first clinical IMRT system and the most common form of IMRT during the mid to late 1990s (Curran 1993). Furthermore, the concept of irregular beam profiles being used to create a nonuniform dose, as originally reported by Brahme (Brahme et al. 1982), referred to a modulated beam rotating around a full circle to create a donut-shaped dose distribution with an organ at risk in its center. Thus tomotherapy in some form was how intensity-modulated delivery was originally envisaged and first implemented.

7.16 Analogy of Tomotherapy to CT Imaging

With CT imaging, a large number of physically created narrow beam attenuation profiles are mathematically projected back along each beam path to reconstruct the subject's density distribution within a narrow slice (a process called *back-projection*). With tomotherapy, the required dose distribution is mathematically projected out of the patient to create narrow beamlet intensity profiles, which are then physically projected back into the patient during slice-based beam delivery. So both processes include a mathematical and a physical projection, with the direction of each being reversed between the imaging and treatment scenarios.

The actual mode of beam delivery used by the Hi-Art system is a modulating fan beam moving around the patient in a helical fashion. The advantages of helical over slice-by-slice CT image acquisition (Kalender and Polacin 1991) apply in a similar manner to tomotherapy beam delivery. In particular, with helical tomotherapy beam delivery there can be no "matchlines" containing hot or cold dose regions as can be present when slice-by-slice delivery is used, especially if couch translation between slices is inaccurate. In CT acquisition, a small object can appear an unrealistic shape, size, or not be seen at all, with errors in couch translation per slice.

Tomotherapy beam optimization is in fact very similar to the process used for conventional IMRT, and the gradient method described earlier in this chapter is equally applicable to tomotherapy. The Hi-Art system actually updates beamlet weights using dose ratios (desired to actual) rather than dose differences. Algorithms describing the type of beamlet optimization used can be found in Olivera et al. (1998) and Shepard et al. (2000).

7.17 The TomoTherapy Hi-Art System®

The Hi-Art System®, including a depiction of the linac waveguide, modulated fan beam, and image detector is illustrated in figure 7.38. The accelerator produces x-rays of energy at 6 MV, which for rotational IMRT is a good balance between rate of attenuation (generally better if lower, as with high energies) and lateral penumbra width (better if smaller, as with lower energies). Source-to-axis distance (SAD) is 85 cm, with the diameter of the bore through which the patient translates also being 85 cm. The detector is designed for single-slice CT imaging and consists of xenon gas–filled ionization chambers encased in tungsten. This detector has high quantum efficiency for the photon energies in the imaging beam. Images are used intrafraction image guidance (see chapter twelve, section 12.4). For imaging, the linac is de-tuned to produce 3.5 MV x-rays and the pulse repetition frequency decreased to keep the dose delivered to the patient low. The data acquisition is fast enough to determine the dose given in individual linac pulses and the detector acquisition system (DAS) files are a useful tool for commissioning and QA of the unit (Balog et al. 2003).

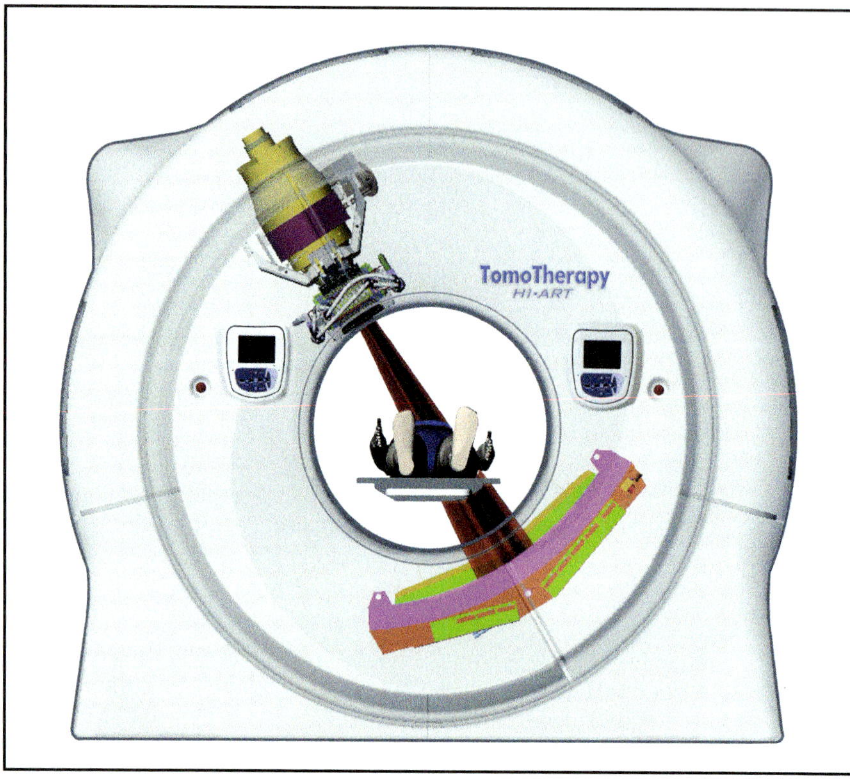

Figure 7.38: The TomoTherapy Hi-Art System®, including the 6 MV linac, a modulated fan beam, and image detector. Source-to-axis distance is 85 cm, with the diameter of the bore through which the patient translates also being 85 cm. The detector is designed for single-slice CT imaging and consists of xenon gas–filled ionization chambers encased in tungsten. (Courtesy of TomoTherapy Inc.)

The x-ray beam is collimated to a fan beam geometry with a width of 40 cm and a fan beam thickness (FBT) that is typically set at either 1.0 cm, 2.5 cm, or 5.0 cm. Symmetrical tungsten jaws are used to define these beam thicknesses. Beneath the jaws is a binary MLC (Swerdloff et al. 1994), where leaves are in one of two states: open or closed. The mode of leaf actuation is pneumatic control via banks of pistons on each side of the MLC assembly. Leaves move across the fan beam, in the superior/inferior direction relative to the patient. The time for the leaves to open and close is approximately 20 ms. The 64 leaves diverge with the beam and project a 6.25 mm width at isocenter. With a binary MLC, leaf end design is not an issue to be taken into account in beam modeling since the end of the leaf is only in the radiation field during transition between states and this time is exceedingly short. Beam output corresponding to a given leaf has a dependence on whether one or both adjacent leaves are open. This is modeled by the dose algorithm, using data collected by the CT detector as part of the beam commissioning process (Balog et al. 1999). Note that adjacent leaves interdigitate, so the actual number of leaves for a given total field width is half that for a MLC where leaves oppose each other.

Since the unit is specifically designed for IMRT, the leaf thickness (10 cm of tungsten) is thicker than in conventional MLCs and the overall shielding of the head is thicker. Jaws provide another 13 cm of tungsten shielding in the beam direction. Therefore total leakage radiation is low despite beam on times that are usually several times longer than non-IMRT treatments and at least as long as for conventional IMRT. Figure 7.39 is a photograph of the MLC with a few central leaves open—this would correspond to a state of the MLC at an instant during treatment.

With all leaves open, fan beam width is 40 cm at isocenter and any point within the diverging beam can be irradiated via opening of the required leaves. Although the fan beam width is 40 cm, points beyond 20 cm from the beam central axis at a given gantry angle can still be irradiated from other angles. Beyond 20 cm from the axis, the number of angles from which a point can be irradiated decreases with distance from the axis.

For delivery and planning a full gantry rotation is divided into 51 projections. Each projection is characterized by its own leaf opening pattern and covers an arc segment of approximately 7 degrees. The gantry rotation period can be anywhere between about 10 and 60 seconds. As such, each projection takes 0.2 to 1 second, during which time each leaf opens for a time proportional to the required intensity for the beamlet to which it corresponds at that angle. All leaves are shut between projections. It is important to note that longest opening time required for *any* leaf in *any* projection for *any* rotation is the determinant of how fast the gantry can rotate in order to allow the required opening time per projection.

The set of leaf opening times per projection for all leaves, at all gantry angles, for all gantry rotations is contained in a *delivery sinogram*. A graphical representation of a sinogram is given in figure 7.40. A delivery sinogram controlling leaf opening times in a tomotherapy treatment is analogous to a map of detector signals in CT imaging, which is also called a sinogram. A single row in the delivery sinogram represents leaf opening times for a single projection, where one pixel in the map corresponds to one leaf. Specifically, the sinogram value gives the fraction of the corresponding arc segment for which the leaf should be open. Each successive row is a new projection. The regular sinusoidal pattern, with one period per gantry rotation, is dominated by pixels that indicate a small or zero leaf opening times—these are leaves whose corresponding beamlets are directed at a sensitive structure.

7.18 Helical TomoTherapy Parameters

Collimator Width

Collimator width (fan beam thickness): This is chosen by the operator to achieve a compromise between fast treatment times and dose modulation in superior/inferior direction. Currently, clinical beams are commissioned for widths of 1 cm, 2.5 cm, and 5 cm and can be chosen for treatment. The 5 cm

Figure 7.39: The binary MLC used by the Hi-Art System® with a few central leaves open, corresponding to a state of the MLC at an instant during treatment. The MLC has 64 leaves, each creating a 6.25 mm wide beamlet at the isocenter. Leaves transit between states (open or closed) in about 25 ms. (Courtesy of TomoTherapy Inc.)

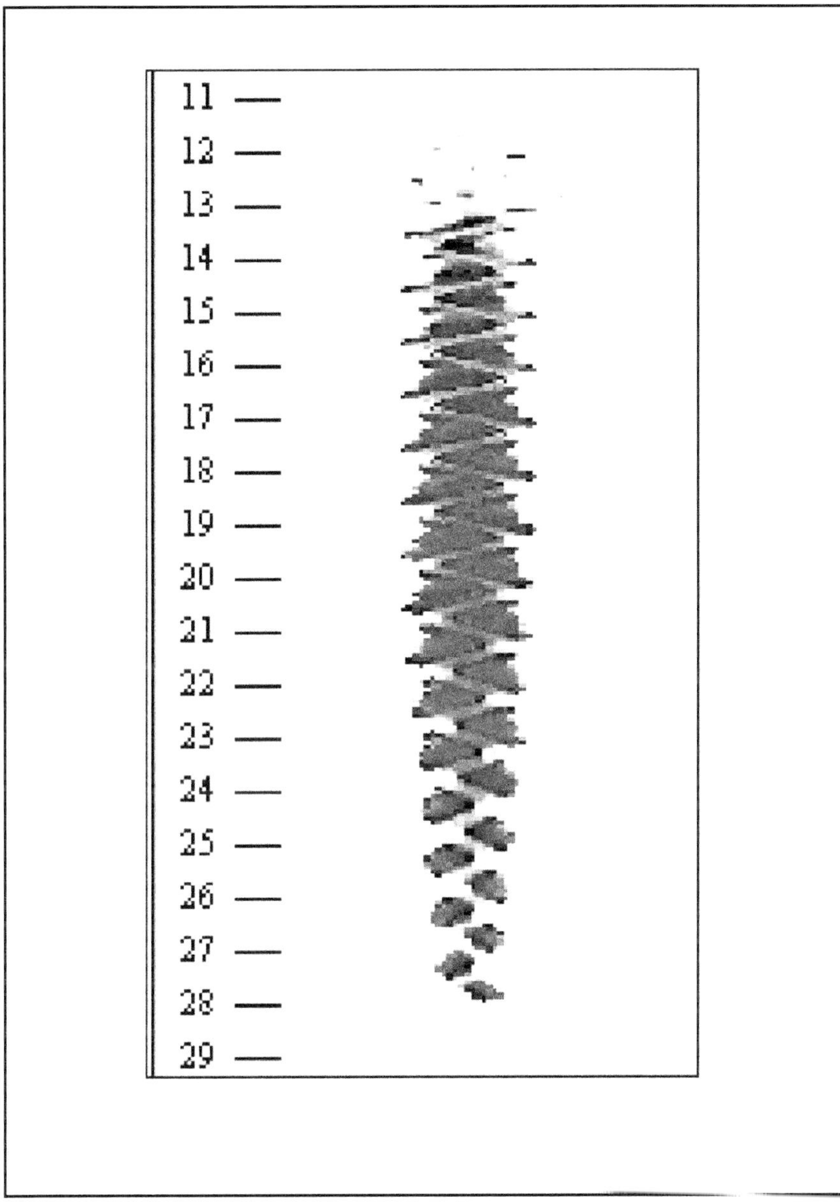

Figure 7.40: Graphical representation of a delivery sinogram. A single row in the delivery sinogram represents leaf opening times for a single projection, where one pixel in the map corresponds to one leaf. Each successive row is a new projection. A regular sinusoidal pattern is seen, with one period per rotation (this is dominated by pixels that indicate small or zero leaf opening times, corresponding to beamlets at a sensitive structure).

width is generally used only for long target volumes (e.g., craniospinal axis or abdominal cavity treatments) and 1 cm is employed for small lesions such as brain metastases. In practice, the 2.5 cm collimator width is used most frequently. Total treatment time is approximately inversely proportional to collimator width (Mackie et al. 2004) since a larger width results in a proportionally longer irradiated volume in any projection. Central axis dose output is also increased, further reducing treatment time with increased collimator width. Collimator width affects superior/inferior resolution and, in particular, the gradient of the dose ramp-up at the ends of the target. With appropriate setting of optimization parameters, however, the impact on extent of this ramp with increase in collimator width can be minimized. This may be preferable to reducing collimator width and incurring additional treatment delivery time.

Pitch

Pitch: A different way to improve the modulation capabilities in the superior/inferior direction is the choice of a small pitch. The pitch is defined as couch movement per rotation in units of the collimator width. For example, with a collimator width of 2.5 cm and a pitch of 0.2, couch translation is 0.5 cm for each gantry rotation. While it is common to use pitch values of 1.0 or higher in diagnostic CT scanning, the pitch in helical tomotherapy is typically between 0.1 and 0.5, resulting in overlap between adjacent rotations during the helical delivery. An interesting property of pitch is that decreasing it does not necessarily increase treatment time (Mackie et al. 2004). This is because a smaller pitch corresponds to a greater number of gantry rotations and greater overlap between fan beams on successive rotations. A smaller dose needs to be deposited per rotation when there are more rotations to contribute to the total dose, so gantry rotations can be faster. This property only holds however while the gantry is free to speed up—if the maximum rotation speed has been reached, then reducing pitch will increase treatment time due to the increased number of gantry rotations.

Modulation Factor

Modulation factor (*MF*): The modulation factor represents the ratio of maximum leaf opening time to the mean leaf opening time for all MLC leaves which open throughout a treatment. Modulation factor is proportional to the overall treatment time and with typical physical constraints for the tomotherapy delivery; modulation factors between 1.0 and approximately 4 are suitable. A smaller modulation factor results in shorter treatment times and is adequate for relatively symmetrical targets close to the central axis of the patient (Grigorov et al. 2003). In fact, the Hi-Art user does not directly select the modulation factor used, but rather specifies a maximum value. During optimization, the system ensures that the final modulation factor is less than or equal to this maximum value.

7.19 Clinical Use of Helical TomoTherapy

Treatment Planning

A course of treatment on a TomoTherapy Hi-Art System begins with acquisition of planning scans and delineation of targets and sensitive structures. Contour creation is mostly done using a conventional RTP system or virtual simulation system. Contours are then transferred to the Hi-Art database via DICOM-RT (Digital Imaging and Communications in Medicine-Radiotherapy) to be used for optimization. The user sets optimization parameters (prescription dose, goal doses, penalties, overlap-priorities, etc.) as for other IMRT planning systems. The specific helical TomoTherapy parameters are collimator width, pitch, and maximum modulation factor.

To accelerate the optimization process, beamlet doses are usually precalculated and stored prior to the interactive optimization session so that only beamlet weights need to be altered during the iterative process (see section 7.10). This beamlet calculation step can require from tens of minutes to several hours, depending on factors such as target size, collimator width, pitch, and selected resolution of the dose calculation matrix. Beamlet calculations are however typically done in a batch mode during nighttime hours—beamlets for several cases can be calculated and ready for optimization by morning. Once optimization begins, the user can monitor progress and adjust some of the parameters to "steer" the result to the desired outcome. Optimization parameters and maximum modulation factor can readily be altered; however a change in collimator width or pitch requires beamlet recalculation. For this reason, a disease-specific protocol known to work well is normally adopted.

A further function of the treatment planning interface is to define the number of fractions to be used. At this stage, it is verified that the plan is deliverable and a final dose is calculated that takes into account mechanical factors such as leaf transit time. In particular, if the optimizer has specified that a leaf must open during an arc segment for a time less than the minimum allowed, this leaf opening will not occur and the final dose will reflect the small effect of this change. The user interface reports at this stage what the final modulation is (less than or equal to the value set), how many gantry rotations will occur, and what the rotation time of each will be.

Treatment Delivery

TomoTherapy Hi-Art treatments almost always include an image-guidance step prior to beam delivery. The paradigm for this is to align the patient's anatomy in a manner that is consistent with its geometry in the planning CT scan. Essentially, the location of the target anatomy with respect to the center of the planning CT volume must be consistent with the location of the same anatomy at treatment time relative to the isocenter of the treatment device. This is a specific example of a general concept in all image-guided radiotherapy setups: there is a planned location of the anatomy and the image guidance system must ensure that the true location is consistent with this.

Where the Hi-Art's version of image-guided setup differs from most is that the planned setup position is not target-specific—instead there is a single, centralized setup for all targets. Prior to the image guidance procedure, laser-based setup is used for initial alignment in a manner similar to that for conventional linacs.

A megavoltage CT scan is acquired for image guidance, at fine, normal, or coarse resolution (Meeks et al. 2005). Dose delivered to the patient during this procedure ranges from about 0.5 to 2.0 cGy depending on resolution (coarse giving the lowest dose). The operator uses an automated image registration algorithm to align the images (Ruchala et al. 1999, Ruchala et al. 2002) and determine any required patient offset.

Once patient setup is complete, beam delivery can begin. The delivery sinogram generated during optimization instructs the pneumatic leaf actuators to open and close the MLC leaves as a function of gantry angle as planned. Precise timing keeps the gantry and couch in synchrony during helical delivery and this is part of a regular QA procedure (Fenwick et al. 2004).

A Plan Example

TomoTherapy beam delivery has value in creating dose distributions that can conform tightly to the target volume and also be multi-level. That is, different dose levels can be created during a single delivery, for example in treating the prostate and seminal vesicles with a different dose per fraction (Grigorov et al. 2003), or whole brain irradiation with a simultaneous infield boost to one or more metastases (Bauman et al. 2007).

A plan is shown in figure 7.41a for a prostate treatment including pelvic nodes. The prescription was for 70.2 Gy in 39 fractions (1.8 Gy per fraction) to the PTV including prostate and seminal vesicles plus a setup margin (PTV1: pink contours) and to simultaneously deliver 52 Gy in 39 1.33 Gy fractions to a second PTV including just the nodes (PTV2: black contours). Figure 7.41b shows the corresponding DVH. Good separation between dose levels for the two PTVs is seen for this case. The prescription called for 98% of PTV1 to receive 70.2 Gy and it can be seen from the corresponding DVH curve that this has been achieved. In fact, the optimizer treats the prescription as a *hard constraint* that must be achieved. PTV2 achieved its dose-volume goal of 52 Gy to 95% of the volume. The rectum abuts PTV1 and the bladder is partially overlapped by PTV so their DVH curves (brown and yellow respectively), as expected, include high-dose levels. A characteristic of a TomoTherapy plan is relatively isotropic dose fall-off outside the target volumes. This is due to the combination of rotational delivery with IMRT.

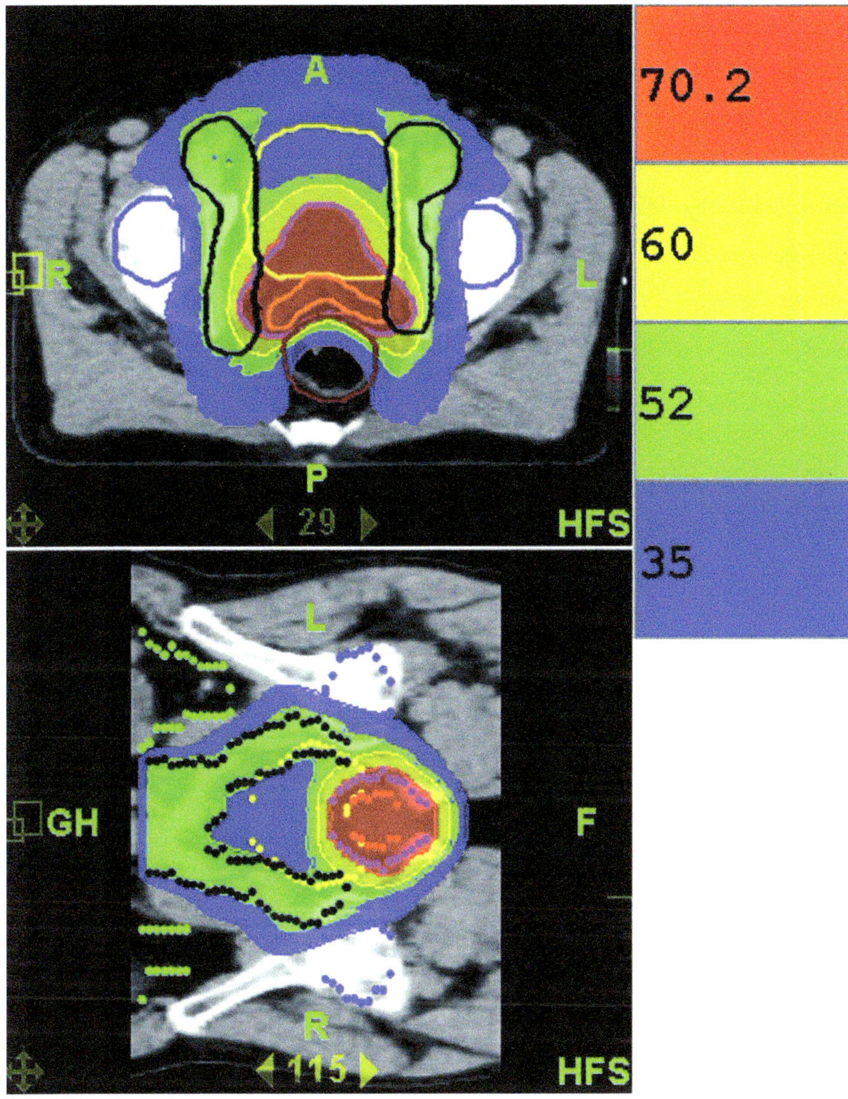

(a)

Figure 7.41: (a) Isodose distribution for the TomoTherapy prostate plan including pelvic nodes. The prescription is to deliver 70.2 Gy in 39 fractions (1.8 Gy per fraction) to the PTV (pink contours) and simultaneously to deliver 52 Gy in 1.33 Gy fractions to a second PTV including just the nodes (black contours).

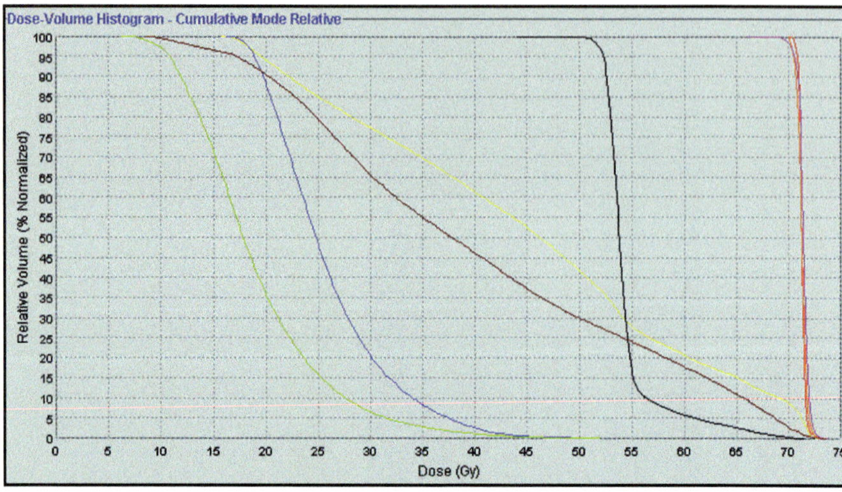

(b)

Figure 7.41: (Continued) (b) The resulting DVH (as well as the isodose levels) shows good separation between the two required dose levels.

References

Adler J.R., Murphy M.J., Chang S.D., Hancock S.L. (1999) Image-guided robotic radiosurgery. *Neurosurg.* 44: 1299–1306.

Alexander E. III, Loeffler J.S., Lunsford L.D. (1993) *Stereotactic Radiosurgery*. McGraw Hill, New York.

Andrews D.W., Scott C.B., Sperduto P.W., Flanders A.E., Gaspar L.E., Schell M.C., Werner-Wasik M., Demas W., Ryu J., Bahary J.P., Souhami L., Rotman M., Mehta M.P., Curran W.J. Jr. (2004) Whole brain radiation therapy with or without stereotactic radiosurgery boost for patients with one to three brain metastases: Phase III results of the RTOG 9508 randomised trial. *Lancet* 363: 1665–1672.

Balog J., Olivera G., and Kapatoes J. (2003) Clinical helical tomotherapy commissioning dosimetry. *Med. Phys.* 30: 3097–3106.

Balog J.P., Mackie T.R., Wenman D.L., Glass M., Fang G., Pearson D, (1999) Multileaf collimator interleaf transmission. *Med. Phys.* 26: 176–186.

Bauman G., Yartsev S., Fisher B., Kron T., Laperriere N., Heydarian M., Van Dyk J. (2007) Simultaneous infield boost with helical tomotherapy for patients with 1 to 3 brain metastases. *Am. J. Clin. Oncol.* 30: 38–44.

Blomgren H., Lax I., Näslund I., and Svanström R. (1995) Stereotactic high dose fraction radiation therapy of extracranial tumors using an accelerator. Clinical experience of the first thirty-one patients. *Acta Oncol.* 34: 861–870.

Bortfeld T., Stein J., and Preiser K. (1997) Clinically relevant intensity modulation optimization using physical criteria. In *Proceedings of the XII International Conference on the Use of Computers in Radiation Therapy*, Salt Lake City, Utah, May 1997. Medical Physics Publishing, Madison, Wisconsin, pp. 27–30.

Bortfeld T., Kahler D., Waldron T., Boyer A. (1994) X-ray field compensation with multileaf collimators. *Int. J. Radiat. Oncol. Biol. Phys.* 28: 723–730.

Bova F.J., Buatti J.M., Friedman W.A., Mendenhall W.M., Yang C.C., Liu C. (1997) The University of Florida frameless high-precision stereotactic radiotherapy system. *Int. J. Radiat. Oncol. Biol. Phys.* 38: 875–882.

Boyer A.L., Geis P., Grant W., Carol M. (1997) Modulated beam conformal therapy for head and neck tumors. *Int. J. Radiat. Oncol. Biol. Phys.* 39: 227–236.

Boyer A., Xing L., Ma C-M., Curran B., Hill R., Kania A., Bleier A. (1999) Theoretical considerations of monitor unit calculations for intensity modulated beam treatment planning. *Med. Phys.* 26: 187–195.
Brahme A. (1988) Optimization of stationary and moving beam radiation therapy techniques. *Radiother. Oncol.* 12: 129–140.
Brahme A., Roos J.E., Lax I. (1982) Solution of an integral equation encountered in rotation therapy. *Phys. Med. Biol.* 27: 1221–1229.
Burman C., Chui C., Kutcher G., Leibel S., Zelefsky M., LoSasso T., Spirou S. et al. (1997) Planning, delivery, and quality assurance of intensity modulated radiotherapy using dynamic multi-leaf collimator: A strategy for large-scale implementation for the treatment of carcinoma of the prostate. *Int. J. Radiat. Oncol. Biol. Phys.* 39: 863–873.
Cadman P., McNutt T., Bzdusek K. (2005) Validation of physics improvements for IMRT with a commercial treatment-planning system. *J. Appl. Clin. Med. Phys.* 6: 74–86.
Cadman P., Bassalow R., Sidhu N.P., Ibbott G., Nelson A. (2002) Dosimetric considerations for validation of a sequential IMRT process with a commercial treatment planning system. *Phys. Med. Biol.* 47: 3001–3010.
Carol M.P. (1995) A system for planning and rotational delivery of intensity-modulated fields. *Int. J. Imaging Syst. Tech.* 6: 56–61.
Cho K.H., Hall W.A., Lo S.S., Dusenbery K.E. (2004) Stereotactic radiosurgery versus fractionated stereotactic radiotherapy boost for patients with glioblastoma multiforme. *Technol. Cancer Res. Treat.* 3: 41–49.
Chui C., LoSasso T., Palm A. (2003) Computational algorithms for independent verification of IMRT. Chapter 5 in *A Practical Guide to Intensity-Modulated Radiation Therapy*. Medical Physics Publishing, Madison, Wisconsin, pp. 83–102.
Convery D., Rosenbloom M. (1992) The generation of intensity-modulated fields for conformal radiotherapy by dynamic collimation. *Phys. Med. Biol.* 37: 1359–1374.
Cozzi L., Fogliata A., Lomax A., Bolsi A. (2001) A treatment planning comparison of 3D conformal therapy, intensity modulated photon therapy and proton therapy for treatment of advanced head and neck tumors. *Radiother. Oncol.* 61: 287–297.
Curran B. (2003) IMRT delivery using serial tomotherapy. In *Intensity-Modulated Radiation Therapy: The State of the Art.* J. Palta and T. R. Mackie (ed.). Medical Physics Publishing, Madison, Wisconsin, pp. 221–245.
Dawson L.A., Anzai Y., Marsh L., Martel M.K., Paulino A., Ship J.A., Eisbruch A. (2000) Patterns of local-regional recurrence following parotid sparing conformal and segmental intensity-modulated radiotherapy for head and neck cancer. *Int. J. Radiat. Oncol. Biol. Phys.* 46: 1117–1126.
Ding M., Newman F., Kavanagh B., Stuhr K., Johnson T., Gaspar L. (2006) Comparative dosimetric study of three-dimensional conformal, dynamic conformal arc, and intensity-modulated radiotherapy for brain tumor treatment using Novalis system. *Int. J. Radiat. Oncol. Biol. Phys.* 66: S82–86.
Dogan N., Glasgow G.P. (2003) Surface and build-up region dosimetry for obliquely incident intensity modulated radiotherapy 6 MV x-rays. *Med. Phys.* 30: 3091–3096.
Ezzell G., Galvin J., Low D., Palta J. (2003) Guidance document on delivery, treatment planning, and clinical implementation of IMRT: Report of the IMRT subcommittee of the AAPM radiation therapy committee. *Med. Phys.* 30: 2089–2115.
Fenwick J.D., Tomé W.A., Jaradat H.A., Hui S.K., James J.A., Balog J.P., DeSouza C.N., Lucas D.B., Olivera G.H., Mackie T.R., Paliwal B.R. (2004) Quality assurance of a helical tomotherapy machine. *Phys. Med. Biol.* 49: 2933–2953.
Grant W., Woo S.Y. (1999) Clinical and financial issues for intensity modulated radiation therapy delivery. *Semin. Radiat. Oncol.* 9: 99–107.
Greer P.B., Popescu C.C. (2003) Dosimetric properties of an amorphous silicon electronic portal imaging device for verification of dynamic intensity modulated radiation therapy. *Med. Phys.* 30: 1618–1627.
Grigorov G., Kron T., Wong E., Chen J., Solazzo J., Rodrigues G, (2003) Optimization of helical tomotherapy treatment plans for prostate cancer. *Phys. Med. Biol.* 48: 1933–1943.

Hartigan J.A. (1975) *Clustering Algorithms*. Wiley, New York.
Holmes T., Mackie T.R., Simpkin D., Reckwerdt P. (1991) A unified approach to the optimization of brachytherapy and external beam therapy. *Int. J. Radiat. Oncol. Biol. Phys.* 20: 859–873.
ICRU (International Commission on Radiation Units and Measurements) (1999) Report 62: Prescribing, recording, and reporting photon beam therapy (Supplement to ICRU report 50). ICRU, Bethesda, Maryland.
Kalender W.A., Polacin A. (1991) Physical performance characteristics of spiral CT scanning. *Med. Phys.* 18: 910–915.
Kestin L.L., Sharpe M.B., Frazier R.C., Vicini F.A., Yan D., Matter R.C., Martinez A.A., Wong J.W. (2000) Intensity modulation to improve dose uniformity with tangential breast radiotherapy: Initial clinical experience. *Int. J. Radiat. Oncol. Biol. Phys.* 48: 1559–1569.
Kooy H., van Herk M., Barnes P., Alexander E. III, Dunbar S.F., Tarbell N.J., Mulkern R.V., Holupka E.J., Loeffler J.S. (1994) Image fusion for stereotactic radiosurgery and radiotherapy treatment planning. *Int. J. Radiat. Oncol. Biol. Phys.* 28: 1229–1234.
Larsson B., Linden K., Sarby B. (1974) Irradiation of small structures through the intact skull. *Acta Radiol. Ther. Phys. Biol.* 13: 512–534.
Lee N., Xia P., Fischbein N., Akazawa P., Quiveley J. (2003) Comment from author on skin toxicity due to intensity-modulated radiotherapy for head and neck cancer. *Int J. Radiat. Oncol. Biol. Phys.* 55: 1150–1151.
Leksell L. (1951) The stereotaxic method and radiosurgery of the brain. *Acta Chir. Scand.* 102: 316–319.
Ling C., Leibel S., Fuks Z., Humm J. (2003) *A Practical Guide to Intensity-Modulated Radiation Therapy*. Medical Physics Publishing, Madison, Wisconsin.
Ling C.C., Humm J., Larson S., Amols H., Fuks Z., Leibel S., Koutcher J.A. (2000) Towards multidimensional radiotherapy (MD-CRT): Biological imaging and biological conformality. *Int J. Radiat. Oncol. Biol. Phys.* 47: 551–560.
Lomax A., Bortfeld T., Goitein G., Debus J. et al. (1999) A treatment planning intercomparison of proton and intensity modulated photon radiotherapy. *Radiother. Oncol.* 51: 257–271.
LoSasso T., Chui C., Ling C. (2001) Comprehensive quality assurance for the delivery of intensity modulated radiotherapy with a multileaf collimator used in dynamic mode. *Med. Phys.* 28: 2209–2219.
LoSasso T., Chui C., Ling C. (1998) Physical and dosimetric aspects of a multileaf collimation system used in the dynamic mode for implementing intensity modulated radiotherapy. *Med. Phys.* 25: 1919–1927.
Low D., Harms W., Mutic S., Purdy J, (1998) A technique for the quantitative evaluation of dose distributions. *Med. Phys.* 25: 656–661.
Mackie T.R. (2006) History of tomotherapy. *Phys. Med. Biol.* 51: R427–453.
Mackie T.R., Hughes J., Olivera G., Kapatoes J., Ruchala K., Ramsey C., Kissick M., Jeraj R. (2004) The delivery time for helical tomotherapy. In *Proceedings of the XII International Conference on the Use of Computers in Radiation Therapy*, Seoul, Korea, 2004, pp 27–30.
Mackie T.R., Holmes T.W., Swerdloff S., Reckwerdt P.J., Deasy J.O., Yang J., Paliwal B.R., Kinsella T.J. (1993) Tomotherapy: A new concept for the delivery of dynamic conformal radiotherapy. *Med. Phys.* 20: 1709–1719.
Maes F., Collignon A., Vandermeulen D., Marchal G., Suetens P. (1997) Multimodality image registration by maximization of mutual information. *IEEE Trans. Med. Imaging* 16: 187–198.
Meeks S.L., Harmon J.F., Langen K.M., Willoughby T.R., Wagner T.H., Kupelian P.A. (2005) Performance characterization of megavoltage computed tomography imaging on a helical tomotherapy unit. *Med. Phys.* 32: 2673–2681.
Metcalfe P., Tangbooduangjit P., White P. (2004a) Intensity modulated radiation therapy: overlapping co-axial modulated fields. *Phys. Med. Biol.* 49: 1–9.
Metcalfe P., Chapman A., Arnold A., Arnold B., Tangboonduangjit P., Capp A., Fox C. (2004b) Intensity-modulated radiation therapy: Not a dry eye in the house. *Australas. Radiol.* 48: 35–44.

Mohan R., Arnfield M., Tong S., Wu Q., Sieber J. (2000) The impact of fluctuations in intensity patterns on the number of monitor units and the quality and accuracy of intensity modulated radiotherapy. *Med. Phys.* 27: 1226–1237.

Olivera G.H., Shepard D.M., Reckwerdt P.J., Ruchala K., Zachman J., Fitchard E.E., Mackie T.R. (1998) Maximum likelihood as a common computational framework in tomotherapy. *Phys. Med. Biol.* 43: 3277–3294.

Perez C. (1997) 3D conformal therapy and potential for IMRT in localized carcinoma of the prostate. In T*he Theory & Practice of Intensity Modulated Radiation Therapy*. E.S. Sternick (ed.). Advanced Medical Publishing, Madison, Wisconsin.

Podgorsak E.B., Oliver A., Pla M. (1988) Dynamic stereotactic radiosurgery. *Int. J. Radiat. Oncol. Biol. Phys.* 14: 115–126.

Preiser K., Bortfeld T., Hartwig K., Schlegel W., Stein J. (1997) A new program for inverse radiotherapy planning. In *Proceedings of the XII International Conference on the Use of Computers in Radiation Therapy*, San Francisco, California, 2007, pp. 425–428.

Price S., Williams M., Butson M., Metcalfe P. (2006) Measurement and comparison of IMRT and conventional skin dose using radiochromic film. *Australas. Phys. Eng. Sci. Med.* 29: 272–277.

Ruchala K.J., Olivera G.H., Kapatoes J.M. (2002) Limited-data image registration for radiotherapy positioning and verification. *Int. J. Radiat. Oncol. Biol. Phys.* 54: 592–605.

Ruchala K.J., Olivera G.H., Schloesser E.A., Mackie T.R. (1999) Megavoltage CT on a tomotherapy system. *Phys. Med. Biol.* 44: 2597–2621.

Shepard D.M., Olivera G.H., Reckwerdt P.J., Mackie T.R. (2000) Iterative approaches to dose optimization in tomotherapy. *Phys. Med. Biol.* 45: 69–90.

Shepard D.M., Olivera G., Angelos L., Sauer O., Reckwerdt P., Mackie T.R. (1999) A simple model for examining issues in radiotherapy optimization. *Med. Phys.* 26: 1212–1221.

Siochi R.A. (1999) Minimizing static intensity modulation delivery time using an intensity solid paradigm. *Int. J. Radiat. Oncol. Biol. Phys.* 43: 671–680.

Stein J., Mohan R., Wang X. (1997) Number and orientations of beams in intensity-modulated radiation treatments. *Med. Phys.* 24: 149–160.

Stein J., Bortfeld T., Doerschel B., Schlegel W. (1994) Dynamic x-ray compensation for conformal radiotherapy by means of multi-leaf collimation. *Radiother. Oncol.* 32: 163–173.

Swerdloff S., Holmes T., Mackie T.R. (1994) Multi-leaf radiation attenuator for radiation therapy. U.S. Patent 5351280.

Tangboonduangjit P., Metcalfe P., Butson M., Quach K.Y., Rosenfeld A. (2004) Matchline dosimetry in step and shoot IMRT fields: A film study. *Phys. Med. Biol.* 49: N287–N292.

Thomas S.J., Hoole A.C. (2004) The effect of optimization on surface dose in intensity modulated radiotherapy (IMRT). *Phys. Med. Biol.* 49: 4919–4928.

Van Esch A., Depuydt T., Huyskins D. (2004) The use of an aSi-based EPID for routine absolute dosimetric pre-treatment verification of dynamic IMRT fields. *Radiother. Oncol.* 71: 223–234.

Williams M., Metcalfe P. (2006) Verification of a rounded leaf-end MLC model used in a radiotherapy treatment planning system. *Phys. Med. Biol.* 51: N65–N78.

Winston K.R., Lutz W. (1988) Linear accelerator as a neurosurgical tool for stereotactic radiosurgery. *Neurosurgery* 22: 454–464.

Wu Q., Manning M., Schmidt-Ullrich R., Mohan R. (2000a) The potential for sparing parotids and escalation of biologically effective dose with intensity-modulated radiation treatments of head and neck cancers: A treatment design study. *Int. J. Radiat. Oncol. Biol. Phys.* 46: 195–205.

Wu Q., Arnfield M., Tong S., Wu Y., Mohan R. (2000b) Dynamic splitting of large intensity-modulated fields. *Phys. Med. Biol.* 45: 1731–1740.

Wu Y., Yan D., Sharpe M., Miller B., Wong J. (2001) Implementing multiple static field delivery for intensity-modulated beams. *Med. Phys.* 28: 2188–2197.

Questions

7.1 Explain briefly how the coordinate system defined by a stereotactic head ring and BRW localizer allows us to position the linac isocenter at an anatomical isocenter as chosen during treatment planning.
(section 7.3)

7.2 Explain the basic principle behind frameless stereotactic targeting using either a camera-based or x-ray image-based system.
(section 7.4)

7.3 For a single isocenter SRS treatment, 5 arcs are to be used with a circular collimator. Arcs 1–5 are respectively 100, 80, 120, 60, and 150 degree arcs and each arc is to use the same "weight" (dose per degree will be the same for all beams). If the total dose to be delivered is 18 Gy, how much dose will be delivered by arc 1?

If the output factor for this beam is 0.9 and the average TMR value for the arc at isocenter is 0.7, how many MU will be required for arc 1?
(section 7.5)

7.4 A CyberKnife planner wants to achieve a conformal dose to an irregular target that is about 2 cm across. What are a few strategies available to achieve this?
(section 7.7)

7.5 A planner wants to keep the minimum PTV dose to a prescribed level (avoid "cold spots" in the target) but at the same time has to contend with an abutting organ at risk that cannot tolerate more than a small volume receiving near the prescription dose. Discuss the pertinent issues and come up with a reasonable strategy in terms of setting dose goals and penalties to achieve an acceptable solution?
(section 7.10)

7.6 Discuss the perceived disadvantages of high monitor to unit dose ratio in IMRT.
(section 7.4)

7.7 List all the main steps in the IMRT radiotherapy treatment planning and dose verification processes required before patient treatment proceeds.
(sections 7.11–7.12 and section 11.4)

7.8 Assume a tomotherapy treatment plan using a 2.5 cm collimator width and a pitch of 0.5 can be delivered in 10 gantry rotations, each taking 40 seconds. If the planner decides to redo the plan with a pitch of 0.3 in order to improve the resolution of the optimization in the hope of coming up with a better plan, what is the likely effect on gantry rotation time and overall treatment time?
(section 7.18)

8 Calibration of Megavoltage Photon and Electron Beams

The nice thing about standards is that there are so many to choose from.
(Attributed to several, last mentioned by G. Sherouse, Global Medical Physics Mailing List, February 23, 2007)

8.1 Introduction

Dosimetry can be divided into two different tasks:

i) Relative dosimetry, where dose measured at a point of interest under certain irradiation conditions is compared with the dose given at a reference point under reference conditions.
ii) Absolute dosimetry, where the dose at this reference point is defined. This is also commonly referred to as *calibration*.

Examples for relative dosimetry are profile and depth dose measurements (varying the point of interest) and field size factor or wedge factor determination (varying the irradiation conditions). This is discussed in detail in chapter three. Since relative dosimetry is just establishing a dose ratio between a point of interest and the reference point, great importance has to be placed on absolute dosimetry; if at the reference point 1 Gy is not 1 Gy, all other measurements will be incorrect as well.

Therefore, a strict set of procedures is followed for the calibration of all radiation beams. A protocol sets a standard that ensures that all centers using the same protocol follow the same procedures and use the same data sets in their dose calibration. Unfortunately, there is not one single calibration protocol, and different centers in different countries follow different protocols. This chapter describes some of the more recent protocols used for the dosimetry of megavoltage (MV) x-rays and electrons from linacs. **It is not intended to be used as a protocol, and the user must follow the appropriate protocol when calibrating a radiotherapy treatment unit.**

In the last few years an important transition has occurred in the calibration of megavoltage radiotherapy beams (AAPM 1999, Andreo et al. 2002, IAEA 2000 [available on the Internet as version 11.b (2004) at http://www-naweb.iaea.org/nahu/dmrp/pdf_files/CoPV11b.pdf]). Older protocols such as the protocols of AAPM task groups 21 and 25 (AAPM 1983, AAPM 1991) were based on calibration factors relating to air kerma (kinetic energy released per unit mass) (or exposure) in cobalt-60 beams. These have been largely replaced by protocols that are based on dose-to-water calibration factors such as the protocol of AAPM TG 51 (AAPM 1999) and the IAEA Technical Report Series (TRS) 398 (IAEA 2000). These protocols are considerably easier to implement for the user and consequently they have found widespread acceptance and the present chapter will focus on them.

The present chapter is largely based on parts of the first edition of Metcalfe, Kron, and Hoban (1997) and chapter 9 in Van Dyk's *The Modern Technology of Radiation Oncology, Volume 2* (2005). It will start by providing some background information on ionization chambers and other equipment used for calibration in radiotherapy departments. This includes a discussion on calibration factors and their traceability and complements the material on ionization chambers in chapter three, sections 3.4 and 3.5. The second section of this chapter is concerned with the historical development of calibration protocols. Despite being more than 20 years old, the AAPM Task Group 21 (TG 21) protocol (AAPM 1983, 1984) is still worth reading as it provides excellent background for understanding the process of absolute dosimetry with ionization chambers. It will therefore be discussed in some depth. After a presentation of the two most important current protocols, the AAPM TG 51 and the IAEA TRS 398 protocol, their differences to previous protocols and each other will be discussed. The chapter is complemented by a discussion of uncertainty and quality assurance (QA) including a brief description of dosimetric intercomparisons.

A few general comments need to be made at the onset of discussing calibration protocols:

- The present chapter describes ideas from many protocols. However, every user must only use one. As a protocol is as much a procedural and legal document as a scientific one, it is inappropriate to try to *improve* on a protocol. All protocols should be used in their entirety and no parts of one protocol should be combined with another.
- Unfortunately, the nomenclature in different protocols is not identical even when it refers to the same physical phenomena. The present chapter aims to use a single nomenclature based on recent protocols. In general, this nomenclature is used even if equations are reproduced from different protocols.

- The term *calibration* can be used in two different meanings: (1) for the determination and certification of a calibration factor for a particular ionization chamber by the National Standards Laboratory (NSL) and (2) for the act of determining the relationship between the dose at a reference point in a given radiation quality using the "calibrated" ionization chamber and relating this dose to a selectable parameter on a treatment unit (e.g., beam-on time or number of monitor units [MU]). The first is the calibration of an instrument, while the second is referred to as calibration of a radiation beam. Protocols assume the former has been done (i.e., a calibration factor is available) and describe how the latter is done.
- This chapter is restricted to the dosimetry of electrons and photons with maximum energies above 1 MeV. As such, photons from cobalt-60 units are included but not orthovoltage and superficial x-ray beams. Similarly, neither brachytherapy sources, neutrons nor heavy ions and protons is discussed. The reader is referred to IAEA (2000) and Ma et al. (2001) for these purposes.
- And finally again: **The contents of the present chapter must not be used for calibration.** Therefore, no data or tables are given which could be used for this purpose.

There are a number of general publications on absolute dosimetry (Mijnheer et al. 1993, Kron 2005), but none of them can replace the actual protocols which mostly offer ample explanation and reasoning on the procedures recommended. Another resource is the webpage of the Radiological Physics Center in Houston, TX (http://rpc.mdanderson.org/rpc/), which has practical tips on implementation of AAPM protocols.

8.2 Absolute Dosimetry with Ionization Chambers

Absolute dosimetry requires the knowledge of a fundamental physical relation between the measured output of a detector, M, and the quantity of interest, the radiation dose, D:

$$D = p\, c\, M, \tag{8.1}$$

with p representing a combination of correction factors (see section 8.2.3) and c a fundamental physical quantity which relates detector response to dose. Examples for this are chemical dosimetry where the yield of a chemical reaction G would be the fundamental quantity and ion chamber measurements where the mean energy extended in air per ion pair created; W/e is this quantity. An even more fundamental process is calorimetry where energy in the form of heat is directly measured as a result of deposition of dose. In this case, only the heat defect for the material in question must be known as well as the respective mass (compare chapter three).

The present chapter does not deal with chemical dosimetry or calorimetry, even if these techniques can in principle be used for beam calibration from first principles. In radiotherapy practice, however, virtually the only technique used to calibrate megavoltage radiation beams is ionization chamber dosimetry and all protocols are focused on this.

8.2.1 Calibration Factors

It would, in principle, be possible to convert the charge collected in the ionization chamber into a dose reading following the discussion in chapter three and the scheme illustrated in figure 3.17. This figure is reproduced and expanded here as figure 8.1. However, absolute dosimetry from first principles is typically not done and the fundamental constant, c, is combined with some correction factors in equation 8.1 to a calibration factor, N, which allows the user to convert their electrometer reading into a dose measure after applying appropriate correction factors.

The calibration factor N:

- Is issued by a National Standards Laboratory (such as NIST [National Institute of Science and Technology] in the United States).
- Is determined by the NSL after the user has sent in the ionization chamber for calibration.
- Is specific for a particular ionization chamber; therefore, not only the chamber type but also its serial number must be recorded on the calibration certificate. In practice, the calibration factor typically pertains to a combination of ionization chamber and electrometer.
- Can link the dose in the user's beam to different physical quantities such as kerma, exposure, or dose. All of these are typically relating to the measurement conditions (radiation quality) in the standards laboratory. The type of calibration factor is identified by a subscript index (D for dose, X for exposure, and K for kerma).

This is illustrated in figure 8.1. It shows quantities used in dosimetry with ionization chambers. An ionization chamber detects charge and therefore mass ionization in air. There are three conceptual steps to convert this to the dose reading in water, which is desired for clinical dosimetry:

i) Conversion of charge to energy deposited: this is the constant c in equation 8.1 with $c = W/e$ for ionometry.
ii) Conversion of dose assessed in air to dose in water: this is at the heart of Bragg-Gray cavity theory. The dose in water (index w) can be determined from the dose deposited in the detector material (dm) using the ratios of the restricted mass stopping power ratios, S/ρ, as:

$$D_w = D_{dm}[(S/\rho)_w/(S/\rho)_{dm}]. \tag{8.2}$$

In most ionization chambers the detector material is air (index a), in which case equation 8.2 is typically abbreviated to $D_w = D_{dm}\, s_{wa}$.

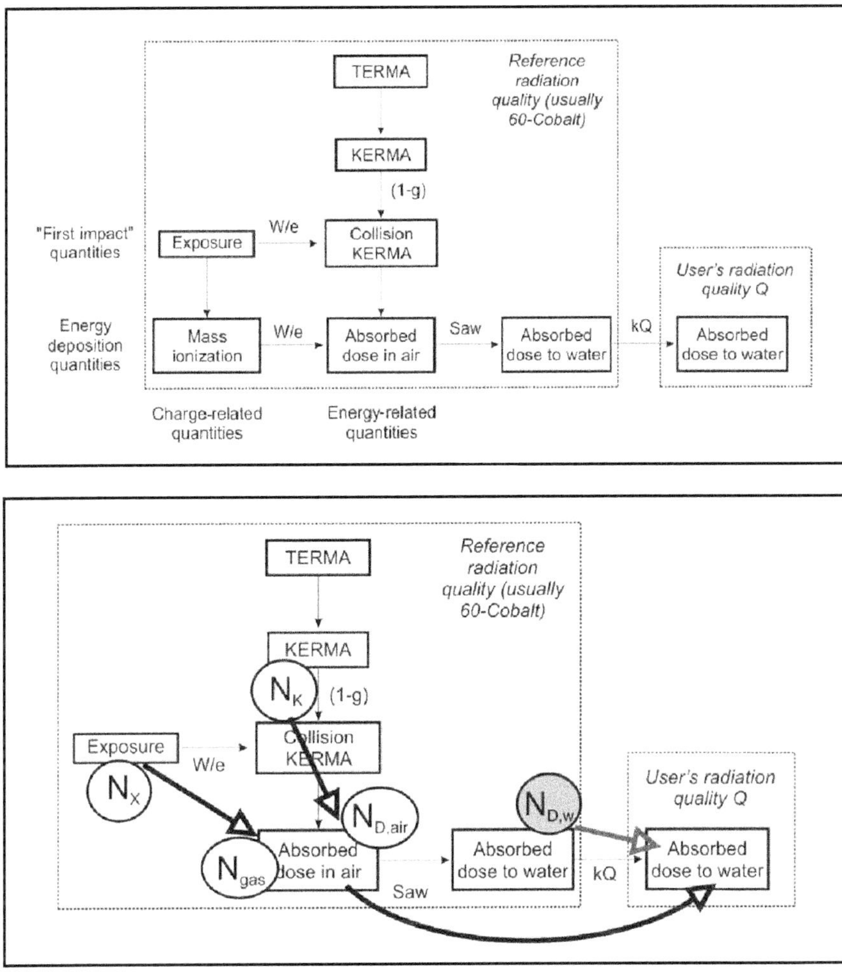

Figure 8.1: Quantities used in radiation dosimetry with ionization chambers: (a) The figure illustrates three steps that may be required during calibration: conversion of charge to energy-related quantities; conversion of a measurement in air to one in water; and correction for a difference in radiation quality between the standards laboratory and the user. (b) Calibration factors associated with the scheme shown in (a). (Adapted from Van Dyk 2005).

This is similar to equation 3.5 in chapter three. More details on the formalism are provided in the context of the description of the AAPM TG 21 protocol (AAPM 1983).

iii) Figure 8.1b shows the variety of calibration factors that may be provided to the user by the NSL. They all refer to a different physical quantity. However, all calibration factors are determined by the NSL for the radiation quality available at the NSL, typically cobalt-60 radiation. As such, a further correction is required to convert for the differences of chamber response in the radiation quality at the NSL and the one of the user. In the most recent protocols (AAPM

TG 51 1999, IAEA TRS 398 2000), which are based on the availability of an absorbed-dose-to-water calibration factor, N_{Dw}; this is the only correction required. All other corrections have been taken into consideration by the NSL, a fact that makes the application of these protocols significantly easier for the user.

It is important to note that the applicability of a particular protocol depends on which calibration factor can be provided by the relevant standards laboratory. If no dose-to-water calibration factor is available, TG 51 or IAEA TRS 398 cannot be used.

8.2.2 The International Framework and Traceability

Calibration of radiation beams shall ensure that every treatment unit delivers the same radiation dose when 1 Gy is prescribed. This requires all calibration factors to be traceable to a primary standard where dose can be measured from first principles. Laboratories that have the resources and expertise to set up and maintain a primary standard for dose are called Primary Standard Dosimetry Laboratories (PSDL). Examples for these are Australian Radiation Protection and Nuclear Safety Agency (ARPANSA), NIST, National Physical Laboratory, UK (NPL), U.S. Navy National Research Laboratory (NRL), and Physikalisch-Technische Bundesanstalt, Germany (PTB) (IAEA 2000). It is important to note though that a PSDL may have established one primary dose standard (e.g., dose for megavoltage radiation using calorimetry) but not another (e.g., exposure for kilovoltage radiation using a free air ionization chamber (Reich 1990).

PSDLs intercompare their dose standard on a regular basis. Most of them are part of the International Metre Convention signed in 1875, which established the Bureau International des Poids et Mesures (BIPM) in Paris. From PSDLs or the BIPM, the calibration can be transferred to a secondary

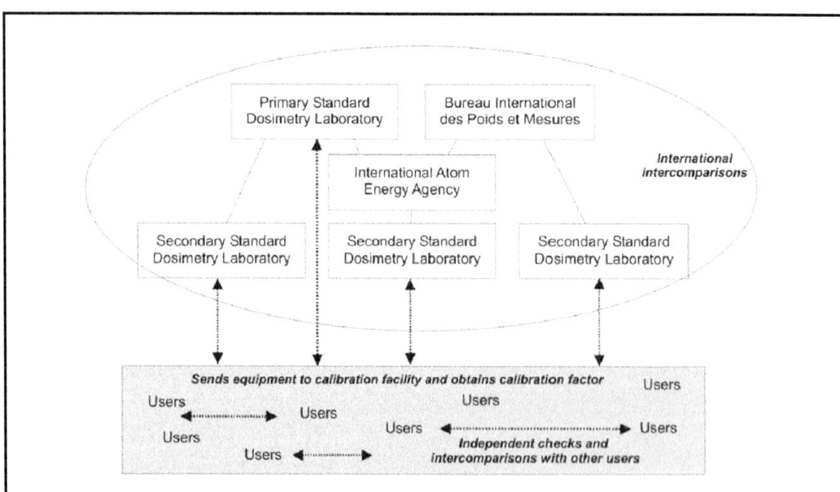

Figure 8.2: International system of standards laboratories.

standard via direct comparison. A secondary standard for dosimetry is maintained by a Secondary Standard Dosimetry Laboratory, or SSDL. In 1976, the World Health Organization (WHO) and the International Atomic Energy Agency (IAEA) jointly established a network for SSDLs in order to disseminate calibrations for dosimetry and ensure traceability. The user of ionizing radiation in radiotherapy can then directly interact with a PSDL or an SSDL as appropriate in their national regulatory framework. Figure 8.2 illustrates this process, which is described in more detail in (IAEA 2000).

8.2.3 Correction Factors

The correction factors p required in equation 8.1 depend on the type of protocol and the basics for the calibration factor. In general, the user has to apply less corrections if he/she already has a calibration factor based on absorbed dose as per TG 51 and TRS 398 protocols. Looking at all possible areas where corrections may be required one can distinguish four different sets of corrections: chamber, air calibration, phantom, and radiation quality correction factors.

Chamber Correction Factors

All protocols require the use of chamber correction factors. They have been discussed in chapter three and correct for environmental influences (such as temperature, pressure, and humidity), recombination losses, and the polarity effect. In addition to this, factors may be applied to account for different settings chosen on the electrometer, P_{range}, and the fact that a chamber may be calibrated separately from an electrometer. In the latter case, a factor, P_{elect}, can be applied to correct for the electrometer reading to charge conversion. Combining all the chamber correction factors the saturation charge collected on an electrometer corrected to standard environmental conditions is:

$$R_{corr} = R_{raw} P_{pT} P_{hum} P_{ion} P_{pol} P_{elect} P_{range} \tag{8.3}$$

A summary of these factors is given in table 8.1 for the three protocols discussed in more depth in this chapter.

When performing calibration measurements it is assumed that all charge recorded has originated in the active volume of the ionization chamber. Charges that may originate from other parts of the measurement assembly in the radiation beam contribute to the so-called "stem effect." By using elongated radiation fields with the chamber in different orientation (Kron 2005) and by placing loops of cable purposely in the radiation beam, it is relatively easy to identify and isolate any stem effect. It should be small for modern equipment. When applying equation 8.3 it is understood that stem effect and other influences on the chamber reading such as the guarding and venting (compare chapter three) have been corrected for.

Air-Calibration Correction Factors

If the user obtains a kerma or exposure calibration factor N_K or N_X as required for the AAPM TG 21 (AAPM 1983) and the IAEA TRS 277 (IAEA 1987) protocols, a set of *air-calibration correction factors* must be

Table 8.1: Chamber correction factors for three calibration protocols

	Symbol	*IAEA TRS 398 (2000) Calibration factor based on dose to water*	*AAPM TG 21 (1983) Calibration factor based on air kerma or exposure*	*AAPM TG 51 (1999) Calibration factor based on dose to water*
Temperature and pressure	P_{pT}	$(273.2+T) / (273.2+T_{ref} \times p_{ref}/p$	$(273.2+T / 295 \times 760/p$	$(273.2+T) / (295.2) \times 101.33/p$
Reference temperature (°C)	T_{ref}	As used by standards lab, generally 20	22	22
Reference pressure	p_{ref}	As used by standards lab, generally 101.3 kPa	760 mmHg	101.33 kPa
Recombination: in all cases determined using a two-voltage method	P_{ion}	$a_0 + a_1 (R_1/R_2) + a_2 (R_1/R_2)^2$ with a_i constants given in the protocol and R_1 and R_2 charge readings at two different voltages; small differences in the constants between TRS 277 and TRS 398 exist. The correction factor is called k_s in the protocol.	Included in the formula for calculation of dose, not the chamber reading correction taken from fig. 4 of the protocol after taking two readings at different voltages.	$(1 - V_H/V_L) / (R^H_{raw} / R^L_{raw} - V_H/V_L)$ with V as the voltage applied against H and L indicating high and low.
Polarity	P_{pol}	$(M^+_{raw} - M^-_{raw}) / 2M_{raw}$ (the protocol uses the terms M_+ and M_-.)	Readings are assumed to be the average of readings obtained with positive and negative polarity.	$(M^+_{raw} - M^-_{raw}) / 2M_{raw}$
Humidity	P_{hum}	Assumed to be 1		
Leakage		Should be <0.1% than the measurement current.		
Electrometer	P_{elect}	Required if chamber and electrometer have been calibrated separately		
Range	P_{range}	Settings to be noted and corrected if needed.	Settings to be noted and corrected if needed.	Settings to be noted and corrected if needed.

applied (Mijnheer et al. 1993). These take into account that the calibration of the chamber at the standards laboratory is usually performed in air using a build-up cap in a cobalt-60 beam; hence an air-kerma or exposure factor. In this case, the calibration factor has been obtained in electron equilibrium conditions while the user is required to perform the measurement under Bragg-Gray conditions. Typical corrections are for absorption and scattering in the wall, build-up cap, and central electrode (k_{att}), for differences between air and wall/build up cap (k_m), and the influence of the central electrode (k_{cel}). In addition to this, potential influences of the stem on the reading (k_{stem}) have to be corrected. It depends on the protocol as to which factors are explicitly used; they may be combined with other factors or deemed not important. Table 8.2 provides a summary of air calibration factors that will be discussed in more detail in the context of the TG 21 protocol discussed in section 8.3.

Phantom Correction Factors

For megavoltage radiation all protocols require the user to perform measurements in a suitable phantom. An exception is the calibration of cobalt units where calibration in air is allowed. This takes advantage of the similar measurement conditions in the standards laboratory and the user's beam and makes it possible to calibrate the beam in air. If the calibration measurement is performed in a phantom, correction factors are needed to account for the measurement conditions in the phantom in the user's beam. As the chamber wall is not perfectly water equivalent as required for Bragg-Gray conditions, a correction, p_{wall}, is applied. Other corrections are for fluence perturbation of the finite size chamber, p_f, a correction for the effective point of measurement which is not necessarily in the center of the chamber, and a material correction for the central electrode, p_{cel}. Particular care must be taken if phantom materials other than water are used. Again the nature and value of these factors that are also listed in table 8.2 will be discussed in more detail in the context of the TG 21 protocol discussed in section 8.3.

Radiation Quality Correction Factors

In protocols based on an absorbed dose to water factor (AAPM 1999, IAEA 2000, DIN 1997) the user is only required to correct for the difference in radiation quality between the standards laboratory and his/her own beam with quality, Q. For photons this factor is commonly termed k_Q. Some standards laboratories (United Kingdom, Switzerland) have commenced to offer calibration factors obtained in reference beams other than cobalt-60. In this case, an additional index Q_0 is used in connection with k_Q to identify this radiation quality. For electron beams the factor k_Q is somewhat more complex because it must also take into account the depth of measurement.

8.2.4 Typical Hardware and Useful Tools

Phantom materials and phantom design for ionization chamber measurements have been discussed in section 3.3, as has been the basic function of

Table 8.2: Corrections to be used in exposure or air-kerma factor-based protocols for the calibration conditions in the standards laboratory. The factors are chamber specific and typically relate to a cobalt-60 beam. In the AAMP TG 21 protocol the symbol used for the in-air calibration factor is A.

	Symbol	*Typical Value*	*Comments*
a) Air calibration correction factors			
Lack of air-equivalence of chamber wall and build-up cap	k_m	0.98 to 1.02	Depends on chamber material and build-up cap; for chambers used in calibration procedures it should be close to 1.
Attenuation in wall and build-up cap	k_{att}	0.97 to 1.00; if not known, the IAEA TRS 277 protocol recommends the use of 0.99.	k_m and k_{att} are often combined and provided as a single factor.
Correction for effect of central	k_{cel}		Significant for chambers with an aluminum central electrode. In the IAEA TRS 277 protocol combined with the correction for the central electrode required in the measurement in the user's beam, p_{cel}.
Correction for stem effect	k_{stem}	Usually assumed to be 1	
Correction for ion recombination in the chamber during calibration at the standards laboratory in Co-60	k_{ion}	Close to unity	Explicitly used in AAPM TG 21 protocol as A_{ion}.
b) Phantom correction factors			
Replacement correction	P_{repl}	Unity for electrons	Includes a correction for gradient effects and effective point of measurement.
Correction for wall material	P_{wall}	Close to unity in MV beams; for electrons usually considered to be 1.	Equations 9.22 and 9.31. With α the fraction of electrons coming from the wall.
Perturbation correction factor (IAEA TRS 277)	p_u	Depends on chamber material and beam quality; for photons 0.985 to 1.015. Significant for large chambers in low-energy electron beams.	Equation 9.28.

Table 8.2: Continued

	Symbol	Typical Value	Comments
b) Phantom correction factors (continued)			
Perturbation correction factor (IAEA TRS 398)	p_Q	0.97 to 1.1	Equation 9.30
Correction for effective depth of measurement	P_d	About 0.99 for MV photons; 1 if chamber is physically displaced to effective point of measurement.	May also be corrected by a physical shift of the chamber (IAEA 1987).
Fluence perturbation correction	P_f	1 in photon beams where secondary electron equibi-rium is assumed to exist.	Significant for electrons.
Correction for central electrode material when a cylindrical chamber with aluminum electrode is used	P_{cel}	In photon beams up to 25 MV unity. In electron beams depending on radius, 1.008 to 1.032.	Includes both the effects during calibration of chamber in-air and the measurement in the user's beam as a global P_{cel}; the IAEA TRS 398 protocol uses p_{cel} only for the measurement in the standard beam.
Cavity correction (IAEA TRS 398)	p_{cav}	Typically assumed to be unity,	
Displacement correction (IAEA TRS 398)	p_{dis}	Around 0.988 for a Farmer-type chamber in Co-60 radiation.	Depends on inner radius of the cylindrical chamber.

ionization chambers and electrometers. A few additional considerations may be useful when extending this to absolute dosimetry.

Ionization Chambers

Thimble-type ionization chambers are recommended in most protocols as the transfer instrument for the calibration factor from the standards laboratory to the user. It is also the chamber of choice for calibration of photon beams and high-energy electron beams. Thimble-type ionization chambers have the principal advantage of no variation in directional response around their axis. In practice, there may be small variations though, and most cylindrical chambers have a fiducial mark to indicate a particular orientation. The mark should be kept in the same orientation for all measurements.

A good list of chambers suitable for calibration is provided in the IAEA TRS 398 protocol (IAEA 2000). It should be noted that chambers may look identical but may not be identical. The materials used in construction determine the correction factors required. A good example is the Farmer-type chamber (compare figure 3.25 in chapter three), which is the most com monly used chamber for calibration purposes. The materials given in the drawing refer to the original design by Aird and Farmer (Aird 1972); however, a variety of thimble materials are available to replace the fragile graphite and potentially provide waterproofing. In many Farmer-type chambers the central electrode is not made from aluminum but from tissue-equivalent conducting plastics. This variation in chamber design is not detectable from outside but will require a correction of the order of 0.7%, which is typically included in protocols as P_{cel} (cel = central electrode).

A correction for thimble chambers not discussed before is the effective point of measurement (IAEA 1987, Andreo 1992). In radiation fields in which the ionizing particles (electrons) have a preferential direction, the air cavity of the chamber will reduce attenuation and the effective point of measurement is moved towards the phantom surface. This is illustrated in figure 8.3. The shift of effective point of measurement, Δ_{effpt}, will depend on the radiation quality and chamber diameter. In practice, small variations in shift of effective point of measurement are largely academic as the chambers that are used for calibration have a diameter less than 8 mm.

This means that the point in the phantom that receives the dose as determined by the chamber is a few millimeters closer to the phantom surface than the center of the cylindrical chamber. Two methods are possible to correct for this:

- The chamber can be displaced for all measurements by the distance of the shift towards the surface: actual depth = stated depth $-\Delta_{effpt}$. This method should be used for all relative measurements required in the calibration process, such as percentage depth dose and tissue-phantom ratio (TPR) measurements to determine the radiation quality.
- A correction can be applied: In the AAPM TG 21 protocol (1983) this is included in a gradient correction factor, P_{grad}, which is part of the replacement correction, P_{repl}. This will be discussed in more depth

below. Both the absorbed-dose factor–based protocols, AAPM TG 51 (1999) and IAEA TRS 398 (2000), use this approach in a slightly modified form. The gradient correction is included in the general quality correction factor, k_Q.

In plane parallel ionization chambers the effective point of measurement can be assumed to be directly behind the entrance window, as can be seen in figure 8.3. Due to their better spatial resolution in beam direction, plane parallel chambers are generally recommended for calibration of electron beams with energies below 10 MeV and required for beams with an incident energy less than approximately 5 MeV. In these beams the dose gradients are so sharp that a cylindrical chamber would not be suitable for

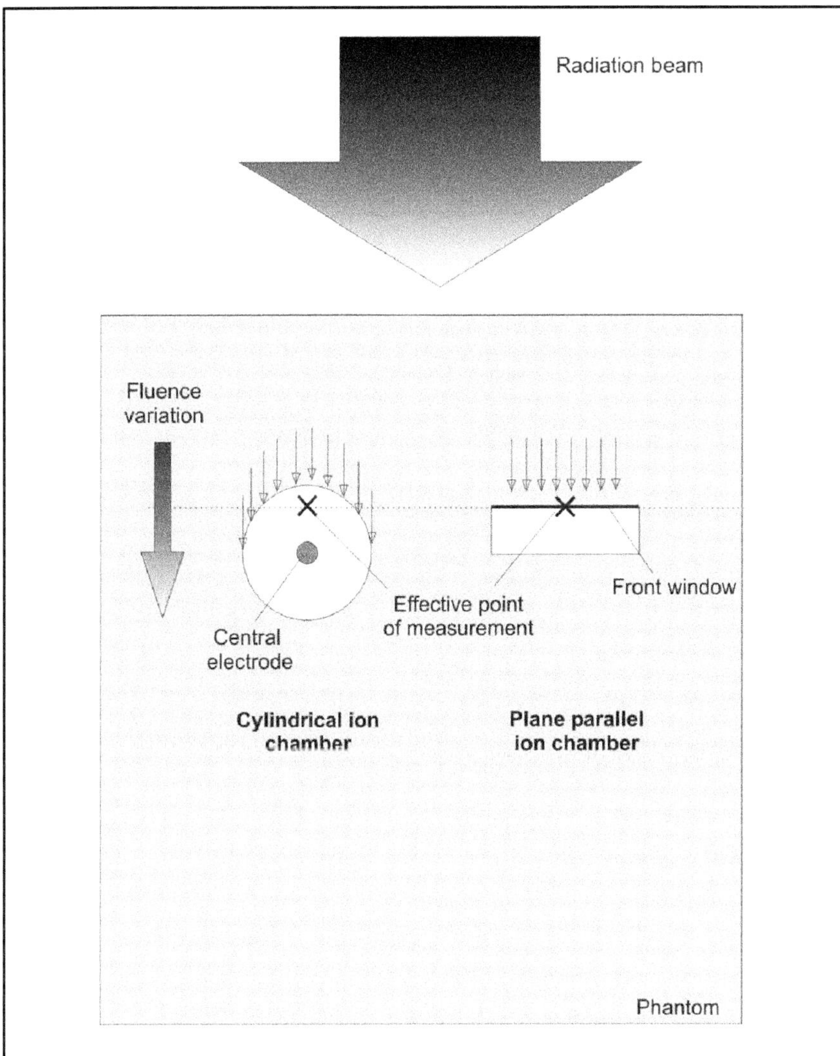

Figure 8.3: Illustration of the concept of effective point measurement. (Adapted from Van Dyk 2005.)

accurate positioning due to its diameter being always larger than the plate separation in plane parallel chambers.

Plane parallel chambers should ideally be waterproof to be suitable for measurements in a water phantom. In this case, the chamber body should also be made from water-equivalent material. However, more commonly, plane parallel chambers are used in combination with a solid slab phantom. It is advisable to use the chamber in a phantom of the same material as the one the chamber is constructed of to avoid interface problems. A lot of additional useful information on plane parallel ionization chambers and their use is given in a technical report of the IAEA devoted to plane parallel chambers (IAEA 1997).

Table 8.3 lists recommendations for characteristics of ionization chambers fit for calibration of megavoltage radiation beams.

Calibration measurements are typically performed less frequently than other QA activities (compare chapter eleven). As such, it is important to pre-irradiate all ionization chambers prior to making calibration measurements—in particular if they have not been used for some time (McCaffrey et al. 2005). It was found that depending on the type of ionization chamber the chamber response changed by up to 1% over the first 15 minutes of irradi-

Table 8.3: Requirements for ionization chambers used for megavoltage beam calibration (adapted from IPEMB 1996, IAEA 1987, 2000, Khan et al. 1991). The table is adapted from Van Dyk (2005) and provides a summary only. The requirements may differ in individual protocols.

	Cylindrical Chambers	*Plane Parallel Chambers*	*Comment*
Volume (cm^3)	0.1 to 1	0.1 to 1	
Dimensions	Inner diameter ≤7 mm Cavity length ≤25 mm	Front window thickness ≤1 mm Plate separation ≤2 mm Collecting electrode ≤20 mm Guard ring ≥3 mm	
Material	Low Z materials	Water equivalent, waterproof	
Polarity effect	≤0.2%	≤1%	
Leakage	<0.1%	≤1%	
Long-term stability	Better than 1%	Better than 1%	Graphite-walled cylindrical chambers typically have better long-term stability than plastic-walled ones.
Recombination	<5%	<5%	Voltage should be chosen high enough to ensure this is met.
Stem effect	<1%	<1%	

ation. Depending on the chamber type this effect can result in both an increase or a decrease of signal from the chamber, and a variation of change in response between different chambers of the same type has been observed.

The volume of a chamber is the most important feature that determines the number of ions that can be collected after irradiation. Uncertainty in determining this volume and variations from chamber to chamber lead to the need for determination of calibration factors. In principle, the standards laboratory determines the chamber volume by counting the number of ions created by a known amount of radiation. Assuming that W/e is constant for all types of radiation, all differences in calibration factor between chambers of the same type reflect primarily small variations in chamber volume. Ionization chambers recommended for calibration typically have a volume between 0.1 and 1 cm^3. This volume is a compromise between adequate spatial resolution and sufficient charge to be collected.

Electrometers

As discussed in section 3.5.7 a high-quality electrometer that can determine charges of a few pC accurately is required for calibration. This requires also a leakage current below 1 pA.

A high-quality electrometer is probably the most expensive item for radiotherapy beam calibration. An Institute of Physics in Engineering and Medicine (IPEM) report (Morgan et al. 2000) provides specifications and guidance as to what electrometers are suitable for calibration purposes. Based on this and other sources, table 8.4 lists some of the features to be expected from an electrometer to be used for calibration. In addition to the requirement for accuracy and high sensitivity, the electrometer needs to be stable over time. Typically, calibrations are performed on an ionization chamber/electrometer combination once every 2 years at an NSL. The electrometer used as a local standard should not change in this time period by more than 0.25%.

Prior to any calibration measurements the electrometer should be warmed up (compare chapter three, section 3.5.7), checked for short-term reproducibility, linearity of charge collection (between 10 and 200 cGy), range factors, leakage (with and without charge accumulated), and the high voltage should be verified (often part of the instrument options). Using different ionization chambers and electrometer combinations allows troubleshooting, as it is possible to distinguish between effects caused by the chamber and by the electrometer. This is particularly useful when leakage has to be checked and minimized.

Other Useful Tools for Calibration of Megavoltage Radiation Beams Using Ionization Chambers

A variety of other items are either required or helpful for calibration of megavoltage beams. As calibration is such an important step, all tools should be of high quality; for example, a good-quality steel ruler is important and a plastic ruler (which may distort with temperature) is not acceptable. Other devices that may be useful are a spirit level (to ensure good alignment of linac and dose measurement equipment), calipers, and a scale to verify the

Table 8.4: Requirements for electrometers to be used in megavoltage beam calibration Please note that these are suggestions only adapted from various calibration protocols and Morgan et al. (2000). (Table reproduced courtesy of J. Van Dyk (Van Dyk 2005).)

Feature		*Comment*
Display	At least 4.5 digits	
Maximum resolution	0.01 pA 0.1 pC	Depends on range setting.
Range	0.01 pA to >100 nA 0.1 pC to >1000 nC	Several range settings may be possible; the conversion must be verified.
Linearity	Better than ±0.1% for all ranges.	
Leakage	Should be <0.1% than the measurement current.	Should be checked for electrometer for zero and a large charge collected.
Polarizing voltage	Maximum voltage ≥400 V 0, + and – voltage must be possible. Variable; at least half voltage, preferably others as well.	Should be accurate better than 1 V. Required to determine polarity correction, zero voltage. Useful for disconnecting chambers. Required to determine recombination correction.
Input impedance (Ω)	$>10^{14}$	
Zero drift	<5 fA	Rate of change in measure mode without chamber attached; may depend on temperature and humidity.
Charge leakage	<5 fA	
Short-term stability	Electrometer change per year <0.1%.	
Long-term stability	Electrometer change per year <0.2%.	

thickness and density of slab phantom materials. If access to a diagnostic x-ray facility or a CT scanner is available, it is useful to verify the integrity of the chamber by imaging it.

A good-quality thermometer is needed as well as access to a high-quality barometer. Provided all equipment is allowed to equilibrate thermally, it is typically not a significant problem to determine temperature accurately. One should take into account though that an open water phantom is typically a degree or more colder than the surrounding air due to evaporation. If calibration is performed in an environment where humidity can change significantly, a hygrometer may be useful. In general, it is good practice to keep ionization chambers and electrometers dry when not in use. A radioactive check source such as the one discussed in chapter three (figure 3.22) can determine the constancy of a chamber electrometer reading prior to performing a calibration.

As mentioned above, cables and connectors can be sources of leakage and electrical noise. As such, it is important to keep the number of adapters to a minimum and to avoid long cables, in particular if they are looped and act as an antenna. Both cables and connectors should be shielded.

Waterproofing of chambers is essential if they are used for calibration in a water phantom. An increasing number of chambers are waterproof, however, the most commonly used Farmer-type chambers usually require protection. This can be a rubber sheath often referred to as "condom" or a thin acrylic cap that can be attached to a watertight rubber sleeve around the cable. It is important to ensure that with waterproofing the chamber must still be open to air. There has been concern that this is not always given when a rubber sheath is used (IAEA 2000). Another disadvantage of the rubber sheath is that it is often coated with a fine powder to make it easier to place and to remove the chamber from the sheath. The powder may find its way into the chamber and can affect its response. Due to these disadvantages, the IAEA TRS 398 protocol does not recommend the use of a rubber sheath for waterproofing of ionization chambers (IAEA 2000).

8.2.5 Chamber Cross Calibration

Most radiotherapy centers will only have one ionization chamber calibrated at a standards laboratory. This chamber, in combination with a calibrated electrometer, is usually referred to as *local standard*. In practice, one would like to perform most measurements not with the local standard and some calibration procedures cannot be performed with the cylindrical chamber usually used as local standard. Therefore, it is necessary to transfer the calibration factor from a local standard to a field chamber/electrometer combination, a process called *cross calibration* or intercomparison.

Most calibration protocols provide guidance and worksheets as to how the cross calibration should be performed. It is usually recommended to perform the intercomparison in a high-energy electron beam (IAEA 1997) as this type of beam does have a relatively large region with constant dose where uncertainty about the effective point of measurement and a small displacement of a chamber would have a negligible effect on the measurement (compare chapter five). If no electron beam of sufficient energy (around 18 MeV or higher) is available, the second best choice may be a cobalt-60 radiation beam due to its continuous radiation output and inherent stability.

In general, two techniques are possible. The first is a replacement technique in which the standard chamber is replaced by the field chamber positioned at exactly the same location in the radiation beam. This technique assumes that it is indeed possible to find an exactly equivalent location for two chambers, which may be different in construction, and it requires that the output of the unit does not change with time. The latter is typically verified by repeating the measurement with the standard after using the field instrument. The comparison is then performed between the field instrument reading and the average of the standard readings prior and after the field instrument reading was taken.

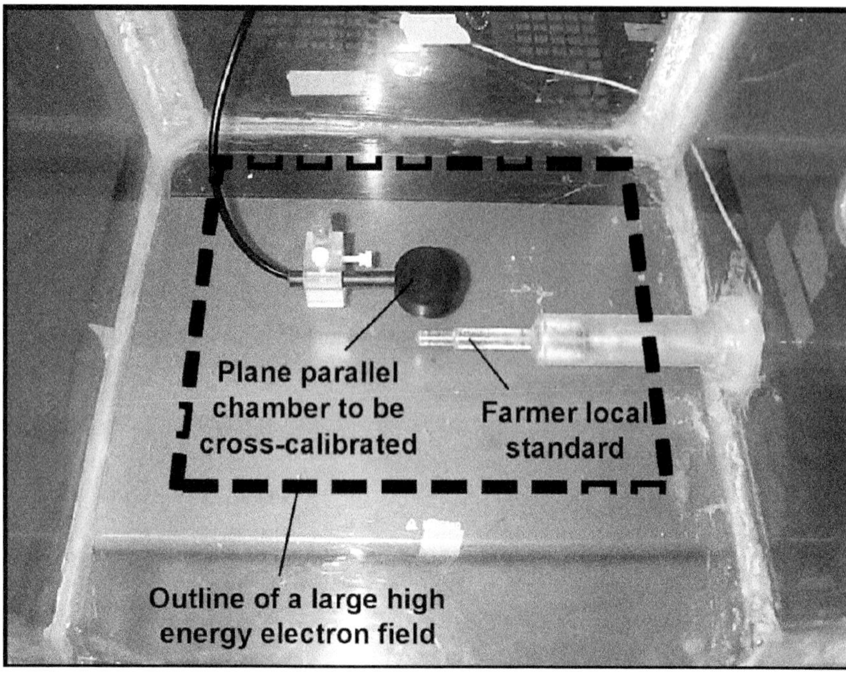

Figure 8.4: A water phantom used for cross calibration of ionization chambers. (Adapted from Van Dyk 2005.)

A different approach is to use both chambers in the same beam at two locations, which are expected to experience the same dose (e.g., at the same depth in a relatively large field). Then the two chambers are exchanged and the same comparison performed. This can easily be done by placing a phantom such as the one shown in figure 8.4 on an isocentrically mounted couch in 90° location with the isocenter between the two chambers. After rotating the couch through 180° to 270°, the two chambers have swapped locations. The average of the two comparisons can be used for the intercomparison. This approach assumes, in addition to the fact that effective point of measurement for the chamber is known, that the radiation beam profile does not change, a less stringent requirement than the assumption that the output does not alter with time.

8.3 Air-Kerma Calibration-based Protocols

8.3.1 History and Background

The history of protocols for dosimetry of radiation was driven by needs of practicing medical physicists and, as such, many protocols were developed by national or international professional organizations of medical physics. A good summary of the historical developments between 1960 and the mid 1990s is given by Nath and Huq (1995). The first scientific unit for quantification of ionizing radiation was the *roentgen* (R). At the 5th Congress of Radiologists in Chicago the International Commission on Radiation Units and Measurements (ICRU) defined that "the roentgen shall be the quantity of X- or gamma-radiation such that the associated corpuscular emission per 0.001293 g of air produces, in air, ions carrying 1 electrostatic unit of quantity of electricity of either sign." (ICRU 1938). The important concept here is that the secondary electrons (i.e., associated corpuscular emission) are not required to deposit their energy locally. This is an essential part of the dosimetry of megavoltage photon radiation, which considers the location of first impact (i.e., exposure, kerma) and the final deposition of energy (i.e., absorbed dose) separately as discussed in chapter three (compare figure 3.17 and the associated discussion).

The concepts of kerma and absorbed dose were introduced in the 1950s, following the need for a distinction between radiation field and radiation effect quantities in megavoltage radiation (Reich 1990). The unit rad (1 rad = 100 erg/g = 0.01 J/kg) was introduced for absorbed dose as a convenient way to maintain a number equivalence for roentgen and rad, in the case of megavoltage radiation interacting in water. For photons above 0.2 MeV one can assume with reasonable accuracy that 1 R = 0.97 rad. As can be seen in figure 8.1, there is a well-defined relationship between kerma and exposure and the respective calibration factors relate as:

$$N_K = N_X\,(W/e)(1-g)^{-1} \tag{8.4}$$

Since 1875 there has been an international conference for definition and maintenance of metrical units. In 1956 this *Conférence Générale des Poids et Mesures* (CGPM) introduced officially the *Système International d'Unités* (SI), which had an effect also on the units used in radiation dosimetry. The CGPM has an advisory committee, the *Comité International des Poids et Mesures* (CIPM) which prepares documents for the CGPM and supervises the *Bureau International des Poids et Mesures* (BIPM) in Paris. In 1958 a subcommittee for the CIPM was formed which specifically consults on ionizing radiation, the *Comité Consultatif pour les Etalons de Mesure des Rayonnements Ionisants* (CCEMRI). As part of these developments also the new units gray (1 Gy = 100 rad) for dose and C/kg (1 R = 2.58 10^{-4} C/kg) for exposure were adopted.

Three important sets of dosimetric protocols are worth discussing in more detail:

- The early C_λ and C_E based protocols of the Hospital Physicists' Association (HPA) in the United Kingdom. These HPA protocols are prescriptive in allowing only two chamber types to be used and as such are simple for the user to apply.
- The development of more protocols in America where a similar approach was taken by the Subcommittee on Radiation Dosimetry of the American Association of Physicists in Medicine (SCRAD). These protocols influenced the work of the ICRU.

- The scientifically more explicit and complex protocols of the Nordic Association of Clinical Physicists (NACP), which laid the framework for the IAEA protocols.

Early EPA Protocols

Considering the available radiation sources at the time a code of practice was published in 1964 for x-rays between 2 and 6 MV and cesium-137 and cobalt-60 sources (HPA 1964). This protocol introduced a process that is still commonly used: a cylindrical ionization chamber is calibrated in air with a suitable electrometer in a reference beam (typically cobalt-60) at the standards laboratory. This is performed with a build-up cap in place and yields an exposure calibration factor, N_X. The user irradiates the same chamber in a phantom at a reference depth in the radiation beam to be calibrated. In the HPA protocols, the x-ray quality is denoted as Ï. The dose in the user's beam can then be calculated from the ionization chamber reading, R_{corr}, corrected for environmental influences as:

$$D = R_{corr} N_X C_\lambda. \tag{8.5}$$

The factor C_λ is an overall correction factor that includes energy, displacement, and perturbation effects. The simplicity of the protocol is a major attraction; however, as a "price" for this simplicity the protocol is only applicable for two types of ionization chamber (Farmer-Baldwin and Victrometer). In 1969, the HPA extended the protocol to higher photon energies (up to 35 MV) and allowed other ionization chambers, provided they had a suitable composition and an internal diameter less than 10 mm (HPA 1969). Unlike the protocol of 1964, the 1969 protocol assumes all chambers of suitable design to be ideal Bragg-Gray cavities. As such, the displacement effect is not explicitly taken into account leading to uncertainties of 3% at higher photon energies. As the C_λ factor is also limited to regions of electron equilibrium, different depths were recommended as reference depth for photons with energies exceeding 10 MV. The method proposed by HPA in 1969 was also with minor modifications recommended by ICRU in 1969 in their report 14 (ICRU 1969) where different methods for megavoltage photon beam calibration were compared.

SCRAD and NACP Protocols

In North America the Subcommittee on Radiation Dosimetry of the American Association of Physicists in Medicine (SCRAD) also published protocols for megavoltage electron (SCRAD 1966) and photon (SCRAD 1971) dosimetry. In contrast to other protocols, the SCRAD protocols recommend the depth of maximum dose as the calibration depth. An advantage of this approach is that the percentage depth dose data, which was used most commonly at the time for patient dosimetry, is normalized at this point. Depth of maximum dose is also the only point in megavoltage depth dose curves where true electron equilibrium exists—the curve is flat,which leads to a minimal influence of small uncertainties in chamber positioning. Another important legacy of these protocols is the SCRAD polystyrene phantom as a convenient environment for calibration measurements.

The SCRAD electron dosimetry protocol from 1966 was extended by ICRU (ICRU 1972) and the Nordic Association of Clinical Physicists (NACP 1972). The NACP protocol included electron beams with energies up to 50 MeV to accommodate betatrons that can produce electrons of this energy. The ICRU and NACP protocols extended the C Ï algorithms familiar from photon calibration to electrons:

$$D = R_{corr} N_X C_E. \tag{8.6}$$

C_E is an absorbed dose conversion factor that can be calculated as

$$C_E = W/e\, s_{w,air} \quad p_{air} A, \tag{8.7}$$

where

- $s_{w,air}$ is the mass stopping power ratio for water to air for electrons,
- A is a correction for the attenuation in the build-up cap,
- p_{air} corrects for reduced scatter in the air cavity, and
- the average energy required to produce an ion pair in air, W/e, was assumed to be 33.7 J/C.

The terms of equation 8.7 follow the process shown in the bottom row of quantities in figure 8.1 (or figure 3.21). The protocol provides values for C_E as a function of electron energy and depth in a phantom.

In the 1970s and 1980s the equipment used for radiotherapy started to change. Fewer cobalt-60 units were installed and betatrons ran out of fashion. Linacs with multiple electron and photon energies up to around 20 MeV became available. This led to a need to revise calibration protocols and the first protocol of a new type was introduced by the Nordic Association of Clinical Physicists (NACP) in 1980 (NACP 1980). This protocol, which is an important predecessor of the IAEA protocol from 1987 (IAEA 1987), is based on an air-kerma calibration of an ionization chamber in a cobalt-60 beam at a standards laboratory. This factor can be linked to exposure as shown above in equation 8.4. The fundamental ideas of the protocol are:

I Obtain an air-kerma calibration factor for cobalt-60 from the standards laboratory.

II Determine a factor for dose to chamber air from air-kerma factor.

III Determine dose to water in a suitable phantom using Bragg-Gray cavity theory.

The NACP protocol also replaces the C-based quality correction with a more explicit way to link dose in water in the user's beam to the air-kerma calibration at the standards laboratory. The advantage of this approach is that the formalism can, in principle, be applied to all ionization chamber types. The air-kerma calibration factor can be converted into an absorbed-dose-to-air calibration factor $N_{D,air}$ according to:

$$N_{D,air} = N_K (1 - g) \, k_{att} \, k_m \tag{8.8}$$

with the correction factors k_{att} and k_{att} accounting for attenuation and scatter of primary photons in wall and build-up cap, and lack of air equivalence, respectively. This is step II above. The knowledge of $N_{D,air}$ allows the derivation of the dose in any medium from the measured dose in the air cavity $D_{air} = N_{D,air} \, R_{corr}$ with R_{corr} as the chamber reading corrected using the chamber correction factors listed in table 8.1. Using Bragg-Gray cavity theory one can determine the dose to water as step III following:

$$D_w = D_{air} \, s_{w,air} \, p_u = N_{D,air} R_{corr} \, s_{w,air} \, p_u \tag{8.9}$$

This equation is identical to equation 8.2 with the exception of the perturbation correction factor, p_u. It is necessary to include this correction to allow for lack of water equivalence in the chamber wall, the perturbation of the charged particle fluence due to the cavity, and thus the shift of the effective point of measurement in the cylindrical chamber. P_u is dependent on the chamber and the index u indicates that it refers to measurements performed in the user's beam. The AAPM TG 21 protocol (1983), which will be discussed in the next section, is based on this work.

In 1983 and 1985, the HPA issued revised protocols for photon (HPA 1983) and electron (HPA 1985) dosimetry. These protocols still maintain the simplicity of the *C*-factor–based protocols. However, the interpretation of the correction factors changed and equation [9.13] of the protocol, for example, can be reformulated as:

$$C_E = (1 - g) \, k_{att} \, k_m \, s_{w,air} \, p_u, \tag{8.10}$$

which reflects a change from exposure to an air-kerma calibration factor and formulation consistent with other protocols. Both HPA protocols are rather restrictive in the type of ionization chamber that can be used for calibration. For photon beams only the NE 2561 cylindrical chamber designed by the National Physics Laboratory (Kemp 1972) was recommended (HPA 1983). For electron beams with an energy of 10 MeV or above the Farmer-type NE 2571 graphite-walled chamber was recommended. For beams of an energy below 10 MeV a plane parallel chamber (Vinten 631) was to be used. The use of only very few chamber types is simple and results in improved precision in all participating radiotherapy centers. While it was possible to achieve this in a national environment, international organizations such as the ICRU or the IAEA have to adopt an approach that allows users from a variety of countries with different resources to participate. This is reflected in data for a large number of ionization chamber types included in these protocols.

Characterization of Beam Quality for Photons

By 1980 it was generally accepted that an x-ray beam cannot be sufficiently described by a "nominal accelerating potential" but must be characterized by measurements in the beam. This concept had already been introduced by SCRAD in 1966 when it was recommended to use ionization chamber readings at different depths in a phantom as a means to measure radiation quality. The NACP protocol uses the ratio of ionization readings at 10 cm depth and at 20 cm depth in a water phantom, termed J_{100}/J_{200}, as the quality index for photon beams. These measurements were to be performed at a fixed source to surface distance (SSD) of 100 cm.

With the changing emphasis from fixed SSD treatment setups to isocentric patient treatment, the beam quality index also changed in later protocols to a ratio of TPRs. Most commonly the ratio of the TPR with 20 cm water over the chamber to 10 cm overlaying water, TPR_{10}^{20} is determined. A field size of 10 × 10 cm^2 defined at isocenter distance is to be used with the chamber at source to axis distance (SAD) of the treatment unit (IAEA 1987, 2000). This eliminates variations of scatter due to effective field size changes at the location of the chamber during percentage depth dose measurements. In addition to being more compatible with common treatment practice, a TPR setup is also easier to achieve when materials other than water are used. In this case, just the same thickness in g/cm^2 is used of any water-equivalent phantom material.

While TPRs have a number of theoretical advantages, commercial scanning water phantoms are better suited to determine percentage depth dose than TPR (compare chapter three, section 3.3). Therefore, an empirical formula given by Followill et al. (1998) provides a useful relationship between TPR_{10}^{20} and the ratio of percentage depth dose values at 20 and 10 cm depths (%dd(20)/%dd(10)):

$$TPR_{10}^{20} = 1.2661\left(\%dd(20) / \%dd(10)\right) - 0.0595. \tag{8.11}$$

It is important to note that this equation only applies for percentage depth dose measurements at an SSD of 100 cm using a field size of 10 × 10 cm^2 from an x-ray beam produced by a clinical linac. As such, a TPR measurement is still preferable.

However, there may be a reason to go back and specify photon beam quality using percentage depth dose. In 1993, Kosunen and Rogers (1993) showed that the percentage depth dose at 10 cm depth, %dd(10), may characterize a wider range of x-ray beams consistently than TPRs, including x-ray beams produced by nonclinical linacs. While these beams are typically not of particular concern to the clinical physicist, experimental linacs may be found in standards laboratories. To take this into account, the AAPM TG 51 (1999) protocol uses the percentage depth dose at a depth of 10 cm, %dd(10), as a photon beam quality measure.

A potential problem with a percentage depth dose measurement is that the dose measurement at depth of maximum dose may be affected by electron contamination in the beam. This can particularly be a concern for high-energy photon beams and the TG 51 protocol recommends the use of a thin, lead foil in the beam for percentage depth dose measurements for x-ray beams of 10 MV or higher. Figure 8.5 shows a lead filter mounted to a tray for a linac. The function of the tray is not to completely remove the electron contamination; it merely replaces an unknown electron contamination with

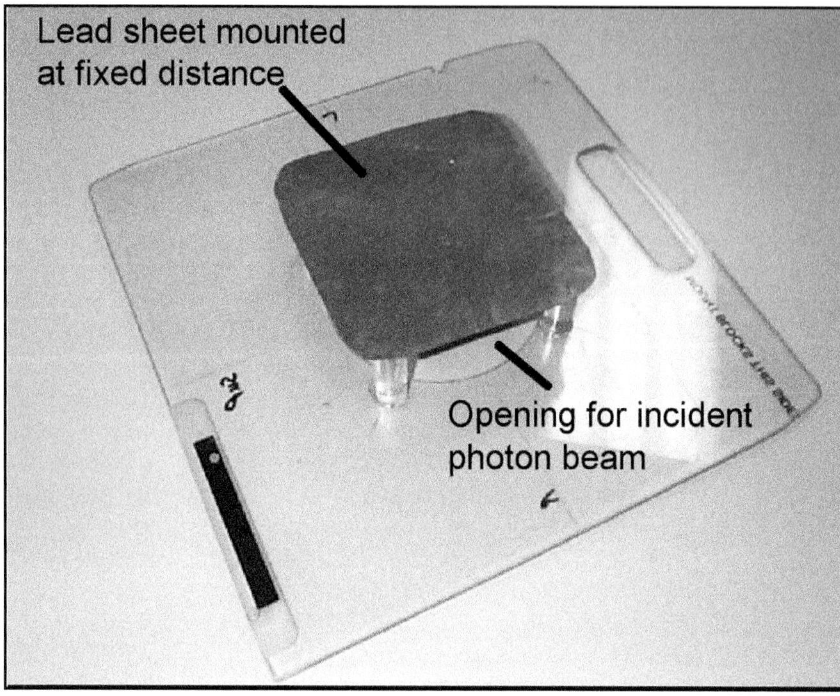

Figure 8.5: Photograph of a lead filter mounted at the recommended distance from the target on a block tray for a linac.

a known one. Therefore, it is important to place the filter at a fixed distance from the phantom surface. AAPM TG 51 (1999) allows two distances: 50 and 30 cm. The advantage of the tray shown in figure 8.5 is that it exactly positions the filter at the correct distance, here 30 cm. In order to ensure that the electron contamination is indeed well defined, this distance must be accurate within 1 cm.

Figure 8.6 shows the percentage depth dose curve measured for an 18 MV beam with and without a lead filter. The figure also shows the difference between the two measurements performed with a cylindrical chamber with 4 mm internal diameter (Scanditronix RK). For better visibility the difference is multiplied by a factor of 10. As can be expected, at the surface of the phantom the dose measured in the unfiltered beam is higher due to electron contamination of the beam. The difference demonstrates that these electrons are noticeable up to a depth of 5 cm and the difference depth dose is similar to the one from a polyenergetic electron beam with a maximum energy around 10 MeV.

The true x-ray percentage depth dose $\%dd(10)_X$ can be calculated according to the AAPM TG 51 protocol (1999) from the depth dose measured with the lead filter, $\%dd(10)_{Pb}$ as

$$\%dd(10)_X = (0.8905 + 0.00150\ \%dd(10)_{Pb})\%dd(10)_{Pb}, \qquad (8.12a)$$

with the lead foil at 50±5 cm from the surface, or

$$\%dd(10)_X = (0.8116 + 0.00264\ \%dd(10)_{Pb})\%dd(10)_{Pb}, \qquad (8.12b)$$

with the lead foil at 30±1 cm from the surface.

The photon beam quality specification and the fact that one of the measurements for photon beam quality determination (the reading at d_{max}) is potentially affected by electron contamination has been subject of a lively debate (Andreo 2000a,b; Rogers 2000) which is summarized also in appendix III of IAEA (2000). However, for the user the difference of protocol in terms of absolute beam calibration is small as long as the protocol is applied consequently. Whatever beam quality specifications are chosen, all of them are related to each other and figure 8.7 illustrates their approximate relationship to nominal beam energy.

Characterization of Beam Quality for Electrons

For electron beams, the NACP 1980 protocol specified three parameters of interest: the most probable energy in front of the accelerator, the most probable energy at the surface of the phantom, $E_{p,0}$, and the mean energy at the phantom surface, E_0. These parameters can be calculated from data acquired in a depth dose measurement in the relevant electron beam as discussed in chapter five and illustrated in figure 5.6. For electron beam energies below 30 MeV the mean energy at the surface can be determined from the half-value depth, R_{50}, as:

$$E_0 = 2.33\ R_{50}\ \text{MeV}. \qquad (8.13) = (5.1)$$

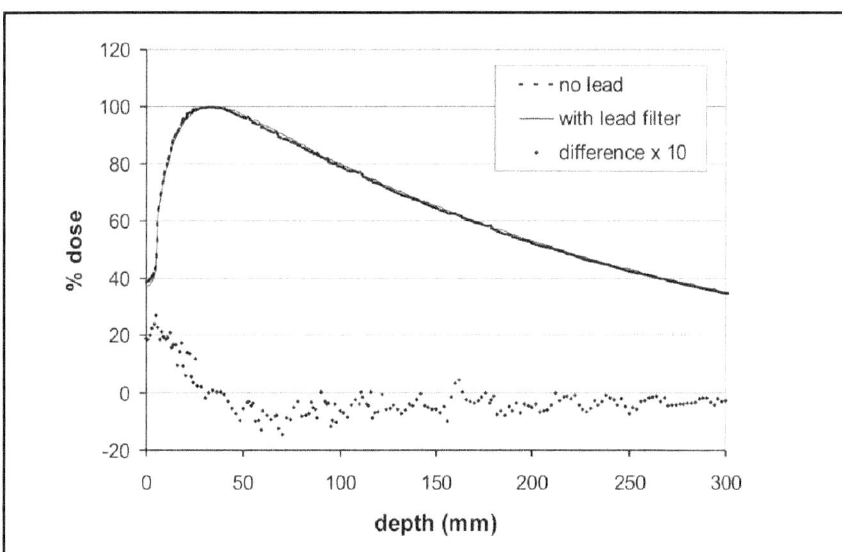

Figure 8.6: Percentage depth dose of an 18 MV photon beam measured with a cylindrical ionization chamber with and without a lead filter in place. The difference is multiplied by a factor of 10 to improve visibility in the graph.

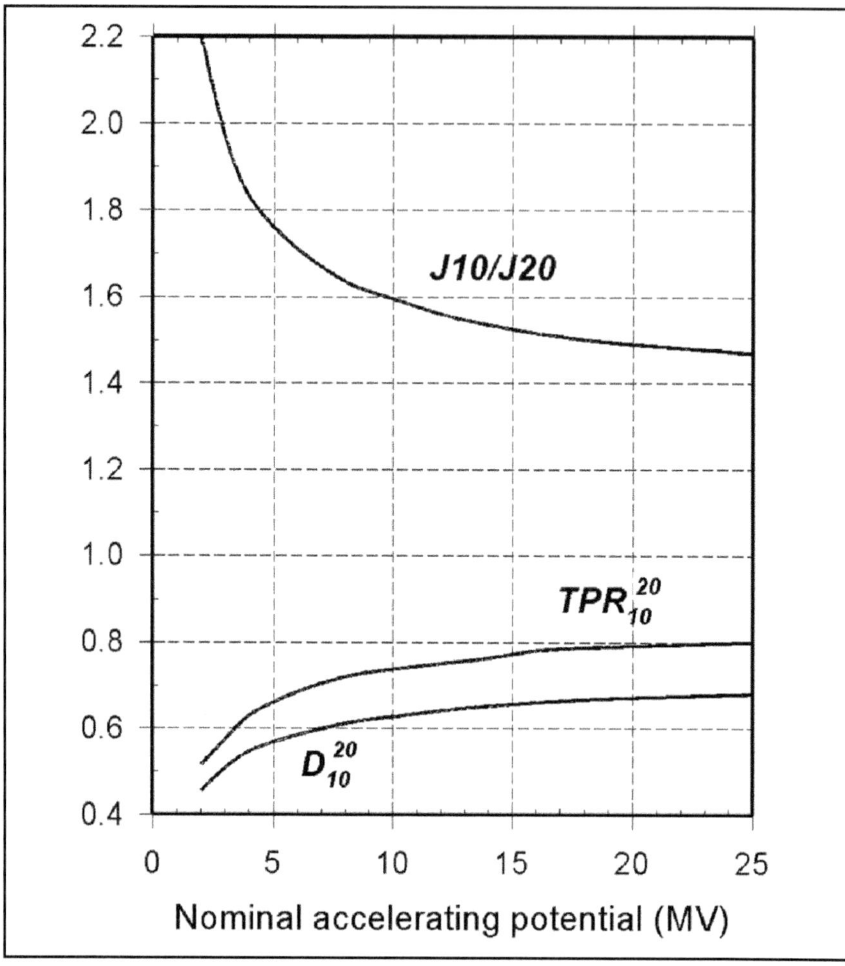

Figure 8.7: Approximate relationship between nominal accelerating potential and different specifications for photon beam radiation quality. This is not a physically accurate relationship as it depends on the actual spectrum of the x-ray beams involved.

This is also discussed in chapter five and equation 5.1 is repeated here as it is an important relationship that can be used to *reconstruct* the percentage depth dose of a given electron beam. Unlike in x-ray beams there is uniform agreement on R_{50} as the appropriate beam quality measure for electron beams produced by linacs.

Protocols for Electron Dosimetry

In general, dosimetry of electron beams is more complicated than photon beam dosimetry. This is due to the variation of electron energy with depth and the relatively steep dose gradients in lower energy electron beams as discussed in chapter five. The variation of electron energy with depth has a very important consequence for dosimetry with ionization chambers including calibration: the conversion of ionization to dose is dependent on the measurement position. This can be easily appreciated from equations 8.2 and 8.9. The stopping power ratio between water and air is dependent on the radiation quality. In photon beams this is more or less constant with depth in a phantom. However, as electrons lose energy with depth, the stopping power ratio between air and water will also change and the conversion of ionization to dose is not straightforward.

This is illustrated in figure 8.8. In electron beams the stopping power ratio between water and air, s^{w}_{air} increases with decreasing energy. Therefore, the relative dose at surface dose is lower than the relative ionization and the relative dose increases at depth, where the stopping power ratio becomes larger. The conversion of ionization to dose requires the knowledge of the stopping power at a particular depth. Unfortunately, due to the random probabilities of interaction and the increased angular spread of the electrons with depth, the electron energy does not change linearly with depth. This has been discussed in chapter five and is illustrated in figure 5.4. Different protocols provide different data sets for the conversion. The software for most computerized beam data analysis systems ("the software driving a scanning water phantom") provide at least one method of converting electron depth ionization curves to depth dose curves. In practice, the variations between the different approaches have a negligible influence on the electron beam quality determination (as long as a correction is performed), for which R_{50} is the most important parameter in most protocols (compare equation 8.13).

As electron beams became more readily available on modern linacs in the 1980s, the need for more accurate electron beam calibration arose. A very comprehensive treatment of electron beam dosimetry is given in report 35 of the ICRU (ICRU 1984). However, it was felt that this document was somewhat difficult to implement for clinical physicists and a number of other protocols were published specifically for beam calibration. Of note is

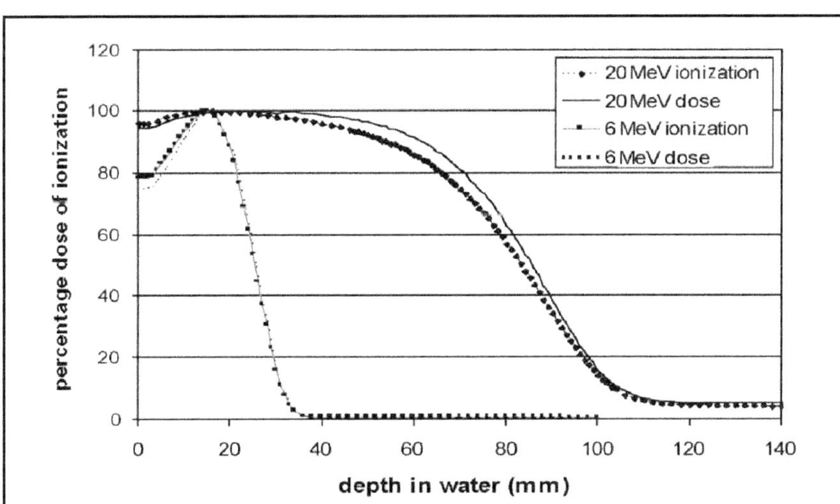

Figure 8.8: Depth dose and depth ionization curves for two electron energies.

the code of practice of the HPA (HPA 1985), which was mentioned before, the IAEA TRS 277 protocol (IAEA 1987), and the report of AAPM TG 25 (Khan et al. 1991) that also covers all aspects of electron dosimetry, including a section of beam calibration based on the AAPM TG 21 protocol.

The complexity of electron dosimetry is also demonstrated in the need to use different types of ionization chambers for the calibration of high (cylindrical chambers) and low-energy (plane parallel chambers) electron beams. The cut-off point for low-energy beams is typically 10 MeV because low-energy electron beams feature strong dose gradients where the high spatial resolution afforded by a thin plane parallel chamber is required. However, cylindrical chambers are normally calibrated by standards laboratories for a number of reasons, such as the amount of surrounding material, directional dependence, polarity effects, waterproofing, and long-term stability. Therefore, cross calibration of chambers is required, as discussed in section 8.2.

Both the AAPM and the IAEA devoted separate publications to the use of plane parallel chambers for calibration (mainly of electron beams). These documents, the AAPM report of TG 39 (AAPM 1994) and the IAEA technical report 381 (IAEA 1997) constitute at the same time extensions and clarifications of the AAPM TG 21 and IAEA TRS 277 protocols. The main issues addressed are an update on perturbation caused by different plane parallel chamber types in electron beams and the way to perform a cross calibration of chambers. In general, it is recommended that users cross calibrate their plane parallel chamber with calibrated cylindrical ionization chambers in the electron beam with the highest energy available.

8.3.2 AAPM TG 21 Protocol

Photons

The AAPM TG 21 protocol is based on the determination of a cavity gas calibration factor, N_{gas} (AAPM 1983). The index *gas* refers to all possible gases in ionization chambers; in practice, gas = air. N_{gas} can be calculated form the exposure calibration factor N_X as follows (equivalent to equations (5) and (6) in the protocol):

$$N_{gas} = \frac{\left[N_X k (W/e) A_{ion} A_{wall} \beta_{wall}\right]}{\left[(L/\rho)_{gas}^{wall} (\mu_{en}/\rho)_{wall}^{air}\right]}, \tag{8.14}$$

where

- $k = 2.58 \times 10^{-4}$ C/(kg R) is a correction for the use of the unit roentgen in the AAPM 1983 protocol. It would not appear in other formalisms that use C/kg as the unit for exposure.
- A_{ion} is the correction for ionization recombination under the reference conditions used for the calibration at the NSL.

- A_{wall} is the correction for attenuation and scattering of photons in the chamber wall.
- β_{wall} is the ratio between absorbed dose and collision kerma in the wall.
- L is the restricted stopping power following the Spencer-Attix theory as discussed in chapter three, section 3.4.3. The cut-off energy for secondary electrons was chosen in the AAPM TG 21 protocol as 10 keV, which matches the energy loss of an electron traversing the ionization chamber cavity.
- μ_{en} is the energy absorption coefficient.
- ρ is the physical density of the material; this converts stopping powers and attenuation coefficients in mass stopping powers and mass attenuation coefficients, respectively. The index order $\left(L/\rho\right)_{gas}^{wall}$ indicates the gas ratio of stopping power in the wall to that of air.

Equation 8.14 is valid if the build-up cap and the chamber wall used in the calibration are of identical material. If this is not the case, it is necessary to extend the numerator to account for the difference in composition between wall plus build-up cap and air. In this case the denominator, called for convenience here A_m (m to indicate materials) can be written as:

$$A_m = \alpha\left(L/\rho\right)_{gas}^{wall}\left(\mu_{en}/\rho\right)_{wall}^{air} + \left(1-\alpha\right)\left(L/\rho\right)_{gas}^{cap}\left(\mu_{en}/\rho\right)_{cap}^{air}, \qquad (8.15)$$

with α as the fraction of ionization in the detector caused by electrons generated in the chamber wall, and $(1 - \alpha)$ as the fraction of ionization in the cavity caused by electrons generated in the build-up cap. The value for α depends on the wall thickness of the chamber and can be taken from figure 1 in the protocol.

The AAPM TG 21 protocol provides the user with worksheets to calculate N_{gas} and to determine the absorbed dose in photon and electron beams. It also provides values for some common chambers in table XVII of the protocol (AAPM 1983). Once N_{gas} is known, the dose to the medium of interest, D_{med}, can be calculated according the AAPM TG 21 protocol as follows: the AAPM TG 21 protocol aims again to be general by not specifying the medium which in practice is usually water:

$$D_{med} = R_{corr}^{*} N_{gas}\left(L/\rho\right)_{gas}^{med} P_{ion} P_{repl} P_{wall}. \qquad (8.16)$$

Equation 8.16 equals equation (9) in the AAPM TG 21 protocol and is similar to equation 3.5 in chapter three, which is at the heart of Bragg-Gray cavity theory.

- R_{corr}^{*} is the meter reading corrected for temperature, pressure, and polarity effects. The asterisk indicates that the AAPM protocol does not include recombination correction, which is accounted for by the term P_{ion} (see below). In the protocol the symbol M is used.

- $(L/\rho)_{gas}^{med}$ is the ratio of the mean restricted collision mass stopping power of the phantom material (*med*) to that of the chamber air (*gas*). They can be obtained from a graph in the original protocol and a clarification document, which was published in 1986 (Schulz et al. 1986).

- P_{ion} is the correction factor for incomplete collection efficiency due to recombination, i.e., $P_{ion} > 1$.

- P_{repl} is the replacement correction factor. It accounts for dose gradient and air cavity effects. The replacement correction depends on radiation quality, the ion chamber used, and the location of the chamber in the phantom. P_{repl} is a combination of two effects: These are the gradient correction, P_{grad}, due to the air cavity, and P_f, which accounts for disturbances in electron fluence caused by the replacement of the phantom material by the detection device. If measurements are performed at depth of maximum dose or deeper in the phantom in megavoltage x-rays, no fluence correction has to be applied since transient electron equilibrium is established. P_{grad} is also small for Baldwin-Farmer probes in photon beams-according to AAPM TG 21 $P_{grad} = 0.99$ for a Baldwin-Farmer thimble chamber (AAPM 1983). The term P_f can have a small effect when measurements in low-density media in disequilibrium situations are undertaken (Rice et al. 1988).

- P_{wall} is a correction for the difference between the material of the wall and the medium of interest. For photon beams it can be calculated as:

$$P_{wall} = \frac{\left[\alpha(L/\rho)_{gas}^{wall}(\mu_{en}/\rho)_{wall}^{med} + (1-\alpha)(L/\rho)_{gas}^{med}\right]}{(L/\rho)_{gas}^{med}}, \tag{8.17}$$

where L/ρ is the ratio of mean restricted mass stopping powers between medium and gas (=air) and (μ/ρ) is the average mass energy absorption coefficient. The term ·is the fraction of ionization in the detector cavity caused by electrons from the chamber wall and $(1 - \alpha)$ is the fraction of ionization in the cavity caused by electrons from the surrounding medium. This works in the same fashion as the ·in equation 8.15. In high-energy photon beams α is generally small since thin-walled chambers should be used for measurements under Bragg-Gray conditions, and the build-up cap is removed for calibration measurements in the phantom.

The measurement of R_{corr} has to be performed under reference conditions as listed in table 8.5. These are a 10×10 cm^2 field size for all photon beams and a depth of measurement depending on the radiation quality. The AAPM TG 21 protocol allows measurements to be made in three phantom materials: water, polystyrene, and acrylic (synonymous with Lucite™, PMMA [polymethyl methacrylate], Perspex™, or Plexiglas™). While the

Table 8.5: Comparison of reference conditions for megavoltage photon beams in three calibration protocols (Adapted from Van Dyk 2005).

	AAPM TG 21	*AAPM TG 51*	*IAEA TRS 398*
Year of publication	1983	1999	2000
Range of energies	Co-60 and x-rays between 2 and 50 MV	Co-60 to 50 MV	< Co-60 to 50 MV; the protocol is also applicable to kV x-rays.
Beam quality specification	TPR_{10}^{20}	$\%dd(10)_X$	TPR_{10}^{20}
Correction for effectve point of measurement in beam quality determination?	No	Yes	No
Recommended dose specification medium	Water	Water	Water
Phantom materials allowed	Acrylic, water, or polystyrene	Water	Water
Cylindrical chamber displacement for calibration	None	$0.6r_{cav}$	None for reference conditions, however if dose reading to be transferred to a different depth, the %dd curve must be shifted by $0.6r_{cav}$.
Reference depth	Co-60 to 15 MV: 5 cm 16 to 25 MV: 7 cm[a]	10 cm	TPR_{10}^{20} ≤0.7: 5 or 10 g/cm TPR_{10}^{20} >0.7: 10 g/cm^2
Reference field size	10 x 10 cm^2	10 x 10 cm^2	10 x 10 cm^2
Restricted mass stopping power ratios, $s_{w,air}$, taken from	Monte Carlo calculations by Berger[b]		Implicit in k_q derivation based on ICRU (1984).

[a] In the AAPM TG 21 protocol (1983) the reference depth depends also on the inner diameter of the chamber. Table XI in the protocol gives the exact details of the reference conditions.

[b] The stopping power ratios used in the AAPM TG 21 protocol (1983) are referenced to a private communication by M. Berger from the National Bureau of Standards in Washington. The cut-off energy was 10 keV.

protocol is restricted to these three phantom materials, it can, in principle, be adapted to others as shown by Reft for the case of solid water (Reft 1989). Stopping power ratios are given in the protocol for water, polystyrene and acrylic and depth scaling is necessary for the latter two to account for different attenuation. Figure 8 in the protocol illustrates the scaling procedure that requires the chamber to be at a distance from the source, which is appropriate for water. Only the amount of material above the chamber is then varied according to its relative attenuation. This avoids the need for an inverse square law correction for changes in the distance between chamber and source.

In the case of acrylic an excess scatter correction, *ESC*, must also be employed, which allows for the small increase in scatter to the chamber in acrylic compared to water. This correction increases with depth of measurement and is listed in the protocol for acrylic. It is not required for polystyrene as its electron density is very close to water. The dose to water can then be calculated for photon beams as

$$D_{water} = D_{med} \left(\mu_{en}/\rho\right)_{med}^{water} ESC. \tag{8.18}$$

Despite the fact that the protocol allows the use of different phantom materials, a point is made that all dose should be expressed as dose to water. This is an important recommendation because it acknowledges that all patients are slightly different and that it is therefore neither appropriate nor necessary to calibrate for dose in "standard" tissue. This variation between patients can be taken into account in the treatment planning stage, in particular if a CT scan of the patient is available (compare chapters nine and ten).

Electrons

The AAPM TG 21 protocol recommends the calibration of electron beams at depth of maximum dose. Table 8.6 provides more information on the reference conditions for electron beam calibration. For cylindrical ionization chambers, the dose to gas calibration factor, N_{gas}, is determined identical to photon beam calibration using equation 8.14. The dose to the medium can then be calculated using an equation equivalent to equation 8.16:

$$D_{med} = R_{corr} N_{gas} \left(L/\rho\right)_{gas}^{med} P_{ion} P_{repl}. \tag{8.19}$$

The difference to photon beam calibration is that no wall correction, P_{wall}, is required as the response of thin-walled ionization chambers in electron beams is largely independent of the wall material. The replacement correction, P_{repl}, is different in electron beams from the one used in photon beams. The protocol has a whole section (IV.D) devoted to this correction. In a typical photon calibration the replacement correction is determined by the gradient correction, which accounts for the shift in effective point of measurement as discussed in the context of equation 8.16. In electron beams calibrated at depth of maximum dose this correction is not required and P_{repl} is equal to the electron fluence correction. In the cavity the number of electrons and the track length is different from the surrounding medium. For a

Table 8.6: Comparison of reference conditions for megavoltage electron beams in three calibration protocols (Adapted from Van Dyk 2005).

	AAPM TG 21	*AAPM TG 51*	*IAEA TRS 398*
Year of publication	1983	1999	2000
Range of energies (MeV)	5 to 50	4 to 50	3 to 50 MV
Beam quality specification from	$E_0 = 2.33\ R_{50}$ MeV	Depth of 50% dose, corrected for effective point of measurement, R_{50}, protocol provides correction from ionization to dose.	Half-value depth of electrons in water, R_{50} determined for $E_0 < 16$ MeV in field size (FS) $10 \times 10\ cm^2$ and for higher energies in FS $20 \times 20\ cm^2$.
Ionization to dose conversion	Stopping power ratios given in tables V to VII for water, polystyrene, and acrylic	(Burns et al. 1996) Not required as an empirical formula is given for determination of R_{50}.	Not required as an empirical formula is given for detemination of R_{50}.
Cylindrical chamber displacement	Chamber to be placed that the front half is within 99% line of maximum dose,	$0.5r_{cav}$	The reference point of the chamber is at central axis, however, this point should be placed $0.5r_{cav}$ below the point of interest.
Phantom material	Acrylic, water, or polystyrene	Water	Only for $R_{50} < 4\ g/cm^2$ (<10 MeV) plastic phantoms instead of water can be used.
Reference SSD	No specification	90 to 110 cm	100 cm
Reference depth	d_{max}	$0.6\ R_{50}$ to 0.1 cm	$0.6\ R_{50}$ to $0.1\ g/cm^2$
Use of plane parallel chambers		Recommended for $R_{50} \leq 4.3$ cm (10 MeV); required for $R_{50} \leq 2.6$ cm (6 MeV).	Required for $R_{50} < 4\ g/cm^2$ (<10 MeV): possible for all other energies.
Field size	No specification; large enough to allow the use of stopping powers derived for "infinitely wide" beams.	For $R_{50} \leq 8.5$ cm (20 MeV) at least $10 \times 10\ cm^2$; For $R_{50} > 8.5$ cm at least $20 \times 20\ cm^2$.	For $R_{50} \leq 7\ g/cm^2$ (16 MeV) at least $10 \times 10\ cm^2$; For $R_{50} > 7\ g/cm^2$ at least $20 \times 20\ cm^2$.

well-guarded plane parallel chamber this correction is negligible; however, for cylindrical chambers the correction factor is tabled in the protocol.

If the measurement medium is not water, the dose to water can be derived using the following equation:

$$D_{water} = D_{med} \left(S_{unrestricted} / \rho \right)_{med}^{water} \Phi_{med}^{water}. \tag{8.20}$$

The unrestricted stopping power must be used here because one is interested in all electron interactions. Φ_{med}^{water} is the ratio of electron fluence at the recommended calibration point (d_{max}) between water and the medium. The unrestricted mass stopping power ratios for acrylic and polystyrene as well as Φ_{med}^{water} water for polystyrene is tabled in the protocol.

8.4 Dose-to-Water Calibration-based Protocols

Conceptually, there are a number of advantages in the use of dosimetric protocols based on an absorbed-dose-to-water factor (Rogers 1992). Compared to air-kerma or exposure-based calibration procedures, the protocols based on absorbed-dose calibration factors are very easy to use for physicists in the clinic. They require the standards laboratory to provide a suitable absorbed dose calibration factor, N_D, and leave less steps to the user (compare figure 8.1). Already in 1979, Reich proposed that therapy beam calibration should be based on calibration of dosimeters in terms of dose to water (Reich 1979). At that time, one of the major problems was that only a few standards laboratories were able to provide the required calibration factors. In 1990, the British HPA (Lillicrap et al. 1990) published a protocol based on availability of absorbed-dose calibration factors. However, this type of approach has only become widely available to users at the end of the century, with standards laboratories acquiring the capability to issue absorbed-dose-to-water calibration factors. The publication of the important protocols of the AAPM TG 51 (AAPM 1999) and the IAEA TRS 398 (IAEA 2000) coincide with this. These protocols will be discussed in more detail in the next sections.

8.4.1 AAPM TG 51 Protocol

Photons

In 1999, the AAPM updated their megavoltage beam calibration protocol. The protocol is based on availability of an absorbed-dose-to-water calibration factor, $N_{D,w}$ (this refers to the right lower corner of figure 8.1b). $N_{D,w}$ is determined by a standards laboratory for a reference radiation quality-in practice, usually cobalt-60 radiation. All standard dosimetry laboratories in North America provide this service.

In Canada this is the National Radiation Council in Ottawa, while in the United States the calibration factors are disseminated by a network of Accredited Dosimetry Calibration Laboratories (ADCLs). ACDLs have been accredited by the AAPM and function as secondary standards laboratories for NIST. The criteria for dissemination of calibration factors are given by an AAPM document, "Guidelines for Accreditation of Dosimetry Calibration Laboratories (for Ionization Chamber Instruments)" published in November 1989.

The central formula for all protocols based on an absorbed-dose-towater calibration factor is:

$$D^{Q}_{water} = R_{corr} N^{Cobalt}_{D,water} k_{Q}, \tag{8.21}$$

where

- D^{Q}_{water} is the dose to water in the user's beam with quality Q.
- R_{corr} is the ion chamber reading corrected for environmental influences, polarity and recombination (e.g., using equation 8.3).
- $N^{Cobalt}_{D,water}$ is the calibration factor provided by the standards laboratory using cobalt-60 as reference radiation.
- k_Q is a correction factor to account for the difference in beam quality in the user's beam and the quality used for deriving the calibration factor $N_{D,water}$ in the standards laboratory.

In order to select the correct k_Q, it is important to specify the photon energy appropriately. Equation 8.21 is identical to equation (3) in the AAPM TG 51 protocol (1999). The equation is similar to equation 8.9 herein; however, Bragg-Gray cavity theory does not need to be applied explicitly by using the stopping power ratio. In equation 8.21 the standards laboratory already provides a calibration factor referenced to water. However, the physics does not change and cavity theory is implicitly included in $N^{Cobalt}_{D,water}$ and k_Q.

Calibration must be performed under reference conditions, which are listed for photon dosimetry in table 8.5. The protocol specifies that all photon calibrations must be performed in water at a depth of 10 cm with a field size of 10 × 10 cm². The only exception is a cobalt-60 beam, where also a depth of 5 cm can be used in line with previous protocols. As in the previous AAPM protocol (AAPM 1983) the reference temperature is 22 °C. However, unlike the AAPM TG 21 protocol (1983), the TG 51 protocol (1999) explicitly includes polarity corrections in all measurements. Correction for recombination is to be performed by the two-voltage method discussed in chapter three and the correction factor can be calculated as:

$$P_{ion} = \frac{\left[1-\left(V_{high}/V_{low}\right)^2\right]}{\left[R_{high}/R_{low}-\left(V_{high}/V_{low}\right)^2\right]} \tag{8.22a}$$

for cobalt-60 radiation, and

$$P_{ion} = \frac{\left[1 - \left(V_{high}/V_{low}\right)\right]}{\left[R_{high}/R_{low} - \left(V_{high}/V_{low}\right)^2\right]} \quad (8.22b)$$

for pulsed or pulsed swept radiation,

with the indexes *high* and *low* referring to high and low voltage (V), respectively, and R the raw electrometer reading.

The photon beam quality specification is done using the x-ray percentage depth dose at 10 cm depth, $\%dd(10)_X$, as discussed in section 8.3. For photon beam quality determination, the AAPM TG 51 protocol recommends that the depth dose curve measured with a cylindrical ionization chamber using the reference point at the central axis of the chamber be shifted by 0.6 times the internal radius of the chamber. This allows for the effective point of measurement of the chamber to be closer to the radiation source as the physical center as discussed above (compare figure 8.3). For nominal beam energies below 10 MV this is the measured depth dose in the open beam. However, at a nominal beam energy of 10 MV or above, electron contamination may affect the dose at depth of maximum dose significantly and a lead filter must be used as discussed above.

The calibration measurement itself is performed with the reference point (i.e., the central axis of the cylindrical chamber) at the point of interest. The values for k_Q required in equation 8.21 are dependent on beam quality and the type of chamber used and include the correction for the dose gradient throughout the chamber. Figure 1 and table I of the TG 51 protocol provide values for several commercially available chamber types. Other chambers may be used at the discretion of the user, provided wall material and dimensions can be matched with a chamber given in the protocol. Like most protocols, the AAPM TG 51 protocol provides an easy-to-use worksheet for photon beam calibration.

Electrons

Unlike in the protocols discussed above where the depth of calibration is recommended to be the depth of maximum dose, the AAPM TG 51 and the IAEA TRS 398 protocols specify the calibration depth as a function of R_{50}:

$$d_{ref} = 0.6\, R_{50} - 0.1 \text{ cm}. \quad (8.23)$$

This overcomes the problem that particularly in higher-energy electron beams the depth of maximum dose is difficult to determine because of the relatively flat region in the depth dose curve around d_{max} (compare chapter five and figure 8.8). As the location of R_{50} is nearly independent of the ionization to dose conversion, the effect of this conversion on the actual location of the reference depth is also negligible.

R_{50} is the main measured parameter that determines the electron beam quality in all protocols. An interesting effect of using R_{50} both as the beam quality index and the factor that determines the calibration depth is the following: If the calibration depth is dependent on the same measurement parameter as the beam quality, the main correction factors for the electron calibration also depend just on one parameter. This is a major simplification in the electron dosimetry process, as can be seen in both the AAPM TG 51 (1999) and the IAEA TRS 398 (2000) protocols. The AAPM TG 51 protocol provides a formula to derive R_{50} directly from an ionization chamber measurement in which the depth of 50% ionization, I_{50}, has been determined.

$$R_{50} = a\, I_{50} - b. \tag{8.24}$$

with a and b constants provided by the protocol as a function of electron energy specified in terms of the value of I_{50}. For 2 cm $\leq I_{50} \leq$ 10 cm, a = 1.029 and b = 0.06. Once R_{50} has been determined, the reference depth can be determined using equation 8.23.

As can be seen in table 8.6, plane parallel ionization chambers must be used for electron beams with $R_{50} \leq$ 2.6 cm and their use is recommended for $R_{50} \leq$ 4.3 cm. For electron beam calibration, equation 8.21 for photon beam calibration is modified to:

$$D_{water,Q} = R_{corr} P_{gr} k'_{R50} k_{ecal} N_{D,water}^{^{60}Co}. \tag{8.25}$$

This is equation (6) of the AAPM TG 51 protocol (1999). By comparing equations 8.21 and 8.25, one can see that the quality conversion factor k_Q in electron beams consists of three factors: P_{gr}, k'_{R50}, and k_{ecal}. P_{gr} is a gradient correction required for cylindrical chambers. It can be determined by two measurements, one with the center of the chamber at the reference depth, d_{ref}, and one with the chamber displaced downstream by 0.5 times the chamber radius to bring the effective point of measurement to the reference depth.

$$P_{gr} = R_{raw}\left(d_{ref} + 0.5 r_{cav}\right) / R_{raw}\left(d_{ref}\right). \tag{8.26}$$

This procedure is equivalent to making the measurement at the effective point of measurement of the chamber; however, the protocol aims to be self-consistent by always using the center of the chamber as the reference point. In plane parallel chambers where the effective point of measurement is directly behind the entrance window, P_{gr} is not required

The AAPM TG 51 protocol is based on an absorbed-dose-to-water calibration factor determined in a cobalt-60 beam, $N_{D,water}^{^{60}Co}$. Therefore, the beam quality correction, k_{R50}, for electrons has the two terms shown in equation 8.25:

$$k_{R50} = k'_{R50} k_{ecal}. \tag{8.27}$$

The factor k_{ecal} converts an absorbed-dose-to-water calibration factor determined for cobalt-60 to a factor applicable to a reference electron quality, *ecal*. This factor is fixed for a given chamber model. Values are given in the protocol for cylindrical and plane parallel chambers; however, a better method for plane parallel chambers is to determine k_{ecal} as part of the chamber cross calibration. The AAPM TG 51 protocol provides a worksheet for the determination of a combined $k_{ecal}N_{D,water}^{^{60}Co}$ through cross calibration.

The factor k'_{R50} corrects the reading for the electron beam to be calibrated. It is dependent on R_{50} and the protocol provides four figures where k'_{R50} can be looked up for a variety of cylindrical and plane parallel chambers. The AAPM TG 51 protocol also provides two empirical formulae to calculate k'_{R50} for cylindrical and plane parallel ionization chambers:

For cylindrical chambers:

$$k'_{R50} = 0.9905 + 0.071\exp\left(-R_{50} / 3.67\right) \tag{8.28a}$$

for plane parallel chambers:

$$k'_{R50} = 1.2239 - 0.145\left(R_{50}\right)^{0.214}, \tag{8.28b}$$

with the half-value depth, R_{50}, expressed in centimeters. This factor is only applicable at the reference depth.

8.4.2 TRS 398 Protocol

Photons

After an extensive consultation process, the IAEA published in 2000 a code of practice for radiotherapy beam calibration based on standards of absorbed-dose to water (IAEA 2000). Unlike the AAPM protocol it is not only applicable to high-energy photons and electrons, but also to low- and medium-energy photons, protons, and heavy ion beams. A separate section in the protocol is devoted to each of the radiation qualities and each section provides an easy-to-use worksheet at the end. The protocol also gives some background information on the network of calibration facilities and uncertainties in the calibration process.

The formalism in IAEA TRS 398 is similar to the one described above for AAPM TG 51 and shown in equation 8.21. The protocol also uses the factor k_Q as correction for the radiation quality of the beam in question. The protocol is more general in terms of allowing any radiation quality Q_0 as reference, even if in practice most standards laboratories use cobalt-60. At time of writing only a few standards laboratories offer calibration factors derived in radiation qualities other than cobalt-60. While it is not possible to reproduce every clinical beam, the quality index allows interpolation between different beams. Of particular interest for users are calibration factors obtained in electron radiation beams. The English NPL is one of the first laboratories to offer this service. Factors can be obtained for a variety of energies between 3 and 17 MeV (http://www.npl.co.uk/ionrad/services/sl25.html). A similar service is available in Switzerland (http://www.metas.ch/). Clearly, these developments make calibration of megavoltage radiation easier for the user and errors less likely. As such, it can be anticipated that in the future an increasing number of standards laboratories will offer absorbed-dose-to-water calibration factors derived in a variety of clinical radiation qualities.

In anticipation of these developments, the IAEA TRS 398 protocol uses k with two indices and provides a general expression for it:

$$k_{Q,Q_0} = \frac{\left[(s_{w,air})_Q (W_{air})_Q \, p_Q\right]}{\left[(s_{w,air})_{Q_0} (W_{air})_{Q_0} \, p_{Q_0}\right]}. \tag{8.29}$$

This is equation (4) in the protocol where

- W_{air} is the mean energy expended in air per ion pair formed. In practice, it is generally assumed that W is independent of radiation quality, which allows cancelling of the two terms in the equation.
- p_Q is a perturbation factor for the chamber irradiated with photons of quality Q. The perturbation factor combines all effects of the ion chamber in the beam and can be calculated as:

$$p_q = p_{cav}\, p_{dis}\, p_{wall}\, p_{cel}. \tag{8.30}$$

The correction factors in this equation are discussed in detail in appendix II of the protocol. The user is not required to determine them all because values for k_Q are provided in the protocol for cobalt-60 as reference radiation quality and a large number of commercial ionization chambers. However, it is important to understand the type of correction needed to appreciate the uncertainty of the k_Q value used. The correction factors are also listed as part of table 8.2b:

- p_{cav} accounts for differences of secondary electron flux in the cavity and water.
- p_{dis} is a displacement correction similar to the gradient correction in the AAPM TG 21 protocol. As the IAEA TRS 398 protocol places the center of the cylindrical chamber and not the effective point of measurement at the point of interest (as is done in IAEA TRS 277), a correction must be made.
- p_{wall} is the wall correction factor. As the IAEA protocol requires all measurements to be performed in a water phantom, equation 8.17 can be modified to include not only the chamber wall but also the protective sleeve to waterproof the chamber. Then

$$p_{wall} = \frac{\left(\alpha\, s_{wall,air} (\mu_{en}/\rho)_{w,wall} + \tau s_{sleeve,air} (\mu_{en}/\rho)_{w,sleeve} + (1-\alpha-\tau)\, s_{w,air}\right)}{s_{w,air}}, \tag{8.31}$$

with $s_{a,b}$ as the restricted stopping power ratio between materials where a and b refer to water, air, wall, or sleeve, respectively, $(\mu_{en}/\rho)_{a,b}$ as the ratio of energy absorption coefficients between materials a and b and α and τ the relative contributions of the wall and the sleeve compared to the ionization in the chamber. This is the same approach as already discussed in the context of equations 8.15 and 8.17.

- p_{cel} is the correction for the central electrode (particularly if made from aluminum). As the protocol does not require an in-air calibration, pcel only accounts for the effect of the electrode in the user's beam. This is different from pcel as it is defined in the IAEA TRS 277 protocol, which includes both the effects during the in-air calibration and the measurement in the user's beam.

The IAEA TRS 398 protocol uses TPR_{10}^{20} as quality index (determined without formally accounting for the shift of the effective point of measurement away from the chamber axis) and allows high-energy photon beam calibration only to be performed in water. The reference conditions are given in table 8.5. In cobalt-60 beams and radiation with $TPR_{10}^{20} < 0.7$, both 5 and 10 cm depth in water can be used while in higher-energy photon beams a reference depth of 10 cm is recommended. The field size for calibration should be 10×10 cm^2 in any case.

Electrons

The protocol of the IAEA TRS 398 handles electron beam calibration very similarly to the AAPM TG 51 protocol. This can be seen in table 8.6 where the reference conditions for electron beam calibration are listed. However, there are several small differences.

The reference depth in both protocols is identical as given in equation 8.23; however, the IAEA TRS 398 protocol specifies all depths in g/cm^2 rather than cm. The IAEA also utilizes the same formulas to convert ionization readings to dose in the determination of R_{50} (equation 8.24). The preferred detector for depth ionization curves in electron beams is a plane parallel chamber; however, if R_{50} is larger than 4 g/cm^2, a cylindrical chamber can also be used with its reference point 0.5 times the radius of the cavity below the measurement point.

The IAEA TRS 398 protocol uses $R_{50} = 4$ g/cm$_2$ as cut-off for the choice of phantom material. Above this energy only water can be used, while below this value several plastic materials can also be employed. For the beam quality determination a field size of at least 10×10 cm^2 shall be used if $R_{50} \leq 7$ g/cm^2. If the beam quality is higher, a field size of 20×20 cm^2 or larger must be employed to ensure that the field size exceeds the lateral range of the electrons.

The dose to water at the reference point in an electron beam can be determined using equation 8.21 discussed in the previous section. This is similar to equation (25) in the protocol:

$$D_{water}^{Q} = R_{corr} N_{D,water}^{Cobalt} k_{Q}. \tag{8.21}$$

The protocol uses the term $N_{D,w,Q0}$ to indicate that the calibration factor has been determined at the reference beam quality Q_0, which in practice is most likely cobalt-60. The factor $k_{Q,Q0}$ corrects for differences between reference beam quality and actual beam quality Q. Values for $k_{Q,Q0}$ are given for cobalt-60 as reference radiation quality for a variety of plane parallel and cylindrical chambers as a function of R50 of the beam in the protocol.

There are three important issues in the context of electron beam calibration, which the IAEA TRS 398 protocol discusses in more depth:

- Cross calibration of ionization chambers
- Use of plastic phantoms
- The transfer of a calibration factor obtained at a depth of (0.6 R_{50} – 0.1 g/cm^2) to another depth, e.g., the depth of maximum dose.

Cross-calibration or intercomparison of ionization chambers has been discussed in section 8.2.4. The IAEA dedicates a section (7.6) of the protocol to this topic. The highest available electron energy should be used.

Plastic phantoms may be used only for low-energy electron beams with a beam quality $R_{50} < 4$ g/cm^2, which refers in practice to a nominal radiation energy of less than 10 MeV at the phantom surface. The protocol provides depth scaling factors, c_{pl}, and fluence scaling factors, h_{pl}, for many plastic materials in its table 21. The depth-scaling factor can be used to convert depth in plastic to depth in water using

$$z_{water} = z_{plastic}\, c_{pl} \text{ (all depths given in g/cm}^2\text{).} \tag{8.32}$$

This is obviously also required when the beam quality is expressed as half-value depth, R_{50}. If the absolute dose is determined in the plastic, a fluence correction must be applied to the ionization chamber reading in addition to this. This fluence correction factor h_{pl} ("*pl*" for plastic) plays a similar role as the factor h_m in the IAEA TRS 277 protocol (1987). However, h_{pl} as provided in the protocol is generally larger and the uncertainty associated with its use is one of the major reasons to discourage the use of plastic phantoms for calibration.

Unfortunately, the recommended measurement depth of 0.6 R_{50} – 0.1 g/cm^2 is often not the depth to where a department wants to specify the dose. This is more commonly done at the depth of maximum dose or a "round" number of centimeters close to d_{max}. The transfer of dose to a different depth is done using a measured percentage depth dose curve in which the effective point of measurement of the chamber has been taken into account.

8.4.3 Comparison of Protocols

Photons

Many authors have compared the different protocols for calibration. An excellent early comparison between several protocols was published by Mattsson in 1985 (Mattsson 1985). Other useful summaries and comparisons of protocols can be found in the book chapters by Mijnheer and Thwaites (Mijnheer et al. 1993), Nath and Huq (1995), and Kron (2005).

The introduction of protocols based on absorbed dose in water calibration factors has prompted several studies comparing old (air-calibration factor) and new (dose-to-water calibration factor) protocols. The AAPM protocols were compared by researchers in Houston connected with the Radiological Physics Center (Tailor and Hanson 2002; Cho et al. 2000). Cho et al. provide a formula for the ratio between the respective calibration factors N_K and $N_{D,water}$:

$$N_{D,water}/N_K = 0.8791^{-1}\left(N_{gas}/N_X\right)P_{repl}P_{wall}\left(S_{restricted}/\rho\right)_{air}^{water}. \quad (8.33)$$

The factor 0.8791 converts from exposure to air-kerma. Equation 8.33 is derived using equations 8.14, 8.16, and 8.21 and does not require any measurement. The ratio between the calibration factors can be calculated directly from factors provided in the AAPM TG 21 protocol (AAPM 1983). By also performing measurements with six ionization chambers, it was found that the measured $N_{D,water}/N_K$ was on average 1.2% higher than the calculated one. This value gives the expected difference for the calibration of a cobalt-60 beam using the two protocols (Cho et al. 2000, Tailor and Hanson 2002).

The change to absorbed-dose calibration factor protocols has not uniformly been welcomed as a significant advance. Khan argued that an update of the AAPM TG 21 protocol would have been sufficient (Khan 2000). He presented a relation similar to equation 8.33, which relates N_{gas} to the absorbed-dose-in-water calibration factor in a cobalt-60 beam, $N_{D,w}^{^{60}Co}$:

$$N_{gas} = \frac{N_{D,w}^{^{60}Co}}{\left[P_{repl}\right]P_{wall}P_{cel}\left(S_{restricted}/\rho\right)_{air}^{water}}, \quad (8.34)$$

with the denominator referring to cobalt-60. The main difference to equation 8.33 is the inclusion of a correction for the central electrode of the cylindrical chamber that has to be applied if it is made from aluminum. By updating the AAPM TG 21 protocol with more recent stopping power data, the differences between AAPM TG 21 and AAPM TG 51 protocols for photon beams were found to be less than 0.5% using a PTW N23333 Farmer-type chamber.

For x-ray beams produced by linacs, the TG 51 protocol based on an absorbed-dose-to-water calibration factor was found to result in megavoltage photon beams in a dose measurement of up to 1.5% higher than the older TG 21 protocol (Cho et al. 2000; Tailor and Hanson 2002). The difference depends on the ionization chamber used and was found to be generally larger for lower-energy beams. The order of magnitude for the difference between air-kerma and absorbed-dose calibration factor based protocols as observed in the AAPM protocols is mirrored in the experience when moving from IAEA TRS 277 protocol to the TRS 398 protocol (Andreo et al. 2002).

There has been an extensive scientific debate about the relative merits of the two new protocols based on absorbed-dose-to-water calibration factors (Andreo 2000a,b; Rogers 2000). This has included open discussions at conferences such as the World Congress on Medical Physics and Biomedical Engineering in 2000 in Chicago and several scientific publications and letters. Probably the most important point of the debate is the quality index for photon beams, which has been discussed here in the context of photon beam quality determination. The issue is that many standards laboratories have nonclinical linacs with a relatively soft x-ray spectrum. If these were to be included in the consideration, $\%dd(10)_X$ appears to be a beam quality index which can be applied to more photon beams (Kosunen and Rogers 1993; Li and Rogers 1994). However, issues with potential electron contamination at depth of maximum dose may make these measurements more difficult for medical physicists in a clinic. Therefore, there are merits in both the TPR_{10}^{20} and the $\%dd(10)_X$ concepts.

In practice, the differences in outcome for calibration of clinical photon beams are small and both protocols result in dose measurements within 1% for most clinical linac beams. However, it is good practice for radiotherapy centers to confirm any differences between protocols when changing. This in itself constitutes a comparison of protocols. In general, differences in dose should be small and the user should try to understand the physical reason for any differences.

Electrons

Table 8.6 provides a summary of the recommendations for electron dosimetry from the four protocols described in this section. Some of the fundamental differences between the protocols have already been discussed in the previous section in the context of photon beam calibration. However, several important differences apply, particularly to electron beams where the effective energy alters with depth of measurement. As the effective energy will affect the stopping power ratio between water and air, there is a fundamental influence of the choice of energy/depth correction and point of measurement on the outcome of the calibration itself.

In general, the older protocols are in line with the clinical practice to choose the calibration point at—or close to—the depth of maximum dose. This process makes dose calculations by hand relatively easy. In addition to this, the AAPM protocols do not prescribe the reference SSD. A radiotherapy center that decides to treat all electron fields at an SSD of 110 cm (e.g., to avoid hitting the patient with the electron applicator) can actually calibrate the beam at this distance.

In comparison, the more recent protocols are more prescriptive. This pertains to the use of plane parallel chambers for low-energy beams as well as for the reference depth for calibration. Both, the AAPM TG 51 protocol and the IAEA TRS 398 protocol use more or less the same criteria.

A comparison between results obtained using the two AAPM protocols (AAPM 1983; AAPM 1999) was made by the Radiological Physics Center in Houston (Cho et al. 2000). For electron beams of 9 MeV and 16 MeV nominal energy they found that the AAPM TG 51 protocol typically resulted in a dose about 1.5% higher at depth of maximum dose. In a theoretical analysis of absorbed-dose ratios between AAPM TG 51 and TG 21 protocols, Tailor and Hanson (2002) showed that the difference typically increased with nominal electron beam energy. The difference depends on the choice of ionization chamber and is similar for cylindrical and plane parallel chambers. In the worst case it was up to 2.5% for a 20 MeV electron beam (Tailor and Hanson 2002).

Andreo compared the IAEA TRS 398 protocol electron contamination procedures with the earlier protocols of the IAEA, TRS 277 and TRS 381

(Andreo et al. 2002). The article provides a good summary of the corrections applied for the central electrode of cylindrical chambers in the different protocols: TRS 277 combines the correction required in the standards laboratory during the cobalt-60 air calibration and the correction in the user's beam in a global correction factor. This is clarified in TRS 381 and in TRS 398 the correction is included in the quality correction factor k_Q. In a variety of electron radiation beams Andreo et al., found that the difference between calibra-tions performed using TRS 277 and TRS 398 was typically around 1% and never exceeded 2%. The absorbed dose to water protocol TRS 398 yielded a larger dose estimate than the air-kerma calibration factor-based protocols. This is in general agreement with the experience in North America where AAPM TG 51 and AAPM TG 21 protocols were compared (Cho et al. 2000, Tailor and Hanson 2002).

8.5 Uncertainty Analysis and Verification

8.5.1 Uncertainty Analysis

Most current approaches to uncertainty analysis in dosimetry follow the procedures based on a publication of a working party of the BIPM (BIPM 1981), which is outlined by the International Standards Institute in 1995 (ISO 1995). Other publications that provide a good overview of the approach are the IAEA protocols (IAEA 1987; IAEA 2000) and an article by Ibbott et al. (1997). This uncertainty analysis distinguishes between uncertainties of type A and B.

Type A uncertainties are a result of random variations, which are assumed to be normally distributed. As such, they can be expressed as a standard error of repeated measurements and they are an objective estimates based on statistical reasoning. Uncertainties of class A can be reduced by performing more measurements of the same quantity; the uncertainty reduces by the square root of the number of measurements. If there are more than one source of uncertainties of type A, they are usually assumed to be independent of each other. In this case they are like two independent geometric dimensions—changing one of them does not alter the value of the other. In other words they are orthogonal and the overall uncertainty can be seen as the geometric diagonal. Thus, the uncertainties can be added in quadrature as per Pythagoras' theorem:

$$U_{total}^2 = \sum \left(U_i^2\right). \tag{8.35}$$

This is often referred to as *propagation of uncertainties, U*.

Type B uncertainties are best estimates of other parameters that can influence the outcome of a measurement. Their estimation requires a thorough understanding of the process, and its determination is commonly subjective. Uncertainties of type B are dependent on the experimental procedure and it is the duty of the operator to estimate these uncertainties to the best of his or her knowledge. Examples are uncertainties in constants used such as W/e and output variations of a linac with warming up due to previous irradiations. While it is strictly speaking not appropriate to use variances (or standard deviation which is the square root of the variance) to quantify uncertainties of class B, this is still commonly done to enable combining uncertainties from more than one parameter. A useful method to derive a standard deviation for type B uncertainties is to estimate the maximum value of uncertainty, U_{max}.

There are several possibilities to describe the distribution of uncertainties of type B ranging from a square to a triangular and Gaussian distributions (IAEA 2000). In case of the latter, the standard deviation can be derived by dividing the estimate for the maximum uncertainty by a suitable value. If the experimenter is "almost certain" about his or her estimation of uncertainty, U_{max} can be assumed to be 3 standard deviations (3SD), representing a 99% confidence limit. If the physicist is only "fairly" sure, U_{max} should be assumed to represent 2SD from the mean (mathematically about 95% confidence level) (IAEA 2000).

If both types of uncertainty, A and B, are expressed with the same level of confidence (e.g., 2SD) and if it can be assumed that all sources of uncertainty are independent of each other, the combined uncertainty can be calculated by taking the quadratic sum as per equation 8.35.

Typically, the uncertainty associated with the calibration in radiotherapy can be separated into three parts:

- Uncertainty of the calibration factor obtained from the standards laboratory
- Uncertainty in the formalism and factors used in the protocol
- Uncertainty in the user's measurement.

The latter two are sometimes combined as the choice of factors from a protocol often depends on a measurement by the user (e.g., the beam quality of the user's beam). Table 8.7 provides a summary of estimates for uncertainties from all sources. This is an illustration of typical values only and the actual uncertainties for a particular calibration can be different from the list. It is the responsibility of the qualified medical physicist to provide an estimate for the accuracy of beam calibration in her/his facility and to communicate this to radiation oncologists.

Uncertainty of the Standard

The uncertainty of the calibration factor given by the standards laboratory to the user depends on the type of calibration factor, the reference beam, and the nature of the standards laboratory. A primary standards laboratory has one less source of uncertainty, as a transfer of calibration from a primary to a secondary standard is not necessary. This additional link in the traceability chain adds approximately 0.5% to the overall uncertainty.

NSLs often compare their dose standards with each other, primary standard dosimetry laboratories, and international organizations such as the IAEA or BIPM. As shown in many publications, the differences between the dose standards are generally very small (IAEA 2000, Shortt et al. 2000, 2001,

Table 8.7: Estimation of uncertainties in the calibration of megavoltage photon and electron beams. The table was compiled from various sources in the literature (AAPM 1983, 1999, IAEA 1987, 2000, Mijnheer et al. 1993) and experience of the authors. All figures are expressed in percent and refer to a 95% confidence range (approximately equivalent to 2SD for a normal distribution of results). The values are an estimate only and can vary considerably depending on local circumstances. (Table reproduced courtesy of J. Van Dyk 2005)

	Photons		*Electrons*		*Comments*
Type of protocol	Based on N_K or N_X	Based on $N_{D,w}$[a]	Based on N_K or N_X	Based on $N_{D,w}$[a]	
Uncertainties associated with the standards library					
Standard relative to primary standard	0.5	0.5	0.5	0.5	Assumes that the standards laboratory maintains a secondary standard which has to be compared to a primary standard; in case of a primary standards library this factor is zero.
N_K or N_X calibration of the user's instrument	1.0		1.0		
$N_{D,w}$ calibration of the user's instrument		0.6		0.6	
Total standards laboratory	**1.1**	**0.8**	**1,1**	**0.8**	
Uncertainties in the user's beam					
Stability of dosimeter	0.6	0.6	0.6	0.6	Can be established using a check source.
Transfer of calibration factor to field instrument	0.5	0.5	0.5	0.5	Depends on chamber type and can be larger for plane parallel chambers.
Set up of the instrument	0.6	0.6	0.6 to 1.0	0.6 to 1.5	Localization of the chamber at reference point particularly

Table 8.7: Continued

	Photons		Electrons		Comments
Stability of radiation source	1.0[b]	1.0[b]	1.0	1.0	Timer error in Co-60 and monitor chamber accuracy in linac.
Chamber correction factors	0.8	0.8	0.8	0.8	Compare table 9.1: P_{ion} usually the most difficult to determine.
Air calibration to absorbed-dose conversion	2.5		3.0		Includes stopping power ratios and correction factors, in particular P_{wall}.
Beam quality correction		1.2		1.5	Largest for plane parallel chambers in electron beams.
Use of phantom materials other than water[c]	1.0		1.5	1.5	As water is the only reference material allowed for photon dosimetry in AAPM TG 51 and IAEA TRS 398, it is not applicable to N_{Dw}-based protocols.
Total users beam	**3.0**	**2.0**	**3.5**	**2.4**	
Combined overall uncertainty	**3.2**	**2.2**	**3.6**	**2.6**	

[a] Assuming the $N_{D,w}$ has been determined in a cobalt-60 beam.

[b] For a cobalt-60 beam this value could be halved.

[c] Not included in the final uncertainty as water phantoms are preferred for use in calibration.

Ibbott et al. 1997). It is interesting to note though that there is a difference between different primary standards. This is particularly noticeable when primary standards laboratories use different absolute dosimetric methods to determine the dose. The largest difference has been observed between calorimetry and Fricke dosimetry.

The value given in table 8.7 for the total uncertainty arising from the calibration of the user's instrument at the standards laboratory is in accordance with a study by Ibbott et al. (1997). These researchers found that the upper bound for the overall uncertainty at the five U.S. ADCLs is about 1.2% (2SD) for an air-kerma calibration. This interesting article also provides a thorough analysis of all the sources of uncertainty in this instance.

The accuracy of the calibration factor is somewhat better if the standards laboratory provides an $N_{D,water}$ (IAEA 2000). Other sources of uncertainty in the standards laboratory are the long-term stability of the radiation source used for calibration (typically a small contribution as cobalt-60 is used) and the physical calibration of the user's instrument as such. In order to minimize the uncertainty during the calibration process, calibration laboratories typically combine a thorough inspection of the ionization chamber and electrometer provided by the user. This includes measurement of the polarization voltage, determination of leakage, or any modifications made by the user.

Some standards laboratories allow the user also to send a reference radiation source with the instruments to be calibrated. This is usually a strontium-90 reference source as the one illustrated in chapter three (figure 3.22). The additional dose readings with the reference source at the standards laboratory and the user's facility reduce the uncertainty of the transfer of the calibration factor.

Uncertainty of the Measurement

The uncertainty arising from the work of the user is typically larger than the one from the instrument calibration at the standards laboratory. The major sources of uncertainty are:

i) *Stability of dosimeter:* The uncertainty here can be reduced if a radioactive check source can be used to determine any variations in detector response.
ii) *Transfer of calibration factor to a field instrument:* This uncertainty arises from an intercomparison of chambers at the user's facility. It is only applicable if the instrument used for calibration is different from the one calibrated at the standards laboratory. In the case of low-energy electrons where a plane parallel chamber should be used, this factor needs to be carefully examined.
iii) *Setup of the measurement:* This uncertainty comprises:
    - Setup of the radiation source
    - Setup of the phantom
    - Placement of the ionization chamber within the phantom.

    Provided sufficient care is taken when positioning the cobalt-60 unit or linac, the first contributes typically little to the overall uncertainty.

The placement of the phantom and adjustment of the water height can be typically achieved with an overall accuracy of better than 2 mm. Due to the inverse square law this results in an uncertainty of some 0.4%. The largest uncertainty arises often from the placement of the chamber within the phantom. In particular in low-energy electron beams, this can be tricky and a small displacement can result in a large difference in dose reading in the steep dose gradients present in these beams. It is interesting to note that the older style protocols that require positioning of the chamber at depth of maximum dose have an advantage over the newer protocols, which require the chamber to be placed at a depth derived from depth dose measurements. It is possible with an ionization chamber placed in water to "hunt" the depth of maximum dose by moving the chamber slightly up and down and taking the final measurement at the location of maximum dose rate. This ensures accurate placement at least in regard to the reference location.

iv) *Stability of radiation source:* This refers to variations of output of the radiation source. This is typically larger in the case of linacs and it is the responsibility of the user to operate the linac during calibration as close as possible to clinical conditions. The uncertainty arising from stability of the source can be reduced through good QA measures as discussed in the next section.

v) *Chamber correction factors:* These factors are determined by the user after determining environmental conditions (temperature, air pressure, and humidity) and performing preparatory measurements (polarity, two voltages). Again, good QA can improve the accuracy of the determination of these factors. The other important factor in these measurements is patience. Unfortunately, accurate calibration requires stable conditions and temperature equilibrium. In addition to this, any change of polarizing voltage requires long waiting times before the measurement has stabilized. The influence of the latter can easily be ascertained by the user for his or her particular instrumentation.

vi) *Air-calibration-to-absorbed-dose conversion:* This is only applicable to protocols based on air-kerma or exposure calibration factors (AAPM 1983; IAEA 1987). The conversion depends on the chamber type and the uncertainty is mainly of type B, as it depends on factors provided by the protocol in use. It also includes the uncertainty of the beam quality determination, as this will determine the stopping power ratio to be used in the Bragg-Gray cavity theory.

vii) *Beam quality correction:* This is the most important correction in protocols based on an absorbed-dose-to-water calibration factor. The IAEA TRS 398 (2000) protocol devotes a lengthy appendix II to the calculation of the correction factor k_Q and its uncertainty. In general, most of the corrections required in the air-kerma factor-based protocols that make explicit use of Bragg-Gray cavity theory are condensed in this factor. Therefore, the uncertainty in k_Q includes components for W/e, the stopping power ratio S^{air}_{water}, and the perturbation correction factors summarized in table 8.2.

There are other sources of uncertainty that the user must take into consideration depending on the protocol used and the equipment available. An example is the measurement in an environment different from water. While most protocols allow the use of plastic phantoms, their use introduces additional uncertainty. There is a component to this from the correction factors required to correct to dose in water and a component due to the quality of the actual phantom used. In any case it is advisable to verify homogeneity, atomic number, and density of the material first, e.g., by taking a CT scan.

Table 8.7 provides a summary of uncertainties in the context of megavoltage beam calibration. It distinguishes between different types of protocols. However, all values given are only "typical" and may not reflect the individual calibration situation. As such, it is important that the user makes her or his own assessment of the overall uncertainty and communicates this to all relevant people in the department.

8.5.2 Dosimetric Intercomparisons

As the calibration is the one measurement that determines the output of a radiotherapy treatment unit, it is advisable to perform an independent check. This is particularly important in the case of a new department where there are no other treatment units from which the magnitude of a dosimeter response can be estimated. In this case, it is recommended to ask someone who is neither responsible nor involved in the local procedures to verify the calibration. This should be done using an independent measurement system. Ideally, the independent check should be performed for all radiation qualities available at a treatment unit.

If no visit of an independent expert can be arranged, there should at least be an independent verification of output using, for example, thermoluminescent (TL) dosimeters which can easily be sent from one radiotherapy center to another (compare chapter three). While the accuracy of this verification procedure is lower than having an independent physicist performing an ionization chamber measurement, it can prevent gross mistakes in the calibration.

External Audits

External audits are a useful tool to ascertain the quality of radiotherapy treatment in a cancer center. They may be required by the regulatory authority or a clinical trials group, or participation can be voluntary. An example of the latter is the IAEA/WHO TLD Postal Progamme for Radiotherapy Hospitals (Izewska and Andreo 2000). This program is aimed particularly at photon beams in developing countries and is considered as a service to users rather than an audit program with punishments for noncompliance.

Figure 8.9 shows the kit for setup of a capsule with thermoluminescence dosimetry (TLD) powder (LiF:Mg,Ti). The simple plastic rod with a hole at 5 cm depth allows positioning the black TLD capsule at reference depth below the water surface in a phantom, which can in its simplest form be just a bucket of water. The water surface is set to the top of the clear rod. The TLD holder can be disassembled and mailed in a small envelope to users all over the world.

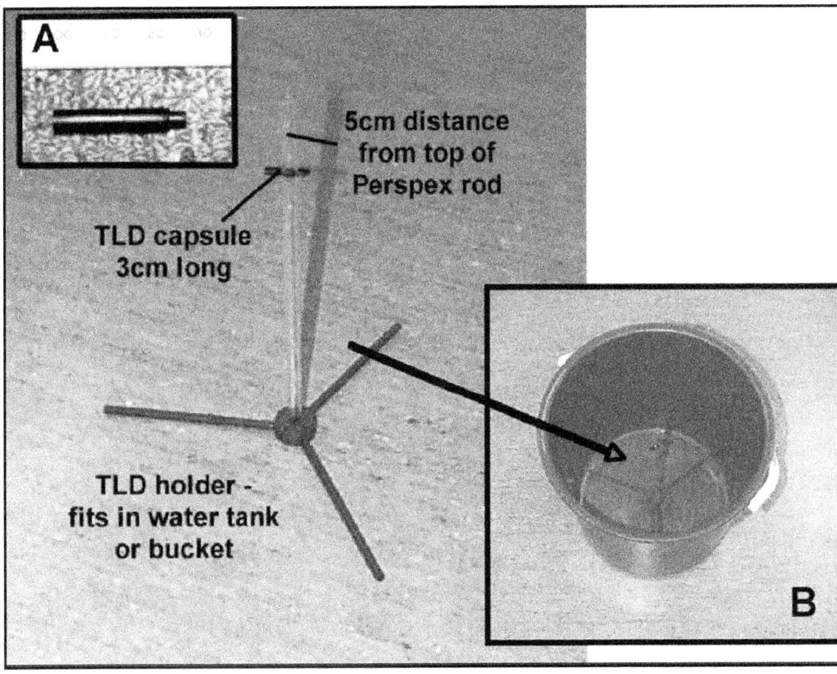

Figure 8.9: IAEA kit for TLD measurements to verify the dose under reference conditions in photon beams. Insert A: Capsule for TLD powder; B: A simple but appropriate water phantom.

The IAEA/WHO TLD program has been in operation for more than 30 years and more than 3500 photon beams have been checked in this period. Izewska and Andreo (2000) reported that a considerable number of hospitals that are participating in the program for the first time were found to be outside the tolerance limits of ±5% of the stated dose. However, it is encouraging to note that participation in the program itself appears to improve the performance: the rate of one-third of hospitals outside the limits reduces to one-fifth if they have participated in previous TL dosimetric audits.

The advantage of a TLD-based service is that the dosimeters can easily be mailed to every location. TL dosimeters are relatively robust and can be used to verify simple output as well as complex dose distributions (Kron et al. 2002, 2003). One of the major drawbacks of TLD measurements is the delay in obtaining the results. In case of the IAEA/WHO service it usually takes more than 1 month before the hospital receives the results; however, in case of significant deviations every effort is made to communicate the results as quickly as possible. At present, the IAEA only offers a dosimetric verification for photon beams; however the group at IAEA is working on a similar service for megavoltage electron beams.

Other dosimetric audit programs exist in many parts of the world. In Europe these range from early programs in the United Kingdom (Thwaites et al. 1992) to the recent EQUAL program. The latter is an initiative of the European Society of Therapeutic Radiation Oncology (ESTRO) and provides a TLD-based dosimetric audit service from the European Quality Assurance

Laboratory (EQUAL) located at the Gustave Roussy Institute in Villejuif, France. This service is available for photon and electron beams (information accessible via the ESTRO webpage: http://www.estro.be/).

In North America check of machine output is verified for many radiotherapy facilities by the Radiological Physics Center (RPC) in Houston, TX (http://rpc.mdanderson.org/rpc/). This group offers an independent check service to radiotherapy centers and is contracted by the Radiation Therapy and Oncology Group (RTOG) to perform QA audits for centers which participate in multicenter clinical trials under the auspices of the RTOG. The RPC offers a TLD output verification service for both electron and photon beams (Aguirre et al. 2002).

Figure 8.10 shows the setup for photon beam dose verification required by the RPC. Three TLD capsules are mounted in a Perspex block, which is positioned on a platform. The platform itself is positioned at 100 cm SSD and can be collapsed to allow easy transport. The size of the Perspex block depends on the photon radiation quality to ensure adequate build-up. The TL dosimeters are to be irradiated to some 3 Gy by the user and sent back to RPC. The physicists at RPC will then correct the dose reading for the incomplete scatter conditions and any other variations in the setup.

For the interpretation of an electron dose measurement it is necessary to have knowledge of the electron energy and therefore its depth dose characteristics. As such, a single measurement point is not sufficient to provide an accurate verification of electron dose, and the RPC TLD holder used for

Figure 8.10: TLD miniphantom and dosimetric setup for RPC photon dose calibration verification.

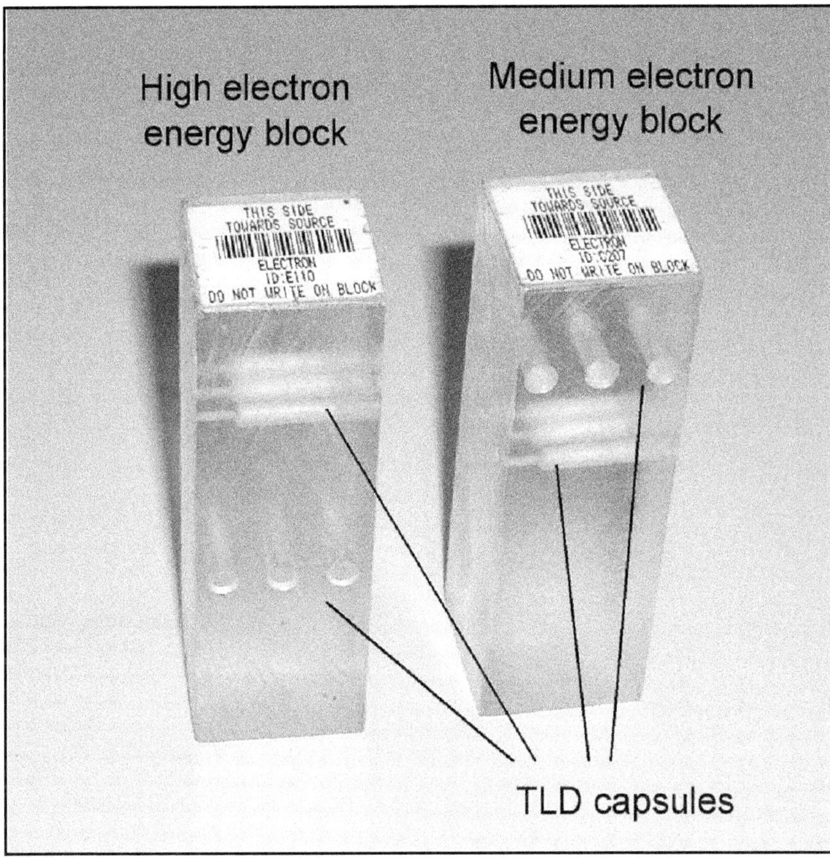

Figure 8.11: Perspex blocks for verification of electron beam dosimetry using TLD-RPC system. As a single measurement is insufficient to interpret the output of an electron beam because of uncertainty of electron energy, a second set of TL dosimeters is placed at expected depth of half maximum dose.

electron dosimetry contains two sets of TL dosimeters as can be seen in figure 8.11. The depth of the TL dosimeters in the Perspex block depends on the electron energy and is chosen to be around the depth of maximum dose and somewhere in the fall-off region to verify the beam energy. The block on the left of figure 8.11 is designed for a nominal energy of 6 MeV while the one on the right is for 12 MeV electrons. The small Perspex block used for electron dosimetry is then placed in the center of a stack of three Perpex squares, as shown in FIgure 8.12. Once the RPC has evaluated the TLD results, they provide the user not only with a dose verification but also with an estimate of the beam energy expressed as a difference in depth for the dose fall-off in the depth dose curve.

To qualify for the participation in multicenter clinical trials under the auspices of the RTOG, a radiotherapy center typically requires an audit of every photon beam in the department as well as approximately half of all electron beams. A good summary of external audits for dosimetry is provided by Izewska et al., during a recent IAEA conference on dosimetry (Izewska et al. 2002).

Dosimetric intercomparisons can be performed on different levels of complexity. Level I dosimetric intercomparisons verify the dose under reference conditions in a single point. The WHO/IAEA photon dose service is a good example for a level I dosimetric intercomparison. A level II intercomparison also verifies a relative dose ratio such as percentage depth dose, wedge angles, or transmission factors. As a meaningful electron beam intercomparison also requires an energy check, the RPC dose verification for electrons is an example for a level II intercomparison. Unfortunately, it is impossible to verify many of the features of modern radiotherapy delivery, such as IMRT, with these types of intercomparisons. A level III dosimetric intercomparison verifies the whole treatment chain in an anthropomorphic phantom such as the one shown in figure 8.13. The phantom can be imaged like a patient and a treatment plan can be developed using the clinical treatment planning system. When the treatment is delivered, dosimeters such as

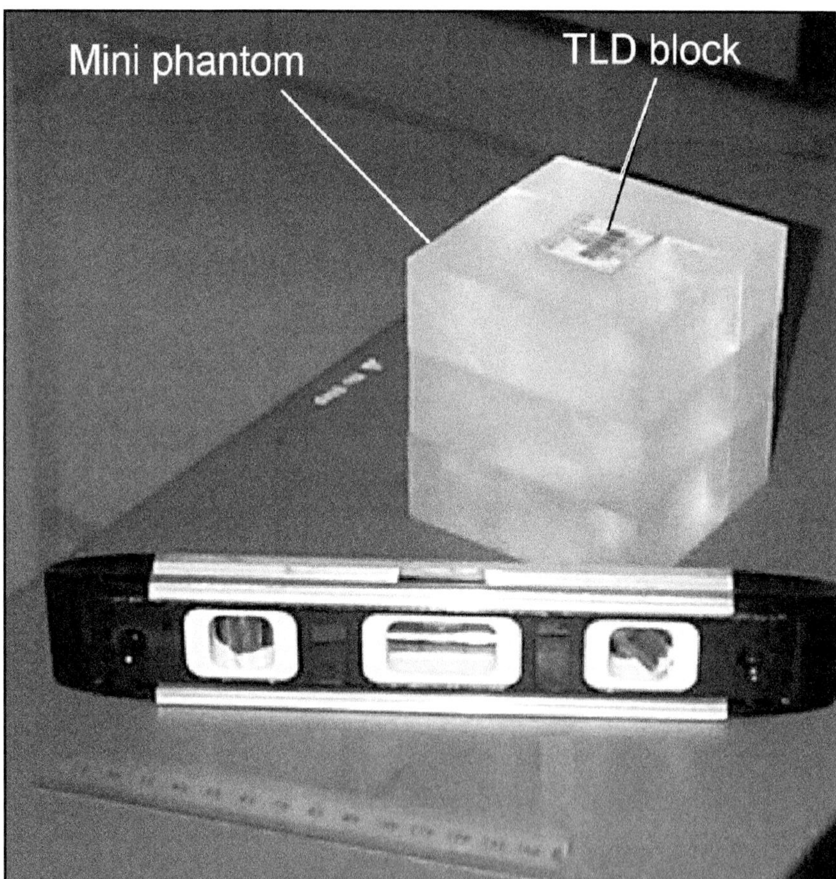

Figure 8.12: Miniphantom used by the RPC for electron beam dosimetry. The TLD block shown in figure 8.11 is in the center of the phantom.

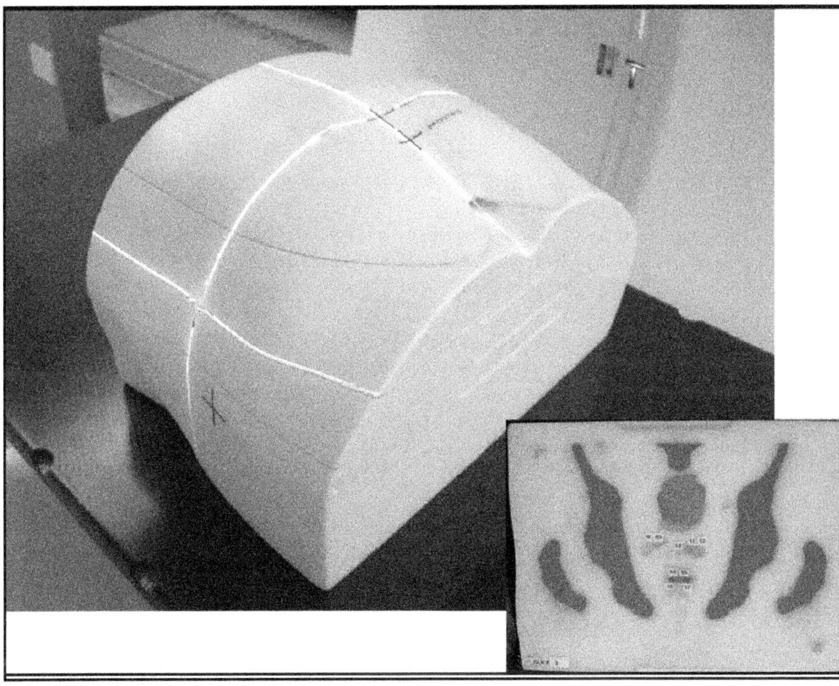

Figure 8.13: Anthropomorphic phantom for level III dosimetric intercomparisons. (Courtesy of M. Ebert.)

ionization chambers, TL dosimeters, or film can be placed in the phantom and the dose verified.

Several groups have performed these measurements (Johansson et al. 1987, Kron et al. 2002, 2003) and the RPC offers this service for verification of stereotactic radiotherapy and IMRT (http://rpc.mdanderson.org/rpc/).

Reporting and Response

If a discrepancy between calibration and independent check has occurred, the reason for it must be established. If necessary, an external expert should be called upon to check the calibration process. The process of investigation is similar for a near miss where a discrepancy has been found prior to treating patients and for a radiation accident where the wrong calibration has been used. In any case, the situation must be rectified before patients are treated (again).

Any radiation accident must also be reported to the regulatory authority. Action levels and limits for the definition of an accident are typically provided by the regulatory authority as part of the licensing process for the radiotherapy facility. Reporting ensures not only that all persons who must be notified are notified, but it also helps to ensure that the situation is rectified and others learn from any mistakes. Therefore, it may also be appropriate to report an incident or a near miss to a relevant professional organization. Dissemination of the detailed information that led to an accident or a near miss can prevent the same from occurring in other institutions.

References

AAPM (American Association of Physicists in Medicine) (1999) AAPM's TG-51 protocol for clinical reference dosimetry of high-energy photon and electron beams. *Med. Phys.* 26: 1847–1870.

AAPM (American Association of Physicists in Medicine) (1994) The calibration and use of plane-parallel ionization chambers for dosimetry of electron beams: an extension of the 1983 protocol. Report of Task Group 39. *Med. Phys.* 21: 1251–1260. Also available as AAPM Report No. 48.

AAPM (American Association of Physicists in Medicine) (1991) Clinical electron-beam dosimetry. Report of Task Group 25. *Med. Phys.* 18: 73–109. Also available as AAPM Report No. 32.

AAPM (American Association of Physicists in Medicine)(1984) Erratum: A protocol for the determination of absorbed dose from high energy photon and electron beams. Report of Task Group 21. *Med. Phys.* 11: 213.

AAPM (American Association of Physicists in Medicine) (1983) A protocol for the determination of absorbed dose from high energy photon and electron beams. Report of Task Group 21. *Med. Phys.* 10: 741–771.

Aguirre J.F., Tailor R., Ibbott G., Stovall M., Hanson W. (2002) Thermoluminescence dosimetry as a tool for remote verification of output for radiotherapy beams: 25 years experience. IAEA, Vienna, pp. 191–199.

Aird E.G., Farmer F.T. (1972) The design of a thimble chamber for the Farmer dosemeter. *Phys. Med. Biol.* 17: 169–174.

Andreo P. (2000a) On the beam quality specification of high-energy photons for radiotherapy dosimetry. *Med. Phys.* 27: 434–440.

Andreo P. (2000b) A comparison between calculated and experimental kQ photon beam quality correction factors. *Phys. Med. Biol.* 45: L25–L38.

Andreo P., Huq M.S., Westermark M., Song H., Tilikidis A., DeWerd L., Shortt K. (2002) Protocols for the dosimetry of high energy photon and electron beams: A comparison of the IAEA TRS-398 and previous international protocols. *Phys. Med. Biol.* 47: 3033–3054.

Andreo P., Nahum A., Hohlfeld K., et al. (1992) Status report on technical report series No. 277 of the IAEA.

BIPM (Bureau International des Poids et Mesures) (1981) Rapport du groupe de travail sur l'expression des incertitudes au CIPM. *BIPM Proc. Verb. Com. Int. Poids et Mesures* 49: Annexe A.

Burns D.T., Ding G.X., Rogers D.W.O. (1996) R50 as a beam quality specifier for selecting stopping-power ratios and reference depths for electron dosimetry. *Med. Phys.* 23: 383–388.

Cho S.H., Lowenstein J.R., Balter P.A., Wells N.H., Hanson W.F. (2000) Comparison between TG-51 and TG-21: Calibration of photon and electron beams in water using cylindrical chambers. *J. Appl. Clin. Med. Phys.* 1: 108–115.

Deutsches Institut für Normung e.V. (DIN) [German Institute for Standardization] (1997) Dosismessverfahren nach der Sondenmethode für Photonen- und Elektronenstrahlung. Beuth, Berlin.

Followill D., Tailor R., Tello V., Hanson W. (1998) An empirical relationship for determining photon beam quality in TG-21 from a ratio of percentage depth doses. *Med. Phys.* 25: 1202–1205.

HPA (Hospital Physicists Association) (1985) Code of practice for electron beam dosimetry in radiotherapy. *Phys. Med. Biol.* 30: 1169–1194.

HPA (Hospital Physicists Association) (1983) Revised code of practice for the dosimetry of 2 to 35 MV x-rays and of caesium-137 and cobalt-60 gamma ray beams. *Phys. Med. Biol.* 28: 1097–1104.

HPA (Hospital Physicists Association) (1969) A code of practice for the dosimetry of 2 to 35 MV x-ray and caesium-137 and cobalt-60 gamma ray beams. *Phys. Med. Biol.* 14: p. 1.

HPA (Hospital Physicists Association) (1964) A code of practice for the dosimetry of 2 to 8 MV x-ray and caesium-137 and cobalt-60 gamma ray beams. *Phys. Med. Biol.* 9: 457.

Ibbott G.S., Attix F.H., Slowey T.W., Fontenia D.P., Rozenfeld M. (1997) Uncertainty of calibrations at the accredited dosimetry calibration laboratories. *Med. Phys.* 24: 1249–1254.

IPEMB (Institution of Physics and Engineering in Medicine and Biology) (1996) The IPEMB code of practice for electron dosimetry for radiotherapy beams of initial energy from 2 to 50 MeV based on an air-kerma calibration. *Phys. Med. Biol.* 41: 2557–2603.

IAEA (International Atomic Energy Agency) (2004) Absorbed dose determination in external beam radiotherapy: An international code of practice for dosimetry based on standards of absorbed dose to water. IAEA Technical Report Series N398 (V.11b). IAEA, Vienna.

IAEA (International Atomic Energy Agency) (2000) Absorbed dose determination in external beam radiotherapy. IAEA Technical Report Series N398. IAEA, Vienna.

IAEA (International Atomic Energy Agency) (1997) The use of plane parallel ionization chambers in high energy electron and photon beams. IAEA Technical Report Series N381. IAEA, Vienna.

IAEA (International Atomic Energy Agency) (1987) Absorbed dose determination in photon and electron beams. IAEA Technical Report Series N277. IAEA, Vienna.

ICRU (International Commission on Radiation Units and Measurements) (2001) Report 64, Dosimetry of high-energy photon beams based on standards of absorbed dose to water. *Journal of the ICRU* 1.

ICRU (International Commission on Radiation Units and Measurements) (1984) Report 35. Radiation dosimetry: Electron beams with energies between 1 and 50 MeV. Bethesda, Maryland, ICRU.

ICRU (International Commission on Radiation Units and Measurements) (1972) Radiation dosimetry: Electrons with initial energies between 1 and 50 MeV. Washington, DC, ICRU.

ICRU (International Commission on Radiation Units and Measurements) (1969) Report 14. Radiation dosimetry: X-rays and gamma rays with maximum energies between 6 and 50 MeV. Bethesda, Maryland, ICRU.

ICRU (International Commission on Radiation Units and Measurements) (1938) Recommendations of the ICRU (Chicago 1937). *Br. J. Radiol.* 39: 295–298.

ISO (International Standards Organisation) (1995) Guide to the expression of uncertainties in measurements. 2nd. Geneva, International Standards Organisation.

Izewska J., Andreo P. (2000) The IAEA/WHO TLD postal programme for radiotherapy hospitals. *Radiother. Oncol.* 54: 65–72.

Izewska J., Andreo P., Vatnitsky S., Shortt K.R. (2003) The IAEA/WHO TLD postal dose quality audits for radiotherapy: A perspective of dosimetry practices at hospitals in developing countries. *Radiother. Oncol.* 69: 91–97.

Izewska J., Svenson H., Ibbott G. (2002) Worldwide QA networks for radiotherapy dosimetry. IAEA, Vienna.

Johansson K.A., Horiot J.C., van der Schueren E. (1987) Quality assurance control in the EORTC cooperative group of radiotherapy. 3. Intercomparison in an anthropomorphic phantom. *Radiother. Oncol.* 9: 289–298.

Kemp L. (1972) The NPL secondary standard therapy-level x-ray exposure meter. *Br. J. Radiol.* 45: 775-778.

Khan F.M. (2000) Comment on "AAPM's TG-51 protocol for clinical reference dosimetry of high-energy photon and electron beams." [*Med. Phys.* 26:1847–1870 (1999)]. *Med. Phys.* 27: 445–447.

Khan F.M., Doppke K.P., Hogstrom K.R., Kutcher G.J., Nath R., Prasad S.C., Purdy J.A., Rozenfeld M., Werner B.L. (1991) Clinical electron-beam dosimetry: Report of AAPM Radiation Therapy Committee Task Group No. 25. *Med. Phys.* 18: 73–109. Also available as AAPM Report No. 32.

Kosunen A., Rogers D.W.O. (1993) Beam quality specification for photon beam dosimetry. *Med. Phys.* 20: 1181–1188.

Kron T. (2005) Calibration of megavoltage radiation beams. Chapter 9 in *The Modern Technology of Radiation Oncology, Volume 2*. J. Van Dyk (ed.). Medical Physics Publishing, Madison, Wisconsin, pp. 301–382.

Kron T., Barnes K., O'Brien P. (2003) Multicentre dosimetric comparison of photon-junctioning techniques in head and neck radiotherapy. *Australas. Radiol.* 47: 289–294.

Kron T., Hamilton C., Roff M., Denham J. (2002) Dosimetric intercomparison for two Australasian clinical trials using an anthropomorphic phantom. *Int. J. Radiat. Oncol. Biol. Phys.* 52: 566–579.

Li X., Rogers D.W.O. (1994) Reducing electron contamination for photon beam quality specification. *Med. Phys.* 21: 791–798.

Lillicrap S., Owen B., Williams J., Williams P. (1990) Code of practice for high energy photon therapy dosimetry based on the NPL absorbed dose calibration. *Phys. Med. Biol.* 35: 1355–1360.

Ma C.-M., Coffey C.W., DeWerd L.A., Liu C., Nath R., Seltzer S.M., Seuntjens J.P. (2001) AAPM protocol for 40-300 kV x-ray beam dosimetry in radiotherapy and radiobiology. *Med. Phys.* 28: 868–893. Also available as AAPM Report No. 76.

Mattsson O. (1985) Comparison of different protocols for the dosimetry of high energy photon and electron beams. *Radiother. Oncol.* 4: 313–318.

McCaffrey J.P., Downton B., Shen H., Niven D., McEwen M. (2005) Pre-irradiation effects on ionization chambers used in radiation therapy. *Phys. Med. Biol.* 50: N121–N133.

Metcalfe P., Kron T., Hoban P. (1997) *The Physics of Radiotherapy X-Rays from Linear Accelerators.* Medical Physics Publishing, Madison, Wisconsin.

Mijnheer B., Thwaites D., Williams J. (1993) Absolute dose determination. *Radiotherapy Physics.* J. Williams D. Thwaites (eds.). Oxford University Press, Oxford, pp. 31–52.

Morgan A.M., Aird E.G., Aukett R.J., Duane S., Jenkins N.H., Mayles W.P., Moretti C., Thwaites D.I. (2000) IPEM guidelines on dosimeter systems for use as transfer instruments between the UK primary dosimetry standards laboratory (NPL) and radiotherapy centres. *Phys. Med. Biol.* 45: 2445–2457.

NACP (Nordic Association of Clinical Physicists) (1980) Procedures in external radiation therapy dosimetry with electron and photon beams with maximum energies between 1 and 50 MeV. *Acta Radiol. Oncol.* 19: 55–79.

NACP (Nordic Association of Clinical Physicists) (1972) Procedures in radiation therapy dosimetry with 5 to 50MeV electrons and roentgen and gamma rays with maximum energies between 1 and 50 MeV. *Acta Radiol. Ther. Phys. Biol.* 2: 603–624.

Nath R., Huq M.S. (1995) Advances in radiation dosimetry. *Radiation Therapy Physics.* A. Smith (ed.). Springer, Berlin, pp. 401–448.

Reft C. (1989) Output calibration in solid water for high energy photon beams. *Med. Phys.* 16: 299–301.

Reich H. (1990) Dosimetrie ionisierender Strahlung. Teubner, Stuttgart.

Reich H. (1979) Choice of measuring quantity for therapy-level dosimeters. *Phys. Med. Biol.* 24: 895–900.

Rice R., Hanson J., Chin L., Mijnheer B., Bjärngard B. (1988) The influence of ionization chamber and phantom design on the measurement of lung doses in photon beams. *Med. Phys.* 15: 884–890.

Rogers D.W.O. (1992) The advantages of absorbed-dose calibration factors. *Med. Phys.* 19: 1227–1239.

Rogers D.W.O. (2000) Comment on "On the beam quality specification of high-energy photons for radiotherapy dosimetry" [*Med. Phys.* 27: 434–440 (2000)]. *Med. Phys.* 27: 441–444.

Schulz R.J., Almond P.R., Kutcher G., Loevinger R., Nath R., Rogers D.W.O., Suntharalingam N., Wright K.A., Kahn F.M. (1986) Clarification of the AAPM Task Group 21 protocol. *Med. Phys.* 13: 755–759.

SCRAD (Sub-Committee on Radiation Dosimetry of AAPM) (1966) Protocol for the dosimetry of high energy electrons. *Phys. Med. Biol.* 11: 505–520.

SCRAD (Sub-Committee on Radiation Dosimetry of AAPM) (1971) Protocol for the dosimetry of X- and gamma-ray beams with maximum energies between 0.6 and 50 MeV. *Phys. Med. Biol.* 16: 379–396.

Shortt K., Ross C.K., Seuntjens J.P., Delaunay F., Ostrowsky A., Gross P., Leroy E. (2001) Comparison of dosimetric standards of Canada and France for photons at 60-Co and higher energies. *Phys. Med. Biol.* 46: 2119–2142.
Shortt K., Shobe J., Domen S. (2000) Comparison of dosimetry calibration factors at the NRCC and the NIST. National Research Council of Canada. National Institute of Standards and Technology. *Med. Phys.* 27: 1644–1654.
Tailor R.C., Hanson W.F. (2002) Calculated absorbed-dose ratios, TG51/TG21, for most widely used cylindrical and parallel-plate ion chambers over a range of photon and electron energies. *Med. Phys.* 29: 1464–1472.
Thwaites D.I., Williams J.R., Aird E.G., Klevenhagen S.C., Williams P. C. (1992) A dosimetric intercomparison of megavoltage photon beams in UK radiotherapy centres. *Phys. Med. Biol.* 37: 445–461.
Van Dyk J. (ed.) (2005) *The Modern Technology of Radiation Oncology, Volume 2.* Medical Physics Publishing, Madison, Wisconsin.

Questions

8.1 Please describe the difference between absolute and relative dosimetry. (Compare chapter three and section 8.1.)

8.2 Which dosimetric technique would be useful for absolute dosimetry? (Compare chapter three and sections 8.1 and 8.2.)

8.3 You have been called as a consultant to check the calibration of a medical linac. The center provides the following information (may or may not be relevant):

- The linac is calibrated to give 1 Gy at depth of maximum dose for 100 monitor units in a 10×10 cm^2 field at 100 cm FSD.
- $TPR_{10}^{20} = 0.74$.
- The percentage depth dose in a 10×10 cm^2 field at 5 cm depth is 0.92.

Using your Capintec-Farmer 0.6cc ionization chamber with graphite wall, you assess the dose at 5 cm depth in a water phantom. The wall thickness of the chamber is 0.046 g/cm^3 and the PMMA cap thickness is 0.625 g/cm^3. You intend to use the AAPM TG 21 protocol (1983) and perform your measurements in a water phantom. The temperature is 24 °C and the pressure is 101.5 kPa (761.5 mm Hg). From your Accredited Dosimetry Calibration Laboratory (ADCL) you have got an N_X (exposure calibration factor) for cobalt-60 exposure of 0.505 R/division for your ion chamber/electrometer combination. For the following applied high voltages you find the following readings for 100 monitor units:

–150V: 188.0, 187.6, 187.6

–300V: 190.2, 190.5, 190.0

Can you confirm the calibration of the center?

You may want to use the worksheet given in the protocol. The following information, which has been adapted from the protocol, may (or may not) be useful (AAPM 1983):

- The ion collection efficiency factor for the NE 2571 chamber A_{ion} can be assumed to be 1 in the cobalt-60 beam used at the ADCL. The collection efficiency in the linear accelerator beam at the center P_{ion} can be determined from the AAPM protocol using the half-voltage method (in the AAPM protocol (1983), section IV.C, figure 4). For the purpose of the present exercise it can be assumed that the factor is equal to the ratio of ionization readings as long as the ratio does not exceed 1.10. Therefore: $P_{ion} = R_{high\ voltage} / R_{half\ voltage}$.
- N_X at the ADCL has been obtained for T = 22 °C, p = 760 mmHg, a relative humidity of 50% and by using a negative voltage of 300V.
- At cobalt-60 gamma rays the fraction of the energy which is lost by the secondary electrons due to bremsstrahlung in air is approximately 0.3%.
- For the NE 2571 chamber the protocol gives the following values for
    - -- the correction for attenuation and scattering of photons in the chamber wall $A_{wall} = 0.99$.
    - -- the fraction of the ionization due to electrons from the chamber wall irradiated by cobalt-60 at the ADCL, $\alpha = 0.42$; therefore the fraction due to the PMMA cap is 0.58.
    - -- the stopping power ratio for graphite walled chambers in a cobalt-60 beam $\left(L/\rho\right)_{gas}^{wall} = 1.01$.
    - -- the mass energy absorption coefficient for graphite walled chambers in a cobalt-60 beam $\left(\mu/\rho\right)_{wall}^{air} = 0.999$.
    - -- the stopping power ratio for a PMMA build-up cap in a cobalt-60 beam $\left(L/\rho\right)_{gas}^{PMMA} = 1.103$.
    - -- the mass energy absorption coefficient for a PMMA build-up cap in a cobalt-60 beam $\left(\mu/\rho\right)_{PMMA}^{air} = 0\ 925$.
- The humidity correction can be assumed to be 1.
- The average energy required to produce an ion pair in air is W/e = 33.7 J/C.
- The conversion factor of roentgens to J/kg, $k = 2.58 \times 10^{-4}$ C/kg R.
- The absorbed dose/collision fraction of KERMA $\beta_{wall} = 1.005$.
- The wall correction factor can be calculated using equation (10) of the protocol (equation 8.17) in the present chapter) as:

$$P_{wall} = \frac{\left[\alpha\left(L/\rho\right)_{gas}^{wall}\left(\mu_{en}/\rho\right)_{wall}^{med} + (1-\alpha)\left(L/\rho\right)_{gas}^{med}\right]}{\left(L/\rho\right)_{gas}^{med}},$$

In the present case this is simplified as α, the fraction of the electrons arising in the wall, is less than 25%. In this case the AAPM protocol suggests to set $\alpha = 0$ and $P_{wall} = 1$.

- The following table gives values for the (restricted) stopping power ratios $s_{w,air}$, $\left(L/\rho\right)_{air}^{water}$ and the replacement correction factor P_{repl} (AAPM 1983). Some of the values were interpolated or transcribed from graphs. The data in the table are not meant to be used for calibration.

Table Q8.1:

TPR_{10}^{20}	Nominal energy (MV)	D_{20}/D_{10}	$\left(L/\rho\right)_{air}^{water}$ (AAPM)	P_{repl} for chamber with 6.5 mm diameter (AAPM)
0.57	cobalt-60	0.5	1.134	0.991
0.62	3.9	0.535	1.131	0.992
0.64	4.4	0.55	1.130	0.992
0.66	5.0	0.57	1.129	0.992
0.68	5.8	0.585	1.127	0.992
0.70	7.0	0.60	1.124	0.992
0.72	8.0	0.615	1.121	0.992
0.74	9.5	0.63	1.118	0.992
0.76	12	0.645	1.113	0.993
0.78	20	0.66	1.096	0.993

(section 8.3)

8.4 A colleague calls you for advice regarding the calibration of their 6 MeV electron beam. She wants to use the AAPM TG 51 protocol for its calibration and would like to use her calibrated Farmer-type chamber (Capintec PR-06) for the calibration. The beam quality has been determined by acquiring a depth ionization curve using a Scanditronix RK chamber with an internal radius of 2 mm. Using the data from these measurements directly without any correction for the effective point of measurement, the depth of 50% ionization was found to be $I_{50} = 25$ mm.

i What type of chamber should be used for the calibration?

ii) What are the reference conditions (field size, depth, SSD) used for the calibration?

(Compare section 8.4.1.)

8.5 The best way to check output calibration is to invite an independent expert (preferably an accredited physicist) using an independent measurement system (electrometer and ionization chamber). Ideally, the independent check should be performed for all radiation qualities available at a treatment unit.

If no visit of an independent expert can be arranged or indeed as a second check anyway, there should at least be an independent verification of output using, for example, thermoluminescent (TL) dosimeters which can easily be sent from one radiotherapy center to another. These devices provide an alert for further investigation if large differences in calibrated output are discovered.
(section 8.3.)

9 Beam Models: Part I

"The essence of the equivalent tissue-air ratio method is the idea that a quantity, analogous to a tissue-air ratio, may be determined for phantoms containing non-water equivalent materials by modeling...."
(Marc R. Sontag et al. 1978)

9.1 Introduction

The effectiveness of radiotherapy for curing cancer depends on the ability to maximize the probability of tumor control while minimizing the probability of complications arising from the irradiation of healthy tissue. Since both these quantities are very sensitive to absorbed dose, accurate knowledge of the respective dose levels is essential if their ratio is to be maximized. Furthermore, the uniformity of dose throughout the tumor is critical, as is the integral dose received by each healthy organ. At treatment time, various techniques are available (such as multiple fields, blocking, wedging, compensating, and, recently, multileaf collimators) to shape the radiation field so that the dose distribution is concentrated within a well-defined volume. However, the effectiveness of these methods is dependent on the treatment planning process being able to correctly predict the dose distribution resulting from a particular treatment configuration. An accurate density map of the patient's internal structures is available from CT data (Brooks and Di Chiro 1975, Geise and McCullough 1977), and, thus, there is minimal uncertainty in the density distribution throughout structures such as lung (Van Dyk et al. 1982, Wegener et al. 1978). ICRU (1976) has recommended an overall dose accuracy of 5%. Assuming an equal allowed error for dose uncertainties in three aspects of patient treatment (patient setup, machine calibration, and the dose calculation algorithm), it is necessary to have an algorithm that can calculate the dose to an accuracy of 3%.

Accurate calculation of an x-ray beam dose distribution within an inhomogeneous medium such as the human body is a complex task since the effect of inhomogeneities on primary and scattered radiation must be treated realistically. The usual way to obtain the dose distribution within an inhomogeneous medium is to adjust the dose measured within a homogeneous water medium (Spiers 1951) in a procedure known as *inhomogeneity correction*. These methods involve varying degrees of complexity for x-ray and electron beams. Some, such as the *effective pathlength* (E-depth) correction (Milan and Bentley 1974, Redpath et al. 1977) for photon beams, discussed in section 9.2, treat all dose as primary dose and thus only take into account the variation in medium density along a path from the radiation source to the point of calculation. More complex photon correction–based algorithms are discussed in section 9.3, starting with the *power law* method that considers the position of inhomogeneities along a path as well as their thickness and density (Batho 1964). This method takes some account of the perturbation in scattered dose. A method that explicitly accounts for scattered dose is the *equivalent tissue-air ratio* (ETAR) method (Cunningham 1972, Sontag and Cunningham 1978). This method considers the spatial distribution of inhomogeneities in determining an equivalent depth and field size for each calculation point. Other methods that attempt to account for scatter perturbations include the *differential scatter-air ratio* (Cunningham 1982) and *delta volume* (Wong and Henkelman 1983) methods. The accuracy with which such methods are able to calculate scattered dose is dependent on the assumptions made about the medium geometry and about the transport of scattered photons. In general, accuracy increases with the computation time required to calculate a dose distribution.

Although these methods are able to calculate the effect of inhomogeneities on photon fluence, the perturbation of *electron fluence* is not considered. Since it is secondary electrons (set in motion by photons) that deposit energy, the effects of inhomogeneities are not predicted accurately unless perturbations in electron fluence as well as photon fluence are taken into account. Electron transport can only be ignored when electronic equilibrium (see chapter two) exists. In this case, the change in dose caused by an inhomogeneity is proportional to the change in photon fluence. Situations of electronic disequilibrium are common for high-energy photon beams, where the range of electrons can be several centimeters. Reviews on inhomogeneity corrections (Battista and Sharpe 1992, Mohan 1989) point out that the non-allowance for electronic disequilibrium can lead to significant errors at high x-ray energies, especially in media containing low-density regions such as lung and air cavities, where the range of electrons is greatly increased from that in water (Mohan and Chui 1985, Hoban et al. 1992). The assumption of electronic equilibrium is implicit in the use of tissue-air ratio measurements (obtained under electronic equilibrium conditions in water) to calculate the primary and scattered dose components. Dose computation methods that do account for electron transport include *convolution* and *Monte Carlo* methods. These models are discussed in chapter ten.

Electron pencil beam models are commonly used in radiotherapy treatment planning (RTP) calculations. The Fermi-Eyges pencil beam model and the Hogstrom implementation of this model are discussed in section 9.4. Dose models for radioactive sources vary significantly from those of external beam models. So as an interesting contrast the brachytherapy source model used in most brachytherapy planning systems is briefly discussed in section 9.5.

Common to most correction-based RTP external beam calculation algorithms is the use of beam data collected in water under reference conditions. These data normally include a depth dose curve and several profile curves stored for a range of field sizes at each available beam energy. A dose distribution for a particular treatment beam and patient configuration is then determined by manipulating these data. Major considerations in the calculation of a dose distribution include the position and shape of the patient external contour with respect to the beam and density variations, or inhomogeneities, within the patient.

9.2 The Milan/Bentley Model

In the early 1970s, an efficient method of calculating dose distributions, which accounts for changes in SSD, surface curvature, and (to some extent) inhomogeneities, was developed by J. Milan and R. E. Bentley at the Royal Marsden Hospital (Milan and Bentley 1974). The Milan/Bentley model is used by some commercial treatment planning systems (for example, the GE Target II) as the basis for their photon beam dose calculation algorithms. The Milan/Bentley model makes use of measured central axis depth dose and profile data for a range of field sizes at each beam energy. Data are collected in water at a reference SSD and manipulated to account for patient and setup variations.

9.2.1 Beam Data Requirements

Beam data are measured for square field sizes covering the range used for treatment. For each field size, these data include (a) 17 central axis depth dose measurements and (b) 47 profile measurements at depths corresponding to d_{max} and central axis points 5, 9, 13, and 17. This arrangement and number of data points is that which was originally chosen by Milan and Bentley, and is also used by the Target system. The spacing of central axis points is user selectable, and the profile point spacing is dependent on field size (such that the tails of the penumbra are included). Profile point spacing increases linearly with depth, in proportion to the distance from the source, thus forming a *fan line* grid. A representation of the beam data is shown in figure 9.1, where the 47 fan lines and 17 central axis points are indicated. Each measured profile is normalized to 1.0 at the central point, so that the profile value gives a dose relative to that at d_{max} when multiplied by the central axis value at that depth. Profiles at several depths are needed because of the variation in the shape of the curves brought about by (a) an increasing scattered dose contribution with depth and (b) an increasing geometric penumbra width. Fine resolution in each profile curve is required because of the rapid dose gradient in the penumbra.

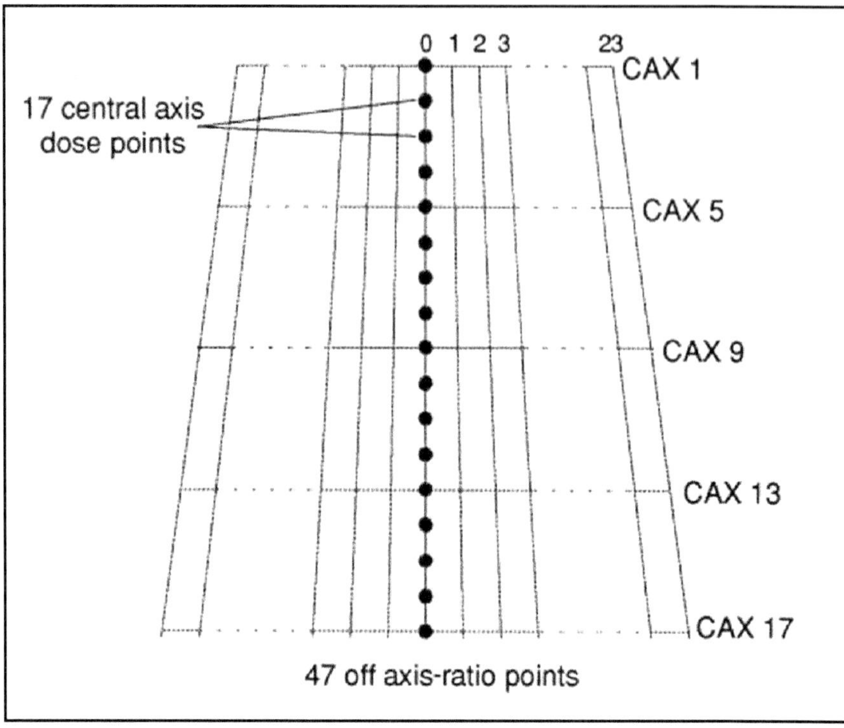

Figure 9.1: Schematic of the fan line grid representing beam data used in the Milan/Bentley model. Seventeen central axis dose ponts are stored, along with 47 off-axis ratios (23 either side of the axis) at the depths of central axis points 1, 5, 9, 13, and 17.

9.2.2 Infinite SSD Central Axis Data

The 17 central axis dose points in the Milan/Bentley model are in fact stored with the inverse square dependence removed. This is known as infinite SSD depth dose data, where the dose fall-off is that which would exist if there were no inverse square falloff. In converting depth dose data collected at a reference SSD, SSD_1, to infinite SSD data, the following transformation is performed on the dose points:

$$C(d,FS_1) = P(d,FS_1,SSD_1)\left(\frac{SSD_1 + d}{SSD_1 + d_{max}}\right)^2. \tag{9.1}$$

In this equation $P(d,FS_1)$ is the measured percentage depth dose at depth d for a field size at the surface of FS_1 and $C(d,FS_1)$ is the corresponding infinite SSD dose.

Central axis data are stored in this format because of the need to convert measured percentage depth doses to percentage depth doses at another SSD, such as in extended SSD treatments or when the SSD varies with surface curvature: For depth dose calculations at any SSD, the appropriate inverse square dependence is reintroduced as described below. Note that $C(d,FS_1)$ is not exactly equivalent to the percentage depth dose for a field size FS_1 at an infinite SSD (a tissue-maximum ratio) because the effect on scatter dose of field size increase with depth is still included.

9.2.3 Obtaining Percentage Dose from Central Axis and Profile Data

The percentage dose (relative to that at d_{max}) is found by converting the infinite SSD depth dose data to diverging field data (reintroducing the inverse square dependence) and multiplying by the off-axis ratio. The off-axis distance is in fact converted to that at the reference SSD by a simple divergence factor, so that all points along a ray line from the source are treated as having an off-axis distance equal to that at the reference SSD. The fan line grid used to store off-axis ratios is thus treated as if it consisted of parallel rays.

For the simple case where SSD is the same as the reference SSD, when field size at the surface is FS, the dose at a depth d and off-axis distance y is given by

$$D(d,y,FS)=C(d,FS)R(d,y',FS)\left(\frac{SSD+d_{max}}{SSD+d}\right)^2, \tag{9.2}$$

where $R(d,y',FS)$ is the off-axis ratio for depth d and off-axis distance y' (projected to that at the reference SSD). The values for $C(d,FS)$ and $R(d,y',FS)$ are obtained by interpolation between measured values.

9.2.4 Correction for SSD Change

If a source surface distance (SSD_2) other than that used for data collection is used, the percentage depth dose (with respect to d_{max} at SSD_1) is given by

$$P(d,FS_2,SSD_2)=P(d,FS_1,SSD_1)\left(\frac{SSD_1+d}{SSD_2+d}\right)^2, \tag{9.3}$$

where FS_2 is the field size at the new SSD, with the same jaw settings: $FS_2 = FS_1(SSD_2/SSD_1)$.

The percentage depth dose can also be written in terms of the central axis dose by substituting equation 9.1 into equation 9.3:

$$P(d,FS_2,SSD_2)=C(d,FS_1)\left(\frac{SSD_1+d_{max}}{SSD_2+d}\right)^2, \tag{9.4}$$

If the change in field size is large enough, there will be a significant change in field size at the surface and hence a change in the dose contribution due to scattered photons. The central axis data obtained for field size FS_1 will thus not be appropriate to use for a field size FS_2, as the shape of the depth dose curve (even with inverse square dependence removed) should be different for the different field sizes. The central axis data used in equation 9.3 should in fact be those obtained for a field size FS_2 at SSD_1:

$$P(d, FS_2, SSD_2) = C(d, FS_2)\left(\frac{SSD_1 + d_{max}}{SSD_2 + d}\right)^2. \tag{9.5}$$

Note that equations 9.4 and 9.5 evaluate faster than equation 9.3 due to the fact that $(SSD_1 + d_{max})$ is a constant. This is another reason for storing infinite SSD data. For SSD_2, the dose at depth d and off-axis distance y is given by including the off-axis term in equation 9.5:

$$D(d, y, FS_2) = C(d, FS_2)R(d, y', FS_1)\left(\frac{SSD_1 + d_{max}}{SSD_2 + d}\right)^2. \tag{9.6}$$

Equation 9.6 includes the change in output brought about by the change in SSD. If a percentage depth dose with respect to d_{max} at the new SSD is required, however, the expression is

$$\begin{aligned} P(d, FS_2, SSD_2) &= P(d, FS_1, SSD_1)\left(\frac{SSD_1 + d}{SSD_1 + d_{max}}\right)\left(\frac{SSD_1 + d_{max}}{SSD_2 + d}\right)^2 \\ &= C(d, FS_2)\left(\frac{SSD_2 + d_{max}}{SSD_2 + d}\right)^2, \end{aligned} \tag{9.7}$$

where the advantage of using infinite SSD data is clearly seen to reduce computations.

9.2.5 Correction for Surface Curvature

The procedure used in dealing with surface curvature is a generalization of that described above for a change in SSD, except that the SSD corresponding to each ray line must be considered separately. Consider a point at a position x,y with respect to the central axis of the beam at the reference SSD (the origin). To obtain the dose at this point, we need (a) the position at which a ray line from the source intersects the surface and (b) the projected off-axis distance at the reference SSD.

From (a), the depth, d, beneath the surface is obtained and also (via the distance of the intersection point from the source) the effective field size to use in looking up the central axis value for depth, d. If the intersection point is a distance x_s from the origin in a direction parallel to the beam axis, then the depth is $d = x - x_s$ and the field size at the surface is $FS_s = FS_1(SSD_1 + x_s)/SSD_1$. The projected off-axis distance is simply $y' = y(SSD_1/(SSD_1 + x))$. The dose at x,y is thus given by

$$D(x, y) = C(d, FS_s)R(d, y', FS_1)\left(\frac{SSD_1 + d_{max}}{SSD_1 + x}\right)^2, \tag{9.8}$$

where $SSD_1 + x$ is equivalent to $SSD_2 + d$ in equation 9.6.

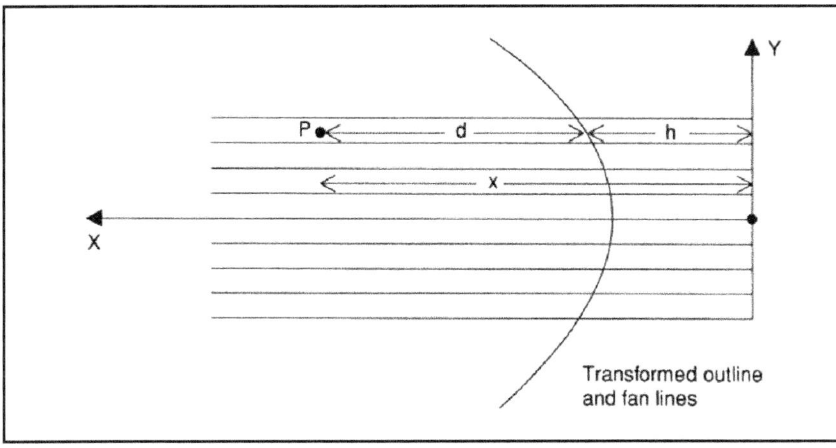

Figure 9.2: Transformed patient outline, such that the beam z-axis (the direction of the beam) is aligned with the x-axis of the patient coordinate system and the patient y-axis is aligned with the beam x-axis at the reference SSD. All outline and dose calculation points are multiplied by a divergence factor to convert each y value to that at the reference SSD.

In practice, this procedure is carried out by first transforming points in the patient outline so that the beam z-axis (the direction of the beam) is aligned with the x-axis of the patient coordinate system and the patient y-axis is aligned with the beam x-axis at the reference SSD. Furthermore, all outline and dose calculation points are multiplied by a divergence factor to convert each y-value to that at the reference SSD (as in $y \to y'$). This makes the intersection coordinates of ray lines with the outline easier to find. The transformed fan lines and patient outline are shown in figure 9.2.

An important need for surface curvature correction is isocentric treatments, where the SSD is always less than the distance from the source to the machine isocenter (the *source-axis distance* [SAD]). For a point in the y-axis in the plane of the isocenter, the dose is given by a special case of equation 9.8, where $SSD_1 + x = SAD$:

$$D(x,y) = C(d, FS_s)\, R(d, y, FS_1) \left(\frac{SSD_1 + d_{max}}{SAD} \right)^2 . \tag{9.9}$$

Furthermore, in isocentric treatments the standard SSD for beam measurements (SSD_1) will normally be equal to the SAD. In this case, in equation 9.9 $SSD_1 = SAD$ and $y' = y$. The dose at isocenter is of course given by equation 9.9 with off-axis term omitted.

9.2.6 Inhomogeneity Correction Using Effective Pathlength

A simple method of correcting for inhomogeneities can be incorporated into the Milan/Bentley method, where the depth used to find the central axis term, C, is the effective depth at the dose calculation point. Effective depth at this point, which has x-coordinate xc with respect to the standard SSD, is found using a ray tracing procedure where increments in effective path-

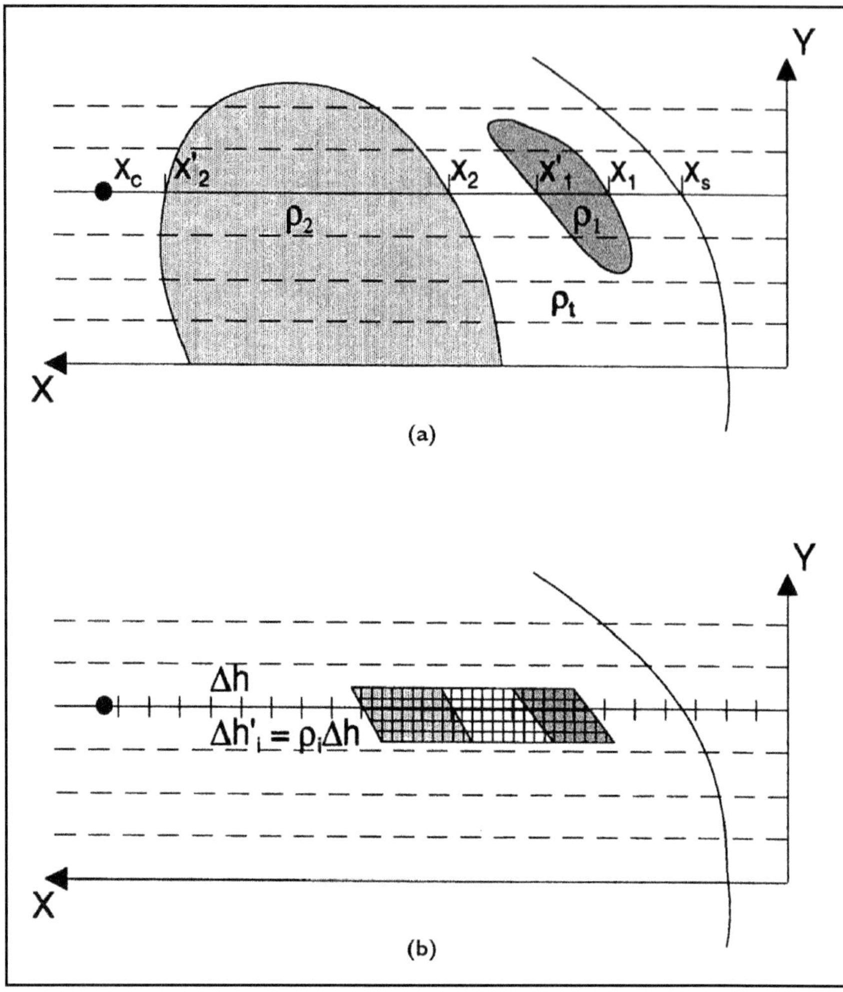

Figure 9.3: (a) Effective depth found from the intersections of ray lines with inhomogeneity boundaries (see equation 9.10). (b) Effective depth found from increments in effective pathlength, $\Delta h'$, for each increment in depth Δh along each ray line (see equation 9.12). Effective pathlength increments are found from pixel values in that depth interval.

length along a ray line from the source are summed. The increments can either be found using densities along the ray line obtained from a CT pixel matrix, or using the density of each outlined region intersected by the ray. In the original Milan/Bentley model, the latter method is used, where first the entry and exit points associated with each outlined structure must be found. Note that each outline point is first transformed as described in the last section, so the rays are considered parallel. If a ray enters the ith structure at an x-coordinate x_i and exits the structure at x_i', where the density of the ith structure is ρ_i, then the effective depth at an x-coordinate x_c is given by (see figure 9.3a)

$$d_{eff} = (x_1 - x_s)\rho_t + (x_1' - x_1)\rho_1 + (x_2 - x_1')\rho_t + \cdots + (x_n' - x_n)\rho_n + (x_c - x_n')\rho_t. \tag{9.10}$$

The calculation point is in this case beyond the nth inhomogeneity, in normal tissue of density ρ_t ($\rho_t = 1.0$ for unit density tissue). If the calculation point is inside the nth inhomogeneity, equation 9.10 is replaced by

$$d_{eff} = (x_1 - x_s)\rho_t + (x_1' - x_1)\rho_1 + (x_2 - x_1')\rho_t + \cdots + (x_c - x_n)\rho_n. \tag{9.11}$$

If a CT pixel matrix is used to find the effective depth, a summation of the products of pixel width and pixel density along ray lines can be performed. In practice, as is done by the GE Target II system, ray lines are divided into segments of length Δh, and the increments in effective pathlength $\Delta h'$ are summed. Each increment in effective pathlength is found by first averaging the densities represented by nearest neighbor pixels along the ray line segment. If the average density of increment i is ρ_i, then the effective pathlength increment is $\Delta h' = \rho_i \Delta h$. For a point at depth d, the effective depth d' is then found from the sum

$$d' = \sum_{i=1}^{n} \Delta h_i' + \Delta h_{n+1}' (d - n\Delta h) / \Delta h, \tag{9.12}$$

where the calculation point lies in the $(n + 1)$th ray line segment. The method of finding effective depth using pixel-by-pixel ray-tracing is shown in figure 9.3b.

Using the effective depth found by either of the methods described above, equation 9.8 can be rewritten to take account of inhomogeneities:

$$D(x, y) = C(d_{eff}, FS_s) R(d, y', FS_1) \left(\frac{SSD_1 + d_{max}}{SSD_1 + x} \right)^2. \tag{9.13}$$

Note that the physical depth, d, is still used in the off-axis ratio term and inverse square term (where $x = x_s + d$).

The methods discussed above provide an accurate and efficient way of predicting dose distributions in situations where the source-surface distance is not that used for data collection, and when the patient surface is curved (as is generally the case). The method of inhomogeneity correction is however not rigorous, as it does not take account of the effects of heterogeneities on scattered photons or secondary electrons. The strength of the Milan/Bentley method is thus in obtaining an accurate dose distribution for a water-equivalent medium, based on data measured under standard conditions. It should be noted that although the theoretical basis of algorithms that account for perturbation of scattered photon dose did exist at the time the Milan/Bentley method was developed, such routines were not generally implemented due to the lack of computing power. Similarly, the more complex inhomogeneity correction methods implemented on treatment planning computers today were generally formulated one to two decades ago.

9.3 Inhomogeneity Corrections

9.3.1 Effective Depth Methods

In equation 6.13, the inhomogeneity correction appears as a modification to the central axis term (by replacing the depth argument with the effective depth). It is also common to express the dose at a point in an inhomogeneous medium as the dose to a water medium (with the same SSD and curvature) multiplied by a correction factor *CF*. In the Milan/Bentley method, the dose to point x,y if the medium were water, $D_W(x,y)$, is given by equation 9.8. Thus, from these two equations, the correction factor is simply a ratio of infinite SSD central axis dose at the effective depth to that at the physical depth, for the field size at the surface. Equivalently, this is a ratio of tissue-air ratios (TARs) at the effective and physical depths, for the field size at the depth of calculation. If the field size at the surface and calculation depths is FS_s and FS_d respectively, the correction factor is given by

$$CF = \frac{C(d_{eff}, FS_s)}{C(d, FS_s)} = \frac{TAR(d_{eff}, FS_d)}{TAR(d, FS_d)}. \tag{9.14}$$

This is the most basic inhomogeneity correction and is known as the *ratio of TARs* (RTAR) method. The correction factor can also be given in terms of percentage depth dose data. Using equation 9.1 (for field size F_s) in equation 9.14 gives

$$CF = \frac{P(d_{eff}, FS_s)}{P(d, FS_s)} \left(\frac{SSD_1 + d_{eff}}{SSD_1 + d} \right)^2. \tag{9.15}$$

Physically, the inverse square term in equation 9.15 accounts for the fact that the percentage depth dose values (which have an inverse square dependence) are from measurements at different distances from the source. Other effective depth methods, which differ in their implementation but are common in that they correct only for primary dose perturbation, include the *isodose shift* and *effective SSD* methods (Cunningham 1982).

The correction factor format is used by many inhomogeneity correction algorithms, both in simple effective pathlength methods (as above) or for more rigorous algorithms as those discussed below. It is important to stress here the independence of the Milan/Bentley model from inhomogeneity corrections. The curvature and SSD corrections provided by the Milan/Bentley method (using equation 9.8) are a platform from which to correct for heterogeneities, using the factor given by either equation 9.14 or another more sophisticated method.

9.3.2 Batho Power Law Method

The power law method of inhomogeneity correction uses tissue-air ratios raised to an exponent based on density differences. This algorithm does not explicitly correct for primary and scattered dose perturbations, but the effect of an increase or decrease in scatter is implicitly considered via the tissue-air ratio terms. The main difference between this and the effective depth method is that the position of an inhomogeneity above the dose calculation point is considered, where the inhomogeneity is assumed to be a slab extending well beyond the beam boundaries (Batho 1964).

The power law method was initially formulated for calculating the effect of an air space on the dose at a point P below the air space. The dose at P will in general be increased by the presence of the air space (due to reduced primary photon attenuation), but the degree of increase will depend on the position of the space due to the varying effect on scattered dose reaching P. In particular, when the air space is close to P, the amount of scattered dose at P will be significantly reduced, somewhat counteracting the increase in primary dose. When the space is further toward the medium surface, the effect on scattered dose at P will be less and the net increase in dose at P will be greater. This phenomenon is not accounted for by effective depth methods, where the correction factor will be the same regardless of the position of the air space.

The power law method takes some account of the position of inhomogeneities by using the distances from a dose calculation point to the surfaces of inhomogeneities as arguments to tissue-air ratios. Consider the very simple case of a point P at a depth d_2 in a homogeneous unit density medium. The ratio of the dose at P to the dose in air at the same point is TAR(d_2). If the depth of unit density material overlying P is now reduced to d_1, where $d_1 < d_2$, then this is equivalent to introducing an air space of thickness $d_2 - d_1$ at a distance d_1 above P. The dose at P is now the dose in air multiplied by $TAR(d_1)$ and is a factor $TAR(d_1)/TAR(d_2)$ greater than was previously the case. Note that this correction factor is the same as would be predicted by the RTAR method. Now consider a further layer of unit density tissue placed above the air space. In this case, the ratio of dose at P with and without the air gap should be almost independent of the thickness of overlying tissue. This is because the required modification to the correction factor for the addition of tissue above the air gap is a multiplication factor to both the numerator and denominator, where this factor should be the same (approximately exponential) function in both cases. The power law method thus makes the assumption that the correction factor is still TAR(d_1)/TAR(d_2). Importantly, the correction factor varies with d_2 and d_1 for a constant air gap thickness $d_2 - d_1$, unlike the case with effective depth methods where the correction is invariant with the position of the air space.

To illustrate the effect on the calculated correction factor for the same air gap at a varying position, consider a point P at 15 cm deep in a medium containing a 5 cm air gap and irradiated with a 10 × 10 cm cobalt-60 beam

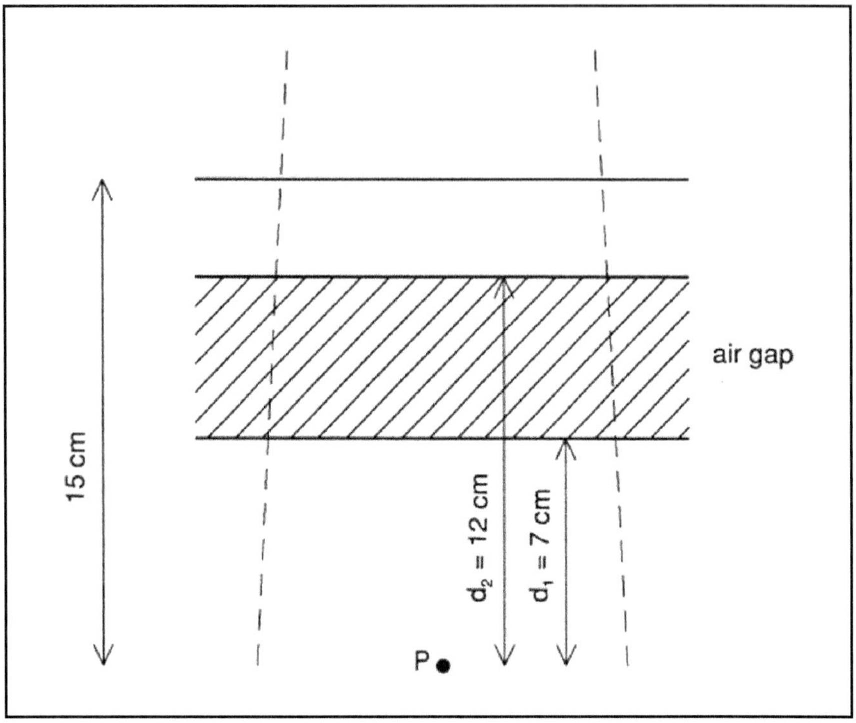

Figure 9.4: Configuration for the example Batho Power Law calculation above. A 10×10 cm Co-60 beam is incident on a phantom containing a 5 cm air gap at 3 cm deep. The dose calculation point is at 15 cm deep.

(see figure 9.4). If the air gap lies between 3 and 8 cm deep, $d_1 = 7$ cm, $d_2 = 12$, cm, TAR$(d_1) = 0.827$, TAR$(d_2) = 0.636$ (from tables), and the correction factor is 1.300.

If the air space is now "slid" down so that $d_1 = 2$ cm and $d_2 = 7$ cm, the correction factor is 1.214. This reflects the less rapid proportional decrease in a TAR curve at shallow depths for a given difference in depth. Note that if the fall-off in a TAR curve were purely exponential (as in the absence of scattered dose), the correction factor would not change. If the air gap began at the surface ($d_1 = 10$ cm and $d_2 = 15$ cm), thus representing 5 cm of missing tissue, the correction would be 1.312, which is also the prediction of the RTAR method (regardless of the position of the air space). Experimental verification of the power law correction for air gaps of varying thicknesses at varying positions with respect to the dose calculation point was carried out by Batho (1964). Results showed that the measured and calculated correction factors in a 12 × 10 cm cobalt-60 agreed to within 1.5% in all cases. The accuracy of the power law method for air gap inhomogeneities indicates that, in the above example, the factor given by the RTAR method is incorrect by about 8% in the case of the 2 cm air gap above the calculation point. The effective depth method is only correct when the air space takes the form of missing tissue at the surface.

The power law correction can be extended to apply to inhomogeneities of arbitrary density by first approximating the tissue-air ratios by exponentials.

Consider now that the air space above is replaced by tissue of density ρ. The ratio of primary beam fluence at P with and without the inhomogeneity is given by the correction factor

$$CF = \frac{\exp\left(-\mu\left(d_1 + \rho\left(d_2 - d_1\right)\right)\right)}{\exp\left(-\mu d_2\right)}, \tag{9.16}$$

which can be rewritten as

$$CF = \left(\frac{\exp\left(-\mu\, d_1\right)}{\exp\left(-\mu\, d_2\right)}\right)^{1-\rho}. \tag{9.17}$$

This expression is rigorous for the primary but not for the scattered component of the beam. The perturbation of scattered dose can be included in an approximate manner by replacing the exponentials with tissue-air ratios. Therefore, the correction becomes a more general case of the expression for the air gap:

$$CF = \left(\frac{TAR\left(d_1\right)}{TAR\left(d_2\right)}\right)^{1-\rho}. \tag{9.18}$$

Note that for the air gap, the exponent is simply unity. The exponential nature of the correction factor is the reason for the name "power law." Again, the correction factor does not take account of any material overlying the inhomogeneity.

In the slightly more general case of the calculation point lying in a medium of density ρ_1, with the inhomogeneity referred to above now having density ρ_2, the derivation of a correction factor becomes (el-Khatib et al. 1984)

$$CF = \frac{TAR\left(d_1\right)^{\rho_1-\rho_2}}{TAR\left(d_2\right)^{1-\rho_2}}. \tag{9.19}$$

It has also been shown that a small modification that appears to improve accuracy slightly in situations involving lung is to replace the tissue-air ratios by tissue-phantom ratios (el-Khatib et al. 1984). Equation 9.19 can, of course, give the correction factor inside an inhomogeneity of density ρ in an otherwise unit density medium ($\rho_1 = \rho, \rho_2 = 1.0$). In this case, if P is a distance d_1 below the top surface of the inhomogeneity, the correction factor is (Sontag et al. 1977),

$$CF = TAR\left(d_1\right)^{\rho-1}. \tag{9.20}$$

In the most general case, the whole medium is considered a stack of layers each with potentially different density. If P lies in the Nth layer and the density of layer n is ρ_n, equation 9.19 is generalized to

$$CF = \frac{\prod_{n=2}^{N} TAR\left(d_n\right)^{\rho_n-\rho_{n-1}}}{TAR\left(d_1\right)^{1-\rho_1}}, \tag{9.21}$$

where the distance from P to the top surface of layer n is d_n. Note that the slab configuration of the medium is particular to each calculation point, such that the N layers are defined by the material encountered by ray lines from the beam source. It is simpler to write equation 9.21 without the denominator. So, by including n = 1 in the product and assigning a value of 1.0 to ρ^0, we have the simple expression

$$CF = \prod_{n=2}^{N} TAR(d_n)^{\rho_n - \rho_{n-1}}. \tag{9.22}$$

Since the density terms in equation 9.22 represent attenuation, and this is only proportional to density for materials of similar composition (such as all soft tissues), density can be replaced by the linear attenuation coefficient relative to water. A final expression is thus

$$CF = \prod_{n=2}^{N} TAR(d_n)^{\frac{(\mu_n - \mu_{n-1})}{\mu_0}}. \tag{9.23}$$

where μ_0 is the linear attenuation coefficient of water and μ_n is that for the nth layer. The extension of the power law correction for multilayer geometries was first presented by Webb and Fox (1980).

As noted by others (Wong et al. 1982), a fundamental derivation of the power law equation has not been presented, even in Batho's original paper. Comparisons with experiment indicate that the method is accurate for at least some geometries, in particular beneath a single inhomogeneous layer (except for large fields and dense inhomogeneities). The multiplicative rule (as in equations 9.21–9.23) is adequate for most small fields but is less acceptable for large fields or large inhomogeneities. Wong and Henkelman (1982) determined that the major weakness of the method is the neglect of the effect from a layer of overlying unit-density tissue. That is, if $\rho_1 = 1.0$ in equation 9.22 then the first term in the product is unity and thus has no effect. Another major limitation is that inhomogeneities are assumed to be slabs (though the slabs are unique to each particular ray line) and as such their finite lateral extent is not accounted for. Furthermore, perturbations in backscattered dose due to underlying inhomogeneities are not considered.

Wong and Henkelman (1982) suggested a modification to the power law method where primary and scattered dose correction factors are calculated separately and summed. The scatter correction uses a sum of scatter-air ratios in a similar manner to the differential scatter-air ratio method developed by Cunningham (1972). This additive method removes the limitation of ignoring the overlying layer of unit density tissue and gives a good approximation in situations of a large field size where the multiplicative method is inadequate.

9.3.3 Equivalent Tissue-Air Ratio Method

The equivalent tissue-air ratio (ETAR) method developed by Sontag and Cunningham (1978) was the first algorithm that attempted to correct for the scattered radiation component by using multiple-slice CT pixel density information. As with the effective pathlength method, a ratio of TARs is used, but now the numerator depends on an *effective beam radius* as well as an effective depth. The effective radius (discussed below) is intended to account for the modification to scattered photon fluence, while the effective depth accounts for the primary fluence.

At present, the ETAR method is still arguably the benchmark to which other algorithms are compared and is a good compromise between dose accuracy and speed of computation. When one considers that this algorithm was developed in an era of limited computing power, it is a tribute to its developers that it performs so well in many situations. The compromises in the method were necessary due to limitations in computing power, and these were clearly identified by the original authors.

This algorithm is based on O'Connor's range scaling theorem (O'Connor 1957), which predicts that the dose at a point in a homogeneous but nonunit density medium is the same as the dose at the same point in space in a unit density medium, provided that all beam dimensions are scaled in proportion to the relative density. These dimensions include the beam radius, *r*, and depth, *d*, of the calculation point. The inhomogeneity correction factor is defined using the ETAR method as

$$CF=\frac{TAR(d',\tilde{r})}{TAR(d,r)}, \tag{9.24}$$

where d' is the effective depth of the calculation point and $\tilde{r}$ is an effective beam radius. The numerator of equation 9.24 is thus the tissue-air ratio for the equivalent configuration in water, hence the name of the method. For rectangular fields, the beam radius can be replaced by the equivalent square field size, and an effective equivalent square is calculated. Depending on the implementation, the effective depth and radius may be used in a tissue-air ratio split into a zero-area tissue-air ratio and a scatter-air ratio to account for the perturbation in primary and scattered dose respectively.

$$TAR(d',\tilde{r})=TAR(d',0)+SAR(d',\tilde{r}). \tag{9.25}$$

The effective depth is simply found as shown previously, and the effective beam radius is found from the product of the actual beam radius and an effective density: $\tilde{r} = r\tilde{\rho}$. The effective density is a weighted average of voxel densities throughout the medium, where the weights depend on the relative scatter contributions from each voxel.

$$\tilde{\rho}-\frac{\sum_{i,j,k} W_{i,j,k}\,\rho_{i,j,k}}{\sum_{i,j,k} W_{i,j,k}} \tag{6.26}$$

Each weight is a function of the distance and angle from the scattering element to the calculation point, as well as the photon fluence at the scattering element. It is assumed that (a) all scattering is Compton scattering and (b) the medium density between the scattering element and the point of dose calculation is unity. In a medium that contains lung, the effective depth and radius will of course be smaller than the physical depth and radius, giving a TAR in the numerator for a smaller depth and field size than that in the denominator. The task of calculating $\tilde{\rho}$ by equation 9.26 is very time consuming as it involves a three-dimensional weighted sum of densities over all volume elements in the irradiated medium.

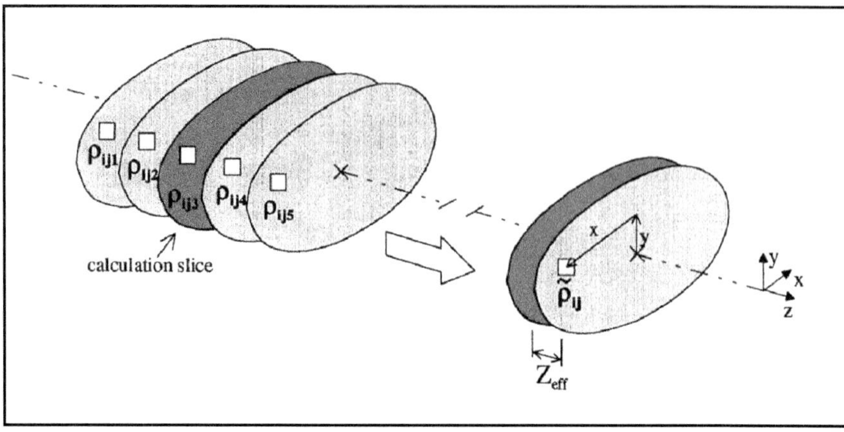

Figure 9.5: On the left, the slice in which dose is to be calculated is shown, with four surrounding slices. A pixel at location *i,j* in each slice is also shown. The five slices are coalesced into an effective slice at a distance Z_{eff} from the calculation slice, as shown on the right, where the effective density of element *i,j* is $\tilde{\rho}_{i,j}$.

To reduce computation time, the three-dimensional volume of scattering elements defined by multiple CT scan slices is coalesced into a single effective slice located at a distance Z_{eff} from the calculation plane, which is in the same x-y plane as the CT scans. Z_{eff} is a weighted average of the distances of each CT slice from the calculation plane, where the weights, W_k, represent the significance of scattering from that slice and are calculated from scatter-air ratios. The density element $\tilde{\rho}_{i,j}$ at each position in the effective slice is calculated by a weighted average over k of all elements at i,j,k, where the weighting factors are the same as those used in calculating Z_{eff}. A weighted average of the density elements in the coalesced slice is then performed as depicted in figure 9.5, with equation 9.26 becoming

$$\tilde{\rho} = \frac{\sum_{i,j} W_{i,j}(Z_{eff})\tilde{\rho}_{i,j}}{\sum_{i,j} W_{i,j}(Z_{eff})}. \tag{9.27}$$

Note that Z_{eff} and $\tilde{\rho}_{i,j}$ correspond to a particular dose calculation slice and are invariant for points within that slice. The weighting factors W_{ij} take account of the contribution to both first scatter and multiple scatter fluence Primary and scattered photon fluence in a density element are approximated using zero-area tissue-air ratios and scatter-air ratios, respectively, while photon transport from the density element to the calculation point assumes Compton scattering and that photons are attenuated as though the medium was all water. Since the weighting factors appear in both numerator and denominator, their magnitudes are not very critical as only relative values are important. Further details of the procedure can be found in the original paper describing the algorithm (Sontag and Cunningham 1978).

The method of primary and scatter dose separation using zero-area tissue-air ratios and scatter-air ratios (such as in equation 9.25) is only valid when electron equilibrium exists. This is because the zero-area TAR represents the primary dose under equilibrium conditions, or more specifically, the dose in the absence of scattered photons. The zero-area tissue-air ratio is thus larger than the true primary dose in situations of electron disequilibrium. This problem has been addressed by Woo et al. (1990), who used Monte Carlo data instead of zero-area TAR. Using Monte Carlo data, the primary dose can be made a field size dependent quantity (as in the case of the scattered dose) that rapidly decreases for field sizes less than that required for equilibrium. This extension has been shown to improve greatly ETAR results in regions of lateral electron disequilibrium (Woo et al. 1990).

9.3.4 Differential Scatter-Air Ratio Method

This method makes good use of the separation of primary and scatter dose by explicitly summing the dose from scatter elements and adding this to the primary dose. Scatter dose can be calculated by either (a) assuming that the medium consists entirely of water, (b) considering the attenuation of the primary beam in reaching each scattering element, or in addition (c) accounting for the attenuation of scattered photons on traveling to the dose calculation site. Primary dose corrections are easily made by using the zero-area tissue-air ratio for the effective depth of the calculation point as in equation 9.27.

The dose can be found by taking a ratio of the effective tissue-air ratio at the calculation point, $\overline{TAR}$, to the tissue-air ratio for the equivalent-square field size at d_{max}, $TAR(d_{max}, r_{eq})$

$$D(P) = D_0 \frac{\overline{TAR}}{TAR(d_{max}, r_{eq})}, \tag{9.28}$$

where D_0 is the dose at d_{max} for the equivalent square. $\overline{TAR}$ is the sum of $TAR(d', 0)$ and an effective scatter-air ratio:

$$\overline{TAR} = TAR(d', 0) + \overline{SAR}, \tag{9.29}$$

where $\overline{SAR}$, may be calculated with one of the three levels of complexity mentioned above. To illustrate this procedure, consider the calculation point P to be on the axis of a cylindrical coordinate system with radial increments Δr, depth increments Δz, and sectors of angular increment $\Delta\varphi$. The scatter dose is due to photons scattered from all voxels within the irradiated volume, where the value of $\overline{SAR}$ is the sum of all *differential scatter-air ratio* values $DSAR(d,z,r,\varphi)$ for elements within the volume. Here, d is the depth of the calculation point, and the scattering element is a distance z above it, at a radius r and at azimuthal angle φ. This situation is represented by figure 9.6, where the scattering element has electron density $\rho(d,z,r,\varphi)$. $\overline{SAR}$ is thus given by

$$\overline{SAR} = \sum_z \sum_r \sum_\varphi \overline{DSAR}(d, z, r, \varphi), \tag{9.30}$$

where if there is no inhomogeneity correction for scatter dose, the effective differential scatter-air ratio value in the summation is just $DSAR(d,z,r,\varphi)$.

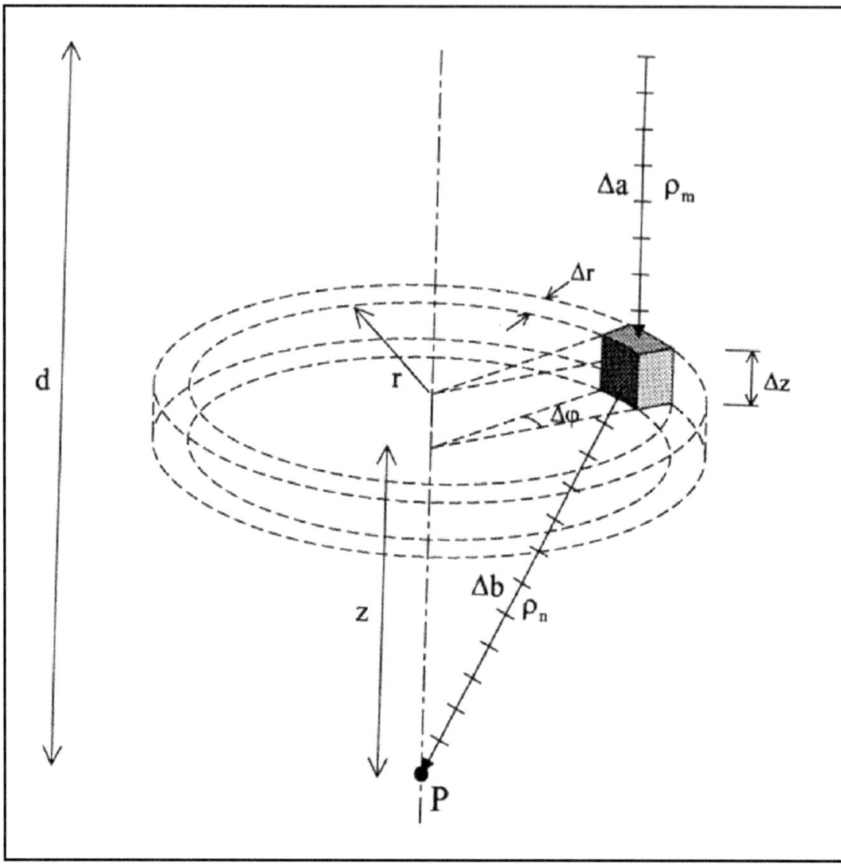

Figure 9.6: Scattering element of volume $r\Delta\varphi\Delta z\Delta r$ at coordinates z,r,φ with respect to the dose calculation point P. Attenuation corrections are made for primary photons reaching the scattering element and for scattered photons traveling to the calculation point, using the density elements along these paths.

On the other hand, if the effect of inhomogeneities on primary fluence at the scattering element is taken into account, an attenuation correction can be made to the effective differential scatter-air ratio value:

$$\overline{DSAR}(d,z,r,\varphi) = DSAR(d,z,r,\varphi) \times \rho(d,z,r,\varphi) \times \exp\left(\mu_0 \Delta a \sum (1-\rho_m)\right) \quad , \qquad (9.31)$$

where μ_0 is the attenuation coefficient of water, Δa is the incremental step size in a ray line tracing from the surface to the calculation point along the direction of the primary beam, and ρ_m is the electron density of elements along this line. Note that the exponent in equation 9.31 is the ratio of attenuation along the inhomogeneous path from the surface to the scattering element to the attenuation along this path if the medium were all water.

A further correction is required to account for perturbation in the attenuation of scattered photons. In this case, if the attenuation coefficient of first-scattered photons in water is μ_1 and the density of element n, of length Δb, along a line from the scattering element to the calculation point is ρ_n, then

$$\overline{DSAR}(d,z,r,\varphi) = DSAR(d,z,r,\varphi) \times \rho(d,z,r,\varphi) \times \exp\left(\mu_0 \Delta a \sum (1-\rho_m)\right) \times \exp\left(\mu_1 \Delta b \sum (1-\rho_n)\right). \quad (9.32)$$

In equation 9.32, scaling of the scatter contribution is performed as though all scatter is first scatter, since only rays directly between scattering elements and the calculation point are considered. It is important to note however that since experimentally derived SAR values are used, all orders of scatter are included in the DSAR values. This approximation will be better as energy increases, and thus the contribution from all orders of scatter decreases, as any error in scatter dose calculation becomes less significant. A further deficiency of this method is that perturbations in backscatter cannot be explicitly accounted for, since the differential scatter-air ratio values represent increments in dose from scattering material *above* the calculation point. As with treating all scatter as first scatter, this will be less of a problem at high energies.

The differential scatter-air ratio method is obviously well suited to situations where the field is of irregular shape, since the primary photon fluence incident on each element in the medium is explicitly considered in the scatter dose calculation. This is especially true if the medium is homogeneous, since the approximations involved in inhomogeneity corrections are avoided. In fact, the summation of differential scatter-air ratios can then be reduced to a summation over sectors, where the DSAR values are replaced by scatter-air ratio values for depth d and the radius from the calculation point to the field edge corresponding to each sector, where the scatter-air ratio values are divided by the number of sectors. In the homogeneous case, the effective SAR value is then given by

$$\overline{SAR} = \frac{\Delta\varphi}{2\pi} \sum_{\varphi} \overline{SAR}\left(d, r(\varphi)\right), \quad (9.33)$$

where $\Delta\varphi$ is the width in radians of each sector and $r(\varphi)$ is the sector radius. The scatter-air ratio in the summation is written as $\overline{SAR}$ since it is in general a sum of SAR values corresponding to each radius at which the line bisecting the sector intersects the field edge. The coefficient for the summation is +1 if the line approaches the edge from inside the field and is –1 if the line approaches from outside. This procedure is also known as sector-integration and was first proposed by Clarkson (1941).

9.3.5 Delta Volume Method

A similar technique to the differential scatter-air ratio method was proposed by Wong and Henkelman (1983) which employs analytical calculations for the primary and first scatter dose. The main advantage of this approach is that first and multiple scatter doses can be calculated separately, instead of all scatter being treated as first scatter when perturbations due to inhomogeneities are considered. This is important, especially for lower energies where multiple scatter dose makes a very significant contribution to the absorbed dose. For example, in a 10 cm radius ^{60}Co beam, multiple scatter is responsible for approximately 10% of the total dose. The algorithm actually treats some second order scatter as first scatter, since for small second-scattering angles this component is perturbed like first scatter dose. Because of this extra component, the first scatter component is known as *augmented first scatter*. This addition to the first scatter component approximately halves the remaining multiple scatter component for ^{60}Co.

In the delta volume method, first scatter is calculated analytically, using Compton scattering cross sections to calculate the scattering angles of primary photons and the density of volume elements along a ray line from the scattering element to the dose calculation point to obtain the perturbation in scattered photon attenuation. Density along a ray from the surface to the scattering element is also used to account for modifications to primary photon fluence. If the calculated augmented first scatter dose in water due to volume element i, with density ρ_i, is $\Delta S_{1,i}$, then the augmented first scatter dose in the medium is given by

$$\Delta S_{1,i}(med) = \rho_i \Delta S_{1,i}(w) \times \exp\left(\mu_0 \Delta d \Sigma\left(1-\rho_k\right)\right) \times \exp\left(\mu_i \Delta r \Sigma\left(1-\rho_l\right)\right) \times \left(\frac{\mu_{ab}}{\rho}\right)_{\mathrm{w_i}}^{\mathrm{med_i}} \tag{9.34}$$

where μ_0, μ_i are the linear attenuation coefficients in water for primary and first scattered photons, Δd and Δr are the incremental step lengths along the paths of primary and first scattered photons, ρ_k and ρ_l are the relative electron densities along these paths, and the last term is the mass energy absorption coefficient ratio of the medium at element i to water at element i.

Multiple scatter dose is obtained via the multiple scatter dose in water, which is determined from the difference in measured SAR values and the calculated first scatter component. In this way, total scatter dose is forced to be correct for a water medium. Perturbations to multiple scatter dose due to inhomogeneities are calculated via the measured perturbation caused by a small void at the position of the inhomogeneity and from the average density of the irradiated medium.

Not only does this method allow for different treatment of first and multiple scatter, but because of the analytical calculation of first scatter perturbation, scatter from *behind* the dose calculation point can be explicitly considered. Thus the first scatter contribution to backscatter can be calculated, unlike in the differential scatter-air ratio method.

One disadvantage of this method is long computation times. A report by the Photon Treatment Planning Collaborative Working Group (Mohan 1991) gives a time of three hours for calculation of dose due to a single beam (with an array of 24 voxels cubed), even with a 30 megaflop array processor.

As with the methods above, electronic equilibrium is assumed, so errors will occur when electronic disequilibrium is brought about by an inhomogeneity.

9.3.6 Fast Fourier Transform (FFT) Convolution Method

This method is a variation of the convolution method, which is discussed in more detail in chapter ten. The convolution method uses the physical principles of primary/scattered photon transport and energy transfer to obtain the dose distribution. It does not rely on measured data as input, except in the sense that the convolution calculation can be used to obtain a correction factor that is then applied to measured dose in water. The components of the convolution vary depending on the implementation, but they are basically (a) energy released in primary photon interactions at an *interaction site* and (b) the distribution of energy deposition about this interaction site. The first of these components is often represented by the terma (see chapter two and chapter ten), or in the case of the FFT convolution discussed here, by the photon fluence. The second component is known as the convolution kernel, which has components representing the primary and scatter dose separately (scatter can be further divided into first and multiple scatter). Kernels may be either analytically formulated (for example, using Compton scattering cross sections, exponential attenuation, and inverse square fall-off) or represented by numerical arrays.

In the FFT implementation referred to here (Boyer and Mok 1986), the time-consuming convolution process is reduced to a multiplication of Fourier-transformed components representing the primary photon fluence and dose kernels. The ability to perform the convolution in Fourier space comes at the expense of requiring that the kernels are spatially invariant, which means that they cannot be scaled to account for density variations. However, in the usual implementation of FFT convolution, an approximation is made that effectively allows a first-order scaling of the first-scatter kernel according to the density at the primary photon interaction site.

The basic equations for primary, first scatter, and multiple scatter dose, respectively using this method are given below, where each can be transformed and multiplied using the convolution theorem of Fourier transforms.

$$D_p(r) = \int_{r'} \Phi_0(r')\rho(r')k_0(r-r')\,dV' \tag{9.35a}$$

$$D_s(r) = \int_{r'} \Phi_0(r')\rho(r')k_1(r-r')\left\{1+\left[1-\rho(r')\right]\mu_1(r-r')\left|r-r'\right|\right\}dV' \tag{9.35b}$$

$$D_m(r) = \int_{r''} \Phi_m(r'')\rho(r'')k_m(r-r'')\,dV'' \tag{9.35c}$$

Here, r′ and r″ are position vectors representing sites of primary and scattered photon interaction respectively; dV' and dV'' are the associated volumes of integration; ρ is the relative electron density; μ_1 is the linear attenuation coefficient of first scattered photons at r'; k_0, k_1, and k_m are the primary, first scatter, and multiple scatter kernels; Φ_0 and Φ_m are the fluence of primary and multiple scattered photons. Note that the term in braces in equation 9.35b represents the first-order correction to the first scatter dose calculation. For the primary dose, the implementation of Boyer and Mok (1986) used Monte Carlo-generated kernels produced by Mackie et al. (1985). Thus, the function k_0 is represented by an array. The primary fluence, Φ_0, is calculated exactly by ray-tracing from the surface to the interaction point r'. Multiple scatter fluence, Φ_m, is calculated by convolving the primary photon fluence with a kernel representing the distribution of interactions of first-scattered photons. The multiple scatter kernel is an isotropic function derived from diffusion theory.

The approximations used in this method are designed to avoid kernel scaling and the associated 3-D ray trace, as must be performed to account for inhomogeneities rigorously. As already mentioned, it is the invariance of the kernels that allows the convolution to be carried out in Fourier space. If the convolution is represented simply by

$$D = \Phi \otimes k \tag{9.36}$$

then in Fourier space

$$\Im(D) = \Im(\Phi) \cdot \Im(k) \tag{9.37}$$

and dose is obtained using an inverse transform

$$D = \Im^{-1}\{\Im(\Phi) \cdot \Im(k)\}. \tag{9.38}$$

Using FFTs to perform the convolution yields a speed increase of more than two orders of magnitude compared with the real space convolution employing 3-D ray tracing. However, the accuracy is decreased. Results of this method are compared with other methods in the next section.

9.3.7 Inhomogeneity Correction Results

Effective Pathlength Method

Since effective pathlength inhomogeneity corrections consider only modification of the primary dose, it is expected that the predicted result will reflect the perturbation of primary photon fluence. This implies that if the effective pathlength along a ray line from the surface to a dose calculation point is less than the geometric pathlength, the predicted dose will *always* be greater than if the medium were unit density (the correction factor will be greater than unity). Similarly, a greater effective than geometric pathlength will give rise to lower dose than for unit density and a correction factor less than unity. These predictions are generally true, but the contribution to total dose from scattered photons means that the magnitudes of the correction factors will be incorrect. For example, a low-density region located in an otherwise unity density medium will cause an increase in primary photon fluence in and beyond that region, causing an increase in primary dose (as long as electronic equilibrium is established). On the other hand, scattered photon fluence in and around the low-density region will generally be reduced (because of a decrease in the number of primary photon interactions), causing a drop in scattered dose. The increase in total dose due to the low-density inhomogeneity is thus not as great as would be predicted from the modification of primary photon attenuation. Similarly, a high-density region gives rise to an increase in scattered dose and can cause a lesser reduction in total dose than would be predicted by the effective pathlength method.

Predictions of correction factors by the RTAR method (equation 9.14), along with other algorithms, have been compared with experimental measurements by Wong and Purdy (1990). These authors constructed layered dosimetry phantoms consisting of polystyrene (ρ = 1.03), cork (ρ = 0.26) and aluminum (ρ = 2.37) slabs (whose positions could be interchanged) that were irradiated by a 15 × 15 cm ^{60}Co (1.25 MeV) beam. In simple configurations consisting of a 10 cm thick layer of cork with either 5 or 10 cm of overlaying polystyrene, the RTAR gave a large overestimate of the correction factor both within and beyond the cork. Mean overestimates were 6.46% and 8.92% for these configurations, respectively, reflecting the considerable scattered dose contribution whose reduction in magnitude is not predicted by the RTAR method. Note that the increase in overestimate of the correction factor with depth of the cork layer corresponds to the increasing significance of scattered dose with depth. When a 5 cm layer of aluminum was placed beyond the cork layer (with 5 cm of overlaying polystyrene), the correction factor in

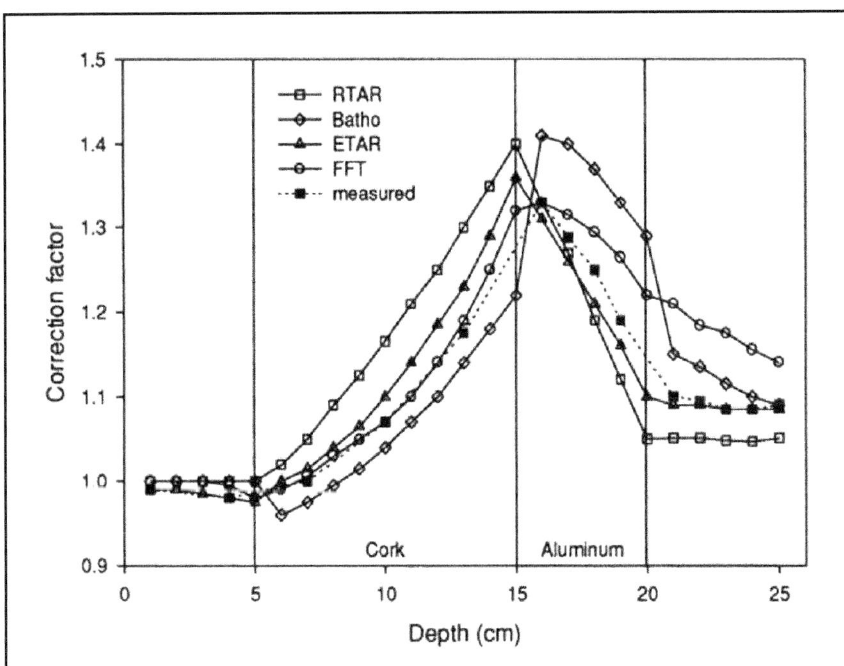

Figure 9.7: Results from a report by Wong and Purdy (1990) comparing inhomogeneity correction methods. Here, RTAR (effective depth), Batho Power Law, ETAR, and FFT methods are compared with experiment for a 15 × 15 cm ^{60}Co beam incident on a phantom containing a 10 cm thick cork layer and a 5 cm aluminum layer. Most notable is the large overestimate and underestimate of the correction in cork and aluminum, respectively, by the RTAR method. Adapted with permission from: Wong and Purdy (1990).

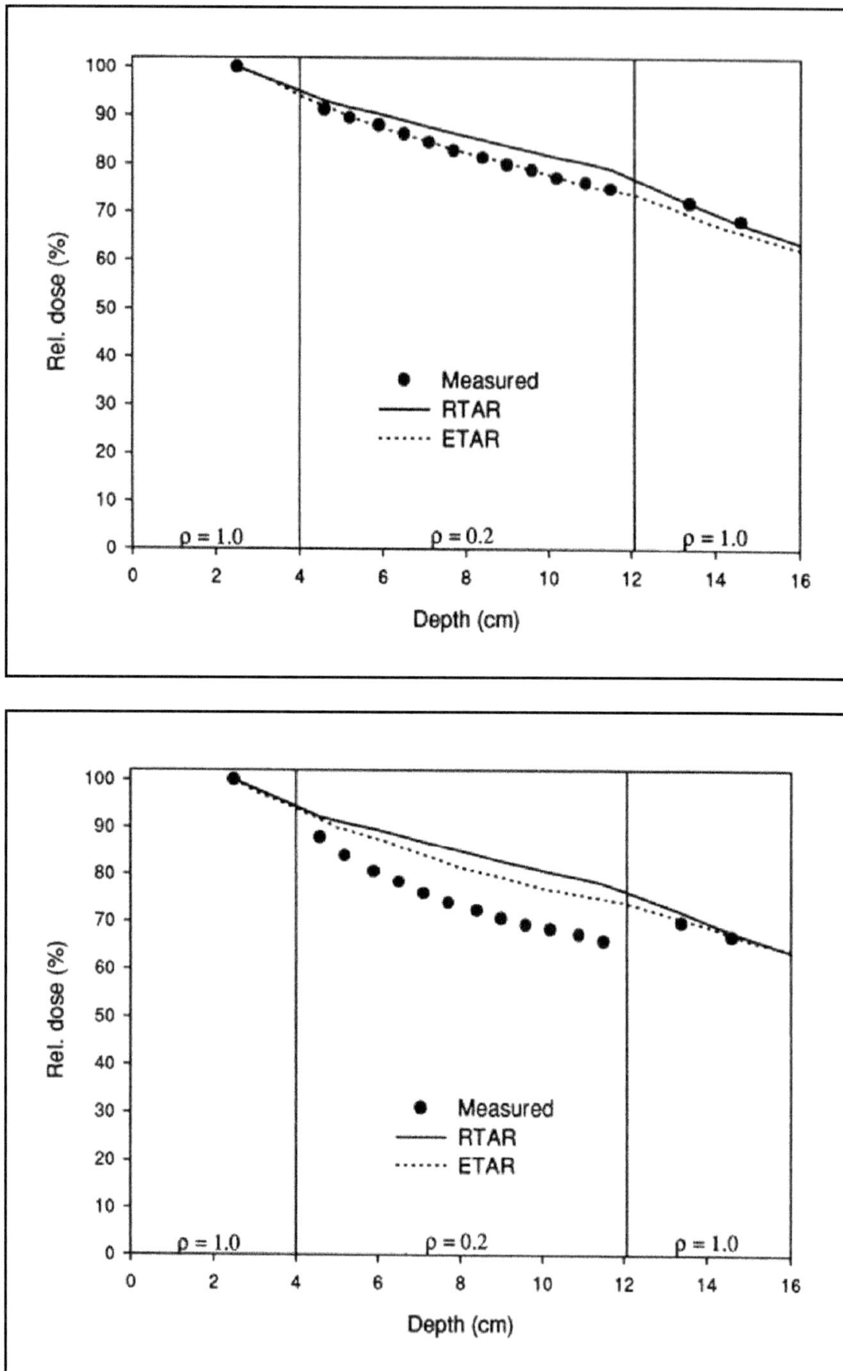

Figure 9.8: Comparison of measured dose for a 10 MV beam incident on a lung phantom with that predicted by the effective depth (RTAR) and equivalent tissue-air ratio (ETAR) methods for (a) 10×10 cm^2 and (b) 5×5 cm^2 fields. The lung is simulated by a cork layer of thickness 8 cm with density 0.2 gcm^{-3}. Note the large overestimate in lung dose for the small field (by both RTAR and ETAR) due to lateral electronic disequilibrium.

cork was again overestimated, while, due to the failure to predict the increase in scattered dose, the correction factor in and beyond the aluminum was underestimated. If the positions of the cork and aluminum layers were swapped, the correction factors were again over- and underestimated in cork and aluminum respectively, with an overestimate in the region beyond the cork. The small mean differences between calculated and experimental correction factors as tabulated by Wong and Purdy (1990) for the cork/aluminum phantom (–0.18% and +1.88% for the two configurations, respectively) do not reflect the large discrepancy in each slab, since differences used to calculate the means are signed and tend to cancel each other (positive in cork and negative in aluminum).

A selection of Wong and Purdy's results for a 10 cm thick cork layer and an underlying 5 cm thick slab of aluminum is shown in figure 9.7. For the moment concentrating on the RTAR results, the overestimate and underestimate of dose in cork and aluminum, respectively, is obvious, particularly in cork where the large reduction in scatter dose is unable to be predicted. In aluminum, the agreement with measurement is better, but this is likely to be due to the fact that the inability to predict the increase in scatter is masked by the high correction factor arrived at due to the overlying cork. Beneath the aluminum, the higher than calculated dose due to increased scatter from aluminum is obvious.

Comparisons of measured and calculated dose for a lung phantom were also carried out by one of the authors (Metcalfe) for a 10 MV, Varian Clinac 18/10 beam. Lung of density 0.2 gcm^{-3} was simulated with a stack of cork of thickness 8 cm. Four-centimeter thick polystyrene slabs were placed on either side of the cork to simulate the chest wall and mediastinum. Dosimetry was performed with a 0.6 cm^3 NE2571 ionization chamber. Calculations were performed using the effective depth and ETAR methods on a Theratronics Theraplan treatment planning computer. Results for the 10×10 cm^2 beam are shown in figure 9.8a, where it can be seen that the RTAR calculation gives an overestimate of the dose in the lung region. This is consistent with the reduction in scatter not being considered (as in the results of Wong and Purdy above). More interesting is the RTAR result for a 5×5 cm^2 beam in figure 9.8b, where there is a large overestimate in lung dose. This is because lateral electronic disequilibrium now exists on the beam axis, whereas it does not if the medium is all unit density (the ratio of TAR values used in the calculation are, of course, both for water media of field size 5×5 cm^2). Lateral electronic disequilibrium is brought about by the lateral electron range being larger than half the beam width, with a consequent reduction in secondary electron fluence from that which would exist if the field size were larger (see chapter two).

To illustrate this point further, an EGS4 Monte Carlo simulation was performed in which a pencil beam of 500 10 MV photons was incident on the lung phantom. The energy spectrum used was the National Research Council 10 MV spectrum supplied with the EGS4 system. In figure 9.9, secondary electron tracks are shown in the first water-lung-water regions, and it can be seen that the maximum lateral range of these electrons in the lung region easily exceeds 2.5 cm. It is expected in fact from this demonstration

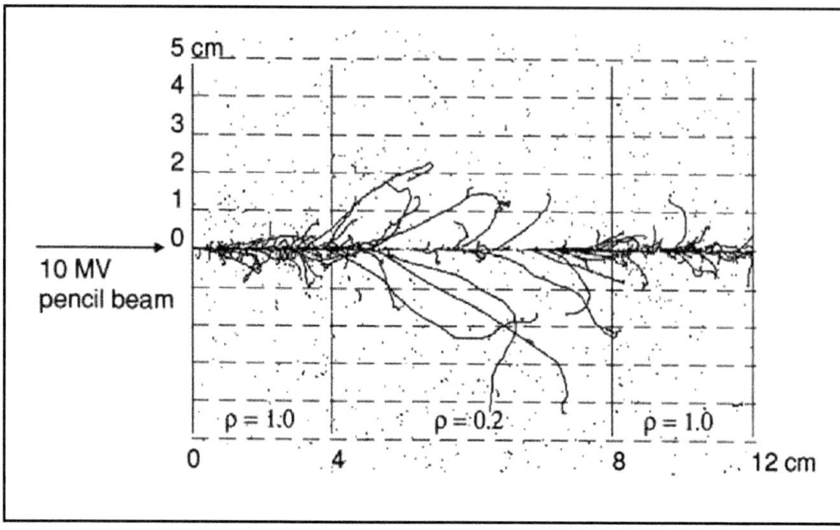

Figure 9.9: EGS4 Monte Carlo simulation showing secondary electron tracks due to a 10 MV photon pencil beam incident on the lung phantom described above (only first lung region shown). The lateral range of the electrons can be seen to be at least 4 cm, well exceeding the 2.5 cm required for disequilibrium at the central axis of a 10×10 cm^2 field.

that disequilibrium in the lung will exist at the central axis of fields up to almost 10×10 cm^2.

Equivalent Tissue-Air Ratio Method

This method is popular as the most rigorous (and slowest) choice of inhomogeneity correction on commercial treatment planning systems. Its popularity is due to its high accuracy resulting from realistically accounting for scatter. In comparisons with experiment, the ETAR method generally gives accurate results when the field size is large enough for electronic equilibrium, indicating that its modeling of scatter dose is adequate. There are approximations in the method that should be mentioned in order to gain some insight into the reason for calculation errors.

First, all inhomogeneities are considered independently. That is, the material between an inhomogeneity and the dose calculation site is assumed to be water in calculating the weighting factors. This can, for example, mean that beneath a lung inhomogeneity the significance of the lung can be underestimated due to water-like attenuation being assumed for scattered photons originating in the lung. This will result in an overestimation of the effective density and consequent overestimation of the effective field size. This is consistent with the results of Kan et al. (1995), where a lung phantom was irradiated with 5×5 and 10×10 cm^2 25 MV beams. In this case there was lateral electronic disequilibrium existing in the lung (extreme in the case of the smaller field) and a continued overestimate of dose on the distal water region after equilibrium was re-established.

A further nonideality that is fundamental to the method is that loss of scatter dose is determined via a smaller effective field size, rather than explicitly. This is a good approach as long as the loss of scatter is largely due to a surrounding low-density medium, in which case O'Connor's density scaling theorem (O'Connor 1957), which ETAR is based on, can be applied. If on the other hand the loss of scatter is due to sharp density changes, such as a water-air interface, the theorem does not apply as well. For example, at the vertical interface of a half-air-half-water geometry, half the scatter dose would be lost. The calculated effective field size of half the geometric size would however underestimate this loss due to the nonlinear shape of the scatter dose versus field size curve. However, another configuration that gives the same effective density is likely to give much better results, since the resulting effective field size in water may then give the correct scatter dose. This is to say that a particular effective density can be obtained for many different configurations, some of which will be better represented by the "equivalent" water configuration than others. There is also the assumption that the same effective depth can be used for the zero-area TAR (for primary dose) and the SAR (for scatter dose) in obtaining the total TAR value (see equation 9.24). This depth is correct for the first of these components but may not be for the second.

In figure 9.7, the ETAR results appear to follow the same trend as those of RTAR, but with much less deviation from measurement (dose is still a little overestimated in cork and underestimated in aluminum). This indicates the dominance of the effective depth in ETAR calculations, but consideration of scatter via the effective field size makes a significant improvement. This is the case with all of the configurations in the study by Wong and Purdy, indicating that in general the effective field sizes calculated by ETAR are a little large for points in a low-density medium and slightly small for points in a high-density medium.

ETAR results for the Metcalfe lung phantom are presented in figure 9.8a for a 10×10 cm^2 field. Calculated results in the lung are now in good agreement with experiment, indicating that the reduction in scatter dose is satisfactorily accounted for in this case (unlike with the effective depth method where dose is overestimated). For the 5×5 cm^2 field however there is still an overestimate of dose in lung due to the assumption of electronic equilibrium, as can be seen in figure 9.8b. Specifically, the error is caused by the use of extrapolated zero-area TAR values to describe the primary dose. The ETAR result in lung is slightly better than the effective depth result due to the correct calculation of the reduction in scatter dose. However, the unaccounted for reduction in primary dose due to disequilibrium is far more significant in this case.

Batho Power Law Method

Analysis of this method by Wong and Henkelman (1982) has revealed that in media with a single-layered inhomogeneity, dose is underestimated within an inhomogeneity of less than unit density and is overestimated when the density is greater than unity. This is the opposite trend to that of the effective path-length method. The largest errors occur near the top of the inhomogeneity, due to the neglect of the overlying (and nearby) water layer.

With a 4.8 cm layer of cork at 2.8 cm deep in a polystyrene phantom, the Batho method was found by Cunningham (1982) to underestimate the dose in cork by approximately 3% for a 20 × 20 cm² beam, and the error was much reduced for a 10 × 10 cm² beam. Even for the larger field, the Batho result was very good for points below the inhomogeneity; this is to be expected since it is for this region that the method was originally derived. For very large field sizes, such as used for half and total body irradiation, Van Dyk et al. (1982) have shown that errors as large as 20% may occur when the Batho method is used. In general it has been found that the method works best for small fields and densities less than unity but is poor for large fields and higher densities. This is presumably because it is based on an exponential characterization of the depth dose, and this is only a good approximation when scatter is minimized, as in the case of small fields and low densities. Thus, unlike the RTAR method, there is an attempt to account for scatter, but this seems to break down in situations where scatter dose makes up a high proportion of the total dose.

In figure 9.7, the Power Law method shows a large overestimate in the correction factor within the aluminum slab and a smaller underestimate in the cork. This is understandable since there is a lot of scatter dose in the aluminum, and this will cause problems for the exponential-based derivation of the method.

DSAR and Delta Volume Methods

The accuracy of the DSAR method depends on its implementation. In the originally developed algorithm, a 3-D scatter ray trace was performed between each elemental scattering volume and the dose calculation point in order to obtain an attenuation correction for the incremental SAR value (see equation 9.32). In some implementations, however, the medium is assumed to be water equivalent in the scatter dose calculation. This is the case for the DSAR-based algorithm used to produce the results reported on by Kan et al. (1995), where the scattering volumes are sectors extending to the field edges (see equation 9.33). As such, the two implementations would be expected to give quite different results. The work of Wong and Purdy shows that the DSAR method gives relatively poor results, even though a scatter ray trace was performed in that case. They state that this is possibly due to the treatment of all scatter as first scatter. Results show a consistent overestimate of dose in cork and an underestimate in aluminum, indicating an underestimate of the effect of gain or loss of scatter due to high- or low-density regions, respectively. Kan's work uses a lung phantom and includes situations where lateral electronic disequilibrium exists. In such regions, all of the algorithms mentioned here will be inaccurate since they assume that equilibrium exists. In regions of equilibrium, however, the method still shows an overestimate of dose, presumably mainly due to the assumption that the scattering medium is water (which produces more scatter than lung).

The delta volume method gives the most accurate results of the algorithms looked at here. This is to be expected due to its 3-D scatter ray tracing, analytical calculation of first scatter (including backscatter), and separate calculation of multiple scatter. A major improvement over the DSAR method, pointed out in the original paper proposing this method

(Wong and Henkelman 1983), is that discontinuities at interfaces are avoided. This is because backscatter is accounted for, therefore including the effects of underlying material in changes to scatter dose.

FFT Convolution Method

This method is the newest of those described here, and, like all convolution algorithms, it is straightforward in principle, consisting of simply a set of superimposed dose kernels, weighted by the photon fluence. FFT convolution requires a number of approximations to be made so that the kernels may be invariant, and these will inevitably cause some inaccuracy in the results. As with most methods, any inaccuracy is most likely to stem from the treatment of scattered photons (particularly the first scatter component, as this is the most significant). In this case, the energy released to first scattered photons is calculated correctly via the primary photon fluence and density at the interaction site. The transport of these photons is modeled as though the medium were water, with only a first order correction based on the density at the interaction site. This will cause an error in scatter dose calculation, especially when considering interaction sites in regions other than those containing the point of dose calculation.

The FFT convolution method was included in the survey by Wong and Purdy (1990), and results are shown in figure 9.7 for the 15×15 cm^2 ^{60}Co beam. Agreement in the cork region is seen to be excellent, but in the aluminum layer a significant overestimate in the correction factor is seen. This is probably due to neglect of the much larger attenuation in aluminum when considering interaction sites in the overlaying cork and polystyrene. The same situation occurs for the underlying polystyrene, where scatter from the upper polystyrene and cork will be overestimated, though scatter from the aluminum should be overestimated and somewhat counter this effect.

Although this method contains approximations that limit its potential accuracy, it is a very important step in the evolution of photon dose calculation algorithms (see chapter ten).

9.4 Electron Pencil Beams

An excellent review of electron beam therapy physics and the planning dose models employed by RTP computers is given by Hogstrom and Almond (2006). Lillicrap (Lillicrap et al. 1975) suggested that broad electron beams could be represented by summing the dose distribution from many smaller beams, i.e., pencil beams. Several authors turned to the Fermi-Eyges theory of thick target multiple coulomb scattering as an appropriate model for how electron beams would behave and be modified in patients (Eyges 1948, Perry and Holt 1980, Brahme 1981).

9.4.1 Fermi-Eyges Pencil Beams

The Eyges equation models multiple coulomb scattering by predicting the probability P of an electron having lateral deviation x and an angular deviation θ_x to depth z (see figure 9.10).

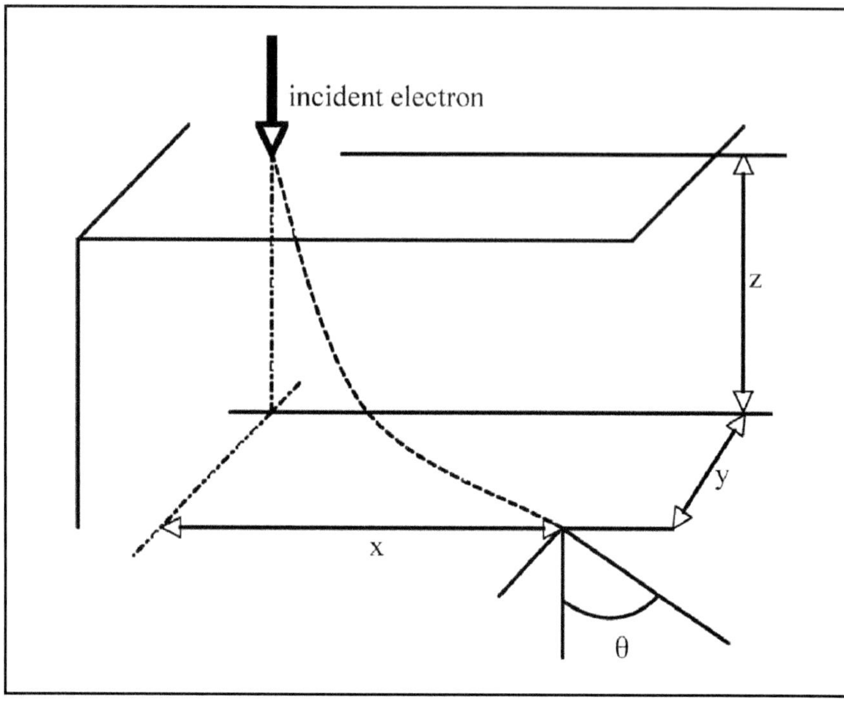

Figure 9.10: Spatial and angular deflection of an electron pencil beam.

$$\frac{\partial P}{\partial z} = -\theta_x \frac{\partial P}{\partial x} + \frac{1}{4}\frac{d\overline{\theta}^2}{dz}\frac{\partial^2 P}{\partial \theta_x^2}, \tag{9.39}$$

where $\frac{d\overline{\theta}^2}{dz}$ is the linear angular scattering power. The solution to mass angular scattering power provided by Fermi was

$$\frac{d\overline{\theta}^2}{dz}(z) = \frac{\overline{\theta}^2}{\rho l}\rho(z). \tag{9.40}$$

However, this solution assumed constant scattering power, which holds reasonably for high-energy electrons. As electrons penetrate tissue, they lose energy at about 2 MeV/cm so the increase in scattering power with decreasing energy becomes important. Eyges provided a solution to this such that using the formalism of Jette (1984)

$$P_E(x,\theta,z) = \frac{1}{2\pi\sqrt{A_0 A_2 - A_1^2}} e^{\left(-A_0 x^2 - 2A_1 x\theta + A_2\theta^2\right)/2\left(A_0 A_2 - A_{1\,1}^2\right)}. \tag{9.41}$$

A_n are scattering moments that depend on z and the varying scattering power of the media prior to reaching z, given by

$$A_n(z) = \frac{1}{2}\int_0^z \frac{d\bar{\theta}^2}{dz}(z-z')^n\,dz'. \tag{9.42}$$

This can be written as the location probability density $P_L(x, z)$ and the angular probability density $P_A(\theta, x, z)$. The location probability density at x is found by integrating the Eyges solution over θ at the point (x, z))

$$P_L(x,z) = \int_{-\infty}^{\infty} P(x,\theta,z)\,d\theta = \frac{1}{\sqrt{2\pi A_2}}e^{-x^2/2A_2}. \tag{9.43}$$

By mathematical substitution using a Gaussian function, as shown in figure 9.11, which is

$$f(t) = \frac{1}{\sqrt{2\pi\sigma}}e^{-t/2\sigma^2}, \tag{9.44}$$

it is possible to show that

$$P_L(x,z) = \frac{1}{\sqrt{2\pi\sigma_x}}e^{-x^2/2\sigma_x^2}, \tag{9.45}$$

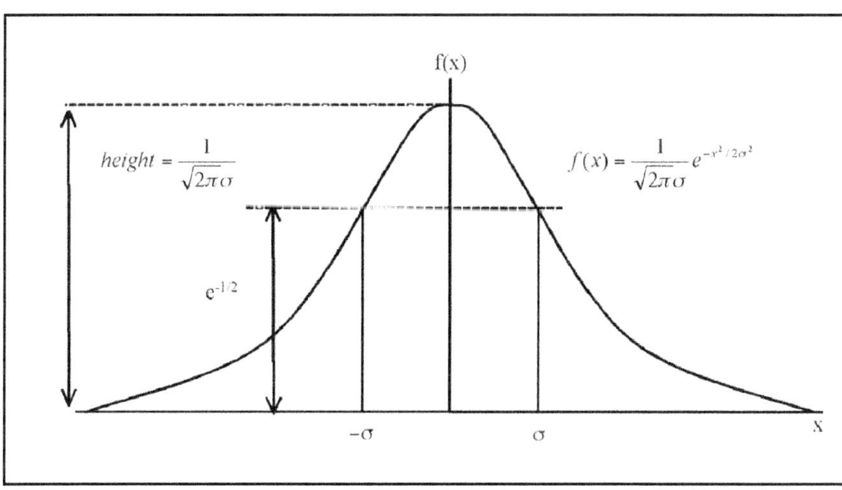

Figure 9.11: A Gaussian curve with standard deviation and the Gaussian function.

where σ_x is the lateral standard deviation

$$P_A(\theta, x, z) = \frac{1}{2\pi\sigma_\theta} e^{-(\theta-\theta_0)^2/2\sigma_\theta^2}, \tag{9.46}$$

where $\theta_0 = \frac{A_1(z)}{A_2(z)} x$, θ_0 is the mean angular deflection and $\sigma_\theta = \sqrt{(A_0 - A_1^2 / A_2)}$ is the angular deviation. Note the quantity PL gives planar fluence and therefore as the electrons are scattered to more oblique paths, the Fermi-Eyges theory becomes less accurate.

$P_L(x, z)$ is the probability distribution for an infinitely thin pencil beam. In order to ascertain the dose for a finite beam, one needs to do a summation or integration of pencil beams. The probability distribution $P(x,y,z)$ that an electron originating in one of the beams will reach unit area (x,y,z)) (see figure 9.12) becomes a very complicated integral; by making the assumption that we are dealing with strip beams, the integral reduces to

$$P(x,y,z) = \sum_i \left[\frac{1}{\sqrt{2\pi}\sigma_x(z)} \int_{x_i-\Delta X/2}^{x_i+\Delta X/2} e^{-(x-X')^2/2\sigma_x^2(z)} dX' \right] \times \frac{1}{\sqrt{2\pi}\sigma_y(z)} \int_{-WY/2}^{WY/2} e^{-(y-Y')^2/2\sigma_y^2(z)} dY' \tag{9.47}$$

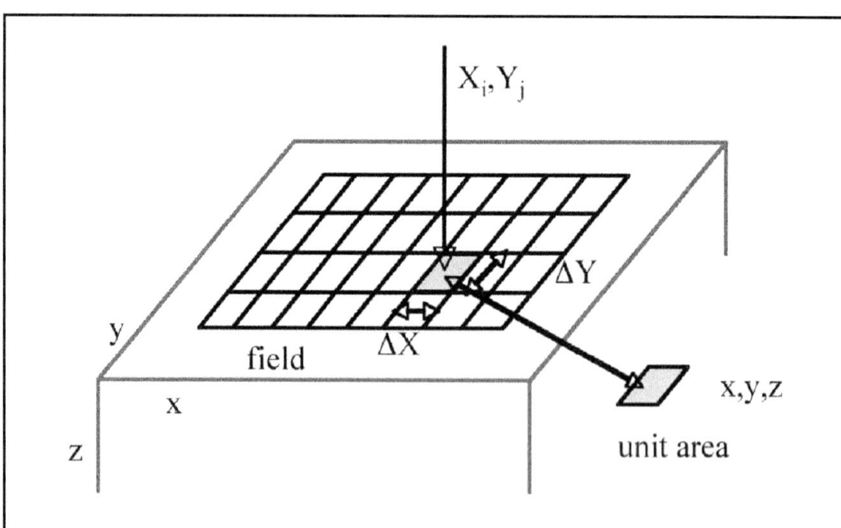

Figure 9.12: Electron incident at X_i, Y_j at the surface ranging to and crossing a unit area at x, y, z.

by substituting the error function where

$$erf(z) = \frac{2}{\sqrt{\pi}} \int_0^z e^{-x} dx, \tag{9.48}$$

then equation 9.47 can be written as

$$P(x,y,z) = \sum_i \frac{1}{2}\left[erf\left(\frac{(X_i + \Delta X/2) - x}{\sqrt{2}\sigma_x(X_i,z)} \right) - erf\left(\frac{(X_i - \Delta X/2) - x}{\sqrt{2}\sigma_y(X_i,z)} \right) \right] \times \frac{1}{2}\left[erf\left(\frac{WY/2 - x}{\sqrt{2}\sigma_y(X_i,z)} \right) + erf\left(\frac{WY/2 + y}{\sqrt{2}\sigma_y(X_i,z)} \right) \right], \tag{9.49}$$

where the first term defines the x scatter factor and the second term is the y scatter factor. This is the generic pencil beam equation and most RTP computer implementations are a variation on this theme.

9.4.2 Hogstrom Pencil Beam Implementation

One particular approach to the Fermi-Eyges pencil beam calculation that of Hogstrom (Hogstrom et al. 1981) developed at the M.D. Anderson Cancer Center and named the *pencil beam algorithm* (PBA) was adapted by several RTP computer vendors. The approach divides the medium into a computational grid consisting of horizontal lines of spacing ΔZ and fan lines diverging from the electron source with spacing of fan lines ΔX at depth z (see figure 9.13). The probability of an electron beam reaching a unit area at

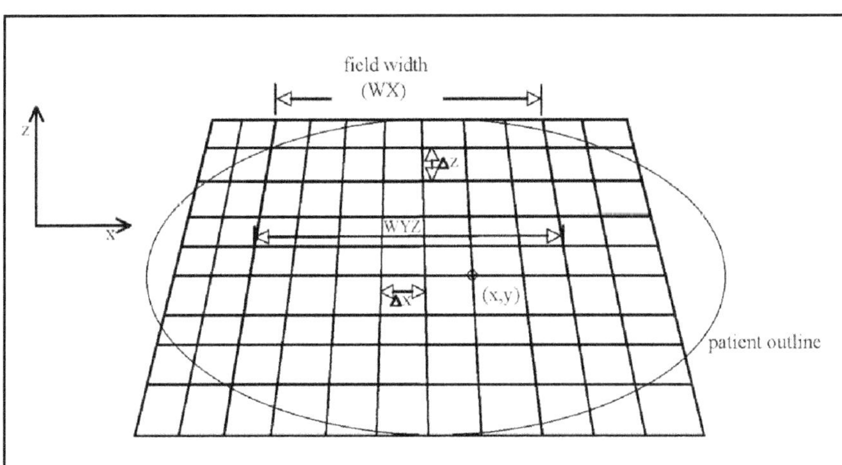

Figure 9.13: Fan line calculation grid used to define the calculation points in the Hogstrom implementation of the Fermi-Eyges pencil beam model.

point x,y,z is as for equation 9.49, except ΔX is the projected width at depth z and beam WY is replaced by projected depth WYZ.

The dose at depth z in water for a rectangular field WX by WY at the surface is calculated as

$$D_{water}^{WX,WY}(0,0,z) = D_{water}^{\infty}(0,0,z)\left[erf\left(\frac{WXZ/2}{\sqrt{2}\sigma_{water}(z)}\right)erf\left(\frac{WYZ/2}{\sqrt{2}\sigma_{water}(z)}\right)\right], \quad (9.50)$$

where WXZ and WYZ are the field dimensions at depth z. D for an infinite field size in water can be found from

$$D_{water}^{\infty}(0,0,z) = D_{water}^{WX,WY}(0,0,z)\left[erf\left(\frac{WXZ/2}{\sqrt{2}\sigma_{water}(z)}\right)erf\left(\frac{WYZ/2}{\sqrt{2}\sigma_{water}(z)}\right)\right]^{-1}. \quad (9.51)$$

Instead of collecting and storing data for a large number of rectangular fields, square field data are used and conversion is made to rectangular fields using a square root method.

Correction for inhomogeneities is done by calculating an effective depth z_{eff} in water for each strip and replacing this in the first term of [9.50]. z_{eff} is determined using the collisional stopping power ratios such that:

$$z_{eff}(z) = \int_{-L_0}^{z}\left(\frac{(dE/dz)^{med}}{(dE/dz)^{water}}\right)_{coll}(z')dz', \quad (9.52)$$

where L_0 is the gap from the applicator to the patient surface. The Hogstrom method also includes formalisms for how to deal with nonconstant fluence across the beam and also how to calculate values for $\sigma_{\theta x}$ at the collimator face and subsequently the x component of the lateral standard deviation σ_x. An effective source collimator distance and virtual source position are also input requirements for most Hogstrom-implemented electron pencil beam codes (see chapter five).

In summary, the approach suggests the mean path of an electron pencil beam can be approximated by the central pencil beam axis. The approach (Hogstrom and Almond 2006) suggests that:

i) Provides sufficient input of square central axis data for irregular electron field dose distribution prediction.
ii) Utilizes CT data on a pixel-by-pixel basis.
iii) Accurately models variable air gaps.
iv) Enables pencil beams to be redefined at surface to account for irregular surface contours.

The 2-D approach which approximated infinite heterogeneities in one lateral dimension has some shortcomings as shown in figure 9.14. This has also been improved to a 3-D implementation (Starkschall et al. 1991) and this

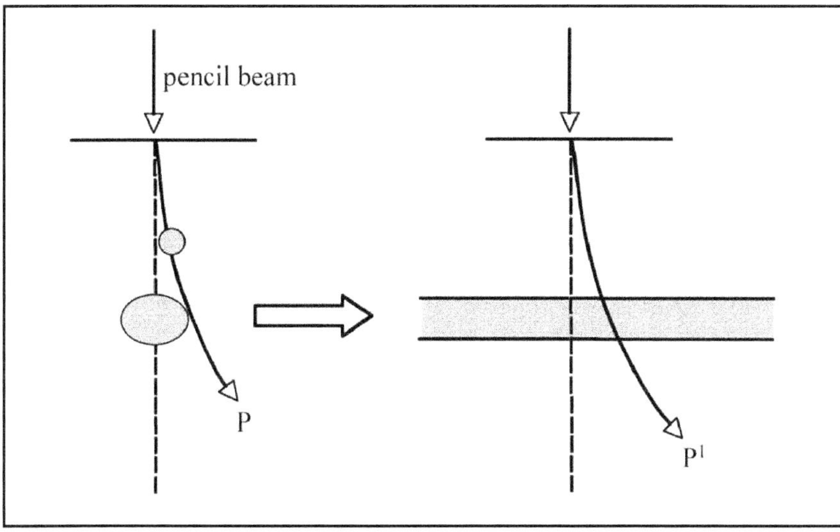

Figure 9.14: Approximation caused by assuming infinite lateral extent of inhomogeneity at central axis where electron expected to reach P reaches P[l].

is the implementation currently used by two planning vendors for electron beam planning (Philips Pinnacle and CMS Focus).

The Fermi-Eyges theory underestimates large angle scattering and Lax et al. (1983) have shown this can be improved by using 3-Gaussian kernel distributions. Another assumption of the theory is that all electrons reach a predicted practical range, whereas electrons near the end of their range undergo scattering through larger angles. To account for this Werner et al. (1982) further modified the σ of the Gaussian distribution for the final third of electron range.

Another shortcoming of the approach as reviewed by Hogstrom and Steadman (1996) was the central axis approximation. This has been addressed by other approaches such as the *phase space evolution* algorithm (e.g., Huizenga and Storchi 1989, Korevaar et al. 1996) and *pencil beam redefinition* algorithm (PBRA) (Shiu and Hogstrom 1991).

Despite these modifications and improvements to the pencil beam algorithms there has been computing processor speed improvements and new variance reduction approaches that have led to Macro Monte Carlo methods becoming competitive alternatives as electron beam calculation algorithms on commercial RTP systems. These new Macro Monte Carlo methods are reviewed in chapter ten.

9.5 Brachytherapy Source Models

This textbook does not focus on brachytherapy in depth; we refer the reader to the following texts for more detailed information about brachytherapy physics (Baltas et al. 2007, Williams and Thwaites 1993) and various reports deal with safe treatment delivery (e.g., AAPM 1997, 1998). Some common definitions and interesting physics pertaining to radioactive sources are presented here as a snapshot. Note how dose models for radioactive sources vary significantly from those of external beam models.

9.5.1 Definitions

Brachytherapy is the use of radioactive sources placed directly into or adjacent to the volume of tissue to be treated. Brachytherapy sources provide a non-uniform irradiation of the treatment volume, with a very steep dose gradient decreasing with distance from the source. By placing brachytherapy sources within the target volume, a high dose is given to the tumor while a low dose is given to surrounding normal tissue. Entrance and exit doses associated with external beam radiotherapy are not present with brachytherapy. These highly sought dosimetric properties allow brachytherapy to be an alternative or complementary modality to external beam radiotherapy for the treatment of localized tumors in sites surrounded by radiosensitive organs, such as the prostate.

Two types of brachytherapy employed are interstitial and intracavitary (Aird and Williams 1993). In *interstitial therapy* the sources are introduced surgically. Usually needles are used to pierce the skin, then plastic tubes are introduced to house the sources, which are introduced by a remote after-loader (e.g., prostate). In *intracavitary therapy* the sources are placed into pre-existing body cavities. The most common intracavitary therapy is for gynecological malignancy such as cancer of the cervix. Other intracavitary sites are bronchus and esophagus. Interstitial therapy uses two different types of implants: removable and permanent.

Removable implants are associated with high-dose rate (HDR) brachytherapy. HDR brachytherapy uses a high-activity, low-energy x-ray source, which is placed inside the target volume using an array of needles. The source is stepped to predetermined locations (dwell points) along each needle for a short period of time (order of minutes). *Remote afterloading* is the preferred method of delivering HDR brachytherapy. These systems use machines consisting of a safe by the bed with complex driving mechanisms that deliver the sources from the safe to within the patient applicator. The device can be controlled from outside the room; hence sources can be removed during nursing procedures and other visits.

Permanent implants are associated with low-dose rate (LDR) brachytherapy. LDR brachytherapy uses a low activity usually contained within an array of seeds. The number of seeds used in LDR brachytherapy varies with the size of the target volume, but can exceed 150 for the treatment of the prostate. Computer-generated treatment plans are used to predetermine the desired location of each seed to be implanted. A seed applicator and guiding grid are used to insert the seeds into the target volume, mimicking the treatment plan. Once inserted, the seeds reside permanently within the patient, delivering dose to the treatment volume over the course of several months.

Brachytherapy sources are usually *sealed sources*, whereby the radioactive material is encapsulated so that the material cannot be lost under any foreseeable degree of physical or chemical stress. Usually the source is encapsulated in a steel casing to prevent isotope leakage and to help absorb beta particles.

9.5.2 Types of Sources

Iridium-192 (^{192}Ir) has a principle gamma ray that is emitted at 370 keV. While ^{192}Ir can be loaded as a relatively low dose rate source it is now most extensively used as an HDR source. ^{192}Ir 10-curie (Ci) sources are used in (HDR) remote afterloading devices such as the microSelectron-HDR as shown in figure 9.15. These devices typically deploy a single 1.1 mm diameter by 3 mm length cylindrical source. They allow computer-controlled, precise position deployment of a single high-activity source into applicators. The source is stepped to various locations for dwell times required to deliver dose at each dwell position consistent with a predetermined dose plan. All staff and operators remain remote from the room. The dose rate of HDR can be up to 5 Gy/minute at 0.5 cm. Extensive use of ultrasound, CT, and MRI techniques have been developed for planning the dose distributions prior to delivery by the source. The use of HDR in four to six large fractions is now a common procedure for advanced prostate boosts and cervical brachytherapy procedures. For prostate, typically, a 45 Gy external beam regime will be augmented with a four-fraction boost of about 6 Gy per fraction. The biologically effective dose (BED) (see chapter fourteen) for the

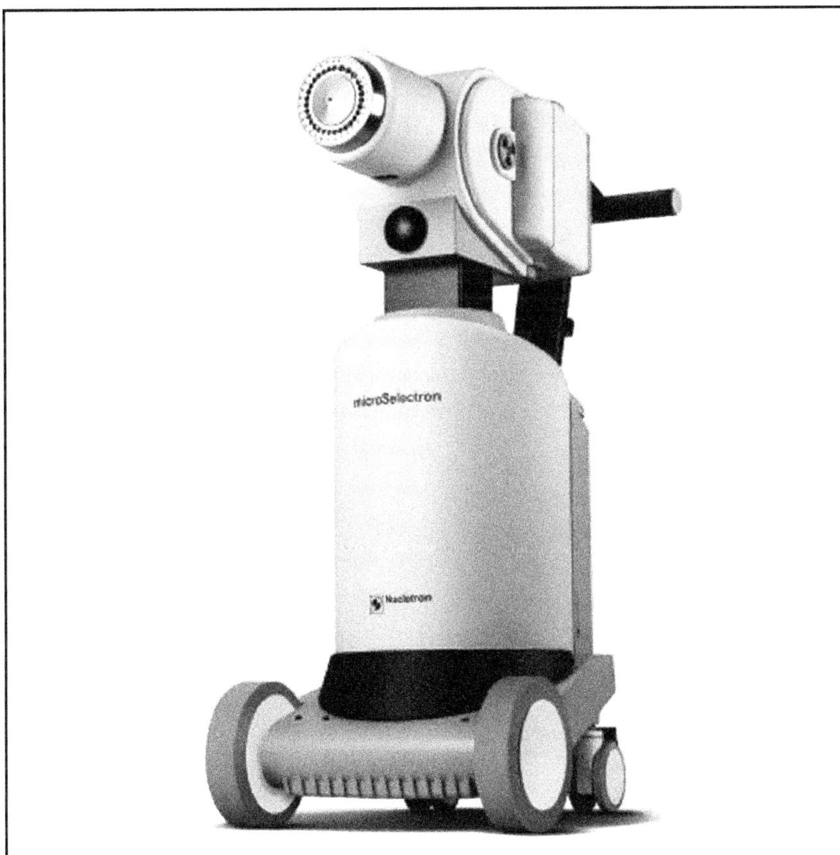

Figure 9.15: MicroSelectron HDR transportable HDR afterloading unit, which uses 10 Ci iridium-131 sources. (Reprinted with permission from Nucletron.)

combined treatment is over 90 Gy. Such a high dose can be achieved because of the localized high dose boost from the brachytherapy procedure. Images are often taken between treatment fractions and the applicator can be repositioned precisely if the applicator has been deemed to move with reference to fiducial markers loaded in the prostate.

Despite the short half-life of 74 days, which required source replacement often, this isotope appears to be gaining favor over the others at present. The energy and high source activity loading into a small source makes insertions possible via thin-gauge needle applicators. This isotope is currently extensively used in prostate treatment as a boost technique, with the current isotope of choice being ^{131}Ir. Monte Carlo techniques which characterize source properties are ideal as accurate dosimetry is always challenging due to the high dose gradients involved. Since the work of Williamson and Li (1995) who characterized iridium sources from pulsed dose rate and HDR devices, various other isotopes have been characterized using Monte Carlo methods (e.g., Russell and Ahnesjö 1996, Karaiskos et al. 1998).

Iodine-125 (^{125}I) has a half-life of 60 days and emits x-rays with energies of 27.4, 31.4, and 35.5 keV. This is a very commonly used LDR isotope. The low dose rate and energy render them reasonably safe for live insertions. There are a few different kinds of constructions, but they are usually in small seeds that can be inserted in connected strands (e.g., RAPID Strand™, Oncura); they can also be individually loaded using a seedSelectron™ (Nucletron) (see figure 9.16). These are permanent implants as the low energy presents a very low radiation hazard to others. The low energy ensures dose to the exit surface of the patient skin is minimal. The low dose rate ensures very high doses can be given with reduced late effects. BEDs (see chapter fourteen) in excess of 145 Gy BED can be delivered because the LDR repair effect can be capitalized upon. The calibration estimates for these sources have been recently revised (AAPM 2004). These implants are generally given as monotherapy to low Gleason score prostate tumors. There is some use of palladium-103 (^{103}Pd) for these treatments. ^{103}Pd has a higher initial dose rate but a shorter half life (AAPM 2000). The seeds (60 to over 100) are left in the prostate permanently as radioactive decay renders them inert. Dose to the urethra from these treatments is reasonably high and urethritus is an early effect which does resolve after some months. There has been progress in development of dosimetric methods to undertake online urethral dosimetry (Rosenfeld et al. 2004, Cutajar et al. 2006).

Radium-226 (^{226}Ra) was used extensively from about 1900 to 1960. This is because it was the only naturally occurring radionuclide that could be refined with a high specific activity. Several disadvantages include decay to radon-222, which is a toxic gas. This gas is also an alpha emitter (high LET), which decays with short half-life to other alpha-emitting radionuclides. Radium also has a very long half-life, 1620 years. *Cesium-137* (^{137}Cs) became more popular from the 1960s to the 1990s. This isotope emits beta particles and gamma rays at energy 662 keV. The beta particles are absorbed in the metal casing (usually stainless steel or iridium-platinum alloy, 0.5 mm thickness). The long half-life of 30.07 years ensures a long period of use without replacement. Predominantly used in LDR continuous source afterloaders for gynecological treatment, the most commonly used applicators for these

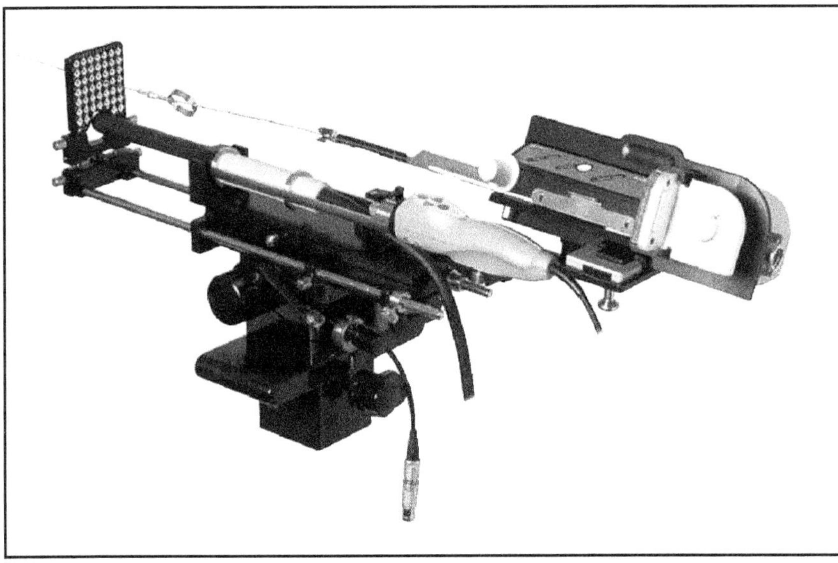

Figure 9.16: SeedSelectron™ LDR afterloading unit which uses iodine-125 sources. Transrectal ultrasound probe below the template is visible. (Reprinted with permission from Nucletron.)

insertions were called Fletcher suit applicators. These consist of a uterine tube and two cervix tubes into which cylindrical sources could be manually afterloaded. A machine called a Selectron (Nucletron) has spherical ^{137}Cs sources that can be automatically remotely afterloaded into Fletcher suit type applicators. LDR for intracavitary insertions is generally regarded as about 2 Gy/hour to point A (i.e., 2 cm from the lower end of the central uterine tube). This isotope is being replaced more and more by HDR techniques with iridium. Cesium can only be produced as an LDR isotope because of its low specific activity. Some properties of the sources discussed here are shown in table 9.1.

Table 9.1: Properties of radionuclides commonly used in brachytherapy

Nuclide	*Energy* Mean (MeV)	Energy Max. (MeV)	Half-Life	Lead HVL (mm)	Lead TVL (mm)
^{192}Ir	0.37	0.61	74 days	4.5	15
^{125}I	0.028	0.035	60 days	0.03	0.1
^{226}Ra[a]	--	2.4	1600 years	16	45
^{137}Cs	0.662	0.662	30.0 years	6.5	22

HVL: half-value layer; TVL: tenth-value layer.
[a]Radium source filtered by 0.5 mm platinum.

9.5.3 A 2-D Source Calculation Method

Historical formalisms (Aird and Williams 1993) failed to account for the difference in shielding at the end of brachytherapy sources compared with source encapsulation shielding at the ends of sources. This creates a significant anisotropy of dose for ^{125}I capsules in particular.

Task group 43 (TG 43) (AAPM 1995) and the TG 43 update (AAPM 2004) have introduced a formalism which accounts for this factor. We present the 2-D formalism here with reference to figure 9.17. It is interesting to note that source orientation cannot always be resolved, so sometimes a 1-D formalism is used in brachytherapy computer planning programs. This overestimates the anisotropy along the source and underestimates this at the source ends. This does not significantly affect the dose calculation, provided a large number of sources are loaded.

The general 2-D calculation is

$$\dot{D}(r,\theta) = S_K \times \Lambda \times \frac{G_L(r,\theta)}{G_L(r_0,\theta_0)} \times g_L(r) \times F(r,\theta). \tag{9.53}$$

where r is distance in centimeters from the center of the active source to the point of interests, r_0 denotes the reference distance (1 cm), and t is the polar angle to the point of interest at $P(r,\theta)$ relative to the source longitudinal axis; θ_0 is the source transverse plane, i.e., 90°.

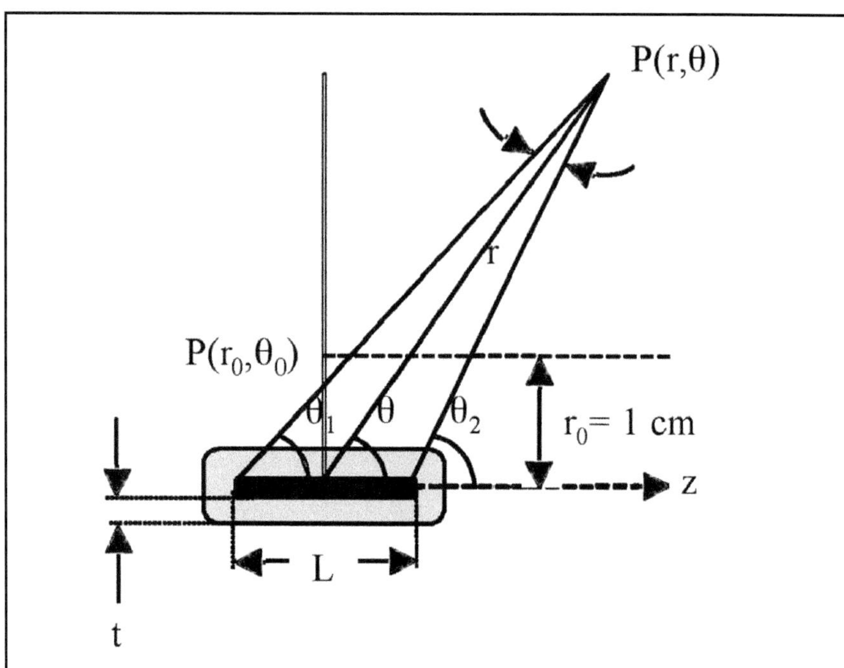

Figure 9.17: Brachytherapy line source parameters in the AAPM 2-D source calculation model. (Reprinted with permission from AAPM 2004.)

S_k is the air-kerma strength, sometimes denoted as U where 1U = 1 μGym^{2}h^{-1}, which is also equal to 1U = 1 cGycm^{2}h^{-1}. *The dose rate constant* in water Λ with units cGyh^{-1}U^{-1} reduces to cm^{-2}, such that:

$$\Lambda = \frac{D(\dot{r}_0, \theta_0)}{S_K}. \tag{9.54}$$

The dose rate constant is measured using a combination of experimental and Monte Carlo simulations. This constant has historically changed from 1.18 to 0.96 for ^{125}I, and this in turn changed the dose coverage on some plans.

The geometry function which accounts for line source is given by

$$G_L(r,\theta) = \frac{\beta}{Lr\sin\theta} \quad \text{if} \quad \theta \neq 0^\circ, \tag{9.55}$$

and

$$G_L(r,\theta) = \left(r^2 - L^2/4\right)^{-1} \quad \text{if} \quad \theta = 0^\circ. \tag{9.56}$$

Also, the point source approximation is

$$G_L(r,\theta) = r^{-2}. \tag{9.57}$$

The radial dose function accounts for fall-off in the transverse plane due to photon scattering, where x is either a line source L or point source P, then

$$g_x(r) = \frac{D(\dot{r}, \theta_0)}{D(\dot{r}_0, \theta_0)} \frac{G_x(r_0, \theta_0)}{G_x(r, \theta_0)}. \tag{9.58}$$

The anisotropy function, which accounts for the difference in source end shielding, is given by

$$F(r,\theta) = \frac{\dot{D}(r, \theta_0)}{\dot{D}(r, \theta_0)} \frac{G_L(r, \theta_0)}{G_L(r, \theta)}. \tag{9.59}$$

This function describes the variation in dose as a function of polar angle relative to the transverse plane. $F(r,\theta)$ on the transverse plane is defined as unity, as encapsulation thickness increases $F(r,\theta)$ is generally less than one.

Because it is difficult to determine the source orientation a 1-D formalism is often used by brachytherapy planning computer systems.

$$\dot{D}(r) = S_K \times \Lambda \times \frac{G_x(r,\theta_0)}{G_x(r_0,\theta_0)} \times g_x(r) \times \phi_{an}(r), \tag{9.60}$$

where in this case the anisotropy function $\phi_{an}(r)$ is the ratio of solid angle weighted dose rate over the entire 4π steradian space. This approximation appears to work accurately for sources such as ^{125}I because multiple sources are loaded.

There are interesting radiobiological differences due to the different dose rates used in LDR and HDR brachytherapy where particularly the late dose complications vary with HDR. Comparison of LDR with external beam (which is high dose rate) has also had implications in explaining some radiobiological parameters such as the suspected low alpha beta ratio for prostate tumors, as discussed further in chapter fourteen.

References

AAPM (American Association of Physicists in Medicine) (2004) Update of AAPM TG 43: A revised AAPM protocol for brachytherapy dose calculations. *Med. Phys.* 31: 633–674.

AAPM (American Association of Physicists in Medicine) (2000). Recommendations of the AAPM on 103Pd interstitial source calibration and dosimetry: Implications for dose specification and prescription. *Med. Phys.* 4: 634–642. Also available as AAPM Report No. 69.

AAPM (American Association of Physicists in Medicine) (1998) High dose-rate brachytherapy treatment delivery. *Med. Phys.* 25: 375–403. Also available as AAPM Report No 61.

AAPM (American Association of Physicists in Medicine) (1997) Code of practice for brachytherapy physics. *Med. Phys.* 24: 1557–1598. Also available as AAPM Report No. 59.

AAPM (American Association of Physicists in Medicine) (1995) Dosimetry of interstitial brachytherapy sources. Report of Task Group 43. *Med. Phys.* 22: 209–234.

Aird E., Williams J. (1993) Brachytherapy. Chapter 9 in *Radiotherapy Physics in Practice.* J. Williams and D. Thwaites (eds.). Oxford University Press. pp. 187–226.

Baltas D., Sakelliou L., Zamboglou N. (2007) *The Physics of Modern Brachytherapy for Oncology*. CRC Press, Taylor Francis Group, Atlanta, Georgia.

Batho H.F. (1964) Lung corrections in cobalt-60 beam therapy. *J. Can. Assoc. of Radiologists* 15: 79–83.

Battista J.J., Sharpe, M.B. (1992) True three-dimensional dose computations for megavoltage x-ray therapy: A role for the superposition principle. *Australas. Phys. Eng. Sci. Med.* 15: 159–178.

Boyer A.L., Mok E.C. (1986) Calculation of photon beam dose distributions in an inhomogeneous medium using convolutions. *Med. Phys.* 13: 503–509.

Brahme A. (1981) Electron beam dose planning using Gaussian beams: Mathematical background. *Acta Oncol.* 20: 147–158.

Brooks R.A., Di Chiro G. (1975) Theory of image reconstruction in computed tomography. *Radiology* 117: 561–572.

Clarkson J. J. (1941) A note on depth doses in fields of irregular shapes. *Br. J. Radiol.* 14: 265–268.

Cunningham J.R. (1982) Tissue inhomogeneity corrections in photon beam treatment planning. *Progress in Medical Radiation Physics*, vol 1. Edited by C.G. Orton. Plenum Publishing, New York.

Cunningham J.R. (1972) Scatter-air ratios. *Phys. Med. Biol.* 7: 45–51.

Cutajar D., Takacs G., Lerch M., Braddock T., et al. (2006) Intraoperative solid-state based urethral dosimetry in low dose rate prostate brachytherapy. *IEEE Trans. Nucl. Sci.* 53: 1408–1412.

el-Khatib E., Battista J.J. (1984) Improved lung dose calculation using tissue-maximum ratios in the Batho correction. *Med. Phys.* 11: 279–286.

Eyges L. (1948) Multiple scattering with energy loss. Phys. Rev. 74: 1534–1535.

Geise R.A., McCullough E.C. (1977) The use of CT scanners in megavoltage photon beam therapy planning. *Radiology* 124: 133–141.

Hoban P. (1988) Electron Beam Modelling in Radiotherapy. M.Sc. Thesis. University of Waikato, New Zealand.

Hoban P.W., Keall P.J., Round W.H. (1992) The effect of density on the 10 MV photon beam penumbra. *Australas. Phys. Eng. Sci. Med.* 15: 113–123.

Hogstrom K., Almond P. (2006) Review of electron beam therapy physics. *Phys. Med. Biol.* 51: R455–R489.

Hogstrom K., Steadman R. (1996) Electron beam dose computation. *Teletherapy: Present and Future*. Proc. Summer School of AAPM. J. Palta and T.R. Mackie (eds.). Advanced Medical Publishing, Madison, Wisconsin, pp. 137–174.

Hogstrom K., Mills M., Almond P. (1981) Electron beam dose calculations. *Phys. Med. Biol.* 26: 445–459.

Huizenga H., Storchi P. (1989) Numerical calculations of energy deposition by broad high energy electron beams. *Phys. Med. Biol.* 34: 1371–1396.

ICRU (1976) Report 24. Determination of absorbed dose in a patient irradiated with beams of x or gamma rays in radiation procedures. ICRU, Washington D.C.

Jette D. (1984) The problem of electron dose calculation: 1 Multiple scattering methods. *A. A. Med. Dosim.* 9: 12–17.

Kan M.W., Wong T.P., Young E.C., Chan C.L., Yu P.K. (1995) A comparison of two photon planning algorithms for 8 MV and 25 MV x-ray beams in lung. *Australas. Phys. Eng. Sci. Med.* 18: 95–103.

Karaiskos P., Angelepoulos A., Sakelliou L., Sandilos P., Antypas C., Vlachos L., Koutsouveli E. (1998) Monte Carlo and TLD dosimetry of an 192Ir high dose-rate brachytherapy source. *Med. Phys.* 25: 1975–1984.

Korevaar E.W., Dabrowski R., Janssen J.J., Storchi P.R., Huizenga H. (1996) Phase space evolution distribution functions for high energy electron beams. *Phys. Med. Biol.* 41: 2079–2090.

Lax I., Brahme A., Andreo P. (1983) Electron dose planning using Gaussian beams. Improved dose profiles. *Acta Radiol. Suppl.* 364: 49–59.

Lillicrap S., Wilson P., Boag J. (1975) Dose distributions in high energy electron beam distributions from narrow beam data. *Phys. Med. Biol.* 20: 30–38.

Mackie T.R., Scrimger J.W., Battista J.J. (1985) A convolution method of calculating dose for 15-MV x rays. *Med. Phys.* 12: 188–196.

Milan J., Bentley R.E. (1974) The storage and manipulation of radiation dose data in a small digital computer. *Br. J. Radiol.* 45: 115–121.

Mohan R. (1991) Three dimensional dose calculations for radiation treatment planning. *Int. J. Radiol. Oncol. Biol. Phys.* 21: 25–36.

Mohan R. (1989) Three dimensional radiation treatment planning. *Australas. Phys. Eng. Sci. Med.* 12: 73–91.

Mohan R., Chui C. (1985) Validity of the concept of seperating primary and scatter dose. *Med. Phys.* 12: 726–730.

O'Connor J.E. (1957) The variation of scattered x-rays with density in an irradiated body. *Phys. Med. Biol.* 1: 352–367.

Perry D., Holt J. (1980) A model of calculating the effects of small inhomogeneities on electron beam dose distributions. *Med. Phys.* 7: 207–215.

Redpath A.T., Vickery B.L., Duncan W. (1977) A comprehensive radiotherapy planning system implemented in FORTRAN on a small interactive computer. *Br. J. Radiol.* 50: 51–57.

Rosenfeld A., Cutajar D., Lerch M., Takacs G. (2004) In vivo dosimetry and seed localization in prostate brachytherapy with permanent implants. *IEEE Trans. Nucl. Sci.* 51: 1–6.

Russell K., Ahnesjö A. (1996) A dose calculation in brachytherapy for ^{192}Ir source using a primary dose separation technique. *Phys. Med. Biol.* 41: 1007–1024.

Shiu A., Hogstrom K. (1991) Pencil beam redefinition algorithm for electron dose distributions. *Med. Phys.* 18: 7–18.

Sontag M.R., Cunningham J.R. (1978) The equivalent tissue-air ratio method for making absorbed dose calculations in heterogeneous medium. *Radiology* 129: 787–794.

Sontag M.R., Cunningham J.R. (1977) Corrections to absorbed dose calculations for tissue inhomogeneities. *Med. Phys.* 4: 431–36.

Spiers F.W. (1951) Radiotherapeutic physics II: Dosage in irradiated soft tissue and bone. *Br. J. Radiol.* 24: 365.

Starkschall G., Shiu A.S., Bujnowski S.W., Wang L.L., Low D.A., Hogstrom K.R. (1991) Effect of dimensionality of heterogeneity correction on the implementation of a three-dimensional electron pencil beam algorithm. *Phys. Med. Biol.* 36: 207–227.

Van Dyk J., Keane T.J., Rider W.D. (1982) Lung density as measured by computerised tomography: Implications for radiotherapy. *Int. J. Radiat. Oncol. Biol. Phys.* 8: 1363–1372.

Webb S., Fox R.A. (1980) Verification by Monte Carlo methods of the power law tissue air ratio algorithm for inhomogeneity corrections in photon beam dose calculations. *Phys. Med. Biol.* 25: 225–240.

Wegener O.H., Koeppe P., Oeser H. (1978) Measurement of lung density by computed tomography. *J. Comput. Assist. Tomogr.* 2: 263–273.

Werner B., Khan F., Deibel F. (1982) A model for calculating electron beam scattering in treatment planning. *Med. Phys.* 9: 180–187.

Williams J., Thwaites D. (1993) *Radiotherapy Physics in Practice.* Oxford University Press. pp. 187–226.

Williamson J., Li Z. (1995) Monte Carlo aided dosimetry of the microselectron pulsed and high dose rate ^{192}Ir sources. *Med. Phys.* 22: 809–819.

Wong J.W., Purdy J.A. (1990) On methods of inhomogeneity corrections for photon transport. Med. Phys. 17: 807–814.

Wong J.W., Henkelman R.M. (1983) A new approach to CT pixel-based photon dose calculations in heterogeneous media. *Med. Phys.* 10: 199–208.

Wong J.W., Henkelman R.M. (1982) Reconsideration of the power law (Batho) equation for inhomogeneity corrections. *Med. Phys.* 9: 521–529.

Woo M.K., Cunningham J.R., Jezioranski J.J. (1990) Extending the concept of primary and scatter separation to the condition of electron disequilibrium. *Med. Phys.* 17: 588–595.

Questions

9.1 A beam of 6 MV x-rays (field size 10×10 cm^2 at 100 cm SSD) penetrates through the chest wall of a patient through rib into lung. Assuming the chest wall thickness is 4 cm, the rib thickness is 2 cm, and the lung thickness is 7 cm; given the %D(8 cm) = 74.3%, %D(9 cm) = 70.4%, %D(13 cm) = 56.7%, then:

i) Calculate the effective depth to the distal side of the lung (ignore surface curvature).

ii) Calculate the correction factor for this point which would be applied to the percentage depth dose value using effective depth correction.

iii) Calculate the new effective percentage depth dose.

(section 9.2)(section 9.3)

9.2 In reference to the effective depth correction method:

i) Should the effective depth method of inhomogeneity correction over estimate or underestimate the correction factor beneath low-density region such as lung? Explain this in terms of the effect of neglecting to consider the perturbation of scatter dose.

ii) Why would you expect this effect to be smaller when the calculation point is further away from the low-density region?

(section 9.3)

9.3 The Batho Power Law inhomogeneity correction method is based on differences in exponential attenuation between an inhomogeneous medium and a homogeneous unit-density (water) medium. The correction factor for a point beneath an inhomogeneity of density ρ, where the distance along a ray line from the point to the upper and lower surfaces of the inhomogeneity is d_2 and d^1, respectively, is given by

$$CF = \left(\frac{TAR(d_1)}{TAR(d_2)} \right)^{1-\rho} . \quad \text{(Q9.1)}$$

i) Starting with a ratio of exponential attenuation factors, derive the correction factor if the point is in a medium of density ρ_1 and the inhomogeneity has a density ρ_2.

ii) Also derive the correction factor if the point is a distance d_2 beneath the top surface of an inhomogeneity of density ρ, within an otherwise unit density medium.

iii) For a dose calculation point beneath a slab of low-density material with a given thickness, the effective depth method predicts a correction factor that is independent of the position of the slab (since effective depth does not change). Considering the less-than-exponential falloff of a TAR curve at shallower depths, why does the Batho power law method yield a correction factor that decreases as the inhomogeneity becomes closer to the calculation point (which should increase the accuracy of the correction)? Why is the TAR falloff less than exponential?

(section 9.3)

9.4 The equivalent tissue-air ratio (ETAR) method of inhomogeneity correction makes use of an effective field size as well as an effective depth.

i) Which important principle of radiation transport is this method attempting to account for, that is not accounted for by methods that only consider effective depth or the medium along ray lines from the source?

ii) Consider a point 10 cm deep in a medium consisting of a water-like substance with relative density 0.3. The field size at the surface is 20×20 cm^2 at 100 cm SSD. Accounting for beam divergence, what is the effective depth and field size at the calculation point? Why should the ETAR method give an almost exact correction factor for this case? Which well-proven theorem applies here?

iii) Consider the situation where a dose calculation point is at the edge of a block of solid water extending half way across a 20×20 cm^2 field (the point is at the central axis of the field). The ETAR method would calculate an effective field size that is the equivalent square of a 20×10 cm^2 field (since half the field is incident on a medium of density zero) and use this reduced field size to account for the reduced scatter resulting from the missing medium. Considering that in reality only half the scatter dose will exist at the calculation point compared to that if the medium extended over the whole field, explain why the ETAR method will give an incorrect answer in this case. Will the correction factor be too high or too low?

(section 9.3)

9.5 Inhomogeneity correction methods relying on measured data, such as tissue-air ratios, to account for the variation in dose with field size are known to overestimate the correction factor in some situations, particularly those involving small field sizes, low-density regions, or high- energy beams. The problem stems from the fact that measurements are not made for field sizes smaller than that required for electronic equilibrium, thus data for smaller field sizes must be obtained by extrapolation.

i) Which characteristic of dose versus field size dependence is not accounted for by this extrapolation and is responsible for the inaccuracy?

ii) Why can the three situations mentioned accentuate the problem?

iii) The true dose for a field size of zero is zero; however, extrapolation of measured data will yield a finite value. What does this finite value really represent?

(sections 9.3)

10 Beam Models: Part II

"There is no hard boundary between correction-based and model-based algorithms. . . . The essential difference is that model-based calculations do not attempt to reconstitute measured data before correcting it for patient treatment situations. The treatment situation is simulated from principles."

(T. Rock Mackie et al. 1996)

10.1 Introduction

Of all the possible methods of predicting a dose distribution due to an incident photon beam, the one that is potentially the most accurate is Monte Carlo modeling. This is particularly true when there are geometrically complex inhomogeneities whose density deviates significantly from unity, in which case the correction methods discussed in chapter six fail to predict the deviation from an in-water dose distribution properly. Indeed, as in the situation of small-field irradiation of a medium containing lung (where electronic disequilibrium is present), low or high-density regions pose problems for correction methods), even if the geometry is simple.

An alternative to the correction methods is to calculate the distribution in an inhomogeneous medium from first principles, using the physics of photon and electron transport. It is in principle possible to do this using analytical methods, where differential equations (in particular the Boltzmann transport equation) are solved for the specific beam and medium conditions. In practice, however, there are so many variables affecting the resulting dose distribution that the equations are rendered almost impossible to solve. Solutions can only be found if many approximations are applied in both formulating and solving the equations. This means, for example, that only simple interaction mechanisms can be accounted for and that solutions for inhomogeneous media are prohibitively difficult to obtain.

Also using photon and electron transport physics, it is possible to consider the trajectories of individual particles and thus the pattern of energy deposition due to each: This is the approach used in a Monte Carlo simulation. The transport of an incident particle, and of particles that it subsequently sets in motion, is referred to as a particle *history*. Each history is uniquely determined by random selection from the probability distributions that control each possible interaction. A dose distribution can be built by summing the energy deposition in each particle history. However, a very large number of histories are required before the uncertainty in the distribution is small enough for it to be used in treatment planning—the lower the uncertainty, the smoother will be the depth dose, profile, or isodose curves obtained from the distribution. A reduction in uncertainty is also often referred to as an increase in reproducibility.

Apart from a Monte Carlo simulation, the most physically realistic method of generating a photon beam dose distribution is using a *convolution* algorithm. Convolution methods use a combination of analytically calculated and Monte Carlo-generated physical quantities to construct the dose distribution. Unlike in a Monte Carlo simulation, the convolution method does not consider individual particles, but rather the macroscopic, or statistically averaged, effects of large numbers of particles. It is like Monte Carlo, however, in that it does not rely on measured dose data; dose distributions are calculated directly from the known characteristics of the incident beam.

In this chapter, Monte Carlo methods will be discussed in section 10.2, with emphasis on how photon/electron transport is handled and the concept of statistical uncertainty. Discussion is mainly based on the EGS4 system (Electron Gamma Shower, version 4), and examples of its use are presented. While there are now new versions, EGS5 and EGSnrc, the simulations done for the first edition of this textbook still show valid demonstrations of the physics. EGSnrc and BEAMnrc simulations, which model linac treatment head effects, are also shown in section 10.3. Another Monte Carlo code finding favor in many applications related to radiotherapy, Geant4, is also discussed in section 10.4.

The convolution/superposition method is also described in section 10.5, focusing on the motivation for the development of this technique, the essential components of convolution, and accounting for inhomogeneities and for the polyenergetic nature of megavoltage photon beams.

A new generation of Macro Monte Carlo codes is being developed for introduction as dose calculation engines for routine radiotherapy treatment planning. Progress with these models is discussed in section 10.6.

10.2 Monte Carlo Simulation

10.2.1 The EGS4 System

The most widely used Monte Carlo code in medical radiation physics today is the EGS4 system. It has been used extensively by the authors and many others for the generation of benchmark dose distributions with which to compare the results of other less-computer-intensive dose calculation methods. It is also a valuable tool in producing data required for algorithms such as the convolution method (see section 10.3).

The early Monte Carlo systems used for simulating radiation transport were mostly applied to higher energies than those used in medical physics; for example, in cosmic ray and high-energy linear accelerator physics. The predecessor to EGS4 was called EGS3 and was developed by Ford (HEPL) and Nelson (SLAC) (Ford and Nelson 1978). This code could be used for simulating particles from 0.1 MeV (photons) and 1.5 MeV (electrons) to several thousand MeV. EGS3 was upgraded to EGS4 by Nelson, Hirayama (KEK), Rogers (NRC), and Bielajew (NRC) (Nelson et al. 1985), in an effort to fix bugs, make the system more flexible, and improve performance at low energies. In particular, a lot of work was done on making the modeling of electron transport more rigorous and efficient.

The EGS4 code is written in a language called MORTRAN, which is based on FORTRAN but has extensions to make it more flexible and easier to use. MORTRAN code is translated into FORTRAN before compilation using a MORTRAN processor supplied with EGS4. The most important difference between the two languages is that MORTRAN uses macro replacement to make the code more readable, whereby a piece of code can be represented by a short string that is replaced by the full code in a preprocessing step before translation to FORTRAN.

The main components of the EGS4 system are:

i) *The usercode.* Only a relatively small part of the EGS4 code is written by the user. The usercode has the function of: (a) specifying the geometry, incident beam, and output requirements of the simulation; (b) initiating each history; (c) recording, or *scoring*, parameters of interest (dose per voxel for example); (d) recognizing when a transport step will take a particle into a new region; and (e) outputting results. Of these functions, (a), (b), and (e) are performed by the MAIN subroutine, (c) is carried out by subroutine AUSGAB, and (d) by subroutine HOWFAR.

ii) *The EGS4 system code.* The simulation of particle transport is carried out by the main system code. Initial parameters for each particle history are passed from subroutine MAIN in the usercode to the system code subroutine SHOWER. This routine in turn calls the particle simulation subroutines, which handle displacement and interaction processes. The particle simulation routines also interface with the routines AUSGAB and HOWFAR in the usercode.

iii) *PEGS4.* Before a simulation is begun, data sets containing physical properties of each material in the simulation must be created using the EGS4 preprocessor PEGS4. This program generates parameters such as scattering cross sections, mean free paths, and electron stopping powers for a user-defined energy range. PEGS4-generated data is read in via the subroutine HATCH in the user code before the simulation begins.

One of the major tasks for the user is to set up the geometry for the problem. In the case of an external beam simulation where a dose distribution is to be obtained, this involves defining the extent and orientation of the radiation field as well as defining the array in which to score energy deposition. It is common to define a scoring array in either cylindrical or cartesian geometry.

Cylindrical geometry is well-suited to obtaining depth dose curves, since energy deposition can be scored in a narrow column of voxels on the beam central axis. Furthermore, if the beam is circular in cross section and its axis coincides with that of the array, radial symmetry exists and energy scored in successive annuli can be used to obtain dose profile curves. Examples of EGS4 user codes that employ cylindrical geometry are INHOM (Rogers 1984), and DOSRZ, both of which are distributed with the EGS4 system.

Cartesian geometry is more flexible than cylindrical geometry since energy can be scored in a three-dimensional array (thereby creating a three-dimensional dose distribution) and one- or two-dimensional dose distributions in all rows, columns, and planes can be obtained. Also, voxels in a three-dimensional cartesian array can each be assigned a unique medium composition, thereby allowing the definition of complex inhomogeneities. Often, however, there is a price for the increased flexibility of cartesian geometry: (a) larger arrays (and hence more memory) are required to define the medium and in which to score the dose; (b) scoring voxels are usually smaller, requiring a longer simulation time to obtain results with a low uncertainty; and (c) each particle history may take longer to simulate because there are more voxel boundaries for particles to cross. Examples of cartesian EGS4 user codes are XYZDOS and RTPCART (Murray 1990), which is adapted from INHOM.

10.2.2 Monte Carlo Particle Transport

The unique history of an incident particle is determined from: (a) medium geometry and composition (controlled by the user, who defines the irradiated medium prior to running the simulation); (b) the initial state of the particle, such as incident position, angle, and energy (determined randomly, with certain user-defined constraints); and (c) random selection from the set of probability distributions governing the possible interactions of photons and electrons. The basic strategy used in Monte Carlo particle transport is outlined in this section, along with an overview of how a beam can be defined and particle histories initiated.

At any instant during a photon history, one or more particles (photons or electrons) have their position, direction, and energy stored on a stack of variables. The transport of a particle to its next position is called a *step,* after which the variables on the stack associated with that particle are updated. The key factors determining the unique sequence of steps constituting a photon history are now outlined.

During a photon step, the following points must be considered:

i) *Distance to the next interaction.* In a photon step, the particle is transported in its current direction for a random distance, where the probability of selecting a particular distance is determined by the mean free path corresponding to (a) the current photon energy and (b) the medium in which the photon currently resides.

ii) *Type of interaction*. After a step is completed, a decision is made as to which type of interaction is to take place at that position (usually a photoelectric, Compton, or pair production event). Interaction type is chosen randomly, with probabilities corresponding to photon energy and medium.

iii) *New angle and energy*. When the type of interaction has been chosen, the resultant angle and energy of the original particle are chosen. These parameters are determined from cross section tables, again taking into account energy and medium.

iv) *New particles*. As a result of some interactions, new particles are created and/or set in motion (as in pair production, for example). The position, direction, and energy of these particles are then added to the data stack.

Photon Interactions

The position and type of each photon interaction in a Monte Carlo simulation is governed by pregenerated distributions along with random selection from these distributions. In particular, the distance from the surface (or from the current interaction site) to the next interaction site is determined by selection from a distribution characterized by the *mean free path* of the photon, which depends on the photon's energy and on the medium composition. Once an interaction site is chosen, an interaction type is chosen based on the known likelihoods of each process. Briefly, the important processes are the following (see chapter two for details):

Compton scattering takes place when a photon of energy $h\nu$ collides with a "free" electron, transferring some of its energy, E_{tr}, to the electron. The photon is scattered, retaining an energy $h\nu - E_{tr}$. The energy transferred, photon scattering angle, and initial direction of the recoil electron are determined by sampling from *Klein-Nishina* cross section data.

The Photoelectric effect takes place when the incident photon loses all its energy in ejecting an orbital electron (sometimes a tightly bound inner electron). The energy left over from that required to free the electron is given to the electron as kinetic energy.

Pair production can occur when the photon has an energy greater than the rest mass of an electron and a positron (1.022 MeV). The electron-positron pair are created in a conversion of energy to mass, and the positron eventually annihilates with an electron, producing a pair of 0.511 MeV photons.

To illustrate typical outcomes of the three main types of photon interaction, EGS4 simulations were performed by the authors and the histories of one or more incident photons were displayed using the EGS_Windows graphics system. The user code employed was a modified version of DOSRZ,

(a)

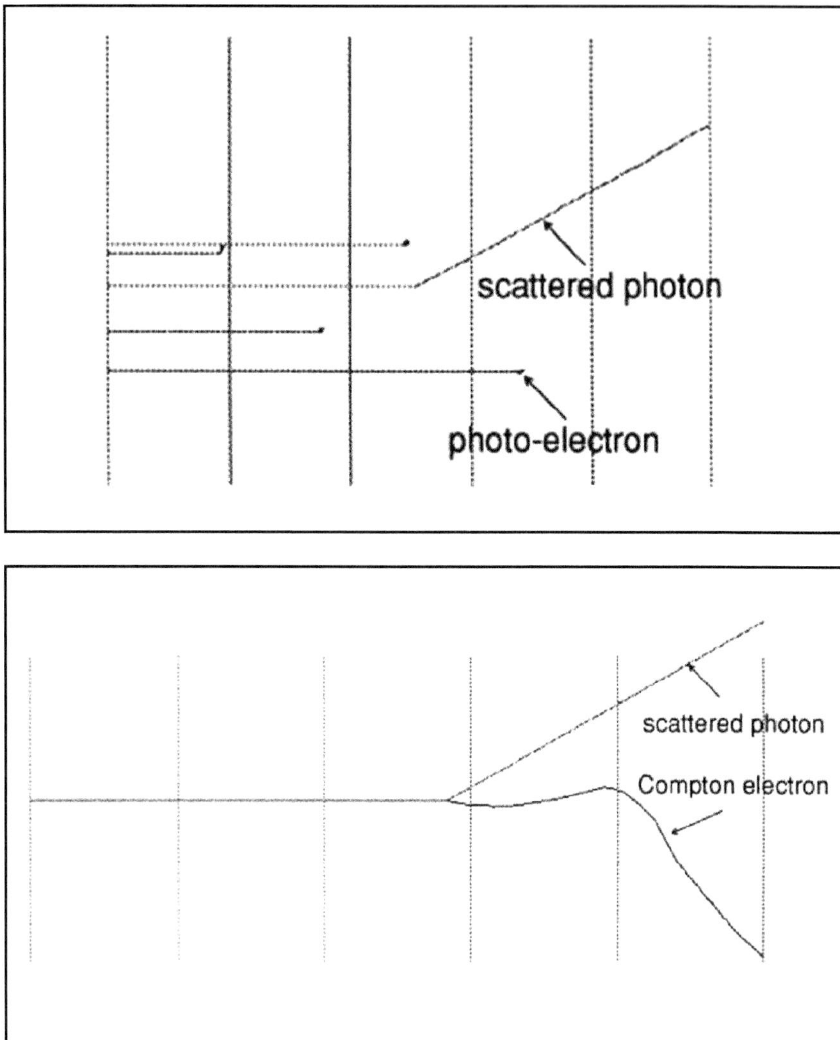

(b)

Figure 10.1: (a) Five 500 keV photons incident on lead, showing four photoelectric interactions (including ejected photoelectrons) and a single Compton scattering event (vertical bars are 1 mm apart). (b) A 10 MeV photon incident on water, showing a Compton scattering event with high-energy recoil electron event (vertical bars are 1 cm apart).

which is supplied with the EGS4 distribution. In each case, 25 incident photons were simulated, where the energy and medium composition was chosen to give an abundance of the desired interaction type. The EGS_Windows package allows interactive 3-D color visualization of particle histories and simulation geometry. The images shown in figure 10.1 are 2-D projections viewed in a direction perpendicular to the incident photon direction.

Photoelectric interactions were produced using 500 keV photons incident on a lead block. From tables (Johns and Cunningham 1983), more than half of all primary photon interactions are expected to be photoelectric events in this

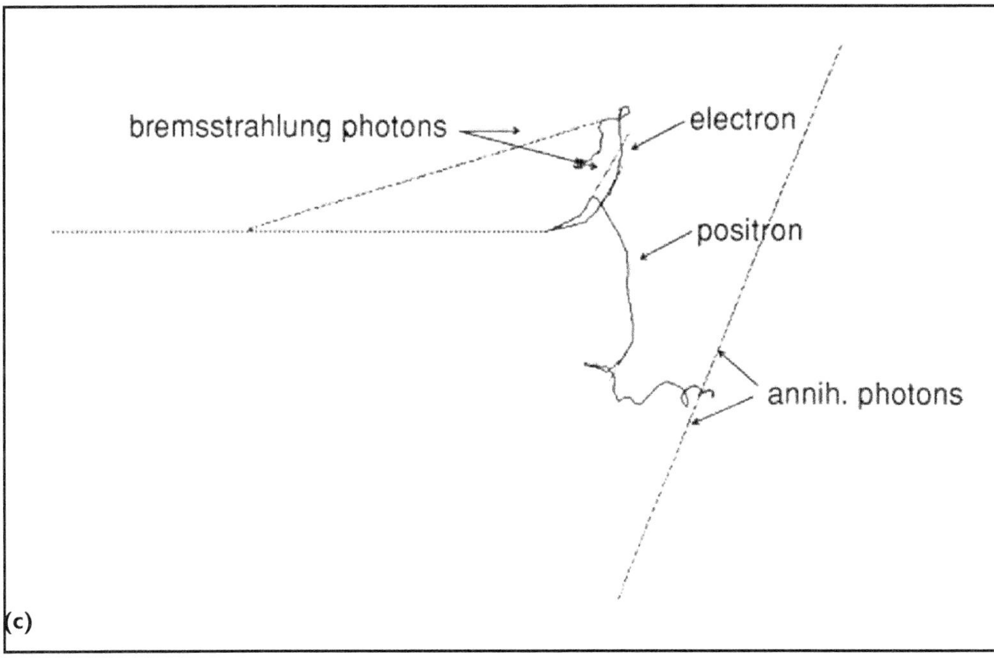

Figure 10.1: (Continued) (c) A 10 MeV photon incident on lead, producing an electron-positron pair, each of which produces a bremsstrahlung photon, with the positron annihilating with an electron to form a pair of annihilation photons.

case. In figure 10.1a, five photons are shown interacting in a 5 mm thick lead block. Four of the photons undergo photoelectric interactions, each of which can be seen to eject a very short range photoelectron. One photon is Compton scattered, and the Compton electron cannot be seen due to its extremely short range. Here the interaction type and distance to interaction have been chosen randomly by EGS4, while the energy and angle of the photoelectron are fixed (in the default version of EGS4).

Figure 10.1b shows a single Compton scattering event undergone by a 10 MeV photon incident on a 5 cm water slab. Approximately 70% of primary photon interactions will be Compton scattering events in this case. This Compton recoil electron has enough energy to exit the medium, after traveling on a curved path determined by multiple scattering steps. Parameters chosen randomly include distance to interaction, interaction type, photon scattering angle, electron recoil angle, photon energy loss (and consequent energy transfer to the electron), and electron multiple scattering angles.

Pair production was simulated using 10 MeV photons incident on a 5 cm lead block, where approximately 80% of primary photon interactions will be pair production events. Some very interesting histories exist, which include electron-positron pairs, bremsstrahlung photons, and annihilation photons. Each of these phenomena are seen in figure 10.1c, where the primary photon produces an electron-positron pair, the electron and positron both produce bremsstrahlung photons, and the positron produces a pair of 0.511 MeV, oppositely directed, annihilation photons.

Electron Transport

While the spatial distribution of energy imparted in photon interactions largely determines the dose distribution, it is electron transport that consumes most of the computing time in a Monte Carlo simulation (at least for high energies where electron range is long). This is because there are usually many short electron transport steps corresponding to each photon step. For example, after a Compton interaction involving a 5 MeV incident photon, a 3 MeV electron could be set in motion that may undergo tens of steps before running out of energy and stopping. In EGS4, electrons are transported until they fall below a user-defined cutoff energy *ECUT*. Unlike in photon transport, electron steps do not usually correspond to the transport of a particle from one discrete interaction to another. The energy loss and multiple scattering interactions of an electron can be condensed, where after each step the electron is considered to have lost a small amount of energy (in many collision and radiative energy losses) and been scattered through a small angle (after undergoing multiple Coulomb scattering). The energy lost in the step is then the product of the stopping power of the medium (which is also dependent on the electron energy) and the length of the step. The angle of deflection after the step is obtained by sampling from an angular distribution characterized by the scattering power (which is medium and energy dependent) and the step length.

In the electron transport procedure described above, energy losses are grouped such that energy is considered to be deposited evenly, or continuously, along each step. This is known as *continuous* energy loss, and a *continuous slowing down approximation* (CSDA) is said to have been used. While being a much faster method of modeling electron transport than considering electron interactions individually, the use of the CSDA means that much of the randomness in electron transport is lost. For example, all electrons with initially the same energy will travel the same distance before stopping. In EGS4, a mixed procedure is used whereby energy losses below a certain user-defined threshold are modeled using the CSDA, while those above the threshold are treated individually in much the same way as photon interactions. The thresholds for discrete collision and radiative energy losses are called *AE* and *AP*, respectively. The lower these thresholds are set, the more randomness will be present in electron transport. The particular phenomenon that is only modeled if discrete energy loss is included is *energy loss straggling*, where the energy lost by electrons of a certain energy in a unit pathlength is not constant (as in the case of CSDA) but has a statistical distribution.

Classes I and II Electron Transport Procedures

The two main Monte Carlo codes used in radiotherapy physics, EGS4 and ETRAN, use different methods of accounting for the large energy loss events that occur due to production of delta rays and bremsstrahlung (Rogers et al. 1988).

ETRAN uses a class I procedure, where the randomness of collision energy loss in a step is accounted for by sampling the energy loss from a distribution such that for a step length t, the energy lost is $\Delta E(t)$. There may also be a delta ray produced of energy E_δ whose energy is included in $\Delta E(t)$ in a statistical sense. Radiative energy loss is randomized via bremsstrahlung photon energies, E_γ, being sampled from a distribution, where the photon is transported if its energy is above the photon transport cutoff. If the energy of the electron at the start of a step is E_0, the energy at the end of the step is given by

$$E_1 = E_0 - \Delta E(t) - E_\gamma, \tag{10.1}$$

noting that the energy of the delta ray can be considered to be included in the $\Delta E(t)$ term. Subtracting the energy given to knock-on electrons and bremsstrahlung from the energy lost in the step, we have the energy *deposited* in the step:

$$E_{dep} = \Delta E(t) - E_\delta. \tag{10.2}$$

In a class II procedure, as used by EGS4, collision energy losses below the energy $\Delta = AE - 0.511$ MeV and radiative energy losses below AP MeV are characterized by the restricted total stopping power $(dE/dS)_{\Delta,AP}$. After a step length t, an electron with initial energy E_0 will have energy

$$E_1 = E_0 - t(dE/dS)_{\Delta,AP} - E_\delta - E_\gamma, \tag{10.3}$$

where the total energy of delta rays (E_δ) and bremsstrahlung photons (E_γ) will decrease with increases in AE and AP, respectively. Energy deposited in the step is simply

$$E_{dep} = t(dE/dS)_{\Delta,AP}, \tag{10.4}$$

since the delta ray and bremsstrahlung energies will be deposited elsewhere. It is useful to have flexibility in the parameters AE and AP, as computation time is increased with a decrease in these values (because of the increase in number of discrete interactions modeled) and it is often not necessary to use small values. For example, if a simulation is performed to determine a photon or electron depth dose curve, quite large values can be used (say $\Delta = AP = 0.2$ MeV); whereas if an electron energy spectrum due to an incident electron beam at a certain depth is required, low values such as 0.01 MeV are best to ensure that energy loss straggling is modeled realistically.

Besides the low energy cutoffs for secondary particle production AE and AP, EGS4 (and ETRAN) have low energy transport cutoffs. When a particle's energy falls below this threshold, its history is terminated and the residual energy is deposited at that site. In EGS4 these parameters are *ECUT* and *PCUT* for electrons and photons, respectively. Appropriate values to choose depend on initial particle energy and dimensions of energy scoring regions. Larger values can be used for higher beam energies (since the same residual energy is then a smaller proportion of the original energy) and for larger scoring regions (since particles can be terminated at higher energies without risk that they would otherwise deposit energy in another scoring region). Lower values of these parameters, of course, slow down the simulation because particles are transported through more steps before being discarded, though the time-dependence is not as great as on

AE and *AP*. In EGS4, *ECUT* and *PCUT* cannot be less than *AE* and *AP*, respectively, as cross section data are not produced for energies less than these values. Note that while discrete energy losses above a certain threshold may be modeled, subsequently produced secondary particles are not necessarily transported. For example, if $AE = 0.521$ MeV and $ECUT = 0.700$ MeV, discrete energy losses above 0.01 MeV are modeled, while only delta rays above 0.189 MeV kinetic energy will be transported.

A further parameter, *ICSDA*, can be set in EGS4 that turns off all discrete energy losses. This forces a pure CSDA approximation, where the only energy loss modeled is that due to sub-*AE*, *AP* collisions. This, of course, has the effect of decreasing the total energy loss per unit pathlength.

The effects of different values of *AE* and *AP*, as well as *ICSDA*, on electron transport are illustrated in figure 10.2, where 25 9-MeV electrons are

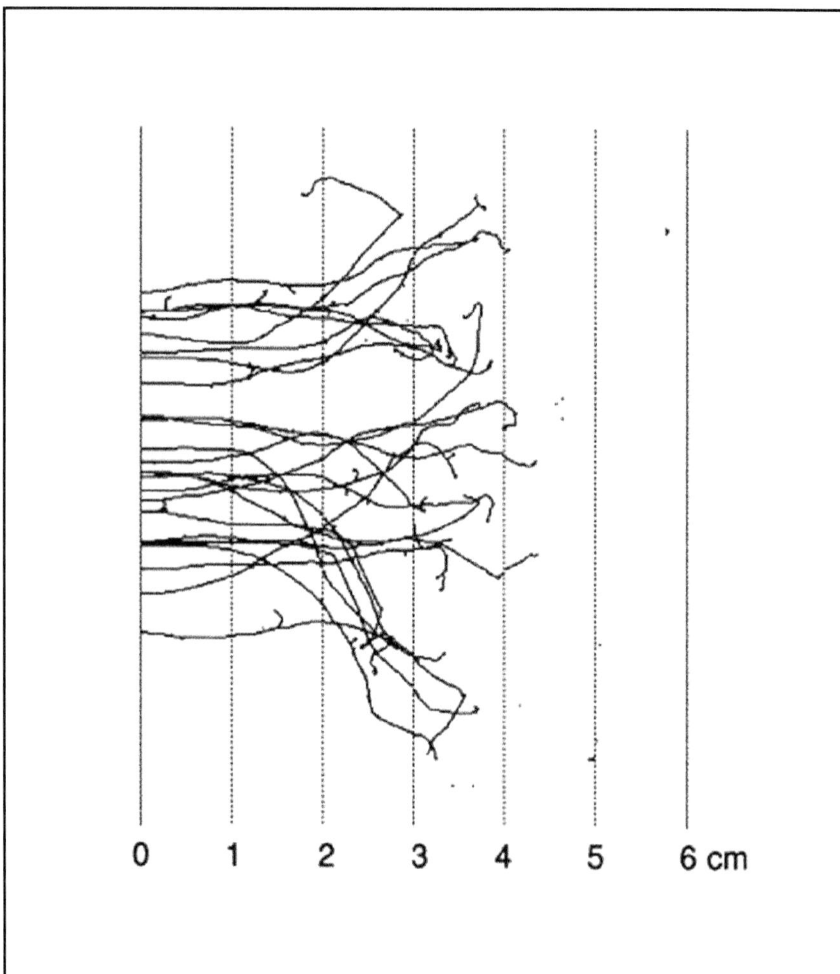

(a)

Figure 10.2: Histories of 25 9-MeV electrons with different values of *AE*, *AP*, and *ICSDA*. (a) $AE = 0.521$ MeV, $AP = 0.01$ MeV, $ICSDA = 0$ (continuous slowing down approximation turned off). Tracks are jagged with abundant delta rays.

incident on a water medium. The electron paths are displayed with the EGS_WINDOWS system. (Photon paths are not displayed.) In each case, *ECUT* and *PCUT* are equal to *AE* and *AP*, respectively. In figure 10.2a, the parameters are set to *AE* = 0.521 MeV, *AP* = 0.01 MeV, *ICSDA* = 0. These

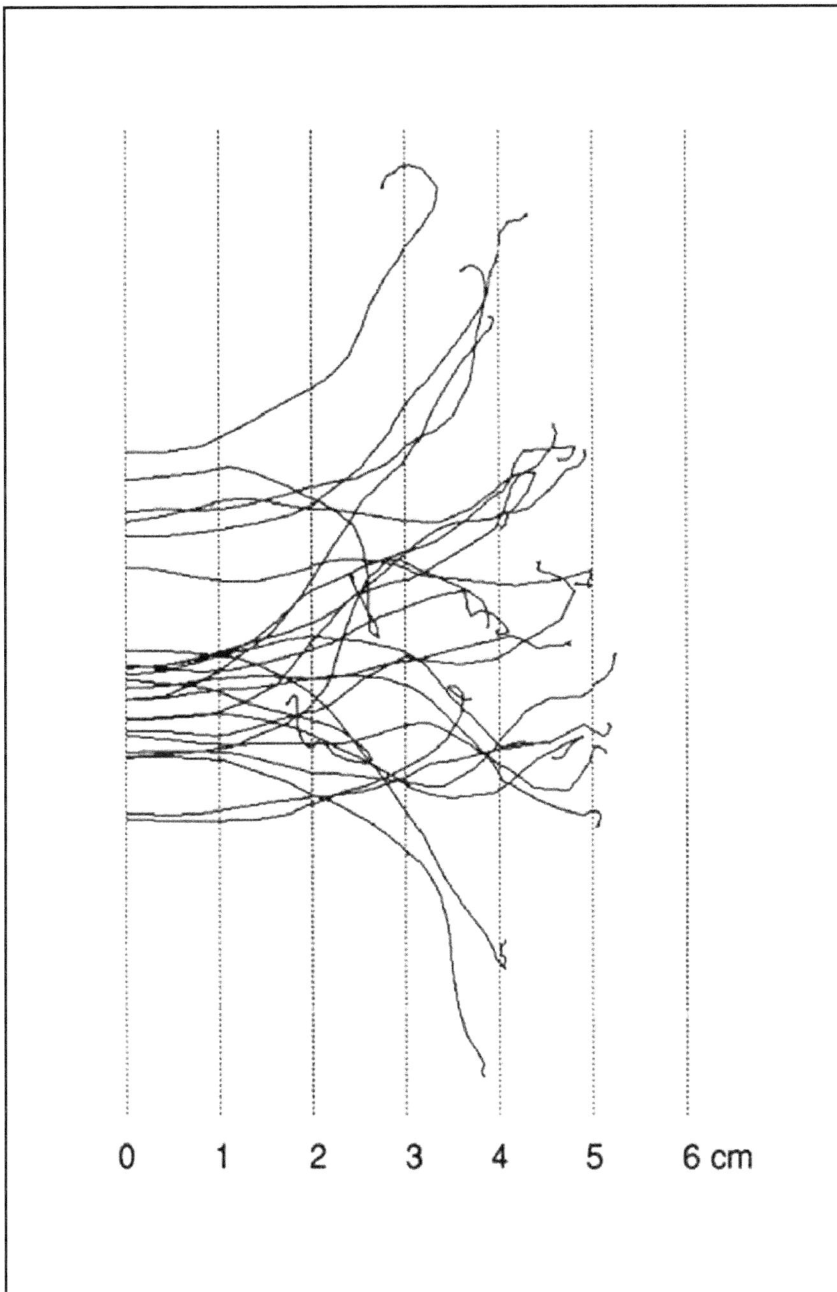

(b)

Figure 10.2: (Continued) (b) *AE* = 0.521 MeV, *AP* = 0.01 MeV, *ICSDA* = 1. Total energy loss is less because only losses of less than 0.01 MeV are accounted for, and tracks are smoother with no delta rays.

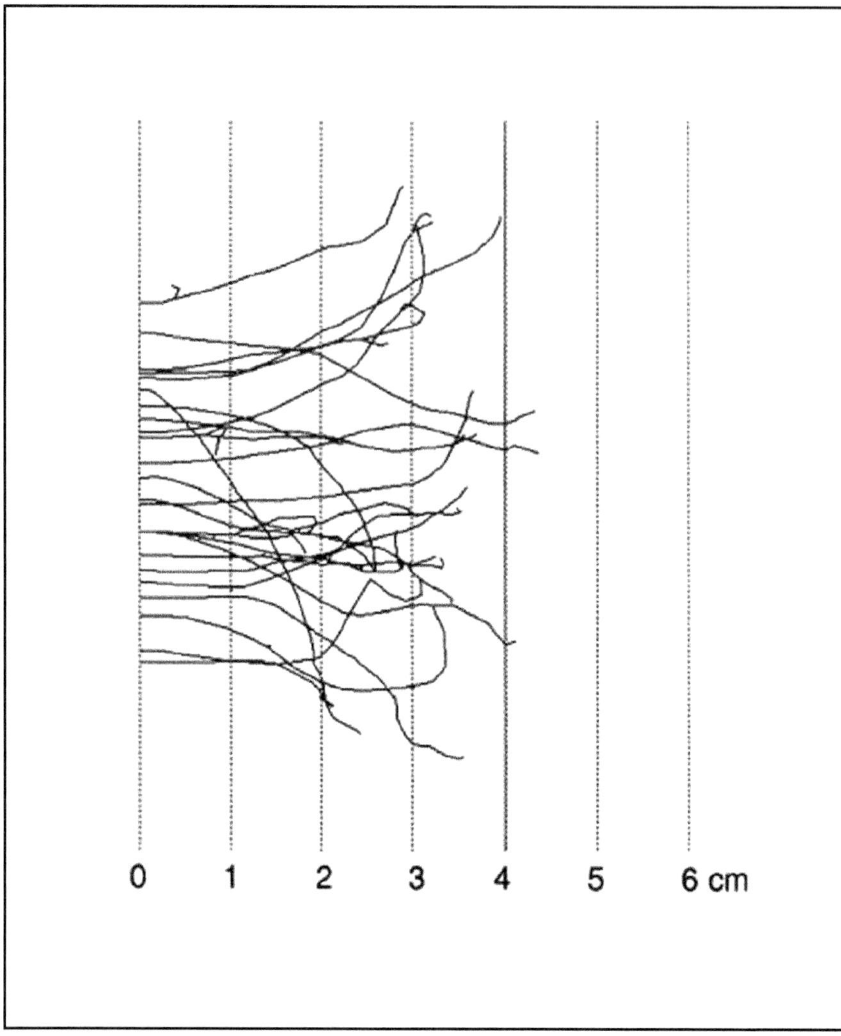

(c)

Figure 10.2: (Continued) (c) AE = 1.011 MeV, AP = 0.5 MeV, $ICSDA$ = 0. Almost all energy is lost continuously, and tracks are smoother than in (a) but have approximately the same penetration.

parameters are typical for situations in which accurate modeling of energy loss straggling and secondary particle transport is required. The electrons can be seen to follow jagged paths, and there are many obvious delta rays produced. Although not shown, there are abundant bremsstrahlung photons produced. Note that the maximum penetration depth is approximately 4 cm. In figure 10.2b, the values of AE and AP are the same, but $ICSDA$ = 1. The electron range is now seen to be larger (due to a lower total energy loss per unit pathlength), the paths are smoother, and there are no delta rays produced (and no bremsstrahlung). It is obviously not acceptable to turn off discrete energy loss events when AE and AP are set to these low values. In figure 10.2c, AE = 1.011 MeV (0.5 MeV kinetic energy) and AP = 0.5 MeV with $ICSDA$ = 0. It is expected that the electron range will now be approximately the same as in figure 10.2a, since the correct total energy loss per unit pathlength is now used (though it is almost all accounted for in continuous energy loss). The electron range is indeed again approximately 4 cm. However, the paths are smoother, and the distribution of penetration depths is narrower than in figure 10.2a since most of the randomness is removed from the energy loss process.

10.2.3 Sampling from a Distribution

Although selection of such parameters as photon step length, electron scattering, etc., is random, the likelihood that a particular value will be chosen is dependent on the probability distribution governing the outcome of the event; the selection is therefore biased. The way in which random numbers are used depends on the type of distribution being sampled. The use of random numbers is best illustrated by example. Below are two random selection schemes: The first for choosing the distance to the next photon interaction and the second for choosing the type of interaction that will occur.

Pathlength Selection

Pathlength selection uses *direct sampling* where the variable (in this case distance to the next interaction) is chosen using a probability density function (PDF) that expresses the relative likelihood that the variable will have a certain value. The PDF is normalized such that its area is unity and then integrated to give a cumulative density function (CDF) that has a maximum value of 1.0. The distance a photon is likely to travel before its next interaction depends on the mean free path of the medium for the photon's current energy. If the mean free path is λ, the linear attenuation coefficient is $\mu = 1/\lambda$ and the PDF is given by

$$f(x) = \mu e^{-\mu x}. \tag{10.5}$$

Integrating this expression gives the CDF

$$F(x) = 1 - e^{-\mu x}, \tag{10.6}$$

where the value of $F(x)$ ranges between 0 and 1 as x goes from 0 to infinity; there is therefore a mapping between a random variable r $(0 \le r < 1)$ and each value of x. Now for any r, the probability that the corresponding value of x will lie between x and $x + dx$ is proportional to the gradient of $F(x)$, which is equal to the PDF, $f(x)$. In the case of pathlength selection, this means that the probability of a pathlength x being chosen decreases exponentially with x. A plot of such PDFs and CDFs is shown in figure 10.3, where the value of μ in this case is 0.1.

Choosing Interaction Type

When a photon interaction type is to be chosen, the relative probabilities or *branching ratios* for each interaction are used. The branching ratio for interaction type i is simply the cross section σ_i, divided by the total cross section σ_t. Now if the range of real numbers between 0 and 1 is divided up into intervals of length equal to each branching ratio, the likelihood of a random variable between 0 and 1 falling in an interval corresponding to interaction i

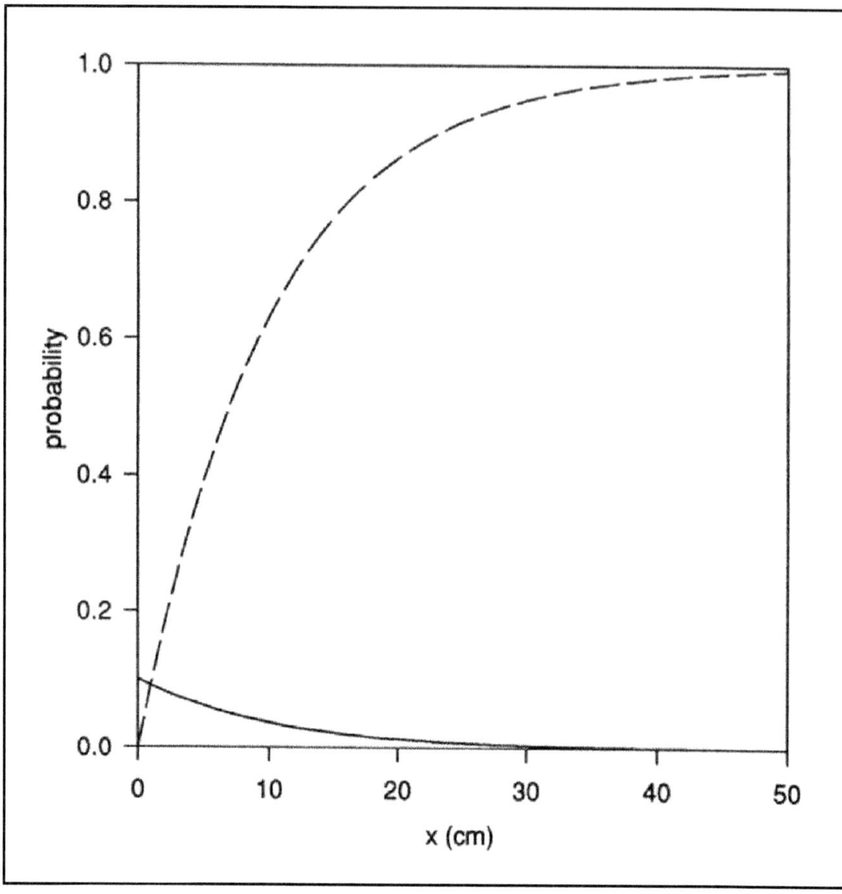

Figure 10.3: Cumulative density function (CDF) used in inversion sampling to obtain distance to next photon interaction (dashed line). The CDF, *F(x)*, is the integral of the probability density function (PDF) *f(x)* (solid line). A value of *x* is chosen by randomly selecting a value of *F(x)* and mapping to the *x* axis. Probability of choosing between *x* and *x* + *dx* is proportional to the slope of *F(x)*.

is equal to the branching ratio σ_i/σ_t. Interaction type is therefore chosen by sampling a random variable between 0 and 1 and finding the interval to which it corresponds.

10.2.4 Uncertainty and Statistics

In using a Monte Carlo simulation for calculation of radiotherapy dose distributions, the objective is to produce a result that is realistic and reproducible. The potential with which the calculated dose can resemble physical reality depends both on the underlying physics in the software and on the ability of the user to characterize the incident beam and to use this information to "drive" the simulation. Reproducibility means that successive simulations (with a different set of incident particles) will give almost identical results, even though the position and magnitude of energy deposition events throughout the simulation will be completely different each time. A completely reproducible distribution also implies that it will be free of statistical uncertainty, since statistical uncertainty is nonreproducible! In reality, it is impossible to achieve complete reproducibility because of the stochastic nature of the energy deposition events.

The idea behind a Monte Carlo simulation is that while the histories of N successive particles incident on a medium (where each particle has identical energy, point of incidence, and direction) will all be different, the average behavior of the particles will be the same as that of other N particles. A simple illustration of this is shown in figure 10.4. In figure 10.4a, the histories of four 3 MeV electrons in a water medium are shown. Each history is allowed by the physics of the simulation, though it is more likely that a fifth electron would follow a path close to that of the second electron from the left than say the rightmost electron (because larger lateral displacements are known to be less likely). In figure 10.4b, 200 electrons are incident, and it can be seen that there is a reasonably symmetrical distribution of electron tracks, forming a forward-directed "lobe." A separate run of 200 histories would produce a set of tracks that resemble closely the pattern shown here, and with many more electrons, the resulting dose distribution would also become quite reproducible. The appearance of successive sets of four electrons would, of course, not closely resemble each other.

In determining the physically realistic value of a particular quantity using a Monte Carlo simulation, the contribution to that quantity from each particle history is summed. This means that the summed value is effectively the average of each contribution. In the case of energy deposition within a voxel, the summed (or average) energy will be different in successive simulations of the same number of incident particles because of the random selection of transport parameters. The amount of energy deposited in a voxel is a stochastic variable obeying Poisson statistics, having a mean λ and variance $\sigma^2 = \lambda$. In successive simulations where the mean summed energy is λ, the standard deviation is thus $\sigma = \lambda^{1/2}$. It can be seen that the fractional error, σ/λ (which is the quantity quoted as the uncertainty) is equal to $1/\lambda^{1/2}$ and thereby decreases with the square root of the total energy deposited. The uncertainty will also decrease with the square root of the number of incident particles. It is a rather unfortunate consequence of this that to halve the uncertainty the number of incident particles must be quadrupled (along with the computation time).

It is essential that the uncertainty in a Monte Carlo result is known, otherwise we do not know the probable error in the result. In EGS4, the uncertainty is calculated by splitting a simulation up into a number of *batches*, where each batch is a separate simulation of an equal number of particle histories. When all batches are completed, the results are averaged (for example, the mean energy deposited in each voxel is calculated), and the standard error in the mean (the uncertainty) is calculated. If the number of batches is n, and the standard deviation in the n results is σ, the uncertainty is $\sigma/n^{1/2}$. This figure represents the standard deviation in the results of successive simulations. Note that quadrupling the number of batches while keeping the number of incident particles in each batch constant has the effect of halving the uncertainty in the same way the uncertainty is halved by quadrupling the

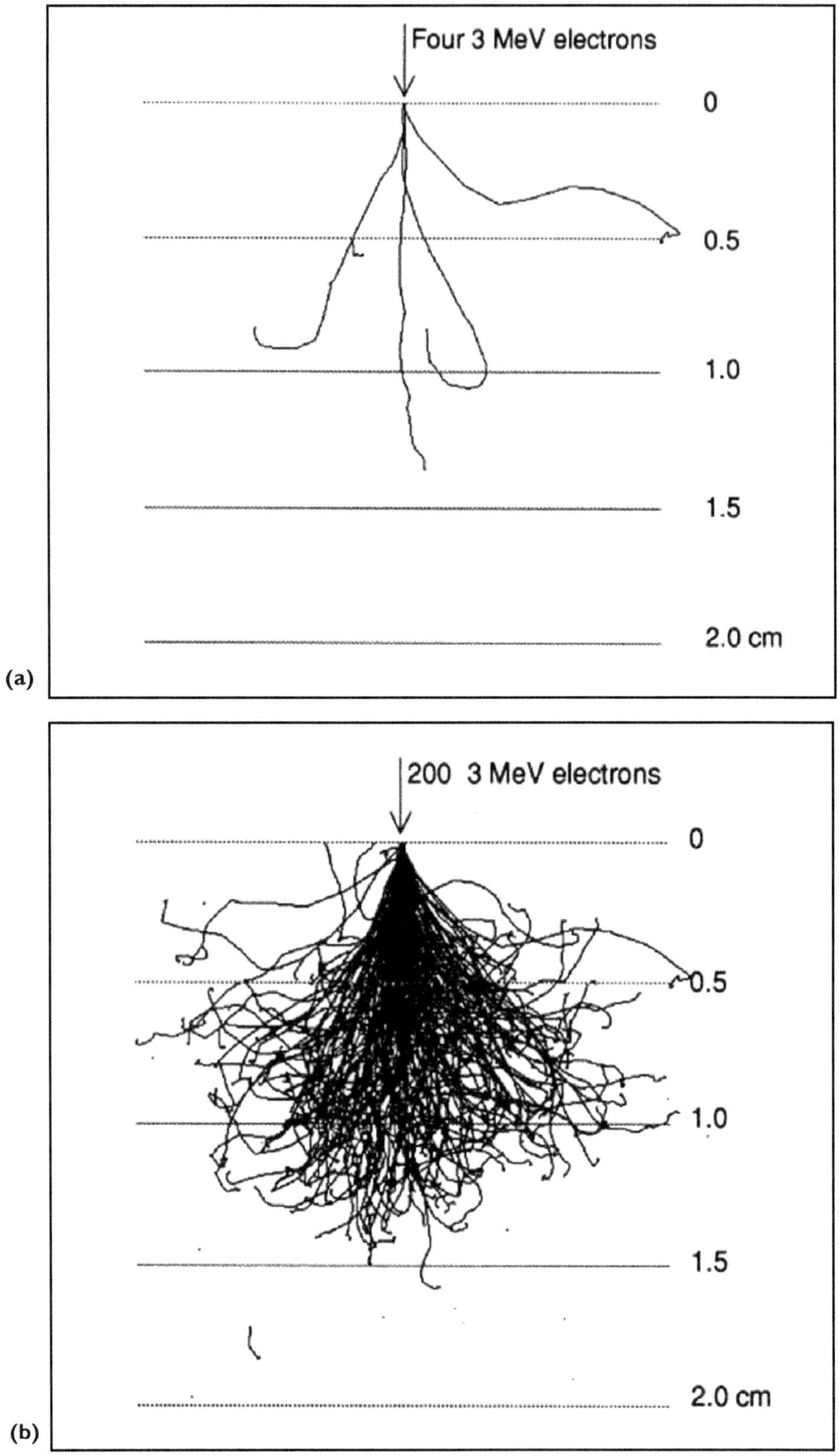

Figure 10.4: (a) Simulation of four 3 MeV electrons incident on a water medium showing different, yet all physically realizable, histories. The appearance of this set of tracks is not reproducible. (b) 200 3 MeV electrons showing a much more reproducible pattern resembling a forward-directed lobe.

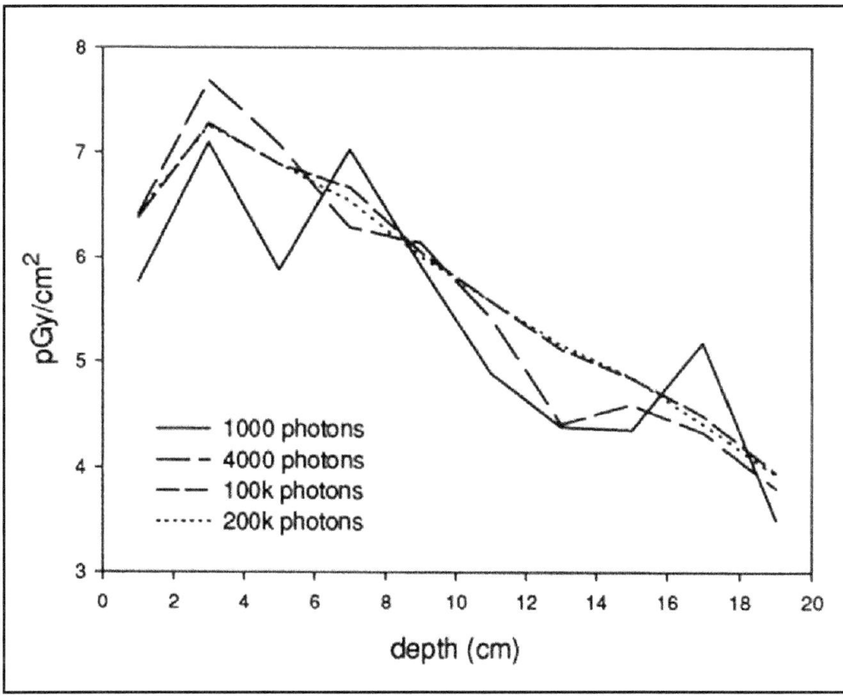

Figure 10.5: Simulations of a 5 cm radius, 6 MV beam where the number of photon histories was 1000, 4000, 100,000, and 200,000. Uncertainty in the voxel centered at 5 cm deep is approximately 17%, 8.5%, 1.7%, and 1.2% respectively.

number of incident particles in each batch. Usually, five to ten batches are used for calculating the uncertainty—the number of batches cannot be too large, since a separate scoring array must be declared for each batch. In EGS4, the uncertainty in a quantity is given as a percentage of the mean of that quantity.

As an example of the reduction in uncertainty with increase in number of histories, separate simulations of a 5 cm radius, 6 MV photon beam were performed, where the number of incident photons was respectively 1000, 4000, 100,000, and 200,000. It is expected that the uncertainty with 4000 photons will be halved from that with 1000, and the uncertainty of the 100,000 photon run should be 1.414 times that with 200,000 photons. The depth dose curves resulting from each simulation are shown in figure 10.5, where dose is normalized to incident photon fluence. Uncertainty of the 200,000 photon run is approximately 1.2% in the voxel centered at 5 cm deep. This means that the predicted uncertainties in this voxel for the other runs (with decreasing photon number) are approximately 1.7%, 8.5%, and 17%, respectively. Figures close to these values were obtained, although there is large variation in the uncertainty values in successive simulations when the uncertainty is high. The apparently small reduction in uncertainty when simulating 200,000 instead of 100,000 particles is obvious, highlighting the difficulty in obtaining good reproducibility by increasing computation time. It should be noted, however, that even for the 200,000 history simulation, computation time was only 15 minutes (on a Silicon Graphics Indigo computer) due to the use of a variance reduction method known as the *reciprocity technique* (see next section).

Efficiency and Variance Reduction

The efficiency of a particular simulation is inversely proportional to both the variance of the results and the time taken for the simulation (Rogers and Bielajew 1988). If the uncertainty of a result is σ and the computing time used in the simulation is T, then the efficiency ε is defined as

$$\varepsilon = \frac{1}{\sigma^2 T}. \tag{10.7}$$

Since the fractional uncertainty is inversely proportional to the square root of the incident particle number, it is also inversely proportional to the square root of the simulation time. This means that efficiency is invariant with simulation time for a particular simulation type.

It is possible to increase efficiency by using *variance reduction* techniques, where either (a) the variance for a particular simulation time is reduced or (b) the time required to obtain results with a certain variance is reduced. It is simple to quantify the advantage gained by employing a variance reduction technique, since the gain is just the ratio of efficiencies obtained using equation 10.7. Examples of variance reduction techniques that can be employed in generating dose distributions are:

i) *Geometry interrogation.* When a particle is being transported, geometry-checking is normally performed (by calling HOWFAR in EGS4) before each step to determine whether the particle would cross a voxel boundary during that step. This is, however, a waste of time if the particle is far from any boundaries. To save time, the current closest distance from a boundary can be stored (the variable DNEAR in EGS4) so that geometry checking is only performed when the particle comes close to the boundary.

ii) *Zonal discard.* If an electron has a lower energy than that required for it to escape from the voxel in which it currently resides, its kinetic energy can be deposited in that voxel instead of transporting the electron any further.

iii) *Reciprocity technique.* The reciprocity technique makes use of the radial symmetry that exists about the axis of a narrow beam. This symmetry means that if a nondivergent beam of circular cross section with radius R is incident on a homogeneous medium, the dose distribution within a coaxial cylindrical scoring region of smaller radius r will be equivalent to that if the beam were of radius r and the scoring region were of radius R. In particular, this means that the central axis depth dose for a broad beam can be obtained by using a very narrow pencil beam and scoring the energy deposition within a radius equal to that of the broad beam. A diagram of this geometry is shown in figure 10.6, where the cylindrical beam on the left is of radius R and the dose scoring region on the central axis is of radius r. The same energy is deposited in the small volume element shown on

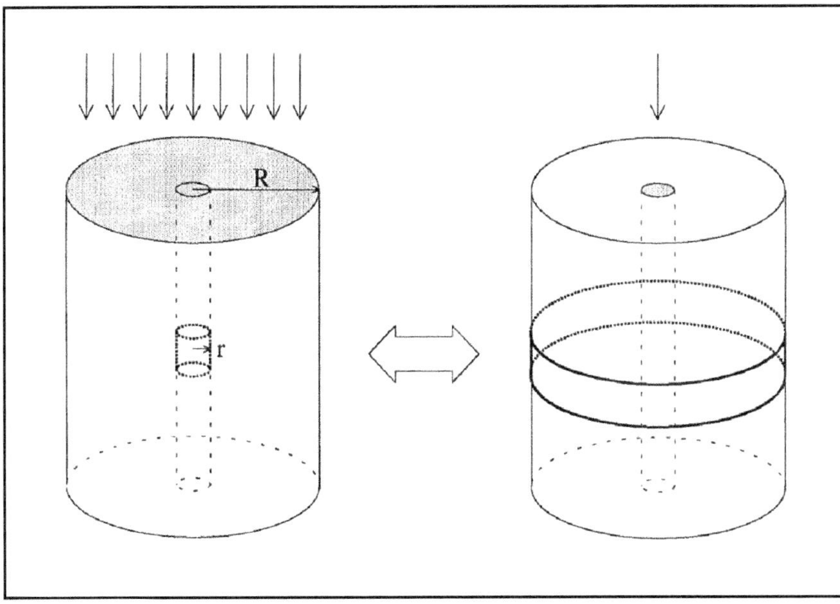

Figure 10.6: Schematic of the reciprocity technique geometry. The cylindrical beam on the left is of radius *R*, and the dose scoring region on the central axis is of radius *r*. The same energy is deposited in the small energy deposition volume element shown on the left as is deposited in the element of radius *R* on the right when the beam is of radius *r*.

the left as is deposited in the region of radius *R* on the right when the beam is of radius *r*. Here, the variance reduction comes about because if particles are incident within a large radius on the surface, most will not contribute to the central axis dose, thus wasting the computing time spent transporting these particles. However, if particles are incident in a very narrow beam, almost all will contribute to the dose within a coaxial region of larger radius. A further bonus in using the reciprocity technique is that several beam diameters can be simulated at once, since energy deposited within multiple radii can be scored. The reciprocity technique has been used to obtain the depth dose results shown in figure 10.5.

10.2.5 Random Number Generation and Sampling

The Random Number Generator

Central to all Monte Carlo systems is the use of random numbers to select initial conditions and events during each history. A significant proportion of the computing time in a simulation is spent generating random numbers, so it is important that the code used for generating these numbers is efficient. Most important, however, is that the sequence of random numbers is very long so that it does not repeat itself during a simulation. If this occurs, new particle histories will eventually be identical to histories earlier in the simulation. The random number generator that is provided with EGS4 is of the multiplicative congruential type—the nth random number is found recursively using the expression

$$X_n = \left(a X_{n-1}\right)_{\mathrm{mod}\, 2^k}, \quad (10.8)$$

where a is a constant multiplier and k is the integer word size of the computer. The first number in the sequence, X_0, is called the *seed.* With two's complement integer arithmetic, the random numbers range from $-2^k/2$ to $2^k/2-1$. This number must be converted into a floating point number in the range 0.0 to 1.0 so it is first multiplied by 2^{-k} and the result added to 0.5. In order for the random number generator to be of long sequence, the choice of a is most important. For 32-bit computers, a good choice for this *magic* number is a = 663608941, as is used by EGS4. The maximum sequence length (with an appropriate constant multiplier such as that given) for this type of generator is $2^{k-2} = 2^{30} \approx 10^9$. This generator has a long enough sequence for most simulations, although a longer sequence may be necessary if, for example, a three-dimensional dose distribution with very low uncertainty is to be produced. In this case, a generator such as the lagged-Fibonacci type is more appropriate (Marsaglia et al. 1990). This generator has a sequence length of 2^{144}, which is long enough for all practical purposes, and it produces identical sequences on all computers.

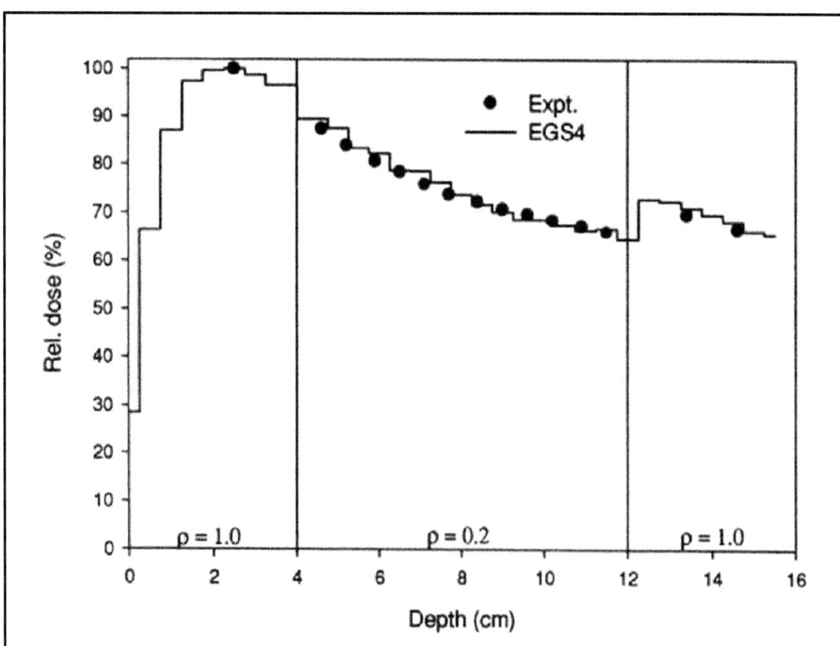

Figure 10.7: EGS4 Monte Carlo results compared with measurement for a 5×5 cm^2 10 MV beam. Measurements were made with a 0.6 cm^2 Baldwin Farmer ionization chamber embedded in polystyrene or between cork tiles.

10.2.6 Monte Carlo Results

Lung Phantom

The irradiation of the lung phantom described in chapter nine with a 5×5 cm^2 10 MV beam was simulated by the authors using EGS4. It was shown in chapter nine that both the effective depth and ETAR algorithms fail to accurately predict the large reduction in lung dose caused by lateral electronic disequilibrium. The Varian 10 MV spectrum of Mohan et al. (1985) was used as input to the simulation, which was controlled by the usercode RTPCART (Murray 1990). Results for the first three regions of the phantom are shown in Figure 10.7, where it can be seen that agreement with experiment is very close. In this simulation, 7.2 million incident photons were used, with $AE = ECUT = 0.561$ MeV, $AP = PCUT = 0.05$ MeV.

Electron Backscatter

A recent and interesting Monte Carlo simulation performed by the authors was for a beam of 9 MeV electrons incident on a water medium containing a lead sheet at 2.2 cm deep (approximately d_{max}). To visualize the electron transport process, a simulation of 25 incident electrons was performed and the particle tracks displayed (see Figure 10.8a). Here the electrons can be seen entering the lead with most rapidly running out of energy and stopping. A significant proportion are scattered back out of the lead, however, and travel back toward the surface. The simulation was also performed using 20,000 electrons to obtain a dose distribution. The reciprocity technique could be used to reduce the required computation time since a circular, non-divergent beam was used. The result is shown in Figure 10.8b, where a standard water-only depth dose curve is plotted for comparison. A dose enhancement of approximately 50% can be seen to occur due to backscattering off the lead.

Energy Spectrum Determination

In an investigation of changing sensitivity of therapy film with depth, a Monte Carlo simulation was carried out to determine the variation with depth of the photon energy spectrum. A cylindrical 6 MV beam 5 cm in radius was simulated, again using the reciprocity technique, where the energies of photons and electrons were scored in 0.2 MeV bins as they passed into a selected region. This is a simple process of reading the current particle's parameters to see if (a) the current region is the region of interest and (b) the particle has entered this region in the current step (new region not equal to old region). If the particle satisfies these conditions, the charge and energy of the particle are determined and the appropriate bin in the photon or electron spectrum array is incremented. At the end of the simulation, the bins are normalized so that their values integrate to 100%. The photon spectra at $d_{\max}$ and 30 cm deep are shown in figure 10.9, where 1 million incident particles were used, with the 6 MV spectrum generated by Mohan used as input. Uncertainty is very low in this case, and the irregularity of the curves is due to that of the incident spectrum. Beam hardening with depth is obvious, with the average energy increasing from 1.72 to 1.95 MeV. An

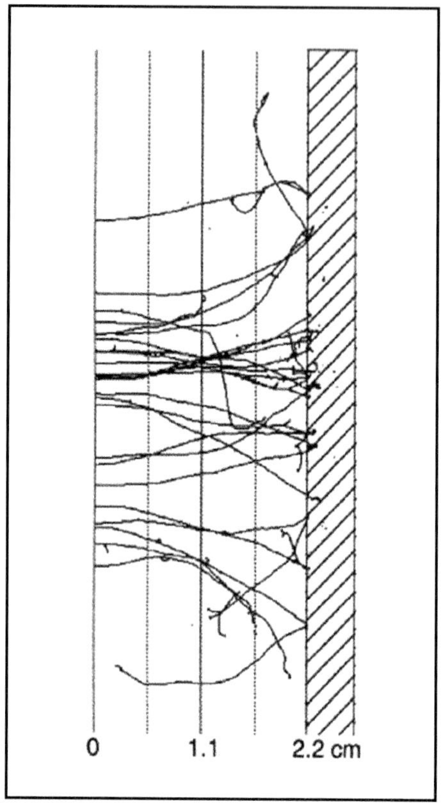

(a)

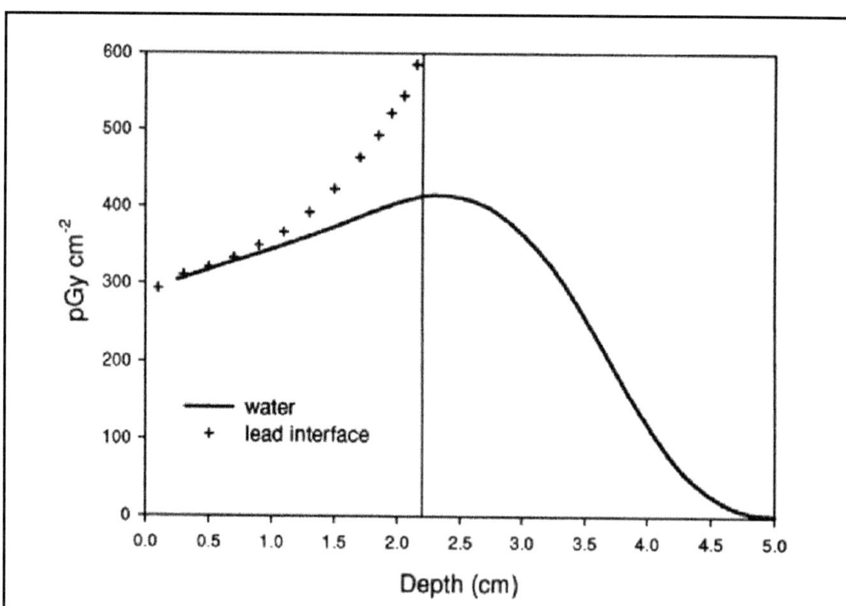

(b)

Figure 10.8: EGS4 simulation of 9 MeV electrons incident on a water medium containing a lead slab. (a) 25 electrons penetrating to the lead with some being scattered back toward the surface. (b) Dose distribution due to 20,000 electrons showing a 50% dose enhancement proximal to the lead compared to a standard water depth dose curve.

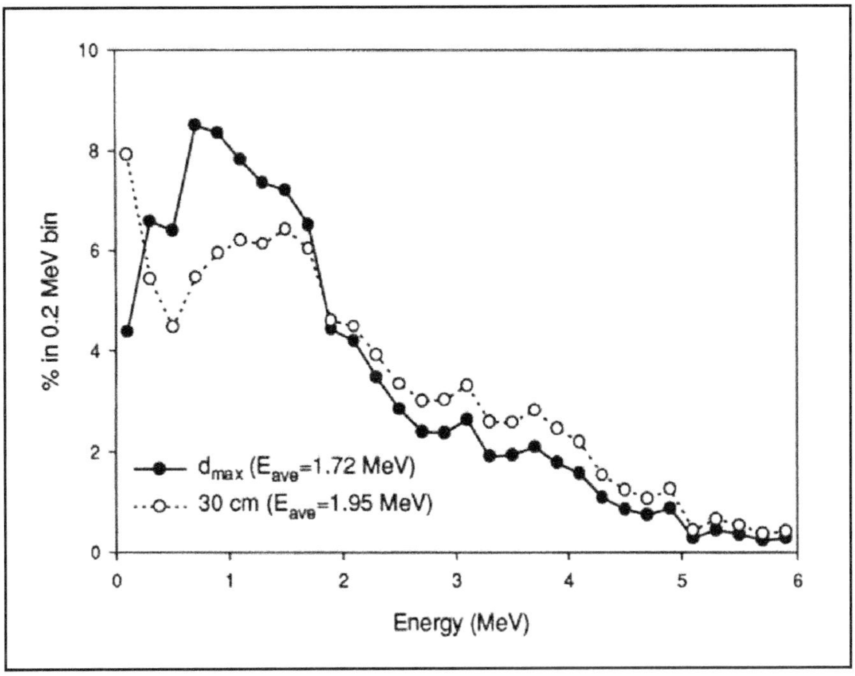

Figure 10.9: Energy spectra determined at d_{max} and 30 cm deep in water for a 6 MV beam. One million incident photons were used in each simulation. Average beam energy increases from 1.72 to 1.95 MeV over this range. There is also an increase in low-energy scattered photons.

increase in low-energy scattered photons is also apparent, which is of interest for the film dosimetry study.

10.3 BEAM

A very sophisticated system for Monte Carlo simulation of linear accelerator treatment heads has been developed as part of the OMEGA project (Rogers et al. 1995), a collaboration between the National Research Council of Canada (NRCC) and the University of Wisconsin (UW). This software, known as BEAM, is based on EGS4 and models electron and photon transport through the treatment head of a linear accelerator, with the result being a data set (called a *phase-space* file) characterizing the beam incident on the patient.

The BEAM code is able to calculate the energy and angular distributions of electrons incident at the patient surface by simulating their transport from the stage when they first enter the treatment head (from the waveguide) until they exit the treatment head. Components of the treatment head (such as the scattering foils, monitor ionization chambers, beam jaws, and electron applicator) are each represented in the simulation by *component modules* (CMs), which are three-dimensional (3-D) structures defined by the user in setting up the simulation. Electrons and photons may enter each component module and will either be absorbed or exit the module. Particles that finally exit the treatment head have their energy, position, and direction written to the phase-space file for later use. Photons produced by interaction of electrons with treatment head elements and air are also included in the phase-space file. In addition, the *history* of each particle is included in the phase-space file so that it is known with which components the incident electrons have interacted.

In order to run a Monte Carlo simulation using BEAMnrc (Rogers et al. 2007), a virtual accelerator is built by combining CMs. These CMs describe various geometrical shapes and are used to model the different components of an accelerator head (e.g., target, collimators). For example, the flattening filter is modeled as a series of conical sections. The dimensions of the CMs are entered by the user; therefore knowledge of the dimensions of the accelerator head is required. Beam properties from several commercial linac treatment machine heads have been benchmarked using BEAMnrc (Sheikh-Bagheri and Rogers 2002). Small discrepancies in the geometric description can lead to dose differences between model and experiment (Chibani and Ma 2007).

Once the accelerator head has been built and compiled, the user chooses the input variables. This starts with defining the incident source. If the accelerator is being modeled from above the target then the incident source is an electron beam incident on the target. A large range of sources with varying geometrical and energy characteristics are available from which to choose.

A common source for a subsequent simulation may be a phase-space file. This is the output file of a previous simulation and contains information (e.g., charge, momentum, energy, latch value) on every particle crossing a chosen plane in the geometry.

If one is running a simulation from above the target to above the jaws, an electron source will be used for this simulation and a phase-space file will be produced at the jaw level. This phase-space file can then be used as a beam input source into accelerator components that contains CMs that define geometry from the jaws to the level of the phantom. This allows a variety of different jaw and MLC configurations to be used without having to restart the simulation from the incident electron beam position each time.

Once the source has been chosen, the user chooses the latch values. These are values that get assigned to each particle that describe the origin of the particle. This allows the user to track from which CM each particle originates, and can be used as a filter when doing dose calculations and analysis of phase-space files.

The user has control of all of the EGS transport parameters and these must be set. Variance reduction techniques can be used to increase the efficiency of simulations. These are particularly useful when creating photon fields from incident electrons. The number of histories to run is also chosen by the user. This number will depend on the accuracy required and the time available.

The output of a BEAM simulation, as stated earlier, is the phase-space file. This can be used as an input into a DOSXYZnrc simulation to calculate dose in a 3-D data set (e.g., a water phantom). It can also be analyzed using

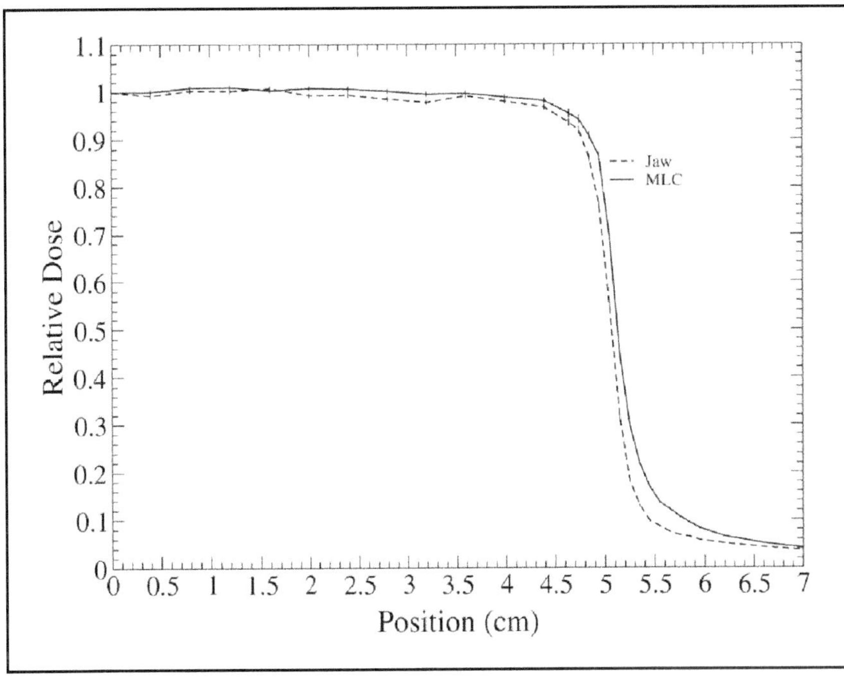

Figure 10.10: EGSnrc simulations using BEAMnrc and DOSExyz modules to model the dose profile for a 10 x 10 cm² MLC compared with a collimated 10 x 10 cm² field across a Varian rounded leaf end MLC at d_{max} in a 6 MV photon field. Note only half of the symmetric profile is shown; also the jaws were set at 30 x 30 cm² for data collection when the MLC were set at 10 x 10 cm². (Courtesy of N. Hardcastle, private communication.)

the utility BEAMDP to obtain spectral, fluence, and statistical information about all of the particles.

Figure 10.1 shows a comparison of MLC and jaw dose profiles using two Monte Carlo simulations. The graph shown in figure 10.10 represents half-dose profiles of dose distributions modeled using the BEAMnrc07 (Rogers et al. 2007) and DOSXYZnrc (Walters et al. 2007) modules within the EGSnrc Monte Carlo code. The graph represents a simulation equivalent to the ionization chamber dose profile graph shown in chapter four (see figure 4.34).

Essentially, the beam has been modeled from electrons generated above the target The electron beam used for creation of the photon beam had an energy of 6.2 MeV with Lorentzian energy spread with full-width at half maximum (FWHM) of 3%. This was a parallel circular beam with a gaussian radial distribution with a 1.3 mm FWHM. The flattening filter density was changed to 18 g/cm³ (Keall et al. 2003).

An ECUT value of 0.7 MeV and PCUT value of 0.01 MeV were used as these match the minimum energies of electrons (AE) and photons (AP) considered in the PEGS data file creation.

The component modules used (in order) were SLABS (target), CONS3R (primary collimator), SLABS (vacuum window), FLATFILT (flattening filter),

CHAMBER (ion chamber), MIRROR (mirror), JAWS (jaws), DYNVMLC (120 leaf Varian Millennium MLC), SLABS (air gap to phantom). A variance reduction technique, *directional bremsstrahlung splitting*, was used (Kawrakow et al. 2004). A splitting number of 1000 was used.

Five million histories were run to created phase-space files at 100 cm SSD from the target. The jaw- and MLC-defined phase-space files had about 1×10^7 particles; this took about 16 hours on a standard specification PC.

The phase-space data were then used as input into a water phantom using DOSXYZnrc07. Two billion histories were run using this phase-space file (recycled 154 times). This was the same for both the MLC-defined field and the jaw-defined field. For the MLC-defined field the jaws were set well back from the MLCs to define a 30×30 cm^2 field at 100 cm SSD. The statdose utility was then used to produce the profiles shown in the figure 10.10.

The added advantage of Monte Carlo simulations is that other data can be collected that is not represented by standard detector systems, in particular beam spectral information. The BEAMDP utility was used to analyze the phase-space files used to generate the data for figure 10.10. The mean energy distribution was plotted in profile with 2 mm resolution. As shown in figures 10.11 and 10.12, the mean beam energy across the profile can also be represented for the jaw- and MLC-defined fields. In this case, as shown in figure 10.12, the mean energy of the 6 MV beam is 1.7 MeV at the surface in the inter umbra and 1.1 MeV in the penumbra. In the MLC-defined beam the mean energy is also 1.7 MeV in the inter umbra; however there is an interesting dip in mean beam energy to about 1.25 MeV in the

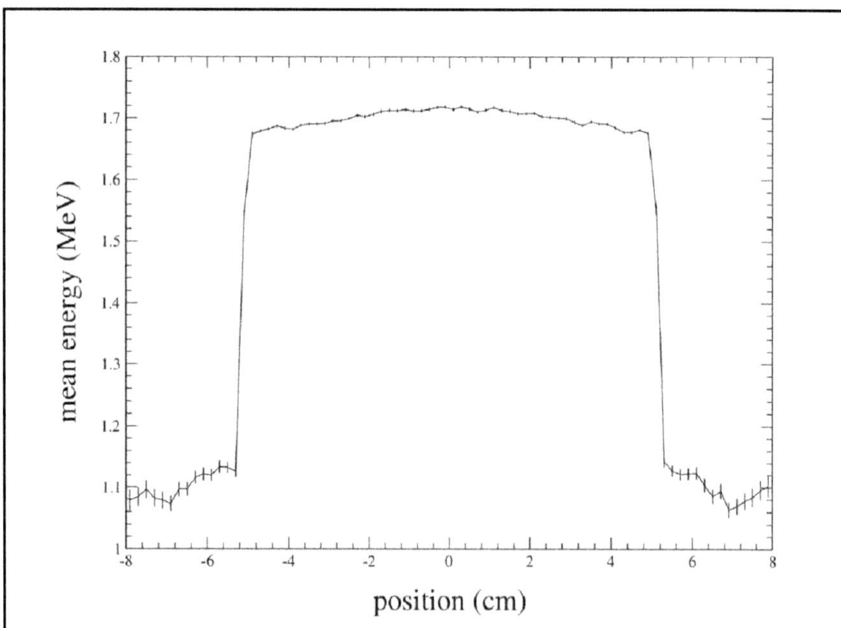

Figure 10.11: EGSnrc simulations using BEAMnrc and DOSExyz modules to model the mean energy profiles in the phase-spaces at 100 cm SSD for a jaw-defined 10×10 cm^2 field. (Courtesy of N. Hardcastle, private communication.)

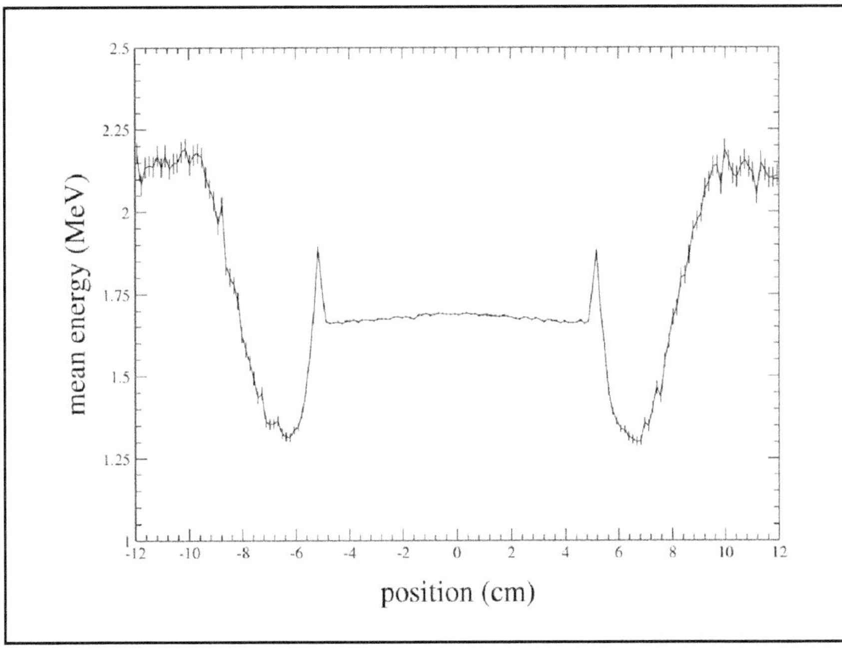

Figure 10.12: EGSnrc simulations using BEAMnrc and DOSExyz modules to model the mean energy profiles in the phase-spaces at 100 cm SSD for a jaw-defined 10 x 10 cm² field with jaws set at 30 x 30 cm². (Courtesy of N. Hardcastle, private communication.)

penumbra just beyond the MLC edge, then the energy increases to 2.15 MeV in the tail of the penumbra. Another advantage of Monte Carlo simulations is that particles that interact with individual components of the linac head can be latched and tracked to show how they contribute to the final result. The mean energy for various linac component modules is shown in figure 10.13, and the fluence contributions are shown in figure 10.14. The fluence contribution from the flattening filter, which remains high near the leaf edge, and the flattening filter lower mean energy give a hint as to why the dip in mean energy is observed. In this textbook the umbra is defined (see chapter four) as the in-field part of the beam. In other historical texts the definition of umbra is different; what we define as the tail of the penumbra in this text is actually termed the *umbra*; i.e., the region of total shadow (e.g., Meredith and Massey 1972). Therefore, in this edition of the text we have defined the i-field region as *inter umbra*.

This interesting effect has also been described by the previous Varian MLC design by Lyddon (2006), who gives the following explanation: "The sharp rise in the mean energy over the edge of the MLC-defined field is due to the preferential attenuation of lower energy photons through inner MLC edges which have shorter attenuation path lengths. The sharp drop-off is due to the fluence of scattered photons from the flattening filter (with lower energy) being greater than that of the primary photons hence the mean energy decreases. The increase in energy again is then due to the primary fluence being greater than the flattening filter-scattered fluence."

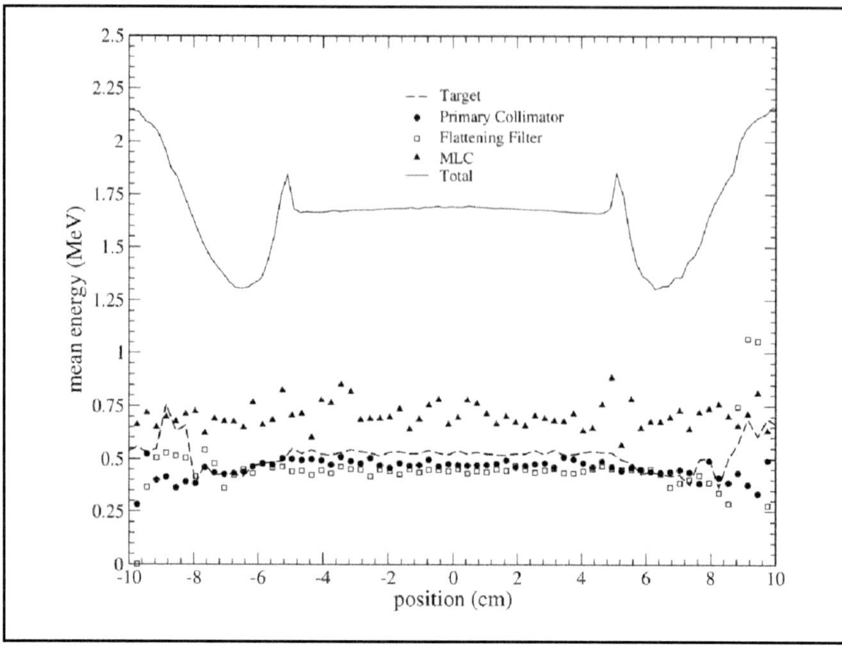

Figure 10.13: EGSnrc simulations using BEAMnrc and DOSExyz modules to model the mean energy profiles in the phase-spaces at 100 cm SSD for an MLC-defined 10 × 10 cm^2 field with jaws set at 30 × 30 cm^2. Showing the contributions from different accelerator head modules. (Courtesy of N. Hardcastle, private communication.)

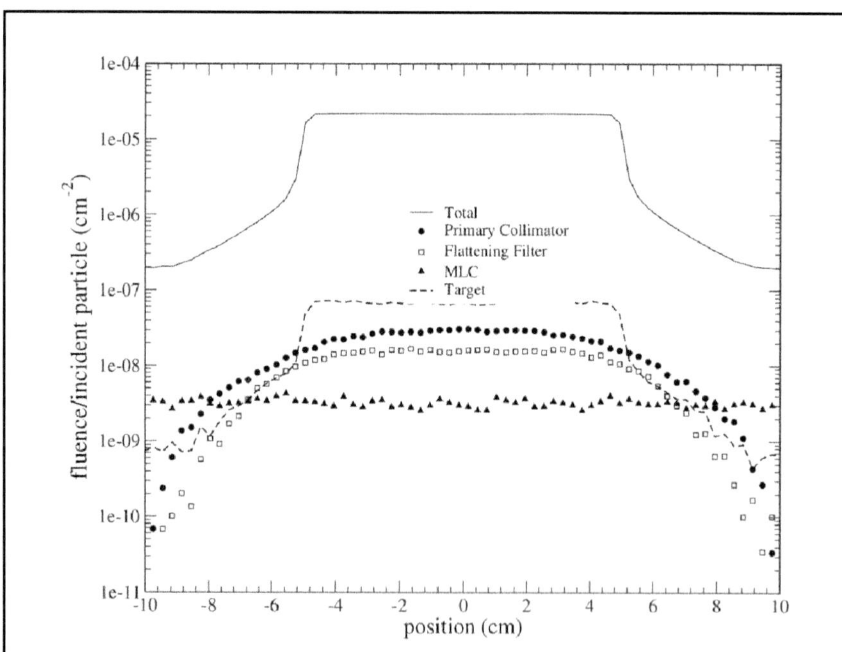

Figure 10.14: EGSnrc simulations using BEAMnrc and DOSExyz modules to model the fluence profiles in the phase-spaces at 100 cm SSD for an MLC-defined 10 × 10 cm^2 field with jaws set at 30 × 30 cm^2. (Courtesy of N. Hardcastle, private communication.)

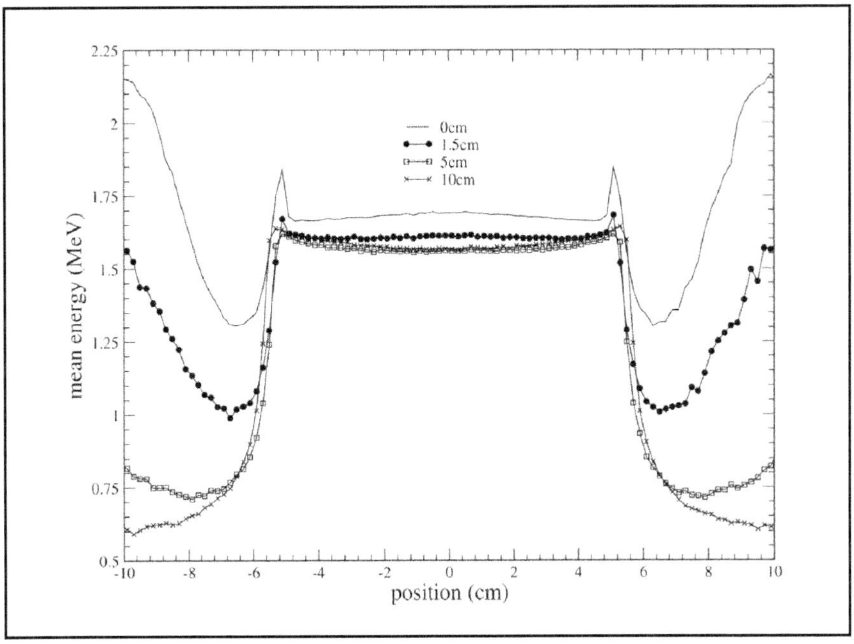

Figure 10.15: EGSnrc simulations using BEAMnrc and DOSExyz modules to model the fluence profiles in the phase-spaces at 100 cm SSD for an MLC-defined 10 × 10 cm^2 field with jaws set at 30 × 30 cm^2. Showing the contributions from different accelerator head modules. (Courtesy of N. Hardcastle, private communication.)

Investigating this effect further at different depths in a water phantom using data from the phase-space as input into DOSXYZnrc as shown in figure 10.15, the mean energy of the beam then starts to be affected more by in-scatter contributions from the beam, and the dip and rise in mean beam energy in the penumbral tail begins to diminish.

These simulations are just one example of information that can be gained from Monte Carlo simulations that is difficult to match by other experimental means.

It is possible to speed up EGSnrc and BEAMnrc Monte Carlo simulations; they have a parallel facility in the code. By using parallel processing capability on a small 14 × 3.2 Ghz × dual-core processor PC cluster, which is shown in figure 10.16, we hope to run simulations at close to 14 times faster than a single dual-core processor with a small communications overhead. The cost of this cluster was only $10,000.

Figure 10.16: Small, low-cost PC cluster used for Monte Carlo simulations. (Courtesy of Center for Medical Radiation Physics, University of Wollongong, Australia.)

10.4 Geant4

Geant4* is a software toolkit with strong links to CERN. It has been developed for the simulation of the passage of various particles through matter (Agostinelli et al. 2003). It has been used for a large number of experiments and projects in a variety of application domains, including high-energy physics, astrophysics and space science, medical physics, and radiation protection. Its functionality and modeling capabilities continue to be extended. An overview of recent developments in diverse areas of the toolkit is presented in Allison et al. (2006).

Medical Physics Application

The toolkit has been used extensively to model proton beams (e.g., Jiang et al. 2005). Its recent use in radiation oncology physics applications has been highlighted by various projects (Verhaegen and Seuntjens 2003); for example, those which report the impact of adding magnetic fields like those used by MRI equipment (Raaijmakers et al. 2005, Raaymakers et al. 2004)). Micro beams and brachytherapy sources have also been modeled using Geant4.

Geant4 has also been used to model positron emission tomography (PET) detector modules (Barret et al. 2005, Moehrs et al. 2006, Lamare et al. 2006). GATE is an application that utilizes the Geant4 libraries and offers a fully scripted simulation for nuclear medicine applications. No coding is required from the user, making the time to develop simulations much less.

*Geant stands for *GEO*metry *AND* *T*racking.

Creating a Geant4 Application

In contrast to other Monte Carlo codes, which are stand-alone applications, Geant4 is a toolkit of classes written in object-oriented C++. In order to create a simulation, one must write and compile the application; this application must utilize several Geant4 classes. Consequently, a reasonable understanding of object-oriented programming and C++ is required; in particular, one should be familiar with pointers, I/O, file I/O, arrays, dynamic memory allocation, static methods, inheritance, polymorphism, and command line arguments. Familiarity with building and running applications in the Linux shell environment is essential.

The toolkit comes bundled with many useful examples. It is advisable to use one of these examples as a starting point when creating simulations for the first time. Geant4 has no default behavior, so the user must define the behavior of the application using the appropriate classes. There are three compulsory user classes that must be created by the user, which inherit behavior from an abstract base class:

i) The first of these is the *detector construction* class. In this class the materials of the simulation are defined, the *geometry* is defined, and attributes of the geometry, such as colors for visualization and which regions of the geometry are deemed to be sensitive, are all specified.
ii) The second, the so-called *primary generator action* class, is used to define the aspects of the *primary beam* such as: particle type, initial kinetic energy, and direction. Any randomization of the primary beam must also be done in this class. Geant4 has in-built libraries for sampling values from uniform, Gaussian, distributions. The user can also specify their own energy distributions of the primary beam by reading in values from a text file.
iii) The third class called the *physics list* is used to define all the particles that will be encountered in the simulation. Along with the physics processes that must be assigned to these particles, a separate in-built class deals with each physics model. Of particular interest to the medical physics community are the low-energy extensions to the electromagnetic physics models, with accurate models of photon and electron transport to low energies. One of the strengths of Geant4 is the ability to change with ease the physics processes for each particle. Geant4 also contains a model for photonuclear reactions, which may be utilized in neutron dose studies and shielding considerations for higher-energy photon beams.

There are several ways to implement scoring in Geant4, depending on the user requirements, and it is up to the user to define explicitly how this is done by implementing a scoring class. The most common method is to create a sensitive detector class. This class carries out a task whenever a particle undergoes a step in one of the volumes of the geometry that has been deemed as a sensitive volume. One can obtain information about this step such as ionization energy loss, particle type and kinetic energy, position of the step. From this information one can tally energy deposition in the sensitive volume for the purpose of dose calculation, or determine the particle type and energy for spectroscopic applications.

Once these classes have been created, the user must build the application using a makefile. The executable file is then run from the command line by typing the program name.

Geant4 applications are generally run in one of two modes: batch mode or interactive mode. The interactive mode involves the creation of a user interface for the simulation. A shell-like environment is created and the user may type a number of in-built commands at the interface. These commands can be used to modify aspects of the geometry, primary beam, and physics models, and to set the *verbosity* of the simulation. For example, the simulation is able to print all tracking information to the standard output. Although this does slow the simulation somewhat, it is useful for debugging purposes. The user is also able to define his/her own user interface commands by creating what are known as *messenger classes*, which essentially modify the behavior of a user-defined class. For example, one may create a Detector Messenger class that is used to modify the geometry between runs. More details on these messenger classes can be found in the Geant4 examples. In batch mode a number of these user interface commands can be placed in a macro file and the filename can be passed as a command line argument to the simulation.

One may visualize the geometry, particle tracks, and energy deposition in the sensitive volume with Geant4. There are several ways to do visualize simulations; the most common uses the OpenGL libraries for visualization. Visualization is essential in the earlier stages of application development to debug the simulation and is generally done in interactive mode.

Further Reading and Useful Information

The Geant4 toolkit is not aimed at the novice application developer (yet makes for great reference material). First timers are often put off by the steep learning curve compared to other Monte Carlo applications. An online user forum exists for those who encounter problems using the Geant4 toolkit. Queries are generally addressed promptly and searching the forum for similar problems to your own can save time. There are also some examples packaged with the toolkit. The Geant4 toolkit provides a very powerful tool for modeling medical physics problems using Monte Carlo. Although the learning curve is steep, knowledge of the toolkit can be applied to a vast range of problems in medical physics, as well as nuclear physics, space physics, and high-energy physics.

10.5 Convolutions/Superposition Methods

Several variations of the convolution algorithm exist (Boyer and Mok 1985, Mackie et al. 1985, Mohan et al. 1986, Ahnesjö 1989, Fogliata et al. 2006), but all have two essential components: one representing the energy imparted to the medium by the interactions of primary photons (usually called the *terma*) and one representing the energy deposited about a primary photon interaction site (the *kernel*). Values in the kernel are the measurements of energy deposited at a vectorial displacement from the interaction site, as a fraction of the terma at that site. The kernel is usually separated into two or three components: one used to calculate the primary dose (the *primary kernel*) and, in addition, either (a) a component used to calculate the first and multiple scatter dose together (the *scatter kernel*) or (b) another two components used to calculate the first and multiple scatter doses separately (the *first-scatter kernel* and *multiple-scatter kernel*). With the different implementations of the convolution method, the kernel has been referred to as either (a) a *dose spread array* (Mackie et al. 1985), (b) a *differential pencil beam* (Mohan et al. 1986), (c) a *point spread function* (Ahnesjö et al. 1987), or (d) an *energy deposition kernel* (Ahnesjö 1989).

The convolution method is often referred to as the superposition method. This stems from the use of scaled kernels to account for inhomogeneities, meaning that the process is not a true convolution because the kernels are not invariant. The process is as such a superposition of (in general) varying kernels, each modulated by the terma. It is common, however, to use the term convolution, even though in a strict mathematical sense this is not the case.

10.5.1 Terma

Terma is an acronym for *total energy released per unit mass* and represents the energy that is (a) imparted to secondary charged particles and (b) retained by the scattered photon when primary photons interact in a unit mass. For each primary photon interaction, the sum of these energies is, of course, equal to that of the primary photon itself. The terma can thus be thought of as the energy *lost* out of the primary beam in a unit mass. A fundamental definition of the terma at a point is the negative of the *divergence of the energy fluence* per unit mass, as this is the energy carried into a unit mass minus the energy carried out. Further discussion of the concept and formulation of terma and its relationship to other quantities such as the energy fluence and kerma can be found in chapter two. Briefly here, if the energy fluence of primary photons at a point r is $\Psi(r)$ then the terma is given by

$$T(r) = -\frac{1}{\rho(r)} \nabla \bullet \Psi(r). \tag{10.9}$$

Since the energy fluence has an exponential falloff with depth, evaluating equation 10.9 gives

$$T(r) = \frac{\mu}{\rho}(r)\Psi(r). \tag{10.10}$$

This expression can be derived by intuition since the terma is just the energy fluence times the fraction of that energy given to the unit mass, and this fraction is obviously the likelihood of a photon interacting (and therefore having all of its energy given up to the medium) in that unit mass.

When performing a terma calculation, the polyenergetic nature of the beam must be taken into account by incrementing the terma for each energy component. Hence, if the energy fluence of primary photons in spectral component n at a point r' is $\Psi_n(r)$, the terma is given by

$$T(r') = \sum_{n=1}^{N} \Psi_n(r') \left(\frac{\mu}{\rho}\right)_n, \qquad (10.11)$$

where $(\mu/\rho)_n$ is the mass attenuation coefficient for energy component n. The energy fluence for each component must in turn be calculated by exponentially attenuating the energy fluence at the surface. For a divergent beam, inverse-square fall-off must also be taken into account. If the energy fluence for component n at the point r_0, where a ray line from the source to point r' intersects the medium surface, is $\Psi_n(r_0)$, the terma is given by

$$T(r') = \left(\frac{|r_0|}{|r'|}\right)^2 \sum_{n=1}^{N} \Psi_n(r_0) e^{-\mu_n|r'-r_0|} \left(\frac{\mu}{\rho}\right)_n, \qquad (10.12)$$

where $|r - r_0|$ is the geometric depth to the point r. For an inhomogeneous medium, the radiological depth to the point r, $d_{eff}(r)$, must be used instead of the geometric depth. This is obtained by ray tracing from r_0 to r. In the ray trace, pathlength increments are multiplied by the corresponding average electron density in order to convert to radiological pathlength increments. Ray tracing can be done either for a medium defined by CT pixels or for a medium where different structures are delimited by contours. A fast and exact method of ray tracing through a 3-D voxel array of CT data has been developed by Siddon (1985). With this method, the intersection of a ray with the three orthogonal sets of planes delimiting the CT voxels is considered. The positions along the ray of the intersection points and the electron densities between each of these points (defined by CT number) are used to increment the radiological depth.

In practice, a 3-D voxel array containing terma values is obtained before the convolution process itself begins. This involves (a) filling an array with electron densities using a set of CT slices, (b) calculating radiological depth for each of these voxels, and (c) calculating terma for each voxel. When the ray trace is performed, radiological depth is found from points at the "top"surface of the array to the center of each voxel, so in fact the intersection point (represented by r_0 above) is merely the first voxel encountered with a density greater than that of air. In cartesian coordinates and for inhomogeneous media, equation 10.12 is replaced by

$$T(x',y',z') = ISQR(x',y',z') \sum_{n=1}^{N} \Psi_n(x_0,y_0,z_0) e^{-\mu_{n,water} d_{eff}(x',y',z')} \left(\frac{\mu}{\rho}\right)_n \qquad (10.13)$$

where $ISQR$ is the inverse-square term.

Even for a polyenergetic beam, a polyenergetic terma such as in equation 10.13 is not always calculated. This is because the convolution calculation may be performed on a component-by-component basis, where terma and kernels

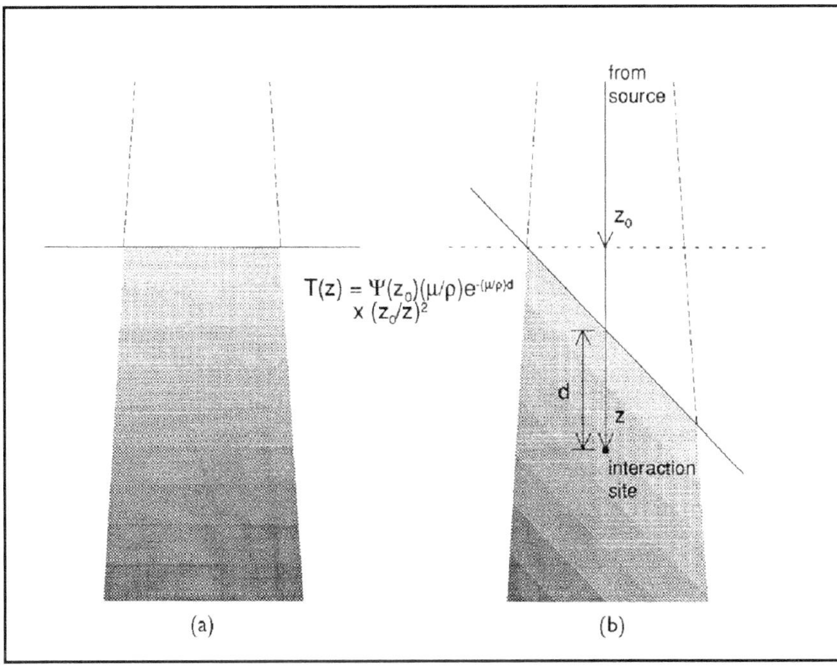

Figure 10.17: Schematic of terma in 2-D for (a) a flat surface normal to the beam direction and (b) a 45 degree surface. The terma calculation for the interaction site at distance *z* from the source is given by the equation shown, where *d* is the depth beneath the surface along a ray line from the source.

are generated for multiple single energies. Such a calculation will be referred to as the *components method* of convolution, and a convolution using a poly-energetic terma and kernels will be called the *polyenergetic method.*

In figure 10.10, simple representations of terma in two dimensions are shown, where in figure 10.17a the surface is normal to the beam direction so the terma varies in the depth direction only. In figure 10.17b, the surface is at 45 degrees to the beam direction, so the terma distribution is tilted. The equation shown is a simplification of equation 10.13, where the medium is homogeneous (so the effective depth is the geometric depth *d* beneath the sloped surface) and the beam is monoenergetic. Note that the shading levels in figure 10.17b do not take account of the inverse square effect.

10.5.2 Energy Deposition Kernels

Energy deposition kernels can be easily obtained using a Monte Carlo simulation, where photons are forced to interact at an appropriate point within an array, and the energy subsequently deposited within each voxel in the array is recorded. Also recorded is the total energy imparted by the primary photons. This is equal to the total primary photon energy since energy imparted includes both that transferred to charged particles and that retained by the scattered photon. The recorded energy in each voxel is then divided by the total energy imparted, giving the *fractional energy* for each voxel.

If separate primary and scattered kernels are required, separate arrays that correspond to primary and scattered fractional energy can be produced. Primary fractional energy is found from energy deposited by (a) electrons ejected by the primary photons in Compton collisions, (b) electrons ejected by primary photons in photoelectric interactions, and (c) charged particles created by primary photons in pair-production events. Scattered fractional energy is from energy deposited by charged particles set in motion by scattered photons and, normally, all bremsstrahlung photons (to ensure that the primary kernel is finite in extent). The distinction between primary and scattered energy can be made by setting the value of a certain "flag" associated with each particle after the primary photon interaction. In EGS4, each flag is a single digit of an integer variable. The nth digit can be incremented and decremented by adding or subtracting 10^n and can be interrogated using modulo and integer arithmetic. For example, FLAG1 is the 100s digit of the variable LATCH, FLAG2 is the 1000s digit, etc.

If the total energy imparted by primary photons in the chosen "interaction" voxel is E_{tot} and the primary energy deposited in a voxel at a vectorial displacement Δr is $E_p(\Delta r)$, the corresponding primary kernel value is

$$H_p(\Delta r) = \frac{E_p(\Delta r)}{E_{tot}}. \tag{10.14}$$

This expression can also be applied to the scattered kernel, where the subscript p is replaced with s. For a cartesian kernel array, Δr is replaced by $\Delta x, \Delta y, \Delta z$.

As with the terma, kernels that account for the energy spectrum of the beam can be generated (for polyenergetic convolution). This can be done in two ways: (a) kernels can be created separately for each energy component and subsequently combined with appropriate weights or (b) a polyenergetic spectrum can be used as the kernels are being generated, with the recorded energy deposited and imparted being weighted. If the method (a) is used, the weight should be equal to the relative total energy imparted by that spectral component. This quantity is, of course, the terma for that component; so, referring to equation 10.10, if the primary kernel for component n is $H_{p,n}(\Delta x, \Delta y, \Delta z)$, the polyenergetic primary kernel is (in cartesian coordinates)

$$H_p(\Delta x, \Delta y, \Delta z) = \frac{\sum_{n=1}^{N} \Psi_n (\mu/\rho)_n H_{p,n}(\Delta x, \Delta y, \Delta z)}{\sum_{n=1}^{N} \Psi_n (\mu/\rho)_n}. \tag{10.15}$$

If method (b) is used, it must be remembered that different photon energies have a different attenuation coefficient, and therefore different probabilities of interacting. This fact must be taken into account when the primary photons are forced to interact at the kernel origin. Simply forcing interactions implies that the probability of interacting is independent of energy. The energy dependence of the attenuation coefficient is accounted for by weighting by the mass attenuation coefficient of that photon, both (a) increments in energy deposited due to a primary photon and (b) the energy imparted by a primary photon. This "corrects" for the forcing procedure and can be shown to be mathematically equivalent to method (a).

If the kernels were infinite in extent, the sum of the fractional energies in the primary and scattered kernels would be unity. However, the finite size of the scattered kernel array means that many scattered photons "escape" from the bounds of the array. This means that a considerable proportion of the scattered photon energy may not be deposited within the array and recorded as fractional energy. In the case of the primary kernel, the total fractional energy is equal to the fraction of primary photon energy that is transferred to charged particles (and not converted to bremsstrahlung). For a single photon energy, this total fractional energy should be equal to μ_{en}/μ (Mackie et al. 1988, Hoban 1995). For a polyenergetic spectrum, total fractional energy in the primary kernel is equal to the ratio of primary collision kerma to terma K_c/T (note that this is equal to μ_{en}/μ for a single energy).

An example of a primary kernel for a 10 MV spectrum, generated in a cylindrical polar array, is shown in figure 10.18. The primary photon interaction

0.25 cm → ←

↓ 0.5 cm ↑

	.0001	.0001	.0001							
×.1970	.0118	.0025	.0008	.0003	.0001	.0001				
.1260	.0468	.0140	.0044	.0015	.0006	.0002	.0001			
.0279	.0313	.0162	.0073	.0031	.0013	.0005	.0002	.0001		
.0071	.0131	.0103	.0063	.0034	.0017	.0008	.0003	.0001	.0001	
.0021	.0049	.0050	.0038	.0024	.0014	.0007	.0004	.0002	.0001	
.0007	.0017	.0021	.0018	.0013	.0009	.0005	.0003	.0001	.0001	
.0002	.0006	.0008	.0007	.0006	.0004	.0003	.0002	.0001		
.0001	.0002	.0002	.0003	.0002	.0002	.0001	.0001			
		.0001	.0001	.0001						

Figure 10.18: Primary kernel generated for a 10 MV spectrum in a cylindrical polar array. The primary photon interaction site is indicated by a cross on the array axis. Total fractional energy contained in the kernel array is 0.574. When the beam is hardened so that the spectrum is as it would be at 20 cm and 40 cm deep, total fractional energy increases to 0.597 and 0.613, respectively.

site is indicated by a cross on the axis of the array. The rapid decrease in fractional energy away from the interaction point can be seen. This kernel was generated for the spectrum that exists at the surface and has a total fractional energy of 0.574. As depth increases, the spectrum hardens, and hence the kernel should ideally be modified to account for the increasing ratio of K_c/T. For the spectra at 20 cm and 40 cm deep, total fractional energies are 0.597 and 0.613, respectively (Hoban et al. 1994). Neglecting this effect causes a small but significant error and can be avoided by either (a) performing the convolution by the components method (Papanikolaou et al. 1993), (b) using collision kerma instead of terma in a polyenergetic convolution, with kernel values now a fraction of K_c (Hoban 1995); or (c) correcting the polyenergetic primary kernel values using the fractional increase in the K_c/T value with depth (Hoban et al. 1994).

10.5.3 The Convolution Process

In a convolution calculation, the dose at each point in the medium is calculated by considering primary photon interactions throughout the irradiated volume. This is done by summing the dose contributions from each irradiated volume element. The vectorial framework for a convolution calculation

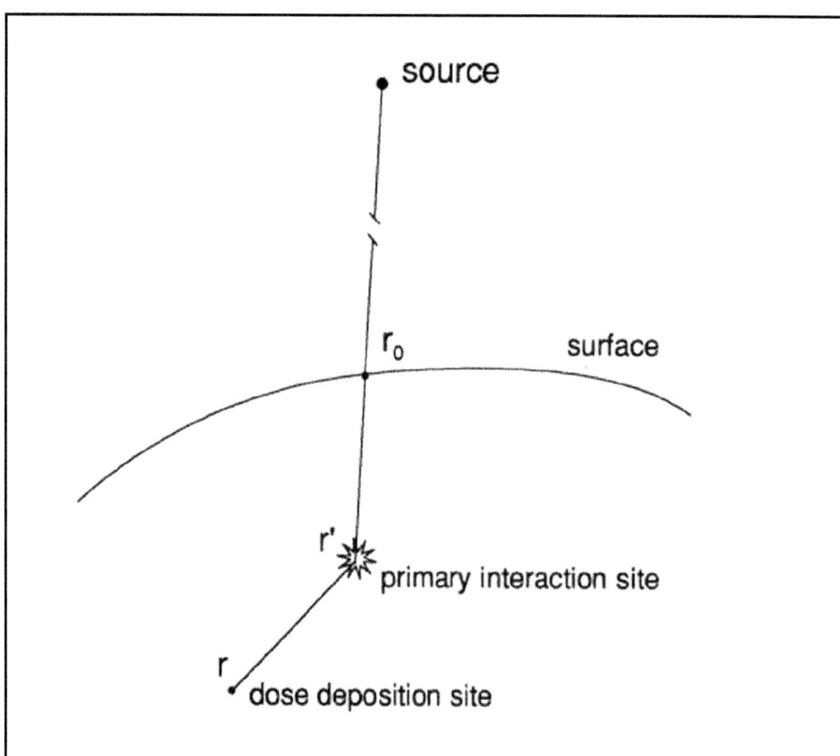

Figure 10.19: Vectorial framework for the convolution process. A ray from the source to the interaction site r' intersects the medium surface at r_0. Dose at r is calculated due to terma at all interaction sites r' via the kernel value corresponding to the vector $r - r'$.

is shown in figure 10.19. If the total energy imparted to a unit mass at r' is the terma $T(r')$, the energy deposited in a unit volume at another point r due to this energy imparted is given by $T(r')H(r-r')$, where $H(r-r')$ is the kernel value (primary plus scattered) for a displacement $r-r'$ from the kernel origin. The total dose at r is given by integrating over all unit masses in the irradiated volume; so, with the kernel separated into primary and scattered components the dose is given by

$$D(r)=\int_{r'} T(r')\left[H_p(r-r')+H_s(r-r')\right]d^3r'. \tag{10.16}$$

Note that the terma is an expression of energy imparted per unit mass, while the integration is over volume elements. This means that the terma in equation 10.16 should be multiplied by the density at r', $\rho(r')$, to convert the terma to energy imparted per unit volume. However, to convert the result of the integration to dose (energy deposited per unit mass) from energy deposited per unit volume, a division by $\rho(r)$ is necessary. Thus, for homogeneous media, where $\rho(r)=\rho(r')$, the density terms cancel and equation 10.16 is correct.

In practice, the convolution can proceed by either (a) integrating the dose deposited at successive dose deposition points due to terma throughout the irradiated medium (as implied in equation 10.16) or (b) calculating the dose contribution throughout the medium due to terma at successive interaction points. The former is known as the *deposition point of view* and the latter as the *interaction point of view* (Mackie et al. 1985). The deposition point of view is best if the dose in only a small subset of voxels is required since time is not wasted calculating dose throughout the whole medium. However, terma in each voxel must be stored in an array since each value is used as many times as there are dose deposition voxels. With the interaction point of view, only a single terma value needs to be stored at any time, so a terma array is not essential. Two-dimensional schematics of each of these methods are shown in figure 10.20, where limits i_{min} and i_{max} are shown, indicating field size-based restrictions on the range of interaction and deposition voxels, respectively.

A convolution calculation such as in equation 10.16, where the kernels are spatially invariant, can be performed in Fourier space (Boyer and Mok 1985, Mohan and Chui 1987, Murray et al. 1989). This saves computation time since the calculation is reduced to a multiplication of values in a Fourier-transformed terma array by corresponding values in transformed kernel arrays. If the convolution of terma with a kernel to obtain dose is written as

$$D=T\otimes K, \tag{10.17}$$

the Fourier transform of the dose is given by

$$\Im(D)=\Im(T)\times\Im(K), \tag{10.18}$$

from which the dose is obtained by taking the inverse Fourier transform. Not only is this computation faster than a "real space" convolution, but the increase in computation time with terma and kernel array sizes is much less rapid. The problem with Fourier space-convolution is that inhomogeneities

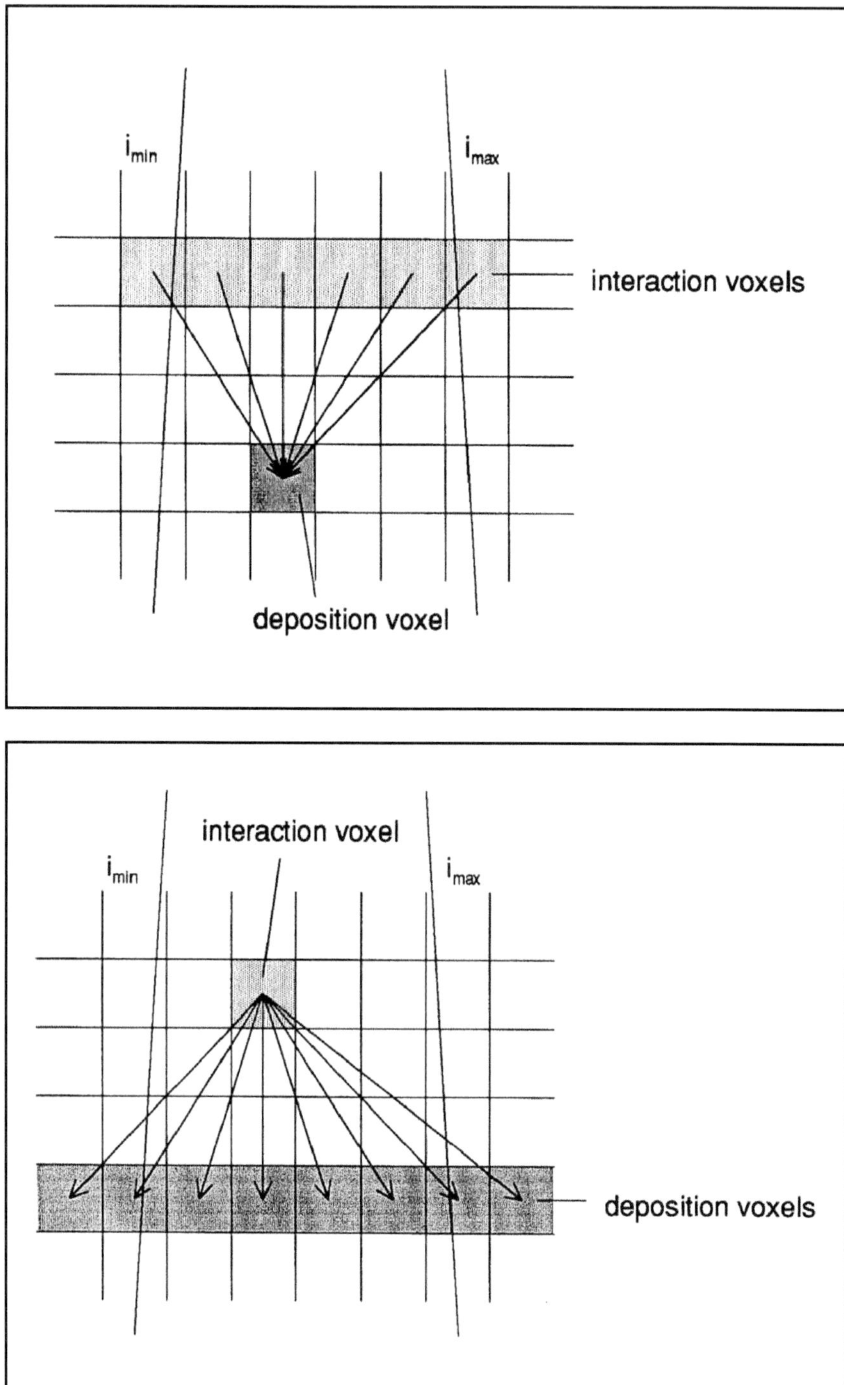

Figure 10.20: (a) Convolution by the *deposition position* point of view, where dose in successive *deposition voxels* is calculated due to all *interaction voxels*. (b) The interaction point of view, where dose throughout the medium is calculated due to successive *interaction voxels*.

cannot be accounted for except in the terma calculation, although a first-order scatter kernel correction has been devised by Boyer and Mok (1986). See chapter nine for further discussion of Fourier space (FFT) convolution.

Convolution in an Inhomogeneous Medium

In equation 10.16 the kernels are functions of the displacement only; they are invariant with r'. For inhomogeneous media, the true fractional energy distribution about each interaction site will depend on the position of that site; hence, the kernels will be a function of r' as well as $r' - r$. Furthermore, the densities at r and r' are in general no longer equal, so the convolution equation for inhomogeneous media becomes

$$D(r) = \frac{1}{\rho(r)} \int_{r'} T(r')\,\rho(r')\left[H_p(r', r - r') + H_s(r', r - r')\right] d^3r'. \tag{10.19}$$

A method must be devised with which to determine the value of $H(r', r - r')$ that is both realistic and without excessive computational overhead. It can be assumed to a first approximation that the energy loss by secondary electrons in traveling from r' to r is dependent on the effective pathlength, and hence on the average density, between the two points (where it is assumed that the medium is all of similar elemental composition). As a consequence, the number and energy of electrons crossing a surface in a given solid angle will also be dependent on the average density. Average density between r' and r can be found by ray tracing along a line through the intervening voxels. This is represented mathematically by the line integral

$$\rho_{ave} = \frac{1}{|r - r'|} \int_{r'}^{r} \rho(r'')\,dr''. \tag{10.20}$$

A representation of the density scaling procedure is shown in figure 10.21.

Because of electron scattering, the use of average density in this manner is strictly incorrect since the number and energy of electrons depends not only on the *average* density but also on the density *distribution* (Woo and Cunningham 1990, Hoban et al. 1990). Scaling based on average density is thus only a fair assumption for the primary kernel. For the scatter kernel, such density scaling is a good approximation because the fluence of once-scattered photons reaching r is exactly proportional to average density between r' and r, and the range of electrons ejected by these photons is very small. For multiply scattered photons, the assumption is invalid, but the dose contribution from multiple scatter is small.

Since the rate of energy deposition in each unit volume is proportional to the density within the volume, the kernel value for the actual density distribution, with density $\rho(r)$ at r, can be obtained from that for the average density by the relation

$$H(r', r - r') = H(\rho_{ave}, r - r')\frac{\rho(r)}{\rho_{ave}}. \tag{10.21}$$

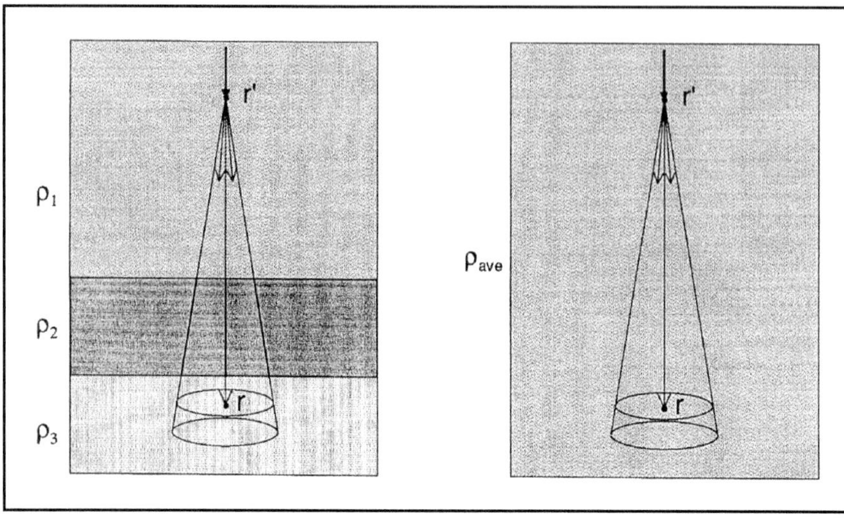

Figure 10.21: Kernel scaling using average density between the interaction and deposition sites. The number and energy of electrons and/or scattered photons crossing a surface into a unit volume at r due to primary photon interactions at r' is intended to be equal for the density distribution in (a) and for the average density in (b).

Substituting equation 10.21 into equation 10.19 gives a final expression for convolution in a heterogeneous medium using density scaling:

$$D(r) = \int_{r'} T(r') \frac{\rho(r')}{\rho_{ave}} \left[H_p\left(\rho_{ave}, r - r'\right) + H_s\left(\rho_{ave}, r - r'\right) \right] d^3r'. \quad (10.22)$$

Note that the $\rho(r)$ cancels in the numerator and denominator in this case. To implement this procedure, kernels must be generated for a range of densities; the value corresponding to the average density can then be found by interpolation between kernels.

In equation 10.21, it was assumed that average *mass density* could be used to characterize electron energy loss and photon attenuation. It is more correct to use average *electron density*, since energy loss is mainly due to electron-electron collisions and most photon interactions are Compton events. If the atomic number varies, such as when regions of bone exist, electron density and mass density are not proportional. The ratio of electron to mass density for bone is less than that for soft tissue because of the lower hydrogen content. This means that the cancellation of $\rho(r)$ in equation 10.22 is not valid. First, assuming that the terma has been calculated using the mass attenuation coefficient of water, the density by which the terma is multiplied in equation 10.19 should be the electron density relative to water $\rho_e^w(r')$. This is because the terma is the energy imparted to a unit mass of water-like material (or to a unit volume if the material is water). So the energy imparted to a unit volume of any other material is the terma multiplied by the number of electrons per unit volume relative to that in water. Second, the numerator in equation 10.21 should be $\rho_e^w(r)$, since the kernel value is the energy deposited in a unit volume of water-like material with density ρ_{ave}, and the energy deposited per unit volume is proportional to number of electrons in that volume. The division by $\rho(r)$ in equation 10.19 remains as this converts energy per unit volume to energy per unit mass. With these modifications, equation 10.22 becomes

$$D(r)=\frac{1}{\rho(r)}\int_{r'}T(r')\rho_e^w(r')\left[H_p(\rho_{ave},r-r')+H_s(\rho_{ave},r-r')\right]\frac{\rho_e^w(r')}{\rho_{ave}}d^3r'. \quad (10.23)$$

If, however, the dose to a water element at r is desired (as is normal), $\rho(r)=\rho_e^w(r)=1.0$ and the only difference to equation 10.19 is that the terma is multiplied by relative electron density instead of mass density. Further experiments which show the relative merits of scaling using mass density or relative electron density are outlined by Seco and Evans (2006).

An interesting illustration of the result of kernel scaling, as well as of the increased electron range in lung, is to perform a dose calculation for the lung phantom shown in figure 9.8 in three stages, with the terma only nonzero in one region at a time. The results of this are seen in figure 10.22 for a 10 × 10 cm², 10 MV beam. A large dose contribution to the lung region from primary photon interactions in the first water region is apparent, which is of course due to the long range of secondary electrons (and also the low attenuation of scattered photons) in lung. This graph can be compared with the electron tracks in figure 9.9, which show the long range of electrons in the lung.

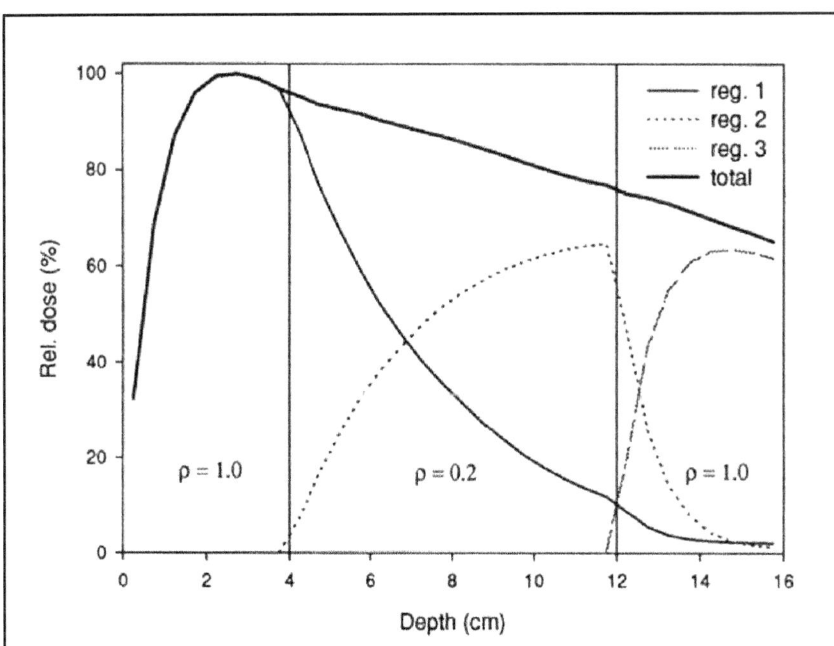

Figure 10.22: Convolution calculation performed in three stages for the lung phantom shown in figure 9.8 (with a 10 × 10 cm², 10 MV beam), where the terma was made nonzero in one region at a time to illustrate the different contributions of primary interaction sites in each region. The legend indicates the region in which the terma is nonzero.

Variation in Incident Beam Intensity

A very important consideration in the convolution process is how incident beam intensity varies across the field and the consequent impact on spatial variation of the terma. In particular, the incident energy fluence will not be constant in a plane perpendicular to the beam axis, and this must be accounted for in obtaining the energy fluence at the surface due to each spectral component (the term $\Psi_0(x_0,y_0,z_0)$ in equation 10.13). The main factors affecting the energy fluence profile are: Intensity profile of bremsstrahlung emitted from the target, variation in attenuation due to nonuniform flattening filter thickness, and extra-focal radiation due to scattering in the flattening filter and primary collimator. Furthermore, the energy spectrum varies with off-axis distance, and this should be accounted for in the terma calculation. In particular, the flattening filter thickness reduces with off-axis distance, causing a softening of the spectrum. Such variations can be determined by Monte Carlo simulation of the treatment head (Mohan et al. 1985, Rogers et al. 1995).

10.5.4 Convolution Results for a Lung Phantom

For the same lung phantom, a convolution calculation was performed for a 5×5 cm^2, 10 MV beam. The results are shown compared with experiment in figure 10.23 and can be seen to give some overestimate in dose throughout the lung. This is consistent with the results of different implementations of the convolution method (Ahnesjö 1989, Battista and Sharpe 1992) and can

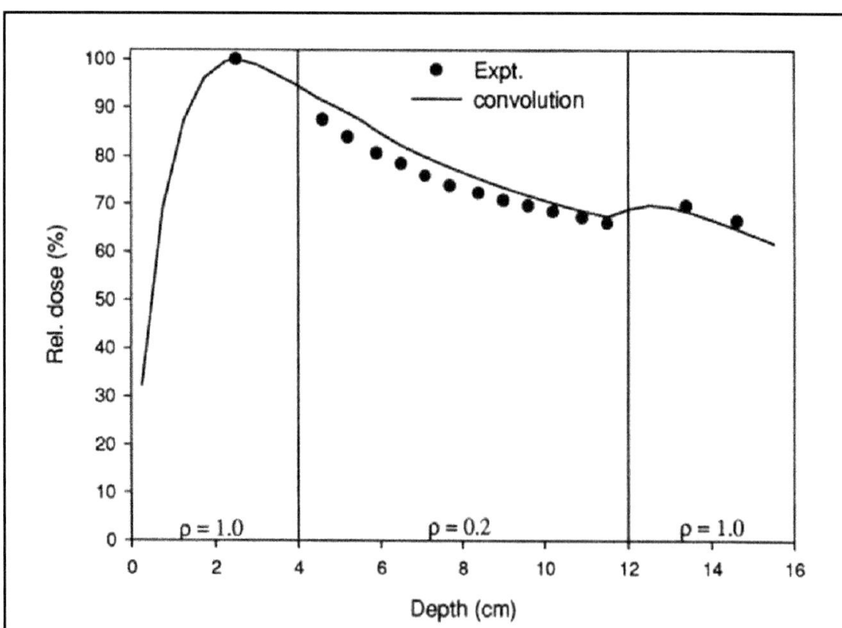

Figure 10.23: Convolution dose calculation compared with experiment for a 5 × 5 cm^2, 10 MV beam in the previously described lung phantom. Accuracy is good, but there is an overestimate of dose in lung due to the assumption of rectilinear secondary electron transport.

be attributed to the assumption implicit in the density scaling method that secondary electrons radiate out from the primary interaction site in straight lines. The accuracy is still much better than that of methods that totally ignore perturbations to electron transport, such as the ETAR method. (See figure 9.8b.)

The specific problem with using the density scaling method for the primary kernel is as follows. As electrons spread out from the primary interaction site, they scatter as well as lose energy. Scattering causes the electrons to be laterally deflected, so electrons initially traveling within a cone of solid angle will quickly exceed the boundaries of that cone. Furthermore, the lateral spread of electrons on reaching an arbitrary plane at a certain distance from the interaction site depends on the *number and position* of scattering events (scattering further from the plane contributes more to lateral deflection than scattering close to the plane). Density scaling assumes that the fluence distribution of electrons at such a plane is only dependent on the average density in reaching the plane and hence not dependent on the position of scattering events. A more rigorous scaling method for the primary kernel thus needs to take account of the density distribution as well as the average density. The dependence of lateral spread on the density distribution is easily proven with the use of the Fermi-Eyges theory of multiple coulomb scattering, as applied to electron pencil beam calculations (Hogstrom et al. 1981). An attempt to modify the density scaling method by calculating an *effective average density* using Fermi-Eyges theory has been made by Keall and Hoban (1995) with improved results. To account for electron scattering properly in a convolution calculation, macroscopic scaling of any kind does not suffice, and the effect of inhomogeneities on the transport of individual electrons must be considered. Such a scheme has been implemented in the Super Monte Carlo method, where a large set of Monte Carlo-generated electron tracks replaces the primary kernel (Keall and Hoban 1996). The nodes in each electron track are stepped through, with the length of the steps being scaled with the inverse of the density of the current voxel, while the energy deposited in each step remains the same. The scatter dose calculation is performed using standard convolution, since rectilinear scaling is a good approximation in this case. A stored electron track is represented on the left in figure 10.24, and the modified track is shown on the right, where a portion of the track passes through a low-density lung region.

10.5.5 Variations of the Convolution Method

In the mid-eighties, several variations on the convolution method emerged, with the original works being presented by Mackie et al. (1985) and Boyer and Mok (1985). Mackie's work was based on real-space convolution, incorporating density-scaling to obtain kernel values in the presence of inhomogeneities, and Boyer and Mok developed a much faster Fourier-space technique that used invariant kernels (a true convolution). The other major contributors to this field were Mohan et al. (1986), who developed the *differential pencil beam* (DPB) method, and Ahnesjö et al. (1987), who later developed the *collapsed-cone convolution* method (Ahnesjö 1989).

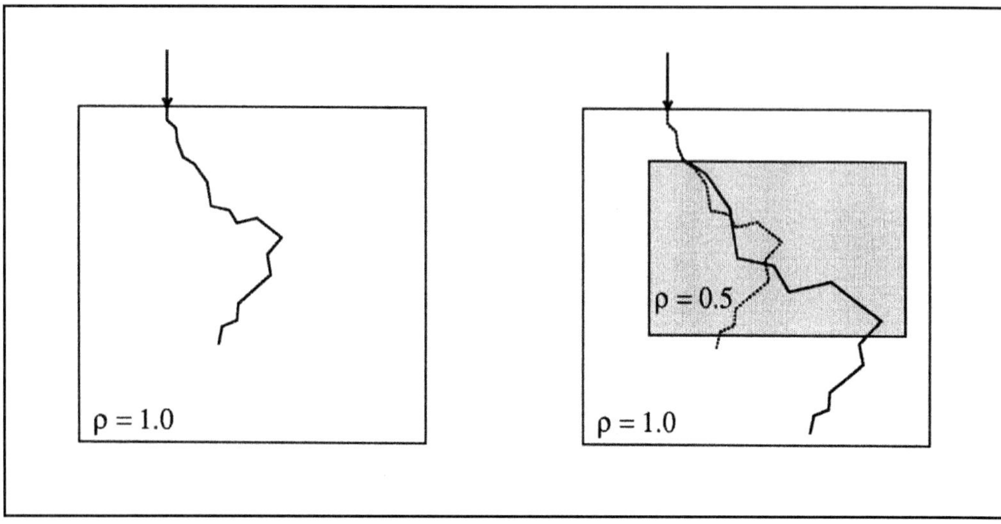

Figure 10.24: A stored (left) and transported (right) electron track representing density scaling of individual transport steps in the Super Monte Carlo method. Each step length is modified by the inverse of the density of the current voxel. Shown is a region of density 0.5, inside which all electron step lengths are doubled. Also shown in the low-density region is the original path.

Original Real-Space Convolution Method

In Mackie's original work (Mackie et al. 1985), the kernels (called dose spread arrays) were separated into primary, truncated first scatter (TFS), and residual first and multiple scatter (RFMS) arrays. The TFS array represents first scatter dose, which is deposited relatively close to the interaction site, and the RFMS array represents multiple-scatter dose as well as first scatter dose, which is not contained in the TFS array. This scatter separation was to allow a smaller, higher resolution array for the TFS component and also to allow a different kernel scaling method. Dose spread arrays were generated for a range of densities (or equivalently, voxel dimensions), and scaling based on average density between interaction and deposition sites was performed for the primary and TFS components. For the RFMS component, average density of the phantom was used in kernel scaling. A FORTRAN-based convolution package is available via anonymous ftp from the University of Wisconsin. This system can be configured to work with arbitrary beam spectra, where polyenergetic kernels are constructed by the user.[1]

Differential Pencil Beam Method

The differential pencil beam method is essentially the same as the components method of convolution. A differential pencil beam refers to an infinitesimal segment of a pencil beam (directed along a ray line from the beam source), within which primary photons interact and give rise to a differential pencil beam dose distribution. The differential pencil beam dose distribution,

[1] Anonymous login to madrad.radiology.wisc.edu: go to directory pub/conv.

denoted as DPB, is equivalent to a convolution kernel except that it is dose deposited in water per unit primary photon collision density (number of primary photons interacting in a unit volume), instead of per unit energy imparted by primary photons. DPBs are generated for a range of primary photon energies, and the spectrum is accounted for by performing a weighted sum of the dose calculated for each energy.

Collapsed Cone Convolution

The collapsed cone convolution method uses a polyenergetic terma and kernel, where the kernel is represented analytically and combines primary and scatter contributions. The functions used to characterize the kernel are of the form

$$H(r,\theta)=\frac{A_\theta e^{-a_\theta r}+B_\theta e^{-b_\theta r}}{r^2}, \qquad (10.24)$$

where there are a finite number of polar angles, θ, with respect to the primary beam along which the function is defined. The interaction site can be considered to be at the apex of a set of radially directed lines spreading out in three dimensions, where each line has a polar angle of θ and azimuthal angle of ϕ. Each of these lines is further considered to be the axis of a cone, where the solid angle covered by the complete set of cones is 4π. The kernel function along each line is actually the energy deposited within the entire cone at radius *r*, collapsed onto the line, hence the name of the method. This means that when the interaction site is at a point in a cartesian voxel array, fractional energy is deposited in voxels along each line, which should in fact be spread among all voxels within the cone. This is not really a problem, however, as those voxels that do not receive energy are compensated by energy deposited due to adjacent lines radiating out from a different interaction site. Furthermore, close to the interaction site (where most fractional energy is deposited) a cartesian voxel will probably enclose more than one cone, so no accuracy is lost.

In practice, there is not a set of lines radiating out from each interaction voxel but a lattice of lines throughout the dose calculation array, in each of the directions of the cone axes. There must be one line of each direction passing through each voxel. Terma is calculated throughout the cartesian array, and this is used to assign terma to points along each cone axis line, where there is a cone of the same orientation originating at each of these points. Inhomogeneities can be accounted for in the terma calculation, and the fractional energy deposited along the cone axes is determined in a manner that is equivalent to the density scaling method already discussed (although with very different implementation). The big advantage of the collapsed cone method over standard convolution is that the computation time only increases with MN^3 (where M is the number of cone axes, and N is the number of voxels along one side of the calculation volume) as opposed to N^6. This is comparable to Fourier convolution, which increases with $(2N)^3\ln 2N$, but does not allow kernel scaling. The collapsed cone convolution method is implemented on a commercially available treatment planning system.

10.6 Macro Monte Carlo

Since the work of Neuenschwander and Born (1992) and Neuenschwander et al. (1995), who described a so-called Macro Monte Carlo technique, there has been great progress in designing faster Monte Carlo codes. Their original approach was based on a data base calculated from Monte Carlo simulations of electron transport through macroscopic spheres consisting of different media. The reduction in the number of electron steps per history using this technique led to a speed increase of about a factor of 10 over conventional Monte Carlo methods.

Since then, another approach called Voxel Monte Carlo (VMC) has been developed. The VMC code developed by Kawrakaw et al. (1996) and tested by Fippel et al. (1997, 1999) has been adapted as an optional electron beam calculation engine by various commercial vendors, e.g., Nucletron. The technique includes several approximations to electron processes and a history repetition technique to reduce the simulation times by about a factor of 30–40 when compared to EGS4 calculation times. A similar approach called XVMC has also been reported by Kawrakow (Kawrakow and Fippel 2000) for photon beams.

Kawrakow et al. (1996) identified the electron multiple scattering models (e.g., Molier's multiple scattering theory) used in EGS require the sampling of many multiple scattering angles. They suggested this as being the main time sink in the EGS Monte Carlo code.

They present a simplified model of the multiple scattering distribution. The multiple angle is sampled from the distribution, such that

$$f(q)qdq = \frac{2}{p^2\bar{\theta}^2\left(1-\exp\left(-4/\bar{\theta}^2\right)\right)}\exp\left(-\frac{q^2}{\bar{\theta}^2 p^2}\right), \tag{10.25}$$

where q denotes momentum transfer and p is the electron momentum. The term q is related to scattering angle by

$$q^2 = 2p^2\left(1-\cos\theta\right) \tag{10.26}$$

and

$$\bar{\theta}^2 = \int_0^t dt' T_s\left(t'\right), \tag{10.27}$$

where t is the path length of the electron step, and T_s is the linear scattering power of the material (ICRU 1992). While the equation does give a better description of large angle electron scattering than Gaussian approximations, there are still large angle scattering simplifications. A detailed analysis by Kawrakow et al. (1996) shows that these are acceptable at radiotherapy beam energies.

These codes are now fast enough to be used on some commercial RTP computer systems to compute electron dose distributions in a few minutes and appear to show superior accuracy to electron pencil beam approaches

(Ding et al. 2006). Dose calculations also seem reasonably fast for current complex intensity-modulated radiation therapy (IMRT) x-ray beam calculations (Laub et al. 2000).

There are other approaches to fast Monte Carlo calculations for RTP; these include the PEREGRINE code (Hartmann-Siantar et al. 2001) developed at the Lawrence Livermore laboratory. Another is the *dose planning method* (DPM) code developed by Sempau et al. (2000). These codes present quite different simplifications to the transport model to VMC or indeed the original Macro Monte Carlo approach.

References

Agostinelli S., Allison J., Amako K., Apostolakis J. et al. (2003) Geant4—A simulation toolkit. *Nucl. Instrum. Methods A* 506: 250–303.

Ahnesjö A. (1989) Collapsed cone convolution of radiant energy for photon dose calculation in heterogeneous media. *Med. Phys.* 16: 577–592.

Ahnesjö A., Andreo P., Brahme A. (1987) Calculation and application of point spread functions for treatment planning with high energy photon beams. *Acta Oncol.* 26: 49–55.

Allison J., Amako K., Apostolakis J., Araujo H., et al. (2006) Geant4 developments and applications. *IEEE Trans. Nucl. Sci.* 53: 270–278.

Barret O., Carpenter T., Clark J., Ansorge R. et al. (2005) Monte Carlo simulation and scatter correction of the GE Advance PET scanner with SimSET and Geant4. *Phys. Med. Biol.* 50: 4823–4840.

Battista J.J., Sharpe M.B. (1992) True three-dimensional dose computations for megavoltage x-ray therapy: A role for the superposition principle. *Australas. Phys. Eng. Sci. Med.* 15: 159–178.

Boyer A. L., Mok E. C. (1986) Calculation of photon dose distributions in an inhomogeneous medium using convolutions. *Med. Phys.* 13: 503–509.

Boyer A. L., Mok E. C. (1985) A photon dose distribution model employing convolution calculations. *Med. Phys.* 12: 169–177.

Carrier J.-F., Archambault L., Beaulieu L., Roy R. (2004) Validation of GEANT4, an object-oriented Monte Carlo toolkit, for simulations in medical physics. *Med. Phys.* 31: 484–492.

Chibani O., Ma C. (2007) On the discrepancies between Monte Carlo dose calculations and measurements for the 18MV Varian photon beam. *Med. Phys.* 34: 1206–1216.

Ding G., Duggan D., Coffey C., Shokrani P., Cygler J. (2006) First macro Monte Carlo based commercial dose calculation module for electron beam treatment planning-new issues for clinical consideration. *Phys. Med. Biol.* 51: 2781–2799.

Emily P., Frank V. (2005) Accuracy of the photon and electron physics in GEANT4 for radiotherapy applications. *Med. Phys.* 32: 1696–1711.

Fippel M., Kawrakow I., Friedrich K. (1997) Electron beam dose calculations with the VMC algorithm and the verification data of the NCI working group. *Phys. Med. Biol.* 42: 501–520.

Fippel M., Laub W., Huber B., Nüsslin F. (1999) Experimental investigation of a fast Monte Carlo photon beam dose calculation algorithm. *Phys. Med. Biol.* 44: 3039–3054.

Fogliata A., Nicolini G., Vanetti E., Clivio A., et al. (2006) Dosimetric validation of the anisotropic analytical algorithm for photon dose calculation: fundamental characterization in water. *Phys. Med. Biol.* 51: 1421–1438,

Foppiano F., Mascialino B., Pia M.G., Piergentili M. (2004) A Geant4-based simulation of an accelerator's head used for intensity modulated radiation therapy. IEEE Nuclear Science Symposium Conference Record 4: 2128–2132.

Ford R. L., Nelson W. R. (1978) *The EGS Code System: Computer Programs for the Monte Carlo Simulation of Electromagnetic Cascade Showers (Version 3)*. Stanford Linear Accelerator Center Report SLAC-210.

Hartmann-Siantar C., Vetter K., DeNardo G., DeNardo S. (2001) Description and dosimetric verification of the PEREGRINE Monte Carlo dose calculation system for photon beams incident on a water phantom. *Med. Phys.* 28: 1322–1337.

Hoban P. W. (1995) Accounting for the variation in kerma/terma ratio in photon beam convolution. *Med. Phys.* 22: 2035–2044.

Hoban P. W., Murray D. C., Round W. H. (1994) Photon beam convolution using polyenergetic energy deposition kernels. *Phys. Med. Biol.* 39: 669–685.

Hoban P. W., Murray D. C., Metcalfe P. E., Round W. H. (1990) Superposition dose calculation for 10 MV photons. *Australas. Phys. Eng. Sci. Med.* 13: 81–92.

Hogstrom K.R., Mills M.D., Almond P.R. (1981) Electron beam dose calculations. *Phys. Med. Biol.* 26: 445–459.

ICRU (International Commission on Radiation Units and Measurements) (1992) Report No. 46. Photon, electron, proton and neutron interaction data for body tissues.

Jiang H., Wang B., Xu X.G., Suit H.D., Paganetti H. (2005) Simulation of organ-specific patient effective dose due to secondary neutrons in proton radiation treatment. *Phys. Med. Biol.* 50: 4337–4353.

Johns H.E., Cunningham J.R. (1983) *The Physics of Radiology*, Fourth edition. Charles C Thomas. Springfield, Illinois.

Kawrakow I., Rogers D., Walters B., (2004) Large efficiency improvements in BEAMnrc using direction bremsstrahlung splitting. *Med. Phys.* 31: 2883–2898.

Kawrakow I., Fippel M., Friedrich K. (1996) 3D electron dose calculation using a voxel based Monte Carlo algorithm (VMC). *Med. Phys.* 23: 445–448.

Kawrakow I., Fippel M. (2000) Investigation of variance reduction techniques for Monte Carlo photon dose calculation using XVMC. *Phys. Med. Biol.* 45: 2163–2183.

Keall P.J., Hoban P.W. (1996) Superposition dose calculation incorporating Monte Carlo generated electron track kernels. *Med. Phys.* 23: 479–486.

Keall P. J., Hoban P. W. (1995) Accounting for primary electron scatter in x-ray beam convolution calculations. Med. Phys. 22: 1413–1418.

Keall P., Siebers J., Libby B., Mohan R. (2003) Determining the incident electron fluence for Monte Carlo based photon treatment planning using a standard measured data set. *Med. Phys.* 30: 574–582.

Lamare F., Turzo A., Bizais Y., Le Rest C., Visvikis D. (2006) Validation of a Monte Carlo simulation of the Philips Allegro/GEMINI PET systems using GATE. *Phys. Med. Biol.* 51: 943–962.

Laub W., Alber M., Birkner M., Nüsslin F. (2000) Monte Carlo dose computation for IMRT optimization. *Phys. Med. Biol.* 45: 1741–1754.

Lyddon J. (2006) Monte Carlo characterization of narrow photon fields for intensity modulated radiotherapy. Ph.D. Thesis, University of Melbourne, Australia.

Mackie T.R., Reckwerdt P., McNutt T., Gehring M., Sanders C. (1996) Photon beam dose computations. *Proceedings of AAPM Summer School—Teletherapy: Present and Future*. Edited by T. R. Mackie and J. R. Palta. Advanced Medical Publishing, Madison, Wisconsin, pp. 103–137.

Mackie T.R., Bielajew A.F., Rogers D.W.O., Battista J.J. (1988) Generation of photon energy deposition kernels using the EGS Monte Carlo system. *Phys. Med. Biol.* 33: 1–20.

Mackie T.R., Scrimger J.W., Battista J.J. (1985) A convolution method of calculating dose for 15-MV x rays. *Med. Phys.* 12: 188–196.

Marsaglia G., Zaman A., Tsang W.W. (1990) Toward a universal random number generator. *Statistics and Probability Letters* 8: 35–39.

Meredith W.J., Massey J.B. (1972) *The Fundamental Physics of Radiology*. 2nd edition. John Wright and Son Publishers, Bristol, UK, p. 211.

Moehrs S., Del Guerra A., Herbert D., Mandelkern M. (2006) A detector head design for small-animal PET with silicon photomultipliers (SiPM). *Phys. Med. Biol.* 51: 1113–1127.

Mohan R., Chui C. (1987) Use of fast Fourier transforms in calculating dose distributions for irregularly shaped fields for three-dimensional treatment planning. *Med. Phys.* 14: 70–77.

Mohan R., Chui C., Lidofsky L. (1986) Differential pencil beam dose computation model for photons. *Med. Phys.* 13: 64–73.

Mohan R., Chui C., Lidofsky L. (1985) Energy and angular distributions of photons from medical linear accelerators. *Med. Phys.* 12: 592–597.

Murray D.C. (1990) Using EGS4 Monte Carlo in medical radiation physics. Aust. *Phys. Eng. Sci. Med.* 13: 132–147.

Murray D.C., Hoban P.W., Metcalfe P.E., Round W.H. (1989) 3-D superposition for radiotherapy treatment planning using fast Fourier transforms. *Australas. Phys. Eng. Sci. Med.* 12: 128–137.

Nelson W.R., Hirayama H., Rogers D.W.O. (1985) The EGS4 Code System. Stanford Linear Accelerator Center Report SLAC-265.

Neuenschwander H., Born E. (1992) A macro Monte Carlo method for electron beam dose calculations. *Phys. Med. Biol.* 37: 107–125.

Neuenschwander H., Mackie T.R., Reckwerdt P.J. (1995) MMC-a high performance Monte Carlo code for electron beam treatment planning. *Phys. Med. Biol.* 40: 543–574.

Nüsslin F. (2000) Monte Carlo dose computations for IMRT optimization. *Phys. Med. Biol.* 45: 1741–1754.

Papanikolaou N., Mackie T.R., Meger-Wells C., Gehring M., Reckwerdt P. (1993) Investigation of the convolution method for polyenergetic spectra. *Med. Phys.* 20: 1327–1336.

Raaijmakers A.J., Raaymakers B.W., Lagendijk J.J. (2005) Integrating a MRI scanner with a 6 MV radiotherapy accelerator: Dose increase at tissue-air interfaces in a lateral magnetic field due to returning electrons. *Phys. Med. Biol.* 50: 1363–1376.

Raaymakers B.W., Raaijmakers A.J., Kotte A.N., Jette D.D., Lagendijk J.J. (2004) Integrating a MRI scanner with a 6 MV radiotherapy accelerator: dose deposition in a transverse magnetic field. *Phys. Med. Biol.* 49: 4109–4118.

Rogers D.W.O. (1984) Low energy electron transport with EGS. *Nucl. Instr. Meth.* A227: 535.

Rogers D.W.O., Faddegon B.A., Ding G.X., Ma C.-M., Wei J., Mackie T.R. (1995) BEAM: A Monte Carlo code to simulate radiotherapy treatment units. *Med. Phys.* 22: 503–524.

Rogers D.W.O., Bielajew A.F. (1988) A comparison of EGS and ETRAN. *Monte Carlo Transport of Electrons and Photons*. Edited by T.M. Jenkins, W.R. Nelson, and A. Rindi. Plenum Publishing Corporation, New York, p. 323.

Rogers D., Walters B., Kawrakow I. (2007) BEAMnrc users manual. National Council of Radiation Canada (NRCC) Report PIRS 509.

Seco J., Evans P. (2006) Assessing the effect of electron density in photon dose calculations. *Med. Phys.* 33: 540-552.

Sempau J., Wilderman S., Bielajew A. (2000) DPM, a fast, accurate Monte Carlo code optimized for photon and electron radiotherapy treatment planning dose calculations. *Phys. Med. Biol.* 45: 2263–2291.

Sheikh-Bagheri D., Rogers D. (2002) Monte Carlo calculations of nine megavoltage photon spectra using the BEAM code. *Med. Phys.* 29: 391–402.

Siddon R. L. (1985) Fast calculation of the exact radiological path for a three-dimensional CT array. *Med. Phys.* 12: 138–144.

Verhaegen F., Seuntjens J. (2003) Monte Carlo modelling of external radiotherapy photon beams. *Phys. Med. Biol.* 48: R107–R164.

Walters B., Kawrakow I., Rogers (2007) DOSXYZnrc users manual. National Council of Radiation Canada (NRCC) Report PIRS 794.

Woo M. K., Cunningham J. R. (1990) The validity of the density scaling method in primary electron transport for photon and electron beams. *Med. Phys.* 17: 187–194.

Questions

10.1 Consider a Monte Carlo simulation of an electron beam incident on a slab of water-equivalent material, where the spectrum of electrons emerging from the slab is recorded. If this spectrum is to be realistic, the true randomness of energy loss events within the slab must be closely approximated. It is generally too time consuming to model all energy loss events discretely, so only those where the energy lost is above a certain threshold are considered individually: small energy losses are grouped and the particle loses energy via a restricted stopping power.

i) What are the two electron energy loss mechanisms that may be modeled discretely and how is the threshold energy loss for modeling each type of discrete energy loss controlled in an EGS4 simulation. What general type of electron transport does EGS4 use?

ii) Explain why values of AE and AP of 0.611 MeV and 0.5 MeV are not suitable for the simulation mentioned above, considering that most knock-on and bremsstrahlung collisions will involve energy losses of less than 0.1 MeV and 0.5 MeV, respectively.

iii) Why is an ECUT value of 0.711 not suitable when the lowest kinetic energy bin in the spectrum to be recorded is 0-0.1 MeV?

(section 10.2)

10.2 In a Monte Carlo simulation, a probability density function (PDF) gives the probability of an event occurring at a certain point in a phase space. For example, the likelihood of a photon interaction being at a particular distance from the previous interaction point, or the probability of each scattering angle in a Compton collision.

i) Given that the probability of photon interaction per unit pathlength is μ, show that the PDF for interaction after traveling a distance x is

$$f(x) = \mu e^{-\mu x} \qquad \text{(Q10.1)}$$

Use the fact that the probability of interaction is the product of the probability that a photon at x will interact and the probability that it has not already interacted before it reaches x.

ii) The quantity used for selecting a variable such as distance to next interaction is the cumulative density function (CDF). This is the indefinite integral of the PDF and for this example is:

$$F(x) = 1 - e^{-\mu x} \qquad \text{(Q10.2)}$$

To choose the distance x, a random number between 0 and 1 is generated and assigned to $F(x)$. The CDF is then inverted (if possible) and solved for x. In this example what does the CDF physically represent, on examination of $F(x)$, and why must this be the case? What is the equation for x as a function of a random number R?

iii) Why is a value of x corresponding to the steepest part of the CDF curve most likely to be obtained? Hint: To which part of the PDF does this correspond?

iv) If the CDF is not invertable (i.e., it is too complex), how might x be found iteratively?

(section 10.2)

10.3 In order for a Monte Carlo simulation to consist of totally independent particle histories, the random number generator must not repeat its sequence. If the sequence does repeat, it does not take long before the incident particle parameters and transport steps of a history are being chosen and controlled using exactly the same random numbers as have been used for a previous history. Once this occurs, the simulation is locked into a repetition of previous histories. This unfortunate occurrence can usually be recognized by the uncertainty not being reduced by increasing incident particle number.

i) Why does such a repetition of particle histories occur, even though it may seem unlikely that a repeated random number being used at the start of a history will have previously been used at the start of a history?

ii) After a sequence begins repeating itself, approximately how many histories are likely to be simulated before previous histories are repeated and why? Does this disprove the theory that history never repeats?

(section 10.2)

10.4 In a Monte Carlo simulation to calculate a photon beam dose distribution, most computation time is spent simulating small electron multiple scattering steps, each of which involves determination of energy loss/deposition and angular deflection; plus there are checks for boundary crossing in some steps. Time spent performing multiple scattering steps is obviously reduced if the number of steps is reduced, and this will occur if either less secondary electrons are transported, they are stopped earlier, or the steps are lengthened.

i) In an EGS4 simulation, how can the threshold for transporting electrons be set to a kinetic energy of 0.1 MeV or above?

ii) Why is it not necessary to set a low value of ECUT when a dose distribution is to be obtained in an array of relatively large voxels?

iii) Given that the stopping power of electrons in water is approximately 2 MeV cm^{-1}, what would be a reasonable value for ECUT when scoring voxels have 0.5 cm sides (allow a residual range of 1/3 the voxel width)?

(section 10.2)

10.5 i) Explain the physical meaning of the terma definition in textbook equation 10.9.

ii) For an exponential fall-off in energy fluence, terma is given by textbook equation 10.10. Show how this expression is obtained from equation 10.9 and explain it in terms of the meaning of each of the two factors. Why is the textbook equation 10.10 obvious from the meaning of terma?

(section 10.5)

11 Quality Assurance in Radiotherapy

"As a good quality system reduces the likelihood of errors, it will also reduce the likelihood of litigation...clearly defined responsibility and accountability is a first requirement of a comprehensive QA program."
(Jacob Van Dyk and James A. Purdy 1999)

11.1 Introduction

Prior to linear accelerator and ancillary equipment arriving in the clinic for acceptance testing it is likely the equipment will have been manufactured by a company adhering to a series of equipment manufacturing quality assurance (QA) standards. A common set of standards are those produced by the International Organization of Standardization (ISO). The ISO-9000 series defines QA as all those planned and systematic actions necessary to provide adequate confidence that a product or service will satisfy given requirements for quality. Within this series ISO-9001 (1994) presents a model for QA in design, development, production, installation, and servicing.

A regulatory organization within the United States is the Federal Drug Administration (FDA), which grants approvals for therapeutic goods and products based on good manufacturing practice. The International Electrotechnical Commission (IEC) also has a long history in publishing standards for electrical, electronic, and related technologies (IEC 1989a,b, 2000). Van Dyk and Purdy (1999) present an excellent review of these standards.

QA in the radiation therapy clinic includes those procedures that ensure a consistent and safe fulfillment of the dose prescription to the target volume, with minimal dose to normal tissue and minimal exposure to personnel, and adequate patient monitoring aimed at determining the end result of treatment (AAPM 1984, WHO 1988).

How does QA assist patients undergoing radiotherapy? Thwaites et al. (2003) have summarized this as follows:

i) QA reduces uncertainties and errors in dosimetry, treatment planning equipment performance, treatment delivery, thus improving dosimetric and geometric accuracy and precision of dose delivery. This improves radiotherapy results (treatment outcomes), raising tumor control rates as well as reducing complication and recurrence rates.
ii) QA not only reduces the likelihood of accidents and errors occurring it also increases the probability that they will be recognized and rectified sooner, thereby reducing their consequences for patient treatment. This is the case for larger incidents and minor incidents.
iii) QA allows a reliable intercomparison of results among different radiotherapy centers, ensuring a more uniform and accurate dosimetry and treatment delivery. This is necessary for clinical trials and also for sharing clinical radiotherapy experience within and between centers.
iv) Improved technology and more complex treatments in modern radiotherapy can only be fully exploited provided a high level of accuracy and consistency is attained.

This chapter is not to be used as a protocol for QA—it aims to provide motivation, rationale, and some background for implementation of QA in radiotherapy. The setup of a QA program depends on local circumstances and the present chapter does not provide a complete, comprehensive guide. QA has been tremendously successful in industry and it comprises a certain disciplined mind-set. Interesting thoughts about QA and its future were recently published by Pawlicki and Mundt (2007).

In general, the need for QA and the complexity and frequency of tests is informed by three parameters:

i) Likelihood of failure.
ii) Severity of the consequence of failure.
iii) Ease of the test.

It is the responsibility of the local QA team to determine the QA activities required to ensure optimum delivery of radiotherapy.

The present chapter provides some definitions, followed by a discussion of QA needs associated with typical equipment in a radiotherapy department. As IMRT QA is one of the most important (and frequently discussed) issues in radiotherapy, section 11.4 is dedicated to this. After some brief comments on QA for image guidance, the chapter concludes with a discussion of in vivo dosimetry, which is an important tool for QA in radiotherapy.

11.2 Definition of Terms

11.2.1 Uncertainty

The terms *accuracy* and *precision* have been outlined in chapter three of this text. See also Kron (1999). Uncertainties are classified as type A—those that can be assessed by statistical means—and type B—those which may be assessed by some other means.

Uncertainties are also referred to as random or as systematic. *Random* uncertainties can be assessed by repeated measurements and these uncertainties can be expressed as having a standard deviation. While the probability distribution may not necessarily be normal in shape, it is generally approximated by a Gaussian distribution. *Systematic* uncertainties can only be assessed by an analysis of the process. A standard deviation may also be estimated for systematic errors. This will estimate where the correct value is likely to lie. Uncertainties are usually added in quadrature this means they are combined as the square root of the sum of squares of the errors. For example if an uncertainty of 3% is expected in the radiotherapy treatment planning (RTP) computer calculations and 3% in the treatment delivery process the combined quadrature uncertainty is approximately 4.2%.

An example of random and systematic uncertainties is the tracking of prostate motion using fiducial markers with an image provided by an electronic portal imaging device (EPID). A random error may be observed due to setup error (typically 3 mm); a systematic error of several millimeters may be observed occasionally due to differences in rectal filling.

11.2.2 Accuracy

What accuracy should we strive for in radiotherapy? While the maximum dose level in the tumor is required for dose reporting consistency, the minimum dose level to the tumor is critical in terms of tumor control. The precision of dose delivery required depends on the steepness of the dose versus the tumor control probability (TCP) response curve. Note that the dose-response curve for tumors is not well characterized and may be quite steep or shallow, depending on tumor type. Therefore the optimum patient treatment accuracy depends on the steepness of the curve as shown in figure 11.1. QA of machine calibration, machine alignment, and dose calculation within the RTP computer are all essential to ensure that the prescribed dose is delivered with the required accuracy. These are the QA issues that are generally addressed by medical physicists in the radiotherapy clinic.

Evidence suggests that a 7% to 10% change in dose to the target volume may result in significant change in the probability of tumor control (ICRU 1976). Other papers indicate that dose variations with accuracy as low as 5% and as high as 15% may be critical (Brahme 1984, Chappell and Fowler 1994). Mijnheer et al. (1987) using normal tissue control probabilities (NTCPs) and Brahme et al. (1988) using TCPs quote required uncertainties of about 7% (2σ). Considering errors are combined in quadrature, this may be a practical aim within the entire radiotherapy treatment process.

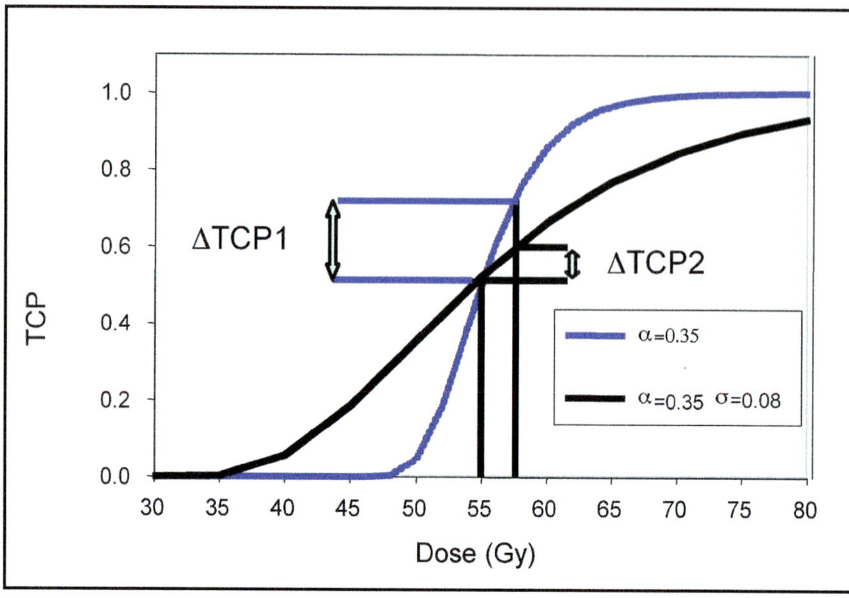

Figure 11.1: Dose versus tumor control probability (TCP) curve showing the same uncertainty in dose can lead to more uncertainty in TCP if the TCP curve is steeper. Note ΔTCP1 is larger than ΔTCP2 with the same dose uncertainty.

11.2.3 Radiation Accidents and Risk Assessments

Most jurisdictions require reporting as a radiation accident under the guideline that 10% difference has occurred to the total fractionated dose prescription. Some jurisdictions also include a 20% difference in a single fraction as reportable.

A number of radiation accidents have been reported. Reviews about what can be learned from these accidents can be found in the literature (e.g., IAEA 2000, ICRP 2001, Cossett 2002, Dixon and O'Sullivan 2003). Accidents usually involve the incorrect use of radiotherapy or ancillary equipment. These events highlight that QA recommendations should be implemented and adhered to in an attempt to minimize the risk. Some of the major categories of accidents include: calculation errors, calibration errors, inadequate review of radiotherapy treatment plans, inadequate review of transcribed patient treatment charts, and error in anatomical area to be treated.

Errors may be the result of an individual's mistake (human error), but flaws in the checking system or nonadherence to some part of the QA system are often highlighted when accidents have been retrospectively analyzed.

With increasing complexity of equipment and more QA tests being required as each new equipment option becomes available, it is likely that more quantitative risk assessment will drive the requirement for QA tests. In industry a risk probability number (RPN) is often calculated:

$$RPN = O \times S \times D\,, \tag{11.1}$$

where O is the percentage probability that a specific cause will result in a failure, S is the severity of the effect, and D is the probability the cause will go undetected. Finding quantitative data to plug into this equation for radiotherapy equipment is not straightforward.

11.2.4 Responsible Professional Groups

It is important to note that QA of systems and how various groups interact within these systems are as important as individual QA equipment tests. Several professional groups are involved in the overall QA measures required to treat the patient correctly. Although there may be some variation depending on the expertise available, in general, the following definitions apply.

The *radiation oncologist* is a medical specialist who ensures the patient treatment is optimal. This specialist's role in the planning and treatment process involves ensuring that the treatment prescription is optimum for the stage and grade of tumor. This includes optimizing dose, fractionation, and duration. The radiation oncologist also ensures that the tumor is covered adequately by the radiation field while radiation-sensitive structures are avoided (target and normal tissue volumes are defined with appropriate margins). Follow-up of patient response and further treatment or referral for other treatment (e.g., chemotherapy) is also an integral part of the radiation oncologist's role.

The *radiation dosimetrist* is responsible for planning the field setup required to deliver the prescribed dose. Note that outside North America the radiation dosimetrist's planning role is covered by a planning *radiation therapist* (historical term: planning radiographer). The treatment radiation therapists (therapy radiographers) ensure that the dose is delivered reproducibly to the patient during each fraction of treatment. In some centers the therapy and planning radiation therapists rotate through planning; in other clinics they specialize in one or the other.

The *radiation oncology medical physicist* (ROMP) is responsible for ensuring that the prescribed dose is accurately given using adequate calibration, machine checks, and checks of RTP calculations and processes. Frequent meetings involving these three expert groups to discuss QA issues are vital. The aim of these meetings is to pursue total quality improvement by reviewing, discussing, and updating the systems in place. The roles of other professional groups, such as linac engineers, should not be overlooked. Linacs are extremely complex machines including electronics, computers, and mechanical devices. Timely communication between linac engineers and ROMPs is essential if any adjustments have potentially affected beam properties.

11.2.5 Quality Audits and Quality Standards

Department QA procedures may also be helped by a *quality audit*. This is a systematic and independent examination to determine if activities and results comply with planned arrangements and if the arrangements are implemented effectively and are suitable to achieve the desired goals (Thwaites et al.

2003). It is important that these audits are performed by independent personnel and are judged against a predetermined set of standards.

Quality standards are a set of accepted criteria against which the quality of an activity in question can be assessed. Various standards recommended for radiotherapy have been agreed upon (e.g., WHO 1988, AAPM 1994, ESTRO 1995). For example, a quality standard might require that "radiation therapy equipment is performing to clearly specified parameters designed to ensure accurate clinical treatment." To meet this standard, there may need to be a detailed criterion, rationale, and indicators listed which encompass the processes of equipment acquisition, commissioning, and QA tests and their frequency.

11.2.6 Commissioning a Linac

Commissioning is the general term used to describe the process whereby a linac is installed and various tests are carried out to take the linac to a clinical treatment-ready state. This process involves a number of steps including, room preparation, installation, acceptance testing, calibration, beam data collections and collation for manual and RTP computer data entry, training, and clinical commissioning. Short descriptions of the processes involved are included in table 11.1. Acceptance test documents supplied by the manufacturer are particularly useful for newly developed equipment, however customer designed tests as specified in the tender document should also be included.

11.3 Equipment That Requires QA

A number of QA documents include details of QA tests for medical linear accelerator equipment (e.g., IEC 1989a,b, IPSM 1988, AAPM 1984, 1994, NCRP 1989, WHO 1988, Constantinou 1993, ESTRO 1995, ACPSEM 1997). Most of these documents were written before the advent of EPIDs and complex record and verify (R&V) systems, so most current documents focus on practical ways to test geometric and radiation tolerances of linacs.

We have identified five categories of devices requiring scheduled rigorous QA programs within a radiotherapy department:

i) Linac geometric and MV beam QA tests.
ii) Linac ancillary high technology devices (Klein et al. 1996), e.g., MLC, EPID.
iii) Record and verify (R&V) computer.
iv) RTP computers (software and hardware, data transfer to R&V).
v) CT simulators and classical simulators.

As shown in figure 11.2, linacs may consist of the conventional MV portal with an EPID flat-panel detector (FPD), and also a linac-mounted imaging device (LMID) consisting of an FPD and a kV tube. Varian calls this device

Table 11.1: Commissioning process for linacs from installation to clinical acceptance

Process Name	*Process Description*	*Comments*
Room preparation	Ensure all services are acceptable and building site is handed over as a safe working environment.	Ensure all services are ready for use including electric, water, and pneumatic air.
Install	Rig in linac in the component breakdowns, base-frame, modulator, linac stand, and gantry. Company linac engineer connects and tests all linac components.	Ensure all OHS and contractor training and certificates are appropriate.
Acceptance	Fill detailed worksheet tests including geometry such as isocenter and beam reproducibility tests.	Radiation survey of the room and surrounding areas. Undertake all acceptance test documents as specified by vendor and include extra customer tests as specified in tender document. Check beam matching if specified. If linac beam matching can be achieved, this makes transfer of patients to different machines much more straightforward.
Calibration	Absolute calibration.	Calibrate all beam energies following calibration protocol. Invite physicist from another center to calibrate same beams independently using his/her own independent equipment. Send TL dosimeter or similar dosimeters to another center for dose comparison.

Table 11.1: (Continued)

Process Name	*Process Description*	*Comments*
Commissioning beam data for RTP	Beam data collected in water phantom is transferred for RTP purposes.	Beam data collected in water phantom is transferred to RTP computer for beam modeling to commence. Beam data are also entered into MU check programs. Paper copy beam data files may also be collated and checked. Ensure all beam data used for RTP purposes are independently checked.
Clinical commissioning	Getting staff ready for clinical treatment of patients.	Staff training for use of equipment. Preparation of treatment protocols. Test setups with anthropomorphic phantom preferably with some dosimetry tests included.

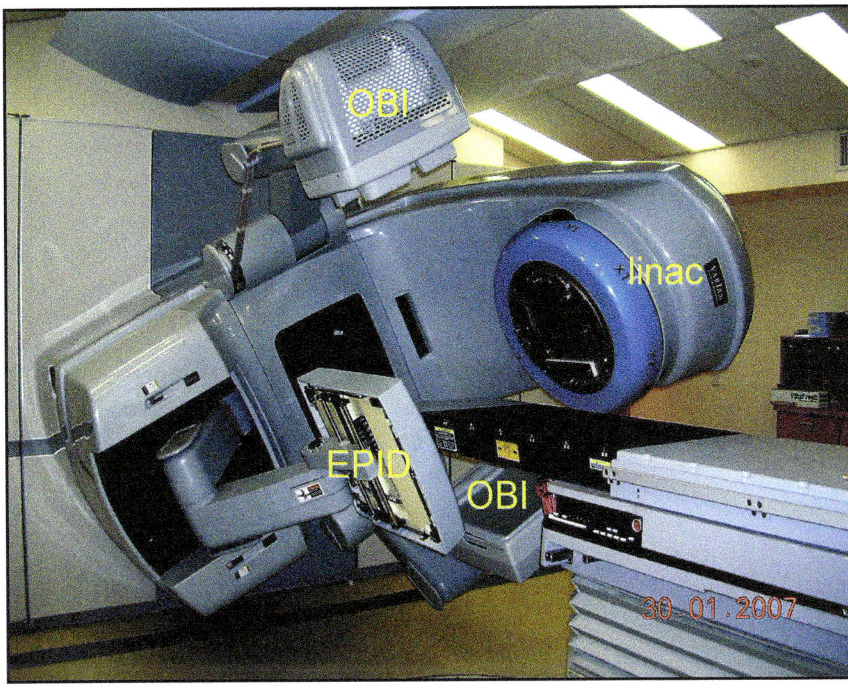

Figure 11.2: Showing the complexity of a modern linac with ancillary equipment extended including: MV treatment head (right); EPID FPD (left); OBI™ kV x-ray tube (top); and OBI FPD (bottom). (Image of Varian Trilogy located at Peter MacCallum Cancer Institute, Melbourne, Australia.)

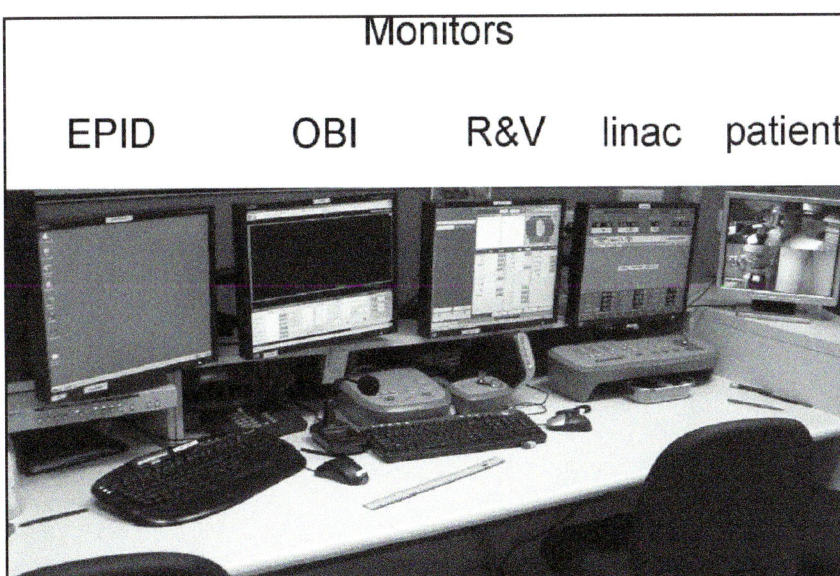

Figure 11.3: Showing the complexity of a modern linac control area with a myriad of VDU screens. From left to right: EPID view monitor, OBI view monitor. R&V and MLC leaf control monitor, linac control monitor, patient view monitor.

an On-Board Imager™ (OBI). The control area located outside the bunker, as shown in figure 11.3, consists of a myriad number of monitors including the R&V monitor, video display unit for patient monitoring, EPID, and/or OBI monitors.

While the software complexity of the RTP computer has increased due to the use of image file transfer from CT and paperless transfer of data to the R&V computer, the RTP computer consists of a set of terminal screens with PC- or Unix-powered workstations linked to a server (see chapter six). There is still the occasional digitizer used for block and electron applicator shape input. Color printers are also used for hardcopy printout of plan data for treatment machines and isodose curves on CT outlines, but these devices are being used less and less as departments move to paperless processes. Note that most of the imaging and treatment parameter data can now be transferred from the RTP computer and viewed directly at the R&V or imaging workstations of the linac.

11.3.1 Linac Geometric and MV Beam QA

Linac Geometric QA Tests

Geometric orientations are defined with reference to figure 11.4a. The axis that runs on a line through isocenter and parallel to the linac waveguide from gun (G) to target (T) in the gantry is referred to as the beam radial axis, and the line that runs perpendicular to this axis through isocenter across the room as defined by the lateral wall lasers is known as the transverse axis. These points may be given an orientation reference A and B.

The *mechanical isocenter* is the point about which the gantry rotates at a distance, usually 100 cm SSD. It is also the point along the axis of rotation about which the collimators rotate and the patient support assembly couch rotates.

Mechanical isocenter in the radial axis for the patient couch is found by gradually moving the linac with a front pointer in place and rotating the couch. The front pointer inscribes a smaller and smaller semicircle until isocenter is achieved. Mechanical isocenter in the vertical axis is achieved by rotating the gantry around a sharp pointer mounted on and protruding from the treatment couch, with its pointer placed at isocenter (i.e., the couch pointer). The front pointer should inscribe a very small arc as it rotates around isocenter. The reference radius of either arc should be less than 1 mm.

During commissioning a very precise front pointer is used to define isocenter. This device is locked in place by the jaws. A front pointer that mounts easily into the block tray slot is used during subsequent routine QA checks.

Once isocenter is obtained, a plumb bob can be used to mark the vertical axis from the ceiling through isocenter to a point on the floor. The position that the plumb bob is held to the ceiling is the position from where the overhead laser should emit its light. When the gantry is set at zero, the crosshair defined by the light field should align directly with the couch pointer and a mark on the floor that defines the center of rotation for the couch.

Ensure that the linac is rotated to an exactly vertical position by placing a level on a machined surface to ascertain the level (horizontal) position. Linac

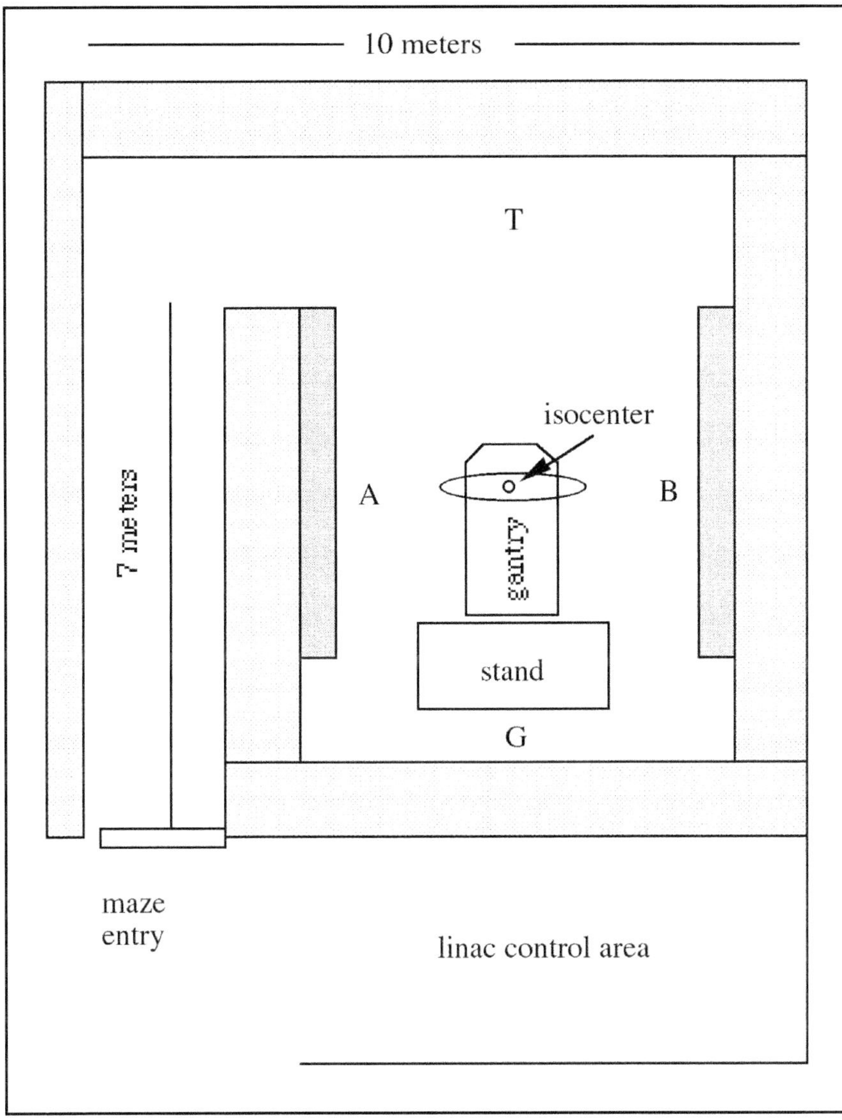

(a)

Figure 11.4: Linac orientation conventions and geometric check devices. (a) Axis orientation used to define profile dimensions and laser orientations in linac bunkers, gun, target, A and B.

(b)

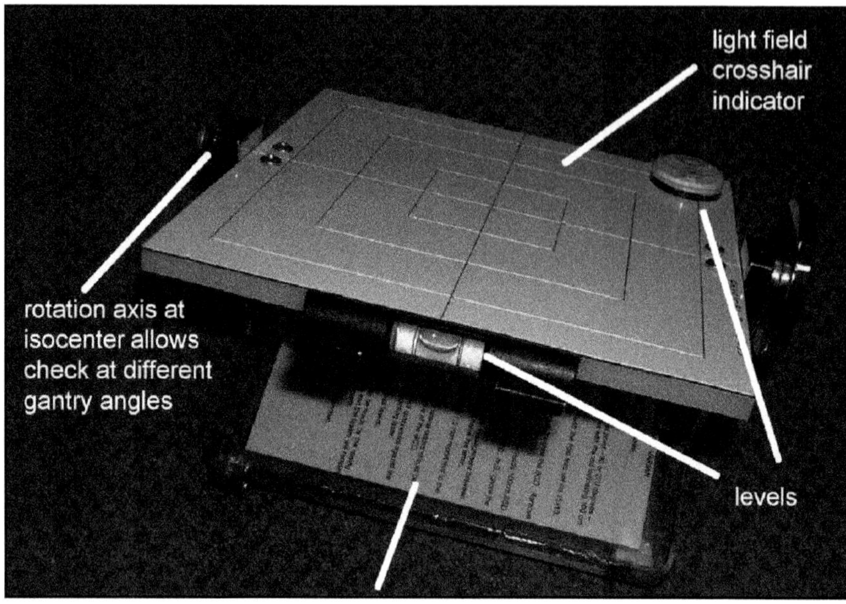

(c)

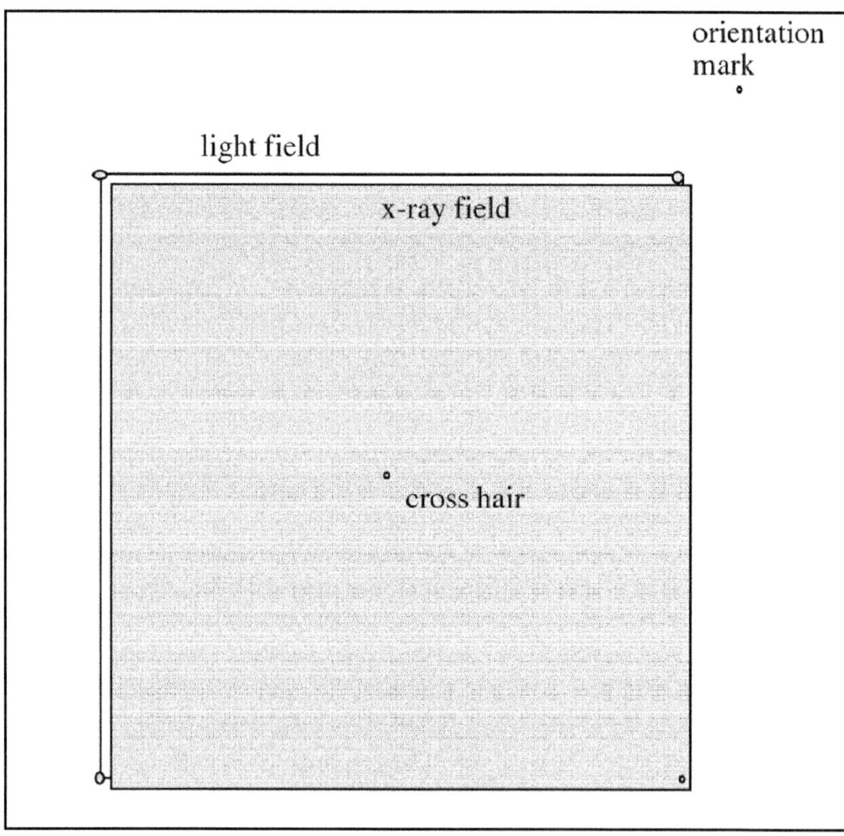

Figure 11.4: (Continued.) Linac orientation conventions and geometric check devices. (b) Linac light field check device: this one is called a flipper because it can be used at a range of gantry angles. (c) X-ray field light field coincidence.

covers generally have to be removed to find a flat machined surface that is not affected by paint finish. The assumption with this test is that jaw mounts are exactly perpendicular to the beam central axis. If this is incorrect, it will show up during subsequent radiation alignment tests.

The *radiation isocenter* is defined as the smallest sphere through which the axes of radiation beams pass in all conditions (IPSM 1988). One way of finding this position is by using spoke shots. To achieve this in the gantry rotation plane an x-ray film is placed upright on the couch, one set of collimators are closed to a narrow slit, and the gantry is rotated to various angles. Four suitable gantry angles are 360, 185, 90, 265 degrees. The film is exposed at different angles, and the spokes of film blackening should coincide. If the spoke lines are too wide, diagonals from entry to exit positions of the spokes can be drawn to get a more exact position of isocenter, i.e., where the diagonals cross.

The process can be repeated with the film placed flat on the couch and the collimators rotated to different angles; once again the spokes should be coincident at the center. This test shows whether the electron beam, the target, and the jaw are aligned correctly.

The tolerance for radiation isocenter is generally 2 mm. Note that this is the more important isocenter, as this is the point to which the tumor volume is referred (IEC 1989b). Mechanical and radiation isocenter should be close to coincident (within about 1 mm on modern linacs).

X-ray versus *optics* alignment tests are outlined as follows. The collimator crosshairs are used as a visual aid to show the beam central axis entry position. The cross hair is generally projected onto the patient through a Mylar™ sheet with crosshairs engraved. Crosshair projection should rotate about isocenter with less than 1 mm drift when the collimators are rotated. Graph paper is placed on the couch top, and the center of the crosshair position is marked every 45 degrees during collimator rotation. The same thing can be done with the front pointer firmly in place. If the front pointer drifts, this may mean the front pointer is bent or that isocenter is not well defined.

If the crosshairs drift but not the front pointer, this may mean the crosshair defining either the decal or the light/mirror assembly is out of adjustment. While the graph paper is present, the edges of the light field can be marked and the collimators rotated; any movement from the marked positions when the collimators are rotated by 90 degree-increments will show jaw alignment asymmetry. If the isocenter drifts with gantry rotation, this generally points to gantry give in the gantry bearing assembly. If isocenter drifts when the table is rotated, this generally points to the linac not being physically located correctly in respect to the table.

The light field size can be checked with graph paper or any etched scale, such as the device shown in figure 11.4b. To routinely test for light field x-ray beam coincidence, a film is placed at the surface of a phantom, and the edges of the light field marked. A schematic representation of the set up is shown in figure 11.4c. This can be achieved by a pin perforating through the film wrap, heavy scoring by a pen, or using lead markers on the edge of the field. Marking the position of the crosshairs on the film with a perforation and one quadrant (e.g., GTR = gun transverse right quadrant) gives a reference position; if any asymmetry is found, the film is then exposed. Because there is a rapid dose fall-off at the surface in the penumbra, an optical assessment of beam edge position with the naked eye is generally adequate, although a densitometer may be employed for further accuracy. Checks at several field sizes are necessary at commissioning. Generally, checks at monthly intervals of one or two field sizes are sufficient after commissioning.

Four room-mounted *lasers* shine their crosshairs through isocenter. These require scheduled alignment checks. With reference to figure 11.4a, these lasers are referred to here as the left transverse, B; right transverse, A; sagittal; and overhead lasers.

Another pair of lasers may also be employed to shine a crosshair that rotates as the gantry rotates. These are known as the back pointer lasers. These lasers intersect to create a crosshair that represents the beam central axis exit position from the patient. This is particularly useful for ensuring that opposed fields will enter and exit at the same location.

The left and right lasers need to be mounted on the walls perpendicular to the radial axis and at the correct height (the same as isocenter) so that they shine through isocenter without diverging off. The appropriate vertical position of a wall laser is located by placing a water tube from the bottom of the linac front pointer to the laser wall position, then marking the point at which the water levels match. The horizontal position is found by shining a correctly aligned crosshair at the wall with the collimator set at 0 degrees and the gantry set at 90 degrees.

Once the laser is mounted, small positional adjustments may be necessary. Angular adjustments in both planes on the laser are also provided. Both lasers should shine through isocenter as set by the front pointer, and both lasers should shine directly into the linac crosshairs when the machine gantry is rotated to 90 degrees and 270 degrees, respectively.

If the lasers are firmly mounted to a metal plate on the wall and are protected from bumping, realignment is rarely required. Often too little attention is paid to the quality of the mounts, particularly in simulator rooms where the luxury of extremely solid concrete walls (as provided in linac bunkers) is not provided. The alignment of some lasers may change after warm up; hence the stability of laser alignment soon after switch on and at a later stage should be tested.

The machine *gantry angles* are normally defined by an analog scale (which generally uses a gravity pendulum). Modern linacs also have a digital display. A 1-degree tolerance is regarded as acceptable throughout the 360-degree gantry rotation. The gantry is generally set to 0, 90, 270, and 360 degrees using a spirit level placed on a flat surface plate near the jaws, and the digital readout potentiometers are adjusted for span and gain to meet the tolerance.

The collimator angles are set by turning the gantry to 90 degrees, i.e., on its side, and using a spirit level placed on the couch. A projection of the light field on the wall can then be compared with the light projection from the top of the spirit level. The top jaw is closed so that a small line of light is visible between the jaw and the top of the spirit level. The jaw is rotated until the top and bottom lines defining the light projection are parallel. The process is repeated for each subsequent jaw to define the 0-, 90-, and 270-degree collimator rotation points.

Make sure that the couch's 0-degree position is parallel to the jaws when the collimators are set to 0 degrees. With the gantry at 0 degrees, close the jaws down to a long rectangular field until the light field and the light projection of the spine of the couch can be seen to be parallel.

Couch rotation can be checked with reference to the gantry and the collimators set to zero, with graph paper aligned on the couch to the edges of the light field. When the couch is rotated to display 90 and 270 degrees, the graph paper lines and the light field edge should coincide.

Couch longitudinal travel and lateral travel can be checked by comparing the digital or analog readout of movements with movements measured using graph paper or a ruler placed on the treatment couch as referenced to the light field crosshair. There are generally analog scales for these also, but the digital scales are the most critical as readouts from these are commonly used as input for patient treatment setup verification computers. Note that as the patient may be placed in slightly different longitudinal and lateral positions on the couch, it is not unusual for digital readout of these positions to vary slightly from day to day. This parameter should vary less from day to day when a patient head cast is used in a fixed couch position.

The couch vertical movement should coincide at zero with the front pointer and at the appropriate offsets when the couch is raised (e.g., to 90 cm SSD) and lowered (e.g., 130 cm SSD). A ruler placed beyond the front pointer is generally adequate to test this if the front pointer is not sufficiently long. The optical distance indicator (ODI) displayed should also agree at these positions. Note that if there is a discrepancy between ODI and lasers or couch readout, always check both by using the physical front pointer.

The integrity of the couch table rigidity and movement can be checked annually by placing weights on the couch as discussed in detail in IEC (1989a). The test involves taking a radiographic film at two table heights 20 cm apart, e.g., at 90 and 110 cm SSD. Drawing diagonals across the film should give an exact match, provided the table travel is true. The test is repeated with the table loaded to 30 kg over 100 cm of its length and then 125 kg over 200 cm of its length. The table should be tested at two angles of rotation, e.g., 0 and 90 degrees, respectively.

Linac MV Beam QA Tests

The *beam flatness* and *symmetry* are carefully aligned during commissioning and periodically checked. It is the symmetry that tends to require periodic beam steering to ensure adherence to the tolerances outlined in IEC (1989a). A photograph of a transverse section through a linac bending magnet in a linac is shown in figure 11.5, this reiterates just how sensitive the beam steering control needs to be to ensure the electrons stay precisely on their trajectory path. The routine checks of beam symmetry and flatness are made easier by using diode or ion chamber arrays that can assess beam symmetry without the time constraint of a full water tank setup (see figure 11.6a).

Usually the *percentage depth dose* properties of each individual linac are unique. The constancy of the depth dose should be checked monthly. Spot checks in solid phantoms are a practical way to streamline the time taken for this QA test. If the depth dose changes, a small degree of freedom for

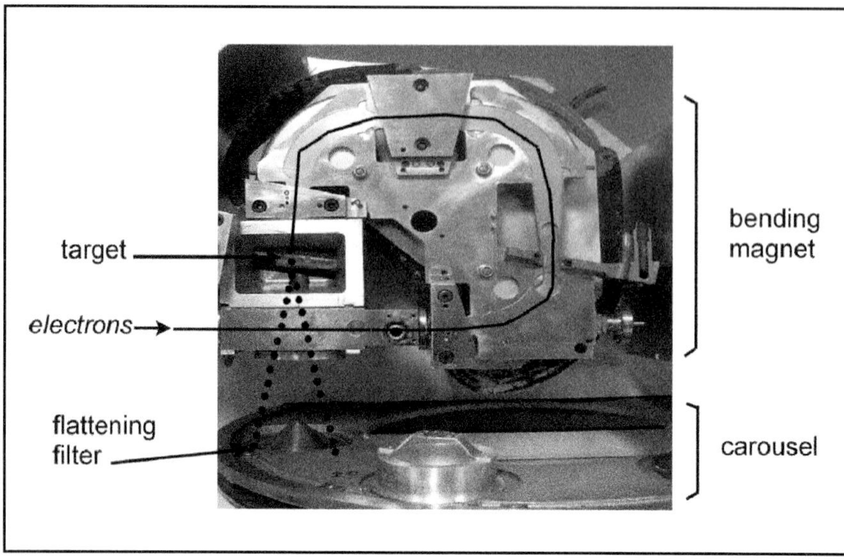

Figure 11.5: Transverse cutaway section of a linac bending magnet showing the complexity of beam steering required to maintain symmetry and output. (Image courtesy of D. Pinchon and G. Sorrel.) SEE COLOR PLATE 50.

linac tuning is available for this parameter to be brought back into specification. However, questions should be carefully asked about what is likely to have caused the change in this parameter. The dosimeter should be checked

(a)

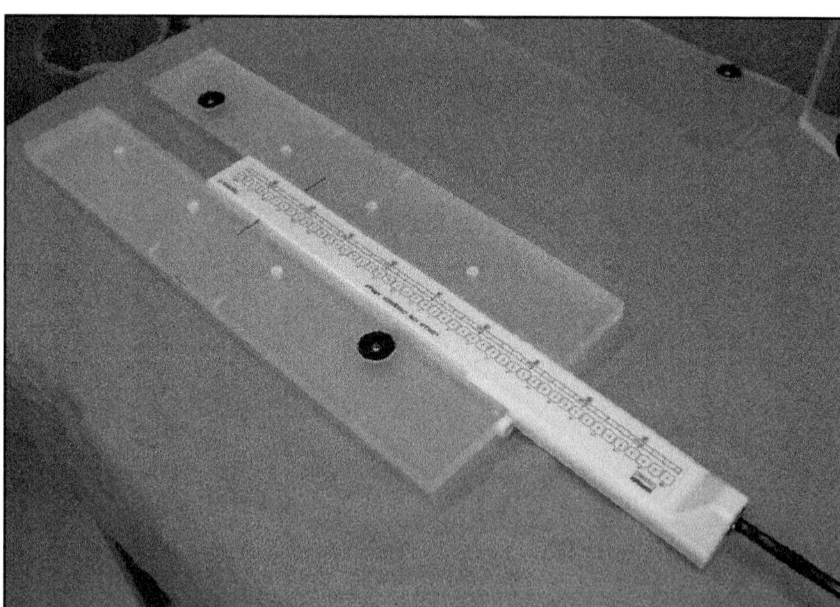

Figure 11.6: Routine MV beam check devices: (a) Symmetry checking dosimetry array.

with a second dosimeter and the equipment and setup (e.g., SSD) should also be checked prior to adjustment.

Block transmission factors may be measured by placing various layers of lead on the block tray with a detector placed at isocenter. A narrow field is usually employed (e.g., 5 × 5 cm^2) to minimize collimator and in-phantom scatter effects. Once again a build-up cap may be employed as measurements in water will include some backscatter. It is important that block thickness is at least 5 half-value layers (HVLs). This reduces the block transmission to 3.125%. However, measured dose beneath blocks may be higher than this,

(b)

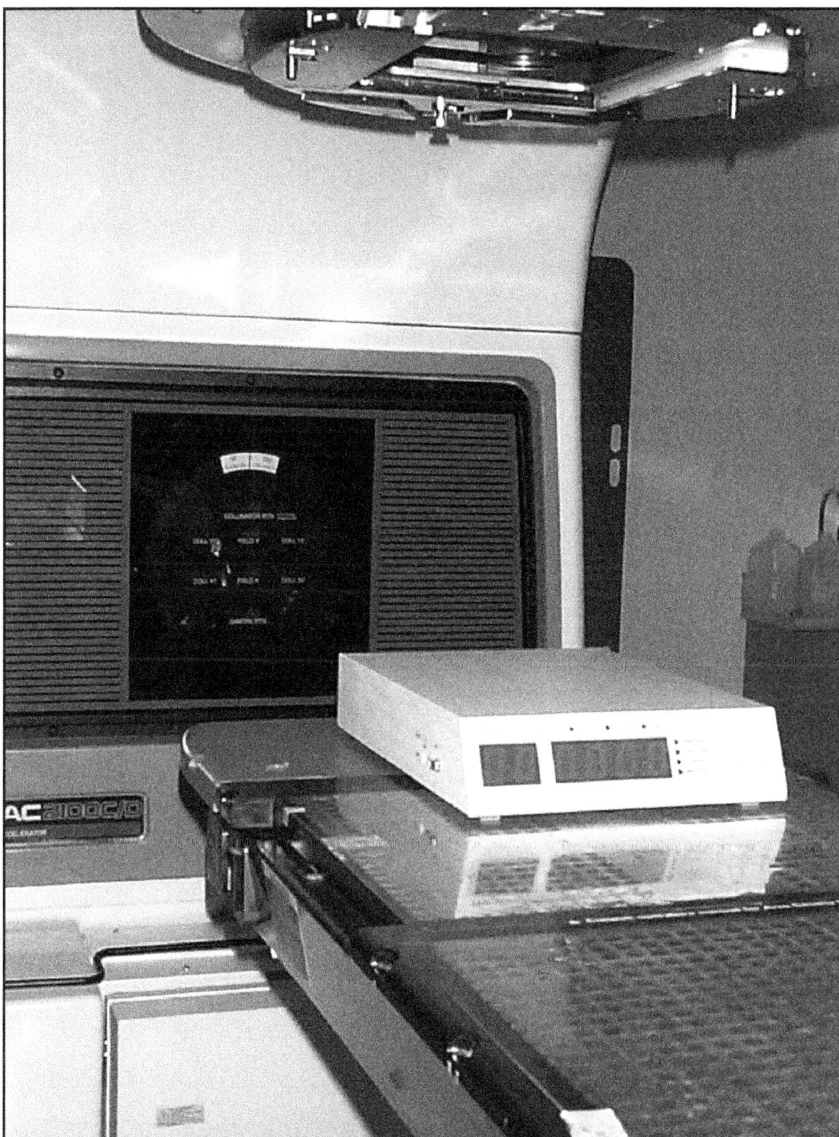

Figure 11.6: (Continued) Routine MV beam check devices: (b) Daily calibration constancy check ion chamber device.

particularly in big fields where in-phantom scatter contributes significantly to the dose. Examples of block transmission factors are given in table 4.1d.

Absolute calibration checks should be undertaken with a field instrument that has been cross calibrated with the department's secondary standard chamber. The secondary standard chamber should be recalibrated by the standards laboratory at least every 2 years. The absolute calibration from first principles in a water tank should be repeated annually. Calibration checks using a standard conversion factor should be checked daily and monthly in a check phantom. Common check phantoms consist of acrylic, polystyrene, solid water, or a small water tank. The consistency of ion chambers should be checked quarterly using a constancy check source. (See chapter eight.) There are also specially designed linac dose consistency check ion chamber devices designed for fast assessment of dose on a daily basis. A typical device is shown in figure 11.6b.

During commissioning an independent calibration check should be undertaken by a physicist visiting from an independent institution. The visiting physicist should measure and calculate the dose independently, and agreement within ±2% should be expected. A separate independent institute should also send thermoluminescent (TL) dosimeters, and these should be exposed and the dose readout quoted by this institute. The agreement here should be within ±3%. Both safeguards are recommended to ensure that a major calculation or measurement error does not occur. The protocol calculations consist of several steps where incorrect data or incorrect calculations could inadvertently be made.

Initial baseline tests at commissioning and future periodic tests should be recorded, signed, and dated on data sheets provided in the clinic. These results should be periodically evaluated to determine the performance level of the equipment.

The exact periodicity of linac geometric and beam QA checks should be tailored to the institute's own needs and capabilities. The important thing is to choose a realistic periodic check schedule that is adhered to rather than an overly optimistic check schedule that is not. In tables 11.2 and 11.3 we group the tests in sequence of importance including tolerances and give a realistic practical guide as to how often these methods might be applied in the clinic. The frequency of these tests and tolerances is based on the recommendations of an AAPM Task Group 40 (TG 40) (1994).

11.3.2 Linac Ancillary High-Technology Devices

Multileaf Collimators

The properties of new devices such as MLCs are dealt with in detail in chapters four and seven. High-technology devices began to proliferate just after the advent of the TG 40 report (AAPM 1994); hence there is one excellent AAPM document dealing with MLC properties (report of AAPM TG 50) (AAPM 2001a) but not yet any AAPM TG article specifically covering recommended QA regimens for these devices. It is understood that as this book goes to press AAPM TG 100 has been formed to update the TG 40 report and to include high-technology devices such as MLCs and EPIDs in this QA document.

Table 11.2: Suggested frequency of linac QA tests (daily and monthly) (Adapted from AAPM 1994)

Frequency	*Procedure Type*	*Procedure Test*	*Tolerance*
Daily	dosimetry	x-ray output constancy	3%
		electron output const.	3%
	mechanical	lasers	2 mm
		ODI	2 mm
	radiation safety	door interlock	functional
		audiovisual monitor	functional
Monthly	dosimetry	x-ray output constancy	2%
		electron output const.	2%
		back-up monitor const.	2%
		PDD/TAR constancy	2%
		beam flatness	per specification
		beam symmetry	3%
	safety interlocks	emergency off switch	functional
		wedge interlocks	functional
	mechanical checks	light/radiation co-inc.	2 mm
		gantry/coll. angle	1 degree
		wedge position	2 mm
		tray position	2 mm
		applicator position	2 mm
		field size indicators	2 mm
		crosshair centering	2 mm diameter
		treatment couch indic.	2 mm/1 degree
		latching of wedges	functional
		jaw symmetry	2 mm
		field light intensity	functional

There are, however, several other documents which outline suggested QA procedures for MLCs (e.g., Klein et al. 1996, Huq et al. 1995, Boyer et al. 1992). These tests generally focus on using film or EPIDs to test leaf position reproducibility in a series of moves, e.g., the picket fence test pattern (Saw et al. 2001). Vieira et al. (2006) have found MLC position reproducibility was 0.1 mm (1σ) for Varian MLCs and 0.2 mm (1σ) for Siemens and Elekta MLCs, although this implies a random distribution. Individual checks on the MLC in question should be carried out as large outliers of leaf position, if found, could be problematic. A thorough understanding of leaf position reproducibility is even more critical for IMRT treatments and any limitations should be taken into account when segment optimization parameters are defined on the RTP computer (Xia and Verhey 2001).

Many MLC QA tests are interlinked with IMRT QA tests because as IMRT was developed, the precision of the MLC device became more critical. This is particularly the case for *dynamic MLC delivery*. The advantage of dynamic sliding window MLC delivery of IMRT is that a smoothly varying intensity profile can be delivered. The delivery is rather complex, however,

Table 11.3: Suggested frequency of linac QA tests (annual) (Adapted from AAPM 1994)

Frequency	*Procedure Type*	*Procedure Test*	*Tolerance*
Annual	dosimetry	x-ray output calibration	2%
		field size dependence	2%
		constancy electron appl.	2%
		central axis constancy	2%
		off-axis constancy	2%
		transmission constancy	2%
		wedge trans. constancy	2%
		monitor chamber linearity	1%
		x-ray output constancy	1%
		electron constancy	2%
		arc mode	per specification
	safety interlocks	follow test procedures	functional
	mechanical checks	collimator rotation	2 mm diameter
		gantry rotat. isocenter	2 mm diameter
		couch rotat. isocenter	2 mm diameter
		coincidence couch axes	2 mm diameter
		coincidence rad. mech	2 mm diameter
		table top sag	2 mm
		vertical travel of table	2 mm

as it involves leaf speed and dose rate modulation. This requires precise control of dose rate and leaf position. LoSasso et al. (1998, 2001) have calculated the effect of leaf offset on dose differences when this technique is employed. Some other material dealing with MLC QA checks as part of pre-patient IMRT QA check is outlined in section 11.4 on IMRT QA.

Electronic Portal Imaging Devices

Images from EPIDs are used in assessing geometric misalignments to bony anatomy or implanted fiducial markers. Maintaining good image quality from these devices is essential to ascertain the best images possible. Image QA tests for these devices have been established as with other imaging modalities around *spatial resolution* and *contrast resolution*. The resolution achievable is often related to the *signal-to-noise ratio* (SNR) that can be achieved by the system. More detailed mathematical analysis of the general principles of SNR and how it is related to x-ray quanta and detector quantum efficiency is given in the AAPM TG 58 report (AAPM 2001b).

Image quality test phantoms generally consist of a line pair per millimeter (lp/mm) device for viewing spatial resolution and another section for analyzing contrast (see figures 11.7 and 11.8). Amorphous silicon (a-Si) EPIDs of the Varian linacs have been reported as being relatively stable over time by Menon and Slodoba (2004). They also introduce a number of

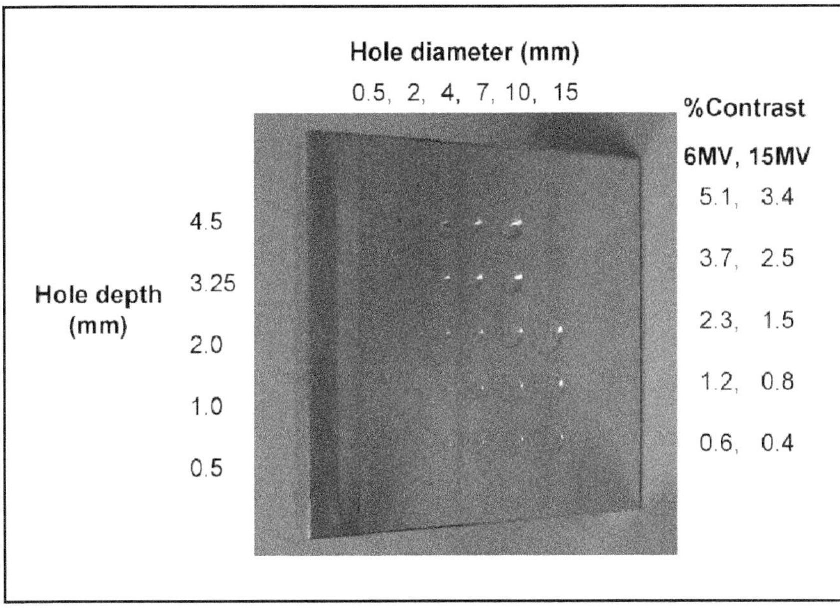

(a)

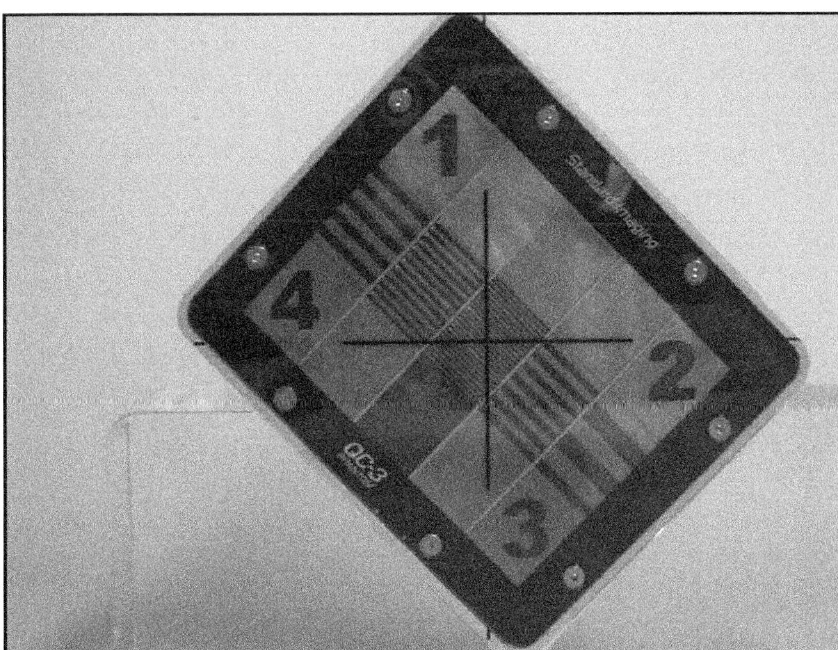

(b)

Figure 11.7: Image quality test phantoms for EPIDs: (a) The Las Vegas phantom, an EPID spatial and contrast resolution image quality test phantom; (b) Spatial resolution image quality test phantom. (Type: QC-3 Standard Imaging Inc., Middleton, WI.)

simple equations which are useful for quantifying and tracking EPID performance over time. These include:

i) Dark and flood field images.
ii) Contrast resolution.
iii) Spatial resolution.
iv) Contrast-to-noise ratio.
v) Dose rate stability.

i) Equations 11.2 to 11.6 follow the formalism of Menon and Slodoba (2004). Dark and flood field calibration images are usually stored in a data base and correct for stationary pixel response non-uniformities, such that for each pixel in the clinical object image taken is corrected by

$$CI = \left[\frac{OI - DF}{FF}\right] \times k_{FF}, \tag{11.2}$$

where CI is the corrected object image, OI is the uncorrected object image, and DF and FF are the dark and flood field images; k_{FF} is the mean of the flood field image values.

ii) Contrast resolution is a measure of object recognition from its background, and this can be quantified as

(a)

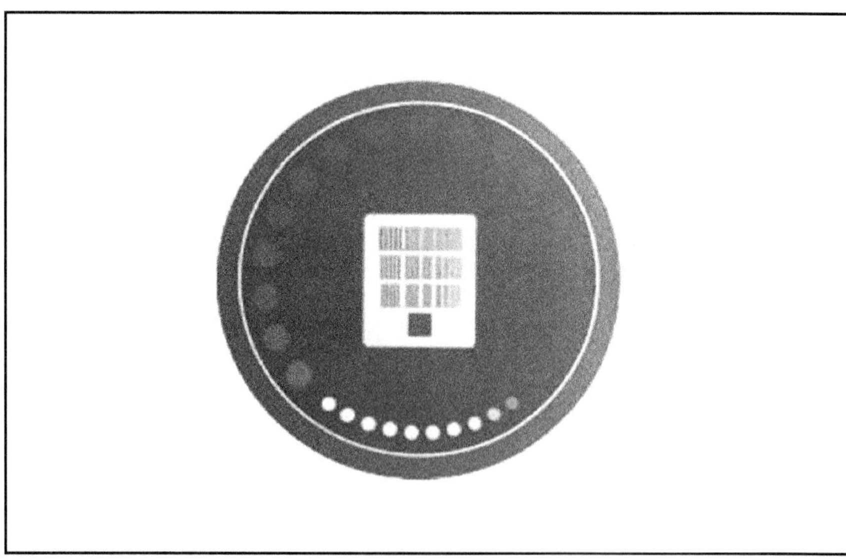

Figure 11.8: Image quality test phantoms for simulators and CT simulators: (a) Resolution phantom for classical simulators (Type: Leeds Test Object). (Reprinted with permission from Leeds Test Objects, Ltd., UK.)

(b)

Figure 11.8: (Continued) Image quality test phantoms for simulators and CT simulators: (b) Resolution phantom for CT simulators (Type: Catphan 500/600). (Reprinted with permission from The Phantom Laboratory, Salem, NY.)

$$C = \left[\frac{I_0}{I_B} - 1\right] \times 100, \tag{11.3}$$

where C is the quantified percentage contrast resolution, I_0 is the intensity of an object, and I_B is the intensity of background.

A more usual test of contrast and spatial resolution for EPIDs is to use a phantom with a number of holes of different diameter and depth. For example, the Las Vegas resolution phantom (Klein et al. 1996) consists of a number of different diameter holes milled to different depths as shown in figure 11.8. Image quality performance is tracked over time by observing how many circles can be visualized by the observer in the imaged phantom.

iii) Spatial resolution may be quantified in imaging systems using the relative modulation transfer function (RMTF) where

$$RMTF = \frac{MTF(f)}{MTF(f_i)}, \tag{11.4}$$

where $MTF(f)$ is the modulation transfer function for the frequency measured and $MTF(f_i)$ is the lowest frequency measured. The phantom Menon and Slodoba (2004) used extensively was the QC-3 phantom shown in figure 11.7b. By setting f_i = 0.1 lp/mm, they found the critical frequency RMTF = 50% lp/mm for a 6 MV beam using a Varian a-Si EPID to be 0.435±0.005 lp/mm.

iv) Contrast-to-noise ratio (CNR) is a measure of how the image deteriorates if the background noise increases. This is given by

$$CNR = \frac{P_b - P_d}{\sigma}, \tag{11.5}$$

where P_b and P_d are the average pixel values of the brightest and darkest irradiated regions and σ is the random image noise calculated by sampling several uniform regions.

v) Dose rate stability can be checked by comparing

$$\eta = \left[\frac{P_d}{P_b} - 1\right] \times 100. \tag{11.6}$$

at different dose rates for a standard (e.g., 20 × 20 cm²) field image collected at different dose rates.

Many EPID QA tests are interlinked with IMRT QA tests because the EPID has been found to be useful as an IMRT dosimetry check device. Further details about this are outlined in section 11.4.

Linac Mounted Imaging Devices

While mounting an x-ray tube offset to the linac treatment head was first reported by Biggs et al. (1985), the commercial proliferation of linac-mounted imaging devices (LMIDs) is relatively recent. The combined development of flat-panel detectors (FPDs), cone beam CT reconstruction, and precision robotic arms seem to have been the three technology improvements that have led to the recent proliferation of these devices.

The first vendor to offer a commercial LMID with cone beam CT was Elekta. The properties of this device are reported by Jaffray et al. (1999). Varian soon followed with a commercial product, which they named an On-board Imager™ (OBI) (Huntzinger et al. 2006). Note that Siemens uses MV cone beam reconstruction by exploiting the resident linac beam and EPID, thus avoiding the extra mounted devices on the gantry (Pouliot et al. 2005). Image quality tests for all three devices should be similar, involving image quality tests of soft tissue contrast and spatial resolution.

Until protocols outlining the best QA tests for these devices are developed, manufacturer acceptance test documents may provide useful guidance. As the imaging components and robotic arms used on LMIDs are an amalgam of those used on classical treatment simulators, QA documents that relate to treatment simulators may be adapted to LMID technology. The LMID FPDs are also similar to EPID FPDs. They are even more similar to kV x-ray tube FPDs used in diagnostic x-ray applications, so similar tests for spatial and contrast resolution apply.

A tight isocenter has always been an important requirement for treatment simulators. This was to ensure setup error due to isocenter drift between simulator imaging and linac treatment was avoided, i.e., the bird's-eye view (BEV) from simulator matched the BEV from linac. It appears that reducing the movement of the tube or the detector during rotation is also critical in avoiding cone beam CT reconstruction artifacts. While cone beam CT is a relatively new technology, researchers have identified and quantified artifacts and performance requirements in detail for this technology (e.g., Jaffray et al. 1999).

11.3.3 Record and Verify (R&V) Computer

R&V systems compare the linac setup parameters with the prescribed parameters. Increasingly, the prescribed parameters are transferred to the R&V computer by file transfer directly from the RTP computer. The file format is usually DICOM-RT (Digital Imaging and Communications in Medicine-Radiotherapy). The data transferred includes: patient ID, dose prescription, MU (monitor units), time, beam angle, MLC leaf positions, and dose per segment.

Other parameters, such as patient couch position, may be acquired during the first fraction as the exact location of the patient on the linac treatment couch is often determined during patient setup for the first fraction. Clinical position tables are set to different tolerances in the R&V system to account for small patient shifts, particularly with regard to the couch positions. Tolerances vary depending on treatment site. For example, a head and neck setup may allow tighter couch tolerances than an abdominal treatment setup.

Many R&V systems now also have an option called *auto field sequence* (*AFS*). This feature allows the semiautomatic drive of the linac to its next control point including gantry angle, collimator settings, and MLC shape. Because these systems are automated, even more vigilance is required to ensure mistakes in parameters set before or during the first fraction setup are not automatically transcribed to all the subsequent fractions.

It is important to ensure there is QA of the data transferred to the R&V system by independent cross-checks. This is particularly important if software upgrades to the R&V or associated RTP computer have occurred. There is little literature available on suitable ways to check these systems. Most ROMPs undertake indirect checks by setting or monitoring parameters on the R&V and checking that they are reproduced in the treatment room by the linac.

More often now, R&V systems have grown into full Radiation Information Systems (RIS), which provide a data base of patient clinical information. This data can, in turn, be retrieved for patient clinical trials. The data storage retrieval from these devices can assist in tracking down any incidents that might have occurred with a patient. They may also assist in tracking down other patients that might have received similar events. They may also provide useful data in the pursuit of quality improvement exercises within the clinic. An analysis by Fraass et al. (1998) showed that complex treatments with R&V gave lower error rates than simpler treatments with no R&V. Another study by Huang et al. (2004) shows the value of analyzing the R&V data to subsequently reduce errors in treatment delivery.

11.3.4 Radiotherapy Treatment Planning (RTP) Computer

Many documents address issues related to RTP computer commissioning and QA tests (e.g., ICRU 1987, AAPM 1995, AAPM 1997, IAEA 2001, ESTRO 2004, Able and Thomas 2005). We have chosen to show accepted tolerances for these systems based on those listed in table 11.4, which is adapted from Van Dyk et al. (1993).

There are also a number of other periodic tests recommended for the RTP computer planning process in the AAPM TG 53 document (AAPM 1997). Table 11.5 shows only a summary of the recommended RTP *systems tests* adapted from this document. There are a large number of other tables outlining *functional tests* in the TG 53 report that the reader is recommended to adopt.

The following is a summary of the aspects of an RTP process that have to be addressed in an RTP QA regimen:

i) Data input from CT or virtual simulator to RTP.
ii) Anatomical structure tests—ensure external and internal contours defined.
iii) Contour tool tests.
iv) Electron density tests.
v) Virtual bolus tests.
vi) Image display tools, e.g., grayscale removal of CT couch.
vii) Beam parameters, e.g., gantry angle readouts agree with linac conventions.
viii) MLC field shaping, geometry, and leaf travel restrictions.
ix) Algorithms and dose display, e.g., grid size and calculation.
x) Dose display and analysis tests, e.g., isocenter, isodoses, DVHs.
xi) Data transfer to MU check program.
xii) Data transfer to R&V computer.

The RTP computer represents the most complex software device in the department with multiple inflows and outflows of data as outlined in chapter six. QA processes surrounding this equipment are therefore extremely important.

Table 11.4: Radiotherapy treatment planning systems acceptance criteria when compared with measured beam data (Adapted from Van Dyk et al. 1993)

Beam and Phantom	*Region*	*Criteria*
(a) Photon beams		
Homogeneous, unblocked	central ray (except build-up region)	2%
	high dose (small dose gradient)	3%
	large dose gradients (>30%)	4 mm
	low dose regions (small dose gradient)	3%
Inhomogeneity	central ray (slab geometry, electronic equilibrium)	3%
Anthropomorphic (off axis, electronic equilibrium)	high dose region (small dose gradient)	4%
	large dose gradient (>30%)	4 mm
	low dose region (small dose gradient)	3%
(b) Electron beams		
Homogeneous, unblocked	central ray (except build-up region)	2%
	high dose (small dose gradient)	4%
	large dose gradients (>30%)	4 mm
	low dose region (small dose gradient)	4%
Inhomogeneity	central ray (slab geometry, electronic equilibrium)	5%
Anthropomorphic (off axis, electronic equilibrium)	high dose region (small dose gradient)	7%
	large dose gradient (>30%)	5 mm
	low dose region (small dose gradient)	5%

Table 11.5: Summary of QA systems tests for RTP computers (Adapted from AAPM 1997)

Recommended	*Item*	*Comments/Details*
Daily	Error log	Review report log listing system failures error messages, hardware malfunctions, and other problems. Triage list and remedy any serious problems that occur during the day.
	Change log	Keep log of hardware/software changes.
Weekly	Digitizer	Review digitizer accuracy.
	Hardcopy output	Review all hardcopy output, including scaling for plotter and other graphics-type output.
	Computer files	Verify integrity of all RTP system data files and executables using checksums or other simple software checks. Checking software should be provided by the vendor.
	Review clinical planning	Review clinical treatment planning activity. Discuss errors, problems, complications, difficulties. Resolve problems.
Monthly	CT data input into RTP system	Review the CT data within the planning system for geometrical accuracy, CT number consistency (also depends on the QA and use of the scanner), and derived electron density.
	Problem review	Review all RTP problems (both for RTP system and clinical treatment planning) and prioritize problems to be resolved.
	Review of RTP system	Review current configuration and status of all RTP system software, hardware, and data files.
Annually	Dose calculations	Annual checks. Review acceptability of agreement between measured and calculated doses for each beam/source.
	Data and I/O devices	Review functioning and accuracy of digitizer tablet, video/laser digitizer, CT input, MR input, printers, plotters, and other imaging output devices.
	Critical software tools	Review BEV/DRR generation and plot accuracy, CT geometry. density conversions, DVH calculations, other critical tools, machine-specific conversions, data files, and other critical data.
Variable	Beam parametrization	Checks and/or recommissioning may be required due to machine changes or problems.
	Software changes, including operating system	Checks and/or recommissioning may be required due to changes in the RTP software, any support/additional software such as image transfer software or the operating system.

11.3.5 Classical Simulator and CT Simulator QA Tests

Classical simulator geometric tests are similar to those for linacs, as they were designed to mimic the geometric movements of the linac. Typical tests are available in most QA documents (e.g., ACPSEM 1997, AAPM 1994), the following should apply to classical treatment simulators:

i) The specifications for the mechanical accuracy of the gantry, couch, and simulated field definition are at least equivalent to those required for the megavoltage treatment machines.
ii) Any geometry of patient position, treatment beam dimensions, and orientation used in a fixed SSD or an isocentric treatment can be reproduced on the simulator.
iii) The couch is of similar design to those used during treatment, including provision for the attachment of treatment immobilization devices.
iv) The laser alignment system is arranged in the same manner as in the megavoltage treatment rooms.

Image quality tests follow the same procedures as those for most imaging devices, such as image intensifiers and, more recently, for FPDs. Spatial and contrast resolution tests are involved. The Leeds test object set shown in figure 11.8a is a phantom designed for such tests.

A *CT simulator* consists of a CT scanner (usually wide bore) with a flat tabletop insert (to mimic the treatment couch) and a laser patient positioning system. If the CT is used to acquire a volumetric data set, then a virtual patient data set can be created. If the CT data can be manipulated to provide virtual BEVs placed over patient digitally reconstructed radiographs (DRRs), then this part of the process is known as *virtual simulation*.

The QA requirements for CT simulators are dealt with in detail in the AAPM TG 66 report (AAPM 2003). This report comprehensively outlines tests covering the electromechanical components of the CT and ancillary equipment (e.g., alignment lasers). The other important tests include QA of the spatial and contrast resolution of the CT images. A Catphan is such a phantom (see figure 11.8b), which has a series of cylindrical inserts with inserted material designed to test aspects of image noise and spatial and contrast resolution.

Another important aspect with CT is the dose given during various scanning procedures. This is quantified using the CT dose index (CTDI). The CTDI_{100} is measured using a 100 mm length ionization chamber and is defined in AAPM (1990) as:

$$CTDI_{100} = \frac{Rdg \times N_x \times C_{TP} \times K_{EL} \times f_{med} \times 100(\mathrm{mm})}{TNBW(\mathrm{mm})}\left[\mathrm{cGy}\right]. \tag{11.7}$$

C_{TP} is the pressure temperature correction, K_{el} is the electrometer calibration factor in C/Rdg, N_x is the chamber calibration factor in R/C, 100 mm is the length of the ionization chamber, and f_{med} is the conversion of exposure in air to absorbed dose in the medium, typically about 0.87 at this energy but this depends on the medium. *TNBW* is total nominal beam width.

When performing a series of scans the dose tails accrue to a higher dose value, known as the multislice absorbed dose (MSDA). This has not been consistently defined, however, so another quantity which essentially defines the same value but more consistently is the $CTDI_{vol}$, where

$$CTDI_{vol} = \frac{N \times T}{I} \times CTDI_{w}, \tag{11.8}$$

where $CTDI_w$ is usually $CTDI_{100}$, N is the number of simultaneous axial scans per x-ray source rotation, T is the thickness of one axial scan in millimeters, and I is the table increment per axial scan in millimeters.

In spiral CT, which is very commonly used, the ratio of the table travel per rotation I to total nominal beam width (N × T) is referred to as the pitch; therefore

$$CTDI_{vol} = \frac{1}{pitch} \times CTDI_{w}. \tag{11.9}$$

The $CTDI_w$ is the average radiation dose over the x and y directions, and $CTDI_{vol}$ is average radiation dose over x, y, and z directions. It is common to have small pitch and oversampling for four-dimensional CT (4DCT) procedures. While these image sets are terrific for defining target coverage, the extra dose due to overscanning should be assessed (Theocharopoulos et al. 2007).

Recent work has shown that polymer gel dosimetry is a useful device for determining CTDI and dose distributions from CT equipment, and this dosimeter may provide additional information not possible with just the use of an ionization chamber (Hill et al. 2005).

Shown in table 11.6 is a summary of the electromechanical tests and image evaluation tests that need to be considered for CT simulators. Suggested frequency of tests is dealt with in AAPM (2003). These tests are aimed at ensuring that placement of virtual treatment beams match real treatment beam positions and that accurate localization of target and critical structures occurs.

An additional requirement for CT simulators is consistent checking of the CT number table, as this has an impact on the dose calculation. Also, consistent checks of the laser system alignment and the laser moves with respect to the CT are important aspects that require QA tests. The process used for laser shifts in virtual simulation are dealt with in chapter six, incorrect laser shifts at virtual simulation can result in geometric misses of the target volume at treatment (see also chapter six).

Table 11.6: Summary of CT simulator QA checks (Adapted from AAPM 2003)

Performance	*Requirement Parameter*	*Tolerance*
Gantry lasers	Coincident, parallel, and orthogonal to scan plane	±2 mm
Wall and ceiling lasers	Accurate spacing from image plane, parallel and orthogonal to scan plane	±2 mm
CT table vertical and longitudinal indexing	Table indexing accurate under CT control	±1 mm
Scout or pilot scans	Positions coincident with CT scans	±1 mm
Radiation profile	Ensure radiation profile does not exceed as specified	Meets manufacturer specifications
Sensitivity profile	Ensure sensitivity profile matches radiation profile	Meets manufacturer specifications
X-ray generator	If generator adjusted or replaced, can have an impact on kV	Meets manufacturer specifications
CT number accuracy	CT number stays consistent from scan to scan and across the scan plane. CT number is consistent with procedure choice	10 HU is approx. equal to 0.01 density change
Image noise	Image noise does not degrade the image	Test daily within manufacturer specifications
Spatial resolution		Test with line pair or MTF phantom
Contrast resolution		Test with low contrast resolution phantom

11.4 Intensity-Modulated Radiation Therapy (IMRT) QA

11.4.1 Action Levels

Two documents deal with recommendations on safe implementation of IMRT into the clinic (Ezzell et al. 2003, Galvin et al. 2004). While not QA protocols per se these are very useful guidance documents.

IMRT QA is rather unique in radiotherapy in that QA checks are carried out on the dose delivery predicted by an RTP computer prior to each patient treatment by taking dose measurements in a surrogate phantom under the linac beam. What action levels should be set to accept an IMRT plan? According to Ezzell et al. (2003) there is a developing consensus that ion chamber measurements in high dose, low gradient regions should agree with the plan to within 3% to 4%.

11.4.2 Steps in the IMRT QA Process

As mentioned for traditional conformal radiotherapy treatments, generally only the plan calculation is checked and the other offline QA procedures are relied on for treatment delivery. Because of the added complexity in IMRT planning and delivery there are three elements that need to be verified for each IMRT plan prior to the first patient treatment fraction commencing. These are (Ezzell et al. 2003):

i) Dose and MU calculation.
ii) Information transfer from RTP to R&V.
iii) Treatment delivery (e.g., beam angles, MU, and segments).

i) Several commercial MU check calculation programs are now commonly used to check IMRT plans in the clinic (e.g., RadCalc®). These programs can satisfy the first requirement to give an estimate of planned dose estimated by the RTP computer. Shown in table 11.7 is a comparison between an RTP calculation and a check by an MU check program for a patient plan. In general, there is less agreement when more segments are planned in each field. The MU check calculation generally does not account for patient contour effect. But these results are certainly adequate in terms of flagging planning errors.
ii) Information transfer from the RTP computer to the R&V computer may be checked using the same protocols outlined in section 11.3.3. An example of such a test would be viewing segment shapes using light fields compared with RTP planning hardcopy templates. Because there are so many segments, often only the first segment for each field is checked this way.
iii) While an MU check program can check the planned MU, the transfer of data and treatment delivery are routinely checked by using dosimeters in an IMRT check phantom on the linac. Several options exist including various ionization chambers, film combinations, and/or an EPID, and collecting single or multiple beam data. The current most commonly used options available are summarized in figure 11.9.

Table 11.7: Patient plan of RTP computer calculation (Pinnacle) vs. MU check calculation (RadCalc)

Beam name	*MU*	*RTP dose (cGy)*	*MU check dose (cGy)*	*% difference*
150	75	29.6	30.5	3.0
118	65	27.0	26.8	-0.7
90	71	26.8	26.5	-1.1
68	77	28.0	27.7	-1.1
30	142	34.8	35.9	3.2
0	61	31.7	32.2	1.6
345	58	32.4	33.8	4.3
Total		210.3	213.4	1.5

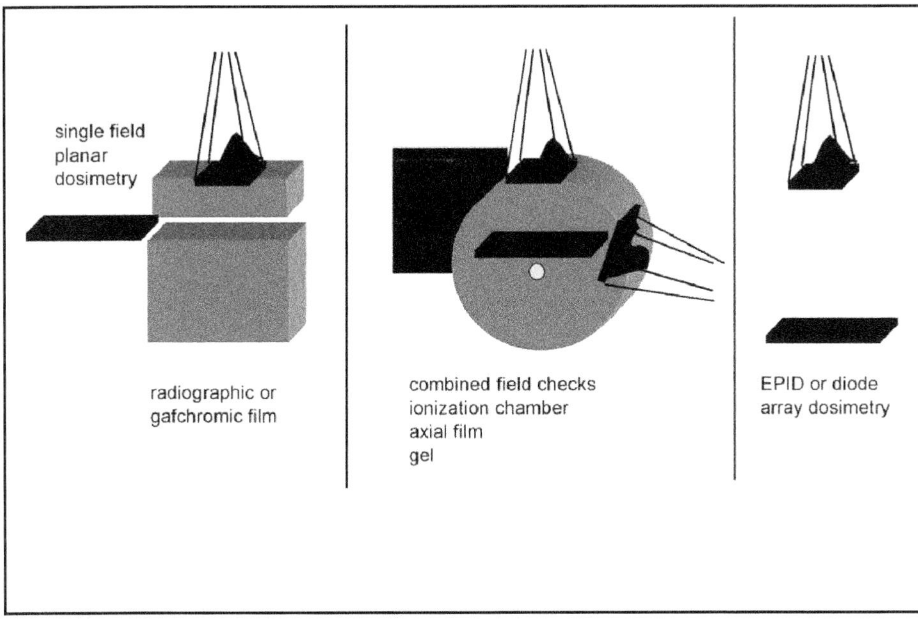

Figure 11.9: Schematic of various dosimetry options and beam combinations available for checking IMRT dose delivery.

The most common check is a point dose check undertaken with an ionization chamber placed in an IMRT phantom to check the treatment dose delivery. This is still the gold standard check for IMRT treatments. The patient planned beam information is usually transferred to a CT contour of an IMRT phantom using an RTP copy-to-phantom feature and doses are re-estimated to isocenter, based on the new effective depths to the phantom dose point.

A patient IMRT planned dose distribution and corresponding copytophantom IMRT planned dose distribution is shown in figure 11.10. The real phantom which dose is delivered to on the linac is also shown. Converting charge to dose in the IMRT phantom is achieved by integrating charge as a ratio to a standard charge measurement taken in the IMRT phantom at isocenter with a standard 10×10 cm^2 field given 100 MU. Shown in table 11.8 is an RTP phantom plan dose compared with measured dose and to RTP computed dose. This table also includes comparison with the MU check calculated dose to the phantom (e.g., RadCalc). This is the type of QA sheet that needs to be retained with each patient IMRT treatment planning documentation.

Shown in figure 11.11 is the cGy prediction using an ion chamber for many individual IMRT fields. This shows an historical improvement in agreement between ionization chamber measured cGy and the RTP predicted cGy per field. Note there was an improvement in agreement when a geometrically more precise IMRT phantom was employed, and a further improvement when a more precise MLC leaf model was employed in a later version of the RTP computer software. While these differences may seem significant, per field the combined dose distribution usually shows a percentage difference that is within 3% of the planned RTP.

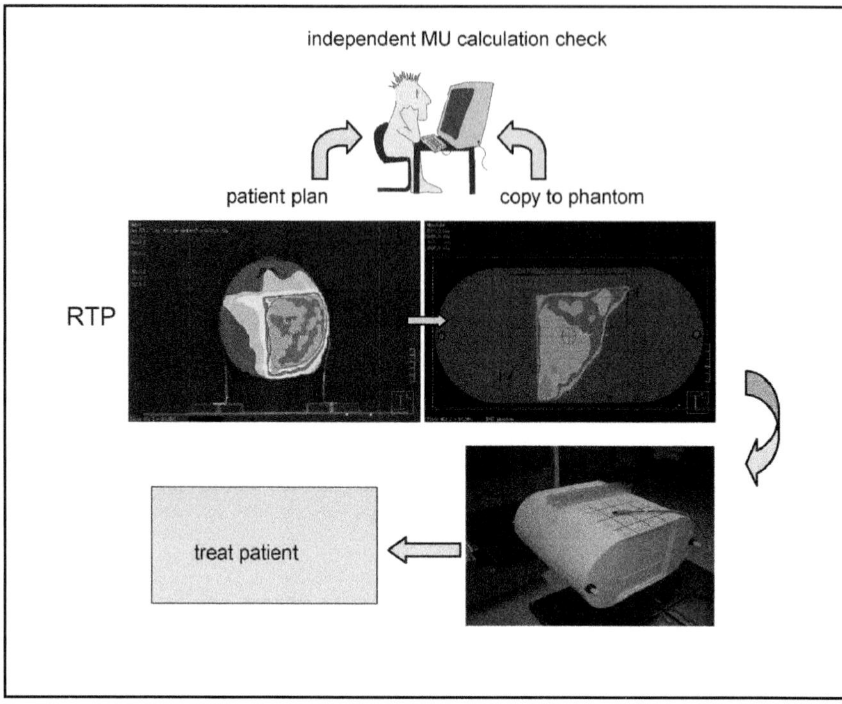

Figure 11.10: The IMRT treatment delivery QA check sequence from RTP patient dose plan transfer to phantom dose plan, data transfer to R&V; then phantom under linac irradiation with dosimeter and finally treatment of patient.

Table 11.8: Dose measured by ionization chamber in IMRT phantom during pre-treatment dose delivery on Varian linac compared with RTP computer (Pinnacle) and MU check calculation (RadCalc)

Beam name	*MU*	*Measured dose (cGy)*	*MU check dose (cGy)*	*RTP dose (cGy)*	*MU check vs. measured % difference*	*RTP vs. measured % difference*
150	75	27.7	27.6	28.2	-0.4	1.8
118	65	19.3	19.0	19.0	-1.6	-1.6
90	71	15.7	15.2	15.4	-3.2	-1.9
68	77	17.3	17.0	17.1	-1.7	-1.2
30	142	27.1	27.2	27.6	0.4	1.8
0	61	27.7	27.6	27.5	-0.4	-0.7
345	58	27.2	27.6	26.9	1.5	1.1
Total		**162.0**	**161.2**	**161.7**	**-0.5**	**-0.2**

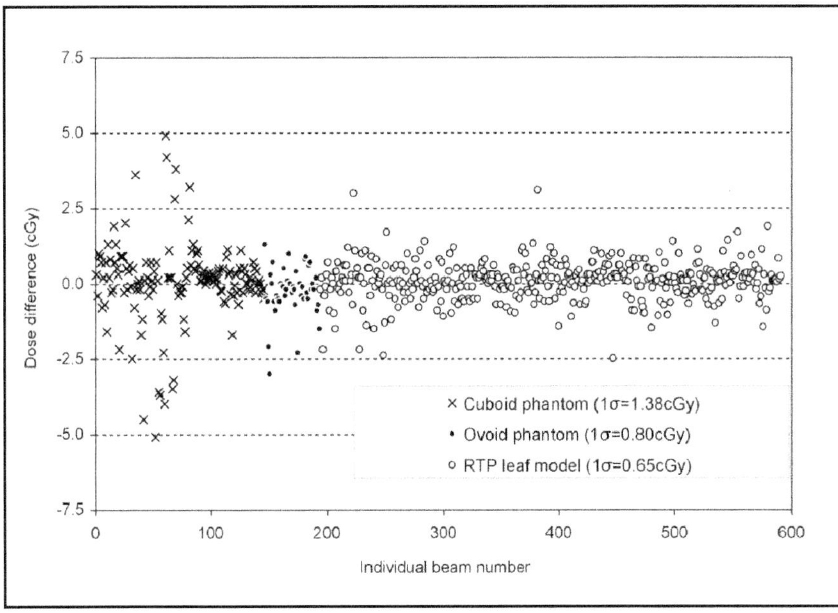

Figure 11.11: Difference in cGy dose between RTP (Pinnacle) calculation and ionization chamber measured dose. Two regions of improvement occurred when (i) the phantom geometric precision was improved and (ii) when the RTP leaf modeling improved. (Data courtesy of M. Williams, Illawarra Cancer Care Centre, NSW, Australia.)

11.4.3 Film Dosimetry

Further validation of planar and axial dose distributions can be undertaken using comparison with film dosimetry. Individual planned IMRT beams can be transferred to a phantom and the dose distribution delivered to film can be compared against the dose predicted using the RTP computer dose distribution. Examples of radiographic and Gafchromic axial dosimetry images that can be collected for IMRT beam validation are shown in figure 11.12. It is also possible to collect individual or combined beam planar dose maps placed in the coronal plane. This is particularly useful for highlighting MLC effects.

One particularly novel and useful approach is to place the film in a spiral pattern in a phantom (Paliwal et al. 2000, Richardson et al. 2003). This provides up to three dose comparison points along the ray of each beam at different distances. By wrapping the film in such a way it is also close to perpendicular to most of the high-dose film contributions.

Film or other 2-D dosimeters can highlight problems that may be directly due to delivery (e.g., MLC calibration). This process may indeed also be useful acting as a beacon that could be tracked back to R&V. Both planar dose maps (film or EPID) and axial dose maps (film) obtained can point out problems that may have occurred in the R&V transfer. Simple geometric orientation data can also be checked with these images. It is important also to check that the treatment fields are deliverable without gantry or beam obstruction before patient treatment. The treatment to the phantom can sometimes point out these problems.

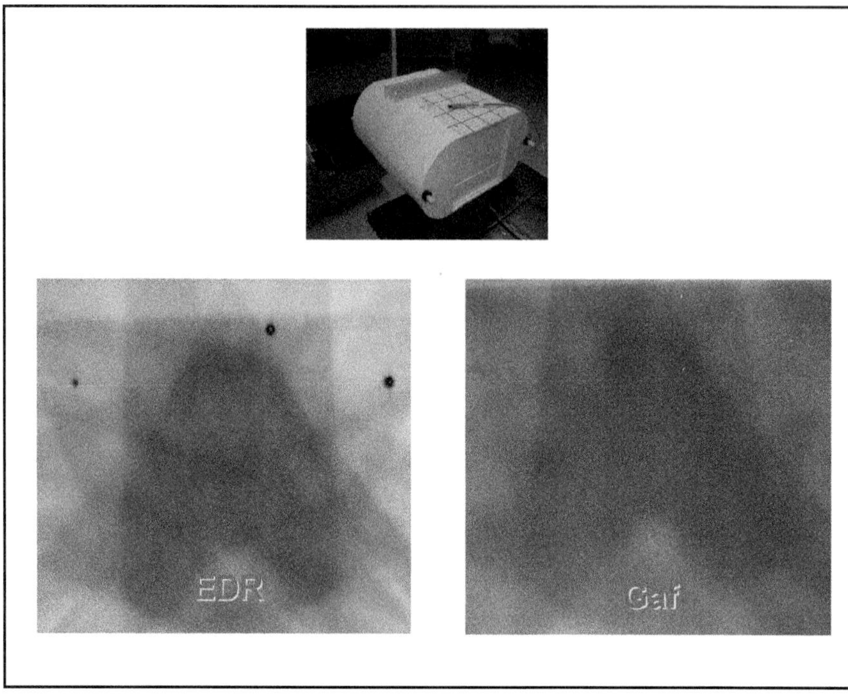

Figure 11.12: Axial density maps showing dosimetry films taken using EDR radiographic film (left) and Gafchromic EDB film (right). Both films were given 1.4 Gy delivered from combined IMRT beams to isocenter. SEE COLOR PLATE 51.

11.4.4 EPID Dosimetry

There has been significant progress toward using electronic portal images to calculate the dose and replace film as a planar dose map analysis tool (see figure 11.13). The conversion from dose to EPID to dose to water can be achieved by using dose deposition kernels coupled with intensity corrections.

The EPID image must be corrected for variations in pixel sensitivity due to dark current, inherent pixel sensitivity, and variations in off-axis response (variations in the spectrum). This is achieved by using a series of off-axis corrections (Greer and Popescu 2003, McDermott et al. 2004, Greer 2005). After corrections have been made to the EPID image a conversion must be performed to obtain a dose to water image. First, a kernel describing the dose deposition in the EPID fluorescent layer is deconvolved out of the image (Steciw et al. 2005, Warkentin et al. 2003, Kirkby and Sloboda 2005a,b). This results in an incident fluence image. A kernel describing the dose deposition in water (Ceberg et al. 1995) is then convolved with the fluence image to obtain the dose to water image.

A unique method of IMRT dosimetry check has been presented by Ansbacher (2006) whereby combined IMRT dose distributions collected with an EPID are reconstructed on a virtual cylindrical water phantom for comparison with copy-to-phantom RTP doses or, indeed, film or other dosimetry collections in a cylindrical phantom. Louwe et al. (2003) and Wendling et al. (2006) take this a step further and backproject EPID data collected from beams exiting the patient during treatment to form a dose map. A similar approach seems to have been first devised using film technology by McNutt et al. (1996). The process is described in figure 11.14. The suggested technique is one of the approaches being evolved for adaptive radiotherapy procedures.

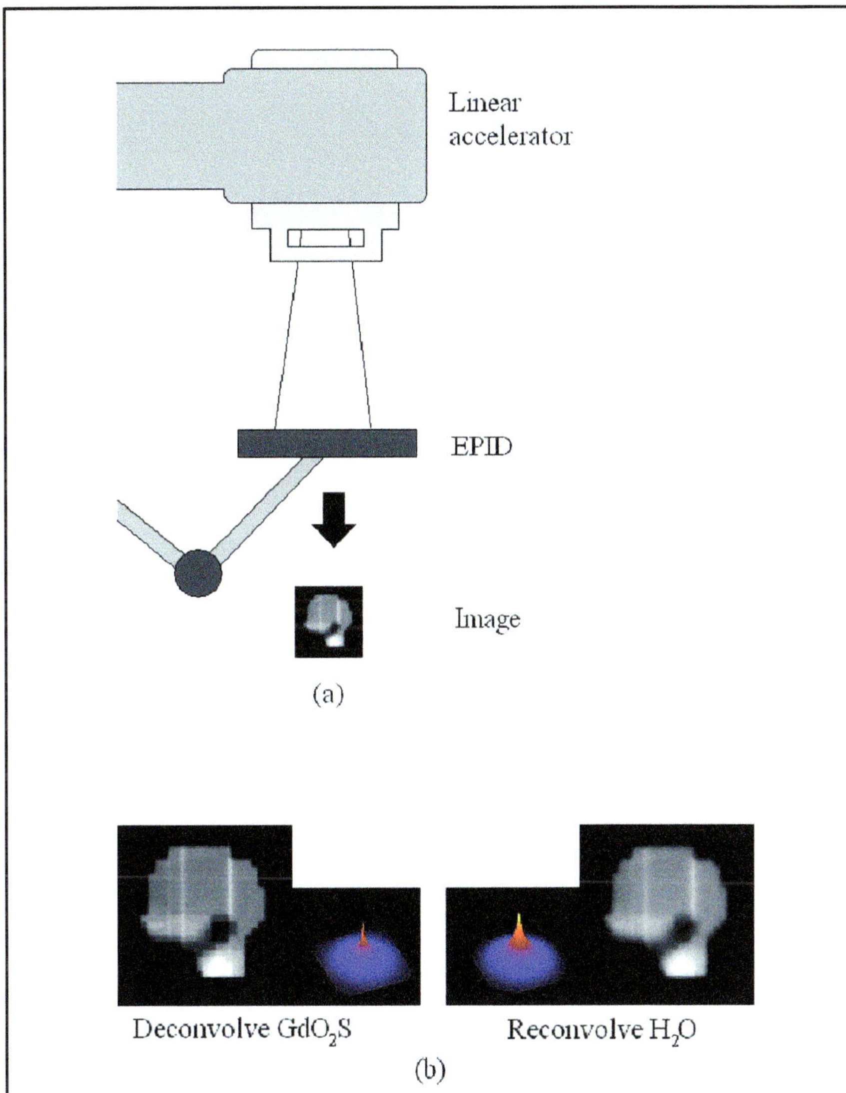

Figure 11.13: Schematic of linac and EPID set up as a pre-treatment dose delivery QA check device. (a) Set up with no patient in the beam as used for IMRT leaf position validation and comparison with RTP calculations of dose-eye views. (b) Dose-eye views taken on an a-Si electronic portal image converted from dose to GdO_2S (left, image and kernel) to dose to water (right, image and kernel).

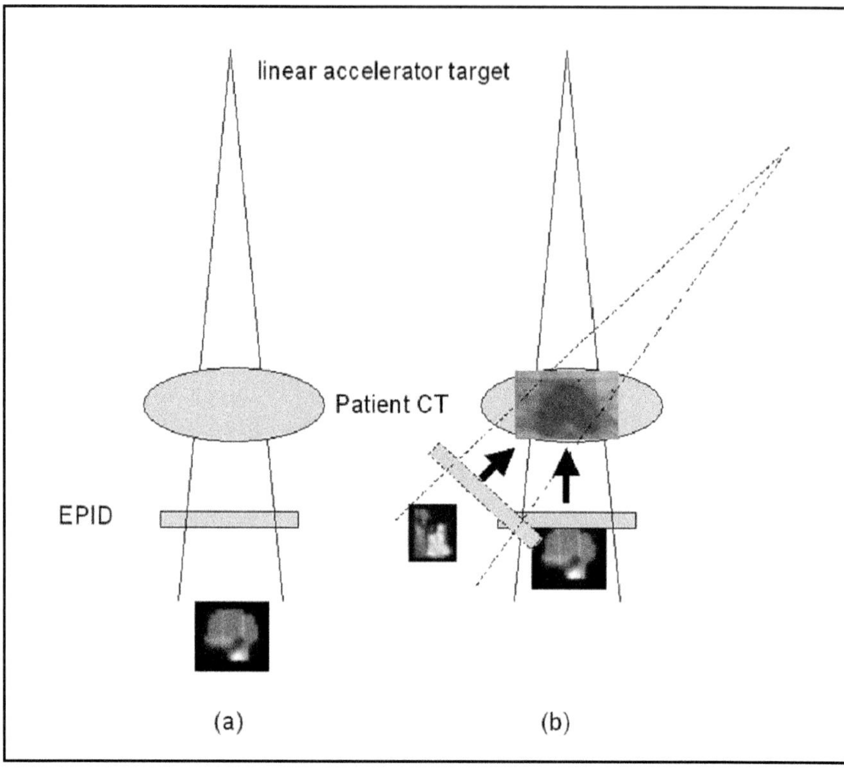

Figure 11.14: Schematic of using an EPID to collect dose information with patient in the beam. (a) Transit dosimetry using an EPID. (b) Schematic of the dose reconstruction process using an EPID. Only two fields have been shown, but EPID images from each beam delivery angle could be used and backprojected.

11.4.5 Gel Dosimetry

In addition to the material provided in chapter three about gel dosimeters a brief review of gel as an IMRT dosimeter is provided in this section. The use of radiation-sensitive gels was first suggested in 1950 when radiation was used to produce color changes in gels containing dyes (Day and Stein 1950). However, radiation-sensitive gels for the purposes of dosimetry, as currently used, are a result of the work undertaken by Gore et al. in the early 1980s, which built upon the earlier development of the Fricke dosimeter in the 1920s (Fricke and Morse 1927).

Gore et al. investigated the nuclear magnetic resonance (NMR) relaxation properties of irradiated Fricke or ferrous sulfate dosimetry solutions (Gore et al. 1984), showing that radiation-induced changes, in which ferrous (Fe^{2+}) ions are converted to ferric (Fe^{3+}) ions, could be quantified using NMR relaxation measurements. It was shown that Fricke dosimetry solutions dispersed throughout a gel matrix could be used to obtain three-dimensional (3-D) spatial information using magnetic resonance imaging (MRI). It was subsequently shown that irradiated Fricke-type gel dosimeters did not retain a spatially stable dose distribution due to ion diffusion within the irradiated dosimeters (Baldock et al. 2001).

In 1993 a polymer gel dosimeter was proposed, which was based on the polymerization of monomers infused in an aqueous gel matrix (Maryanski et al. 1993). In 1994 this formulation was patented (Maryanski et al. 1994) and became commercially available and known as BANG®. Subsequently, due to the commercialization of this product, PAG (polyacrylamide gel) (Baldock et al. 1998a,b) became the polymer gel dosimeter acronym used by most authors. Although polymer-type dosimeters did not have the diffusion limitations of Fricke-type gel dosimeters, there was another significant limitation in their use. Due to the nature of their free radical chemistry, polymer gel dosimeters were susceptible to atmospheric oxygen inhibition of the polymerization processes. As a result, these gel dosimeters had to be manufactured in an oxygen-free environment (Baldock et al. 1998a), which was a significant limitation in the introduction of gel dosimetry into the clinic.

A significant development in the field of gel dosimetry occurred when results of using an alternative polymer gel dosimeter formulation were published (Fong et al. 2001). This new type of polymer gel dosimeter, known as MAGIC gel, bound atmospheric oxygen in a metallo-organic complex, thus removing the problem of oxygen inhibition and enabling polymer gels to be manufactured on the bench-top in the laboratory. This created what was to be known as a *normoxic* polymer gel dosimeter, compared with the PAG formulation which subsequently became known as a *hypoxic* polymer gel dosimeter. It was subsequently shown that other formulations of normoxic polymer gel could be manufactured (De Deene et al. 2002).

Since the work of Gore in 1984, the majority of investigations have been undertaken with MRI. However, alternative techniques for evaluation have been introduced including optical computed tomography (CT) (Gore et al. 1996), x-ray CT (Audet et al. 2002), vibrational spectroscopy (Baldock et al. 1998b), and ultrasound (Mather et al. 2002).

Another new class of polymer dosimeter is PRESAGE™ (Heuris Pharma, Skillman, New Jersey) (Guo et al. 2006). PRESAGE is a solid dosimeter, which is based on clear polyurethane combined with leuco-dye leucomalachite green. Although not suitable for MRI evaluation, the leuco dyes have a maximum absorbance at a wavelength of 633 nm and are therefore suitable for evaluation with a helium-neon (He-Ne) laser-based optical scanning system.

Several studies (De Deene et al. 1998, 2000, Gustavsson et al. 2003, Duthoy et al. 2003, Vergote et al. 2003, 2004) have reported this form of radiological tissue equivalent dosimeter (Keall and Baldock 1999; Venning et al. 2005) to be advantageous over other IMRT dosimetry methods. This is because unlike other 2-D dosimeters polymer gel dosimetry offers a complete 3-D dose data set for QA in IMRT.

11.5 Image-Guided Radiation Therapy (IGRT) QA

As this textbook goes to print there has not yet been a document outlining recommendations for appropriate implementation of the image-guided radiation therapy (IGRT) process. There is however an AAPM TG 75 committee currently looking into recommendations. The recommendations can be summarized as follows:

i) Compile a list of all imaging procedures to be undertaken before, during, and after treatment.
ii) Identify image-guided steps that do not require ionizing radiation (e.g., Tomé et al. 2002).
iii) Minimize dose outside the treatment field.
iv) Ensure imaging technique gives sufficient image quality.
v) Estimate the total effective dose from imaging that the patient will receive.
vi) Evaluate total dose patient-by-patient.
vii) Estimate potential stochastic and deterministic effects.

It is beyond the scope of this book to cover QA of diagnostic procedures. However, it will be important for the QA team to solicit expertise from other professionals for the determination and quantification of image quality, which is something not traditionally done by staff in radiotherapy. It is important to note though that QA of images in radiotherapy typically has additional requirements over the ones used in radiology. In particular, the requirements for spatial accuracy are usually more stringent.

Similarly, image guidance goes beyond the quality of images. A QA program must include the decision-making process and any actions performed as a result of the image guidance. For example, if a system allows automatic couch movement after a difference between reference position and actual position of the target has been identified, it is essential to verify that direction and magnitude of the shift is correct.

11.6 In Vivo Dosimetry

Useful information is provided in the first book of a new series of the European Society for Therapeutic Radiology and Oncology (ESTRO), which is dedicated to the subject of in vivo dosimetry (van Dam and Marinello 1994).

The International Commission on Radiation Units and Measurements (ICRU) stated in 1976 that "All the procedures involved in the planning and execution of radiotherapy can contribute to a significant uncertainty in the absorbed dose delivered to the patient...an ultimate check of the actual treatment given can only be made by using in vivo dosimetry." In the 30 years since then radiotherapy techniques of increasing complexity and sophistication have become available. This makes treatment verification even more critical. In addition to verification of the patient's position by means of portal imaging, the given dose can be assessed using in vivo dosimetry. Assessment of dose in individual patients is increasingly required due to:

i) More complex equipment. In a dual-modality linac with MLC and dynamic beam modification more things can go wrong than in a cobalt-60 unit.
ii) More difficult techniques. The dose distributions in three dimensions are more difficult to predict than in conventional radiotherapy. Similarly, closer tumor margins and potentially gated radiotherapy delivery demand more treatment verification.
iii) Tighter dose rationale. With improving knowledge of the effects of ionizing radiation, the dose to a particular tumor can be more accurately prescribed. This requires better control of the dose actually given.
iv) Less dose to normal tissues. With increasing follow-up of radiotherapy patients and improved survival rates, side effects of radiotherapy become more visible. These include effects such as secondary cancers due to scattered radiation at large distances from the field edge. Most planning systems are not designed to predict these doses. In vivo dosimetry can help.
v) More awareness of patient's rights. The possibility of and the increasing costs associated with litigation make in vivo dosimetry an important part of the treatment verification record.

In ICRU report 24 (1976) it is also specified what in vivo dosimetry might include:

i) Entrance dose measurements.
ii) Exit dose measurements.
iii) Transmission measurements.
iv) Intracavitary absorbed dose measurements.

In general, most in vivo dosimetry is limited to measurements outside of the patient—mostly on his or her surface. This is realized in the first three in vivo dosimetry aims listed above. However, it is possible to obtain information on the dose to the tumor or critical structures also by intracavitary measurements (Lindskoug and Lundberg 1984), by using the combined information of exit and entrance dose measurements for an estimation of the tumor absorbed dose (Leunens et al. 1990) or potentially by dose reconstruction using the information obtained by an EPID (Ansbacher 2006, van Elmpt et al. 2006).

One should also add assessment of dose in a particular organ to the ICRU list. This may be the tumor if located at the skin. This could be a critical organ. This typically encompasses lens dose measurements, scrotal dose measurements, and skin dose determination.

An impact on clinical practice by in vivo dosimetry has been demonstrated in many studies (Leunens et al. 1990, 1993, Heukelom et al. 1991, Kron et al. 1993). There are obvious advantages and improvements in the treatment for some patients (Kron et al. 1993):

i) Shielding and bolus may been altered.
ii) Field spacing and incident angles may be adjusted.
iii) Determination of boost dose (e.g., excision scars).
iv) Less reproducible treatment techniques may be modified.
v) Data used in planning may be reviewed.

11.6.1 Suitable Dosimetric Techniques

Various measurement techniques have been developed for in vivo dosimetry. Details of most of these devices have been discussed in chapter three. Here are some brief comments on how these devices seem to be shaping up as in vivo dosimeters:

i) Ionization chambers have a number of disadvantages that make them less well suited for in vivo dosimetry. Ionization chambers are usually bulky and fragile and therefore not easily used on patients. In addition, they require a connection to an electrometer, and a high voltage is applied, which can be an electrical hazard for the patient. Therefore, ionization chambers are currently only recommended for a very limited number of applications such as total body irradiation (AAPM 1986b).

ii) Presently, scintillation dosimeters do not play an important role in x-ray dosimetry. However, scintillation crystals can be built to small dimensions, and the rapid development of optical fibers may render them more suitable in the future (compare chapter three, section 3.9.3).

iii) Chemical dosimetry was one of the first means of in vivo dosimetry. In the 1920s skin dose was often determined using bariumplatinumcyan ($BaPt(CN)_43H_2O$), which changes its color with increasing dose from green to brown (Reich 1990). The color change was compared with color charts. The major problem with this method was that the dosimeters exhibited a large dependence on the radiation quality because of the heavy metal contents of the chemical compound. Today, chemical dosimetry is not used for in vivo dosimetry.

iv) Port films have been proposed as an easy and nonobstructive method of dose verification (van Dam et al. 1992, Ebert and Kron 1994). These efforts are now nearly completely replaced by the use of electronic portal imaging for in vivo dosimetry (Wong et al. 1990, Ying et al. 1990). EPID images, unlike film, do not require post irradiation development. The data are also stored electronically and are immediately viewable after exposure. This and the fact that EPIDs are permanently connected to the linac make this new technique a promising option for in vivo dosimetry, as the work of the group in Amsterdam demonstrates (Essers et al. 1996, Nijsten et al. 2007, McDermott et al. 2007).

v) However, film as a media has maintained some attention. This is largely due to the introduction of radiochromic film (compare chapter three, section 3.8.2). As this film does not need to be protected from light and is self developing, it has some attraction for determination of 2-D dose distributions on the skin of patients undergoing radiotherapy. As the measurements are confined to skin, applications in electron treatments are more common (Butson et al. 2000, Ciocca et al. 2003, Gamble et al. 2005). Radiographic film may still be useful in some circumstances where dose information from large areas is required. A recent example is the independent in vivo verification of complex treatments delivered in helical tomotherapy (Kron et al. 2005).

vi) Thermoluminescence dosimetry (TLD) is one of the most commonly used techniques for in vivo dosimetry (Kron et al. 1993). Its typical applications are the assessment of the dose to critical organs and measurements in difficult geometries. In clinical practice, LiF:Mg,Ti is usually used. This material has relatively close tissue equivalence ($Z_{eff} = 8.3$) and can be produced in the desired forms such as powder, ribbons, or rods with small dimensions of the order of a few millimeters. The typical accuracy reached in clinical practice of better than $\pm 5\%$ (2σ) is sufficient for the requirement of in vivo dosimetry.

The major advantages of TLD for in vivo dosimetry are the small size of the detectors, their stand-alone measurement characteristic. TL materials typically consist of one material compound only and not a series of different layers. Therefore, the reading of the TL dosimeter is generally independent of the angular distribution of the radiation. This is an important feature for measurements in treatment geometries where incident beam angles vary. TLD detectors are typically prepackaged prior to placing them on the patient's skin in appropriate locations after the setup. This causes minimal interference with patient setup and treatment. A major disadvantage of TLD is the delayed readout. However, in fractionated radiotherapy it is feasible to obtain the results of measurements after the first fraction of treatment and before the next fraction is delivered to the patient.

vii) The most commonly used detectors for in vivo verification of entrance dose are p-type diodes (Rikner and Grusell 1987, Heukelom et al. 1991, Howlett et al. 1999, Marre and Marinello 2004). Diodes are convenient to use because in commercially available semiconductors, "adequate" build-up is already included. This build-up also acts as a filter for low-energy x-rays to which diodes would over respond. However, the built-in build-up usually requires the user to purchase a set of diodes for each radiation quality in use. The optimum setup is one dedicated diode system for each radiation unit. In this case, no cables have to be moved or connections made prior to the measurement, and virtually no interference with the patient's treatment is necessary. Most commercial systems consist of five or more diodes, which enable entrance and exit dose measurements at the same time. Due to their build-up material, these detectors have dimensions of some 5 to 10 mm. This and their directional dependence render them less suitable than TL dosimeters for measurements in complicated geometries, such as lens dose determinations. The accuracy of clinically used diodes is of the order of ±3% (2σ) or better in particular if the detectors are recalibrated on a regular basis (e.g., weekly or monthly). The major advantages of an immediate readout and the possibility of obtaining multiple readings make diodes particularly useful for in vivo dosimetry of total or half-body irradiations. Possible disadvantages of diodes are that cables are required and that their response depends on the beam energy and direction. Temperature compensation is also required for these devices.

viii) For in vivo dosimetry, MOSFET detectors have similar properties to diodes. They have the additional advantage of being integrative dosimeters. This allows the dose record for a particular patient to be recorded as patient dose fractions are accrued. In high-dose applications MOSFETs may be operated without bias, which precludes the requirement for cables.

In the last years MOSFET detectors have been demonstrated to be useful in vivo dosimeters for a variety of applications. Their main use has been for skin dose measurements in field (Scalchi et al. 2005, Rosenfeld et al. 2002, Bloemen-van Gurp et al. 2006) and out of field (Butson et al. 2005).

Being very small, there have also been attempts to encapsulate MOSFETs entirely and to make them stand-alone detectors, which can be implanted into tissues of interest (Beddar et al. 2005). These prototype detectors were powered by radiofrequency telemetry for the dose measurements this makes a power supply within each capsule unnecessary. With further miniaturization it may be possible to implant the detector directly in the target or critical structure, which would allow for the ultimate in vivo dosimetry.

Whatever technique is used for in vivo dosimetry, it is essential to accurately determine and document the position of the detector. Image guidance may be a tool to not only position radiation beams but also to determine the location of the detector (Engstrom et al. 2005, Wertz et al. 2007).

11.6.2 Applications of In Vivo Dose Verification

In vivo dosimetry workload can be separated into two different groups:

(i) Routine verification on every patient.
(ii) Verification in special cases on request.

Routine Verification

In the first type of in vivo dosimetry, central axis measurements are performed during the first patient treatment session, or as soon after as practical, on every patient in every treatment field. This can be achieved using TLD (Ruden 1976), but most commonly semiconductors are employed for these measurements as they offer a real-time dose reading (Leunens et al. 1990, Alecu et al. 1999, Fiorino et al. 2000). Entrance dose measurements are a verification of the performance and the settings of the treatment unit itself. In addition to this, the entrance dose measurement can be used to check the treatment distance and verify the dose delivered in IMRT fields (Higgins et al. 2003). If entrance dose measurements are combined with exit central axis dose measurements (Leunens et al. 1990, Heukelom et al. 1991, Huyskens et al. 1994), one can also obtain information on the patient parameters such as separation and tissue inhomogeneity effect on dose measured versus dose planned on the RTP computer. More recently, it has been demonstrated that the central axis exit dose can also be reliably determined using the central portion of an a-Si EPID (Piermattei et al. 2006). The ease of use and ready availability of EPI has generated renewed interest for in vivo dosimetry (Pasma et al. 1998), in particular in the context of IMRT (Partridge et al. 2002, Wendling et al. 2006). In a very interesting development this type of dosimetry has been shown to have the potential to replace the normal QA procedures required for IMRT (McDermott et al. 2007). If this proves to be widely applicable, QA time for IMRT could be reduced and the time from planning to treatment significantly shortened. (See also figures 11.13 and 11.14.)

In vivo dosimetry as routine verification is practiced on every patient in some centers in Europe and Scandinavia. In order to reduce the workload, some centers verify the dose only in a certain percentage of patients.

Special Cases

In vivo dosimetry in special cases is typically only requested for patients where there is an indication of possible benefits. The measurements are normally requested from the radiation oncologist or staff in treatment planning. In a typical radiotherapy center between 5% and 10% of the patients may require these types of measurements. The applications can be broadly divided in three different categories:

i) Determination of the delivered dose in difficult geometries.

   In vivo dosimetry is useful for patients whose dose was difficult to predict from the treatment plan. An example is the skin dose in breast treatment (Heukelom et al. 1991). This is of particular interest at the site of excision. Among other variables, the skin dose depends on the use of shadow trays, blocks, positioning devices, and the obliquity of the incident beam (Kron et al. 1996). Most RTP computer calculations still only use empirical models for dose in the build-up region so skin dose is more accurately measured than modeled. This application group also includes a wide variety of head and neck treatments.

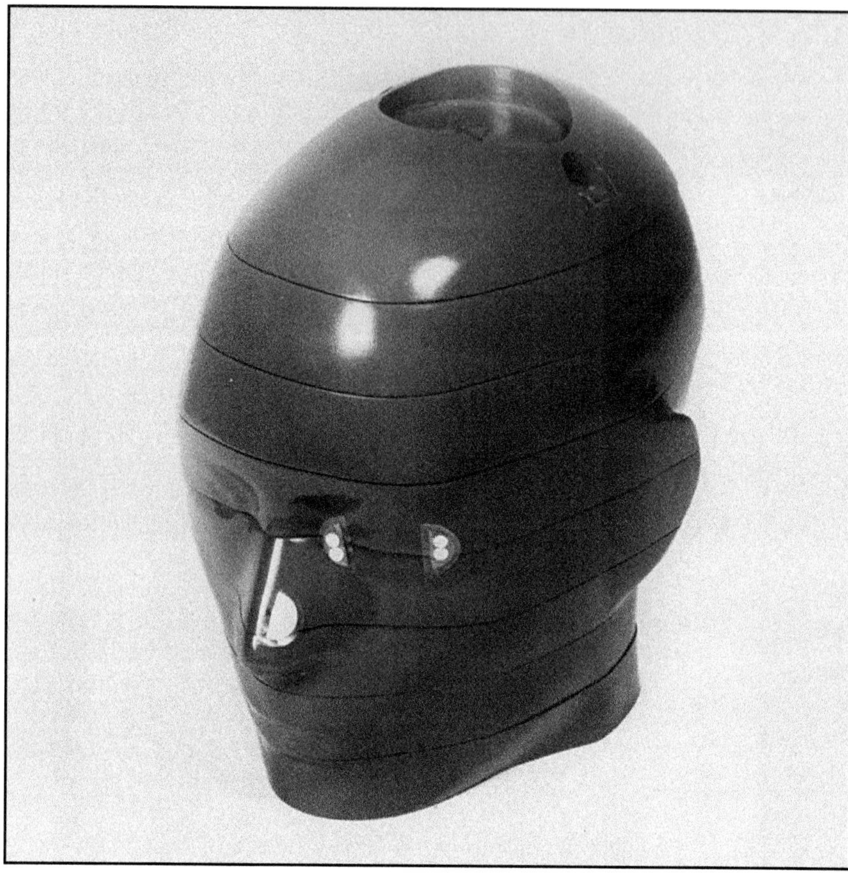

Figure 11.15: TLD measurement of lens dose on an anthropomorphic RANDO phantom. The packs with two TL dosimeters each would be placed similarly for in vivo dosimetry on patients.

ii) Assessment and record of the dose to critical structures.

The most common request for in vivo dosimetry in this category is to record the dose to a critical organ, such as the lens of the eye or the testis. An example of a lens dose assessment in lateral photon beams is shown in figure 11.15, whereby TL dosimeters are placed on a phantom in the same positions as would be used for patient in vivo measurement. The assessment of the dose to a fetus is another scenario where at least the transmission and scatter components from the collimator can be determined using in vivo dosimetry. Other measurements would be the assessment of the rectal dose using intracavitary measurements.

iii) Monitoring of the dose in specialized treatments.

For certain treatment techniques the delivered dose may be monitored over the whole course of treatment. This includes x-ray total body irradiations (TBI) (AAPM 1986b) and total body skin electron (TBSE) irradiation (Desai et al. 1988). RTP computer calculations are generally not optimized for these long SSD, large geometry treatments. In both circumstances treatment planning is difficult, and the dose prediction may be affected by many uncertainties that are hard to control (e.g., scatter from floor and walls).

QA and quality improvement is an ongoing process in radiotherapy departments and covers all aspects of treatment planning, treatment delivery, and treatment verification. Linac QA machine checks are a small but important subset of this overall process. When new techniques are developed, such as IMRT, then new QA procedures and protocols start to evolve (Williams 2003). As new technology matures, the literature also seems to evolve from various clinically based groups characterizing some properties of the device. There is often then a document outlining appropriate definitions and implementation of the device to the wider community; then finally comprehensive QA documents are produced.

References

AAPM (American Association of Physicists in Medicine) (2003) Quality assurance for computed-tomography simulators and the computed-tomography simulation process. Report of Task Group 66. *Med. Phys.* 30: 2762–2792. Also available as AAPM Report No. 83. Medical Physics Publishing, Madison, Wisconsin.

AAPM (American Association of Physicists in Medicine) (2001a) Report No. 72: Basic applications of multileaf collimators. Report of Task Group 50. Medical Physics Publishing, Madison, Wisconsin.

AAPM (American Association of Physicists in Medicine) (2001b) Clinical use of electronic portal imaging. Report of Task Group 58. *Med. Phys.* 28: 712–737. Also available as AAPM Report No. 75. Medical Physics Publishing, Madison, Wisconsin.

AAPM (American Association of Physicists in Medicine) (1997) Quality assurance for clinical radiotherapy treatment planning. Report of Task Group 53. *Med. Phys.* 25: 1773–1829. Also available as AAPM Report No. 62.

AAPM (American Association of Physicists in Medicine) (1995) Report No. 55: Radiation treatment planning dosimetry verification. Report of Task Group 23.

AAPM (American Association of Physicists in Medicine) (1994) Comprehensive QA for radiation oncology: Report of AAPM radiation therapy committee task group 40. *Med. Phys.* 21: 582–618. Also available as AAPM Report No. 46.

AAPM (American Association of Physicists in Medicine) (1990) Report No. 31: Standardized methods for measuring diagnostic x-ray exposures. Report of Task Group 8. American Institute of Physics, New York.

AAPM (American Association of Physicists in Medicine) (1986) Report No. 17: The physical aspects of total and half body photon irradiation. American Institute of Physics, New York.

AAPM (American Association of Physicists in Medicine) (1984) Report No. 13: Physical aspects of quality assurance in radiation therapy. American Institute of Physics, New York.

Able C.M., Thomas M.D. (2005) Quality assurance: Fundamental reproducibility tests for 3-D treatment-planning systems. *J. Appl. Clin. Med. Phys.* 6: 13–22.

ACPSEM (Australasian College of Physical Scientists and Engineers in Medicine) (1997) Position Paper: Recommendations for the safe use of external beams and sealed brachytherapy sources in radiation oncology. *Suppl.* 20: 1–35.

Alecu R., Loomis T., Alecu J., Ochran T. (1999) Guidelines on the implementation of diode in vivo dosimetry programs for photon and electron external beam therapy. *Med. Dosim.* 24: 5–12.

Ansbacher W. (2006) Three-dimensional portal image-based dose reconstruction in a virtual phantom for rapid evaluation of IMRT plans. *Med. Phys.* 33: 3369–3382.

Audet C., Hilts M., Jirasek A., Duzenli C. (2002) CT gel dosimetry technique: Comparison of a planned and measured 3D stereotactic dose volume. *J. Appl. Clin. Med. Phys.* 3: 110–118.

Baldock C., Harris P.J., Piercy A.R., Healy B. (2001) Experimental determination of the diffusion coefficient in two-dimensions in ferrous sulphate gels using the finite element method. *Australas. Phys. Eng. Sci. Med.* 24: 19–30.

Baldock C., Burford R.P., Billingham N., Wagner G.S., Patval S., Badawi R.D., Keevil S.F. (1998a) Experimental procedure for the manufacture and calibration of polyacrylamide gel (PAG) for magnetic resonance imaging (MRI) radiation dosimetry. *Phys. Med. Biol.* 43: 695–702.

Baldock C., Rintoul L., Keevil S.F., Pope J.M., George G.A. (1998b) Fourier transform Raman spectroscopy of polyacrylamide gels (PAGs) for radiation dosimetry. *Phys. Med. Biol.* 43: 3617–3627.

Beddar A.S., Salehpour M., Briere T.M., Hamidian H., Gillin M.T. (2005) Preliminary evaluation of implantable MOSFET radiation dosimeters. *Phys. Med. Biol.* 50: 141–149.

Biggs P.J., Goitein M., Russell M.D. (1985) A diagnostic x-ray field verification device for a 10 MV linear accelerator. *Int. J. Radiat. Oncol. Biol. Phys.* 11: 635–643.

Bloemen-van Gurp E.J., Minken A.W., Mijnheer B.J., Dehing-Oberye C.J., Lambin P. (2006) Clinical implementation of MOSFET detectors for dosimetry in electron beams. *Radiother. Oncol.* 80: 288–295.

Boyer A.L., Ochran T.G., Nyerick C.E., Waldron T.J., Huntzinger C.J. (1992) Clinical dosimetry for implementation of a multileaf collimator. *Med. Phys.* 19: 1255–1261.

Brahme A. (1984) Dosimetric precision requirements in radiation therapy. *Acta Oncol.* 23: 379–391.

Brahme A., Chauvaaudra J., Landberg T., McCullough E., Nusslin F., Rawlinson A., Svensson G., Svensson H. (1988) Accuracy requirements and quality assurance of external beam therapy with photons and electrons. *Acta Oncol.* 27: *Suppl.* 1.

Butson M.J, Cheung T., Yu, P. (2005) Peripheral dose measurement with a MOSFET detector. *Appl. Radiat. Isot.* 62: 631–634.

Butson M., Cheung T., Yu P.K., Metcalfe P.E. (2000) Assessment of large single-fraction, low-energy X-ray dose with radiochromic film. *Int. J. Radiat. Oncol. Biol. Phys.* 46: 1071–1075.

Ceberg C.P., Bjärngard B.E., Zhu T.C. (1995) Experimental determination of the dose kernel in high-energy x-ray beams. *Med. Phys.* 23: 505–511.

Chappell R., Fowler J.F. (1994) Steepness of dose-response curve for larynx cancer. *Radiother. Oncol.* 30: 90.

Ciocca M., Orecchia R., Garibaldi C., Rondi E., Luini A., Gatti G., Intra M., et al. (2003) In vivo dosimetry using radiochromic films during intraoperative electron beam radiation therapy in early-stage breast cancer. *Radiother. Oncol.* 69: 285–289.

Constantinou C. (1993) *Protocol and Procedures for Quality Assurance of Linear Accelerators*. Constantinou, Brockton, Massachusetts.

Cosset J.M. (2002) ESTRO Breur Gold Medal Award Lecture 2001: Irradiation accidents—lessons for oncology? *Radiother. Oncol.* 63: 1–10.

Day M.J., Stein G. (1950) Chemical effects of ionizing radiation in some gels. *Nature* 166: 146–147.

De Deene Y., Hurley C., Venning A., Vergote K., Mather M., Healy B.J., Baldock C. (2002) A basic study of some normoxic polymer gel dosimeters. *Phys. Med. Biol.* 47: 3441–3463.

De Deene Y., De Wagter C., Van Duyse B., Derycke S., Mersseman B., De Gersen W., Voet T., Achten E., De Neve W. (2000) Validation of MR-based polymer gel dosimetry as a preclinical three-dimensional verification tool in conformal radiotherapy. *Magn. Reson. Med.* 43: 116–125.

De Deene Y., De Wagter C., Van Duyse B., Derycke S., De Neve W., Achten E. (1998) Three-dimensional dosimetry using polymer gel and magnetic resonance imaging applied to the verification of conformal radiation therapy in head-and-neck cancer. *Radiother. Oncol.* 48: 283–291.

Desai K.R., Pezner R.D., Lipsett J.A., Vora N.L., Luk K.H., Wong J.Y., Chan S.L., et al. (1988) Total skin electron irradiation for mycosis fungoides: Relationship between acute toxicities and measured dose at different anatomic sites. *Int. J. Radiat. Oncol. Biol. Phys.* 15: 641–645.

Dixon P., O'Sullivan B. (2003) Radiotherapy quality assurance: Time for everyone to take it seriously. *Eur. J. Cancer* 39: 423–429.

Duthoy W., De Gersem W., Vergote K., Coghe M., Boterberg T., De Deene Y., De Wagter C., Van Belle S., De Neve W. (2003) Whole abdominopelvic radiotherapy (WAPRT) using intensity modulated arc therapy (IMAT): first clinical experience. *Int. J. Radiat. Oncol. Biol. Phys.* 57: 1019–1032.

Ebert M., Kron T. (1994) Portal films for relative exit dose determination. *Australas. Phys. Eng. Sci. Med.* 17: 71–78.

Engstrom P.E., Haraldsson P., Landberg T., Sand H.H., Aage E.S., Nystrom H. (2005) In vivo dose verification of IMRT treated head and neck cancer patients. *Acta Oncol.* 44: 572–578.

Essers M., Boellaard R., van Herk M., Lanson H., Mijnheer B. (1996) Transmission dosimetry with a liquid-filled electronic portal imaging device. *Int. J. Radiat. Oncol. Biol. Phys.* 34: 931–941.

ESTRO (European Society for Therapeutic Radiology and Oncology) (1995) Advisory report to the Commission of the EU for the Europe Against Cancer Program. Quality Assurance in Radiotherapy. *Radiat. Ther. Oncol.* 35: 61–73.

ESTRO (European Society for Therapeutic Radiology and Oncology) (2004) Quality assurance of treatment planning systems: Practical examples for non IMRT photon beams. ESTRO, Brussels, Belgium.

Ezzell G.G., Galvin J.M., Low D., Palta J.R., Rosen I., Sharpe M.B., Xia P., et al. (2003) Guidance document on delivery, treatment planning, and clinical implementation of IMRT: Report of the IMRT Subcommittee of the AAPM Radiation Therapy Committee. *Med. Phys.* 30: 2089–2115. Also available as AAPM Report No. 83.

Fiorino C., Corletto D., Mangili P., Broggi S., Bonini A., Cattaneo G.M., Parisi R., Rosso A., Signorotto P., Villa E., Calandrino R. (2000) Quality assurance by systematic in vivo dosimetry: Results on a large cohort of patients. *Radiother. Oncol.* 56: 85–95.

Fong P.M., Keil D.C., Does M.D., Gore J.C. (2001) Polymer gels for magnetic resonance imaging of radiation dose distributions at normal room atmosphere. *Phys. Med. Biol.* 46: 3105–3113.

Fraass B.A., Lash K.L., Matrone G.M., Volkman S.K., McShan D.L., Kessler M.L., Lichter A.S. (1998) The impact of treatment complexity and computer-control delivery technology on treatment delivery errors. *Int. J. Radiat. Oncol. Biol. Phys.* 42: 651–659.

Fricke H., Morse S. (1927) The chemical action of roentgen rays on dilute ferrous sulphate solutions as a measure of radiation dose. *AJR Radium Ther. Nucl. Med.* 18: 430–432.

Galvin J.M., Ezzell G., Eisbrauch A., Yu C., Butler B., Xiao Y., Rosen I., et al. (2004) Implementing IMRT in clinical practice: A joint document of the American Society for Therapeutic Radiology and Oncology and the American Association of Physicists in Medicine. *Int. J. Radiat. Oncol. Biol. Phys.* 58: 1616–1634.

Gamble L.M., Farrell T.J, Jones G.W., Hayward J.E. (2005) Two-dimensional mapping of underdosed areas using radiochromic film for patients undergoing total skin electron beam radiotherapy. *Int. J. Radiat. Oncol. Biol. Phys.* 62: 920–924.

Gore J.C., Ranade M., Maryanski M.J., Schulz R.J. (1996) Radiation dose distributions in three dimensions from tomographic optical density scanning of polymer gels: I. Development of an optical scanner. *Phys. Med. Biol.* 41: 2695–2704.

Gore J.C., Kang Y.S., Schulz R.J. (1984) Measurement of radiation dose distributions by nuclear magnetic resonance (NMR) imaging. *Phys. Med. Biol.* 29: 1189–1197.

Greer P.B. (2005) Correction of pixel sensitivity variation and off-axis response for amorphous silicon EPID dosimetry. *Med. Phys.* 32: 3558–3568.

Greer P.B., Popescu C.C. (2003) Dosimetric properties of an amorphous silicon electronic portal imaging device for verification of dynamic intensity modulated radiation therapy. *Med. Phys.* 30: 1618–1627.

Guo P., Adamovics J.A., Oldham M. (2006) Characterization of a new radiochromic three-dimensional dosimeter. *Med. Phys.* 33: 1338–1345.

Gustavsson H., Karlsson A., Back S.A., Olsson L.E., Haraldsson P., Engstrom P., Nystrom H. (2003) MAGIC-type polymer gel for three-dimensional dosimetry: Intensity-modulated radiation therapy verification. *Med. Phys.* 30: 1264–1271.

Heukelom S., Lanson J.H., van Tienhoven G., Mijnheer B.J. (1991) In vivo dosimetry during tangential breast treatment. *Radiother. Oncol.* 22:269–279.

Higgins P.D., Alaei P., Gerbi B.J., Dusenbery K.E. (2003) In vivo diode dosimetry for routine quality assurance in IMRT. *Med. Phys.* 30: 3118–3123.

Hill B., Venning A.J., Baldock C. (2005) A preliminary study of the novel application of normoxic polymer gel dosimeters for the measurement of CTDI on diagnostic x-ray CT scanners. *Med. Phys.* 32: 1589–1597.

Howlett S., Duggan L., Bazley S., Kron T. (1999) Selective in vivo dosimetry in radiotherapy using P-type semiconductor diodes: a reliable quality assurance procedure. *Med. Dosim.* 24: 53–56.

Huang G., Medlam G., Lee J., Billingsley S., Bissonnette J.P., Ringash J., Kane G., Hodgson D.C. (2005) Error in the delivery of radiation therapy: Results of a quality assurance review. *Int. J. Radiat. Oncol. Biol. Phys.* 61: 1590–1595.

Huntzinger C., Munro P., Johnson S., Miettinen M., Zankowski C., Ahlstrom G., Glettig R., et al. (2006) Dynamic targeting image-guided radiotherapy. *Med. Dosim.* 31: 113–125.

Huq M.S., Yu Y., Chen Z.P., Suntharalingam N. (1995) Dosimetry characteristics of a commercial multileaf collimator. *Med. Phys.* 22: 241–247.

Huyskens D., Van Dam J., Dutreix A. (1994) Midplane dose determination using in vivo dose measurements in combination with portal imaging. *Phys. Med. Biol.* 39: 1089–1101.

IAEA (International Atomic Energy Agency) (2004) Commissioning and quality assurance of computerized planning systems for radiation treatment of cancer. Technical Report Series 430. IAEA, Vienna, Austria.

IAEA (International Atomic Energy Agency) (2000) Lessons learned from accidental exposures in radiotherapy. Safety Report Series: N17. IAEA, Vienna, Austria

ICRP (International Commission on Radiological Protection) (2001) Prevention of accidental exposures to patients undergoing radiation therapy. ICRP Report 86. Oxford: Pergamon Press, UK.

ICRU (International Commission on Radiation Units and Measurements) (1987) Use of computers in external beam radiotherapy procedures with high energy photons and electrons. Report 42. Bethesda, Maryland.

ICRU (International Commission on Radiation Units and Measurements) (1976). Determination of absorbed dose in a patient irradiated by beams of X or gamma rays in radiotherapy procedures. ICRU Report 24. Bethesda, Maryland.

IEC (International Electrotechnical Commission) (2000) Document 62083. Medical electrical equipment-requirements for the safety of radiotherapy treatment planning systems.

IEC (International Electrotechnical Commission) (1989a) Document 976. Medical electron accelerators-functional performance characteristics.

IEC (International Electrotechnical Commission) (1989b) Document 977. Medical electron accelerators in the range 1 MeV to 50 MeV - guidelines for functional performance characteristics.

IPSM (Institute of Physical Sciences in Medicine) (1988) *Commissioning and Quality Assurance of Linear Accelerators*. IPSM Report 54. IPSM, York, England.

ISO (International Organization of Standards) (1994) Quality systems-model for quality assurance in design, development, production, installation and servicing. Report series ISO 9000. Report ISO 9001.

Jaffray D.A., Drake D.G., Moreau M., Martinez A.A., Wong J.W. (1999) A radiographic and tomographic imaging system integrated into a medical linear accelerator for localization of bone and soft-tissue targets. *Int. J. Radiat. Oncol. Biol. Phys.* 45: 773–789.

Keall P., Baldock C. (1999) A theoretical study of the radiological properties and water equivalence of Fricke and polymer gels used for radiation dosimetry. *Australas. Phys. Eng. Sci. Med.* 22: 85–91.

Kesteloot K. Dutreix A., van der Schueren E. (1993) A model for calculating the costs of in vivo dosimetry and portal imaging in radiotherapy departments. *Radiother. Oncol.* 28: 108–117.

Kirkby C., Sloboda R. (2005a) Comprehensive Monte Carlo calculation of the point spread function for a commercial a-Si EPID. *Med. Phys.* 32: 1115–1127.

Kirkby C., Sloboda R. (2005b) Consequences of the spectral response of an a-Si EPID and implications for dosimetric calibration. *Med. Phys.* 32: 2649–2658.

Klein E.E., Low D.A., Maag D., Purdy J.A. (1996) A quality assurance program for ancillary high technology devices on a dual energy accelerator. *Radiother. Oncol.* 38: 51–60.

Kron T. (1999) Calibration of megavoltage radiation beams. Chapter 9 in *The Modern Technology of Radiation Oncology*. J. Van Dyk (ed.). Medical Physics Publishing, Madison, Wisconsin, pp. 287–312.

Kron T., Yartsev S., Mackie T.R. (2005) Verification dosimetry during treatment for helical tomotherapy using radiographic film. *Australas. Phys. Eng. Sci. Med.* 28: 232–237.

Kron T., Butson M., Hunt F., Denham J. (1996) TLD extrapolation for skin dose determination in vivo. *Radiother. Oncol.* 41: 119–123.

Kron T., Schneider M., Murray A., Mameghan H. (1993) Clinical thermoluminescence dosimetry: How do expectations and results compare? *Radiother. Oncol.* 26:151–161.

Leunens G., Van Dam J., Dutreix A., van der Schueren E. (1993) Importance of in vivo dosimetry as part of a quality assurance program in tangential breast treatments. *Int. J. Radiat. Oncol. Biol. Phys.* 28: 285–296.

Leunens G., Van Dam J., Dutreix A., van der Schueren E. (1990) Quality assurance in radiotherapy by in vivo dosimetry. 2. Determination of the target absorbed dose. *Radiother. Oncol.* 19: 73–87.

Lindskoug B., Lundberg L. (1984) Clinical applications of thermoluminescent dosimetry. In: *Thermoluminescence and Thermoluminescent Dosimetry* by Y. Horowitz. CRC Press, Boca Raton, Florida, pp. 49–132.

LoSasso T., Chui C.S., Ling C.C. (2001) Comprehensive quality assurance for the delivery of intensity modulated radiotherapy with a multileaf collimator used in dynamic mode. *Med. Phys.* 28: 2209–2219.

LoSasso T., Chui C.S., Ling C.C. (1998) Physical and dosimetric aspects of a multileaf collimation system used in the dynamic mode for implementing intensity modulated radiotherapy. *Med. Phys.* 25: 1919–1927.

Louwe R.J., Damen E.M., van Herk M., Minken A.W., Torzsok O., Mijnheer B.J. (2003). Three-dimensional dose reconstruction of breast cancer treatment using portal imaging. *Med. Phys.* 30: 2376–2389.

Marre D., Marinello G. (2004) Comparison of p-type commercial electron diodes for in vivo dosimetry. *Med. Phys.* 31: 50–56.

Maryanski M.J., Gore J.C., Schulz R.J. (1994) Three-dimensional detection, dosimetry and imaging of an energy field by formation of a polymer in a gel. U.S. Patent 5,321,357.

Maryanski M.J., Gore J.C., Kennan R.P., Schulz R.J. (1993) NMR relaxation enhancement in gels polymerized and cross-linked by ionizing radiation: a new approach to 3D dosimetry by MRI. *Magn. Reson. Imaging* 11: 253–258.

Mather M.L., Whittaker A.K., Baldock C. (2002) Ultrasound evaluation of polymer gel dosimeters. *Phys. Med. Biol.* 47: 1449–1458.

McDermott L.N., Wendling M., Sonke J.J., van Herk M., Mijnheer B.J. (2007) Replacing pretreatment verification with in vivo EPID Dosimetry for prostate IMRT. *Int. J. Radiat. Oncol. Biol. Phys.* 67: 1568–1577.

McDermott, L.N., Louwe R.J., Sonke J.J., van Herk M., Mijnheer B.J. (2004). Dose-response and ghosting effects of an amorphous silicon electronic portal imaging device. *Med. Phys.* 31: 285–295.

McNutt T.R., Mackie T.R., Reckwerdt P., Paliwal B.R. (1996) Modeling dose distributions from portal dose images using the convolution/superposition method. *Med. Phys.* 23: 1381–1392.

Menon G.V., Sloboda R.S. (2004) Quality assurance measurements of a-Si EPID performance. *Med. Dosim.* 29: 11–17.

Mijnheer B.J., Battermann J.J., Wambersie A. (1987) What degree of accuracy is required and can be achieved in photon and neutron therapy? *Radiother. Oncol.* 8: 237–253.

Nijsten S.M., Mijnheer B.J., Dekker A.L., Lambin P., Minken A.W. (2007) Routine individualised patient dosimetry using electronic portal imaging devices. *Radiother. Oncol.* 83: 65–75.

Paliwal B., Tomé W., Richardson S., Mackie T.R. (2000) A spiral phantom for IMRT and tomotherapy treatment delivery verification. *Med. Phys.* 27: 2503–2507.

Partridge M., Ebert M., Hesse B.M. (2002) IMRT verification by three-dimensional dose reconstruction from portal beam measurements. *Med. Phys.* 29: 1847–1858.

Pasma K.L., Dirkx M.L., Kroonwijk M., Visser A.G., Heijmen B.J. (1999) Dosimetric verification of intensity modulated beams produced with dynamic multileaf collimation using an electronic portal imaging device. *Med. Phys.* 26: 2373–2378.

Pasma K.L., Kroonwijk M., de Boer J.C., Visser A.G., Heijmen B.J. (1998) Accurate portal dose measurement with a fluoroscopic electronic portal imaging device (EPID) for open and wedged beams and dynamic multileaf collimation. *Phys. Med. Biol.* 43: 2047–2060.

Pawlicki T., Mundt A. (2007) Quality in radiation oncology. *Med. Phys.* 34: 1529–1534.

Piermattei A., Fidanzio A., Stimato G., Azario L., Grimaldi L., D'Onofrio G., Cilla S., et al. (2006) In vivo dosimetry by an aSi-based EPID. *Med. Phys.* 33: 4414–4422.

Pouliot J., Bani-Hashemi A., Chen J., Svatos M., Ghelmansarai F., Mitschke M., Aubin M., et al. (2005) Low-dose megavoltage cone-beam CT for radiation therapy. *Int. J. Radiat. Oncol. Biol. Phys.* 61: 1552–1560.

Reich H. (1990). Dosimetric ionisirender Strahlung. Teubner, Stuttgart, Germany.

Richardson S.L., Tomé W.A., Orton N.P., McNutt T., Paliwal B.R. (2003) IMRT delivery verification using a spiral phantom. *Med. Phys.* 30: 2405–2411.

Rikner G., Grusell E. (1987) Patient dose measurements in photon fields by means of semiconductor detectors. *Med. Phys.* 17: 870–873.

Rosenfeld A.B. (2002) MOSFET dosimetry on modern radiation oncology modalities. *Radiat. Prot. Dosimetry* 101: 393–398.

Ruden B.I. (1976) Evaluation of the clinical use of TLD. *Acta Radiol. Ther. Phys. Biol.* 15: 447–464.

Saw C.B., Ayyangar K.M., Zhen W., Thompson R.B., Enke C.A. (2001) Commissioning and quality assurance for MLC based IMRT. *Med. Dosim.* 26: 125–133.

Scalchi P., Francescon P., Rajaguru P. (2005) Characterization of a new MOSFET detector configuration for in vivo skin dosimetry. *Med. Phys.* 32: 1571–1578.

Scalchi P., Francescon P. (1998) Calibration of a mosfet detection system for 6-MV in vivo dosimetry. *Int. J. Radiat. Oncol. Biol. Phys.* 40: 987–993.

Steciw S., Warkentin B., Rathee S., Fallone B.G. (2005) Three-dimensional IMRT verification with a flat-panel EPID. *Med. Phys.* 32: 600–612.

Theocharopoulos N., Damilakis J., Perisinakis K., Gourtsoyiannis N. (2007) Energy imparted-based estimates of the effect of z overscanning on adult and pediatric patient effective doses from multi-slice computed tomography. *Med. Phys.* 34: 1139–1152.

Thwaites D., Mijnheer B., Mills J. IAEA (2003) Chapter 12 in *Review of Radiation Oncology Physics: A Handbook for Teachers and Students*. E. Podgorsak (editor). IAEA Educational Series, Vienna, Austria, pp. 335–366.

Tomé W.A., Meeks S.L., Orton N.P., Bouchet L.G., Bova F.J. (2002) Commissioning and quality assurance of an optically guided three dimensional target localization system for radiotherapy. *Med. Phys.* 29: 1781–1989.

Van Dyk J., Purdy J. (1999) Clinical implementation of technology and the quality assurance process. Chapter 2 in *The Modern Technology of Radiation Oncology*. J. Van Dyk (ed.). Medical Physics Publishing, Madison, Wisconsin, pp. 19–51.

Van Dyk J., Barnett R.B., Cygler J.E., Schragge P.C. (1993) Commissioning and quality assurance of treatment planning computers. *Int. J. Radiat. Oncol. Biol. Phys.* 26: 261–273.

Van Dam J., Vaerman C., Blanckaert N., Leunens G., Dutreix A., van der Scheuren E. (1992) Are port films reliable for exit dose measurements? *Radiother. Oncol.* 25: 67–72.

Van Dam J., Marinello G. (1994) *Methods for In Vivo Dosimetry in External Radiotherapy*. Garant Publishers, Leuven, Belgium.

van Elmpt W.J., Nijsten S.M., Schiffeleers R.F., Dekker A.L., Mijnheer B.J., Lambin P., Minken A.W. (2006) A Monte Carlo based three-dimensional dose reconstruction method derived from portal dose images. Med. Phys. 33: 2426–2434.

Venning A.J., Nitschke K.N., Keall P.J., Baldock C. (2005) Radiological properties of normoxic polymer gel dosimeters. *Med. Phys.* 32: 1047–1053.

Vergote K., De Deene Y., Duthoy W., De Gersem W., de Neve W., Achten E., De Wagter C. (2004) Validation and application of polymer gel dosimetry for the dose verification of an intensity-modulated arc therapy (IMAT) treatment. *Phys. Med. Biol.* 49: 287–305.

Vergote K., De Deene Y., Claus F., De Gersem W., Van Duyse B., Paelinck L., Achten E., De Neve W., De Wagter C. (2003) Application of monomer/polymer gel dosimetry to study the effects of tissue inhomogeneities on intensity-modulated radiation therapy (IMRT) dose distributions. *Radiother. Oncol.* 67: 119–128.

Vieira S.C., Bolt R.A., Dirkx M.L., Visser A.G., Heijmen B.J. (2006) Fast, daily linac verification for segmented IMRT using electronic portal imaging. *Radiother. Oncol.* 80: 86–92.

Warkentin B., Steciw S., Rathee S., Fallone B.G. (2003). Dosimetric IMRT verification with a flat-panel EPID. *Med. Phys.* 30: 3143–3155.

Wendling M., Louwe R.J., McDermott L.N., Sonke J.J., van Herk M., Mijnheer B.J. (2006) Accurate two-dimensional IMRT verification using a back-projection EPID dosimetry method. *Med. Phys.* 33: 259–273.

Wertz H., Boda-Heggemann J., Walter C., Dobler B., Mai S., Wenz F., Lohr F. (2007) Image-guided in vivo dosimetry for quality assurance of IMRT treatment for prostate cancer. *Int. J. Radiat. Oncol. Biol. Phys.* 67: 288–295.

Williams P.C. (2003) IMRT: delivery techniques and quality assurance. *Br. J. Radiol.* 76: 766–776.

Wong J.W., Slessinger E.D., Hermes R.E., Offutt C.J., Roy T., Vannier M.W. (1990) Portal dose images I: Quantitative treatment plan verification. *Int. J. Radiat. Oncol. Biol. Phys.* 18: 1455–1463.

WHO (World Health Organization) (1988) Quality assurance in radiotherapy. WHO Geneva, Switzerland.

Xia P., Verhey L.J. (2001) Delivery systems of intensity-modulated radiotherapy using conventional multileaf collimators. *Med. Dosim.* 26: 169–177.

Ying X.G., Geer L.Y., Wong J.W. (1990) Portal dose images II: Patient dose estimation. *Int. J. Radiat. Oncol. Biol. Phys.* 18: 1465–1475.

Questions

11.1 Outline the main stages in commissioning a new linear accelerator. (section 11.2)

11.2 Define the terms "mechanical isocenter" and "radiation isocenter". (section 11.3)

11.3 Name the five common imaging QA tests for an EPID. (section 11.3)

11.4 Describe the challenges of using EPIDs as IMRT dosimetry check arrays. (section 11.4)

11.5 Because there are a large number of beam segments delivered during IMRT, manual check of the patient treatment plan is impossible. MU checking programs can now check IMRT planned doses. Why is this not adequate? Outline some options for dose delivery QA for IMRT prior to patient treatment. (section 11.4)

12 Patient Immobilization and Image Guidance

"There was a general acknowledgment that anatomic motion was abhorrent, but change is on the way with 4-D and adaptive therapy during imaging, planning and treatment."

(Paul Keall, 2007 private communication)

12.1 Introduction

The ultimate goal of radical radiotherapy is to deliver a high radiation dose to a target while minimizing the dose to surrounding healthy tissues. In order to fulfill this aim, it is essential to consider what constitutes the *target* and how to ensure that the target is in the correct location for treatment. The present chapter (section 12.2) commences with a discussion of target volumes as they are developed in the highly influential reports 50 and 62 of the International Commission on Radiological Units and Measurements (ICRU) (ICRU 1993, 1999). An important aspect of target definition is the concept of margins to ensure that the anatomical volume that must be irradiated is actually receiving the correct dose of radiation on every treatment day (or at least on an acceptable number of treatment days). To optimize radiotherapy delivery, it must be the aim to minimize margins as much as possible. This can be achieved by improving patient setup and immobilization, as will be discussed in section 12.3.

However, knowledge also of the target location on every day of treatment will reduce the uncertainty of dose delivery and therefore the margin required for the delivery. This interfraction motion management can be achieved by using image-guided radiation therapy (IGRT), which is discussed in section 12.4. Another important aspect of uncertainty in delivery is the possible motion of the target during delivery. Accounting for this intrafraction motion is currently an area of intensive research and some aspects of this will be presented in section 12.5.

12.2 The ICRU Target Definitions

The ICRU was founded in 1925 with the brief to develop quantities and units for radiation as well as to establish methods for their measurements and provide the physical data required for this process. The ICRU works now in several technical areas, one of them being radiotherapy. There are important ICRU publications on dosimetry such as the report 64 on radiation beam calibration (ICRU 2001; compare also chapter eight). However, the most influential publications of the ICRU for radiotherapy have been concerned with "Prescribing, Recording and Reporting." This information was covered briefly in chapter six, section 6.9. We delve into it in more detail in this chapter.

These publications provide a uniform system of volumes and concepts for dose reporting in brachytherapy (ICRU 1985, 1998) and external beam radiotherapy (ICRU 1993, 1999, 2004). These publications are centered on the concept of several target volumes as first introduced by ICRU report 50 and illustrated in figure 12.1, such that:

i) The *gross tumor volume* (GTV) is the detectable extent of a tumor. Metastases and lymph nodes, which are demonstrably involved, e.g., by PET imaging, are part of the GTV. The GTV is considered the volume of highest density of malignant cells.
ii) The GTV is surrounded by a *clinical target volume* (CTV) which contains local sub-clinical spread. The density of the tumor cells is typically assumed to reduce with distance from the GTV as demonstrated in the case of breast cancer by Holland et al. (1985). Unlike the GTV, the CTV is determined anatomically not by imaging the tumor itself. If the tumor has been removed by surgery, no GTV exists and the CTV becomes the primary target in adjuvant radiotherapy.
iii) The *planning target volume* (PTV) is a geometric volume around the CTV. The expansion from CTV to PTV involves a margin that depends on the uncertainty of being able to direct the radiation beam to the target. Uncertainties arise from two distinct sources: motion and deformation of the target and the ability to position the target reproducibly in the radiation beam. The ICRU coins the terms internal margin and setup margin for the uncertainties attributable to the target and patient setup, respectively (ICRU 1999). This concept is illustrated in figure 12.2. It is important to note that if the magnitude of these margins change randomly, then the two margins should be added in quadrature, hence

$$IMSM_{combined} = \sqrt{IM^2 + SM^2}. \tag{12.1}$$

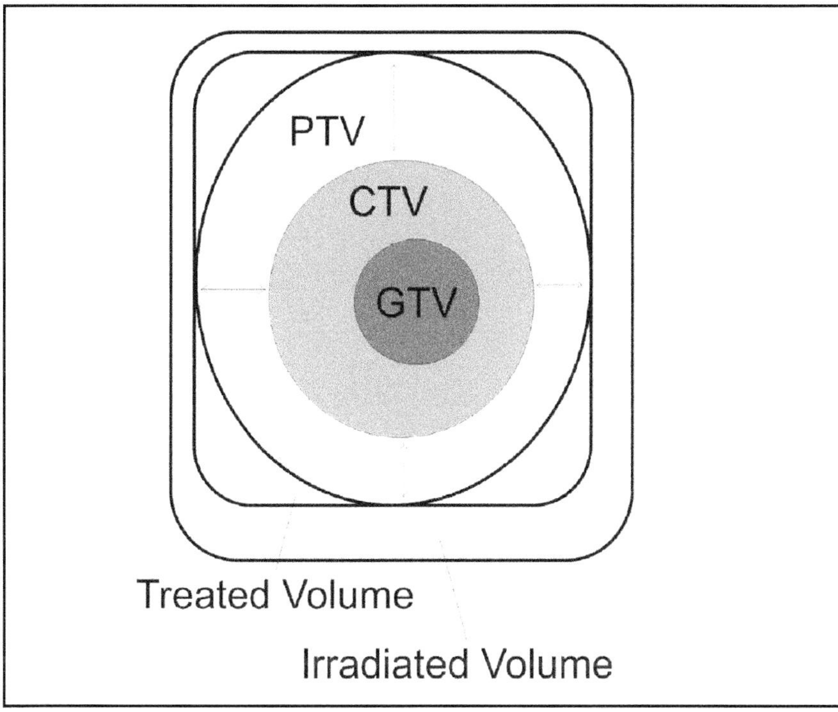

Figure 12.1: Schematic illustration of different volumes used for prescribing and reporting dose. PTV = planning target volume; CTV = clinical target volume; GTV = gross tumor volume. [Adapted from ICRU report 50 (1993).]

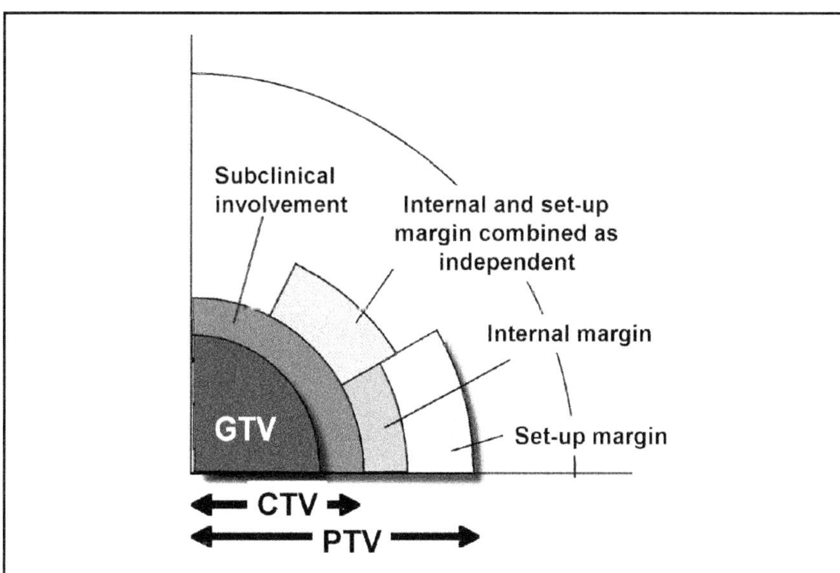

Figure 12.2: Schematic illustration of different target volumes and margins. PTV = planning target volume; CTV = clinical target volume; GTV = gross tumor volume. [Modified from ICRU (1999).]

iv) While PTV margins are typically applied without distinguishing between internal and setup margin, it is at least in principal possible to determine the internal margin by using time-resolved imaging, such as four-dimensional CT (4DCT) scanning (Rietzel et al. 2005, Hugo et al. 2007). This allows the creation of an *internal target volume* (*ITV*) (ICRU 1999). The concept has been used successfully for lung (Jin et al. 2006) and esophageal cancer (Lorchel et al. 2006). It is important to note that short time-resolved imaging does not account for changes that occur over longer time intervals, such as organ deformation due to swelling or organ filling (e.g., bladder and rectum in prostate radiotherapy). Serial scans over intervals of hours or days can be used for this purpose—in particular for determining systematic and random error in organ position. It is important to note that the margins around the different target volumes do not need to be symmetric; however, they always have to be considered in 3-D. For example a margin may be created by expanding the CTV using directions normal to its surface, but with an expansion along each direction that depends on the direction.

12.2.1 Irradiation Volumes

In addition to these target volumes, the ICRU defines two volumes that indicate which volume in the patient is actually receiving dose. Even with the best IMRT delivery techniques it is not possible to confine the high-dose region entirely to the PTV. As such, the *treated volume* (TV) encompasses the total volume that receives at least a dose required to achieve the treatment aim (best possible tumor control probability while maintaining acceptable complication probability). This dose level could be, for example, the prescription dose or the minimum dose to the target volume. Once a reference dose level is chosen, it is easy to calculate a conformity index such as that described by Knoos et al. (1998):

$$CI = \frac{V_{PTV}}{TV}. \tag{12.2}$$

A conformity index (CI) of 1.0 would indicate that the same total volume is covered by the prescription dose as by the PTV. Note however that the actual region in space corresponding to each volume is, in general, not identical even if $CI = 1.0$ (the prescription dose volume may be a different shape to, or offset from, the PTV), so the dose distribution must be examined along with taking note of the CI value.

In addition to this, the ICRU defined the *irradiated volume* (IV) as the volume receiving a dose that is considered significant in relation to normal tissue tolerance (ICRU 1999). The dose level used to define the irradiated volume will vary with treatment scenario.

12.2.2 Organs at Risk

The more recent ICRU reports on prescribing, recording, and reporting radiotherapy, ICRU report 62 (1999) and ICRU report 71 (2004) extend the original concept of target volumes explicitly to *organs at risk* (OAR). The OARs—or critical normal structures as they are often referred to—are "normal tissues...whose radiation sensitivity may significantly influence treatment planning and/or prescription dose" (ICRU 1999). The actual tolerance doses and volume effects concerning these tissues are discussed in more detail in chapter fourteen. The ICRU extends the idea of margins to the OARs. Therefore, a *planning organ at risk volume* (PRV) is derived from the OAR in the same way as the PTV is derived from the CTV.

The concept of PRVs has not yet been generally accepted into radiotherapy treatment planning. While several groups found it useful for spinal cord (Breen et al. 2006) and intestines (Muren et al. 2005, Hysing et al. 2006), there is still some discussion about its usefulness in clinical practice (Stroom and Heijmen 2006). It is also likely that the approach needs to be varied depending on the type of critical structure.

As both PTV and PRV are geometrical volumes they may overlap with each other and other structures.

12.2.3 Margins

When considering figures 12.1 and 12.2 it becomes obvious that one of the aims must be to reduce the PTV margin as much as possible. It is therefore useful to consider all aspects that will impact on these margins:

Evidence suggests that a 7% to 10% change in dose to the target volume may result in significant change in the probability of tumor control (ICRU 1976). Other papers indicate that dose variations with accuracy as low as 5% and as high as 15% may be critical (Brahme 1984, Chappell and Fowler 1994). Mijnheer et al. (1987) using normal tissue control probabilities (NTCPs) and Brahme et al. (1988) using TCPs quote required uncertainties of about 7% (2σ). Considering errors are combined in quadrature, this may be a practical aim within the entire radiotherapy treatment process.

Internal margins account for:

i) Intrafraction organ motion.
ii) Variations in site, size, and shape of the organs and tissues contained in or adjacent to the CTV (ICRU 1999). A very important element of the internal margin to account for is systematic error in target position. This is the difference between the mean position over the course of treatment and the position determined during planning scans. An inadequate margin for systematic error will mean that part of the CTV will be systematically underdosed.

Setup margins account for:

i) Patient positioning.
ii) Dosimetric uncertainties (e.g., positioning of shielding).
iii) Mechanical uncertainties (e.g., gantry sag, couch sag).
iv) Differences between planning equipment (CT, simulator) and treatment units.
v) Human factors (ICRU 1999).

Unfortunately, the distinction between the two margins is not perfect. Setup variations may lead to variations in organ site and shape. However, setup margins are mostly a result of interfraction variations, while internal margins include also a large component of intrafraction variation. Modern image guidance also has the capability to reduce some uncertainties affecting both, internal and setup margin (but not all!). For example, daily CT acquisition of the patient in treatment position allows reduction of internal margin due to determination of organ shape and location while at the same time reducing the uncertainty of setup and patient positioning. As such, it may be increasingly difficult to apply the two margins independently of each other.

12.2.4 A Margin Recipe

A so-called *margin recipe* was derived by van Herk et al. (2000). This is used to derive the magnitude of margin to be placed beyond a CTV to obtain a PTV. Derivation of this was based on the random and systematic geometric target position variations and the effects they have on dose coverage of the CTV. It also takes into account the penumbral width of the beam. For a penumbral width of 3.2 mm the margin recipe was expressed as:

$$IMSM_{combined} \approx 1.5\sum + 0.7\sigma, \tag{12.3}$$

where Σ is the systematic error and is the σ random error. This margin should only be considered as a *lower limit* for safe radiotherapy as it does not take into account rotational positional variations or shape distortion.

12.3 Patient Setup and Immobilization

12.3.1 Patient Support Assembly: Couch

Modern linacs such as the one shown in figure 12.3a have a couch top that includes indexing (see figure 12.3b). This consists of simple recesses on the edge of the couch that immobilization devices clip into. This ensures they can be fixed in the same superior, inferior, and lateral locations on the patient couch for each treatment fraction.

Immobilization Devices

While immobilization aims to provide reproducible patient positioning with minimal patient movement, patient comfort should also be considered; there needs to be a balance between the two aims. Various immobilization devices for use on the linac couch are available; these include:

i) Head cast and support systems provide fixation predominantly for patients undergoing brain and head and neck radiotherapy treatments. The casts may also be used for upper mantle treatments. The

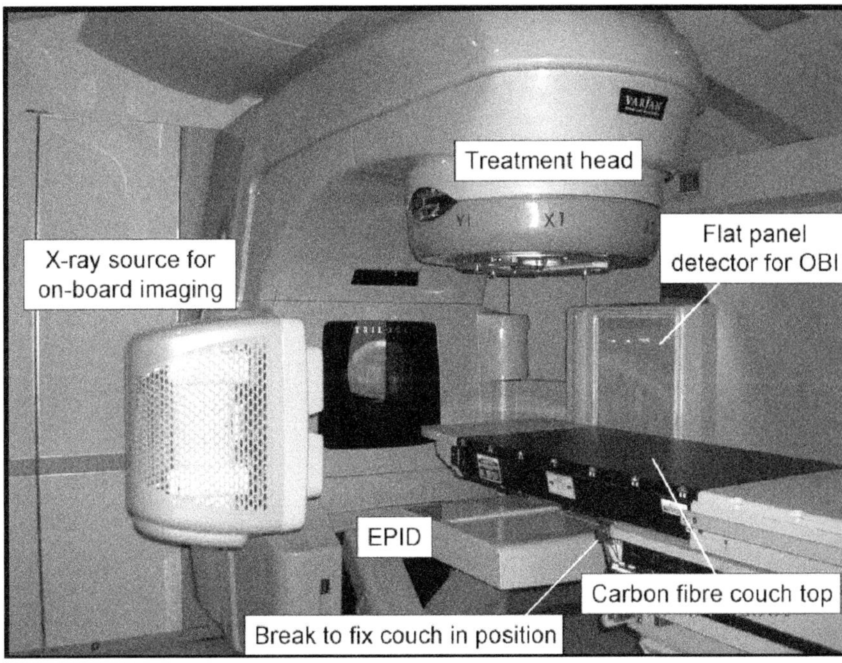

(a)

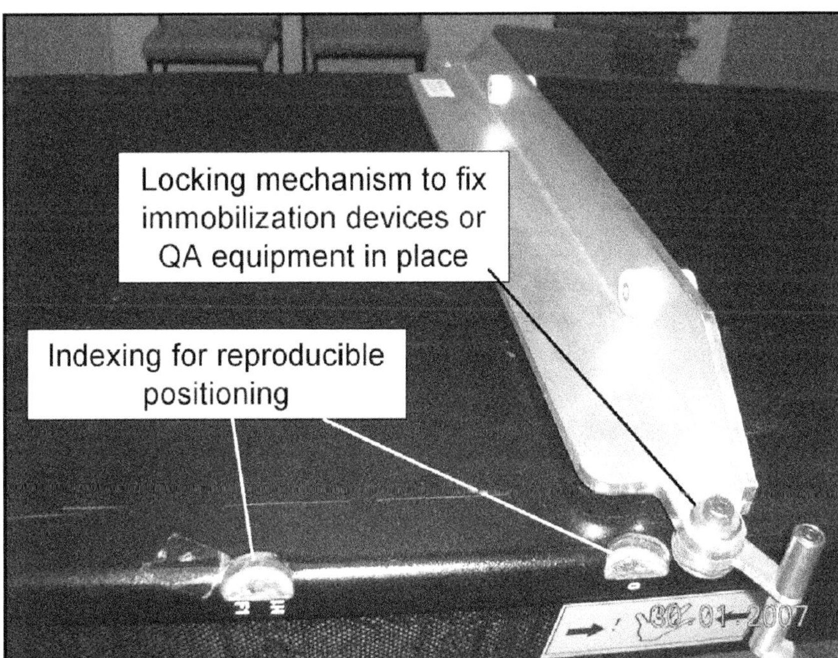

(b)

Figure 12.3: Linac and couch top indexing. (a) Photograph of a linac with carbon fibre couch top shown in relation to linac treatment head, EPID, On-Board Imager (OBI)™ tube, and flat-panel detector (FPD). (b) Close-up view of couch top and indexing device.

head and chin position may be critical for field matching. The head mask can have various points of attachment to the head mask fixation device and sometimes the mask extends to the shoulder region. Molds are commonly made from thermoplastic Orfit™ material, which molds when heated (see figures 12.4a ,12.4b). Figure 12.4c shows a head and shoulder mask mounted to a head board that extends beyond the end of the linac couch, ideal for multi-field IMRT.

ii) Vacuum bags and body casts are filled with small polystyrene beads that mold to the patient's shape when evacuated. These reduce pelvic tilt, which can otherwise occur after the first few fractions when the patients tend to be more relaxed during treatment (see figure 12.5a). Body casts are sometimes used for further immobilization (see figure 12.5b).

iii) Bellyboards for prone positioning (see figure 12.5c) are often used for patients having rectum or other pelvic treatments. The advantage is that the small bowel falls into the gap in the board, away from the treatment volume.

iv) Breast jigs similar to that shown in figure 12.5d are used to help immobilize the patient and to create a sternum that aligns with the field edge without collimator rotation, thus minimizing the lung irradiated. It is important that patients are CT simulated then treated in the identical jig arrangement; e.g., jig wedge angle needs to be identical.

(a)

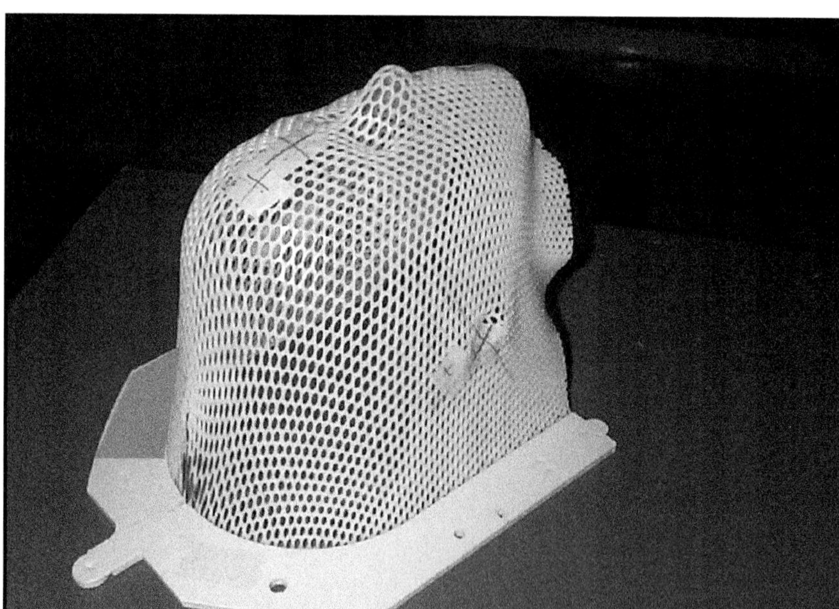

Figure 12.4: Patient head and neck immobilization devices. (a) Patient head cast. (b) Different head supports. (c) Head and shoulder mask mounted to a head board that extends beyond the end of the linac couch, ideal for multi-field IMRT.

v) Knee and ankle supports, often used in combination with other fixation devices, ensure reproducible leg position, which seems to help with reproducible setup (see figure 12.5e).

This is a snapshot of the many devices available. The implementation of immobilization strategies into each clinic usually follows a detailed examination of portal (or other) images collected in the linac bunker on the patient just prior to or just after treatment. If the immobilization method is shown to improve setup reproducibility, then it is implemented for routine use on the cohort group of patients for whom it has been shown to be useful.

A higher level of precision of immobilization can be obtained by using stereotactic frames (these are outlined in chapter seven); however they are not always practical for all patients having day-to-day fractionated treatment.

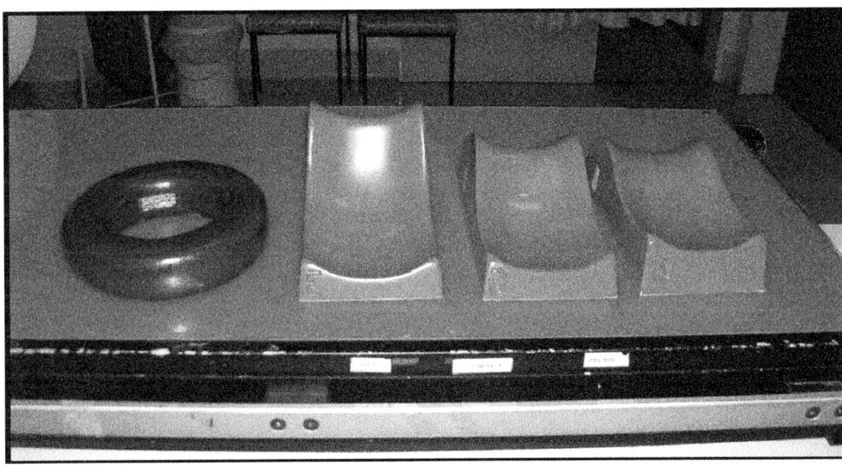

(b)

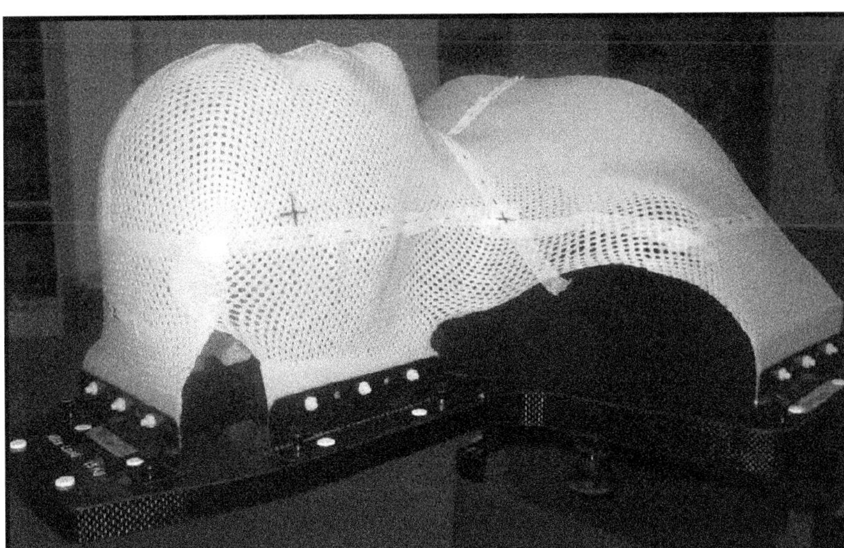

(c)

Figure 12.4: (Continued)

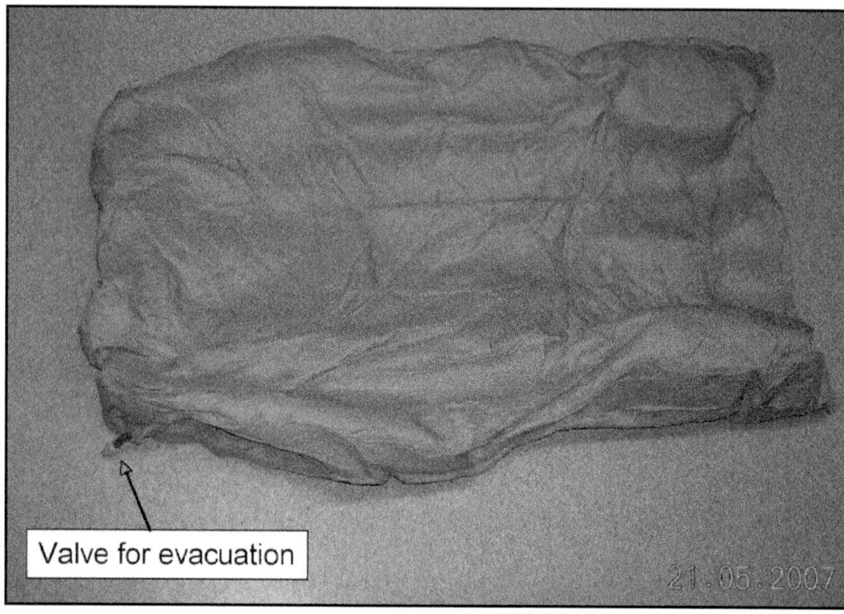

(a)

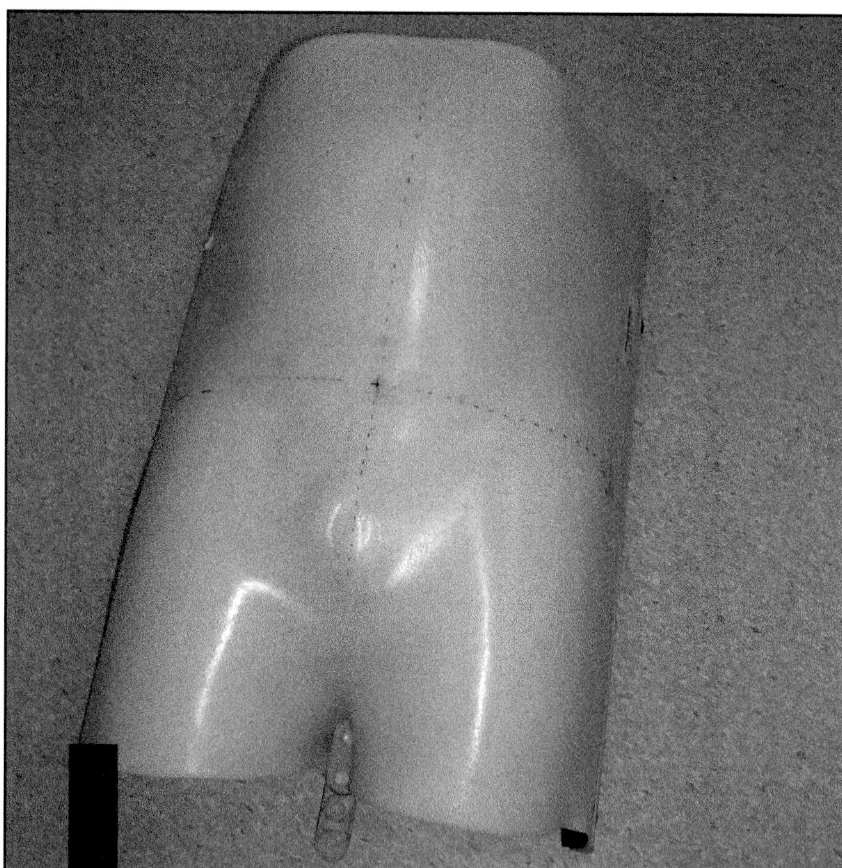

(b)

(c)

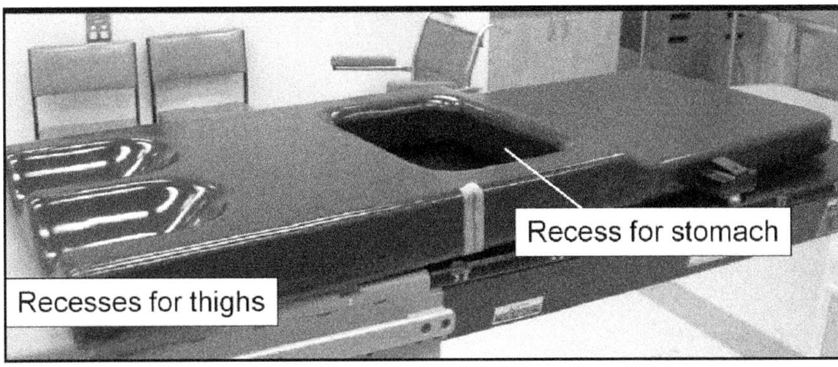

(d)

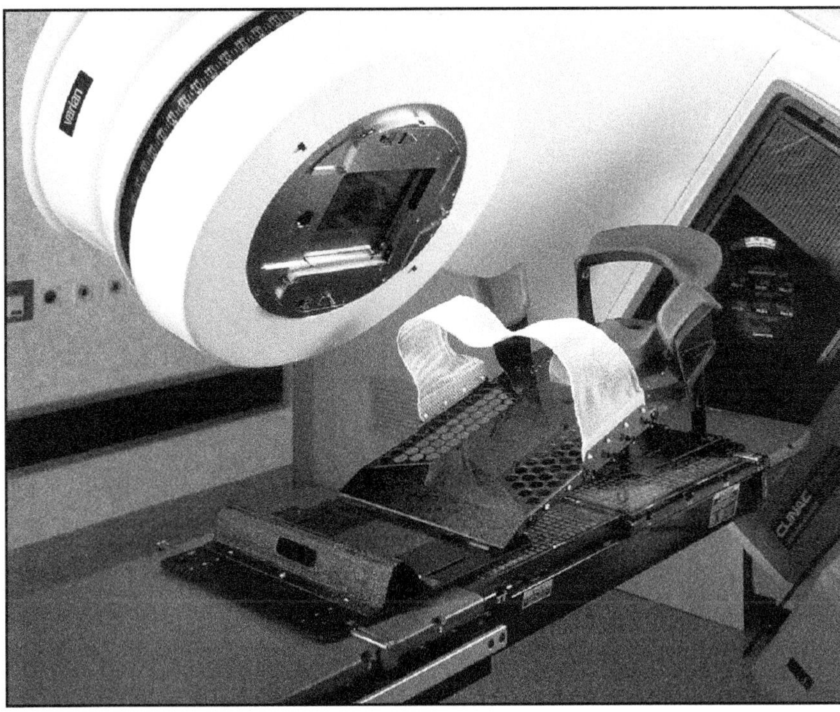

Figure 12.5: Patient body immobilization devices: (a) Vac fix bags: These are filled with small polystyrene balls. When the air is removed from the bag while placed under the patient trunk, a solid impression is made of the patient's shape, then the same device with the fixed impression is used as a custom fit for immobilization of patient during each treatment fraction. (b) Body casts, for abdominal immobilization, are not used at all centers but may aid in further immobilization. (c) Bellyboards for prone positioning are often used for patients having rectum or other pelvic treatments. The advantage is that the small bowel falls into the gap in the board, away from the treatment volume. (d) A typical breast jig. This is used to help immobilize the patient and to create a sternum that aligns with the field edge without collimator rotation.

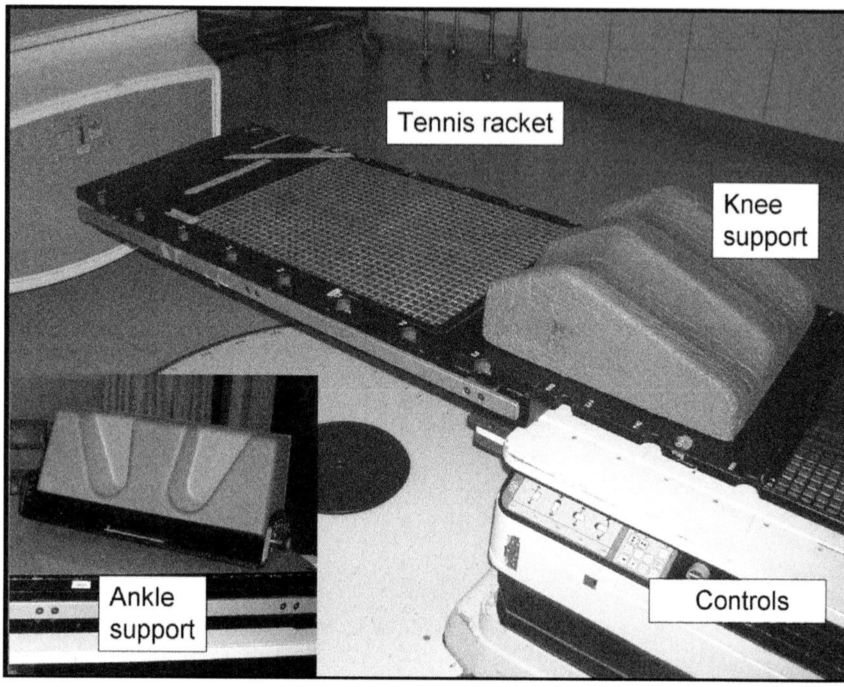

(e)

Figure 12.5: (Continued) (e) Patient body immobilization devices: (e) Knee and ankle supports used to ensure reproducible leg and foot position, which, in turn, stabilizes setup in the trunk region.

12.3.2 External and Internal Markers

Patients are often aligned to external markers either placed on the head cast or positioned at patient tattoo position. External marker positions are usually defined at CT simulation time, whereby external fiducial markers are placed with reference to isocenter. Unfortunately, there is internal organ mobility so external markers can at best only guide the initial treatment setup. While some structures such as the pelvic lymph nodes may follow the bony anatomy quite nicely, other organs such as the prostate undergo significant motion with respect to the bony anatomy. The extent of this motion is summarized in table 12.1.

It is the verification views using electronic portal imaging devices (EPIDs) or linac-mounted imaging devices (LMIDs) that more and more frequently guides *on-line* correction of position. This can be achieved by small repositioning moves of the linac couch in the (x, y, z) axis, essentially guiding the center of each treatment beam to isocenter.

Because the images from EPIDs are not ideal for showing soft tissue anatomy, it is possible to place internal fiducial markers in some organs to guide the treatment, as shown in figure 12.6a and 12.6b (e.g., for prostate typically three gold markers are inserted). In this case, isocenter positioning is being based on soft tissue anatomy surrogates. This is a major mind shift

Table 12.1: Mean and standard deviation (SD) of prostate movements from various reports in the literature. Movements are in centimeters (cm) and axis moves are classified as anterior-posterior, superior-inferior, and left-right

		Ant.-Post (cm)		*Sup.-Inf. (cm)*		*Left-right (cm)*			
Patients	*Method*	*Mean*	*SD*	*Mean*	*SD*	*Mean*	*SD*	*Reference*	*Comments*
55	Fiducial markers	0.56 (post)	0.41	0.59 (inf.)	0.5	0.05	0.15	Crook et al. (1995)	Treated supine
50	CT	0.2 (post)	0.29	0.05 (inf.)	0.33	0.06 (left)	0.08	Zelefsky et al. (1999)	Large deviations linked to large bladder and rectal volumes
6	Ultrasound	0.5	0.4	0.3	0.4	0.3	0.3	Chinnaiyan et al. (2003)	Post prostectomy
10	Ultrasound	0.4	0.3	0.4	0.4	0.4	0.3	Chinnaiyan et al. (2003)	No prostectomy
23	Fiducial markers		0.44		0.37		0.24	Nederveen et al. (2003)	Off-line target position verification allowed reduced margins
13	Fiducial markers		Base 0.29 (ant.) Apex 0.21 (sup.)		Base 0.21 (sup.) Apex 0.21 (sup.)			Wu et al. (2001)	
35	BAT and Portal images	0.13	0.57	0.16	0.64	0.09	0.33	Little et al. (2003)	Comparison between BAT + Portal imaging

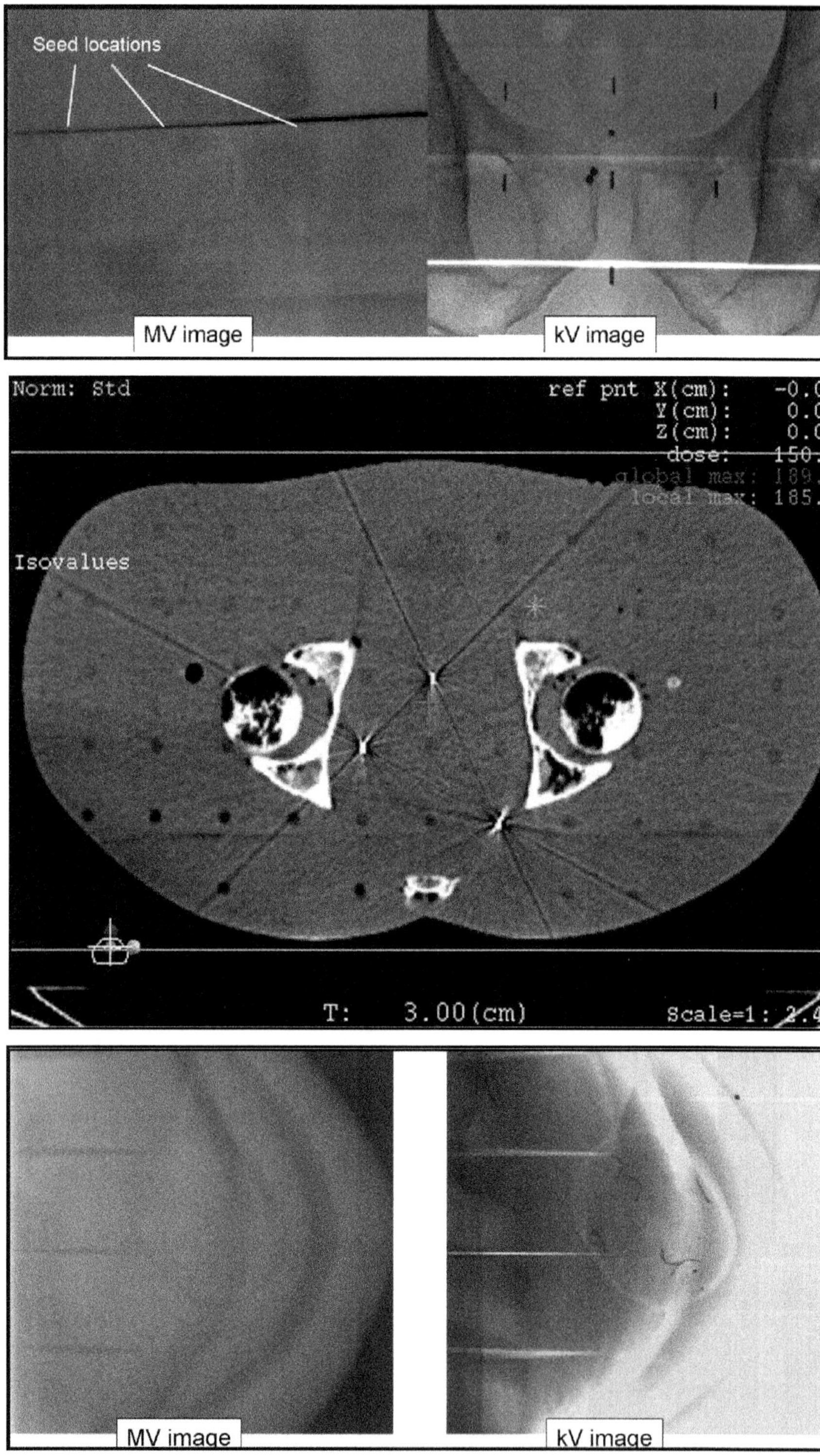

Figure 12.6: Internal fiducial markers. (a) Rando phantom with fiducials showing MV compared with kV beam's-eye view image. (Courtesy of R. Amner.) (b) Rando phantom kV CT image showing three internal gold fiducials. (Courtesy of R. Amner.) (c) Breast phantom showing visibility of surgical clips. (Courtesy of D. Willis.)

away from alignment to bony landmarks, which many centers have traditionally used.

Surgical clips in breast cancer patients can also act as good potential fiducial markers (see figure 12.6c). In other organs it is more difficult to place internal markers without risk of other complications (e.g., lung).

12.4 Interfraction Motion Management: Image Guidance

Several new developments have helped us monitor movement of bony anatomy and soft tissue surrogates within the patient in the linac room. With the exception of EPIDs or port films, which can gather images during treatment, all the other more recent linac add-on devices can monitor patient position just prior to treatment. Image guidance tools that are available include:

i) *EPIDs*. Because they collect image data using the linac's megavoltage beam, with a detector that is relatively insensitive to these energies, they generally give the lowest quality images of this group. They do however work reasonably well with fiducial marker surrogates (see figure 12.6a). While they are able to acquire images through the course of treatment, they are more routinely used to take a snapshot beam's-eye view (BEV) requiring about 2 monitor units just prior to treatment.

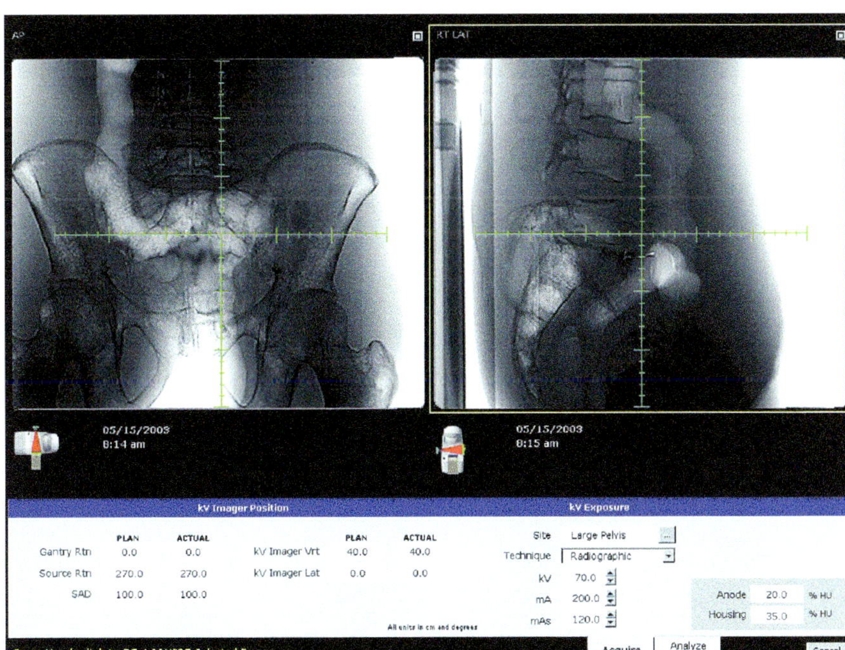

Figure 12.7: Two orthogonal kV images (i.e., two images taken 90° gantry rotation separating the image) acquired using a Varian OBI™ to be used for 3-D patient positioning.

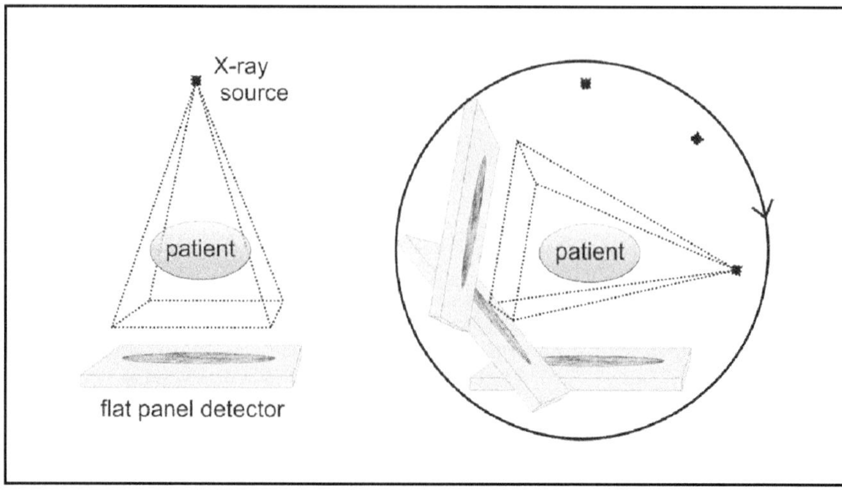

Figure 12.8: Schematic illustration of cone beam CT acquisition principle. The x-ray source and FPD rotate 360° and a pyramidal cone beam is used to acquire the information, whereas in conventional fan beam CT a narrow slit (in the superior-inferior direction) is used. This reduces scatter effects.

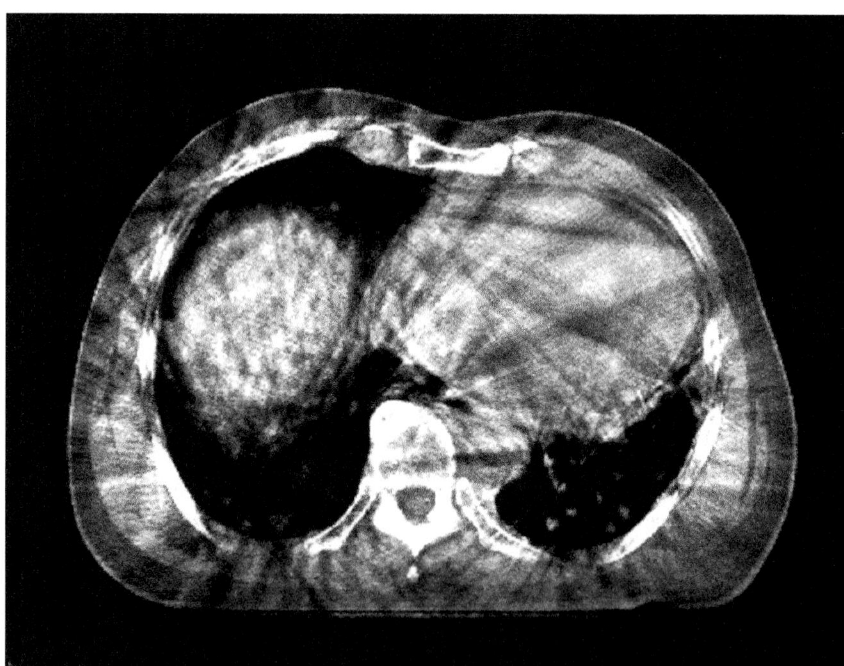

Figure 12.9: Cone beam CT image affected by motion artifacts; lung acquisitions show movement artifacts due to the slow rotation required for the open gantry to acquire the data during several breathing cycles.

ii) *LMIDs*. kV x-ray BEV images taken orthogonal to each other (90-degree gantry rotation between images) are shown in figure 12.7. With these images 3-D patient repositioning using couch moves can be undertaken using *on-line* fraction-by-fraction correction just prior to treatment. The other option is *off-line* correction after analyzing the data for several fractions (See et al. 2000). The bony landmarks show up well, but soft-tissue surrogates such as gold seeds are still probably required for prostate positioning.

iii) *kV cone-beam CT imaging*. This is achieved by rotation of the LMID around the patient using up to a 360-degree gantry rotation. The principle is explained in figure 12.8. Reasonable images are achievable in prostate (see figure 1.23) and head and neck sites. Because of the open gantry design rotation is legally restricted to between 45 and 60 seconds, so there are motion artifacts in some sites, especially in lung as shown in figure 12.9. Vendors that supply options ii and iii are Varian Medical Systems and Elekta.

iv) *CT-on-Rails*. As shown in figure 12.10, a CT scanner is mounted at the end of the linac couch and a 180-degree couch rotation, with the patient on couch, is then required before treatment. The device offers fan beam kilovoltage (diagnostic) quality CT scans. The vendor that offers this option is Siemens Medical Solutions.

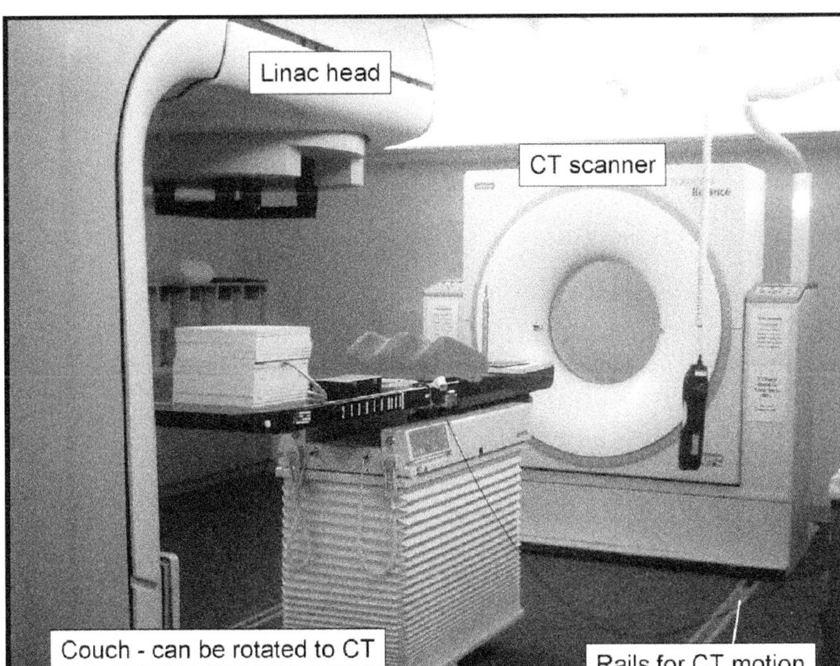

Figure 12.10: CT-on-Rails in linac room, mounted at end of linac couch. Patient has CT just before treatment, then a 180° rotation of couch enables patient treatment by linac.

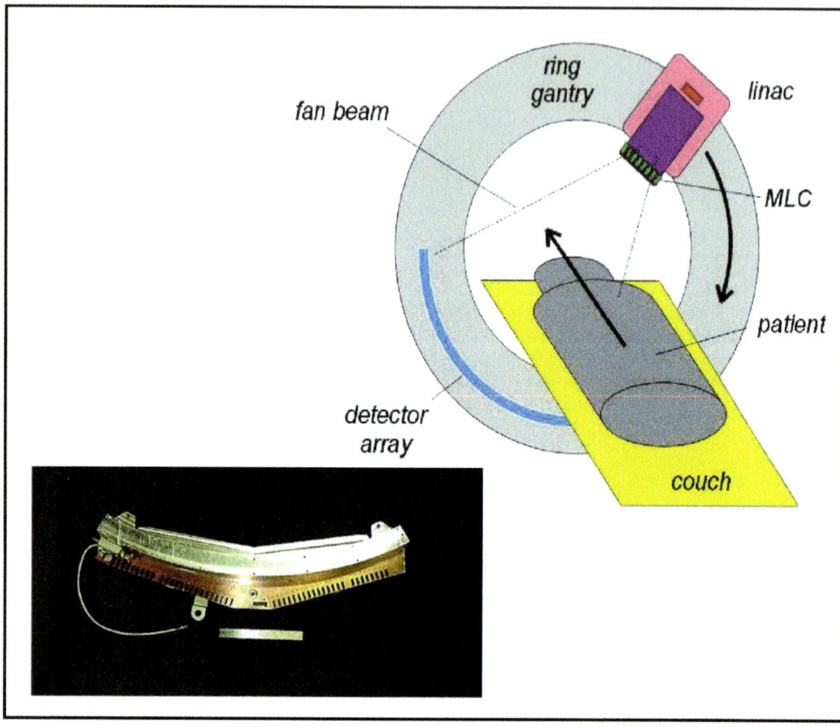

Figure 12.11: Illustration of a fan beam CT acquisition principle as used by the TomoTherapy device. The x-ray source and detector array rotate 360° because in a conventional fan beam CT a narrow slit (in the superior-inferior direction) is used; this reduces scatter effects. The MV beam reduces bone contrast. The closed gantry design ensures no restriction on rotation time.

v) *MV cone-beam CT imaging*. Images are acquired using the linac beam and the existing EPID rotated around the patient. Reasonable quality images can be obtained (see Pouliot et al. 2005), though generally with a higher dose than with kilovoltage cone-beam imaging. The vendor that offers this option is Siemens.

vi) *Fan-beam MV CT*. The TomoTherapy Hi-Art System® uses a megavoltage beam and a conventional CT detector to collect CT images before each treatment (see figure 12.11). Unlike the methods described in (iii) and (v) a fan-beam geometry rather than a cone-beam geometry is used for imaging.

Both MV CT devices offer an advantage when scanning patients with hip prostheses or dental fillings because the high Z material does not produce an artifact at megavoltage energies (see figure 12.12).

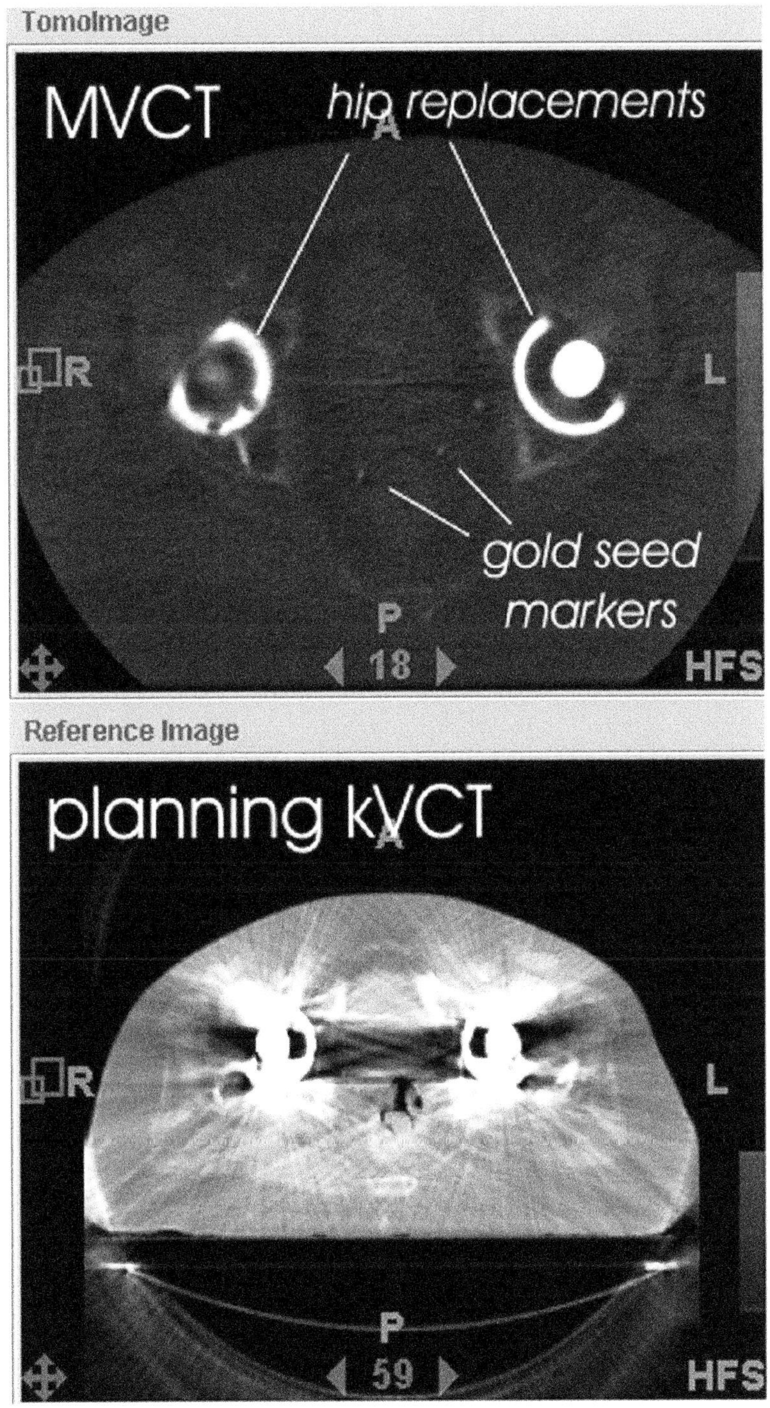

Figure 12.12: Comparison of a MV CT image acquired using the TomoTherapy device and a kV CT image collected using a conventional kV CT of a patient with dual hip replacements. The patient had gold fiducial markers implanted in the prostate and seminal vesicles.

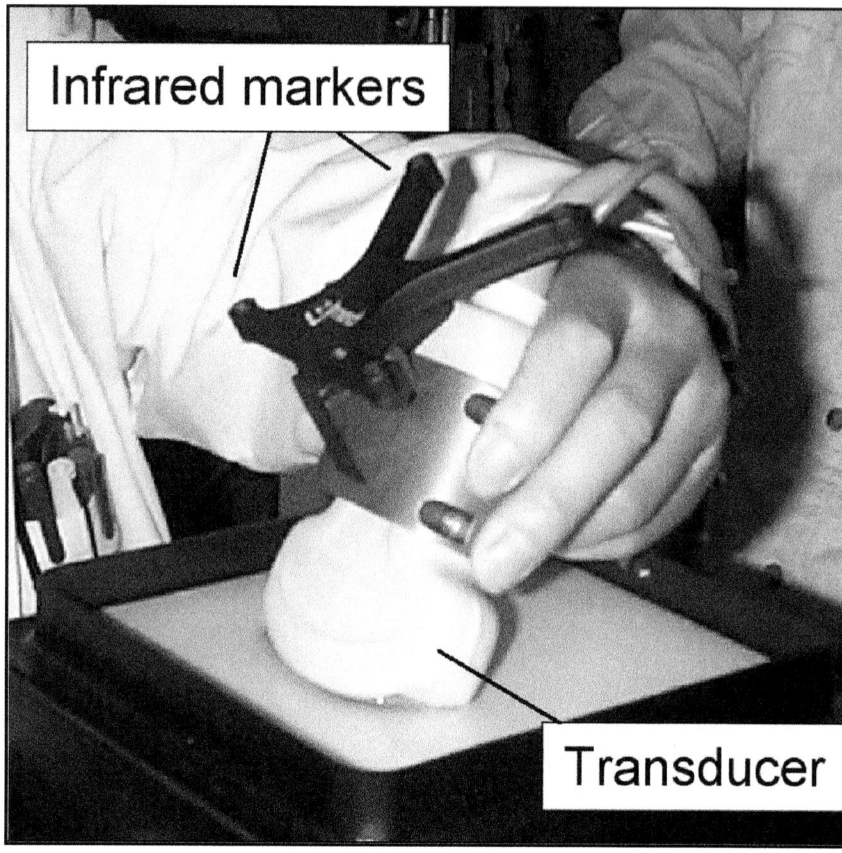

Figure 12.13: Ultrasound probe used for positioning of prostate in the treatment room prior to linac treatment delivery. The exact location of this probe is determined using two infrared cameras. The combination of many 2-D ultrasound images also allows reconstruction of 3-D images. (Courtesy of London Regional Cancer Program, Canada.)

There are two other modalities that also offer imaging of the internal anatomy with no extra ionizing radiation due to their imaging mode; these are:

i) *Ultrasound*, which has been used for in-room verification for prostate position for some time (see also chapter six). An example of such a device is shown in figure 12.13. Ultrasound has been shown to be extremely useful in tracking prostate shifts. The types of images which can be collected using ultrasound are shown in figure 12.14.

   The position of the prostate, as determined by the transabdominal ultrasound imaging, is compared to the previously CT-defined prostate location relative to isocenter. The position of the ultrasound transducer is referenced to the coordinate system of the linac's isocenter, either through physical connection to the gantry, or alternately via an infrared localizing system (Tomé et al. 2002). Typically it is the interface regions between the bladder and prostate and the rectum and prostate which show on ultrasound that are used as reference landmarks. The locations of these interfaces relative to isocenter are compared to the planned location, and the necessary calculated couch shifts relocate the prostate to the correct position.

There have been several studies into the use of transabdominal imaging for tracking prostate motion (e.g., Jani et al. 2005, Lattanzi et al. 2000). The technique allows for the daily relocation of the prostate in an attempt to better cover the mobile prostate organ, leading to potential reduction in the target margin. At the very least ultrasound has pioneered our knowledge base about prostate movements. Some user training is required to use these devices effectively. These devices produce no hazard from additional dose from ionizing radiation for image collection, which is an advantage. Tomé has also suggested prostate treatment of post-prostatectomy patients can also be better located using one of these devices; they used the visualization of the bladder neck as a localization reference for the prostatic fossa (Chinnaiyan et al. 2003).

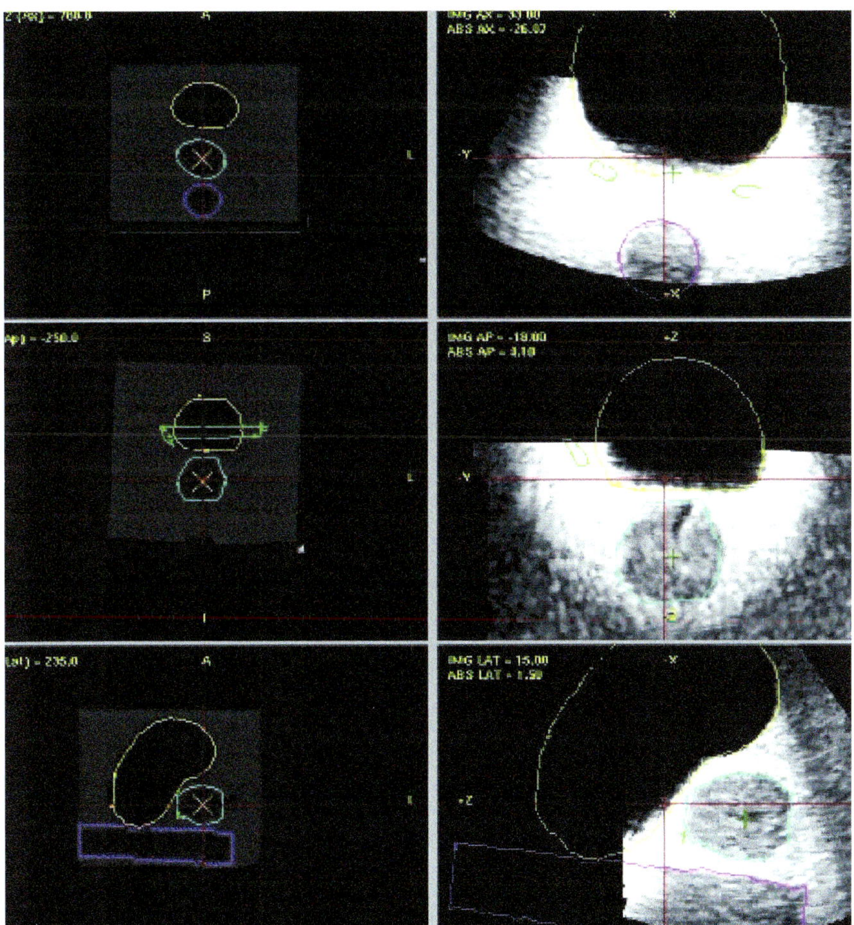

Figure 12.14: Three-dimensional reference image (CT, on the left) and ultrasound image of a prostate phantom. (Courtesy of London Regional Cancer Program, Canada.)

ii) *Magnetic resonance imaging* (MRI) is really the most exciting new modality that is being investigated as an in-linac room or MRI LMID option. There are devices that have been suggested such as the one shown in figure 12.15. This design envisages two cobalt-60 sources, each with a binary multileaf collimator (MLC) sandwiched between high magnetic field strength magnets for collection of MRI information during or just after radiotherapy treatment.

A similar machine called the Renaissance™ (ViewRay, Inc.) is being planned by Dempsey and collaborators. This has three cobalt-60 beam sources, each mounted in its own treatment head, with conventional MLC. The multiple source cobalt-60 heads ensure dose rates approaching those currently provided by linacs. It is suggested the simplicity of a cobalt-60 source drive would reduce the impact of the magnetic field from the MRI on the beam steering.

There are also linac-mounted designs using lower strength magnetic fields. Images from these devices have been reported in the literature (Raaymakers et al. 2004). Electrons generated from x-rays are obviously changed in the tracks they follow and this can have large effects on the dose distribution if large magnetic fields are employed (Bielajew 1993). However at these low magnetic field strengths smaller effects on the linac beam dose distributions have been characterized (Raaijmakers et al. 2005).

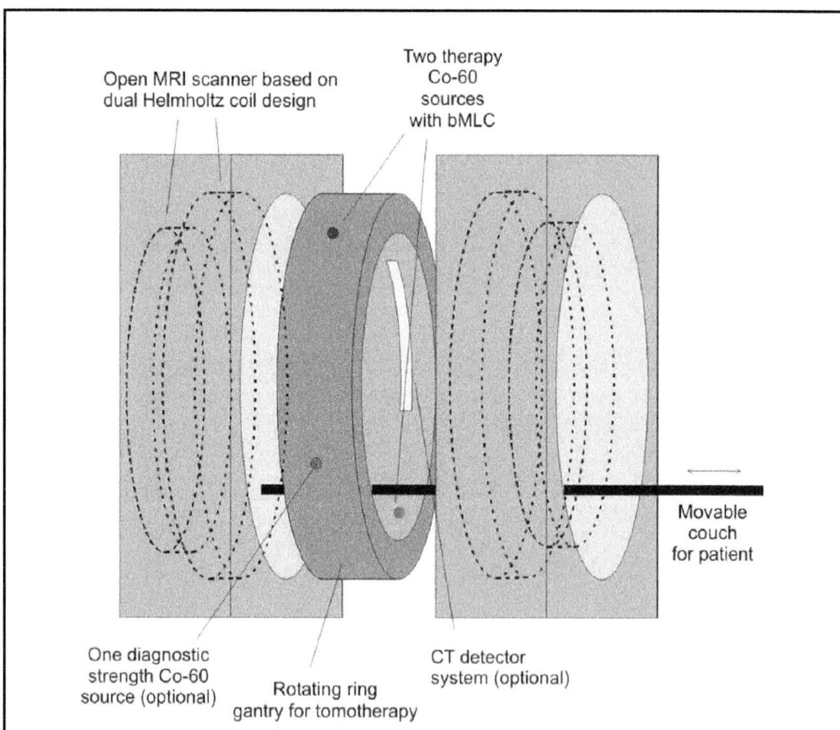

Figure 12.15: Schematic illustration of a suggested design for a cobalt-60 unit combined with an MRI scanner (adapted from Kron et al. 2006). This unit design had two cobalt-60 sources, each with a binary MLC sandwiched between MRI scanner magnets.

Table 12.2: Comparison of image guidance modalities. Note this is a suggestion for illustration purposes only, as a more tailored evaluation depends on the environment and needs of the user.

	Ultra-sound	*EPI*	*kV Imaging*	*CT (in room)*	*CBCT*	*MVCT*	*MRI*
Imaging dimensions	2 or 3	2	2 plus two orthogonal	3	3	3	3
Image quality	+	+	++	++++	+++	+++	++++
Soft tissue contrast	+++		+	++	++	+	+++
Field verification	no	yes	no	no	no	no	no
Dose reconstruction		+		+	+	+++	
Dose (cGy)	none	3	0.3	2	2	2	none
Time per patient (min.)	5	1	3	5	5	5	not known
Feasibility/ ease of use	training required	easy	easy	large bunker	--	--	not tested
Cost	low	low	medium	high	high	medium	very high

Most of these various competing linac-mounted, or in-linac room, imaging modalities are very new to the market; some are still just designs. Early comparisons such as in table 12.2 give estimates only of the relative merits of each device.

12.5 Intrafraction Motion Management

Intrafraction variation is due to a change in patient geometry during the dose delivery within a treatment fraction, often caused by involuntary internal organ motion. Skeletal, gastrointestinal, and cardiac systems all contribute to intrafraction motion; however, it is respiratory motion towards which significant research and development have been directed (Keall et al. 2006a,b). This may be attributed to the extent of tumor movement caused by respiratory motion; lung and thoracic tumors moving as much as 5 cm during respiration have been reported (Keall et al. 2006a,b). More typical movements may be in the order of 1 to 2 cm; however the actual extent of motion depends very much on where the tumor is located in the lung—in general, peripheral tumors move less.

Large margins are necessary in the PTV to account for the limits of tumor motion. These increased margins in turn lead to the volume of normal tissue irradiated being larger, and this may be associated with a necessary reduction in prescribed dose. It is also suggested that breathing leads to a spreading out of the dose distribution during radiation delivery, resulting in a deviation between the intended dose and the dose delivered (Li et al. 2006). Methods which specifically account for respiratory motion include:

i) Motion encompassing volume, this may involve 4DCT of respiratory volume or inhale-exhale breath-hold (see figure 12.16a).
ii) Respiratory gating (see figure 12.16b).
iii) Motion adaptive therapy known as tumor tracking (see figure 12.16c).

12.5.1 Gating

Respiratory gating during imaging allows selected portions of the breathing waveform to be reconstructed between specified time intervals or specified amplitudes. Gated delivery only treats in a selected portion, or *gate*, thus reducing tumor movement when the beam is on by providing a series of snap shots of tumor treatment tuned to the respiratory cycle.

Optimally, because the gate has finite and necessary time width to achieve a reasonable dose rate, this would be chosen where tumor motion is the least,

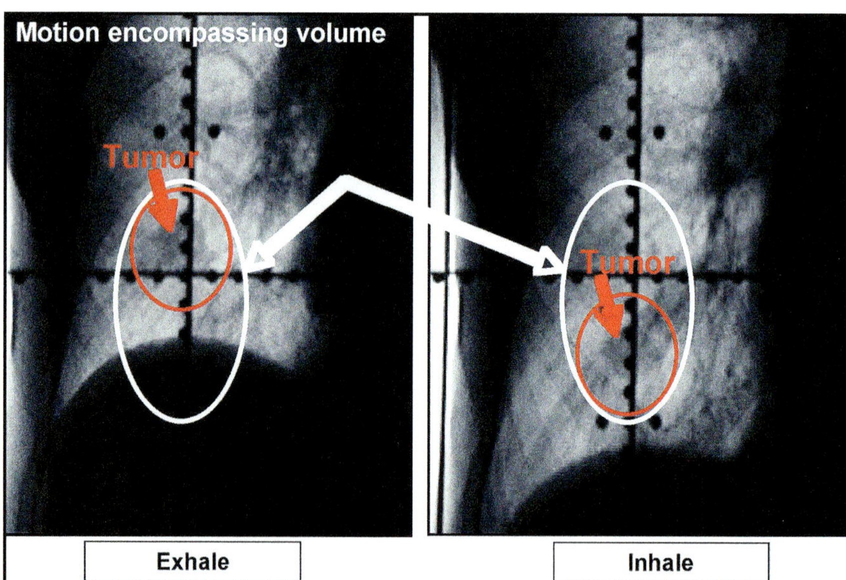

(a)

Figure 12.16: Strategies for dealing with intrafraction motion: (a) Motion encompassing volume; this may involve 4DCT of respiratory volume or inhale-exhale breath-hold; (b) Respiratory gating involving beam-hold off and beam-on during selected phases of the respiration cycle; (c) Motion adaptive therapy known as tumor tracking, whereby a device such as a DMLC follows (tracks) the tumor target motion during beam-on. (Courtesy of P. Keall, Stanford University.)

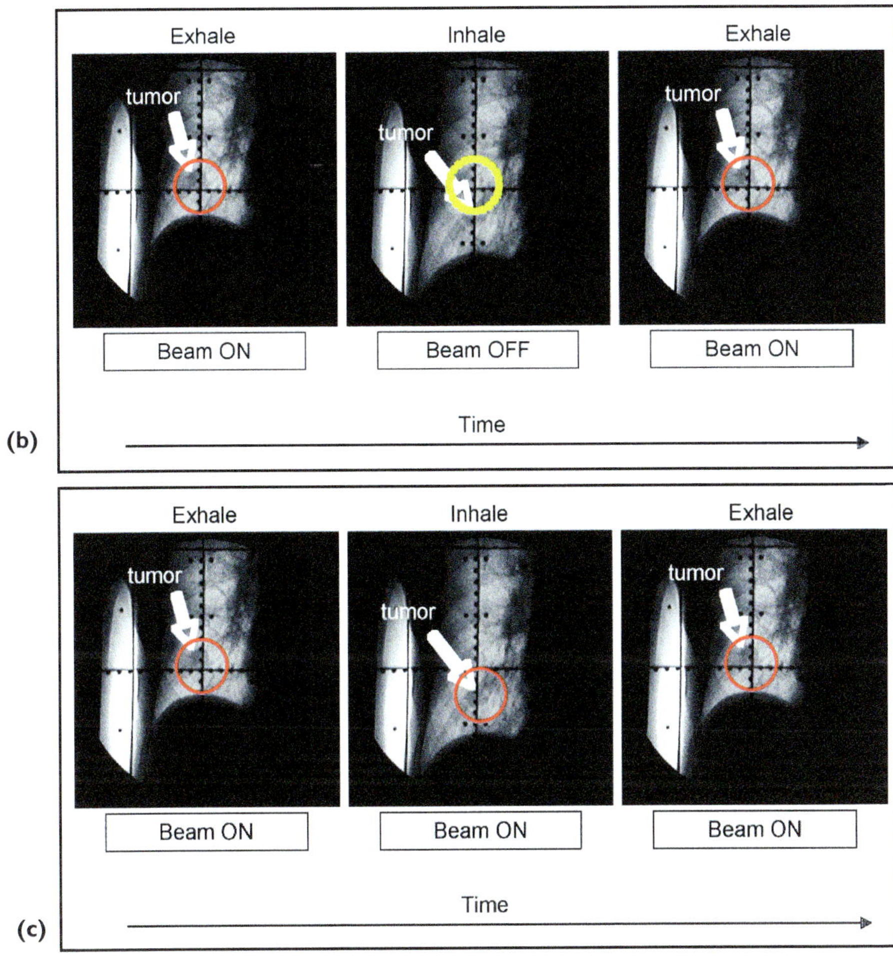

Figure 1.16: (Continued)

perhaps end expiration. Another option might be at end inspiration where lung volume is maximal, often separating the tumor from critical structures such as the heart and spinal cord (Keall et al. 2006a,b). The choice of gate width must be a compromise between the time the beam is on as a ratio of the total treatment time, the *duty cycle*, and the amount of tumor motion within the gate, *tumor residual motion*. Treatment delivery time is typically increased up to threefold using respiratory gating due to the average duty cycle for treatment ranging from 30% to 50% (Jiang 2006).

Real-time knowledge of the tumor position is necessary in all methods accounting for intrafraction motion. Respiration monitoring techniques include internal fiducial-based methods, and externally based surrogate methods. Surrogate methods rely on external surrogates such as reflective skin markers or lung air volume measurements corresponding to internal tumor motion. Internal fiducial-based methods rely on on-line fluoroscopic or electronic portal imaging to directly detect the location of the tumor through implanted fiducials.

There are several commercially available respiratory gating systems that utilize external surrogates for tumor motion. One such system is the *real-time position management* (RPM) system. The RPM system (Varian Medical Systems) uses a video camera to detect the motion of external infrared reflecting markers Figure 12.17a shows the RPM respiratory gating system coupled with a Philips CT. *Four-dimensional CT* (4DCT) is produced by oversampled data acquisition at each slice. Multiple images can then be reconstructed per slice and are evenly distributed over the acquisition time. Each of these images represents a different anatomical state of the breathing cycle. Reconstructed images can then be sorted into spatio-temporally coherent volumes based on respiratory phase as discerned by the external surface marker (Reitzel et al. 2005). Figure 12.17b shows the plastic marker box with two infrared-reflecting markers placed on the patient's chest between the umbilicus and xiphoid. The infrared markers are illuminated by infrared light-emitting diodes, and images of the markers are captured by

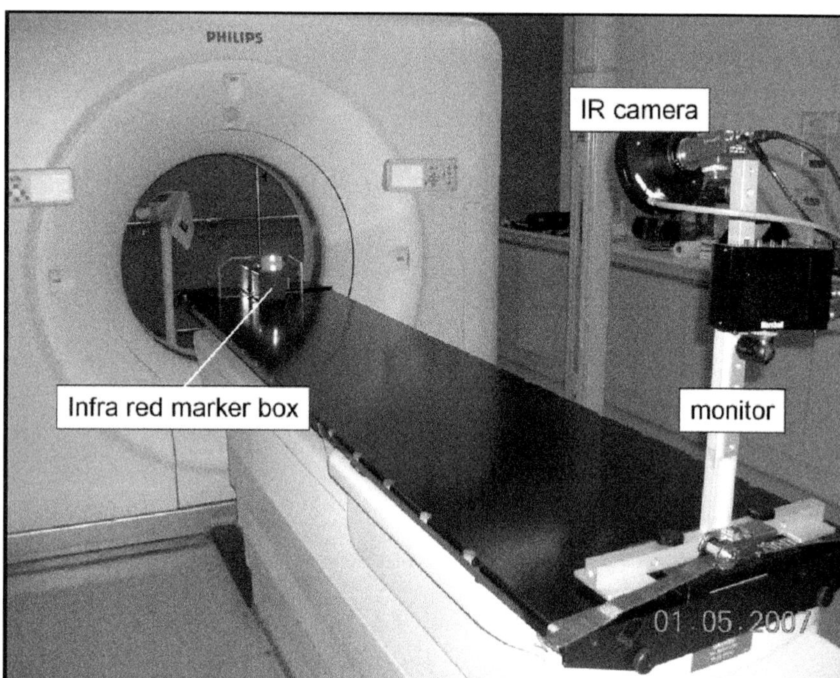

(a)

Figure 12.17: The real-time position management (RPM) system (vendor, Varian Medical Systems). The RPM system uses a video camera to detect the motion of external infrared reflecting markers: (a) RPM respiratory gating system coupled with a Philips CT. 4DCT images are produced by oversampled data acquisition at each slice; (b) Plastic marker box with infrared markers placed on the abdomen of patient for motion detection by infrared camera; (c) The image in the top left-hand corner displays the real-time video footage of marker motion, while the image in the bottom right-hand corner displays the motion track. The gated portions of the breathing cycle, or beam-on pulses, can be seen below the respiratory waveform. During treatment on the linac, a signal to the gun allows treatment only when motion is within the gate.

(b)

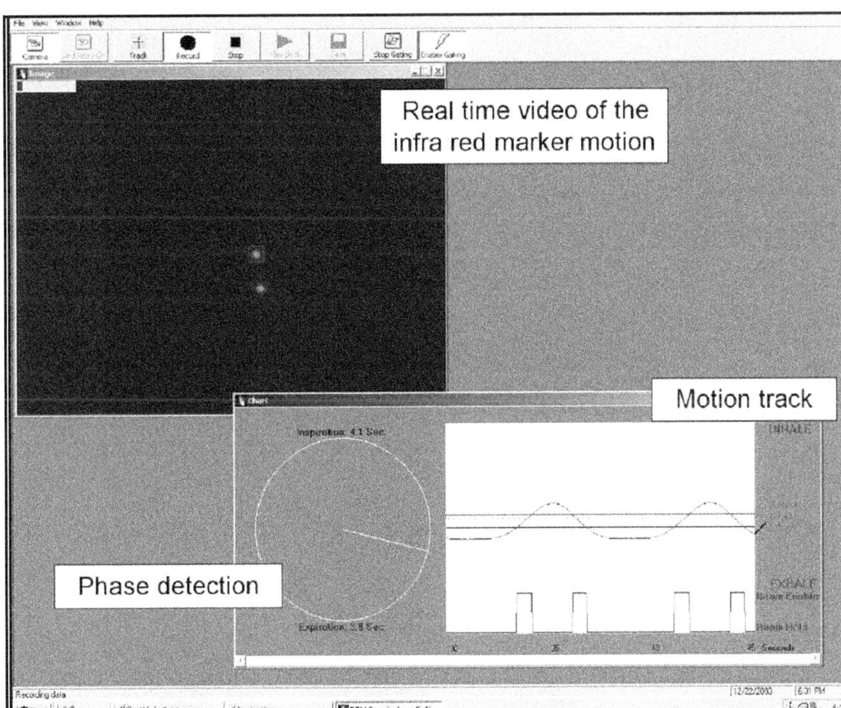

(c)

Figure 12.17: (Continued)

the camera at 30 frames per second. The camera output is directed to a PC running the RPM software; a tracking algorithm establishes the amplitude and period of the motion. The software designates end-inhale and end-exhale positions and assigns phases by linear interpolation (Rietzel et al. 2005). The user interface can be seen in figure 12.17c. The image in the top left-hand corner displays the real-time video footage of marker motion, while the image in the bottom right-hand corner displays the motion track. The gated portions of the breathing cycle, or beam-on pulses, can be seen below the respiratory waveform. During treatment on the linac, a signal from the PC to the grid gun allows treatment only when motion is within the gate (Yorke et al. 2005). Patient internal anatomy positioning during treatment should be verified using in-room imaging before treatment, and marker block position and patient breathing waveform should be as close as possible to that obtained in planning.

Another commercial gating system is based on a pressure belt feedback mechanism; the Anzai belt (AZ-733V by Anzai Medical) is used by Siemens 4DCTs and linacs. This is depicted in figure 12.18a and consists of an elastic belt containing a load cell which detects external respiratory motion in real-time through changes in abdominal motion. The signal from the load cell is amplified and fed into the scanner. As the patient breathes in, the belt tight-

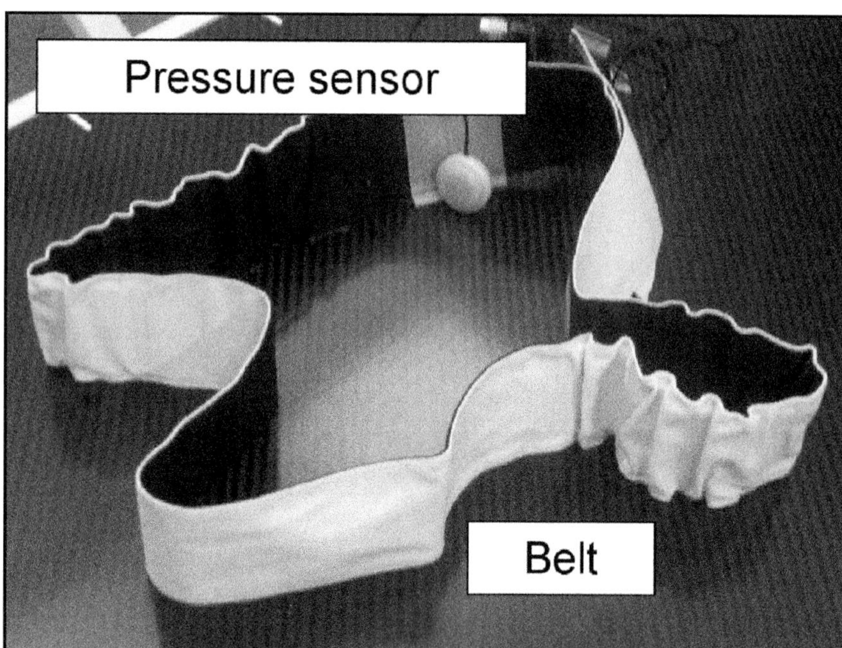

(a)

Figure 12.18: Two other commercial system approaches to monitoring the breathing cycle: (a) The Anzai belt (model AZ-733V, vendor Anzai Medical), used by Siemens 4DCT and linac devices. This consists of an elastic belt containing a load cell which detects external respiratory motion in real-time through changes in abdominal motion. As the patient breathes in, the belt tightens and pressure is exerted on the load cell; thus a higher amplitude signal is produced. A decreasing in amplitude of the signal corresponds to the exhalation respiratory phase.

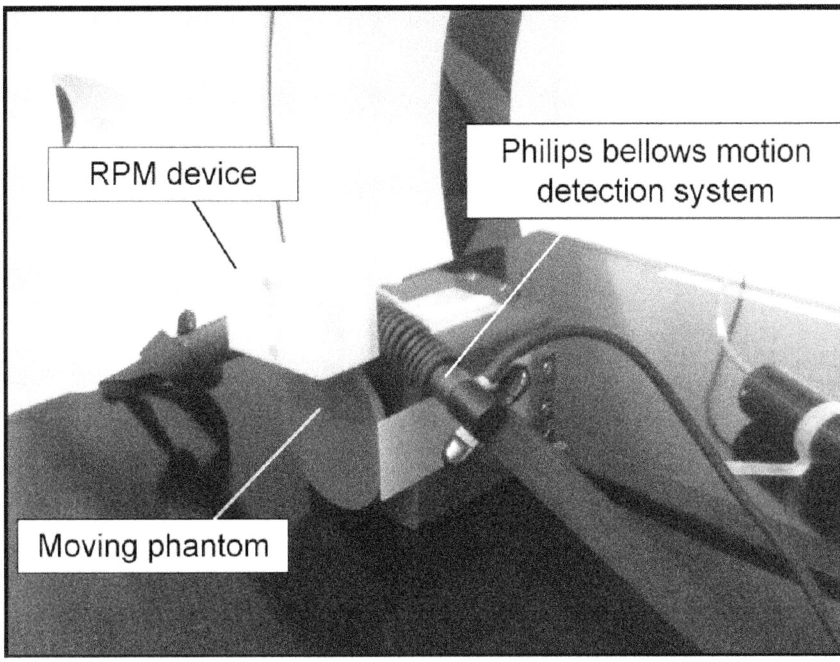

(b)

Figure 12.18: (Continued.) Two other commercial system approaches to monitoring the breathing cycle: (b) A bellows system (Philips) shown with the RPM system (Varian) on a phantom; it is important to correlate two different systems if one is to be used for 4DCT acquisition and the other for treatment delivery gating.

ens and pressure is exerted on the load cell; thus a higher amplitude signal is produced. A decreasing in amplitude of the signal corresponds to the exhalation respiratory phase.

The belt is connected to the sensor port, which is in turn connected to the wave deck. The system is run by a PC which offers a graphical user interface for monitoring and recording the signal. During spiral CT mode respiratory data are acquired during a whole helical CT scan and stored. Projections are integrated over a 250 ms window, starting from a given phase. The Anzai belt recognizes the lowest amplitude phase point (0% inhalation) and the highest amplitude phase point (100% inhalation). In-between phases for inhalation are calculated by linear scaling. Similarly, the peak of the curve (maximum amplitude) is recognized as 0% exhalation and the trough, 100% exhalation with in-between phases calculated by linear scaling. Yet another system using a bellows approach is implemented by Philips (see figure 12.18b).

Another issue is that although motion artifacts are reduced in 4DCT, they still exist. Each phase of retrospective re-binning is a snapshot of regular respiration and is subject to artifacts due to incorrect phase determination or an irregular breathing pattern. Motion intraphase will also produce artifacts due to the partial projection effect and this must be taken into account in planning (Rietzel et al. 2005).

Gated linac treatment using the Anzai belt follows the same principle as the RPM respiratory gating system, where the beam is only triggered at predetermined portions of the respiratory cycle. If at any stage, breathing becomes erratic, such as the patient coughing, the system is designed to automatically switch off. Design of phantoms which test the effectiveness of these devices during the treatment cycle are really in their infancy; usually an existing phantom is modified to move in 1-D or 2-D (as an example see figure 12.19).

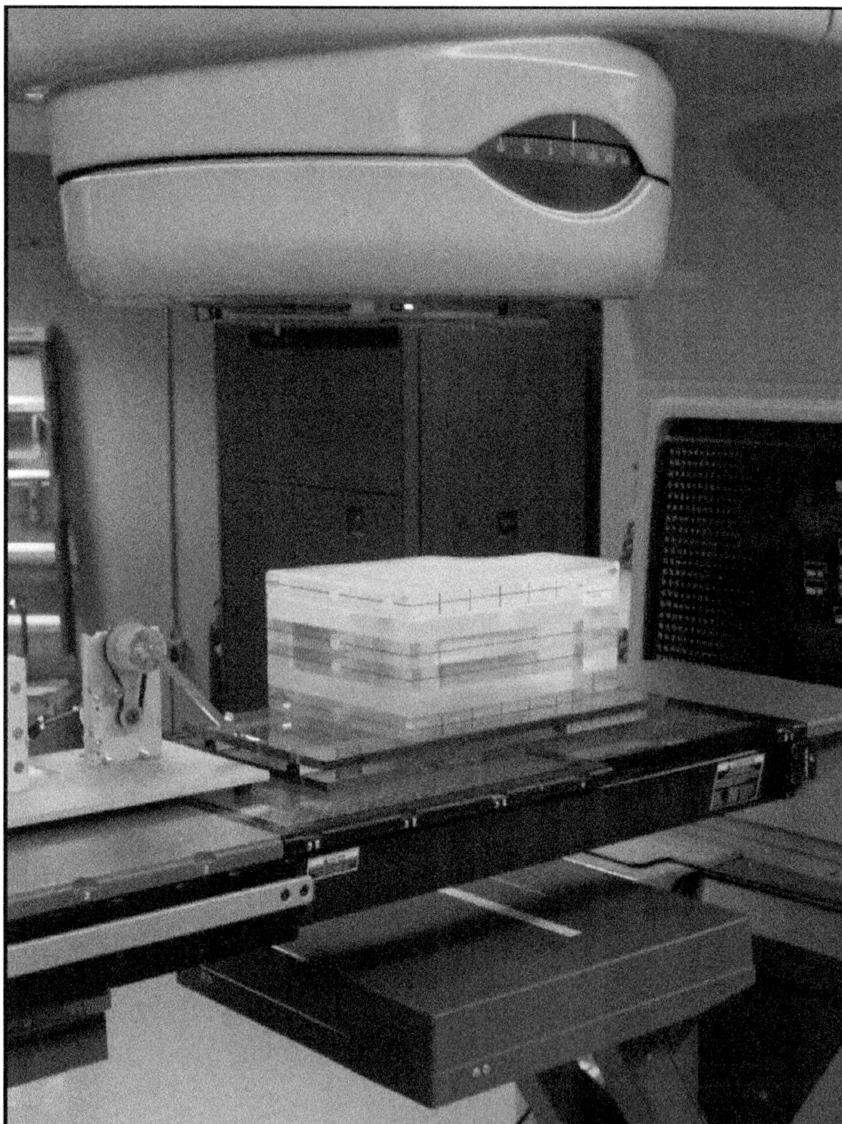

Figure 12.19: Commercial dosimetry phantom (Standard Imaging Ltd.) moving on rails driven by a rotating arrangement. (Courtesy of P. Keall, Stanford University.)

Respiratory gating using external surrogates is noninvasive, and allows the patient to breathe freely during both planning and treatment. Patients chosen should have fairly stable breathing, and reasonably large intrafraction tumor motion, so as to benefit from treatment because a reduced target volume can be specified. Studies have shown that respiratory training and visual and audio feedback can improve respiratory reproducibility for gating purposes (e.g., Jiang 2006). External surrogates rely heavily on internal tumor motion correlating with the motion of the external marker, which may not always be the case.

Internal-based fiducial gating utilizes fiducial markers detected by radiographic means to gate radiation delivery (Keall et al. 2006a,b). The technique was jointly developed by Hokkaido University and Mitsubishi (Seppenwoolde et al. 2002). Gold seeds are implanted close to the target location, and the fiducial position is tracked in 3-D using a pair of stereotactic kilovoltage x-ray imaging systems with automatic detection software (Keall et al. 2006a,b).

During treatment, the linac is activated when the desired position of the fiducial is obtained for both stereotactic x-ray cameras. Internal-based fiducial methods have the advantage of directly measuring the tumor position by fluo-roscopic imaging. However, placement of fiducials for internal-based gating of lung tumors is associated with a high risk of pneumothorax (Jiang 2006).

Patients must also be able to tolerate the implant procedure and remain motionless on the couch for up to 45 minutes (Keall et al. 2006a,b). Fiducial markers can also be loaded by bronchoscope; however fixation and migration of markers in the bronchial tree also pose problems due to the relationship of the markers and the tumor shifting over time (Imura et al. 2005).

12.5.2 Breath-hold

Breath-hold methods involve the patient holding his or her breath at certain phases of the breathing cycle to reduce tumor movement. Breath-hold methods consist of deep inspiration, active breathing control, or self-held breath-hold.

Deep inspiration breath-hold involves coaching the patient to a reproducible maximum inhalation breath-hold during simulation and treatment (Keall et al. 2006a,b). The aim is to immobilize the tumor and to expand healthy lung tissue out of the high-dose region. Control devices used include spirometry and external markers; however they are not used to trigger the linac. The therapist or physicist will observe and record the patient breathing waveforms during simulation and assess multiple breath-holds to establish the reproducibility of the *deep inspiration breath-hold* (DIBH). CT scans are obtained during free breathing, DIBH, end-inspiration and end-expiration. The free-breathing scan is used for reference, and serves as an alternate treatment plan if DIBH is not feasible for the duration of treatment. End-inspiration and expiration scans give an estimate of tumor motion during normal tidal breathing and can be used for QA purposes. During treatment, the therapist will turn the beam on only when the target breath-hold has been attained and turn the beam off if the breath-hold level falls outside a preset tolerance (Keall et al. 2006a,b).

Active breathing control (ABC) is a method used to assist reproducible breath-hold. The ABC spirometry system (Elekta Oncology Systems) enforces patient breath-holds at preselected volumes of respiration during which dose delivery occurs (Wong et al. 1999). The volume of air entering and leaving the lungs is monitored by a spirometer. The changing air volume corresponds to the patient's respiratory signal. The patient breathes normally through the apparatus. The ABC system acquires and controls the desired respiration flow-volume. The operator can activate the software to force a breath-hold for a predetermined period of time by closing the valve of the breathing tube and restricting air flow to and from the patient. The breath-hold is then maintained for 15 to 30 s during treatment, depending on the patient's tolerance (Keall et al. 2006a,b).

There is a *self-held breath-hold* where the patient voluntarily holds his/her breath at some point in the breathing cycle while the prescribed dose is administered. This can occur with or without respiratory monitoring.

Without respiratory monitoring this technique relies on the patient determining when an adequate breath-hold is in place and clearing the interlock by pressing a hand-held switch. This allows the therapist to activate the beam. Releasing the switch asserts the interlock, disabling the beam. Both the therapist and the patient have the power to turn the beam off. The drawback of this method is that it is reliant on a patient's ability to maintain a breath-hold in which internal anatomy is stable for an adequate period of time (>10 s) while controlling the switch. The AAPM report on the management of respiratory motion in radiation oncology (Keall et al. 2006a,b) advises that PTV margins must take into account breath-hold reproducibility, patient setup reproducibility, and internal motion. Due to lack of significant statistical data, margins for internal motion should be tailored to the individual patient by measuring the reproducibility at the simulator session.

Respiratory monitored self-breath-hold utilizes an external surrogate such as the RPM system to monitor the breathing waveform and to control dose delivery, while the patient is required to perform a voluntary breath-hold. Similar to the method discussed previously, it is required that the patient can perform a reproducible breath-hold for an adequate period of time and respond to verbal instructions from the therapist. Portal imaging is performed prior to treatment to verify patient positioning. Respiratory monitored self-breath-hold has the safeguard that the waveform is monitored; thus if the breath-hold at any time deviates from predetermined breath-hold margins, the beam will automatically turn off.

Berson et al. (2004) found that breath-hold techniques were beneficial in eliminating possible time lag between tumor and external surrogates, improved diaphragm positional reproducibility and that there was an efficiency gain in CT simulation and treatment (longer duty cycle). Breath-hold methods cannot be used on every patient and are reliant on patient compliance or involuntary restriction of airflow. Treatment is prolonged using breath-holds to account for intrafraction motion as additional time has to be added to enable to patient to recover from each breath-hold.

12.5.3 Motion Adaptive Therapy

Motion adaptive therapy consists of respiration-synchronized techniques that shift dose in space to track a tumor's changing position. A dynamic couch, or for small targets, a robotic arm (CyberKnife®, Accuray Inc.) can be used to target a moving tumor's position (Tewatia et al. 2006). 4-D radiotherapy by target tracking has also been proposed (Keall et al. 2004) whereby *dynamic multileaf collimators* (DMLCs) are adjusted to *track* tumor motion.

CyberKnife® is an image-guided radiosurgery system in which the linac beam is moved by a robotic arm that is coupled through a real-time loop to an imaging system that monitors tumor location. The robotic arm can move with six degrees of freedom and track the target through a combination of infrared tracking and synchronized x-ray imaging (Schweikard et al. 2003).

Prior to treatment, gold fiducial markers are placed in the vicinity of the target organ. Stereo x-ray imaging is used during treatment to determine the precise spatial location of these markers. The position of internal fiducials is computed repeatedly during treatment through x-ray image processing. In addition to this, infrared emitters are attached to the chest and the abdominal surface of the patient and the infrared tracking system records the motion of these emitters. The software updates and correlates external body surface movement with the movement of internal tumor fiducials. From this correlation model the placement of the internal target can be devised during time intervals where no x-ray images are taken. The beam can be constantly repositioned to account for tumor motion, and dosimetry adapted to allow for changing lung volume.

DMLC-based motion tracking is achieved as follows:

i) 4DCT is used to create separate CT images at discrete phases of the respiratory cycle, which allows temporal anatomic changes over time to be observed.
ii) Select respiratory phases are reconstructed, and deformable image registration is then performed to map each CT image data set. This allows a representative tumor trajectory to be obtained.
iii) IMRT treatment planning is simultaneously performed on each CT set reconstructed, and a leaf sequence based on optimized fluence maps is generated.
iv) The radiation beam defined by the DMLC then tracks the respiration-induced target motion. This is based on a feedback loop including the breathing signal to a real-time MLC controller (Keall et al. 2004). A bench top version of this DMLC device is shown in figure 12.20.

Tewatia et al. (2006) have also researched *breathing synchronized delivery* (BSD) for DMLC-based motion tracking in which the patient is guided to breathe in accordance with target motion obtained from 4DCT data sets. In the case of IMRT, a breathing guide is developed for each patient from his or her normal breathing patterns (obtained from spirometry). The

Figure 12.20: A bench top version of a DMLC tracking device. The MLC leaves move during beam-on to encompass the target volume which is changing. DMLC tracking is achieved by feedback from the infrared marker. (Courtesy of P. Keall, Stanford University.)

patient is then required to reproduce this breathing cycle by following a visual prompt. 4DCT data sets are then obtained and DMLC planning performed as described previously. It is assumed that the patient maintains a constant breathing cycle and the dose rate from the treatment machine is constant so that the treatment process correlates with the breathing phase. The instantaneous average tumor displacement is overlaid on the DMLC position at corresponding phases and the leaf speed is evaluated to ensure accurate treatment delivery.

The advantages of motion adaptive therapy include:

i) Accounts for 3-D motion of the tumor.
ii) The duty cycle for treatment is 100%.

Time delay compensation is necessary for both the CyberKnife and conventional DMLC; a 200 ms delay between acquisition of tumor coordinates and repositioning of the linac is present, in addition to image acquisition, read-out, and processing times (Tewatia et al. 2006). This necessitates prediction algorithms such that the beam can be synchronized at the position of the tumor, which may be difficult if the patient breathes irregularly. Murphey et al. (2002) have found that by using breathing predictions with a variety of adaptive filters, tumor position can be predicted with up to 80% accuracy in the presence of a 200 ms system time delay, but that accuracy degrades rapidly at longer delay intervals. The report also highlights that dosimetric considerations should be taken into account in treating a changing volume. The anatomy and air volume of the lung are continually changing during breathing, which perturbs the attenuation of the treatment beam and alters the position of the tumor with respect to normal tissue and critical structures.

References

Berson A., Emery R., Rodriguez L., Richards G.M., Nq T., Sanghavi S., Barsa J. (2004) Clinical experience using respiratory gated radiation therapy: Comparison of free breathing and breath-hold techniques. *Int. J. Radiat. Oncol. Biol. Phys.* 60: 419–426.

Bielajew A.F. (1993) The effect of strong longitudinal magnetic fields on dose deposition from electron and photon beams. *Med. Phys.* 20: 1171–1179.

Breen S.L., Craig T., Bayley A., O'Sullivan B., Kim J., Jaffray D. (2006) Spinal cord planning risk volumes for intensity-modulated radiation therapy of head-and-neck cancer. *Int. J. Radiat. Oncol. Biol. Phys.* 64: 321–325.

Chinnaiyan P., Tomé W., Patel R., Chappell R., Ritter M. (2003) 3D-ultrasound guided radiation therapy in the post-prostectomy setting. *Technol. Cancer Res. Treat.* 2: 455–462.

Crook J.M., Raymond Y., Salhani D., Yang H., Esche B. (1995) Prostate motion during standard radiotherapy as assessed by fiducial markers. *Radiother. Oncol.* 37: 35–42.

Holland R., Veling S.H., Mravunac M., Hendriks J.H. (1985) Histologic multifocality of Tis, T1-2 breast carcinomas. Implications for clinical trials of breast conserving surgery. *Cancer* 56: 979–990.

Hugo G.D., Yan D., Liang, J. (2007) Population and patient-specific target margins for 4D adaptive radiotherapy to account for intra- and inter-fraction variation in lung tumour position. *Phys. Med. Biol.* 52: 257–274.

Hysing L.B., Kvinnsland Y., Lord H., Muren L.P. (2006) Planning organ at risk volume margins for organ motion of the intestine. *Radiother. Oncol.* 80: 349–354.

ICRU (International Commission on Radiation Units and Measurements) (2004) Report 71, Prescribing, recording, and reporting electron beam therapy. ICRU, Bethesda, Maryland.

ICRU (International Commission on Radiation Units and Measurements) (2001) Report 64, Dosimetry of high-energy photon beams based on standards of absorbed dose to water. ICRU Bethesda, Maryland.

ICRU (International Commission on Radiation Units and Measurements) (1999) Report 62: Prescribing, recording, and reporting photon beam therapy (Supplement to ICRU Report 50). ICRU, Bethesda, Maryland.

ICRU (International Commission on Radiation Units and Measurements) (1998) Report 58: Dose and volume specification for reporting interstitial brachytherapy. ICRU, Bethesda, Maryland.

ICRU (International Commission on Radiation Units and Measurements) (1993) Report 50: Prescribing, recording, and reporting photon beam therapy. ICRU, Bethesda, Maryland.

ICRU (International Commission on Radiation Units and Measurements) (1985) Report 38: Gynaecological brachytherapy. ICRU, Bethesda, Maryland.

Imura M., Yamazaki K., Shirato H., Onimaru R., Fujino M., Shimizzu S., Harada T. et al. (2005) Insertion and fixation of fiducial markers for setup and tracking of lung tumors in radiotherapy. *Int. J. Radiat. Oncol. Biol. Phys.* 63: 1442–1447.

Jani A., Gratzle J., Muresan E., Martel M. (2005) Impact on late toxicity of using transabdominal ultrasound for prostate cancer patients treated with intensity modulated radiotherapy. *Technol. Cancer Res. Treat.* 4: 115–120.

Jiang S.B. (2006) Technical aspects of image-guided respiration-gated radiation therapy. *Med. Dosim.* 31: 141–151.

Jin J.Y., Ajlouni M., Chen Q., Yin F.F., Movsas B. (2006) A technique of using gated-CT images to determine internal target volume (ITV) for fractionated stereotactic lung radiotherapy. *Radiother. Oncol.* 78: 177–184.

Keall P.J., Mageras G.S., Balter J.M., Emery R., Forster K.M., Jiang S.B., et al. (2006a) The management of respiratory motion in radiation oncology report of AAPM Task Group 76. *Med. Phys.* 33: 3874–3900.

Keall P.J., Vedam S., George R., Bartee C., Siebers J., Lerma F., Weiss E., Chung T. (2006b) The clinical implementation of respiratory gated intensity-modulated radiotherapy. *Med. Dosim.* 31: 153–162.

Keall P.J., Joshi S., Vedam S.S., Siebers J.V., Kini V.R., Mohan R. (2004) Four-dimensional radiotherapy planning for DMLC-based respiratory motion tracking. *Med. Phys.* 32: 942–951.

Knoos T., Kristensen I., Nilsson P. (1998) Volumetric and dosimetric evaluation of radiation treatment plans: radiation conformity index. *Int. J. Radiat. Oncol. Biol. Phys.* 42: 1169–1176.

Kron T., Eyles D., Schreiner J., Battista J. (2006) Magnetic resonance imaging foradaptive cobalt tomotherapy: A proposal. *J. Med. Phys.* (India) 31: 238–250.

Lattanzi J., McNeeley S., Donnelly S., Palacio E., Hanlon A., Schultheiss T.E., Hanks G.E. (2000) Ultrasound-based stereotactic guidance in prostate cancer-quantification of organ motion and set-up errors in external beam radiation therapy. *Comput. Aided Surg.* 5: 289–295.

Li X.A., Stepaniak C., Gore E. (2006) Technical and dosimetric aspects of respiratory gating using a pressure-sensor motion monitoring system. *Med. Phys.* 33: 145–153.

Little D.J., Dong L., Levy L.B., Chandra A., Kuban D.A. (2003) Use of portal images and BAT ultrasonography to measure setup error and organ motion for prostate IMRT: Implications for the treatment margins. *Int. J. Radiat. Oncol. Biol. Phys.* 56: 1218-1224.

Lorchel F., Dumas J.L., Noel A., Wolf D., Bosset J.F., Aletti P. (2006) Esophageal cancer: determination of internal target volume for conformal radiotherapy. *Radiother. Oncol.* 80: 327–332.

Muren L.P., Karlsdottir A., Kvinnsland Y., Wentzel-Larsen T., Dahl O. (2005) Testing the new ICRU 62 ‘Planning Organ at Risk Volume’ concept for the rectum. *Radiother. Oncol.* 75: 293–302.

Murphey M., Jalden J., Isaksson M. (2002) Adaptive filtering to predict lung tumor breathing motion during image-guided radiation therapy. *Proceedings of the 16th International Congress on Computer-Assisted Radiology and Surgery (CARS)*. pp 539–544.

Nederveen A.J., Dehnad H., van der Heide U.A., van Moorselaar R.J., Hofman P., Lagendijk J.J. (2003) Comparison of megavoltage position verification for prostate irradiation based on bony anatomy and implanted fiducials. *Radiother. Oncol.* 68: 81–88.

Pouliot J., Bani-Hashemi A., Chen J., Svatos M., Ghelmansarai F., Mitschke M., Aubin M., Xia P., et al. (2005) Low-dose megavoltage cone-beam CT for radiation therapy. *Int J. Radiat. Oncol. Biol. Phys.* 61:1 552–560.

Raaijmakers A.J., Raaymakers B.W., Lagendijk J.J. (2005) Integrating a MRI scanner with a 6 MV radiotherapy accelerator: Dose increase at tissue-air interfaces in a lateral magnetic field due to returning electrons. *Phys. Med. Biol.* 50: 1363–1376.

Raaymakers B.W., Raaijmakers A.J., Kotte A.N., Jette D., Lagendijk J.J. (2004) Integrating a MRI scanner with a 6 MV radiotherapy accelerator: dose deposition in a transverse magnetic field. *Phys. Med. Biol.* 49: 4109–4118.

Rietzel E., Liu A.K., Doppke K.P., Wolfgang J.A., Chen A.B., Chen G.T., Choi N.C. (2006) Design of 4D treatment planning target volumes. *Int. J. Radiat. Oncol. Biol. Phys.* 66: 287–295.

Rietzel E., Pan T., Chen G.T. (2005) Four-dimensional computed tomography: image formation and clinical protocol. *Med. Phys.* 32: 874–889.

Schweikard A., Shiomi H., Adler J. (2004) Respiration tracking in radiosurgery. *Med. Phys.* 31: 2738–2741.

See A., Kron T., Johansen J., Hamilton C., Bydder S.A., Hawkins J., Roff M., Denham J.W. (2000) Decision-making models in the analysis of portal films: a clinical pilot study. *Australas. Radiol.* 44: 72–83.

Seppenwoolde Y., Shirato H., Kitamura K., Shimizu S., van Herk M., Lebesque J.V., Miyasaka K. (2002) Precise and real-time measurement of 3-D tumor motion in lung due to breathing and heartbeat, measured during radiotherapy. *Int. J. Radiat. Oncol. Biol. Phys.* 53: 822–834.

Stroom J., Heijmen B.J. (2006) Limitations of the planning organ at risk volume (PRV) concept. *Int. J. Radiat. Oncol. Biol. Phys.* 66: 279–286.

Tewatia D., Zhang T., Tomé W., Paliwal B., Metha M. (2006) Clinical implementation of target tracking by breathing synchronized delivery. *Med. Phys.* 33: 4330–4336.

Tomé W.A., Meeks S.L., Orton N.P., Bouchet L.G., Bova F.J. (2002) Commissioning and quality assurance of an optically guided three dimensional ultrasound target localization system for radiotherapy. *Med. Phys.* 29:1781–1789.

van Herk M., Remeijer P., Rasch C., Lebesque J.V. (2000) The probability of correct target dosage: dose-population histograms for deriving treatment margins in radiotherapy. *Int. J. Radiat. Oncol. Biol. Phys.* 47:1121–1135.

Wong J.W., Sharpe M.B., Jaffray D.A., Kini V.R., Robertson J.M., Stromberg J.S., Martinez A.A. (1999) The use of active breathing control (ABC) to reduce margin for breathing motion. *Int. J. Radiat. Oncol. Biol. Phys.* 44: 911–919.

Wu J., Haycocks T., Alasti H., Ottewell G., Middlemiss N., Abdolell M., Warde P., Toi A., Catton C. (2001) Positioning errors and prostate motion during conformal prostate radiotherapy using on-line isocentre set-up verification and implanted prostate markers. *Radiother. Oncol.* 61:127–133.

Yorke E., Rozenweig K.E., Wagman R., Mageras G.S. (2005) Interfractional anatomic variation in patients treated with respiration-gated radiotherapy. *J. Appl. Clin. Med. Phys.* 6: 19–32.

Zelefsky M.J., Crean D., Mageras G.S., Lyass O., Happersett L., Ling C.C., Leibel S.A., Fuks Z., Bull S., Kooy M., van Herk M., Kutcher G.J. (1999) Quantification and predictors of prostate position variability in 50 patients evaluated with multiple CT scans during conformal radiotherapy. *Radiother. Oncol.* 50: 225–234.

Questions

12.1 If the internal margin is 3 mm and the setup margin is 4 mm, calculate the margin between CTV and PTV.
(section 12.2)

12.2 Using the van Herk recipe (equation 12.3) calculate the margin if the systematic movement error is assessed as 2 mm and the random movement error as 3 mm.
(section 12.2)

12.3 Name two considerations when considering implementing a patient immobilization device.
(section 12.3)

12.4 Name three treatment strategy options now that 4DCT images showing surrogate tumor movement during the respirator cycle are available.
(section 12.5)

12.5 Describe some of the technical challenges facing the implementation of tumor tracking of lung tumors.
(section 12.5)

13 Radiation Protection and Room Shielding

"Anything worth doing is worth doing well."
(T.A.J. Metcalfe, private communication)

13.1 Introduction

Modern society has made use of ionizing radiation in many ways. Some of the most important applications are in medicine where radiation is used for diagnostic purposes in nuclear medicine and radiology, and for therapy in the context of radiotherapy. The present chapter is concerned with radiation protection in radiotherapy only and restricted to external beam radiotherapy with an emphasis on applications of linacs. For general radiation protection the reader is referred to books such as Cember (1983) and Kathren (1985).

Hazards from ionizing radiation can broadly be divided in two groups:

i) *External exposure:* This is typically from a radiation source that can be turned off like an x-ray unit. The radiation from this source may cause cell damage while it is turned on. In general, nothing radioactive is left in the body. Hazards of this type may occur in radiology or radiotherapy departments.
ii) *Internal exposure:* This occurs most commonly after the incorporation (e.g., breathing in, consuming with food) of radioactive isotopes. The radioactivity remains in the organism until the isotope has decayed (physical half-life) or until it is excreted (e.g., in urine or during exhalation). These hazards may be present in laboratories of nuclear medicine departments.

In the context of the present chapter we will deal with external exposures only as internal exposure is not of concern in the context of linacs. However, as discussions of accidents in radiotherapy show there are significant risks also involved in the use of iosotopes for cancer treatment (IAEA 2000, ICRP 2001) and the present chapter cannot be seen as a comprehensive treatment of radiation protection in radiotherapy.

The chapter will initially deal with the practical and philosophical background of radiation protection. This is largely based on the recommendations of the International Commission on Radiological Protection (ICRP). The recommendations published in the ICRP report 60 (1991) build the foundation of most radiation protection regulations in the world. While the publication of a new set of recommendations is expected soon (compare www.icrp.org), the discussion here is based on ICRP report 60, the framework of which is discussed in section 13.2. Practical aspects of radiation protection in the context of linacs in radiotherapy are then briefly discussed in section 13.3 and the chapter concludes with a section of bunker design and shielding, probably one of the most important tasks of a medical physicist working in a radiotherapy department.

13.1.1 Background Radiation

Exposure of humans to ionizing radiation is part of normal life. This is schematically illustrated in figure 13.1. External radiation sources are cosmic (e.g., from the sun) and terrestic radiation (from ground and buildings). Internal radiation (excluding radon) stems from ingested radioactive nuclides with food, such as potassium-40. A significant contribution also comes from radon emanating from the ground. It is inhaled and as an alpha particle emitter can contribute significantly to the radiation burden. Human activity contributes at present approximately 15% to the overall radiation

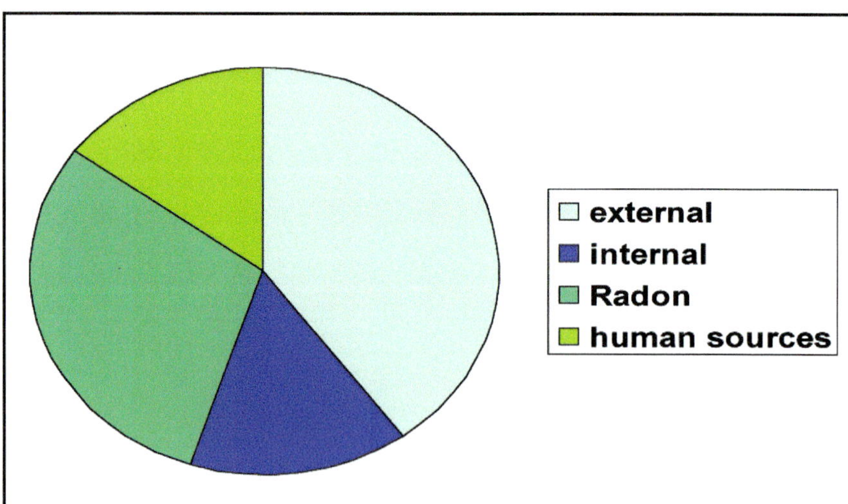

Figure 13.1: Typical human exposure from different sources. The contributions will vary significantly depending on life style and location.

exposure. This is almost entirely made up by medical exposures that account for more than 90% of the exposure due to human activity, with doses from diagnostic procedures being the largest contribution.

The overall *normal* exposure in most countries is approximately 3 mSv per year (for a discussion of the unit sievert, abbreviation Sv, see section 13.3). However, depending on where one lives and works, variations between 1.5 mSv and 5 mSv are normal.

13.1.2 Radiation Effects

Radiation protection must be an integral consideration of all applications of ionizing radiation. Radiation protection in this context has two main objectives:

i) Prevention of deterministic effects such as accidental medical exposure.
ii) Reduction of the probability of stochastic effects.

The difference between the two types of effect is illustrated in figure 13.2a.

Deterministic Effects

Deterministic effects are generally due to cell killing. They have a dose threshold below which no effect can be observed. Above this threshold, the severity of harm increases with dose. The effect is specific for a particular type of tissue. Examples for deterministic effects are erythema of the skin and damage to nerve cells (e.g., leading to paralysis in the case of spinal cord damage). For most deterministic effects, the threshold dose is of the order of several gray (Gy). The actual threshold dose also depends on dose delivery mode with single exposures being typically more detrimental than protracted exposure. However, in general, deterministic effects are of primary concern at high radiation dose levels.

Total body x-ray irradiation (TBI) effects are an example for this as well. Radiation sickness may occur at doses around 2 Gy, where a significant number of intestinal cells are killed. At higher doses the cells of the bone marrow that produce blood cells, and therefore the immune system, are the most important consideration. In TBI for therapeutic purposes (Metcalfe et al. 1997) immune suppression is actually the desired effect, and bone marrow transplants are undertaken to rescue the patient.

Some patients (usually those with chronic myelogenous leukemia) who require alogeneic bone marrow transplants are administered TBI as the method of bone marrow conditioning. Historical regimens of 5 Gy in 1 fraction have been replaced by typically 12 Gy in 6 fractions (delivered in 3 days) in the interests of reduced effect to lung (Lawton et al. 1989). Because all the marrow needs to be irradiated, some complications to radiosensitive organs such as the lens of the eye and lung may occur (Peters et al. 1979, Ozsahin et al. 1994).

In order to encompass the patient with the treatment field extended (usually 300 cm SSD), treatment fields that employ laterals or laterals combined

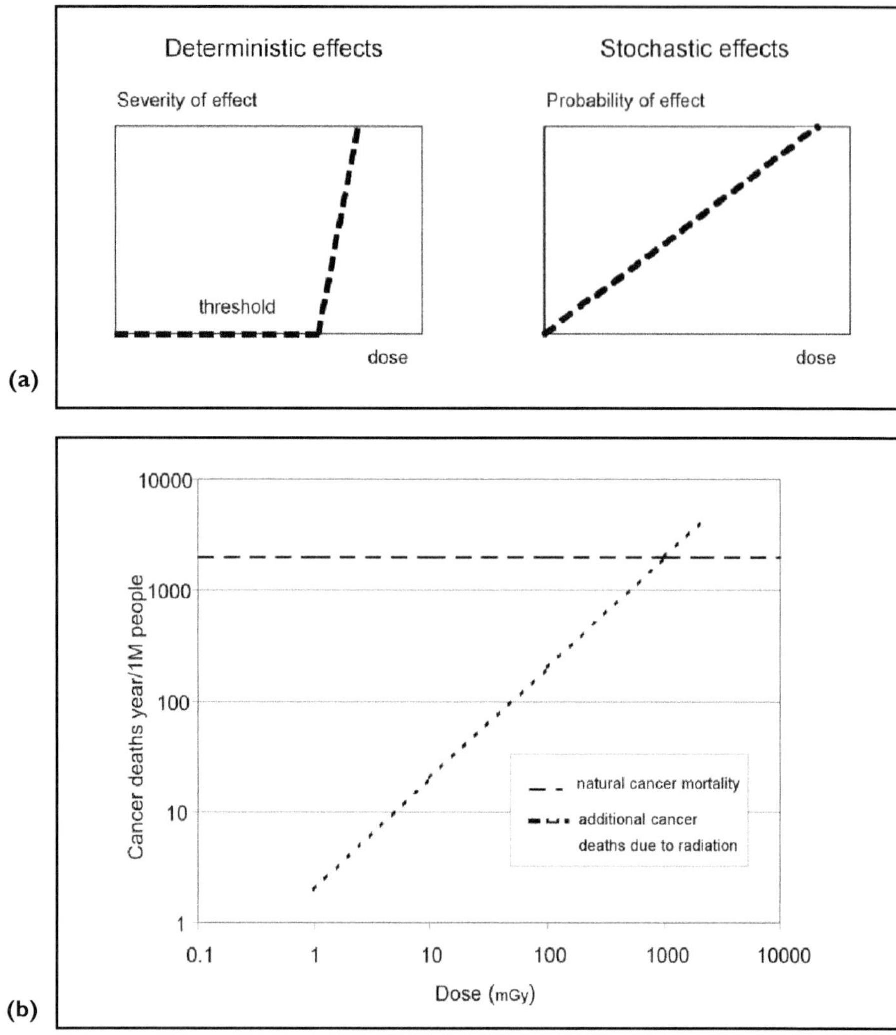

Figure 13.2: Deterministic and stochastic radiation effects and the linear no threshold (LNT) model. (a) Showing the threshold and no threshold concept. (b) Linear relationship between the log dose and the log of cancer induction. There is generally good data to confirm this approach at dose levels above some 200 mSv. The LNT model is only applicable to stochastic effects (i.e., cancer induction or genetic effects).

with AP/PA fields are often used (AAPM 1986, Doughty et al. 1987). A dynamic bed method has sometimes been employed (Quast 1985, Connors et al. 1988, Gerig et al. 1994).

In cases of extreme radiation accidents *radiation syndromes* may occur. However, these patients require intensive care, and for healthy individuals the exposure to several gray of radiation can lead to death due to the shut-down of the immune system within weeks after irradiation. The delay is due to the relatively slow turnover of blood cells. Without intervention (antibiotics, transfusions, bone marrow transplant) this *hematopoietic syndrome* (3 to 8 Gy) leads to death in 3 to 8 weeks.

Once the acute dose to the body exceeds 10 Gy, internal bleeding follows and death may occur within a few days due to gastrointestinal (GI) tract depopulation of epithelium (*gastrointestinal syndrome*). At even higher dose levels, the *central nervous system* (CNS) is directly affected. The *cerebrovascular syndrome* (100 Gy) is characterized by an increase in small vessel permeability and leads to death in 1 to 2 days. Texts that provide further information about CNS syndrome, GI syndrome, or bone marrow syndrome include Dalrymple et al. (1973) and Selman (1983).

Another important (and often discussed) deterministic effect is cataract formation. According to Hall and Giarccia (2005) it is defined as an opacifica-tion of the normally transparent lens of the eye. As a radiation effect it constitutes a failure of the pre-equatorial epithelial dividing cells to differentiate properly, an effect which has been reported to occur after single exposures of 2 Gy or a fractionated exposure of 10 Gy. With increasing dose, the latency period decreases.

Stochastic Effects

Stochastic effects on the other hand are considered to be applicable to all dose levels. The most important stochastic effect from a radiation protection point of view is cancer induction. At least from a conceptual point of view, a single radiation-induced event can lead to cell changes in the DNA, which can result in transformation of a normal cell to a malignant cell. In this case, no threshold dose exists and the severity of the effect is independent of the dose, as cancer induction is equally detrimental if induced by small or large radiation doses. However, the probability of the effect increases with dose.

In practice, the transformation of a normal cell to a malignant cell is not a single event process and consists of many steps. The hallmarks of cancer have been described by Hanahan and Weinberg (2000) as evading apoptosis, self-sufficient growth stimulation, insensitivity to growth inhibition, metastasis, limitless proliferation (*immortality*) and sustained angiogenesis. All these characteristics are associated with changes in the DNA and it requires several (at least three) events to cause the mutation of a normal cell to a cancer.

Non-Cancer Effects

A lot of the risk analysis for stochastic effects are from the survivors of the nuclear bomb explosions in Japan in 1945 (Pierce et al. 1996). It should be noted that the data have also been analyzed to determine the non-cancer mortality for those who died between 1950–1990 (Shimizu et al. 1999). A statistically significant increase with radiation dose has been shown for stroke, heart, respiratory, and digestive tract diseases.

Other Effects

Finally, it is important to also consider genetic/hereditary effects due to exposure. While intensive studies of 70,000 offspring of the atomic bomb survivors have failed to identify an increase in congenital anomalies, cancer, chromosome aberrations in circulating lymphocytes or mutational blood protein changes (Neel et al. 1990), it must be assumed that mutations occur, and there is ample evidence of ionizing radiation causing heritable mutations in many plants and animals.

Genetic effects can be deterministic and stochastic in nature. The radiation effects on embryos depend on the time of gestation. In the pre-implantation period (0 to 9 days) exposure typically leads to death of the embryo (LD_{50} 1 Gy). Growth retardation and malformations are typically not seen. In the organogenesis/embryonic period (10 days to 6 weeks) malformation of organs, small head size, and intrauterine growth retardation are observed. In the early fetal period (6 to 25 weeks) mental and growth retardation are the most significant effect. Increased carcinogenesis must be expected as a result of radiation exposure throughout the fetal phase.

Due to the complexity of radiation exposures during pregnancy, the ICRP has devoted a separate report to the issue (ICRP 1999).

Effects are also categorized as early, which are within one year of irradiation, and late, which appear after this period. Somatic effects are those that affect the human body, including embryo, whereas genetic effects are due to chromosome mutations, which produce heritable effects. If a recessive gene is affected, then mutation of offspring will only be apparent if two recessive genes are linked and the recessive trait shows through.

13.2 The Framework of ICRP Report 60

"The International Commission on Radiological Protection, ICRP, is an independent Registered Charity, established to advance for the public benefit the science of radiological protection, in particular by providing recommendations and guidance on all aspects of protection against ionising radiation." (ICRP, www.icrp.org). The ICRP has been charged with developing guidelines and units for radiation protection since 1928. It also has the task to publish data which can further the field of radiation protection. A historical overview was recently been given by Clarke (Clarke and Valentin 2005), the long-time chairman of the ICRP.

As the webpage of the ICRP states, it "is composed of a *main commission* and five standing Committees: on *radiation effects*, on *doses from radiation exposure*, on *protection in medicine*, and on the *application of ICRP recommendations*, and on *protection on the environment*, all served by a small scientific secretariat. The Main Commission consists of twelve members and a Chairman (currently Dr. L.-E. Holm, Sweden). Like other scientific academies, the Commission elects its own members, under rules that are subject to the approval of ISR. Renewal is assured in that 3 to 5 members must be changed every fourth year. Committees typically comprise 15–20 members. Biologists and medical doctors dominate the current membership; physicists are also well represented." (www.icrp.org).

Recommendations of the ICRP form the basis of radiation protection regulations in most countries of the world. As such, the publication of report 60: "The 1990 recommendations of the ICRP" published in 1991 found their way over the following years into legislation in many nations. An important publication in this context is the *Basic Safety Standards* published in 1996 by the International Atomic Energy Agency (IAEA 1996) and adopted, amongst others, by the World Health Organization (WHO) and the International Labor Organization (ILO). The *Basic Safety Standards* is an internationally binding document based on ICRP report 60 (1991).

ICRP report 60 (1991) defines the philosophy underlying radiation safety, gives extensive rationale for the scientific basis, and provides practical recommendations, e.g., in terms of limit values and dose concepts. In March 2007, the commission approved a new set of recommendations (which was not published at time of writing). However, a press release and versions circulated for comments indicate that the basic concepts and framework of the 1991 recommendations are likely to be still valid. New scientific evidence is incorporated in the new document as well as a novel emphasis on protection of the environment (www.icrp.org).

13.2.1 Quantities for Radiation Protection

The following units are commonly used in radiation protection. The *absorbed dose*, D, is the energy deposited per unit mass in any target material (compare also chapter three). It is the fundamental definition of physical dose and applies to any type of radiation. As discussed in chapter three, absorbed dose is measured in gray (Gy) = 1 Joule kg^{-1}. The old unit for absorbed dose, which is sometimes still used in the context of radiation protection, is the rad = 0.01 Gy.

The *equivalent dose*, H, takes into account the effect of the radiation on tissue by using a *radiation weighting factor*, W_R.

$$H = D \times W_R. \tag{13.1}$$

A summary of weighting factors recommended by the ICRP (1991) is shown in table 13.1. The equivalent dose is measured in *sievert* (*Sv*) and can be used to quantify the dose to individual organs. The old unit for equivalent dose is the Rem = 0.01 Sv.

Finally, the *effective dose*, E, takes into account the varying sensitivity of different tissues to radiation using a *tissue weighting factor*, W_T, according to:

$$E = \sum_{all\ organs} \left(W_T, H\right) = \sum_{all\ organs} \left(W_T, W_R, D\right). \tag{13.2}$$

The effective dose is used to describe the biological relevance of a radiation exposure where different tissues/organs receive varying absorbed doses potentially from different radiation sources. As such, effective dose is essentially a measure of the risk associated with exposure to radiation.

The effective dose is measured in sievert, and the tissue weighting factors are summarized in table 13.2. As can be seen from the weighting factor given to the gonads (0.2) the overall weight of genetic/heredity effects is considered to be of the order of one quarter of the risks due to cancer induction in the person receiving the exposure.

Table 13.1: Radiation weighting factors (Adapted from ICRP 1991)

Type of Radiation	W_R
beta	1
alpha	20
x-rays	1
gamma rays	1
neutrons <10 keV	5
neutrons (10 keV – 100 keV)	10
neutrons (100 keV – 2 MeV)	20
neutrons (2 MeV – 20 MeV)	10
neutrons >20 MeV	5

Table 13.2: Tissue weighting factors (Adapted from ICRP 1991)

Type of Radiation	W_R
gonads	0.2
red bone marrow	0.12
colon	0.12
lung	0.12
stomach	0.12
bladder	0.05
breast	0.05
liver	0.05
esophagus	0.05
thyroid	0.05
skin	0.01
bone surfaces	0.01
remainder	0.05
Total	1

It is important to note that the concept of effective dose and the tissue weighting factors given are only applicable to stochastic effects.

In the context of radiation protection, sometimes the concept of collective dose is also useful. It is used to measure the total impact of a radiation practice or source on all the exposed persons. The collective dose is measured in man-sievert (man-Sv).

13.2.2 Principles of Radiation Protection

Report 60 of the ICRP (1991) defines three underlying principles of radiation protection: justification of practice, optimization of protection, and individual dose limits.

Justification of Practice

No practice involving exposures to radiation should be adopted unless it produces sufficient benefit to the exposed individuals or to society to offset the radiation detriment it causes. This thinking is based on the assumption discussed above that stochastic radiation effects may occur at any dose level and even very small radiation doses are potentially harmful. This concept is captured in the so-called *linear no threshold* (LNT) model which is illustrated in figure 13.2b.

According to this model, there is a linear relationship between the log dose and the log of cancer induction as illustrated in figure 13.2b. There is generally good data to confirm this approach at dose levels above some 200 mSv. The LNT model is only applicable to stochastic effects (i.e., cancer induction or genetic effects).

As a rule of thumb, the total lifetime risk of fatal cancer for general population = 5%/Sv. In terms of equivalent dose, the lifetime fatal cancer risk for cancer of bone marrow is estimated to be 0.5 %/Sv, while breast, lung, and thyroid are 0.2%, 0.85%, and 0.08%, respectively. The relatively low value for thyroid cancer induction does not reflect a low cancer incidence but a good cure rate.

In medical exposures such as occurring to patients undergoing radiotherapy, the justification is given by the prescription of the radiation oncologist. As such the prescription is a very important document from a radiation protection point of view.

The Optimization of Protection

This principle requires that all exposures must be optimized to maximize the benefit and to minimize the risks. In practice, this is the most important principle for radiotherapy physics practice. The optimization of protection obviously does not apply easily to the dose given to the tumor, where the dose delivery is deliberately an objective of the treatment. However, as this is the case, optimization of protection will require the planner and operator to minimize the dose to normal structures as much as compatible with the therapeutic aim. This is good-practice radiotherapy. This may be an area of future improved minimization by more detailed analysis of low- and medium-level dose zones for example in IMRT dose delivery.

For exposures where dose is not the therapeutic agent, optimization of protection can be described by the so-called ALARA principle (As Low As Reasonably Achievable). This means radiation exposure should be limited as much as possible keeping in mind the risk-benefit relation of radiation and its applications. For example, it is unreasonable to refuse an x-ray after a bone fracture because statistically this may shorten your life expectancy by one day. The benefits of the x-ray with its diagnostic value by far outweigh the risk associated with the radiation exposure.

ALARA also implies consideration of the societal and economic context. As such, it cannot be applied to all situations in the same way and professional judgment may be required to achieve an outcome that is optimal for a particular situation.

Individual Dose and Risk Limits

In addition to the two principles discussed above, the ICRP recommends limit values for persons exposed to ionizing radiation. These limit values are meant to restrict the risk due to exposure to the same order of magnitude as other risks we generally accept (e.g., driving a car) or are expected to take as part of employment (e.g., driving a taxi). The limit values recommended by the ICRP in its report 60 (1991) are 1 mSv year^{-1} for the general public and 100 mSv per 5 years for occupationally exposed persons with the effective dose in any single year not exceeding 50 mSv. In addition to this, the commission provides specific annual limits for equivalent dose to the lens of the eye (15 mSv public, 150 mSv occupational), the skin (50 mSv public, 500 mSv occupational), and the hands and feet (only defined for occupationally exposed persons who may need to handle radioactivity, 500 mSv) (ICRP 1991). It is important to note that this dose is in addition to the dose received by any person from natural radiation sources.

13.2.3 Types of Exposure

The ICRP (1991) and subsequently the IAEA Basic Safety Standards (IAEA 1996) define three general categories for exposure (quotes from the glossary of IAEA 1996): *Occupational exposure:* "All exposures of workers incurred in the course of their work, with the exception of exposures excluded from the Standards and exposures from practices or sources exempted by the Standards. *Public exposure* as: "Exposure incurred by members of the public from radiation sources, excluding any occupational or medical exposure and the normal local natural background radiation but including exposure from authorized sources and practices and from intervention situations." *Medical exposure* as: "Exposure incurred by patients as part of their own medical or dental diagnosis or treatment; by persons, other than those occupationally exposed, knowingly while voluntarily helping in the support and comfort of patients; and by volunteers in a program of biomedical research involving their exposure."

The first two concepts are easy to appreciate and the three general principles of radiation protection discussed above apply directly. For medical exposures a limit value as such cannot apply. The exposure always has to be seen in the context of the benefits to the patient (ICRP 1996) and the prescription provides justification for the exposure. However, medical exposures also apply to persons providing support for others who undergo a medical exposure (e.g., parents supporting children) and volunteers in medical research. In both these cases, limit values must be considered and national legislation/regulation usually provides for this.

13.3 Dose Limits

To ensure general radiation standardization of those workers who may be exposed to neutrons or x-rays or other particles, the unit millisievert (mSv) is used for radiation monitoring reporting. For radiotherapy workers using x-rays the weighting factor is nominally one, so a sievert equals a gray.

No practice involving exposures to radiation should be adopted unless it produces sufficient benefit to the exposed individuals or to society to offset the radiation detriment it causes. Currently the assessment of risk is based on the linear (or linear quadratic) no threshold (LNT) model. Obviously, diagnostic medical x-ray procedures are not regarded within occupational or general public dose limits. In these cases the benefit (e.g., x-ray for suspected disease or injury diagnosis) outweighs the risk. In more severe cases where cancer is being treated with radiotherapy to high-dose levels, the benefit of treating to locally control the existing cancer out-weighs the statistical chance of secondary cancer formation.

As a radiation exposure is a small risk to workers using irradiating apparatus, the *radiation safety officer* (RSO) usually issues a *personal radiation monitor* that measures the amount of ionizing radiation received by a person (i.e., natural radiation plus occupational exposure). The weekly radiation limit for the general public is 0.02 mSv and for designated workers is nominally 0.4 mSv. The natural radiation received during any week adds 0.05 mSv (which cannot be avoided), approximately 0.03 mSv of which is external radiation. There historically have been different monitors including film, TL dosimeters, and more recently optically stimulated luminescence (OSL) monitors have become available. For short-term, high-risk procedures ionization chamber-based pocket devices are also available. These usually have a dose threshold alarm. The advantage of these is that dose can be assessed immediately; historically they have not been stable over long time periods of time, hence

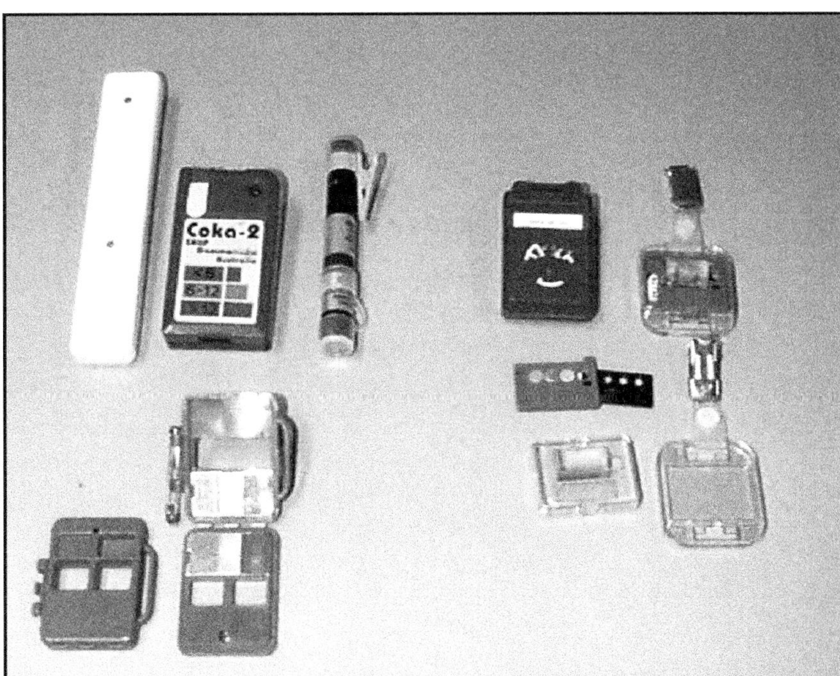

Figure 13.3: Personal dosimeters showing film badges (lower left) with various filters to ensure film response to different energies can be properly assessed. Also TL dosimeters (middle right) and pocket ionization detectors (upper left).

film, TL dosimeters, or OSL is preferred for routine occupational exposure monitoring. A selection of these monitors is shown in figure 13.3.

13.3.1 Practical Radiation Protection

In this section a few issues concerning radiation protection specifically in a medical environment such as a radiotherapy department are discussed. From a practical perspective, radiation exposure can be reduced in three very basic ways: time, distance, and shielding. *Time* refers to the exposure time. Exposure time is directly proportional to the dose received. For radiation workers, this implies keeping the exposure as brief as possible, something of great relevance when working with radioactive isotopes. In the context of linacs, there is typically no option to reduce the exposure time for staff. However, *distance* can be increased in the construction phase of a linac bunker. As distance is related to dose in an inverse-square fashion, distance is a very effective way to improve protection. This is discussed in more detail in the next section.

Also, *shielding* is discussed extensively in the next section. It is an obvious measure to reduce exposure.

13.3.2 A Radiation Protection Program

The concept of a radiation protection program includes a number of measures to reduce the likelihood of adverse effects due to the use of ionizing radiation. A useful publication for the implementation of such a program is Technical Document (TECDOC) 1040 of the IAEA (1998).

A radiation protection program considers a facility as a whole. Figure 13.4 illustrates the relationships between all parties involved in a radiation safety program. The regulatory authority will license the use of ionizing radiation for a particular purpose such as provision of radiotherapy services. The licensee is then the person (in a legal sense) who is responsible for radiation safety. This already implies that it must be of interest to managers to provide sufficient resources to radiation protection. Within the institution, radiation protection affects many parties as indicated in figure 13.4.

Typical elements of a radiation protection program are:

i) Assignment of responsibilities.
ii) The registrant or licensee is the legal person responsible for radiation protection. In practice though, some responsibilities can be delegated for to the *Radiation Protection Officer* (RPO).The RPO is a person technically competent to provide advice and oversight of the local radiation safety program; often (but not necessarily) this may be a radiation oncology physicist. He/she is a crucial component of the program, and should be given the resources and authority necessary. The role would typically include responsibility for designation of controlled and supervised areas, responsibility for ensuring preparation of local rules, the training of new staff in safe radiation work practices, liaison with the regulatory authority on radiation protection matters, supervision of the personnel monitoring program, and

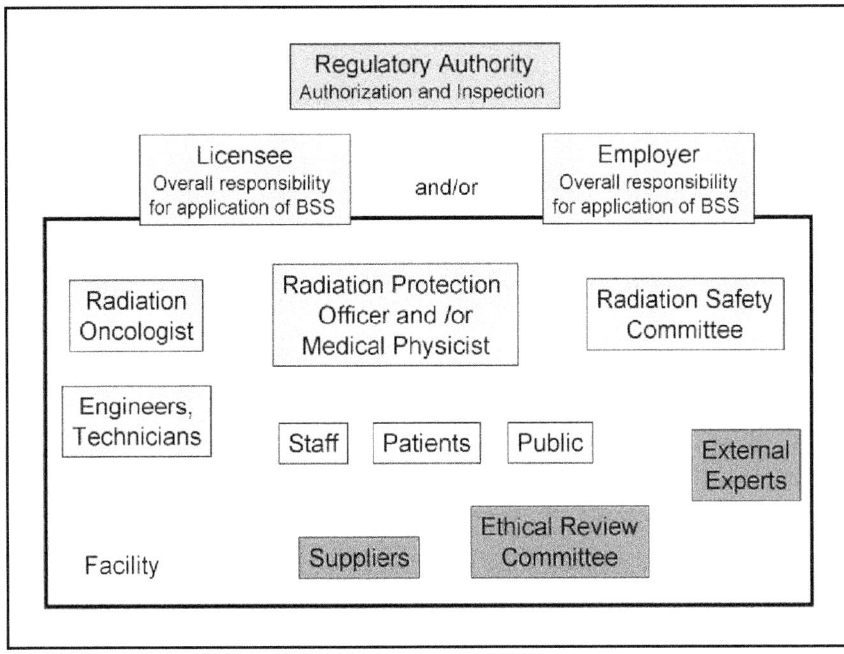

Figure 13.4: Structure chart showing participants in a radiation protection program.

maintenance of records, especially worker radiation histories. In addition to this, she/he would perform routine surveillance of radiation areas and respond to radiation accidents.

iii) A radiation safety committee (RSC): the typical roles of the RSC are to oversee the institutional radiation safety program and advise and review local rules relevant for radiation protection. Members may require special training.

iv) Designation of radiation areas or zones that monitored personnel only should administer.

v) Local rules are intended to provide adequate levels of protection and safety through the establishment of common work procedures and other systems to be followed by all workers in the area. They should be set down in writing and include all information required for work in the area, and be made known to all workers.

vi) Education and training. All staff in radiotherapy must have appropriate education to perform their duties. Staff on whom radiation protection depends must receive training as well as persons who could be subject to irradiation. Technology in radiation therapy is fast developing. It is therefore essential for all staff to have regular updates on radiation protection aspects.

vii) Planning for accidents and emergencies.

viii) Health surveillance and monitoring.

ix) Review and audit of radiation safety procedures.

x) System of recording and reporting. In each radiation facility there should be a system instituted where all relevant information relating to radiation work is recorded, documented, and, when necessary, reported to management. This is a key factor in control of exposures and maintenance of a safe working environment and may depend on national regulations.

Most of the procedures concerning radiation protection can be summarized in a *radiation protection manual*. The manual is ideally a local reference book, and made available to (and possibly mandatory reading for) all radiation workers. In general, it can be expected that the radiation safety manual includes sections on:

i) Basics of radiation safety.
ii) Sources, risks, and effects of radiation.
iii) Local radiation safety organization.
iv) National and/or state regulations.
v) Personnel monitoring rules and requirements.
vi) Emergency procedures.
vii) Sections covering local rules in radiation-user departments include procedures to deal with radioactive cadavers (deceased patients containing radioactive materials).
viii) Radiation and pregnancy (for radiation workers, but also including information relating to patients who may be pregnant).
ix) Incident/accident procedures.
x) Research and radiation, such as guidelines if ethics proposal are sought for clinical trials that include the use of radiation.

13.4 Room Shielding

13.4.1 Primary, Secondary, and Leakage Radiation

Primary radiation is defined as radiation that emanates directly from the target (also termed *useful beam*). The maximum field extension at the shielding wall can be calculated by multiplying the distance across the diagonal of the collimators as light projected to the shielding wall, which for a 40×40 cm^2 field is 56.6 cm (see figure 13.5a), by a divergence factor equal to the distance from target to wall; let this distance be X. A margin of 0.305 m beyond where the primary beam can be directed with fully open collimators is extended to the width of each side of the primary barrier to take account of the beam's divergence through the wall thickness, therefore

$$W = 0.566X + 0.61. \qquad (13.3)$$

Secondary radiation (also termed *scatter*) is that produced by scattering of the primary radiation beam off scatter volumes such as the patient. *Leakage radiation* is that which escapes from the linac head in any direction other than that of the useful beam. Excess leakage may occur if insufficient lead shielding is placed around the head of the machine, but even with correct shielding some leakage will always be present when the beam is activated. A schematic showing the various sources of primary, leakage, and scattered radiation emanating from the patient during linac treatment is shown in figure 13.5b.

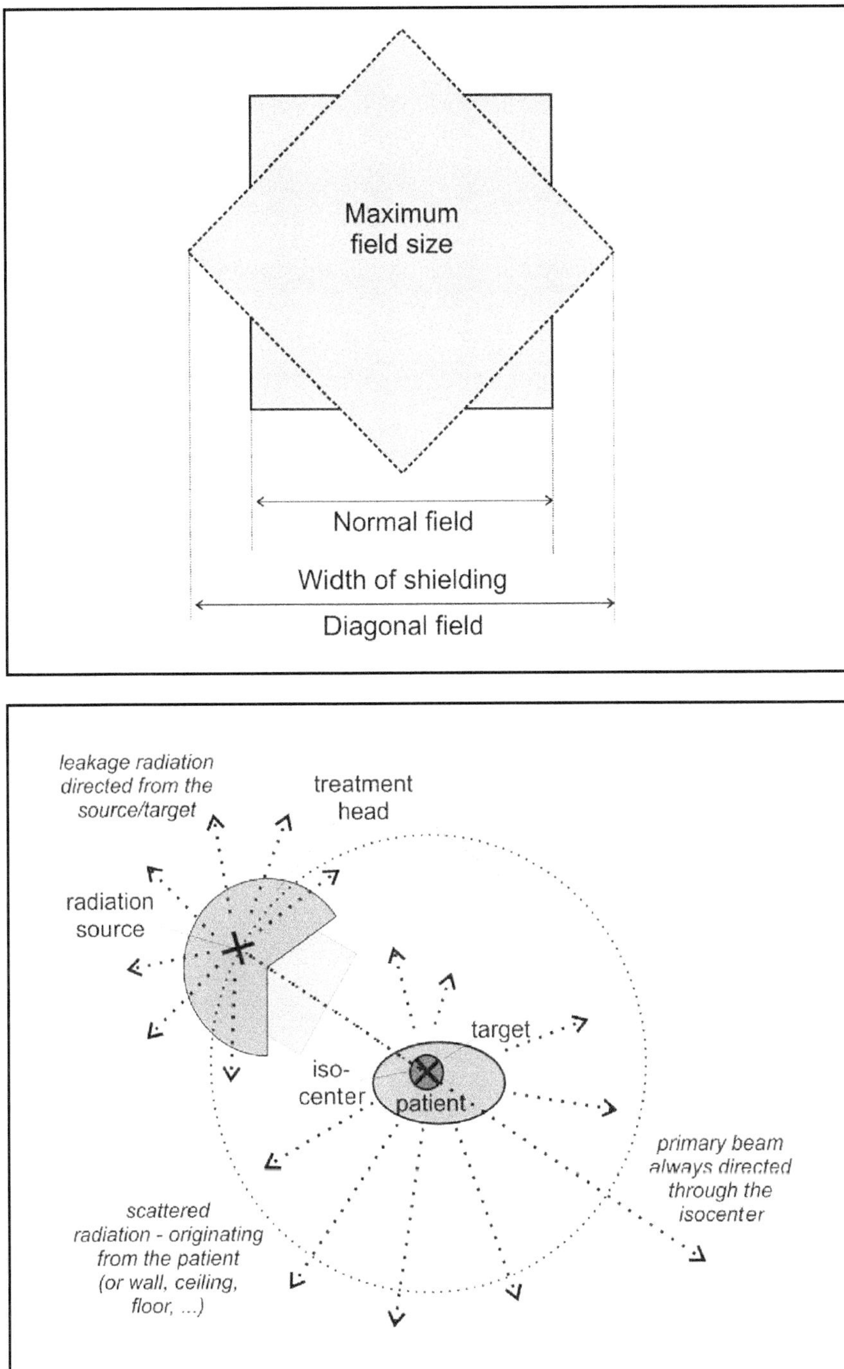

(a) (b)

Figure 13.5: The collimator rotation effect on room shielding extension and the various sources of radiation requiring shielding protection. (a) Demonstrates that the 45° rotation of linac collimator extends the width of room shielding required. An extra safety margin of primary barrier shielding is also provided beyond the diagonal beam width. (b) Schematic showing the sources of primary, leakage, and scattered radiation that needs to be considered in room shielding calculations.

13.4.2 Workload and Dose Limits

Workload, *W*, is the likely maximum dose output from the x-ray machine for linacs; it is usually calculated in units of Gy per year at isocenter. *Use factor*, *U*, is the fraction of the workload for which the radiation is directed at the space for which protection is being estimated. *Occupancy factor*, *T*, is the fraction of the workload that the area is occupied by personnel or members of the general public.

Typical bunker plans are shown in figures 13.6a and b. Note the treatment room is often referred to as a *bunker*. Further details of barrier thickness calculation methods are included elsewhere (Khan 1984, NCRP 1976, NCRP 1977).

There are different ways of carrying out barrier design calculations, and they can be based on various different assumptions. In general, the calculations should use a *conservative* approach to ensure maximum radiation safety for hospital staff and the general public.

Prior to any calculations the assumptions made should be clearly stated. These include physical parameters such as tenth-value layer (TVL) of shielding material used for a particular beam energy. TVLs for different materials at different beam energies are given in table 13.3, angles of the walls, distances to be used in ISL calculations, assumptions concerning *workload*, *W*, (treatment, physics QA, research, and maintenance), *use factor*, *U*, and *occupancy factor*, *T*, of any area to be protected from the radiation.

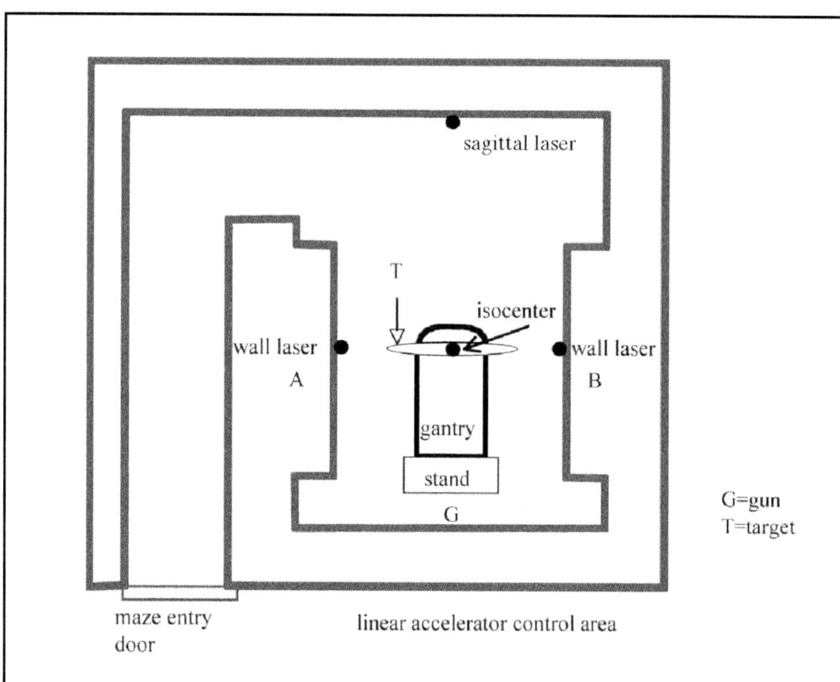

(a)

Figure 13.6: Different linac room design plan layouts: (a) Linac gantry rotating perpendicular to maze; (b) Linac gantry rotating parallel to maze; (c) Plan of linac in bunker showing the distances defined as used in calculating dose contributions to the entry of the maze. (Adapted from Van Dyk 1999.)

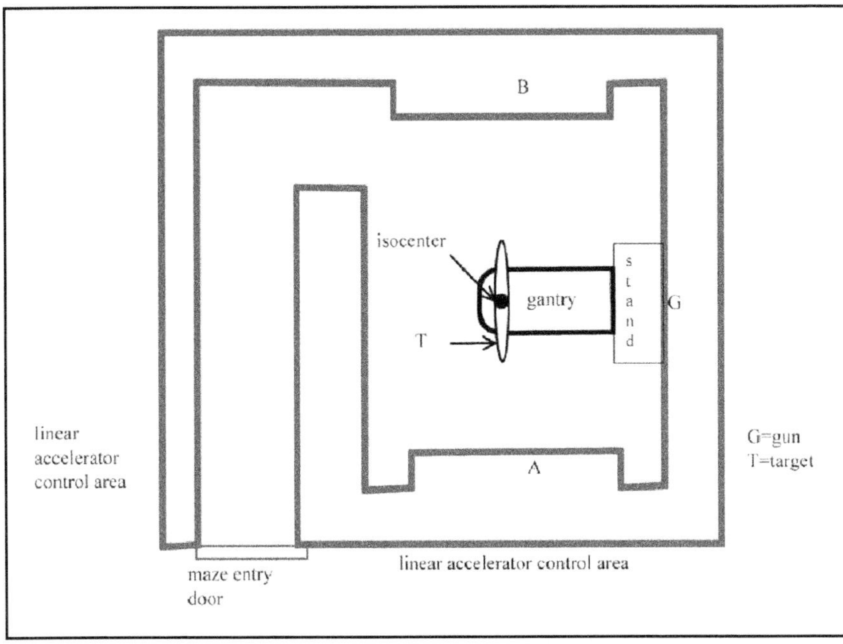

(b)

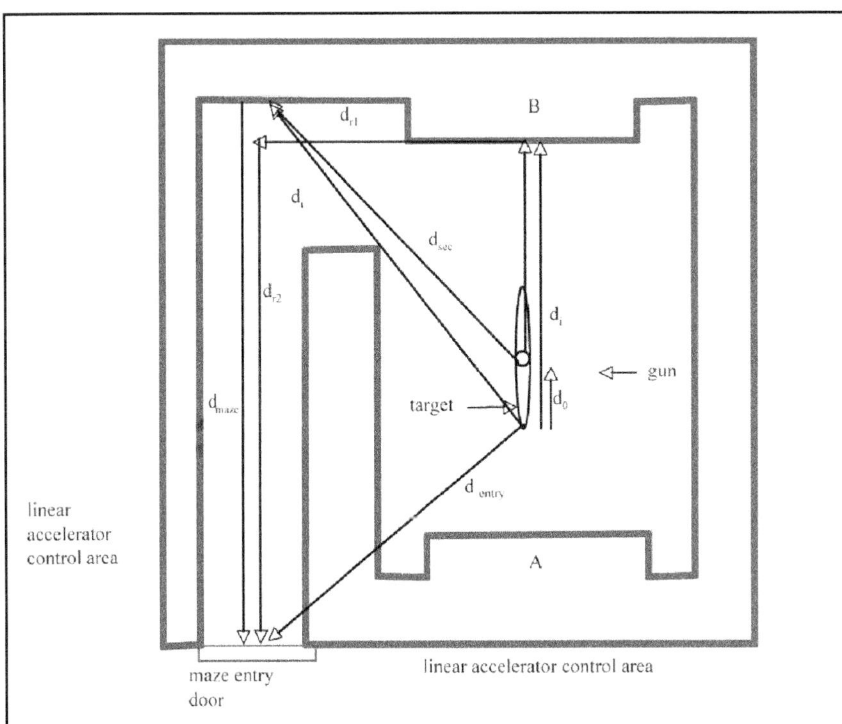

(c)

Figure 13.6: (Continued)

Assumptions should also clearly state the linac type including beam parameters (e.g., energy, maximum collimator settings). In this section it should also be stated which legal (or other) limit values are taken for the exposure of staff and/or general public and patients.

For primary barriers the following use factors can be assumed for a linac (Khan 1984, NCRP 1977): U (floor) = 1, U (walls) = 0.25, U (ceiling) = 0.25, depending on equipment and techniques. Following a conservative approach, these numbers add up to more than 1. Obviously, U = 1 for all secondary barriers.

Following the NCRP (1976) typical occupancy factors are: T (work areas, offices, nurses stations) = 1, T (corridors, rest rooms) = 0.25, and T (waiting rooms, toilets, stairways, elevators, outside areas for pedestrians and traffic) = 0.125 or less.

Protective barriers are designed to ensure that doses to personnel/patients/general public adhere to the ALARA principle. It is always envisaged that these levels are lower than the permissible dose limits. A nominal 1 mSv year^{-1} dose limit or less, H_{limit}, is usually chosen for dose calculations to any room positions that personnel (designated or otherwise) are likely to occupy. In some cases the relevant authority may require the design limit to be lower than this (e.g., 0.3 mSv year^{-1} is becoming a commonly used value).

A check with the authority to ensure that the walls are designed to the correct H_{limit} is advisable. In the United States the state radiological health programs require that an area assigned a full occupancy should not receive more than 0.02 mSv hour^{-1} and 0.5 mSv year^{-1}. As an alternative if it can be demonstrated by measurement, calculation, or simulation model that the dose to an individual that is likely to receive the highest dose will not exceed the annual dose limit of 1 mSv, this is also deemed acceptable.

Calculation of Workload

A typical linac machine workload value may be ascertained as follows:.

i) Clinical workload for 50 patients per working day, 3.3 Gy of radiation delivered at isocenter per patient, 5 working days per week, and 52 working weeks per year. Therefore the total clinical workload $W_{clin} = 3.3 \times 50 \times 5 \times 52 = 42{,}900$ Gy/year at isocenter (Kron et al. 1996).
ii) Physics and electronics workload includes routine quality assurance (QA), phantom measurements, and monthly and annual QA. A conservative estimate of this is 7100 Gy/year. This is likely to depend mainly on the amount of beam data collection and symmetry testing required because both procedures require a lot of beam-on time.
iii) Therefore the total accelerator workload per year is $W = 42{,}900$ Gy + 7100 Gy = 50,000 Gy at isocenter.

Different workload values may be estimated. All shielding calculations should be based on conservative data. In this sense the maximum likely machine workload should be used for shielding calculations. If a workload for a similar machine and environment are available, this is the optimum way to obtain a realistic workload value (Kron et al. 1996). A typical linac workload value can vary and there are several different values quoted in the literature (NCRP 1976, NCRP 1977).

For superficial and orthovoltage units, assuming 32 patients treated per day, the weekly workload is assumed to be 3000 mA min for superficial units up to 150 kVp and some 20,000 mA min for orthovoltage units of 200 to 500 kVp. From this, the dose per mA min can be derived (NCRP 1976).

13.4.3 Shielding Calculations

If a new treatment room has to be designed, a number of issues need to be considered:

i) If more than one bunker is to be built, a common wall reduces the costs for shielding.
ii) Distance to bunker wall barriers should be maximized due to the ISL effect. (Note this in turn increases area to be shielded).
iii) Avoiding high occupancy areas in the primary beam zone reduces primary shielding thickness.
iv) A reasonably large control area is required to house computers and personnel. Usually four but up to seven personnel may be present in the control area during treatment.
v) The bunker needs to be designed with installation of the linac in mind. The linac is generally installed into the bunker via the entry corridor. An atrium is occasionally used as the entry point.
vi) The entry and surrounding areas need to provide easy access for beds and wheelchairs, and easy access to the linac patient couch should be possible for mobile beds and wheelchairs.
vii) The treatment couch needs to be able to rotate in extended position and still have some margin of distance between couch and walls.
viii) More than one scatter event should have taken place prior to radiation reaching the entry door.
ix) The bunker is designed to a full workload at the maximum x-ray energy available, unless the regulatory body allows for percentage workload at each energy to be taken into consideration.

Orientation of Linac

The two common orientations of linacs in the treatment room are shown in plan view in figures 13.6a and 13.6b. Also shown in figure 13.6c are the various distances involved when considering scatter and leakage. Geometric orientations are defined with reference to these figures. The axis which runs on a line through isocenter and parallel to the linac wave guide from gun (G) to target (T) along the gantry at 0 degrees is referred to as the beam radial axis and the line which runs perpendicular to this axis through isocenter across the room as defined by the lateral wall lasers is known as the transverse axis. These may be given an orientation reference A and B. The two most common orientations are with the radial axis parallel to the entry corridor, and perpendicular to the entry corridor.

A linac-to-wall diagonal orientation is sometimes employed to ensure a large distance can be obtained to the nearest walls when the gantry is rotated 90 degrees, this ensures a large field is available for total body irradiation.

Primary Barrier

Following an adaptation of the formalism of McGinley (1997) the transmission of the *primary* barrier B required to reduce the primary beam to an acceptable level is given by

$$B = \frac{H_{limit} \times \left(d/d_0\right)^2}{W \times U \times T}, \tag{13.4}$$

where W is the workload. This is the likely maximum dose output from the x-ray machine. For linear accelerators this can be calculated in units of Gy year^{-1}, Gy week^{-1}, or Gy hour^{-1} at isocenter. Use factor, U, is the fraction of workload for which radiation is directed at the space for which protection is being estimated. Occupancy factor, T, is the fraction of the workload that the area to be protected is occupied by personnel or members of the general public. The distance d is that from the target to a position 0.305 m outside the wall to be shielded. Note 0.305 m, which is 1 ft, is the traditional minimum distance that a person is expected to stand away from a shielding barrier; d_0 is the distance from the target to isocenter.

Tenth-Value Layer

Once the barrier transmission factor has been calculated the thickness of required shielding material can be derived from broad beam transmission data plotted as log of barrier transmission versus barrier thickness as shown in figure 13.7 and elsewhere (NCRP 1976, NCRP 1977). The thickness of a material required to reduce radiation to one tenth of its original intensity is called a *tenth-value layer* (*TVL*). As these plots are often limited in barrier thickness scale, an alternative is to calculate the number n of TVLs where

$$n = \log_{10}\left(1/B\right). \tag{13.5}$$

The thickness of shielding required, s, is then calculated using

$$s = TVL_1 + \left(n-1\right) \times TVL_e. \tag{13.6}$$

TVL_1 is the thickness of the first TVL and TVL_e is the thickness of subsequent TVLs as given in table 13.3 for various materials. TVL_1 may be a little shorter sometimes due to the softer beam components being absorbed in the first TVL. Note for some energies an increase in the value of TVL_1 to TVL_e may be indistinguishable.

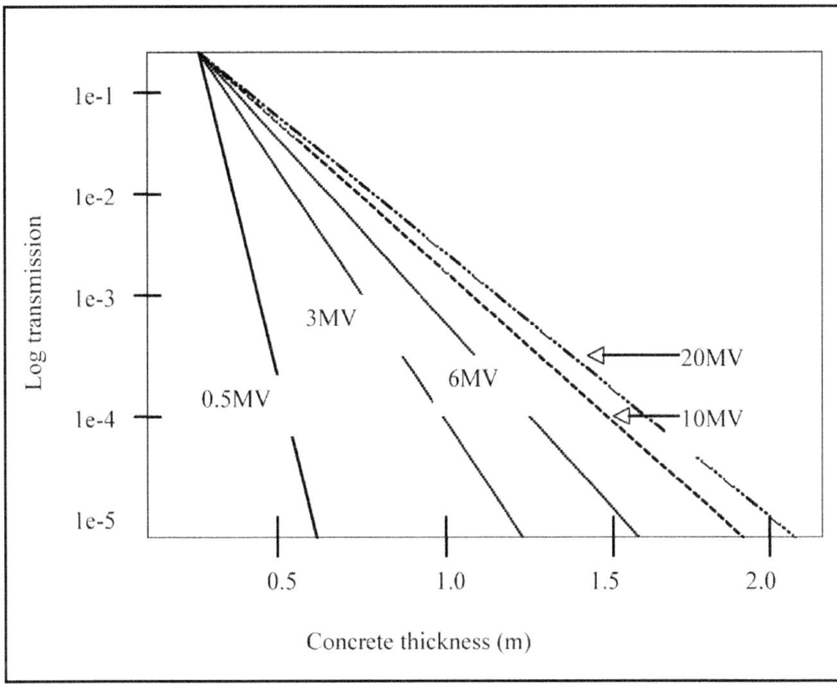

Figure 13.7: Tenth-value layers (TVLs) for different linac beam energies transmitting through concrete barriers.

Secondary Barriers

The secondary barrier calculations incorporate calculations due to *scatter* from the patient and from other scattering surfaces whereby (McGinley 1997):

$$B = \frac{H_{limit} \times (d_1)^2 \times (d_2)^2 \times (F/F_0)}{a \times W \times T}, \tag{13.7}$$

where d_1 is the distance from the patient to the position to be protected beyond the barrier and d_2 is the distance from the target to the patient; a is the ratio of scattered radiation at 1 m from the scattering object to the primary radiation at 1 m (see table 13.4); F_0 is the average field area (400 cm^2) and F is the beam field area at the patient.

The TVL for scattered radiation is less than for primary radiation. The TVL decreases as the angle of scatter increases due to reduction in energy of the scattered radiation. Some scatter TVLs versus scattering angle are given in table 13.5 for a 6 MV x-ray beam.

The calculation of head leakage barrier is done assuming an attenuation due to accelerator shielding of A = 0.001 (NCRP 1976); therefore B due to leakage is given by

$$B = \frac{(1/A) \times H_{limit} \times (d_3)^2}{a \times T}, \tag{13.8}$$

Table 13.3: Tenth-value layer (TVL) for different materials. (Data from McGinley 1997, Metcalfe et al. 1997, Varian 1991, Hendee and Ibbott 1992, NCRP 1976, 1977, Cember 1983, Seelman-Eggbert et al. 1981, Van Dyk 1999)

Electron Energy (MeV)	*Shield Material*	*TVL_l (m)*	*TVL_e (m)*	*TVL_L (m)* 90°*
0.2	Concrete	--	0.084	--
	Lead	--	0.0017	--
0.25	Concrete	--	0.074	--
	Lead	--	0.0029	--
0.5	Concrete	--	0.117	--
	Lead	--	0.012	--
3	Concrete	--	0.245	--
	Lead	--	0.0485	--
4	Concrete	--	0.292	--
	Steel	--	0.091	--
	Lead	--	0.053	--
	Leadite	--	0.14	--
6	Concrete	0.35	0.35	0.279
	Earth	--	0.572	--
	Steel	0.099	0.099	0.080
	Lead	0.055	0.057	0.045
10	Concrete	0.41	0.39	0.305
	Earth	--	0.648	--
	Steel	0.104	0.104	0.085
	Lead	0.057	0.056	0.046
15	Concrete	0.46	0.43	0.330
	Earth	--	0.720	--
	Steel	0.108	0.108	0.087
18	Concrete	0.47	0.43	0.330
	Steel	0.108	0.108	0.087
20	Concrete	0.48	0.44	0.343
	Steel	0.108	0.109	0.088
24	Concrete	0.51	0.46	0.356
	Steel	0.109	0.109	0.089

* The TVLs for this column were defined using BJR (1972). There is a difference in energy above 15 MV between BJR (1972) and BJR (1983).

where d_3 is taken from isocenter. This is not completely rigorous (most leakage is from the bend magnet/target assembly). Since all different source positions due to gantry rotation have to be taken into account, this does however represent a reasonable assumption for large distances compared to the isocenter distance.

There is a special case for d_3 with horizontal wave guide linacs whereby the distance is measured from the gun to the point of interest for the shielding wall directly behind the gun. This is because these accelerators are subject to leakage from the vicinity of the gun.

Table 13.4: Scatter-primary ratio (a) at 1 meter from a human-size phantom for a size 400 cm^2 at the phantom, target to phantom distance of 1 meter (Adapted from NCRP 1976)

Scattering Angle (deg)	*6 MV a*	*10 MV a*
30	0.007	0.0030
45	0.0018	0.0010
60	0.0011	0.0005
90	0.0006	0.0003
135	0.0004	0.0002

Table 13.5: TVL for scatter at various angles from a primary 6 MV beam (Adapted from ICRP 1976)

Angle (°)	*15*	*30*	*45*	*60*	*90*	*135*
TVL (m)	0.33	0.27	0.24	0.21	0.18	0.15

Note that for a simple conservative calculation of leakage primary TVL data can be used (Metcalfe et al. 1997). Alternately, specific TVLs for leakage radiation as shown in the last column of table 13.3 can be used (Varian 1991). As with scattered radiation the energy and hence TVL of leakage depends on the angle of exit from the linac head. Details of different TVL with leakage angle are available (Nelson and LaRiviere 1984), and the TVL increases when the angle of reflected leakage is reduced. The maximum difference in TVL is about 2 cm over an angle range from 35° to 135° for energies from 6 to 25 MV.

The secondary barrier calculated for leakage is usually greater than the calculated scatter barrier thickness. One rule of thumb is that if the leakage barrier exceeds the scatter barrier calculated by more than 3 half-value layers (HVLs), then just the leakage barrier thickness is applied; but if they are within 3 HVLs then 1 extra HVL is added to account for the slight increase in dose exiting from the scatter workload component (Hendee and Ibbott 1992).

To simplify the calculation above 10 MV, the scattered component of radiation is negligible compared with the head leakage component and hence only the leakage calculation is required to estimate the thickness of secondary barriers at high energies (McGinley 1997).

Oblique Surfaces

The radiation may approach a barrier at an oblique angle; for example, if the linac is placed diagonally in the room, it is likely to strike wall barriers at an angle. In this case, the slant thickness can be calculated as radiation path length t of the incident beam, which is given by

$$t = \frac{s}{\mathrm{Cos}\,\theta}, \tag{13.9}$$

where θ is the angle of the beam with respect to a line perpendicular to the barrier and s is the perpendicular thickness of the barrier. One must remember that some scatter will not stay on this path; hence, as a rule of thumb, 1 or 2 HVLs are added to the barrier calculated thickness to allow for this effect; for further details see Biggs (1996).

The Floor

The floor is generally only constructed to be thick enough to support the linac's weight. Therefore, care should be taken to ensure an angled beam path through the equivalent thickness of the floor; the base thickness of the support wall is equivalent to the required shielding wall thickness. The space beneath the floor may need to be temporarily accessible for services such as cables leading from the modulator to the klystron. This area should be entry interlocked to stop the machine if any port to the area is opened. The entry port should also be physically locked and marked with a radiation warning sign.

The Roof

It is generally advisable to ensure the roof of the accelerator bunker is a non-occupied space. A reasonable thickness of concrete is however advisable to reduce the dose from *sky shine*. This phenomenon occurs due to scattered photons reaching accessible areas at ground level due to photon scatter from the air volume above the roof. The effect can be calculated (Borak 1975). If the roof is posted with appropriate warning signs and is well isolated from general access, a nominal dose rate limit may be set. Another approach is to assign a quasi occupancy factor to the roof of 6%. The advised method depends on the regulations in force in the particular country or state.

Calculation of Maze Scatter Protection

Previous reports by Geise (1984) and Cember (1983) have calculated the scattered radiation in the maze but these calculations failed to account for all contributions of dose to the dose at the maze entry. The extra components are included in a more complete analysis by McGinley (1997), which was followed by Podgorsak et al. (1999) as follows. The primary beam scattered off walls to the maze entry (S_s)

$$S_s = \frac{D_0 \times \alpha_1 \times A_1 \times \alpha_2 \times A_2}{\left(d_i \times d_{r1} \times d_{r2}\right)^2}, \tag{13.10}$$

where D_0 is the workload with the machine facing one wall (see figures 13.6). S_s is the dose at the maze entry from primary beam being scattered down the maze; note that as the primary is attenuated by the patient, then an attenuation factor f can be incorporated to calculate the final dose (for 6 MV f = 0.27). α_1 is the reflection coefficient at first reflection based on an energy one-half the primary, α_2 is the reflection coefficient at second reflection based on an energy of 0.5 MeV. A small sample of reflection coefficients is included in table 13.6; a more detailed series of reflection coefficients are graphically plotted against energy up to 10 MeV in NCRP report 51 (1977). A_1 is the beam

Table 13.6: Scattering reflection coefficients for x-rays of normal incidence (Adapted from NCRP 1977)

Angle of Reflection (°)	*4 MV*	*6 MV*	*10 MV*
0	4.5×10^{-3}	3.3×10^{-3}	3×10^{-3}
30	4×10^{-3}	3×10^{-3}	2.7×10^{-3}
60	3.3×10^{-3}	2.6×10^{-3}	2×10^{-3}
75	2×10^{-3}	1.7×10^{-3}	1.5×10^{-3}

area at first reflection; A_2 is the cross section of the maze; and d_i, d_{r1}, d_{r2} are the distances from the target to the wall, across to the center of the maze, and down the maze, respectively.

Scatter dose off the patient then scattered off walls to the maze entry, S_p, is given by

$$S_p = \frac{a \times D_0 \times (F/F_0) \times \alpha_1 \times A_3}{\left(d_0 \times d_{sec} \times d_{maze}\right)^2}, \tag{13.11}$$

where in this case A_3 is the area of the maze back wall that can be seen from the maze entrance, d_0 is the distance from target to patient, d_{sec} is the distance from patient to center of maze back wall, and d_{maze} is the distance down the maze. The scatter of leakage component at the maze entry, L, is calculated from

$$L = \frac{L_0 \times D_0 \times \alpha_1 \times A_3}{\left(d_i \times d_{maze}\right)^2}, \tag{13.12}$$

where L_0 is the ratio of dose due to head leakage to that at isocenter (usually taken as 0.001), and d_i is the diagonal distance from the target to the maze end wall.

Transmission dose from leakage radiation through the maze barrier to the door, T, is

$$T = \frac{L_0 \times D_0 \times B}{\left(d_{entry}\right)^2}, \tag{13.13}$$

where d_{entry} is the distance from the target to the maze entry.

Because the workload D_0 has been calculated with the beam facing the back wall only, the result needs to be multiplied by some factor to account for the other workloads when the linac is in other orientations, i.e., facing up, down, and to the other wall. McGinley (1997) recommends a multiplication factor of 2.64; hence the dose at the door is given by:

$$D = 2.64 \times \left[S_p + f(S_s) + L + T\right]. \tag{13.14}$$

If the scatter dose is significant at the maze exit, then a shielded door may be required. The maze is designed to ensure more than one scatter event (and preferably several) should have taken place prior to radiation reaching the entry door. This ensures the energy of any residual scatter at this position is low and hence the door can be constructed with less shielding material. The radiation energy at the maze entry is much lower than the primary beam (nominally 200 keV) therefore a 3 mm lead door is often sufficient—this is more than 1 TVL at this energy. The thickness of lead required in the entry door to reduce the dose to an acceptable level can be looked up using a barrier transmission graph of log attenuation versus lead thickness for this energy. A mechanical mechanism to open and close this door is advisable for occupational health and safety reasons as these doors are opened and closed about 150 times per day.

Note that any calculation used to estimate the dose at the end of the maze should be checked with a radiation survey instrument. If the dose from scatter is sufficiently low at the exit of the maze (as determined by calculation and verified by survey), then no door is required. In such cases the entry is interlocked by using a door entry sensor (e.g., infrared curtain). It is important that the range of such a sensor is tested to floor level.

If the room is designed without a maze (e.g., floor space may be limited), a mechanically driven door with shielding equivalent to the adjoining wall (scatter or primary barrier thickness) is required in the entry to the room. Suitable designs are listed elsewhere (McGinley 1997).

Shielding considerations for tomotherapy are dealt with elsewhere (Robinson et al. 2000, Balog et al. 2005). Interesting differences are the narrow beam (5 cm) for primary shielding, beams stopper, and low leakage.

13.4.4 Protection Against Neutrons

If an accelerator with a photon peak energy higher than or equal to 10 MV is to be installed, additional consideration must be made regarding neutrons produced by gamma neutron (γ,n) reactions. A number of publications deal with this problem for different accelerator types and energies (e.g., AAPM 1986, Gur et al. 1978, Nath et al. 1980, Sanchez et al. 1989). Neutron contamination is produced by high-energy photons and electrons incident on the target, primary collimator, flattening filter, and jaws. The cross section for gamma neutron reactions is about 10 times larger than electron neutron reactions. Hence the neutron activation from high-energy electron beams is lower than from high-energy x-ray beams. This is the reason why the maximum x-ray energy and not the maximum electron energy is considered the most significant contributor to the neutron dose. X-ray beams that have energies above 10 MV are those worth examining for neutrons.

The most likely source of neutrons is the treatment head, where neutrons can be produced in the target, flattening filter, collimators, and shielding material (Khan 1984). However, neutrons can also be produced in all other materials (McGinley et al. 1984). Note that devices in the primary beam such as wedges and blocks can be activated. Surveys of this equipment and careful selection of appropriate materials for these devices are recommended.

The typical energy spectrum of the neutrons has a peak at 1 MeV, which is similar to the energy of neutrons produced from uranium fission. Detection of neutrons can be made using rem-meters (Khan 1984), TL dosimeters (see chapter three) or activation foils. For the latter, gold or indium (Sanchez et al. 1989) are particularly well suited.

Neutrons can activate other elements that remain radioactive and will contribute to the exposure of radiation therapy staff entering the room after a high-energy photon treatment. A similar problem is posed by the direct activation of elements in (γ,n) reactions (McGinley et al. 1984). The most likely elements to be activated are oxygen (^{16}O (γ,n) ^{15}O, t_H = 2 minutes) and nitrogen (^{14}N (γ,n) ^{13}N, t_H = 10 minutes).

Neutrons can undergo more scatter processes than photons before they are absorbed. Therefore neutron scatter down the maze is a particular problem (Kersey 1979). Typically this is dealt with by a specially designed neutron protection door.

A neutron door is generally provided to ensure the absorption of neutrons that would reach the maze entry. This usually consists of a borated paraffin compound filling the door, which causes a neutron to slow by multiple collisions with the nuclei and then capture gamma ray reactions. The steel case of the door absorbs most of these gammas. Some gammas may escape the steel in the door; therefore, neutron doors are often placed some way down the maze, so that the rest of the maze acts as a shield from the control area. The only alternative to a neutron door is a long and/or baffled maze.

13.4.5 Radiation Surveys

Once a linac has been installed, a radiation survey is essential. Prior to starting the survey ensure available plan data about barrier wall and door thicknesses approximately match the real situation in and around the bunker. Define all areas that are within 1 m of the primary beam as primary barriers and designate other areas as secondary barriers. Clearly define data you are going to use for calculations in terms of use factor, workload, and occupancy factor.

Check that beam energies available on the equipment correspond with the bunker design brief. Be familiar with the likely workload at each energy. Have a suitably sensitive and calibrated x-ray survey meter available, preferably with a secondary back-up monitor. Wearing a personnel monitor and also a pocket ionization dosimeter with an audible alarm is advisable. A preliminary machine calibration and energy check should be undertaken.

Be familiar with rapid ways to disengage the beam if required. For example, be familiar with machine-off mechanisms both at the control console and in the room. The control area should be surveyed with a sensitive radiation survey instrument. All emergency off switches in the room and outside should be tested to ensure that electrical supply to the machine is stopped when these switches are activated.

The door interlocks, entry room interlocks, and/or neutron door interlocks can be tested by opening these devices during beam-on. This should immediately deactivate the radiation beam. For survey measurements away from the control area a second person is required to operate the beam-on and beam-off commands at the control console. Always ensure rapid, clear communication with this person, e.g., voice, intercom, or mobile phone. A distance from door or wall to survey instrument of about 30 cm is usually maintained.

Once this initial checklist has been completed, it should be followed immediately by a detailed radiation survey of all areas surrounding the room. Typical survey meters vary from large ionization volume devices to other variations on the theme. Some examples are shown in figure 13.8. Geiger counters are not suitable as radiation survey instruments for linac room surveys. The areas to be surveyed include the exterior bunker walls, exterior roof, maze at various points, physics port, and particularly the control area where designated staff will be spending most of their working time. Attempt to monitor likely high-dose areas by placing the monitor remotely with the operator well away from it at the time of exposure and either observing dose by sight, video, or recording integral dose.

The linac primary beam is generally pointed at the primary wall barrier to be tested without interposing material, and some interposing material is generally placed in the primary beam when scatter measurements are taken. For scatter measurements the gantry angle is generally less critical. However, several gantry angles should be tested for the special situation of monitoring scatter down the maze.

Dose measurements with a neutron survey meter should be undertaken at the treatment room maze entry with the neutron door both open and closed. The dose from the survey meter in the bunker at various times (e.g., 10-second increments) after beam-off should be recorded. This enables the

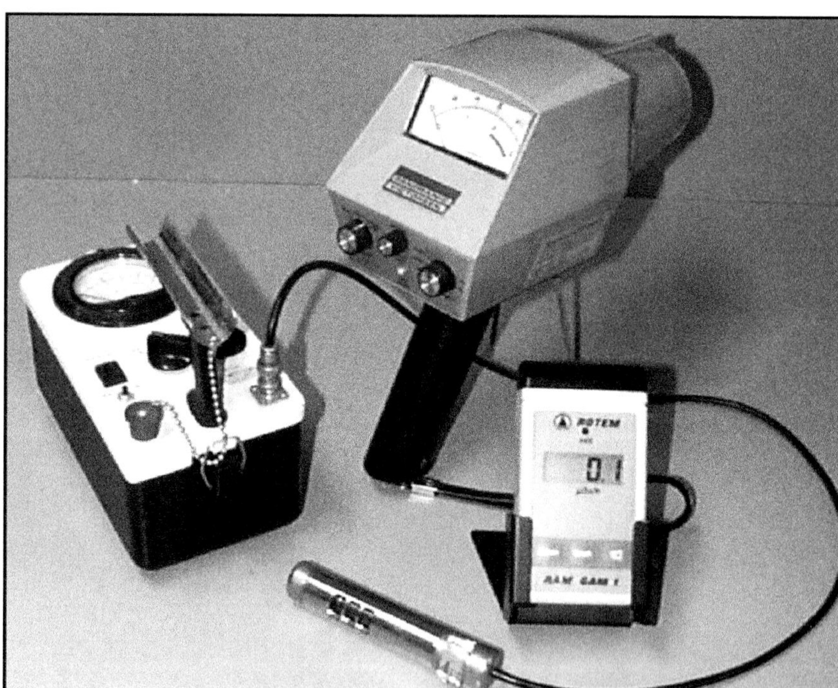

Figure 13.8: A variety of survey dosimeters.

calculation of a safe time interval for staff to enter the room after the use of high-energy x-rays.

Head leakage tests should also be undertaken during the radiation survey. This can be done by wrapping x-ray film around the bend magnet lead shielding area. Develop the film after some beam on-time. Dark patches will show up if there are any gaps in the bend magnet shielding. Alternatively a survey meter placed remotely in various positions beyond the treatment shielding around the back of the gantry may be sufficient (this should be monitored remotely).

References

AAPM (American Association of Physicists in Medicine) (1986) Report 19. Neutron measurements around high energy x-ray radiotherapy machines. AAPM, New York.

Balog J., Lucas D., DeSouza C., Crilly R. (2005) Helical tomotherapy radiation leakage and shielding considerations. *Med. Phys.* 32: 710–715.

Biggs P. (1996) Obliquity factors for cobalt 60 and 4, 10 and 18 MV x-rays for concrete, steel, and lead for angles of incidence between 0 and 70 degrees. *Health Phys.* 70: 527–536.

BJR (British Journal of Radiology) Supplement 17 (1983) Central Axis Depth Dose Data for Use in Radiotherapy. British Institute of Radiology, London.

BJR (British Journal of Radiology) Supplement 11 (1972) Central Axis Depth Dose Data for Use in Radiotherapy. British Institute of Radiology, London.

Borak J. (1975) A simple approach to calculating gamma ray sky-shine for reduced shielding calculations. *Health Phys.* 29: 423–425.

Cember H. (1983) *Introduction to Health Physics*, Second edition. McGraw-Hill, New York.

Clarke R., Valentin J. (2005) A history of the international commission on radiological protection. *Health Phys.* 88: 717–732.

Connors S., Scrimger J., Logus W., Johnson L., Schartner E. (1988) Development of a translating bed for total body irradiation. *Med. Dosim.* 13: 195–199.

Dalrymple G.V., Gaulden M.E., Kollmorgen G.M., Vogel H.H. (1973) *Medical Radiation Biology.* W.B. Saunders Company, Philadelphia, Pennsylvania.

Doughty D., Lambert G., Hirst A., Marks A. et al. (1987) Improved total body irradiation dosimetry. *Br. J. Radiol.* 60: 269–278.

Geise R. (1984) Shielding calculations for megavoltage gamma and x-ray therapy facilities. AAPM Symposium Proceedings 4: *Radiotherapy Safety*. B. Thomadsen (ed.). American Institute of Physics, New York, pp. 35–44.

Gerig L., Szanto J., Bichay T., Genest P. (1994) A translating-bed technique for total-body irradiation. *Phys. Med. Biol.* 39: 19–35.

Gur D., Rosen J., Bukovitz A., Gill A. (1978) Fast and slow neutrons in an 18 MV photon beam from a Philips SL/75-20 linear accelerator. *Med. Phys.* 5: 221–222.

Hall E., Giaccia A. (2005) *Radiobiology for the Radiologist*. Lippincott Williams and Wilkins, Baltimore, Maryland.

Hanahan D., Weinberg R. (2000) The hallmarks of cancer. *Cell* 100: 57–70.

Hendee W., Ibbott G. (1992) *Radiation Therapy Physics*, Second edition. Mosby - Yearbook, St. Louis, Missouri.

IAEA (International Atomic Energy Agency) (2000) Lessons learned from accidental exposures in radiotherapy. Safety Report Series: N17. IAEA, Vienna, Austria

IAEA (International Atomic Energy Agency) (1998) Design and implementation of a radiotherapy programme: clinical, medical physics, radiation protection and safety aspects. IAEA-TECDOC 1040. IAEA, Vienna, Austria.

IAEA (International Atomic Energy Agency) (1996) International basic safety standards for protection against ionizing radiation and for the safety of radiation sources. IAEA, Vienna, Austria.

ICRP (International Commission on Radiological Protection) (2001) Prevention of accidental exposures to patients undergoing radiation therapy. Report 86. Pergamon Press, Oxford.

ICRP (International Commission on Radiological Protection) (1999) Radiological Protection and Safety and pregnancy, ICRP Report 77. Pergamon Press, Oxford.

ICRP (International Commission on Radiological Protection) (1996) Radiological Protection and Safety in Medicine, ICRP Report 73. Pergamon Press, Oxford.

ICRP (International Commission on Radiological Protection) (1991) Recommendations of the International Commission on Radiological Protection. Report 60, Annals of the ICRP, Pergamon, Oxford.

ICRP (International Commission on Radiological Protection) (1960) Report of committee III on protection against x-rays up to energies of 3 MeV and Beta and Gamma rays from sealed sources. Pergamon Press, New York.

ICRU (International Commission of Radiological Units) (1976) Determination of absorbed dose in a patient irradiated by a single beam of x or gamma rays in radiation procedures. Bethesda, Maryland.

Kathren R.L. (1985) *Radiation Protection, Medical Physics Handbook 16*. Adam Hilger, Bristol, United Kingdom.

Kersey R. (1979) Estimation of neutron and gamma radiation doses in the entrance mazes of SL 75-20 linear accelerator rooms. *Medicamundi* 24: 151–155.

Khan F.M. (1984) *The Physics of Radiation Therapy*, First edition. Williams and Wilkins, Baltimore, Maryland.

Kron T., Aldrich B., Jovanovic K., Howlett S., Hamilton C. (1996) Workload and use factor of medical linear accelerators in radiotherapy. *Health Phys.* 69: 971–975.

Lawton C., Barber-Derus S., Murray K., Casper J. et al. (1989) Technical modifications in hyperfractionated total body irradiation for T-lymphocyte depleted bone marrow transplant. *Int. J. Radiat. Oncol. Biol. Phys.* 17: 319–322.

Loshek D.D., Keller K.A. (1988) Beam profile generator for asymmetric fields. *Med. Phys.* 15: 604–610.

McGinley P. (1997) *Shielding Techniques for Radiation Oncology Facilities*. Medical Physics Publishing, Madison, Wisconsin.

McGinley P., Wright B., Meding C. (1984) Dose to radiotherapy technologist from air activation. *Med. Phys.* 11: 855–858.

Metcalfe P., Kron T., Hoban P. (1997) *The Physics of Radiotherapy X-Rays from Linear Accelerators*. Medical Physics Publishing, Madison, Wisconsin.

Nath R., Price K., Holeman G. (1980) An intercomparison of neutron measurements for a 25 MV x-ray radiotherapy accelerator. *Med. Phys.* 7: 545–548.

NCRP (National Commission of Radiation Protection) (1989) NCRP Report 102. Medical x-ray, electron beam and gamma ray protection for energies up to 50 MeV (equipment design performance and use). NCRP, Bethesda, Maryland.

NCRP (National Commission of Radiation Protection) (1977) NCRP Report 51. Radiation protection design guidelines for 0.1-100 MeV particle accelerator facilities. NCRP, Bethesda, Maryland.

NCRP (National Commission of Radiation Protection) (1976) NCRP Report 49. Structural shielding design and evaluation for medical use of x-rays and gamma rays of energies up to 10 MeV. NCRP, Bethesda, Maryland, reprinted 1991.

Neel J., Schull W., Awa A., Satoh C., Kato H. et al. (1990) The children of parents exposed to atomic bombs: Estimates of the genetic doubling dose of radiation for humans. *Am. J. Hum. Genet.* 46: 1053–1072.

Nelson W., LaRiviere P. (1984) Primary and leakage radiation calculations at 6, 10 and 25 MeV. *Health Phys.* 47: 811–818.

Ozsahin M., Belkacemi Y., Pene F., Dominique C. et al. (1994) Total-body irradiation and cataract incidence: A randomized comparison of two instantaneous dose rates. *Int. J. Radiat. Oncol. Biol. Phys.* 28: 343–347.

Peters L., Withers H.R., Cundiff J.H., Dicke K.A. (1979) Radiobiological considerations in the use of total body irradiation for bone-marrow transplantation. *Radiology* 131: 243–247.

Pierce D., Shimizu Y., Preston D., Vaeth M. et al. (1996) Studies of the mortality of atomic bomb survivors. Report 12, Part I. Cancer: 1950-1990. *Radiat.Res.* 146: 1–27.

Podgorsak E., Metcalfe P., Van Dyk J. (1999) Medical linear accelerators. Chapter 11 in *The Modern Technology of Radiation Oncology*. J. Van Dyk (ed.). Medical Physics Publishing, Madison, Wisconsin, pp. 349–435.

Quast U. (1985) Physical treatment planning of total-body irradiation: Patient translation and beam-zone method. *Med. Phys.* 12: 567–574.

Robinson D., Scrimger J., Field G., Fallone B. (2000) Shielding considerations for tomotherapy. *Med. Phys.* 27: 2380–2384

Sanchez F., Madurga G., Arrans R. (1989) Neutron measurements around an 18 MV linac. *Radiother. Oncol.* 15: 259–265.

Seelmann-Eggebert W., Pfennig G., Muenzel H., Klewe-Nebenius H. (1981) Karlsruher Nuklidkarte. Gersbach, Munich.

Selman J. (1983) *Elements of Radiobiology*. Charles C Thomas, Springfield, Illinois.

Shimizu Y., Pierce D., Preston D., Mabuchi K. (1999) Studies of the mortality of atomic bomb survivors. Report 12, part II. Non cancer mortality: 1950–1990. *Radiat. Res.* 152: 374–389.

Van Dyk J. (1999) (ed.). *The Modern Technology of Radiation Oncology*. Medical Physics Publishing, Madison. Wisconsin.

WHO (World Health Organization) (1988) *Quality Assurance in Radiotherapy*. World Health Organization Publications, Albany, New York.

Questions

13.1 Define stochastic and deterministic effects.
(section 13.2)

13.2 Describe typical radiation dose levels for medical procedures and how they compare to background radiation levels.
(section 13.3)

13.3 For the linac bunker plan shown in the sketch plan (figure 13.6a) calculate the wall thicknesses required at a position just outside wall B. The linac is capable of producing 6 MV x-rays with an annual workload of 50,000 Gy per year at isocenter, and the tenth-value layer (TVL) for concrete at 6 MV is 35 cm. The distance to the outer wall exterior is 5 m; hence the calculation distance is 5.305 m from isocenter. Assume the occupancy outside this wall to be 1/16th.
(section 13.4)

13.4 For the linac bunker plan shown in the sketch plan (figure 13.6a) calculate the wall thicknesses required for secondary barrier walls if scatter from patient is considered using a TVL for 90° scatter of 0.17 m. Assume the same workload factors and beam properties as for question 13.4. Assume distance to the outer wall exterior scatter barrier is 5 m; hence the calculation distance is 5.305 m from isocenter. Assume the occupancy outside one of the scatter walls may be 1.
(section 13.4)

13.5 For the linac bunker plan shown in the sketch plan (figure 13.6a) calculate the secondary barrier wall thicknesses required if leakage is considered. Assume the same workload factors and beam properties as for

question 13.3. Assume distance to the outer wall exterior scatter barrier is 5 m; hence the calculation distance is 5.305 m from isocenter. Assume the occupancy outside one of the scatter walls may be 1. Assume the TVL for leakage is the same as for primary.
(section 13.4)

14 Tumor and Normal Tissue Response

"Anything worth doing is worth doing badly
...If there is no other choice."
(John R. Cameron)

14.1 Introduction

The need for accurate prediction of dose distributions in treatment planning is best understood in light of the biological response of tumors and normal tissues to radiation. The whole objective of radiotherapy is to deliver a lethal dose to the tumor while incurring minimal damage to surrounding normal tissues. The primary requirement for ensuring that this is the case is to spatially conform the high dose region to a volume that includes only the tumor plus a margin for microscopic disease and setup error. There is also the rather fortunate fact that by delivering the course of treatment in multiple small-dose fractions, tumor cells are more efficiently eradicated than those of normal tissues. With the combination of spatial dose conformity and the preferential killing of tumor cells by fractionated treatment, radiotherapy can be very effective, but the design of a biologically optimal treatment schedule cannot be achieved by assessment of the dose distribution alone. In this chapter, the conversion of dose to more biologically relevant quantities such as cell survival fraction, biologically effective dose (BED), tumor control probability (TCP), and normal tissue complication probability (NTCP) is discussed, along with the effects of fractionation and duration of treatment.

The mechanism of cell kill by radiation is discussed in section 14.2. In section 14.3 the linear quadratic model for cell survival is introduced, as are the concepts of BED and standard effective dose (SED). The four Rs of radiobiology are discussed in section 14.4. The linear quadratic model including cell *proliferation* is presented in section 14.5. TCP and NTCP models are discussed in section 14.6 with reference to serial and parallel structures. In section 14.7 the concept of equivalent uniform dose (EUD) is introduced as a suitable IMRT optimization tool.

Sometimes findings from radiobiological models guide us to suggested changes in dose fractionation regimes. These changes need to be assessed by applying strict clinical trial guidelines, and these are briefly outlined in section 14.8.

There are many still to be resolved questions in radiotherapy that need further examination using radiobiological modeling as a first pass then backed up or not by clinical trial evidence. Some of the questions that are yet to be fully resolved, include:

i) Low α/β ratio for prostate tumors.
ii) Low radiation doses and secondary cancer induction.
iii) Inhomogeneity of the temporal dose distribution (e.g., IMRT).
iv) Molecular biology and genetic drugs.

These topics are discussed in section 14.9 of this chapter.

It is very important to realize that the success of radiotherapy is to some extent dependent on chance. For example, the dose that is required to give a reasonable (say 60%) probability of tumor control is known to be about 70 Gy for many tumors. A dose of 70 Gy is not some magical threshold dose required to kill cancer cells—it is a dose that gives a reasonable chance that all cancer cells will be killed. The survival probability of any particular cell when a dose of 70 Gy is delivered is extremely low (say 10^{-9}), but because there are so many cancer cells, there will still be a significant chance that one or more remains capable of replication. Some patients receiving a dose of 55 Gy would be cured, while some tumors still will regrow after a dose of 75 Gy; there is no certainty of tumor control, only probability. Unfortunately, there is also a probability of complications to healthy organs, which increases with dose and extent of the organ irradiated, and this must always be balanced against the probability of tumor control. So, while the proportion of patients having a recurrence of the tumor can be reduced by increasing the dose level, this may increase the NTCP to an unacceptable level. It would obviously be an advantage to be able to predict with some accuracy the tumor dose that can be delivered while keeping the NTCP to an acceptable level. It is also useful to know whether the dose throughout the tumor can be nonuniform without unacceptably reducing the TCP, as this may allow a significant reduction in NTCP by changing the beam intensity profiles or directions. Simple statistical models exist that can be used to predict these quantities, and although not rigorous, they are very useful in predicting at least relative changes in TCP and NTCP with variations in beam parameters and tumor characteristics.

14.2 Mechanism of Cell Killing

Ionizing radiation deposits energy in the medium it penetrates, but this energy is very small in terms of thermal effects. For example, consider a dose of 10 Gy (10 J/kg) delivered to water (or soft tissue). Because 1 Gy = 2.39×10^{-4} kcal, 10 Gy will give a temperature elevation of only 0.00239 degrees C! Damage to cells is caused by ionization events occurring as electrons lose their kinetic energy, where DNA strands are broken either via direct action or by indirect action by free radicals. The most important free radical produced is the hydroxyl radical (OH•), which is produced by the ionization of the water molecule.

$$\begin{aligned} H_2O + \text{ionization energy} &\rightarrow H_2O^+ + e^- \\ &\rightarrow H^+ + OH^-. \end{aligned}$$

For direct action, ionization must occur within the DNA molecule, and for indirect action it must occur within about 100 Angstroms of the molecule (1 Angstrom = 10^{-10} m). The site of interaction within the DNA has an effect on the biological damage. The radiochemical lesion in the DNA may be labile for a short time and may revert to the undamaged state (repair).

The number of cells given a lethal dose depends on the total dose, rate of energy loss along the track of each ionizing particle, and the dose rate. There is a certain probability of survival of each cell as a function of dose, and this is an exponentially based curve. A simple exponential equation for cell survival probability, $S = e^{-\alpha D}$, where α is the radiosensitivity and D is total dose, can be derived from single-target theory. This can be easily proven by considering a tumor that initially contains N_0 living cells. When the tumor is irradiated, a cell can be said to die when it is "hit" once. Figures 14.1a and 14.1b are presented to help explain this concept. Now as cells die, the number of living cells still available to be killed decreases, as of course does the *rate* of cell kill with dose. The rate is in fact proportional to the number of cells left alive, N, so the differential equation for the number of cells alive after a dose D is

$$\frac{dN}{dD} = -\alpha N, \tag{14.1}$$

where α is a constant of proportionality. Solving this equation gives

$$N = N_0 \exp(-\alpha D), \tag{14.2}$$

where N/N_0 is the survival probability. An elementary but very important observation for radiotherapy is that although the proportion of cells killed is constant as the original number increases (for the same dose), the probability that *all* cells are killed decreases. This means that a large tumor receiving the same dose as a small tumor is less likely to be eradicated.

Mammalian cell survival curves are characterized by a shoulder region. The initial shallower curve at low dose is thought to be due to single hit killing, where a single ionizing particle causes enough damage to the DNA

(a)

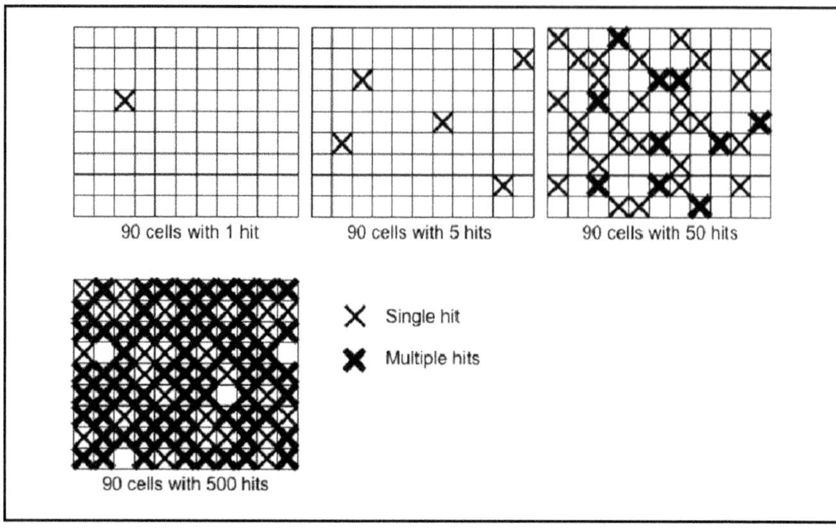

(b)

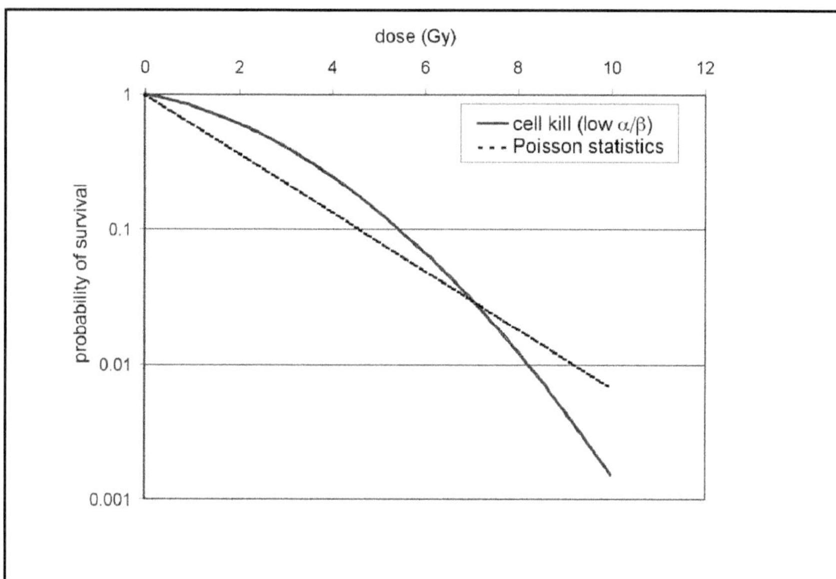

Figure 14.1: The target theory Poisson distribution: (a) the way target hits are gradually building up; (b) the Poisson log plot of survival versus dose compared with a typical mammalian tumor cell kill shaped distribution.

to kill the cell. This damage is usually a double-strand break. At higher doses, cells that have undergone some damage to the DNA (a single-strand break) but remain viable may suffer one or more extra DNA strand breaks (within a short distance from the first break), which will then kill the cell. It has been found that the second break must occur within about four base pairs of the first (Bryant 1989). The rapid increase in multiple breaks as dose increases causes a steepening of the gradient of the cell survival curve (on a logarithmic scale). Because the DNA has the capacity to repair, it seems likely that a second break occurring before the damage from the first has been repaired is less likely at low dose rate. It is also true that densely ionizing (high LET) or high dose rate radiation is more likely to cause multiple, nearby, breaks concurrently, thus reducing the probability of repair.

14.3 Introduction to Radiobiological Models

To improve the effectiveness of treatment, the radiation oncologist has a number of parameters to adjust: treatment technique, treatment volume, radiation modality and energy, total dose, dose per fraction, number of fractions, time interval between fractions, overall treatment time, adjuvant chemotherapy, radiosensitizers, hypothermia, and other factors that influence the treatment outcome. Radiobiological models should help the reader understand the influence of the different factors on treatment outcome and their interrelationships. They can also assist the radiation oncologist in clinical decision making. However, radiobiological calculations should never be used without consideration of clinically derived results and should be taken merely as a guide for each individual patient by the radiation oncologist.

A wide range of different models has been developed to describe the effect of ionizing radiation in living cells or organisms. Some of them were primarily designed to model the response of human skin to radiation based on ideas developed for the response of film to light (Thames and Hendry 1987). This typically resulted in power law type models such as the *nominal standard dose* (NSD) model (Ellis 1967, Orton 1988). The other type of model is based on the biological observation of cell kill after radiation. Examples for these models are the incomplete repair (IR) model (Thames 1985); lethal, potentially lethal (LPL) model (Curtis 1986); and the linear quadratic model (Douglas and Fowler 1976). Of these models, only two (one of each type) have had significant impact on clinical practice. These two models shall be described in more detail.

14.3.1 Nominal Standard Dose and Time Dose Fractionation

The concept of the NSD was introduced by Ellis (1967) for the calculation of the tolerance dose D_T for normal tissues from the overall treatment time T (in days) and the overall number of fractions, N, delivered to the patient:

$$D_T = NSD \times T^{0.11} \times N^{0.24}. \tag{14.3}$$

This mathematical formulation was empirically derived from isoeffect curves for skin and describes the increase in tolerance dose if the number of fractions and/or the treatment time is increased.

Equation 14.3 is valid for 4 to 30 fractions and can only be applied for completed courses of radiotherapy with normal tissues treated to tolerance level (Orton and Ellis 1973). To overcome this limitation, the concept was extended to *partial tolerance* (*PT*) values (Ellis 1969):

$$PT = NSD \times \frac{n}{N}, \tag{14.4}$$

with n the number of fractions given out of N scheduled fractions.

After combination of equations 14.3 and 14.4:

$$\begin{aligned} PT &= NSD^{-0.536} \times n \times d^{1.538} \times x^{-0.169} \\ &= NSD^{-0.536} \times TDF \times 1000 \quad , \end{aligned} \tag{14.5}$$

with d the dose per fraction and x the average time per fraction in days (for example, 5 fractions per week gives x = 1.4 days per fraction). TDF is the time-dose-fractionation factor (Orton and Ellis 1973).

Since NSDs are a constant for a given type of tissue and PTs are additive, different fractionation schedules can be directly compared using equation 14.5. This comparison has been further simplified by the creation of look-up tables for TDFs (Orton and Ellis 1973) for treatment schedules with 1 to 5 fractions per week. These tables have been widely used for the comparison of different fractionation schedules.

In order to account for treatment interruptions and split course radiotherapy, a decay factor (DF) for a radiation dose given prior to a treatment interruption of duration R (days) was proposed (Winston 1969):

$$DF = \left(\frac{t}{t+R} \right)^{0.11}, \tag{14.6}$$

where t is the duration of the radiation given prior to the interruption in days. The total TDF can then be calculated from the dose given before and after the treatment break, D_1, and after the treatment break, D_2:

$$TDF_{total} = TDF_1 \times DF \times TDF_2. \tag{14.7}$$

Various critiques on the NSD/TDF models have shown the following shortcomings (Thames and Hendry 1987, Fowler 1989):

i) NSD/TDF is a mathematical approximation without any biological foundations.
ii) Fraction size d, not the fraction number N, is the important parameter that should be used in equation 14.3. $N^{0.24}$ does not predict the severe late damage that occurs for larger fraction sizes (Overgaard et al. 1988).
iii) Isoeffect curves in a log-log plot of total dose versus number of fractions are curved, not straight as predicted by these equations.
iv) The time factor exponent (0.11) does not distinguish between different types of tissue. It is overestimated for late responding tissues where no proliferation takes place and underestimates the time effect for tumors and early/acute reactions. This problem could be overcome by specifying different exponents. However, it appears

that the exponent also changes during treatment, which makes the model difficult to use in clinical practice.

v) The time factor does not allow for changes of the time effect during treatment (accelerated proliferation).

vi) Time, dose, and fractionation tables are perhaps too easy to use without thinking about late/early reactions or proliferation rates (Fowler 1989). The NSD model is no longer used in clinical practice and should not be used.

The search for a less empirical model with roots closer to radiobiological principles has led to greater acceptance of the linear quadratic model, which is outlined below.

14.3.2 The Linear Quadratic Model

Single Fraction

The linear quadratic (LQ) model is based on the cell survival curves after radiation, assuming a target model (Thames and Hendry 1987). Figure 14.2 shows typical curves for tumor and normal tissue response for irradiation of cells with low LET radiation (x-rays or electrons). This curve in a semi-logarithmic plot of cell survival fraction, S, versus dose, D, can mathematically be described as:

$$S = \exp\left(-\left(\alpha D + \beta D^2\right)\right), \tag{14.8}$$

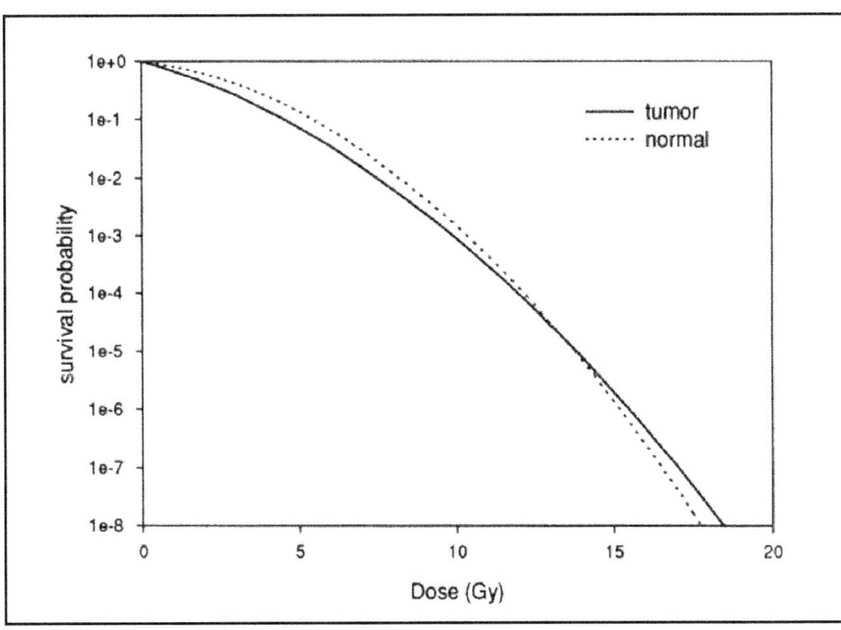

Figure 14.2: Survival fraction for tumor and normal tissue calculated using the linear quadratic model (calculated using equation 14.8). Parameters chosen for tumor are α = .035 Gy^{-1}, β = 0.035 Gy^{-2} (α/β = 10), and for normal tissue are α = 0.15 Gy^{-1}, β = 0.05 Gy^{-2} (α/β = 3).

where $S = 1$ means all cells survive and $S = 0.1$ means 1 in 10 cells survive. Alpha (units Gy^{-1}) and β (units Gy^{-2}) are constants, characteristic for a given type of tumor or tissue. Note that to obtain this curve, the dose is given in a single irradiation and any repair of cells is allowed before survival fraction is measured. In these curves, the values of α,β for the tumor and normal tissue are 0.35 Gy^{-1},0.035 Gy^{-2} and 0.15 Gy^{-1},0.05 Gy^{-2} respectively. These are typical values, but a wide range exists for both tumor and normal tissue. Whatever the individual values of α and β, the ratio of α/β is usually approximately 10 Gy for malignant tumors and 3 Gy for late responding normal tissue. For some normal tissue, the α/β ratio is very low (for example, it is approximately 1 for the spinal cord). Note that the units of this ratio are Gy. The coefficient α gives the initial slope (the linear portion) of the survival curve. It relates to cell kill after a single radiation event and governs the initial slopes of the curves in figure 14.2. The coefficient β describes the quadratic component of the curve where cell kill is the result of multiple hits. Notice that the tumor curve has the largest initial slope, but at larger doses the normal tissue survival falls off most rapidly. This is because (in this case) the tumor has a larger α value, but normal tissue has a larger β. The crossing of the curves is typical and means that small single doses of radiation have a larger effect on tumors than on normal tissue, but large single doses are more damaging to normal tissue than to tumors.

Tumors are often referred to as *early responding* tissues because they respond to radiation after a short time. Normal tissues are mostly late responding, meaning that they do not show the effects of radiation damage for a longer time (maybe several months). However, some normal tissues, such as the skin and mucosa, are early responding. (Skin erythema occurs rapidly after treatment.) Early responding tissues correspond to those having a large α/β ratio in the linear quadratic model, and hence a large initial slope of the cell survival curve. This has to do with faster cell cycles, and hence their faster response to radiation (as cells try to divide).

The cell survival can be considered to be a product of two components:

$$S = \exp(-\alpha D) \times \exp(-\beta D^2), \tag{14.9}$$

referred to as the *linear* and *quadratic* terms respectively. The separate components are plotted for $\alpha = 0.035$, $\beta = 0.035$ (tumor) in figure 14.3a, where it can be seen that the linear term causes a faster fall-off initially but the curves cross at a dose of 10 Gy. It is easily shown by equating the terms that this crossing should occur at a dose equal to α/β, which in this case is equal to 10 Gy. Similarly, for normal tissue with $\alpha/\beta = 3$ Gy (see figure 14.3b), the curves cross at a dose of 3 Gy. It is obvious then that the quadratic term begins to dominate at lower doses for normal tissue (because of the lower value of α/β) than for tumors. Since the influence of the linear and quadratic terms is equal at a dose of α/β Gy, then by extrapolating the initial slope of the survival curve to determine the linear component at any dose, the α/β ratio can be determined by finding the dose at which the linear component is responsible for half of the cell kill.

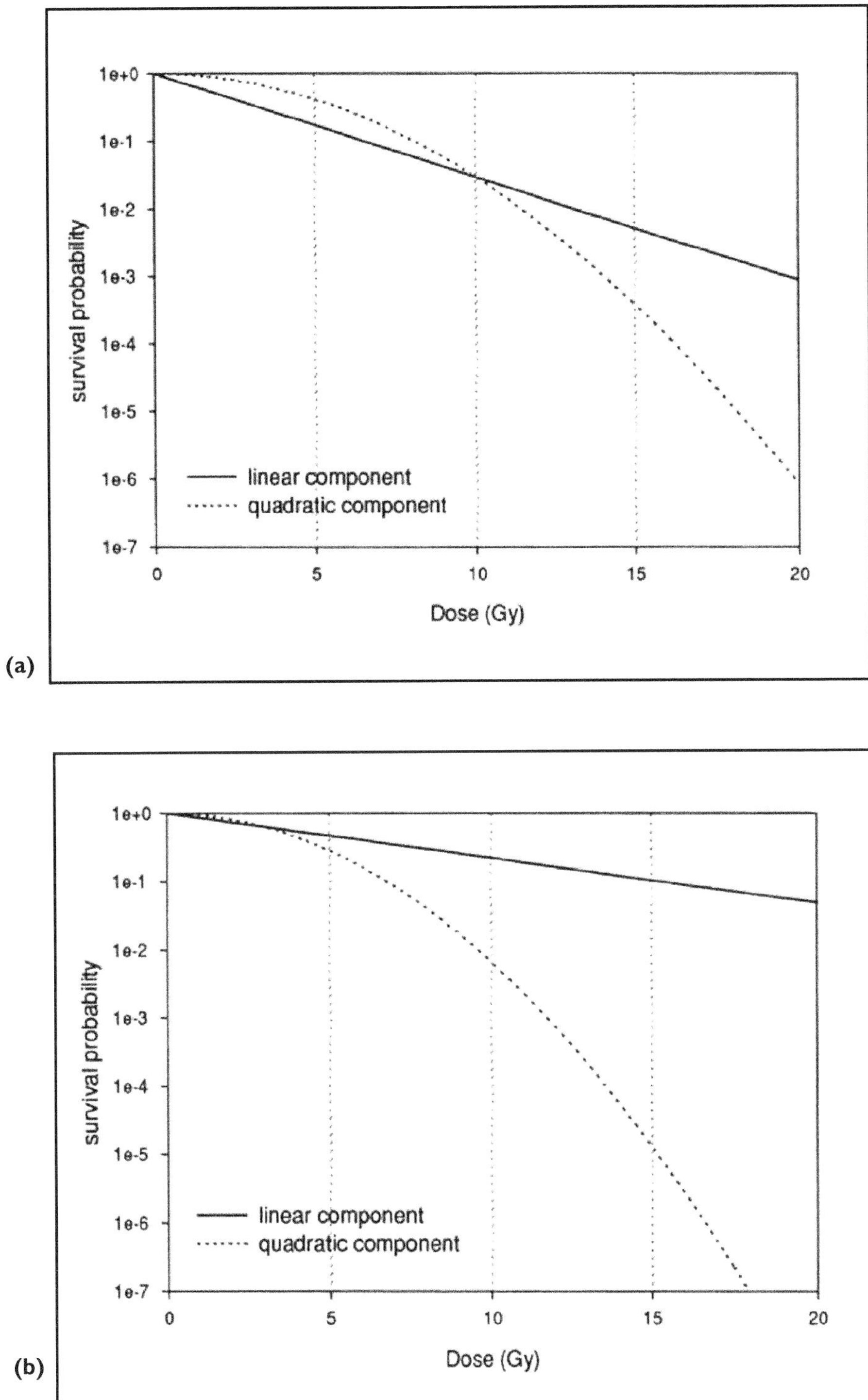

Figure 14.3: Linear (α) and quadratic (β) components plotted separately for (a) tumor and (b) normal tissue. Note that curves cross when dose is equal to the ratio α/β.

Cell Survival with Fractionated Dose Delivery

The fact that small single doses damage tumors more than normal tissue but large single doses are more damaging to normal tissue is one of the reasons why radiotherapy is normally administered in *fractions*. If multiple small dose fractions of say 2 Gy are given, with time for repair between fractions, the shape of the cell survival curve for each fraction always resembles the shape of the curve between 0 and 2 Gy. Thus the quadratic term in the cell survival curve never has a large effect, and the cumulative effect of multiple fractions is larger for tumors than for normal tissue. Mathematically, if the dose per fraction is d Gy and n fractions are delivered, the cell survival is found from the product of cell survivals for each fraction. Thus, using equation 14.8,

$$\begin{aligned} S &= \left[\exp\left(-\left(\alpha d + \beta d^2\right)\right)\right]^n \\ &= \exp\left(-nd\left(\alpha + \beta d\right)\right). \end{aligned} \tag{14.10}$$

It is obvious that for any given total dose nd, the cell kill will be larger as d increases, since then the quadratic term has more effect (mathematically, a larger d makes the exponent larger in equation 14.10 since nd is constant). Furthermore, the increase in cell kill with larger d is more significant for smaller α/β ratios (normal tissue); hence, the cell kill is reduced by fractionation to a larger extent for normal tissue than for tumors. This means that the late effects that occur in normal tissues are reduced by fractionation.

The comparative effects of dose fractionation on tumors and normal tissue are shown in figure 14.4 for different values of dose per fraction, d. In each case, the cell survival due to single dose and fractionated dose is shown, with the upper curves corresponding to fractionated dose delivery. Note that full repair of sublethal damage is assumed between fractions, since this is the case for survival fraction values on the original single-dose survival curves. In figure 14.4a, where $d = 1$ Gy (and the quadratic term has minimal effect on cell kill), it can be seen that the cell kill in the fractionated case is always greater for the tumor, reflecting the larger linear component for tumors. Because the amount by which the gradient of the tumor cell survival curve exceeds that of the normal curve is greatest as dose approaches zero, the optimal fraction size is zero (with an infinite number of fractions)! If this were the case, the βd term in equation 14.10 would disappear and the cell survival would be given by the linear component only. As fraction size increases then, the cell survival curves become closer together and also become closer to the single dose curves. This trend can be followed in figure 14.4b–d where the dose per fraction is 2 Gy, 4 Gy, and 8 Gy, respectively. Obviously from these curves, small fractions require larger total dose for any given tumor cell kill. Hence, the number of fractions that must be delivered can become very large. Typically, fraction size is 2 Gy, which gives good separation of the tumor and normal tissue cell survival curves without the number of fractions and total treatment time being excessively large. The calculation of d,n pairs for a given effect is discussed below.

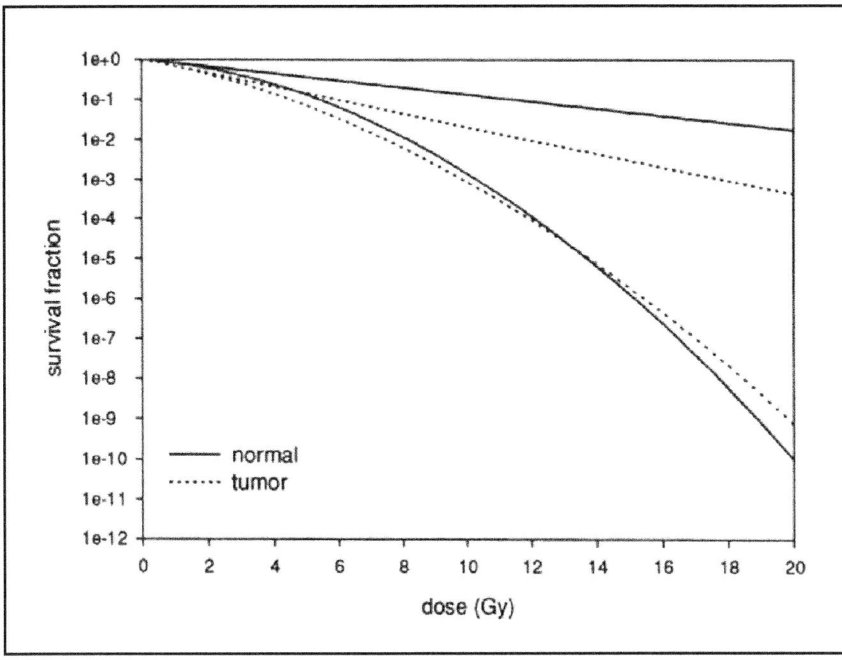

(a)

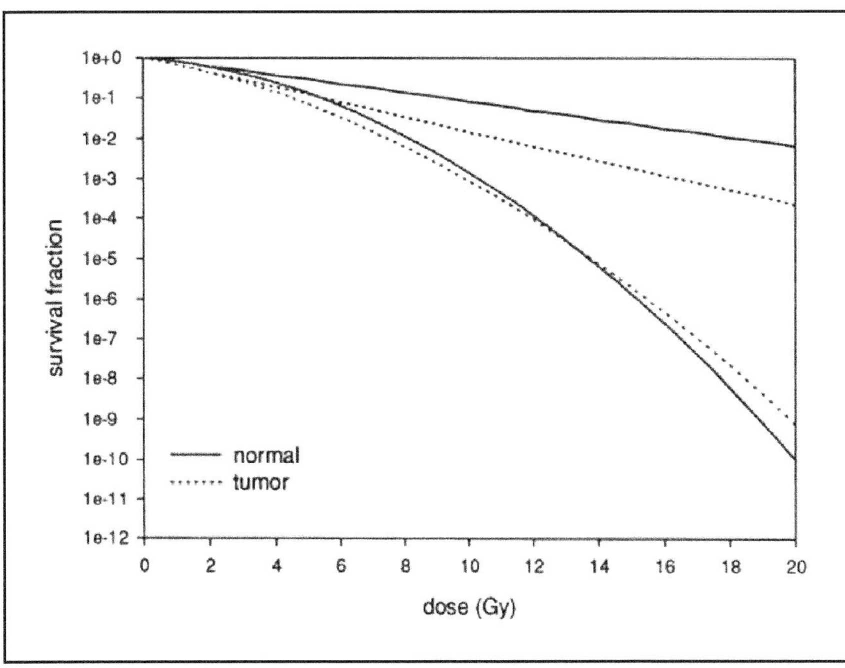

(b)

Figure 14.4: Survival fractions for single dose (lower curves) and fractionated dose (upper curves) for tumor ($\alpha = 0.35$ Gy^{-1}, $\beta = 0.035$ Gy^{-2}) and normal tissue ($\alpha = 0.15$ Gy^{-1}, $\beta = 0.05$ Gy^{-2}). Fraction sizes of 1 Gy, 2 Gy, 4 Gy, and 8 Gy are used in (a), (b), (c), and (d), respectively.

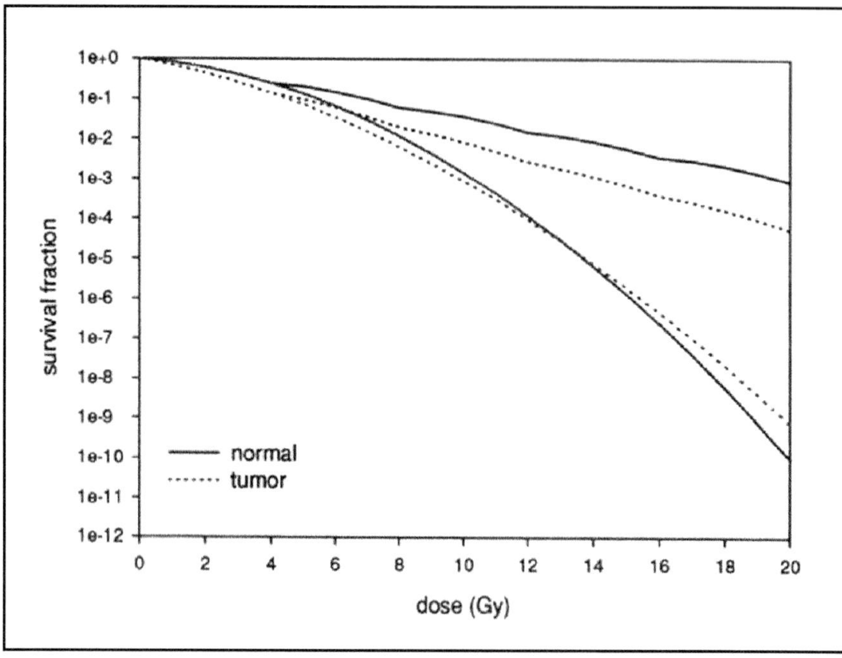

(c)

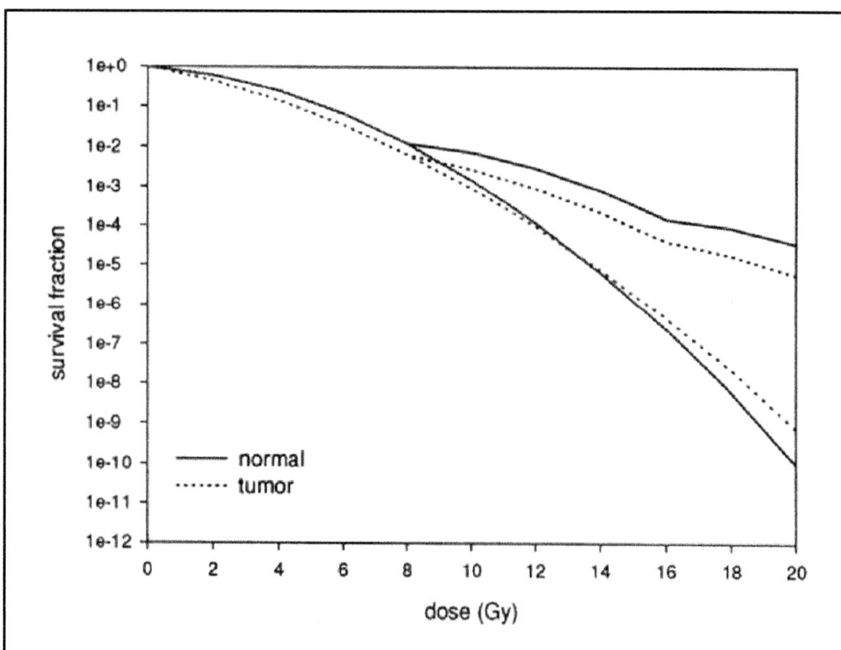

(d)

Figure 14.4: (Continued)

Biologically Effective Dose (BED) and Standard Effective Dose (SED)

Similar to the calculation of a tolerance dose using the TDF model, pairs of values for dose per fraction and fraction number that give a constant biological effect can be calculated using the linear quadratic model. This is commonly done when the fraction size is to be altered from a standard dose, and it is required to determine the number of fractions that will retain the same effect as the standard regimen.

A common method of quantifying the effect of a fractionation scheme is to calculate the equivalent dose in infinitely small fractions. This is known as the biologically effective dose (BED) (Fowler 1989), extrapolated tolerance dose (ETD) (Barendsen 1982), or the extrapolated response dose (ERD) (Dale 1985).

It is simple to calculate the BED by equating the survival fraction for the current fraction size and number with the survival fraction for when the fraction size approaches zero. The model assumes that the biological effect of ionizing radiation is only dependent on the cell survival. We have then, using the exponent in equation 14.10 and the fact that nd is the total dose,

$$\begin{aligned} nd(\alpha + \beta d) &= BED(\alpha + \beta \times 0) \\ &= \alpha\, BED, \end{aligned} \tag{14.11}$$

which after rearranging gives

$$BED = nd\left(1 + \frac{d}{\alpha / \beta}\right). \tag{14.12}$$

Another way of deriving the BED is to consider the semi-log plot of the survival curve. From equation 14.10,

$$\ln S = -nd(\alpha + \beta d), \tag{14.13}$$

which has an intercept at $\ln S = 0$ and an initial slope of $-\alpha$. This initial slope defines the (linear) cell survival curve for infinitely small fractions. If we then choose values of n and d, calculate the corresponding value of $\ln S$ and divide by α, this must be the dose required (the BED) to give the same effect ($\ln S$) in infinitely small fractions. It is easy to show that dividing $-\ln S$ by α gives the BED as in equation 14.12. Another definition of the BED is thus

$$BED = \ln S / \alpha, \tag{14.14}$$

where $-\ln S$ is termed the *effectiveness, E*.

Note that the biologically effective dose will always be larger than the actual total dose, nd, since the gradient of the single fraction cell survival curve continually increases with dose.

The BED can be used to compare the likely effects of different fractionation schemes, giving each in terms of the equivalent dose in infinitely small dose fractions. Expressions for BED with current and new values of fraction size and number can be equated to solve for an unknown variable (usually the new fraction number). It is of course not necessary to first calculate the BED for the current scheme, as survival fraction can be equated directly.

Another useful quantity is the standard effective dose (SED). This is very similar to the BED, but it is the equivalent dose in 2 Gy fractions (instead of infinitely small fractions). Similar to equation 14.11, but setting $d = 2$ Gy we have

$$nd(\alpha + \beta d) = SED(\alpha + \beta \times 2), \tag{14.15}$$

which on solving for SED gives

$$\begin{aligned} SED &= n\,d\left(1 + \frac{d}{\alpha/\beta}\right) \bigg/ \left(1 + \frac{2}{\alpha/\beta}\right) \\ &= BED \bigg/ \left(1 + \frac{2}{\alpha/\beta}\right). \end{aligned} \tag{14.16}$$

The SED is always less than the BED (because a given dose in 2 Gy fractions is more effective than the same dose in infinitely small fractions) and will be greater than nd if $d > 2$ Gy and less than nd if $d < 2$ Gy. This is a useful quantity because 2 Gy is the most common fraction size and it is informative to know the dose in this standard scheme to which the planned fractionation regime is equivalent. Note that the SED to BED ratio is independent of dose and constant for a particular α/β ratio. (If BED remains constant with a change in fractionation schedule, so does SED.) The SED is a particularly useful concept for clinicians as it gives them a feeling for effects relative to a standard fractionation regime that they are familiar with.

The tissue in which it is desired to keep a constant effect may be the tumor or a normal tissue structure, each of which has different α/β ratios, as discussed above. Hence when an isoeffective fractionation schedule is calculated for either tumor or normal tissue, only that structure under consideration will experience the same biological effect. It must therefore be ensured that the overall effect is still acceptable.

Example 1(a): Tumor Effect Is Constant

In this case, the reference fractionation scheme is 60 Gy delivered in 30×2 Gy fractions, and it is required that as the fraction size is changed, the BED and SED remain unchanged for the *tumor*. The SED is of course 60 Gy for both the tumor and normal tissue, since the equivalent dose in 2 Gy fractions is equal to the actual dose. The BED will be higher than the actual dose and the SED since infinitely small fraction sizes always require the largest total dose. BED is easily calculated for the reference scheme using equation 14.12 (with $n = 30$, $d = 2.0$ Gy and $\alpha/\beta = 10$ Gy) to be equal to 72 Gy. We will also calculate the BED and SED for the normal tissue so that the fractionation scheme is not changed without ensuring that the effect on normal tissue is still tolerable. The BED and SED of the tumor are denoted as BED_{10} and SED_{10} to indicate that the α/β ratio is 10 Gy. Similarly, for the normal tissue, these parameters are BED_3 and SED_3.

The procedure is as follows:

i) Calculate the number of fractions required to maintain a BED_{10} of 72 Gy by using equation 14.12, solving for n for a range of values of d. Equivalently we could use equation 14.16 with SED_{10} set to 60 Gy.
ii) Using the values of n obtained in (i), calculate BED_3 and SED_3 for normal tissue as a function of d, via equations 14.12 and 14.16, respectively.

Results are shown in figure 14.5a, where it can be seen that required total dose reduces as dose per fraction increases, reflecting the greater effect of larger fraction sizes for a given total dose. As expected, SED_{10} is equal to total dose for a 2 Gy fraction size, is less than total dose when fraction size is less than 2 Gy (because 2 Gy fractions are more effective than fraction sizes of less than 2 Gy), and is greater than the total dose when dose per fraction is greater than 2 Gy. BED_3 is large since infinitely small fractions are the least effective. Just as the SED_3 and SED_{10} are equal to total dose when the fraction size is 2 Gy, BED_3 and BED_{10} are equal to total dose when the curves are extrapolated to zero fraction size. The SED_3 curve is particularly useful, as the effect on normal tissue receiving the modified total dose can be seen. Note that in this case, BED_3 and SED_3 increase with fraction size, even though total dose decreases. This is because the effect on cell kill due to increased fraction size is larger than that due to the reduction in total dose and is a consequence of the cell survival curve being more rounded for normal tissue than for tumors (α/β is smaller).

Example 1(b): Normal Tissue Effect Is Constant

The reference fractionation scheme is again 60 Gy delivered in 30 × 2 Gy fractions, but it is now required that the BED and SED remain unchanged for the *normal tissue* (the late effects are constant). BED is calculated for the reference scheme using equation 14.12 (with n = 30, d = 2.0 Gy and α/β = 3 Gy) to be equal to 100 Gy. SED is of course 60 Gy, since the reference fraction size is 2 Gy. We will also calculate the BED and SED for the tumor, so that the fractionation scheme is not changed without ensuring that the effect on the tumor is still adequate.

Results are shown in figure 14.5b, where it can be seen that the reduction in total dose with fraction size, for equivalent effect on the normal tissue, is more rapid than is the case when tumor effect is required to be constant. This is because normal tissue response is more sensitive to fraction size than is tumor response (due to the smaller α/β ratio and hence more rounded survival curve). As expected, SED_3 is equal to total dose for a 2 Gy fraction size, is less than total dose when fraction size is less than 2 Gy, and is greater than the total dose when dose per fraction is greater than 2 Gy. Also as expected, BED_{10} and SED_{10} are equal to total dose when the dose per fraction is zero and 2 Gy, respectively. The SED_{10} curve can be used to see the effect on the tumor as

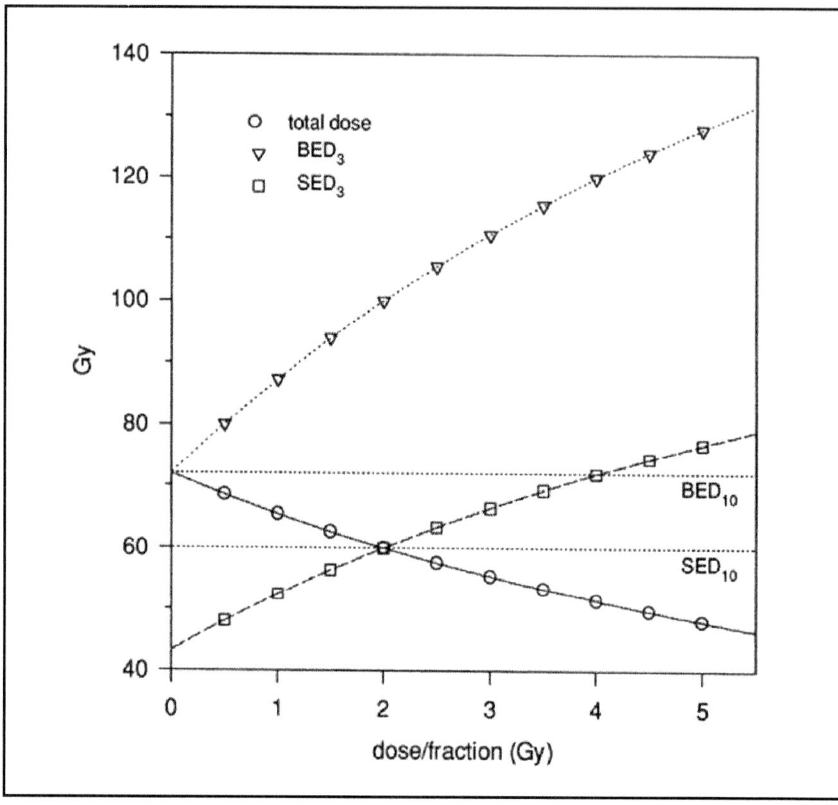

(a)

Figure 14.5: Total dose is plotted as a function of dose per fraction when BED (and SED) is kept the same as for a standard fractionation scheme (30 fractions of 2 Gy). In (a) BED and SED are constant for the tumor (BED_{10} = 72 Gy, SED_{10} = 60 Gy), and BED and SED are plotted for normal tissue (BED_3, SED_3). In (b) BED and SED are constant for normal tissue (BED_3 = 100 Gy, SED_3 = 60 Gy), and BED and SED are plotted for the tumor (BED_{10}, SED_{10}).

fraction size, and hence total dose, is changed. In this case, BED_{10} and SED_{10} decrease with an increase in fraction size, along with the total dose. This is because the effect on cell kill due to increased fraction size is now less than that due to the reduction in total dose, highlighting the fact that tumor response is less dependent than normal tissue response on dose per fraction (due to the smaller α/β for normal tissue).

In both these examples, the gains that can be obtained by increasing fraction number can be seen. The benefit lies either in reducing effects on late responding normal tissue while keeping the same tumor cell kill (as in the first example) or in increasing the tumor cell kill for the same late effects. This is the motivation for *hyperfractionation*, where more than the standard number of fractions is delivered (in smaller fraction sizes), thus exaggerating the effects of a conventional fractionation regime.

(b)

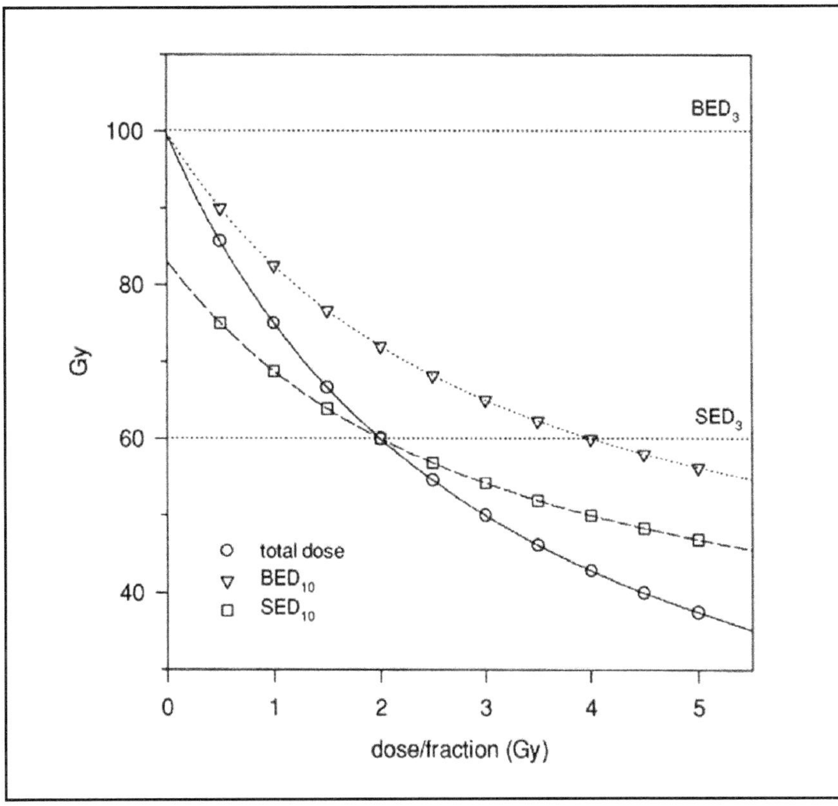

Figure 14.5: (Continued)

Relative Effectiveness and Standard Relative Effectiveness

Other interesting quantities are the *relative effectiveness* (RE) and *standard relative effectiveness* (SRE). These are used to express the effectiveness of a fractionation scheme compared to when infinitely small or 2 Gy fractions are used, respectively. Specifically, it is the ratio of (a) the dose required in 0 or 2 Gy fractions to give the same effect as the current scheme (the BED or SED) to (b) the actual dose (Lebesque et al. 1991, Yaes et al. 1991):

$$\begin{aligned} RE &= \frac{BED}{n\,d} \\ &= \left(1 + \frac{d}{\alpha/\beta}\right) \end{aligned} \tag{14.17}$$

and

$$\begin{aligned} SRE &= \frac{SED}{n\,d} \\ &= \frac{\left(1 + \frac{d}{\alpha/\beta}\right)}{\left(1 + \frac{2}{\alpha/\beta}\right)}. \end{aligned} \tag{14.18}$$

Looking at the SRE for example, it is obvious that the value increases (the scheme becomes more effective) when SED increases; that is, when a larger dose in 2 Gy fractions is required to have the same effect. SRE is, of course, greater than unity when $d > 2$ Gy.

Isoeffect

An isoeffect line is a curve along which the x,y coordinates (factors in some function describing effect) give equal effect. Using the effectiveness ($E = -\ln S$) from equation 14.13, we can write

$$\frac{1}{nd} = \frac{\alpha}{E} + \left(\frac{\beta}{E}\right)d. \tag{14.19}$$

Hence a plot of dose per fraction versus the reciprocal of total dose gives a line with gradient β/E with y intercept α/E. The coordinates for the plot are found from data indicating sets of n,d pairs that had the same biological effect or cell survival probability. Such a plot can be used to derive α/β ratios, using the product (Withers and Peters 1973, Thames and Hendry 1987).

$$\frac{\alpha}{\beta} = \frac{\alpha}{E} \times \frac{E}{\beta}. \tag{14.20}$$

The advantage of this method (as opposed to the use of cell survival curves) is that effect may be measured by clinical outcome and cell survival data are not essential. Because of the meaning of E in equation 14.20, the method of course presumes that equivalent effect means equivalent cell kill.

14.4 The Four Rs of Radiobiology

The benefits of fractionation do not only stem from the difference in shape of the tumor and normal tissue cell survival curves. In fact, fractionated treatment began long before it was appreciated that there was a difference in these shapes. Originally, the motivation for giving a course of radiotherapy in many small doses was to avoid acute reactions of early responding normal tissues, particularly the skin and mucosa. An early application of fractionation was in the sterilization of rams in France in the 1920s (Hall 1988), where it was found that in delivering a single dose large enough to sterilize the testes, acute reaction of the skin of the scrotum was excessive. Multiple small doses over a period of approximately ten days were thus delivered, and it was found that sterilization could be achieved while sparing the scrotum. Fractionated treatment of human tumors followed this discovery, with the ram's testes and scrotum being models for the tumor and skin, respectively. As well as there being a sparing of normal tissue such as skin by fractionation, there is more efficient killing of tumor cells due to changes in oxygenation and in redistribution of cells throughout the cell cycle. The conventional explanation and justification for fractionation (quite apart from the different shapes of cell survival curves) is contained in the so-called *four Rs of radiobiology* (Withers 1975), which are discussed here.

Reoxygenation

Oxygenation of tumor cells is important to assure high radiosensitivity. For low LET radiation (photons and electrons), oxygenated cells require about one-third of the dose of hypoxic cells to achieve the same cell kill (Thames and Hendry 1987). This translates to an oxygen enhancement ratio (OER) of three. Certain types of tumors contain hypoxic cells, which are at a range of distance from a blood vessel such that they are resistant to radiation yet still receive enough oxygen to be viable. These quiescent cells may make up about 15% of the total cell population (Hall 1988). With fractionated radiotherapy, the quiescent cells in the tumor reoxygenate (Withers 1985, Sephton 1991), making them more sensitive to subsequent fractions. That is, the well-oxygenated cells are killed, leaving blood supply available to previously hypoxic cells. In this way, the population of well-oxygenated (and, hence, radiation sensitive) cells is kept to a constant proportion.

Redistribution

Cell cycle redistribution could be an important factor for "self sensitization" (Withers 1985), since the radiosensitivity of cells varies during the cell cycle. Late G2/M and late G1/early S phase are the most sensitive phases of the cell cycle. Fractionation ensures that all cells will be in their sensitive phase at the time that one or more fractions are delivered.

Repair

It is important to allow enough time for normal cells to repair cellular injury. It has been shown that it is essential to allow at least six hours between two fractions to give normal cells that have suffered sublethal damage time to repair. From early results of the continuously hyperfractionated accelerated radiotherapy (CHART) regime (Saunders et al. 1988) it seems to be necessary to allow even longer repair times for the spine to avoid radiation myelitis (Saunders 1991). Repair has particular significance for brachytherapy treatments that are undertaken at different dose rates (Williams and Thwaites 1993).

For fractionated therapy where n fractions of dose d are separated by at least six hours to allow full repair, equation 14.10 is valid; otherwise the fractionated cell survival curve cannot be obtained from multiple applications of the initial part of a single dose survival curve (for which full repair is achieved).

For low-dose rate continuous radiation, account must be taken of the repair of sublethal damage during the time period of treatment such that:

$$\frac{E}{\alpha} = RT\left(1 + \frac{KR}{\frac{\alpha}{\beta}}\right), \tag{14.21}$$

whereby T is the total time of the continuous radiation insertion, R is the dose rate in Gy/hour, and K is a constant related to sublethal repair and is given by:

$$K = \frac{2}{\mu}\left(1 - \frac{1-\exp(-\mu T)}{\mu T}\right), \tag{14.22}$$

in which μ is the time constant for sublethal repair; this is generally taken to be approximately 0.5 h^{-1}, which corresponds to a repair half-time of $T_{½} = 1.4$ hours.

Regeneration

Repopulation of cells is one of the most important factors for treatment outcome and is important for both early responding normal tissue and for tumors. For early responding normal tissue such as the skin, the rate of cell division begins to increase after a certain time since treatment began, such that more and more dose is required to give the same biological effect. This signifies an advantage in prolonging treatment, since the rate of repair becomes faster with time. Protracting the course of radiotherapy therefore spares normal, early responding tissues but is not beneficial to late responding tissues.

The same effect, however, can occur for tumors, where the accelerated repopulation of cells means that significantly more dose is required to achieve the same total cell kill, if total treatment time is longer than the time after which tumor cells begin to divide more rapidly. In head and neck tumors it has been observed that tumors start to grow more rapidly between two to four weeks after the start of radiotherapy (Withers et al. 1988). This time is often referred to as the *time for kickoff*, T_k. Prolonging treatment is therefore detrimental to tumor control, while being beneficial to normal early responding tissue. This causes a dilemma, of course, and the best method of treatment depends on the relative merits of tissue sparing and avoiding tumor proliferation. This requires particular consideration of the tumor type. For example, it may be important to minimize the treatment time for very fast proliferating tumors. An indication of the proliferative potential of a tumor can be gained from the potential doubling time (T_p) of the cells. T_p is not necessarily the volume doubling time of the tumor: It is the time required to double the cell population assuming no cell loss has occurred. Times of three to four days have been recorded for head/neck squamous cell carcinomas. The times for breast carcinomas are typically between about eight and 30 days (mean 12 days). For prostate carcinoma T_p is typically 60 days. If T_p is short, accelerated fractionation regimes may be worthwhile, whereas if T_p is long any advantage gained in tumor cell kill will be outweighed by the increase in early normal tissue reactions.

Typical values of repair half-times, α/β ratios, and potential doubling times are given in table 14.1a–c.

The Fifth R—Radiosensitivity

Radiosensitivity has been identified as another important radiobiological variable (Steel et al. 1989). Cells from different types of tumors differ in

Table 14.1: Tissue repair half-times for lung and spinal cord

(a)

Tissue (Reference)	*Dose Delivery*	*$T_{½}$ (Hours)*
Lung (Vegesna et al. 1985)	fractionated	1.5 (1.2–1.9)
(Down et al. 1986)	LDR	0.68 (0.55–0.89)
(Travis et al. 1987)	fractionated	0.74 (0.68–0.81)
Spinal cord (Ang et al. 1987)	fractionated	1.55 (1.40–1.73)

(b) α/β parameters for various tissues (from Thames et al. 1987)

Tissue	*α/β (Gy)*
1. Epidermal healing	16 (15–18)
2. Skin atrophy	29 (15–130)
3. Spinal cord	3.4 (2.7–4.3)
4. Lung	2.5 (2.0–3.2)
5. Lung	3.7 (2.8–4.5)
6. Lung	3.7 (3.2–4.3)

(c) T_p values for different tumor types (from Trott et al. 1985)

Tumor	*T_p (days)*
1. Hodgkin's	3
2. Myeloma	22
3. Small cell lung	4
4. Breast	12 (8–50)
5. Prostate	60
6. Melanoma	15 (8–20)

their radiosensitivity. Human tumor cell lines have sufficiently different cell survival curves in vitro to explain why there is a difference in curability of patients due to radiosensitivity of the tumor.

14.5 Linear Quadratic Model Including Tumor Proliferation

The regeneration effects discussed in the previous section made it imperative for the linear quadratic model to consider time effects—in particular the impact of overall treatment time, T. This is especially true for tumors that are capable of rapid proliferation. Several attempts have been made to account for time factors in the model. Most commonly, equation 8.10 is extended with a factor reducing the overall effect of the treatment with time. If the cell population grows exponentially with time, where the coefficient of time in the exponent is γ, equation 8.10 can be modified to

$$S = \exp(-nd(\alpha + \beta d) + \gamma T). \tag{14.23}$$

Now if for zero dose the population doubles in a time T_p, by setting S equal to 2 and $T = T_p$, it is easy to show that $\gamma = \ln 2/T_p$. Also, if the cells do not begin active proliferation until a time T_k (time for kickoff) after treatment begins, then

$$S = \exp\left(-n\,d\left(\alpha + \beta d\right) + \frac{\ln 2\left(T - T_k\right)}{T_p}\right) \qquad (14.24)$$

and the effectiveness is

$$E = -\ln S = n\,d\left(\alpha + \beta d\right) - \frac{\ln 2\left(T - T_k\right)}{T_p}. \qquad (14.25)$$

The time factor in these equations is only defined when $T > T_k$ otherwise it is omitted. An example of a survival curve that includes proliferation is shown for a tumor with $\alpha = 0.35$ Gy^{-1} and $\beta = 0.035$ Gy^{-2} in figure 14.6a–c, where in each case a 2 Gy fraction is given each day and $T_k = 20$ days. In figures 14.6a, b, and c, $T_p = 6, 4$, and 2 days, respectively. The time, T, in days after treatment began was first simply taken to be equal to the number of fractions, n. Also in these plots, the time in days was modified to include the effects of treatment breaks over each weekend. The time is then given by $T = n + 2 \times \text{int}[(n\text{–}1)/5]$, where the extra term gives the number of weekend days passed. The reduction in cell kill due to proliferation can be seen to increase with a reduction in T_p as expected. Repopulation during weekend treatment breaks is obvious and is, of course, most significant for small values of T_p. Note that with weekend breaks, the delivered dose at which proliferation begins is less than when there are no interruptions. It must be remembered that the tumor cells are not in fact repopulating as rapidly as described by the time factor in equation 14.24 since many cells are lost. The curves including repopulation therefore show the potential (an upper limit) of the effect of accelerated repopulation, and a cell loss term should be included in the above equations.

Note that for tumors that have a large T_p, the effect of proliferation is small and effectiveness is reduced by a lesser amount than for tumors with a smaller T_p. Furthermore, these tumors will also in general have a value of T_k that is larger than the total treatment time; hence, proliferation will not be an important factor.

Since the effectiveness of radiation on the late responding tissue is not reduced by prolongation of treatment time, whereas early responding tissues are spared, it is the late responding tissues that limit the dose that can be delivered in normal treatment regimes (they become *dose limiting*). For short treatment times, where early responding tissue is not spared by proliferation, it is often these tissues that are dose limiting.

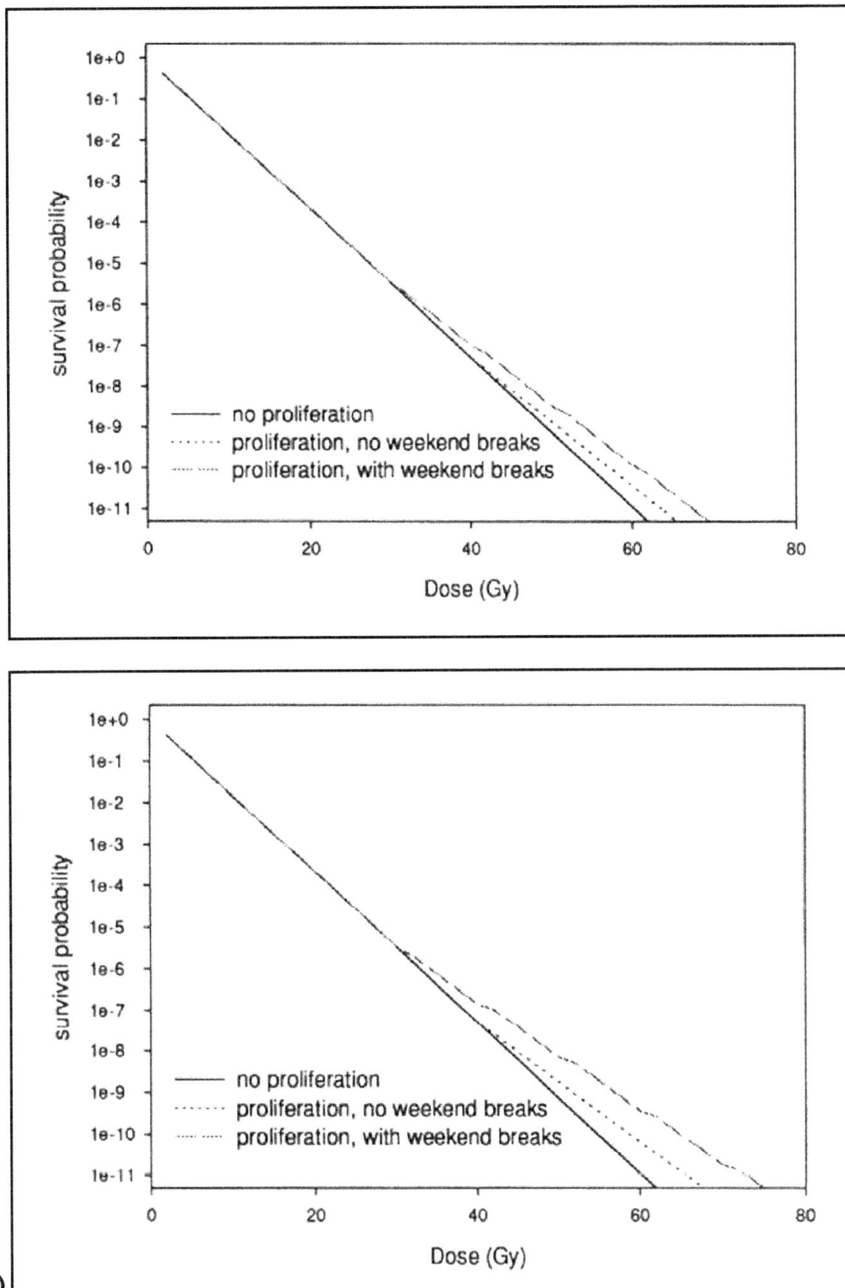

Figure 14.6: Survival curves for a tumor with $\alpha = 0.35\ Gy^{-1}$ and $\beta = 0.035\ Gy^{-2}$, dose per fraction = 2 Gy, T_k = 20 days, and one fraction is delivered per day. The effects of proliferation after time T_k is shown, with and without weekend breaks in treatment. In (a), (b), and (c) the potential doubling time, T_p, is 6, 4, and 2 days, respectively. The significant reduction in cell kill due to proliferation can be seen, with a further reduction when weekend breaks are included.

(c)

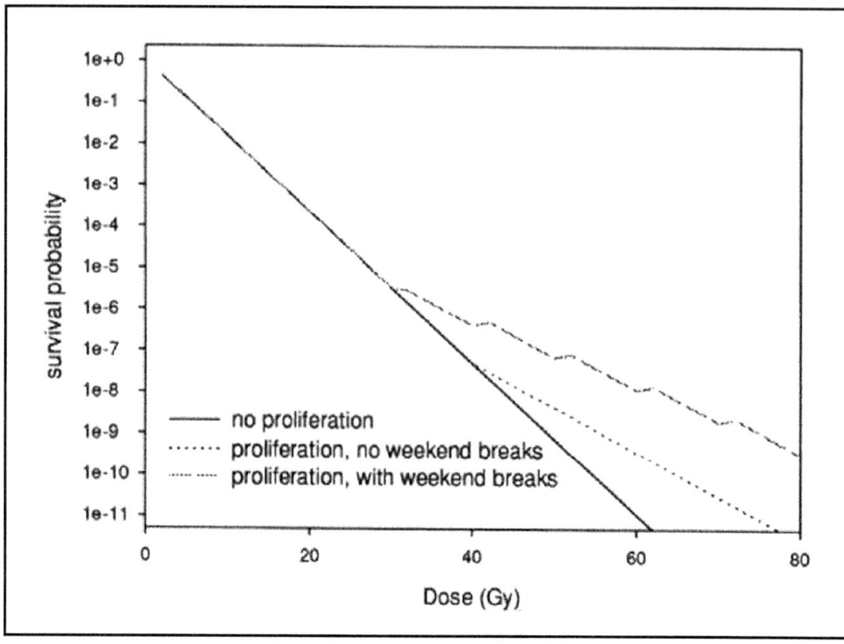

Figure 14.6: (Continued)

14.5.1 Biologically Effective Dose Including Tumor Proliferation

Using equations 14.14 and 14.25, the BED for a proliferating tumor is given by (Fowler 1989):

$$BED = \frac{E}{\alpha} = n\,d\left(1 + \frac{d}{(\alpha/\beta)}\right) - \frac{\ln 2\left(T - T_k\right)}{\alpha\,T_p}. \tag{14.26}$$

This is the BED allowing for proliferation and is physically the equivalent dose in infinitely small fractions when proliferation does not have an effect. For normal late responding tissue, the time factor is not required since $T < T_k$ throughout the treatment, and furthermore, T_p is large enough such that tissue growth is not important (equation 14.12 still holds).

It is obvious from equation 14.26 that while the BED delivered to a tumor increases with total dose *nd*, it *decreases* with time when $T > T_k$. The real effect of increments in dose after the kickoff time is reached is therefore reduced. It may therefore be beneficial to deliver a smaller dose in a shorter time, rather than a larger dose over a time greater than T_k. This is the motivation behind accelerated treatment, such as the procedure used in the CHART program (Saunders et al. 1988), where 36 fractions of 1.5 Gy delivered in 12 days (3 fractions per day) were used. The BED for tumor and normal tissue, using both CHART and standard schedules, is calculated below as an example. In the standard schedule, 30 fractions of 2 Gy are delivered over 40 days.

Example: BED for Tumor and Normal Tissue Using CHART and Standard Schedules

Use of equation 14.26 can be demonstrated for a head and neck tumor with a potential doubling time of $T_p = 4$ days, $\alpha/\beta = 10$ Gy, $\alpha = 0.3$ Gy^{-1}, and $T_k = 10$ days.

For the CHART regime, tumor BED is

$$\begin{aligned} BED &= \frac{E}{\alpha} = n\,d\left(1 + \frac{d}{(\alpha/\beta)}\right) - \frac{\ln 2(T - T_k)}{\alpha T_p} \\ &= 36 \times 1.5\left(1 + \frac{1.5}{10}\right) - \frac{\ln 2(12 - 10)}{0.3 \times 4} \\ &= 60.9\ \text{Gy}_{10}, \end{aligned} \tag{14.27}$$

where the subscript 10 after Gy is to indicate that the BED is for tissue with an α/β ratio of 10 Gy.

For the standard regime

$$\begin{aligned} BED &= \frac{E}{\alpha} = n\,d\left(1 + \frac{d}{(\alpha/\beta)}\right) - \frac{\ln 2(T - T_k)}{\alpha T_p} \\ &= 30 \times 2\left(1 + \frac{2}{10}\right) - \frac{\ln 2(40 - 10)}{0.3 \times 4} \\ &= 54.7\ \text{Gy}_{10}. \end{aligned} \tag{14.28}$$

This shows that the tumor BED is higher for CHART than the standard regime due to the fact that the reduction in effect caused by proliferation is less. This is despite the total dose being reduced to 54 Gy from 60 Gy.

Now examining the late effects (note that we don't include proliferation), for CHART we have

$$\begin{aligned} BED &= \frac{E}{\alpha} = n\,d\left(1 + \frac{d}{(\alpha/\beta)}\right) \\ &= 36 \times 1.5\left(1 + \frac{1.5}{3}\right) \\ &= 81\ \text{Gy}_3, \end{aligned} \tag{14.29}$$

and for the standard regime

$$\begin{aligned} BED &= \frac{E}{\alpha} = n\,d\left(1 + \frac{d}{(\alpha/\beta)}\right) \\ &= 30 \times 2\left(1 + \frac{2}{3}\right) \\ &= 100\ \text{Gy}_3. \end{aligned} \tag{14.30}$$

The predicted late effects are less for the CHART regime than the standard regime since both the total dose and fraction size are significantly lower.

14.5.2 Accounting for Treatment Interruption

Accounting for treatment interruptions and resulting prolongation of the treatment time is difficult since it depends on many factors:

- When did the interruption occur in the treatment schedule (was the tumor reoxygenated and had it started accelerated proliferation)?
- What is the potential cell doubling time T_p and kickoff time T_k?
- How radiosensitive is the tumor (mainly what is the value of α)? Although the α/β ratios of many tumors are well-known, the individual values for α and β are less readily available as their derivation is difficult from the experimental data (Thames and Hendry 1987).

Therefore, various rules of thumb have been proposed to account for a prolongation of treatment time (if there was a correction made at all). Liversage (1971) mentioned 30 cGy per day; Fowler (1989) and Barendsen (1982) suggested adding 50 cGy per day to the BED as defined above.

One can attempt to make use of the potential doubling times, T_p, known for several tumors (Trott et al. 1985). Assume n_1 fractions of the planned n fractions (size d) have been given. After a break of R days the treatment shall be resumed. Using equation 14.26 one can calculate the remaining biologically effective dose, BED_2, which needs to be given (since BED for parts of the treatment are additive):

$$BED_2 = BED_{planned} - BED_1. \tag{14.31}$$

Two solutions can be explicitly solved for when either (a) the dose per fraction after the break is to be kept the same as that before the break or (b) the prescribed number of fractions is to be retained. In deriving the following equations, the treatment break occurs after T_k.

i) *The dose per fraction shall be kept*

If the dose per fraction d is kept the same for the remaining part of the treatment, additional fractions have to be given. The new total number of fractions can be calculated according to:

$$n_{new} = n_{old} + \frac{R}{T_p} \times \frac{\ln 2/\beta}{d \times (\alpha/\beta + d)}. \tag{14.32}$$

Values of T_p and α/β can be looked up in table 14.1, and a fairly typical value of β for most tumor types is 0.03 Gy^{-2} (Fowler 1989). In arriving at equation 14.32, the new total treatment time is the intended time, T, plus the interruption time, R, plus one day for each additional required fraction $(n_{new} - n_{old})$.

With values for T_p of four days and $\alpha/\beta = 10$ Gy one calculates two additional 2 Gy fractions for a treatment interruption of seven days (and two additional treatment days for the additional fractions); more precisely, the calculation gives $n_{new} = n_{old} + 2.2$ days.

ii) *The overall number of fractions shall be kept*

If the overall number of fractions shall be kept to $n = n_1 + n_2$, the dose per fraction has to be increased from d to d_2 after the break where d_2' can be calculated according to:

$$d_2 = -\frac{\alpha}{2\beta} + \sqrt{\left(\frac{\alpha}{2\beta} + d\right)^2 + \frac{R}{T_p} \times \frac{\ln 2/\beta}{n_2}}. \qquad (14.33)$$

With values for T_p of four days and $\alpha/\beta = 10$ Gy, a dose for the remaining 10 fractions of a 2 Gy per fraction schedule, after a seven-day break, is 2.3 Gy per fraction. Note that the efficacy of increasing the dose per fraction as suggested by this calculation is the decision of the radiation oncologist who must balance the benefit of maintaining the same tumor control probability against the possible side effects of a higher total dose and dose per fraction.

Retreatment

The effect of an earlier treatment on the tolerance doses in retreatment depends on the tissue type, the damage done in the first treatment, and the time interval. In general, there is not sufficient data available to make firm recommendations about retreatment. Results from animal experiments—in particular with rodents—cannot easily be transferred to humans since tolerance doses are different (for example, 120 Gy for a rat spinal cord), and the laboratory animals do not have a long enough natural life span to judge late effects after retreatment. For nerve cells and other late responding tissue there is hardly any recovery from damage. In a retreatment, a different technique should be used to avoid the late responding tissues treated before. Skin (acute reactions), mucosa, and epithelia do recover to some degree from radiation-induced damage.

Metastasis

Radiotherapy can be effectively used for relief of pain caused by metastasis. Sadly however, cure is rare with radiotherapy due to the systemic nature of the disease. It is probably fair to say that surgery combined with radiotherapy leads to long-term local control of most primary tumors, which of course reduces the likelihood of metastatic disease. New therapies that control or inhibit the growth of metastasis are a major challenge faced by researchers in the future. If these agents can be developed, the circle for successful management of the cancer will be closed. Agents that are being investigated include new chemotherapy agents, genetic agents, and many other potentially useful agents.

14.6 Models of Tumor and Normal Tissue Response

Models with which to calculate TCP and NTCP have been developed that are based on clinical data, in vitro cell survival curves, and appropriate statistical theory. There are well-known dependencies on dose and irradiated volume for both tumors and normal structures. For example, it is known that a region of low dose in a tumor will increase the risk of recurrence (since the chance of killing all cells in this region is reduced). Indeed, a region of zero dose will reduce the TCP to zero. Likewise, increasing the irradiated volume of some structures (such as the spinal cord) causes a sharp increase in the likelihood of complications. On the other hand, some organs can tolerate quite a large volume being irradiated to a high dose without their function being unduly impaired.

14.6.1 Functional Sub-Units (FSUs)

These phenomena can be explained with appropriate modeling of the structure, taking into account both the response of cells to radiation and the organization of the cells into *functional sub-units* (FSUs). These FSUs are the elemental components of an organ or tumor, such as nephrons in the kidney, alveoli in the lung, or groups of oligodendrocytes in the spinal cord. For cells where there are no standard elemental components of the organ, for instance skin, a cell number is chosen as the sub-unit size. In a tumor, the FSUs are single clonogenic cells. The organization of the FSUs is referred to as the *architecture* of the structure. It is assumed that if any cell within an FSU is left alive, the FSU can repopulate, but the number of FSUs that must be inactivated for complications to occur depends on the architecture of the structure. The basic mathematics of cell killing and of FSU-based theory of TCP and NTCP will be discussed here. Figure 14.7 shows schematically the irradiation of a structure composed of FSUs and the cells within a single FSU.

14.6.2 Inactivation of One FSU

As already mentioned, to inactivate an FSU requires that all cells within it are killed. This can be calculated easily if the probability of a cell being killed (or surviving) after receiving a certain dose is known. The survival probability of a cell can be described by the linear-quadratic model, where, for a fractionated treatment, the parameters are the dose per fraction and number of fractions, as well as the α and β values corresponding to the type of cell in the tumor or organ. For the simple case of a dose D in a single fraction, the cell survival probability is given by equation 14.8. The cell survival probability after n fractions of d Gy is given by equation 14.10. From that expression it can be seen that if α is large compared to βd, the quadratic term is insignificant and can be ignored, giving a linear fall-off of the logarithm of survival probability with dose. This approximation can generally be made for tumors, where we may have for example $\alpha = 0.30\ \text{Gy}^{-1}$, $\beta = 0.03\ \text{Gy}^{-2}$, and $d = 2$ Gy ($\beta d = 0.06$). Such a simplification cannot be made

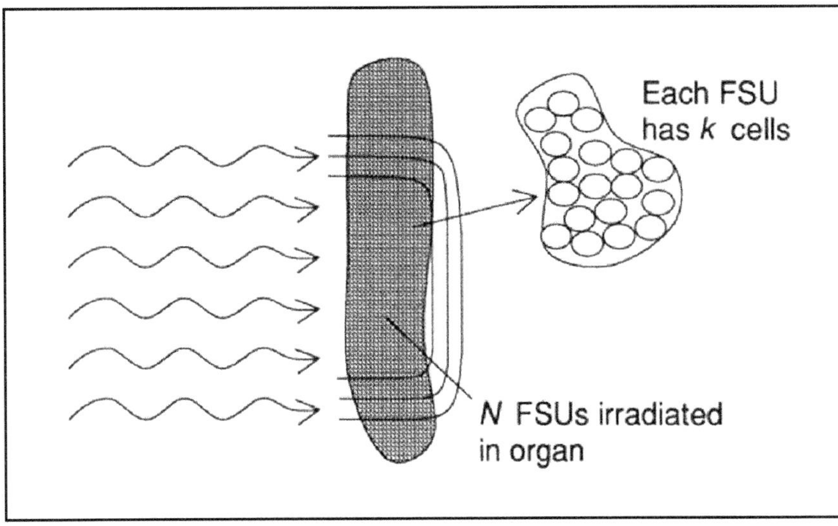

Figure 14.7: Irradiation of a structure containing N FSUs, each containing k cells. To inactivate an FSU requires that all k cells are killed.

for normal tissues, however, where the values will be more like α = 0.15 Gy^{-1}, β = 0.05 Gy^{-2}, and (βd = 0.10). For tumors then,

$$S \approx \exp(-\alpha D) \tag{14.34}$$

where $D = nd$.

Using the survival probability as a function of dose, the likelihood of killing all cells in a single FSU can be calculated. The chance of a cell *not* surviving is obviously equal to $1 - S$, so the probability of *all* cells not surviving is simply

$$P = (1 - S)^k, \tag{14.35}$$

where there are k cells in the FSU.

It is apparent that the more cells there are in the FSU, the lower is the likelihood of inactivation. Now, recalling that there is only one cell per FSU in tumors, substituting equation 14.34 into 14.35 gives the probability of eradicating a tumor FSU (a clonogenic cell):

$$P_{tum} = (1 - \exp(-\alpha D)). \tag{14.36}$$

For a normal organ, where cell survival must take account of the β term, we have, using equations 14.10 and 14.35,

$$P_{norm} = (1 - \exp(-n\,d(\alpha + \beta d)))^k. \tag{8.37}$$

These probabilities can now be used in expressions for the TCP and NTCP, where for the latter the type of expression depends on the architecture of the organ.

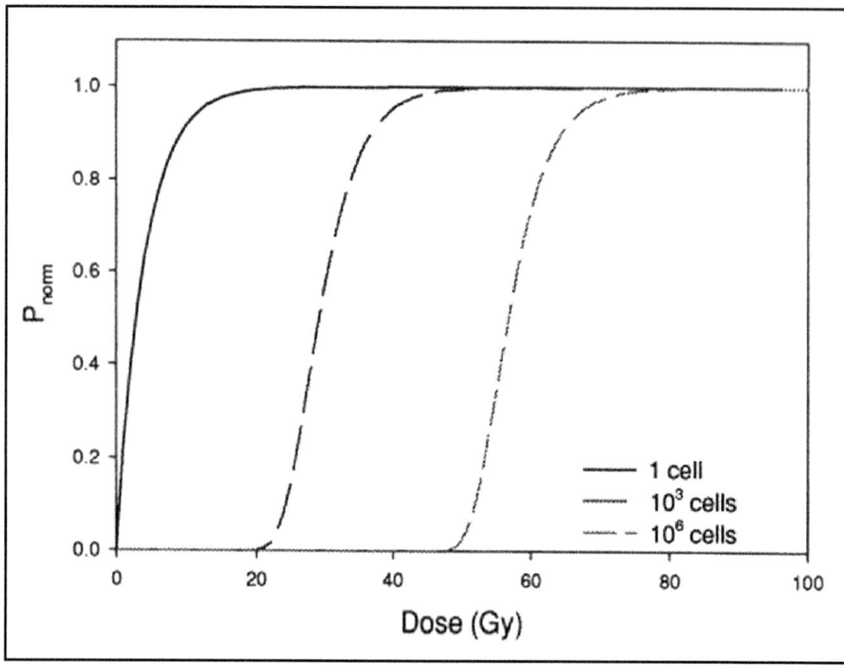

Figure 14.8: Probability of inactivation of one FSU versus dose, when each FSU contains 1, 10^3, or 10^6 cells, all of which must be killed. Survival probability of each cell is described by $\alpha = 0.15\ Gy^{-1}$, $\beta = 0.05\ Gy^{-2}$. Dose required for the same probability of inactivation varies with the logarithm of the cell number.

To illustrate the effect of the number of cells per FSU on the probability of inactivation, equation 14.37 was evaluated and plotted as a function of total dose for k equal to 1, 10^3, and 10^6. In this example, dose per fraction is $d = 2$ Gy, $\alpha = 0.15, \beta = 0.05$, and number of fractions, n, is varied from 0 to 50. The results are shown in figure 14.8 where it can be seen that the dose required for any given probability is proportional to the logarithm of the number of cells per FSU. Thus, the radioresistance of an FSU increases with the log of number of cells per FSU.

14.6.3 Serial and Parallel Structures

In structures where the FSUs act as a "chain gang," such as in the spinal cord, the organ architecture is said to be *serial*. If any one FSU is rendered inoperable, the function of the whole organ suffers because of this weak link. Note that all FSUs may still perform the same job, but they take input from, and pass output to, each other. Complications will thus arise when all cells in *one or more* FSUs are killed. As each FSU is critical, such organs are known as *critical element* structures.

In a parallel structure, each FSU performs the same task independently of each other, as is the case with the nephrons in the kidney. The inactivation of some (but not all) FSUs may make the function of the organ less efficient, but still allow it to perform its task. However, if too many FSUs are

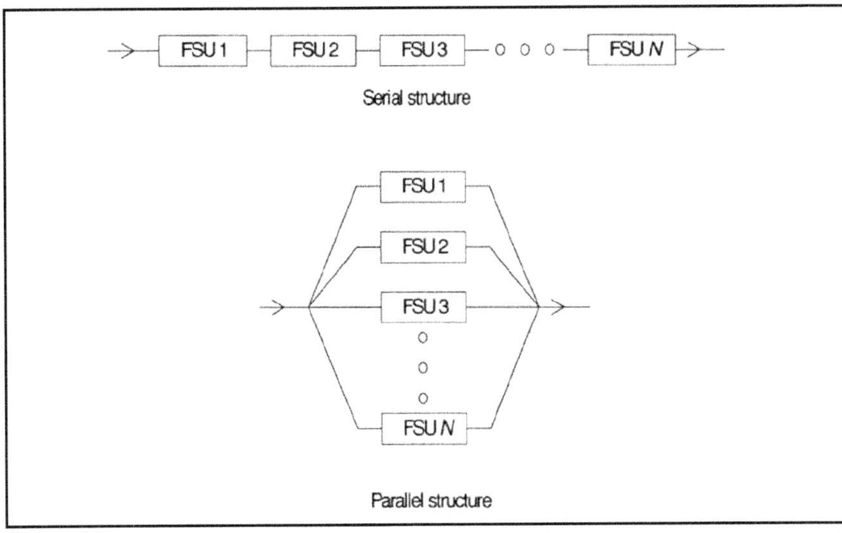

Figure 14.9: FSUs arranged in a serial or parallel architecture. In a serial structure, one or more FSUs must be inactivated to cause complications (since this breaks the chain); whereas in a parallel structure, all, or a certain proportion, must be inactivated (since FSUs work independently).

taken out of the workforce, the organ will not be able to cope with its workload and complications will set in. Parallel organs each have a critical proportion μ of FSUs that can be destroyed such that the organ will still function. This represents the threshold fractional volume for complications; hence parallel organs are known as *critical volume* structures.

One can make the analogy to electrical circuits, where there are N components either connected in series or parallel as shown schematically in figure 14.9.

14.6.4 NTCP for a Serial Organ

Simple expressions for the likelihood of complications in series and parallel organs are obtained from probability theory. For serial structures the NTCP is the probability that any one of the FSUs is damaged. If I_j denotes the event "inactivation of FSU j," then, when there are N irradiated FSUs in the organ, the logical expression for when complications will occur is

$$I_1 + I_2 + \cdots + I_N, \tag{14.38}$$

where + is the logical OR operator. In order to convert this into a mathematical expression, a logical expression involving the AND operator must be obtained. Using negation operators and simple identities, equation 14.38 can be rewritten as

$$\begin{aligned} &\overline{\overline{I_1 + I_2 + \cdots + I_N}} \\ &= \overline{\overline{I_1} \bullet \overline{I_2} \bullet \cdots \bullet \overline{I_N}} \\ &= 1 - \left(\overline{I_1} \bullet \overline{I_2} \bullet \cdots \bullet \overline{I_N}\right) \end{aligned} \tag{14.39}$$

where • is the logical AND operator and a bar is the logical NOT. The quantity in braces is, of course, the event that no FSU is inactivated. Now if the probability of inactivation of FSU j (event I_j) is P_j, equation 14.40 is obtained:

$$NTCP = 1 - \left(\left(1 - P_1\right) \cdot \left(1 - P_2\right) \cdots \left(1 - P_N\right)\right). \tag{14.40}$$

Each probability can be calculated using the cell survival probability and the number of cells per FSU as in equation 14.35. Now for the moment assuming that the probability of FSU inactivation has a constant value P throughout the organ, equation 14.40 becomes

$$\begin{aligned} NTCP &= 1 - \left(1 - P\right)^N \\ &= 1 - \left(1 - \left(1 - S\right)^k\right)^N . \end{aligned} \tag{14.41}$$

This equation forms the basis for the volume effect relationship for serial organs, since the number of FSUs irradiated, N, is proportional to the volume irradiated. Using equations 14.41 and 41.37, NTCP can be plotted against total dose for different values of N as in figure 14.10, where k = 1000, d = 2 Gy, α = 0.15 Gy^{-1}, β = 0.05 Gy^{-2}, and n is varied up to 50. The most obvious feature of these plots is the increase in steepness as the number of FSUs irradiated increases. Interestingly, the threshold dose for nonzero NTCP changes very little. This is because for very small values of P in equation 14.41 (where it is very unlikely that all cells in an FSU will be killed), $(1 - P)^N$ is very close to unity even for large N.

If the complication probability is known to be $NTCP_1$ for whole organ irradiation, it is easily shown using equation 14.41 that when a fractional volume v is irradiated the complication probability is equal to

$$NTCP_v = 1 - \left(1 - NTCP_1\right)^v . \tag{14.42}$$

It is also interesting to allow the probability of FSU eradication in different subvolumes to vary due to, say, dose inhomogeneity. The organ can be divided into subvolumes, each of volume v_{sub}, which contain N_{sub} FSUs. Now if the probability of FSU eradication in subvolume i is P_i, from equation 14.41 the NTCP for M irradiated subvolumes is

$$NTCP = 1 - \prod_{i=1}^{M} \left(1 - P_i\right)^{N_{sub}} . \tag{14.43}$$

Furthermore, if the whole-organ complication probability is $NTCP_{1,i}$ when the dose in volume i is D_i, then for M irradiated subvolumes, each of volume v_{sub}, the complication probability is given by

$$NTCP = 1 - \prod_{i=1}^{M} \left(1 - NTCP_{1,i}\right)^{v_{sub}} . \tag{14.44}$$

Hence the NTCP for a serial organ can be calculated for inhomogeneous irradiation.

It is common to use clinical data in calculating NTCP instead of using cell survival curves based on in vitro data. The clinical data used is normally

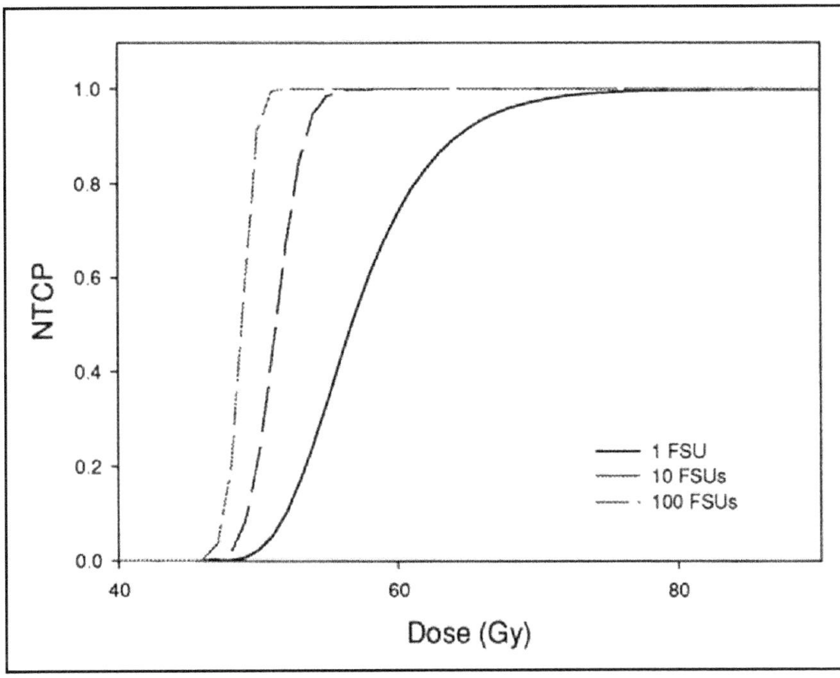

Figure 14.10: NTCP versus dose for a serial organ, when there are 1, 10, or 100 FSUs irradiated. In each case there are 1000 cells per FSU, $\alpha = 0.15, \beta = 0.05$, and the dose per fraction is 2 Gy. The probability of inactivation of at least one FSU increases with the number irradiated. Note that the minimum dose required for complications is relatively constant, since the probability of complications in all cases only becomes significant when the probability of killing all cells in an FSU is not infinitesimally small.

the dose that will cause complications in an organ in 50% of cases, where this quantity, D_{50}, normally refers to whole organ irradiation. Using the logistic model (Schultheiss et al. 1983), the whole-organ complication probability is given by

$$NTCP_{1,i} = 1/(1 + (D_{50}/D)^{1/n}, \tag{14.45}$$

in which n is the volume effect parameter (giving the degree to which the probability of damage is proportional to the fractional volume irradiated). Evaluation of equation 14.44 is thus simple if the absorbed dose in each fractional volume of the organ is known. Extension of this method to multiple organs is straightforward; the product in equation 14.44 is simply used as a factor in a product of like terms for each organ.

14.6.5 NTCP for a Parallel Organ

As already stated, complications to an organ with FSUs operating in parallel will only set in if the number of FSUs inactivated is greater than a certain threshold. If there are N FSUs irradiated in the organ and the threshold number for complications is M, the complication probability is given by

$$NTCP = \sum_{l=M}^{N} Q(l). \tag{14.46}$$

where $Q(l)$ is the probability that *l and only l* FSUs will be eradicated. Now if the probability of any FSU being eradicated is P (as in the case of serial organs), then using the binomial distribution, the likelihood of eradicating exactly l FSUs is

$$Q(1) = \binom{N}{l} P^{1} (1-P)^{N-1}. \tag{14.47}$$

The NTCP is thus given by a *cumulative binomial distribution*. Because of the summation in equation 8.46, for a constant critical volume (and hence M), NTCP increases with the number of FSUs irradiated (N) and hence the volume of the organ irradiated. Also, for a given irradiated volume, NTCP increases as the critical volume (and hence M) decreases.

For inhomogeneous irradiation, the situation is more complex and Q must be calculated using a distribution of FSU eradication probabilities corresponding to different dose levels, where different subvolumes of the organ each receive a discrete dose level and hence their FSUs have a discrete probability of inactivation. The reader is referred to Jackson et al. (1993) and Niemierko and Goitein (1993) for a full explanation. There are several different elegant models, some which include clinical data, for describing NTCP and the reader is referred to some of these as follows (e.g., Kutcher and Burman 1989, Lyman 1985, Burman et al. 1991).

14.6.6 Tumor Control Probability

The FSU organization of a tumor is parallel, but this is a special case where (a) the critical volume is the total volume of the tumor and (b) an FSU consists of a single clonogenic cell. This means that the tumor will still perform its (sinister) function as long as one clonogenic cell survives (as it will repopulate the tumor). As already pointed out, this makes k equal to 1 in equation 14.35. Also, we can think of the tumor control probability (TCP) as the complication probability for the tumor and thus use equation 14.46, with M equal to N (and hence a single term), to obtain TCP. Since $Q(N) = P^N$ from equation 14.47 we have

$$TCP = P^{N}, \tag{14.48}$$

which, using equation 14.35 with k equal to 1 along with equation 14.34 gives

$$\begin{aligned} TCP &= (1-S)^{N} \\ &= \left(1-\exp(-\alpha D)\right)^{N}. \end{aligned} \tag{14.49}$$

This is, of course, the probability of killing all N clonogenic cells in the tumor and will decrease as N increases. It will also decrease as the sensitivity α decreases. These trends are intuitive; however, an expression such as equation 14.49 is required to deduce the actual functional dependence of TCP on N (and hence tumor volume) and α. It is useful to write the TCP in terms of clonogen density, ρ, and tumor volume, V, where $N = \rho V$. Also, for small values of $\exp(-\alpha D)$ in equation 8.49, $1-\exp(-\alpha D)$ is approximately equal to $\exp(-\exp(-\alpha D))$, so we have

$$TCP = \exp\left(-\exp\left(-\alpha D\right)\right)^{\rho V}$$
$$= \exp\left(-\rho V \exp\left(-\alpha D\right)\right). \tag{14.50}$$

This is the usual formula for TCP (Webb and Nahum 1993), which can also be derived using Poisson statistics using the fact that TCP is the probability of zero surviving clonogens, and the *expected* number surviving is $\rho V\exp(-\alpha D)$.

The relationship of TCP to tumor volume is seen by plotting TCP versus dose for different volumes as in figure 14.11, where the clonogen density ρ is 10^7 cm^{-3} and sensitivity α is 0.35 Gy^{-1}. It is important to understand the reason behind the volume dependence of TCP. In particular, it is interesting (and elementary) that the probability of controlling a tumor *decreases* as volume increases, whereas the probability of complications in a serial organ *increases* with volume (where in both cases the endpoint is loss of function). At first thought this may seem contradictory, as in one case a large volume is harder to "kill," and in the other case it is easier. The reason is, of course, that to control a tumor requires that *all* clonogenic cells are killed (and this becomes harder to achieve when there are more cells to kill), but with a serial organ, complications occur when *one or more* FSUs are inactivated (which is easier to achieve with more FSUs).

It can be seen from figure 14.11 that the dose required for any given TCP is proportional to the logarithm of the volume. This is easily shown by solving equation 14.50 for D:

$$D = \frac{1}{\alpha}\left(\ln\left(\rho V\right) - \ln\left(\ln\left(\frac{1}{TCP}\right)\right)\right). \tag{14.51}$$

The logarithmic dependence of required dose on density and volume means that it should not be assumed that small tumors do not require as much dose. A reduction in tumor size by a factor of two and four corresponds to a reduction in required dose of approximately 2 Gy and 4 Gy, respectively, with α = 0.35 Gy^{-1}. Perhaps more importantly, the iso-TCP dose is also proportional to the logarithm of the clonogen density. This means that the required dose for a certain cell kill does not reduce rapidly as cell density reduces. For example, if we demand a TCP of 0.5, with α = 0.35 Gy^{-1}, $\rho = 10^7$ cm^{-3}, and $V = 100$ cm^3, the required dose is 60.3 Gy. Now if the density is reduced by a factor of 100, the reduction in dose required to maintain the same TCP is only 13 Gy. Note that this figure is the same for any value of V or TCP (since $\partial D/\partial \rho = 1/\alpha\rho$).

Also from equation 14.51 it can be seen that the dose for equivalent TCP is inversely proportional to the sensitivity α. This dependence is illustrated in figure 14.12, where the density and volume are 10^7 cm^{-3} and 100 cm^3, respectively. Also note from figure 14.12 that the slope of the TCP curve increases only slowly with sensitivity.

In order to account for dose inhomogeneity, it is convenient to treat the tumor as being made up of multiple *tumorlets*, each of which has its own *local TCP*. Because all tumorlets must be controlled in order to control the tumor,

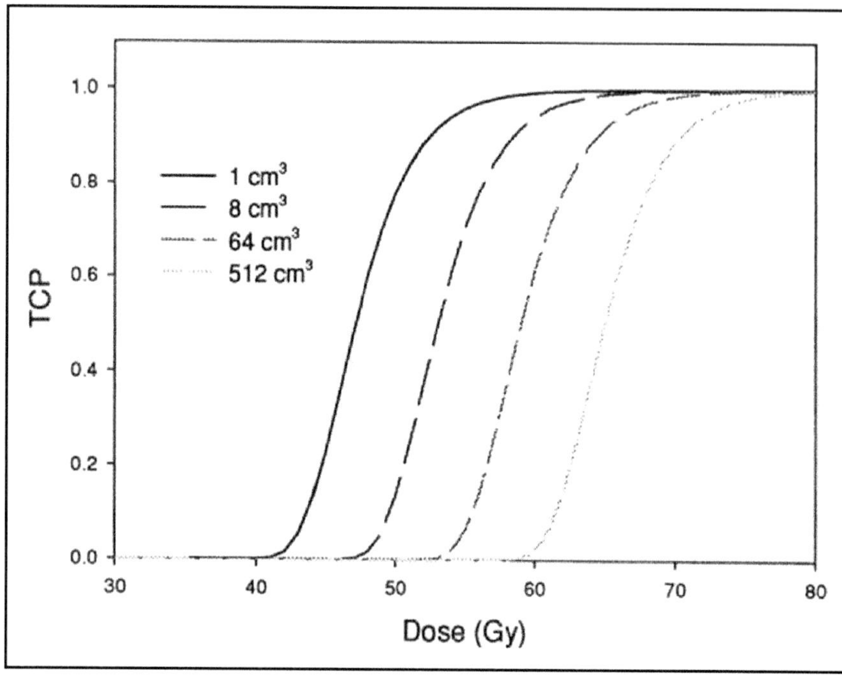

Figure 14.11: TCP versus dose for tumor volumes of 1, 8, 64, and 512 cm^3 (side lengths of 1, 2, 4, and 8 cm). Clonogen density, ρ, is 10^7 cm^{-3}, and sensitivity, α, is 0.35 Gy^{-1} in all cases. Dose required for a constant TCP increases with the logarithm of the tumor volume, as can be seen from equation 14.51.

the total TCP is simply the product of local TCPs for all tumorlets. If the volume of each tumorlet is v_t and the dose to tumorlet i is D_i, total TCP is

$$TCP = \prod_{i=1}^{N} TCP_i = \prod_{i=1}^{N} \exp\left(-\rho v_t \exp\left((\alpha D)\right)\right) \qquad (14.52)$$

when there are N tumorlets. Note that it is possible to also account for variations in clonogen density and sensitivity using the tumorlet model. These parameters are difficult to ascertain, but it is interesting for example to plot local TCP versus the estimated clonogen density across the tumor to highlight the requirement for high dose even in regions of low density.

A TCP versus dose curve calculated with equation 14.50 or 14.52 is known to have a slope that is much greater than a clinically derived TCP curve. This can be explained by the fact that curves derived from actual treatment outcomes are indicative of the population as a whole and do not reflect the sensitivity of any particular patient's tumor. Thus, if curves obtained from the TCP equations above are to match these data, the equations should be averaged over a range of α values. Using equation 14.50 for a term in the averaging, we have

$$TCP = \frac{1}{K}\sum_{k=1}^{K} \exp\left(-\rho V \exp\left(-\alpha_k D\right)\right), \tag{14.53}$$

and in the case of a product of tumorlet TCPs,

$$TCP = \frac{1}{K}\sum_{k=1}^{K}\left[\prod_{i=1}^{N} \exp\left(-\rho v_i \exp\left(-\alpha_k D_i\right)\right)\right]. \tag{14.54}$$

This procedure has been investigated by Webb and Nahum (1993) and Webb (1994), who showed that a good match to clinical TCP curves for bladder tumors was obtained by averaging equation 14.50 over a Gaussian distribution of α values with a mean of $\alpha_0 = 0.35$ Gy^{-1} and a standard deviation of $\sigma_\alpha = 0.08$ Gy^{-1}. Values of α_0 and σ_α have been determined by Webb (1994) by fitting clinically derived TCP data compiled by Brenner (1993). These data covered a range of tumor types including breast, malignant melanoma, squamous cell carcinoma of the upper respiratory and digestive tracts, and affected nodes in squamous cell carcinoma of the upper respiratory and digestive tracts. The α value distributions obtained for these and other tumor types (stored in a look-up table) can be used in equation 14.54, along with the calculated dose distribution (to find values of D_m) to yield a TCP value. It

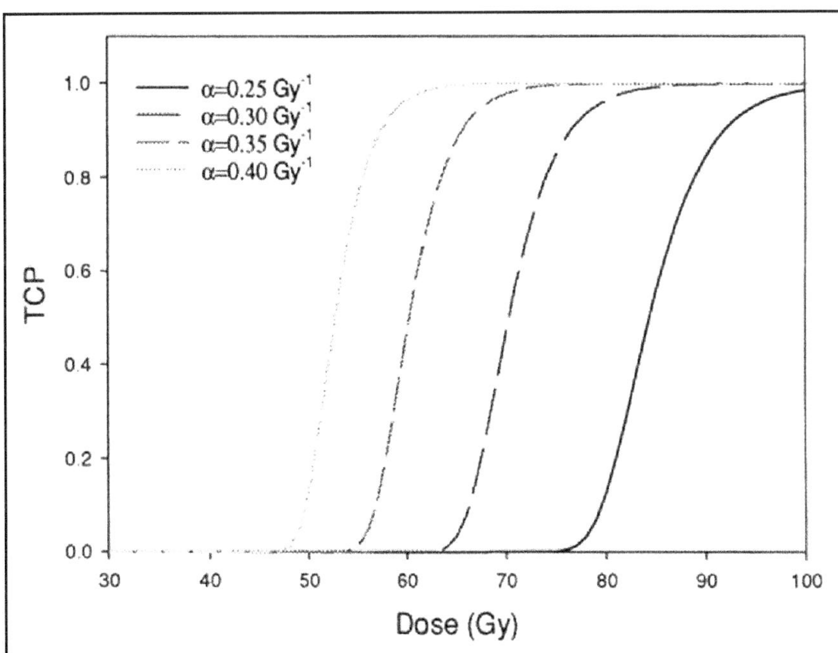

Figure 14.12: TCP versus dose for sensitivities α of 0.25, 0.30, 0.35, and 0.40 Gy^{-1}. Clonogen density, ρ, is 10^7 cm^{-3} in all cases. Dose required for a constant TCP increases with the logarithm of the tumor volume, as can be seen from equation 14.51.

is important to note that the flattening of the TCP curve on averaging is because of the displacement of curves for different sensitivities along the dose axis (see figure 14.12)—it is **not** due to the variation in slope (which is small). Such population averaging means that if clinically derived TCP curves are used to prescribe a dose, this dose will be a good choice for the patient population as a whole—not necessarily for the patient being treated. Note that an average TCP curve is not the same as a TCP curve for an average patient (which will be much steeper). It would be a great advantage to know the shapes and positions of TCP (and NTCP) curves for individual patients, as then doses could be appropriately tailored. Predictive assays of radiosensitivity, oxygen status, and proliferative potential are the "holy grail" of radiobiology (Hall 1994).

14.6.7 Use of TCP and NTCP in Treatment Planning

As discussed above, models with which to calculate TCP and NTCP have been developed based on clinical data, in vitro cell survival curves, and appropriate statistical theory (Brahme 1984, Withers 1985, Nahum and Tait 1992, Webb and Nahum 1993). It is known that for a given mean dose throughout the tumor volume, where the density of clonogenic cells is constant, a uniform dose gives the highest TCP for a given total energy deposited in the tumor (Brahme 1984). Furthermore, Ebert and Hoban (1996) have shown that maximum TCP is obtained by keeping the localized TCP constant throughout the tumor, where clonogen density varies arbitrarily.

Deciding on the best beam configuration requires consideration of both TCP and NTCP, where the goal is to either (a) maximize TCP while keeping NTCP to an acceptable limit or (b) minimize NTCP while keeping TCP acceptably high. For example, a significantly higher TCP may be obtained at the cost of only a slight increase in NTCP (by a boost treatment to a small region of the tumor for example), or a considerable drop in NTCP may be possible for only a small drop in TCP (for example, allowing a more nonuniform dose distribution across the tumor). In both of these cases, the overall benefit to the patient may be increased, but this cannot be revealed or quantified by visual inspection of a dose distribution. An example of this lies in the design of a plan for prostate treatment, where the NTCP of the rectum, bladder, and femoral heads was considered (Mohan et al. 1992). It was found that by increasing the dose to as much of the target volume as possible, with the constraint that the NTCP must be kept to less than or equal to 5%, a TCP of 59% was obtained. In an alternative plan, concentrating simply on dose uniformity across the tumor and low dose to surrounding organs, where the allowable limit of 5% NTCP was not taken advantage of, a TCP of 47% (a 12% drop) with an NTCP of 1% (a 4% drop) was obtained. The former plan is clearly better, given that 5% NTCP was considered acceptable in this case, but this conclusion could not have been drawn without the aid of a TCP/NTCP calculation.

As a simple example of a TCP/NTCP calculation, consider the treatment of a squamous cell carcinoma (SCC) of the digestive tract, where a portion of the esophagus is irradiated. The volume of the tumor is 300 cm^3, clonogenic cell density is 10^7 cm^{-3}, the mean radiation sensitivity (α) to small dose fractions is $\alpha_0 = 0.31$ Gy^{-1}, and the interpatient variation in sensitivity is $\sigma_\alpha = 0.07$ Gy^{-1} (Webb 1994). This gives the "population averaged" TCP curve shown in figure 14.13, which is representative of clinical response over a large number of patients. Now the esophagus is considered to be uniformly irradiated over a fractional volume of 0.3 and unirradiated elsewhere. From clinical data, $D_{50/5}$ (the dose giving 50% likelihood of complications after five years when the whole organ is irradiated) for this organ is 68 Gy. The NTCP curve in figure 14.13 is obtained for this situation. From these curves, the dose corresponding to a 50% TCP is 60.2 Gy, at which dose the NTCP is 4%. Whereas this may be considered an acceptable balance, the NTCP increases to 8% for only a 5% increase in dose (the TCP increases to 58%). A 5% decrease in dose reduces the TCP to 41% and the NTCP to 2%.

These figures indicate the sensitivity of biological outcome to dose and hence the requirement for accurate dose delivery. Furthermore, the *uncomplicated tumor control probability* (UTCP) is also plotted in figure 14.13, and this is seen to have a narrow peak about which its value falls off rapidly.

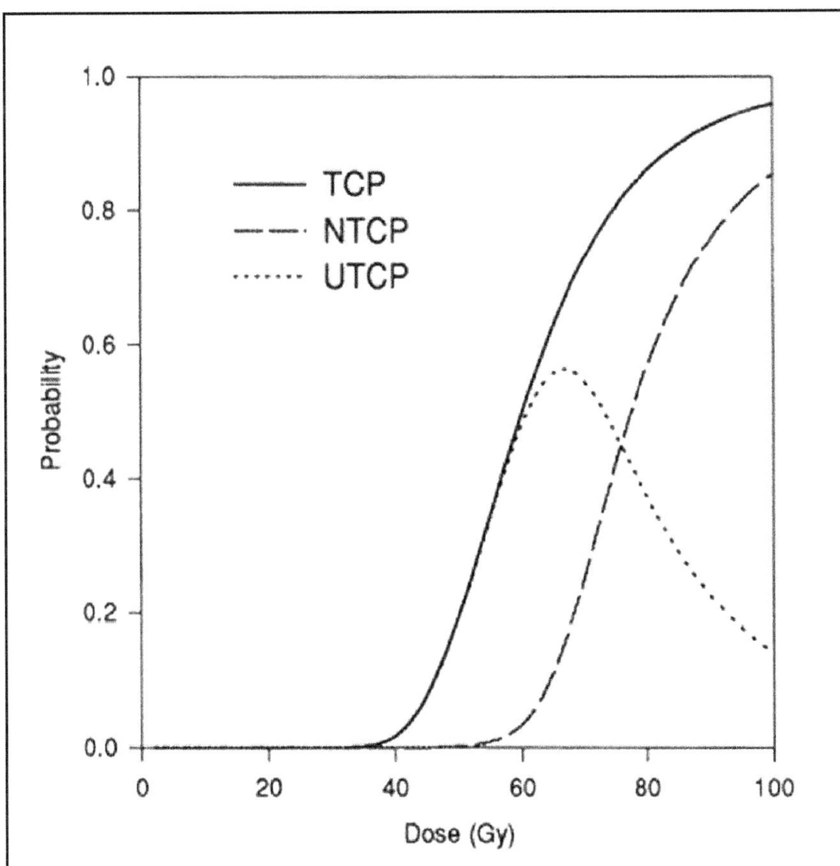

Figure 14.13: Tumor control probability (TCP) calculated for treatment of squamous cell carcinoma of the upper respiratory tract and normal tissue complication probability (NTCP) for partial irradiation of the esophagus. Also shown is the probability of uncomplicated tumor control (UTCP).

UTCP is equal to TCP × (1–NTCP) and is the probability of tumor control while having no complications. It should be noted that the curves in figure 14.13 are for population-averaged radiosensitivity data, which yields less steep curves than would be the case for an individual patient. Thus, when it becomes possible to routinely characterize the sensitivity of individual tumors, the dose tolerances to adhere to will be even tighter. The strong dependence of tumor response on dose is supported by calculations of Brahme (1984), who states that for common dose response characteristics, a variation in dose of 10% leads to a variation in TCP of 30%–50% when on the steep portion of the response curve.

Therapeutic Ratio

The curves relating tumor control probability and normal tissue complication probability when plotted against dose should give curves of similar shape due to the similar statistical basis for tumor and normal tissue damage. As outlined in chapter three, the optimum favorable therapeutic ratio occurs when the NTCP curve is maximally displaced to the right of the TCP curve. The ratio of doses at which the NTCP and TCP curves, respectively, are equal is one definition of the therapeutic ratio. In figure 14.13, the therapeutic ratio is approximately 1.4. The position and gradient of the dose response curves depend on the radiosensitivity of the cell populations, the size of the tumor/organ and, in the case of tumors, on the density of clonogen cells. Higher sensitivity brings the curves to the left, as does a smaller volume or clonogen density for tumors. The volume dependence of normal structures is more complex due to the different organ architectures, as outlined earlier. A higher sensitivity also steepens the curves, meaning that the difference in dose between zero and 100% TCP/NTCP is reduced. There is a greater risk of recurrence with tumors that have steeper dose-response curves, since it is easier to give a dose that does not give a high enough TCP, especially if the clinician is trying to minimize the dose in order to avoid complications (though a steeper TCP curve is also likely to be further to the left and thus have more separation from the NTCP curve).

In radiotherapy the total dose is usually escalated to a tolerance dose (TD) that has a 5% probability of a severe complication (but not lethal) within a five-year period when taken over the patient population. The tolerance dose is an attempt to express the maximum dose acceptable to the clinician. This requires the assignment of an arbitrary but useful percentage of complications. A 5% severe complication rate up to five years after therapy could be assigned as the minimum level, and a 50% severe complication rate as the maximum level. In favorable circumstances the radiation oncologist would not be willing to exceed the minimum rate, and in unfavorable conditions he or she would be unwilling to exceed the maximum. The minimum tolerance dose is denoted as the $TD_{5/5}$. Nominal TD levels for different organs are presented in table 14.2.

Table 14.2: Suggested radiation tolerance doses for human organs and structures

Organ (Reference)	*Injury at Five Years*	*(TD5/5)*	*(TD50/5)*	*Volume or Length*
Skin*	ulcer, severe fibrosis	55	70	100 cm^2
Skin**	teleangectasia	50	65	100 cm^2
Oral Mucosa*	ulcer, severe fibrosis	60	75	50 cm^3
Testes*	permanent sterilization	5–15	20	whole
Ovary*	permanent sterilization	2–3	6–12	whole
Fetus*	death	2	4	whole
Kidney**	clinical nephritis	23	28	whole
		30	40	2/3 rds
Rectum**	ulcer, stricture	55	80	100 cm^3
Bladder**	ulcer, contracture	65	80	whole
Liver**	liver failure,	30	40	whole
	ascites	35	45	2/3 rds
CNS-brain*	necrosis	50	>60	whole
Lung*	pneumonitis, fibrosis	40	60	lobe
Lung**		17	24	whole
		30	40	2/3 rds
Heart**	pericarditis,	40	50	whole
	pancarditis	45	55	2/3 rds
Eye lens**	cateract requiring intervention	10	18	whole
Eye retina**	blindness	45	65	whole
Ear**	acute serous otitis	30	40	mid/external
Spinal cord**	myelitis necrosis	50	70	20 cm length

* Data from Dalrymple et al. 1973.
** Data from Emami et al. 1991.

14.7 Equivalent Uniform Dose

Because most dose distributions across a target volume are inhomogeneous, a parameter that gives equivalent uniform dose (EUD) has been developed (Niemierko 1997). This is a hybrid model which includes dose and an empirical biological parameter, loosely based on the seriality or parallel nature of the tumor or critical structure, to estimate the equivalent dose if given uniformly.

The EUD formula is usually written as:

$$EUD = \left(\frac{1}{N} \sum_{i=1}^{N} D_i^a \right)^{\frac{1}{a}}, \tag{14.55}$$

where N is the number of voxels in the structure, D_i is the dose in the ith voxel, and a is the tumor or normal tissue parameter that characterizes the dose-volume effect (Wu et al. 2005). The EUD equation can be cast in slightly different forms (Moiseenko et al. 2005), but in general, as a approaches infinity, EUD approaches the maximum D; as a approaches negative infinity, EUD approaches the minimum of D; as a approaches 1, the EUD approaches the mean of D. Of course the dose-volume histogram (DVH) established depends on the a parameter chosen for the tissue. Negative a values are usually (but not always) chosen for tumor targets and positive values for organs at risk with the value of a increasing slightly for serial organs. The way EUD is affected by dose and the value of a is shown in figure 14.14. Figure 14.14a shows an inhomogeneous dose given over five voxels, and the graph in 14.14b shows how EUD varies with a for this simple scenario.

For IMRT planning a good cost function is important even if the models the cost functions are based on are not perfect; they do allow for good plan trial production and comparison. The EUD formalism can be easily differentiated, which makes it a good candidate as an inverse planning objective function tool for IMRT.

$$F = \prod_j f_j, \tag{14.56}$$

where for tumors f_j is given by

$$f_j = \frac{1}{1 + \left(\frac{EUD_0}{EUD} \right)^n}, \tag{14.57}$$

and for organs at risk

$$f_j = \frac{1}{1 + \left(\frac{EUD}{EUD_0} \right)^n}, \tag{14.58}$$

where n is the weight which indicates the structure-dependent endpoint (Moiseenko et al. 2005).

By setting an EUD objective, usually only one EUD objective is required. This avoids having to define several dose objectives; for example, with the comparative five-field IMRT prostate plan first presented in chapter six (figure 6.21). However, this time three plan trials have been undertaken; they were:

i) Conformal forward planned dose to rectum.
ii) IMRT with four rectal dose objectives for rectum.
iii) IMRT with one EUD objective for rectum.

Figure 14.15 shows the lowest DVH to rectum using a single maximum EUD objective to 50 Gy, with a positive a value. A homogeneous target dose using EUD is difficult because excess high doses are not penalized (Jones

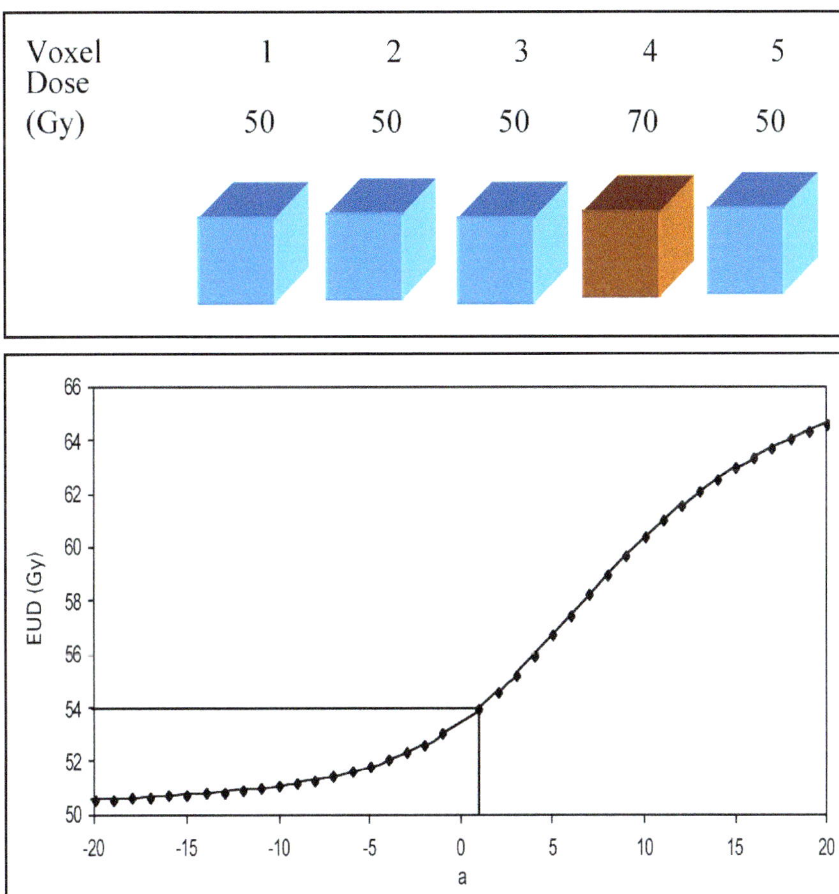

Figure 14.14: The dependence of EUD on dose and a. (a) A simple five-voxel inhomogeneous dose distribution. (b) The way the EUD varies for this distribution depending on the value of a selected. EUD is less than mean dose when a is less than one and more when a is greater than one.

1999). However, for this trial a homogeneous target dose has also been met using two EUD objectives using the technique suggested by Wu et al. (2002). The trick is to set a minimum EUD with negative value *a* and maximum EUD with positive value *a*.

As also shown in figure 14.15, all three planned DVHs were well below trial dose thresholds (e.g., Pollack et al. 2002). All three plan trials also had an efficiently low monitor unit/dose ratio (nominally 400 MU/200 cGy); hence the time overheads between conformal and IMRT dose delivery are converging. However, there are other issues to consider and one of these may be the different temporal distribution between conformal and IMRT dose delivery.

14.8 Clinical Trials Definitions

Radiobiological models are just that, models. These predictive models need to be backed up by clinical trial data.

A clinical trial (also clinical research) is a research study in human volunteers to answer specific health questions. Carefully conducted clinical trials are regarded as the fastest and safest way to find treatments that work in people and ways to improve health. Interventional trials determine whether experimental treatments or new ways of using known therapies are safe and effective under controlled environments (see U.S. National Institutes of Health (NIH) web site: www.nih.gov).

14.8.1 Different Phase Trials

Clinical trials are conducted in phases. The trials at each phase have a different purpose and help scientists answer different questions. *Phase I trials* test

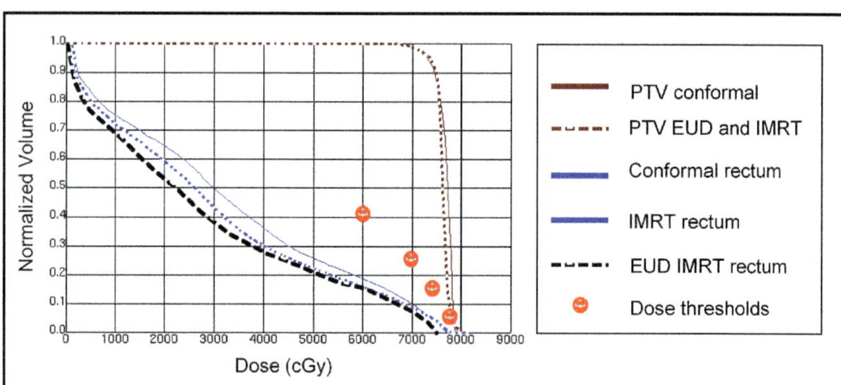

Figure 14.15: DVH for rectum using single maximum EUD objective of 50 Gy (with $a = 6$). Target volume objectives were a minimum EUD of 72 Gy (with $a = 10$) and maximum EUD of 76 Gy (with $a = -10$). The DVH is compared with standard IMRT dose plan objectives whereby four dose objectives needed to be specified for the rectum. Note the IMRT and conformal DVH plots are the same as those presented in figure 6.21 (SEE COLOR PLATE 23). The dose threshold points have also been included from the Pollack et al. (2002) trial data.

experimental treatment in a small group of people (20 to 80) for the first time to evaluate its safety, to determine a safe dosage range, and to identify side effects. In *phase II trials* experimental treatment is given to a larger group of people (100 to 300) to see if it is effective and to further evaluate its safety. In *phase III trials* experimental treatment is given to large groups of people (1000 to 3000) to confirm its effectiveness, to monitor side effects, and to compare it to commonly used treatments.

Phase III trials compare one treatment with another, to find out which is better. Most phase III trials are randomized. This means that there are at least two different groups in the trial, and those taking part are put into one or other group at random. Typically, one group will have the new treatment being tested and the other the standard treatment that they would have to treat their cancer if not in the trial. Those having the standard treatment are called the control group. A randomized trial that has a control group is called a randomized controlled trial (see U.K. Cancer Help web site: www.cancerhelp.org.uk).

14.8.2 Different Levels of Evidence

The credence given to a trial can be grouped into various levels of evidence. *Level I evidence* is arranged from a systematic review (meta analysis) of all randomized controlled trials; there may be three or more phase III clinical trials. *Level II evidence* is obtained from at least one properly designed randomized phase III control trial. *Level III-1 evidence* is obtained from a well-designed pseudo-randomized controlled trial (alternate allocation or some other method). *Level III-2 evidence* is obtained from comparative studies not randomized (cohort studies). *Level IV evidence* is obtained from case series.

14.9 Current Issues in Radiobiology

It is beyond the scope of the present book to cover radiobiology in depth. There are many general textbooks that cover this (Hall and Giaccia 2005, Steel 2002, Wigg 2001). However, many radiobiological issues are currently subject of intense debate and investigation. Several of them have also direct implications for the use of the technology covered in the present book. Four of them are mentioned here as they feature prominently in the literature. The reader is referred to the references provided for more information.

14.9.1 Low α/β Ratio

Recent interest in the α/β ratio has focused on prostate tumors. Clinical outcome comparisons between external beam and brachytherapy treatments have revealed α/β for prostate to be about 1.5, which is much lower than previously expected (Duchesne and Peters 1999, Brenner and Hall 1999, Fowler et al. 2001).

There is some counter argument that hypoxia may be masking the effect and that hypofractionation may lead to less reoxygenation of tumors. An excellent point/counterpoint article on the issue has been debated in the literature

(Fowler, Nahum, and Orton 2006), but overall the low α/β evidence has been sufficiently compelling to lead to several clinical trials based on the hypofractionation of prostate tumors. The hypothetical gains from hypofractionation were explored by Fowler et al. (2003). Clinical trails for external beam treatment of prostate with early reported toxicity outcomes for both arms include those of Pollack et al. (2006) where 70.2 Gy in 26 fractions was tested against 76 Gy in 38 fractions for the control arm. There are several other similar trials currently being reported in the literature.

Dose fractionation is the realm of the radiation oncologist and is increasingly becoming the realm of the radiobiologist and physicist (in implementing models as outlined in this chapter). Improved understanding of dose fractionation and the development of models to quantify its effects have led to the investigation of hyperfractionation (delivering dose in smaller fractions separated by smaller time intervals) for proliferative tumors such as head and neck and lung.

The low α/β ratio for prostate tumors has turned previous assumptions about optimum fractionation regimes for prostate on their head. This has led to introduction of hypofractionation trials for prostate radiotherapy treatment. The bonus here, if clinical trial evidence is shown to support this, is less fractions for patients and more patients treated on each linac per year.

Radiobiological models have so far offered guideline predictions if fractionation and treatment duration are altered. These calculations at present are predictors that are applied to whole populations. There may, however, be variations in tumor and normal tissue sensitivity from patient to patient.

14.9.2 Low Radiation Doses and Secondary Cancer Induction

Modern cancer therapy has had many successes resulting in a significant number of long-term survivors. In these patients the risk of late effects to normal tissues must be carefully evaluated, as discussed in section 14.5. However, there is also an increasing awareness of the potential risk of low radiation doses to large volumes of the body (Nelson 1997). Clinical evidence for secondary cancer induction due to radiotherapy is available for many cancers including prostate (Moon et al. 2006) and particularly lymphomas (Hoppe 1997, Tward et al. 2006). Models for secondary cancer induction have been developed and must take into consideration the balance between the probability of cell mutation and cell kill (Dasu et al. 2005, Suit et al. 2007).

This requires a paradigm shift in terms of dose calculation. At present, dose calculation algorithms are commissioned to predict the dose in the target and closely located critical structures accurately. However, there is an increasing need to also predict the dose in organs distant to the tumor with reasonable accuracy. Of particular concern are IMRT treatments where there is a risk that dose is spread somewhat arbitrarily throughout the body in order to improve the dose in the target and selectively spare identified critical structures (conformal avoidance). In addition to this, there may be increased leakage due to the longer beam-on times. Consequently, it has been estimated that the risk of radiation-induced malignancies may be significantly increased in these treatments (Hall 2006a,b).

14.9.3 Inhomogeneous Temporal Dose Distribution

With conformal radiotherapy and fixed wedges the temporal dose rate within each field was relatively constant and similar from voxel to voxel, with of course a varying time gap between fields as the gantry rotated to a new position. With step and shoot IMRT in particular there is potential for dose to be delivered in discrete time stamps within each treatment field. This is because segments open, too, and shield the dose accrual point.

While the on-off effect will be less with sliding window IMRT delivery and even less with dynamic wedge, the dose rate will certainly vary during these delivery techniques. There has been some modeling work to try to optimize the delivery using the LQ model with a time parameter accounting for the on-off nature of dose within a field and the time gap between dose delivery (Altman et al. 2006), such that

$$S = \exp\left(-\alpha D + \beta G(t) D^2\right), \tag{14.59}$$

where $G(t)$ is a dose protraction factor. Then creating a metric $G(t_{eff})$ depending on this and the linear quadratic model, they calculate

$$BED = nD\left(1 + \frac{G_{eff}(t) D}{(\alpha/\beta)}\right). \tag{14.60}$$

This model predicts fields with increasing beam weight toward the middle of the time cycle; then decreasing beam weight in a triangular pattern with decreasing dose at each end is optimal for cell kill. Caveat: this is early modeling work only. There have also been in vitro cell studies showing more cell survival for IMRT multi-field delivery versus conformal two-field dose delivery (i.e., typical scenario for head and neck treatments). However, this article also reiterates there is no clinical trial evidence for this (Moiseenko et al. 2007).

14.9.4 Molecular Biology and Genetic Drugs

In the last 20 years cancer treatment has developed to a multidisciplinary approach. Radiotherapy and/or surgery as localized treatment modalities are often combined with chemotherapy to reduce the risk of distant failure. Unfortunately, combination therapy is usually more toxic despite better survival (e.g., DeAngelis et al. 2002, Forastiere et al. 2003). Therefore, the development of molecular agents which target specific features of tumor cells without affecting normal tissue is of great interest for radiotherapy. A good example is inhibitors of epidermal growth factor receptors (EGFR) such as Cetuximab. In a recent study by Bonner et al. (2006) it was shown that mortality in locoregionally advanced head and neck cancer can be improved by addition of Cetuximab to radiotherapy without increasing toxicity. It is interesting to note that a treatment such as this would allow radiotherapy to be optimized virtually independently of the drug. However, even more important may be the fact that the drug is targeting one of the possible pathways that tumor cells may activate in reaction to radiation.

As the name indicates, Cetuximab is a monoclonal antibody (MAB). There are a large number of other drugs available based on this concept. Of particular interest is trastuzumab (trade name Herceptin®), which specifically targets human epidermal growth-factor receptors 2 (HER2). This gene is overexpressed in many cancers and some 25% of breast cancer cases express this. The drug specifically targets this receptor and has been shown to improve disease-free survival for patients with advanced breast cancer (Piccart-Gebhart et al. 2005). The interesting part is that this type of adjuvant therapy can be targeted better to individual patients. It will be one of the challenges of the future to ensure that radiation therapy can work optimally with these individualized schedules.

We leave the final word to Bentzen (2005) "*Theragnostic imaging* for radiation oncology is the use of molecular and functional imaging to prescribe radiation in the four dimensions of space and time. *Theragnostics* is derived from the Greek *theropea* (to treat medically) and *gnosis* (knowledge)." The translation of imaging information into meaningful tumor response information is in its infancy but is progressing (Ballegeer et al. 2006). One day we may be able to use biological information derived from imaging and knowledge of molecular biology in prescribing therapy for the individual patient (Bentzen 2005).

References

Altman M.B., Chmura S.J., Deasy J.O., Roeske J.C. (2006) Optimization of the temporal pattern of radiation: An IMRT based study. *Int. J. Radiat. Oncol. Biol. Phys.* 66: 898–905.

Ang K.K., Landuyt W., Xu F.X., Vanuytsel L., van der Schueren E. (1987) The effect of small radiation doses per fraction on mouse lip mucosa assessed using the concept of partial tolerance. *Radiother. Oncol.* 8: 79–86.

Ballegeer A., Forrest L.J., Jeraj R., Mackie T.R., Nickles R.J. (2006) PET/CT following intensity-modulated radiation therapy for primary lung tumor in a dog. *Vet. Radiol. Ultrasound* 47: 228–233.

Barendsen G.W. (1982) Dose fractionation, dose rate, and iso-effect relationships for normal tissue responses. *Int. J. Radiat. Oncol. Biol. Phys.* 8: 1981–1987.

Barton M. (1995) Tables of equivalent dose in 2 Gy fractions: A simple application of the linear quadratic formula. *Int. J. Radiat. Oncol. Biol. Phys.* 31: 371–378.

Bentzen S.M. (2005) Theragnostic imaging for radiation oncology: Dose-painting by numbers. (Abstract). *Lancet Oncol.* 6: 112–117.

Bonner J.A., Harari P.M., Giralt J., Azarnia N., Shin D.M., Cohen R.B., Jones C.U., et al. (2006) Radiotherapy plus cetuximab for squamous cell carcinoma of the head and neck. *N. Engl. J. Med.* 354: 567–578.

Brahme A. (1984) Dosimetric precision requirements in radiation therapy. *Acta Radiol. Oncol.* 23: 379-391.

Brenner D.J. (1993) Dose, volume and tumor control predictions in radiotherapy. *Int. J. Radiat. Oncol. Biol. Phys.* 26: 171–179.

Brenner D.J., Hall E.J. (1999) Fractionation and protraction for radiotherapy of prostate cancer. *Int. J. Radiat Oncol. Biol. Phys.* 43: 1095–1101.

Bryant P.E. (1989) Mechanism of repair of DNA damage induced by ionizing radiation. *The Biological Basis of Radiotherapy*, Second edition. Elsevier Publishers, Amsterdam, pp. 15–28.

Burman C., Kutcher G.J., Emami B., Goitein M. (1991) Fitting of normal tissue tolerance data to an analytical function. *Int. J. Radiat. Oncol. Biol. Phys.* 21: 123–135.

Curtis S.B. (1986) Lethal and potentially lethal lesions induced by radiation—a unified repair model. *Radiat. Res.* 106: 252–270.

Dale R.G. (1985) The application of the linear-quadratic dose-effect equation to fractionated and protracted radiotherapy. *Br. J. Radiol.* 58: 515–528.

Dalrymple G.V., Gaulden M.E., Kollmorgen G.M., Vogel H.H. (1973) *Medical Radiation Biology*. W.B. Saunders Company, Philadelphia, Pennsylvania.

Dasu A., Toma-Dasu I., Olofsson J., Karlsson M. (2005) The use of risk estimation models for the induction of secondary cancers following radiotherapy. *Acta Oncol.* 44: 339–347.

DeAngelis L.M., Seiferheld W., Schold S.C., Fisher B., Schultz C.J., RTOG Study Group 93-10 (2002) Combination chemotherapy and radiotherapy for primary central nervous system lymphoma: Radiation Therapy Oncology Group Study 93-10. *J. Clin. Oncol.* 20: 4643–4648.

Douglas B.G., Fowler J.F. (1976) The effect of multiple small doses of x-rays on skin reactions in the mouse and a basic interpretation. *Radiat. Res.* 66: 401–426.

Down J.D., Easton D.F., Steel G.G. (1986) Repair in mouse lung during low dose-rate irradiation. *Radiother. Oncol.* 6: 29–42.

Duchesne G.M, Peters L.J. (1999) What is the α/β ratio for prostate cancer? Rationale for hypofractionated high-dose-rate brachytherapy. *Int. J. Radiat. Oncol. Biol. Phys.* 44: 747–748.

Ebert M.A., Hoban P.W. (1996) Some characteristics of tumour control probability for heterogeneous tumours. *Phys. Med. Biol.* 41: 2125–2133.

Ellis F. (1969) Dose, time and fractionation. A clinical hypothesis. *Clin. Radiol.* 20: 1–7.

Ellis F. (1967) *Modern Trends in Radiotherapy*, Vol. 1. T.J. Deeley and C.A.P. Wood (eds.). Butterworth, London, pp. 34–51.

Emami B., Lyman J., Brown A., Coia L., Goitein M., Munzenrider J.E., Shank B., Solin L.J., Wesson M. (1991) Tolerance of normal tissue to therapeutic irradiation. *Int. J. Radiat. Oncol. Biol. Phys.* 21: 109–123.

Forastiere A.A., Goepfert H., Maor M., Pajak T., Weber R., Morrison W., Glisson B., et al. (2003) Concurrent chemotherapy and radiotherapy for organ preservation in advanced laryngeal cancer. *N. Engl. J. Med.* 349: 2091–2098.

Fowler J.F. (1993) The volume effect in radiotherapy. *J. Jpn. Soc. Ther. Radiol. Oncol.* 5: 75–87.

Fowler J.F. (1989) The linear-quadratic formula and progress in fractionated radiotherapy. *Br. J. Radiol.* 62: 679–694.

Fowler J.F., Nahum A.E., Orton C.G. (2006) The best radiotherapy for the treatment of prostate cancer involves hypofractionation. Point/Counterpoint. *Med. Phys.* 33: 3081–3084.

Fowler J.F., Ritter M.A., Chappell R.J., Brenner D.J. (2003) What hypofractionation protocols should be tested for prostate cancer? *Int. J. Radiat. Oncol. Biol. Phys.* 56: 1093–1104.

Fowler J.F., Chappell R.J., Ritter M.A. (2001) Is α/β for prostate tumors really low? *Int. J. Radiat. Oncol. Biol. Phys.* 50: 1021–1031.

Gelber R.D., Goldhirsch A., Cole B.F., Wieand H.S., Schroeder G., Krook J.E. (1996) A quality-adjusted time without symptoms or toxicity (Q-TWiST) analysis of adjuvant radiation therapy and chemotherapy for resectable rectal cancer. *J. Natl. Cancer Inst.* 88: 1039–1045.

Hall E.J. (2006a) The inaugural Frank Ellis Lecture: Latrogenic cancer the impact of intensity-modulated radiotherapy. *Clin. Oncol.* 18: 277–282.

Hall E.J. (2006b) Intensity-modulated radiation therapy, protons, and the risk of second cancers. *Int. J. Radiat. Oncol. Biol. Phys.* 65: 1–7.

Hall E.J. (1994) Predictive assays of radiosensitivity, oxygen status and proliferative potential. (abstract). In *Proceedings of 21st Scientific Meeting of the Clinical Oncology Society of Australia.*

Hall E.J. (1988) *Radiobiology for the Radiologist*, Third edition, J.B. Lippincott, Philadelphia, Pennsylvania.

Hall E.J. (1983) *Radiobiology for the Radiologist*, Second edition, J.B. Lippincott, Philadelphia, Pennsylvania.

Hall E.J., Giaccia A.J. (2005) *Radiobiology for the Radiologist*, Sixth edition. Lippincott Williams & Wilkins, Baltimore, Maryland.

Hoppe R.T. (1997) Hodgkin's disease: Complications of therapy and excess mortality. *Ann. Oncol.* 8 *Suppl* 1: 115–118.

Jackson A., Kutcher G.J., Yorke E.D. (1993) Probability of radiation-induced complications for normal tissues with parallel architecture subject to non-uniform irradiation. *Med. Phys.* 20: 613–625.

Jones L. (1999) Radiotherapy optimization: Biological and physical concepts. Ph.D. Thesis. University of New South Wales. Sydney. Australia.

Jones L. (1996) The physics of dose fractionation. B.Sc. (honors) Thesis, University of Wollongong, New South Wales, Australia.

Kutcher G.J., Burman C. (1989) Calculation of complication probability factors for non-uniform normal tissue irradiation: The effective volume method. *Int. J. Radiat. Oncol. Biol. Phys.* 16: 1623–1630.

Lebesque J.V., Keus R.B. (1991) The simultaneous boost technique: The concept of relative normalized total dose. *Radiother. Oncol.* 22: 45–55.

Liversage W.E. (1971) A critical look at the ret. *Br. J. Radiol.* 44: 91–100.

Lyman J. (1985) Complication probability as assessed from dose-volume histograms. *Radiat. Res. Suppl.* 104: S13–19.

Mohan R., Mageras G.S., Baldwin B., Brewster L.J., Kutcher G.J., Leibel S., Burman C.M., Ling C.C., Fuks Z. (1992) Clinically relevant optimization of 3-D conformal treatments. *Med. Phys.* 19: 933–944.

Moiseenko V., Deasy J., Van Dyk J. (2005) Radiobiological modeling for treatment planning. Chapter 5 in *The Modern Technology of Radiation Oncology,* Volume 2. J. Van Dyk (ed.). Medical Physics Publishing, Madison, Wisconsin, pp. 185–220.

Moiseenko V., Duzenli C., Durand R. (2007) In vitro study of cell survival following dynamic MLC intensity-modulated radiation therapy dose delivery. *Med. Phys.* 34: 1514–1520.

Moon K., Stukenborg G.J., Keim J., Theodorescu D. (2006) Cancer incidence after localized therapy for prostate cancer. *Cancer* 107: 991–998.

Nahum A.E., Tait D.M. (1992) Maximizing control by customized dose prescription for pelvic tumors. *Advanced Radiation Therapy: Tumor Response Monitoring and Treatment Planning*. A. Breit. (ed.). Springer, Heidelberg.

National Radiological Protection Board (1989) *Handbook #13, Living with Radiation*, Fourth edition, London.

Nelson N.J. (1997) Solutions bring new problems: Secondary cancers are a risk for some survivors. *J. Natl. Cancer Inst.* 89: 1483–1485.

Niemierko A. (1997) Reporting and analyzing dose distributions: A concept of equivalent uniform dose. *Med. Phys.* 24: 103–110.

Niemierko A., Goitein M. (1993) Modeling of normal tissue response to radiation: The critical volume model. *Int. J. Radiat. Oncol. Biol. Phys.* 25: 135–145.

Niemierko A., Goitein M. (1991) Calculation of normal tissue complication probability and dose-volume histogram reduction schemes for tissue with a critical element architecture. *Radiother. Oncol.* 20: 166–176.

Orton C.G. (1988) A unified approach to dose-effect relationship in radiotherapy. II: Homogeneous dose distributions. *Int. J. Radiat. Oncol. Biol. Phys.* 14: 557–560.

Orton C.G., Ellis F. (1973) A simplification in the use of the NSD concept in practical radiotherapy. *Br. J. Radiol.* 46: 529–537.

Overgaard J., Hjelm-Hansen M., Johansen L.V., Anderson A.P. (1988) Comparison of conventional and split-course radiotherapy as primary treatment in carcinoma of the larynx. *Acta Oncol.* 27: 147–152.

Piccart-Gebhart M., Procter M., Leyland-Jones B., Goldhirsch A., Untch M., Smith I., Gianni L. et al. (2005) Trastuzumab after adjuvant chemotherapy in HER2-positive breast cancer. *N. Engl. J. Med.* 353: 1659–1672.

Pollack A., Hanlon A.L., Horwitz E.M., Feigenberg S.J., Konski A.A., Movsas B., Greenberg R.E. et al. (2006) Dosimetry and preliminary acute toxicity in the first 100 men treated for prostate cancer on a randomized hypofractionated dose escalation trial. *Int J. Radiat. Oncol. Biol. Phys.* 18:115–120.

Pollack A., Zagars G.K., Starkshall G., Antolak J.A., Lee J.J., Huang E.E., von Eschenbach A.G., Kuban D.A, Rosen I. (2002) Prostate cancer radiation dose

response: Results of the M.D. Anderson phase III randomized trial. *Int. J. Radiat. Oncol. Biol. Phys.* 53: 1097–1105.
Saunders M.I. (1991) *Br. J. Radiol.* 64 (Congress suppl.): 43.
Saunders M.I., Dische S., Fowler J.F., Denekamp J., Dunphy E.P., Grosch E., Fermont D., Ashford R., Maher J., Des Rochers C. (1988) Radiotherapy employing three fractions on each of twelve consecutive days. *Acta Oncol.* 27: 163–167.
Schultheiss T.E., Orton C.G., Peck R.A. (1983) Models in radiotherapy: Volume effects. *Med. Phys.* 10: 410–415.
Selman J. (1983) *Elements of Radiobiology*, First edition. Charles C Thomas Publisher, Springfield, Illinois, pp. 140–150.
Sephton R.G. (1991) Current concepts in radiobiology applied to radiotherapy. Australas. *Phys. Eng. Sci. Med.* 14: 130–140.
Steel G.G. (2002) *Basic Radiobiology*, 3rd edition. Hodder Arnold, London.
Steel G.G., McMillan T.J., Peacock J.H. (1989) The 5Rs of radiobiology. *Int. J. Radiat. Biol.* 56: 1045–1048.
Stupp R., Mason W., van den Bent M., Weller M., Fisher B., Taphoom M.J., Belanger K, et al. (2005) Radiotherapy plus concomitant and adjuvant temozolomide for glioblastoma. *N. Engl. J. Med.* 352: 987–996.
Suit H., Goldberg S., Niemierko A., Ancukiewicz M., Hall E., Goitein M., Wong W., Paganetti H. (2007) Secondary carcinogenesis in patients treated with radiation: A review of data on radiation-induced cancers in human, non-human primate, canine and rodent subjects. *Radiat. Res.* 167: 12–42.
Thames H.D. (1985) An incomplete repair model for survival after fractionated and continuous radiation. *Int. J. Radiat. Oncol. Biol. Phys.* 54: 319–339.
Thames H.D., Hendry J.H. (1987) *Fractionation in Radiotherapy*. Taylor and Francis, London.
Travis E.L., Thames H.D., Watkins T.L., Kiss I. (1987) The kinetics of repair in mouse lung after fractionated irradiation. *Int. J. Radiat. Biol. Relat. Stud. Phys. Chem. Med.* 52: 903–919.
Trott K.R., Kummermehr J. (1985) What is known about tumor proliferation rates to choose between accelerated fractionation or hyperfractionation? *Radiother. Oncol.* 3: 1–9.
Tward J.D., Wendland M.M., Shrieve D.C., Szabo A., Gaffney D.K. (2006) The risk of secondary malignancies over 30 years after the treatment of non-Hodgkin lymphoma. *Cancer* 107: 108–115.
Vegesna V., Withers H.R., Thames H.D. Jr., Mason K. (1985) Multifraction radiation response of mouse lung. *Int. J. Radiat. Biol. Relat. Stud. Phys. Chem. Med.* 52: 459–468.
Webb S. (1994) Optimum parameters in a model for tumour control probability including interpatient heterogeneity. *Phys. Med. Biol.* 39: 1895–1914.
Webb S., Nahum A.E. (1993) A model for calculating tumor control probability in radiotherapy including the effects of inhomogeneous distributions of dose and clonogenic cell density. *Phys. Med. Biol.* 38: 653–666.
Wigg D. (2001) *Applied Radiobiology and Bioeffect Planning*. Medical Physics Publishing, Madison, Wisconsin.
Williams J.R., Thwaites D.I. (1993) *Radiotherapy Physics in Practice.* Oxford University Press, Oxford, U.K.
Winston B.M. (1969) Correcting for radiotherapy treatment interruptions. *Clin. Radiol.* 20: 8–10.
Withers H.R. (1985) Biological basis for altered fractionation schemes. *Cancer* 55: 2086–2095.
Withers H.R. (1975) The four Rs of radiotherapy. *Adv. Radiat. Biol.* 5: 241–271.
Withers H.R., Taylor J.M., Maciejewski B. (1988) The hazard of accelerated clonogen repopulation during radiotherapy. *Acta Oncol.* 27: 131–146.
Withers H.R., Peters L.J. (1973) Basic principles of radiotherapy: Biological aspects of radiation therapy. *Textbook of Radiotherapy*, Third edition. G.M. Fletcher (ed.). Lea and Feabiger, Philadelphia, Pennsylvania, pp. 105–173.

Wu Q., Xing L., Ezzell G., Mohan R. (2005) Inverse treatment planning, Chapter 4 in *The Modern Technology of Radiation Oncology*, Volume 2. J. Van Dyk (ed.). Medical Physics Publishing, Madison, Wisconsin, pp. 1–30.

Wu Q., Mohan R., Niemierko A., Schmidt-Ullrich R. (2002) Optimization of intensity-modulated radiotherapy plans based on the equivalent uniform dose. *Int. J. Radiat. Oncol. Biol. Phys.* 51: 224–235.

Yaes R.J., Patel P., Maruyama Y. (1991) On using the linear-quadratic model in daily clinical practice. *Int. J. Radiat. Oncol. Biol. Phys.* 20: 1353–1363.

Questions

14.1 Calculate the standard effective dose SED (equation 14.16) for early/tumor effects; also late and spinal cord effects for 30 Gy delivered in 3 Gy/F.
(section 14.3)

14.2 Calculate the biological effective dose (BED) for early and late effects for a LDR brachytherapy treatment given at 0.5 Gy/hour over 150 hours to a dose of 75 Gy.
Assume $\alpha/\beta = 10$ for early effects and $\alpha/\beta = 3$ for late effects.
(section 14.4)

14.3 There are two fractionation regimes:

Regimen 1 is hyperfractionation of 2 fractions per day 1.15 Gy/ fraction in 70 fractions to 80.5 Gy.

Regimen 2 is a standard one fraction per day 2 Gy/fraction in 35 fractions to 70 Gy.

The duration of both regimens is 47 days (including weekend treatment breaks). Assume T_k is 14 days and T_p is 4 days.

i) Calculate and compare the fractional clonogen survival.
ii) Calculate the early/tumor and late effects using the BED calculation and explain what the results mean in terms of the favorability of each regimen.
iii) Once T is greater than T_k, what is the BED loss in tumor control dose (Gy_{10}) per day.
iv) Describe how clonogen repair time affects the implementation of HF in the clinical environment.
v) Name some tumor types that would be good candidates and bad candidates for this HF schedule and explain why.

(section 14.5)

14.4 Tumor control probability (TCP) can be defined as the probability of killing all clonogenic cells in the tumor. The cell survival probability for a given dose can be used to calculate the expected number of surviving clonogens for a given initial number of clonogens, and from this expected number, a Poisson probability distribution can be used to obtain the probability that the surviving number is zero.

i) Give the Poisson equation for the TCP in terms of initial number of clonogens, N_0, and the survival probability of each clonogen, S. From a mathematical perspective, why is a large tumor harder to kill?
ii) What does the survival probability need to be for a tumor containing 10^9 clonogens to have a TCP of 0.6? What must the expected number surviving be?
iii) If the survival probability of a clonogen is given by $S = e^{-\alpha D}$, show that the dose D required to give a certain TCP when the initial clonogen number is N_0 is given by

$$D = \frac{1}{\alpha}\left(\ln N_0 - \ln\left(\ln\left(1/TCP\right)\right)\right). \qquad \text{(Q14.1)}$$

Why does this make it clear that dose should not be reduced for small tumors?
iv) Show that according to the above equation, for $\alpha = 0.35\ \text{Gy}^{-1}$,the change in dose required to maintain the same TCP as the logarithm of initial clonogen number changes by 1 is 6.6 Gy. Note that $\ln N = 2.3\log_{10} N$.

(section 14.6)

14.5 A new hypofractionation regime (Pollack et al. 2002) uses 70.2 Gy in 26 fractions with the control arm being 76 Gy in 38 fractions. Calculate the SED advantage of the hypofractionation regimen if α/β =1.5 and disadvantage if $\alpha/\beta = 10$.
(section 14.8)

Answers to Chapter Questions

1: Medical Linear Accelerators

1.1 In contrast to a traveling wave accelerator the microwave power is not absorbed at the end of the waveguide; instead it is reflected back to create a standing wave.

1.2 An analogy for a load line is that if an electric device is rotating, when pressure (extra load) is put on the device, then more voltage needs to be applied to retain the same revolutions per minute. In an accelerator structure the guide becomes more loaded as the gun current is increased. Hence a higher voltage needs to be applied to the gun to

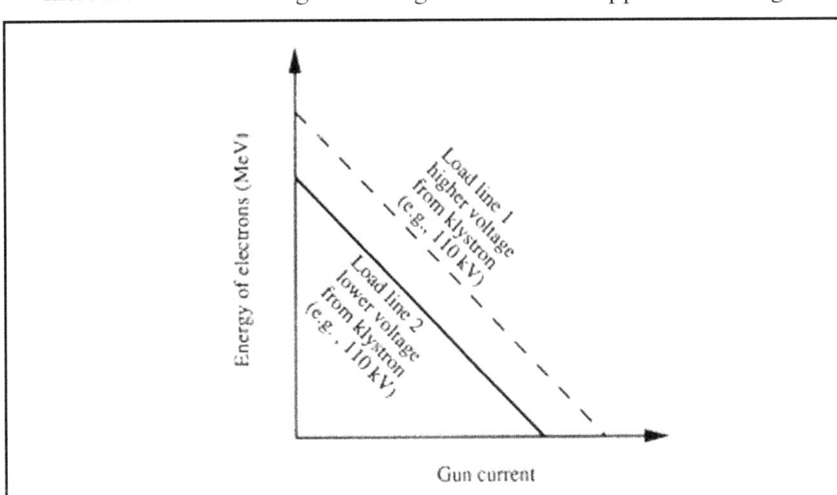

Figure A1.1: Schematic showing load lines of gun current versus beam energy.

maintain the same beam energy. The lines which describe the relationship between gun current and beam energy are referred to as load lines (see figure A1.1).

1.3 The force on an electron of charge q in a magnetic field is $F = qvB$, where v is the velocity of the electron and B is the magnetic field strength. As $F = (mv^2)/r$, where r is the radius of curvature, then $qvB = (mv^2)/r$ as the momentum is $p = mv$, then rearranging gives $B = p/(qr)$.

The momentum p can be calculated using equation 1.3

$$p = \left[E_k\left(\frac{E_k}{c^2} + 2m_0\right)\right]^{\frac{1}{2}} \tag{A1.1}$$

or by

$$p = mv \tag{A1.2}$$

then

$$\begin{aligned} m &= m_0 + \frac{E_k}{c^2} \\ &= 9.11\times10^{-31} + \frac{6\times(1.602\times10^{-13})}{(3\times10^8)^2} \\ &= 1.159\times10^{-29} \end{aligned} \tag{A1.3}$$

and using equation 1.5

$$\begin{aligned} v &= c\left(1 - \frac{m_0^2}{m^2}\right)^{\frac{1}{2}} \\ &= 3\times10^8\times\left[1 - \frac{(9.11\times10^{-31})^2}{(1.159\times10^{-29})^2}\right]^{\frac{1}{2}} \\ &= 2.99\times10^8\,\text{ms}^{-1}. \end{aligned} \tag{A1.4}$$

Therefore

$$\begin{aligned} B &= \frac{mv}{qr} \\ B &= \frac{(1.159\times10^{-29})\times(2.99\times10^8)}{(1.6\times10^{-19})\times(1.5\times10^{-1})} \\ B &= 0.1444\ \text{Tesla}. \end{aligned} \tag{A1.5}$$

1.4 Shielding blocks need to be positioned in the room by the radiation therapist, whereas the multileaf collimator (MLC) fields can be set up remotely from outside the treatment room; this reduces the treatment time between fields. Shielding blocks are heavy and may cause back injury to staff while being placed in position or, if dropped, may cause injury to patients. In addition, some shielding blocks are made from a low melting point alloy (LMA) that contains cadmium, a hazardous material.

1.5 Having remote programming with MLCs is somewhat negated if the radiation therapist still has to enter the room to locate and remove the physical wedges between fields. This is a particular problem if the MLC is being used to deliver many preprogrammed treatment portals, as is usually required for conformal treatments. In this situation then dynamic wedges that can also be programmed to deliver dose from the control console become a useful feature in combination with MLC. One disadvantage is that sometimes only one set of jaws has the motion range to produce dynamic wedge and this may not be in the required orientation for optimum MLC shielding. This is the case for the Varian linac when concave chest wall shielding is required for breast tangent fields. Creating a pseudo wedge compensator by using segmented MLC tangent fields is another alternative to shape the dose cloud in the chest wall region.

2: Interaction Properties of X-Rays and Electrons

2.1 The cross section is the area about the interaction center within which an incident particle must pass in order to interact. This area is thus the probability that a particle incident on a unit area will interact with a particular interaction center that is somewhere within that unit area.

2.2 The relationship between the linear attenuation coefficient μ and cross section σ can also be expressed as in equations 2.10 and 2.11, where N_A is Avogadro's number (6.022×10^{-23} mol^{-1}), ρ is the mass density (for oxygen $\rho = 1.332 \times 10^{-3}$ gcm^{-3}), and A is the mass number (for oxygen A = 16). The conversion from σ to μ using the equation 2.11 for oxygen is therefore

$$\mu = \frac{N_A \rho}{A}\sigma$$

$$\mu = \frac{(6.022 \times 10^{23}) \times (1.332 \times 10^{-3})}{16}\sigma \tag{A2.1}$$

$$\underline{\mu = \left(5.01 \times 10^{19}\right)\sigma \qquad \left[cm^{-1}\right]}$$

$$\frac{\mu}{\rho} = \frac{5.01 \times 10^{19}}{\left(1.332 \times 10^{-3}\right)}$$

$$\frac{\mu}{\rho} = 3.76 \times 10^{22}\sigma \quad \left[cm^2 g^{-1}\right]. \tag{A2.2}$$

2.3 Using the classical scattering expression from equation 2.31:

$$\frac{d\sigma_0}{d\Omega} = \frac{r_0}{2}(1 + \cos^2\theta)$$

$$\frac{d\sigma_0}{d\Omega} = \frac{(2.818 \times 10^{-15})}{2} \times \left(1 + \cos^2 10^0\right) \tag{A2.3}$$

$$\underline{\frac{d\sigma_0}{d\Omega} = 7.82 \times 10^{-30} \qquad [m]}$$

$$F_{KN} = \left(\frac{1}{1 + \alpha(1 - \cos\theta)}\right)^2 \left(1 + \frac{\alpha^2 (1 - \cos\theta)^2}{\left[1 + \alpha(1 - \cos\theta)\left(1 + \cos^2\theta\right)\right]}\right)$$

$$F_{KN} = \left(\frac{1}{1 + 5.87 \times (1 - \cos\theta)}\right)^2 \left(1 + \frac{5.87^2 (1 - \cos\theta)^2}{\left[1 + 5.87 \times (1 - \cos\theta)\left(1 + \cos^2\theta\right)\right]}\right) \tag{A2.4}$$

$$\underline{F_{KN} = 0.873.}$$

$$\frac{d\sigma_e}{d\Omega} = \frac{d\sigma_0}{d\Omega} F_{KN}$$
$$\frac{d\sigma_e}{d\Omega} = 6.83 \times 10^{-30}. \qquad \text{(A2.5)}$$

Note that r_0 and α are calculated using $r_0 = \frac{kq^2}{m_0c^2}$ $\alpha = \frac{h\nu}{m_0c^2}$.

2.4

(i) According to equation 2.16, the ratio of the Compton component of the mass attenuation coefficient for lead to that for water is the ratio of numbers of electrons per gram $\langle Z / A \rangle$. From table 2.1, this ratio is 0.396/0.555 = 0.714. The Compton component of the mass attenuation coefficient for 3 MeV photons in lead is thus 0.0385 $cm^2g^{-1} \times 0.714 = 0.0275\ cm^2g^{-1}$.

ii) A water molecule contains 10 electrons and 18 nucleons (since hydrogen contains no neutrons), giving it a $\langle Z / A \rangle$ value of 0.555. Most other organic substances (e.g., muscle or bone) have a lower ratio of electrons to nucleons because they have a lower hydrogen content by weight. Heavier elements have less electrons and protons than neutrons, hence their $\langle Z / A \rangle$ value is less than 0.5.

2.5 The increase in mean energy with depth occurs because, at radiotherapy energies in soft tissue, mass attenuation coefficient decreases with an increase in photon energy. Lower energy photons are thus removed from the beam by scattering and absorption more rapidly than higher energy photons. This phenomenon is known as beam hardening.

3: Dosimetry of Megavoltage X-Rays and Electrons

3.1 From a number of points it is advantageous if a dosimeter consists of only one material:

- It is easier to correct for nontissue (or water) equivalence.
- No interface problems may occur within the detector. In particular, secondary electron disequilibrium can affect the properties of a detector.
- Directional dependence is reduced (or can easily be predicted).
- The same detector material can be used in different shapes and sizes, thereby allowing extrapolation to infinitesimal small detectors (compare e.g., Kron et al. 1996).
- The detector shape can be made to suit particular holes in phantoms.

The following detectors consist of one material only:

1. Thermoluminescence dosimeters (TLDs, compare section 3.7): Different shapes and forms are available for dosimetry which make TLDs ideal for measurements in phantoms.
2. Gel dosimetry (compare section 3.9.5): The particular advantage of this technique is that the phantom and the dosimeter are identical.

A disadvantage of the above detectors is that their evaluation is typically time-consuming. For clinical dosimetry no technique is available which would allow the direct (in situ) readout of a detector without introducing different materials (e.g., cables) into the radiation beam. However, developments in scintillator dosimetry and optically stimulated luminescence may lead to a detector that allows direct readout via the emitted light in situ (Jones et al. 1993, Aznar et al. 2004).

3.2 See table A3.1.

3.3 The aim of the design shown in figure 3.72 is to minimize wall effects of the chamber. In addition to a thin front window and a small plate separation, the chamber offers a relatively wide guard ring and angled side walls. Rawlinson et al. (1992) recommended that the chamber diameter should be at least 25 times the chamber height, preferably more.

In addition to this, they recommend angled side walls as a useful feature to reduce wall effects from electrons created in the wall scattered into the chamber. Similarly, a low-density material may be employed for the construction of the wall. Other features include that no flange is used within the active window and the construction of a thin collecting electrode to minimize polarity effect.

Table A3.1: Ionization chamber characteristics

	Typical Diameter (mm)	*Typical Height (mm)*	*Advantages and Disadvantages for Measurements in a Water Phantom*	*Typical Applications (Please list two per chamber type)*
Parallel plate ionization chamber	5-20 (plus guard	1-2	+ good spatial resolution for depth dose measurements + perturbation small + effective point of measure- well defined – waterproofing – adjusting chamber position in relation to water surface – creates waves when moving	• surface dose • measurements in build-up region • electron dosimetry and calibration
Cylindrical/ thimble ionization chamber	4-7	4-25 (= "length")	+ can be symmetric (diameter = length) + relatively smooth movement in water – spatial resolution – effective point of measurement may change	• calibration • profiles • Depth dose measurements

3.4 Some advantages of semiconductor diode detectors for radiotherapy dosimetry are:

1. Semiconductors (i.e., their active volumes) are small (typically about 2 mm diameter); therefore they are better suited than ionization chambers for water-tank measurements in strong dose gradients such as a photon beam penumbra.
2. Due to their small physical size they are one of the most versatile in vivo dosimeters.
3. Diodes provide an immediate readout (unlike thermoluminescence dosimeters, film or MOSFET detectors).
4. Diodes are robust and can easily be made waterproof.
5. Modern p-type diodes have a long useful life span and their dose response varies only slightly with total absorbed dose.
6. The electron stopping power ratio to water for silicon is relatively constant with energy; as such, they are well suited for relative electron dosimetry.

(Compare section 3.6.1.)

3.5 The glow curve is the graphical representation of the light output of thermoluminescence dosimeters (TL) dosimeters during readout. This can be plotted against temperature or time. As for glow curve evaluation typically a constant heating rate is applied; the shape of the glow curve does not alter much when displayed as a function of time or temperature. Figure 3.74 shows a glow curve for LiF:Mg,Ti. Using a constant heating rate of between 1° and 5 °C per second, one can clearly distinguish the different peaks that belong to different traps. There are a variety of models, which lead to mathematical expressions that can be used to fit individual glow peaks. This can be useful for glow curve deconvolution. Some models, mathematical expressions and glow peak fitting algorithms are discussed by Horowitz and Yossian (1995).

In many dosimetric applications the readout is performed using a so-called "stepped heating cycle." In this heating cycle the temperature of the TL dosimeter is raised very quickly to the readout temperature, which is then maintained for a couple of seconds. In this case, glow peaks can typically not be resolved. Glow curves are also a feature of optically stimulated dosimetry.
(Compare section 3.7.)

3.6 Both types of dosimeters offer the possibility to acquire a two dimensional dose distribution in a single measurement based on chemical changes due to irradiation. Both film types employ a sensitive layer of some 10 μm thickness on a polyester base. However, there are a number of differences, which are summarized in table A3.2. It should be noted that the information given should only provide guidance. Individual film types may have slightly different dosimetric features. The table also attempts to give some information on potential applications.
(Compare section 3.8.)

3.7
- NE 2571 Farmer type ionization chamber

  The radius of the active volume of the chamber is 3.15 mm (Aird and Farmer 1972). Therefore, the spatial resolution is of the order of 6 mm. However, care should be taken if one uses the chamber in a water tank; its length is 24 mm, which will be reflected in the spatial resolution if scanning in the same direction as the chamber axis.

- TLD100 ribbon

  The typical TLD100 ribbon has a size of $3.1 \times 3.1 \times 0.9$ mm^3. Therefore, one can expect a resolution of some 3 mm if the ribbon is properly aligned with the radiation field. In the diagonal direction its resolution is worse. If the TLD detector is irradiated side on, the spatial resolution can be improved. However, for this purpose, other TLD geometries such as micro cubes and rods are commercially available.

- Radiographic film evaluated with a typical densitometer

  The spatial resolution of the film is typically limited by the densitometer to somewhere between 1 and 2 mm.

- A silicon diode detector for water tank measurements (e.g., Scanditronix)

  The size of a silicon wafer in a typical diode for water tank measurements is of the order of 2.5 mm. This is also the approximate spatial resolution.

Table A3.2: Comparison of dosimetric features of radiographic and radiochromic films

	Radiographic film	*Radiochromic film*
Manufacturer	Kodak, Agfa, CEA, DuPont.	GAF Chemicals.
Examples for radiotherapy dosimetry	X-Omat V (slow), Ektomat V (fast).	MD-55 (single and dual emulsion) compare e.g., Meigooni et al. 1996, Klassen et al. 1997.
Sensitive layer	Silver in silver bromide is reduced to metallic silver due to ionizing radiation.	Radiochromic material turns blue on irradiation.
Developing required	Yes.	No.
Stability	After developing, good.	May exhibit changes over a couple of days (fading).
Dose required to produce an optical density (OD) of 1	Slow film: 0.25 Gy; fast film: <0.01 Gy (depends on developing).	Single emulsion low sensitivity: 100 Gy. Dual emulsion high sensitivity: 20 Gy.
Dose linearity	Conventional film: only over a limited range. CEA film (Cheng and Das 1996): linear.	Linear, reaches saturation.
Photon energy dependence, response at 30 kVp compared to 6 MV	Conventional film: 16. CEA film: 25.	0.5 (Muench et al. 1991).
Cost	Relatively low.	High.
Applications in radiotherapy	Treatment verification (portal films). Assessment of dose distribu- in an anthropomorphic phantom. Assessment of dose distribution in dynamic treatments (Elder et al. 1995). Quality assurance: X-ray vs. light field verification.	Brachytherapy dosimetry ^{125}I). Small field dosimetry (e.g., stereotactic fields). Surface dose measurements (Butson et al. 1996b).

In principle, it is possible to correct for the spatial resolution of any detector by using a detector response function (Higgins 1995). This describes the variation of response throughout the detector. It will be different eg. for a square or a round detector. The measured dose distribution has to be de-convolved with the detector response function.

3.8 Assume a 6 MV x-ray beam with a radiation output of 4 Gy per minute at isocenter and depth of maximum build-up. For the purpose of this argument we can neglect the field size dependence of the dose. Similarly, one can neglect the pulsed nature of the radiation delivery as the pulse length is considerable shorter (3 ms) than the acquisition time (200 ms), and most scanning systems operate with a reference chamber which integrates over the same number of pulses.

The dose delivered at isocenter in 200 ms is $4/60 \times 0.2 = 0.0133$ Gy or 13.3 mGy. Under standard conditions 1 cm^2 of air weighs 1.293×10^{-3} g. The weight of the gas in the chamber of 0.2 cm^2 is therefore 2.586×10^{-7} kg. As 1 Gy = 1 J/kg, the energy delivered in 200 ms to the air in the chamber is 3.45 nJ. The mean energy required to produce a charge of 1 C of one polarity is 33.97 J (this is W/e). Therefore, approximately 0.1 nC is produced of either charge in the chamber.

If one wants to deduct the collimator transmission of 0.5% from such a measurement, the ion chamber, cable, and electrometer combination must be able to detect 0.5 pC with sufficient accuracy.

3.9 The film has been exposed without build-up with a screw placed in the field center, as indicated in figure 3.73b. The top of the screw offers just enough build-up to maximize the dose in a ring around the center. However, the radiation is attenuated sufficiently in the long part of the screw to reduce the radiation in the center of the film again. If the screw is long enough, this can even produce less film blackening than the field without build-up.

3.10 There are, in general, three different approaches:

1. The first approach would be to ensure that the treatment plan includes the pacemaker and specifically calculating a point dose to the pacemaker. The accuracy of the planning system in predicting the dose in this point correctly depends on the dose calculation algorithm, the commissioning data, and the distance from the field edge. Significant uncertainties exist. However, by carefully placing the beams and making use of asymmetric collimation to avoid unnecessary exit dose due to beam divergence from a posterior beam, one may be able to increase the margin (Riley 2004).
2. Second, measurements in an anthropomorphic phantom can be made to assess the dose directly in the location of the pacemaker. It appears necessary to have a phantom with lung-equivalent material as this will alter the attenuation and may affect the range of scattered photons and electrons. Allowance has to be made for the physical differences between the patient and the anthropomorphic phantom. These phantom measurements will typically be performed using TL dosimeters. It may be necessary to drill additional holes into the phantom to approximate the location of the pacemaker.

   To improve the quality of the phantom measurements several strategies can be employed:

i) The measurement can be used to distinguish between the contribution to the pacemaker dose from different radiation fields.
ii) The measurement and phantom setup can be repeated several times. This will give an idea about the reproducibility of the setup and the correctness of the chosen margins in an "ideal" patient. As equipment tolerances and treatment techniques can vary from place to place; this will give additional information about the reproducibility that can be achieved.
iii) Additional TL dosimeters can be used in the vicinity of the pacemaker location. This will allow estimating dose gradients around the pacemaker. From this, the effect of a variation of pacemaker position from day to day (due to pacemaker movement or patient treatment setup) can be estimated.
iv) It is good practice to include additional detectors in the phantom measurements that are placed on the phantom surface. These can be repeated in in vivo measurements to validate the phantom measurements.

3. In vivo verification should include two parts:

i) The verification of the patient and pacemaker position using portal films or electronic portal imaging. One should keep in mind though that each port film, in particular when taken in double-exposure technique, contributes additional dose to the pacemaker.
ii) The dose can be estimated from in vivo dosimetry measurements. These can directly establish collimator transmission from entrance dose measurements in a ray-line through the pacemaker. However, if the pacemaker is only about 1 cm from the field edge, it is unlikely that external radiation will contribute considerably in a treatment using a modern linac. A secondary shielding block close to the patient's skin will reduce the pacemaker dose, though not significantly as could have happened in a cobalt-unit due to its larger radiation source size.

The direct assessment of internal scatter that is likely to be the main contribution to the pacemaker dose is difficult. It may prove more viable to include in the computerized treatment plan a *verification* point, which is both accessible to in vivo dosimetry and includes a major contribution from internal scatter. The exit dose 1 cm out of the field may be a suitable position. If measurement and calculation agree at the verification point, it is likely that the calculation is also correct at the point of interest in the pacemaker.

In any case, repeated in vivo dosimetry measurements will pick up any accidental irradiation of the pacemaker as well as establish the reproducibility of the field positioning. The knowledge of both is of importance in estimating the risk of pacemaker failure during radiotherapy.

The report of AAPM task group 34 (Marbach et al. 1994) provides more information on the problems raised here. Other more recent publications on effects of radiation on pacemakers are Mouton (2002), Solan (2004) and Hurkmans (2005). In any case one should check the pacemaker after radiotherapy as the battery may deplete more rapidly under irradiation as discussed by Wilm et al. 1994.

4: X-Ray Beam Properties

4.1 Sources of electron contamination in x-ray beams include: electron contamination produced by x-ray collisions with the treatment head and/or beam modifying devices, and electron contamination produced by interactions in the air column between treatment head and patient.

4.2

(i) The tissue-phantom ratio (TPR) is defined as the ratio of absorbed dose at any given point to the absorbed dose at the same distance from the source but with the surface of the phantom moved so that this point is at a specified reference depth of d_{ref}. The tissue-maximum ratio (TMR) is a special case of the TPR where the specified reference depth is d_{max}

$$TMR(d_{ref}, S, E) = \frac{D_{tissue}(d_{ref2}, S, E)}{D_{tissue}(d_{\max}, S, E)} \quad . \tag{A4.1}$$

ii) The ratio of the dose at a point in tissue to the dose at the same point in air is known as the tissue-air ratio (TAR).

(iii) If the TAR values are taken at d_{max}, these values have the special term, peak scatter factors (PSF), such that

$$PSF(S) = TAR(d_{\max}, S) \quad . \tag{A4.2}$$

If all the PSF(S) values are renormalized so that PSF(10×10 cm^2) = 1, these values are called normalized peak scatter factors (NPSF). This is approximately equal to the in-phantom scatter factor S_p, which, as previously stated, can be measured directly with a build-up cap and does not require the extrapolation step. Note: PSF is usually measured at 100 cm source surface distance (SSD) whereas S_p and TARs are usually measured at 100 cm source axis distance (SAD).

4.3 Using equation 4.17a, then

$$TAR(d,S) = \frac{\%D\left(d, f, \dfrac{Sf}{(f+d)}\right)}{100} \times PSF\left(\frac{Sf}{f+d}\right) \times \left(\frac{f+d}{f+d_{\max}}\right)^2$$

$$TAR(10,22) = \frac{\%D\left(10, 100, \dfrac{22 \times 100}{(100+10)}\right)}{100} \times PSF\left(\frac{22 \times 100}{100+10}\right) \times \left(\frac{100+10}{100+1.5}\right)^2$$

$$TAR(10,22) = \frac{\%D(10,100,20)}{100} \times PSF(20) \times \left(\frac{100+10}{100+1.5}\right)^2$$

$$TAR(10,22) = \frac{70}{100} \times 1.06 \times 1.17$$

$$TAR(10,22) = 0.87. \tag{A4.3}$$

4.4 Using equation 4.17b and as the PSFs are ratios, the NPSF values have been looked up using *BJR Supplement 17* or *25*, and as we are calculating a TMR for a 6 MV beam, we will assume $d_{ref} = d_{max}$, which is approximately 1.5 cm.

$$TMR(d,S) = \frac{\%D\left(d,f,\frac{Sf}{(f+d)}\right)}{100} \times \frac{NPSF\left(\frac{Sf}{(f+d)}\right)}{NPSF\left(\frac{Sf}{(f+d_{ref})}\right)} \times \left(\frac{f+d}{f+d_{\max}}\right)^2$$

$$TMR(10,22) = \frac{\%D(10,100,20)}{100} \times \frac{NPSF(20)}{NPSF(21.6)} \times \left(\frac{100+10}{100+1.5}\right)^2$$

$$TMR(10,22) = \frac{70}{100} \times \frac{1.023}{1.026} \times 1.17$$

$$TMR(10,22) = 0.82. \tag{A4.4}$$

From rearranging equation 4.21 but including INVSQ because calibrated at 100 cm SSD (also see equation 6.5 in chapter 6, given that $FF = S_{cx}S_p$):

$$MU = \left(\frac{D \times INVSQ}{\left(\frac{D}{MU}\right)_{std,SSD,water} \times S_c \times S_p \times TMR}\right)$$

$$MU = \left(\frac{200 \times 0.97}{1 \times 1.033 \times 0.82}\right) = 229MU. \tag{A4.5}$$

4.5 The most common quantitative measures of symmetry and flatness follow the IEC (1989) guideline. The percentage flatness %F is assessed by finding the maximum D_{max} and minimum D_{min} dose point values on the profile within 80% of the beam width, such that

$$\%F = \left(\frac{D_{\max} - D_{\min}}{D_{\max} + D_{\min}}\right) \times 100 \quad . \tag{A4.6}$$

Dose profiles are taken to test flatness at a typical treatment depth, generally at a depth of 10 cm where the horns are reduced and the flatness is generally better than 5%. The scan at depth is used to avoid the horn effect, which is greatest at d_{max}. Dose *horn* is the term given to describe the profile shape if the dose at the beam off-axis position exceeds the dose in central axis.

Symmetry may be assessed at dmax. This is the most sensitive position to test symmetry. The beam symmetry %S is assessed by comparing the area under each side of the profile, such that

$$\%S = \left(\frac{Area_L - Area_R}{Area_L + Area_R} \right) \times 100 \quad . \tag{A4.7}$$

The area under the profiles can often be obtained using an automatic option on the water tank scanning device. If this is not available, then counting squares on graph paper with a hard copy of the profile is an alternative. With current linac beam steering the beam profiles should be able to be tuned close to symmetry in both the transverse and radial axes. The symmetry should be within ±2%.

5: Linear Accelerator Electron Beam Properties

5.1 Electrons show only modest skin sparing since (unlike photons) the electrons will start interacting with the medium and depositing dose as soon as the surface is reached. In megavoltage photons the build-up is a result of the indirect energy deposition where the secondary electrons are mostly forward directed and therefore deposit their energy beyond the actual photon interaction point. Typical surface doses in electron beams from a 15 × 15 cm field are in the range of 75% to 95% of maximum dose (increasing with increasing electron energy). This compares with surface doses around 20% in photon beams of similar size. In electron beams the surface dose increases with increasing electron energy. This is opposite to photon beams, where the surface dose decreases with increasing photon energy as the secondary electrons have longer range, scatter less, and become more forwardly directed. On entry, the electron paths are approximately parallel to the beam direction. With depth the electron path becomes progressively more oblique as the electrons scatter through the medium. This increases the electron fluence and therefore the rate of energy deposition.

5.2

(i) The range of a high-energy electron in water is many centimeters. Its essential for calibration purposes to ensure that lateral electron equilibrium exists at central axis, i.e., the lateral range of electrons must not exceed half the field width. In this case R_{50} is independent of the field size. This is reflected in the statement by the IAEA that smaller field sizes may be used if it can be proven that R_{50} does not change more than 0.1 g/cm^2 by going to a larger field size (IAEA 2000, p. 87).

ii) 16.3 MeV (compare equation [5.1]).

5.3 The inverse square law in the case of electron beams is based on the use of the effective SSD. Therefore oldSSD = ESSD. The ESSD depends on field size and electron energy and is typically tabulated after measurements are performed during commissioning. Table 5.3 in this chapter gives a set of sample data. From this, the closest field size is 6 × 6 cm^2 and the oldSSD is therefore around 55 cm. NewSSD is 70 cm and the output at the extended SSD is more than 1.6 times smaller than the original one. This compares to a reduction of dose by a factor of 1.32 if the nominal SSDs would have been used.

5.4 To start with, all electrons reaching air will travel further. Due to the low density of the air, the number of interactions per centimeter is nearly 3 orders of magnitude lower than in tissue and the track should have been shown extending out of the page. This does not alter the message of the figure as the direction is not reversed and no electrons generated in air are included in the figure.

Another interesting fact can be noticed. As discussed in section 5.4, the distance traveled laterally by electrons is a random distribution that depends on electron energy (lower energy electrons scatter more), the scattering power of the material, and penetration depth in the material (the larger the depth in matter the more scattering events have occurred already and the lower the energy of the electrons). In principle, the largest distance traveled laterally would be close to the maximum remaining electron range; after an initial large-angle scattering event, the electron moves more or less straight. In practice, this is very unlikely and the effective lateral travel is significantly smaller than the average penetration depth in material as the electrons are mostly forwardly directed.

With increasing depth in a phantom the mean angle of electron movement away from incident beam direction increases. However, the remaining range also becomes smaller.

Scatter in air and secondary photon production are ignored.

5.5 Using equation 5.1:

$$E_0 = 2.33 \times d_{50}$$

$$E_0 = 2.33 \times 5 = 11.65 \text{ MeV}.$$

The practical range can be estimated using equation 5.3:

$$R_p = 1.193\, d_{50} + 0.154.$$

Thus

$$R_p = 6.1 \text{ cm}.$$

The nominal energy of the electron beam is likely to be 12 MeV as can be seen in table 5.1.

6: Radiotherapy Treatment Planning X-Rays

6.1 As follows:

i) Using equation 6.1 and tables 4.1 and 4.4 for data:

$$MU = \frac{D \times 100}{\left(\frac{D}{MU}\right)_{std,SSD,Water} \times F \times \%D} = \frac{200 \times 100}{1 \times 1.056 \times 70.5} = 269 \quad MU \quad . \tag{A6.1}$$

ii) Using equation 6.5 and table 4.5 for data:

$$MU = \left(\frac{D \times INVSQ}{\left(\frac{D}{MU}\right)_{std,SSD,Water} \times F \times TMR} \right) = \left(\frac{200 \times 0.97}{1 \times 1.056 \times 0.819} \right) = 224 \quad MU \quad . \tag{A6.2}$$

6.2 Using equation 6.10 with data from tables 4.1 and 4.5 then:

$$MU = \frac{D}{\left(\frac{D}{MU}\right)_{std,SAD,Water} \times TMR(d, s_{phant}) \times S_c(s_{col}) \times S_p(s_{phant}) \times WF \times BF} = \frac{300}{(1.0) \times (0.698) \times (1.022) \times (1.033) \times (0.97) \times (0.63)} = 667 \tag{A6.3}$$

Note: S_p was calculated using $S_p = \left(\frac{F_c}{S_c}\right)$.

6.3 It was generally believed that backscatter into the ion chambers (which occurs less with blocked fields because blocks are further away) may be the cause of the difference. This point has been argued in the literature with conflicting views. The effect is important for some linear accelerators (Tatcher and Bjärngard 1994, Hounsel 1998). Experiments which involve the turning off of the ion chamber feedback and the use of a timer (á la cobalt) show that the effect due to ion chamber over-response is negligible for some linacs.

The term *collimator scatter* includes scatter contributions from all sources, excluding in-phantom scatter. Perhaps a better term for this is *extra focal radiation*, and a significant proportion of this is due to scattering off the flattening filter. Though some scatter is off the collimators themselves, they also act as blocks to the scatter created from the flattening filter (Chaney et al. 1994).

Due to divergence when the field is blocked, the in-phantom scatter component will be reduced as the effective field size is reduced. However scatter from multiple foci (non point source) such as the flattening filter will have a similar extra focal scatter component reaching the patient for both the collimated only and blocked field.

6.4 The relation between CT number and electron density for common tissues as shown in figure 6.9a should be closely reproduced for this answer.

The relation between CT number and electron density is given by rearranging equation 6.11 such that

$$N_{CT} = \frac{1000}{R_\gamma}(\rho_e^w - 1)\,. \qquad \text{(A6.4)}$$

Table A6.1:

Tissue	*Element and (%f_i)*
Human bone	H(8.67), C(13.00), N(3.6), O(66.40), S(0.46), P(2.43), Ca(4.96), Trace(0.48)
Water	H (12.0), O (88.0)

6.5 As follows:

i) Following the calculations in section 6.5, page 403 and example 6.8, the appropriate data needed from table 3.1 are:

Using equation 6.21

$$\rho_e^w = \frac{N_A \rho}{\rho_{ew}}\left[\sum_{j=1}^{N} f_j\left(\frac{Z}{A}\right)_j\right]_s\,, \qquad \text{(A6.5)}$$

where N_A is Avogadro's number (6.023 x 1023 mol^{-1}) and ρ_{ew} is 3.343 x 1023 cm^{-3}. Substituting values for N_A and ρ_{ew} into equation 6.21 gives

$$
\begin{aligned}
\rho_e^w &= 1.81\rho\left[\sum_{j=1}^{N} f_j\left(\frac{Z}{A}\right)_j\right]_s \\
\rho_e^w &= 1.81\times 1.12\times\left[0.0867\times 1.0+0.9133\times 0.5\right] \\
\rho_e^w &= 1.10 \quad .
\end{aligned}
\tag{A6.6}
$$

ii) To calculate R_σ for water

$$
\begin{aligned}
\left[\sum_{i=1}^{N} f_i\left(\frac{\mu}{\rho}\right)_i\left(\frac{A}{Z}\right)_i\right]_w &= 0.12\times 1.008\times 0.3165+0.88\times 2.0\times 0.1736 \\
&= 0.3426.
\end{aligned}
\tag{A6.7}
$$

Note: A slightly different elemental ratio of H:O was used in example 6.8 in the text.

For human soft bone (ignore trace elements in the calculation:

$$
\begin{aligned}
\left[\sum_{i=1}^{N} f_i\left(\frac{\mu}{\rho}\right)_i\left(\frac{A}{Z}\right)_i\right]_s &= 0.0867\times 0.3165\times 1.0+0.13\times 0.1666\times 2.0 \\
&+0.036\times 0.1706\times 2.0+0.664\times 0.1732\times 2.0 \\
&+0.005\times 0.3063\times 2.0+0.024\times 0.2706\times 2.0 \\
&+0.05\times 0.4597\times 2.0 \\
&= 0.3746\,.
\end{aligned}
\tag{A6.8}
$$

Then from equation 6.20 the ratio of these two gives $R_\sigma = 0.915$.

Then using equation 6.19:

$$
\begin{aligned}
N_{CT} &= \frac{1000}{R_\sigma}(\rho_e^w - R_\sigma) \\
&= \frac{1000}{0.915}(1.10-0.915) \\
&= 287HU \quad .
\end{aligned}
\tag{A6.9}
$$

7: Special Procedures: Stereotactic Radiosurgery and IMRT

7.1 Pixel locations in the planning image coordinate system are transformed into the stereotactic coordinate system via the detected positions of the localizer rods on each CT slice (see figure 7.3). From this transformation, we know the stereotactic coordinates of the desired isocenter. The stereotactic targeting device (e.g., figure 7.4a and 7.4b) can then be set to the isocenter position in stereotactic coordinates and this point can be made to coincide with the linac isocenter during patient setup.

7.2 Reflectors on an array in a camera-based system, or bony anatomy features in an x-ray image-based system, have a planned location and orientation relative to the linac isocenter. During treatment-room imaging (camera or x-ray), the location of reflectors or bony anatomy can be detected. The system can inform the user of any required changes in position and orientation to achieve consistency between the planned and actual position/orientation.

7.3 There are 510 degrees of arc in total with which to deliver the 18 Gy. Arc 1 will deliver 100/510 of the total dose so approximately 3.5 Gy (see equation 7.3).

The number of MU required is the dose to be delivered divided by dose per MU. We know the dose is 3.5 Gy (350 cGy) so we must calculate the dose per MU. Using equation 7.1, average dose per MU = 0.9 × 0.7 = 0.63 cGy/MU. MU required is thus 350/0.63 = 556 MU.

7.4

(i) Use either multiple isocenters or a conformal (non-isocentric) plan.
(ii) Use a small collimator diameter (e.g., 1 cm diameter), or a small collimator along with a larger collimator to help keep treatment time down while achieving a conformal plan.
(iii) Use of a large number of beam "nodes."

7.5 A minimum dose goal for the PTV would be set along with a minimum dose penalty. This penalty should be kept high to minimize the volume of PTV receiving a dose less than the goal, but this may cause excess spillage of high dose into the OAR. Careful checks of both minimum PTV dose and the volume of high dose in the OAR will need to be performed during optimization. The maximum OAR dose will be approximately equal to the minimum PTV dose so it is reasonable and realistic to set the same dose for the maximum OAR dose. The best we can aim for in terms of OAR dose is a rapid dose fall-off away from the PTV. This can be encouraged by setting a DVH goal such that only a small volume of the OAR receives a dose higher than a dose somewhat lower than the maximum (minimum PTV dose). Adjustment of the DVH goal during optimization to successively lower d_{max} or Vmax values, as each goal is achieved, can be effective.

7.6 Many small segments can often create a more intricate conformality to re-entrant shaped target volumes. When many small segments are used to create IMRT dose distributions, the number of monitor units increases, because small segments have low field ratio outputs. This increases the leakage component of total dose from blocked segments and also increases the treatment delivery time. There is also less confidence in the accuracy of dosimetry of small segments and leaf positional precision becomes more critical.

7.7 In summary, the main sequence of events which occur for a patient IMRT plan and treatment delivery process include:

i) Set beam angles and beam energy as you would for normal planning. Note: no opposing beam angles seem better for the optimizer, e.g., 7 or 9 equally spaced angles for prostate.
ii) Set tumor and normal tissue constraints and objectives, e.g., dose to PTV between a lower limit dose (e.g., 60 Gy) and a higher limit dose (e.g., 61 Gy). Most systems now include a tool which enables the user to draw the desired DVH shape and the computer optimization engine is set to try to match this desired shape.
iii) Use inverse planning software to iterate to a solution for intensity-modulated beams.
iv) Create an MLC leaf sequence which approximates the intensity-modulated field (or design a compensator).
v) Forward-compute the dose distribution for the leaf sequence to ensure the accurate dose model is used to compute the final dose distribution. By including this step it is as though one has used the inverse planning software as a tool to arrive at a good plan, then one has computed the dose using the forward plan computation as with any other plan.
vi) Export the leaf sequence in a control sequence to the linac leaf controller software, usually via the oncology information system.
vii) Check the dose by applying the plan to a known phantom and checking planar doses and axial doses using a combination of dosimeters. (Boyer et al. 1999, Chui et al., chapter 5 in Ling et al. 2003).
viii) Check using MU check software for IMRT segmented fields; this does not check the dose delivery.
ix) Check that the treatment fields are deliverable without gantry or beam obstruction, then deliver the patient treatment.
x) Target verification is essential and improvements in electronic portal imaging are complementary to this process.
xi) Rescan with CT and re-plan if excessive tumor shrinkage during treatment course.

7.8 Reducing the pitch from 0.5 to 0.3 means that there will be 5/3 times the number of rotations. Consequently, only 3/5 as much dose needs to be deposited per rotation. This should speed up the rotations by a factor of 5/3, so the rotation will become 24 seconds. Overall treatment time will remain unchanged because the greater number of rotations is offset by each rotation being faster.

8: Calibration of Megavoltage Photon and Electron Beams

8.1 Absolute dosimetry yields information directly about absorbed dose in Gy. These measurements are typically done under standard (or "reference") conditions. Absolute dosimetry involves a number of factors, which relate an instrument reading to absorbed dose. For each radiation quality of a linac normally only one absolute measurement is performed which relates the number of monitor units (or the exposure time) to an absorbed dose in Gy at the reference point. Only few dosimetric techniques are suitable to do absolute dosimetry-see next question.

Relative dosimetry relates the dose measured under any conditions or performed with any dosimeter type to the dose at the reference location. These measurements can relate the dose at different locations to the reference point such as percentage depth dose measurements or beam profiles. Other types of relative dosimetry establish the ratio of the dose measured at the reference point in different radiation conditions to the dose under standard conditions. Examples for this are collimator, tray, or wedge factor measurements. In general, no factors are required for relative dosimetry since it is only the comparison of two dosimeter readings, one of them being performed under reference conditions.

(Compare chapter three and sections 8.1 and 8.2.)

8.2

i) Calorimetry: Heat is a direct measure of the energy absorbed in matter-as long as the energy lost in chemical reactions (i.e., the heat defect) can be accounted for, the temperature change in a medium is a direct measure for the absorbed dose. The specific heat cp,m has to be known of the material in use. (Compare chapter three, section 3.9.1.)
ii) Chemical dosimetry: If the yield of a chemical reaction per absorbed dose is known, the quantification of end products of a radiation induced reaction can be converted into absorbed dose. The yield is called G factor (compare chapter three, section 3.9.4)
iii) Ion chamber measurements: Ideally, an ionization chamber collects all ions produced by ionizing radiation in the active volume. If recombination is accounted for and if the average energy for the production of an ion pair W/e is known, this can be used to calculate the absorbed dose. This is in practice only possible for low-energy radiation where a free-air ionization chamber can be used (compare chapter three, section 3.4.2)

In all techniques that are used for absolute dosimetry, a property of the detector has to be known which relates the measurement quantity (e.g., temperature, the amount of a certain chemical produced, or the collected charge) to the absorbed energy. To be suitable for absolute dosimetry this factor should be:

- independent to environmental influences
- invariant from one detector to a similar one
- constant over time
- known with good accuracy.

While the light emitted by a TLD or the charge collected in a diode under irradiation is theoretically linked in an absolute way to the absorbed dose in the detectors, the correction factor depends on many environmental influences such as temperature and it would be difficult to manufacture multiple detectors with identical behavior due to the strong dependence of the reading on the actual doping of the detectors. Similarly, the blackening of radiographic film depends on the developing process that is difficult to maintain stable over long periods of time. Therefore, these detectors are generally not suitable for absolute dosimetry.

8.3 Can you confirm the calibration of the center? The first step is the calculation of N_{gas}, the cavity-gas calibration factor. This can be done according to equation (6) of the protocol (equation 8.14):

$$N_{gas} = N_X k\left(W/e\right) A_{ion} A_{wall} \beta_{wall} A_m^{-1}, \qquad \text{(A8.1)}$$

with A_m as the correction for the difference in composition between wall plus build-up cap and air. Generally, A_m can be written as (equation 8.15):

$$A_m = \alpha\left(L/\rho\right)_{gas}^{wall}\left(\mu/\rho\right)_{wall}^{air} + \left(1-\alpha\right)\left(L/\rho\right)_{gas}^{cap}\left(\mu/\rho\right)_{cap}^{air}. \qquad \text{(A8.2)}$$

Using the figures given in the question this can be calculated as:

$$A_m = 0.42 \times 1.01 \times 0.999 + 0.58 \times 1.103 \times 0.925 = 1.015 \qquad \text{(A8.3)}$$

and

$$N_{gas} = 2.58 \times 10^{-4} \times 33.7 \times 1 \times 0.99 \times 1.005 \times 1.015^{-1} \times N_X$$
$$= 8.52 \times 10^{-3}\, Gy / R \times N_X. \qquad \text{(A8.4)}$$

This is the same relation as given in table XVII of the protocol. Therefore, the cavity-gas calibration factor for the Capintec 0.6cc chamber in use is $N_{gas} = 4.303 \times 10\text{-}3$ Gy/div.

In the next step, the reading of the chamber acquired in the linac beam has to be converted to standard conditions. The temperature and pressure correction factor can be calculated as follows (equation (3.7) in chapter three):

$$\left(\text{T in}^{\circ}C + 273\right) / 295 \times \left(760\text{mmHg}\right) / p = 1.005, \tag{A8.5}$$

with reference conditions for the AAPM protocol being T = 22 °C (=295K) and p = 760 mmHg. As no polarity, correction is required for photon beams and the humidity correction can be taken as 1; the corrected reading per monitor unit MU of the chamber at the measurement depth is

$$\begin{aligned} M / MU &= 1.005 \times < 190.2, 190.5, 190.0 > /100 \\ &= 1.005 \times 1990.2 / 100 = 1.912. \end{aligned} \tag{A8.6}$$

Please note, that this does not yet include the correction for recombination losses in the user's beam.

The nominal energy of the radiation beam under consideration can be taken from the given TPR_{10}^{20} and the table in the question as 9.5 MV. The dose to water has been measured at a depth of 5 cm. This is in accordance with the protocol, which states in table XI that, the calibration depth for this energy and chambers with an internal diameter of more than 6 mm should be 5 cm in water. The dose at this depth in the measurement performed can be calculated as:

$$D_{med} \left(M / MU\right) N_{gas} \left(L/\rho\right)_{gas}^{med} P_{ion} P_{repl} P_{wall}. \tag{A8.7}$$

This is equation (9) in the AAPM protocol (1983) and (8.16) here. The medium for the measurement is water in the example. The recombination P_{ion} can be calculated from the mean reading at the two different voltages. With the values given above, this results in:

$$\begin{aligned} D_{water} &= 1.912 \times 4.303 \times 10^{-3} \times 1.118 \times \left(190.2 / 187.7\right) \times 0.992 \times 1 \\ &= 9.246 \times 10^{-3}\,\text{Gy/MU}. \end{aligned} \tag{A8.8}$$

Therefore, the dose to water at depth of maximum dose can be calculated using the percentage depth dose at d = 5 cm to 0.01005 Gy per MU or 1.005 cGy/MU. Within the uncertainty of the measurement (see also section 8.5), this would confirm the calibration of the linac.

(Compare section 8.3.2.)

8.4

a) The AAPM protocol TG 51 (1999) is applicable to electron beams with nominal energies of 4 to 50 MeV. In the protocol the beam quality is specified by the depth in water where the depth dose reaches 50% of the maximum dose, R_{50}. R_{50} can be calculated using equations 16 and 17 in the protocol from a depth ionization curve after shifting the curve by the chamber displacement 0.5 r_{cav} for a cylindrical chamber with radius r_{cav}.

In the example, the depth ionization curve has been acquired using a Scanditronix RK chamber with an internal radius of 2 mm. After shifting the curve by 1 mm ($r_{cav}/2$) to account for the effective point of measurement, the depth of 50% ionization is I_{50} = 24 mm. Then according to equation (16) of the protocol (equation 8.24) one can determine R_{50} as:

$$R_{50} = a\,I_{50} - b = 1.029\,I_{50} - 0.06\text{ cm}\left(\text{for}2 \le I_{50} \le 10\text{cm}\right) = 24.1\text{mm}. \quad (A8.9)$$

According to the AAPM TG 51 protocol, a plane parallel ionization chamber should be used for an electron beam with an $R_{50} \le 2.6$ cm. Therefore, your colleague needs to cross calibrate her Farmer type chamber with a suitable plane parallel ionization chamber in order to calibrate the electron beam.

b) The reference conditions for calibration of electron beams are given in table 8.6. Following the AAPM TG 51 protocol, clinical reference dosimetry for electrons can be performed at an SSD between 90 and 110 cm. For the low-energy electron beam of the example, the reference field size at this distance is 10 × 10 cm^2.

The reference depth for dosimetry is according to equation (18) of the protocol (equation 8.23):

$$d_{ref} = 0.6\,R_{50} - 0.1\text{ cm} = 13.5\text{ mm},$$

a depth which is very close to d_{max}.

(Compare section 8.4.1.)

8.5 The best way to check output calibration is to invite an independent expert (preferably an accredited physicist) using an independent measurement system (electrometer and ionization chamber). Ideally, the independent check should be performed for all radiation qualities available at a treatment unit.

If no visit of an independent expert can be arranged or indeed as a second check anyway, there should at least be an independent verification of output using, for example, thermoluminescent (TL) dosimeters which can easily be sent from one radiotherapy center to another. These devices provide an alert for further investigation if large differences in calibrated output are discovered.

9: Beam Models: Part I

9.1

(i) The effective depth assuming the relative electron densities of tissue,bone, and lung are: 1.0, 1.2, and 0.3, respectively is

$$\begin{aligned} d_{eff} &= \sum_i \rho_i d_i \\ &= (1.0 \times 4.0) + (1.2 \times 2.0) + (0.3 \times 7.0) \\ &= 8.5 \ \text{gcm}^2 . \end{aligned} \tag{A9.1}$$

(ii) From equation 9.15 and interpolating to find the percentage depth dose at 8.5 cm,

$$\begin{aligned} CF &= \frac{P(d_{eff}, FS_s)}{P(d, FS_s)} \left(\frac{SSD_1 + d_{eff}}{SSD_1 + d} \right)^2 \\ &= \frac{P(8.5)}{P(13)} \left(\frac{100 + 8.5}{100 + 13} \right)^2 \\ &= 1.18 . \end{aligned} \tag{A9.2}$$

(iii) The new effective percentage depth dose is given by

$$\begin{aligned} P(d_{eff}, FS_s) &= P(d, FS_s) \times CF \\ &= 56.7 \times 1.18 = 66.9\% . \end{aligned} \tag{A9.3}$$

9.2

(i) The effective depth method accounts for perturbation of the primary fluence due to non-unit density material. Only material along a ray- line from the source to the dose calculation point is considered. Primary fluence and hence primary dose will be increased when a low-density region lies above the calculation point. However, due to the reduced density and number of scattering centers in this region, scatter dose will be reduced. There is a reduction in scatter dose, which offsets the increase in primary dose. The effective depth method will thus overestimate the dose beneath a low-density region such as lung.

(ii) When the dose calculation point is further from the low-density region, the contribution of scatter dose to the total dose decreases. This means that the significance of not accounting for perturbation of scatter dose will be reduced.

9.3

(i) The dose calculation point P is in a medium of density ρ_1 (relative to water) and there is an overlying region of density $\rho 2$, where the distance (along a ray-line to the source) from P to the top and bottom of the region with density ρ_2 is d_2 and d_1, respectively. Using an attenuation coefficient μ, the ratio of exponential attenuation from the top of the overlying region to P to the exponential attenuation if the medium were all of unit density is

$$R = \frac{\exp(-\mu\rho_1 d_1 - \mu\rho_2(d_2 - d_1))}{\exp(-\mu d_2)}$$
$$= \frac{\exp(-\mu d_1(\rho_1 - \rho_2}{\exp(-\mu d_2(1-\rho_2))}$$
$$= \frac{\exp(\mu d_1)^{\rho_1-\rho_2}}{\exp(\mu d_2)^{1-\rho_2}}. \quad \text{(A9.4)}$$

Now assuming that tissue-air ratios have an exponential falloff with depth, where the coefficient in the exponent is μ, using the above ratio as the correction factor gives

$$CF = \frac{TAR(d_1)^{\rho_1-\rho_2}}{TAR(d_2)^{1-\rho_2}}. \quad \text{(A9.5)}$$

ii) When the calculation point is a distance d_2 below the top surface of a medium of relative density ρ, the ratio in part (i) is replaced by

$$R = \frac{\exp(-\mu\rho d_2)}{\exp(-\mu d_2)}$$
$$= \exp(-\mu d_2)^{\rho-1}. \quad \text{(A9.6)}$$

Again assuming that a TAR is exponential with coefficient μ, the correction factor is given by

$$CF = TAR(d_2)^{\rho-1}. \quad \text{(A9.7)}$$

(iii) Because the proportional contribution of scatter dose increases with depth, the fall-off in a TAR curve with depth (past d_{max}) is slower than the pure exponential that would exist if only primary dose were important. For a pure exponential, the ratio of values at depths d_1 and d_2 is constant for constant d_2 - d_1. For a TAR curve however, this ratio will decrease as d_1 and d_2 decrease, causing the ratio in equation A9.5 to be reduced when the low-density region is closer to the calculation point.

9.4

(i) The effective field size used in the ETAR method is intended to represent the field size in water that gives rise to the same scatter dose as at the current dose calculation point in an arbitrary medium for an arbitrary field shape. It is accounting then for the increase in scatter dose that occurs with an increase in scattering material, and hence with field size.

(ii) The effective depth at a point 10 cm deep in a medium of relative density 0.3 is $d_{eff} = 10 \times 0.3 = 3.0$ cm of water. To find the effective field size r_{eff}, first calculate the physical field size at 10 cm deep, then multiply by the medium density:

$$r_{eff} = 20 \times ((100+10)/100) \times 0.3 = 6.6 \text{ cm}.$$

Since the medium is homogeneous, the scatter dose at a point P due to primary photon interactions in any volume V at a vectorial displacement r from P is the same as the scatter from a volume in water whose displacement from P and volume are scaled by the relative density. Depth and field dimensions can therefore be scaled by density to obtain an equivalent in-water geometry. This is an application of O'Connor's theorem.

(iii) In this case, the ETAR method will calculate a correction factor using the TAR for a field size of half the physical field size. In halving the field size the remaining scatter dose is greater than half the original amount because of the sub-linear increase in scatter dose with field size, whereas in reality there will be half the scatter dose that would exist for a continuous medium. The method will therefore underestimate the reduction in scatter dose and the correction factor will be too high.

9.5

(i) For field sizes larger than that required for lateral electronic equilibrium, central-axis dose decreases gradually with field size as scat ter dose is reduced. For field sizes whose half-width is less than the maximum lateral electron range however, the reduction in dose is much more rapid (due to lateral electronic equilibrium) and in fact approaches zero as field width approaches zero. The extrapolation of dose versus field size to smaller field sizes assumes a continuation of the larger-field curve and thus overestimates the true dose.

(ii) Lateral electronic disequilibrium will occur anywhere that is closer to the field edge than the lateral secondary electron range. Because of the increased electron range when beam energy is increased or medium density is reduced, and since the average distance to the field edge is less for small fields, lateral electronic disequilibrium is more likely to occur for high energies, low densities, and small field sizes.

(iii) When a dose versus field size curve is extrapolated to zero field size, the value obtained approximates the dose in the absence of scatter. This is the primary dose component on the axis of a field large enough for lateral electronic equilibrium to exist on the axis.

10: Beam Models: Part II

10.1

(i) The two types of electron energy loss mechanisms modeled discretely in an EGS4 simulation are knock-on (electron-electron) collisions and bremsstrahlung-producing interactions. In a knock-on collision, the orbital electron is ejected from its shell, becoming a delta ray which will deposit energy in ionizations and excitations. Bremsstrahlung is produced when the incident electron is decelerated in the field of a nucleus and the energy lost is taken by the bremsstrahlung photon. The threshold energy losses above which these interactions are modeled discretely in EGS4 are the user-defined parameters *AE* and *AP*, respectively, where *AE* is given in total energy (rest mass plus kinetic) and *AP* is in kinetic energy (MeV). This type of electron transport mechanism is known as class II.

(ii) Values of AE = 0.611 MeV and AP = 0.5 MeV mean that only knock-on collisions involving an energy loss (transfer to the delta ray) of greater than 0.1 MeV and bremsstrahlung collisions with an energy loss of greater than 0.5 MeV will be modeled. Since most collisions involve energy losses of less than these amounts, most of the true randomness in energy lost per unit pathlength (energy loss straggling) will not be simulated. Therefore, the spectrum of electrons exiting the slab will not be realistic.

(iii) When the total energy of an electron falls below ECUT, its remaining kinetic energy is deposited on the spot and the electron is no longer transported. An ECUT value of 0.711 MeV means that no electrons with kinetic energy of less than 0.2 MeV will be transported, so there will be no electron fluence scored in the kinetic energy bin from 0 to 0.1 MeV.

10.2

(i) The probability per unit pathlength that a photon will interact at a distance x from its current location is the product of the likelihood that is has not already interacted and the probability per unit pathlength that it will interact at that point. From simple exponential attenuation it is known that the first of these factors is $\exp(-\mu x)$. The second factor is by definition the attenuation coefficient, μ. The probability density function (PDF) for a photon interacting after adistance x is therefore $f(x) = \chi\exp(-\mu x)$.

(ii) On examination of the CDF $F(x)$ it is obvious that it represents the probability that the photon has interacted at a distance less than x, since it is one minus the probability that it has not interacted. This must be the case because the CDF is, in this case, the sum of interaction probabilities up to distance x. The value of x at which the photon will interact is found by randomly assigning a value R to $F(x)$, where $0.0<R<1.0$, then obtaining the corresponding value for x. Inverting equation Q10.2 gives

$$x = -\frac{1}{\mu}\ln(1-R) \tag{A10.1}$$

or equivalently, since R is distributed the same as $1 - R$,

$$x = -\frac{1}{\mu}\ln(R). \tag{A10.2}$$

(iii) The most likely value of x to choose is that corresponding to the steepest part of the CDF curve because a given increment x then corresponds to the largest range of R. This must be the case because the steepest part of the CDF corresponds to the peak of the PDF (since the integral is increasing most rapidly at this value of x).

(iv) The CDF is not in general a function that can be inverted as in equation A.10.2. In such a case, the random value R is chosen and x is incremented until the CDF exceeds R. The exact value of x can be found by interpolation within the final increment.

10.3

(i) When a random number sequence begins repeating, each time a history is initiated, there is a chance that the initial random used was previously used as an initial random number. Once this occurs, the remaining sequence of histories will be exactly the same as the previous sequence that was begun with this initial random number.Furthermore, this sequence will repeat over and over again until all histories have been completed.

(ii) The chance that a repeated random number being used to initiate a history was used to initiate a previous history is of course equal to the inverse of the average amount of random numbers used per history. The expectation value for the number of histories to be run (after the random number sequence repeats) before repetition occurs is thus the average number of random numbers used per history. No, history does not repeat itself because nature has a true random number generator that never repeats, unlike the pseudo-random number generators used in Monte Carlo simulations!

10.4

(i) In EGS4, the cutoff energy for electron transport is set to a kinetic energy of 0.1 MeV by specifying an ECUT value of 0.611 MeV (0.511 + 0.1).

(ii) If an electron has a kinetic energy, that means it has no chance of escaping the voxel it is currently in, and there is no point in transporting it further as its remaining energy will all be scored in that voxel. It is thus more efficient to terminate its transport and deposit all remaining kinetic energy in the voxel. Note that there is a reason not to terminate the electron's transport if we want to obtain accurate information about bremsstrahlung photons produced in the medium, as the electron may still produce bremsstrahlung (though unlikely at very low electron energies).

(iii) If we want to terminate electron transport when the residual range is 1/3 of the voxel width (0.5 cm) and the stopping power is 2 MeV cm^{-1}, an appropriate kinetic energy at which to set ECUT is $E_k = 0.33 \times 0.5 \times 2.0 = 0.33$ MeV. This corresponds to a total energy of $0.511 + 0.33 = 0.841$ MeV. Since this calculation is approximate, an ECUT value of 0.6 MeV is reasonable.

10.5

(i) Terma at a point r is the total energy imparted by primary photons at that point, including both energy transferred to secondary charged particles and retained by scattered photons. Equation 10.9 includes the divergence of primary photon energy at r. This is physically the energy emerging from a small volume at r minus the energy entering that volume. The negative sign is because we are interested in the energy entering minus that emerging, and this net energy is divided by the density because the units of terma are energy per mass (like dose).

(ii) For an exponential fall-off in energy fluence in the medium, where the point r is at a depth r along a ray-line from the source, the divergence of energy fluence is found from its derivative with respect to d, so

$$\begin{aligned} -\frac{1}{\rho(r)}\nabla\cdot\psi(r) &= -\frac{1}{\rho(r)}\frac{\partial}{\partial d}\psi_0 e^{-\mu d} \\ &= \frac{\mu}{\rho}(r)\psi_0 e^{-\mu d} \\ &= \frac{\mu}{\rho}(r)\psi(r). \end{aligned} \quad \text{(A10.3)}$$

The factors on the right hand side of equation A10.3 show us that the terma is the probability that photons will interact in a unit equivalent pathlength at r, multiplied by the energy fluence at r. This product is obviously the terma because it gives the total energy of photons involved in interactions, and this is the energy of ejected charged particles plus scattered photons.

11: Quality Assurance in Radiotherapy

11.1 Tender, acquisition, installation, acceptance tests, calibration, commissioning for RTP, pre patient phantom tests, patient treatment.

11.2 The mechanical *isocenter* is the point about which the gantry rotates at a distance, usually 100 cm SSD. It is also the point along the axis of rotation about which the collimators rotate and the patient support assembly couch rotates.

The *radiation isocenter* is defined as the smallest sphere through which the axes of radiation beams pass in all conditions (IPSM 1988). One way of finding this position is by using spoke shots.

11.3 The five common imaging QA tests for an EPID as outlined by Menon and Sloboda (2004) are:

- Dark and flood field images
- Contrast resolution
- Spatial resolution
- Contrast to noise ratio
- Dose rate stability.

11.4 The EPID is a device designed for imaging. The device has high Z material and non-uniform scattering material within its design. Off-axis corrections are required to correct the device's energy response, and kernel corrections are required to correct from dose to EPID material to dose to tissue.

11.5 While an MU check program can check the plan MU, the transfer of data and treatment dose delivery is checked by using dosimeters in an IMRT check phantom on the linac.

A point dose check can be undertaken with an ionization chamber placed in an IMRT phantom at isocenter. The patient plan information is usually transferred to the IMRT phantom using a copy-to-phantom calculation and doses are re-estimated to isocenter, based on the new effective depths to the phantom dose point by integrating charge using a standard in the phantom, then measuring charge collected in the IMRT delivery using planned MUs and segments, and converting this to dose for each field, which is then summed up over the entire treatment of combined multiple IMRT fields.

Usually, validation of planar and axial dose distributions may be undertaken using comparison of film and ionization chamber readings. Each individual planned IMRT beam is transferred to a phantom and the dose delivered to a film is compared against the dose predicted using the RTP computer.

Then all beams are transferred to a phantom, which may be a cylinder or a simple block shape; then the dose is delivered and dose checks are made using film placed parallel to the beam axis to get an axial dose comparison. The films also act as a good qualitative tool for viewing the dose-planned and delivered-dose maps. There has been progress toward using EPIDs to calculate the dose and to replace film as the planar dose map analysis tool. Gel dosimetry may be another useful option, in particular as a complete snap shot of 3-D dose distribution is obtained. This may be particularly useful when commissioning an IMRT treatment program at a new clinic.

12: Patient Immobilization and Image Guidance

12.1

$$IMSM_{combined} = \sqrt{IM^2 + SM^2} \tag{A12.1}$$

$$IMSM_{combined} = \sqrt{3^2 + 4^2} = 5 \text{ mm}. \tag{A12.2}$$

12.2

$$IMSM_{combined} \approx 2.5\sum +0.7\sigma \tag{A12.3}$$

$$IMSM_{combined} \approx 2.5(2) + 0.7(3) \approx 7.1 \text{ mm}. \tag{A12.4}$$

12.3 i) Can it significantly reduce patient movement?
ii) Is patient comfort acceptable?

12.4

i) Define motion encompassing volume.
ii) Respiratory gating.
iii) Tumor tracking.

12.5

i) Delay in communication between movement tracking device and equipment (e.g., DMLC) response.
ii) Correcting for dose in dynamically changing inhomogeneity.
iii) Viewing the tumor volume clearly and using a combination of external and perhaps internal fiducials (implantation of which carries a risk). This problem also applies to gating and volume encompassing methods.
iv) Synchronizing infrared external markers perhaps with x-ray imaging and predictive mapping of the breathing cycle. Phase differences between external markers and internal tumor movement may exist.
v) For some patients breathing cycles are irregular.

13: Radiation Protecton and Room Shielding

13.1 Deterministic effects are generally not governed by probability, i.e., the effect will happen. These effects are due to cell death and they have a dose threshold that can typically be assessed in terms of dose Gy and volume of tissue exposed. The deterministic dose threshold is specific to particular tissues and severity of harm is dose dependent.

Stochastic effects may be due to cell changes (DNA) and proliferation towards a malignant disease. Severity (example cancer) is independent of the dose and there may be no dose threshold. Stochastic effects may be applicable to very small doses; however the probability of effect increases with dose.

13.2 Medical diagnostic x-rays vary from about 0.1 mSv for a chest x-ray to 10 mSv for a CT scan; this varies greatly depending on scan protocol. The dose values for radiotherapy are generally 50 to 70 Gy (equivalent to 50 to 70 Sv). Natural background is usually about 2 mSv year^{-1} but can be higher than this in some countries due to terrestrial component of radiation varying due to radiation background in surrounding rocks.

13.3 Calculating the thickness of concrete for primary barrier using equations 13.4, 13.5, and [13.6], then:

$$B = \frac{H_{\text{limit}} \times ((d/d_0)^2}{W \times U \times T}$$

$$B = \frac{\left(3 \times 10^{-4}\right) \times (5.305/1)^2}{\left(5 \times 10^4\right) \times 0.25 \times 0.0625} = 1.08 \times 10^{-5} \tag{A13.1}$$

$$n = \log_{10}\left(\frac{1}{B}\right) = 4.97 \text{ TVLs.} \tag{A13.2}$$

If 1st and nth TVL is 35 cm concrete at 6 MV, then primary barrier concrete thickness required is 4.97 TVL × 0.35 = 1.74 m of concrete.

13.4 Calculating the thickness of concrete of secondary barrier from scatter from equation 13.7, for a 40 × 40 cm^2 field (F/F_0) = 4.

$$B = \frac{H_{limit} \times (d_1)^2 \times (d_2)^2 \times (F/F_0)}{a \times W \times T}$$

$$B = \frac{(3 \times 10^{-4}) \times (5.035)^2 \times 1^2 \times 4}{(6 \times 10^{-4}) \times (5 \times 10^4) \times 1} = 1.014 \times 10^{-3} \tag{A13.3}$$

$$n = \log_{10}\left(\frac{1}{B}\right) = 3 \text{ TVLs.} \tag{A13.4}$$

The TVLs for 90° scatter is 17 cm concrete so the thickness is 0.51 m.

13.5 Calculating the thickness of concrete of secondary barrier from leakage using equation 13.8:

$$B = \frac{(1/A) \times H_{\text{limit}} \times (d_3)^2}{W \times T}$$

$$B = \frac{1000 \times (3 \times 10^{-4}) \times (5.035)^2}{(5 \times 10^4) \times 1} = 7.6 \tag{A13.5}$$

$$n = \log_{10}\left(\frac{1}{B}\right) = 3.8 \text{ TVLs.} \tag{A13.6}$$

Because the TVL for leakage is the same as for primary, then the 3.8(TVL) × 35 cm concrete is required to shield leakage; note this is 1.33 m of concrete and because this thickness exceeds the scatter thickness calculated in the previous answer (answer 13.3), then secondary barriers should be constructed to the leakage barrier thickness specification; note the lesser scatter barrier calculated thickness.

14: Tumor and Normal Tissue Response

14.1 Assume $\alpha/\beta = 10$ for early effects and $\alpha/\beta = 3$ for late effects and $\alpha/\beta = 1.0$ for spinal cord.

Tumor and early

$$SED = nd\left(\frac{1+\frac{d}{\alpha/\beta}}{1+\frac{2}{\alpha/\beta}}\right)$$

$$SED = 10\times 3\left(\frac{1+(3/10)}{1+(2/10)}\right) = 32.5\ \mathrm{Gy}_{10/2} \quad \text{(A14.1)}$$

Late effects

$$SED = nd\left(\frac{1+\frac{d}{\alpha/\beta}}{1+\frac{2}{\alpha/\beta}}\right)$$

$$SED = 10\times 3\left(\frac{1+(3/3)}{1+(2/3)}\right) = 36\ \mathrm{Gy}_{3/2} \quad \text{(A14.2)}$$

Spinal cord

$$SED = nd\left(\frac{1+\frac{d}{1}}{1+\frac{2}{1}}\right)$$

$$SED = 10\times 3\left(\frac{1+(3/1)}{1+(2/1)}\right) = 40\ \mathrm{Gy}_{1/2} \quad \text{(A14.3)}$$

14.2 Using text book equations 14.21 and 14.22

$$K = \frac{2}{\mu} = \frac{2}{0.5} = 4 \quad \text{(A14.4)}$$

$$\frac{E}{\alpha} = RT\left(1+\frac{KR}{\alpha/\beta}\right) = 75\left(1+\frac{4\times 0.5}{10}\right) = 90\ \mathrm{Gy}_{10} \quad \text{(A14.5)}$$

$$\frac{E}{\alpha} = RT\left(1+\frac{KR}{\alpha/\beta}\right) = 75\left(1+\frac{4\times 0.5}{3}\right) = 125\ \mathrm{Gy}_{3} \quad \text{(A14.6)}$$

Calculate the BED for early and late effects for a HDR $d = 6.8$ Gy per fraction given in 6 fractions to 40.8 Gy.

$$\frac{E}{\alpha} = nd\left(1+\frac{d}{\alpha/\beta}\right) = 6\times 6.8\left(1+\frac{6.8}{10}\right) = 68.5\ \mathrm{Gy}_{10} \qquad \text{(A14.7)}$$

$$\frac{E}{\alpha} = nd\left(1+\frac{d}{\alpha/\beta}\right) = 6\times 6.8\left(1+\frac{6.8}{3}\right) = 133\ \mathrm{Gy}_{3} \qquad \text{(A14.8)}$$

Note this shows that to obtain similar late effects which are dose limiting in brachytherapy, the dose given for tumor control is lower for the HDR unit.

14.3

(i) Using equation 14.24:

Regimen 1

$$S = \exp\left(-nd(\alpha+\beta d) + \frac{\ln 2(T-T_k)}{T_p}\right)$$

$$S = \exp\left(-70\times 1.15(0.35+0.035\times 1.15) + \frac{\ln 2(47-14)}{4}\right)$$

$$S = 6.56\times 10^{-12} \qquad \text{(A14.9)}$$

Regimen 2

$$S = \exp\left(-nd(\alpha+\beta d) + \frac{\ln 2(T-T_k)}{T_p}\right)$$

$$S = \exp\left(-35\times 2(0.35+0.035\times 2) + \frac{\ln 2(47-14)}{4}\right)$$

$$S = 5.22\times 10^{-11} \qquad \text{(A14.10)}$$

(ii) Using equation 14.26:

Regimen 1 - Tumor/Early

$$BED = nd\left(1+\frac{d}{(\alpha/\beta)}\right) - \frac{\ln 2(T-T_k)}{\alpha T_p}$$

$$BED = 70\times 1.15\times\left(1+\frac{1.15}{(10)}\right) - \frac{\ln 2(47-14)}{0.35\times 4}$$

$$BED = 73.5\ \mathrm{Gy}_{10} \qquad \text{(A14.11)}$$

Regimen 2 - Tumor/Early

$$BED = nd\left(1+\frac{d}{(\alpha/\beta)}\right)-\frac{\ln 2(T-T_k)}{\alpha T_p}$$
$$BED = 35\times 2\times\left(1+\frac{2}{(10)}\right)-\frac{\ln 2(47-14)}{0.35\times 4}$$
$$BED = 67.7\ \mathrm{Gy}_{10} \tag{A14.12}$$

Regimen 1 - Late

$$BED = nd\left(1+\frac{d}{(\alpha/\beta)}\right)$$
$$BED = 70\times 1.15\times\left(1+\frac{1.15}{(3)}\right)$$
$$BED = 111.3\ \mathrm{Gy}_{3} \tag{A14.13}$$

Regimen 2 - Late

$$BED = nd\left(1+\frac{d}{(\alpha/\beta)}\right)$$
$$BED = 35\times 2\times\left(1+\frac{2}{(3)}\right)$$
$$BED = 116.7\ \mathrm{Gy}_{3} \tag{A14.14}$$

As regimen 1 gives a higher tumor BED_{10}, then tumor control should be higher. However early effects will be greater also.

Late effects should be less for regimen 1 as the BED_3 is less.

(iii) Once T is greater than T_k then:

$$\frac{\ln 2(1)}{0.35\times 4} = 0.5\ \mathrm{Gy}_{10}\ /\ \mathrm{day} \tag{A14.15}$$

(iv) The repair half time is long enough to require a minimum of 6 hours' gap between each fraction of dose delivered. This means patients need to be scheduled morning and afternoon with this appropriate time gap. This is even more of a challenge when CHART is applied as CHART is 3 fractions per day.

(v) Regimen 1 seems suitable for tumors that are fast proliferators, e.g., squamous cell carcinoma head and neck and lung tumors. It does not appear that regimen 1 would be advantageous for slowly proliferating tumors, e.g., prostate, if proliferation is the test.

Table A14.1: Calculated BEDs and SEDs for 70.2 Gy in 26 fractions

Pollack et al. (2002) regime with varying α/β values					
α/β	1.5	α/β	3	α/β	10
d	2.7	d	2.7	d	2.7
n	26	n	26	n	26
BED	196.56	BED	133.38	BED	89.154
SED	84.24	SED	80.028	SED	74.295

14.4

(i) In textbook equation 14.50, TCP is given by a Poisson probability distribution for the likelihood of there being no surviving clonogens in a tumor which is initially of clonogen density ρ, volume V and receives a dose D. This simply says that the probability of zero surviving clonogens is the exponential of the expected number of surviving clonogens. For an initial number of clonogens N and a uniform survival probability S, the TCP is then

$$TCP = e^{-NS}. \qquad \text{(A14.16)}$$

From a purely statistical point of view, a large tumor is harder to kill because there is a lower probability of killing all clonogens (the expected number surviving is larger).

(ii) Substituting TCP = 0.6 and $N = 10^9$ into the equation gives a survival probability of 5.11×10^{-11}. The expected number of surviving clonogens is 0.51 (note that this is independent of the initial number).

(iii) For an initial clonogen number N_0 and a survival probability of $S = e^{-\alpha D}$, TCP is given by

$$TCP = e^{-N_0 e^{-\alpha D}}. \qquad \text{(A14.17)}$$

Taking the logarithm of both sides of this equation twice, then rearranging to make dose the subject gives the required expression. The logarithmic dependence of required dose on initial clonogen number indicates that the clonogen number must vary by orders of magnitude before a significantly different dose is required. A tumor that is half the size of another tumor will require almost as much dose [see also the answer to part (iv) below].

(iv) For a constant *TCP*, the change in dose required for a change in clonogen number from N_1 to N_2 is:

$$\Delta D = \frac{1}{\alpha} \ln \frac{N_2}{N_1}$$
$$= \frac{1}{\alpha} 2.3 \log \frac{N_2}{N_1}. \quad \text{(A14.18)}$$

For one logarithm of difference in clonogen number, the change in dose required when $\alpha = 0.35\ \text{Gy}^{-1}$ is $\Delta D = (1/0.35) \times 2.3 \times 1 = 6.6$ Gy.

14.5 The benefit of using the Pollack et al. (2002) regime compared to the standard 2 Gy/fraction is dependent on α/β ratio for the prostate. When SED is calculated for a prostate with α/β value of 1.5, 3, and 10 as shown in table A14.1 it becomes clear that it is beneficial to use the Pollack regime if the α/β value for a prostate is as low as 1.5.

In the case where α/β equals 1.5 for the prostate tumor, there is an increase in the tumor cell kill with the Pollack regime and it would require a dose increase of 14 Gy to achieve the same tumor response using the standard 2 Gy/fraction. This would be associated with a small increase in tissue effect for ($\alpha/\beta = 3$) of 4 Gy. In the case if $\alpha/\beta = 10$ for prostate, which is unlikely, then there is a reduction in SED by 2.3 Gy compared with the standard regime.

Index